The Genus Citrus

The Genus Citrus

Edited by

Manuel Talon
Marco Caruso
Fred G. Gmitter Jr.

ELSEVIER
WOODHEAD PUBLISHING
An imprint of Elsevier

Woodhead Publishing is an imprint of Elsevier
The Officers' Mess Business Centre, Royston Road, Duxford, CB22 4QH, United Kingdom
50 Hampshire Street, 5th Floor, Cambridge, MA 02139, United States
The Boulevard, Langford Lane, Kidlington, OX5 1GB, United Kingdom

Notices

Knowledge and best practice in this field are constantly changing. As new research and experience broaden our understanding, changes in research methods, professional practices, or medical treatment may become necessary.

Practitioners and researchers must always rely on their own experience and knowledge in evaluating and using any information, methods, compounds, or experiments described herein. In using such information or methods they should be mindful of their own safety and the safety of others, including parties for whom they have a professional responsibility.

To the fullest extent of the law, neither the Publisher nor the authors, contributors, or editors, assume any liability for any injury and/or damage to persons or property as a matter of products liability, negligence or otherwise, or from any use or operation of any methods, products, instructions, or ideas contained in the material herein.

Library of Congress Cataloging-in-Publication Data
A catalog record for this book is available from the Library of Congress

British Library Cataloguing-in-Publication Data
A catalogue record for this book is available from the British Library

ISBN: 978-0-12-812163-4(print)
ISBN: 978-0-12-812217-4 (online)

For information on all Woodhead publications
visit our website at https://www.elsevier.com/books-and-journals

Publisher: Charlotte Cockle
Acquisition Editor: Nancy Maragioglio
Editorial Project Manager: Vincent Gabrielle
Production Project Manager: Sojan P. Pazhayattil
Cover Designer: Victoria Pearson

Typeset by SPi Global, India

Contents

Contributors

Numbers in parentheses indicate the pages on which the authors' contributions begin.

Manuel Agustí (193, 219), Mediterranean Agroforestry Institute, Polytechnic University of Valencia, Valencia, Spain

Pablo Aleza (171), Centro de Citricultura y Producción Vegetal, Instituto Valenciano de Investigaciones Agrarias (IVIA), Valencia, Spain

Fernando Alferez (411), University of Florida, Institute of Food and Agricultural Sciences, Southwest Florida Research and Education Center, Immokalee, FL, United States

Vicent Arbona (271, 291), Ecophysiology and Biotechnology, Department of Agricultural and Environmental Sciences, Universitat Jaume I, Castelló de la Plana, Spain

Graham H. Barry (83), XLnT Citrus Company, Cape Town, South Africa

Ozgur Batuman (349), Department of Plant Pathology, Southwest Florida Research and Education Center, University of Florida, Immokalee, FL, United States

Manosh Kumar Biswas (33), Key Laboratory of Horticultural Plant Biology (Ministry of Education), Huazhong Agricultural University, Wuhan, China

Rodrigo Marcelli Boaretto (311), Sylvio Moreira Citrus Research Center, Instituto Agronômico (IAC), Cordeirópolis, Brazil

Kim D. Bowman (105), U.S. Horticultural Research Laboratory, USDA-ARS, Fort Pierce, FL, United States

Liliana M. Cano (349), Department of Plant Pathology, Indian River Research and Education Center, University of Florida, Fort Pierce, FL, United States

Marco Caruso (83, 129), CREA Research Centre for Olive, Citrus and Tree Fruit, Acireale, Italy

Vittoria Catara (349), Department of Agriculture, Food and Environment, University of Catania, Catania, Italy

Yijing Cen (333), South China Agricultural University, Guangzhou, China

José M. Colmenero-Flores (291), Institute of Natural Resources and Agrobiology, Spanish National Research Council (CSIC), Seville, Spain

Paul J.R. Cronje (421), Citrus Research International, Department of Horticultural Sciences, Stellenbosch University, Stellenbosch, South Africa

Jose Cuenca (171), Centro de Citricultura y Producción Vegetal, Instituto Valenciano de Investigaciones Agrarias (IVIA), Valencia, Spain

Franck Curk (57), French National Institute for Agricultural Research (INRA), AGAP Research Unit, San Giuliano, France

John V. da Graça (391), Citrus Center, Texas A&M University-Kingsville, Weslaco, TX, United States

Xiuxin Deng (33), Key Laboratory of Horticultural Plant Biology (Ministry of Education), Huazhong Agricultural University, Wuhan, China

Nuria Duran-Vila (391), Centro de Protección Vegetal y Biotecnología, Valencian Institute of Agrarian Research (IVIA), Valencia, Spain

Manjul Dutt (171), Citrus Research and Education Center, Department of Horticultural Sciences, University of Florida, Lake Alfred, FL, United States

Simona Fabroni (447), CREA, Research Center for Olive, Citrus and Tree Fruit, Acireale, Italy

Shi Feng (447), Citrus Research and Education Center, Lake Alfred, FL, United States

Waldir Fernandes, Jr. (471), Economics, Business Administration and Education Department, São Paulo State University, Jaboticabal; College of Technology, State Center of Technological Education, São Jose do Rio Preto, Brazil

Juliana Freitas-Astúa (391), Embrapa Cassava and Fruits, Cruz das Almas, Brazil

Yann Froelicher (129), Unité Mixte de Recherche Amélioration Génétique et Adaptation des Plantes (UMR Agap), Centre de Coopération Internationale en Recherche Agronomique pour le Développement (CIRAD), San Giuliano, France

Dean Gabriel (371), Plant Pathology Department, University of Florida, Gainesville, FL, United States

Zhifeng Gao (471), Food and Resource Economics Department, University of Florida, Gainesville, FL, United States

Maria Antonietta Germanà (171), Department of Agricultural, Food and Forest Sciences (SAAF), University of Palermo, Palermo, Italy

Frederick G. Gmitter, Jr. (1, 9, 83, 129), Citrus Research and Education Center (CREC), Institute of Food and Agricultural Sciences (IFAS), University of Florida, Lake Alfred, FL, United States

Aurelio Gómez-Cadenas (271, 291), Ecophysiology and Biotechnology, Department of Agricultural and Environmental Sciences, Universitat Jaume I, Castelló de la Plana, Spain

Timothy R. Gottwald (371), US Department of Agriculture, Agricultural Research Service, Fort Pierce, FL, United States

Santin Gravena (333), Paulista State University (UNESP, Retired) and Member of GCONCI-Consultant Citrus Group, Jaboticabal, Brazil

Jude W. Grosser (171), Citrus Research and Education Center, Department of Horticultural Sciences, University of Florida, Lake Alfred, FL, United States

Tim G. Grout (333), Citrus Research International, Nelspruit, South Africa

Jae-Wook Hyun (349), Citrus Research Institute, National Institute of Horticultural and Herbal Science, Jeju, South Korea

Johan Joubert (105), Citrus Research International, Nelspruit, South Africa

Davie M. Kadyampakeni (311), Citrus Research and Education Center, University of Florida, Lake Alfred, FL, United States

Masaya Kato (495), Department of Bioresource Sciences, Faculty of Agriculture, Shizuoka University, Shizuoka, Japan

Prabhjot Kaur (171), Citrus Research and Education Center, Department of Horticultural Sciences, University of Florida, Lake Alfred, FL, United States

Robert Krueger (57), United States Department of Agriculture-Agricultural Research Service National Clonal Germplasm Repository for Citrus and Dates, Riverside, CA, United States

Irene Lavagi (391), University of California, Riverside, CA, United States

Hongye Li (349), Institute of Biotechnology, Zhejiang University, Hangzhou, People's Republic of China

Concetta Licciardello (245), CREA Research Centre for Olive, Citrus and Tree Fruit, Acireale, Italy

Silvio A. Lopes (371), Research and Development Department, Fundecitrus, Araraquara, Brazil

Gang Ma (495), Department of Bioresource Sciences, Faculty of Agriculture, Shizuoka University, Shizuoka, Japan

Haijie Ma (349), Institute of Biotechnology, Zhejiang University, Hangzhou, People's Republic of China; Department of Horticultural Science, Citrus Research and Education Center, University of Florida, Lake Alfred, FL, United States

Dirceu Mattos, Jr. (311), Sylvio Moreira Citrus Research Center, Instituto Agronômico (IAC), Cordeirópolis, Brazil

Kelly T. Morgan (311), Southwest Florida Research & Education Center, University of Florida, Immokalee, FL, United States

Raphaël Morillon (271, 291), Equipe "Amélioration des Plantes à Multiplication Végétative", UMR AGAP, Département BIOS, CIRAD, Guadeloupe, France

Ana Quiñones Oliver (311), Centro de Desarrollo de Agricultura Sostenible, Instituto Valenciano de Investigaciones Agrarias (IVIA), Valencia, Spain

Patrick Ollitrault (57), French Agricultural Research Centre for International Development (CIRAD), AGAP Research Unit, San Giuliano, France

Lluís Palou (421), Centre de Tecnologia Postcollita, Institut Valencià d'Investigacions Agràries (IVIA), Valencia, Spain

Eduardo Primo-Millo (193, 219), Centro de Citricultura y Producción Vegetal, Instituto Valenciano de Investigaciones agrarias (IVIA), Valencia, Spain

Jose Antonio Quaggio (311), Center of Soils and Environmental Resources, Instituto Agronômico (IAC), Campinas, Brazil

Paolo Rapisarda (447), CREA, Research Center for Olive, Citrus and Tree Fruit, Acireale, Italy

Mark Ritenour (349), Department of Horticultural Science, Indian River Research and Education Center, University of Florida, Fort Pierce, FL, United States

María J. Rodrigo (245), Instituto de Agroquímica y Tecnología de Alimentos (IATA)—Consejo Superior de Investigaciones Científicas (CSIC), Valencia, Spain

Daniel S. Rokhsar (1, 9), US Department of Energy Joint Genome Institute, Walnut Creek; Department of Molecular and Cell Biology and Center for Integrative Genomics, University of California, Berkeley, Berkeley, CA, United States; Molecular Genetics Unit, Okinawa Institute of Science and Technology Graduate University, Onna, Japan

Russell Rouseff (447), Citrus Research Institute of China, Chinese Academy of Agriculture Sciences, Southwest University, Chongqing, People's Republic of China

Giuseppe Russo (129), CREA Research Centre for Olive, Citrus and Tree Fruit, Acireale, Italy

Avi Sadka (245), Department of Fruit Tree Sciences, Agricultural Research Organization, The Volcani Center, Rishon Le'zion, Israel

Tokurou Shimizu (149), Division of Citrus Research, Institute of Fruit Tree and Tea Science, National Agriculture and Food Research Organization, Shizuoka, Japan

Malcolm W. Smith (129), Department of Agriculture and Fisheries, Bundaberg Research Station, Bundaberg, QLD, Australia

Thomas H. Spreen (471), Food and Resource Economics Department, University of Florida; Economic and Marketing Research Department, Florida Department of Citrus, Gainesville, FL, United States

Philip A. Stansly (333), University of Florida—IFAS, Southwest Florida Research & Education Center, Immokalee, FL, United States

Minoru Sugiura (495), Department of Food Science and Nutrition, Doshisha Women's College of Liberal Arts, Kyoto, Japan

Francisco R. Tadeo (245), Centro de Genómica, Instituto Valenciano de Investigaciones Agrarias (IVIA), Valencia, Spain

Manuel Talon (1, 9), Centro de Genómica, Instituto Valenciano de Investigaciones Agrarias (IVIA), Valencia, Spain

Javier Terol (245), Centro de Genómica, Instituto Valenciano de Investigaciones Agrarias (IVIA), Valencia, Spain

Alberto Urbaneja (333), Centro de Protección Vegetal y Biotecnología, Instituto Valenciano de Investigaciones Agrarias (IVIA), Valencia, Spain

Antonio Vicent (349), Centro de Protección Vegetal y Biotecnología, Instituto Valenciano de Investigaciones Agrarias (IVIA), Valencia, Spain

Georgios Vidalakis (391), University of California, Riverside, CA, United States

Christopher Vincent (271), University of Florida Citrus Research and Education Center, Lake Alfred, FL, United States

Nian Wang (171), Citrus Research and Education Center, Department of Microbiology, University of Florida, Lake Alfred, FL, United States

Siyu Wang (447), Citrus Research and Education Center, Lake Alfred, FL, United States

Yu Wang (447), Citrus Research and Education Center, Lake Alfred, FL, United States

Fengnian Wu (333), South China Agricultural University, Guangzhou, China

Guohong Albert Wu (1, 9), US Department of Energy Joint Genome Institute, Walnut Creek, CA, United States

Nelson A. Wulff (371), Research and Development Department, Fundecitrus, Araraquara, Brazil

Masashi Yamamoto (33), Faculty of Agriculture, Kagoshima University, Kagoshima, Japan

Xiaoming Yang (33), Key Laboratory of Horticultural Plant Biology (Ministry of Education), Huazhong Agricultural University, Wuhan, China

Lorenzo Zacarias (421), Instituto de Agroquímica y Tecnología de Alimentos, Consejo Superior de Investigaciones Científicas (IATA-CSIC), Valencia, Spain

Marisa L. Zansler (471), Economic and Marketing Research Department, Florida Department of Citrus, Gainesville, FL, United States

Lancui Zhang (495), Department of Bioresource Sciences, Faculty of Agriculture, Shizuoka University, Shizuoka, Japan

Changyong Zhou (391), Southwest University, Chongqing, People's Republic of China

Chapter 1

The citrus genome

Frederick G. Gmitter, Jr.[a], Guohong Albert Wu[b], Daniel S. Rokhsar[b,c,d], Manuel Talon[e]

[a]*Citrus Research and Education Center (CREC), Institute of Food and Agricultural Sciences (IFAS), University of Florida, Lake Alfred, FL, United States,* [b]*US Department of Energy Joint Genome Institute, Walnut Creek, CA, United States,* [c]*Department of Molecular and Cell Biology and Center for Integrative Genomics, University of California, Berkeley, Berkeley, CA, United States,* [d]*Molecular Genetics Unit, Okinawa Institute of Science and Technology Graduate University, Onna, Japan,* [e]*Centro de Genómica, Instituto Valenciano de Investigaciones Agrarias (IVIA), Valencia, Spain*

1.1 Concept of genome

The genome of any living organism is indeed the foundation upon which every aspect of that organism's life is built. The organism's shape, size, color, mode of acquisition, and requirements for nutrients, interactions within the environment where the organism may be found, the timely and appropriate regulation of expression of fundamental coding information leading to form and function of the organism and its distinctive organs and tissues, and even the mechanisms by which the genome sequence content is passed forward to future generations—all these and many other fundamental aspects of the life of any organism are determined by the information content of the genome. The genome is the fundamental basis of all living things. Plants of the genus *Citrus* are certainly no exception. In this sense, the genome underlies or influences every aspect of citrus trees and fruits, and therefore certainly the next few chapters, and nearly every chapter topic that follows in this book, may best be understood when viewed through the lens of genomic comprehension. From this perspective, the editors have chosen to open the book with an introductory chapter recognizing and describing something of the foundational role of the citrus genome.

DNA sequencing and analysis seems ubiquitous in contemporary biology; however, it is worth recognizing that this technology arose in the last 25 years of the previous century. The first method of DNA sequencing, based on primers for extension that were location specific and coupled with nucleotide-specific labeling, were reported first in the early 1970s (Wu, 1972; Padmanabhan et al., 1974). These reports were followed by the classic work of Sanger et al. (1977), which remained the gold standard for genome sequencing through the first decade of the current century. The earlier methods were applied to viruses, bacteriophage, and soon to a bacterium *Haemophilus influenzae*, using a whole-genome shotgun (WGS) approach (Fleischmann et al., 1995). Perhaps one of the most significant developments in the earliest days of the new millennium, the first draft sequence of the human genome was published (Venter et al., 2001). Since then, sequencing technology experienced exponential growth in capacity and quality that continues unabated until today (summarized in Goodwin et al., 2016). It is as a consequence of this technological revolution that agricultural scientists, and citrus researchers specifically, have been able to make the progress and lead to incredible new insights to objectively address biological and evolutionary questions.

1.2 Citrus genomes

Within the context of the rapid evolution of genomic science and associated bioinformatics tools, the global citrus community came together in Valencia, Spain in 2003 to establish a framework for international collaboration, referred to as the International Citrus Genome Consortium (ICGC). The historical background of the ICGC as well as the fundamental principles, objectives, and goals were described previously (Talon and Gmitter, 2008; Gmitter et al., 2012). Initially, the ICGC estimated the cost of producing "the citrus genome" to be about US$16 million, and so individual research groups went to their home industries and countries to seek support for the ambitious international project. Finally, researchers from the United States, France, Italy, Spain, and Brazil acquired financial resources and engaged primarily with the US Department of Energy-Joint Genome Institute (USDOE-JGI), Genoscope in France, and Istituto Genomica Applicata (IGA) in Italy to work together for producing and making publicly available through the USDOE-JGI web portal Phytozome in January 2011, the first ever reference citrus genome sequence of a haploid clone derived from Clementine mandarin (Aleza et al., 2009;

The Genus Citrus. https://doi.org/10.1016/B978-0-12-812163-4.00001-2

Germana et al., 2013). This was accomplished by Sanger sequencing technologies in the three institutes, coupled with a reference genetic linkage map (Ollitrault et al., 2012), community EST resources, and contributions from other entities (see Wu et al., 2014 for full details of the work, and a complete listing and description of the contributions of collaborating individuals and organizations). In parallel, a different approach using what then was the new pyrosequencing methodology of Roche 454 was applied to sequence the diploid sweet orange Ridge Pineapple, which had originally been selected by the ICGC for the global sequencing project. Though substantially more fragmented, this diploid orange assembly covered much of the same predicted gene space as the haploid. It was made available at the same time as the haploid Clementine through Phytozome in January 2011 to the global research community (Wu et al., 2014).

In 2012, a draft assembly of the Valencia sweet orange genome was published (Xu et al., 2013). A dihaploid line derived from anther culture was sequenced using a WGS paired-end strategy with Illumina GAII (genome analyzer II) technology. Then, shotgun Illumina sequence reads from the diploid source Valencia tree were generated and mapped to the de novo haploid reference assembly to provide more complete sequence information. The shotgun sequencing approach was applied to three additional presumed accessions of mandarin (*Citrus reticulata*) and three pummelo accessions (*Citrus maxima*); it was concluded that sweet orange arose as a simple backcross hybrid of (mandarin × pummelo) × mandarin, though this conclusion was subsequently challenged by Wu et al. (2014).

As sequencing technology advanced, citrus genome sequence studies followed suit. Relatively expensive Sanger sequencing technology was followed by pyrosequencing, which was less expensive but yielded technological challenges resulting in lower-quality assemblies. Illumina-based sequencing came next and allowed for greater depth of coverage for less cost, but again there were compromises in terms of final quality of the genome assemblies. The advent of long read single molecule sequencing approaches such as PacBio addressed some of these challenges and led to assemblies with better quality indicated by increased contig and scaffold N50 and N90 values compared with previous citrus genome assemblies. Two research groups, one in China and the other in Japan, adopted new sequencing technology and published in 2017 the pummelo and Satsuma mandarin reference genomes, respectively (Wang et al., 2017; Shimizu et al., 2017). The former research group used a haploid clone of pummelo (*C. maxima*) to produce the first de novo sequence assembly of this species based on the single-molecule sequencing approach provided by the PacBio RS II system (Wang et al., 2017). They first used 56.8× coverage of long reads to assemble and followed this with 307.3× coverage of Illumina short reads to correct errors and fill in the gaps. They have also used a similar approach to produce de novo sequence assemblies of three other citrus species representing wide variation within the genus, including *Citrus ichangensis* (Ichang papeda), *Citrus medica* (citron), and a Rutaceous relative *Atalantia buxifolia* (the Chinese box orange) The Satsuma draft genome was produced from the widely grown commercial diploid cultivar, Miyagawa Wase, also by a hybrid sequencing and assembly approach. In this case, sequencing consisted of Illumina-based short reads, three mate-pair libraries, long read PacBio sequences, and a hybrid assembly approach. Pseudomolecules were constructed following alignment to three SSR (simple sequence repeat)- and SNP (single nucleotide polymorphism)-based genetic linkage maps from Satsuma hybrid families. Very recently the de novo sequence of *Fortunella hindsii* was constructed from a plant produced by three generations of selfing (Zhu et al., 2019). A summary of these various de novo assemblies is illustrated in Table 1.1.

As sequencing and assembly technologies continue to evolve and improve, better assemblies of important citrus genomes will be produced in the near future. Improved sequencing technologies also will impact the quality of associated transcriptome information, leading to better annotations as well. But the availability of rapid and inexpensive WGS and the consequent generation of deep sequence coverage of many genomes have enabled the exploration of important questions relating to the phylogeny and evolution of the very broad range of phenotypes found among citrus accessions and commercial cultivar groups, as described below.

1.3 Genomes: Pure, admixed, and domesticated

1.3.1 Inference of pure and admixed genome regions

Significant insight has been gained in our understanding of the origin and evolution of the genus *Citrus* based on the analysis of WGS sequencing of a wide collection of citrus species, hybrids, and admixed varieties (Wu et al., 2018). The phylogenomic analysis, in combination with biogeographic data and the use of citrus leaf fossil calibration, revealed that citrus diversified during the late Miocene through a rapid Southeast Asian radiation, giving rise to at least seven progenitor species. This was followed by a second radiation in the early Pliocene of Oceania resulting in several Australian citrus species (Wu et al., 2018). Please refer to Chapter 2 for a detailed treatment of citrus evolution and dispersal, including a chronogram of citrus speciation.

TABLE 1.1 Summary of de novo citrus genome sequences cited above and their characteristics.

Genome	Ploidy	Platform	Total number of contigs	Contig N50 (*L50) (kb)	Longest contig (Mb)	Total number of scaffolds	Scaffold N50 (*L50) (Mb)	Longest scaffold (Mb)	Size of assembled scaffolds (Mb)
Haploid *C.* × *clementina*	Haploid	Sanger	8692	118.9*	1.23	1,398	31.4*	30.50	301.4
C. × *sinensis* v1.0 Ridge Pineapple	Diploid	Sanger/454	53,536	6.6*	0.119	12,574	0.2505*	5.93	319.2
Citrus sinensis	Dihaploid	Illumina GAII	17,140	51*		4,811	8.4*	8.16	301.02
C. grandis	Haploid	PacBio RS II/ Illumina	2602	2,183	10.62	1,612	4.21		346
C. medica	Diploid	PacBio RS II/ Illumina	51,023	46.50	0.41	32,731	0.367	2.44	405.0
C. ichangensis	Diploid	PacBio RS II/ Illumina	28,551	76.56	0.77	14,915	0.504	2.98	357
Atalantia buxifolia	Diploid	PacBio RS II/ Illumina	44,724	23.89	0.20	25,600	1.073	7.16	316
Citrus unshiu	Diploid	Illumina HiSeq 2000, PacBio RS II	N/A	N/A	N/A	20,876	3.864	5.22	359.7
Citrus reticulata (Mangshan wild mandarin)	Diploid	Illumina	N/A	24.76	N/A	42,714	1.705	7.20	334
Fortunella hindsii	Diploid	PacBio/Illumina	1226	2209	12.00	N/A	5.156	N/A	N/A

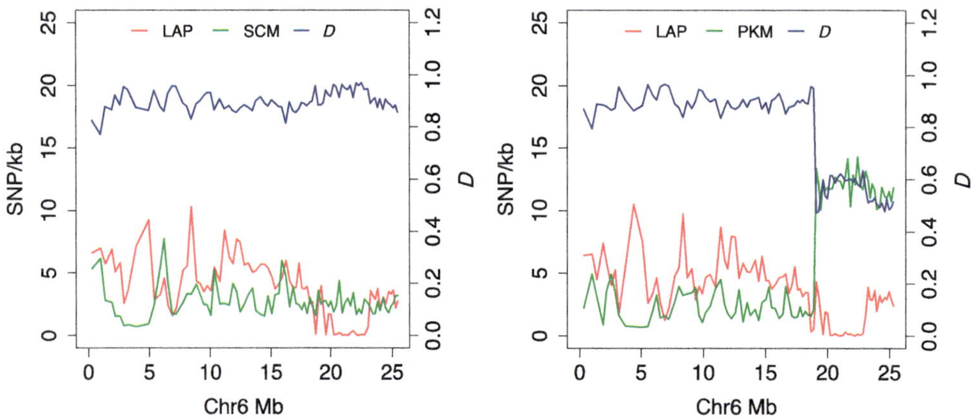

FIG. 1.1 Segmental admixture in the citrus genome. Plotted are average nucleotide diversity and genome divergence D between the low acid pummelo [also known as Siamese Sweet (CES 2240) Hodgson, 1967] and a mandarin in sliding windows of 100 kb. In the left panel, the divergence D between Siamese Sweet and Sun Chu Sha mandarin stays around 0.9, consistent with no interspecific admixture in either genotype on chromosome 6. By contrast, the divergence D between Siamese Sweet and Ponkan mandarin shown in the right panel displays a sharp drop at position 19.0 Mb, with a corresponding sharp rise in the nucleotide diversity of Ponkan at the same location, indicating a transition for Ponkan to interspecific admixture in the region chr6: 19.0–25.6 Mb [coordinates based on the reference Clementine sequence (Wu et al., 2014)].

The ancient speciation processes resulted in deep sequence divergence between different species (~1%–2%) compared to within-species variation (~0.1%–0.6%) (Wu et al., 2018). The degree of divergence between two citrus species can be quantified by the genetic differentiation between two diploid genomes representative of the two species (Wu et al., 2014):

$$D = 1 - 0.25 * (\pi_1 + \pi_2) / \pi_{12}$$

where π_1 and π_2 are the nucleotide diversity (i.e., heterozygosity) of the two diploid genomes and π_{12} is their sequence divergence (i.e., probability that two randomly chosen alleles from the two diploids are different). The value of *D* ranges from 0 to 1, with two unrelated individuals from a panmictic population having *D*=0.5 and two deeply divergent genomes having D approaching one. Between the two Asian species *C. maxima* (pummelos) and *C. reticulata* (wild pure mandarins), *D* ~ 0.9, whereas among the later diversifying Australian citrus species, *D* is in the range 0.75–0.85.

Genomic regions with interspecific admixture can be inferred based on a sliding window estimate of the pairwise divergence (*D*) and the local nucleotide diversity of each genome. As an illustration, consider the genomes of a pummelo (*C. maxima*) and a mandarin with pummelo admixture. Pure genomic regions are characterized by a deep divergence of the two species with a high *D* value (~0.9), whereas regions with interspecific admixture will see a significant drop in *D* value to ~0.5 accompanied by a sharp increase in the local heterozygosity of the mandarin (Fig. 1.1). A more complete description of the various admixture patterns throughout citrus and their associated signatures are given by Wu et al. (2014).

1.3.2 Identification of the ancient progenitor citrus species

With WGS, progenitor citrus species (i.e., accessions without interspecific admixture) can be identified based on a combined analysis of genetic divergence *D* and heterozygosity as outlined above. A recent comparative genomic analysis of 60 accessions of diverse citrus and related varieties identified 10 progenitor citrus species, with most commercial varieties (e.g., oranges, grapefruits, lemons, and limes) having admixed genomes derived from two or more progenitor species (Wu et al., 2018). Of the 10 progenitor species 7 resulted from the first phase of speciation in the late Miocene of East and Southeast Asia (see Chapter 2), and include *C. medica* (citrons), *C. maxima*, *C. reticulata*, *Citrus micrantha*, *C. ichangensis*, *Citrus japonica* (a.k.a Kumquat), and *Citrus mangshanensis*. The other three species arose from the second phase of speciation in the early Pliocene of Oceania and include three Australian species *Citrus glauca* (Australian desert lime), *Citrus australis* (Australian round lime), and *Citrus australasica* (Australian finger lime). Phylogenomic analysis (Wu et al., 2018) shows that *Poncirus trifoliata* has high affinity to *Citrus* and is an outgroup to the citrus clade.

Evidence for additional progenitor citrus species exists from chloroplast, SSR, and other molecular marker studies (Bayer et al., 2009; Schwartz et al., 2016; Nagano et al., 2018). These phylogenetic analyses show that the genus citrus forms a monophyletic group, though discrepancy was noted between the chloroplast tree and nuclear marker-derived phylogeny. Whole-genome phylogenetic study revealed a single origin for citrus, with Oceanic citrus evolving from an Asian species that has close affinity to kumquat (Wu et al., 2018). WGS sequencing of additional citrus genotypes in Oceania and Asia will provide a more detailed picture of the evolutionary trajectory of the genus.

1.3.3 A whole-genome perspective of citrus hybrids and admixtures

Genome-wide admixture analyses for citrus have recently been performed with both WGS sequencing (Wu et al., 2018; Wang et al., 2018) and genotyping by sequencing technology (Oueslati et al., 2017; Ahmed et al., 2019). The ancestry of some notable hybrids (oranges, lemons, and limes) has been confirmed at the genomic level, and the origins of Clementine mandarin and some other important cultivars have been verified (Wu et al., 2014, 2018). The admixture process can be complex as in the case of sweet orange, and one Australian finger lime was found to be admixed with round lime alleles (Wu et al., 2018). The most surprising finding, however, came from the highly heterogenous group of mandarins (Fig. 1.2).

(1) Unsuspected pummelo admixture was initially identified in some traditional mandarins including Ponkan and Willowleaf that formerly were considered to be pure *C. reticulata* (Wu et al., 2014). Later analyses of a large collection of mandarins showed that pummelo admixture is widespread among mandarins with varying degrees of admixture (Oueslati et al., 2017; Wu et al., 2018; Ahmed et al., 2019).
(2) Analysis of the pummelo admixture pattern among mandarins indicates two phases of pummelo introgression, with the first phase involving possibly only a single pummelo tree, resulting in mandarins with a small amount of pummelo admixture (1%–10%). Later, additional pummelo introgressions gave rise to sweet orange and mandarins with higher proportions of pummelo alleles (Wu et al., 2018).

A list of various citrus hybrids and admixtures can be found in Chapter 4, along with detailed descriptions.

1.3.4 Comparative analyses of citrus genomes: The example of evolution and domestication of mandarins

The availability of increasing numbers of plant genomes resulting from the rapid development of sequencing technologies and sequence data analysis platforms are generating a huge number of comparative genomic data opportunities that are significantly transforming the way we approach the study of plants, with strong impact on many relevant agricultural crops. While our understanding of the citrus genome is currently in its infancy, many examples in the following chapters of this book clearly show that this knowledge is becoming a very powerful tool to bring to light unsuspected relationships and propose new and original hypotheses, influencing many scientific disciplines in the citrus community.

To finish this introductory chapter, we would like to illustrate these novel approaches, briefly summarizing as an evocative example the latest findings on the processes of mandarin evolution and domestication, a fascinating story that is emerging from very recent genomics developments. During the last century and until very recently, the taxonomy of citrus and especially that of mandarins has been a matter of major discussion and controversy. Historically prestigious citrus scholars

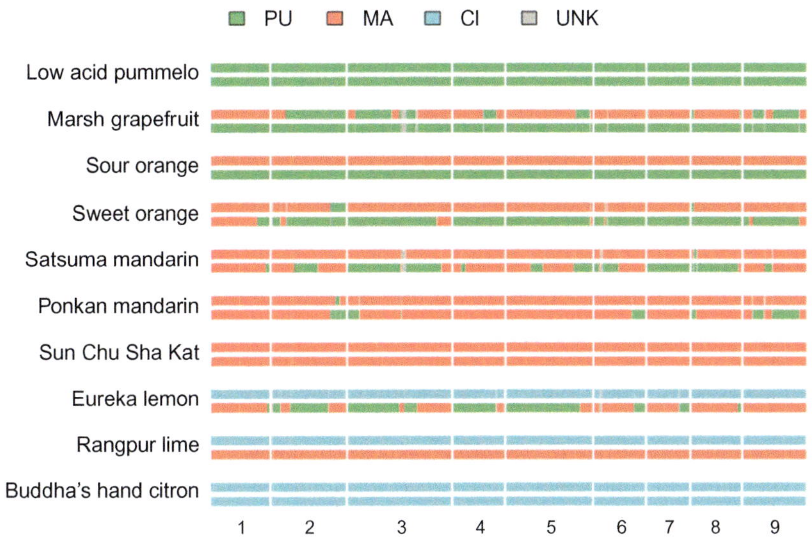

FIG. 1.2 Local ancestry along nine chromosomes of some citrus cultivars as revealed by interspecific admixture analysis. Each bar represents one chromosome of a diploid genome, colored by its local species ancestry. PU=*C. maxima*, MA=*C. reticulata*, CI=*C. medica*, and UNK=unknown. Note that the pair of bars of a genome may not correspond to phased chromosomes.

and botanists such as Swingle, Tanaka, Webber, and Hodgson did not fully agree on the number of mandarin species that could be distinguished, ranging from 3 to no less than 36.

By carefully analyzing the tremendous volume of data resulting from next-generation sequencing, we know today that the term "mandarin" actually comprises a few though still undetermined number of pure species, as well as a vast collection of heterogenic admixtures with different proportions of pummelo introgression into a maternal mandarin genome (Wu et al., 2018; Wang et al., 2018). What ties all these different genotypes together consequently are their strong phenotypic resemblances.

Among the "pure mandarins," genome analyses in these last two works suggest that different species are included in this group. For instance, "wild mandarins" collected from Mangshan (a northern branch of the Nanling Mountains, an east to west mountain range that separates the southern and central subtropical regions of South China), actually represent two distinct species, *C. mangshanensis* and *C. reticulata* (Wang et al., 2018; Xu et al., 2013; Wu et al., 2014). The mandarin *C. mangshanensis* (Mangshanyegan) was initially identified and described in 1990 (Liu et al., 1990), while an independent study also recognized Mangshanyegan as a species distinct from Mangshanyeju (*C. reticulata*) and proposed the species name *Citrus nobilis Lour* for Mangshanyegan (Li et al., 2009). As *C. nobilis* has been used to refer to other mandarin oranges (such as King, Kunenbo, and other tangors), we prefer *C. mangshanensis* to refer to Mangshanyegan. Importantly, both Mangshanyegan and Mangshanyeju are pure species without interspecific admixture signatures and can thus be used as species reference accessions as proposed earlier (Li et al., 2009). Whereas the former is monoembryonic with inedible fruit and considered to be a primitive form of citrus (Fig. 1.3), the latter is polyembryonic and bears edible fruit (Li et al., 2009). Indeed, the wild mandarin under the binomial name of *C. mangshanensis* lies at the basal position of the citrus phylogenetic tree (Wu et al., 2018), predating the emergence of Mangshanyeju mandarin (*C. reticulata*) and other citrus species (e.g., citrons, pummelos, and papedas).

Both studies (Wang et al., 2018; Wu et al., 2018) also suggest that there are other so-called "wild mandarins" that can be considered pure species or that contain very little interspecific admixture from pummelo, such as Sun Chu Cha Kat. This group differs widely from the huge collection of mandarin admixtures that carry variable pummelo introgression in their genomes. Several introgressions appear to have played an important role in the domestication of mandarins and sweet orange, based on the association of pummelo admixture proportion with mandarin fruit size and palatability (Wu et al., 2018) and other traits (Wang et al., 2018).

The analysis of WGS sequencing of wild and cultivated mandarins collected around the Nanling Mountains also identified two groups of cultivated mandarins located to the north and south of the Nanling range (Wang et al., 2018). Phenotypically, the two cultivated mandarin groups were characterized by distinct peel color, fruit size, and acidity. Distinct pummelo admixture pattern was also noted for two mandarin groups that were proposed to have resulted from two independent domestication events involving different mandarin progenitors (Wang et al., 2018).

Mangshanyegan (*C. mangshanensis*) represents a genetic isolate without gene exchange with other citrus types. By contrast, Mangshanyeju (*C. reticulata*) represents ancient wild mandarins and was proposed to be at the origin of the evolution and domestication of cultivated mandarins (Wang et al., 2018), though it may be an ancestor that diverged from the progenitors of the two populations of cultivated mandarins. The subpopulation of cultivated mandarins that evolved in the area south of Nanling not only includes native mandarins such as Qingtianju or Bingtangju but is also genetically related to many popular mandarins developed or well known in the Western world such as Willowleaf, Ponkan, Orah, Murcott, and Wilking. In contrast, the subpopulation of the north region is composed of native admixtures generally bearing intense colored fruit such as Changsha mandarin or the red tangerine (Hongju), and is clearly related to the group of Satsuma mandarins mostly developed in Japan.

An extensive network of relatedness was found among the cultivated mandarins and sweet orange, due to the sharing of common *C. reticulata* haplotypes presumably resulting from the human selection process for large palatable fruits, a clear signature of the domestication process (Wu et al., 2018).

Our understanding of the process of domestication of mandarin, as we know it today, has many mysterious aspects that will undoubtedly be answered in the near future. One striking and unexplained observation is why the mandarin phenotype has been recurrent over time, that is, how very similar combinations of traits emerged in different epochs and locations. As mentioned above, pure wild mandarins have a strong phenotypical resemblance, outstandingly more or less lanceolate leaves, and small, inedible, and seedy, carotenoid-colored fruits, a fact that was highlighted and recognized by earlier generations of citrus botanists. While *C. mangshanensis* was unknown to Swingle and Tanaka, they did agree that *C. reticulata* (a citrus type native to continental Asia), *Citrus tachibana* (considered native to Japan), and *Citrus indica* (considered native to India) were unquestionably three species of mandarins. Based on genomic evidence, we know today that *C. mangshanensis* is not a direct ancestor of *C. reticulata* and that both are separate species, while *C. tachibana* is currently diverging from *C. reticulata* and therefore it may be more correct to consider it a subspecies of *C. reticulata*. Furthermore, the wild Indian mandarin, *C. indica*, is nested in none of the Mangshan mandarin clades, indicating again that it is genetically unrelated to *C. mangshanensis* or *C. reticulata*. Thus, the mandarin phenotype apparently has emerged in

FIG. 1.3 (A) Typical immature fruit (early August) of *C. mangshanensis* on the tree along with a stem, leaves, and small thorns. In addition to the rough and irregular fruit surface, note also the conspicuous wax found on most of the fruit. This and the following two photos were provided by Dr. Xiaochun Zhao of the CAAS Citrus Research Institute, Southwest University, Chongqing, China. (B) Cross-section of immature fruit of *C. mangshanensis* with abundant and large monoembryonic seeds; much of the space within the segment walls is occupied by seeds. Also included is a typical leaf with narrowly winged petiole, showing typical size and venation. (C) Cross-section of immature fruit of *C. mangshanensis* with the seeds and some pulp removed, revealing the conspicuous thick and rigid segment walls; vascular tissues in the central core; large green oil glands on the edge of the fruit surface; and the firm, globular stalked juice vesicles, similar to those found in fruit of the Australian native species *C. australasica* (Finger lime).

different epochs in the citrus clade, a situation that recalls a question that Tanaka raised regarding the possibilities of nature to recreate citrus species.

Today, with the increasing availability and quality of the citrus genome sequence assemblies, this question and many others represent new approachable subjects that will enable better understanding and shed a brighter light on the origin, evolution, and diversification of this most unique and important group of plants, the genus *Citrus*. Sequencing technology and associated informatics will continue to improve in the coming years, and the application of the new fundamental information to be generated will support and inform future advances in all the subject areas that follow in this book.

References

Ahmed, D., et al., 2019. Genotyping by sequencing can reveal the complex mosaic genomes in gene pools resulting from reticulate evolution: a case study in diploid and polyploid citrus. Ann. Bot. 123, 1231–1251.

Aleza, P., et al., 2009. Recovery and characterization of a *Citrus clementina* Hort. ex Tan. 'Clemenules' haploid plant selected to establish the reference whole *Citrus* genome sequence. BMC Plant Biol. 9, 110.

Bayer, R.J., Mabberley, D.J., Morton, C., Miller, C.H., Sharma, I.K., Pfeil, B.E., et al., 2009. A molecular phylogeny of the orange subfamily (Rutaceae: Aurantioideae) using nine cpDNA sequences. Am. J. Bot. 96 (3), 668–685.

Fleischmann, R.D., et al., 1995. Whole-genome random sequencing and assembly of Haemophilus influenzae Rd. Science 269, 496–512.

Germana, M.A., et al., 2013. Cytological and molecular characterization of three gametoclones of *Citrus clementina*. BMC Plant Biol. 13, 129.

Gmitter Jr., F.G., Chen, C., Machado, M.A., de Souza, A.A., Ollitrault, P., Froehlicher, Y., Shimizu, T., 2012. *Citrus* genomics. Tree Genet. Genomes 8, 611–626. https://doi.org/10.1007/s11295-012-0499-2.

Goodwin, S., McPherson, J.D., McCombie, W.R., 2016. Coming of age: ten years of next-generation sequencing technologies. Nat. Rev. Genet. 17, 333–351. https://doi.org/10.1038/nrg.2016.49.

Hodgson, R.W., 1967. Horticultural varieties of citrus. In: Reuther, W., Webber, H.J., Batchelor, L.D. (Eds.), The Citrus Industry, pp. 431–588.

Li, R.T., Zhang, Y.N., Chen, M.L., Jiang, K.J., 2009. Taxonomy analysis on wild mandarins originating from Mangshan Mountain. Guangdong Agric Sci. 8, 11–13 (in Chinese).

Liu, G.F., He, S.W., Li, W.B., 1990. Two new species of citrus in China. Acta Bot. Yunn. 12, 287–289.

Nagano, Y., et al., 2018. Phylogenetic relationships of Aurantioideae (Rutaceae) based on RAD-Seq. Tree Genet. Genomes 14, 6. https://doi.org/10.1007/s11295-017-1223-z.

Ollitrault, P., et al., 2012. A reference genetic map of *C. clementina* hort. ex Tan.: citrus evolution inferences from comparative mapping. BMC Genomics 13, 593.

Oueslati, A., Salhi-Hannachi, A., Luro, F., Vignes, H., Mournet, P., Ollitrault, P., 2017. Genotyping by sequencing reveals the interspecific *C. maxima/C. reticulata* admixture along the genomes of modern citrus varieties of mandarins, tangors, tangelos, orangelos and grapefruits. PLoS One 12 (10), e0185618.

Padmanabhan, R., Jay, E., Wu, R., 1974. Chemical synthesis of a primer and its use in the sequence analysis of the lysozyme gene of bacteriophage T4. Proc. Natl. Acad. Sci. U. S. A. 71, 2510–2514. https://doi.org/10.1073/pnas.71.6.2510.

Sanger, F., Nicklen, S., Coulson, A.R., 1977. DNA sequencing with chain-terminating inhibitors. Proc. Natl. Acad. Sci. U. S. A. 74 (12), 5463–5477. Bibcode:1977PNAS...74.5463S. https://doi.org/10.1073/pnas.74.12.5463.

Schwartz, T., Nylinder, S., Ramadugu, C., Antonelli, A., Pfeil, B.E., 2016. The origin of oranges: a multi-locus phylogeny of Rutaceae subfamily Aurantioideae. Syst. Bot. 40 (4), 1053–1062.

Shimizu, T., Tanizawa, Y., Mochizuki, T., Nagasaki, H., et al., 2017. Draft sequencing of the heterozygous diploid genome of Satsuma (*Citrus unshiu* Marc.) using a hybrid assembly approach. Front. Genet. https://doi.org/10.3389/fgene.2017.00180.

Talon, M., Gmitter Jr., F.G., 2008. *Citrus* genomics. Int. J. Plant Genomics. https://doi.org/10.1155/2008/528361.

Venter, J.C., Adams, M.D., et al., 2001. The sequence of the human genome. Science 291, 1304–1351. https://doi.org/10.1126/science.1058040.

Wang, X., Xu, Y., Zhang, S., Cao, L., Huang, Y., Cheng, J., Wu, G., Tian, S., et al., 2017. Genomic analyses of primitive, wild and cultivated citrus provide insights into asexual reproduction. Nat. Genet. 49, 765–772.

Wang, L., et al., 2018. Genome of Wild Mandarin and Domestication History of Mandarin 5-772. Mol. Plant 11, 1024–1037.

Wu, R., 1972. Nucleotide sequence analysis of DNA. Nat. New Biol. 236, 198–200. https://doi.org/10.1038/newbio236198a0.

Wu, G.A., Prochnik, S., Jenkins, J., Salse, J., Hellsten, U., Murat, F., et al., 2014. Sequencing of diverse mandarin, pummelo and orange genomes reveals complex history of admixture during citrus domestication. Nat. Biotechnol. 32 (7), 656–662.

Wu, G.A., Terol, J., Ibanez, V., Lopez-Garcia, A., Estela, P.-R., Carles, B., et al., 2018. Genomics of the origin, evolution and domestication of citrus. Nature 544, 311–316. https://doi.org/10.1038/nature25447.

Xu, Q., et al., 2013. The draft genome of sweet orange (*Citrus sinensis*). Nat. Genet. 45, 59–66.

Zhu, C., Zheng, X., Huang, Y., Ye, J., Chen, P., Zhang, C., Zhao, F., Xie, Z., Zhang, S., Wang, N., Li, H., Wang, L., Tang, X., Chai, L., Xu, Q., Deng, X., 2019. Genome sequencing and CRISPR/Cas9 gene editing of an early flowering mini-citrus (*Fortunella hindsii*). Plant Biotechnol. J. https://doi.org/10.1111/pbi.13132.

Further reading

Lander, E.S., Linton, L.M., Birren, B., Nusbaum, C., Zody, M.C., et al., 2001. Initial sequencing and analysis of the human genome. Nature 409, 860–921. https://doi.org/10.1038/35057062.

Chapter 2

The origin of citrus

Manuel Talon[a], Guohong Albert Wu[b], Frederick G. Gmitter, Jr.[c], Daniel S. Rokhsar[b,d,e]

[a]*Centro de Genómica, Instituto Valenciano de Investigaciones Agrarias (IVIA), Valencia, Spain,* [b]*US Department of Energy Joint Genome Institute, Walnut Creek, CA, United States,* [c]*Citrus Research and Education Center (CREC), Institute of Food and Agricultural Sciences (IFAS), University of Florida, Lake Alfred, FL, United States,* [d]*Department of Molecular and Cell Biology and Center for Integrative Genomics, University of California, Berkeley, Berkeley, CA, United States,* [e]*Molecular Genetics Unit, Okinawa Institute of Science and Technology Graduate University, Onna, Japan*

2.1 The mythological origin of citrus

Citrus are part of the Rutaceae, one of the 21 families that form the order Geraniales. This family is composed of about 160 genera including more than 1600 species divided in 7 subfamilies and 12 tribes, comprising essentially trees or bushes adapted to tropical and subtropical regions. One of the subfamilies of the Rutaceae, the Aurantioideae, groups a number of genera including the genus *Citrus*, exhibiting a form of fructification that gives rise to a peculiar berry known as hesperidium. These fruits originate from single ovaries that present a cavity divided into loculi (carpels) corresponding to the segments that enclose the juice vesicles. Citrus are fruit crops very much appreciated worldwide although their origins have been unknown until very recently and, therefore, their birthplace has been a matter of controversy for the past centuries. The oldest references to the origin of citrus that we are aware of come from the world of fables and are associated with the golden apples of the garden of the Hesperides. Legend has it that Heracles had to perform 12 titanic works as a penitence imposed by Eurystheus. Among these labors, the mythological Greek hero had to steal the golden apples that granted immortality from the garden of the Hesperides, the three daughters of Hesperis. For several centuries, many early authors have believed that they recognized the garden of the famous women to different places, such as Syria or the coasts of Libya, although the most important authorities agreed to locate it in the western part of the known world at that time, that is, in the west of Mount Atlas (Risso and Poiteau, 1818). Actually, the Hesperides in ancient Greek means "the West," in addition to "daughter of Hesperis." This legend began to spread in the eighth-century BC, long before the arrival of the first citrus, the citrons, to the West, which occurred approximately in the fourth-century BC. Since the first bibliographic references relating the yellow fruit of citrons with the golden apples did not occur until the third-century AD, it is clear that citrus could not have been the original golden fruits growing in the garden of the Hesperides. Nevertheless, after the publication in 1646 of the Ferrari masterpiece on citrus culture, "*Hesperides sive de Malorum Aureorum Cultura et Usu*" (Ferrari, 1646), a long-standing association between the legendary fruits of the garden of the Hesperides and citrus fruits was established. This idea rooted so much in other later authors of great renown and citrus erudition as Volkamer (1708), for example, that the name of Hesperides was definitively linked to citrus. Such an association is already a part of our culture and tradition and its traces are perceived in several terms of our current botanical definitions that refer to this mythological origin, such as hesperidium, *Citrus aurantium*, Aurantioideae, etc. However, the legend locates the origin of citrus in the West, while the real origin of citrus must be situated in the East, as shown below.

2.2 The origin of citrus

2.2.1 The concept of citrus

Many decades of dedicated research and developments around the world have produced a rich legacy of knowledge addressing fundamental issues supporting the concept of citrus. The current notion of the genus *Citrus*, however, includes an elusive number of species because its boundaries and composition have been subjects of controversy during the past centuries and are still not well defined. Hundreds or even thousands of years of citrus cultivation have generated myriads of cultigens, plants that have been artificially crossed or selected by humans and, therefore, show complex mixtures of

The Genus Citrus. https://doi.org/10.1016/B978-0-12-812163-4.00002-4

botanical characters that are very difficult to classify. Furthermore, the taxonomic delimitation of species in *Citrus* is also greatly complicated by the high degree of polyembryony, that is, the formation of nucellar embryos as exhibited by both intergeneric and intrageneric citrus hybrids and that basically reproduce the phenotype of the mother variety. In fact, the two basic current taxonomic citrus systems, proposed by Tanaka (1954) and Swingle and Reece (1967), are so different that they are considered to represent two extremes or visions of the same genus. While the first author, for instance, identifies 36 different species of mandarins the last one only enumerates 3. Between these two extremes, other propositions have been added with particular nomenclatures and classifications. During the past decades, this lack of definition or vagueness has hampered many scientific advances and developments that are essential to delimit and circumscribe the own concept of species and genus. Under these circumstances, the elucidation of the place of birth of citrus can only be properly answered by appealing to solid fundaments supporting the concept of citrus. In this chapter, we will revise our knowledge regarding the phylogeny, chronology, and biogeography of citrus to locate its center of origin and speculate then on the evolution and natural dispersal of the genus *Citrus*. The availability of the citrus genomes (Xu et al., 2013; Wu et al., 2014; Wang et al., 2017, 2018) and in particular for our purposes, the comparative genomic analysis presented in Wu et al. (2018) are of special relevance because they offer original information based on the new approaches opening unprecedented hypothesis, challenges, and suggestions. The identification of a set of "pure" true citrus species, principally, has provided solid insights into the phylogeny of citrus, and their origins, evolution, and dispersal. Taken together, the information presented in this and in Chapter 1 directly points to a reformulation of the definition of the genus *Citrus*.

2.2.2 Phylogeny of citrus pure species

Citrus phylogeny has been particularly controversial due to the difficulty in identifying pure wild progenitor species in the face of a substantial interspecific hybridization that has resulted in several clonally propagated and cultivated species or varieties. Traditional studies on *Citrus* phylogenetic relationships have been mostly based on the investigations with a few chloroplastic or nuclear gene sequences. According to these approaches, there is an agreement in the monophyletic origin of all citrus despite the topologies reported vary considerably. Many studies of this type reported a superior clade including at least five genera: *Citrus*, *Microcitrus*, *Eremocitrus*, *Fortunella*, and *Poncirus*. On the other hand, the genus *Citrus* has usually been divided into two subgenera: *Citrus* and *Papeda* (Swingle and Reece, 1967). The subgenus *Citrus* also has invariably been split into three major phylogenetic groups coincident with the traditional classification of the three main true citrus species (Scora, 1975; Barrett and Rhodes, 1976): citrons, pummelos, and mandarins. However, even the grouping of these three clusters is rather polemic because the different studies used different DNA sequences (see Nicolosi et al., 2000; Pfeil and Crisp, 2008; Bayer et al., 2009; Schwartz et al., 2015; Carbonell-Caballero et al., 2015 for a detailed account). This last report that was based on the complete chloroplast sequence concluded that the genus *Citrus* was composed of three main clades including the citrons/Australian species, the pummelos/*Micranthas*, and the mandarins/*ichangensis*. Other citrus-related genera found as wild species in New Guinea and New Caledonia (*Clymenia* and *Oxanthera*) nested also in the first clade together with *Citrus medica*, traditionally believed to be native in India, and the Australian limes (Pfeil and Crisp, 2008; Bayer et al., 2009; Schwartz et al., 2015). According to these data, citrus ancestors of current species were probably generated in a succession of speciation events that separated citrons from *Clymenia*, *Oxanthera* and the Australian species, *Micrantha* from pummelos, and *C. ichangensis* from mandarins.

The most comprehensive phylogenetic study in citrus was presented by Wu et al. (2018), who reported a nuclear genome-based phylogeny of the pure citrus species, derived from more than 300,000 SNPs located in intergenic and nonpericentromeric genomic regions. These authors resequenced the genome of representative citrus genotypes and applied combining diversity analysis, multidimensional scaling, and chloroplast genome phylogeny to discriminate citrus pure species from hybrids and admixtures. The results identified 10 ancestral citrus species including 7 Asian species, *C. medica* (citrons), *C. maxima* (pummelos), *C. reticulata* (mandarins), *C. micrantha*, *C. ichangensis*, *Fortunella margarita* (Nagami kumquat), and *C. mangshanensis* and 3 Australian accessions, *Eremocitrus glauca* (Australian desert lime), *Microcitrus australis* (Australian round lime), and *Microcitrus australasica* (Australian finger lime). On the other hand, Tachibana has been traditionally assigned its own species in most classification systems, *C. tachibana* (Swingle and Reece, 1967; Tanaka, 1954). However, sequence analysis reveals its close affinity to *C. reticulata* and, therefore, does not support its taxonomic position as a separate species, although both chloroplast and nuclear genome phylogenies clearly distinguish Tachibana from the mainland Asian mandarins. This suggests that Tachibana should be designated a subspecies of *C. reticulata* (see Chapter 3 for a deep description of this citrus). By contrast, the wild Mangshan "mandarin" (*C. mangshanensis*) clearly represents a distinct species, with comparable distances to *C. reticulata* or other true species (Wu et al., 2018; Wang et al., 2018). This listing of true citrus species cannot be considered a definitive list of pure species since there are still many genomes that need to be sequenced in order to definitively clarify their genetic status.

The phylogenomic analyses performed on these sets of pure citrus species revealed that citrus form a monophyletic group and that all these 10 species should be considered members of the genus *Citrus* although for clarity we will keep the traditional names of *Fortunella*, *Eremocitrus*, and *Microcitrus* in the chapter. In analyzing this set of data, two important points are worth mentioning. The first one is that although the topology of the phylogenetic tree presented is the one that had higher probability, the sudden split of the first branches of the tree affecting mostly *C. mangshanensis* and *C. ichangensis* still might allow alternative topologies. The reliability of the tree, however, appears to be optimal for the two main clades, one of which establishes phylogenetic relationships between mandarins, kumquats, and Australian limes, while the other clade groups together *Micrantha*, pummelos, and citrons. The second observation is related to the differences observed between the nuclear genome-derived phylogeny and the phylogeny based on the chloroplast genome that show several discrepancies. This is not unexpected, as the cpDNA is a single, nonrecombining unit, and is unlikely to show perfect lineage sorting during a rapid radiation. One of the most important differences is related to the location of *Poncirus*. In the nuclear genome tree, *Poncirus* is as an out-group of citrus, while in the chloroplast phylogeny it is nested inside a major superclade that also contains mandarins, pummelos, and *Micrantha*. This fact is compatible with the view that *Poncirus* is likely an admixture descendent of an ancient hybrid of citrus with an unknown parent (Carbonell-Caballero et al., 2015), although the low heterozygosity level of this genotype may imply a more complex alternative mechanism such as chloroplast capture, for instance. Another inconsistency between both trees is associated with the position of citrons and pummelos. In the chloroplast tree, citrons cluster with Australian species while in the nuclear phylogeny, citrons form a clade with pummelos while Australian limes cluster with *Fortunella*. The link between citrons and pummelos is more credible than that with the Australian limes since under a morphological point of view, citrons and pummelos share several significant characteristics that are usual in the subgenus *Citrus,* the true citrus. These two species are characterized by a complex floral vascular anatomy with large flowers that carry joined stamens and ovaries with many loculi. The flowers of both species develop large fruits with yellow or pale yellow peels, thick rinds, a higher number of segments, and larger columellas (Swingle and Reece, 1967). Australian limes in no way resemble the citrons, since they first develop small leaves, fruits, and flowers that carry free stamens and fewer loculi. Australian limes, on the other hand, show characteristic patterns of dimorphic foliage, with coriaceous strongly veined leaves that are quite unique among another citrus. The geographical distribution of wild genotypes also favors a citron-pummelo relationship since both share overlapping regions in mainland South East Asia, and observation that might be related to the occurrence of a common ancestor. The association of citrons with Australian limes, in contrast, requires a convincing hypothesis linking the area of origin of citrons with that of the Australian limes (see below). Among the pure species of citrus, the fruits of *Fortunella* and Australian desert lime that share the same clade are the only ones that contain a low number of fruit loculi and ovules, one more independent observation reinforcing the reliability of the nuclear phylogenetic tree.

2.2.3 Genealogy of cultivated citrus

Chapter 4 presents a brief historical recapitulation of the efforts dedicated to the elucidation of the genealogy and parentage of the cultivated citrus during several decades of the past century (see references therein). Investigations and studies in these matters were initially focused on biochemical markers that were soon substituted by nuclear and organelle DNA markers such as RFLPs and SSRs. As mentioned above, most of these studies recognized three ancestral species, *C. medica* (citrons), *C. maxima* (pummelos), and *C. reticulata* (mandarins). With these premises, an important breakthrough in this direction was provided by Nicolosi et al. (2000), who correctly predicted the parentage of a number of commercial citrus types. These authors concluded that sweet and sour oranges and grapefruits had the same origin, while lemon, Palestine sweet lime, bergamot, and Volkamer lemon were hybrids of sour orange and citron. Rough lemons and Rangpur limes, however, were crosses between mandarin and citron, while Mexican lime was a hybrid of *C. micrantha* and citron.

The identification of citrus species based on genome analyses as defined in the previous section, provided hundreds of thousands of diagnostic SNPs of pure genomes opening in principle the possibility of determining the genetic constitution of any citrus genotype. These valuable informative markers allowed genome-wide admixture analysis of a number of citrus types discriminating very precisely among citrus hybrids and admixtures. Pioneer observations presented in Wu et al. (2014) revealed the occurrence of pummelo introgressions among the mandarins analyzed, indicating that mandarins were actually mandarin-pummelo admixtures. Interestingly, oranges and grapefruits that contain pummelo chloroplasts also derived their genetic ancestry from the same species, *C. maxima* and *C. reticulata*, although the forms or varieties involved in both crosses were not the same. In general, sour oranges are just hybrids, while the origin of sweet orange certainly involves more complex backcrosses. Besides, grapefruits arose from a cross between pummelo and sweet orange validating reports that located the grapefruit origin in the Barbados Island during the mid-18th century.

It was confirmed later that the pattern of pummelo introgression into a mandarin genetic background was a general rule for the type of citrus popularly called "mandarins," since the bulk of mandarins studied, except for a few of them that can be considered pure *C. reticulata* varieties, contained a pummelo admixture as determined by nuclear SNP markers (Curk et al., 2015), genotyping by sequencing (Oueslati et al., 2017), and through whole genome analyses (Wu et al., 2018; Wang et al., 2018). The analyses of the set of mandarin genomes reported in Wang et al. (2017, 2018) and Wu et al. (2018) suggest that there are several mandarins that, in addition to Sun Chu Cha Kat and Tachibana, contained pure *C. reticulata* genomes and, therefore, represent distinct varieties of this species.

In addition to the different forms of *C. reticulata*, which can be considered an ancestral taxa by itself, the extension of the pummelo introgression into the mandarin genome differentiated two groups of mandarins, one of them containing small amounts of pummelo admixture (1%–10%) and only two pummelo haplotypes, and the other showing significantly higher proportions of pummelo alleles (12%–38%). The first group includes what appear to be older or more traditional mandarins such as Cleopatra, Sunki, Huanglingmiao, Changsha, Dancy, Willowleaf, and Ponkan, while more modern mandarins, for example, W. Murcott, King, Wilking, Fallglo, Kiyomi, Satsuma, and Clementine are ascribed to a second group. A number of additional unnamed Chinese mandarins reported in Wang et al. (2017) also contained pummelo introgression. The admixture pattern of the traditional mandarins is compatible with a unique ancient introgression event, possibly involving a single pummelo ancestor, whose haplotypes were broken into shorter segments after repeated backcrossing with mandarins. This suggestion explains why the initial pummelo introgression is currently widespread among the mandarins, including the younger mandarins of the second group that in addition contain other pummelo introgressions. Estimations based on the smallest sizes of the pummelo segments currently found in these older mandarins indicate that such an admixture pattern can be only generated after five or six generations of backcrosses, suggesting that fewer sexual generations were probably involved in other mandarins with longer segments. Wu et al. (2018) also reported an extensive network of relatedness among mandarins and sweet oranges, an association that appears to form part of the basic axioms of the domestication of the edible citrus. In this admixture-based classification framework, breeding between sweet orange and mandarins or admixture mandarins would produce additional admixture mandarins. Therefore, admixture mandarins, oranges, grapefruits, and all hybrids derived from them such as those known as tangors (mandarin × sweet orange), tangelos (mandarin × grapefruit), and orangelos (sweet orange × grapefruit) are basically admixtures (Oueslati et al., 2017) originated from initial interspecific crosses involving *C. reticulata* and *C. maxima*. However, it should be noted that genetically they can readily be separated into two groups according to the mandarin or pummelo nature of the organelle genomes.

Similarly, whole genome analyses and the maternal parentage provided by the chloroplast genome (Carbonell-Caballero et al., 2015) rendered the genealogy of other relevant cultivated citrus confirming previous results (see Chapter 4). Thus, Rangpur lime and red rough lemon are both hybrids (*C. reticulata* × *C. medica*) of citron, a citrus species that always acts as the parental father because of its cleistogamous flowers. The genome of lemon contains the three species *C. medica* (50%), *C. maxima* (18%), and *C. reticulata* (31%) and exhibits a pummelo chloroplast, indicating that lemons very likely originated from a cross between sour orange and citron. Finally, analyses of nuclear and chloroplast genomes revealed that the Mexican lime is a hybrid originated from a *C. micrantha* × *C. medica* cross, while Calamondin is the result of a hybridization of *Fortunella* with *C. reticulata* pollen (Fig. 2.1). The above conclusions are essentially in line with those provided in Curk et al. (2016) and Penjor et al. (2016) that, based on nuclear and cytoplasmic markers and RAD-Seq respectively, presented comprehensive analyses of the parentage of the lemon and lime horticultural groups. These works further established that *C. medica* was the direct male parent of the main subgroups of acidic citrus. Thus, a citron cross with *C. micrantha* or a close papeda species originated the hybrids named *C. aurata*, *C. excelsa*, and *C. macrophylla*. On the other hand, crosses between *C. reticulata* × *C. medica* also generated Volkamer lemon. Curk et al. (2016) also suggested that among triploid limes, *C. latifolia* resulted from the fertilization of a haploid ovule of *C. limon* by a diploid gamete of *C. aurantifolia*. Finally, *C. amblycarpa*, some lumias (*C. lumia*), and the bergamot apparently resulted from hybridizations between *C. micrantha* × *C. reticulata*, *C. maxima* × *C. medica*, and *C. limon* × *C. aurantium*, respectively.

In a comprehensive work studying the relevant Japanese indigenous citrus, Shimizu et al. (2016) identified 12 varieties consisting of Kishu, Kunenbo, Yuzu, Koji, sour orange, Dancy, Kobeni Mikan, sweet orange, Tachibana, Cleopatra, Willowleaf, and pummelo that played pivotal roles in the generation of these Japanese varieties. According to this work, the inferred parentage confirmed the hybrid origins of the indigenous varieties.

It is worth mentioning that most hybrids and admixtures of these commercial/cultivated citrus exemplified in this section exhibit phenotypes frozen in time that are perpetuated through clonal propagation by grafting or apomixis via nucellar polyembryony.

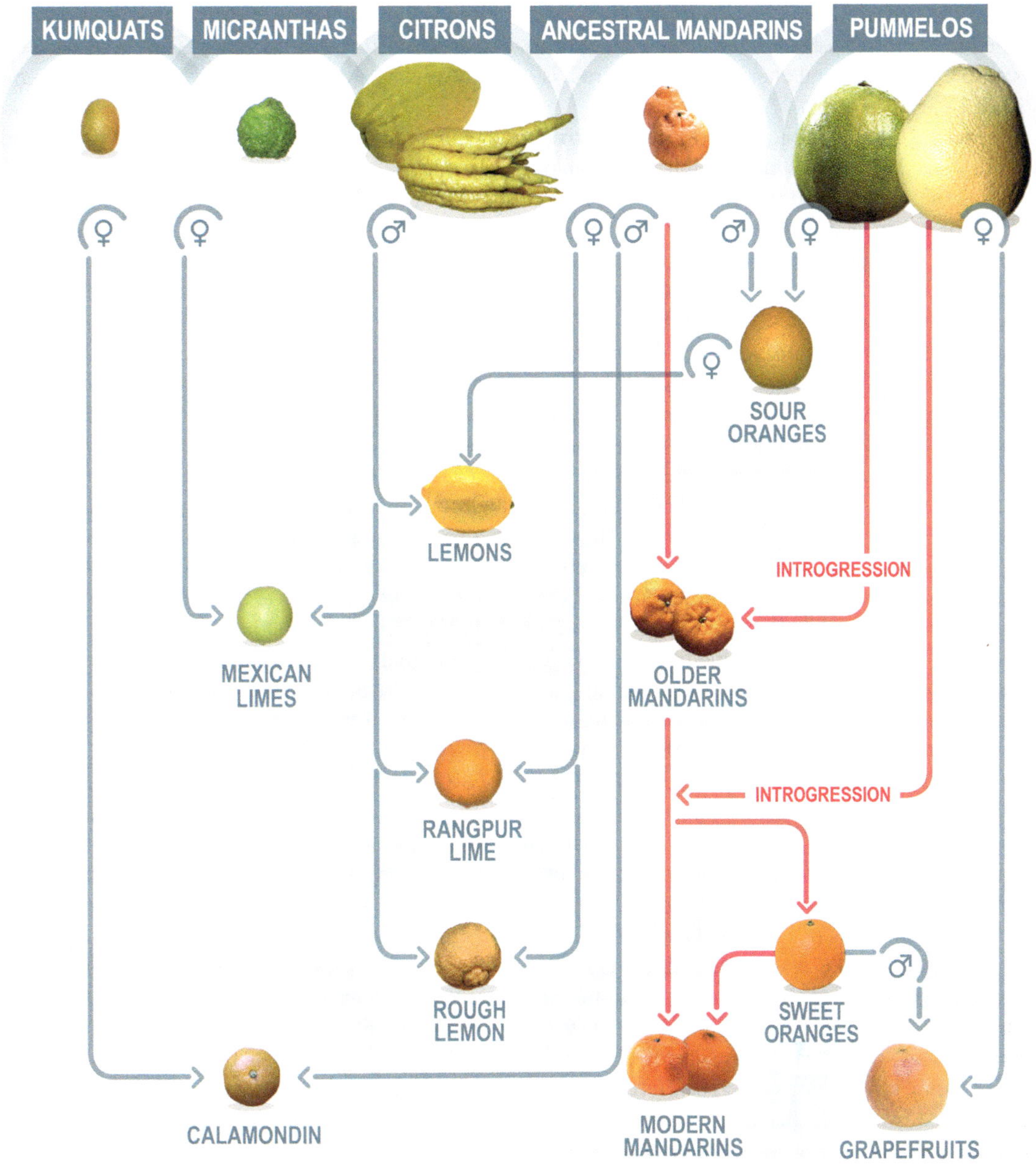

FIG. 2.1 Genealogy of cultivated citrus. Genealogy of major citrus types according to their genomic structures. Five progenitor pure species are shown at the *top*. *Blue lines* indicate simple crosses generating hybrids between two parental genotypes. *Red lines* represent more complex processes involving multiple individuals, generations, and/or backcrosses. The extension of the pummelo introgression differentiates two groups of mandarins. One of them contains small amounts of pummelo admixture and only two pummelo haplotypes and includes what appear to be older or more traditional mandarins. The other group shows higher proportions of additional pummelo alleles and is represented by more modern mandarins. Fruit images are not to an exact scale. *Adapted from Wu, G.A., Terol, J., Ibanez, V., Lopez-Garcia, A., Estela, P.-R., Carles, B., et al., 2018. Genomics of the origin and evolution of Citrus. Nature 544, 311–316. https://doi.org/10.1038/nature25447.*

2.2.4 Paleontology of citrus

Among the few number of specimens comprising the paleontological record of *Citrus* only two accounts categorically reported fossil species resembling extant citrus. These specimens that correspond to fossilized leaves are *Citrus meletensis* from the Pliocene (1.8–5.3 Ma) of Valdarno, Italy (Fischer and Butzmann, 1998) and *Citrus linczangensis* from the Late Miocene (5.3–11.6 Mya) of Lincang, Yunnan (China) (Xie et al., 2013). Other studies initially reporting older citrus-like leaf remains were later cataloged as doubtful since their phylogenic relationship with extant *Citrus* was weakly supported.

Leaves of *C. meletensis* and *C. linczangensis* share at least three significant characteristics such as intramarginal veins well developed in the lower part of the blade, entire margins, and articulated winged petioles. The occurrence of intramarginal veins, rather than a general characteristic of citrus, is occasionally seen in only some citrus such as *C. reticulata* and *C. aurantifolia* (Xie et al., 2013).

Margins of leaves from the species of the genus *Citrus* are mostly crenate or bluntly toothed and only sporadically are entire. Typically, the same species may exhibit more than one type of leaf margin and among the few species that may show entire margins are *C. maxima* and its derivatives (grapefruits, sour oranges and a few sweet oranges, tangors, and mandarin admixtures) and some species of *Fortunella* and some of its hybrids, for instance, Calamondin.

The articulation of the petiole and the occurrence of wings are present in a number of different genera of the Aurantioideae and even in other subfamilies of the Rutaceae (Swingle and Reece, 1967; Fischer and Butzmann, 1998). While the articulation of the petiole takes place in genera distantly (e.g., *Galipea, Burkillanthus, Paramignya, Glycosmis, Merrillia, Murraya*, etc.), or more closely related to citrus (*Pleiospermium, Citropsis*, and *Atalantia*), the presence of a perceptible wing is more exclusive of citrus and other related species from the genera *Citropsis* and *Pleiospermium.* The genus *Citropsis* is native to tropical and subtropical Africa whereas the species of *Pleiospermium* are found in the islands of the Malay Archipelago, in Sri Lanka and in India. The similarity among the leaves of these three genera was evidenced by Swingle (Swingle and Reece, 1967) who reported that the "large unifoliolate leaves of *Citropsis gabunensis* (Gabon cherry-orange) look astonishingly like those of *Citrus* in almost every character" … including narrow … "winged petioles very like those of some forms of *Citrus aurantium* and even *C. maxima*." These leaves also closely resemble those of the unifoliolate species of *Pleiospermium*, which are similar to citrus leaves, as observed, for example, in *P. alatum* that exhibits a petiole narrowly winged. Since the species of *Citropsis* graft readily on *Citrus*, and vice versa, these authors suggested that *Citropsis* and *Pleiospermium* are the two genera that most nearly represent today the remote ancestors of citrus. Current molecular determinations based on chloroplast genome fragments suggest, indeed, that these two genera are included in a cluster that shares a common ancestor with the cluster containing citrus (Pfeil and Crisp, 2008; Bayer et al., 2009). According to the divergence estimations (7.0–20.7 Ma) presented by these authors, the age of the *C. meletensis* and *linczangensis* fossils is in principle compatible with their belonging to any of the three genera in dispute, *Citrus, Citropsis*, and *Pleiospermium.*

The two fossil specimens show distinct leaf dimensions since *C. meletensis* exhibits a much smaller leaf size and a higher length-to-width ratio than *C. linczangensis*. The size and shape of the petiole wings that are considerably wider and bigger in *C. linczangensis* is, however, the principal distinctive characteristic between both species. The leaves of *C. linczangensis* carry a prominent articulated, subcordate, and broadly winged petiole that although of different morphology is to a great extent very similar to those of *C. maxima,* a species that may show narrower, medium, or wide petioles. However, the relation between the sizes of wings and leaf blades is clearly higher in the fossil specimen, as observed in *C. junos*. On the contrary, *C. ichangensis, C. micrantha,* and several citrus included in the papeda group, such as *C. macroptera* and *C. hystrix,* show wide petioles and a much higher wing-to-blade ratio than *C. linczangensis.*

Furthermore, *C. meletensis* exhibited much narrower petiole wings, similar to those typical of *C. aurantium* (*C. maxima* × *C. reticulata*). Other extant citrus such as *C. macrophylla* (*C. micrantha* × *C. medica*), *C. sudachi*, and *Fortunella* also may develop narrow petiole wings. Therefore, the shape and size of the petiole can be considered a diagnostic feature discriminating these two fossil species. According to the longer and wider wings, *C. linczangensis* should be considered a member of the genus *Citrus,* while *C. meletensis* with narrower wings might also belong to this genus or alternatively to *Citropsis* or *Pleiospermium* since these two genera generally also develop narrow wings. However, the most intriguing remark of *C. meletensis* is unquestionably related to its localization in the Pliocene of Italy, a region far removed from the native geographic distribution of these three genera. While this finding is still not properly contextualized, the identification of *C. linczangensis* from the Late Miocene of Lincang (Xie et al., 2013) provides definite evidence of the existence of *Citrus* within the province of Yunnan (China) since ca. 8.0 Ma.

The two leaf specimens that define *C. linczangensis* appear to combine characteristic traits of current major citrus groups nested in distinct clades as defined in the citrus phylogeny described in previous sections and, therefore, do not exactly resemble any particular extant citrus species. The original report by Xie et al. (2013) established the presence of intramarginal veins in *C. linczangensis, C. aurantifolia*, and *C. reticulata.* We now know that *C. aurantifolia* is a hybrid between *C. micrantha* and *C. medica*, while most of the *C. reticulata* assignments actually are mandarin admixtures (*C. reticulata* × *C. maxima*). The fossil also has a clear resemblance to *C. maxima* and to other citrus included in the papeda group because of the broadly winged petioles (Fig. 2.2). In addition, it shows an entire margin as seen in both *C. maxima* and *Fortunella,* two groups that phylogenetically are nested in two different clades. The specimen, therefore, might well represent a common ancestor or a form related to an ancestor of the citrus groups. This finding implies that southwestern China (Gmitter and Hu, 1990) in the Late Miocene was a native habitat of citrus and, therefore, a potential region of early diversification.

FIG. 2.2 Leaf characteristics of *C. linczangensis* and extant citrus. *C. linczangensis* (1) appears to combine characteristic traits of current major citrus groups nested in distinct clades and, therefore, do not exactly look like any particular extant citrus species. The fossil has a clear resemblance to *C. maxima* (2) and to other citrus included in the papeda group (3, *C. ichangensis*; 4, *C. hystrix*; 5 *C. macroptera*) because of the broadly winged petioles. In addition, it shows an entire margin as seen in *Fortunella obovata* (6), a species closer to *C. reticulata* (7) than to papedas. Margins of leaves from the species of the genus *Citrus* are mostly crenate or bluntly toothed and only sporadically are entire. The specimen, therefore, might well represent a common ancestor or a form related to an ancestor of the major citrus groups. Leaf drawings are not to an exact scale. *Fossil adapted from Xie, S.P., Manchester, S.R., Liu, K.N., Wang, Y.F., Sun, B.N., 2013. Citrus Linczangensis sp n., a leaf fossil of Rutaceae from the Late Miocene of Yunnan, China. Int. J. Plant Sci. 174, 1201–1207. https://doi.org/10.1086/671796.*

2.2.5 Chronology of citrus speciation

The scant record of citrus fossils has also limited the estimations of citrus divergence times that are, therefore, similarly scarce. Another difficulty hampering the achievement of developments in this area is related to indeterminate definition of pure citrus species as discussed above. Using molecular dating based on partial chloroplast sequences and fossil calibrations, the crown of citrus was initially estimated to be 18–22 Ma (Muellner et al., 2007). In a further detailed contribution, Pfeil and Crisp (2008), who used the same approach to date the age of Rutaceae and Aurantioideae found that citrus radiated at 7.1 Ma (3.7–11.8 Ma) while the estimated average age of the Australasian clade was 5.8 Ma (2.9–9.7 Ma), time intervals consistent with the dating of the most relevant citrus fossils described until today, *C. linczangensis* and *C. meletensis*. Additional molecular phylogenies grounded also on chloroplast sequences also suggested that *Citrus* radiated in the Late Miocene or Pliocene (Schwartz et al., 2015). Recent phylogenomics works based on whole genome sequences have likewise examined the chronology of citrus speciation, providing new information complementing these assessments. Estimation times reported in Pfeil and Crisp (2008), for instance, were later used to calibrate a phylogenetic tree generated with the whole sequence of 34 chloroplast genomes that provided the maternal phylogeny of most important groups of citrus and a number of cultivated varieties, hybrids, and admixtures (Carbonell-Caballero et al., 2015). These authors reported that the estimations of divergence times in relatively recent varieties such as cultivated hybrids and admixtures were clearly overestimated. This increase in variability was obviously due to hybridization although the contribution of bud grafting and the propagation of chimeric (Terol et al., 2015) or heteroplasmic (Carbonell-Caballero et al., 2015) genotypes cannot be ruled out.

The phylogenetic study presented by Wu et al. (2018) reports a nuclear genome-based phylogeny of the pure citrus species calibrated with *C. linczangensis*, the leaf fossil from the Late Miocene of Yunnan. According to the approaches developed in these reports, there is a general agreement that the emergence of the crown of citrus emerged ca. 7.0–8.0 Ma during the Late Miocene, while the Australian species were dated at the transition of this geological epoch to the Early Pliocene.

The analyses of the nuclear genomes established the occurrence of a relatively rapid radiation in the Late Miocene (Wu et al., 2018) that expanded only during 2 million years (6–8 Ma) but gave rise to the most important species of extant citrus such as *Citrus mangshanensis* (8 Ma), *C. ichangensis* (7.5 Ma), mandarins (*C. reticulata*, 6.9 Ma), micranthas (*C. micrantha*, 6.6 Ma), kumquats (*F. margarita*, 6.5 Ma), and citrons and pummelos (*C. medica* and *C. maxima*, 6.2 Ma). The ancestors of the Australian species experienced a second radiation during the Early Pliocene (4.0–4.6 Ma) that separated *E. glauca*, the unique species of this genus from the species of the genus *Microcitrus*. The results also showed that the ancestor of the Tachibana mandarin, naturally found in Taiwan, the Ryukyu Archipelago, and Japan (Tanaka, 1931) split from mainland Asian mandarins during the Early Pleistocene (2.4 Ma).

2.2.6 Biogeography of citrus

The genus *Citrus* and related genera (*Fortunella, Eremocitrus, Microcitrus*, and *Poncirus*) belong to the Aurantioideae, a subfamily of Rutaceae found exclusively in the Old World that generally occurs in humid evergreen forests from the lowland to the montane belt. This subfamily is widely distributed in the Monsoon region from West Pakistan to north-central China and thence south through Malay Archipelago to New Guinea and Bismarck Archipelago, northeastern Australia, New Caledonia, Melanesia, and the western Polynesian islands. Of the 203 species that comprise the 33 genera of this subfamily, most of them are mainly Indomalayan, only 21 are native to Africa and 15 are found in Australia and New Caledonia. Thus, the center of diversity of Aurantioideae extends from tropical Africa through the monsoon region of Southeast Asia and eastern Australasia to Polynesia. The natural habitats of citrus and related genera, although they do not include Africa, roughly extend through this broad area, where wild genotypes of the several species of the genus *Citrus* have been reported to grow freely in this wide region. The following description mostly relies on the compilation texts of Tanaka (1954) and Swingle and Reece (1967) who illustrate the pattern of the geographical distribution of citrus, providing essential information on their birthplace and dispersal, although the wild nature of the genotypes reported cannot be always warranted.

Citrons (*C. medica*) have been reported to be native in an extensive area comprising India, Bhutan, Bangladesh, Myanmar, and China (Swingle and Reece, 1967; Gmitter and Hu, 1990; De Candolle, 1883; Rajput and Hari Babu, 1985; Zhang, 1981; He et al., 1984; Zhou, 1991). In India, wild citron plants have been found in the northeast in Assam as well as in the warm areas of the foothills of the Himalayas, between Garhwal and Sikkim, in Punjab and Uttar Pradesh in northern India, in the Satpura mountains in the central part of the country and in southern India, in the Nilgiri Hills in the Western Ghats. In virgin areas of the south of the Himalayas, wild plants of *C. medica* are apparently in excellent vegetative conditions (Chen, 1997). In the east of Bhutan, in the Daphla Hills in the state of Arunachal Pradesh in India, *C. nana,* a primitive form of citron grows freely (Tanaka, 1961; Scora, 1988). In China, wild and semi-wild varieties of citron such as *C. medica* var. *sarcodactylis* and var. *yunnanensis* have been found in Yunnan while *C. medica* var. *muliensis* has been reported in Sichuan. Citrons also grow in some places in the mountain ranges of central China mostly in Hubei, Shaanxi, and Henan. In South China, there are two clearly diversified regions for citrons (Yang et al., 2015). In spite of the extensive area that citrons occupy today, it is widely accepted that these citrus are native of the foothills of the Himalayas in India and perhaps Myanmar.

Pummelos (*C. maxima*), which are widely distributed in an extensive area, are found wild in Malaysia, the peninsula of Indochina and nearby islands as well as in Yunnan, Hainan, and the south of China (Gmitter and Hu, 1990; De Candolle, 1883; Zhou, 1991). Its occurrence in the province of Assam in northeast India has been also described (cited in Swingle and Reece, 1967; Tanaka, 1961), although the wild nature of the specimens found there remains doubtful. In China, two distinct gene pools from Southeast and Southwest China have been recently described (Yu et al., 2017).

Mandarins, in its popular meaning, includes a varied set of forms and genotypes with strong morphological resemblance between them that groups both pure *C. reticulata* as well as numerous admixtures of this species with *C. maxima* (see below), and even other distinct species exhibiting related phenotypes such as *C. mangshanensis, C. daoxianensis*, or *C. indica*. Mandarins are extensively distributed in South China, where it is possible to find some wild forms growing in ancient forests. Several genotypes of "pure" *C. reticulata* as well as the wild mandarins Mangshan (*C. mangshanensis*) and Daoxian (*C. daoxianensis*) have been discovered in the mountain forests of Mangshan and in the Nanling Mountains of the southern Hunan province in China (He et al., 1984, 1988; Liu et al., 1990; Wang et al., 2018). On the other hand, it is generally accepted that *C. indica* (Indian wild orange) is a distinct very primitive mandarin (Swingle and Reece, 1967; Tanaka, 1961). *C. indica* was observed growing in a wild state in the province of Assam, in Khasi Hills (Tanaka, 1954), in Naga Hills' primeval forests and in dense undisturbed forests in the Garo Hills (Singh, 1981). The extent of variability encountered in these forests in the wilderness indicates that this area is the native place of the Indian wild orange (see Chapter 3 for further information on *C. indica*).

In connection with the citrus-related genera (*Fortunella, Eremocitrus*, and *Microcitrus*), the genus *Fortunella* is composed of four species known as kumquats (*F. margarita, F. japonica, F. polyandra*, and *F. hindsii*), natives of regions located in the southeast of China including the provinces of Sichuan, Yunnan, Guizhou, Hunan, Jiangxi, Fujian, and Zhejiang (Zhang, 1981; He et al., 1984; Zhou, 1991). *F. hindsii* has also been reported in the mountain ranges of central China that traverse the provinces of Hubei, Shaanxi, and Henan. *F. polyandra* is distributed in Thailand, continental Malaysia, and in Hainan (Tanaka, 1954). References to their fruits are also found in very ancient Chinese writings (cited in Cooper, 1990). The Calamondin (*Fortunella* × *C. reticulata*) is widely distributed in the Philippines and occurs wild as well as cultivated (Wester, 1915).

Australian limes have been traditionally assigned the distinct generic names of *Eremocitrus* and *Microcitrus*, and have been classified as citrus-related genera, in the same way as *Fortunella* (Swingle and Reece, 1967; Tanaka, 1954). Australian citrus are diverse and found native in both dry and rainforest environments in northeast Australia, depending on species. The single *Eremocitrus* species, *E. glauca*, known as the Australian desert lime, is the only known xerophytic citrus. Plants of this species are grown in dry and solid soils in northeastern Australia (Sykes, 1997). Out of the six species of the genus *Microcitrus*, *M. warbugiana* is the only one originated in southeastern New Guinea (Tanaka, 1954) while the other ones, *M. australasica* (finger lime), *M. australis* (round lime), *M. garrowayi*, *M. inodora*, and *M. maideniana* are thought to be native to Queensland and northern New South Wales in eastern Australia (Swingle and Reece, 1967; Sykes, 1997).

On the other hand, wild hybrids and admixtures between the species of the genus *Citrus* have also been reported to grow in general in areas and habitats occupied or shared by the parental genotypes. Thus, natural colonies of *C.* × *limonia* (Rangpur lime; *C. reticulata* × *C. medica*) have been found east of the mountain range that crosses the Chinese province of Guangxi but mostly in the South of the Tibet (Gmitter and Hu, 1990; Zhang, 1981; He et al., 1984; Chen, 1997). Its presence in Karen Hills, in Eastern Myanmar, Sikkim, Khasi Hills, Thailand, Laos, and Annam (Vietnam) has been also cited in Tanaka (1954). Similarly, *C.* × *jambhiri* (Rough lemon; a different *C. reticulata* × *C. medica* hybrid) has been reported to grow in the semi-wild state in the northeastern Himalayan region (Singh, 1981) and Bonavia (1888) indicated that at the end of the 19th century it was not known outside India.

The sour orange tree [*C.* × *aurantium*, (*C. maxima* × *C. reticulata*)] has been reported to grow spontaneously in the warm zones of the northern slopes of the Himalayas between Nepal, Garhwal, and Sikkim, in the northeast of India (De Candolle, 1883; Tanaka, 1961; Hooker, 1872). Its presence in other areas such as Thailand or Annam has only been sporadically described. It has been reported to exist in India from ancient times and is considered to be of Indian origin (Singh, 1981). There are also several accessions from China described as sour orange (Wang et al., 2017).

Regarding the relevant cultivated hybrids, plants of sweet oranges [*C.* × *sinensis*, (*C. maxima* × *C. reticulata*)] were found in the wild in the tropical forests of Northern Myanmar and along the Salween River in the east, in the Khasi Mountains in Assam (Cooper, 1990) and in Chandu, in Uttar Pradesh. It is common in herbarium sheets collected in the Salween region, Thailand, Tonkin, and Yunnan (Tanaka, 1954). Thus, for these authors, sweet orange is believed to be native to northeastern India and Myanmar although De Candolle (1883) claimed that sweet orange was not indigenous to India.

According to several authors, the natural region of lemons (*C.* × *limon* (*C.* × *aurantium* × *C. reticulata*) × *C. medica*) is unknown since lemon trees have not been found growing freely in noncultivated areas. However, Singh (1981) based on previous studies cited that lemons are native to Eastern Himalaya, because a number of high-quality lemons growing in a wild state were collected in Assam, Sikkim, and nearby regions. Similarly, Tanaka (1954) reported its presence at 1000 m in Ketah, the northeastern corner of Myanmar and in Dudya in central India. For most authors, it is assumed that it appeared in the foothills of the Himalayas, in India (Swingle and Reece, 1967; Bonavia, 1888). Interestingly, there are a number of lemon-like fruits, still awaiting detailed studies, that are thought to be indigenous to Eastern Himalaya since no such varieties have been found elsewhere (Singh, 1981).

C. aurantifolia (Mexican lime, sour lime; *C. micrantha* × *C. medica*) shows great extension in East Indies and the Pacific Islands (Tanaka, 1954). There is a general agreement that *C. aurantifolia* comes from the Southeast Asian Archipelago (Swingle and Reece, 1967) although Singh (1981) claims that sour lime, which grows all over India, is indigenous to this country since it is found in a wild state in the outer Himalaya, from Garhwal and Sikkim to Khasi or Garo Hills and Madhya Pradesh. Its occurrence in Gujranwala and in Nagpore was also cited in Tanaka (1954). However, other authors have stated that there are no untamed plants of Mexican lime.

It is well known that grapefruit [*C. paradisi*, (*C. maxima* × *C. sinensis*)] originated in the 18th century, as a chance seedling hybrid between pummelo and sweet orange, on the island of Barbados, in the Caribbean Sea (cited in Swingle and Reece, 1967; Tanaka, 1954).

Mandarin admixtures, that is, introgression of pummelo on *C. reticulata*, are present in a wide area from the Himalaya region, in northeastern India and Myanmar to southern and southeast China (Swingle and Reece, 1967; Tanaka, 1961; Hodgson, 1967) including the Nanling region (Wang et al., 2018). Most authors believed that mandarins developed along the Yangtze River and the Pacific coast of China although reports of *C. reticulata*, probably mandarin admixtures, have informed on their occurrence in the Gujranwala district (Punjab Province of Pakistan), Karen Hills, in Eastern Myanmar and at 1200 m in the Lower Pulnys (Palani Hills) in southern India (cited in Tanaka, 1954). However, Bonavia (1888) mentioned that the true mandarin, which was certainly found in Sri Lanka, did not exist in India at that time, suggesting that many if not all these reports are related to mandarins derived from cultivated areas.

The distribution of papedas has been described in detail in Swingle and Reece (1967), Tanaka (1954), Zhou (1991), Bonavia (1888), and Ghosh (1997). The authors have reported that the papeda or pseudopapeda, *C. ichangensis*, was discovered in the wild in several regions including Northeastern India and Northern Myanmar. In Assam, it is found growing freely in Shillong and other parts of the Khasi and Naga Hills at 1800 m. In China, it has been reported in central and southwestern provinces of Hubei, Sichuan, Yunnan, and Hunan, with several wild variants of this species found in Hunan. *C. ichangensis* has a low level of genetic diversity, a circumstance that has been attributed to the mountainous regions and the Yangtze River that probably provided stable habitats that resulted in restricted gene flow, genetic drift, and population bottlenecks (Yang et al., 2017). *C. junos* ("yuzu"), probably a hybrid of *C. ichangensis* × *C. reticulata*, shows a wide distribution from almost the Indian frontier and the Yunnan province through the valleys of the Yangtze River to Anhui and Hangzhou in the East Coast of China (Tanaka, 1954). Another wild or semi-wild papeda of wide distribution is *C. macroptera* that is found in Northeast India abundantly growing in wild state in the Khasi Hills of Meghalaya, North Cachar, Karimganj, and Karbi Anglong districts of Assam, Mizoram, and Tripura and Manipur (Singh, 1981; Bhattacharya and Dutta, 1956). It grows also in northern Myanmar, Thailand, and in the peninsula of Indochina. This citrus also spreads all over the Islands of Borneo and the Philippines and crosses the Wallace Line through the Sulawesi, Maluku, and the Lesser Sunda Islands reaching New Guinea, the Bismarck Archipelago, New Caledonia, and Melanesia including the Fiji and Samoa islands (Tanaka, 1954). In contrast, wild *C. micrantha* appears to be more restricted to the Philippines (Wester, 1915) while *C. celebica* is extended to the northeast of the Sulawesi Islands and also to the Philippines. Tanaka (1954), however, already informed of the presence of these two citrus in the islands beyond the Wallace Line. *C. hystrix* also shows a wide area of expansion since it can be found over Yunnan, Sri Lanka, Upper Myanmar and Malaysia in the continent, as well as in the islands of Sumatra, Java, Timor Sulawesi, and the Philippines (Tanaka, 1954). This report also mentions the occurrence of *C. annamensis* (*C. combara*) in the Kochin Hills in the upper region of Myanmar and of *C. latipes* in the Khasi Hills of Meghalaya in northeastern India and in northern Myanmar (Hooker, 1872). According to Singh (1981), this papeda may be considered endemic to Eastern Himalaya and is found in a semi-wild state also in Naga Hills. In the Khasi and North Cachar Hills, *C. assamensis*, another papeda with same similarities to *C. hystrix* has also been described growing freely. In contrast, *C. hystrix* does not occur in India.

Of the two species of *Poncirus*, *P. trifoliata* has been reported to be native to central and southeastern China (Zhang and Mabberley, 2008; Fang et al., 2011), while *P. polyandra* has been identified in Yunnan Province (Ding et al., 1984). There are also very ancient references to the presence of *Poncirus* in China (Cooper, 1990).

It is worth mentioning that *Clymenia polyandra* that strongly resembles a true citrus is found in Papua New Guinea and close islands as New Ireland (Tanaka, 1954) while the genus *Oxanthera* is a native of New Caledonia.

2.2.7 The center of origin of citrus

Native habitats of citrus and related genera roughly extend through a broad area mostly comprising India, Southeast Asia, Australasia, Melanesia, and Polynesia. It is generally believed that the papedas (subgenus *Papedas*, Swingle) developed in the Pacific Islands of these areas, while the origin of the true citrus (subgenus *Eucitrus*, Swingle) is associated with the continental Southeast Asia region, a biodiversity hotspot with a climate largely influenced by monsoons. In their master review, Swingle and Reece (1967) suggested, for example, that most important species of true citrus appeared in an extensive region of Southeast Asia comprising northeast India, Myanmar and southeast China.

Pioneering works of Vavilov (1927) identified two centers of the origin of citrus, the Indo-Burma center (Assam and Myanmar; oranges, mandarins, and citrons) and the Indo-Malayan center (Indochinese Peninsula and the Malay Archipelago; pummelo). This vision was modified by Tolkowsky (1938) who stressed the relevance of the northeast region of India (Chapter 3) and also aimed directly at the mountains of southern China. Tanaka (Swingle and Reece, 1967; Tanaka, 1959, 1961) in subsequent work reformulated these ideas and concluded that the primary center of citrus was placed within northeast India and northern Myanmar, from where citrus dispersed to secondary centers located in Indochina and southeast China. Tanaka (1954) noticed that most citrus types were well represented in wild forms in the Eastern Himalaya as exemplified by the presence not only of true species as we know today, but also of intergenic hybrids. Thus, primitive citrons (*C. nana*) from one side have been found in eastern Bhutan, while the Indian wild mandarin (*C. indica*) and a papedocitrus (*C. ichangensis*) have been discovered in Naga Hills, in Assam. The occurrence of pummelos, *C. maxima*, on the other hand, was also reported in Garo Hills, Assam, and in the north of Bangladesh, although the development of untamed pummelos in these areas appears to be debatable (Bhattacharya and Dutta, 1956). In Assam, relevant citrus intergenic hybrids such as both sour and sweet oranges (*C. aurantium* and *C. sinensis*) and Rangpur limes and rough lemons (*C. limonia* and *C. jambhiri*) have been found presumably in a wild state although the presence of undomesticated sweet oranges is not generally accepted. Other reports on untamed *C. aurantifolia* and *C. limon* growing freely in the Himalayan regions have been presented although again these accounts might be also questionable. Tanaka (1954) and/or Singh (1981) also reported in Assam, Myanmar, and nearby regions, the occurrence of a number of papedas, such as *C. latipes, C. macroptera, C. combara, C. assamensis*, and *C. hystrix* (see Chapter 3 for further information). This last species, however, has not been ever reported in India. Unfortunately, the genomic status of these papedas is currently unknown since their genomes have not yet been studied in detail and their categorization as true species or intergenic hybrids is still pending.

From the above description, northeast India and Myanmar appear to be devoted to indigenous species of the most representative Chinese citrus, that is, the Chinese mandarins, the hybrids of wide distribution in China, *C. junos,* and the members of the genus *Fortunella* and *Poncirus.* Mandarin admixtures and Sun Chu Cha Kat (*C. reticulata*), a true species of mandarin, are certainly present in many areas in northeastern India, although it is believed that many of these forms are in fact apparently related to the Chinese mandarins introduced in India. Consequently, Tanaka (1954) proposed a theoretical dividing line (Tanaka's line), running from the northeast border of India through Yunnan until the island of Hainan, conceptually similar to the Wallace Line, that geographically separated the citrus of the foothills of the Himalaya from those found mostly in China. According to Tanaka (1954), therefore, *C. junos*, admixture mandarins, mandarins nonrelated to the Indian wild mandarin and the members of the citrus-related genera *Poncirus* and *Fortunella*, citrus types that can be found today in China along the Yangtze River and the Pacific coast, originated east of this line. This division establishes a distinction among the two kinds of species found to each side of the geographical line. This was previously noticed by Tolkowsky (1938) who suggested the occurrence of two major centers of citrus origin, as mentioned above. Thereafter, other propositions have been advanced emphasizing the importance of the Chinese center and suggesting, for example, that primary centers for citrus took place in the southwestern mountains of China (Zhang, 1981) or in Yunnan and adjacent areas (Gmitter and Hu, 1990).

As explained in the preceding sections, new evidence has been presented in recent years in different disciplines that may contribute to elucidation or resolution of questions revolving around the origin of citrus. A pivotal observation in this direction is related to the identification of a new species of citrus, *C. linczangensis* (Xie et al., 2013), from the Late Miocene of Lincang, Yunnan. The leaf fossil, which possesses broadly winged petioles as the current papedas, provides definite evidence that in Late Miocene, western Yunnan was a native habitat of citrus.

On the other hand, the chronogram of the citrus phylogenetic tree presented in Fig. 2.3 shows that the citrus studied are essentially grouped into two main clades roughly corresponding to those two regions previously identified as the centers of citrus origin, the northeast region of India and the mountains of southern China. Thus, the topology of this phylogenetic tree appears to follow to some degree the geographic distribution of the native places of birth of current citrus.

The analyses of the native habitats of citrus based on the documented reports on the presence of wild genotypes growing freely in noncultivated areas reveal that the triangle limited by Northeastern India, Northern Myanmar, and Western Yunnan concentrated the highest number of wild citrus species and hybrids found in any region and, therefore, exhibits the highest citrus diversity. In this area, true species of citrus such as citrons (*C. medica*), pummelos (*C. maxima*), Indian wild oranges (actually, Indian wild mandarins, *C. indica*), and pseudopapedas (*C. ichangensis*) have been found growing in the wild state. Moreover, relevant intergenic hybrids such as sour and sweet oranges (*C. aurantium* and *C. sinensis*), Rangpur limes (*C. limonia*), and rough lemons (*C. jambhiri*) are not uncommon in these regions. In addition, there are also reports on the presence of the wild pure mandarin Sun Chu Cha Kat (*C. reticulata*) and of other important hybrids, namely, lemons (*C. limon*) and Mexican limes (sour lime, *C. aurantifolia*). As mentioned above, the wild state of some of these citrus forms maybe doubtful. It is worth mentioning that out of the seven continental true citrus species identified to this day according to the genome analyses, only *C. micrantha* (Pacific Islands) and *F. margarita* (South China) are not found in this triangle, in contrast to most important papedas such as *C. junos* (yuzu), *C. macroptera*, *C. hystrix*, *C. annamensis*, *C. latipes*, and *C. assamensis* that can be unquestionably found in the region.

The estimations of divergence times (Fig. 2.3) indicated that the ancestral citrus definitively experienced a relatively fast radiation giving rise to all major citrus species, except the Australian ones, in ca. 2.0 Ma, an observation consistent with a unique initial area of diversification that argues against the occurrence of several centers of citrus origin. Therefore, current evidence indicates that most important species of citrus originated in a region located between the province of Assam, in Northeast India, Northern Myanmar, and the province of Yunnan in China. While our proposal offers partial support to some of the previous formulations conferring geographic accuracy to the otherwise broad and vague propositions, overall, it clearly challenges other ideas based on centers located in Australia or nearby islands (Swingle and Reece, 1967; Beattie et al., 2009), the Malay Archipelago (Webber, 1967), Thailand (Scora, 1988), or southeast China (cited in Zhang, 1981).

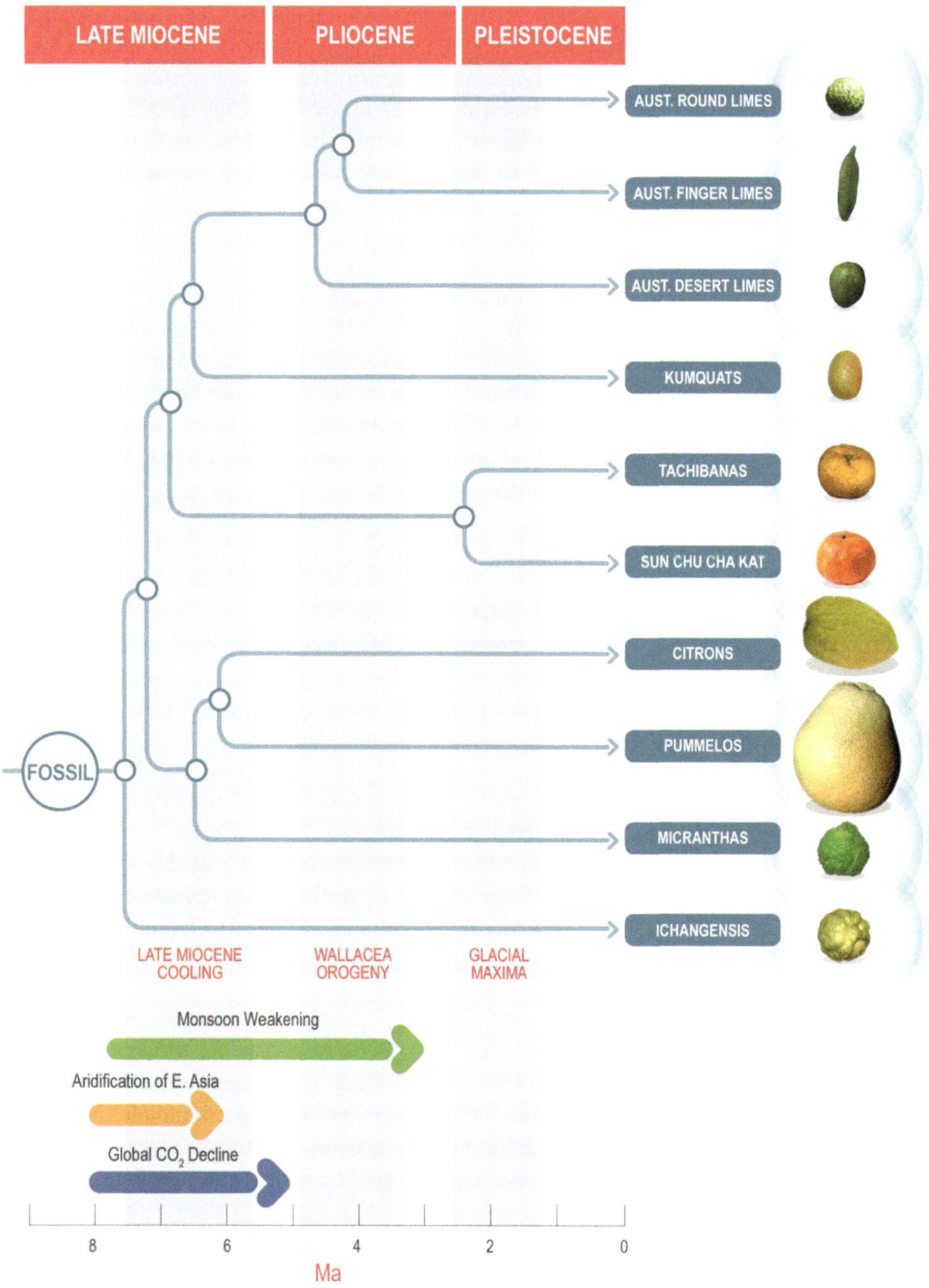

FIG. 2.3 Chronology of citrus speciation. The analyses of the nuclear genomes suggest that the ancestral citrus evolved during Late Miocene through a rapid radiation (6–8 Ma) coincident with a dramatic weakening of the summer monsoon. This drastic transition from wetter monsoonal conditions to a drier climate was probably triggered by a global reduction of CO_2 levels that brought about a period of global cooling accompanied by a strong phase of aridification. The Australian citrus species and the native Japanese Tachibana mandarin split later from mainland citrus. The oceanic dispersal of citrus to Australia occurred during the Early Pliocene (4.0–4.6 Ma) while Tachibana probably arrived to the Pacific Islands during the Early Pleistocene (2.4 Ma). Age calibration is based on the citrus fossil *C. linczangensis* from the Late Miocene. Fruit images are not to an exact scale. *Adapted from Wu, G.A., Terol, J., Ibanez, V., Lopez-Garcia, A., Estela, P.-R., Carles, B., et al., 2018. Genomics of the origin and evolution of Citrus. Nature 544, 311–316. https://doi.org/10.1038/nature25447.*

2.3 Citrus radiation and evolution

2.3.1 Citrus radiation

It is widely accepted that true citrus originated in continental Southeast Asia from older papedas although there are different opinions regarding the birthplace of papedas, and hence on the own direction of the dispersion and evolution of citrus.

The most complete hypothesis on the dispersal and evolution of citrus was developed by Tanaka (1954, 1959, 1961), and although we know now that contains several inconsistences it is worth summarizing its basic premises. According to his view, the Eastern Himalayas that contained at least a representative member of the most important citrus types was the major center of dispersion of the true citrus. He also identified two major routes of dispersion along the Yangtze River and the coast of southern China. Regarding the ancestor of citrus, Tanaka (1959) suggested that this should be a species similar to those of the African genus *Citropsis* that exhibit winged petioles and may develop simple leaves very much resembling those of citrus. He also related the origin of the true citrus in the Eastern Himalayas to the loss of the inflorescence from the papeda *C. latipes*, or from other similar papedas.

The loss of the inflorescences in favor of a solitary flower is a characteristic observed, for instance, in *C. ichangensis* and mandarins. According to Tanaka (1959), the acclimation of the papedas to the tropical climate in continental mainland during their southward dispersion to the Pacific Ocean gave rise to limes and pummelos. These two kind of citrus experienced major fruit changes such as either thinning or thickening of pericarp, considerable elongation of pulp vesicles and major changes in their content, for example, the loss of acrid oil. The similarities between some fruit characteristics of limes, lemons, and citrons led him to propose that limes gave rise to lemons and citrons through the gain of strong citral aroma, the enhancement of anthocyanin pigmentation, the enlargement of the flower and fruit, and the complete loss of petiole wings. These citrus types, citrons and lemons, advanced then westward along the Himalaya range. On the other hand, he claimed that pummelos gave rise to both sour and sweet oranges, two citrus types that maintained the original globose fruit shapes of pummelos but developed new and attractive carotenoid pigmentations. Finally, Tanaka (1959) thought that mandarins probably evolved from *C. indica* to *C. reticulata* (Ponkan) via *C. erythrosa* (actually Sun Chu Cha Kat or similar) and *C. tangerina*.

Swingle, on the other hand, thought that the genus *Citrus* originated in the New Guinea-Melanesia region although a few species occur in New Caledonia or in neighboring islands to the east and southeast (Swingle and Reece, 1967). He speculated that most papedas certainly developed in the East Indian Archipelago, in the Philippines, New Guinea, and Melanesia and that only a few of these papedas reached the Asiatic continent to give rise to the true citrus that we know today.

A third hypothesis proposed from Australian researchers (Beattie et al., 2009) claims that the genus *Citrus* originated in Australia and dispersed westward from Australasia to Asia via the Halmahera route. The authors suggest that these primitive citrus moved in equatorial currents on island terranes to the Southeast Asia Archipelago and later on from there to continental Southeast Asia to evolve to modern citrus. According to the current knowledge of plate tectonics, these equatorial currents and island terranes departed from the north of Papua New Guinea and moved westward across the same area. Thus, the ascendants of the well-known commercial *Citrus* species and hybrids and the species of the genus *Fortunella* and *Poncirus* originated from the ascendants of the citrus and related genera that are found today in the Australasian region, that is, *Clymenia* and *Microcitrus* in New Guinea, *Eremocitrus* and *Microcitrus* in Australia, and *Oxanthera* in New Caledonia. Part of the support for this hypothesis is provided by the phylogenetic relationships generated with chloroplast markers that suggest, as explained above (Bayer et al., 2009) that *C. medica* is nested with the Australian limes and with *Clymenia polyandra*.

In light of new discoveries it seems relatively simple to point out the inconsistencies of these proposals, although this circumstance in no way detracts from the contribution of such proposals to the development of the ideas that we present here. Citrus speciation as explained by Tanaka (1959) for instance is rather improbable because several critical ancestors in his scheme are in fact intergeneric hybrids, as we now know. However, the identification of the northeast India region as the primary center of citrus diversification is outstanding. On the other hand, the Melanesian origin of the genus *Citrus* suggested by Swingle and Reece (1967) does not appear to be supported for a high degree of diversity in the area, although again, his vision on the evolution of true citrus in continental South East Asia is also meritorious. The Melanesian and Australian hypotheses are both based on a preponderant and relatively old (more than 20 Ma) east to west dispersal. The nuclear genome phylogeny and the timing of speciation events, however, point to an Asian origin of citrus. In this scheme, Australian limes evolved later from an ancient form that certainly derived from the continental Southeast Asia (Wu et al., 2018). The chloroplast genome phylogeny, although different from the nuclear genome phylogeny, is also consistent with an Asian origin of citrus (see citations in the previous section) and, therefore, both phylogenies reject the proposed Australasian origin. Critical comments to these formulations were also advanced in previous works (Pfeil and Crisp, 2008; Schwartz et al., 2015), that reported that the inferred age, ancestral area reconstruction, and lack of plausible vicariant breaks were incompatible with a hypothetical citrus movement from Australasia to Asia. These authors instead suggested that *Citrus* radiated from an East Asian origin, most probably in the Late Miocene or Pliocene and dispersed from western to eastern Malesia.

Recent evidence based on the whole genome sequencing and the dating of citrus speciation (Wu et al., 2018) revealed two well-separated phases of radiation associated with the diversification, first, of the continental citrus and then of the Australian species (Fig. 2.3). The Asian speciation was a relatively rapid radiation that took place in the Late Miocene (6–8 Ma) and spanned a period of almost 2 million years. This radiation generated at least seven species of which we are aware (*C. mangshanensis*, *C. ichangensis*, *C. micrantha*, *C. medica*, *C. maxima*, *C. reticulata*, and *F. margarita*) as well as the ancestor species that later evolved into the Australian limes. Although this listing is still awaiting the incorporation of more genomes as explained in previous sections, most of the major citrus species that we know today emerged during the period of this radiation. The second phase of citrus speciation consisted of the diversification of the Australian limes, *E. glauca*, *M. australis*, and *M. australasica*, during the Early Pliocene. It spanned the period 4.0–4.6 Ma and was separated from the Asian citrus radiation epoch by about 1.5 Ma. The closest relative to Australian citrus is *Fortunella*, a genus whose species have been reported to grow wild along the coast of southern China and Hainan. Therefore, the data suggest that the Australian citrus progenitor likely migrated via transoceanic dispersal across the Wallace Line from Southeast Asia to Australia and later adapted to these diverse climates.

The analyses of the whole genome also show that that Tachibana mandarin, naturally found in Taiwan, the Ryukyu Archipelago, and Japan (Tanaka, 1931), split from mainland Asian mandarin (*C. reticulata*) species during the Early Pleistocene (ca. 2.4 Ma). *C. tachibana* probably evolved in these islands as a genetic isolate from the mainland Asian gene pool, as revealed by both the chloroplast phylogeny and haplotype sharing analysis.

In conclusion, based on the genomic, phylogenetic, and biogeographic analyses, we propose that citrus diversified during the Late Miocene epoch through a rapid Southeast Asian radiation. In fact, most of the major citrus species that are currently known emerged during the period of this radiation. A second radiation in the Early Pliocene giving rise to the Australian limes was enabled by migration across the Wallace Line of an ancestor of the citrus extant species in Australia and New Guinea. Finally, Tachibana mandarins originated from mainland mandarins in the Early Pleistocene.

2.3.2 Late Miocene: Global cooling and the Southeast Asian radiation

As explained in the previous sections, citrus phylogenies and the corresponding estimation of divergence times (Pfeil and Crisp, 2008; Bayer et al., 2009; Schwartz et al., 2015; Carbonell-Caballero et al., 2015; Wu et al., 2018) indicated that citrus underwent a rapid and extensive radiation during the Late Miocene, approximately 6.0–8.0 Ma. This radiation gave rise to the ancestors of the extant citrus species and types that we know today. On the other hand, the geographic distribution of the citrus species growing wild in noncultivated areas point to the Himalayan Eastern foothills, an area dramatically influenced by the Indian monsoon, as the primary center of citrus dispersion. These two independent conclusions are directly connected with the identification of *C. linczangensis* from the Late Miocene of Lincang (Xie et al., 2013), a citrus leaf fossil specimen providing definite evidence of the existence of *Citrus* within the province of Yunnan since this epoch. Through the review of the geological and climatological conditions of this area during the Late Miocene, in this section we will provide a plausible hypothesis regarding the causes that triggered this initial citrus radiation, and hence the speciation and the evolution of the genus *Citrus*.

It is well accepted that diversification can be induced by large-scale reconfigurations of the landscape due for instance to the separation of land masses, the formation of rivers, or the uplift of mountains. There are a large number of studies correlating high diversity with the uplift of mountains, an observation reported in practically all important mountain ranges including the Himalayas, Tianshan, Caucasus, and Andes (Luebert and Weigend, 2014). In principle, several mechanisms driving this kind of rapid plant radiation have been proposed (Wen et al., 2014). In the Himalaya region, the shaping of the geographic genetic structure of the area has been traditionally correlated with the climatic oscillations provoked by monsoonal seasonality and the formation of new ecological niches associated with the Himalayan uplift. Many studies, actually, agree that the emergence and uplift of the Himalaya had a determinant influence over the monsoonal seasonality of Southeast Asia. Therefore, it has been widely accepted during decades that the Himalaya rise has determined monsoon activity and that the climate changes associated with the uplift of these mountain ranges were the pivotal factors facilitating speciation and diversification. However, contrary to this general and extended belief, current evidence clearly indicates that monsoons in this region, first, arose at different times and, second, arose in a way certainly unrelated to the Tibetan uplift (Renner, 2016). Since the Early Miocene, extensive uplifts of the Tibetan plateau occurred in at least four major periods: 17–25, 13–15, 7–8, and 1.6–3.5 Ma (cited in Wen et al., 2014). Monsoon regimes in South and East Asia were probably established in these regions in Early Miocene and showed an extended period of intensification with strong summer monsoons in the Middle Miocene reaching a maximum between 10 and 18 Ma (Valdiya, 1999; Zhisheng et al., 2001; Clift et al., 2014). At the onset of the Late

Miocene, climate was largely controlled by the Indian summer monsoon that brought heavy summer rainfall to the southern part of the Himalayas (Zhisheng et al., 2001). This phase of monsoon intensification abruptly ended around 10 Ma in East Asia and about 8 Ma in South East Asia, while a period of strong monsoon weakening started thereafter in these regions (Clift et al., 2008, 2014). It is accepted that the sudden weakening of the monsoon provoked a drastic climate transition from wetter conditions to a drier climate with seasonal heavy rains (Clift et al., 2008, 2014). Thus, the rapid radiation of citrus that took place in the southeastern foothills of the Himalaya during the Late Miocene coincided with the onset of a weakening phase of the monsoon that started around 8.0 Ma and lasted about 5 Ma in the southern Himalaya. This dramatic alteration obviously did not only affect citrus but also caused major and severe biota changes in the whole region. In the surrounding areas of the Himalaya, an elevated number of spectacular rapid radiations in various groups of plants, such as the eudicot genera *Caragana*, *Rheum*, *Pedicularis*, *Saussurea*, *Rhododendron*, *Primula*, *Meconopsis*, *Rhodiola*, and many lineages of gymnosperms, for example, the conifer genus *Juniperus*, has been documented. These species also evolved surprisingly in a short time period from a common ancestor, as observed in citrus. Moreover, rapid biological radiations have also been reported for practically all other major groups of organisms including insects, fishes, crabs, amphibians, reptiles, birds, or ferns (Wen et al., 2014; Wang et al., 2012; Favre et al., 2015). In the Siwalik region, it has been documented that the humid climate that prevailed earlier supported thick rain forests and that the weakening of the monsoon intensity brought about drastic changes in the vegetation. These changes in the paleoflora provoked major and conspicuous substitutions of evergreen C_3 tropical trees to tall C_4 grasses, transitions that took place around 7.0–8.0 Ma in southeastern Nepal (Valdiya, 1999, 2002). Taken together, these observations indicate that the sudden and drastic changes in the amount and distribution of precipitation are in the core of the rapid radiations of various plant lineages (Wen et al., 2014) comprising the unique biodiversity of the southeastern margin of the Tibetan plateau, a region that acted as a hotspot or center of evolution (Favre et al., 2015).

Interestingly, the physiological and phenotypic adaptations of current citrus (Iglesias et al., 2007; Tadeo et al., 2008) appear to still carry the signature of that old transition from wetter to dryer conditions since continental citrus that can be defined as mesophytes exhibit intriguing adaptations to cope with periods of water stress. Citrus, for instance, possesses an efficient waxy coating on leaves and fruit peels that may help to reduce water losses; the fruit also develops individual juice sacs, which may play also a similar role, while the tree shows low photosynthetic and transpiration rates associated with slower growth patterns. In contrast, citrus exhibits lush foliage, high chlorophyll content, and shallow root systems as found in tropical understory bushes with lower light availability and poor organic material soils. Citrus also develop a unique spongy fruit albedo that may help also to cushion rapid volume alterations produced by sudden water inputs. These observations are compatible with the assumption that ancestral citrus were native of regions with tropical monsoon climates but evolved in a "humid subtropical climate" with more pronounced dry seasons.

Recent evidence has positioned the local climatic changes associated with the monsoon intensity in the South East Asia in a global context. While the Miocene epoch has been described as "the making of the modern world" (Potter and Szatmari, 2009), the influence of the Late Miocene appears to be crucial. The Late Miocene was strictly characterized by a cooling trend that inaugurated a period of aridification in the interior of the continents, generating dryer vegetation types and even propitiated the appearance of major deserts. A comprehensive study that conveniently resumes this thinking was recently presented in Herbert et al. (2016). These authors found the sustained Late Miocene cooling that considerably reduced ocean temperatures occurred in a global scale between about 7.0 and 5.0 Ma. The global cooling brought about aridification in large areas of the continents enhancing seasonality, shaping modern ecosystems, and restructuring terrestrial plant and animal communities. The authors also indicated that the magnitude of the changes in temperature, climate, and biota that can be observed all along the subtropics during the Late Miocene cannot be attributed to regional tectonic forces and concluded that such drastic changes were an ultimate consequence of the decline in atmospheric CO_2 levels that started about 8.0 Ma (Fig. 2.3). A recent work has provided independent evidence to this proposal revealing that this long-term cooling trend and the weakening of the summer monsoon were synchronous with intensification of the Asian winter monsoon (Holbourn et al., 2018). According to these data, the climate shift occurred from 7.0 to 5.5 Ma starting at the end of a global ^{13}C decrease. These observations also suggest that changes in the carbon cycle involving the terrestrial and deep ocean carbon reservoirs were instrumental in driving Late Miocene climate cooling.

Therefore, the changes in the climate and ecosystems described above in the South East Asia and specifically in the areas close to the center of origin of citrus had not a local character but were part of a global phenomenon that started at the onset of the Late Miocene epoch. Thus, the extreme climate transition from wetter conditions to the drier climate detected in this region at this time was just one more event among all drying episodes that modified the landscape of the subtropical regions, a statement that can be reliably exemplified with the establishment of the Sahara Desert in northwest Africa. Similarly, the expansion of C_4 grasslands at the expense of C_3 plants observed in the boundaries of the Himalaya (Valdiya, 1999, 2002) in the Late Miocene was not an exception but a trend in tropical regions, coincident with the global radiation of succulent plant lineages (Arakaki et al., 2011) and the changes in the accompanying mammalian fauna (Valdiya, 1999, 2002; Badgley et al., 2008). The reduction in CO_2 levels and the expansion of arid environments during the Late Miocene favored the global radiation and evolution of C_4 and succulent plants observed between 8.0 and 6.0 Ma (Cerling et al., 1997), since the C_4 photosynthetic strategy is more efficient under carbon limitation than the normal C_3 pathway, in addition to the better adaption of succulent plants to water deficit.

Regarding the causes of the citrus radiation, it is also worth noticing that Herbert et al. (2016) clearly stated that the vegetation changes observed in the region were part of a much larger global temperature decline, a statement that complements the absence of available evidence linking the Tibetan uplift with the Southeast Asian monsoon intensification (Renner, 2016).

Therefore, we propose that citrus evolved in the southeastern foothills of the Himalaya from an ancestral species in a global frame of CO_2 level decline at the onset of Late Miocene. The reduction of available carbon brought about a period of global cooling that resulted in an enhancement of aridity in the subtropical regions. These changes very likely contributed many other rapid radiations and major migrations, turnovers, and substitutions of the terrestrial biota. The process of citrus speciation, therefore, cannot be considered a single or isolated episode, but on the contrary is part of a global trend that occurred in tropical regions during the Late Miocene.

2.3.3 Dispersal routes of ancestral citrus

The drastic climate change that occurred at the onset of the Late Miocene probably triggered adaptive radiations forcing long-distance dispersal of the incipient citrus species. Thus, we propose that from the putative center of origin, the region located between the provinces of Assam in India and Yunnan in China, citrus dispersed essentially in two directions, south- and eastward, as many groups of plants and animals did in continental South East Asia (Wen et al., 2014; Favre et al., 2015). Migration to the north was certainly arrested by the Himalaya range while only citrons appear to have journeyed westward. Support for this proposal comes from studies on citrus biogeography and phylogeny, as explained above and on the paleogeography of the region, especially the geological history of the Wallacea area and Japan, as discussed below.

Citron is native in a wide area that extends from India to central and southwest China but the abundant number of reports informing of its excellent conditions in many virgin areas at the foothills of the Himalaya indicates that citron very likely evolved there and that only later or even in more recent times arrived to China. In this regard, it is interesting to mention that ancient Chinese writings appear to suggest that citron was a fruit not known in China in the second century, while in Rome it was already highly appreciated (Han, 1979), implying that *C. medica* was possibly spread earlier in the West than in the Orient. From the region of origin, other citrus types followed predominant ancient routes to the south and east, although similar to the citron dispersal, additional cores of dispersion can also be distinguished in these paths. Previous researchers recognized as centers of citrus origin three areas chiefly located in Assam and Myanmar, the Indochinese Peninsula, and the mountains of southeast China (Vavilov, 1927; Tolkowsky, 1938). Tanaka (1954), however, separated a primary center of origin in northeast India and northern Myanmar from the other two centers, located in Indochina and southeast China. He thought that these played subsidiary roles as secondary centers of origin. Our proposal, on the other hand, defines a single center of origin from which citrus dispersed in several directions and under different environments through a few additional main cores of dispersion.

The pummelo probably migrated from the center of origin, through a southern route, to more tropical monsoonal lands in the Indochinese Peninsula and nearby islands. A high proportion of the Sulawesi taxa with an origin in Asia, for instance, also dispersed across this route (Lohman et al., 2011). Scarce and indirect evidence (Yu et al., 2017) suggests that its irruption in South China might have occurred later. Nested in the same phylogenetic clade as pummelos, the papeda *C. micrantha* may have followed the same route to reach the Philippines and the islands beyond the Wallace Line.

In addition to the west- and southward routes, a major ancestral path opened too to the east. It is accepted that during the Late Miocene, there were not orogenic barriers between Myanmar and Yunnan, a geological configuration that may have favored eastward dispersal of biota. There is compelling evidence that the eastern edge of the Himalaya, the Hengduan range, comprising several mountain subranges in western Chinese provinces, underwent rapid uplift only after the Miocene and that they reached maximum elevations shortly before the Late Pliocene, ca. 3.6 Ma (Golonka et al., 2006; Sun et al., 2011, and references therein). Additional evidence based on the reconstruction of paleovegetation and paleoclimate in the Late Pliocene of west Yunnan indicates that uplift of Gaoligong and Nu Mountains (Hengduan range) and the eastern portion of the Tibetan Plateau (Western Yunnan) must have occurred during or after the Late Pliocene (Kou et al., 2006; Su et al., 2013). Through this eastward route probably evolved kumquats and mandarins, two citrus types that are phylogenetically related and found today in China along the coast of southern China and the Yangtze River, respectively. The kumquats are native in regions mostly located in the southeast of China, although they can be found also in central China, Thailand, continental Malaysia, and in Hainan. The "wild mandarins" Mangshan, Daoxian, and several genotypes of *C. reticulata* have been discovered in the Mangshan and Nanling Mountains of southern Hunan province in China, whereas mandarin admixtures are present in wide areas all along southern and southeast China. Furthermore, this route was probably also followed by the pseudopapeda *C. ichangensis* that in China has been reported in the central and southwestern provinces of Hubei, Sichuan, Yunnan, and Hunan, in addition to Assam in India.

Therefore, current evidence suggests that after the initial radiation citrus dispersed in three main directions (Fig. 2.4), westward (citrons), southward (pummelos and *Micrantha*), and eastward (*ichangensis*, mandarins, and kumquats). The phylogenetic relationships among these citrus grouped citrons, pummelos, and *Micrantha* in a clade separated from a second one composed of mandarins, kumquats, and Australian limes. Available information also appears to suggest that there were two major cores of secondary dispersion. One of them, from where pummelos and *Micrantha* evolved, was roughly located in the Thai-Malay Peninsula, an area that appears to have acted as a biogeographic crossroads (Lohman et al., 2011). The other hub, in the confluence of the Yunnan province with the southern mountains of China probably was prolonged within three corridors, one to the south through the Pacific coast (kumquats), a second one along the Yangtze River (mandarins), and the third one to the northeast, to central China (*ichangensis*).

2.3.4 Early Pliocene: Wallacea orogeny and the dispersal of Australian limes

The diversification of Australian limes was not driven by the Late Miocene cooling, since speciation of these citrus and their dispersal to Australia occurred at the beginning of the Pliocene epoch, around 4.6 Ma. This statement agrees with previous estimations that dated the emergence of a citrus Australasian clade about 5.8 Ma (Pfeil and Crisp, 2008; Carbonell-Caballero et al., 2015). Under the generic name of Australian limes are included seven species of the genera *Eremocitrus* and *Microcitrus* native to Australia and Papua New Guinea. The genus *Eremocitrus* includes one single species, *E. glauca*, known as Australian desert lime, the only citrus that shows strong xerophytic adaptations. This species has been adapted

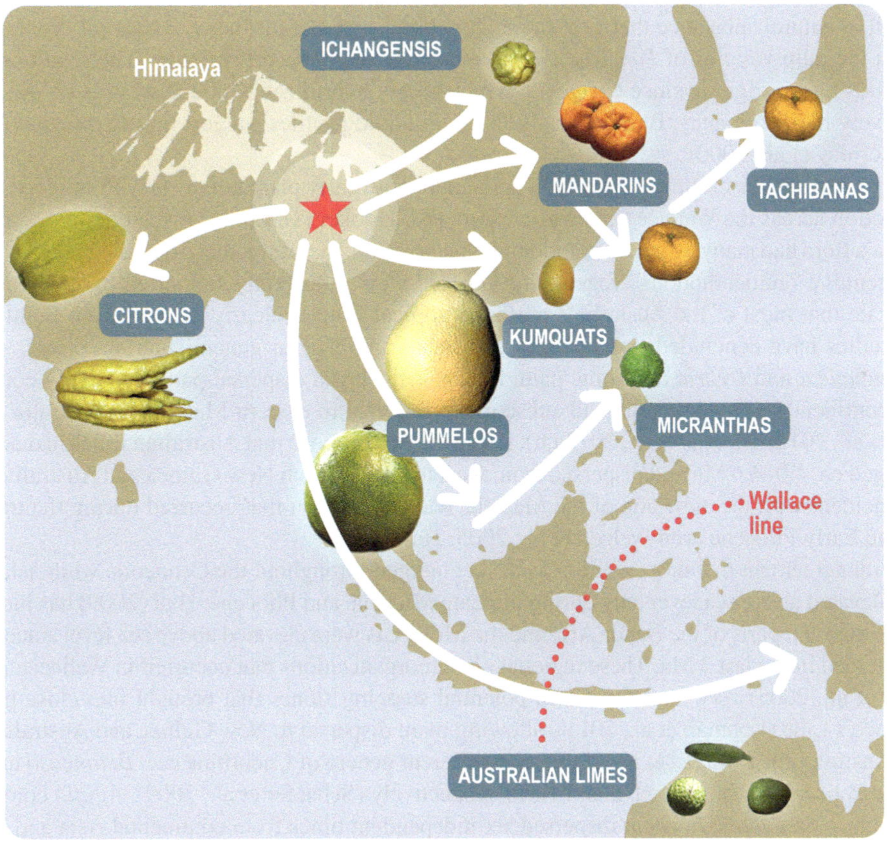

FIG. 2.4 Ancestral dispersal routes of citrus. *Arrows* suggest plausible migration directions of the ancestral citrus species from the center of origin, the triangle formed by northeastern India, northern Myanmar, and northwestern Yunnan. After the initial radiation citrus dispersed in three main directions, westward (citrons), southward (pummelos and *Micrantha*), and eastward (*ichangensis*, mandarins, and kumquats). The phylogenetic relationships among these citrus group citrons, pummelos, and *Micrantha* in a clade separated from a second one composed of mandarins, kumquats, and Australian limes. The oceanic dispersal of citrus to Australia during the Early Pliocene was enabled by the occurrence of stepping stones that allowed overcoming the Wallace Line during a profound reshuffling of the orogeny of the region. Tachibana arrived to the Pacific Islands favored by the reduction of the sea levels and the emergence of land bridges between mainland and the islands, promoted by the expansion of ice sheets during the Early Pleistocene. The *red star* denotes the location of the *C. linczangensis* fossil in Yunnan Province. Fruit images are not to an exact scale. *Adapted from Wu, G.A., Terol, J., Ibanez, V., Lopez-Garcia, A., Estela, P.-R., Carles, B., et al., 2018. Genomics of the origin and evolution of Citrus. Nature 544, 311–316. https://doi.org/10.1038/nature25447.*

to dry and solid soils and is a native of northeastern Australia (Sykes, 1997). The other genus, *Microcitrus* that can endure long periods of droughts and show semi-xerophytic adaptations, contains six species: *M. warbugiana*, the only one originated in southeastern New Guinea (Tanaka, 1954), *M. australasica* (Australian finger lime), *M. australis* (Australian round lime), *M. garrowayi*, *M. inodora*, and *M. maideniana* that are thought to be native to Queensland and northern New South Wales in eastern Australia (Swingle and Reece, 1967; Sykes, 1997). There are two additional citrus-related genera that according to the chloroplastic markers (Bayer et al., 2009; Oueslati et al., 2016) cluster with the Australian limes, *Clymenia* (New Guinea wild lime), native to New Guinea and nearby islands, and *Oxanthera* (false orange) that is endemic to New Caledonia. While *Clymenia* is strictly acclimated to tropical environments, *Oxanthera* on the contrary, exhibits several xerophytic characteristics based on its leaf morphology (Swingle and Reece, 1967). Therefore, the species of the genera *Eremocitrus*, *Microcitrus*, *Clymenia*, and *Oxanthera* that have adapted to tropical environments and dry conditions are found native in Australia, New Guinea, and New Caledonia, and may form a monophyletic clade with the rest of members of the genus *Citrus*, implying that they adapted in these islands and regions after the migration of an common citrus ancestor from continental Southeast Asia.

We speculate that this citrus ancestor probably was closely related to the species of the genus *Fortunella* for several converging reasons. Kumquats that are widespread along the Pacific southern coast of China, phylogenetically group with the Australian limes (Fig. 2.3) and with *Clymenia* and *Oxanthera* according to chloroplast markers. Among the pure species of citrus, the fruits of *Fortunella* and *Eremocitrus* are the only ones that show a surprisingly similar low number of fruit loculi and ovules. These two genera exhibit 3–5 locular ovaries with only 2 collateral ovules in each loculus, while other citrus types show higher number of loculi and with a minimum of 4–12 ovules (Swingle and Reece, 1967). Furthermore, the small fruits of *Fortunella* and the Australian desert lime also share many characters including a low content in the pulp vesicles of acrid oil, a natural substance that give the juice a bitter, unpleasant flavor. These oil droplets, on the contrary, are very profuse in the pulp vesicles of *Poncirus*, *Microcitrus*, and papedas, for instance. The relationship between kumquats and *Clymenia* is also intriguing since *Clymenia*, which barely hybridizes with a few citrus types, generates kumquat hybrids with relatively high efficiency. Besides, analyses of limonoid profiles suggest that *Clymenia* is closely associated with *Fortunella* (Berhow et al., 2000).

This migration of citrus to Australia is basically in agreement with early plant botanists who suggested that the predominant dispersal direction across the Wallace Line was west to east. Hooker (1859), for instance, suggested that a substantial part of the Australia's flora had many affinities to Indomalayan taxa and, therefore, these plants can be considered "intrusive" in contrast to the genuine "autochthonous" vegetation. This old view is currently supported by molecular and geographic analyses that indicate that most of the Australia's northern tropical flora is clearly derived from Southeast Asia. Thus, a high number of studies have concluded that lineages of several angiosperm genera such as *Aglaia*, *Alocasia*, *Begonia*, *Pseuduvaria*, *Neonauclea*, and *Uvaria* and some palm lineages displayed dispersal patterns largely consistent with initial diversification in continental Southeast Asia and subsequent dispersal to eastern Malesia or Australia (Richardson et al., 2012; Nauheimer et al., 2012; Thomas et al., 2012a,b). Likewise, we propose that Australian limes had an Asian continental ancestor that diverged ca. 4.0–4.6 Ma and dispersed from west to east to reach New Guinea and Australia. Interestingly, this migration was coincident with the elevation of the Malesia/Wallacea region that occurred during the transition during the Late Miocene to the Early Pliocene (van Welzen et al., 2005; Hall, 2009).

There is a general agreement that most western Malesia emerged throughout the Cenozoic, while islands and lands east of Wallace's Line elevated above sea level only during the Late Miocene and Pliocene. Hall (2009) has indicated that most of Sumatra and Java, Sulawesi, parts of the Banda Arc, and the Moluccas were elevated above sea level since 5 Ma while Seram and Timor have emerged in the last 3 Ma. These regional plate reorganizations that occurred in Wallacea in the last 5 million years (van Welzen et al., 2005) very likely provided potential stepping stones that brought into close proximity biotas on either side of Wallace's Line (Lohman et al., 2011), allowing plant dispersal to New Guinea and Australia. Thus, the genera *Bridelia,* reached Australia ca. 2 Ma (Li et al., 2009), two different genera of Cucurbitaceae, *Benincasa* and *Neoachmandra*, arrived to Australia from Southeast Asia ca. 5 and 1 Ma, respectively (Schaefer et al., 2009), *Aglaia* colonized Australia ca. 5 Ma (Muellner et al., 2008), while *Begonia* dispersed six independent times from continental Asia and western Malesia to Wallacea and New Guinea dating from the Late Miocene to the Pleistocene (Thomas et al., 2012a). Other species that radiated from west to east in this region during the last 20 Ma belong to the genera *Rhododendron*, the subtribe Livistoninae or the tribe Isonandreae (cited in Schwartz et al., 2015). As the initial citrus radiation triggered by the Late Miocene cooling, the migration of citrus from Southeast Asia through Australia was not unique, but one more of the numerous overwater dispersal events across Wallace's Line into Australasia, a circumstance that was naturally followed by citrus diversification.

An alternative to the west to east dispersal of citrus was proposed by Beattie et al. (2009) that suggested that citrus originated in Australia and dispersed westward in equatorial currents to Southeast Asia. This view that is not supported by the preceding data was also formerly questioned (Pfeil and Crisp, 2008; Bayer et al., 2009; Schwartz et al., 2015). For instance, the inferred age of the terrane movement (Muellner et al., 2008) during the Eocene (about 37 Ma) via the Halmahera route, as Beattie et al. (2009) suggested, appears to be too old to have allowed the rafting of the ancestral citrus species (Pfeil and Crisp, 2008). The age estimations presented in this last work dated the crown of *Citrus* and the Australasian radiation at maximums of 11.8 and 9.7 Ma, respectively, time intervals more compatible with transoceanic dispersal rather than with movements in island terranes, as postulated in previous epochs. Besides, Schwartz et al. (2015) reconstructed the ancestral area of citrus to test the hypothesis that citrus may have originated in Australasia and then migrated to mainland Asia. They found that the ancestral area of *Citrus* was west of Wallace's Line, an observation contradicting an east to west hypothesis. These authors, therefore, speculated that *Citrus* dispersed from western to eastern Malesia across Banda Arc islands connected through stepping stones.

Current evidence, therefore, suggests that citrus radiated in the Late Miocene from mainland Southeast Asia and that during the Early Pliocene an ancestor species related to the extant *Fortunella* crossed the Wallace Line from western to eastern Malesia and then diversified, especially in Australia (Wu et al., 2018). This proposal is compatible with the citrus phylogenetic relationships and the inferred timing of Australian limes diversification (Fig. 2.3). The proposal is also concordant with the predominant west to east dispersal trend observed in the region (Hooker, 1859; Richardson et al., 2012; Thomas et al., 2012a) and with the spatiotemporal diversification patterns reported in other genera. Lastly, our idea agrees with the paleogeography of the region (Hall, 2002, 2009), especially the geological history of the Wallacea that postulates that during Late Miocene onward extensive land masses and islands emerged providing potential stepping stones and allowing island-hopping dispersal (van Welzen et al., 2005). It has also been indicated that some islands of this area have had no connection to surrounding lands and, therefore, land bridges could not have played a significant role in the demarcation of their biota. For these islands, their biota necessarily arose via transoceanic dispersal (Lohman et al., 2011). Lastly, New Guinea was very likely an intermediate stop on the way to Australia, as all the five known *Microcitrus* species found native in eastern Australia are very probably derived from *M. warburgiana*, a species exclusively found in New Guinea (Swingle and Reece, 1967).

2.3.5 Early Pleistocene: Glacial maxima and the diversification of mandarins

Tachibana mandarin is naturally found in Taiwan, the Ryukyu Archipelago, and Japan (Tanaka, 1931). Recent work in Japan suggests that there are at least three types of Tachibana that arose from hybridization of the same unknown parents (Shimizu et al., 2016). Although Tachibana has been assigned its own species, *Citrus tachibana* (Swingle and Reece, 1967; Tanaka, 1954), sequence analysis (Shimizu et al., 2016; Wu et al., 2018) does not support its taxonomic position as a species separate from *C. reticulata*. According to Wu et al. (2018), Tachibana separated from mainland Asian mandarins during the Early Pleistocene (around 2.4 Ma), a period of glacial maxima. This mandarin very probably arrived in these islands from the adjacent mainland during the drop in the water level of the South China Sea promoted by the expansion of ice sheets (Gibbard and Cohen, 2008).

It has been well documented that Taiwan, the Ryukyu Archipelago, and Japan attained their flora (Chiang and Schaal, 2006; Huang and Lin, 2006; Huang, 2014) and fauna (Dobson and Kawamura, 1998) from the adjacent mainland through the emergence of land bridges that occurred mostly during the Pleistocene. Current paleoceanographic evidence indicates that the extension of glaciers led to drastic reductions in the levels of the South China Sea (Wang and Sun, 1994; Voris, 2000) creating land bridges and providing major corridors between the islands. The bridges connecting the mainland with the islands occurred many times throughout the Quaternary (Gibbard and Cohen, 2008) including the period of the Tachibana split.

2.4 A new evolutionary framework for the genus *Citrus*

The advent of new information on the genetic structure of the genome of citrus has provided for the first time, solid bases on the taxonomy, phylogeny, and genealogy of this fruit crop, a fundament allowing unprecedented insights on the origin and evolution of the genus *Citrus*. These findings draw a revolutionary framework for the development of these fruit trees, a scenario that challenges current thoughts and points directly toward a reformulation of this genus. A major consequence of these ideas first affects undoubtedly the taxonomy of the genus since the new framework challenges prior proposals for citrus classification. The current chaotic citrus taxonomy, based on long-standing, conflicting proposals, is certainly awaiting a solid reformulation in line with a full understanding of the hybrid/admixture nature of cultivated citrus. It is now obvious that several "citrus-related genera" (*Fortunella*, *Eremocitrus*, *Microcitrus*, *Clymenia*, and *Oxanthera*) that are in fact nested within the citrus clade should, therefore, more properly be considered species within the genus *Citrus*, while *Poncirus*, a subject of continuous controversy is clearly an out-group to citrus.

During hundreds or even thousands of years, domestication and cultivation of citrus have artificially produced many new types of citrus, a circumstance that has profoundly misled researchers in the recent past and has arrested the reliable development of basic theories and hypotheses on citrus evolution. The discrimination among hybrids, admixtures, and pure citrus species provided by the irruption of comparative genomics, has allowed us to trace back the genealogy and genetic origin of the major commercial citrus cultivars, separating them from the ones that are considered to be pure citrus species.

The identification of citrus species on the other hand has enabled the proposal that the center of origin of citrus was located in the Southeast foothills of the Himalayas, in a region including the eastern area of Assam, northern Myanmar, and western Yunnan, and that from there, citrus dispersed through ancient routes to continental and maritime South East Asia regions. We also propose that the ancestral citrus evolved during Late Miocene through a rapid radiation coincident with a dramatic weakening of the summer monsoons. This drastic transition from wetter monsoonal conditions to a drier climate was probably triggered by a worldwide reduction of CO_2 levels which brought about a period of global cooling that in the tropics was accompanied by a strong phase of aridification. The Australian citrus species and Tachibana, a native Japanese mandarin, apparently split later from mainland citrus. The oceanic dispersal of citrus to Australia occurred during the Early Pliocene and was probably enabled by the occurrence of stepping stones that allowed overcoming the Wallace Line during a profound reshuffling of the orogeny of the region. Tachibana, on the other hand, probably arrived to the Pacific Islands during the Early Pleistocene, a period of glacial maxima that provoked the reduction of the sea levels in the area and the emergence of land bridges between mainland and the islands. These three speciation processes, the Southeast Asian, Australian, and Japanese radiations, were not exclusive episodes of citrus, since similar radiations are also observed in the same areas in many other plant and even animal lineages.

Undoubtedly, many details of this proposal will be nuanced and modified in the coming years as soon as new information becomes available, but today, the proposal has the immense value of providing new suggestions and ideas that will help to renew stimulating challenges and new guidelines in the field of citrus research.

Acknowledgments

The authors acknowledge financial support from the following institutions and companies: Instituto Nacional de Investigación y Tecnología Agraria y Alimentaria, INIA (Ministerio de Economía, Industria y Competitividad e Innovación, Spain) through grant no. RTA-00071-C06-01; the Citruseq-Citrusgenn consortium (Anecoop S. Coop., Eurosemillas S.A., Fundación CajaMar Valencia, GCM Variedades Vegetales A.I.E., Investigación Citrícola Castellón S.A. and Source Citrus Genesis); Florida Citrus Production Research Advisory Council (FCPRAC), Florida Department of Agriculture and Consumer Services grant no. 013646, Florida Department of Citrus (FDOC) and Citrus Research and Development Foundation grant no. 71, on behalf of the Florida citrus growers. The work conducted by the US Department of Energy Joint Genome Institute, a DOE Office of Science User Facility, is supported under Contract No. DE-AC02-05CH11231. Help and expertise of Isabel Sanchis and Angel Boix (IVIA) in lab and field tasks are also gratefully acknowledged.

References

Arakaki, M., Christin, P.A., Nyffeler, R., Lendel, A., Eggli, U., Ogburn, R.M., et al., 2011. Contemporaneous and recent radiations of the world's major succulent plant lineages. Proc. Natl. Acad. Sci. U. S. A. 108, 8379–8384.

Badgley, C., Barry, J.C., Morgan, M.E., Nelson, S.V., Behrensmeyer, A.K., Cerling, T.E., et al., 2008. Ecological changes in Miocene mammalian record show impact of prolonged climatic forcing. Proc. Natl. Acad. Sci. U. S. A. 105, 12145–12149.

Barrett, H., Rhodes, A., 1976. A numerical taxonomic study of affinity relationships in cultivated *Citrus* and its close relatives. Syst. Bot. 1, 105–136.

Bayer, R.J., Mabberley, D.J., Morton, C., Miller, C.H., Sharma, I.K., Pfeil, B.E., et al., 2009. A molecular phylogeny of the orange subfamily (Rutaceae: Aurantioideae) using nine cpDNA sequences. Am. J. Bot. 96 (3), 668–685.

Beattie, G.A.C., Holford, P., Mabberley, D.J., Haigh, A.M., Broadbent, P., 2009. On the origins of *Citrus,* Huanglongbing, Diaphorina citri and Trioza erytreae. In: Gottwald, T.R., Graham, J.H. (Eds.), Plant Management Network. International Conference of Huanglongbing, pp. 25–57.

Berhow, M.A., Hasegawa, S., Kwan, K., Bennett, R.D., 2000. Limonoids and the chemotaxonomy of *Citrus* and the Rutaceae family. In: Berhow, M.A., Hasegawa, S., Manners, G.D. (Eds.), *Citrus* Limonoids: Functional Chemicals in Agriculture and Food. American Chemical Society, Washington, DC, pp. 212–228.

Bhattacharya, S.C., Dutta, S., 1956. Classification of Citrus Fruits of Assam. New Delhi. Monogr. No. 20, Indian Council of Agricultural Research.

Bonavia, E., 1888. The Cultivated Oranges and Lemons, etc. of India and Ceylon. Applewood Books.

Carbonell-Caballero, J., Alonso, R., Ibanez, V., Terol, J., Talon, M., Dopazo, J., 2015. A phylogenetic analysis of 34 chloroplast genomes elucidates the relationships between wild and domestic species within the genus *Citrus*. Mol. Biol. Evol. 32 (8), 2015–2035.

Cerling, T.E., Harris, J.M., MacFadden, B.J., Leakey, M.G., Quade, J., Eisenmann, V., et al., 1997. Global vegetation change through the Miocene/Pliocene boundary. Nature 389, 153–158.

Chen, Z., 1997. Citrus germplasm in China. In: Broadbent, P., Sykes, S.R., Bevington, K.B., Hailstones, D. (Eds.), Proceedings of the Citrus Germplasm Conservation Workshop, Brisbane, Australia, pp. 85–95. October.

Chiang, T.Y., Schaal, B.A., 2006. Phylogeography of plants in Taiwan and the Ryukyu Archipelago. Taxon 55, 31–41.

Clift, P.D., Hodges, K.V., Heslop, D., Hannigan, R., Long, H.V., Calves, G., 2008. Correlation of Himalayan exhumation rates and Asian monsoon intensity. Nat. Geosci. 1, 875–880.

Clift, P.D., Wan, S.M., Blusztajn, J., 2014. Reconstructing chemical weathering, physical erosion and monsoon intensity since 25 Ma in the northern South China Sea: a review of competing proxies. Earth Sci. Rev. 130, 86–102. https://doi.org/10.1016/j.earscirev.2014.01.002.

Cooper, W.C., 1990. Odyssey of the Orange in China. EO Painter Printing Company.

Curk, F., Ancillo, G., Ollitrault, F., Perrier, X., Jacquemoud-Collet, J.P., Garcia-Lor, A., et al., 2015. Nuclear species-diagnostic SNP markers mined from 454 amplicon sequencing reveal admixture genomic structure of modern citrus varieties. PLoS One 10 (5), e0125628.

Curk, F., Ollitrault, F., Garcia-Lor, A., Luro, F., Navarro, L., Ollitrault, P., 2016. Phylogenetic origin of limes and lemons revealed by cytoplasmic and nuclear markers. Ann. Bot. 117, 565–583.

De Candolle, A., 1883. In: Alcan, F. (Ed.), Origine des plantes cultivées. Anncienne Librairie Germer Baillière, Paris, p. vol. 43.

Ding, S.Q., et al., 1984. A new species of Poncirus from China (in Chinese with Latin summary), Acta Bot. Yunnanica 6, 292–293.

Dobson, M., Kawamura, Y., 1998. Origin of the Japanese land mammal fauna: allocation of extant species to historically-based categories. Quat. Res. 37, 385–395.

Fang, J., Wang, Z., Tang, Z., 2011. Atlas of Woody Plants in China: Distribution and Climate. vol. 1 Springer Science & Business Media.

Favre, A., Päckert, M., Pauls, S.U., Jähnig, S.C., Ingo, D.U., Muellner-Riehl, A.N., 2015. The role of the uplift of the Qinghai-Tibetan Plateau for the evolution of Tibetan biotas. Biol. Rev. 90, 236–253. https://doi.org/10.1111/brv.12107.

Ferrari, G.B., 1646. Hesperides sive de malorum aureorum cultura et usu. Sumptibus Hermanni Scheus, Rome.

Fischer, T.C., Butzmann, R., 1998. *Citrus meletensis* (Rutaceae), a new species from the Pliocene of Valdarno (Italy). Plant Syst. Evol. 210, 51–55.

Ghosh, S.P., 1997. Indigenous citrus germplasm in India and conservation strategy. In: Broadbent, P., Sykes, S.R., Bevington, K.B., Hailstones, D. (Eds.), Proceedings of the Citrus Germplasm Conservation Workshop, pp. 96–105. Brisbane, Australia.

Gibbard, P., Cohen, K.M., 2008. Global chronostratigraphical correlation table for the last 2.7 million years. Episodes 31, 243–247.

Gmitter, F.G., Hu, X.L., 1990. The possible role of Yunnan, China, in the origin of contemporary *Citrus* species (Rutaceae). Econ. Bot. 44, 267–277. https://doi.org/10.1007/Bf02860491.

Golonka, J., Krobicki, M., Pajak, J., Nguyen, V.G., Zuchiewicz, W., 2006. Global Plate Tectonics and Paleogeography of Southeast Asia.

Hall, R., 2002. Cenozoic geological and plate tectonic evolution of SE Asia and the SW Pacific: computer-based reconstructions, model and animations. J. Asian Earth Sci. 20, 353–431.

Hall, R., 2009. Southeast Asia's changing palaeogeography. Blumea 54, 148–161. https://doi.org/10.3767/000651909X475941.

Han, C., 1979. Nan fang cao mu zhuang: A Fourth Century Flora of Southeast Asia: Introduction, Translation, Commentaries by Li, Hui-Lin. Chinese University Press, Hong Kong. (it was written in AD 304 by Ji Han).

He, S., Liu, G., Xiang, D., 1984. On the three problems in the study of the germplasm resources of citrus. In: Studart-Montenegro, H.W., Moreira, C.S. (Eds.), Proceedings of the International Society of Citriculture. vol. 1. International Society of Citriculture, Sao Paulo, pp. 100–102.

He, S., Liu, G., Wenpin, L., 1988. Wild mandarin oranges in China. In: Goren, R., Mendel, K. (Eds.), Proceedings of the International Society of Citriculture. vol. 1. Balaban Publishers, Rehovot, pp. 113–121.

Herbert, T.D., Lawrence, K.T., Tzanova, A., Peterson, L.C., Caballero-Gill, R., Kelly, C.S., 2016. Late Miocene global cooling and the rise of modern ecosystems. Nat. Geosci. 9, 843–847. https://doi.org/10.1038/ngeo2813.

Hodgson, R., 1967. Horticultural varieties of citrus. In: Reuther, W., Webber, H., Batchelor, L. (Eds.), The Citrus Industry. vol. I. University of California Press, Berkeley, pp. 431–591.

Holbourn, A.E., Kuhnt, W., Clemens, S.C., Kochhann, K.G.D., Jöhnck, J., Lübbers, J., Andersen, N., 2018. Late Miocene climate cooling and intensification of southeast Asian winter monsoon. Nat. Commun. 9, 1584.

Hooker, J.D., 1859. On the flora of Australia. In: Reeve, L. (Ed.), Its Origin, Affinities, and Distribution, Being an Introductory Essay to the Flora of Tasmania. John Edward Taylor. London.

Hooker, J.D., 1872. In: Reeve, L. (Ed.), The Flora of British India. Savill, Edwards and Co., London.

Huang, S.F., 2014. Hypothesizing origin, migration routes and distribution patterns of gymnosperms in Taiwan. Taiwania 5, 139–163.

Huang, S.F., Lin, T.P., 2006. Migration of *Trochodendron aralioides* (Trochodendraceae) in Taiwan and its adjacent areas. Bot. Stud. 47, 83–88.

Iglesias, D.J., Cercós, M., Colmenero-Flores, J.M., Naranjo, M.A., Ríos, G., Carrera, E., et al., 2007. Physiology of citrus fruiting. Braz. J. Plant Physiol. 19, 333–362.

Kou, X.Y., Ferguson, D.K., Xu, J.X., Wang, Y.F., Li, C.S., 2006. The reconstruction of paleovegetation and paleoclimate in the Late Pliocene of West Yunnan, China. Clim. Change 77, 431–448.

Li, Y., Dressler, S., Zhan, D., Renner, S.S., 2009. More Miocene dispersal between Africa and Asia—the case of *Bridelia* (Phyllanthaceae). Syst. Bot. 34, 521–529.

Liu, G., He, S., Li, W., 1990. Two new species of citrus in China. Acta Bot. Yunnanica 12, 287–289.

Lohman, D.J., de Bruyn, M., Page, T., von Rintelen, K., Hall, R., Ng, P.K.L., et al., 2011. Biogeography of the Indo-Australian Archipelago. Annu. Rev. Ecol. Evol. Syst. 42, 205–226.

Luebert, F., Weigend, M., 2014. Phylogenetic insights into Andean plant diversification. Front. Ecol. Evol. 2, 27. https://doi.org/10.3389/fevo.2014.00027.

Muellner, A.N., Vassiliades, D., Renner, S., 2007. Placing Biebersteiniaceae, a herbaceous clade of Sapindales, in a temporal and geographic context. Plant Syst. Evol. 266, 233–252.

Muellner, A.N., Pannell, C.M., Coleman, A., Chase, M.W., 2008. The origin and evolution of Indomalesian, Australasian and Pacific island biotas: insights from Aglaieae (Meliaceae, Sapindales). J. Biogeogr. 35, 1769–1789.

Nauheimer, L., Boyce, P.C., Renner, S.S., 2012. Giant taro and its relatives: a phylogeny of the large genus *Alocasia* (Araceae) sheds light on Miocene floristic exchange in the Malesian region. Mol. Phylogenet. Evol. 63, 43–51.

Nicolosi, E., Deng, Z.N., Gentile, A., Malfa, S.L., Continella, G., Tribulato, E., 2000. *Citrus* phylogeny and genetic origin of important species as investigated by molecular markers. Theor. Appl. Genet. 100 (8), 1155–1166.

Oueslati, A., Ollitrault, F., Baraket, G., Amel, S.H., Luis, N., Ollitrault, P., 2016. Towards a molecular taxonomic key of the Aurantioideae subfamily using chloroplastic SNP diagnostic markers of the main clades genotyped by competitive allele-specific PCR. BMC Genet. 17, 118. https://doi.org/10.1186/s12863-016-0426-x.

Oueslati, A., Salhi-Hannachi, A., Luro, F., Vignes, H., Mournet, P., Ollitrault, P., 2017. Genotyping by sequencing reveals the interspecific *C. maxima/C. reticulata* admixture along the genomes of modern citrus varieties of mandarins, tangors, tangelos, orangelos and grapefruits. PLoS One 12 (10), e0185618.

Penjor, T., Mimura, T., Kotoda, N., Matsumoto, R., Nagano, A.J., Honjo, M.N., et al., 2016. RAD-Seq analysis of typical and minor *Citrus* accessions, including Bhutanese varieties. Breed. Sci. 66 (5), 797–807.

Pfeil, B.E., Crisp, M.D., 2008. The age and biogeography of *Citrus* and the orange subfamily (Rutaceae: Aurantioideae) in Australasia and New Caledonia. Am. J. Bot. 95, 1621–1631. https://doi.org/10.3732/ajb.0800214.

Potter, P.E., Szatmari, P., 2009. Global Miocene tectonics and the modern world. Earth Sci. Rev. 96 (4), 279–295. https://doi.org/10.1016/j.earscirev.2009.07.003.

Rajput, C.B.S., Hari Babu, R., 1985. Citriculture. Kalyani Publishers, Ludhiana, New Delhi, p. 368.

Renner, S.S., 2016. Available data point to a 4-km-high Tibetan Plateau by 40 Ma, but 100 molecular-clock papers have linked supposed recent uplift to young node ages. J. Biogeogr. 43, 1477–1489.

Richardson, J.E., Costion, C.M., Muellner, A.N., 2012. The Malesian floristic interchange: plant migration patterns across Wallace's Line. In: Gower, D., Johnson, K., Richardson, J., Rosen, B., Rüber, L., Williams, S. (Eds.), Biotic Evolution and Environmental Change in Southeast Asia. Cambridge University Press, New York, pp. 138–163.

Risso, A., Poiteau, A., 1818. Histoire Naturelle des Orangers. Imprimerie de Mme Hérissant Le Doux, Imprimeur ordinaire du Roi et des Musées Royaux, Paris.

Schaefer, H., Heibl, C., Renner, S.S., 2009. Gourds afloat: a dated phylogeny reveals an Asian origin of the gourd family (Cucurbitaceae) and numerous oversea dispersal events. Proc. R. Soc. Lond. B Biol. Sci. 276, 843–885.

Schwartz, T., Nylinder, S., Ramadugu, C., Antonelli, A., Pfeil, B.E., 2015. The origin of oranges: a multi-locus phylogeny of Rutaceae subfamily Aurantioideae. Syst. Bot. 40 (4), 1053–1062.

Scora, R.W., 1975. On the history and origin of *Citrus*. Bull. Torrey Bot. Club 102, 369–375.

Scora, R.W., 1988. Biochemistry, taxonomy and evolution of modern cultivated citrus. In: Goren, R., Mendel, K. (Eds.), Proceedings of the International Society of Citriculture. vol. 1. Balaban Publishers, Rehovot, pp. 83–92.

Shimizu, T., Kitajima, A., Nonaka, K., Yoshioka, T., Ohta, S., Goto, S., et al., 2016. Hybrid origins of citrus varieties inferred from DNA marker analysis of nuclear and organelle genomes. PLoS One 11 (11), e0166969.

Singh, B., 1981. Establishment of First Gen Sanctuary in India for *Citrus* in Garo Hills. Concept Publishing Company, New Delhi, p. 182.

Su, T., Jacques, F.M.B., Spicer, R.A., Liu, Y.S., Huang, Y.J., Xing, Y.W., et al., 2013. Post-Pliocene establishment of the present monsoonal climate in SW China: evidence from the late Pliocene Longmen megaflora. Clim. Past 9, 1911–1920.

Sun, B.N., Wu, J.Y., Liu, Y.S., Ding, S.T., Li, X.C., Xie, S.P., et al., 2011. Reconstructing Neogene vegetation and climates to infer tectonic uplift in western Yunnan, China. Palaeogeogr. Palaeoclimatol. Palaeoecol. 304, 328–336.

Swingle, W.T., Reece, P.C., 1967. History, world distribution, botany, and varieties. In: Reuther, W., Webber, H.J., Batchelor, L.D. (Eds.), The Citrus Industry. Revised second ed. vol. 1. University of California Press, Berkeley, pp. 190–430.

Sykes, S.R., 1997. Citrus germplasm in Australia with special reference to indigenous members of the sub-family Aurantioideae. In: Broadbent, P., Sykes, S.R., Bevington, K.B., Hailstones, D. (Eds.), Proceedings Citrus Germplasm Conservation Workshop. NSW Agriculture, Brisbane, pp. 76–84. 1999.

Tadeo, F.R., Cercós, M., Colmenero-Flores, J.M., Iglesias, D.J., Naranjo, M.A., Ríos, G., et al., 2008. Molecular physiology of development and quality of citrus. Adv. Bot. Res. 47, 147–223.

Tanaka, T., 1931. The discovery of *Citrus tachibana* in Formosa, and its scientific and industrial significance. Stud. Citrol. 5, 1–20.

Tanaka, T., 1954. Species Problem in *Citrus*. Japanese Society for Promotion of Science.

Tanaka, T., 1959. Origin of Citrus fruits with reference to Himalaya region. Hortic. Soc. J. 28 (2), 71–75.

Tanaka, T., 1961. Citologia: Semi-Centennial Commemoration Papers on *Citrus* Studies. Citologia Supporting Foundation, Osaka.

Terol, J., Ibañez, V., Carbonell, J., Alonso, R., Estornell, L.H., Licciardello, C., et al., 2015. Involvement of a citrus meiotic recombination TTC-repeat motif in the formation of gross deletions generated by ionizing radiation and MULE activation. BMC Genomics 16, 69. https://doi.org/10.1186/s12864-015-1280-3.

Thomas, D.C., Hughes, M., Phutthai, T., Ardi, W.H., Rajbhandary, S., Rubite, R., et al., 2012a. West to east dispersal and subsequent rapid diversification of the mega-diverse genus *Begonia* (Begoniaceae) in the Malesian archipelago. J. Biogeogr. 39 (1), 98–113. https://doi.org/10.1111/j.1365-2699.2011.02596.x.

Thomas, D.C., Surveswaran, S., Xue, B., Sankowsky, G., Mols, J.B., Keßler, P.J.B., Saunders, R.M.K., 2012b. Molecular phylogenetics and historical biogeography of the Meiogyne-Fitzalania clade (Annonaceae): generic paraphyly and late Miocene-Pliocene diversification in Australasia and the Pacific. Taxon 61 (3), 559–575.

Tolkowsky, S., 1938. Hesperides: A History of the Culture and Use of Citrus Fruits. J. Bale, Sons and Curnow, London.

Valdiya, K.S., 1999. Rising Himalaya: advent and intensification of monsoon. Curr. Sci. 76, 514–524.

Valdiya, K.S., 2002. Emergence and evolution of Himalaya: reconstructing history in the light of recent studies. Prog. Phys. Geogr. 26 (3), 360–399. https://doi.org/10.1191/0309133302pp342ra.

van Welzen, P.C., Slik, J.W.F., Alahuhta, J., 2005. Plant distribution patterns and plate tectonics in Malesia. Biol. Skr. 55, 199–217.

Vavilov, N.I., 1927. Botanical-geographical principles for selection. Bull. Appl. Bot Plant Breed. 16 (2), 420–428.

Volkamer, J.C., 1708. Nürnbergische Hesperides, oder gründliche Beschreibung der Edlen Citronat, Citronen, und Pomerantzen-Früchte. 2. Sohn and Erben editors, Nuremberg, pp. 1708–1714. Bei Johan Andrea Endters feel.

Voris, H., 2000. Maps of Pleistocene sea levels in Southeast Asia: shorelines, river systems and time durations. J. Biogeogr. 27, 1153–1167.

Wang, P., Sun, X., 1994. Last glacial maximum in China: comparison between land and sea. Catena 23, 341–353.

Wang, L., Schneider, H., Zhang, X.C., Xiang, Q.P., 2012. The rise of the Himalaya enforced the diversification of SE Asian ferns by altering the monsoon regimes. BMC Plant Biol. 12, 210. https://doi.org/10.1186/1471-2229-12-210.

Wang, X., Xu, Y., Zhang, S., Cao, L., Huang, Y., Cheng, J., et al., 2017. Genomic analyses of primitive, wild and cultivated citrus provide insights into asexual reproduction. Nat. Genet. 49 (5), 765–772.

Wang, L., He, F., Huang, Y., He, J., Yang, S., Zeng, J., Deng, C., Jiang, X., Fang, Y., Wen, S., Xu, R., Yu, H., Yang, X., Zhong, G., Chen, C., Yan, X., Zhou, C., Zhang, H., Xie, Z., Larkin, R.M., Deng, X., Xu, Q., 2018. Genome of wild mandarin and domestication history of mandarin. Mol. Plant 11 (8), 1024–1037. https://doi.org/10.1016/j.molp.2018.06.001.

Webber, H.J., 1967. History and development of the citrus industry. In: Reuther, W., Webber, H.J., Batchelor, L.D. (Eds.), The Citrus Industry. Revised second ed. vol. 1. University of California Press, Berkeley, pp. 1–39.

Wen, J., Zhang, J.Q., Nie, Z.L., Zhong, Y., Sun, H., 2014. Evolutionary diversifications of plants on the Qinghai-Tibetan Plateau. Front. Genet. 5, 4. https://doi.org/10.3389/fgene.2014.00004.

Wester, P.J., 1915. Citrus fruits in the Philippines. Philipp. Agric. Rev. III (1), 5–29.

Wu, G.A., Prochnik, S., Jenkins, J., Salse, J., Hellsten, U., Murat, F., et al., 2014. Sequencing of diverse mandarin, pummelo and orange genomes reveals complex history of admixture during citrus domestication. Nat. Biotechnol. 32 (7), 656–662.

Wu, G.A., Terol, J., Ibanez, V., Lopez-Garcia, A., Estela, P.-R., Carles, B., et al., 2018. Genomics of the origin and evolution of *Citrus*. Nature 544, 311–316. https://doi.org/10.1038/nature25447.

Xie, S.P., Manchester, S.R., Liu, K.N., Wang, Y.F., Sun, B.N., 2013. *Citrus linczangensis* sp n., a leaf fossil of Rutaceae from the Late Miocene of Yunnan, China. Int. J. Plant Sci. 174, 1201–1207. https://doi.org/10.1086/671796.

Xu, Q., Chen, L.L., Ruan, X., Chen, D., Zhu, A., Chen, C., et al., 2013. The draft genome of sweet orange (*Citrus sinensis*). Nat. Genet. 45 (1), 59–66.

Yang, X., Li, H., Liang, M., Xu, Q., Chai, L., Deng, X., 2015. Genetic diversity and phylogenetic relationships of citron (*Citrus medica* L.) and its relatives in southwest China. Tree Genet. Genomes 11, 129. https://doi.org/10.1007/s11295-015-0955-x.

Yang, X., Li, H., Yu, H., Chai, L., Xu, Q., Deng, X., 2017. Molecular phylogeography and population evolution analysis of *Citrus ichangensis* (Rutaceae). Tree Genet. Genomes 13, 29. https://doi.org/10.1007/s11295-017-1113-4.

Yu, H., Yang, X., Guo, F., Jiang, X., Deng, X., Xu, Q., et al., 2017. Tree Genet. Genomes 13, 58. https://doi.org/10.1007/s11295-017-1133-0.

Zhang, W., 1981. Thirty years achievements in citrus varietal improvement in China. In: Matsumoto, K. (Ed.), Proceedings of the International Society of Citriculture. vol. 1. International Society of Citriculture, Tokyo, pp. 51–55.

Zhang, D., Mabberley, D.J., 2008. CITRUS Linnaeus, Sp. Pl. 2: 782. 1753. Fl. China. 11, pp. 90–96.

Zhisheng, A., Kutzbach, J.E., Prell, W.L., Porter, S.C., 2001. Evolution of Asian monsoons and phased uplift of the Himalayan Tibetan plateau since Late Miocene times. Nature 411, 62–66.

Zhou, J., 1991. Exploration on the original region of the citrus plants. In: Bangyan, H., Qian, Y. (Eds.), Proceedings of the International Citrus Symposium Guangzhou, China. International Academic Publishers, Beijing, pp. 70–91.

Chapter 3

Domestication and history

Xiuxin Deng[a], Xiaoming Yang[a,*], Masashi Yamamoto[b], Manosh Kumar Biswas[a]

[a]*Key Laboratory of Horticultural Plant Biology (Ministry of Education), Huazhong Agricultural University, Wuhan, China,* [b]*Faculty of Agriculture, Kagoshima University, Kagoshima, Japan, *Present address: College of Forestry, Nanjing Forestry University, Nanjing, China*

The domestication of plants and animals is the most important development in the past 13,000 years of human history (Diamond, 2002). Undoubtedly, domestication is an ongoing process operating under the influence of human activities and nature (Miller and Gross, 2011; Gross and Olsen, 2010). Compared with the most studied annual crops, less is known about how domestication has imprinted the variation of fruit trees (Meyer et al., 2012). *Citrus* and its relatives are the most economically important perennial horticultural crops in the world. In general, *Citrus,* like other long-lived perennials, has a long juvenile phase, extensive outcrossing, widespread hybridization, and limited population structure. Most of the cultivated citrus was domesticated from its wild ancestors or underwent cross-breeding though artificial selection for important agronomic traits, resulting in the unprecedented improvements in citrus flavor, quality, yield, etc. Under long domestication, these outstanding agronomic characteristics, combined with clonal propagation, multiple origins, and frequent gene flow between wild and cultivated species or varieties, contributed to the abundant genetic diversity through the long-term evolution under a complicated environment. Recently, an increasing number of studies have focused on the domestication of *Citrus* during the past years. This not only helps us understand the establishment and origin of cultivated species, but also offers a comprehensive utilization of wild resources to improve the existing varieties and even create new germplasms for *Citrus* genetic improvement. The rapid development of molecular biology and the increasing availability of various germplasms enabled reconstruction of the domestication and evolution history of *Citrus*. Different molecular markers have been successfully applied to shed light on the evolutionary history and domestication of *Citrus* and its relatives (Barkley et al., 2006; Li et al., 2010; Yang et al., 2010; Garcia-Lor et al., 2012; Ollitrault et al., 2012a, b; Snoussi et al., 2012; Ramadugu et al., 2013; Garcia-Lor et al., 2015; Yang et al., 2015; Barbhuiya et al., 2016; Curk et al., 2016; Yang et al., 2017; Yu et al., 2017). Amazingly, the application of whole genome sequences, phylogenomics, genome scans, and selective sweep mapping have greatly expanded the pool of available evidence for *Citrus* (Xu et al., 2013; Wu et al., 2014a, 2018; Carbonell-Caballero et al., 2015; Wang et al., 2017, 2018). This has deeply bolstered and reshaped the current understanding of the pattern and process of domestication and the evolution history of *Citrus* and its relatives.

3.1 The taxonomy, cultivars, and genetic origin of citrus

3.1.1 The taxonomy and the true citrus group

As we know, the name "citrus" derives from the Latin form of "Kedros," a Greek word denoting trees such as cedars, pines, and cypress. Citrus fruits are regarded as hesperides in Greek mythology. Based on the standard of classification the famous botanist Linnaeus ordained, all citrus species known to him belonged to the genus *Citrus*. *Citrus* and related genera are primarily evergreen species belonging to the order Geraniales and family Rutaceae, according to the Engler system. Rutaceae along with other 11 families constituted the suborder Gernaiineae. Rutaceae was further subdivided into six subfamilies including the subfamily Aurantioideae, of which true citrus and related genera were a part. Generally, there were four typical phenotype characteristics for species within the family Rutaceae (Swingle and Reece, 1967). First, the fruit and leaf contained oil glands; second, the ovary was raised on a floral (nectary) disc; third, leaves commonly had pellucid dots; and fourth, axile placentation was present in the fruit. Additionally, many species had polyembryonic seeds that contained both zygotic and nucellar embryos (Davies and Albrigo, 1994).

Unfortunately, the taxonomic situation was still controversial, complex, and sometimes confusing in Rutaceae, especially for the tribes, subtribes, and genera within the Aurantioideae. *Citrus* and related genera hybridize readily and this process has lasted up to now in the wild for thousands of years (Swingle and Reece, 1967). No rigorous reproductive separation

The Genus Citrus. https://doi.org/10.1016/B978-0-12-812163-4.00003-6

among different species, and the greater vigor of the nucellar compared with the zygotic embryos in many species, together increased the challenge of properly classifying the species of the Aurantioideae. In general, species in the Aurantioideae could be characterized as small trees, shrubs, or lianas that produced fruit with a leathery rind or hard shell, and often with a juicy pulp. The leaves and fruits were characterized by schizolysigenous oil glands that released an aroma when touched; the flowers were typically white or purple with fragrance. The fruits were hesperidium berries with specialized structures and juice vesicles, which resulted in plants within the Aurantioideae with their own unique characteristics. Most species in Aurantioideae were evergreen trees, except for three monotypic genera (*Poncirus*, *Aegle*, and *Feronia*), three species of *Clausena*, and one species of *Murraya* (Morton et al., 2003).

The primitive taxonomic systems of the Aurantioideae were artificial systems based just on the morphological characteristics (Bentham and Hooker, 1867; Engler and Rutaceae, 1896; see Chapter 4 for a historical perspective and a modern proposal). Hooker and Engler were the pioneer taxonomists who carried out the research on Aurantioideae, but both of them ignored the existence and importance of nucellar embryony when developing their classifications. During the mid-1900s, Swingle (1948) and later Swingle and Reece (1967) developed a taxonomic system from a practical standpoint, based on several morphological characteristics, that was quite useful and functional. The tribe Clauseneae with five genera, together with the tribe Citreae with 28 genera, for a total of 33 genera, were classified into the subfamily Aurantioideae in Rutaceae (Swingle and Reece, 1967). The Clauseneae was divided into three subtribes: Micromelinae, Clauseninae, and Merrillinae. The Citreae was further subdivided into three subtribes: the Triphasiinae, Balsamocitrineae, and Citrinae. Nearly all the species developed axillary spines in the Citreae, which was totally different from Clauseneae. The simple leaves of the tribe Citreae were also easily distinguished from those of the tribe Clauseneae. The presence of a pulp vesicle in the fruit in the subtribe Citrinae was common, which made them distinct from all the other subtribes in the Aurantioideae. The Citrinae contained a primitive citrus group, a near citrus group, and a true citrus group that included six genera: *Citrus*, *Poncirus*, *Fortunella*, *Eremocitrus*, *Microcitrus*, and *Clymenia*. A description of different species in the true citrus group is given below.

The genus *Fortunella* was named after the English collector for the London Horticultural Society, Robert Fortune. It resembles *Citrus* in the general aspect of the plant and fruits, including four species: *F. margarita*, *F. japonica*, *F. polyandra*, and *F. hindsii*. Hodgson (1967) adds another two species—*F. crassifolia* and *F. obovata*—into this genus. Species in *Fortunella* may be edible and have small fruits, and the number of locules is relatively low. They are slow-growing evergreen shrubs or short trees with dense branches, sometimes displaying small thorns. The leaves are unifoliate with dark glossy green and distinctive silver coloration on their underside, which is another obvious characteristic. The flowers are white, similar to other *Citrus* flowers, and borne singly or in clusters in the leaf-axils. The flowering time is much later than other common *Citrus* species, from one month to two months. The size of the fruit is relatively small and the shape of the fruit ranges from round to ovate. The entire fruit including the peel may be eaten, which is completely different from other species in *Citrus*. Out of these species, only *Fortunella hindsii* is still in a wild state; along the Nanling Mountains of south China, wild populations of this species are commonly found.

Poncirus, also called trifoliate orange or hardy orange, consists of two trifoliolate species. But this genus was thought to be monotypic, with *P. trifoliata* being the only described species, until Ding et al. (1984) discovered and described *P. polyandra* as a new species from Fuming County in Yunnan, China. It is characterized by small trees with trifoliate, deciduous leaves and pubescent fruit. The tree is thorny and marked by being cold hardy. Three-lobed leaves that have a similar size and oval leaves emerge yellowish-green in spring, turn glossy dark green in summer, and fade to yellow in autumn. The flowers are white with pink stamens, and the scent is much less pronounced than that of other species in *Citrus*. The fruits have a very bitter taste due to the presence of ponciridin, which is not palatable. Seeds are plump with many embryos, which means breeders prefer *Poncirus* as a male parent in citrus breeding programs. In addition, it is mainly used as rootstock for cultivated citrus as well as to produce an excellent protective hedge due to the extreme thorniness. Its hybrids, including citrange, citrumelo, and some other varieties, are also used as citrus rootstocks.

Eremocitrus is a xerophytic tree native to Australia, and along with *Microcitrus* has long been geographically isolated from other citrus. Its long and drooping branches guarantee that it would be well-adapted to their xerophytic environment and be able to withstand severe drought and hot dry winds. Leaves are gray-green, very thick, and unifoliate. Flowers are single or in groups in the axils of the leaves on slender pedicels. The fruits are oval to pyriform in shape with fresh peel, similar to all true species and they contain small and hard seeds with wrinkled coats. Although the commercial value of *Eremocitrus* is not significant, it is a good rootstock germplasm and has been used in citrus breeding programs to improve citrus tolerance to salinity, drought, and freezes.

Microcitrus is a delicate rainforest tree that naturally occurs as an understory tree in Australia and consists of six species. This genus has several distinctive characteristics separating it from all of the species in *Citrus,* because of the long period of geographical isolation from other citrus species for millions of years. The special phenotypic traits of *Microcitrus* reflect in its long and slender fruits. Fruits also contain considerable amounts of acidic oil droplets, which is unique in the Citrinae. The thin skin of *Microcitrus* ranges in color from green, yellow, red, and purple to even black. The oil cells on the rind are small giving the fruit a glossy appearance. The inner juice vesicles are cylindrical balls filled with juice. Seeds are white and ovate in shape. Flowers borne in leaf axils and petals are white, also having purplish coloration at the base. Just as the indication of the prefix "micro," the sizes of leaves, flowers and fruits are smaller than that of other citrus species. Occasionally, it has been used as a rootstock in breeding programs.

Clymenia was indigenous in Australia and differed in many aspects of its appearance for the special geographical isolations from the other genera in Citrinae. *Clymenia* is a shrub or small tree without spines. A short, narrow petiole is the typical characteristic that sets them apart from most other citrus. It is usually used as interstock, but it has no commercial importance. It is characterized by a single species with flowers borne singly in the leaf axils on short, stout pedicels and ovoid fruit containing numerous oil glands as well as a large number of polyembryonic seeds.

The *Citrus* are small trees or large shrubs with spiny shoots and alternately arranged evergreen leaves. Branches change from angular to cylindrical as they grow from young to mature. Most species have numerous thorns in the leaf axils, but the thorns may become less prominent as trees mature. The leaves are usually unifoliate with laminae and the petiole size ranges from large (pummelos and its relatives, some citrons) to moderate (oranges and lemons) to small (mainly mandarins and some citrons). Citrus flowers are pentamerous and usually hermaphroditic. The flowers are borne as solitary or in small corymbs. Each odoriferous flower has five white petals and numerous stamens. Citrus fruits are modified berries known as hesperidia. The fruit shapes are various, ranging from spheroid (orange) to oblate (grapefruit and mandarin) to prolate (lemons and limes). Citrus fruits are either seedy or seedless. The shape of seeds varies from obovoid to roundish, and seeds contain one or many embryos. The outstanding difference in the formation of a seed is the polyembryony characteristic, while seeds contain several pairs of cotyledons, including one zygotic embryo and several nucellar embryos as well.

The taxonomy and systematics of the *Citrus* are complex and the precise number of natural species is unclear. This is because some citrus are hybrids clonally propagated through apomixis, and there is genetic evidence that even some wild citrus are of hybrid origin. A high frequency of bud mutation, a long history of cultivation, wide dispersion, wide cross-compatibility, and the paucity of remaining wild citrus stands have joined together to complicate the taxonomy of the genus *Citrus* (Barkley et al., 2006; Scora, 1975; Nicolosi et al., 2000). It is a common phenomenon that bud sports would be vegetatively propagated and maintained, which could lead to subtle differences at the level of genomics among different varieties. Beyond that, the degree of apomictic seed production also caused reduced variability within the species.

Previous studies on relationships between genera and species were carried out based mainly on morphological characteristics. Numerous classification systems have been formulated during the past years of taxonomic research. There is no agreement on what degree of difference justifies species status, and whether apparent hybrids among naturally occurring forms should be assigned species rank (Roose et al., 1995). Most of the modern cultivars have an interspecific origin for the reason that all citrus species are fully sexually compatible, capable of producing fertile interspecific hybrids. Thus, it is common that some of them should be considered as separate races instead of different species.

The genus *Citrus* has variously been described as consisting of from 1 to 162 species. The most widely accepted taxonomic systems today are those of Swingle and Reece (1967) and Tanaka (1977), who recognized 16 and 162 species, respectively. The core arguments between Swingle and Tanaka are that the Swingle taxonomic system has combined too many dissimilar plants into single species while the Tanaka taxonomic system has included many hybrids that do not warrant species status. Another major conflict concerned the treatment of mandarins between the Swingle and Tanaka systems. Swingle placed all mandarins except *C. tachibana* and *C. indica* in *C. reticulata*, where Tanaka recognized 36 species. Swingle and Reece (1967) also divided the genus *Citrus* into two subgenera, *Papeda* and *Citrus*. The subgenus *Citrus* contains edible citrus fruits while *Papeda* consists of various papedas that contain inedible fruits because the fruit has a high concentrations of acrid oil, causing a bitter and unpleasant flavor. In general, researchers prefer to adopt the classification system of Swingle and Reece (1967) that appears to be more popular.

Traditionally, taxonomic determination based on morphological traits at various developmental stages is easily influenced by the environment. According to the opinion of Scora (1975), and Barrett and Rhodes (1976), there are only three basic species within the subgenus *Citrus*: *C. medica*, *C. reticulata,* and *C. maxima,* based on the biochemical and morphological polymorphism analyses. All other species within the subgenus *Citrus* might have derived from hybridization among these three basic species or between them and species of the subgenus *Papeda* or closely related genera. A series of molecular marker studies using restriction fragment length polymorphism (RFLP; Federici et al., 1998), random amplified polymorphic DNA (RAPD), sequence characterized amplified regions (SCARs; Nicolosi et al., 2000), simple sequence repeat (SSR; Barkley et al., 2006), insertion-deletion (indels; Garcia-Lor et al., 2012) and single nucleotide polymorphism (SNP; Ollitrault et al., 2012b) generally supported the idea that these three taxa were certainly ancestors of cultivated citrus. These studies also proposed *C. micrantha* as the fourth basic species, and highlighted that it was the ancestor of lime (*C. aurantifolia*). Although these molecular studies have provided some insight to citrus phylogeny and the three-species concepts were generally accepted, uncertain cases still remained. One of these cases is the state of *Citrus ichangensis,* a very cold-hardy evergreen wild citrus distributed in the mountains of Wuling, Qingba, and Wumeng of China. Molecular marker analysis verified that it is not a hybrid of these three species (see Chapter 2 for further details on the genetic status of *C. ichangensis*). Even today, there exist wild populations in these mountains (Yang et al., 2017).

More recently, Mabberley (1997) also proposed a new classification system for edible citrus that recognized three species and four hybrid groups. This classification system was as similar as a pioneering numerical taxonomic study (Barrett and Rhodes, 1976), and all of them consistently considered that there were only three main taxa as the ancestors of cultivated *Citrus*. However, the subdivision into four hybrid groups remained controversial and agreement could not be widely reached, which limited the potential applications of this system in the field of *Citrus* taxonomy (refer to Chapter 4 for a classification proposal based on genomics).

3.1.2 The genetic origin of some hybrid citrus

Citrus is one of the most important fruit crops in the world, and its diversity and origin have been widely studied (Webber, 1967; Calabrese, 1992; Krueger and Navarro, 2007). The phylogeny and taxonomy for citrus cultivars have been somewhat debatable in the past, as illustrated in Chapters 2 and 4. As mentioned previously, most modern cultivars have an interspecific origin and their genomes can be considered mosaics of large DNA fragments inherited from the basic taxa. The precise contribution of each of these basic species to the genomes of some secondary cultivated species, such as *C. sinensis* (sweet orange), *C. limon* (lemon), *C. aurantium* (sour orange), *C. aurantifolia* (lime), *C. paradisi* (grapefruit), *C. clementina* (Clementine), and *C. unshiu* (Satsuma), is gradually being understood. A series of studies has gained support to clarify some of these relationships in recent years from data derived from morphological characteristics (Swingle and Reece, 1967; Scora, 1975; Tanaka, 1977; Ollitrault et al., 2003), metabolites (Fanciullino et al., 2007), isoenzymes (Ollitrault et al., 2003; Herrero et al., 1996), RFLP (Federici et al., 1998), RAPD and SCAR (Nicolosi et al., 2000), AFLP (Pang et al., 2007), SSR (Barkley et al., 2006; Yang et al., 2015; Curk et al., 2016), chloroplast SSR (Cheng et al., 2005), chloroplast DNA (Li et al., 2010; Carbonell-Caballero et al., 2015; Penjor et al., 2013), and nuclear DNA (Ramadugu et al., 2013; Xu et al., 2013; Wu et al., 2014a, 2018; Wang et al., 2018). All these studies generally supported the three basic species and clarified the complex genetic origin of some hybrid citrus (see Chapters 2 and 4 for a report on the genetic constitution of several citrus hybrids and admixtures through genome analyses).

C. aurantium (sour orange) was related with *C. reticulata* but displayed introgressed traits and markers of *C. maxima.* Based on the analyses of chloroplast genes (Yang et al., 2015; Carbonell-Caballero et al., 2015; Nicolosi et al., 2000) and mitochondrial genes (Froelicher et al., 2011), the results indicated that *C. maxima* acted as the maternal phylogeny. After analyses with nuclear markers and genes, the sour oranges were thought to be a natural hybrid of a mandarin and a pummelo (Barkley et al., 2006; Yang et al., 2015; Scora, 1975; Nicolosi et al., 2000; Barrett and Rhodes, 1976; Uzun et al., 2009). Recently, based on the whole genome resequencing, Xu et al. (2013) and Wu et al. (2014a) consistently supported that sour orange (*C. maxima*×*C. reticulata*) was indeed the direct result of a simple interspecific F1 cross between *C. maxima* as the seed parent and *C. reticulata* as the pollen parent.

C. sinensis (sweet orange) was a natural hybrid that inherited predominately mandarin traits with some pummelo. Although the chloroplast genome of the sweet orange was derived from pummelo, the nuclear analyses suggested that sweet orange had a majority of its genetic makeup derived from mandarin and only a small proportion from pummelo. There was credible evidence that the chloroplast of sweet orange was probably derived from pummelo (Carbonell-Caballero et al., 2015; Nicolosi et al., 2000). Both the whole genome SNP and 202 pairs of polymorphic SSR markers consistently showed that the number of pummelo-hereditary markers and mandarin-hereditary markers fit well with a 1:3 ratio, which comprehensively verified the genome composition of sweet orange (Xu et al., 2013). Therefore, the sweet orange ((*C. maxima*×*C. reticulata*)×*C. reticulata*) originated from an interspecific hybridization with pummelo as the female parent and mandarin as the male parent, followed by a backcross with a male mandarin (Fig. 3.1) (Xu et al., 2013). Roose et al. (2009) and Garcia-Lor et al. (2012) independently provided a similar hypothesis with a different set of molecular markers. However, Wu et al. (2014a) set forth a different argument about the genetic origin of sweet orange (((*C. maxima*×*C. reticulata*)×*C. maxima*)×*C. reticulata*).

The scientific name of grapefruit (*C. paradisi*) was given by Macfadyen in 1830 to a citrus tree known as "forbidden fruit" in Jamaica. The origin of *C. paradisi* was attributed to a natural hybridization between pummelo and sweet orange in the West Indies, where the hybridization would have occurred more than 200 years ago (Nicolosi et al., 2000; Barrett and Rhodes, 1976). Thus, grapefruit might be more correctly classified as an interspecific hybrid than as a species (Gmitter

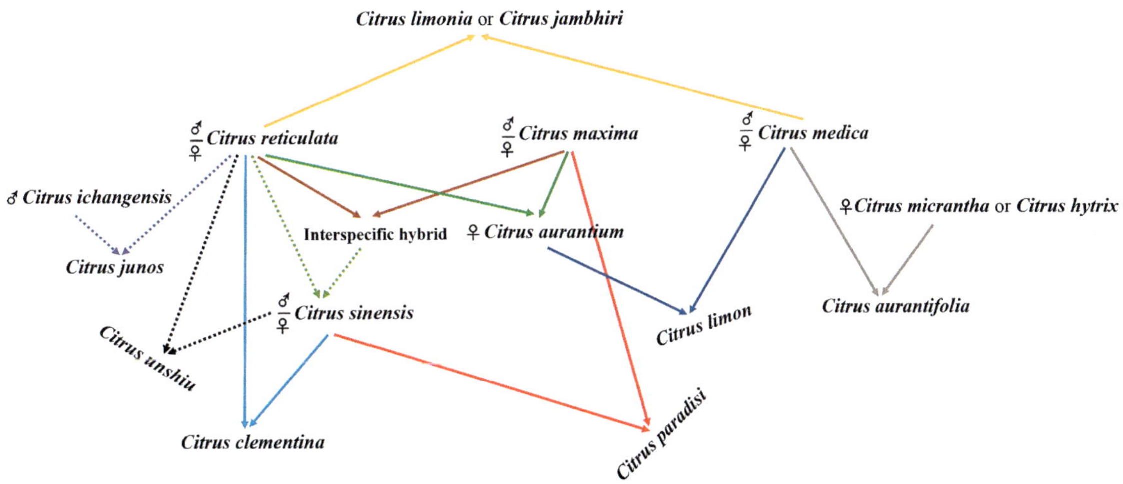

FIG. 3.1 The possible genetic origin of some hybrid citrus, such as *C. sinensis* ((*C. maxima* × *C. reticulata*) × *C. reticulata*), *C. clementina* (*C. reticulata* × *C. sinensis*), *C. limon* (*C. aurantium* × *C. medica*), *C. paradisi* (*C. maxima* × *C. sinensis*), *C. aurantifolia* (*C. hystrix* × *C. medica*), *C. junos* (*C. reticulata* × *C. ichangensis*), *C. limonia* (*C. reticulata* × *C. medica*), and *C. unshiu* (*C. reticulata* × *C. sinensis*). The *dotted lines* represent that there is still some controversies about the origin of these hybrid species, and the *solid lines* represent that the phylogeny relationships have strong evidence to support, such as *C. sinensis* ((*C. maxima* × *C. reticulata*) × *C. reticulata*), *C. clementina* (*C. reticulata* × *C. sinensis*), *C. limon* (*C. aurantium* × *C. medica*), *C. paradisi* (*C. maxima* × *C. sinensis*), *C. aurantifolia* (*C. hystrix* × *C. medica*), *C. junos* (*C. reticulata* × *C. ichangensis*), *C. limonia* (*C. reticulata* × *C. medica*), and *C. unshiu* (*C. reticulata* × *C. sinensis*). The *dotted lines* represent that there is still some controversies about the origin of these hybrid species, and the *solid lines* represent that the phylogeny relationships have strong evidence to support.

Jr., 1995). Most current grapefruits were apparently derived from a stock of wild plants and successfully selected different varieties, such as Duncan, Marsh, Walters, etc. Karyotype analysis suggested that grapefruit possibly resulted from a cross between pummelo and sweet orange. Chloroplast SSR (Cheng et al., 2005) and cpDNA (Carbonell-Caballero et al., 2015; Nicolosi et al., 2000) analyses on the other hand, revealed that *C. maxima* was the female parent of *C. paradisi*. Finally, analyses with nuclear SNP (Ollitrault et al., 2012b), SSR (Barkley et al., 2006), RAPDs, and SCARs (Nicolosi et al., 2000), indicated that *C. paradisi* was a hybrid of the pummelo and sweet orange (*C. maxima*×*C. sinensis*). Although *C. paradisi* was considered a natural hybrid in the Caribbean, another grapefruit called "Chang Shan Hu You" (*Citrus changshan-huyou* Y. B. Chang, sp. Nov.) was also found in Zhejiang Province, China, where it has been cultivated for more than 500 years. Its supposed ancestors were native there from where other natural hybrids of pummelo were known (Hodgson, 1967; Bowman and Gmitter Jr., 1990).

The origin of lemon (*C. limon*) has been a source of conflicting taxonomic opinion. Different varieties, such as Eureka lemon, Lisbon lemon, Femminello lemon, and Villafranca lemon, could not be distinguished by 77 nSSR markers (Yang et al., 2015), which suggested that they originated from a single clone parent via a series of mutations (Gulsen and Roose, 2001). Therefore, different lemons were highly heterozygous and were very similar to each other. In the initial phase, *C. limon* was conventionally considered to be a species by the two most widely cited taxonomic systems (Swingle and Reece, 1967; Tanaka, 1961). However, Nicolosi et al. (2000) were the first to propose by molecular markers that lemons arose from hybridization between *C. aurantium* and *C. medica*. Chloroplast DNA analyses showed that the *C. limon* displayed a contribution from the *C. aurantium* genome. Cytogenetic studies also demonstrated that *C. medica* was a true parental species of lemons (Carvalho et al., 2005). Comparing chromosome types between lemon and citron, Carvalho et al. (2005) showed that *C. medica* was probably a cytogenetically homozygous accession, but the lemon was a cytogenetically heterozygous accession, while all the chromosome types of *C. medica* were clearly presented in lemons. Furthermore, an increasing number of studies based on the nuclear markers supported the opinion proposed by Nicolosi et al. (2000). The results suggested that lemon was likely to be of hybrid origin, with sour orange being the maternal parent and citron being the paternal parent (Garcia-Lor et al., 2012; Ollitrault et al., 2012a, b; Yang et al., 2015; Curk et al., 2016; Garcia-Lor et al., 2013; Gulsen and Roose, 2001).

Rough lemon (*C. jambhiri*) was completely different from the typical lemon mentioned above. Although Webber (1943) thought *C. jambhiri* was a hybrid of lemon and citron, this opinion seemed to have erred from truth. The chloroplast phylogeny analysis verified *C. reticulata* as the maternal parent of *C. jambhiri* (Li et al., 2010) while the citron participated in the genetic origin of *C. jambhiri*, a theory that gained support by RAPD, RFLP, SCAR (Nicolosi et al., 2000), and SSR analyses (Barkley et al., 2006; Yang et al., 2015). Therefore, *C. reticulata* and *C. medica* participated in the origin of *C. jambhiri* (*C. reticulata*×*C. medica*), with the former as the female parent and the latter as the male.

Limonia (*C. limonia*) was an interspecific hybrid and the genetic origin was confused. *C. limonia* was reported to be closely related with citron in the previous studies, and it could be a result of a cross between citron and rough lemon (*C. jambhiri*) or mandarin (Li et al., 2010; Nicolosi et al., 2000). After a series of studies based on the nSSR and chloroplast datasets, there was a great possibility that both *C. reticulata* and *C. medica* participated in the formation of *C. limonia* (*C. reticulata*×*C. medica*), with the former as a female parent and the later as the male (Yang et al., 2015; Curk et al., 2016).

The origin of lime (*C. aurantifolia*) was controversial and several hypotheses were proposed for its genetic origin. Barrett and Rhodes (1976) thought that it probably arose from a tri-hybrid cross involving the *C. medica*, *C. maxima*, and *Microcitrus* species, whereas molecular data supported the hypothesis that the lime was a hybrid between *C. medica* and a Papeda species (Nicolosi et al., 2000; Federici et al., 1998; Torres et al., 1978). Several studies demonstrated that it is a direct interspecific cross between *C. medica* and *C. micrantha*, which was confirmed by the analyses of chloroplast, mitochondria, and nuclear markers (Garcia-Lor et al., 2012; Ollitrault et al., 2012a, b; Nicolosi et al., 2000; Garcia-Lor et al., 2013; Froelicher et al., 2011; Wu et al., 2018). *C. hystrix* also might be another potential female parent to participate in the genetic origin of *C. aurantifolia* (Ollitrault et al., 2012a, b; Yang et al., 2015; Curk et al., 2016).

Clementine mandarin (*C. clementina*), also known as the Algerian tangerine, is believed to be a chance seeding from a Mediterranean mandarin that appeared a little more than one century ago in Algeria. Molecular studies have demonstrated that the Clementine was likely to be a hybrid of mandarin and sweet orange (Ollitrault et al., 2012a; Nicolosi et al., 2000; Samaan, 1982). Wu et al. (2014a) further confirmed this hypothesis at the whole genome level by definitively identifying that the Willowleaf mandarin and sweet orange were both parents of Clementine (*C. reticulata*×*C. sinensis*), with the former as the female parent and the latter as the male.

The taxonomy of *C. junos* is still controversial. *C. junos* was called "Xiang Cheng" in Chinese or "Yuzu" in Japanese. Swingle and Reece (1967) believed that *C. junos* was an intersubspecific hybrid of *C. ichangesis* and *C. reticulata*, and therefore did not grant it species status. However, Tanaka (1954) claimed that Yuzu was a species instead of a hybrid. Hirai et al. (1986) suggested that Yuzu was not a hybrid between *C. ichangesis* and *C. reticulata* by isozyme analysis. Bayer et al. (2009) suggested that *C. ichangensis* hybridized easily with *C. reticulata* and the hybrids, called ichandrins (ichang lemon), were probably of the same origin as the long-cultivated *C. junos*. Based on the cpDNA analysis, Bayer et al. (2009) further inferred that *C. junos* was a hybrid of *C. reticulata* and *C. ichangensis*. However, mitochondrial markers revealed that *C. junos* would have the same maternal origin as *C. ichangensis* and *C. micrantha*, suggesting that the maternal origin of *C. junos* was papeda (Froelicher et al., 2011).

The genetic origin of Satsuma mandarin (*C. unshiu*) was full of mysteries. Satsuma mandarin was similar to "Bendiguangju" in the tree habit and morphological characteristics of fruit and flower. Based on the isoenzyme and nuclear DNA marker analyses, both Bendiguangju and the Satsuma mandarin have identical geographical origins in Zhejiang province, China, and similar nucleotide composition, which were not only morphologically similar but also clustered together (Fang, 1990; Hu, 1998). In addition, the results analyzed by cpSSR data indicated that these two genotypes differed in chloroplast genome composition: the chloroplast of Satsuma mandarin came from the *C. reticulata*, whereas that of Bendiguangju came from *C. sinensis* (Cheng et al., 2005). In other words, it was possible that the Japanese Buddhists took a mandarin that had gotten pollen from sweet orange back to Japan and successfully bred the primitive Satsuma (*C. reticulata*×*C. sinensis*) through seed propagation (Cheng et al., 2005). However, the controversy about the origin of the Satsuma mandarin still exists and different hypotheses have been provided during the past years. Satsuma mandarin was a putative hybrid between Kunenbo (*C. nobilis*) and Kishu mikan (*C. kinokuni*) (Shimizu et al., 2016). According to the opinion from Fujii et al. (2016), the parental diagnosis of the Satsuma mandarin by nuclear and cytoplasmic markers showed that it appeared to derive from an occasional cross with the Kishu mikan (*C. kinokuni*) type of mandarin, which would be a derivative or synonym of Nanfengmiju, and the pollen parent of the Kunenbo (*C. nobilis*) type of mandarin, which would be a derivative or synonym of Benduguanju.

3.2 The cultivation history and distribution of citrus

3.2.1 Ancient Chinese citrus

3.2.1.1 The citrus taxonomy and varieties in ancient China

There was a long domestication, cultivation, and management history for *Citrus* in China. The time could trace back to at least more than 4000 years ago. The Chinese people started to domesticate wild citrus in the ancient eastern Asian continent. During the natural evolution and culture, many types or cultivars with important or potential values have been formed. Throughout the ages, the ample available literature and references have recorded some relevant information about plentiful citrus species or varieties as well as citrus cultivation management obtained from daily experience and practices in great detail.

Ancient dynasties in China regarded citrus as a highly valued tribute (Webber, 1967; Tolkowsky, 1938). Citrus as fruit offerings first appeared in the oldest Chinese reference, "*Shih Ching*" (Si Jing) ("*Book of Odes*"), under the section titled "*Tribute to Yu*" of the Chinese Imperial Encyclopedia, which could date back as far as the early 8th century BCE or the late 9th century BCE (Cooper and Chapot, 1977). It was written down that *Citrus* was one of the most important tributes to be sent to the imperial court at An-Yang, a territory near the big bend in the Yellow River, China, to commemorate the god of irrigation and mythical Emperor Yu (2205–2197 BCE) who was regarded as the founder of the equally ancient and mysterious Xia dynasty. *Chu* (Ju) and *Yu* (You) were selected as the main tributes by the primitive tribes in the slave society from the not yet completely reunified areas of central and south coastal China. According to the records in *Si Jing*, the term *Ju* refers to small mandarins and the kumquat (*Fortunella*) while the term *You* indicated the pummelo (*C. maxima*) and the Yuzu (*C. junos*) (Spiegel-Roy and Eliezer, 1996). In the list of tribes, there are other names such as *Luan*, *Pao*, and *Wendan* that represented some common pummelo varieties that have been cultivated for millennia in China. Pummelo was probably multiplied by air layering (marcottage) in traditional citrus cultivation history in ancient China.

But there was no record of *Gan* in the list of tributes, which might include large mandarin-type fruits and possibly also oranges. During the Zhou dynasty (1046–256 BCE), literature flourished and lots of poems and books were written in this period. And it was no surprise that *Ju* and *You* as two types of citrus were mentioned most often in the "*The Five Canons,*" which was one of the most famous mythological, historical, philosophical, and literary documents in China for a millennium. In ancient China, *Zhi* represented *Poncirus*, which is a close genus to *Citrus*, and it was first mentioned by the poet Yu Song in the "*Fu on the wind*" at the end of the Zhou Dynasty as that was a favorite tree of birds to build their nest according to descriptions.

Some other citrus were mentioned in the Han dynasty (202 BCE–220 CE). A Chinese poet in the Western Han period described *Cheng*, *Lu Ju*, and *Huang Gan* or *Kan* in his prose poem. The *Cheng* in ancient China usually would be the sour orange (*Citrus aurantium*), *Lu Ju* might represent the kumquat, and the *Huang Gan* referred to the yellow mandarin, probably including large mandarin-type fruits and sweet oranges. Apparently, there was no specific mention of sweet orange in the ancient Chinese literature. However, more and more scholars suggest that the term *Gan* corresponded to sweet orange in some cases. In fact, this was a generic name for all citrus in China. From the time of the Han dynasty, Guandong selected yellow- and red-skinned *Gan* orange-type fruits to pay tribute to the courts and emperors for centuries. The Tai race, who were once dominant in Southern China and then migrated from the west and southward into Burma and Assam, would promote the propagation and distribution of citrus in some cases because the species or varieties termed *Kan* were propagated by seeds in the Khasi hills of Assam until now.

In the Jin Dynasty (263–306 CE), there are numerous historical records referring to citrus in this period. The first description of citron appeared only later by Han Ji in the Jin Dynasty in "*Records of the Plants and Trees of the Southern Region,*" which was the earliest regional flora report in the world and has been handed down from ancient times. In the book, the term *Kuo Han* was used for the first time to represent citron and its relatives, such as Finger citron, called the hand of Buddha. Citron and finger citron were considered a sort of talisman and were exchanged for good luck in ancient China. Additionally, the book also first pointed out that fruit farmers could effectively control fruit fly damage to citrus fruits by using the yellow ant as a method of biological control. This indicated that the ancient Chinese had acquired scientific management measures for citrus cultivation.

There were a great number of citrus species or varieties in ancient China. The "*Collection of Citrus,*" written in 1178 by Yanzhi Han in the Southern Song Dynasty, was the first monograph on citrus taxonomy and variety in China and the world as well. It records 27 species or varieties of citrus in total, mainly grown in Wenzhou, Zhejiang province, east China. It was considered to be the most influential and famous book on citrus ever written. Han developed a taxonomic system and citrus was first divided into three categories based on the morphological characteristics: the orange, the mandarin, and the tangerine. Han's system further divided the orange, mandarin, and tangerine into 5, 14, and 8 species, respectively. Based on today's knowledge, most of the classification is reasonable, as only the kumquat was wrongly classified as a tangerine. Recent studies have suggested that some of the citrus varieties recorded in this book are still under cultivation. Besides, in the book, Han also summarized common citrus propagation, cultivation, and pest control techniques, and showed the curative effects of citrus fruit on health. Shizheng Li, a prominent doctor in the Ming Dynasty (1368–1644 CE), also took full advantage of the characteristics of the fruit peel to make the distinction among different citrus, such as mandarin, tangerine, and pummelo; this was recorded in the well-known traditional Chinese medicine book "*The Compendium of Materia Medica.*" These classification criteria had a profound effect on the present citrus taxonomy and still have a reference value up to now.

3.2.1.2 The distribution of citrus in ancient China

The book "*The History Records*" clearly documented that *Citrus* was widely cultivated in the Sichuan, Hunan, and Hubei provinces in the southwest and center of China and achieved high economic benefits in the Han Dynasty. But beyond that, citrus seeds were excavated in the Ma Wang Dui tomb of the Western Han dynasty in Changsha, Hunan province, China. Both history records and archeological evidence consistently showed that citrus was cultivated extensively and formed a large-scale industrial cluster in the Western Han dynasty. Meanwhile, according to some relevant records, with the development and booming of the citrus industry, the imperial governments in ancient China set up management officials in the main production areas in the Sichuan and Guangdong provinces of China.

Even as early as the Xia Dynasty (2100–1600 BCE), *Citrus* had been listed as one of the main tributes and taxes in major Chinese citrus producing areas, such as Sichuan, Jiangsu, Anhui, Jiangxi, Hunan, and Hubei provinces. Although Nanchong is a small city that lies in Sichuan province in the southwest of China, people named the city the "Hometown of the orange" because its high citrus production and the local cultivation history could data back to the Zhou Dynasty. The time when citrus (possible the Ruju tangerine, a small citrus-like Nanfeng tangerine) first left the original habitats in South China and was introduced to the north part of China would be at the end of the Kaiyuan Reign period (713–741 CE) in the Tang dynasty (618–907 CE) using the seed propagation system according to the records in the book "*Unofficial Biography of Yang Taizheng.*" There was also a long history regarding citrus in Yunnan province and *C. verrucosa* was one of the well-known endemic varieties characterized by its crumpled pericarp. *C. verrucosa* would be at least 1700 years old there. Both Fuzhou city and Zhangzhou city in Fujian province in the southeast coastal areas of China were inextricably linked to the cultivation of citrus for the suitable climatic conditions. The primary outline of the citrus industry in ancient China did not form until the Tang Dynasty and the distribution trend continues to this day. This has brought significant impact to the current Chinese citrus industry and remains meaningful.

3.2.1.3 The spiritual symbol, medicinal value, and management of citrus in ancient China

Citrus has been a symbol of the distinguished spirit and moral integrity since ancient times. "*Ode to the orange*" was the most famous and first romantic collection of poems in the history of Chinese literature by Yuan Qu in the Warring States Period (403–221 BCE). Qu highly eulogized the holy and pure symbol of citrus and wanted to use citrus to dilute the reality of darkness and pain, and to show his big and bold ambitions for the other lands. Recent investigations showed that he might possibly have written this poem during his exile for a few years in the south of Chu, the current location of Huaihua of Hunan province. In this area, a great diversity of sweet orange varieties can still be found along the Yuan River, which is connected to the Yangtze River. In ancient times, the river was the most convenient way to travel by boat. The citrus trees also have religious significance and are thought to ward off evil. In Fujian province, Chinese ancestors preferred to plant citrus fruit trees around the temples, which indicated that citrus was endowed with a sense of holiness to some extent during the process of cultural heritage.

For thousands of years, the ancient Chinese civilization highly prized citrus so much for its medicinal values. Some relevant medical mechanism was described in detail and valuable viewpoints of its medical function were put forward in multiple medicinal works. "*Shennong's Classic of Materia,*" a medical book from the Qin dynasty (221–207 BCE), mentioned that some citrus could keep breathing unobstructed, clear internal heat, and make us fresh. The book was the first to summarize and emphasize the medicinal value of citrus in daily life. "*Compendium of Materia Medica,*" the best-known medical book of traditional Chinese medicine by Shizheng Li in the Ming Dynasty (1368–1644 CE), addressed the medicinal functions of some citrus in a much more systematic manner than before. The treatment concept has been held in high esteem since it was first published. *P. trifoliata*, *C. aurantium*, *C. wilsonii,* and *C. maxima* had been widely utilized as natural medicines and successfully applied to treat a variety of ailments and injuries, such as coughing, phlegm, dyspepsia, chronic tracheitis, emphysema, chronic gastritis, and even blood hypertension.

As early as the Tang Dynasty (618–907 CE), people had mastered fruit storage technology. Farmers usually put a thin layer of wax onto the surface of citrus fruit to maintain freshness, which made the fruit's appearance better. In the late stage of the Northern Song Dynasty (960–1127 CE), people made full use of the different natures of citrus (hot in nature) and mung beans (mild in nature) to mix them together to prolong the shelf life of citrus.

3.2.1.4 Wild citrus in China

During the past 60 years, Chinese scientists have been investigating the citrus germplasm. During the 1970s and 1980s, several wild mandarins including Daoxian and Mangshan mandarins as well as Chongyi wild mandarin, have been located in the northern slopes of Nanling mountains along the 25°N of latitude. Later, a primitive type of Mangshan wild mandarin was also found in this area. Recent genome sequencing indicated that this Mangshan mandarin (Fig. 3.2) that shows a lot of

FIG. 3.2 Mangshan wild mandarin tree and its fruit.

seeds and a few juice sacs, is the most primitive type of loose skin mandarin (Xu Qiang, unpublished data). Besides the wild mandarin, wild papeda including *Citrus hongheensis* was exploited in Honghe of Yunnan province. A recent investigation of the citrus germplasm further showed that a wild *C. ichangenesis* population could be found in the mountains of a large area, including the Three Gorges area, Wuling and Wumeng mountains, more than hundreds of kilometers from the south to the north (Yang et al., 2017). Also, wild kumquat (*Fortunella hindsii*) populations with diversity in fruit and leaf shape have been exploited recently in the mountains of Nanling and Wuyi, covering an area of 800 km (from east to west) along the 25–26°N latitude (Deng Xiuxin, unpublished data).

3.2.2 Ancient citrus in Japan

Citrus is one of the most important fruit crops in Japan. Although various accessions of *Citrus* species are cultivated in the southwest of Japan, almost all of them are nonnative, that is, they were introduced from abroad (e.g., *Citrus maxima*, *C. nobilis*, and *C. aurantium*), arose as chance seedlings (e.g., *C. unshiu* and *C. iyo*), were selected from bud sports (e.g., many early maturing *C. unshiu*), or were bred by artificial pollination (e.g., Kiyomi, Shiranuhi, and Setoka). Only two species, *C. tachibana* and *C. depressa*, were present in Japan before the start of recorded history.

3.2.2.1 Citrus tachibana *(Tachibana)*

C. tachibana is a small mandarin (Fig. 3.3) mainly distributed on the Pacific side of west Honshu, Hikoku, and Kyushu, the main islands of Japan. In addition, it was discovered on the Ryukyu Islands and Taiwan (Tanaka, 1931; see Chapter 2). Most *C. tachibana* trees are now grown in yards of houses or shrines. Wild trees also exist (Hirai et al., 1990). Although this species is not edible, the Japanese have deified it. Because the Imperial Household placed a high value on *C. tachibana*, its trees were planted in the garden of the Imperial Palace. Some traditional poems described this species in the oldest existing collection of poems, called "Manyo-shu." Because it was the only evergreen citrus in ancient Japan, the Japanese have focused much reverent attention on it. Trees are highly resistant to frost and snow. Its petiole wing is small. The fruit is somewhat flattened and very small. The rind color and texture are reddish yellow and smooth, respectively. It has polyembryonic seeds.

C. tachibana has not been distributed on the Asian continent and can be clearly distinguished from mandarins that originated there. Hirai et al. (1990) demonstrated unique isozyme genotypes of *C. tachibana*. Almost all wild trees of *C. tachibana* had CC at peroxidase (Px) while the allele Px-D was common in the Chinese mandarin. The Allele A at the glutamate oxaloacetate transaminase (Got)-2 was also a characteristic allele in *C. tachibana*. Some *C. tachibana* had M at the Got-2 allele, which was widely found in the Chinese and Indian mandarins. However, this M allele of *C. tachibana* was not considered to have originated from the Chinese mandarin because there were no indications of an influence of the Chinese mandarin on the Px genotype. Hence, M at Got-2 is considered to have originally developed in *C. tachibana*. In addition, DNA analyses revealed the individual characteristic of *C. tachibana*. It could be clearly distinguished from

FIG. 3.3 Fruits and a leaf of *C. tachibana*.

other *Citrus* species regarding the nuclear genome (Shimizu et al., 2016). Both the mitochondrial DNA (mtDNA) and the chloroplast DNA (cpDNA) of *C. tachibana* were different from those of other species (Araújo et al., 2003; Nagano et al., 2014; Yamamoto et al., 1993; Yamamoto et al., 2013; Carbonell-Caballero et al., 2015). The chromosome configuration also demonstrated the individuality of *C. tachibana*. When citrus chromosomes are stained with chromomycin A3 (CMA), each chromosome shows a characteristic CMA-positive band and each species possesses a unique chromosome configuration. Among various *Citrus* species, the Japanese mandarin, including *C. tachibana,* only has the type F chromosome (Yamamoto, 2007). These findings indicate that *C. tachibana* in Japan has been isolated from the citrus grown in other regions and has differentiated in a unique way.

There is diversity in the isozyme genotypes in the *C. tachibana* population of Japan, although no variation in morphological traits has been observed (Hirai et al., 1990). In contrast, there is no diversity of genotypes found in introduced species (Hirai and Kajiura, 1987). Although *C. tachibana* is polyembronic, not only nucellar seedlings but also zygotic ones appear. Thus genetic variation exists in the *C. tachibana* population. This is one reason why *C. tachibana* is considered to be indigenous to Japan.

Some natural hybrids of *C. tachibana* were discovered (Hirai et al., 1990). They were considered to be chance seedlings that arose from cross-combination between *C. tachibana* and introduced mandarins. Some old Japanese mandarins such as Koji and Yatsushiro may have been selected among natural hybrids (Hirai et al., 1990; Hirai and Kajiura, 1987).

3.2.2.2 Citrus depressa *(Shiikuwasha)*

C. depressa is indigenous to the Ryukyu Islands (Tanaka, 1926). The Ryukyu Islands are located between approximately 24 and 31° north and 122 and 131 degrees east (Fig. 3.4). Its wild trees were also discovered in Taiwan (Tanaka, 1931). The largest difference in habitat between *C. depressa* and *C. tachibana* is that the former has not been found on the main islands (Honshu, Shikoku, and Kyushu) of Japan. It can be said that *C. depressa* is an indigenous citrus growing on subtropical Japanese islands (Ryukyu Islands) and in Taiwan. Most of the *C. depressa* trees are grown in yards of houses or small orchards. Wild trees also exist. This species is edible and has been used for several purposes. Premature fruits are popular acid citrus on the Okinawa Islands. Mature fruits are used for fresh consumption and making juice. Various fruit-processed products such as candy and dressing have been developed.

It is a small mandarin and shows an oblate and peelable fruit like *C. tachibana* (Fig. 3.4). In contrast to *C. tachibana* (Fig. 3.3), highly diverse morphological traits are shown by *C. depressa*. There are many accessions of *C. depressa* and each of them shows various fruit sizes, rind texture, rind color, and juice quality. All forms of *C. depressa* have polyembryonic seeds. Almost all of them are seedy, but a seedless cultivar, Nakamoto Seedless was selected (Medoruma et al., 2011).

Citrus fruits are essential sources of some phytonutrient components. Among them, polymethoxyflavonoids (PMFs), unique components of citrus, show efficacy against lifestyle-related diseases such as cancer and diabetes (Kawai et al., 1999; Lee et al., 2010; Miyata et al., 2008). The PMF content of *C. depressa* was much higher than that of *C. unshiu*, which is a leading citrus species in Japan (Yamamoto et al., 2008). Products that are rich in PMFs have been produced using the fruits of *C. depressa*.

FIG. 3.4 Fruits and a leaf of *C. depressa*.

C. depressa has not been distributed on the Asian continent and can be clearly distinguished from mandarins that originated there by isozyme analysis. All *C. depressa* accessions analyzed had A at Got-2 (Yamamoto et al., 2011). As mentioned above, A at Got-2 is a characteristic allele of the Japanese mandarin. This result indicates the individuality of *C. depressa*. According to the isozyme genotypes, there is a possibility of relationships between *C. depressa* and *C. tachibana*. However, differences between them were detected. In comparison with *C. tachibana*, all *C. depressa* showed heterozygous genotypes at the Got-2 and Px allele except for Ishikunibu. It is uncertain whether the heterozygous genotypes are due to a hybrid origin of this species or mutation.

It is well known that wide phenotypic diversity exists within *C. depressa*. DNA analyses have shed some light on the existence of genetic diversity within *C. depressa*. First, Yamamoto et al. (1998) reported diversity within this species using a limited number of accessions on the basis of random amplified polymorphic DNA (RAPD) analysis. Then, more accessions of *C. depressa* derived from several of the Ryukyu Islands were used for sequence-related amplified polymorphism (SRAP) analysis (Yamamoto et al., 2017). This revealed the wide diversity of the species. Although *C. depressa* shows polyembryony, its differentiation may have been mainly caused by natural hybridization. A mutant usually cannot be distinguished from the original accession by SRAP analysis. Almost all accessions of *C. depressa* could be differentiated from each other by SRAP. In addition, a given accession could be easily propagated by nucellar seedlings after a new genotype arose. No clear relationships between our SRAP results and geography were evident.

DNA analysis of the cytoplasmic genome also provides useful information on the genetic characteristics of *C. depressa*. Some cpDNA analyses of different regions of the genome revealed that there were two types of cpDNA in *C. depressa* (Nagano et al., 2014; Yamamoto et al., 2013; Yamamoto et al., 2017; Urasaki et al., 2005). One type was identical to that of *C. tachibana* and the other type was identical to that of the Chinese mandarin *C. sunki* and Indian mandarin *C. reshni* (Yamamoto et al., 2013; Yamamoto et al., 2017). Both types were different from those of major mandarins such as *C. reticulata*, *C. unshiu*, *C. kinokuni*, *C. medica*, *C. maxima*, and papeda. The divergence of the cpDNA genome of *C. depressa* indicates a polyphyletic origin of this species. In addition, there is a possibility that hybrids derived from accessions with both cpDNA types would appear because some *C. depressa* accessions are self-incompatible (Yamamoto et al., 2006) and accessions with both cpDNA types are distributed on the same islands. In addition, the type of cpDNA, namely that of maternal origin, influences the constitution of the nuclear genome of *C. depressa* (Yamamoto et al., 2017). The type of mtDNA of *C. depressa* was identical to that of *C. sunki* and *C. reshni* and different from that of major mandarins (Froelicher et al., 2011).

Although genetic diversity existed in *C. depressa*, it could be clearly distinguished from the other species (Fig. 3.4) (Yamamoto et al., 2017). *C. depressa* was related to *C. tachibana* but both species could be distinguished from one another. The fruit characteristics of *C. sunki*, such as the size, shape, skin color, texture, and polyembryony, resemble those of *C. depressa* (Tanaka, 1948). However, *C. sunki* was distinguished from *C. depressa* by SRAP markers, most closely aligned instead with *C. reshni* and more distantly with *C. kinokuni* and *C. reticulata*. Japanese mandarins, *C. depressa* and *C. tachibana*, could be discriminated from other citrus species in SRAP analysis. Studies on DNA (Barkley et al., 2006; Federici et al., 1998) and isozyme profile (Hirai et al., 1986) as well as the chromosome configuration (Yamamoto and Tominaga, 2003) also reported characteristic features of those Japanese mandarins. From these results, Japanese mandarins are considered to be genetically isolated from other citrus, suggesting that they evolved and differentiated in a unique way. They may possess a unique genome within the genus *Citrus*.

The indigenous *C. depressa* and introduced *C. nobilis* may have played a role in the origin of many local citrus on the Ryukyu Islands. *C. keraji* and *C. oto* are closely related to *C. nobilis* based on the results of isozyme and DNA analyses (Yamamoto et al., 2011).

3.2.3 Ancient citrus in India

Citrus is one of the vital fruit crops that has both economic and social impacts in Indian society. It's not only important for the human diet in India, but also very important from a religious point of view because citrus is one of the most important items in many Hindu customs. According to old literature that was documented in 800 BCE in Chinese and Sanskrit, citrus fruits have a long history of use on the Indian subcontinent (Upadhaya et al., 2016). From then to today, many tribes have used citrus species to prepare traditional medicine. In northern India, the indigenous tribal community conserved and maintained indigenous wild and semiwild citrus species in the forest and also homestead gardens. In this section, we mainly demonstrate the past, present, and future prospects of the ancient citrus variety on the Indian subcontinent.

3.2.3.1 History, distribution, and origin of ancient citrus in India

Citrus has a long domestication history in the Southeast Asian region. The use of citrus in the Indian subcontinent started thousands of years back, as described in the ancient book "Ayurveda." For example, the peels, seeds, and pulp of citron fruit were used in Ayurvedic treatment in the subcontinent for treating abdominal colic, digestive disorders, piles, etc. History books, story books and literature demonstrated the cultivation and use of citrus in Indian society. During the archeology excavation of the Mesopotamian civilization, citrus seeds were found that were from 4000 BCE. It is believed that there existed a road connecting the Mesopotamian area with the Indian subcontinent and the west area of China. The archeology in Chengdu, Sichuan of China has proved that there existed commercial activity between Sichuan and the Middle East in about 1000 BCE. Recent studies have shown that the area that includes Yunnan of China, northern Burma, and northeast India harbors the most diversified citron germplasm (Yang et al., 2015).

During the regime of Akbar the Great, citrus fruits were sold in the local market. At that time, the most common citrus varieties were limes (*Citrus aurantifolia*, Kagzi Neembu); Galgal (*Citrus limon*, Galgal, Jamir); Ghep (citrus like fruits); Bijaura (*Citrus medica*, Bijora or Pahari Nimbu); Narangi (*Citrus reticulata* mandarin, Santra); Jambhiri (*Citrus limon* Jambhir, Jamir, Kagzi nimbu) etc. Through today, these varieties remain common in the local markets in India. From the above discussion, it is clear that Indians started using citrus a long time ago. It also proved the availability of citrus in this region since ancient times.

The Indian subcontinent is a land of plant and animal diversity and its civilization was one of the most ancient civilizations on Earth. This region is the origin of many plant species. Although citrus grows in almost all states of India, the states from the northeast are known as the "Treasure House of Citrus Genetic Wealth." It is claimed that *Citrus* and its relative species originated in south or southeast Asia, and the main center of origin is northeastern India, including the Assam, Arunachal Pradesh, Meghalaya, Manipur, Mizoram, Nagaland, Tripura, and Sikkim states (Bhattacharya and Dutta, 1951; Bhattacharya and Dutta, 1956; Dutta, 1958; Singh, 1967; see also Chapter 2 for a recent proposal). These regions are located in the northeastern Himalayas and foothills of the central and western Himalaya tracts. A large amount of citrus diversity is found in this region in both wild and cultivated forms (Chadha, 1995; Singh and Singh, 2001). This northeastern region, globally known as the paradise of citrus genetic diversity, is the reservoir of megabiodiversity of citrus and its relative species. This is the natural home of many citrus species; therefore, many citrus explorers recognized this region as the hot spot of citrus diversity (Singh and Singh, 2006; see Chapter 2). In the south Indian regions, there are many other types of indigenous and wild mandarins (Singh and Singh, 2006).

Vavilov in the early 1950s (Vavilov, 1950) reported that mandarin (*C. reticulata*), sweet orange (*C. sinensis*), citron (*C. medica*), lemon (*C. limon*), sour lime (*C. aurantifolia*), Rangpur lime (*C. limonia*), and Jenerutenga (*C. nobilis*) were found in both the cultivated and wild forms in the northeast Himalayan region in India. Bhattacharya and Dutta (1951, 1956) reported that the wild sweet orange "Soh Nairiang," the wild Indian mandarin (*C. indica*), *C. ichangensis*, *C. latipes*, *C. assamensis*, and *C. microptera* were widely distributed in the various parts of northeastern India. They are even found up to 2000 M in elevation (Ghosh, 1977). Verma (1999) and Singh and Singh (2001) found that the Hill lemon (Galgal), *C. pseudolimon*, and Attani (*C. rugulosa*) were common in the foothills of the Himalayas in the northwest. Singh (1981) recommended the northeastern Himalayan region and the northwestern region of India as the best places for collecting primitive germplasms of citrus. Many Indian citrus researchers believed and proved that the progenitors of commercial citrus varieties originated in India (Bhattacharya and Dutta, 1951; Bhattacharya and Dutta, 1956; Dutta, 1958; Singh, 1967). Hore et al. (1997) and Singh and Singh (2001) collected several types of wild and semiwild citrus from the Khasi, Jaintia, and Garo Hills of Meghalaya, Mizoram, and Tripura Hills.

3.2.3.2 Citrus indica

Citrus indica is a wild endangered *Citrus* species native to northeastern India that is mainly found in the Tura range of the Garo Hills region of Meghalaya. Locally, it is known as Memong Narang (Memong = Ghost, Narang = Citrus) or as the Indian wild orange (Hazarika, 2012). According to Tanaka's opinion, it is one of the most primitive *Citrus* species and is perhaps the progenitor of many cultivated citrus (Singh, 1981; Malik et al., 2006). This species is well adapted in a wide range of elevations. *C. indica* bears flowers and fruits more abundantly in dense forests than those found in the periphery of human habitats. The fruits are small and weigh 15–20 g with a spherical shape. The surface is a deep orange red to almost scarlet; it is also smooth, juicy, and inedible. The fruits of *C. indica* are used in traditional medicines against kidney stones, many stomach problems, and viral infections. This species has some unique features that are very special among the citrus species, for example, its flowering time starts from September and lasts until January, a phenomenon that is very rare in other citrus species. These features may contribute to keeping this species' purity because this time period is the resting period for other citrus species. *C. indica* is highly resistant to biotic and abiotic stresses (Malik et al., 2006). Molecular marker analysis verified that *Citrus indica* is the hybrid of citron with *Citrus reticulata* instead of pure wild citrus.

3.2.3.3 Citrus macroptera

Citrus macroptera is a wild citrus species found in the evergreen and moist deciduous forests of the northeast region, including the Meghalaya, Manipur, and Mizoram states and the north Himalayas, Assam (Cachar, Karbi Anglong, Nagaon and Sivsagar) in India (Nair and Nayar, 1997). This species, commonly known as Satkara or Hatkara, is also known by its local name in different states in the northern region of India such as being called the Heiribob in Manipur. Its fruits are edible and commonly used in culinary preparations as well as in making squashes. The fruits are harder than other citrus species and their storage life is about two months. Its fruits are also used as traditional medicine for the treatment of stomach pains and alimentary disorders. *C. macroptera* is tolerant to greening and exocortis virus diseases. It is widely used as rootstock.

3.2.3.4 Citrus latipes

C. latipes is a wild nonedible citrus species mostly distributed in the Shillong plateau, Pynursha, Mawphlang, and Cherrapunji area of the Khasi hills of Meghalaya (Singh, 1981), Assam, and Nagaland (Dutta, 1958) in the northeastern states of India. Locally, it is known as Soh-Shyrkhoit in the Khasi language (Soh = fruit, Shyrkhoit = monkey), meaning the fruit of a monkey. *C. latipes* is commonly called "Khasi papeda." Its size and shape are very similar to the Kaffir lime (*C. hystrix*). Therefore, sometimes *C. latipes* is mistakenly identified as the Kaffir lime. It grows like a small, thorny tree and looks similar to kaffir limes and Ichang papedas. The tree grows to 30 ft, the foliage is dense, and the petioles are winged. The fruits are nonedible, big in size with large and bold seeds. This species is endangered and facing the threat of extinction.

3.2.3.5 Citrus assamensis

C. assamensis is locally known as Soh- Sying (Soh = fruit, Sying = ginger) and Ada Jamir (Ada = ginger, Jamir = Citrus) in the Khasi hills and Assam, respectively. This noneconomical wild citrus species is threatened with extinction. They are found in the home stated garden in Assam, Meghalaya and Bangladesh. It is native to the Cachar area of Assam and Khasi hills in India. *C. assamensis* is very closely related to the sour pummelo (*C. megaloxycarpa*). The tree grows from medium to tall sizes and is thorny with pear-shaped fruits. The leaves and fruits of this species have a strong aroma like ginger.

3.2.3.6 Citrus megaloxycarpa *Lushaigton (Sour Pummelo)*

C. megaloxycarpa, also known as the Sour pummelo, is one of the ancient and indigenous citrus species in India. It is widely found in the Tripura and Jaintia hills in Meghalaya, Mizoram, Assam, and Meghalaya. According to the IUCN list of categories, *C. megaloxycarpa* faces the threat of extinction. Fruits of this species are nonedible, large and seedy, and the fruit juice is highly acidic. *C. megaloxycarpa* apparently resembles the *C. maxima* (true pummelo), but is distinct in the presence of marginate or slightly winged petioles. The fruits are bitterly sour, and the purplish tinged petals and seeds have purple colored chalazal spots.

3.2.3.7 *Citrus limettioides*

C. limettioides originated in India and is known as the Indian sweet lime; it is also locally known as Mithanimbu or Sharbati or the Palestine sweet lime. *C. limettioides* is commonly found in the Garo and Jaintia hills of Meghalaya. The tree is medium to large size with an irregular spreading form. The flowers are pure white, and the fruits are small and round to slightly oblong with a thin smooth rind with prominent oil glands. Also, the juice is less acidic while the seeds are few and highly polyembryonic.

3.2.3.8 *Citrus limetta*

It is commonly known as mousambi, musambi, sweet lime, sweet lemon, and sweet limetta. It is native to South Asia and Southeast Asia, especially northeast India, and is commonly found in the Baghmara area of the Garo hills, Meghalaya. It is a small tree growing up to 8 m (26 ft) in height, with irregular branches with numerous thorns. Its bark is relatively smooth and brownish-gray. The petioles are narrow but the distinctly winged leaves are compound with acuminate leaflets. The flowers are white, and the fruits are oval and green with a greenish pulp. *C. limetta* begins bearing fruit at 5–7 years old and its peak production often occurs at 10–20 years. It is propagated by seeds.

3.2.3.9 *Ancient member of* Citrus sinensis

There are three wild citrus species: the Tasi, Sohbitara, and Soh nairiange belonging to *Citrus sinensis*. These are native to the northeastern hilly region in India, including Arunachal Pradesh and Meghalaya. The fruits are medium in size with a skin that is smooth to rough and juice that is highly acidic.

3.2.4 The origin, spread, and introduction of citrus

Plenty of hypotheses about the history and geographical origin of *Citrus* have been suggested during the last century, but the origin of *Citrus* is still identified with a history full of controversy and legends (see Chapter 2). It would appear that *Citrus* and its related genera were generally believed to be native to the subtropical and tropical areas of southeastern Asia, including southern China, northeast India, the Indo-Chinese peninsula, and the Malay Archipelago, and then spread to other continents (Webber, 1967; Gmitter Jr and Hu, 1990; Khan, 2007). Within this large region, as early as the 1930s, Tolkowsky (1938) raised the idea that the possible origin center would be mountainous parts of southern China and northeastern India. This is where sheltered valleys and southern slopes as geographic barriers were protected from cold and dry winds, and extremely influenced by the warm rains of the summer monsoon. Although Tanaka (1954) acknowledged that some species started in China and spread to Indo-China, Malaysia, northeastern Asia, and finally Japan, Tanaka insisted that *Citrus* might have originated in northeastern India and Burma, and China was just a secondary center of distribution. Tanaka (1954) also provided a line of demarcation called the "Tanaka line" running from the northwestern border of India, above Burma, to Yunnan province of China, to south of the island of Hainan in the South China Sea. Some citrus species such as citron, lemon, pummelo, and sour and sweet orange originated south of this line while mandarin, trifoliate orange, and others originated north of the line. But, Swingle and Reece (1967) and Jackson and Ziegler (1991) did not support Tanaka's opinion and suggested that the introduction of citrus into cultivation and the origin of several species were more likely to have started in China. Owing to complicated and unique landforms and various climate types, the local area contained a diversity of terrestrial ecosystems, and was one of the places enjoying the richest biodiversity (Huang et al., 2011). Interestingly, conclusive evidence and detailed distribution information of the wild citrus in these areas were more and more unambiguous. With the development of the economy in China and improvement of science supported by the Chinese government, an increasing number of wild, semiwild, and garden-yard citrus germplasms has been explored in succession during the past decades. Interestingly, some of these rare valuable citrus germplasms were sporadically distributed as individuals, but most of them were widely distributed as populations in different scales. There is no doubt that these new discoveries of old, wild, indigenous, representative, and diverse citrus germplasms, such as *C. hongheensis* (Yang et al., 2010), *C. ichangensis* (Yang et al., 2017), *C. junos* (Tanaka, 1954; Yu, 1979), *C. aurantium* (Gmitter Jr and Hu, 1990), *C. sinensis* (Gmitter Jr and Hu, 1990), *C. medica* (Yang et al., 2015; Guo, 1981; Ramadugu et al., 2015), *C. limonia* (Gmitter Jr and Hu, 1990), *C. hystrix* (Yu, 1979; Guo, 1981), *P. trifoliata* (Pang et al., 2007; Hu, 2015), *P. polyandra* (Ding et al., 1984), *F. hindsii* (Chen, 2011), and various wild mandarins including *C. nobilis*, *C. reticulata* Daoxian, *C. reticulata* Chongyi, and *C. reticulata* Mangshan (Li et al., 2007) will challenge the traditional hypothesis of the genetic origin center of *Citrus*. Both the specific climate and complex geomorphology have profound effects on local climate and consequently on *Citrus* and its relative's distribution. In addition, the enormous river channels promoted citrus distribution and large-scale valley networks provided a natural and stable ecosystem for generations of propagation. Calabrese (1998) indicated that China was the origin center and citrus passed from its original location to other Asian regions accompanied by civilization. Gmitter Jr and Hu (1990) also emphasized that southwest China, especially Yunnan province, acted as a joining link between the Indo-Burma and central China-Japan regions of citrus diversification and evolution instead of a dividing zone devoid of significant citrus germplasms. Abundant citrus germplasms and high genetic diversity, coupled with the available natural system for citrus dispersal, totally supported the hypothesis that southwest China was one part of the primitive center origin of contemporary citrus species, along with nearby areas of India, Burma, and southern China (Gmitter Jr and Hu, 1990).

It is well known that China has a magnificent culture with a long history and has domesticated many fruit plants, including citrus, for production purposes and aesthetic reasons since the second millennium BCE. In many cases, water streams and human activities promoted citrus seeds to go through a long-distance spread from the sites of origin and culture to new habitats. Many types of citrus were more likely to move to the west through the Arabian areas before Christ (Tolkowsky, 1938). However, the concrete time that citrus spread from the general origin area to other regions following the paths of civilization and extensive movement of multiple types of citrus would probably occur before the existing literal records.

Citron (*C. medica*) was native to northeastern India, Burma, and southwest China, where it was found in the valleys at the foot of the Himalaya mountains and the adjacent zone (Lim, 2012). Although the concrete origin area of citron was still mysterious, the seeds were found in Mesopotamian excavations dating back to 4000 BCE. However, Isaac (1959) thought that the cultivation history of citron in Mesopotamia was relatively later and the possible time would begin at 600 BCE. When the armies of Alexander the Great entered Asia (about 330 BCE), they found citron in Media, also called the Persian apple, and citron was subsequently introduced into the Mediterranean region. Citron first appeared in the Latin literature in the Cloanzius Verus under "citreum" in the second century BCE. One side of the Jewish coin, coined in 136 BCE, clearly bore the portrait of citron. The great Roman poet Virgil (70–19 BCE) gave a poetic description of the citron fruit and its properties in the book "*The Fruit of Media*." Controversy about whether citron was mentioned in the Bible did not seem to calm down, but the citron was used for worship by Jews in the feast of Tabernacles (50–150 CE), which was supported by ample archeological as well as literature records (Reuther et al., 1967; Nicolosi et al., 2005). It could be determined that citron was the first citrus fruit that was familiar to Europeans and remained the only representative of citrus there until the 7th century CE. Although the genetic origin of citron was once considered to be in India, some reports provided different opinions that the citron would be indigenous and come from China based on molecular marker analyses (Yang et al., 2015; Gmitter Jr and Hu, 1990; Ramadugu et al., 2015). Since the 1970s, Chinese researchers have discovered many wild citrons in the valleys and virgin forests in southwest China, which provided solid evidence for the genetic origin of citron associated with this area. Also, there was a trade route about 2000 km long that was well-known for its silk trade, dubbed the "Southern Silk Road" by historians, and extended from south China to India. Moreover, the engraved symbols on utensils and bronze relics with the Middle East's unique culture and history unearthed at the Sanxindui ruins in Sichuan province, China, were strong proof that the Southern Silk Road boosted trade links between China and India, even countries in the Middle East. With the increase of trade contacts and cultural exchanges, for the long-storage stability and pharmacological functions, citron, one of the most possibly selected fruits, was introduced to India or the Middle East by Chinese traders during the long and drawn out journey (see Chapter 2 for alternatives to this proposal).

The origin of the lemon (*C. limon*) is unknown, although the lemon was thought to have first been grown with a long history of cultivation in southwest China, northeast India, or northern Burma (Morton, 1987). It was possible that a hybrid between citron and the sour orange created an intermediate species (Yang et al., 2015; Curk et al., 2016). Although lots of sculptures, mosaics, and frescoes were regarded as evidence to support the idea that lemon arrived in Europe no later than the second century CE (Swingle and Reece, 1967), the images of that in ancient Rome were too vague to be solid evidence to support this opinion. The lemon was used as an ornamental plant in early Islamic gardens and first recorded in an Arabic treatise on farming near the 10th century CE (Morton, 1987). The wide distribution of lemon would occur between 1000 and 1500 CE through the Arab world and the Mediterranean region in connection with the expansion of the Arabian Empire (Morton, 1987). The famous navigator Christopher Columbus brought lemon seeds to Hispaniola on his voyages and then the lemon was later introduced to the New World near the 16th century. Coincidently, Spanish conquest throughout the New World helped to spread lemon seeds further.

The sour orange (*C. aurantium*) was a hybrid of mandarin (*C. reticulata*) and pummelo (*C. maxima*) (Yang et al., 2015; Wu et al., 2014a), generally considered to be native to southeastern Asia, also possibly China and India. Arabs were thought to promote the distribution of the sour orange to the west associated with society activities and warfare, progressively. According to Swingle's opinion (Swingle and Reece, 1967), Arabs would carry the sour orange to Arabia in the 10th century CE and the spreading range was greatly enhanced, reaching its peak in the heyday of the Arabic dynasty (750–1258 CE). For 500 years, it was the only orange in Europe and was the first orange to reach the New World.

The sweet orange (*C. sinensis*) probably originated as a natural hybrid between pummelo and mandarin with multiple hybrids (Xu et al., 2013). It has been grown in Southeast Asia since ancient times, possibly originating in Southern China,

North Burma, Northeast India, and as far as south Indonesia (Webber, 1967; Tanaka, 1954). Although Tanaka (1954) thought that sweet oranges originated in the tropical rainforests of upper Burma and Assam, Cooper (1989) suggested that the sweet orange migrated from southwest China to upper Burma rather than by the reverse path. Actually, south China was the most possible origin area of the sweet orange because of the existence of wild sweet oranges in the mountains or mixed village gardens there. The apomictic characteristic of the sweet orange provided more chances to make it naturalize later in the tropical rainforest region. The Western countries were unaware of the existence of the sweet orange until 1500 CE. The Portuguese traders were known to bring superior selections of sweet oranges from China; the sweet orange was also known as the Portugal orange in many Mediterranean languages. As we know, the actual commerce of sweet oranges took place with the discovery of a new trade to the Cape of Good Hope and at last reached the destination, the West Indies. On his second voyage in 1493, Columbus brought the seeds of oranges, lemons, and citrons to Haiti and the Caribbean. The sweet orange was introduced to America by Spanish explorers and conquerors. The Washington navel orange originated in Bahia, Brazil, from where it was imported into the United States in 1870, where it apparently received its name. It has been suggested to be a mutation of the Seleta sweet orange. Among varieties of oranges in the world, the Washington orange was the most widely planted variety for its exceptionally delicious, seedless, easy-peeling fruits; many other cultivars arising from mutations of it have promoted its distribution all around the world.

The pummelo (*C. maxima*), also known as shaddock after the sea captain Captain Shaddock, was almost certainly in the region of the Malaysia and Indonesia archipelagos and was also widely distributed in Fiji. Calabrese (1992) stated that pummelo was of tropical origin (Malayan archipelago), which was similar with other researchers' opinions that pummelo might have spread from the Malayan and Indonesian archipelagos to China instead of spreading in the opposite direction. High-quality pummelo cultivars and culture developed in Thailand during the 8th and 13th centuries CE with the mode of propagation being air layering. However, the pummelo also grew widely in parts of China and has been gathered and cultivated by the Chinese for thousands of years. Some hybrids of the shaddock were found by crusaders in Palestine in 900 CE and were brought to Europe and then to the Caribbean through navigation (Webber, 1967). In the mid-17th century, pummelo seeds were introduced to the West Indies from the Malay archipelago, where citrus cultivation was already well established. Pummelo would follow a similar path to that of sour orange and sweet orange to arrive in Europe. One form of shaddock, named Adam's apple, grew in the Holy Land around 1187 CE, and was introduced to Spain by the Arabs at the same period (Tolkowsky, 1938). There were indications that pummelo seeds were brought to Barbados by Captain Shaddock, who also promoted the spread of pummelo to Latin America.

Mandarins (*C. reticulata*) were native to the tropical and subtropical regions containing Indo-China and south China, with traders carrying selections to eastern India. Mandarins were introduced to the West at the turn of the 19th century, which was much later than other citrus. In 1805, two varieties of mandarin were first introduced to England from Guangdong, China. From England, it was then introduced to the Mediterranean region. The Willowleaf (*C. deliciosa*) was transported from China to the Mediterranean region after 1805 and soon become the major species there. *C. unshiu* was known as "Wenzhou migan" in China, but in Japan it was known as "mikan" or formally "unshu mikan," which was similar as the local reading of the same characters used in Chinese. Besides, one of the English names for the fruit was Satsuma, which was derived from the ancient state of Satsuma in the Kagoshima area of Japan. The highly distinctive Satsuma mandarin was considered to have been cultivated in Japan prior to 1600 CE, which was the time of its earliest known reference. Tanaka (1932) suggested that it probably originated as a chance seedling in Japan during the Tang dynasty (618–907 CE). Actually, according to Chinese history records, the Satsuma was a seedless and easy-peeling citrus mutant of Chinese origin and was introduced to Japan by a Japanese Buddhist priest from Unshu, China, who gave rise to Satsuma; however, it was introduced to the West via Japan.

The center of origin of *C. junos*, also called "Xiangcheng" in Chinese or "Yuzu" in Japanese, was presumed to be the upper region of the Yangtze river and was spread widely into the southern areas of China, and as a result, there were many landraces in China. The *C. junos* peel was buried in Jingzhou, Hubei Province, in the Tsu Dynasty (450 BCE) to perfume a lady's suitcase (Zhang, 1990). About the 8th century CE, it was first introduced to Japan from China via Korea (Taninaka et al., 1981). Moreover, after introduction, it has produced various progenies and formed a special group of acid citrus through self- or cross-pollination with other citrus in Japan. The cultivation of *C. junos* was more popular and commonly performed in Japan than in China.

The kumquat (*Fortunella*) was native to the south Asia region, mainly in south China. The earliest historical reference to kumquats appeared in Chinese literature in the 12th century CE. They were introduced to Europe in 1846 by Robert Fortune and shortly thereafter into North America. The trifoliate orange (*P. trifoliata*), a cold-hardy tree that was native to China, was introduced to many countries without clear time records and used primarily as hardy rootstocks for citrus.

The history of the spread and introduction of citrus in China is an epitome of Chinese culture and history merging into the whole world. Up to now, the names of some citrus species or varieties still retained the original Chinese place names or personal names. A kind of sweet orange named "Lue Gim Gong," which was a result of selective breeding from the common sweet orange, was used from the memory of a Chinese-American immigrant named Liu Jingguang (in Mandarin; in Cantonese the name is Lue Gim Gong), who helped local people to plant citrus fruit trees in Florida in the United States. Satsuma was still called unshiu (Wenzhou) with Chinese characters and unshiu was also used as a species name. Besides, "ponkan" was a kind of mandarin and its name also came directly from the Chinese characters.

3.2.5 The genetic diversity of *Citrus*

Genetic diversity is essential for any species to underwrite its ability to adapt and survive in the face of environmental change. Citrus germplasm collections and in-depth characterization of citrus genetic resources were increasingly appreciated as a repository for genetic improvement, and their evaluation was an essential prerequisite for their utilization in citrus breeding. Recently, a series of citrus germplasm studies has received much attention with the development of modern biotechnology, which provided new capabilities to characterize the genetic diversity of the citrus germplasm including wild, semiwild, garden-yard, and cultivated citrus species and varieties. Trends of genetic diversity usage should be of benefit to design more efficient breeding programs and management for citrus, and then promote the effective utilization and conservation of citrus.

3.2.5.1 The genetic diversity of semiwild, cultivated, and garden-yard citrus

As one of the pioneers who focused on the genetic diversity studies on *Citrus*, Barkley et al. (2006) selected SSR markers to detect molecular polymorphisms among 370 mostly sexually derived citrus accessions from the collection of citrus germplasms maintained at the University of California, Riverside. A total of 275 alleles were detected with 11.5 alleles per locus and the value of the observed heterozygosity ranged from 0.171 to 0.710. Most of accessions were initially acquired from other collections around the world and these germplasms have appropriate representativeness to reflect the possible genetic diversity of *Citrus* around the world.

There was a high genetic similarity within the mandarin group, which was a single species composed of several genetically different individuals and great number of hybrids, rather than a large number of species as proposed by some taxonomic studies. The cultivated mandarins had a narrow genetic base after evaluation of the genetic similarity among 35 mandarin accessions, including 10 species and 7 hybrids with RAPD markers (Filho et al., 1998). The values of observed heterozyosity and expected heterozygosity of the wild mandarin in the northern mountain area of Vietnam were from 0.03 to 0.96 and 0.03 to 0.92, respectively, by SSR markers (Froelicher et al., 2008). The phylogenetic analyses of mandarin landrace, wild mandarins, and related species in China were performed using nuclear *LEAFY* second intron and plastid *trn*L-*trn*F sequence (Li et al., 2007). Based on the molecular-based evolutionary tree of relationships among mandarin landraces and wild mandarins, there was an obvious difference between the landrace and wild mandarins. Interestingly, both the phylogenetic analysis of nrDNA and cpDNA sequence data showed that wild mandarins were not a monophyletic group. Garcia-Lor et al. (2015) analyzed the genetic diversity and population structure of the mandarin, mainly obtained from American, Mediterranean, and Asiatic sources, by nuclear, chloroplastic, and mitochondrial markers to point out that the mandarin was highly polymorphic and many genotypes believed to be pure mandarins were really hybrids and had introgressions from other basic taxa in their genomes.

The genetic variability in 38 grapefruits and three pummelos was assessed by RAPD and SSR markers (Corazza-Nunes et al., 2002). The majority of grapefruit accessions showed a narrow genetic base, suggesting that the observed morphological polymorphism within the group must be associated with somatic mutations, which were not easily detected by molecular markers. Yu et al. (2017) used SSR markers to evaluate the genetic diversity and population structure of 274 pummelo accessions in China and the value of the observed heterozygosity and genetic differentiation (F_{st}) was 0.325 and 0.077, respectively. Pummelo germplasms in China could be divided into three main subpopulations (the southeast, southwest, and central regions of South China). Higher genetic diversity was observed in southwest China and frequent gene flow was detected between the southeast and southwest gene pools, which indicated that southwest China was one of the most important origin centers of the pummelo. Similarly, Wu et al. (2014b) used the high resolution melting analysis of SNPs and sequencing of DNA segments to genotype 260 pummelo germplasms, including 55 accessions from Myanmar. Based on the field investigation, high levels of haplotype diversity and apparent population structure were identified among different pummelo populations. Pummelos were accidentally brought downstream by rivers flowing through Yunnan province, China, in ancient times, and then dispersed to surrounding regions by animals and human activities, which eventually developed isolated populations (Wu et al., 2014b).

Luro et al. (2012) studied 24 citron varieties preserved in the citrus germplasm of INRA-CIRAD, San Giuliano, France, with nuclear and cytoplasmic markers and noted a higher-than-expected polymorphism rate among Mediterranean citron varieties, likely due to cross-fecundation. This idea was backed up by further observation from Ramadugu et al. (2015), who also analyzed the genetic diversity of 15 accessions of citron from the Mediterranean and also thought there was a relatively high genetic diversity there with SSR and SNP markers. Yang et al. (2015) evaluated the genetic diversity of 56 accessions of citron and relatives mainly from southwest China by SSR markers and the averaged values of observed and expected heterozygosity of citron and relatives were 0.36 and 0.49, respectively. Coincidentally, Yang et al. (2015) and Ramadugu et al. (2015) got a similar conclusion that the genetic diversity of the citron in southwest China was higher than that of the citron in the Mediterranean. But beyond that, considerable heterozygosity was observed in certain citrons, contrary to previous reports (Yang et al., 2015; Ramadugu et al., 2015). Similarly, Barbhuiya et al. (2016) analyzed 219 individuals of citron collected from four wild and eight domesticated populations using SSR markers to assess the genetic structure and diversity of the citron in the wild and domesticated populations in India. The mean values of observed and expected heterozygosity ranged from 0.222 to 0.540 and 0.438 to 0.733, respectively, among the wild and domesticated populations. However, compared with the genetic diversity of the citron from the east, China and India, the possible origin center of the citron, the citron varieties from the west, the Mediterranean, showed relatively lower genetic diversity (Yang et al., 2015; Barbhuiya et al., 2016; Ramadugu et al., 2015; Luro et al., 2012).

Many lemons and limes have similar morphological and biochemical characteristics and some were known to have originated by mutation from representative varieties. The genetic diversity of a large sample of lemon and lime cultivars from a wide range of geographic locations with molecular markers was reported. With ISSR (intersimple sequence repeats), SSR, and isozyme, Gulsen and Roose (2001) suggested that most lemons had nearly identical marker phenotypes, suggesting they originated from a single clonal parent via a series of mutations. Citrons contributed the largest part of the lemon genome and a major part of the genomes of rough lemons, sweet lemons, and sweet limes. Uzun et al. (2011) also carried out the evaluation of genetic diversity in lemons and some of their relatives based on SRAP and SSR markers. Although nearly all accessions could be distinguished, there was a low level of genetic diversity detected among lemon cultivars. Curk et al. (2016) performed extended analyses of the diversity, genetic structure, and origin of 133 citrus accessions (limes and lemons) on a large scale by cytoplasmic and nuclear makers. The results showed that all lemon and lime accessions were highly heterozygous, with interspecific admixture of two, three, and even four ancestral taxa genomes in *Citrus*. *C. medica* participated in the genetic origin of all limes and lemons, acting as the direct male parent of almost all of them due to the multiflowering times of citron, which would increase the possibility of pollination to other citrus types.

3.2.5.2 The genetic diversity of wild and semiwild citrus

Knowledge regarding the distribution of wild and semiwild citrus (such as *Poncirus*, *C. hongheensis*, *C. ichangensis*, and *C. aurantium*) is relatively limited, especially because they were hard to find and collect, let alone explore the genetic diversity of these accessions. However, some of them were selected as rootstocks, and only a little was known about these valuable citrus germplasms.

The genetic diversity of the Tunisian citrus rootstock germplasms was predominantly due to high heterozygosity and differentiation between four facultative apomictic species (*C. aurantium*, *C. sinensis*, *C. limon*, and *C. aurantifolia*) that contained a total of 201 local accessions, which was assessed with nuclear SSR and mitochondrial Indel markers (Snoussi et al., 2012). The identification of a core sample of rootstock accessions could be used to conduct research on the physiology and agronomical evaluations, and that will be integrated into citrus rootstock breeding programs in the future.

Twenty-eight accessions of trifoliate orange and hybrids and seven accessions of its relatives were analyzed for genetic diversity and phylogenetic relationship using nuclear and chloroplast SSR markers (Gong et al., 2008). There was a rich genetic diversity of the trifoliate orange germplasm in China and the "Fumin" trifoliate orange was a true species that was genetically distant from common trifoliate orange, its hybrids, and its relatives (Gong et al., 2008). With AFLP markers, Pang et al. (2007) analyzed the genetic diversity of *Poncirus* accessions, including accessions from repositories in China. Considerable levels of genetic variation existed in *Poncirus* accessions and the UPGMA cluster analysis showed no clear association between the clustering of accessions and their geographical origin. Accessions from Hubei and Henan provinces exhibited greater genetic diversity than other regions, and that from Hunan also had greater genetic diversity when all accessions were considered, which could be regarded as the prior regions of exploration and collection for *Poncirus*.

Long before *C. hongheensis* was systematically described, it was widely cultivated around villages and houses by the local people of Hani nationality in Yunnan province, China. Actually, it was considered to be practically extinct in the wild. Its initial origin and distribution were still unknown until Yang et al. (2010) collected 138 individual *C. hongheensis* trees in seven populations and performed genetic variation analyses based on ISSR fingerprinting. At the species level, the average expected heterozygosity and Shannon diversity index were 0.3520 and 0.5195 across all populations, respectively. A high G_{st} (gene differentiation coefficient) value (0.6247) indicated that there was significant differentiation among populations, which would be the result of habitat fragmentation and limited gene flow ($N_m = 0.1502$). In the context of in situ conservation and restoration genetics, high outcrossing was particularly important to maintain the historically significant processes of *C. hongheensis*.

C. ichangensis, a wild and endemic perennial plant, was characterized by the existence of wild and natural populations in southwestern and middle-west China. A total of 231 individuals across 16 natural populations were analyzed with chloroplast SSR markers, nuclear SSR markers, and single-copy nuclear genes (Yang et al., 2017). The mean effective number of chloroplast haplotypes per population was 2.061. The value of observed heterozygosity ranged from 0.100 to 0.341, and the value of expected heterozygosity ranged from 0.096 to 0.459 among populations. The genetic variation was mostly within populations (44.14% for the F_{st} and 46.26% for the R_{st}) and among groups (44.89% for F_{st} and 43.56% for R_{st}). However, there was a relatively low level of variation among populations within groups (10.96% for F_{st} and 10.18% for R_{st}). Clear signals of recent bottlenecks and strong patterns of isolation by distance were detected among different subpopulations, indicating a low extent of historical gene flow for *C. ichangensis* and that genetic drift would occur after population differentiation.

References

Araújo, E.F., Queiroz, L.P., Machado, M.A., 2003. What is citrus? Taxonomic implications from a study of cp-DNA evolution in the tribe *Citreae* (Rutaceae subfamily Aurantioideae). Org. Divers. Evol. 3, 55–62.

Barbhuiya, A.R., Khan, M.L., Selvadurai, D., 2016. Genetic structure and diversity of natural and domesticated populations of *Citrus medica* L. in the Eastern Himalayan region of Northeast India. Ecol. Evol. 6, 3898–3911.

Barkley, N.A., Roose, M.L., Krueger, R.R., Federici, C.T., 2006. Assessing genetic diversity and population structure in a citrus germplasm collection utilizing simple sequence repeat markers (SSRs). Theor. Appl. Genet. 112, 1519–1531.

Barrett, H.C., Rhodes, A.M., 1976. A numerical taxonomic study of affinity relationships in cultivated *Citrus* and its close relatives. Syst. Bot. 1, 105–136.

Bayer, R.J., Mabberley, D.J., Morton, C., Miller, C.H., Sharma, I.K., Pfeil, B.E., Rich, S., Hitchcock, R., Sykes, S., 2009. A molecular phylogeny of the orange subfamily (Rutaceae: Aurantioideae) using nine cpDNA sequences. Am. J. Bot. 96, 668–685.

Bentham, G., Hooker, J.D., 1867. Genera Plantarum. vol. I Reeve & Co., London.

Bhattacharya, S., Dutta, S., 1951. *Citrus* varieties of Assam. Indian J. Genet. Pl. Breed. 11, 57–62.

Bhattacharya, S., Dutta, S., 1956. Classification of Citrus Fruits of Assam. Indian Council of Agricultural Research (ICAR).

Bowman, K.D., Gmitter Jr., F.G., 1990. Caribbean forbidden fruit: grapefruit's missing link with the past and bridge to the future? Fruit Var. J. 44, 41–44.

Calabrese, F., 1992. The history of citrus in the Mediterranean countries and Europe. Proc. Int. Soc. Citricult. 1, 35–38.

Calabrese, F., 1998. La Favolosa Storia Degli Agrumi. L'Epos editore, Palermo, Italy.

Carbonell-Caballero, J., Alonso, R., Ibañez, V., Terol, J., Talon, M., Dopazo, J., 2015. A phylogenetic analysis of 34 chloroplast genomes elucidates the relationships between wild and domestic species within the genus *Citrus*. Mol. Biol. Evol. 32, 2015–2035.

Carvalho, R., Soares, F.W.S., Brasileiro-Vidal, A.C., Guerra, M., 2005. The relationships among lemons, limes and citron: a chromosomal comparison. Cytogenet. Genome Res. 109, 276–282.

Chadha, K., 1995. Status report on tropical fruit species in South Asia. In: Expert Consultation on Tropical Fruit Species of Asia, Kuala Lumpur (Malaysia), 17–19 May 1994. IPGRI.

Chen, P., 2011. The Investigation and Genetic Diversity Evaluation of Wild Hongkong Kumquat (*Fortunella hindsii* Swingle) in China. M.Sc. Thesis, Huazhong Agricultural University, Wuhan, P.R. China, p. 11.

Cheng, Y.J., de Vicente, M.C., Meng, H.J., Guo, W.W., Tao, N.G., Deng, X.X., 2005. A set of primers for analyzing chloroplast DNA diversity in *Citrus* and related genera. Tree Physiol. 25, 661–672.

Cooper, W.C., 1989. Odyssey of the Orange in China. Natural history of the Citrus fruits in China. EO Painter Printing Company, DeLeon Springs, Florida, p. 122.

Cooper, W.C., Chapot, H., 1977. Fruit production with special emphasis on fruit for processing. In: Nagy, S., Shaw, P.E., Veldhuis, M.K. (Eds.), *Citrus* Science and Technology, first ed. vol. 2. AVI Publishing Co., Westport, CT, USA, pp. 1–127.

Corazza-Nunes, M.J., Machado, M.A., Nunes, W.M.C., Targon, M.L.P.N., 2002. Assessment of genetic variability in grapefruits (*Citrus paradisi* Macf.) and pummelos (*C. maxima* (Burm.) Merr.) using RAPD and SSR markers. Euphytica 126 (2), 169–176.

Curk, F., Ollitrault, F., Garcia-Lor, A., Luro, F., Navarro, L., Ollitrault, P., 2016. Phylogenetic origin of limes and lemons revealed by cytoplasmic and nuclear markers. Ann. Bot. 117, 565.

Davies, F.S., Albrigo, L.G., 1994. *Citrus*. CAB International, Wallingford, United Kingdom.

Diamond, J., 2002. Evolution, consequences and future of plant and animal domestication. Nature 418 (6898), 700–707.

Ding, S.Q., Zhang, X.N., Bao, Z.R., Liang, M.Q., 1984. A new species of *Poncirus* from China. Acta Bot. Yunnanica 6, 292–293.

Dutta, S., 1958. Origin and history of Citrus fruits of Assam. Indian J. Hort. 15, 146–153.

Engler, A., Rutaceae, 1896. In: Engler, A., Prantl, K. (Eds.), Die Naturlichen Pflanzenfamilien. Engelmann, Leipzig, pp. 95–201. Teil III, Abt. 4–5.

Fanciullino, A.L., Dhuique-Mayer, C., Luro, F., Morillon, R., Ollitrault, P., 2007. Carotenoid biosynthetic pathway in the *Citrus* genus: number of copies and phylogenetic diversity of seven genes. J. Agric. Food Chem. 55, 7405–7417.

Fang, D.Q., 1990. Studying the Origin and Evolution of *Citrus* by Isoenzymes. Ph.D. Thesis, Huazhong Agricultural University, Wuhan, P.R. China, p. 164.

Federici, C.T., Fang, D.Q., Scora, R.W., Roose, M.L., 1998. Phylogenetic relationships within the genus *Citrus* (Rutaceae) and related genera as revealed by RFLP and RAPD analysis. Theor. Appl. Genet. 96, 812–822.

Filho, H.D.C., Machado, M.A., Targon, M.L.P.N., Moreira, M.C.P.Q.D.G., Jr, J.P., 1998. Analysis of the genetic diversity among mandarins (*Citrus* spp.) using RAPD markers. Euphytica 102, 133–139.

Froelicher, Y., Dambier, D., Bassene, J.B., Costantino, G., Lotfy, S., Didout, C., Beaumont, V., Brottier, P., Risterucci, A.M., Luro, F., Ollitrault, P., 2008. Characterization of microsatellite markers in mandarin orange (*Citrus reticulata* Blanco). Mol. Ecol. Resour. 8 (1), 119–122.

Froelicher, Y., Mouhaya, W., Bassene, J.B., Costantino, G., Kamiri, M., Luro, F., Morillon, R., Ollitrault, P., 2011. New universal mitochondrial PCR markers reveal new information on maternal citrus phylogeny. Tree Genet. Genomes 7, 49–61.

Fujii, H., Ohta, S., Nonaka, K., Katayose, Y., Matsumoto, T., Endo, T., Yoshioka, T., Omura, M., Shimada, T., 2016. Parental diagnosis of satsuma mandarin (*Citrus unshiu* Marc.) revealed by nuclear and cytoplasmic markers. Breed. Sci. 66, 683–691.

Garcia-Lor, A., Luro, F., Navarro, L., Ollitrault, P., 2012. Comparative use of InDel and SSR markers in deciphering the interspecific structure of cultivated citrus genetic diversity: a perspective for genetic association studies. Mol. Gen. Genet. 287, 77–94.

Garcia-Lor, A., Curk, F., Snoussi-Trifa, H., Morillon, R., Ancillo, G., Luro, F., Navarro, L., Ollitrault, P., 2013. A nuclear phylogenetic analysis: SNPs, indels and SSRs deliver new insights into the relationships in the 'true citrus fruit trees' group (Citrinae, Rutaceae) and the origin of cultivated species. Ann. Bot. 111, 1–19.

Garcia Lor, A., Luro, F., Ollitrault, P., Navarro, L., 2015. Genetic diversity and population structure analysis of mandarin germplasm by nuclear, chloroplastic and mitochondrial markers. Tree Genet. Genomes 11, 1–15.

Ghosh, S., 1977. Citrus industry of North East India. Punjab Hort. J. 1/2, 13–21.

Gmitter Jr., F.G., Hu, X.L., 1990. The possible role of Yunnan, China, in the origin of contemporary *Citrus* species (Rutaceae). Econ. Bot. 44, 267–277.

Gmitter Jr., F.G., 1995. Origin, evolution, and breeding of the grapefruit. Plant Breed. Rev. 13, 345–363.

Gong, G.Z., Hong, Q.B., Peng, Z.C., Jiang, D., Xiang, S.Q., 2008. Genetic diversity of Poncirus and its phylogenetic relationships with relatives as revealed by nuclear and chloroplast SSR. Acta Hort. Sinica 35 (12), 1742–1750 (in chinese).

Gross, B.L., Olsen, K.M., 2010. Genetic perspectives on crop domestication. Trends Plant Sci. 15 (9), 529–537.

Gulsen, O., Roose, M.L., 2001. Lemons: Diversity and relationships with selected *Citrus* genotypes as measured with nuclear genome markers. J. Am. Soc. Hortic. Sci. 126, 309–317.

Guo, T., 1981. A Survey Report About the Citrus Germplasm in Xishuang Bana of Yunnan. Research Reports on Special Topics No. 81-03, Citrus Research Institute, Chinese Academy of Agricultural Sciences, Chongqing, Sichuan, People's Republic of China.

Hazarika, T., 2012. *Citrus* genetic diversity of north-east India, their distribution, ecogeography and ecobiology. Genet. Resour. Crop. Evol. 59, 1267–1280.

Herrero, R., Asins, M.J., Carbonell, A.E., Navarro, L., 1996. Genetic diversity in the orange subfamily Aurantioideae. I. Intraspecies and intragenus genetic variability. Theor. Appl. Genet. 92, 599–609.

Hirai, M., Kajiura, I., 1987. Genetic analysis of leaf isozymes in citrus. Jpn. J. Breed. 37, 377–388.

Hirai, M., Kozaki, I., Kajiura, I., 1986. Isozyme analysis and phylogenic relationship of *Citrus*. Jpn. J. Breed. 36, 377–389.

Hirai, M., Mitsue, S., Kita, K., Kajiura, I., 1990. A survey and isozyme analysis of wild mandarin, tachibana (*Citrus tachibana* (Mak.) Tanaka) growing in Japan. J. Jpn. Soc. Hort. Sci. 59, 1–7.

Hodgson, R.W., 1967. Horticultural varieties of citrus. In: Reuter, W., Webber, H.J., Batchelor, L.D. (Eds.), The Citrus Industry. History, World Distribution, Botany and Varieties. vol. I. University of California Press, pp. 431–591.

Hore, D., Govind, S., Singh, I., 1997. Collecting Citrus Germplasm From the Mizoram and Tripura Hills of India. Plant Genetic Resources Newsletter (IPGRI/FAO) Bulletin des Ressources Phytogenetiques (IPGRI/FAO) Noticiario de Recursos Fitogeneticos (IPGRI/FAO).

Hu, C.G., 1998. Genetic Diversity and Phylogenetic Relationships Among *Citrus* as Revealed by Molecular Markers. Ph.D. Thesis, Huazhong Agricultural University, Wuhan, P.R. China, p. 117.

Hu, Y.B., 2015. Studies on the Genetic Diversity of Trifoliate Orange (*Poncirus trifoliata* [L.] Raf) and Genetics of Its Hybrids With Red Tangerine (*Citrus reticulata* Blanco). Ph.D. Thesis Huazhong Agricultural University, Wuhan, P.R. China12.

Huang, G.L., Zhou, W.Q., Ali, S., 2011. Spatial patterns and economic contributions of mining and tourism in biodiversity hotspots: a case study in China. Ecol. Econ. 70, 1492–1498.

Isaac, E., 1959. Influence of religion on the spread of citron. Science 129, 179–185.

Jackson, L., Ziegler, L.W., 1991. *Citrus* Growing in Florida. University of Florida Press, Gainesville.

Kawai, S., Tomono, Y., Katase, E., Ogawa, K., Yano, M., 1999. Antiproliferative activity of flavonoids on several cancer cell lines. Biosci. Biotechnol. Biochem. 63, 896–899.

Khan, I., 2007. *Citrus* Genetics, Breeding and Biotechnology. CABI Publishing, Wallingford.

Krueger, R.R., Navarro, L., 2007. Citrus germplasm resources. In: Khan, I. (Ed.), *Citrus* Genetics, Breeding and Biotechnology. CABI Publishing, CAB International, Wallington, pp. 45–140.

Lee, Y.S., Cha, B.Y., Saito, K., Yamakawa, H., Choi, S.S., Yamaguchi, K., Yonezawa, T., Teruya, T., Nagai, K., Woo, J.T., 2010. Nobiletin improves hyperglycemia and insulin resistance in obese diabetic ob/ob mice. Biochem. Pharmacol. 79, 1674–1683.

Li, Y.Z., Cheng, Y.J., Tao, N.G., Deng, X.X., 2007. Phylogenetic analysis of mandarin landraces, wild mandarins, and related species in China using nuclear *LEAFY* second intron and plastid *trn*L-*trn*F sequence. J. Am. Soc. Hortic. Sci. 132 (6), 796–806.

Li, X.M., Xie, R.J., Lu, Z.H., Zhou, Z.Q., 2010. The origin of cultivated citrus as inferred from internal transcribed spacer and chloroplast DNA sequence and amplified fragment length polymorphism fingerprints. J. Am. Soc. Hortic. Sci. 135, 341–350.

Lim, T.K., 2012. Edible Medicinal and Non-Medicinal Plants. Springer, Netherlands, pp. 682–689.

Luro, F., Venturini, N., Costantino, G., Paolini, J., Ollitrault, P., Costa, J., 2012. Genetic and chemical diversity of citron (*Citrus medica* L.) based on nuclear and cytoplasmic markers and leaf essential oil composition. Phytochemistry 77, 186.

Mabberley, D.J., 1997. A classification for edible *Citrus* (Rutaceae). Telopea 7, 167–172.

Malik, S., Chaudhury, R., Dhariwal, O., Kalia, R.K., 2006. Collection and characterization of *Citrus indica* Tanaka and *C. macroptera* Montr.: wild endangered species of northeastern India. Genet. Resour. Crop. Evol. 53, 1485–1493.

Medoruma, K., Higa, A., Kinjyo, H., Zukeyama, H., Awaguchi, Y., Miyagi, T., Arasaki, M., Inoue, H., Onda, S., Kawano, S., Urasaki, N., Shimabuku, S., Hanashiro, S., 2011. Characteristics of seedless *Citrus depressa*, Nakamoto seedless. Bull. Okinawa Pref. Agric. Res. Center 5, 5–10 (In Japanese with English abstract).

Meyer, R.S., DuVal, A.F., Jensen, H.R., 2012. Patterns and processes in crop domestication: an historical review and quantitative analysis of 203 global food crops. New Phytol. 196 (1), 29–48.

Miller, A.J., Gross, B.L., 2011. From forest to field: perennial fruit crop domestication. Am. J. Bot. 98 (9), 1389–1414.

Miyata, Y., Sato, T., Imada, K., Dobashi, A., Yano, M., Ito, A., 2008. A citrus polymethoxyflavonoid, nobiletin, is a novel MEK inhibitor that exhibits antitumor metastasis in human fibrosarcoma HT-1080 cells. Biochem. Biophys. Res. Commun. 366, 168–173.

Morton, J.F., 1987. Lemon in Fruits of Warm Climates. Purdue University, pp. 160–168.

Morton, C.M., Grant, M., Blackmore, S., 2003. Phylogenetic relationships of the Aurantioideae inferred from chloroplast DNA sequence data. Am. J. Bot. 90, 1463–1469.

Nagano, Y., Inafuku-Teramoto, S., Hashimoto, M., Mimura, T., Matsumoto, R., Yamamoto, M., 2014. Characterization of chloroplast matK sequencesof *Citrus tachibana* and *Citrus depressa*, two indigenous species in Japan. Adv. Hortic. Sci. 28, 95–99.

Nair, K.N., Nayar, M.P., 1997. Rutaceae. In: Hajra, P.K., Nair, V.J., Daniel, P. (Eds.), The Flora of India. Botanical Survey of India, Calcutta, pp. 259–408.

Nicolosi, E., Deng, Z.N., Gentile, A., La Malfa, S., Continella, G., Tribulato, E., 2000. *Citrus* phylogeny and genetic origin of important species as investigated by molecular markers. Theor. Appl. Genet. 100, 1155–1166.

Nicolosi, E., Malfa, S.L., El-Otmani, M., Negbi, M., Goldschmidt, E.E., 2005. The search for the authentic citron (*Citrus medica* L.): historic and genetic analysis. HortSci. 40, 1963–1968.

Ollitrault, P., Jacquemond, C., Dubois, C., Luro, F., 2003. Genetic diversity of cultivated tropical plants. In: Hamon, P., Seguin, M., Perrier, X., Glaszmann, J. (Eds.), Montpellier. Science Publishers and Cirad, Paris, pp. 193–217.

Ollitrault, P., Terol, J., Chen, C., Federici, C.T., Lotfy, S., Hippolyte, I., Ollitrault, F., Berard, A., Chauveau, A., Cuenca, J., Costantino, G., Kacar, Y., Mu, L., Garcia-Lor, A., Froelicher, Y., Aleza, P., Boland, A., Billot, C., Navarro, L., Luro, F., Roose, M.L., Gmitter, F.G., Talon, M., Brunel, D., 2012a. A reference genetic map of *C. clementina* hort. ex Tan.; citrus evolution inferences from comparative mapping. BMC Genomics 13, 593.

Ollitrault, P., Terol, J., Garcia-Lor, A., Bérard, A., Chauveau, A., Froelicher, Y., Belzile, C., Morillon, R., Navarro, L., Brunel, D., 2012b. SNP mining in *C. clementina* BAC end sequences; transferability in the *Citrus* genus (Rutaceae), phylogenetic inferences and perspectives for genetic mapping. BMC Genomics 13 (1), 13.

Pang, X.M., Hu, C.G., Deng, X.X., 2007. Phylogenetic relationships within *Citrus* and its related genera as inferred from AFLP markers. Genet. Resour. Crop. Evol. 54, 429–436.

Penjor, T., Yamamoto, M., Uehara, M., Ide, M., Matsumoto, N., Matsumoto, R., Nagano, Y., 2013. Phylogenetic relationships of citrus and its relatives based on *matK* gene sequences. PLoS One 8, e62574.

Ramadugu, C., Pfeil, B.E., Keremane, M.L., Lee, R.F., Maureira-Butler, I.J., Roose, M.L., 2013. A six nuclear gene phylogeny of *Citrus* (Rutaceae) taking into account hybridization and lineage sorting. PLoS One 8, e68410.

Ramadugu, C., Keremane, M.L., Hu, X., Karp, D., Federici, C.T., Kahn, T., Roose, M.L., Lee, R.F., 2015. Genetic analysis of citron (*Citrus medica* L.) using simple sequence repeats and single nucleotide polymorphisms. Sci. Hortic. 195, 124–137.

Reuther, W., Batchelor, L.D., Webber, H.J., 1967. History and development of the citrus industry. In: Reuther, W., Webber, H.J., Batchelor, L.D. (Eds.), The Citrus Industry. University of Califorlia Press, Riverside, pp. 1–39.

Roose, M.L., Soost, R.K., Cameron, J.W., 1995. *Citrus*. In: Smart, J., Simmonds, N.W. (Eds.), The Evolution of Crop Plants, second ed. Longman, Essex, pp. 443–449.

Roose, M.L., Federici, C.T., Mu, L., Kwok, K., Vu, C., 2009. Map-based ancestry of sweet orange and other citrus variety groups. In: Second International Citrus Biotechnology Symposium, Catania, Italy, pp. 28.

Samaan, L.G., 1982. Studies on the origin of Clementine tangerine (*Citrus reticulata* Blanco). Euphytica 31, 167–173.

Scora, R.W., 1975. On the history and origin of citrus. Bull. Torrey Bot. Club 102, 369–375.

Shimizu, T., Kitajima, A., Nonaka, K., Yoshioka, T., Ohta, S., Goto, S., Toyoda, A., Fujiyama, A., Mochizuki, T., Nagasaki, H., 2016. Hybrid origins of citrus varieties inferred from DNA marker analysis of nuclear and organelle genomes. PLoS One 11, e0166969.

Singh, D., 1967. Nakoor Lime—a new *Citrus*. Indian J. Hort. 24, 84–86.

Singh, B., 1981. Establishment of First Gene Sanctuary in India for *Citrus* in Garo Hills. Concept Publishing Company, New Delhi, p. 182.

Singh, I., Singh, A., 2001. Citrus germplasm and its utility. In: *Citrus*. International Book Distributing Co., Lucknow, pp. 45–66.

Singh, I., Singh, S., 2006. Exploration, collection and characterization of citrus germplasm—a review. Agric. Rev. 27, 79.

Snoussi, H., Duval, M.F., Garcia-Lor, A., Belfalah, Z., Froelicher, Y., Risterucci, A.M., Perrier, X., Jacquemoud-Collet, J.P., Navarro, L., Harrabi, M., 2012. Assessment of the genetic diversity of the Tunisian citrus rootstock germplasm. BMC Genet. 13 (1), 16.

Spiegel-Roy, P., Eliezer, E.G., 1996. The Biology of *Citrus*. Cambridge University Press.

Swingle, W.T., 1948. Botany of citrus and its relatives of the orange subfamily. In: Webber, H.J., Batchelor, L.D. (Eds.), The Citrus Industry. University of California Press, California, pp. 129–174.

Swingle, W.T., Reece, P.C., 1967. The botany of *Citrus* and its wild relatives. In: Reuther, W., Webber, H.J., Batchelor, L.D. (Eds.), The Citrus Industry. vol. 1. University of California, Berkeley, pp. 190–430.

Tanaka, T., 1926. Wild Citri of the Japanese territories. Bul. Sci. Fakultato Terkultura Kjusu Imp. Univ. 2, 51–58 (In Japanese with English abstract).

Tanaka, T., 1931. The discovery of *Citrus tachibana* in Formosa, and its scientific and industrial significance. Stud. Citrol. 5, 1–20 (In Japanese with English abstract).

Tanaka, T., 1932. A monograph of the Satsuma orange: with special reference to the occurrence of new varieties through bud variation. Mem. Fac. Sci. Agr. Taihoku Univ. 4, 1–626.

Tanaka, Y., 1948. An Iconographyof Japanese Citrus Fruits. vol. 2 Yokendo, Tokyo (in Japanese).

Tanaka, T.Z., 1954. Species Problem in *Citrus*; A Critical Study of Wild and Cultivated Units of *Citrus*, Based Upon Field Studies in Their Native Homes. Japanese Society for the Promotion of Science, Tokyo.

Tanaka, T., 1961. Citologia: Semi-Centennial Commemoration Papers on *Citrus* Studies. Citrologia Supporting Foundation, Osaka, p. 114.

Tanaka, T., 1977. Fundamental discussion of *Citrus* classification. Stud. Citrol. 14, 1–6.

Taninaka, T., Otoi, N., Morimoto, J., 1981. Acid citrus cultivars related to the Yuzu (*Citrus junos* Sieb. ex Tanaka) in Japan. In: Matsumoto, K., Oogaki, C., Kozaki, I. (Eds.), Proceedings of International Society of Citriculture. Aiko Printing Co., Tokyo, Japan, pp. 73–76.

Tolkowsky, S., 1938. Hesperides: A History of the Culture and Use of Citrus Fruits. John Bale, Sons and Curnow, London, p. 371.

Torres, A.M., Soost, R.K., Diedenhofen, U., 1978. Leaf isozymes as genetic markers in *Citrus*. Am. J. Bot. 65, 869–881.

Upadhaya, A., Chaturvedi, S., Tiwari, B., 2016. Utilization of wild citrus by Khasi and Garo tribes of Meghalaya. Indian J. Tradit. Knowl. 15 (1), 121–127.

Urasaki, N., Yoshida, K., Uehara, T., Inoue, H., Onda, S., Kinjyo, H., Kawano, S., 2005. Single nucleotide polymorphism in shiikuwasha (*Citrus depressa* Hayata) chloroplast DNA, trnL-trnF. Jpn. J. Trop. Agr. 49, 246–251.

Uzun, A., Yesiloglu, T., Akakacar, Y., Tuzcu, O., Gulsen, O., 2009. Genetic diversity and relationships within *Citrus* and related genera based on sequence related amplified polymorphism markers (SRAPs). Sci. Hortic. 121, 306–312.

Uzun, A., Yesiloglu, T., Polat, I., Aka-Kacar, Y., Gulsen, O., Yildirim, B., Tuzcu, O., Tepe, S., Canan, I., Anil, S., 2011. Evaluation of genetic diversity in lemons and some of their relatives based on SRAP and SSR markers. Plant Mol. Biol. Report. 29, 693–701.

Vavilov, N.I., 1950. The origin, variation immunity and breeding of cultivated crops. Chron. Bot. 13, 34.

Verma, S.K., 1999. In: Singh, S., Ghosh, S.P. (Eds.), Hi-Tech Citrus Management. ISC, NRCC, Nagpur, pp. 54–620.

Wang, X., Xu, Y.T., Zhang, S.Q., Li, C., Huang, Y., Cheng, J.F., Wu, G.Z., Tian, S.X., Chen, C.L., Liu, Y., Yu, H.W., Yang, X.M., Lan, H., Wang, N., Wang, L., Xu, J.D., Jiang, X.L., Xie, Z.Z., Tan, M.L., Larkin, R.M., Chen, L.L., Ma, G.B., Ruan, Y.J., Deng, X.X., Xu, Q., 2017. Genomic analyses of primitive, wild and cultivated citrus provide insights into asexual reproduction. Nat. Genet. 49, 765–772.

Wang, L., He, F., Huang, Y., He, J., Yang, S., Zeng, J., Deng, C., Jiang, X., Fang, Y., Wen, S., Xu, R., Yu, H., Yang, X., Zhong, G., Chen, C., Yan, X., Zhou, C., Zhang, H., Xie, Z., Larkin, R.M., Deng, X., Xu, Q., 2018. Genome of wild mandarin and domestication history of mandarin. Mol. Plant 11 (8), 1024–1037. https://doi.org/10.1016/j.molp.2018.06.001.

Webber, H.J., 1943. Cultivated varieties of citrus. In: Webber, H.J., Batchelor, D.L. (Eds.), The Citrus Industry. vol. 1. University of California, Berkeley, pp. 475–668.

Webber, H.J., 1967. History and development of the citrus industry. In: Reuther, W., Webber, H.J., Batchelor, L.D. (Eds.), The Citrus Industry. vol. 1. University of California Press, Berkeley, pp. 1–39.

Wu, G.A., Prochnik, S., Jenkins, J., Salse, J., Hellsten, U., Murat, F., Perrier, X., Ruiz, M., Scalabrin, S., Terol, J., Takita, M.A., Labadie, K., Poulain, J., Couloux, A., Jabbari, K., Cattonaro, F., Del Fabbro, C., Pinosio, S., Zuccolo, A., Chapman, J., Grimwood, J., Tadeo, F.R., Estornell, L.H., Munoz-Sanz, J.V., Ibanez, V., Herrero-Ortega, A., Aleza, P., Perez-Perez, J., Ramon, D., Brunel, D., Luro, F., Chen, C., Farmerie, W.G., Desany, B., Kodira, C., Mohiuddin, M., Harkins, T., Fredrikson, K., Burns, P., Lomsadze, A., Borodovsky, M., Reforgiato, G., Freitas-Astua, J., Quetier, F., Navarro, L., Roose, M., Wincker, P., Schmutz, J., Morgante, M., Machado, M.A., Talon, M., Jaillon, O., Ollitrault, P., Gmitter, F., Rokhsar, D., 2014a. Sequencing of diverse mandarin, pummelo andorange genomes reveals complex history of admixture during citrusdomestication. Nat. Biotechnol. 32, 656–662.

Wu, B., Zhong, G.Y., Yue, J.Q., Yang, R.T., Li, C., Li, Y.J., Zhong, Y., Wang, X., Jiang, B., Zeng, J.W., Zhang, L., Yang, S.T., Bei, X.J., Zhou, D.G., 2014b. Identification of pummelo cultivars by using a panel of 25 selected SNPs and 12 DNA segments. PLoS One 9 (4), e94506.

Wu, G.A., Terol, J., Ibanez, V., Lopez-Garcia, A., Estela, Perez-Roman, Carles, B., et al., 2018. Genomics of the origin, evolution and domestication of citrus. Nature 544, 311–316. https://doi.org/10.1038/nature25447.

Xu, Q., Chen, L.L., Ruan, X.A., Chen, D.J., Zhu, A.D., Chen, C.L., Bertrand, D., Jiao, W.B., Hao, B.H., Lyon, M.P., Chen, J.J., Gao, S., Xing, F., Lan, H., Chang, J.W., Ge, X.H., Lei, Y., Hu, Q., Miao, Y., Wang, L., Xiao, S.X., Biswas, M.K., Zeng, W.F., Guo, F., Cao, H.B., Yang, X.M., Xu, X.W., Cheng, Y.J., Xu, J., Liu, J.H., Luo, O.J., Tang, Z., Guo, W.W., Kuang, H., Zhang, H.Y., Roose, M.L., Nagarajan, N., Deng, X.X., Ruan, Y.J., 2013. The draft genome of sweet orange (*Citrus sinensis*). Nat. Genet. 45, 59–66.

Yamamoto, M., 2007. Application of fluorescent staining of chromosomes to genetic studies in citrus. Jpn. J. Plant Sci. 1, 12–19.

Yamamoto, M., Tominaga, S., 2003. High chromosomal variability of mandarin (*Citrus* spp.) revealed by CMA banding. Euphytica 129, 267–274.

Yamamoto, M., Kobayashi, S., Nakamura, Y., Yamada, Y., 1993. Phylogenic relationships of citrus revealed by RFLP analysis of mitochondrial and chloroplast DNA. Jpn. J. Breed. 43, 355–365.

Yamamoto, M., Matsuo, Y., Kuniga, T., Matsumoto, R., Yamada, Y., 1998. Isozyme and RAPD analyses of shiikuwashas (*Citrus depressa* Hayata). Bull. Natl. Inst. Fruit Tree Sci. 30/31, 39–51.

Yamamoto, M., Kubo, T., Tominaga, S., 2006. Self- and cross-incompatibility of various citrus accessions. J. Jpn. Soc. Hort. Sci. 75, 372–378.

Yamamoto, M., Matsumoto, R., Uechi, Y., Ijichi, T., Kubo, T., Tominaga, S., 2008. The polymethoxy flavones content of local citrus accessions on the island of Kikaijima in Kagoshima Prefecture. Jpn. Bull. Fac. Agr. Kagoshima Univ. 58, 1–7. (In Japanese with English abstract).

Yamamoto, M., Kouno, R., Nakagawa, T., Usui, T., Kubo, T., Tominaga, S., 2011. Isozyme and DNA analyses of local Citrus germplasm on Amami Islands, Japan. J. Jpn. Soc. Hort. Sci. 80, 268–275.

Yamamoto, M., Tsuchimochi, Y., Ninomiya, T., Koga, T., Kitajima, A., Yamasaki, A., Inafuku-Teramoto, S., Yang, X., Yang, X., Zhong, G., Nasir, N., Kubo, T., Tominaga, S., 2013. Diversity of chloroplast DNA in various mandarins (*Citrus* spp.) and other citrus demonstrated by CAPS analysis. J. Jpn. Soc. Hort. Sci. 82, 106–113.

Yamamoto, M., Takakura, A., Tanabe, A., Teramoto, S., Kita, M., 2017. Diversity of *Citrus depressa* Hayata (Shiikuwasha) revealed by DNA analysis. Genet. Resour. Crop. Evol. 64, 805–814.

Yang, Y., Pan, Y.Z., Gong, X., Fan, M., 2010. Genetic variation in the endangered Rutaceae species *Citrus hongheensis* based on ISSR fingerprinting. Genet. Resour. Crop. Evol. 57, 1239–1248.

Yang, X., Li, H., Liang, M., Xu, Q., Chai, L., Deng, X.X., 2015. Genetic diversity and phylogenetic relationships of citron (*Citrus medica* L.) and its relatives in southwest China. Tree Genet. Genomes 11, 129.

Yang, X.M., Li, H., Yu, H.W., Chai, L.J., Xu, Q., Deng, X.X., 2017. Molecular phylogeography and population evolution analysis of *Citrus ichangensis* (Rutaceae). Tree Genet. Genomes 13, 29.

Yu, D., 1979. China's Fruit Taxonomy. Agriculture Publishing House, Bejing, China.

Yu, H.W., Yang, X.M., Guo, F., Jiang, X.L., Deng, X.X., Xu, Q., 2017. Genetic diversity and population structure of pummelo (*Citrus maxima*) germplasm in China. Tree Genet. Genomes 13, 58.

Zhang, W.C., 1990. Scientific collaboration for the development of citrus industry of the world. In: Huang, B.Y., Yang, Q. (Eds.), Proceedings of the International Citrus Symposium. International Academic Publishers, Beijing, China, pp. 3–6.

Chapter 4

Citrus taxonomy

Patrick Ollitrault[a], Franck Curk[b], and Robert Krueger[c]

[a]*French Agricultural Research Centre for International Development (CIRAD), AGAP Research Unit, San Giuliano, France,* [b]*French National Institute for Agricultural Research (INRA), AGAP Research Unit, San Giuliano, France,* [c]*United States Department of Agriculture-Agricultural Research Service National Clonal Germplasm Repository for Citrus and Dates, Riverside, CA, United States*

4.1 The genus *Citrus* definition

4.1.1 The botanical treatment of the genus *Citrus*

Taxonomists now agree that the *Citrus* L. genus is part of the Sapindales Berchtold and J. Presl, order in the Rutaceae Jussieu family (Stevens, 2017; NCBI, 2017), while it was before included in the Geraniales order (Swingle and Reece, 1967). The family Rutaceae, whose name comes from the genus *Ruta* L., includes herbaceous and woody plants with essential oil glands. According to Swingle and Reece (1967), *Citrus* belongs to the subfamily Aurantioideae, which is divided into two subtribes: the Clauseneae (5 genera) and the Citreae (28 genera). The Clauseneae tribe is considered more primitive than the Citreae tribe. Citreae are divided into three subtribes: Triphasiinae, Balsamocitrinae, and Citrinae. Swingle and Reece (1967) subdivided the Citrinae into three groups including the "true citrus" one composed of six genera: *Citrus*, *Clymenia*, *Eremocitrus*, *Fortunella*, *Microcitrus*, and *Poncirus*. Chloroplast molecular studies (Morton et al., 2003; Bayer et al., 2009) confirmed the monophyly of the Aurantioideae. However, the study by Bayer et al. (2009) based on nine cpDNA gene regions, suggested nonmonophyly for several subtribes and resulted in proposals for revision of the Swingle and Reece (1967) Aurantioideae classification, even though some clade remained poorly resolved. In the future, whole chloroplast genome sequencing data of Aurantioideae members should better resolve their phylogeny and, therefore, should provide the key to a definitive classification of tribes and subtribes.

4.1.2 Phenotypical traits of the true *Citrus*

Fruits of *Citrus* are apparently some of the first to be domesticated and exploited by humans. The center of origin and diversity for *Citrus* is southeastern Asia (Chapter 2), particularly northeast India, Myanmar, and southern China (Tolkowsky, 1938; Tanaka, 1954). In those areas, citrus were apparently exploited and consumed during ancient times (Chapter 3), later spreading into the Middle East, Europe, and ultimately the Western Hemisphere (Scora, 1975; Webber, 1943, 1967). Ramon-Laca (2003) and Mabberley (2004) rather interestingly trace the movement of citrus from the center of origin and diversity through the Middle East and into Europe by tracing the etymology of the names of the cultivated citrus. The various peoples and languages through which the several *Citrus* species passed undoubtedly recognized their similarities in the form of fruits and trees. This is supported by the fact that Linnaeus (Linnaeus, 1753) combined the previously named *Aurantium* (orange), *Citreum* (citron), and *Limon* (lemon) into the genus *Citrus*, a name previously applied to an entirely different and unrelated species, *Tetraclinis articulata* (Vahl) Mast. (Mabberley, 2004). This author recognized the importance of Swingle (1943) in making sense of some of the taxonomic confusion within the Aurantioideae, while noting that many taxonomic issues still need clarifications.

4.1.2.1 Morphological characteristics of the genus Citrus (sensu Swingle)

The classification proposed by Swingle in (1943) was only slightly modified in Swingle and Reece (1967); henceforth, reference will be made to Swingle (1943) with the understanding that the same information is available in the more easily accessed Swingle and Reece (1967). Thus, we may develop a working botanical description of *Citrus* by studying and interweaving the description of Swingle (1943) and the more recent and shorter one of Zhang and Mabberley (2008).

The Genus Citrus. https://doi.org/10.1016/B978-0-12-812163-4.00004-8

The fruits of *Citrus* are berries, that is, fleshy, indehiscent, many-seeded fruits containing no hard parts except the seeds (Fig. 4.1A; Chapter 12). More specifically, *Citrus* fruits are hesperidia, in which the fleshy parts of the fruit are divided into segments and are surrounded by a separable skin (Fig. 4.1A and B). Hesperidia are confined to the fruits of Aurantioideae (Webber, 1943). The obovoid or flattened seeds (Fig. 4.1C and D) are attached adaxially (near the central axis or core, Fig. 4.1A), have smooth or ridged seed coats, and contain one to many embryos (Fig. 4.1E–J). The segments are filled with stalked fusiform pulp vesicles, which contain very watery, large-celled tissue (Fig. 4.1A and B); this is the economic part of the fruit. The segments are surrounded by a white endocarp, outside of which is the peel, which contains numerous oil glands (Fig. 4.1A, K, and L). The peel is generally green during the early stages of fruit development and turns yellow or orange at maturity. The fruit arises from the fragrant flowers, which are borne singly or in small racemes in the axils of the leaves. The flowers of *Citrus* are perfect or staminate, the latter condition being due to abortion of the pistil. The calyx is cup shaped with three to five lobes, and is subglabrous. There are four to eight petals (usually five), which are white (Fig. 4.1M) or pink (Fig. 4.1N) outside, imbricate, and thick. There are usually four times as many free or basally coherent stamens as petals (Fig. 4.1M, O, and P), although there may be up to 10 times as many. The disk is annular or short, with nectary glands. The ovary contains 3–18 locules (generally 10–14), each of which contains two to eight ovaries in two collateral rows (Fig. 4.1Q–S). The style is large and cylindrical, expanding abruptly into the subglobose or oblate spheroid stigma (Fig. 4.1T).

Members of Citrus are evergreen shrubs or small trees, generally 3–10 m in height. Young branches are often flat and angled, becoming cylindrical with age, usually with solitary (rarely paired) spines at the axils. Leaves are generally unifoliolate, with petioles that are usually articulated at the base of the blade and conspicuously winged (Fig. 4.1U). The leaf blade is subleathery to leathery with crenulate (rarely entire) margins, and contains numerous fragrant oil glands.

FIG. 4.1 Botanic traits of the *Citrus* species. A: cross-section through a citrus fruit, B: longitudinal section through a citrus fruit; C: semi-deltoid citrus seed; D: obovoid citrus seed; E: longitudinal section through a citrus seed; F: citrus seed with seed coats, G: cross-section through a citrus seed; H: polyembryonic citrus seed; I and J: polyembryonic citrus seedling; K and L: outside citrus peel section with oil glands; M: open citrus flower; N: open lemon flower; O: open orange flower; P: citrus flower stamens; Q: longitudinal section through a citrus flower; R and S: cross-section through a citrus ovary; T: pistil (ovary, style and stigma) of a citrus flower; U: unifoliate citrus leaves. *Modified from "Histoire Naturelle des orangers" (Risso A, Poiteau A. Histoire Naturelle des Orangers. Paris: Imprimerie de Mme Hérissant Le Doux, Imprimeur ordinaire du Roi et des Musées Royaux; 1818).*

4.1.2.2 Traits of the related species of the true citrus

Although recent taxonomic treatments, such as Zhang and Mabberley (2008), have synonymized the corresponding genera with *Citrus*, they are described in this section as distinct genera following Swingle (1943). The true citrus fruit trees, including *Citrus*, share certain characteristics but are clearly differentiated according to the morphological taxonomic key of Swingle (1943), as presented in Fig. 4.2. The fruit generally resembles those described above in the botanical description of *Citrus*. The pulp vesicles contain droplets of oil, which are more abundant in *Poncirus, Microcitrus*, and the papedas. The vesicles differ from those of other members of Aurantioideae in that they narrow into slender stalks with a point at the apex, except for those in *Clymenia*, which has pyriform vesicles. The fruits of the true citrus fruit trees are segmented and the fruit of the genera other than *Citrus* is smaller than those of *Citrus* itself. *Fortunella* and *Eremocitrus* have ovaries with three to five locules, each of which has only two ovules, whereas *Citrus*, *Microcitrus*, and *Poncirus* have ovaries with six to eight locules, each of which contains many ovules. Members of the true citrus fruit trees are generally cross and graft compatible with other members of the group (Krueger and Navarro, 2007; Siebert, 2016; Siebert et al., 2015).

Fortunella (Kumquat) closely resembles *Citrus* (Swingle, 1943), but (in addition to the distinctions described above) has a much larger stigma containing a few large oil glands; pale green abaxial leaf surfaces; and small, angular flower buds. The trees, leaves, flowers, and fruits are generally smaller than those of *Citrus*. The flowers are 1.0–1.5 cm diameter. The fruit is 1.5–2.5 cm in diameter, round to ovoid-ellipsoid, with a peel that is orange and sweet at maturity and acidic flesh. Kumquats are adapted to climates that are marginally cool for most other Aurantioideae, they require less heat to achieve fruit maturity and have a certain level of winter dormancy (Swingle, 1943).

Eremocitrus and *Microcitrus* are both endemic to the Oceania region. Both differ from *Citrus* in having dimorphic foliage and free stamens; however, *Microcitrus* has an ovary with four to eight locules, whereas *Eremocitrus* has an ovary with three to five locules. Both have rather small, coriaceous leaves; however, the leaves of *Eremocitrus* are thick and

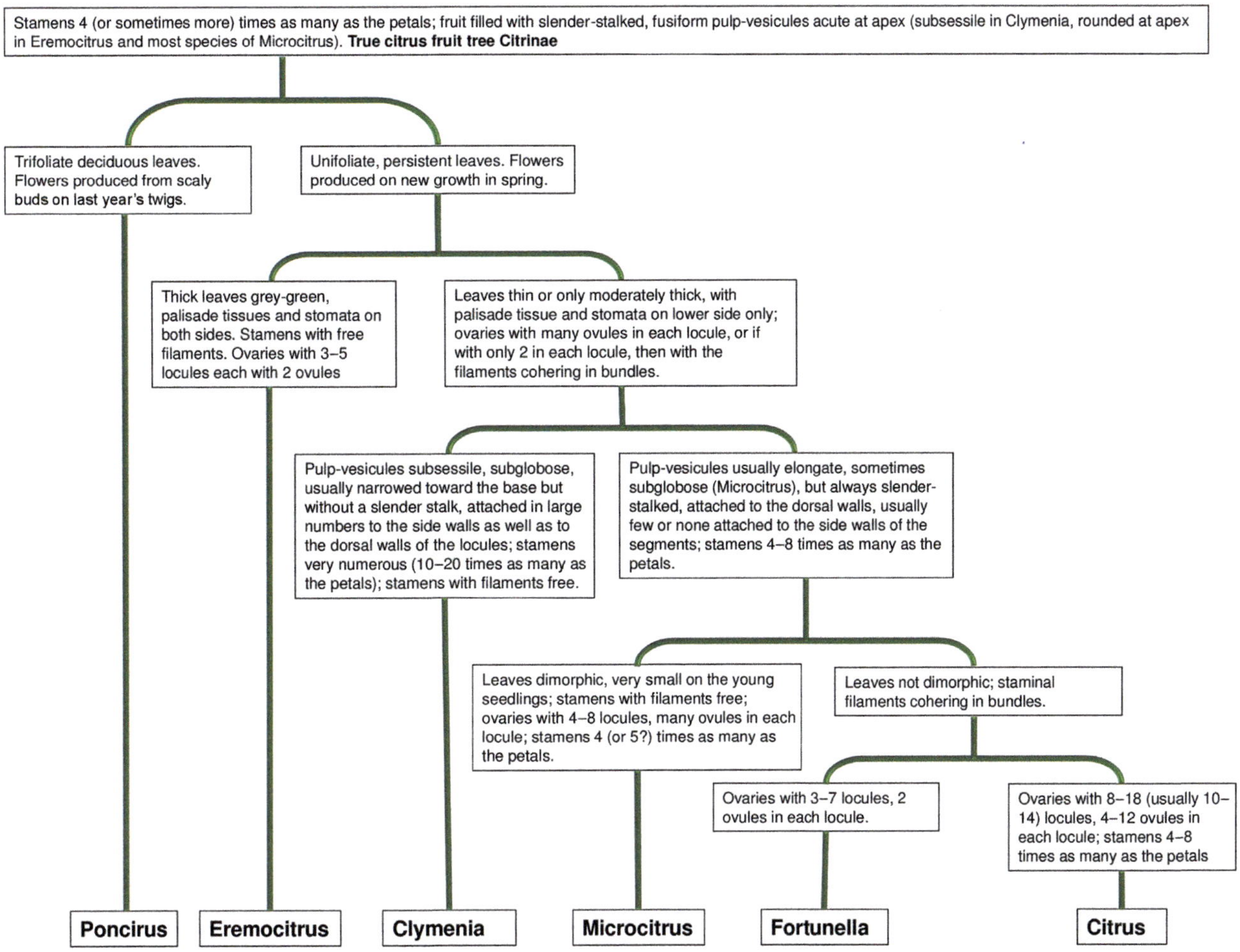

FIG. 4.2 Key to the genera of the true citrus fruit trees of the subtribe Citrinae according to Swingle and Reece (1967).

have a thick palisade layer in the cuticle with stomata on the upper and lower leaf surface. The subglobose or obovoid fruit of the monotypic *Eremocitrus* is small (0.7–1.2×0.8–1.0cm) and is yellow at maturity, whereas fruit of *Microcitrus* is larger (4–5×6–7cm), more variable in form (globose-ovoid or cylindrical, sometimes curved), and varies in color at maturity from greenish-yellow to black. Swingle (1943) describes the xerophytic adaptation of *Eremocitrus*, noting the thick cuticle and this genus's ability to withstand prolonged droughts and extremes of heat and cold (as compared to other Aurantioideae). The cold hardiness of *Eremocitrus* stated in Swingle (1943) and Swingle and Reece (1967) is in error; *Eremocitrus* can probably tolerate temperatures to about −5.5°C, consistent with the original description of the genus in 1914 (Krueger and Navarro, 2007; Swingle, 1914). *Microcitrus*, on the other hand, is considered semixerophytic and able to withstand prolonged periods of drought (Swingle and Reece, 1967; Swingle, 1943). *Eremocitrus* shows some unusual graft relationships (Siebert, 2016; Siebert et al., 2015).

Trifoliate orange was for many years considered a mono-typic genus, represented by *Poncirus trifoliata* (Swingle, 1943), with distinctive trifoliate leaves (unique among the true citrus fruit trees) and deciduous growth habit. The small leaf buds and larger-scale-covered flower buds form in the summer and over winter on leafless terminal twigs, flowering the following spring. This gives to trifoliate oranges the highest degree of cold hardiness among the true citrus fruit trees, surpassing that of kumquats. *Poncirus* flowers are nearly sessile, with petals that open flat, entirely free stamens, and an ovary with six to eight locules. The fruit is smaller than those of *Citrus* (3–5cm diameter), densely and finely pubescent, with many oil glands, and is very seedy. The adaptation of *Poncirus* to cold conditions led Swingle (1943) to speculate that the remote ancestor of the true citrus fruit trees originated in a tropical or semitropical climate. While the other genera of the true citrus fruit trees remained in these climates, *Poncirus* (or its ancestors) "migrated" to the temperate climate of Northeastern Asia, during which time it developed the adaptations to colder winters mentioned previously. In addition to cold tolerance, *Poncirus* exhibits many other characteristics that have been and continues to be used in citrus rootstock breeding, notably disease tolerance (including citrus tristeza virus immunity) and dwarfing. For a more complete treatment of *Poncirus*, the reader is referred to Krueger and Navarro (2007). Relatively recently, a new species, *Poncirus polyandra*, was published (Ding et al., 1984; Duan, 1990), which differs from *P. trifoliata* by its larger leaves, some floral differences, and, most notably, being evergreen. Perhaps, this latter characteristic is related to its habitat in Yunnan, the southernmost province in China.

Clymenia is a very distinctive member of the other true citrus fruit trees. *Clymenia* was separated from *Citrus* by Swingle (1943) based upon the structure of the pulp vesicles, which are short, plump, blunt, ovoid or subglobose, sessile, or very short stalked, and attached to the side walls of the 14–16 locules. In addition, the leaves of *Clymenia* differ from those of the other true citrus fruit trees, and the flowers have enlarged disks with 10–20 times as many stamens as petals. The fruits resemble sweet limes and are edible.

As we will discuss below, a chloroplastic phylogeny (Bayer et al., 2009) integrated *Oxanthera* into the true citrus phylogenetic cluster and, therefore, we will describe it here. According to Swingle and Reece (1967), *Oxanthera* is a very distinct group in having "glabrous, glaucous leaves bluntly rounded or retuse at the tip and cuneate at the base, borne on spineless twigs. The leaves are thick and coriaceous in all the species except one. All the *Oxanthera* types agree in having large, orange-like flowers, and glabrous, more or less glaucous leaves, which usually are rather blunt at the tip and cuneate at the base. All species are thornless, and apparently all have more or less elongate fruits that are longitudinally ribbed at least when young." Therefore, the *Oxanthera* group is easily differentiated from other taxa even though *Oxanthera* flowers are similar to those of cultivated *Citrus* species. However, this group displays an unusual range of variation for characters having high taxonomic value in other plant species. According to Swingle and Reece (1967): "Three of the four species of *Oxanthera* have unifoliolate leaves with clearly articulated petioles that are usually wingless, but are plainly but narrowly winged in one species. One of the species has a hypomerous ovary with only two locules, another has an isomerous ovary with five or six locules, whereas a third species has a hypermerous ovary with seven locules." *Oxanthera* is a specialized xerophytic group like *Microcitrus* but less so than *Eremocitrus* (Swingle and Reece, 1967).

4.1.3 Reproductive biology, cytogenetics and molecular data, and the definition of the genus *Citrus*

4.1.3.1 The genus Citrus and the biological concept of species; genus or species

According to the biological species concept (BSC) developed by Mayr (1942), "species are groups of actually or potentially interbreeding natural populations, which are reproductively isolated from other such groups." This is a popular concept among biologists; however, the debate around the species concepts is still very active and many other concepts have been proposed based, in part, on different biological properties (reviewed in Mayden (1997) and de Queiroz (2007)). For example, the BSC emphasizes the property of reproductive isolation (Mayr, 1942; Dobzhansky, 1970), while the ecological species concept is based on the occupation of a distinct niche or adaptive zone (Andersson, 1990); one approach of the phylogenetic species concept emphasizes diagnosability (Nixon and Wheeler, 1990), and another, monophyly (Donoghue, 1985; Mishler, 1985).

None of the actual citrus classifications fits with the BSC because all *Citrus* species are sexually compatible producing hybrids with moderate-to-high fertility and should, therefore, be joined in a single species according to the BSC. However, nuclear genome sizes of *Citrus* species display differentiation up to 10% between *C. reticulata* (360 Mb per haploid genome) and *C. medica* (398 Mb) (Ollitrault et al., 2003), the two taxa with the smallest and largest genomes in the genus *Citrus*. The differentiation of nuclear genome sizes agrees with cytogenetic observations of chromosome morphology differentiation between the ancestral taxa of cultivated citrus (Raghuvanshi, 1969; Nair and Randhawa, 1969; Guerra, 1993; Hynniewta et al., 2011). In addition, the intermediate and preferential disomic inheritance observed in some doubled diploids of interspecific origin such as for "Volkamer" lemon (*C. reticulata* × *C. medica* (Dirceu et al., 2016)) or "Mexican" lime (*C. micrantha* × *C. medica* (Rouiss et al., 2018)) and allotetraploid somatic hybrids (Kamiri et al., 2011; Xie et al., 2015) attest to preferential chromosome pairing between chromosomes of the same ancestral taxon. All these results reveal a significant genomic differentiation between the four ancestral taxa of the cultivated *Citrus* (*C. maxima*, *C. medica*, *C. micrantha*, and *C. reticulata*). The monophyly of each ancestral taxon has been demonstrated by nuclear markers analysis (Herrero et al., 1996a,b; Nicolosi et al., 2000; Barkley et al., 2006; Garcia-Lor et al., 2013a) and maternal phylogenetic studies (Bayer et al., 2009; Nicolosi et al., 2000; Yamamoto et al., 1993; Froelicher et al., 2011; Carbonell-Caballero et al., 2015; Curk et al., 2016). Recently, nuclear phylogenomic studies have revealed a huge number of diagnostic (discriminant) single-nucleotide polymorphisms (SNPs) for each of these four ancestral taxa (Wu et al., 2014, 2018; Curk et al., 2015; Oueslati et al., 2017). Moreover, an important part of the phenotypic diversity of the cultivated *Citrus* results from the allopatric evolution of the four ancestral taxa (see Section 4.3 for more details). Therefore, recognition of the four ancestral taxa of most cultivated citrus at species rank is supported by the phylogenetic species concept based on diagnosability and monophyly as well as the ecological species concept, considering the past allopatric evolution of the ancestral taxa under different environmental contexts (see below).

4.1.3.2 Sexual compatibility and phylogenetic relationships with related genera of the true citrus; toward a new definition of the genus Citrus?

Two elements disagree with the circumscription of the genus *Citrus* as proposed by Swingle and Reece (1967). The first is the demonstrated sexual compatibility (Fig. 4.3) of the different species of the other "true citrus" genera with the species of *Citrus* as defined by Swingle and Reece (1967). Many fertile hybrids have been produced between *P. trifoliata* and several *Citrus* species and *Poncirus* is a very important genetic resource for rootstock breeding by "intergeneric" hybridization. The so-called citrange, citrumelo, citremon, citradia, and citrandarin result, respectively, from hybridization between sweet orange, grapefruit, lemon, sour orange, and mandarin with *P. trifoliata*. Some of these hybrids were involved in a second round of hybridization producing trigeneric hybrids with *Fortunella* (citrangequat) and *Eremocitrus* (Citrangeremo) or backcrosses in *Citrus* (citrangor) and *Poncirus* (cicitrange). Several hybrids between *Citrus* and *Fortunella* were also created during the 20th century (mandarinquat, limequat) or identified in the germplasms (calamondin) and involved in trigeneric hybridization with *Microcitrus* (faustrimedin, faustrime) and backcrossed in *Fortunella* (procimequat) and *Citrus* to develop triploid hybrids (Viloria et al., 2004). Several hybrids have been created between *Citrus* and *Microcitrus* (e.g., "Australian blood" lime). Hybrids between *Citrus* and *Eremocitrus glauca* were also obtained (eremorange and eremolemon, respectively, with sweet orange and lemon) as well as hybrids between *Fortunella* and *Poncirus* (citrumquat). An accession (CRC 4109) derived from the open pollination of a *Clymenia polyandra* × procimequat hybrid was described by the University of California (UCR, 2017a).

The second discordant element is the nonmonophyly of the chloroplast genomes of the Swingle *Citrus* species, revealed first by Bayer et al. (2009) and more recently from whole genome sequencing (WGS) resequencing data by Carbonell-Caballero et al. (2015). Indeed, while *C. maxima, C. reticulata*, and species of the subgenus *Papeda* form a well-supported clade, *C. medica* is in a separate well-supported clade with Australian citrus (*Microcitrus* and *Eremocitrus*). The Bayer et al. (2009) study also revealed that *Clymenia* and the New Caledonian citrus *Oxanthera* are part of this last clade. Moreover, *P. trifoliata* and *Fortunella* spp. join the *C. maxima*/*C. reticulata*/*Papeda* clade before the Australian citrus/*C. medica* clade (Bayer et al., 2009; Carbonell-Caballero et al., 2015). The "true citrus" group plus *Oxanthera* form a strongly supported clade, highly differentiated from the other Citreae genera (Bayer et al., 2009).

These elements, and the very high synteny and collinearity observed between genetic maps of *Poncirus* and *Citrus* species (Chen et al., 2008; Bernet et al., 2010) and cytogenetic maps (da Costa Silva et al., 2015) strongly support the proposal of Mabberley (1998, 2004) and Zhang and Mabberley (2008) to integrate *Poncirus, Fortunella, Microcitrus, Eremocitrus,*

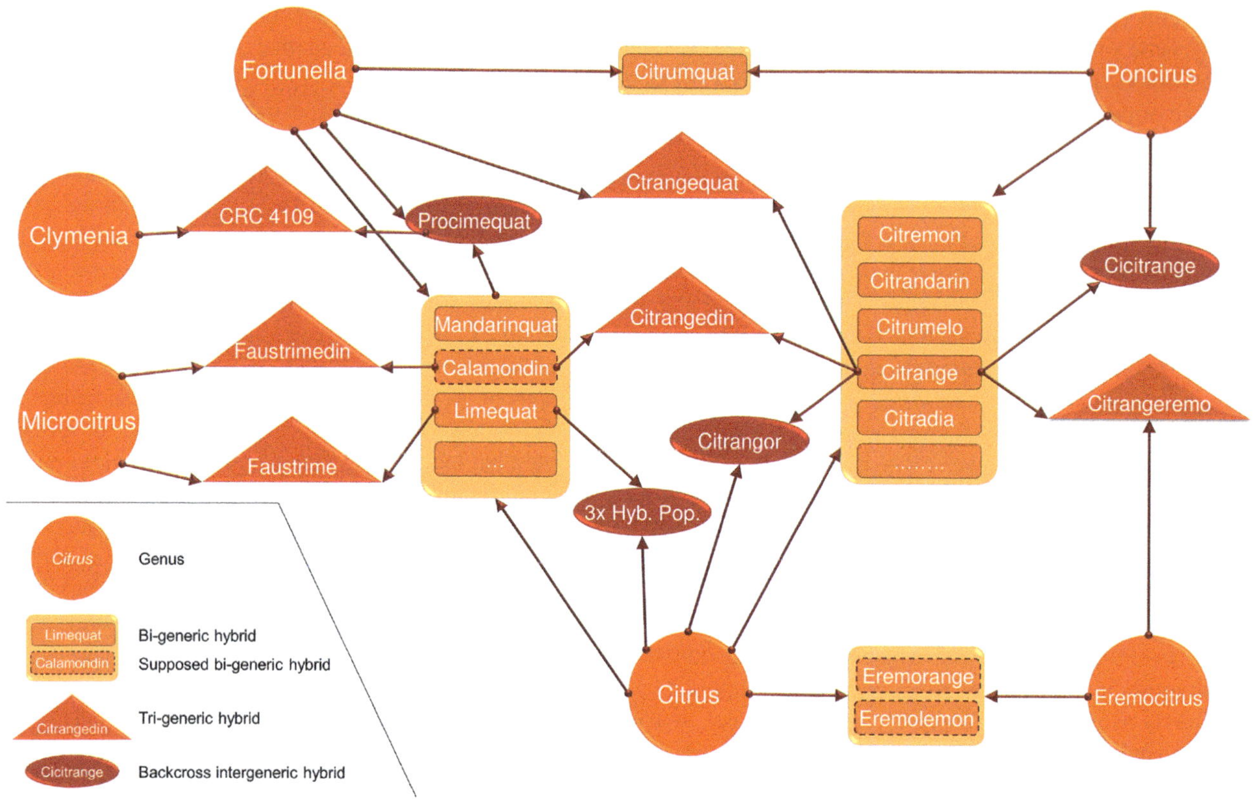

FIG. 4.3 Sexual hybrids obtained between the six genera of Swingle True citrus group.

and *Clymenia* into the genus *Citrus*. According to the results of Bayer et al. (2009), *Oxanthera* may also be integrated into *Citrus*. Some other aspects regarding the specific subdivisions delimitations within the Citrus genus and the origin of admixture types proposed by Mabberley (1997, 2004) are not in agreement with recent molecular studies (see below) and its classification system is still incomplete. This could explain why the Swingle and Reece (1967) classification of the true citrus group still remains popular in the citrus scientific community.

4.2 The genus *Citrus* classifications; an historical, biological, genetic, and phylogenomic perspective

4.2.1 The history of citrus botanical classifications

The extraordinary ability of citrus plants to hybridize with many species and close genera, frequent morphological mutations, and apomixis have complicated matters for early citrus taxonomists, who often did not have the opportunity to observe the plants in their natural environment. We inherited a natural history of citrus rich in folk classifications and in many ancient texts dealing with citrus fruits, the various acidic citrus (lime, lemon, and citron) and the sour and the sweet orange are readily confused.

The ancient descriptions: In China, citrus have been grown for >4000 years. The earliest known mention of citrus fruit is in a text from one of the "Shu Jing" books, also called the Book of Documents or Classic of History, revealing that two types of citrus fruits, large ones and smaller ones (probably pummelos and mandarins), were offered to the emperor (23rd-century BC) as high-value fruits (Medhurst, 1846). Other descriptions can be found in the early Mediterranean literature of Virgil (the Georgics, probably before 37 BC) and Dioscorides (*De Materia Medica*, between AD 50 and 70) who described the citron and its traditional uses. The first certified monograph on citrus fruits, the *Ju Lu*, dating from 1178 was written in China by Han Yanzhi. In this book, 27 citrus fruits were described as well as the different stages of their cultivation, from the propagation to harvest (Dioscorides, 2000).

4.2.1.1 *The early classifications*

Ferrari (1646) was the first to make a citrus classification in his *Hesperides, siue, de malorum aureorum cultura et usu*, a classification followed and elaborated by Johann Christoph Volkamer in his two volumes of *Nurenbergische Hesperides* published in 1708 and 1714 (Volkamer, 1708, 1714). Ferrari and Volkamer are between what Magnin-Gonze (2009) named the *Descriptive Botany* and the *Biological Botany* periods. Linnaeus, with his new *Fundamenta botanica*, established our current binomial nomenclatural system and organized plants by their reproductive features. In his "*Genera plantarum*" (Linnaeus, 1737), basing his work for *Citrus* on that of de Tournefort (1700), Linnaeus included the genus *Citrus* with three species: *Citrus aurantium* Tournef., *Citrus citreum* Tournef., and *Citrus limon* Tournef. according to flower and leaf descriptions. In "*Species plantarum*" (Linnaeus, 1753), Linnaeus recognized only two species subdivided into several varieties: *C. aurantium* var. *aurantium* (the sour oranges), *C. aurantium* var. *sinensis* (the sweet oranges), *C. aurantium* var. *grandis* (renamed *C. aurantium* var. *decumana* by Linnaeus = *Citrus maxima*; the pummelos), *Citrus medica* var. *limon* (the lemons), and *C. medica* var. *medica* (the citrons). In the second edition, a third species, *Citrus trifoliata* (the trifoliate orange) was added.

Between 1700 and 1800, citrus classification transitioned through the *Classificatory Botany* period to the *Evolutionary Botany* period (Magnin-Gonze, 2009) with illustrious taxonomists such as Burman (1768), de Loureiro (1790), and Blanco (1837). Blanco (1837), using the Linnaean sexual system, described seven *Citrus* species and included the first mention of mandarin (*Citrus reticulata* Blanco). The best illustration of what Magnin-Gonze (2009) called *Evolutive Botany* is the incredibly modern "*Tableau synoptique du Genre* Citrus" that Gallesio (1811) published in 1811, in his *Traité du Citrus*. Another work coming from the *Evolutive Botany* period is still considered as an important reference: the "*Histoire naturelle des orangers*" of Risso and Poiteau (1818). Many other taxonomists including Blume, Macfadyen, Tenore, Fortune, Oliver, Pasquale, Hooker, Bonavia, Engler, and Bailey tried to organize and classify citrus taxa (Nicolosi, 2012).

4.2.1.2 The 20th-century classifications

In his "Flore Générale de l'Indo-Chine" based on morphological traits, Guillaumin (Citrus, 1911) included six *Citrus* species (*Citrus decumana* Murr., *C. aurantium* L., *C. medica* L., *Citrus nobilis* Lour., *Citrus japonica* Thumb., and *C. trifoliata* L.).

During the 20th century, two important, but very different taxonomic systems were established by Tyôzaburô Tanaka (1954, 1961, 1977) in Japan and Walter T. Swingle (Swingle and Reece, 1967; Swingle, 1943) in the United States. Both the Tanaka and Swingle systems are still widely used by the citrus scientific community.

The Tanaka system recognizes 157 species of *Citrus*, including 35 species of mandarins in its first publication (Tanaka, 1954). It was expanded to 162 species in the 1977 version (Tanaka, 1977). With the current knowledge on cultivated citrus interspecific admixture, it is clear that the Tanaka classification has too many species, corresponding essentially to varietal groups of clonal origin (due to facultative apomixis) resulting from different reticulation events.

Swingle (1943) was the first citrus taxonomist to propose the use of biochemical markers (glycosides) for taxonomy. The Swingle classification (Swingle, 1943) and the revised version of Swingle and Reece (1967), based on history, morphological, and biochemical characters, also takes into account the vegetative reproduction and particularly the facultative apomixis present in several citrus taxa. Swingle and Reece were aware of the problem of the species concept in agamic complexes as discussed by Stebbins (1950) and Lawrence (1951), who stated "The perpetuation of apomictic hybrids has resulted in some descriptive taxonomists treating each biotype as a morphologically distinct and seed producing species." Taking this into account, the Swingle (1943) classification displays a spectacular reduction of species (Krueger and Navarro, 2007) compared with the Tanaka (1954) classification (157 species). The reduction in species number is particularly important for mandarin with three (*C. reticulata*, *C. tachibana*, and *C. indica*) and 36 species in the Swingle and Tanaka classifications, respectively. Swingle recognized two subgenera and sections: (i) subgenus *Papeda*, with two species in section *Papeda* (*C. latipes* and *C. ichangensis*) and four species in section *Papedocitrus* (*C. hystrix*, *C. macroptera*, *C. micrantha*, and *C. celebica*); and (ii) subgenus *Citrus* (formerly *Eucitrus*), with 10 species (*C. aurantiifolia*, *C. medica*, *C. limon*, *C. grandis*, *C. paradisi*, *C. aurantium*, *C. sinensis*, *C. reticulata*, *C. tachibana*, and *C. indica*). Modifications of the Swingle (1943) classification include 17 species (Bhattacharya and Dutta, 1956), 36 species (Hodgson, 1961), or 31 species (Singh and Nath, 1969). Despite its strengths, the Swingle system does not recognize the hybrid nature of very important horticultural groups such as sweet orange, sour orange, grapefruit, lemon, and lime classified, respectively, as *C. sinensis*, *C. aurantium*, *C. paradisis*, *C. lemon*, and *C. aurantiifolia*. More recently, Mabberley (1997, 1998, 2004) was the first taxonomist to try to integrate new phylogenetic knowledge into the *Citrus* classification. Mabberley (1997) proposed three main species for commercial fruits of *Citrus*: *C. medica*, *C. reticulata*, and *C. maxima*. He also proposed four hybrids: *C.* × *aurantium* for sweet oranges, sour oranges, grapefruits, tangelos, and tangors; *C.* × *jambhiri* for rough lemon; *C.* × *aurantiifolia* for "Mexican" lime types; and *C.* × *limon* for lemons (considered by Mabberley (1997) as backcrosses of a lime by a citron). In 2004, Mabberley extended its proposal for the treatment of hybrids. However, several doubts remained on the origin of admixture taxa and some hypotheses made by Mabberley (1997, 2004) are now clearly inappropriate given recent phylogenomic data (Curk et al., 2016; Wu et al., 2018; Oueslati et al., 2017). In his first classification, Mabberley (1997) listed the following cultivars of *C. reticulata*: "tangerine, mandarin, satsuma, clementine; cultivars include "Clementine," "Dancy," "Emperor," "Fina," "Imperial," "Nova," and "Owari"." However, recent phylogenomic studies (Wu et al., 2014, 2018; Oueslati et al., 2017) revealed that all cited cultivars are not pure *C. reticulata* but are introgressed by *C. maxima*. These introgressions are close to 20% for satsuma and clementine and 30% for Nova tangelo. These varieties should be treated as *C.* × *aurantium* rather than *C. reticulata*. In 2004, Mabberley treated rough lemon (formerly *Citrus* × *jambhiri* in Mabberley, 1997) as *C.* × *taitensis* with two potential origins: *C. reticulata* × *C. medica* or *C. reticulata* × *C. limon*. Recent data (Curk et al., 2016; Wu et al., 2018) agree with the first hypothesis. The origin of *C.* × *aurantiifolia* (lime) proposed in Mabberley (2004) (*C. ichangensis* × *C. maxima*) is erroneous. As proposed by Nicolosi et al. (2000) and confirmed recently (Curk et al., 2016; Wu et al., 2018), *C.* × *aurantiifolia* = *C. micrantha* × *C. medica*. Even if Zhang and Mabberley (2008) split *papeda* into two species *C. cavaleriei* and *C. hystrix* in agreement with the Swingle classification of sections *papeda* and *papedocitrus* (Swingle, 1943), the treatment of *papeda* types is still too limited. Therefore, the conceptual framework proposed by Mabberley (1997) for the classification of edible *Citrus* represents a good foundation to which robust phylogenomic data can now be applied.

Mabberley (1998, 2004) and Zhang and Mabberley (2008) proposed to include all the "true citrus" taxa in the genus Citrus with the following names: *C. australasica* (finger lime), *C. australis* (Australian lime), *C. cavaleriei* and *C. hystrix* (papeda), *C. glauca* (desert lime), *C. japonica* (kumquat), *C. maxima* (pummelo), *C. medica* (citron), *C. reticulata* (mandarin), and *C. trifoliata* (trifoliate orange). As discussed previously, this is fully justified by sexual compatibility and chloroplastic phylogenetic data. According to the chloroplastic phylogenetic study of Bayer et al. (2009), *Oxanthera* spp. from New Caledonia should also be included in *Citrus*. However, by not providing a subgeneric classification Mabberley (1998, 2004) and Zhang and Mabberley (2008), did not fully convey phylogenetic relationships within *Citrus*. The close relationships between the Australian citrus could be reflected in a sectional classification.

4.2.2 1967–2017, from traditional taxonomy to phylogenomy: 50 years to clarify the genetic organization of the genus *Citrus* and the origin of modern citrus varieties

During the 1970s, numerical taxonomy resulted in a better understanding of citrus domestication and of the relationships between the various cultivated species of *Citrus*. Barrett and Rhodes (1976) were the first to propose, based on morphological descriptors, that three basic taxa (*C. maxima, C. medica*, and *C. reticulata*) gave rise to all cultivated *Citrus*. During the 1980s, essential oils and polyphenols were the first molecular markers used for taxonomic purposes. Chemotaxonomic studies revealed four true Citrus species (*C. halimii* B.C. Stone, *C. maxima, C. medica,* and *C. reticulata*) (Scora, 1988). During the same period, the importance of *C. maxima, C. reticulata,* and *C. medica* was also emphasized by total protein analysis (Handa and Ishizawa, 1986). The development of codominant isozyme markers (Herrero et al., 1996a,b; Torres et al., 1982; Hirai et al., 1986) opened the modern era of citrus phylogenic studies. Indeed, codominant markers allow revealing the high heterozygosity of admixture taxa and their haplotype sharing with the ancestral taxa. Restriction fragment length polymorphisms (RFLPs) (Federici et al., 1998; Fanciullino et al., 2007) significantly increased the number of useful codominant markers; however, RFLP assays are time consuming and labor intensive. Since the second part of the 1990s, the development of polymerase chain reaction (PCR) markers and particularly simple sequence repeats (SSRs) (Barkley et al., 2006; Kijas et al., 1995; Chen et al., 2006; Luro et al., 2008; Froelicher et al., 2008; Ollitrault et al., 2010; Liu et al., 2013a; Biswas et al., 2014; Liang et al., 2015; Ramadugu et al., 2015; Shimizu et al., 2016) strongly reinforced the *Citrus* phylogenetic studies. Mitochondrial (Froelicher et al., 2011) and plastome data (Bayer et al., 2009; Nicolosi et al., 2000; Yamamoto et al., 1993; Carbonell-Caballero et al., 2015) provided important information on *Citrus* maternal phylogeny. During the last 5 years, with the availability of the first complete reference sequences of the citrus genome (Wu et al., 2014; Xu et al., 2013), the era of phylogenomics began. WGS and genotyping by sequencing (GBS) data provided a huge number of SNPs and allowed the identification of discriminant polymorphisms of the different ancestral taxa, covering the whole genome (Wu et al., 2014, 2018; Curk et al., 2015; Oueslati et al., 2017). Efficient SNP genotyping methods have been developed for scalable experiments using competitive allele amplification (KASPar© technology (Curk et al., 2015; Garcia-Lor et al., 2013b; Cuenca et al., 2013)). Cleaved amplified polymorphic sequence approaches were successfully developed in Japan (Shimada et al., 2014; Omura and Shimada, 2016). SNP arrays have been developed for high-throughput studies (Ollitrault et al., 2012; Fujii et al., 2013) and recently, in California, two Affymetrix Axion SNP arrays with about 1.5 million and 56,000 SNPs were developed (Eck et al., 1996). GBS (Oueslati et al., 2017) and its variant DArTseq (Penjor et al., 2014, 2016; Curtolo et al., 2017) approaches were also recently developed in citrus. The diagnostic polymorphisms of the ancestral species were successfully used to identify the origin of admixture (Curk et al., 2015, 2016; Wu et al., 2014, 2018) and to infer the phylogenomic karyotypes all along their genomes (Wu et al., 2014, 2018; Oueslati et al., 2017).

Recent phylogenetic and phylogenomic studies validated most of the hypotheses of interspecific hybrids origin proposed in the important paper of Nicolosi et al. (2000). Curk et al. (2016) proposed clarifications for the lemon and lime horticultural groups on the basis of nuclear and cytoplasmic diagnostic markers of the ancestral taxa. The conclusions of these two papers and the phylogenomic studies by Wu et al. (2014, 2018) from WGS data, by Oueslati et al. (2017) from GBS data for the *C. reticulata/C. maxima* complex, and by Penjor et al. (2016) from RAD-Seq for several lime and lemon types are summarized in Fig. 4.4. It provides a good framework to lay the foundation of a classification based on the reticulate evolution of citrus and the resulting phylogenomic structures.

4.2.3 The ancestral and admixture taxa

We propose a new trinomial concept for *Citrus* admixture classification. As illustrated in Fig. 4.5 for the admixture types between the four ancestral taxa of the cultivated *Citrus,* the species name is determined by the phylogenomic admixture revealed by recent phylogenetic and phylogenomic data. Three potential admixture combinations implying *C. micrantha* have not been yet revealed by phylogenomic studies. The variety rank corresponds to the groups of modern cultivars diversified, by mutations, transposable element mobilities, or stable epigenetic variations, without further sexual recombination, from each ancestral reticulation events. An example within *C.* × *aurantium* is given in Fig. 4.6. Sweet oranges and willow leaf mandarins are two *C. reticulata/C. maxima* admixtures groups with unknown origins (Wu et al., 2014, 2018) but both deriving from a single hybrid. A little more than one century ago, in Algeria, Father Clement selected Clementine as a chance seedling from 'Mediterranean' willow leaf mandarin. Phylogenetic and phylogenomic studies demonstrated that it resulted from *C. reticulata* var. *deliciosa* × *C.* × *aurantium* var. *sinensis* hybridization (Nicolosi et al., 2000; Wu et al., 2014; Curk et al., 2015; Oueslati et al., 2017; Ollitrault et al., 2012). Numerous cultivars with significant phenotypic diversity were selected from this initial hybrid. They are treated as *C. aurantium* var. *clementina*. In the same way, the first grapefruit resulted from a spontaneous hybridization in the Carribean between *C. maxima* and *C.* × *aurantium* var. *sinensis* (Wu et al., 2018; Curk et al., 2015; Oueslati et al., 2017; Penjor et al., 2016) followed by an important asexual diversification leading to the actual grapefruit cultivars. They are treated as *C. aurantium* var. *paradisi*. The older is the ancestral reticulation event, the higher is the within-variety diversity, particularly under human selection of phenotypical variants. For sweet oranges, it resulted in a huge amount of phenotypical diversity generally organized in common oranges, navel oranges, blood oranges, and acidless cultivars. They are all treated as *C.* × *aurantium* var. *sinensis*.

The actual citrons, pummelos, and "small flower" *papeda* are mostly pure representatives of, respectively, *C. medica*, *C. maxima*, and *C. micrantha* (Wu et al., 2018; Curk et al., 2014, 2015). The situation is even more complex for *C. reticulata*. Indeed many of the mandarins included in *C. reticulata* by Swingle and Reece (1967) and Mabberley (1997) display introgressions of *C. maxima* (Wu et al., 2014, 2018; Curk et al., 2014, 2015; Oueslati et al., 2017). Recent phylogenomic studies (Wu et al., 2014, 2018; Oueslati et al., 2017) revealed a continuum of *C. reticulata/C. maxima* admixture when including modern mandarins, tangors, tangelos, sweet and sour oranges, orangelos, and grapefruits (Fig. 4.7).

Sweet orange and grapefruit horticultural groups are ideotypes, each arised from a single reticulation event that have been very successful and spread all over the world, but they are fully part of this continuum of *C. reticulata/C. maxima* admixture. Moreover, sweet orange genome share a significant proportion of haplotypes with modern mandarins (Wu et al., 2018). We propose to treat all modern varieties of mandarins classified as type 3 by Wu et al. (2018), due to *C. maxima* introgression, as *C.* × *aurantium*. The species names of the Tanaka (1954, 1961) classification may be used for variety rank when they have taxonomic priority. Indeed, even if the Tanaka classification erroneously gave species rank to hybrids with their structure fixed by apomixis, it had the advantage of recognizing among mandarins many of the different reticulation events. This definition of *C.* × *aurantium* extends the one proposed by Mabberley (1997) to all *C. reticulata/C. maxima* admixtures including tangors, tangelos, and some mandarins. It also retains the variety concept proposed by Linnaeus (1753) for *C. aurantium* var. *sinensis* and *C. aurantium* var. *aurantium* and extends it to all admixture genotypes resulting from independent reticulation events.

In the same way for other *Citrus* admixture species, when *Citrus* types sharing the same kind of phylogenomic admixture result from independent reticulation events, we propose to use the former Tanaka species names (Tanaka, 1954; Tanaka, 1961; Tanaka, 1977), when appropriate (priority), or the priority name for this type, for variety rank.

From the pylogenetic/phylogenomic data actually available (Nicolosi et al., 2000; Curk et al., 2015, 2016; Wu et al., 2014, 2018; Oueslati et al., 2017; Penjor et al., 2016), a revised classification of citrus based on the identified phylogenomic structures could be as follows. Table 4.1 summarizes the correspondence between the proposed classification and the former most important ones of Tanaka (1961), Swingle and Reece (1967), and Mabberley (2004) revised by Zhang and Mabberley (2008).

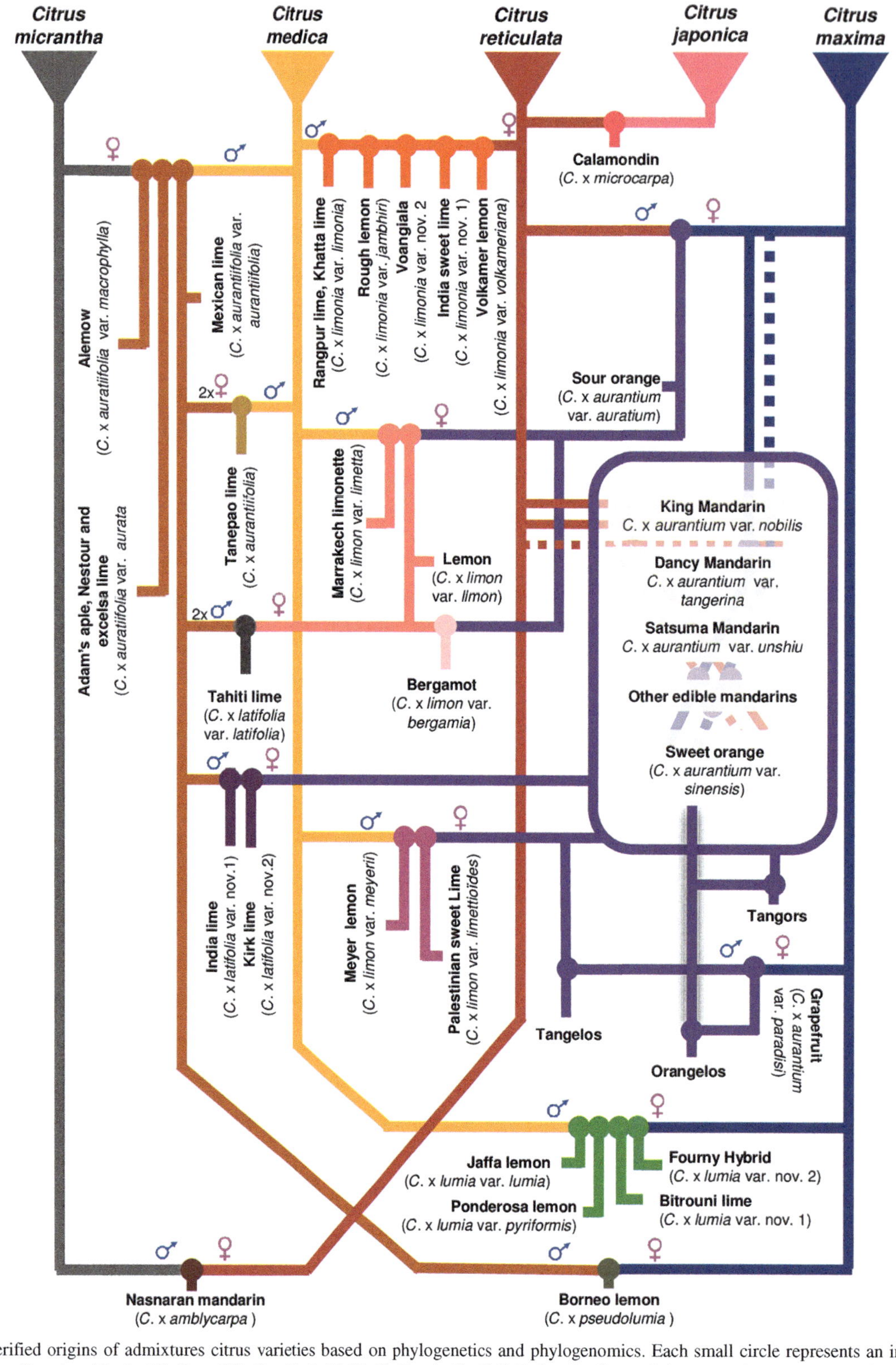

FIG. 4.4 Verified origins of admixtures citrus varieties based on phylogenetics and phylogenomics. Each small circle represents an independent reticulation event. *Based on Nicolosi E, Deng ZN, Gentile A, Malfa Sl, Continella G, Tribulato E. Citrus phylogeny and genetic origin of important species as investigated by molecular markers. Theor. Appl. Genet. 2000;100(8):1155–1166, Wu GA, Prochnik S, Jenkins J, Salse J, Hellsten U, Murat F, et al. Sequencing of diverse mandarin, pummelo and orange genomes reveals complex history of admixture during citrus domestication. Nat. Biotechnol. 2014;32(7):656–662, Wu GA, Terol J, Ibanez V, Lopez-Garcia A, Perez-Roman E, Carles B, et al. Genomics of the origin, evolution and domestication of citrus. Nature 2018., Curk F, Ollitrault F, Garcia-Lor A, Luro F, Navarro L, Ollitrault P. Phylogenetic origin of limes and lemons revealed by cytoplasmic and nuclear markers. Ann. Bot. 2016;117(4):565–583, Curk F, Ancillo G, Ollitrault F, Perrier X, Jacquemoud-Collet JP, Garcia-Lor A, et al. Nuclear species-diagnostic SNP markers mined from 454 amplicon sequencing reveal admixture genomic structure of modern citrus varieties. PLoS One 2015;10(5):e0125628, Penjor T, Mimura T, Kotoda N, Matsumoto R, Nagano AJ, Honjo MN, et al. RAD-Seq analysis of typical and minor Citrus accessions, including Bhutanese varieties. Breed. Sci. 2016;66(5):797–807, and Oueslati A, Salhi-Hannachi A, Luro F, Vignes H, Mournet P, Ollitrault P. Genotyping by sequencing reveals the interspecific C. maxima/C. reticulata admixture along the genomes of modern citrus varieties of mandarins, tangors, tangelos, orangelos and grapefruits. PLoS One 2017;12(10): e0185618.*

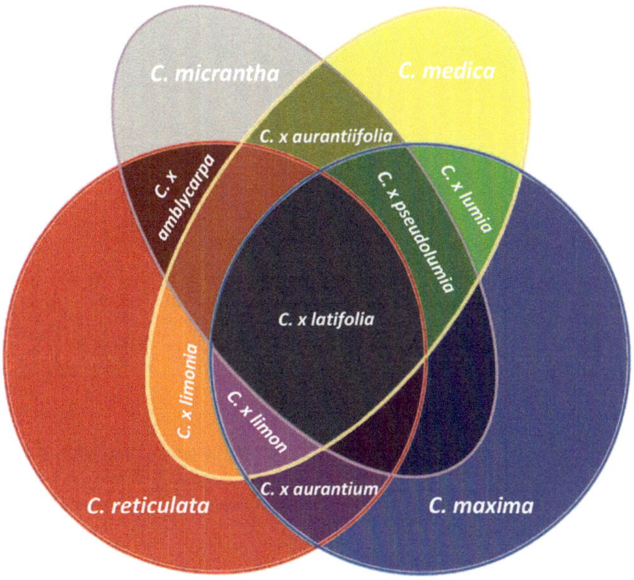

FIG. 4.5 Proposal for an unambiguous *Citrus* classification based on phylogenomic admixture; for example, from the *C. maxima/C. medica/ C. micrantha/C. reticulata* gene pool.

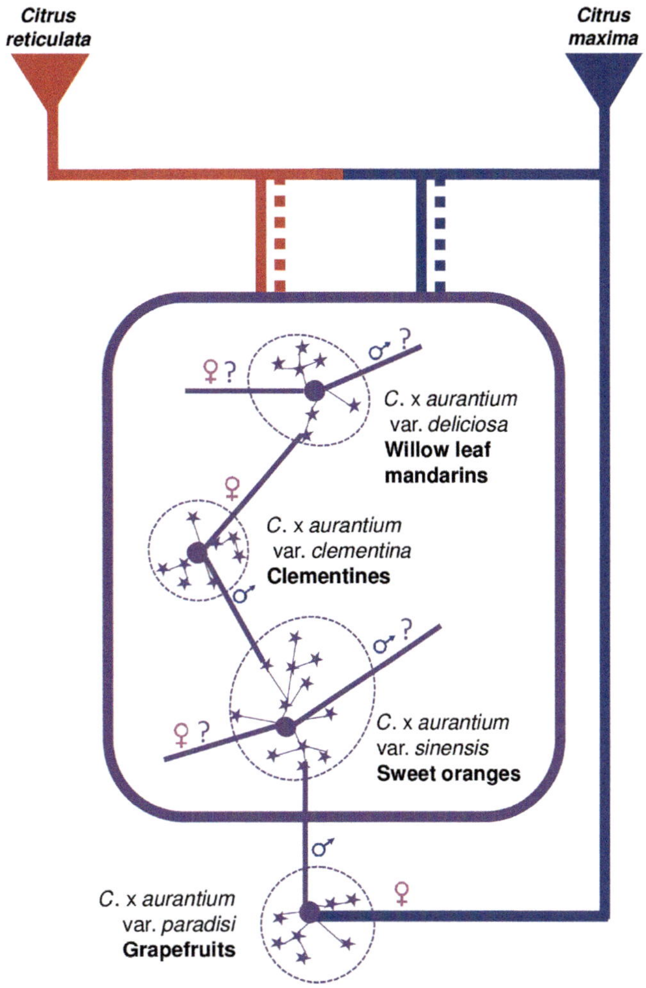

FIG. 4.6 Illustration of the proposed treatment for variety rank in admixture species; an example in *C.* × *aurantium*. *Small circle*: ancestral reticulation event, *stars*: modern cultivars diversified, by mutation, transposable element mobility or stable epigenetic variations, from the ancestral hybrid, without further sexual recombination.

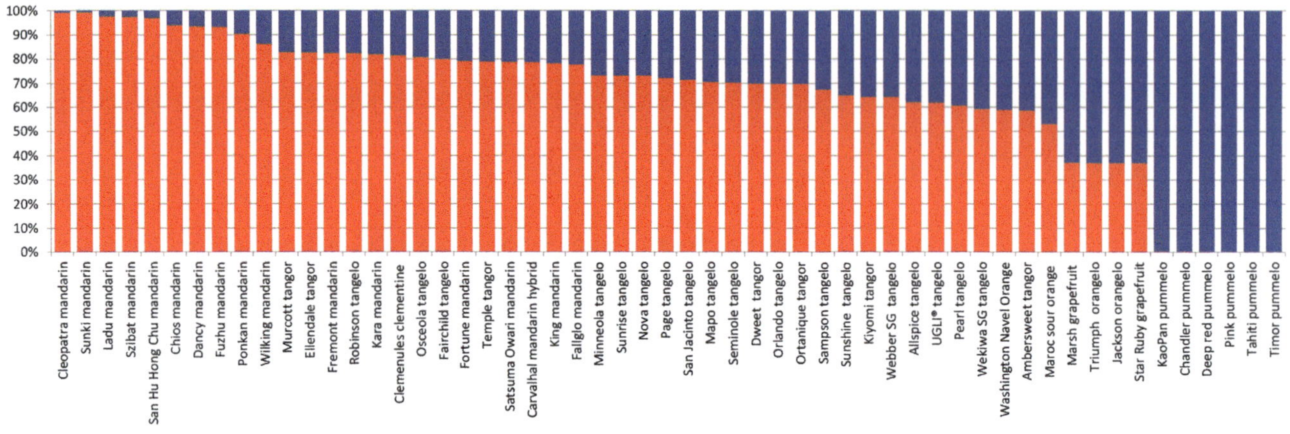

FIG. 4.7 Relative admixtures of *C. reticulata* and *C. maxima* in mandarins, tangors, tangelos, orangelos, grapefruits, and pummelos. *Red*: *C. reticulata* contribution; *blue*: *C. maxima* contribution. *Modified from Oueslati A, Salhi-Hannachi A, Luro F, Vignes H, Mournet P, Ollitrault P. Genotyping by sequencing reveals the interspecific C. maxima/C. reticulata admixture along the genomes of modern citrus varieties of mandarins, tangors, tangelos, orangelos and grapefruits. PLoS One 2017;12(10): e0185618. GBS data.*

4.2.3.1 Pure Citrus species

Given the recent phylogenomic data and biological characteristics (particularly sexual compatibility), we agree with the inclusion of the former genera *Microcitrus, Eremocitrus, Clymenia, Poncirus,* and *Fortunella* in the *Citrus* genus as proposed by Mabberley (1997, 1998, 2004). We also propose to include the New Caledonian genus *Oxanthera* in *Citrus*. Phylogenetic (Bayer et al., 2009) and phylogenomic studies (Wu et al., 2018) do not validate the subgenera *Citrus* and *Papeda* proposed by Swingle (1943). Indeed, they are not monophyletic. Only a few species classified in subgenus *Papeda* by Swingle Swingle (1943) have been deeply characterized by genomic studies (Wu et al., 2018; Wang et al., 2017).

The genomic diversity within trifoliate orange, kumquat, *Microcitrus*, *Eremocitrus*, *Clymenia*, and *Oxanthera* has not yet been established. Therefore, at this point, we do not have a definitive phylogenomic basis to classify the former Swingle species of these genera and the *Papeda* group as all belonging to true species of *Citrus*. Future WGS analysis of the different papedas, trifoliate orange, kumquats, *Microcitrus*, *Eremocitrus*, *Clymenia*, and *Oxanthera* taxa and particularly the clarification of the phylogenetic relationships between the Australian/Oceanian taxa will allow an exhaustive proposal of pure *Citrus* taxa and correspondence with former classifications. As of 2017, the phylogenomically confirmed pure species are as follows:

***C. cavaleriei* H. Lév. ex Cavalerie** (*C. ichangensis* Swingle; *C. latipes* (Swingle.) Tanaka) has, as its natural area of distribution, West-central and Southwestern China. WGS data (Carbonell-Caballero et al., 2015; Wu et al., 2018) reveal a clear differentiation between *C. ichangensis* and *C. micrantha* both for chloroplastic and nuclear phylogenies and a low level of nuclear heterozygosity. These data are coherent with the subdivision in sections *Papeda* and *Papedocitrus* of Swingle classification (Swingle, 1943). It is, therefore, justified to rank *C. cavaleriei* as a species as proposed by Zhang and Mabberley (2008). It should be a parent of *C.* × *junos* (Yuzu).

***C. maxima* (Burm.) Merr.** is the species described as *C. grandis* (L.) Osbeck by Swingle (1943) that originated in the Malay Archipelago and Indonesia. It should include all nonintrogressed pummelos. It is involved in several interspecific hybrid taxa: *C.* × *aurantium, C.* × *lemon, C.* × *latifolia*, and *C.* × *lumia*.

***C. medica* Linnaeus** fits with the species of the same binomial described by Swingle (1943) that initially evolved in Northeastern India and the nearby region of Burma and China. It should include all nonintrogressed citrons. Swingle and Reece (1967) distinguished the standard *C. medica* type, *C. medica* var. ethrog Engl. (= Citrus limonimedica Lush. for Tanaka (1961)) and *C. medica* var. sarcodactylis (Hoola van Nooten) Swingle. The molecular analysis done with citrus germplasm of the INRA/CIRAD collection (Corsica) does not justified this subdivision (Curk et al., 2015; Luro et al., 2012). However, a structure analysis based on the SSR markers and including Mediterranean and Chinese *C. medica* accessions revealed a differentiation between the two regions and a substructuration of Chinese accessions in two clusters (Ramadugu et al., 2015). All fingered lemons were in one of these clusters but associated with some nonfingered varieties. Therefore, at this stage, we do not retain the subdivision proposed by (Swingle and Reece (1967). A deeper phylogenomic analysis of *C. medica* diversity in its main areas of diversification (Yunan, India, Mediterranean Basin) will be necessary to subdivide, or not, this species in different varieties. *C. medica* is involved in several interspecific hybrid taxa: *C.* × *lemon, C.* × *limonia, C.* × *aurantiifolia, C.* × *latifolia, C.* × *lumia*, and *C.* × *pseudolumia*.

TABLE 4.1 **Correspondences between the new phylogenomic classification and the former classifications of Tanaka (1961), Swingle and Reece (1967), and Mabberley (2004) revised by Zhang and Mabberley (2008).**

Phylogenomic classification	Tanaka (1961)	Swingle and Reece (1967)	Zhang and Mabberley (2008)	Common names (examples)	Phylo-genomic references
Citrus cavaleriei H. Lév. ex Cavalerie	*C. ichangensis* Swingle	*C. ichangensis*	*C. cavaleriei*	Adsae	Wu et al. (2018)
C. maxima (Burm.) Merr.	*C. maxima*	*C. maxima*	*C. maxima*	Pummelos (Pink, Deep Red, Timor, …)	Curk et al. (2015) and Wu et al. (2018)
C. medica L.	*C. limonimedica* Lush.	*C. medica*	*C. medica*	Etrog citron	Curk et al. (2015), Curk et al. (2016), and Wu et al. (2018)
	C. medica			Citrons (Corsican, Diamante, Buddha's hand, Humpang, …)	Curk et al. (2015), Curk et al. (2016), and Wu et al. (2018)
C. micrantha Wester	*C. micrantha*	*C. micrantha*	*C. hystrix* DC.	Small-flowered papeda, small-fruited papeda	Curk et al. (2015), Curk et al. (2016), and Wu et al. (2018)
C. reticulata var. *austera* Swingle		*C. reticulata* var. *austera*	*C. reticulata* Blanco	Sun-Chu-Sha-Kat mandarin	Wu et al. (2018)
C. reticulata var. *tachibana* ined.	*C. tachibana* (Makino) Tanaka	*C. tachibana*	*C. reticulata*	Tachibana mandarin	Wu et al. (2018)
C. × *amblycarpa*	*C. amblycarpa*	*C. reticulata* hybrid		Nasnaran mandarin	Curk et al. (2015)
C. × *aurantiifolia* var. *aurantiifolia*	*C. aurantiifolia*	*C. aurantiifolia*	*C.* × *aurantiifolia*	Mexican, Key, West Indies limes…	Curk et al. (2016), Wu et al. (2018), and Penjor et al. (2016)
C. × *aurantiifolia* var. *macrophylla* ined.	*C. macrophylla* Wester	*C. aurantiifolia* (Christm.) Swingle		Alemow	Curk et al. (2016)
C. × *aurantiifolia* var. *aurata* ined.	*C. aurata* Risso	*C. limon* (L.) Burm. f.	*C.* × *aurantium* L.	Adam's apple	Curk et al. (2016)
	C. excelsa Wester	*C. aurantiifolia*		Excelsa and Nestour lime	Curk et al. (2016)
C. × *aurantium* L. var. *aurantium*	*C. aurantium*	*C. aurantium*	*C.* × *aurantium*	Sour orange, Bouquetier	Wu et al. (2014), Curk et al. (2015), Oueslati et al. (2017), Wu et al. (2018), and Penjor et al. (2016)
	C. myrtifolia Raf.			Myrtle-leaf orange, Chinoto	Curk et al. (2015)
C. × *aurantium* var. *clementina* ined.	*C. clementina* hort. ex Tanaka	*C. reticulata*		Clementine	Wu et al. (2014), Curk et al. (2015), Oueslati et al. (2017), and Wu et al. (2018)
C. × *aurantium* var. *deliciosa* ined.	*C. deliciosa* Ten.	*C. reticulata*	*C. reticulata*	Willowleaf, Chios mandarins	Wu et al. (2014), Curk et al. (2015), Oueslati et al. (2017), and Wu et al. (2018)
C. × *aurantium* var. *erythrosa* ined.	*C. erythrosa* hort. ex Tanaka	*C. tachibana*	*C. reticulata*	Fuzhu and San hu hong chu mandarins	Oueslati et al. (2017)

Continued

TABLE 4.1 **Correspondences between the new phylogenomic classification and the former classifications of Tanaka (1961), Swingle and Reece (1967), and Mabberley (2004) revised by Zhang and Mabberley (2008)—cont'd**

Phylogenomic classification	Tanaka (1961)	Swingle and Reece (1967)	Zhang and Mabberley (2008)	Common names (examples)	Phylo-genomic references
C. × *aurantium* var. *kinokuni* ined.	*C. kinokuni* hort. ex Tanaka	*C. tachibana*		Kinokuni, Kishu, Huanglingmiao mandarins	Oueslati et al. (2017) and Wu et al. (2018)
C. × *aurantium* var. *nobilis* ined.	*C. nobilis* Lour.	*C. reticulata* hybrid	*C.* × *aurantium*	King mandarin	Curk et al. (2015), Oueslati et al. (2017), and Wu et al. (2018)
C. × *aurantium* var. *paradisi* ined.	*C. paradisi* Macfad.	*C. paradisi*	*C.* × *aurantium*	Star Ruby, Marsh, Duncan, etc.	Curk et al. (2015), Oueslati et al. (2017), Wu et al. (2018), and Penjor et al. (2016)
C. × *aurantium* var. *paratangerina* ined.	*C. paratangerina* hort. ex Tanaka	*C. reticulata*		Ladu Mandarin	Oueslati et al. (2017)
C. × *aurantium* var. *sinensis* L.	*C. sinensis* (L.) Osbeck	*C. sinensis*	*C.* × *aurantium*	Sweet oranges (Valencia, Washington Navel, Tarroco, etc.)	Wu et al. (2014), Curk et al. (2015), Oueslati et al. (2017), and Wu et al. (2018)
C. × *aurantium* var. *suhuiensis* ined.	*C. suhuiensis* hort. ex Tanaka	*C. reticulata*	*C. reticulata*	Szibat and Se Hui Gan mandarins	Oueslati et al. (2017)
C. × *aurantium* var. *tangerina* ined.	*C. tangerina* hort. ex Tanaka	*C. reticulata*	*C. reticulata*	Dancy, Beauty mandarins	Curk et al. (2015), Oueslati et al. (2017), and Wu et al. (2018)
C. × *aurantium* var. *temple* ined.	*C. temple* hort. ex Yu. Tanaka	*C. sinensis*		Temple tangor	Oueslati et al. (2017)
C. × *aurantium* var. *unshiu* ined.	*C. unshiu* Marcow.	*C. reticulata* clone	*C. reticulata*	Satsuma mandarins	Curk et al. (2015), Oueslati et al. (2017), Wu et al. (2018), and Penjor et al. (2016)
C. × *latifolia* var. nov. 1				India lime	Curk et al. (2016)
C. × *latifolia* var. nov. 2				Kirk lime	Curk et al. (2016)
C. × *latifolia* var. *latifolia*	*C. latifolia*	*C. aurantiifolia*	*C.* × *latifolia*	Bears, Tahiti, Persian limes	Curk et al. (2016)
C. × *limon* var. *bergamia* ined.	*C. bergamia* Risso and Poit.	*C. aurantiifolia*	*C.* × *limon*	Fantastico, Femminello, Castagnaro bergamots	Curk et al. (2016), Penjor et al. (2016)
C. × *limon* var. *meyerii* ined.	*C. meyerii* Yu. Tanaka	*C. limon*	*C.* × *limon*	Meyer lemon	Curk et al. (2016)
C. × *limon* var. *limettioides* ined.	*C. limettioides* Tanaka	*C. aurantiifolia*		Palestinian and Brazil sweet limes and Butnal sweet lemon	Curk et al. (2016), Penjor et al. (2016)
C. × *limon* var. *limetta* ined.	*C. limetta* Risso	*C. limon*		Marrakech limonette	Curk et al. (2016)

TABLE 4.1 Correspondences between the new phylogenomic classification and the former classifications of Tanaka (1961), Swingle and Reece (1967), and Mabberley (2004) revised by Zhang and Mabberley (2008)—cont'd

Phylogenomic classification	Tanaka (1961)	Swingle and Reece (1967)	Zhang and Mabberley (2008)	Common names (examples)	Phylo-genomic references
C. × *limon* var. *limon* (L.) Burm. f.	*C. limon* (L.) Burm. f.	*C. limon*	*C.* × *limon*	Lemons (Lisbon, Eureka, Verna, Luminciana, Interdonato, etc.)	Curk et al. (2016), Wu et al. (2018)
C. × *limonia* var. nov. 1				India sweet lime, Indian lemon	Curk et al. (2016)
C. × *limonia* var. *jambhiri* ined.	*C. jambhiri* Lush.	*C. limon*	*C.* × *taitensis* Risso	Rough lemon	Curk et al. (2016), Wu et al. (2018), and Penjor et al. (2016)
C. × *limonia* Osbeck var. *limonia*	*C. limonia*	*C. limon*		Rangpur lime	Curk et al. (2016), and Wu et al. (2018)
	C. karna Raf.			Khatta Kharna lime	Curk et al. (2016)
C. × *limonia* var. nov. 2				Voangiala	Curk et al. (2016)
C. x limonia var. *volkameriana* Pasquale	C. limonia Osbeck	*C. limon*		Volkamer lemon	Curk et al. (2016)
C. × *lumia* var. nov. 1				Bitrouni lime	Curk et al. (2016)
C. × *lumia* var. nov. 2				Fourny hybrid	Curk et al. (2016)
C. × *lumia* var. *lumia*	*C. lumia* Risso and Poit.	*C. limon*		Jaffa lemon	Curk et al. (2016)
C. × *lumia* var. *pyriformis* ined.	*C. pyriformis* Hassk.	*C. limon*	*C. maxima*	Ponderosa lemon	Curk et al. (2016)
C. × *microcarpa*	*C. madurensis* Lour.	*C. reticulata* hybrid	*C.* × *microcarpa*	Calamondin, Calamansi	Curk et al. (2016)
C. x *pseudolumia* ined.				Borneo, Barum, Baboon lemons	Curk et al. (2016)

***C. micrantha* Wester** originated from the Southern Philippines. According to Swingle and Reece (1967), it should include *C. micrantha* var. *micrantha*, the "small flowered papeda" locally called Biasong and *C. micrantha* var. *microcarpa*, the "small-fruited papeda" with the native name Samuyao. Some chloroplast (Bayer et al., 2009; Nicolosi et al., 2000), mitochondrial (Froelicher et al., 2011), and nuclear phylogenetic studies (Nicolosi et al., 2000; Curk et al., 2015; Ollitrault et al., 2012) suggest that *C. micrantha* and *C. histrix* are closely related. They may eventually be treated as a single species after deeper genomic analysis of *C. hystrix*. *C. micrantha* is involved in several admixture taxa: *C.* × *amblycarpa*, *C.* × *aurantiifolia*, and *C.* × *latifolia*.

***C. reticulata* Blanco** is proposed to include only nonintrogressed mandarins. According to Wu et al. (2018) WGS data, it includes two mandarins. One is classified by Swingle (1943) and Tanaka (1931) as *C. tachibana* (Mak.) Tanaka and is widespread in southern Taiwan, the Ryukyu Islands, and southern Japan (Tanaka, 1931). The second one is the Sun-Chu-Sha-Kat Chinese mandarin treated as *C. reticulata* var. *austera* by Swingle (1943) but confusingly treated (UCR, 2017b) as *C. erythrosa* by Tanaka (1954). Regarding *C. tachibana*, these conclusions for pure *C. reticulata* concern only the type classified as *C. tachibana* (Mak.) Tanaka by Tanaka; indeed, four other Tanaka species *C. erythrosa* hort. ex Tanaka, *C. kinokuni* hort. ex Tanaka, *C. ponki* hort. ex Tanaka, and *C. oleocarpa* hort. ex Tanaka are included in *C. tachibana* Swingle and

Reece (1967). Among these four species, molecular data suggest that at least *C. kinokuni* ("Nanfeng Miju" mandarin) and *C. erythrosa* ("Fuzhu," "San hu hong chu" mandarins) are introgressed at low level by *C. maxima* (Curk et al., 2015; Oueslati et al., 2017). WGS data clearly differentiated *C. tachibana* (Mak.) Tanaka from "Sun-Chu-Sha-Kat" (Wu et al., 2018). Moreover, "Sun-Chu-Sha-Kat" displays much more relatedness with other mandarins than Tachibana does (Wu et al., 2018). Therefore, Wu et al. (2018) suggest "it may be more useful to consider *C. tachibana* (*Mak.*) as a subspecies of *C. reticulata* arising from allopatric isolation." To keep a similar system for pure and hybrid species, we propose to treat them as *C. reticulata* var. *tachibana* and *C. reticulata* var. *austera*, respectively. The treatment of "Cleopatra" and "Sunki" mandarins (respectively, *C. reshni* and *C. sunki* in Tanaka and *C. reticulata* var. *austera* in Swingle), which have only one minor *C. maxima* putative introgression (Wu et al., 2018; Oueslati et al., 2017) needs deeper analysis. A phylogenomic characterization of *C. daoxianensis* S. W. He and G. F. Liu, a wild Chinese mandarin, found to be pure *C. reticulata* in a discrete diagnostic SNP study (Curk et al., 2015), is also necessary to determine whether it is synonymous with *C. reticulata*. *C. reticulata* is involved in several admixture taxa: *C.* × *aurantium, C.* × *amblycarpa, C.* × *limonia, C.* × *microcarpa, C.* × *lemon*, and *C.* × *latifolia.*

4.2.3.2 Admixture types

Bispecific admixture

C. × *amblycarpa* should include all admixtures between *C. micrantha* and *C. reticulata* such as "Nasnaran," an Indonesian citrus considered to be a direct *C. micrantha* × *C. reticulata* hybrid (Curk et al., 2015; Ollitrault et al., 2012). Indeed, it shares the *C. micrantha* mitochondrial genome (Froelicher et al., 2011) and displays interspecific heterozygosity (*C. micrantha/C. reticulata*) for nuclear markers all along its genome (Curk et al., 2015; Ollitrault et al., 2012).

C. × *aurantiifolia* includes all *C. micrantha/C. medica* admixtures and particularly direct hybrids between *C. micrantha* and *C. medica* such as *C.* × *aurantiifolia* var. *aurantiifolia* ("Mexican" lime, "West Indies" lime, and "Thornless" lime) according to Nicolosi et al. (2000), Curk et al. (2016), Penjor et al. (2016), and Wu et al. (2018). According to Curk et al. (2016), *C.* × *aurantiifolia* var. *aurata* ("Adam's apple"; "Excelsa" and "Nestour" limes) and *C.* × *aurantiifolia* var. *macrophylla* are also direct hybrids between *C. micrantha* and *C. medica.* Kaghzi and New Caledonian limes displaying a more complex phylogenomic structure with homozygous areas (Curk et al., 2016) should be classified as *C.* × *aurantiifolia.* The triploid Tanepao, Ambilobe, Coppenrath, and Mothasseb limes, and Madagascar lemon share a similar *C. micrantha/ C. medica* structure and probably derive from an interspecific backcross ((*C. micrantha* × *C. medica*) × *C. medica*) involving a diploid ovule of *C.* × *aurantiifolia* (Curk et al., 2016). They should also be classified as *C.* × *aurantiifolia.*

C. × *aurantium*, as stated before, includes all *C. reticulata/C. maxima* admixtures. If we refer to the demonstrated admixture (Wu et al., 2014, 2018; Curk et al., 2015; Oueslati et al., 2017), it should concern: *C.* × *aurantium* var. *aurantium* (sour oranges, "Bouquetiers"); *C.* × *aurantium* var. *sinensis* (sweet oranges), *C.* × *aurantium* var. *paradisi* (grapefruits); *C.* × *aurantium* var. *tangerina* ("Dancy," "Beauty" mandarins), *C.* × *aurantium* var. *unshiu* (satsuma mandarins), *C.* × *aurantium* var. *clementina* (clementines), *C.* × *aurantium* var. *nobilis* ("King" mandarin), *C.* × *aurantium* var. *temple* ("Temple" mandarin), *C.* × *aurantium* var. *deliciosa* ("Willowleaf" and "Chios" mandarins), *C.* × *aurantium* var. *erythrosa* ("Fuzhu" and "San hu hong chu" mandarins), *C.* × *aurantium* var. *paratangerina* ("Ladu" mandarin), and *C.* × *aurantium* var. *suhuiensis* ("Szibat" and "Se Hui Gan" mandarins). All others mandarins classified as type 3 by Wu et al. (2018) from WGS data, should be treated as *C.* × *aurantium*. Among these varieties, sour orange appears to be the only direct hybrid, while the other ones display more complex genomic structure with phylogenetically homozygous fragments (*C. maxima/ C. maxima* or/and *C. reticulata/C. reticulata*) in addition to *C. reticulata/C. maxima* heterozygosity. Recent hybrids from breeding programs (mandarin hybrids, tangors—mandarin × sweet orange-, tangelos—mandarin × grapefruit-, orangelos—sweet orange × grapefruit) as well as natural tangors and tangelos should also be classified in *C.* × *aurantium.*

C. × *limonia* includes all *C. reticulata/C. medica* admixture types and particularly according to Curk et al. (2016) and Wu et al. (2018) the direct hybrids between these two species: *C.* × *limonia* var. *limonia* ("Rangpur," "Karna," "Khatta," "Khatta Karna" limes); *C.* × *limonia* var. *volkameriana* ("Volkamer" lemon; "Kaghi" lime); *C.* × *limonia* var. *jambhiri* ("Rough" lemon). Cytogenetic studies also provide evidence for a mandarin × citron origin of "Volkamer" lemon, "Rough" lemon, and "Rangpur" lime (Carvalho et al., 2005), while mitochondrial markers (Froelicher et al., 2011) and chloroplast sequences (Carbonell-Caballero et al., 2015) revealed that the female mandarin parent was close to *C. reticulata* var. *austera*. Unclassified cultivars such as the "Voangiala" lemon on one hand and India lemon and Indian sweet lime on the other hand, represent two other *C. reticulata* × *C. medica* independent reticulation events (Curk et al., 2016).

C. × *lumia* corresponds to a *C. medica/C. maxima* admixture. According to Curk et al. (2016), it may include *C.* × *lumia* var. *lumia* ("Jaffa" lemon), *C.* × *lumia* var. *pyriformis* ("Ponderosa" lemon), and the previously unclassified "Bitrouni" lime and "Hybride de Fourny" lemon. The "Bitrouni" lime displays a *C. aurantium* var. *aurantium* cytoplasm, while the others have the *C. maxima* cytoplasm shared with sweet orange (Curk et al., 2016). The 'Hybride de Fourny' appears to be a direct hybrid, while the others have more complex structure with *C. maxima* or *C. medica* homozygosity.

C. × *microcarpa* includes all the kumquat/*C. reticulata* admixtures and particularly the calamondin or calamansi *C.* × *microcarpa* var. *microcarpa* treated as a *C. reticulata* hybrid by Swingle (1943) and *C.* × *microcarpa* by Mabberley (2004). Indeed, according to Wu et al. (2018), the calamondin is a direct hybrid between kumquat and mandarin with a kumquat cytoplasm (Carbonell-Caballero et al., 2015).

Complex tri and tetraspecific admixtures

C. × *limon* includes all *C. reticulata/C. maxima/C. medica* admixtures. Molecular and cytogenetic studies (Nicolosi et al., 2000; Garcia-Lor et al., 2013a; Curk et al., 2016; Ollitrault et al., 2012; Carvalho et al., 2005; Gulsen and Roose, 2001; Ramadugu et al., 2013) suggested that the "yellow lemon" types originated from a *C.* × *aurantium* var. *aurantium* × *C. medica* hybridization and this was definitively proved by WGS data (Wu et al., 2018). *C.* × *limon* var. *limon* should include all lemon types derived by mutation ("Lisbon," "Eureka," "Vern" or "Berna," "Fino," "Santa Theresa," "Adamopoulos," "Luminciana," "Interdonato," etc.) of the original hybrid. *C.* × *limon* var. *limetta* ("Marrakech" limonette) had the same parents as *C.* × *limon* var. *limon* but resulted from an independent reticulation (Curk et al., 2016). *C.* × *limon* should also include *C.* × *limon* var. *limettioides* ("Palestinian" and "Brazilian" sweet limes and "Butnal" sweet lemon) and *C.* × *limon* var. *meyerii* ("Meyer" lemon). These two types probably resulted from hybridization between a *C.* × *aurantium* female parent different than var. *aurantium* (with *C. maxima* cytoplasm) pollinated by *C. medica* (Curk et al., 2016). The Bergamot is also included in this tri-specific group as *C.* × *limon* var. *bergamia* (Curk et al., 2016; Penjor et al., 2016). It probably resulted from hybridization between *C.* × *limon* var. *limon and C.* × *aurantium* var. *aurantium.* Several genotypes, erroneously named citrons, may also be classified as *C.* × *limon*: the "Damas," "Mak Nao Si," and "Rhobs el Arsa" "citrons" that share the *C. aurantium* var. *aurantium* cytoplasm (Curk et al., 2016). The "Milam" lemon and the "Alikioti" lime also display the tri-specific structure of *C.* × *limon* with a *C. reticulata* cytoplasm (Curk et al., 2016).

C. × *pseudolumia* is proposed for admixtures between *C. maxima*, *C. medica*, and *C. micrantha*. Such constitution was revealed by Curk et al. (2016) for Borneo and Barum lemons of the INRA-CIRAD Corsican collection. The two accessions were identical with about 50% *C. maxima*, 38% *C. medica* and 12% *C. micrantha* nuclear genome contributions and a *C. maxima* cytoplasm shared with *C.* × *aurantium* var. *sinensis*. These two varieties may result from a *C. maxima C.* × *aurantiifolia* natural cross. Borneo lemon was morphologically described by Chapot (1964) and considered close to the *Lumia* but different of the *Lumia* cultivars previously described, with serious similarities with the Indian "Gulgul" or "Galgal" fruits. It display profiles of leaf and peel oils very different than other lemons and *lumia* types with high content in linalool/linalyl acetate and α-terpineol and linalool/linalyl acetate, respectively (Lota et al., 2002). Despite its name, Chapot (1964) states that the Borneo lemon was not known in Indonesia but probably originated in India. It was cultivated in North Africa during the 20th century and introduced in the United States under the Baboon lemon name.

C. × *latifolia* includes the genotypes with admixtures of the four ancestors, *C. reticulata/C. maxima/C. medica/C. micrantha*. According to Curk et al. (2016), it may include *C.* × *latifolia* var. *latifolia* (the triploid "Tahiti," "Bears," "Persian" limes) and two diploid limes (Kirk and India) with complex admixture of the four ancestral taxa. Kirk lime and India lime share the *C. micrantha* and the *C. reticulata* var. *austera* cytoplasm, respectively. The triploid *C.* × *latifolia* var. *latifolia* limes may results from a (*C.* × *limon* var. *limon*) × (*C.* × *aurantiifolia* var. *aurantiifolia*) hybridization with a diploid *C.* × *aurantiifolia* pollen (Rouiss et al., 2018; Curk et al., 2016).

4.3 Phenotypic diversity structure strongly reflects evolutionary history

The limitation of gene flow between populations is, with selection, one of the main driving factors for genetic and phenotypic differentiation and can lead to speciation. For *Citrus* s.l. (true citrus plus *Oxanthera*), allopatric evolution has been a clear determinant of the gene pool structure. This is apparent for endemic species of Australia and the Oceanic Islands. There is also evidence in the ancestral Asian species of cultivated *Citrus*, as explained below. Differences in flowering season were also probably a key component for the parapatric differentiation of mandarins, trifoliate oranges, and kumquats in China.

We discuss in this part the phenotypic diversity of the edible Asian citrus classified in the genus *Citrus* by Swingle (1943), not including the kumquats and their hybrids. Indeed, the genomically proven contribution of kumquat to admixtures in citrus germplasm is limited to calamondin (mostly an ornamental type and a condiment in the Philippines cooking) and few studies of phenotypical diversity structure of edible citrus have included kumquat. A description of morphological characteristics of kumquat was provided in Section 4.1.2.2.

4.3.1 Reticulate evolution, apomixis, and the correlation between the structures of genetic and phenotypic diversities in the Asian edible *Citrus* species

The differentiation between the four ancestral taxa of Asian edible *Citrus*, which are sexually compatible, can be explained by a founder effect in four geographic zones and by initial allopatric evolution (Swingle and Reece, 1967; Scora, 1975; Webber, 1967; Wu et al., 2018). *C. maxima* originated in the Malay Archipelago and Indonesia, *C. medica* evolved in Northeastern India and the nearby region of Burma and China, *C. reticulata* was originally found over a region including Vietnam, Southern China, and Japan, and *C. micrantha* is native to the southern Philippines, particularly islands of Cebu and Bohol. As described before, the other edible *Citrus* ideotypes resulted from admixture of these taxa. In addition, vegetative propagation occurred immediately or a few generations after the reticulation events owing the facultative apomixis present in most admixture ideotypes. Therefore, the number of interspecific meiosis and recombination events was limited and large parts of the genome of modern citrus remain in interspecific heterozygosity (Wu et al., 2014, 2018; Oueslati et al., 2017). This reticulate evolution coupled with apomixis also led to generalized linkage disequilibrium when considering the global gene pool of the genus *Citrus*, sensu Swingle (Garcia-Lor et al., 2012). As a consequence: (i) an important part of the actual phenotypic diversity of edible citrus should be related to the differentiation between species before reticulation and introgression processes and (ii) the structures of the phenotypic and genetic diversities are closely correlated. Such correlations were observed for morphological and pomological characters (Ollitrault et al., 2003; Barrett and Rhodes, 1976), flavone constitution (Mizuno et al., 1991), peel oil volatile compounds (Liu et al., 2013b), carotenoid contents (Fanciullino et al., 2006), coumarin and furanocoumarin constitution (Dugrand-Judek et al., 2015), and fingerprinting of secondary metabolites (Matsukawa and Nito, 2017). Recently, Wu et al. (2018) found a relationship between the proportion of *C. maxima* genome and fruit size in the *C. maxima/C. reticulata/C.* × *aurantium* gene pool.

4.3.2 Traits of the four Asian ancestral taxa of the edible *Citrus* (Fig. 4.8)

4.3.2.1 C. maxima *(Burm.) Merri*

Citrus maxima is widely distributed and cultivated in Southeastern Asia and the East Indian Archipelago with the English common name of pummelo. It was introduced into the Caribbean during the discovery period of the New World, where it is named shaddock. A natural hybridization with sweet orange occurring in the Caribbean region produced the grapefruit (see more detail below). According to the description made by Swingle and Reece (1967), *C. maxima* has the biggest flowers (with five sepals and petals and 20–25 stamens, with large linear anthers) and produces the biggest fruits in *Citrus*, which are oblate-spheroid or subpyriform with large, thick, wrinkled seeds. The fruit usually has a thick peel and very large pulp vesicles compared with other *Citrus* species. The membranes enclosing the segments are very strong and can easily be peeled. The weakly adherent pulp vesicles can then be separated. *Citrus maxima* presents additional distinctive characteristics compared with other *Citrus* species. Young angular twigs, leaf midribs, and large veins and petioles are often pubescent. Leaves are "large or very large, oval or elliptic-oval, with a blunt point at the tip and a broadly rounded base, often subcordate and even slightly overlapping the winged petiole… the petiole is broadly winged, and more or less cordate" (Swingle and Reece, 1967). *C. maxima* produces a high level of several secondary metabolites such as naringin (Swingle and Reece, 1967) and coumarins and furanocoumarins (Dugrand-Judek et al., 2015). It is a monoembryonic species with a gametophytic self-incompatibility system (Soost, 1968).

4.3.2.2 C. medica *L.*

C. medica is now widespread in northeastern, central and southern India, Bangladesh, Myanmar, Bhutan and Yunnan Province, and China (Swingle and Reece, 1967; Hodgson, 1967; Gmitter and Hu, 1990; Hazarika, 2012). It was the first species introduced to the Mediterranean Basin following the invasion of Persia by Alexander the Great around 325 BC. It is monoembryonic, self-compatible, and mainly cleistogamous, which led to the high homozygosity of modern cultivars owing to endogamy (Curk et al., 2016; Wu et al., 2018; Curk et al., 2015) and observation that probably explains why it is systematically found as the male parent in admixtures (Nicolosi et al., 2000; Curk et al., 2016; Wu et al., 2018). The following description is adapted from Swingle and Reece (1967). Citron trees are shrubs or small. Leaves are glabrous, elliptic-ovate or ovate-lanceolate, bluntly pointed or rounded at the tips, cuneate or rounded at the base with stout, short, single spines in the axils. Their petioles are short, wingless, or narrowly margined. Inflorescences are short with few-flowered racemes. The flowers are large with generally purplish buds with five petals. They are perfect or male with very numerous stamens (Ollitrault et al., 2003; Raghuvanshi, 1969; Nair and Randhawa, 1969; Guerra, 1993; Hynniewta et al., 2011; Dirceu et al., 2016; Rouiss et al., 2018; Kamiri et al., 2011; Xie et al., 2015; Herrero et al., 1996a,b). The ovary has a height of 18 (usually

FIG. 4.8 Illustrations of morphological traits of the four ancestral taxa of Asian edible *Citrus*. (A) *C. maxima* (©F. Curk-Inra); (B) *C. medica* (© F. Curk-Inra); (C) *C. micrantha* var. microcarpa (photo: Courtesy UC Riverside Citrus Variety Collection); and (D) *C. reticulata* var. *austera* Sun Chu Sha Kat' (photo: Courtesy UC Riverside Citrus Variety Collection).

10–14) locules with—four to eight or more ovules in each locule in two collateral rows. Fruits are medium (10cm) to very large (50cm) according to Ramadugu et al. (2015), oblong, oval, or fingered with smooth or more often rough and bumpy surface. The fruit is very fragrant and yellow when ripe. The rind is very thick with small segments, filled with pale greenish pulp vesicles. They generally contain few seeds (Webber, 1943, 1967) with one white embryo.

4.3.2.3 C. micrantha *Wester*

Future WGS studies of the different Tanaka and Swingle papeda species could lead to *C. micrantha* and other species being synonymized as discussed above for *C. hystrix* (Combava, kaffir lime). The description provided here is from Wester in 1915. *C. micrantha* is cultivated on a small scale as a hair wash in the southern Philippine Islands (Cebu, Bohol, Negros, and Mindanao). It is not eaten and is of no economic importance. Wester (1915) recognized two varieties: *C. micrantha* var. *micrantha* Wester (the small-flowered papeda locally named "Biasong") and *C. micrantha* var. *microcarpa* Wester (small-fruited papeda, locally named "Samuyao"). The two types present similar traits but *C. micrantha* var. *microcarpa* displays a global reduction of morphological characters. According to the original description of the species by Wester (1915), the characteristics of both *C. micrantha* var. *micrantha* (and *C. micrantha* var. *microcarpa* in parentheses), are as follows: a tree attaining a height of 7.5–9 (4.5)m; leaves 9–12 (5.5–8)cm long, 27–40mm broad (Ding et al., 1984; Duan, 1990; Mayr, 1942; Mayden, 1997; de Queiroz, 2007; Dobzhansky, 1970), broadly elliptical to ovate, petioles 35–60mm long (Ding et al., 1984; Duan, 1990; Mayr, 1942; Mayden, 1997; de Queiroz, 2007; Dobzhansky, 1970; Andersson, 1990; Nixon and Wheeler, 1990; Donoghue, 1985; Mishler, 1985; Ollitrault et al., 2003), broadly winged, up to 40 (about 14) mm wide; flowers small,

12–13 mm (Bayer et al., 2009; Tolkowsky, 1938; Tanaka, 1954; Scora, 1975; Webber, 1943) in diameter, white, with a trace of purple on the outside, petals 4 (Swingle and Reece, 1967; Morton et al., 2003; Bayer et al., 2009), stamens 15–18 (Bayer et al., 2009; Tolkowsky, 1938; Tanaka, 1954; Scora, 1975; Webber, 1943, 1967; Ramon-Laca, 2003; Mabberley, 2004; Linnaeus, 1753; Swingle, 1943; Zhang and Mabberley, 2008; Krueger and Navarro, 2007; Siebert, 2016), ovary obovoid with 6–8 locules (Tanaka, 1954; Scora, 1975; Webber, 1943); fruits obovate to oblong-obovate, 5–7 cm long and 3–4 cm in transverse diameter (roundish in outline, 1.5–2 cm in diameter). *C. micrantha* is probably monoembryonic (UCR, 2017c).

4.3.2.4 C. reticulata *Blanco*

Current WGS data provide evidence of only *C. reticulata* var. *tachibana* ined. and *C. reticulata* var. *austera* Swingle as being not introgressed by *C. maxima*. However, few of the primitive mandarin types have been re-sequenced until this moment. *Citrus daoxianensis,* a Chinese wild mandarin without evidence of admixture in a discrete molecular marker study (Curk et al., 2015), is an example of potentially pure *C. reticulata*. We provide here a description of *C. reticulata* var. *tachibana* and *C. reticulata* var. *austera* based on Swingle (1943) and Tanaka (1954). Both types produce small, highly seedy, and very acidic fruits characteristic of undomesticated types. Their fruits are orange at maturity. Both are polyembryonic.

C. reticulata var. *austera* is frequent in the Swatow region of Kwangtung where it is used as rootstock. It is naturally found in Assam (India), China, and Japan (Wu et al., 2018). It differs from the sweet mandarins by its small, intensely acidic fruits. Swingle and Reece, (1967) described it as follows: "Fruits slightly depressed globose, 2.9–3.3 cm long, 3.3–3.6 cm diameter, with smooth, loose peel about 4 mm thick, capucine yellow when ripe; oil glands small, round, far apart, fragrant; segments 9, easily separated; segment walls thin, tender, white; core 6–8 mm diameter, soft; pulp deep chrome yellow, composed of small, short, pulp vesicles, clinging together but irregularly arranged and easily broken; juice reddish yellow, very sour; seeds about 9, rounded at one end, pointed at the other, showing white parallel lines from base to tip; leaves lanceolate–elliptical, blades 6.8 × 2.5 cm, rather acutely cuneate at the base and narrowed to a blunt apex, with about 10 pairs of lateral veins; petioles nearly wingless."

C. reticulata var. *tachibana* is widespread from southern Taiwan to the southwestern province of the main island of Japan. Swingle and Reece (1967) considered it a "wild species that has persisted since prehistoric times." Many of its characteristics are close to *C. reticulata* var. *austera.* Tachibana is self-compatible (Yamamoto et al., 2006). A description of this species by Makino, as translated by Katsura, and reported by Swingle and Reece (1967) reads: "Tree stands over 10 feet. Branches and leaves grow thickly. Strongly resistant to frost or snow… Leaves long, ovate-elliptical, subcoriaceous, broadly acuminate, obtuse and incised at the tip, somewhat broad and convex at the base, indistinctly dentate at the margin, midrib slender, straight and distinct beneath, veins almost indistinct, oil glands indistinct; petiole short, small, with linear wings which seem to be on the verge of degeneration. Flowers axillary, solitary, small. Pedicels 2 mm long, slender, glabrous; scales at the base triangular, ciliate at the margin. Calyx 3 mm in diameter; sepals somewhat recurved outward, densely ciliate at the margin, etc. Ovary almost globular, attenuate at the base, about 2 × 2 mm in size, etc. Fruit somewhat flattened, 2–3 cm lateral diam. Skin smooth, oil glands scattered beneath the skin. Segment cases 6–7; juice bitter and almost inedible. Seeds 1–2 in a segment, and rather large in size, etc. Flowers the same as other *Citrus* plants in time of blooming, shape and color."

4.3.3 Traits of some modern citrus taxa resulting from admixture

With the proposed classification concept based on admixture, the morphological and phenological characteristics within admixture taxa may vary a lot, not only from one ancestor to the other one, but also with transgressive forms. Therefore, an exhaustive description of the pattern of variation is difficult to provide, as new hybrid combinations could produce new transgressive forms. We present here the descriptions of some of the most economically important admixture varieties, synthesized from Swingle (1943) and Zhang and Mabberley (2008).

Sour orange (*C.* × *aurantium* var. *aurantium*) and sweet orange (*C.* × *aurantium* var. *sinensis*) are believed to have arisen from the admixture of *C. reticulata* and *C. maxima*. As per the common names, these two taxa have much in common. The mutually coherent pulp vesicles are free from oil droplets and never contain acrid oils; the medium-sized fruits (5–9 cm diameter) have adherent peels and contain numerous segments (Tolkowsky, 1938; Tanaka, 1954; Scora, 1975; Webber, 1943, 1967; Ramon-Laca, 2003; Mabberley, 2004; Linnaeus, 1753; Swingle, 1943; Zhang and Mabberley, 2008); the flowers are large (2.5–4.5 cm in diameter); and the leaves have winged petioles less than half as long as the leaf blade. The chief differences between the sweet and sour oranges are mostly concerned with the fruit, although the petioles of sour oranges are broader and longer than those of sweet oranges. The fruits of the sour orange have a brighter and rougher peel than those of the sweet orange; the oil glands are in sunken areas of the peel, whereas sweet oranges have oil glands in convex areas of the peel; and, of course, they are not as sweet as those of the sweet orange (lower brix:acid ratio).

Fruits of sour orange and its hybrids have a tendency to have hollow cores and a flattened form, although this is not a consistent characteristic. Sweet orange is one of the most important cultivated forms of *Citrus* due to its palatability and nutritional value. Sour orange is cultivated for flavoring, marmalade, and perfumery and historically it was used as a rootstock.

The grapefruit (*C.* × *aurantium* var. *paradisi*) is believed to have resulted from a further cross of sweet orange with pummelo. In contrast to the origin of its parents, in Southeast Asia, the origin of *C.* × *aurantium* var. *paradisi* is fairly well established as having occurred in historical times in the Caribbean (Kumamoto et al., 1987; Bowman and Gmitter, 1990). It differs from sweet orange chiefly in having larger (9–13 cm diameter) fruits with large, coherent juice vesicles. The yellow fruits have a lower brix:acid ratio than sweet orange, with a distinctive flavor and sometimes with pinkish or pink-tinged flesh. Grapefruit is cultivated commercially as an edible fruit, although its importance has been decreasing in the recent years.

The lemon, *C.* × *limon* var. *limon*, has resulted from a cross between *C.* × *aurantium* var. *aurantium* and *C. medica*. As with sweet and sour oranges, *C.* × *limon* var. limon has mutually coherent pulp vesicles that are free from oil droplets and never contain acrid oils; medium-sized fruits (5–9 cm diameter) having adherent peels and containing numerous segments (Tolkowsky, 1938; Tanaka, 1954; Scora, 1975; Webber, 1943, 1967; Ramon-Laca, 2003; Mabberley, 2004; Linnaeus, 1753; Swingle, 1943; Zhang and Mabberley, 2008); and large flowers (2.5–4.5 cm in diameter), generally with a pink tinge in the common acid types. The fruit shape is more or less oval, with a low apical papilla. The thick peel is yellow when ripe, with fairly prominent oil glands. The lemon is generally an acidic fruit, although low-acid selections occur. It is cultivated for use as fresh fruit and for flavoring.

Small fruited limes, *C.* × *aurantiifolia* var. *aurantiifolia*, have originated as a cross between *C. micrantha* and *C. medica*. As with the admixtures previously described, *C.* × *aurantiifolia* var. *aurantiifolia* has mutually coherent pulp vesicles that are free from oil droplets and never contain acrid oils. The fruits are small (4–6 cm diameter), ovoid, or subglobose, often with a small apical papilla, with 9–12 segments. The thin peel is yellow-green when mature and has prominent oil glands. The flowers are small (<2.5 cm diameter) and have a pinkish tinge in the common acid types. Although both acid and low-acid selections exist, the acid types are the more commonly cultivated.

The large-fruited acid limes, *C.* × *latifolia* var. *latifolia*, were proposed by Curk et al. (2016) to have resulted from admixture of *C. limon* var. limon × *C. aurantiifolia* var. *aurantiifolia*. The fruit is similar to those of *C.* × *aurantiifolia* var. *aurantiifolia*, but is larger (6–9 cm diameter), and seedless, probably due to the triploïdy. Their taste and aroma are similar to *C.* × *aurantiifolia* var. *aurantiifolia* but are not as strong. The trees of *C.* × *latifolia* are also more vigorous and robust than those of *C.* × *aurantiifolia* var. *aurantiifolia*. Interestingly, this triploid lime is described to be one of the less susceptible cultivar to Huanglongbing, a very damaging bacterial disease due to *Liberibacter* spp. (Folimonova et al., 2009).

4.4 Conclusion

The definition and classification of the *Citrus* genus has long been debated by taxonomists. The reticulate evolution combined with partial apomixis has led to very different classification systems. Recently, phylogenomic data revealed the origins and admixtures of modern cultivars and wild types. Coupled with reproductive biology, phylogenomy supports the inclusion of all true citrus of the Swingle system plus Oxanthera in the genus *Citrus*. Considering biological and genomic data, the recognition at species rank of the four ancestral taxa of most cultivated citrus is supported by the phylogenetic species concept based on diagnosability and monophyly. Taking into account the implication of reticulate evolution, partial apomixis and asexual diversification from ancestral hybrids we propose, for the genus *Citrus*, a trinomial concept of classification. For admixture taxa, the species rank is defined by their phylogenomic constitution. The variety rank is defined by the old independent reticulation events from which groups of cultivars were differentiated by asexual mechanisms. It provides an unambiguous conceptual framework for *Citrus* classification based on the phylogenomic and genetic data. However, today, the available genomic data remain parcellar and further WGS studies are needed to establish a definitive classification of the genus *Citrus*.

References

Andersson, L., 1990. The driving force: species concepts and ecology. Taxon 39 (3), 375–382.

Barkley, N.A., Roose, M.L., Krueger, R.R., Federici, C.T., 2006. Assessing genetic diversity and population structure in a citrus germplasm collection utilizing simple sequence repeat markers (SSRs). TAG Theor. Appl. Genet. 112 (8), 1519–1531.

Barrett, H.C., Rhodes, R.A., 1976. A numerical taxonomic study of affinity relationships in cultivated *Citrus* and its close relatives. Syst. Bot. 1, 105–136.

Bayer, R.J., Mabberley, D.J., Morton, C., Miller, C.H., Sharma, I.K., Pfeil, B.E., et al., 2009. A molecular phylogeny of the orange subfamily(Rutaceae: Aurantioideae) using nine cpDNA sequences. Am. J. Bot. 96 (3), 668–685.

Bernet, G.P., FernandezRibacoba, J., Carbonell, E.A., Asins, M.J., 2010. Comparative genome-wide segregation analysis and map construction using a reciprocal cross design to facilitate citrus germplasm utilization. Mol. Breed. 25 (4), 659–673.

Bhattacharya, S.C., Dutta, S., 1956. Classification of Citrus Fruits of Assam. Scientific monograph. Indian Council of Agricultural Research, Delhi. 20.

Biswas, M.K., Xu, Q., Mayer, C., Deng, X., 2014. Genome wide characterization of short tandem repeat markers in sweet orange (*Citrus sinensis*). PLoS One 9 (8):e104182.

Blanco, F.M., 1837. Flora de Filipinas. Segun el Sistema sexual de Linneo. Sto. Tomas por D. Candido Lopez, Manila (Filipinas).

Bowman, K.D., Gmitter Jr., F.G., 1990. Caribbean forbidden fruit: grapefruit's missing link with the past and bridge to the future? Fruit Var. J 44 (1), 41–44.

Burman, N.L., 1768. Flora Indica, cui accedit series zoophytorum indicorum, nec non prodromus florae capensis. Apud Cornelium Haek ed. Amsterdam, Netherlands, Apud Cornelium Haek.

Carbonell-Caballero, J., Alonso, R., Ibanez, V., Terol, J., Talon, M., Dopazo, J., 2015. A phylogenetic analysis of 34 chloroplast genomes elucidates the relationships between wild and domestic species within the genus *Citrus*. Mol. Biol. Evol. 32 (8), 2015–2035.

Carvalho, R., Soares Filho, W.S., Brasileiro-Vidal, A.C., Guerra, M., 2005. The relationships among lemons, limes and citron: a chromosomal comparison. Cytogenet. Genome Res. 109 (1/3), 276–282.

Chapot, H., 1964. Le citron de Borneo. Al Awamia 13, 137–153.

Chen, C.X., Zhou, P., Choi, Y.A., Huang, S., Gmitter Jr., F.G., 2006. Mining and characterizing microsatellites from citrus ESTs. TAG Theor. Appl. Genet. 112 (7), 1248–1257.

Chen, C.X., Bowman, K.D., Choi, Y.A., Dang, P.M., Rao, M.N., Huang, S., et al., 2008. EST-SSR genetic maps for *Citrus sinensis* and *Poncirus trifoliata*. Tree Genet. Genomes 4 (1), 1–10.

Citrus, G.A., 1911. In: Gagnepin, F. (Ed.), Flore Générale de l'Indo-Chine Paris: Masson et Cie. pp. 676–682.

Cuenca, J., Aleza, P., Navarro, L., Ollitrault, P., 2013. Assignment of SNP allelic configuration in polyploids using competitive allele-specific PCR: application to citrus triploid progeny. Ann. Bot. 111 (4), 731–742.

Curk, F., Ancillo, G., Garcia-Lor, A., Luro, F., Perrier, X., Jacquemoud-Collet, J.P., et al., 2014. Next generation haplotyping to decipher nuclear genomic interspecific admixture in *Citrus* species: analysis of chromosome 2. BMC Genet 15. 152–014–0152-1.

Curk, F., Ancillo, G., Ollitrault, F., Perrier, X., Jacquemoud-Collet, J.P., Garcia-Lor, A., et al., 2015. Nuclear species-diagnostic SNP markers mined from 454 amplicon sequencing reveal admixture genomic structure of modern citrus varieties. PLoS One 10 (5):e0125628.

Curk, F., Ollitrault, F., Garcia-Lor, A., Luro, F., Navarro, L., Ollitrault, P., 2016. Phylogenetic origin of limes and lemons revealed by cytoplasmic and nuclear markers. Ann. Bot. 117 (4), 565–583.

Curtolo, M., Cristofani-Yaly, M., Gazaffi, R., Takita, M.A., Figueira, A., Machado, M.A., 2017. QTL mapping for fruit quality in *Citrus* using DArTseq markers. BMC Genomics 18 (1), 289 (-017-3629-2).

da Costa Silva, S., Mendes, S., Soares Filho, W.d.S., Pedrosa-Harand, A., 2015. Chromosome homologies between *Citrus* and *Poncirus*—the comparative cytogenetic map of mandarin (*Citrus reticulata*). Tree Genet. Genomes 11 (1), 811.

de Loureiro, J., 1790. Flora Cochinchinensis: Sistens Plantas in Regno Cochinchina Nascentes: Quibus Accedunt Aliae Observatae in Sinensi Imperio, Africa Orientali, Indiaeque Locis Variis: Omnes Dispositae Secundum Systema Sexuale Linnaeanu. Lisbon, Ulyssipone Typis, et expensis academicis.

de Queiroz, K., 2007. Species concepts and species delimitation. Syst. Biol. 56 (6), 879–886.

de Tournefort, J.P., 1700. Intitutiones Rei Herbariae, Editio Altera. Typographia Regia Ed. Typographia regia, Paris.

Ding, S.Q., Zhang, X.N., Bao, Z.R., Liang, M.Q., 1984. Poncirus polyandra. Acta Bot. Yunnanica 6 (3), 292.

Dioscorides, P., 2000. Dioscorides. In: De materia medica—Five Books in One Volume: A New English Translation by T.A. Osbaldeston. IBIDIS Press, Johannesburg, South Africa.

Dirceu, M., Fermino, C.E., Moreira, N.V., Alves de Azevedo, F., Della Coletta, F.H., Vicente Contador, Z.P. (Eds.), 2016. Usefulness of phylogentic diagnostic SNP markers of citrus ancestral taxa for genetics and breeding. International Citrus Congress, 2016-09-18/2016-09-23; Foz do Iguaçu, Brésil: IAPAR.

Dobzhansky, T., 1970. Genetics of the Evolutionary Process. Columbia University Press, New York.

Donoghue, M.J., 1985. A critique of the biological species concept and recommendations for a phylogenetic alternative. Bryologist 88 (3), 172–181.

Duan, Z., 1990. Primary study of fuming evergreen trifoliate orange (*P. trifoliata* Raf). In: Proceedings of the International Citrus Symposium Guangzhou, China, pp. 166–169.

Dugrand-Judek, A., Olry, A., Hehn, A., Costantino, G., Ollitrault, P., Froelicher, Y., et al., 2015. The distribution of Coumarins and furanocoumarins in *Citrus* species closely matches *Citrus* phylogeny and reflects the organization of biosynthetic pathways. PLoS One 10 (11):e0142757.

Eck, Y., Moragues, M., Ferrante, S., Roose, M., 1996. Development and application of affymetrix SNP arrays for citrus Book of Abstract. In: Sustainable Citriculture, The role of applied knowledge. International Citrus Congress; 18–23 September.

Fanciullino, A.L., Dhuique-Mayer, C., Luro, F., Casanova, J., Morillon, R., Ollitrault, P., 2006. Carotenoid diversity in cultivated citrus is highly influenced by genetic factors. J. Agric. Food Chem. 54 (12), 4397–4406.

Fanciullino, A.L., Dhuique-Mayer, C., Luro, F., Morillon, R., Ollitrault, P., 2007. Carotenoid biosynthetic pathway in the citrus genus: number of copies and phylogenetic diversity of seven genes. J. Agric. Food Chem. 55 (18), 7405–7417.

Federici, C.T., Fang, D.Q., Scora, R.W., Roose, M.L., 1998. Phylogenetic relationships within the genus *Citrus* (*Rutaceae*) and related genera as revealed by RFLP and RAPD analysis. Theor. Appl. Genet. 96 (6/7), 812–822.

Ferrari, G.B., 1646. Hesperides siue de malorum aureorum cultura et usu Libri quator Io. Baptistae Ferrarii Senensis e Societate Iesu. Réédition de Romae: Sumptibus Hermanni Scheus, Milano (Italy). 1646; Milano: F. Motta, 1992.

Folimonova, S.Y., Robertson, C.J., Garnsey, S.M., Gowda, S., Dawson, W.O., 2009. Examination of the responses of different genotypes of citrus to huanglongbing (citrus greening) under different conditions. Phytopathology 99 (12), 1346–1354. 2017/12.

Froelicher, Y., Dambier, D., Bassene, J.B., Costantino, G., Lotfy, S., Didout, C., et al., 2008. Characterization of microsatellite markers in mandarin orange (*Citrus reticulata* Blanco). Mol. Ecol. Resour. 8 (1), 119–122.

Froelicher, Y., Mouhaya, W., Bassene, J.B., Costantino, G., Kamiri, M., Luro, F., et al., 2011. New universal mitochondrial PCR markers reveal new information on maternal citrus phylogeny. Tree Genet. Genomes 7 (1), 49–61.

Fujii, H., Shimada, T., Nonaka, K., Kita, M., Kuniga, T., Endo, T., et al., 2013. High-throughput genotyping in citrus accessions using an SNP genotyping array. Tree Genet. Genomes 9 (1), 145–153.

Gallesio, G., 1811. Traité du citrus. Louis Fantin ed. Chez Louis Fantin Libraire, Paris.

Garcia-Lor, A., Luro, F., Navarro, L., Ollitrault, P., 2012. Comparative use of InDel and SSR markers in deciphering the interspecific structure of cultivated citrus genetic diversity: a perspective for genetic association studies. Molecular Genetics and Genomics 287 (1), 77–94.

Garcia-Lor, A., Curk, F., Snoussi-Trifa, H., Morillon, R., Ancillo, G., Luro, F., et al., 2013a. A nuclear phylogenetic analysis: SNPs, indels and SSRs deliver new insights into the relationships in the 'true citrus fruit trees' group (Citrinae, Rutaceae) and the origin of cultivated species. Ann. Bot. 111 (1), 1–19.

Garcia-Lor, A., Ancillo, G., Navarro, L., Ollitrault, P., 2013b. *Citrus* (Rutaceae) SNP markers based on Competitive Allele-Specific PCR; transferability across the Aurantioideae subfamily. Appl. Plant Sci. 1 (4), https://doi.org/10.3732/apps.1200406. eCollection 2013 Apr.

Gmitter Jr., F.G., Hu, X.L., 1990. The possible role of Yunnan, China, in the origin of contemporary *Citrus* species (Rutaceae). Econ. Bot. 44 (2), 267–277.

Guerra, M., 1993. Cytogenetics of Rutaceae. V. High chromosomal variability in *Citrus* species revealed by CMA/DAPI staining. Heredity 71, 234.

Gulsen, O., Roose, M.L., 2001. Chloroplast and nuclear genome analysis of the parentage of lemons. J. Am. Soc. Hortic. Sci. 126 (2), 210–215.

Handa, T., Ishizawa, Y., Ogaki, C., 1986. Phylogenetic study of fraction I protein in the genus *Citrus* and its close related genera. Jpn. J. Genet. 61, 15–24.

Hazarika, T.K., 2012. *Citrus* genetic diversity of north-east India, their distribution, ecogeography and ecobiology. Genet. Resour. Crop. Evol. 59 (6), 1267–1280.

Herrero, R., Asíns, M.J., Carbonell, E.A., Navarro, L., 1996a. Genetic diversity in the orange subfamily Aurantioideae. I. Intraspecies and intragenus genetic variability. Theor. Appl. Genet. 92 (5), 599–609.

Herrero, R., Asins, M.J., Pina, J.A., Carbonell, E.A., Navarro, L., 1996b. Genetic diversity in the orange subfamily Aurantioideae. II. Genetic relationships among genera and species. Theor. Appl. Genet. 93 (8), 1327–1334.

Hirai, M., Kozaki, I., Kajiura, I., 1986. Isozyme analysis and phylogenic relationship of *Citrus*. Jpn. J. Breed. 36 (4), 377–389.

Hodgson, R.W., 1961. Taxonomy and nomenclature in citrus. In: Proceedings of the Second Conference of the International Organization of Citrus Virologists. University of Florida Press, Gainesville, Florida.

Hodgson, R., 1967. Horticultural varieties of citrus. In: Reuther, W., Webber, H., Batchelor, L. (Eds.), The *Citrus* Industry. vol. I. University of California Press, Berkeley, pp. 431–591.

Hynniewta, M., Malik, S.K., Rao, S.R., 2011. Karyological studies in ten species of *Citrus*(Linnaeus, 1753) (Rutaceae) of North-East India. Comp. Cytogenet. 5 (4), 277–287.

Kamiri, M., Stift, M., Srairi, I., Costantino, G., El Moussadik, A., Hmyene, A., et al., 2011. Evidence for non-disomic inheritance in a *Citrus* interspecific tetraploid somatic hybrid between *C. reticulata* and *C. limon* using SSR markers and cytogenetic analysis. Plant Cell Rep. 30 (8), 1415–1425.

Kijas, J.M., Fowler, J.C., Thomas, M.R., 1995. An evaluation of sequence tagged microsatellite site markers for genetic analysis within *Citrus* and related species. Genome 38 (2), 349–355.

Krueger, R.R., Navarro, L., 2007. Citrus germplasm resources. In: *Citrus* Genetics, Breeding and Biotechnology. CAB International, Wallingford, UK, pp. 45–140.

Kumamoto, J., Scora, R.W., Lawton, H.W., Clerx, W.A., 1987. Mystery of the forbidden fruit: historical epilogue on the origin of the grapefruit, *Citrus paradisi* (*Rutaceae*). Econ. Bot. 41 (1), 97–107.

Lawrence, G., 1951. Taxonomy of vascular plants. The Macmillan Co., New York.

Liang, M., Yang, X., Li, H., Su, S., Yi, H., Chai, L., et al., 2015. De novo transcriptome assembly of pummelo and molecular marker development. PLoS One 10 (3):e0120615.

Linnaeus, C., 1737. Genera Plantarum. Leiden.

Linnaeus, C., 1753. Species Plantarum: Vol. II. Holmiae (Stockholm). Impensis Laurentii Salvii [Printed for the Ray Society, London. in 1959.

Liu, S.R., Li, W.Y., Long, D., Hu, C.G., Zhang, J.Z., 2013 Oct 28. Development and characterization of genomic and expressed SSRs in citrus by genome-wide analysis. PLoS One 8 (10):e75149.

Liu, C., Jiang, D., Cheng, Y., Deng, X., Chen, F., Fang, L., et al., 2013b. Chemotaxonomic study of *Citrus*, *Poncirus* and *Fortunella* genotypes based on peel oil volatile compounds—deciphering the genetic origin of Mangshanyegan (*Citrus nobilis* Lauriro). PLoS One 8 (3):e58411.

Lota, M., de RS, T.F., Jacquemond, C., Casanova, J., 2002. Volatile components of peel and leaf oils of lemon and lime species. J. Agric. Food Chem. 50 (4), 796–805.

Luro, F.L., Costantino, G., Terol, J., Argout, X., Allario, T., Wincker, P., et al., 2008. Transferability of the EST-SSRs developed on Nules clementine (*Citrus clementina* Hort ex tan) to other *Citrus* species and their effectiveness for genetic mapping. BMC Genomics 9, 287.

Luro, F., Venturini, N., Costantino, G., Paolini, J., Ollitrault, P., Costa, J., 2012. Genetic and chemical diversity of citron (*Citrus medica* L.) based on nuclear and cytoplasmic markers and leaf essential oil composition. Phytochemistry 77, 186–196.

Mabberley, D.J., 1997. A classification for edible *Citrus* (Rutaceae). Telopea 7 (2), 167–172.

Mabberley, D.J., 1998. Australian Citreae with notes on other Aurantioideae (Rutaceae). Telopea 7 (4), 333–344.

Mabberley, D.J., 2004. *Citrus* (Rutaceae): a review of recent advances in etymology, systematics and medical applications. Blumea 49 (2/3), 481–498.

Magnin-Gonze, J., 2009. Histoire de la botanique, Référence ed. Delachaux et Niestlé, Paris.

Matsukawa, T., Nito, N., 2017. Metabolic fingerprinting of *Citrus* cultivars and related genera using HPLC and multivariate analysis. J. Plant Stud. 7, 1.

Mayden, R., 1997. A hierarchy of species concepts: the denouement in the saga of the species problem. In: Caridge, M., Dawah, H., Wilson, M. (Eds.), Species: The Units of Biodiversity. Chapman & Hall, pp. 381–424.

Mayr, E., 1942. Systematics and the Origin of Species. Columbia University Press, New York.

Medhurst, W.H., 1846. The Shoo King, or the Historical Classic: being the most Ancient Authentic Record of the Annals of the Chines Empire. The Mission Press, Shanghae.

Mishler, B.D., 1985. The morphological, developmental, and phylogenetic basis of species concepts in bryophytes. Bryologist 88 (3), 207–214.

Mizuno, M., Iinuma, M., Ohara, M., Tanaka, T., Iwamasa, M., 1991. Chemotaxonomy of the genus *Citrus* based on polymethoxyflavones. Chem. Pharm. Bull. 39, 945–949.

Morton, C.M., Grant, M., Blackmore, S., 2003. Phylogenetic relationships of the Aurantioideae inferred from chloroplast DNA sequence data. Am. J. Bot. 90 (10), 1463–1469.

Nair, P.K.R., Randhawa, G.S., 1969. Chromosome morphology of the pachytene stage with respect to different *Citrus* types. In: Chapmn, H. (Ed.), Proc. First Int Citrus Symp (Univ California, Riverside). pp. 215–223. 1.

NCBI, 2017. Taxonomy/Browser. http://www.ncbi.nlm.nih.gov/Taxonomy/Browser/wwwtax.cgi.

Nicolosi, E., 2012. Inquadramento tassonomico. In: Tribulato, E. (Ed.), gli agrumi. Coltura & Cultura ed. Bayer CropScience, Milano, Italy, pp. 1–15.

Nicolosi, E., Deng, Z.N., Gentile, A., Sl, M., Continella, G., Tribulato, E., 2000. *Citrus* phylogeny and genetic origin of important species as investigated by molecular markcrs. Thcor. Appl. Genet. 100 (8), 1155–1166.

Nixon, K.C., Wheeler, Q.D., 1990. An amplification of the phylogenetic species concept. Cladistics 6 (3), 211–223.

Ollitrault, P., Jacquemond, C., Dubois, C., Luro, F., 2003. *Citrus*. In: Genetic Diversity of Cultivated Tropical Plants. CIRAD, Montpellier, pp. 193–217.

Ollitrault, F., Terol, J., Pina, J.A., Navarro, L., Talon, M., Ollitrault, P., 2010. Development of SSR markers from *Citrus clementina* (Rutaceae) BAC end sequences and interspecific transferability in *Citrus*. Am. J. Bot. 97 (11), e124–e129.

Ollitrault, P., Terol, J., Garcia-Lor, A., Berard, A., Chauveau, A., Froelicher, Y., et al., 2012. SNP mining in *C. clementina* BAC end sequences; transferability in the *Citrus* genus (Rutaceae), phylogenetic inferences and perspectives for genetic mapping. BMC Genomics 13, 13.

Omura, M., Shimada, T., 2016. *Citrus* breeding, genetics and genomics in Japan. Breed. Sci. 66 (1), 3–17.

Oueslati, A., Salhi-Hannachi, A., Luro, F., Vignes, H., Mournet, P., Ollitrault, P., 2017. Genotyping by sequencing reveals the interspecific *C. maxima/C. reticulata* admixture along the genomes of modern citrus varieties of mandarins, tangors, tangelos, orangelos and grapefruits. PLoS One 12 (10):e0185618.

Penjor, T., Mimura, T., Matsumoto, R., Yamamoto, M., Nagano, Y., 2014. Characterization of limes (*Citrus aurantifolia*) grown in Bhutan and Indonesia using high-throughput sequencing. Sci. Rep. 4, 4853.

Penjor, T., Mimura, T., Kotoda, N., Matsumoto, R., Nagano, A.J., Honjo, M.N., et al., 2016. RAD-Seq analysis of typical and minor *Citrus* accessions, including Bhutanese varieties. Breed. Sci. 66 (5), 797–807.

Raghuvanshi, S.S., 1969. Cytological evidence bearing on evolution in *Citrus*. In: Chapmn (Eds.), Proc First Int Citrus Symp (Univ California, Riverside). pp. 207–214. 1.

Ramadugu, C., Pfeil, B.E., Keremane, M.L., Lee, R.F., Maureira-Butler, I.J., Roose, M.L., 2013. A six nuclear gene phylogeny of *Citrus* (Rutaceae) taking into account hybridization and lineage sorting. PLoS One 8 (7), e68410.

Ramadugu, C., Keremane, M.L., Hu, X., Karp, D., Federici, C.T., Kahn, T., et al., 2015. Genetic analysis of citron (*Citrus medica* L.) using simple sequence repeats and single nucleotide polymorphisms. Sci. Hortic. 195, 124–137.

Ramon-Laca, L., 2003. The introduction of cultivated *Citrus* to Europe via Northern Africa and the Iberian Peninsula. Econ. Bot. 57 (4), 502–514. 2017/12.

Risso, A., Poiteau, A., 1818. Histoire Naturelle des Orangers. Imprimerie de Mme Hérissant Le Doux, Imprimeur ordinaire du Roi et des Musées Royaux, Paris.

Rouiss, H., Bakry, F., Froelicher, Y., Navarro, L., Aleza, P., Ollitrault, P., 2018. Origin of *C. latifolia* and *C. aurantiifolia* triploid limes: the preferential disomic inheritance of doubled-diploid 'Mexican' lime is consistent with an interploid hybridization hypothesis. Ann. Bot. 121, 571–585.

Scora, R.W., 1975. On the history and origin of *Citrus*. Bull. Torrey. Bot. Club 102, 369–375.

Scora, R.W., 1988. Biochemistry, taxonomy and evolution of modern cultivated citrus. In: Proceedings of the Sixth International Citrus Congress. vol. 1, pp. 277–289.

Shimada, T., Fujii, H., Endo, T., Ueda, T., Sugiyama, A., Nakano, M., et al., 2014. Construction of a citrus framework genetic map anchored by 708 gene-based markers. Tree Genet. Genomes 10 (4), 1001–1013.

Shimizu, T., Kitajima, A., Nonaka, K., Yoshioka, T., Ohta, S., Goto, S., et al., 2016. Hybrid origins of *Citrus* varieties inferred from DNA marker analysis of nuclear and organelle genomes. PLoS One 11 (11):e0166969.

Siebert, T.J., 2016. Cross-Compatibility, Graft-Compatibility, and Phylogenetic Relationships in the Aurantioideae: New Data From the Balsamocitrinae. University of California, Riverside, California.

Siebert, T., Kahn, T., Krueger, R., 2015. Observations of graft compatibility between *Citrus* spp and related Aurantioideae taxa. Acta Hortic. 1065, 173–179.

Singh, R., Nath, N., 1969. Practical approach to the classification of citrus. In: Proceedings of the First International Citrus Symposium.

Soost, R.K., 1968. The incompatibility gene system in citrus. In: Chapmn, H. (Ed.), Proc First Int Citrus Symp. 1. University of California, Riverside, pp. 189–190.

Stebbins, G., 1950. Variation and Evolution in Plants. Columbia University Press Ed. Columbia University Press, New York.

Stevens, P.F., 2017. Angiosperm Phylogeny Website. Available at: http://www.mobot.org/MOBOT/research/APweb/.

Swingle, W.T., 1914. *Citrus* and related genera. In: Bailey, L.H. (Ed.), Standard Cyclopedia of Horticulture. The Macmillan CO., New York, pp. 779–785.

Swingle, W.T., 1943. The botany of citrus and its wild relatives of the orange family. In: Webber, H.J. (Ed.), The Citrus Industry: Berkeley and Los Angeles. University California Press, pp. 129–474.

Swingle, W.T., Reece, P.C., 1967. The botany of *Citrus* and its wild relatives. In: Reuther, W., Webber, H.J., Batchelor, D.L. (Eds.), The Citrus Industry. University of California Press, Berkeley, CA, USA, pp. 190–430.

Tanaka, T., 1931. The discovery of *Citrus tachibana* in Formosa, and its scientific and industrial significance. Studia Citrologica 5, 1–20.

Tanaka, T., 1954. Species Problem in *Citrus* (Revisio Aurantiacearum IX). Tokyo, Japan, Japanese Society for Promotion of Science.

Tanaka, T., 1961. Citologia: Semi-Centennial Commemoration Papers on *Citrus* Studies. Osaka, Japan, Citologia Supporting Foundation.

Tanaka, T., 1977. Fundamental discussion of *Citrus* classification. Studia Citrologia 14, 1–6. Osaka.

Tolkowsky, S., 1938. Hesperides: A History of the Culture and Use of Citrus Fruits. J. Bale, Sons & Curnow, Ltd., London.

Torres, A., Soost, R., Mau-Lastovicka, M., 1982. *Citrus* isozymes: genetics and distinguishing nucellar from zygotic seedlings. J. Hered. 73 (5), 335–338.

UCR, 2017a. Coleara F2, *Citrus* Variety Collection. Available at: http://www.citrusvariety.ucr.edu/citrus/clymenia_procimequat_4109.html.

UCR, 2017b. Sun Chu Sha Kat mandarin. *Citrus* variety collection. University of California, Riverside. http://www.citrusvariety.ucr.edu/citrus/sunchushakat.html.

UCR, 2017c. Samuyao papeda. *Citrus* variety Collection University of California Riverside. Available at: http://www.citrusvariety.ucr.edu/citrus/samuyao.html.

Viloria, Z., Drouillard, D.L., Graham, J.H., Grosser, J.W., 2004. Screening triploid hybrids of 'Lakeland' limequat for resistance to citrus canker. Plant Dis. 88 (10), 1056–1060.

Volkamer, J.C., 1708. Nürnbergische Hesperides, Vol 1. Johann Andreä Endters seel. Sohn & Erben, Nuremberg.

Volkamer, J.C., 1714. Continuation der Nürnbergischen hesperigum. Nuremberg.

Wang, X., Xu, Y., Zhang, S., Cao, L., Huang, Y., Cheng, J., et al., 2017. Genomic analyses of primitive, wild and cultivated citrus provide insights into asexual reproduction. Nat. Genet. 49 (5), 765–772.

Webber, H.J., 1943. Cultivated varieties of citrus. In: The Citrus Industry. History, World Distribution, Botany and Varieties. University of California Press, California, pp. 475–668.

Webber, H.J., 1967. History and development of the citrus industry. In: Reuther, W., Webber, H.J. (Eds.), The Citrus Industry Berkeley. University of California Press, pp. 1–39.

Wester, P.J., 1915. Citrus fruits in the Philippines. Philipp. Agric. Rev. 8 (1), 5–28.

Wu, G.A., Prochnik, S., Jenkins, J., Salse, J., Hellsten, U., Murat, F., et al., 2014. Sequencing of diverse mandarin, pummelo and orange genomes reveals complex history of admixture during citrus domestication. Nat. Biotechnol. 32 (7), 656–662.

Wu, G.A., Terol, J., Ibanez, V., Lopez-Garcia, A., Perez-Roman, E., Carles, B., et al., 2018. Genomics of the origin and evolution of citrus. Nature 554, 311–316.

Xie, K., Xia, Q., Wang, X., Liang, W., Wu, X., Grosser, J.W., et al., 2015. Cytogenetic and SSR-marker evidence of mixed disomic, tetrasomic, and intermediate inheritance in a citrus allotetraploid somatic hybrid between "Nova" tangelo and "HB" pummelo. Tree Genet. Genomes 11 (6), 112. 10/06.

Xu, Q., Chen, L.L., Ruan, X., Chen, D., Zhu, A., Chen, C., et al., 2013. The draft genome of sweet orange (*Citrus sinensis*). Nat. Genet. 45 (1), 59–66.

Yamamoto, M., Kobayashi, S., Nakamura, Y., Yamada, Y., 1993. Phylogenic relationships of citrus revealed by RFLP analysis of mitochondrial and chloroplast DNA. Jpn. J. Breed. 43 (3), 355–365.

Yamamoto, M., Kubo, T., Tominaga, S., 2006. Self and cross-incompatibility of various *Citrus* accessions. J. Jpn. Soc. Hortic. Sci. 75 (5), 372–378.

Zhang, Z., Mabberley, D.J., 2008. *Citrus* Linnaeus. Sp. Pl.2. 782 1753. Fl. China, 90–96.

Chapter 5

Commercial scion varieties

Graham H. Barry[a], Marco Caruso[b], Frederick G. Gmitter, Jr.[c]

[a]*XLnT Citrus Company, Cape Town, South Africa,* [b]*CREA Research Centre for Olive, Citrus and Tree Fruit, Acireale, Italy,* [c]*Citrus Research and Education Center (CREC), Institute of Food and Agricultural Sciences (IFAS), University of Florida, Lake Alfred, FL, United States*

5.1 Pummelos/shaddocks (*Citrus maxima*)

Modern pummelo varieties are likely to differ from their ancient ancestors in that they have been selected over the centuries according to their palatability and appearance. Although bitter- and sour-tasting pummelos exist, very few of the major commercially important pummelo varieties exhibit the trace of bitterness characteristic of grapefruit, and they have considerably lower naringin content than grapefruit. Despite many similarities with grapefruit, pummelos nevertheless have several distinguishing characteristics. The distribution of superior pummelo varieties and the environmental conditions under which good eating quality is attained appear to be restricted to warm subtropical to tropical growing conditions. The greater phenotypic diversity of pummelo varieties compared with grapefruit is a consequence of their ability to produce monoembryonic seeds and thus unique individual seedling genotypes.

The pummelo varieties of commercial importance consist of clones of the common and pigmented groups, the fruits of which are oblate, round, or broadly pyriform, with relatively thin rinds, and sweet to mildly acidic in flavor. Flesh color varies from nonpigmented, ranging from pale green to straw-colored, to pigmented by lycopene, ranging from pale pink to deep red; some of these pummelo varieties are highly attractive and excellent in flavor.

Additionally, pummelos have both slightly acidic and acidless (or sweet) variants. The acidless or sweet pummelos, a group analogous to the sugar oranges, are similar to the common pummelos in all respects except that they are virtually devoid of acid. As a result, the flavor of pummelos ranges from insipidly sweet to acidic, though many are pleasantly flavored.

The accepted common name, pummelo, is of Dutch origin from the then East Indies, modern-day Indonesia, whereas the alternative common name, Shaddock, has persisted in certain countries and cultures. About 10 principal Chinese pummelo varieties, irrespective of variety, are often marketed simply as "Honey pomelo."

5.1.1 Principal commercial pummelo varieties

Numerous pummelo varieties are produced in China and Southeast Asian countries; they vary in flesh and rind color, size, shape, and other characteristics. For convenience, pummelo varieties are divided into two groups based on the pigmentation of their flesh, viz., white-fleshed and pigmented. Many if not most of the commercially important pummelo varieties of Thailand, South China, and Taiwan belong to a highly distinctive group, obviously of common ancestry and presumably of Siamese/Thai origin. Varieties in this group include Kao Pan and Kao Phuang of Thailand, and Mato of China, Taiwan, and Japan. Most pummelo varieties fall in the nonpigmented group, the fruits of which are highly variable in nearly all respects except pigmentation and acid content; they are all nonpigmented, moderately to highly acidic, and typically seedy, although there are varieties that are virtually seedless or nearly so in the absence of cross-pollination.

Some of the better known and locally important pummelo varieties are listed below; a full description of their characteristics is available in Citrus Varieties in China by Deng et al. (2008). The suffix "-you" is used in the Mandarin language to indicate "pummelo," for example, Shatianyou, and will not be used here. Many of the Chinese pummelo varieties are named after the region where they were discovered or where they are produced.

5.1.1.1 White-fleshed pummelo varieties

Shatian is one of the major and the most famous pummelo varieties in China, and is mainly cultivated in its region of origin, Guangxi, as well as in Guangdong, Hunan, and Chongqing. While Shatian is sexually self-incompatible and produces few

The Genus Citrus. https://doi.org/10.1016/B978-0-12-812163-4.00005-X

seedless fruit, it is weakly parthenocarpic and requires cross-pollination to achieve high yields resulting in highly seeded fruit (60–120 seeds per fruit). Fruits are large (±15 cm in diameter and ±17 cm long) with pyriform or gourd shape and typically with a neck and flattened stylar end. The rind is orange-yellow in color, medium thick, and somewhat difficult to peel. The flesh is crisp and tender with rich sweet flavor due to high Brix (12–16°Brix) and moderately low acid content (0.4%–0.5%). Fruit mature from mid- to late-November, and store and ship well.

Guanximi originated in Fujian Province and is mainly grown in Fujian but has been introduced into other pummelo-producing regions in China. Guanximi is strongly parthenocarpic and requires no artificial pollination, producing seedless fruit. Fruits are large (±15 cm diameter and ±17 cm high) and slightly pyriform in shape with a smooth, light green-yellow color when harvested around the mid-autumn or mooncake festival (mid- to late-September), and golden yellow at full maturity (from October to mid-November). The flesh is sweet (9–12°Brix) with moderate acidity (0.6%–1.0%). Fruits do not store as well as Shatian due to juice vesicle drying. Due to its precocity, early maturity, and productivity, Guanximi has become one of the major pummelo varieties in China. Goliath pummelo of the Western Hemisphere is possibly the same variety as Guanximi. Hongroumi, a red-fleshed variant of Guanximi, has been selected in Fujian.

Wanbai originated in Taiwan and is grown to some extent in Taiwan, Fujian, Chongqing, and Sichuan. Wanbai is relatively precocious (early fruiting) and productive, and is one of the latest maturing pummelo varieties, maturing from December through January due to higher acidity than most other commercial pummelo varieties (1.0% acidity).

Other locally important Chinese pummelo varieties include Jinxiang from Hunan Province; Feicui and Sijipao or Siji, chance seedlings mainly grown in Zhejiang Province; Zaoxiang, also called Yongjiaxiangpao, a chance seedling of Buntan/Wendan in Zhejiang Province; Anjiangxiang, cultivated for about 300 years and one of the principal fine local white pummelo varieties in Hunan Province; and Sangma, a local ancient variety first reported in 1761 from the Zhujiang Delta area of Guangdong.

Banpei is of unknown Malayan origin and was introduced into Taiwan in 1920, named in 1925, and taken to Japan soon thereafter. It is thought to be the same one known as *Pai You* in Taiwan and ranked first in Japan and second in Taiwan among pummelo varieties. It develops high-quality fruit only in the hottest regions of Southern Japan.

Hirado is the second most popular pummelo in Japan after Banpei and has white flesh. Hirado originated as a chance seedling in Nagasaki around 1910.

Sarawak or Tahitian was likely taken from Tahiti to Hawaii and then to California. It has a rind thinner than most other pummelos with pale green to amber color flesh and a unique lime-like flavor.

5.1.2 Pigmented pummelo varieties

5.1.2.1 Lightly pigmented pummelo varieties

Duweiwendan, also called Xianyouwuhe (seedless), originated in Fujian Province where it is mainly grown. Fruit characteristics are typical of a pummelo but with an indented stylar end, and the presence of lycopene in the segment walls results in a light pink coloration, whereas the pulp is not pigmented.

Pingshan is a well-known pummelo variety in China. It was widely cultivated in the Song Dynasty, about 600 years ago, and is now grown mainly in Fujian Province where it originated, and to a small extent in Guangdong and Zhejiang. Its segment walls are dark pink in color due to the presence of lycopene. Fruit quality is good and typical of a pummelo, but it is excessively seedy due to self-compatibility with about 100 seeds per fruit. It is an early-maturing variety (late-September to early-October).

Siamese Sweet was introduced by the USDA (United States Department of Agriculture) to the United States from Thailand in 1930 and renamed 'Siamese Sweet.' It is not of commercial relevance but has been involved in numerous hybridizations to create pummelo hybrids with Ruby Red grapefruit and Ruby blood orange.

5.1.2.2 Highly pigmented pummelo varieties

Dianjiang originated in Chongqing and is mainly grown there. Dianjiang has red pigmentation in the juice vesicles.

Chuhong is an early-maturing variety (late-September) selected from a chance seedling in Zhejiang Province. The segment walls and the pulp have red pigmentation, and fruits have few seeds.

Hongroumi, a red-fleshed variant of Guanximi, was selected in Fujian and has deep red colored flesh; fruits frequently are covered with bags to exclude light and thus enhance the red pigmentation of the rind. In the past decade, it has superseded the white-fleshed Guanximi as the main variety produced in Pinghe County of Fujian Province, because of the consumer preferences for red flesh.

Chandler originated as a hybrid of Siamese Sweet and Siamese Pink pummelos at Indio, California, and was released in 1961. Chandler is pink-fleshed like the pollen parent. Fruits are seedless when not pollinated, but otherwise they can be very seedy, and mature in mid-November.

Hirado Buntan Seedling (sometimes referred to as Hirado Buntan or HB) was selected in Florida from a population of Hirado pummelo seedlings by the USDA, Florida. It is the most commonly grown pummelo in Florida where fruit is more oblate in shape than most pummelos with a thin rind, juicy, and a deep pink to pale red colored flesh. It was introduced into China from the United States in 1990 where it is also known as Jianghuwendan or HB.

Pomelit is a probable hybrid from the Indonesian Djeroek Deleema Kopjor pummelo and was selected in 1978 by de Lange of the Citrus and Subtropical Fruit Research Institute and developed by Burdette, Outspan Citrus Center, Nelspruit, South Africa. The tree is vigorous with weak and narrow crotch angle resulting in limb splitting under a typically heavy crop load if not supported. Fruits are smaller in size than most other pummelo selections, and round to slightly oval in shape with smooth rind texture. The rind is extremely thin and slightly yellowish-green to yellow in color, often with pink blush when grown in hot regions. The flesh is pink—sometimes more intensely so adjacent to the segment walls—extremely tender, very juicy, and less coarse and ricey in texture than Chandler or Goliath. Significantly, there are fewer segments than the latter two varieties—usually only 12 or 13—most of which are regular in shape. It is seedless when grown in isolation from other pummelo varieties. The flavor early in the season is mildly acidic combined with very high sugar levels, giving a well-balanced, delicious taste. Later, the acidity drops and sweetness increases to a sugar to acid ratio of around 16:1 resulting in good flavor. Overall, pomelit has an attractive appearance, acceptable size, and good flavor.

5.1.3 Pummelo hybrids

In addition to pummelo varieties, there are various pummelo hybrids that have been produced with varying but limited commercial success.

Hassaku has been around since 1860 in Hiroshima Prefecture, Japan, and is probably a natural pummelo × mandarin hybrid with pummelo-like characteristics. The fruit resembles a Marsh grapefruit from a mild or cool growing region but with orange rind and flesh color. Hassaku is produced in Wakayama Prefecture, Japan.

Hyuganatsu is a pummelo-like hybrid produced in Miyazaki Prefecture, Japan, and is commonly referred to as Summer orange.

Poorman orange or New Zealand grapefruit, with Golden Special, Cutler Red, and Morrison clones, is thought to be a pummelo × mandarin hybrid, and produced in limited quantities in New Zealand.

Melogold and Oroblanco (syn. Sweetie) were the first triploid citrus hybrids developed through interploid hybridization. They are white-fleshed, low-acid, seedless triploid hybrids of Siamese Sweet acidless pummelo and tetraploid Marsh grapefruit developed by Soost and Cameron at the University of California in 1958. Oroblanco fruit more closely resembles a white grapefruit, whereas Melogold fruits are larger and more pummelo-like in appearance. Rind color of Oroblanco remains greener far longer, only achieving full yellow several weeks after being internally mature; the slow rind color development was made a feature of the variety and by highlighting its bright green color and naming the variety "Sweetie," consumer awareness of its outstanding features was created. The higher sugar to acid ratio, combined with a lack of bitterness which often characterizes Marsh when grown in cooler areas, was the initial impetus behind Oroblanco's success. However, fruit yields of Oroblanco fall short of Marsh. Initially Melogold was preferred by growers in California's Central Valley and Oroblanco in Israel, but after a decade or more they were regarded as being commercial failures and trees were removed or topworked to other varieties.

Dawn and Valentine pummelo hybrids are unique in that they are the only pummelo types containing anthocyanins. Dawn is a triploid hybrid of Siamese Sweet acidless pummelo and tetraploid Ruby blood orange, and Valentine is a diploid hybrid of Siamese Sweet acidless pummelo and Ruby blood orange × Dancy mandarin hybrid.

UF-914 (Fig. 5.1) is a recently developed triploid pummelo × grapefruit hybrid with grapefruit-like characteristics from the University of Florida, from a cross between Siamese Sweet acidless pummelo and tetraploid Ruby Red grapefruit. UF-914 produces nearly uniform red-fleshed, seedless, and relatively large fruit, with extremely low levels of furanocoumarins, the chemicals in grapefruit that are responsible for the grapefruit-drug interaction. The most unique and potentially valuable attributes of the fruits are its similarity to grapefruit in flavor and aroma, its sweeter taste (a consequence of slightly higher sugar and lower acidity than normal grapefruit varieties), and the extremely low levels of the furanocoumarin compounds. Though not yet commercially produced, it is performing well under protected structures in Florida.

FIG. 5.1 Fruits of UF-914.

5.2 Grapefruit (*Citrus paradisi*)

Although grapefruit's exact origin is uncertain, there is strong evidence to support the notion that grapefruit originated in the West Indies. The earliest mention of grapefruit was in Barbados (West Indies) in 1750 under the name "forbidden fruit," and received its species designation, *paradisi*, in 1830 by James Macfadyen from the name forbidden fruit. The first known use of the term "grapefruit" occurred in 1814 in Jamaica, probably due to the fruit setting in small grape-like clusters. However, grapefruit were initially known by the Dutch-derived name "pomello," but a group of Florida citrus growers convinced the American Pomological Society to officially adopt the name "grapefruit." Compared to other citrus types, grapefruit is a relatively recent commercial citrus type.

Grapefruit was introduced into Florida from the Caribbean region in 1823, most likely as seed, from where commercial development of grapefruit as a mainline citrus product initially occurred. The first shipments of fruit took place in 1885, while commercial production in California, Arizona, and Texas only started some decades later.

A seedling tree of grapefruit either developed directly from the initial introduction or from subsequent seedlings was later named Duncan grapefruit in 1892. It is from this original parent tree that all subsequent grapefruit varieties have developed. Two principal grapefruit lines of Florida origin were developed in the late 1800s, initially as seeded then as low-seeded, nonpigmented (so-called "white") grapefruit. Subsequently, Texas has contributed meaningfully to the development of numerous grapefruit cultivars containing lycopene pigmentation.

Walters grapefruit, derived from a seedling of Duncan, was never of any commercial importance. Nevertheless, Walters is of historical significance in the grapefruit variety development lifeline since it led to the mutation with lycopene pigmentation in the membranes separating the pulp vesicles and was named Foster in 1907; only a little over 100 years ago the first pink-fleshed grapefruit variety was discovered. In turn, Foster itself never attained commercial acclaim since it was highly seeded (with >50 seeds per fruit) and shortly after its discovery the pink-fleshed and low-seeded Thompson Pink (or Pink Marsh) mutation was discovered (see the following paragraph). Furthermore, the mutation of the pink-fleshed Foster from Walters led to a further spontaneous mutation in 1930 with dark-red flesh, named Hudson, but it was excessively seeded like its predecessors Foster and Walters. The father of deeply pigmented and truly "red" grapefruit, Dr. Richard Hensz, irradiated thousands of Hudson seeds in 1959 to accelerate the mutation process and to create low-seeded or seedless mutations. As a result of this work, Star Ruby grapefruit was released in 1970 and has become the standard for flesh color against which all other pigmented grapefruit varieties are compared. However, Star Ruby has some production problems, possibly related to the mutations induced by thermal neutrons. A nucellar line of Star Ruby was developed in South Africa from seeds imported from Florida in 1972 and appears to be an improvement on the original variety according to Hensz.

The other main line of grapefruit variety led to the development of numerous commercially important varieties. Initially, Marsh Seedless (commonly known as Marsh) originated as a chance mutation of Duncan around 1860 in Lakeland, Florida. At the time of its discovery and commercial development it was considered "seedless", but usually has two or three seeds per fruit; compared to its progenitor, Duncan, with 30–60 seeds per fruit, one can understand why Marsh was referred to as "seedless." While Marsh has remained the principal white-fleshed grapefruit variety in commercial production, it is also the predecessor of a pink-fleshed mutation, initially referred to as Pink Marsh and later named Thompson Pink, which quickly superseded Foster as a pink-fleshed grapefruit variety due to its low seed content, and became the pivotal point from which all other pigmented grapefruit varieties would develop, except Star Ruby. Until this point all grapefruit variety developments had taken place in Florida.

In 1929, a mutation from Thompson, Henninger's Ruby Red, was discovered in Texas and was the first citrus variety to be patented in the United States, in 1934. It became known as Ruby Red (or Ruby) and was also the first grapefruit with an attractive pink blush in the rind; although Thompson was pink fleshed, externally it was indistinguishable from Marsh. This progression represents chimeric expression of the mutation, with Thompson exhibiting the mutation only in the L-2 histogen, but Ruby Red expressing it in both L-1 and L-2. After Henninger's discovery, Webb discovered a similar mutation of Thompson in Texas in 1931, first named as Webb's Redblush Seedless and now known as Redblush. Subsequently, Ruby Red and Redblush were considered to be identical since they were indistinguishable. Ruby Red or Redblush and several other branch mutations were discovered in the Lower Rio Grande Valley of Texas at about the same time, but comparative studies could not determine any significant differences between them and Ruby Red.

Ruby Red (and similar selections) subsequently led to the development of deeper pigmented mutations such as Henderson and Ray Ruby. Budwood from a nucellar Ruby Red source was irradiated in 1963, from which A&I 1–48 was developed, an unreleased, darker pigmented selection. It produced a budsport which was discovered in 1976 and released and named Rio Red in 1984. Flame was in turn developed from Henderson seedlings in Florida and released in 1987, and Nelruby from Ray Ruby seedlings in South Africa in the mid-1980s. Two low-naringin mutations were developed by gamma irradiation from Henderson in South Africa and are in their early stages of commercial development as SweetHeart and RedHeart (initially called Flamingo), and a Rio Red branch mutation (Texas Red) with an intensely deep-red rind coloration is being developed in Texas.

5.2.1 Principal commercial varieties

Grapefruit varieties are divided into two groups, viz., white-fleshed and pigmented grapefruits.

5.2.1.1 White-fleshed grapefruit varieties

From the original grapefruit seedlings in Florida, numerous selections were made and named, and the most important of these was Duncan, regarded as the finest flavored grapefruit variety, although excessively seedy, having 30–60 seeds per fruit. Subsequently, all other commercially important grapefruit varieties have descended from Duncan. However, Duncan is no longer of commercial importance.

Marsh (syn. Marsh seedless)

Initially propagated in Florida as Marsh Seedless and named in 1890, it is now commonly known simply as Marsh. Marsh grapefruit trees are vigorous with a spreading growth habit producing large trees if left unpruned, up to 5–6 m in height by year 10. Under suitable growing conditions, Marsh trees are productive, producing fruits medium to large in size with round to slightly flattened shape. Fruit maturity starts from the fall (October in the Northern Hemisphere) and fruit can hang well on the tree for 3 or more months. However, fruit quality is at its peak from late-October through December. At that stage, rind color is pale to light yellow, and the flesh becomes buff colored. Initial rind color development is typically delayed in warm, subtropical growing environments. The rind is medium thin (8–12 mm thick depending on the fruit size and growing conditions), with a smooth texture and even surface. Although initially named "Marsh Seedless," Marsh fruits rarely have any seeds, but typically have two to four seeds per fruit. At maturity, the flesh is tender and juice content is high, often exceeding 50%. Soluble solid content (Brix) of Marsh grapefruit usually ranges from <8 to almost 12°Brix but is more typically between 8.5 and 10.5°Brix. Acid content ranges from 1.5% to 1.8% early in the season, and from 1.2% to 1.4% later in the season when the flavor is considered best. The maturity ratio of Brix-to-acidity ranges from 5:1 early in the season prior to maturation to over 7:1 later in the season. The fruit ship and store well but are sensitive to low-temperature shipment (<2 °C) for extended periods, developing chilling injury symptoms on the rind.

5.2.1.2 Pigmented grapefruit varieties

The number of pigmented grapefruit varieties increased rapidly following the discovery of the first pigmented grapefruit, Foster, in 1907. The pigmented grapefruit group can be split into "pink" and "red" varieties based on intensity of internal pigmentation as a consequence of lycopene accumulation. Variations in internal color may occur depending on the climatic conditions under which the fruits are produced, although the red pigmentation develops preferably under hot, subtropical conditions.

Pink-fleshed grapefruit varieties

The origins of Foster, Thompson, and Ruby (also known as Henninger or Ruby Red) grapefruits have been described above. Though attractive in appearance and excellent quality, Foster never achieved commercial importance due to its seediness. Ruby Red's fruit characteristics are similar to Thompson in all respects, except for the deeper pink flesh color uniformly distributed throughout and a red blush on the rind. Subsequently, Redblush was selected, which is so similar to Ruby that it is difficult to distinguish these varieties by physical appearance as Redblush also has a pink blush on the rind and deep pink flesh. Tree characteristics of Redblush are similar to those of Marsh, but productivity tends to be slightly lower than Marsh producing fruits of medium-large size. Their fruit mature at the same time as Marsh although commercial harvesting usually starts a few weeks later. At peak maturity, rind color is pale yellow with a pink blush on one cheek, and the flesh is pink throughout. Internal pigmentation tends to fade from a uniform pink to pale pink color when fruits are harvested late. Seed content typically ranges from one to four seeds per fruit. At maturity, juice content ranges from 50% to 55%, soluble solids content from 10.5 to 11.0°Brix, acidity from 1.1% to 1.3%, and maturation ratio from 7 to 8:1.

Ruben Pink, also known as Ruben Seedless, was discovered as a limb sport of Foster in Northwestern Argentina in the late 1960s. Fruits have smooth, thin rinds, and deeper flesh color and fruit size larger than Ruby grapefruit. External appearance is similar to Ruby, but the fruits are seedless compared to the seeded Foster, and productivity in Argentina is very good.

Henderson originated in Texas in 1973 as a limb sport of Fawcett, which in turn originated as a limb sport of Thompson. Henderson's notable difference is a better rind blush and deeper flesh color than Redblush.

Ray Ruby, of uncertain origin, was discovered in Texas in 1970 probably as a limb sport of Redblush. Ray Ruby trees are indistinguishable from those of Redblush, and the principal differences are deeper rind blush and overall redder color than Redblush, a deeper pink flesh coloration but with a prominent "halo" within individual segments in the cross section, and slightly higher soluble solid content resulting in a sweet flavor.

Red-fleshed grapefruit varieties

Star Ruby is the predominant red grapefruit variety worldwide, although Rio Red is also important in selected grapefruit producing regions. Star Ruby originated in Texas as a seedling grown from Hudson grapefruit seed that had been irradiated with thermal neutrons in 1959, named and released in 1970. Star Ruby trees have a compact, relatively bushy growth habit, bearing its fruit on the inside and mainly the lower half of the tree. The compact growth also results in a dense tree with poor light penetration resulting in dieback of bearing wood within the tree, poor internal fruit quality, yield and fruit size, and alternate bearing. Star Ruby leaves tend to be lighter green than other grapefruit varieties, and often show blotchy chlorotic symptoms. Erratic and inconsistent bearing has been a problem of Star Ruby in most production regions, but the nucellar Star Ruby selection in South Africa does not display this characteristic. Star Ruby trees are precocious with an early production peak at 7–9 years after planting. The fruits are slightly smaller than most other grapefruit varieties, and at maturity, rind color has a deep pink blush with yellow and/or red cheeks, resulting in an attractive reddish-orange appearance. Flesh color is deep red, and the rinds are thin and smooth. The fruits are virtually seedless, with 0.1–0.6 seeds per fruit on average, rarely exceeding two seeds per fruit. Juice content is typically high (52%–55%), as is soluble solid content (9.5–11.5°Brix), with moderate acidity (1.1%–1.4%), and good ratio (7–8:1).

Rio Red was discovered as a natural mutation on a limb of A&I 1–48 in Texas in 1978 and was released in 1984. A&I 1–48 originated from irradiated budwood of a nucellar Ruby seedling in 1971. Rio Red has become the preferred red grapefruit variety in Texas due to inconsistent bearing problems of Star Ruby. Tree of Rio Red are vigorous with a more open growth habit than other grapefruit varieties. Fruit maturity is slightly later than that of Star Ruby mainly due to delayed rind color development, although rind color is similar to that of Ray Ruby and Star Ruby once fully developed. Flesh pigmentation is less intense than that of Star Ruby, similar to Nelruby and Flame with the presence of a slight halo due to a lack of uniformity in pigmentation across the segments. Fruit shape tends to be more elongated than that of other grapefruit varieties.

Flame originated as a nucellar selection from Henderson; the seed was collected from Texas and planted in Florida in 1973, and this selection was released by the USDA in 1987. Fruit maturity is slightly later than that of Star Ruby, coinciding with that of Ray Ruby and Nelruby, but before Rio Red. Flame displays a slight rind blush similar to that of Henderson, darker than Redblush but not closely as deep as Rio Red. Flesh color is comparable with that of Rio Red and Nelruby, but paler than Star Ruby.

Nelruby is a nucellar selection of Ray Ruby that was developed in South Africa in the 1980s. Nelruby has deeper red flesh pigmentation than Ray Ruby, but has a prominent halo resulting in flesh color being less intense than Star Ruby.

Oran Red, also known as Rouge Latoma in Northwestern Argentina where it was discovered as a limb sport of Henninger Ruby in the 1970s, has earlier but similar rind color development as Star Ruby, but lacks deep, uniform red flesh color, being more like Rio Red and Flame.

5.2.1.3 Grapefruit-like hybrids

Since all current commercial and true grapefruit varieties have been developed via mutagenesis, the probability of breeding (or rebreeding) new "true" grapefruit varieties is implausible. In fact, few grapefruit hybrids resembling true grapefruit and with grapefruit-like characteristics have been developed. These grapefruit-like hybrids have only enjoyed moderate commercial success, and none are produced in large volumes. Examples include the white-fleshed Oroblanco and Melogold, triploid hybrids of Siamese Sweet acidless pummelo, and a tetraploid Marsh grapefruit.

Triumph grapefruit originated in Florida presumably from a seedling of the initial introduction of grapefruit to Florida and was one of the first named grapefruit varieties (1884). Due to its lack of bitterness typical of a true grapefruit, Triumph is suspected to be an orange-grapefruit hybrid, a natural orangelo.

Jackson grapefruit is thought to be a low-seeded limb sport of Triumph that originated in South Africa, and Jackson LS is a branch mutation of Jackson. Jackson LS has a unique flavor due to low naringin levels combined with lower acidity than other grapefruit varieties resulting in a unique, refreshing fruit-punch-like flavor. Fruits are seedless to low-seeded with a deep yellow, glossy rind color and tawny flesh color. Fruit size is typically smaller than that of Marsh grapefruit, but otherwise appears similar if not with a smoother rind texture.

5.3 Lemons (*Citrus limon*)

The lemon we know today was not known in the wild; modern-day lemons probably arose some millennia ago under cultivation between the foothills of the Himalayas and the Middle East (in Media of ancient Persia, now Northwestern Iran). Arab traders distributed lemons along the "silk routes" and Columbus took seeds of lemons (and oranges and citrons) with him from the Canary Islands on his second voyage (1493). Commercial lemon culture developed first in Italy, mainly in Sicily, and until a few decades ago Italy was leading in global lemon producer. With nearly a millennia of lemon culture in Sicily, it is not surprising that Sicily was traditionally the source of many commercial lemon varieties.

The production of good-quality lemons is mostly restricted to the Mediterranean-type climates, for example, the coastal and semi-coastal locations of Southern California, Sicily, Greece, Spain, and South Africa, and drier, subtropical citrus-producing regions of the world. Nevertheless, Argentina is the largest lemon producer worldwide, accounting for ~25% of global production (mainly from Tucuman Province in Northwest Argentina). In more humid environments higher pest and disease pressure affects both trees and fruits. With minor exceptions, the principal commercial lemon growing regions of the world are characterized by mild winters and cool summers.

The most distinctive fruit traits of the acidic citrus types are high acidity (although sweet or acidless mutants also occur) and a distinctive form with an oval to elliptical shape with a nipple (sometimes suppressed), tightly adhering and highly fragrant rind, and pale, straw-colored flesh. Polyembryonic seeds are typically present. Most lemon varieties are more-or-less ever-flowering due to low floral induction requirements, and consequently more-or-less everbearing depending on cultural practices. Lemon trees are highly sensitive to cold conditions. The distinct flavor and aroma of lemons are due to the presence of limonin and D-limonene.

Due to the lemon seed's polyembryonic nature, it is extremely difficult to breed lemon varieties by hybridization with true lemon characteristics, and most current commercial lemon varieties have been developed through natural or induced mutations from the original ancestral form.

5.3.1 Principal commercial varieties

For most lemon varieties it is difficult to cite accurate and meaningful fruit descriptions, and it is difficult to identify lemon varieties from the fruit alone. Fruit characteristics from the same bloom are highly variable, and fruits on the same tree but from different blooms typically have greater differences than between varieties of the same bloom. As far as fruit characteristics are concerned, the varieties grown in California, Argentina, and many of those in the Mediterranean basin are remarkably alike, if not identical. The principal exceptions are those with a distinctive form, such as Interdonato, Arancino, and Lunario of Italy, some of which are citron-like in certain traits.

The Femminello group (syn. Femminello Comune, Femminello Ovale) represents the oldest Italian lemon varieties and accounts for most of the Italian lemon production. The trees are characterized by an ever-blooming, ever-bearing habit which makes them very responsive to the *verdelli* or the forced summer production of Sicilian lemon culture. The Femminello lemon group comprises several selections each of which has its own distinct maturity periods, producing four crops per year with different names according to their flowering periods or time of harvest, rather than the particular selection from which they have originated.

Crop	Period of harvest
Primofiore	September to November (autumn)
Limoni	December to May (winter)
Bianchetti	April to June (spring)
Verdelli	June to September (summer)

The distribution of the crop over the different harvest periods is determined by the climatic conditions and other factors as well as the characteristic of each of the Femminello selections. Seed content also varies depending on the crop. The juice content of the Limoni and Bianchetti crops are similar, whereas the Verdelli crop usually has less juicy fruit. The Femminello lemon selections are productive and produce fruits of medium size, with moderately thick rind and lower juice content than other lemon varieties. Almost all Femminello selections are susceptible to mal secco disease, which is widespread throughout most Italian lemon orchards, but the selections Santa Teresa and Zagara Bianca (or Fior d'arancio; Fig. 5.2A) show some tolerance to mal secco. While Femminello is essentially an Italian-grown cultivar the Santa Teresa selection of Femminello is also grown in a small area in Tucuman Province in Northwestern Argentina where it is revered for its high rind oil quality.

Femminello Siracusano 2Kr (Fig. 5.2B; Starrantino et al., 1988) is the most widely planted lemon variety in Italy during the past decade, and has outstanding performance in terms of productivity, yield precocity, and fruit quality, but it is not suitable in areas of high mal secco pressure, where Zagara Bianca is preferred.

Italian lemon is a Turkish variety that closely resembles Femminello Santa Teresa in both tree tolerance to mal secco and its roundish fruit shape, suggesting that they are possibly the same selection. Fruits are very seedy with a moderately thick rind and particularly firm condition, more so than most other lemon varieties.

Kütdiken is an old Turkish lemon variety and is believed to have originated from a Femminello selection; Kütdiken means blunt thorn. The trees are medium to large in size and as with most Femminello types are susceptible to mal secco. Kütdiken is incompatible on trifoliate-type rootstocks. The trees are reliable bearers of average sized fruit whose shape is typical of the Femminello Comune selection with its characteristic prominent nipple. Maturing in mid-November and harvested in early February the rind is moderately thick, juice yield is moderate to low and there are often 10 or more seeds per fruit.

Interdonato (syn. Speciale; Fig. 5.2C) was considered a lemon-citron hybrid and is said to have originated around 1875 on the property of Colonel Interdonato in Sicily. More recently, it has been suggested that Interdonato is a true lemon that underwent significant genomic deletion events (Curk et al., 2016). Interdonato lemon is the earliest maturing lemon variety, being at least 4 weeks ahead of Femminello. Like Monachello, it is also tolerant to mal secco disease, and is planted mainly for that reason. Trees are moderately vigorous and nearly thornless, with the leaves resembling to some extent those of citron and produce a low percentage of out-of-season blooms. The fruit is much larger, longer, and smoother than other lemon varieties, with a thin rind and few seeds. The rind is thinner than most Mediterranean lemons but has a low juice content.

FIG. 5.2 Fruits of Femminello Zagara Bianca (A), Femminello Siracusano 2Kr (B) and Interdonato (C) lemons. *(Photo courtesy of Giuseppe Russo, CREA).*

Monachello is an Italian cultivar that was considered to be a lemon-citron hybrid due to peculiar phenotypic features such as low vigor, a distinctive round-topped fruit shape, thick peel, and low juice percentage, but molecular data suggests that Monachello is a lemon variety. Overall fruit quality is low, juice has an acid level lower than most lemons, although still acceptable. The outstanding characteristic of Monachello lemon is its tolerance to mal secco disease, and has been planted in areas of Italy where the disease is most severe; however, in recent years it has been replaced by Femminello Zagara Bianca.

Lamas is a well-known Turkish lemon variety named after the town of Lamas and forms a significant proportion of Turkey's early-season lemon production. Lamas trees are vigorous, high yielding, and come into production at an early age and are harvested immediately after the Interdonato crop is finished, usually in early November. Fruit is of medium size, almost round in shape with only the smallest nipple and neck similar to Genoa lemon in overall appearance, with a smooth thin rind and juicy pulp containing about five seeds per fruit.

Lisbon is one of the most widely grown lemon varieties and is planted extensively throughout the lemon-growing regions of the world. Lisbon lemon originated in Australia and was first described in Sydney in 1824, and is thought to be of Portuguese origin, possibly from the Gallego seedling clonal group. However, given that the English colonialists only arrived, together with some lemon seeds, in the Sydney region via the Canary Islands, Rio de Janeiro, and Cape Town in 1788, it is equally possible that Lisbon lemon originated from this initial introduction of citrus into Australia. A lemon selection known as "Portugal" in Morocco and Algeria is said to be indistinguishable from the Lisbon lemon variety subsequently introduced from California. The earliest reference to Lisbon in California appears to be its listing in an 1853 nursery catalogue having been brought to California in 1849 or soon thereafter from Boston, Massachusetts, where it was listed in a nursery catalogue as early as 1843. Two importations of Lisbon lemon into California from Australia were made in 1874 and 1875. It is almost certain that this variety ultimately reached Southern California as budwood from an old Lisbon lemon tree growing in Alameda (near San Francisco) and sent to Goleta (near Santa Barbara) in 1883.

Lisbon lemon is more cold-tolerant than Eureka lemon as well as more productive, more vigorous with denser foliage, and thornier. Fruits tend to be produced on the inside of the tree where they are protected to some degree from sun, wind, and cold damage. Fruit maturity and harvest time is restricted mainly to the winter and early spring in California, but slightly earlier in Argentina and South Africa. Fruit characteristics do not differ substantially between Lisbon and Eureka, although Lisbon fruit tend to be slightly rougher or less smooth. Unlike Eureka, Lisbon is compatible with trifoliate orange, citrumelo, and citrange rootstocks. Numerous selections of Lisbon lemon have been made in California, some of which are clones and others nucellar selections, for example, Frost nucellar Lisbon.

Limoneira 8A is a Lisbon selection from Limoneira Ranch, Ventura County, California. It is more productive, vigorous, and thornier than Lisbon, with higher rind oil yield.

Yen Ben was selected as a branch mutation of Lisbon lemon in Queensland, Australia, in the late 1930s. Fruit size is smaller than that of standard lemons, but rinds are thinner and juice content is higher.

Eureka (syn. Garey's Eureka) originated in California in 1858 from a group of seedlings grown in Los Angeles from a fruit of Sicilian origin, possibly Lunario. Eureka was selected in 1877 and propagated by Garey who named it Garey's Eureka, but soon became simply known as Eureka. Eureka is a major lemon variety in many important lemon-growing regions except Italy, Spain, and a few other Mediterranean production regions. Outside the Mediterranean basin Eureka is the most widely grown lemon variety, forming a major part of the crop in California, Australia, South Africa, Chile, and a significant proportion of Argentina's production. Most of Israel's lemons are also Eureka.

The Eureka lemon tree has a spreading growth habit and is less vigorous than Lisbon with sparser leaf cover and markedly less cold tolerant. Eureka is well adapted to the coastal areas in Australia, California, and South Africa where frost damage is rarely experienced. Being a slightly smaller tree, Eureka is often less productive than Lisbon but has a well-distributed harvest season, an almost ever-bearing fruiting habit. Eureka shoots are less thorny than Lisbon making picking easier than most other lemon varieties, and Eureka is precocious with high production in its early years. Eureka fruit tend to be borne on terminal shoot positions thereby exposing fruit to greater wind blemish and occasional sunburn. Eureka fruit tend to be smaller than the other important lemon varieties, but they have a smooth and relatively thin rind and high juice content. Eureka is incompatible with trifoliate orange rootstock and many of its hybrids, with the exceptions of Benton and Cox citranges, X639 citrandarin, and Minneola × Trifoliate hybrid. Numerous Eureka clones, mostly of Californian origin, have been developed and distributed around the world.

Seedless Eureka lemon varieties, ARC Eureka SL and 2PH Eureka SL, were developed by induced mutation through gamma irradiation in South Africa and Australia, respectively; currently these are being promoted for commercialization.

Variegated pink-fleshed Eureka lemon originated in a home garden in California as a sport of the conventional Eureka lemon and was selected by Shamel in 1932. The leaves are variegated green and white, making the tree quite ornamental, and the tree grows less vigorously than a typical Eureka lemon tree. The rind is striped green and cream and is sometimes rougher than conventional Eureka. When fully ripe, the stripes fade and the rind turns yellow with distinct pink oil glands. The flesh is pale pink at full maturity due to the presence of lycopene, low-seeded, and very acidic. Fruits are smaller than conventional Eureka fruit.

Genoa was initially described by Webber (1943) as being a Eureka-type, yet there are distinguishing fruit characteristics, and Genoa's compatibility with Swingle citrumelo leads the interpretation that Genoa does not belong in the Eureka group. Genoa was introduced into California in about 1875 from Genoa, Italy, and is therefore also of Italian origin like Villafranca. The Genoa lemon tree is virtually thornless and smaller than Eureka but is more cold resistant and has denser foliage. Genoa has achieved commercial importance only in Argentina and Chile, and to a lesser extent in South Africa, where the Argentine nucellar selection, Genoa EEAT, has proven to be more productive than old-line Genoa. Genoa fruits are less elongated and more spherical than Eureka and Lisbon with a small pointed neck and nipple, and fruit maturity is more concentrated in late autumn/early winter than in winter itself.

Villafranca was introduced into Florida from Sicily in 1875 and was taken to California soon thereafter. The tree is more vigorous than Eureka and is thornier when young, but it is almost impossible to differentiate the two varieties by mature tree habit and fruit characteristics, although the seasonal distribution of the crop is more like Lisbon, being mainly in the winter. Villafranca is of minor importance in most lemon-producing regions but is important in Israel where it produces more seedy fruit than Eureka. Corona Foothill Eureka selection is a Villafranca selection.

Fino (syn. Mesero, Primofiori) is of unknown origin but possibly originated from an old Spanish lemon variety. Fino trees are vigorous, thorny, and productive. Fino has regular-shaped fruit, slightly oval, with a smooth thin rind and a relatively short nipple compared with Verna. Fruits are smaller but juice percentage is higher than that of Verna. Fino has more seeds than Verna, averaging about five per fruit. Unlike Verna, Fino does not produce a secondary crop, but like Eureka, Fino's harvest extends from October through February. Fino is highly susceptible to mal secco disease.

Verna (syn. Berna) is Spain's principal late-season lemon variety and is also of unknown origin. Initially only planted in its native Spain, Verna has recently also been planted in Italy and Argentina. Verna lemon trees are large, spreading, and virtually thornless. They usually flower twice but, in some years, a third crop is produced. The second crop—the "secundus"—is of inferior quality and of little commercial importance. Growers sometimes force the tree to produce a larger third crop, known as "verdelli" or "rodrejos". The main crop or "cosecha" comprises medium to large fruit with a pronounced nipple and usually a well-developed neck. The rind is medium thick to thick, and somewhat rough and uneven in texture. The juice yield is correspondingly lower than other lemon varieties, but the pulp is tender, there are few seeds and the juice has good acidity. The "cosecha" crop extends from mid-February until late July or early August. Verna trees tend towards alternate-bearing, particularly following the 'verdelli treatment,' and are outyielded by Fino.

5.4 Limes (*Citrus aurantiifolia*[a] and *Citrus latifolia*)

The most distinctive fruit traits of limes are their high acidity, flavor, and aroma, although sweet or acidless lime mutants also occur. Lime trees are highly sensitive to cold, and productivity and fruit quality are typically highest in semitropical growing conditions than in milder conditions.

Small-fruited lime (*C. aurantiifolia*) is commonly referred to as West Indian lime (syn. Mexican lime, Key lime). Of the two acid limes commercially produced and traded, it is the longest known and was initially the most widely cultivated. It is a substitute for lemons in the more humid production environments and is better adapted to semitropical conditions. Initially the commercial lime industry was restricted to the small-fruited lime which is highly sensitive to cold and has a high heat requirement to produce fruits of adequate size. Consequently, the lime industry is principally developed in hot semitropical, subtropical, and tropical regions. The climatic conditions most favorable for limes are poorly suited for lemon production, whereas where good quality limes are produced, they are generally preferred for lemons. Hence, the main lime-producing regions are Mexico, India, Egypt, Brazil, Peru, as well as other countries in Central America.

Small-fruited lime production in Mexico only began to gain in importance in the 1920s and large-fruited limes have only been grown in Mexico for <50 years. The small-fruited lime was planted on some of the Florida Keys as early as 1838 and ultimately became naturalized there, hence the name "Key lime." Although at one time "Key lime" was grown in the Florida Keys, <10 ha remain in production, and large-fruited lime became the preferred variety in Florida. Following World War I, it became a well-established commercial crop in Southern Florida, growing to ± 3000 ha in its heyday of the 1980s. After recovering from the devastation of Hurricane Andrew in 1992, Florida's lime production was subsequently wiped out by the citrus canker eradication effort in the late 1990s.

---

a. The original spelling "*aurantifolia,*",being derived from the generic name *Aurantium*, is correctable under Shenzhen ICN Art. 60.10 (see Ex. 35).

There are few true selections of West Indian lime since it is highly polyembryonic, and there is little to no variation in the wild or under cultivation. Furthermore, there are numerous synonyms and local names (Kagzi in India, limón sutil in much of South America), often of the same variety. Nevertheless, besides the classic small-fruited sour lime, there are also some thornless or near thornless selections as well as seedless variants.

West Indian lime trees have a somewhat bushy growth habit with many slender or fine-stemmed branches bearing numerous short spines or thorns and smaller leaves with distinctly paler color than large-fruited limes and lemons; the thorniness nature is one of the distinguishing traits of the tree when compared with large-fruited lime. The trees are also extremely sensitive to frost and cold conditions, requiring a semitropical to tropical growing environment. In contrast to the large-fruited lime, it is highly susceptible to citrus tristeza virus, for which it is currently the most widely used indicator plant.

Fruits are small (typically >20 to <50mm in diameter), slightly elongated with a small nipple, a very slight neck, and a smooth, thin rind. When the fruit mature on the tree the rind becomes greenish-yellow to yellow, though fruits are often harvested earlier while still dark green in color. The rind oil has a distinct aroma which typifies West Indian limes. The flesh is light greenish-yellow in color, very juicy with a pronounced acidity, and has strong lime flavor. Fruits are typically seedy with numerous seeds (8 to >16 seeds per fruit) which are highly polyembryonic allowing seedling trees to be propagated for commercial production.

West Indian limes are produced on an extensive scale in Mexico, Brazil, Egypt, India, and Peru, but are largely unknown in many importing countries, particularly throughout Europe. Besides being consumed fresh as a condiment to meals, lime juice, lime oil, and pectin are important commercial products of West Indian limes.

Large-fruited lime (*C. latifolia*) is commonly referred to as Tahiti lime (syn. Persian lime, Bearss lime) and is comparatively recent in terms of its discovery and commercialization, since it only originated after the development of lemons. Large-fruited lime probably came from its place of origin in Central Asia (possibly Northeastern India or Northern Myanmar, formerly Burma) via Iran (formerly Persia), through the Mediterranean, then taken to Brazil by Portuguese traders, and from there to Australia in 1824, and onto Tahiti. It then reached California through San Francisco from Tahiti between 1850 and 1880, and Florida by 1883. Despite its name, this variety is not currently grown commercially in either Tahiti or in Iran (formerly Persia), where only the small-fruited lime is produced.

The large-fruited seedless Tahiti lime is much more familiar on export markets and is now part of the mainstream group of citrus fruit sold in supermarkets around the world, though in much smaller volume than lemons. Tahiti limes are produced mainly in Mexico, especially in the Martinez de la Torre region of Veracruz, and Brazil, although small consignments also originate from Guatemala, Belize, Honduras, Colombia, Venezuela, and Israel.

The Tahiti lime tree is relatively vigorous with dark green leaves and almost thornless shoots. The fruits are considerably larger than West Indian limes, typically ranging from 40 to 60mm in diameter, and since Tahiti lime is a naturally occurring triploid and produces no viable pollen, fruits are almost always seedless, although seeds occasionally occur. The flavor of Tahiti lime is acidic, but lacks the strong tartness of West Indian lime, and the classic lime aroma although obvious is less pronounced. Stylar-end breakdown is a peculiar physiological disorder of Tahiti lime.

Commercial culture of Tahiti lime is much more recent than that of West Indian lime and was initially less important, originally being confined to the United States where, because of its cold hardiness and lower heat requirement, it was much better adapted than the better known and generally preferred West Indian lime. As limes perform better than lemons in humid climates the US lime industry was almost entirely confined to Dade County, Southern Florida, where its comparatively low heat requirement allowed the fruit to attain acceptable maturity during the summer months. In Florida, Tahiti lime was therefore an acceptable substitute for the preferred West Indian lime. Because Tahiti lime fruits are the size of small lemons, it was initially necessary to educate the markets and consumers to accept this fruit as a true lime rather than a small lemon, but this sentiment is no longer relevant.

Tahiti lime fruit, particularly the smaller ones, can be mistaken for those of West Indian lime since externally they are virtually indistinguishable; furthermore, both may be mistaken for immature or small lemons. Therefore, Tahiti limes are harvested and marketed when they are still dark green as fully mature fruit become greenish-yellow to pale yellow, like the West Indian lime. If left on the tree until past peak maturity the fruit turn yellow and are unmarketable.

5.5 Oranges (*Citrus sinensis*)

As oranges were moved from their site of origin, the primitive, ancestral forms were selected, resulting in a wide range of the currently available cultivars and selections of cultivars. Selection criteria differed according to local preference, and the perception of fruit quality differs according to regional preferences. Due to sweet orange seed's polyembryonic nature and the narrow cultural definition of what is sweet orange, it is virtually impossible to breed "true" orange varieties by hybridization. Consequently, all current commercial orange varieties have been selected and developed from natural or induced mutations (Fig. 5.3).

FIG. 5.3 Fruits of different sweet orange varieties: Vaniglia Sanguigno (A), an acidless variety with lycopene pigmentation; Lempso, a Tarocco clonal selection with an intense rind pigmentation (B); Cross section of fruit of the OLL series, late maturing somaclones with very deep orange color and superb juice quality (C); Lane Late Navel orange (D); Cara Cara Navel orange (E). *(C, photo courtesy of Jude Grosser, UF-CREC; D and E, photo courtesy of Giuseppe Russo, CREA).*

The sweet orange group is typically divided into four types according to their primary distinguishing characteristics, namely:

- Sugar or acidless orange: Characterized by extremely low acid levels resulting in insipid flavor.
- Pigmented or blood orange: Characterized by the presence of anthocyanin pigments in the pulp.
- Navel orange: Characterized by the presence of a secondary fruitlet and small stylar opening.
- Common orange: The remaining sweet oranges are designated to this group, and they may be further distinguished based on their time of maturity, that is, early, midseason, and late.

5.5.1 Sugar or acidless orange varieties

Sugar or acidless oranges are known under various names throughout the world yet are hardly known in many of the western supermarkets, since only small volumes enter international trade and the varieties are typically only appreciated in their regions of production. There are only slight differences between varieties, but they are essentially very similar except for the time of maturity of some selections, and many are local names of the same variety. Their peculiar flavor is relished in many Arabic-speaking countries and to some extent in Spain and Portugal and their former colonies. In the Mediterranean area, their culture was most important in Spain and Egypt, and in the Western Hemisphere they are popular in Brazil and Mexico. In Spain, Sucreña Real or Imperial Grano de Oro is less popular now than it was in the past and production has virtually ceased. The same applies in Italy with the Vaniglia variety (Fig. 5.3 A).

This small group of oranges is characterized by extremely low acid levels resulting in insipid flavor since acid levels are of the order of 0.1% or about one-tenth of the acidity of sweet oranges with sugar to acid ratios of 80–100. However, it is not simply the extraordinarily high ratio that results in an insipid flavor, but the lack of the familiar "orange" aroma which characterizes acidless orange varieties.

5.5.2 Blood or pigmented orange varieties

Blood oranges are characterized by the presence of anthocyanin pigments in their pulp, and sometimes the rind too. Pigmented or blood oranges were once thought to have originated in the Mediterranean area, but it has recently been suggested that they, like navel oranges, are indigenous to China. New molecular findings revealed that all blood orange cultivars have a common origin, although a clone called 'Jingxian' showed a different insertional event (Butelli et al., 2012). The history and the origin of blood oranges have been extensively reviewed by Karp (2007).

Blood oranges have been cultivated in the Mediterranean region since the early 19th century, particularly in the Western Mediterranean countries, in Italy where they are known as Sanguigno and Sanguinello, Spain (Sanguina), Morocco, Algeria, and Tunisia (Sanguine). The Spanish variety, Doble Fina, is of unknown origin, and later gave rise to Sanguinelli which originated as a bud mutation of Doble Fina discovered in 1929 at Almenara in Castellon Province. The Spanish Sanguinelli should not be confused with the Italian Sanguinello. Washington Sanguine (Grosse Sanguine or Doublefina Amelioree) also originated as a bud mutation of Doble Fina discovered by B. Ferrer at Sagunto, Valencia Province. An important, high-quality semi-blood variety, Maltaise Sanguine, is of unknown origin, but Malta is most probable. It is grown extensively in Tunisia and to a lesser extent in Morocco. In addition to the standard Maltaise Sanguine, two derivatives have been identified: Maltaise Sanguine Bokhodza—an early-maturing, lower acid selection which is reportedly more prone to decay and does not ship well, and Maltaise Tardive Barlerin—a late-maturing selection but otherwise similar to the standard Maltaise Sanguine. Fruit of Maltaise Sanguine are slightly elongated or oval-shaped, of medium size with a pebbly rind, and few to no seeds. The rind occasionally has a light red blush and the flesh is lightly pigmented by anthocyanins with excellent flavour and soft, melting texture.

Blood oranges are mostly cultivated in Italy, and the Italian varieties have been distributed to other production regions where the climatic conditions are favorable for their cultivation. The main Italian blood orange varieties over time were Sanguinello, Moro, and more recently Tarocco. The origin of Sanguinello is uncertain, but it was most likely a seedless mutation of Sanguigno, which is an old seedy variety. Sweet and rich-flavored Sanguinello matures a bit later than the other Italian varieties, and hangs well on the tree, so it is generally marketed from February to April. One common selection is named Sanguinello Moscato for its supposed flavor of Muscat grape.

Moro is prized for its reliable and often intense pigmentation, which begins to develop early during the season. The rind is sometimes completely red, and the flesh ranges from burgundy-streaked to deep purple. Maturing early in the season, from December to March, Moro is popular on export markets for its dramatic color, but is comparatively tart, and in Italy it is primarily used to make thick, tangy, dark-hued juice.

Tarocco is the most important orange variety produced in Sicily where it is predominantly consumed as a fresh fruit. In the last two decades, Tarocco clones spread to other citrus growing areas, like Sichuan Province in China, California, South Africa, and Australia. Tarocco is a mutation of Sanguinello discovered around 1900, reputedly in Francofonte, and named for the resemblance of the original strain to a toy top—it can have a "neck," like a Minneola tangelo. Fruits are medium to large in size, with tender, rag-free flesh, and a sweet, well-balanced flavor. Often the pulp is attractively bicolored, blonde at the stem end and red at the blossom end. The variety has given rise to several dozen named mutations, differing in season, fruit size, color, and growing habit (Caruso et al., 2016). Fruits of different Tarocco selections mature from late November until May. Tarocco trees are typically vigorous with long upright scaffold branches creating an upright growth habit. Thorniness is often a feature of most Tarocco selections.

Flowering of Tarocco can be variable resulting in light crops of excessively large fruit. In addition, multiple blooms frequently occur resulting in two crops setting, of which the second crop has unacceptably poor fruit quality. Tarocco fruits are large (200–250 g), with the shape varying from roundish to slightly elongate and often with a pronounced neck. The rind is finely pebbled, moderately thick, and spongy. The rind is loosely adhering and easily peeled. The segment walls are tender, and the flesh is melting. Due to the tender rind and overall softness, Tarocco fruits have a short shelf life and do not store well. Tarocco fruit tend to have a poor tree-holding ability and preharvest fruit drop is a problem in some seasons.

Pigmentation of the pulp of Tarocco orange selections varies from deeply pigmented, for example, TDV, Ippolito, and Rosso, to lightly or unpigmented, for example, Scirè, Gallo, and Sant' Alfio. Rind pigmentation is variable among selections and within a given selection, and is independent of pulp pigmentation. Lempso (Fig. 5.3B) is the Tarocco cultivar with the highest rind pigmentation. In addition, pigmentation of Tarocco selections vary among fruit within the tree, and pulp pigmentation in the market varies throughout the season. Pulp pigmentation is affected by various factors: (i) genetic; the selection of Tarocco plays an overriding role in determining whether fruits of that selection will be pigmented or not; (ii) climatic; cool nights and warm days in the winter are required to stimulate anthocyanin synthesis, however, the actual cold accumulation required is not known; and (iii) the rootstock; *Poncirus* confers more pigmentation, while lemon-like rootstocks such as *Citrus macrophylla* and *Citrus volkameriana* negatively influence anthocyanin accumulation in the pulp.Tarocco fruit tend to be tart due to relatively slow acid degradation combined with rapid fruit softening. As a result, fruit rend to become soft and hint of over-maturity before acid levels have dropped adequately. In addition, Brix levels are typically not high and range from 9 to 11 °Brix, but Tarocco fruit have a sweet, well-balanced flavor.

5.5.3 Navel oranges

Navel oranges are characterized by the presence of a secondary fruitlet and a small stylar opening, the "navel." Navel orange trees, in general, are not particularly vigorous, and have a round, somewhat drooping canopy and grow to a moderate size. The flowers lack viable pollen so the Navel orange does not pollinate other citrus flowers, and due to the lack of functional pollen and viable ovules, Navel orange fruits are seedless. Their large round fruits have a slightly pebbled orange rind that is relatively easily peeled, and the navel-end, really a small secondary fruit, sometimes protrudes from the apex of the fruit. Washington Navel orange, the original commercial Navel orange selection, is at its best in the late autumn into the winter, but holds on the tree for several weeks beyond maturity and stores well.

Although the exact origin of Navel orange is not known, it is believed to have developed from a branch mutation in a Selecta orange tree in the early 1800s in Bahia, Brazil. However, there were also oranges with navels being grown in Spain and Portugal prior to the 1820s and it is likely that oranges with navels originated sometime earlier, possibly even in China. Nevertheless, the initial development that led to the name "Washington Navel" and the worldwide commercialization of the present-day Navel orange originated from the importation of 12 budded trees in 1870 which were sent from Bahia, Brazil by Mr. and Mrs. Schneider, Presbyterian missionaries, to Mr. Saunders, superintendent of gardens and grounds for the USDA in Washington, DC.

The delicious, seedless, easy-peeling fruit quickly attracted the attention of local citrus growers, and within a decade the "Washington Navel" orange, as it came to be known, was the most widely planted variety in California. From California, Washington Navel orange rapidly spread to other parts of the citricultural world and soon became a major variety, now second only to Valencia orange. Washington Navel orange is also known as Bahia Navel in the Brazilian city from which it was imported into the United States in 1870.

Navel oranges are genetically more unstable than most other orange varieties and because of their cultivation globally, countless mutations have been discovered and numerous Navel orange selections have been made in many parts of the world. Many of these Navel orange selections have fruit characteristics which are almost indistinguishable from Washington Navel, except for the time of maturity, and it is now possible in many navel-producing regions to extend the harvest season to 6 months or longer. Some of the most widely planted Navel clones are the following: Fukumoto, Navelina, and Newhall among the early-maturing varieties; Fischer, Washington, among the mid-ripening ones; and Lane Late (Fig. 5.3D), Chislett, Powell, Barnfield, and Cambria among the late-maturing varieties. Of interest was the discovery in 1976 of a red-fleshed mutation of Washington Navel containing lycopene pigmentation, Cara Cara (Fig. 5.3E) from Venezuela. Subsequently, other red-fleshed Navel mutations have been discovered, namely Kirkwood Red and Red Lina from South Africa.

5.5.4 Common orange varieties

Common oranges are a diverse group of sweet oranges with a range of fruit traits and were mostly derived from seedlings. Previously, they were sometimes referred to a "blond" or "white" oranges in the Mediterranean basin (*blanca* in Spain and *biondo* in Italy) to differentiate them from blood oranges and Navel oranges. Many common orange varieties contain seeds, but as seed content has increasingly become important in the fresh fruit markets, especially throughout the European Union, the production of these varieties has decreased in the past few decades in favor of less seedy or seedless types, especially Navel oranges. Nevertheless, seeded oranges still remain important for fresh juicing purposes as well as for processing into juice concentrate or not-from-concentrate orange juice products. This group of sweet oranges can be further differentiated based on their time of maturity, that is, early, midseason, and late, and include some well-known orange varieties such as Valencia, Pera, Hamlin, Pineapple, Salustiana, and Shamouti.

1. Valencia sweet orange is the most abundantly produced orange variety in the world. An earlier assumption that this variety was of Spanish origin is not supported by the evidence, though the name is of Spanish origin. The original Valencia selection, now referred to as Valencia Late, may have originated in Portugal as a chance seedling of Selecta sweet orange, an old Portuguese variety which matures earlier than Valencia. The variety was imported into England from the Azores Islands and named Excelsior in 1865, and thereafter sent to Long Island, New York, then onto California in 1876 where it was renamed Rivers late in California and Hart's Tardiff in Florida. In 1887 the variety name was changed to Valencia Late as suggested by a Spanish visitor who considered it to be similar to a late-maturing variety produced in the Valencia region of Spain. Don João, an orange variety from the Don João orchard near Beja, Portugal, was sent from Portugal to California in

1963 and comparisons showed no difference between the two varieties. Today, numerous selections of Valencia orange are produced in various citrus-producing regions of the world, some which only have local importance, while some others are cultivated worldwide. Natal is a Brazilian orange variety of unknown origin and resembles Valencia in most tree (vigorous, large, and productive) and fruit (medium size fruit, few seeds) characteristics, but matures even later (usually harvested around Christmas in Brazil) and is more elongated than Valencia. Other commonly cultivated Valencia clones include Campbell, Olinda, Delta, Midknight, and Rhode Red. More recently, new selections derived from Valencia somaclones have been developed and are being commercialized in Florida, including Valquarius (maturing 4–6 weeks earlier than Valencia), and EV1 and EV2 that mature in the same season or even earlier than Hamlin sweet orange in Florida. Valencia orange selections containing lycopene pigmentation have been discovered, including Puka Valencia from Argentina and Ruby Valencia from South Africa. Finally, the OLL (Orie Lee Late) series of late maturing oranges (Fig. 5.3C) has become of commercial interest to the Florida juice business. These selections (OLL-4, OLL-8, and OLL-20) are somaclones selected from the original OLL tree. They have very deep orange color and superior flavor profiles, even compared with Valencia orange; this has led to new plantings in Florida that are expected to increase in size over the next few years.

2. Shamouti sweet orange was once an important orange variety grown mainly in Israel and Cyprus, and also in Turkey and eSwatini (formerly Swaziland), but with limited success in Florida, California, and Texas. Despite its fine fruit quality traits, Shamouti production has declined since the advent of improved Navel orange and mandarin varieties. Shamouti originated as a branch mutation on a common orange tree in the mid-1800s near Jaffa, Palestine (now Israel). Shamouti trees have a distinctive upright growth habit, thornless shoots, and large broad leaves. Its productivity and external and internal fruit quality are largely affected by growing conditions, specifically climate and rootstock. Although juice content is relatively low due to the thick rind, and rind texture is pebbly to rough, particularly at the stem end of the fruit, its ease of peeling and rich, sweet flavor once made Shamouti one of the finest fresh oranges in its heyday. It is medium to large in size and oval in shape with a slightly flattened stem end.
3. Pineapple is thought to have originated from seedlings from fruit brought to South Carolina, USA from China and planted in Florida around 1860, and named either due to its delicate fragrance or due to the shape of the tree. Pineapple is very productive and produces medium-sized, round fruit with a smooth, thin rind. At maturity in early- to mid-winter, rind color is good, and fruits are juicy and have a sweet and rich flavor. Pineapple was primarily planted as a midseason processing variety but was also planted on a small scale for the fresh fruit export trade; however, its excessive seediness, having 20 or more seeds per fruit, resulted in it losing favor to superior low-seeded orange varieties.
4. Hamlin sweet orange arose as a chance seedling in 1879 in Florida where it was planted on a large scale and became Florida's leading early-maturing orange variety, but was planted on a more limited scale elsewhere, for example, Brazil and Argentina. Hamlin is highly productive, but fruit size is small, and typically too small for the fresh fruit market. Although Hamlin develops early rind color, maturing at almost the same time as Washington Navel, it has pale-orange rind color and pale flesh and juice color.
5. Jincheng, meaning "glorious orange," is a midseason sweet orange variety, selected in 1939 as a chance seedling from a local common sweet orange in Chongqing, China. Maturity ranges from late-November to early-December when rind color is deep orange and flavor is sweet but lacks richness due to moderate acidity. Fruits are oval with a smooth thin rind that has prominent oil cells and tender flesh. Previously, Jincheng was China's predominant orange variety, but it has been superseded by Washington Navel.
6. Salustiana was discovered as a branch mutation of Comuna sweet orange in the 1950s in a convent garden in Valencia Province, Spain. It was the second most important common orange variety in Spain (after Valencia), but has declined in popularity. Salustiana trees are vigorous and very productive, and optimal fruit maturity in Mediterranean production regions is from mid- to late-January when the fruits have a rich, sweet flavor and smooth, orange rind. The fruits are seedless to very low-seeded and medium in size, with a fine, tender flesh. Salustiana is also produced to a more limited extent in Morocco and Argentina.
7. Pera orange is the principal processing orange in Sao Paulo, Brazil due to its exceptional productivity. Fruits are smaller than those of Valencia orange and have a slightly oval shape and smooth rind. Pera trees are vigorous, upright in habit, large, and very productive.

5.6 Mandarins (*Citrus reticulata*)

Mandarin is one of the three original progenitor species of all modern citrus varieties. The modern-day mandarin group is highly complex due to cross-breeding with compatible citrus species, specifically pummelos, and interbreeding among other mandarins. As a result, "pure" mandarins do not exist among contemporary commercial mandarin varieties (see Chapters 1 and 2 for further details).

Previous attempts to categorize various mandarin types into groups or species (Hodgson, 1967; Saunt, 2000) were largely based on the classifications of Swingle (1943) and Tanaka (1954). With the recent recategorization into Type 1, Type 2, and Type 3 mandarins based on their relative proportion of pummelo admixture (Wu et al., 2018), the classification system followed here will be based on that concept.

The current classification of mandarins is summarized as follows:

- Type 1 mandarins: pure mandarins with no interspecific admixture, for example, Indica, Tachibana, and Sun Chu Sha Kat mandarins (typically small-fruited and sour/acidic, practically primitive types).
- Type 2a mandarins: early-admixture mandarins containing a small amount of pummelo alleles traced back to a common pummelo ancestor, represented by the sour or acidic mandarins, for example, Cleopatra and Sunki mandarins.
- Type 2b mandarins: early-admixture mandarins containing additional pummelo alleles, represented by the nonacidic or sweet mandarins, for example, Ponkan, Dancy, and Mediterranean mandarins.
- Type 3 mandarins: late-admixture mandarins containing additional pummelo introgressions in Type 2 mandarins, and further crossing between sweet orange and mandarins or among late-admixture mandarins produced the modern mandarin varieties, for example, King, Satsuma, Clementine, and Nadorcott. Furthermore, higher level hybridizations have subsequently resulted in both natural and man-made tangors (mandarin × orange hybrids) and tangelos (mandarin × grapefruit hybrids).

5.6.1 Principal commercial mandarin varieties

Ponkan: synonyms include Batangas (Philippines), Chinese Honey orange (China), Nagpur suntara or santra (India), and Swatow orange (Hong Kong).

Ponkan is well suited to tropical climates and is produced in Asia (Central India, South China, the warmer prefectures of Southern Japan, Taiwan, the Philippines, Indonesia, and Malaysia) and São Paulo, Brazil, making Ponkan the most extensively produced mandarin variety worldwide. Trees have a compact, upright growth habit; fruits are large for mandarins (70–80 mm, 120–140 g), slightly flattened with a short neck and depressed stylar end often with a small navel; the rind is smooth but slightly bumpy, loosely adherent, and very easy to peel, yellow-orange in color with a distinct aroma, but distinct from Mediterranean (syn. Willowleaf) mandarin. The flesh is crisp, orange with a mildly sweet flavor and aroma, but tough segment walls, and contains several polyembryonic seeds. The fruits have early-midseason maturity (November), poor tree-holding ability, and fruit tend to puff when overmature. Trees are prone to irregular and biennial bearing and are sensitive to *Alternaria* brown spot.

Numerous closely related selections or clones of Ponkan are known and are nucellar clonal budlines or chance bud sport mutations of Ponkan. They include a seedless Ponkan called Qianyang; Warnuca tangerine, the original introduction of Ponkan into the United States; Oneco (syn. Cravo Tadia in Brazil), a seedling selection Ponkan-type from Florida which has a rougher rind texture, more seeds, and later maturity than Ponkan; Emperor (syn. Emperor of Canton) from Australia; Empress from South Africa; and many others from China and elsewhere.

Ponkan has rarely been used as a breeding parent in the Western Hemisphere, except to produce Fremont in Florida, and Fremont was subsequently used as a breeding parent to produce Daisy, but Ponkan has been used as a breeding parent in Japan. Depending on the parentage, many of the first or higher order hybrids of Ponkan are classified as Type 3 mandarins due to the late and additional admixture of other mandarins. Those varieties of commercial significance are discussed below.

Nadorcott (syn. Afourer, W. Murcott; Fig. 5.4A; Nadori, 2004), a putative spontaneous hybrid of Murcott × Mandalina (Mandalina possibly being a synonym for Ponkan in some Middle Eastern countries from where it was derived), was developed by INRA (Institut National de la Recherche Agronomique), Morocco. Trees are vigorous with an upright growth habit, less dense than Clementine, shoots are generally not thorny. Nadorcott's pollen is highly fertile and is a strong pollinator; furthermore, its ovules are fertile and very sensitive to cross-pollination. Nadorcott is very precocious and productive, but prone to alternate bearing. Fruits are of medium size with deep reddish-orange rind color and pinkish albedo, very smooth rind texture, flattened shape, and slight ribbing at the stem end. Maturity is from early to mid-January when internal quality is excellent although it can mature earlier in subtropical regions. The very high sugars (10–14°Brix) and moderate acidity (1.0%–1.4%) result in a well-balanced ratio and rich flavor, and fruit can be harvested over a 6-week period without loss of quality. Tango and Nadorcott LS are induced mutations of Nadorcott developed by the University of California and the Agricultural Research Council, South Africa, respectively.

Shiranuhi (Kiyomi × Ponkan; Matsumoto, 2001) was developed by NARO-IFTS breeding program in Okitsu, Japan. The vigorous tree has a growth habit like Satsuma and Kiyomi, with shoots that can be thorny to thornless. Lower flowering than Satsuma is required to set a good crop; it produces very little pollen but is not pollen sterile. Fruit set is parthenocarpic and

FIG. 5.4 Fruits and canopy of Nadorcott (A) and Orri (B); fruits of Neufina (C) and Sugar Belle (D). *(C, photo courtesy of Mauel Talon, IVIA).*

generally good with little manipulation required. The large fruits have variable morphology, very necked with a "snout" if hot at bloom or under large day/night temperature differences. The rind can be rough and/or coarse and is sensitive to sunburn and fruit splitting, but it is easy to peel and matures late (February/March) with high Brix, rich flavor, and acidity higher than other varieties in Japan (1%–1.2% vs 1% for Kiyomi). The production of Shiranuhi, once limited to Japan, is expanding in China, South Korea, California, Brazil, and Australia. Fruits are marketed as Shiranuhi, Dekopon, Hallabong, and Sumo Citrus among other names.

Dancy: synonyms include Obeni-mikan (Japan), Ladu, Keonla (India), and Hongju (China).

Dancy is one of the oldest mandarin varieties introduced into Florida and originated in 1867 as a seedling tree of Moragne "tangierine," reportedly introduced from Tangiers (Morocco). The variety name Dancy was first used in 1877 and Dancy became the leading mandarin variety in Florida in the 1890s but declined in the 1980s due to the advent of superior varieties. Although introduced into other parts of the citricultural world, Dancy has failed to compete successfully with other, better adapted varieties. Recent molecular research shows that Ponkan is one of Dancy's parents. Dancy and some of its descendants are very sensitive to *Alternaria* brown spot. Worldwide Dancy has been an important parent of many recently bred mandarin varieties. For example, Dancy is part of the pedigree of the USDA developed tangelos and subsequent hybrids, as well as the Volcani Center's breeding program in Israel where Michal (syn. Mikhal), believed to be a natural hybrid of Clementine and an unknown pollen parent, although it is widely accepted as Dancy mandarin, has been used to produce numerous second-generation Dancy hybrids, for example, Temple × Michal (Rishon, Tami), Wilking × Michal (Shani, Mama, Merav, Edit), Satsuma × Michal (Vered), and higher order, third-generation Dancy hybrids, for example, Orah × Shani (Odem) and Edit × Nova (Dina).

Man-made or "synthetic" tangelos (mandarin × grapefruit hybrids) were first developed by the USDA breeding program in Florida in the early 20th century using Duncan (also known as Bowen) grapefruit as the seed parent and Dancy mandarin as the pollen parent. Numerous named tangelos were released by the USDA, firstly in 1904, but particularly in the 1930s. The flavor of the tangelos are more-or-less intermediate between grapefruit and mandarin, but are too grapefruit-like for modern tastes, that is, too tart. Tangelos are generally characterized by relatively high acidity and have a more pleasant flavor when produced in semitropical and tropical climates than in milder climates. Tangelos are generally weakly parthenocarpic and sexually cross-incompatible. In some cases, tangelos have characteristics more like those of the grapefruit or pummelo parent than the mandarin, for example, Sampson tangelo inherited its bitterness from Duncan grapefruit.

Mandarin-like tangelos include Thornton, Orlando (Lake), Minneola, Seminole, Suwannee, Yalaha, Webber, San Jacinto, Clement, and Broward. Of these, only Orlando and Minneola have gained some commercial significance. Orlando is rarely produced currently, but Minneola is still grown in Florida (marketed under the name Honeybell, due to its round fruit shape with a conspicuous neck), California, South Africa, and Peru.

Second-generation Dancy mandarin hybrids include several hybrids made using Clementine as seed parent such as Nova, Fairchild, and Robinson (Orlando pollen parent); Page and LB8–9 (Sugar Belle; Fig. 5.4 D) Minneola as seed and pollen parent, respectively; and Winola (triploid from a Wilking × Minneola cross; Spiegel-Roy et al., 1992). LB8–9 plantings and production have increased in recent years because this variety has been found to be field tolerant to Huanglongbing in Florida. Fortune mandarin was initially thought to be a Clementine × Dancy hybrid, but recent evidence suggests that it is Clementine × Orlando tangelo hybrid. It produces late maturing, highly colored, and acidic fruit. Although declining in global importance, Fortune has been used (extensively) as a female parent in breeding programs to produce triploid hybrids in Spain (Garbi, Safor; Aleza et al., 2010; Cuenca et al., 2010) in Sicily (Mandalate; Reforgiato Recupero et al., 2005) and California (Daisy, TDEs). Gold Nugget (syn. Pixie-like; Roose et al., 2000) (Wilking × Kincy) was developed by the University of California-Riverside (UCR) and is being commercialized to a limited extent in California. Others with some commercial importance include a series of triploids developed at UCR from the cross of [4n (Temple × Dancy) × 2n Encore] (Williams and Roose, 2004): Shasta Gold (TDE #2), Tahoe Gold (TDE #3), and Yosemite Gold (TDE #4). Also the USDA Florida released two varieties that were commercially important, but have declined substantially in importance because they are seeded: Sunburst (Robinson × Osceola) and Fallglo (Bower × Temple). A low-seeded mutant of the latter known as Early Pride (McCollum and Hearn, 2011) has been planted to some degree in Florida.

Mediterranean: The common mandarin of the Mediterranean basin is known by many different names, most of which are local place names that refer to its origin or are synonyms of the word "common." The Mediterranean or Willowleaf mandarin is thought to be the first mandarin introduced into the Mediterranean basin from China in the early 1800s possibly via England, to Malta and finally to Italy, before being distributed throughout the region and later worldwide.

Mediterranean mandarin is adapted to hot, dry climates, but has major disadvantages including alternate bearing and poor holding and shipping qualities. Its most distinctive characteristics are (i) the small size and narrow leaves, (ii) the aroma of the leaves, (iii) the mild and pleasantly aromatic flavor of the juice, (iv) the distinctive fragrance of the rind oil, (v) spreading-drooping growth habit of the shoots, and (vi) the very high degree of seed polyembryony.

Fruits are medium in size and moderately flattened, with a thin, loosely adherent rind which is yellowish-orange at maturity. Fruit typically have a strongly furrowed neck, and sometimes with a small concealed navel, but no areola. Fruit could be peeled very easily. Seeds are numerous (>20 seeds per fruit) and highly polyembryonic. The flesh is light orange, and the juice is rich in flavor and aroma. Most Mediterranean mandarin selections mature from early to early-midseason, although there are several selections with later maturity. As fruit become overmature, rind separation increases sharply and "puffing" occurs, accompanied by a marked loss of acidity and juiciness. Fruit loses quality if not picked promptly and has a poor tree-holding ability, and poor storage ability. Besides being consumed as a fresh fruit, there are two important by-products, viz., rind oil and oil of petit grain which is distilled from the leaf prunings.

Numerous selections and mutations of Mediterranean mandarin have been discovered and developed, including Canton mandarin (Szinkom, Szekan, or September mandarin) of China from which it is believed that the first importation of mandarins to Europe was made and Canton mandarin is also the presumed progenitor of Clementine, Avana, Avana Apireno, Avana Tardivo di Ciaculli in Sicily, Montenegrina in Brazil, and Clemend'or in Uruguay. Mediterranean mandarin has also been an important parent or grandparent for subsequent generation of hybrids, including Clementine, Wilking, and Kinnow. Of these, Clementine is by far the most significant globally, although Kinnow is also commercially produced in the Punjab regions of India and Pakistan. Imperial mandarin from Australia is believed to be a chance hybrid of Mediterranean mandarin and possibly Emperor (a Ponkan-type), and Mandalate (D8811) (Fortune × Avana 4n) is a triploid variety from CREA (Council for Agricultural Research and Economics), Italy.

An important higher-generation hybrid originating from Mediterranean mandarin is Orah (and Orri) mandarin which was initially thought to be a Temple × Dancy hybrid (Spiegel-Roy et al., 1992), but Dancy has been shown to be an unlikely pollen parent, and Orah is probably a Temple × Kinnow hybrid (Barry et al., 2014) from The Volcani Center, Israel. Orri (= Or 4; Fig. 5.4B), an induced mutation of Orah, is extremely vigorous with an upright growth habit similar to that of Murcott; when in bearing, the top half of the tree comprises long, vigorous watershoots, which are thorny. It has very low pollen viability, but is not completely pollen sterile, and Orri can set seeds (0–2 seeds per fruit), although it is considered to be seedless or low-seeded. Because Orri trees are vigorous they are unproductive in their early bearing years, and although production improves, Orri has a tendency for alternate and erratic bearing. The fruit size of Orri is generally good, but can be too small when trees crop excessively. At maturity, rind color is orange, fruit shape is slightly flattened, but the rind tends to be coarse when fruits are produced in relatively arid conditions, compared with more humid coastal conditions. It can be easily peeled but leaves an oil residue on the fingers. Maturity starts from mid- to late-January when internal quality and flavor are excellent. Because of its eating quality, it has seen consumer demand increasing in recent years.

King mandarin is likely a natural tangor, that is, mandarin Type 2 introgressed with sweet orange. It was discovered in Saigon, Vietnam, 1880, and is late-maturing selection (March through May) producing large fruit with a slightly flattened to round in shape and rough, bumpy rind texture and thick, and an easily-peeled rind. Outside of Southeast Asia, there is little or no commercial production. Kunenbo of Japan is similar to King though fruit size is smaller and has a smoother rind. King was parent of many cultivars and breeding parents of cultivars like Kara, Kincy, Kinnow, Wilking, and Encore.

Satsuma: The origin of Satsuma mandarin or unshiu mikan has been traced to the early 15th century in Japan, although there remains controversy over the exact site of origin, perhaps coming to Japan from Wenzhou in China. Sequence analysis cannot positively answer the question of its origin, but it supports the parentage as being Huanglingmiao (a somatic mutant of Kishu) × Kunenbo (a *Citrus nobilis* selection found in Japan).

Numerous mutations have resulted in the many Satsuma selections currently described, a situation analogous to the proliferation of sweet oranges, grapefruit, and Clementine mandarins; they range in maturity from very early (August–September) to very late (January). Most of these Satsuma selections have been identified and commercialized in Japan, as well as other countries where Satsumas are produced, including the United States, Spain, Argentina, Uruguay, and of course, China. These include the ultra-early selections [Iwasaki, Hashimoto, Miyamoto, Temprana, Wakayama (also known as Xie Shan)], the early selections (Miyagawa Wase, Okitsu Wase, Miho Wase, Guoching No. 1, and several other local selections from Louisiana, USA), midseason maturity (Owari, the most widely grown variety, Clausellina from Spain, Silverhill from the United States, and Dobashi Beni), and the late selections (Aoshima, Ohtsu No.4, Picaña, and Bela/Lola from Spain).

Satsuma trees are among the most cold-tolerant varieties of all citrus (particularly when grown on trifoliate orange rootstocks). In cool autumn/winter climatic regions, they produce the highest quality fruit. Satsuma fruits are almost always seedless and easy to peel, with attractive external and very deep orange internal color. Their flavor is simple and mild but quite pleasant when grown in appropriate environments. Trees of the earlier maturing selections are less vigorous than late-maturing types, but all tend to be precocious bearing and very productive. Satsumas grown in subtropical regions with warmer autumn temperatures, such as the Southeast United States, tend to produce larger fruit with poor rind color, are puffy, and frequently with lower sugars and acids.

Until the 1940s, mainly Owari Satsuma, the progenitor of most modern selections, was produced in Japan. Currently, in Japan, the largest Satsuma mandarin production region in the world, Miyagawa Satsuma is the principal early-maturing selection, Nankan No. 20 Satsuma is the principal mid-maturing selection, and Aoshima Satsuma is the main late-maturing selection.

Kiyomi is a hybrid of Miyagawa Satsuma × Trovita orange developed by the breeding program at Okitsu, Japan. Fruit can resemble a large Satsuma with pyriform fruit shape or an orange when smooth and round; they are late–maturing selection and similar in size to a Navel orange, with a relatively thick and spongy rind that could be easily peeled, though some albedo usually remains on the segment walls. The rind is deep yellow-orange to orange at maturity, and the tender flesh is orange and mildly flavored. Kiyomi is pollen sterile so when planted in solid blocks the fruits are seedless; however, they can produce abundant monoembryonic seeds when cross-pollinated and so Kiyomi has been used in breeding (see Shiranuhi, above, and Setoka, below).

Primosole (Tribulato and La Rosa, 1993)is a hybrid of Miho Satsuma × Carvalhais mandarin from the University of Catania, Italy. It produces large fruit that mature from early October through early November, and are moderately easy to peel, though some peel oil on the fingers can be found. They are nearly seedless but tend to have dry and ricey flesh, particularly in hot production areas. They are produced mainly in Sicily, but also in Peru.

Queen is a seedling selection from an open-pollinated Satsuma developed by the Teresa Bros., Spain. Queen is very late-maturing selection with few to no seeds with cross-pollination and large-fruited and a flatter shape than Satsuma.

Clementine: Genome sequence analysis has shown Clementine to be a natural hybrid of Mediterranean mandarin × sweet orange (see Chapter 2), that is, mandarin Type 2 introgressed with sweet orange. This natural tangor was named by Father Clement Rodier, hence the name Clementine, who discovered it in the garden of an orphanage in Algeria, North Africa, at the end of the 1890s. The original Clementine variety introduced into Spain from Algeria in 1925 was named Fina (syns. SRA 63, Cadoux). Monreal Clementine, a sexually self-compatible selection, was introduced in 1940. Until the 1950s, only Fina Clementine was produced in Spain and is the progenitor of all modern Clementine selections. Numerous mutations have developed from Fina, now allowing harvest for a 5-month period (September–January). Nules Clementine (syn. Clemenules) is the main Clementine selection produced in Spain, the principal Clementine mandarin production region in the world.

Algerian Clementine is one of the original Clementine selections, and is probably the original Clementine selection introduced into the United States in 1909 and taken to California in 1914, and it is probably the Clementine selection that was used in the early USDA breeding program to produce Nova, Fairchild, Page, etc. Algerian has excellent flavor, though smaller fruit size than other Clementine selections. In the 1960s, numerous Moroccan and Algerian Clementine strains were selected and evaluated at SRA, San Giuliano, Corsica.

As has been the case with Satsumas and other clonally propagated citrus, numerous mutations have been found over time and have led to the extension of Clementine season from September through January (in Spain, the major Clementine-producing country in the world). There are some ultra-early selections (September to October) derived from Oronules, including the current earliest Basol reaching internal maturity in early September; this is followed by Clemensoon, Clemenrubi, Cultifort, Orogos, and Ripetri. However, many of these ultra-early-maturing Clementine selections develop severe galls or sphaeroblasts on their trunks and limbs and, combined with multiple bud sprouting, lead to stunted trees. Early-maturing Clementines (October to November) include the attractive and delicious Oronules (syn. Caffin); Spinoso VCR, a mutation of Comune (Fina) found in Sicily; Marisol (declining because of inferior flavor); and several others. The heart of the Clementine season runs from November through December and is dominated by Nules (syn. Clemenules, Reina). Nules is by far the principal Clementine section produced in most of the world's Clementine-producing regions, and is the standard for productivity, fruit size, and quality. Other midseason selections include Oroval (the first mutant from Fina described first in 1950); Sidi Aissa from Morocco; and more recently two selections produced by gamma irradiation of Nules: Nulessin (lower fertility but greater productivity than Nules) and Nero (both pollen and ovule sterile). Finally, the late-maturing selections include Hernandina (once the major late-maturing variety, but declining because of erratic and alternate bearing, and delayed peel color development); Nour from Morocco which likewise bears crops erratically; Clemenlate, a vigorous but low yielding tree; Sando, a Fina mutant with excellent flavor and color and very late maturity (into March in some seasons); and Neufina (López-García et al., 2015; Fig. 5.4C), from the IVIA program with a long potential harvest period and good postharvest behavior.

5.6.2 Other mandarin hybrids of current or potential commercial importance

There are a handful of natural tangelos and tangors which originated as chance seedlings that have either been important commercially (globally or locally) and/or for breeding new potentially useful commercial mandarin varieties. For example, Ugli from Jamaica (likely a pummelo × mandarin hybrid), which is produced in Jamaica where it was first found, has a limited but consistent market in the Eastern United States. Temple "orange" (syn. Sue Linda, Thoro, Royal mandarin, Scarlett, Magnet) is presumably a natural tangor (mandarin × orange), was produced in large quantities in Florida because of its deep reddish-orange rind color and rich, spicy flavor; certain clones are still produced in other production regions, and has been useful in breeding new varieties because it produces monoembryonic seeds with zygotic embryos. Ortanique [syn. Tambor, Topaz, Mandora, Urunique, Ortaline, and Mystique (nucellar)] is a presumed mandarin × common orange hybrid, though it remains in production because it produces large and late-maturing (March–April) fruit that remain on the tree in good condition for extended periods, its importance is declining because it is not easy to peel, has a very oily rind, and will be seedy with cross-pollination. Ellendale arose as a natural tangor in Australia, and although it produces large and easy to peel fruit, it is also in decline because of poor tree structure, fruit splitting, and seediness. Ellendale has been used in breeding because of its monoembryony, and new varieties such as African Sunset and Valley Gold (Ellendale × Robin; Sippel et al., 2015) have been released and are being planted to some extent in South Africa, and Royal Honey (Ellendale × Murcott?) in Australia. Ellendale also has been exploited by the breeding program in Uruguay, at INIA. Murcott (syn. Honey tangerine, Smith), a tangor of unknown parentage, has late maturity (February through April), and was an important variety in Florida because of its excellent eating quality; however, it has all but vanished in Florida because it is excessively seedy, not a true "easy peeler," and has great sensitivity to Huanglongbing. However, several low-seeded mutants have been developed by irradiation, including Mor from Israel; IrM1 and IrM2 (Smith, 2006), and 2PH (Phoenix) from Australia; and Murcott SL from Spain. None of these are completely seedless, and in some cases seed numbers in individual fruit, though they average less than two seeds per fruit, can sometimes exceed 10, depending on the production conditions and location. Furr (syn. C54-4-4, Clemcott) is a Clementine × Murcott hybrid produced by the USDA in Indio, California in 1954. It was then moved to Rio Farms, Texas, and taken to Florida by Dr. Heinz Wutscher (USDA) in 1975, irradiated in 1990 by Wutscher and then again in 1996, and planted on the farm of Mr. Orie Lee, Florida. It has very large fruit with very good flavor and is reasonably easy to peel. Taylor-Lee was developed by 2PH Farms (Australia) via seed from Orie Lee's Clem × Murcott; trees are vigorous and very productive producing very heavy crops of large fruit size, though packout can be low because of cosmetic appearance. Like its Murcott parent, Furr and Taylor Lee are susceptible to *Alternaria* brown spot. Another selection from irradiation of Furr in South Africa, Leanri, was found with earlier fruit maturity and deeper orange rind color. Setoka (Matsumoto et al., 2003) is a hybrid of (Kyomi × Encore) × Murcott developed by the breeding program at Okitsu, Japan. It is seedless and large, with very attractive appearance internally and externally, and very high fruit quality; most frequently it is grown under cover in Japan and Jeju Island, South Korea, where it performs very well and earns high premiums in the marketplace.

A series of triploid mandarin hybrids have been produced to meet the market requirement of seedlessness. This work was pioneered first in California but was exploited more thoroughly by the work of the breeding team at CREA in Sicily. Tacle, a hybrid of Monreal Clementine by tetraploid Tarocco, produces large fruit that are easily peeled, and with deep rind color and occasional red flesh due to the presence of anthocyanins. This has sparked an interest in producing additional red-fleshed mandarins, and breeders in many countries are pursuing this goal. Mandared is another variety developed by the same breeding program, from the cross between Nules and a tetraploid Tarocco, and has been marketed under the name of RedNules. Mandalate is a hybrid of Fortune by Avana 4n (Avana is a clone of Mediterranean mandarin), but is not pigmented by anthocyanins. Other triploid breeding programs have been likewise focused on the development of seedless mandarins including one at IVIA, Spain, and at the University of Florida. Triploid mandarin trees tend to be thorny, not always highly productive, and the fruit generally have large oil glands leading to a coarse appearance and undesirable traces of peel oil on the fingers upon peeling. Garbi and Safor have been released in Spain, and Kid's Favorite (C4-15-19) was released in Florida.

Finally, mandarin breeders have attempted to exploit the genetic seedless characteristic of Seedless Kishu (Mukaku Kishu). This mutant of seedy Kishu from Japan is ovule sterile but pollen fertile, and the trait appears to be monogenic and dominant, so this selection has been used in breeding programs in Japan, Florida, and Australia. Bingo mandarin was first released in Florida from the Kishu background that has been planted on a modest commercial scale, but fruits have not yet entered the marketplace, so its future remains to be seen.

References

Aleza, P., Cuenca, J., Juárez, J., Pina, J.A., Navarro, L., 2010. 'Garbí' mandarin: a new late-maturing triploid hybrid. HortScience 45 (1), 139–141.

Barry, G., Gmitter, F., Chen, C., Roose, M., Federici, C., 2014. Investigating the parentage of 'Orri' and 'Fortune' mandarin hybrids. Acta Hort. 1065, 449–456.

Butelli, E., Licciardello, C., Zhang, Y., Liu, J., Mackay, S., Bailey, P., et al., 2012. Retrotransposons control fruit-specific, cold-dependent accumulation of anthocyanins in blood oranges. Plant Cell 24, 1242–1255.

Caruso, M., Ferlito, F., Licciardello, C., Allegra, M., Strano, M.C., Di Silvestro, S., et al., 2016. Pomological diversity of the Italian blood orange germplasm. Sci. Hortic. 213, 331–339.

Cuenca, J., Aleza, P., Juárez, J., Pina, J.A., Navarro, L., 2010. 'Safor' mandarin: a new citrus mid-late triploid hybrid. HortScience 45, 977–980.

Curk, F., Ollitrault, F., Garcia-Lor, A., Luro, F., Navarro, L., Ollitrault, P., 2016. Phylogenetic origin of limes and lemons revealed by cytoplasmic and nuclear markers. Ann. Bot. 117, 565–583.

Deng, X.X., Peng, C., Chen, Z., Deng, Z., Xu, J., Li, J., 2008. *Citrus* Varieties in China. China Agric. Press, Beijing. 203 pp. (in English and Chinese).

Hodgson, R.W., 1967. History, world distribution, botany and varieties. In: Reuther, W., Webber, H.J., Batchelor, L.D. (Eds.), The Citrus Industry. vol. I. University of California, Berkeley, CA, pp. 431–591.

Karp, D., 2007. The bloods: mutations, chemistry and flavor. Fruit Gardener 39, 16–22.

López-García, A., Terol, J., Tadeo, F.R., Herrero-Ortega, A., Ibañez, V., Talón, M., 2015. Three new cultivars of clementine: 'Clemenverd', 'nero' and 'neufina'. Acta Hort. (1065)239–243.

Matsumoto, R., 2001. 'Shiranuhi', a late-maturing citrus cultivar. Bull. Fruit Tree Res. Stn. 35, 115–120. (in Japanese with English abstract).

Matsumoto, R., Yamamoto, M., Kuniga, T., Yoshioka, T., Mitani, N., Okudai, N., Yamada, A., Asada, K., Ikemiya, H., Yoshinaga, K., Uchihara, S., Oiyama, I., Murata, H., 2003. New citrus cultivar 'Setoka'. Bull. Natl. Inst. Fruit Tree Sci. 2, 25–31. (in Japanese with English abstract).

McCollum, T.G., Hearn, C.J., 2011. 'US Early Pride', a very low-seeded, early-maturing mandarin hybrid. HortScience 46, 1695–1697.

Nadori, E., 2004. Nadorcott mandarin: a promising new variety. Proc. Int. Soc. Citricult 1, 356–359.

Reforgiato Recupero, G., Russo, G., Recupero, S., 2005. New promising *Citrus* triploid hybrids selected from crosses between monoembryonic diploid female and tetraploid male parents. HortScience 40 (3), 516–520.

Roose, M.L., Williams, T.E., Cameron, J.W., Soost, R.K., 2000. 'Gold Nugget' mandarin, seedless, late maturing hybrid. Hortscience 35, 1176–1178.

Saunt, J., 2000. *Citrus* Varieties of the World. Sinclair International Limited, England.

Sippel, A.D., Bijzet, Z., Froneman, I.J., Combrink, N.K., Maritz, J.G.J., Hannweg, K.F., Severn-Ellis, A.A., Manicom, B.Q., 2015. *Citrus* breeding in South Africa: the latest developments in the programme run by the ARC-Institute for Tropical and Subtropical Crops. Acta Hort. 2015, 397–403.

Smith, M.W., 2006. Variety 'IrM2'. Plant Var. J. 19, 260–264.

Spiegel-Roy, P., Vardi, A., 1992. `Shani,' 'Orah' and 'Winola': Three new selections from our breeding program. Proc. Int. Soc. Citricult. 1, 72–73.

Starrantino, A., Russo, F., Donini, B., Spina, P., 1988. Lemon mutants obtained by gamma irradiation of the nucellus cultured in vitro. Proc. Int. Soc. Citricult. 1, 231–235.

Swingle, W., 1943. History, botany and breeding. In: Webber, H.J., Batchelor, L.D. (Eds.), The Citrus Industry. Vol. I. Univ. California Press, pp. 129–474.

Tanaka, T., 1954. Species Problem in *Citrus*. Japanese Society for Promotion of Science.

Tribulato, E., La Rosa, G., 1993. Primosole e Simeto: Due nuovi ibridi di mandarino. Italus Hortus 1, 21–25. In Italian.

Webber, H.J., 1943. Cultivated varieties of citrus. In: Webber, H.J., Batchelor, L.D. (Eds.), The Citrus Industry. vol. I. Univ. California Press, California, pp. 475–668.

Williams, T.E., Roose, M.L., 2004. `TDE2' Mandarin hybrid (Shasta Gold Mandarin), 'TDE3' Mandarin hybrid (Tahoe Gold Mandarin) and 'TDE4' Mandarin hybrid (Yosemite Gold Mandarin): three new mid and late-season triploid seedless mandarin hybrids from California. Proc. Int. Soc. Citricult. 1, 394–398.

Wu, G.A., Terol, J., Ibanez, V., López-García, A., Pérez-Román, E., Borredá, C., et al., 2018. Genomics of the origin and evolution of *Citrus*. Nature 554, 311–316.

Further reading

Davies, F.S., Albrigo, L.G., 1994. *Citrus*. CAB Intl.

Goldenberg, L., Yaniv, Y., Porat, R., Carmi, N., 2014. Effects of gamma-irradiation mutagenesis for induction of seedlessness, on the quality of mandarin fruit. Food Nutr. Sci. 5, 943–952.

Jackson, L.K., Davies, F.S., 1999. *Citrus* Growing in Florida, fourth ed. University Press of Florida.

Martin, F.W., Cooper, W.C., 1977. Cultivation of Neglected Tropical Fruits with Promise. 3. The Pummelo [*Citrus grandis*]. U.S. Agricultural Research Service.

Morton, J., 1987. Fruits of Warm Climates. Miami, FL.

Tolkowsky, S., 1937. Hesperides: A History of the Culture and Use of Citrus Fruit. Staples and Staples, Westminster.

Chapter 6

Citrus rootstocks

Kim D. Bowman[a], Johan Joubert[b]
[a]*U.S. Horticultural Research Laboratory, USDA-ARS, Fort Pierce, FL, United States,* [b]*Citrus Research International, Nelspruit, South Africa*

6.1 Introduction

In present day citriculture, trees are almost always propagated by grafting the fruiting cultivar, or the scion, onto a second cultivar that will form the lower trunk and root system. The second cultivar is termed the rootstock, and in most cases, the rootstock cultivar has been created or selected, and evaluated over time for the specific purpose of being used as a rootstock. The current repertoire of citrus rootstocks includes many that appeared naturally, in the distant past or more recently, as well as others that were the result of specialized citrus rootstock breeding programs over the past 100 years. There are some exceptions to the use of rootstocks in growing citrus. But in most cases, trees will be healthier, grow better, fruit at a younger age, and produce more and better quality fruit if they are grafted on a suitable rootstock.

6.2 Reasons for a rootstock

Probably the most important reason to use a rootstock (rather than growing a seedling) for citrus, is that grafting a branch or bud from a fruiting citrus tree onto a rootstock gives the opportunity to quickly produce an unlimited quantity of the original fruiting tree, each one of which will be genetically identical and have essentially the same fruit characteristics. For some types of citrus, including many mandarins, use of grafting (or cuttings) is the only way the trees of a particular desired fruiting type can be propagated, since they do not grow true to type from seed. In theory, the replication of the fruiting variety by grafting the scion onto a separate plant (to be used as the rootstock) can make use of any healthy growing citrus plant as the rootstock. The rootstock can be nearly any age, or any type of citrus, even a seedling of the same variety as the scion. In practice, this rootstock plant is typically a young seedling 4–8 months old of a variety that is known to have good characteristics for use as a rootstock.

Of additional importance, the use of grafting to replicate trees of the fruiting variety will substantially shorten the time to fruiting of the propagated plants, as compared with planting seeds of that fruiting variety. Although cultivars like sweet orange and grapefruit can usually be propagated through seed, and more or less recover the same fruit traits, seedlings of sweet orange and grapefruit typically must reach at least 5 years of age before they begin to fruit, while grafted trees with a scion of that mature fruiting variety on a suitable rootstock will begin to bear good fruit of that variety within 1–3 years. This extended time period that seedlings require to begin fruiting is termed "juvenility," while varieties that have begun to fruit are considered "mature." The importance of this shortening of time to fruiting (avoiding juvenility) cannot be understated in commercial application, as growers can rarely afford to wait many years to obtain the fruit needed to recover a return on their investment. One variant of propagation that may be used to recover the particular fruiting variety with good fidelity, avoid the long juvenile period, and not require a rootstock, is to propagate the scion variety by making stem cuttings. Procedures for the successful use of stem cuttings in citrus have been described (Bowman and Albrecht, 2017), but with only a few exceptions (such as in lime production), graft propagation onto a rootstock is preferred to cutting propagation.

Most citrus trees are grown for the fruit they produce, and the scion variety to be planted is chosen on the basis of the particular combination of fruit traits that is desired for consumption or profit. Most of the fruiting cultivars in high demand do not have the best combination of traits affecting root health, such as tolerance to salinity, high pH, or flooding, and resistance to biotic threats like *Phytophthora* spp. diseases, nematodes, and *Candidatus* Liberibacter asiaticus (*C*Las). Extremes of heat, cold, and wind can also severely damage citrus trees, and choosing the cultivar of the root system separately from the cultivar of the fruiting variety gives the opportunity to optimize the durability and health of the tree root system independently of the characteristics of the fruit. Planting many of the common scion varieties on their own roots often produces a tree with inferior health and durability, as compared with the same scion grafted onto a good rootstock cultivar.

The Genus Citrus. https://doi.org/10.1016/B978-0-12-812163-4.00006-1

For example, sweet orange (*Citrus sinensis*) is notably susceptible to *Phytophthora* spp. damage to the lower trunk and root system, often resulting in consequent early tree death when a sweet orange tree is grown on its own roots.

6.3 Important rootstock attributes

The potential benefit of using a cultivar that has been specifically selected for outstanding lower trunk and root traits can be of major importance for the successful growth and production of the resulting trees. The particular traits that are most useful for the rootstock will vary greatly by environmental factors, pest and disease threats, and the management systems that will be employed during the lifetime of the tree. All of these may change over time, even at the same location. In this section, we will briefly discuss some of the more important of the rootstock attributes.

Anchorage. Good physical anchorage of the tree to the ground is a critical trait for good citrus tree health and survival. This may be an especially important rootstock trait in growing areas affected by regular or periodic strong winds (Fig. 6.1). The length, number, and strength of both scaffold and fibrous roots contribute to this physical anchorage, and the root development of all rootstock cultivars is strongly influenced by soil characteristics, tree health (pests and disease effects), as well as the irrigation and fertilization patterns. Even in deep soils and with minimal disease pressure, root systems will develop primarily in the top soil layers, if irrigation water does not penetrate more deeply. Limited information is available that compares root development and anchorage of different rootstock cultivars under a range of biotic and abiotic conditions (Albrecht et al., 2017; Eissenstat, 1991; Eissenstat and Achor, 1999), and typically most rootstocks are considered to provide adequate anchorage if growing conditions provide for good tree health. In the absence of other compromising factors, rootstocks that support more vigorous growth of the scion will typically make the largest and strongest root systems.

Flood/drought. Citrus is grown in a diverse range of environmental conditions, including the extremes of low and high rainfall and soil moisture. In many growing areas, management systems include drainage structures and irrigation systems

FIG. 6.1 Trees blown over by hurricane Jeanne in Vero Beach, Florida: (A) fourteen-year-old Nova mandarin trees on Cleopatra rootstock; and (B) two-year-old flame red grapefruit tree on Cleopatra rootstock.

to moderate and shorten the duration of flood and drought stress. However, even in areas with drainage and irrigation systems in place, flood or drought may sometimes affect trees. Field observations and studies of the impacts of flooding and drought stress on rootstocks have provided some information about relative rootstock tolerance to these factors (Bhusal et al., 2002; Garcıa-Sanchez et al., 2007; Vu and Yelenosky, 1991). Based on long-term field performance, sour oranges (*Citrus aurantium*) and selections of *Poncirus trifoliata* are typically considered to have good tolerance to wet soil and flooding, while Cleopatra has poor tolerance to flooding. In contrast, *P. trifoliata* and Swingle have relatively poor drought tolerance, while Rangpur and Sunki have good tolerance, and Carrizo is intermediate. Although reliable information on tolerance to flood and drought may be concluded from nonexperimental but long-term field observations, methods for specialized study have been developed and employed to study tree responses and more rapidly evaluate rootstock tolerance under controlled conditions (Romero et al., 2006). The different drought survival strategies of Rangpur and Sunki, two drought-tolerant rootstocks (Santana-Vieira et al., 2016; Dutra de Souza et al., 2017), were contrasted individually and in graft combinations with scions. The rootstock Forner-Alcaide 5 was observed to be more tolerant to drought stress than either of its parents, Cleopatra and *P. trifoliata* (Rodríguez-Gamir et al., 2010).

Temperature extremes. Citrus rootstocks vary in their tolerance of high- and low-temperature extremes, with tolerance of low temperatures having received more research attention (Yelenosky, 1985; Zandalinas et al., 2016). All citrus is sensitive to freeze damage when it is actively growing, and rootstock effects on citrus tolerance of freezing conditions is primarily an influence to increase or reduce active growth of the tree prior to when it is exposed to freezing temperatures (Yelenosky, 1996). It is the different responses of citrus rootstocks to temperature patterns and other conditions in the days and weeks preceding the freeze event (preconditioning) that primarily determine whether a rootstock will induce more or less freeze tolerance in the tree. Because of this large effect of conditions in the weeks preceding a freeze, damage from low temperatures (and relative rootstock effects) can differ significantly between regions and freeze events. However, rootstock often has a significant effect on tree tolerance of freezes, with the rootstocks rough lemon and Volkamer lemon considered to yield trees most sensitive to freeze damage, Carrizo and sour orange intermediate, while trifoliate orange, Swingle, and some of the other trifoliate hybrid rootstocks provide the best resistance to freeze damage. Under conditions likely to have the most freeze risk and good cool temperature preconditioning, trifoliate orange rootstock is typically observed to provide the best freeze tolerance (Ben Yahmed et al., 2016; Ebel et al., 2008; Kawase et al., 1987; Yelenosky, 1985).

Salinity. Citrus is considered highly sensitive to salinity in the soil and irrigation water (Maas, 1993), although there are large differences in tolerance of salinity among rootstocks. Salinity sensitivity in citrus is thought to be primarily a response to the uptake by the rootstock of large amounts of chloride, but excessive sodium and boron can also be problems (Ruiz et al., 1999; Simón-Grao et al., 2018). Selection of rootstocks for salinity tolerance has received considerable research effort (Forner-Giner and Ancillo, 2013; Gallasch and Dalton, 1989; Raga et al., 2014), and among the new major rootstocks, FA-5 is a hybrid rootstock reported to have useful levels of salinity tolerance (Forner-Giner et al., 2009). Clonal selections of *P. trifoliata* introduced to Australia from China were also identified as having good salt-excluding ability (Sykes, 2011a), in contrast to the sensitivity of other trifoliate orange clones that have been studied. There is some variability in the results from assessment of rootstock tolerance to salinity, depending on scion and other factors (Levy and Syvertsen, 2004; Ream and Furr, 1976). In general, Cleopatra, Sunki, and Rangpur are considered the most tolerant, sour orange and rough lemon are intermediate, and the most sensitive are trifoliate orange, Carrizo, and Swingle (Castle, 1987; Castle et al., 2015).

pH. High pH conditions and calcareous soils are a problem in many citrus areas, and use of rootstocks that are tolerant is critical for successful production (Al-Jaleel et al., 2005; Castle et al., 2004; Louzada et al., 2008; Wutscher, 1979). *P. trifoliata* and many of its hybrids (such as Carrizo and Swingle) have been noted to be especially sensitive to high carbonate soils, and suffer from resultant micronutrient deficiencies, especially iron (Castle and Manthey, 1998; Manthey et al., 1994). Significant effort has been devoted to the study of citrus rootstock tolerance of high pH conditions, and methods to evaluate rootstocks for this trait (Castle et al., 2009; Licciardello et al., 2013; Llosá et al., 2009). In general, sour orange and rough lemon are considered most tolerant of high pH, Cleopatra and US-812 intermediate, and *P. trifoliata* and many of its hybrids intolerant (Castle et al., 2015)

Phytophthora. Three species of *Phytophthora* [*P. nicotianae* (*P. parasitica*), *P. citrophthora*, and *P. palmivora*] cause damage to citrus root systems, young trees, and the lower trunk. A key component of minimizing disease damage is the use of a rootstock which is resistant or tolerant to infection. Many evaluations of resistance have focused on individual *Phytophthora* species, especially *P. nicotianae* (Boava et al., 2011; Carpenter and Furr, 1962; Graham, 1995), but it is known that rootstock resistance to one species of the pathogen often differs markedly from its response to the other species (Graham et al., 2003; Graham and Feichtenberger, 2015; Widmer et al., 1998). For example, Swingle and most clones of *Poncirus trifoliata* are considered highly tolerant of *P. nicotianae*, but are susceptible to damage from *P. palmivora* (Fig. 6.2) (Albrecht and Bowman, 2004; Bowman et al., 2002). Since the species of Phytophthora that most affect trees differ by climate and region, identifying which of the pathogens is found locally will be key to identifying the appropriate

FIG. 6.2 Young Swingle seedlings grown in pots in the greenhouse. (A) Swingle grown in clean soil; (B) Swingle grown in soil inoculated with *Phytophthora palmivora* and developing severe root damage; (C) close up of root system of Swingle seedling shown in (B).

rootstock (Bowman et al., 2007). Soil type, irrigation, and drainage conditions can also have a large effect on potential for damage from *Phytophthora*, as excessive moisture and reduced aeration can exacerbate disease development. Physical damage to roots, such as from feeding of nematodes and insects (including Diaprepes weevil), can also circumvent resistance mechanisms and significantly increase *Phytophthora* problems.

Diaprepes weevil. The root weevil *Diaprepes abbreviatus* is a serious pest of citrus, currently found in many parts of Florida, as well as much of the Northern and Eastern Caribbean region. Adults lay eggs on the leaves of citrus and other plant species, but it is the larval stage that drops to the ground after hatching, which causes the most damage to citrus. As the larvae feed on citrus roots, they cause severe fibrous root loss, as well as tissue wounding, which often results in invasion by *Phytophthora* spp. The combination of *Diaprepes* and *Phytophthora* damage has been termed the Phytophthora-Diaprepes Complex (PDC; Graham et al., 2003, 2007) because of the extensive tree decline and death that results in trees on sensitive rootstocks. Although all sexually and graft-compatible citrus relatives have found to be susceptible to damage by the *Diaprepes* weevil (Bowman et al., 2001; Lapointe and Bowman, 2002; Lapointe et al., 1999), large differences are observed between rootstocks in tolerance to *Diaprepes* under field conditions (Bowman et al., 2003). Soil type can significantly affect rootstock reaction to *Diaprepes* infestation, especially because of the consequent effects from *Phytophthora*. In poorly drained soils, Swingle and Carrizo were found to be severely affected by PDC, while other rootstocks such as US-802, US-897, and US-942 were found to be more tolerant (Fig. 6.3). The UF-CREC rootstock breeding program has used a two-phase assay (damage assay/recovery assay) to screen several hundred diploid and tetraploid rootstock candidates for the Phytophthora-Diaprepes Complex (Grosser et al., 2003, 2007).

CTV. Citrus tristeza virus (CTV) is considered a major disease threat to citrus. Although some strains of the virus can cause severe stem pitting and other damage to scions, the most significant rootstock-specific effect of the virus is the stunting or quick decline reaction suffered by sweet orange trees grafted on sour orange rootstock. It is estimated that CTV led to the death of about 16 million trees on sour orange rootstock in Argentina in the 1940s, and smaller but significant number of trees on sour orange were also destroyed by CTV in Brazil and California (Wallace, 1968). In addition to being spread in infected

FIG. 6.3 One-year-old red grapefruit trees in area severely damaged by Phytophthora-Diaprepes weevil complex. (A) Red grapefruit/Swingle rootstock; (B) Red grapefruit/US-802 rootstock.

budwood, CTV is readily spread from tree to tree by several species of aphids. Because of this sensitivity to CTV and difficulty in controlling spread, sour orange is currently a relatively minor rootstock in much of the world, despite many otherwise excellent rootstock traits. Among common rootstocks, *Citrus macrophylla* is the other rootstock besides sour orange that suffers from a CTV quick decline-type reaction. In the development and testing of new rootstocks, determining resistance or tolerance to CTV is desired and often relies on natural infection in field trials where sour orange can be an effective susceptible experimental control (Wutscher and Bowman, 1999). Although methods have been described to evaluate CTV tolerance of rootstocks more rapidly and under controlled conditions (Bowman and Garnsey, 2001), a large portion of citrus germplasm and hybrids typically show field tolerance and little tendency toward a quick decline-type reaction. Resistance to CTV has been described from *P. trifoliata* and that trait is transmitted to many hybrids (Garnsey et al., 1987). Significant progress has been made in mapping the locus for CTV resistance from *P. trifoliata* (Gmitter Jr. et al., 1996; Deng et al., 1997, 2001).

Huanglongbing. In many citrus production areas, Huanglongbing or citrus greening disease, is considered to be the most serious disease affecting citrus production, and is potentially the most serious looming threat that could devastate the global citrus industries. Huanglongbing is believed to be caused by several bacterial pathogens identified as *Candidatus* Liberibacter spp., although Koch's postulates have not been fulfilled because the bacteria has not been cultured successfully (Wang and Trivedi, 2013) (see Chapter 18). Generally, three species of *Candidatus* Liberibacter (differentiated by 16S rDNA sequence) are recognized as causing Huanglongbing disease, *Candidatus* L. *asiaticus*, *Candidatus* L. *africanus*, and *Candidatus* L. *americanus*, although *Candidatus* L. *asiaticus* (*C*Las) is by far the most widespread among citrus production areas. All citrus species and near relatives are susceptible to infection by *C*Las (Bové, 2006), but significant differences in tolerance can be found, with sweet orange and grapefruit being severely damaged by infection (Folimonova et al., 2009), while other material including *P. trifoliata* and some hybrids with citrus (Albrecht and Bowman, 2011, 2012) are highly tolerant and virtually free of symptoms. Although some rootstocks are tolerant to *C*Las infection as seedlings, generally this tolerance to infection does not result in a large improvement in the *C*Las tolerance of trees, when the rootstock is grafted with a *C*Las-sensitive scion. Consequently, using a *C*Las-tolerant rootstock will only partially improve the overall tree tolerance to *C*Las in the field (Albrecht et al., 2012; Bowman et al., 2016a,b), possibly the result of only a significantly healthier root system and the consequential direct improvements those healthier roots provide for overall tree health. Delaying infection with *C*Las until field trees are established and a larger size will usually result in better tree tolerance to the infection than when trees become infected at a very young age. Although controversial, there is some indication that enhanced nutrition can improve field performance of trees infected with *C*Las (Morgan et al., 2016; Stansly et al., 2014), and this nutritional effect may be somewhat modulated by rootstock. Numerous field trials in Florida have been conducted with trees naturally infected by *C*Las over multiple years. Generally, under Florida conditions and in field plantings infected by *C*Las, US-942 is one rootstock that has provided field performance superior to many other rootstocks, including Swingle, Carrizo, Kuharske, Ridge sweet orange, sour orange, and Cleopatra (Bowman et al., 2016a,b).

Other diseases. Many other pathogens affect citrus, including those caused by viruses, viroids, fungi, and bacteria (Timmer et al., 2000). In some cases, devastating unidentified diseases are believed to be caused by pathogens, but research has yet to elucidate the cause, such as with citrus blight (Young et al., 1984; Derrick and Timmer, 2000). Often, rootstocks differ significantly in resistance or tolerance to pathogens, but limited information is available, or only applicable to local biotypes or isolates of the pathogen where the research was conducted. Some information is available on rootstock tolerance or sensitivity to viroids and other diseases (Vernière et al., 2004; Timmer et al., 2000). Several publications have provided ratings of all available rootstocks for the most widespread and important pathogens (Bitters, 1986; Castle et al., 2015; Lacey, 2017; Lee et al., 2009; Roose, 2014), and these provide a good resource for broad comparisons.

Nematodes. Numerous species of nematodes affect citrus roots, although many of these are mainly of regional or local importance, rather than having worldwide significance. The 'citrus nematode' *Tylenchulus semipenetrans*, is one nematode that has an impact on citrus production in all regions of the world (Duncan and Cohn, 2005). Rootstocks differ in their tolerance to *T. semipenetrans*, although there are known to be some differences by regional biotypes of the nematode. For most biotypes, *P. trifoliata*, Swingle, and some other hybrids of *P. trifoliata* with citrus have good resistance or tolerance, while Cleopatra, sour orange, and rough lemon are susceptible (Duncan et al., 1994; Verdejo-Lucas and Kaplan, 2002; Verdejo-Lucas et al., 1997, 2000).

Tree size. Citrus tree size is strongly influenced by rootstock, with rough lemon rootstock typically producing very large trees with most scions, and Flying Dragon trifoliate orange producing very small (dwarf) trees (Mademba-Sy et al., 2012; Reforgiato Recupero et al., 1992; Roose, 1986; Stuchi et al., 2003). Terminology for rootstock influence on tree size varies, but we suggest useful categories are very vigorous (rough lemon), standard (Carrizo or Swingle), semidwarf, and dwarf (Flying Dragon). In addition to rootstock, many other factors influence tree size, including scion, soil, management practices, and diseases that affect the tree. In the absence of compatibility issues, a group of rootstocks usually have the same relative size effects on a common scion; so, for example, all scions on rough lemon will make larger trees than all scions on Flying Dragon. In one sweet orange field trial in central Florida (Wutscher and Hill, 1995) with good management and in the absence of any severe disease pressure, sweet orange trees at 18 years old and on US-802 rootstock (very vigorous) were about 6.1 m tall, on Swingle (standard) were about 4.5 m tall, and on US-897 (semidwarf to dwarfing) were about 2.7 m tall (Fig. 6.4). Some rootstocks are particularly responsive to extremes of fertilization, so that rootstock effect on tree size may

FIG. 6.4 Hamlin sweet orange on two rootstocks with a large difference in effect on tree size. (A) Hamlin/US-802 rootstock at 6.1 m tall; (B) Hamlin/US-897 rootstock at 2.7 m tall. Trees with both rootstocks are 18 years old, and representative of the size for all trees on that rootstock in the trial. The field trial is located in Osceola County, Florida, and did not have prominent disease effects at the time the photos were taken.

be altered significantly by management practices. The rootstock effect on tree size may also be modulated by disease. For example, some strains of CTV may severely reduce the size of sweet orange trees on sour orange, as compared with trees on other rootstocks more tolerant of the infection (Wutscher and Bowman, 1999), and Huanglongbing has been noted to reduce the size of trees with sweet orange scion on all rootstocks (Bowman et al., 2016b). Some viroids have been proposed for use as size-controlling agents for citrus (Hutton et al., 2000; Vidalakis et al., 2011). It should be noted that the rootstocks which promote the highest yield per tree will often not provide the highest yield per hectare if tree densities are not adjusted to optimize yield per area. In addition, optimum spacing to maximize yields per hectare will be different if based on tree size and yield at 5 years or at 15 years.

Yield and fruit quality. Numerous studies have identified large rootstock effects on fruit yield and quality (Barry et al., 2004; Bowman et al., 2016a,b; Castle et al., 2011; Wutscher and Bistline, 1988; Wutscher and Bowman, 1999). Effects of rootstock on yield are strongly associated with rootstock adaptation to growing conditions, and tolerance to any biotic diseases or abiotic challenges that are present. In the field, there are often many different factors interacting in a complex fashion to determine rootstock effects on yield. In the absence of strong biotic or abiotic factors that limit tree health, trees that grow the strongest may often be those that yield the most fruit on a per tree basis. There is a strong association between tree size and yield per tree, so it is important to consider tree size in assessing potential for good yield per unit area; however, as mentioned above, yield efficiency and planting density should be optimized for maximum yield potential per unit area. Fruit quality is a combination of many different fruit traits, including size, shape, external and internal color, Brix, and acid, that may have differing importance, depending on the scion cultivar and the commercial use (Wutscher, 1988). For sweet orange used in the juice trade, Brix is probably the most important fruit quality trait. But with other scions, like grapefruit and mandarins, fruit size, shape, flavor, and color can be equally important (McCollum and Bowman, 2017). Generally, *P. trifoliata* and many of its hybrids, as well as sour orange, are considered to have strong favorable influences on fruit quality, while the effects of rough lemon and Rangpur are less favorable.

Compatibility. Although most citrus rootstocks are graft compatible with most citrus scions, there are exceptions. For example, incompatibilities have been observed with Roble, Pera, and Shamouti sweet oranges, Murcott tangor, and Eureka lemon with Swingle and some other trifoliate hybrid rootstocks (Garnsey et al., 2001; Weathers et al., 1955), and sometimes the incompatibility takes many years to become apparent. Consequently, caution should be used in choosing new scion-rootstock combinations until a longer history of use is developed. Citrus rootstock and scion cultivars vary in relative growth rate and metabolism, and this often results in different trunk diameter and bark texture which is apparent at the graft union (Fig. 6.5). The difference in trunk diameter is sometimes termed overgrowth or undergrowth. In particular, overgrowth of rootstocks with *P. trifoliata* parentage is often observed when they are grafted with other citrus species or cultivars as scions. It can be noted that the relative growth difference of two cultivars is often similar, regardless of which is in the scion and rootstock position. For example, US-802 will grow to a much larger trunk diameter than Valencia grafted to it, whether it is Valencia scion on US-802 rootstock, or US-802 scion on Valencia rootstock. In extreme cases, overgrowth of the rootstock can result in a weakened graft union or other physiological problems. However, the moderate rootstock overgrowth typically seen with sweet orange on Carrizo or Swingle normally does not seem to present any hazard for trunk strength or tree physiology.

Nutritional utilization and deficiency. Rootstocks can differ markedly in the ability to uptake and use nutrients from the soil. Focused studies have been conducted on several nutrients, including iron (Albano et al., 2005, 2013; Martinez-Cuenca et al., 2017) and boron (Zhou et al., 2014). Differences in nutrient use efficiency and toxicity may have important effects on rootstock use and performance under differing climatic, soil, and management conditions, but are poorly understood (Lea-Cox and Syvertsen, 1993; Zambrosi et al., 2011).

Nursery characteristics. Propagation of citrus rootstocks in the nursery is traditionally accomplished using genetically uniform nucellar seedlings (Anderson et al., 1991; Kepiro and Roose, 2007; Moore and Castle, 1988). The most widely used rootstocks usually produce a high proportion of uniform seedlings, and consequently, seedling uniformity has generally been an important selective factor in efforts to develop promising new rootstocks (Bowman et al., 2016a; Albrecht et al., 2017). Notably, a few rootstocks with a relatively high frequency of off-types from seed have been used commercially (Bowman et al., 1995; Castle et al., 1993; Stover and Castle, 2002), but up until now, none of these have become of wide importance. Over the past 20 years, propagation of citrus rootstocks by cuttings has become common in research efforts (Bowman and Albrecht, 2017; Bowman and McCollum, 2015) and used somewhat in commercial nurseries. Micropropagation of citrus rootstocks has also become common in some regions, to make up for seed deficiencies (Albrecht et al., 2017), and provides the potential to allow mass propagation of new rootstock clones lacking nucellar seed, which would previously have been impractical (Bowman, 1995). In Florida, about 600,000 nursery trees were produced on micropropagated US-942 rootstock between July 2018 and June 2019 (Fig. 6.6).

FIG. 6.5 Trees illustrating different types of natural rootstock overgrowth and undergrowth relative to the scion. (A) Minneola scion/sour orange rootstock; (B) Minneola scion/Swingle rootstock; (C) Valencia scion/Swingle rootstock; (D) Clementina Fina scion/US-897 rootstock; (E) US-897 scion/US-802 rootstock; (F) = US-802 scion/Valencia interstock/sour orange rootstock. All trees are 13–14 years old and canopies appear healthy.

FIG. 6.6 Nursery plants of US-942 rootstock produced by micropropagation. *(Photo courtesy of Mireia Bordas, Agromillora.)*

6.4 Rootstock use by region

Rootstock use varies greatly from one country to another, and sometimes differs dramatically between production regions in the same country. Most citrus production regions do not have good data available to quantify the number of trees propagated on different rootstocks. In addition, there are sometimes rather large shifts in rootstocks used for new propagations over a period of 5 years. Consequently, we have summarized the citrus rootstock usage in new plantings based (for most countries) on the authors' professional opinions, and the insight of regional citrus experts who were willing to provide their estimates in response to an email survey. Rootstocks were only identified in Tables 6.1 and 6.2, when there was data or expert opinion indicating the rootstock was used for at least 5% of new propagations in at least one country (or large region) during the past 5 years. In the following table, we have identified the frequency of use in new propagations, from 2012 to 2017, for the major rootstocks in each of the countries for which good information was available. When there were data or expert opinions that assigned a percentage by rootstock, those numeric values were included in the table. When data were not available and the best expert opinion could not assign a percentage value, rootstocks were ranked by frequency of use, with A = the rootstock that was used for the most propagations. The authors apologize for countries not included, and accept the likelihood of some inaccuracies in the table, but believe that it is broadly a reasonable estimation of worldwide use.

There were 21 rootstocks that were determined to be used for at least 5% of propagations in at least one citrus-producing country. Some major rootstocks, such as Carrizo and sour orange, were used in many countries, while other important rootstocks like Ziyang Xiangcheng are significantly used in only one country. It is apparent that rootstock use in an area is determined by many factors, including tradition, availability of reliable information about the performance of different rootstocks, and the availability of seed. In many areas, probably better rootstocks exist than those currently being used, but there is not enough clear evidence of which rootstocks would be the best or plant material is limited; so, those rootstocks most familiar remain the most used.

6.5 The major rootstocks

Twenty-one rootstocks were identified as major world rootstocks because of their use for at least 5% of recent propagations in at least one significant citrus-producing area. Brief descriptions of these rootstocks are given below, along with references for further information. In some cases, the major rootstocks identified in our list are, in fact, groups of closely related rootstocks rather than individual clones. One example of this is trifoliate orange, which includes many different cultivars (such as Flying Dragon and Rubidoux), some of which may differ in one or more attributes. In other cases, there is controversy about whether named clones that seem to have derived from the same source are, in fact, genetically the same or different. The two similar cultivars, Carrizo and Troyer, are an example of this, and for the purposes of this chapter, we have combined them together. Exploring the diversity of genotypes in these cases was beyond the scope of this chapter, but some references are provided to guide further study. Additional rootstocks that did not meet the criteria we have used for selection as a "major" rootstock now are mentioned in the following section, if we felt they were noteworthy and likely to increase in use during the coming years.

Benton is a citrange (the term applied to hybrids of *C. sinensis* × *P. trifoliata*) rootstock bred in New South Wales, Australia, in an attempt to produce dwarfing rootstocks compatible with Eureka lemon. This rootstock is a hybrid of Ruby Blood orange and trifoliate that has performed well in trials, especially in Australia. Benton has been successfully used for commercial Eureka plantings in Australia since 1990. Benton has also been used successfully as a rootstock for grapefruit, Valencia, navel orange, and mandarins. In South Africa, trees on Benton are more compact than trees on rough lemon, Swingle and Carrizo, resulting in more manageable trees, although trees on Benton have been very large in other areas. When propagating Benton in the nursery, the tree has a "bushy" growth habit with a shrub-like appearance. Consequently, to prepare the rootstock for the budding process, nurserymen plant seeds relatively close together and select the strongest, most upright seedlings. For Valencia types, Benton produces very good quality fruit, with high internal quality early in the season. Medium-to-larger fruit size remains one of the more favorable traits of this rootstock. Midknight Valencia trials were planted on Benton in the Letsitele area of South Africa and produced an average crop of over 80 kg per tree (Lee et al., 2009). The main advantages of Benton are intermediate resistance to *Phytophthora* root rot, nematodes and CTV, good internal fruit quality, and good external color development compared with Swingle. Australia is the only country where Benton is commercially used to a significant extent.

C35 is a semidwarfing rootstock developed by the University of California and is a cross between Ruby sweet orange and *P. trifoliata* (Cameron and Soost, 1986). C35 is tolerant to *Phytophthora*, citrus nematode, and CTV, but sensitive to the citrus exocortis viroid. In South Africa, this rootstock has shown to be sensitive to citrus blight in trials conducted in the Letsitele region (Lee et al., 2009). The growth of trees on C35 is similar to that of trees on Carrizo up to about 6 years

TABLE 6.1 Rootstock use for new plantings in Americas 2012–17.

Country	Benton	Carrizo	Cleopatra	C35	F-A 5	Kuharske	Macrophylla	MXT	*P. trifoliata*	Rangpur	Rough lemon	Sour orange	Sunki	Swingle	US-802	US-812	US-897	US-942	Volkamer	X639	Ziyang X.
USA, Calif.		30	4	24			10		10		4	1		7					9	1	
USA, Fla.			5			18						12		16	8	5	7	7	2	12	
Mexico		E		F			D		G			A		C					B		
Domin. Rep.				B					E	C				D					A		
Costa Rica		15	2	1					20	1				50					10		
Nicaragua		10							30					50					10		
Honduras		10		1					4		1	62		14					7	1	
Panamá		10	2						15					65					8		
Brazil, Parana		1	7							80			5	7							
Brazil, Sao P.			2						1	80			7	10							
Argentina	G	C	F	D					A			E		B							
Colombia		B	C	F					E					A					D		

Estimated percent of trees, or rank for number of trees: A=#1 most used, B=#2, C=#3, D=#4, E=#5, F=#6, G=#7.

TABLE 6.2 Rootstock use for new plantings in Europe, Asia, and Africa 2012–17.

Country	Benton	Carrizo	Cleopatra	C35	F-A 5	Kuharske	Macrophylla	MXT	*P. trifoliata*	Rangpur	Rough lemon	Sour orange	Sunki	Swingle	US-802	US-812	US-897	US-942	Volkamer	X639	Ziyang X.
Spain		21	1	7	29		25					1		15					1		
Italy		60		10	2		5		3			5		10					5		
Greece		C							D			A		B					E		
Turkey		16		25								48							11		
Morocco		B		D			A					E							C		
Tunisia		8		4								80							8		
Egypt		2					2					35							60		
Saudi Arabia			C									A							B		
Israel		25										40				D			30		
Iran		15		8			1		2			63		3					7		
India										B	A										
Pakistan											97	3									
China									80	C		D									10
Japan									99												
Indonesia										90	5										
Thailand		B												A							
Australia	12	34	2	10			1		34					4					1		
New Zealand	2		15						82			1									
South Africa		47		10				5			15			14					1	7	

Estimated percent of trees, or rank for number of trees: A=#1 most used, B=#2, C=#3, D=#4, E=#5, F=#6, G=#7.

of age, at which stage growth is slower with the eventual mature tree size being 15%–25% smaller (mature trees 10 years old). C35 performs well in soils suitable for Carrizo, but it does not adapt well to saline and high pH soils. When planted on very sandy or high clay content soils, C35 performs below average with small and stunted tree size. C35 produces very good quality fruit with high internal quality early in the season and larger fruit size remains one of the better qualities of this rootstock. C35 has performed well in combination with grapefruit, navel orange, Valencia, and mandarin types. There have been reports of tree dieback and incompatibility-like symptoms with Nules Clementine and Fukumoto Navel on C35 (Lee et al., 2009). The main positive attributes for C35 rootstock are intermediate resistance to *Phytophthora* root rot, tolerance to nematodes and CTV, good internal quality fruit, earlier external color compared with Swingle citrumelo, and small-to-medium tree size. C35 is the second-most popular rootstock for new propagations in California, and is widely used to a lesser extent in many other growing regions.

Carrizo (and Troyer) is a hybrid of Washington navel orange and *P. trifoliata*, originating from a cross made at Riverside California in 1909, within the USDA citrus breeding program. Carrizo is often considered identical with Troyer, although separate clones are maintained in many locations, and this is still somewhat controversial (McCarty et al., 1974; Savage and Gardner, 1965). Characteristics of Carrizo and Troyer are very similar; so, for our purposes here, we are considering them the same. With most scions, Carrizo provides excellent early bearing of young trees, internal fruit quality is excellent, and fruit size is usually medium to large. Carrizo is vigorous, and develops medium-to-large tree size (depending on the scion cultivar), or what is often considered "standard" tree size. Carrizo is sensitive to alkaline-induced chlorosis and, in some cases, trees have severely declined as a result of iron and other trace element deficiencies on saline, calcareous, and especially high pH soils. Another major disadvantage of this rootstock in some areas is the tendency to induce a higher incidence of creasing. Carrizo induces higher acid levels than rough lemon, but while this has occasionally created problems with Valencia, Minneola tangelo, and grapefruit in the cooler production areas, it is advantageous in hot areas (Lee et al., 2009). In the absence of limiting disease problems, Carrizo is an excellent rootstock for sweet orange, grapefruit, and most mandarin hybrids. The main advantages of Carrizo citrange are intermediate resistance to *Phytophthora* root rot, nematode and CTV tolerance, good internal quality, earlier external color compared with Swingle citrumelo, and medium tree size. Carrizo is not compatible with Eureka and other Eureka-type lemon selections. Carrizo remains the most-used rootstock in Spain, South Africa, Italy, Australia, and California, and is widely planted in many other countries.

Cleopatra is a small-fruited mandarin that has been often identified as *Citrus reshni*, and is thought to have originated in India where it is known as chota or billi kichili (Hodgson, 1967). The overall field performance of trees on Cleopatra is good in most regions, but trees on this rootstock typically have low yield until the trees are at least 8–10 years old. Cleopatra usually produces a standard to very large tree size with most scions. Fruit of the grafted scion is of high quality, with high sugar content, but sometimes fruit size is somewhat small. Cleopatra induces good cold tolerance and drought tolerance in the scion. Trees are unaffected by CTV and some viroids, as well as having tolerance to some species of *Phytophthora* root rot. It is susceptible to citrus nematode. Cleopatra has been described as having the highest salinity tolerance of all the commercial rootstocks through having the ability to exclude sodium and chloride from being taken up by the root system (Lee et al., 2009). Cleopatra performs well on deep and loamy, well-drained soils but is adversely affected by soils with high water tables. In trials conducted in Florida to compare various rootstocks for blight tolerance, Cleopatra mandarin only began to show decline symptoms after about 12 years. Many years of experience in Florida has shown it to be an excellent rootstock for all mandarin cultivars that are sufficiently large fruited. Cleopatra is used widely as a rootstock in several citrus-production regions, but there are no countries where it is the dominant rootstock. Numerous similar mandarins are sometimes used as rootstocks, and some may have potential advantages over Cleopatra (Castle, 1987; Castle et al., 2004, 2011; Sykes, 2011a,b).

Forner-Alcaide 5 (F-A 5) was the product of a 1978 cross between Cleopatra mandarin and *P. trifoliata* Rubidoux completed in the Instituto Valenciano de Investigaciones Agrarias (IVIA) in Valencia, Spain. Testing in Spain showed that F-A 5 is resistant to CTV, citrus nematode, and *Phytophthora*, and has tolerance to salinity, flooding, and lime-induced chlorosis (Forner et al., 2003; Forner-Giner et al., 2003, 2010, 2011). In trials with Navelina orange and Okitsu satsuma, trees on F-A 5 produced good fruit quality, and higher fruit yields than trees on Carrizo. The rootstock F-A 5 produced a tree that was smaller in size than trees on Carrizo rootstock. F-A 5 has become an important rootstock in Spain, and has begun to gain interest from growers in some of the neighboring countries.

Kuharske was a natural seedling variant of Carrizo identified in a Groveland (Florida) citrus nursery in about 1986, as notably resistant to a locally important problem, burrowing nematode (*Radopholus similis*). Field performance of this rootstock in trials over the past 20 years has generally indicated relatively good influence on tree health and fruit quality, but only average in effects on fruit yield (Bowman et al., 2016b; Castle et al., 2011). Kuharske has been noted to be widely used in Florida because it is similar to Carrizo citrange, but otherwise is exceptional only in having very good tolerance to the burrowing nematode (Castle, 2010). Kuharske is currently of commercial significance only in Florida, where it has been the most widely used rootstock for new propagations over the past 5 years.

Macrophylla, also known as Alemow, was thought to be a hybrid of citron (*Citrus medica*) and pummelo (*Citrus maxima*) originating on the island of Cebu, Philippine Islands (Saunt, 2000), but a recent study revealed that it resulted from an hybridization between *Citrus micrantha* (or a close papeda species) and citron (Curk et al., 2016). Lemon performs well on this rootstock and trees are precocious and produce good-quality fruit, although they are more sensitive to the mal secco disease (*Plenodomus tracheiphilus*). Trees on this rootstock grow vigorously for a few years, with growth thereafter being considerably less vigorous due to heavy fruit loads. Fruit on the trees are large, but the fruit on young trees generally have a low juice and brix content, and tend to dry out early. In rootstock trials, trees on *Macrophylla* typically have the lowest juice quality. Trees on *Macrophylla* are very susceptible to cold damage, CTV (when used with other scions besides lemon), xyloporosis, citrus blight, and citrus blackspot (Castle et al., 1993). *Macrophylla* is resistant to root rot as well as being tolerant to salt and boron. The insensitivity of lemons as a scion to CTV makes this a rootstock option for lemons under South African conditions, and it also appears one of the best options for Tahiti lime in South Florida. *Macrophylla* has been extensively used as a lemon rootstock in California and, in recent years, has become popular in Spain and Morocco, where water with high salinity levels is used for irrigation (Lee et al., 2009).

Minneola × Trifoliate (M×T) is the product of a cross between the parents indicated, within the USDA-Florida breeding program. Although it is not known commercially in Florida, field performance in Florida has been reported as HRS/US-934 (Wutscher and Bowman, 1999). At present, this rootstock is only of commercial significance in South Africa, where it was introduced from Florida in 1992. The rootstock (known as M×T in South Africa) performs very vigorously in the nursery, but once the scion is budded, it induces some size-controlling effect in the scion. In two trials, one in the Eastern Cape and one in Limpopo Province (South Africa), in combination with Palmer navel as the scion, tree volumes were about 55% of trees budded to rough lemon or Troyer citrange. Excellent per tree yields and tree productivity (expressed as yield/unit tree size) were obtained with M×T, significantly higher than that obtained for the standard rootstocks (Lee et al., 2009). The tree size of Clementine on M×T in the Eastern Cape (South Africa) has been slightly smaller, although quality and production have been similar to Swingle and Carrizo. Several scion cultivars budded to this rootstock have performed outstandingly in a planting at Malelane (Lee et al., 2009). No incompatibility problems have been identified with Eureka lemon on M×T, but severe incompatibility signs were visible in combination with kumquat (*Fortunella* spp.). The adaptation of this rootstock to various soil types is not well known, but early indications are that it does not adapt well to soils with a high pH, and MxT is sensitive to some citrus viroids. Trials done to date indicate M×T has good tolerance to CTV, *Phytophthora*, and nematodes (Lee et al., 2009).

Poncirus trifoliata (L.) Raf. (trifoliate orange) is regarded as a separate genus from citrus, and has been grown in China for thousands of years (Swingle and Reece, 1967). Many different selections of this species exist, and have been often categorized either as large-flowered types which grow upright and develop fewer branches (e.g., Pomeroy), or small-flowered types which are less vigorous and with more branches (e.g., Rubidoux and Australian), although it is not clear that these morphological differences have a strong association with horticultural traits when used as a rootstock. Specific selections of this species have been described, which differ from many of the other selections of *P. trifoliata* in one or more important horticultural attributes (Lapointe and Bowman, 2002; Sykes, 2011a,b). In general, trees on trifoliate orange have a low to moderate vigor, but bear a good crop of fruit for the tree size. Fruit quality on trifoliate orange is excellent, generally with high Brix and acid contents, and developing a smooth, thin rind. The fruit size may be smaller compared with other rootstocks, but fruit has been noted to hang well on the trees. For the most common selections of trifoliate orange, they are noted to be resistant or tolerant to citrus nematode, some *Phytophthora* species, CTV, flooding, and cold, but sensitive to high pH, salinity, drought, and exocortis. Trifoliate orange rootstocks seem more favorably suited to areas with cool winter temperatures, and are often regarded as a good choice for cases where significant tree dwarfing is desired, especially the *P. trifoliata* cultivar Flying Dragon (Castle et al., 2005; Mademba-Sy et al., 2012). Trifoliate orange rootstocks may also have utility in some situations because of their particularly strong ability to improve internal fruit quality of the grafted scion. Trifoliate orange is currently heavily favored as a rootstock in Japan, China, Australia, New Zealand, and Argentina, and is significantly used in many other citrus-producing regions.

Rangpur (*Citrus limonia* Osbeck) is a natural selection that probably originated in India, and was previously thought to be a hybrid of acid mandarin (*Citrus reticulata*) with either the rough lemon or sour orange (Hodgson, 1967; Saunt, 2000); a recent study suggested an acidic mandarin as the female parent and citron as the male parent (Curk et al., 2016), and this has more recently been confirmed by genome sequence analysis (Wu et al., 2018; see Chapter 2). Trees on this rootstock are very vigorous and productive, similar to rough lemon. Trees typically bear fruit at an early stage and are tolerant to CTV, but are susceptible to citrus and burrowing nematodes, exocortis, *Phytophthora*, and citrus blight. Rangpur has good tolerance to drought as well as calcareous soils, and for these reasons has popularity in areas with possible drought conditions and lacking supplemental irrigation. Trees on Rangpur rootstock typically produce large-size fruit, with internal quality similar to rough lemon, and notably inferior to the internal fruit quality of fruit on many other rootstocks, such as Carrizo or Swingle. Rangpur is the major rootstock in Brazil and Indonesia, but is relatively little used in most other countries.

Rough lemon (*Citrus jambhiri* Lush) is presumed to be indigenous to north-east India, where it still grows wild (Lee et al., 2009); like Rangpur, it has been shown to be a hybrid of *C. reticulata* × *C. medica* (Wu et al., 2018; see Chapter 2). Many rough lemon selections are known, and clone choice differs between countries. In South Africa, where rough lemon is common and is currently the second-most-used rootstock, the selection that has been favored is Cairn. Rough lemon is one of the most vigorous rootstocks used with citrus, under favorable conditions producing a very large tree, and has seen widespread use around the world. Typically, grafted scions on rough lemon produce large crops and large fruit size. Other advantages of rough lemon are the resistance to high pH and drought, and good growth in deep sandy soils. One of the disadvantages of rough lemon is the internal quality of the fruit is lower in Brix and percent juice than other higher fruit quality-inducing rootstocks. Rough lemon is not tolerant of either citrus or burrowing nematodes, and is not sufficiently tolerant to *Phytophthora* root and foot rot to be used on wet, poorly drained soils, even though it is relatively tolerant to flooding. Trees on rough lemon are more sensitive to cold and frost damage, but young trees planted on rough lemon and damaged by cold recover more rapidly than trees on less vigorous rootstocks that are severely damaged. Rough lemon has additional negative effects in combination with mandarins, as the fruit tend to dry out (granulate), reducing the already naturally short harvest season of most mandarin cultivars, and also tends to increase the alternate-bearing tendency of many mandarin cultivars. Rough lemon is the most used rootstock for new plantings in India and Pakistan, and important in South Africa, but is currently relatively unimportant in most other countries.

Sour orange (*Citrus aurantium* L.) is an F1 hybrid between mandarin and pummelo (Wu et al., 2014; see Chapter 2). The origin of this natural hybrid is speculated as from northeastern India and nearby regions in neighboring countries (Hodgson, 1967), but it had spread to most other citrus-growing areas very early in recorded history. Sour orange is close to the ideal rootstock in many horticultural respects, inducing high yields and superb fruit quality, along with good tolerance to root rot, high pH, salinity, flooding, cold, and blight. However, in many areas and scion combinations, sour orange cannot be used without extreme risk, due to its susceptibility to a tree stunting or quick decline induced by infection of the scion with CTV. At some point in time, sour orange was the most important rootstock in almost every citrus growing region of the world. However, millions of trees of sweet orange, grapefruit, or mandarin were lost on sour orange in South America, Central America, the Caribbean, Florida, Spain, Italy, and South Africa after trees became infected with CTV. At present, virulent forms of CTV and the prevalence of the aphid vector have discouraged use of sour orange as a rootstock for new plantings in many regions. Scions that are less susceptible to CTV, such as lemons, can be successfully used in combination with sour orange. Sour orange is a heavily used rootstock for new propagations in Greece, Turkey, Tunisia, Saudi Arabia, Israel, Iran, Honduras, and Mexico, and sees significant use in many other regions. Many different regional clones of sour orange exist, although all seem to have similar sensitivity to CTV (Bowman and Garnsey, 2001). Efforts to create a sour orange-like rootstock without sensitivity to CTV have received considerable attention in Florida (Bowman, 2007; Bowman and McCollum, 2018a,b,c; Grosser et al., 2004).

Sunki (*Citrus sunki* hort ex. Tanaka) is small-fruited acidic mandarin that is considered native to China, and has some similarity to Cleopatra (Hodgson, 1967). Sunki was used historically as a rootstock in China, and has been included in rootstock trials in several other areas. Sunki is reported to be tolerant of CTV, xyloporosis, blight, and salinity, but is only moderately tolerant of high pH and cold, and susceptible to *Phytophthora* and exocortis. Fruit yield and fruit quality are probably similar to that of trees on Cleopatra (Castle, 1987; Castle et al., 1993, 2015; Wutscher, 1979). There is interest in replacing Rangpur with Sunki mandarin in Brazil because of good salinity and drought tolerance along with better blight tolerance (Girardi et al., 2017). Brazil is the only country where Sunki is significantly used for new tree propagations.

Swingle citrumelo is a hybrid of Duncan grapefruit (*Citrus paradisi*) and trifoliate orange (*P. trifoliata*) created in 1907 in Eustis (Florida) by the USDA pioneering citrus breeder for which it is named, Walter T. Swingle. The Swingle rootstock was officially released by USDA in 1974 (Hutchison, 1974; Wutscher, 1979). In the field, Swingle rootstock produces large, vigorous, and productive trees with fruit that is similar to or slightly larger than fruit on Carrizo, and good internal quality. It has been noted that trees on Swingle have an average of 3 weeks delayed rind color development and later fruit maturity as compared with Carrizo, due to higher acidity (Lee et al., 2009). Swingle has good tolerance to drought, cold, *P. nicotianae* (but not *P. palmivora*), citrus nematode, blight, and CTV. The rootstock is suitable for most soils except highly calcareous conditions, salinity, and heavy clay soils, with a clay content greater than 25%–30%, or heavy clay layers. Swingle shows trunk overgrowth with most scion cultivars, and this overgrowth has been considered a significant problem in some cases with mandarins and with early slow-growing navel sweet orange selections, when it can result in compression girdling (Garnsey et al., 2001; Lee et al., 2009). Swingle has been observed as not compatible with the scions Eureka lemon, Tomango and Shamouti oranges, and Murcott tangor. Swingle is the dominant rootstock for new propagations in Colombia, Panama, Nicaragua, Costa Rica, and Thailand, and is used for a significant portion of new propagations in Argentina and several other regions. In Florida, Swingle was used for nearly half of all propagations from 1988 to 2010, but since then, it has declined significantly because of better performance from other rootstocks showing broader soil adaptability, better cropping, and improved tolerance to Huanglongbing.

US-802 citrus rootstock was released by the US Department of Agriculture in 2007 (Bowman, 2007a), and has subsequently been widely used in Florida, especially with sweet orange scions. Florida Department of Agriculture Annual Budwood Reports puts the number of propagations on US-802 from 2011 to 2018 at 1,782,501 trees. This rootstock is from a cross of Siamese pummelo (*C. maxima*) × Gotha Road Trifoliate Orange (*P. trifoliata*) and was tested primarily by USDA with industry partners in Florida prior to release, but has subsequently been widely planted and evaluated. The rootstock induces strong vigor in grafted scions, similar to the effects of rough lemon, and in most cases trees will become large. With sweet orange scion in Florida, US-802 was the most productive rootstock in some trials (on a per tree basis) both before *C*Las, and afterwards as well when trees were affected by Huanglongbing. Other positive attributes of US-802 are resistance or tolerance to high pH, CTV, citrus nematode, *P. nicotianae* and *P. palmivora* foot and root rot, Diaprepes root weevil, and citrus blight. Most notable faults with this rootstock are very high vigor, and a corresponding tendency to induce lower Brix in the fruit of grafted scions. Field performance has been described in several publications (Bowman et al., 2003, 2016a,b; Castle et al., 2004; Wutscher and Hill, 1995). US-802 is virtually unknown in other areas outside Florida.

US-812 citrus rootstock was released by the US Department of Agriculture in 2001 (Bowman, 2001), and has subsequently been widely used in Florida. Florida Department of Agriculture Annual Budwood Reports puts the number of propagations on US-812 from 2011 to 2018 at 1,377,171 trees. This rootstock originated from a cross of Sunki mandarin × Benecke trifoliate orange (*P. trifoliata*) and was tested primarily by USDA with industry partners in Florida prior to release, but has subsequently been widely planted and evaluated. Prior to the introduction of Huanglongbing in Florida, US-812 was often among the most productive rootstocks with a range of scions in field trials (Bowman and Rouse, 2006; Castle et al., 2011; Wutscher and Bowman, 1999), and typically induced good fruit quality. Trees on US-812 are similar in vigor to trees on Carrizo and Swingle, and US-812 has good tolerance to CTV, *P. nicotianae*, citrus blight, and high pH. In trials strongly affected by Huanglongbing, field performance of trees on US-812 has generally been about average, but not outstanding (Bowman et al., 2016a,b; McCollum and Bowman, 2017). US-812 is used to a limited extent in other countries outside the US, including Israel and South Africa, where it is often referred to as Sunki-Benecke.

US-897 citrus rootstock was released by the US Department of Agriculture in 2007 (Bowman, 2007b), and has subsequently been widely used in Florida, especially with the goal of achieving tree size control for close set plantings and intensive management. Florida Department of Agriculture Annual Budwood Reports puts the number of propagations on US-897 from 2011 to 2018 at 1,646,569 trees. This rootstock originated from a cross of Cleopatra mandarin × Flying Dragon trifoliate orange (*P. trifoliata*) and was tested primarily by USDA with industry partners in Florida prior to release, but has subsequently been widely planted. Calculated on a per tree basis, fruit productivity of trees on US-897 is typically poor, but because of the small tree size, calculations of potential yield per hectare at predicted optimum spacing may be very high. US-897 usually induces good internal fruit quality on grafted scions, but fruit size may often be smaller than average. In addition to the tendency for US-897 to produce small or semidwarf trees, this rootstock was also noted to be tolerant of CTV, citrus nematodes, the PDC, and high pH. US-897 exhibits a susceptibility to citrus blight that is intermediate, as compared with other rootstock cultivars. Because of its effects to control tree size, US-897 has been a preferred rootstock in many new attempts at very high-density plantings, and plantings in protected structures (Ferrarezi et al., 2017). US-897 has reportedly been introduced to many other countries, but as of yet, commercial use is only of significance in Florida. Field performance has been described in several publications (Bowman et al., 2003, 2016a,b; Castle et al., 2004; McCollum and Bowman, 2017; Wutscher and Hill, 1995).

US-942 citrus rootstock was released by the US Department of Agriculture in 2010 (Bowman and McCollum, 2010), and since then has rapidly gained popularity in Florida. Records from Florida Department of Agriculture Annual Budwood Reports puts the number of propagations on US-942 from 2011 to 2018 at 1,784,729 trees. This rootstock originated from a cross of Sunki mandarin × Flying Dragon trifoliate orange (*P. trifoliata*) and was tested primarily by USDA with industry partners in Florida prior to release, but has subsequently been widely planted and evaluated. Field performance has been studied primarily with sweet orange scion, but trials have also indicated excellent performance with grapefruit, mandarin, pummelo, and Tahiti lime scions. The major positive attributes of US-942 are ease of uniform seed propagation, resistance or tolerance to CTV, citrus blight, *Phytophthora* root diseases, Diaprepes weevil, and high pH soils, as well as a semidwarfing effect on scion tree size, induction of good fruit quality, and induction of good fruit productivity per tree and per canopy volume. Performance has been outstanding in trials on both well-drained deep soils and heavy soils with a high water table. More recently, US-942 has been a top performer in field trials affected by Huanglongbing disease (Bowman et al., 2016a,b), and the nature of its Huanglongbing tolerance has been the subject of detailed greenhouse studies (Albrecht and Bowman, 2012; Albrecht et al., 2016; Bowman and Albrecht, 2015). Major faults are, as yet, unidentified. Although relatively unknown in other citrus production areas, US-942 will likely be the most used rootstock for new propagations in Florida during 2018–19, surging ahead of the previously most used rootstocks Kuharske, Swingle, sour orange, and X639. Characteristics and field performance of US-942 have been described in additional publications (Albrecht and Bowman, 2019; Bowman and Albrecht, 2017; Bowman et al., 2003; Castle et al., 2004, 2011; McCollum and Bowman, 2017; Wutscher and Bowman, 1999).

Volkamer (*Citrus volkameriana*) is likely a mandarin × citron hybrid, and of Italian origin. The Volkamer lemon was named in honor of J.C. Volkamer, the 18th-century German botanist (Saunt, 2000). Volkamer resembles rough lemon in many respects, including the induction of high vigor and poor internal fruit quality on grafted scions. In combination with lemon selections, Volkamer is considered an excellent choice in many areas, including southern Africa. Volkamer is susceptible to *Phytophthora* root rot and cold damage, although slightly less sensitive to both than rough lemon. In South Africa, Volkamer is considered sensitive to CTV and citrus nematodes (Lee et al., 2009), and it is susceptible to citrus blight in Florida. However, the rootstock performs well on a fairly wide range of soil conditions, including sandy, high pH, and areas affected by salinity. Volkamer is not susceptible to some citrus viroids. In some cases, performance of Volkamer is considered superior to rough lemon, but this is not uniformly the case. Volkamer is a major rootstock in Mexico, Dominican Republic, Egypt, and Israel, and has widespread use in many countries with lemon scions.

X639 rootstock arose from a cross between Cleopatra and *P. trifoliata* made at the CSFRI (now ARC-ITSC) South Africa, by Dr. Hojby in the early 1950s (Lee et al., 2009). The overall characteristics appear similar to those of Carrizo, but it may be more sensitive to *Phytophthora* root rot in replant conditions. X639 develops a fairly vigorous medium-sized tree in combination with most scion cultivars. In susceptibility tests for saline and calcareous conditions conducted in Australia, X639 was found to be similar to or slightly less susceptible than Swingle citrumelo (Lee et al., 2009). X639 has performed well in the higher pH soils of the Tshipise (Limpopo) and Lower Orange River (Northern Cape) areas in South Africa. Trees on this rootstock generally are productive, and the rootstock is considered to induce good internal fruit quality and good-sized fruit in South Africa (Lee et al., 2009), although some studies in Florida have indicated particularly low Brix for grapefruit on X639 (McCollum and Bowman, 2017). X639 is considered to have good promise in South Africa for Valencia, navel, grapefruit, and mandarin scions (Lee et al., 2009). Trials in Florida indicated that X639 behaved similarly to Cleopatra mandarin with regard to citrus blight tolerance, that is, trees were susceptible, but took about 12 years to show decline and die-back. The main advantages of X639 are *Phytophthora* root rot tolerance, nematode and CTV tolerance, good internal quality fruit (under some conditions), medium tree size, good performance on higher pH soils, good cold hardiness, and compatibility with Eureka lemon. X639 been used for a significant number of propagations over the last 5 years in both Florida and southern Africa.

Ziyang Xiangcheng (Fragrant citrus) is a local citrus rootstock that originated in southwest China. It is sometimes identified as *Citrus junos* (Tan et al., 2015), and has been suggested as a putative hybrid of *Citrus ichangensis* and *C. reticulata* (Liu et al., 2008; Zhou et al., 2014); in fact, it is a local selection of what is commonly known outside of China as Yuzu. This appears to be similar to, but not the same as Zhique or *Citrus wilsonii* Tanaka (*C. junos*) (Fu et al., 2016; Hancevic et al., 2013). Ziyang Xiangcheng has been described as having excellent tolerance to biotic and abiotic stresses, including high soil pH. It has been used in Sichuan and Chongqing regions of China, primarily in citrus orchards planted on wet and heavy paddy soils. Although not widely used, it is increasingly being used for new plantings of Orah mandarin trees in Guangxi and Yunnan provinces. Ziyang Xiangcheng is as yet untested in other citrus growing regions, and limited descriptive information is available.

6.6 Rootstock trends and future prospects

The common use of rootstocks in citriculture began in many countries during the mid-1800s (Castle, 2010), and initial rootstocks used included sour orange and other naturally occurring citrus types. A series of disasters affecting the citrus industries in different countries followed, often driven by epidemic spread of CTV strains that destroyed vast areas of trees on sour orange rootstock, and prompting a search for other, better rootstocks. During this phase, rough lemon, Cleopatra mandarin, Rangpur, trifoliate orange, and sweet orange enjoyed popularity for use as rootstocks in some areas, and guided breeding of citrus began in earnest within the USDA in Florida under the leadership of Walter T. Swingle. Although they took many years to reach commercial use, out of those early USDA efforts came the first hybrids from controlled crosses to see worldwide use as citrus rootstocks, Carrizo/Troyer citrange and Swingle citrumelo (Webber et al., 1967). During the 1960s and after, hybrid rootstocks began to increase in popularity through many citrus-producing regions, although the timetable and particular rootstocks involved varied by country. During the 1970s–80s, Carrizo and Swingle displaced sour orange and rough lemon as the dominant rootstocks in Florida (2004 Florida Department of Agriculture Citrus Budwood Annual Report). It can be noted that among the 21 major world rootstocks, 10 are naturally occurring species or hybrids, and 11 are F1 hybrids from breeding programs. Among all the F1 hybrids, *P. trifoliata* is one of the parents. None of the major rootstocks are more advanced hybrids, somatic hybrids, the product of induced mutation, or from any other type of genetic manipulation. As there are many efforts underway to create more complex hybrids and somatic hybrids to broadly improve performance, as well as genetically engineer citrus rootstocks to obtain improved tolerance to disease, it may be anticipated that the future of citrus rootstocks will include more diversity of germplasm and other types of genetic combinations. It should also be recognized that all of the 21 major rootstocks produce a high proportion of nucellar seedlings and are thus easily and uniformly propagated by seed. This ease of seed propagation has been a major selective factor for citrus rootstocks until now, but with the commercialization of citrus rootstock micropropagation, it may become of considerably less importance in the years to come. It is easy to see how a rootstock with outstanding field performance might become a major world rootstock in the future, even if it cannot be propagated by seed.

Numerous citrus rootstock breeding and testing programs exist, and in many cases, these programs have developed and released new or improved rootstocks. In all, 11 of the 21 rootstocks listed as major world rootstocks were the product of past citrus breeding efforts in the United States (7), South Africa (2), Australia (1), and Spain (1). Typically, new citrus rootstocks are only slowly adopted for widespread use because of the many years required to collect reliable field performance information, and the long life of a commercial planting. In the past, reliance on seed for rootstock propagation also slowed the adoption of new rootstocks to an area, and the production of seed often has greatly limited the use of new rootstocks even where they are desired. Sometimes vast regional differences in soils, pests and pathogens, scion cultivars, and management practices contributes to the long delay in determining suitability of new rootstocks for use in multiple regions. Nevertheless, the history of Carrizo and Swingle rootstocks has shown that new rootstocks developed in one region can eventually become widely used throughout citrus production regions of the world. There are many new rootstocks from regional programs that did not make the list of 21 major rootstocks now, but appear to have promising characteristics, and could become major rootstocks in the coming years. In the following is mentioned some of these new rootstocks, poised to possibly move into more significant roles for the future.

The University of Florida citrus breeding program in Lake Alfred, Florida, has developed a series of new rootstocks, many of them tetraploids, and the product of somatic hybridization or the subsequent sexual hybridization of two somatic hybrids (Grosser et al., 2011, 2015). UFR-4 rootstock [(Nova mandarin + Hirado Buntan pummelo seedling) × (Cleopatra mandarin + Argentine trifoliate orange)], UFR-5 rootstock [(Nova + Hirado Buntan pummelo seedling) × (Succari sweet orange + Argentine trifoliate orange)], and UFR-6 rootstock (Changsha mandarin + Trifoliate orange 50-7) are three examples that have evidenced good field performance and are seeing considerable commercial interest. These were reported by Florida Department of Agriculture as having been used for the propagation of 71,231 trees, 16,735 trees, and 28,617 trees, respectively, during the period July 2017 to June 2018. UFR-4 and UFR-5 performed well in the University of Florida two phase assay for PDC, and then have demonstrated good tolerance to PDC and HLB, along with good yield, through 5 years of field testing. UFR-6 is a cold-hardy somatic hybrid rootstock that produces small trees which bear fruit with high soluble solids, thus making it a good candidate for new production systems featuring high-density plantings. Additional new hybrids derived from Flying Dragon maternal parentage are showing promise for tree size control as well as tolerance of Huanglongbing and good yields.

The citrus rootstock breeding program at IVIA in Valencia, Spain, was responsible for development of F-A 5, one of the major world rootstocks from our list. Two other promising rootstocks from this program, Forner-Alcaide 517 and Forner-Alcaide 418, have more recently been described (Forner-Giner et al., 2014). As these two new rootstocks combine tree size control, high yield efficiency, and good fruit quality with good tolerance to high pH, traits of great current interest, it might be expected that at least one of these new rootstocks will also grow in popularity.

The USDA rootstock breeding program in Ft. Pierce, Florida, was responsible for the development of four rootstocks from our major rootstocks list, released between 2001 and 2010, and all having names that begin with "US." Over the past 15 years, the USDA rootstock breeding program has focused efforts on developing an improved sour orange-type rootstock, targeting a rootstock that combines all the best traits of sour orange with good tolerance to CTV and outstanding productivity. This genetically broad group of hybrids created and being evaluated under this effort includes hybrids of sour orange itself, as well as other hybrids that include the two parental species of sour orange, *C. reticulata* and *C. maxima*. The first of these "SuperSour" rootstocks were released in 2018, based on field performance considerably superior to both sour orange and Swingle in tree health and productivity (Bowman and McCollum, 2018a,b,c). The three SuperSour rootstocks released in 2018 are US SuperSour 1 rootstock = *C. maxima* × *C. reticulata,* US SuperSour 2 rootstock = *P. trifoliata* × (sour orange × *C. ichangensis*), and US SuperSour 3 rootstock = Cleopatra mandarin × (*C. maxima* × *P. trifoliata*). There has been considerable commercial interest in these new rootstocks.

The citrus breeding program at the University of California, Riverside, was responsible for testing and releasing of one of the major rootstocks on our list, C35. In addition, several other rootstocks from the University of California program, including C22, C54, C57, and C146 (all of Sunki × Swingle trifoliate orange parentage), show promise for some citrus production areas (Federici et al., 2009; Roose, 2007), especially in areas of high pH soils (Louzada et al., 2008), and seem likely to increase in use over the coming years.

Several other citrus-breeding programs have significant efforts directed toward development of improved rootstocks, including work in Brazil (Fadel et al., 2018), Australia (Smith, 2017; Sykes, 2011a,b), Italy (Reforgiato Recupero et al., 2009), and China (Guo and Deng, 2001; Guo et al., 2002; Liu and Deng, 2007). In addition, numerous programs have invested strongly in genetic transformation or other methods of genetic manipulation, that offer the opportunity to introduce major new types of disease resistance or other important traits to citrus rootstocks. Although most of these efforts are probably more than a decade away from a product that will become a major world rootstock, it seems likely these efforts will eventually yield new rootstocks with commercial impact.

Although creating the best rootstock through sexual hybridization, somatic hybridization, genetic transformation, or mutation is itself a huge challenge, it may be that the greatest difficulty in development of a new rootstock and widespread commercial use, is in the sufficiency and accuracy of field evaluation that can provide clear information about which rootstock is the best choice for each environment, scion, and production system. Probably many excellent rootstocks already exist, but only require thorough multiyear evaluation at multiple sites, to determine which few of the many hundreds of promising rootstocks are truly worthy of widespread commercial use in each area. Through the many efforts underway now, the field of rootstocks will evolve greatly over the coming 10 years, and should provide growers in every citrus region better facts to inform which of the many new and existing rootstocks provide the best opportunity to improve production and profitability.

Acknowledgments

The authors are grateful to the following individuals for information provided to help in estimating rootstock usage by country: Najar Asma, Marzieh Shirdast Bahnamiri, Marco Caruso, Berken Cimen, Xiuxin Deng, Juan Francisco Delgado Fernández, Walter dos Santos Soares Filho, Maria Angeles Forner-Giner, Paul Fourie, Magripis Gerasimos, Francisco J. de la Osa González, Snoussi Hajer, Nathan Hancock, Casimiro Humberto, Yaser Ibrahim, Muhammad Jafar Jaskani, Jorgino Pompeu Junior, Simon Jurisich, Tahir Khurshid, Hussein Marei, Jackie Maxwell, Emilio Arévalo Peñaranda, Ron Porat, Teófilo Ramírez, Jonathan Roberts, Mikeal Roose, Giuseppe Russo, Takehiko Shimada, I.P. Sing, Beatriz Stein, Zuleide Hissano Tazima, Agorastou Theodora, Sri Widyaningsih, Masashi Yamamoto, and Shiping Zhu.

References

Albano, J.P., Bausher, M.G., Evens, T.J., Bowman, K.D., 2005. Dynamics of nutrient uptake during root and shoot growth in three diverse citrus rootstock cultivars. Fla. Sci. 68, 7–8.

Albano, J.P., Bowman, K.D., Wilson, P.C., 2013. Differential response of citrus rootstocks to CuEDTA concentration in sand culture. J. Appl. Hortic. 15 (2), 81–86.

Albrecht, U., Bowman, K.D., 2004. Growth comparison of citrus roostocks after artificial infection with Phytophthora. Proc. Fla. State Horticult. Soc. 117, 156–160.

Albrecht, U., Bowman, K.D., 2011. Tolerance of the trifoliate citrus hybrid US-897 (*Citrus reticulata* Blanco x *Poncirus trifoliata* L. Raf.) to huanglongbing. HortScience 46 (1), 16–22.

Albrecht, U., Bowman, K.D., 2012. Tolerance of trifoliate citrus rootstock hybrids to Candidatus Liberibacter asiaticus. Sci. Hortic. 147, 71–80.

Albrecht, U., McCollum, G., Bowman, K.D., 2012. Influence of rootstock variety on huanglongbing disease development in field-grown sweet orange (*Citrus sinensis* [L.]) Osbeck trees. Sci. Hortic. 138, 210–220.

Albrecht, U., Fiehn, O., Bowman, K.D., 2016. Metabolic variations in different citrus rootstock cultivars associated with different responses to Huanglongbing. Plant Physiol. Biochem. 107, 33–44.

Albrecht, U., Bordas, M., Lamb, B., Meyering, B., Bowman, K.D., 2017. Influence of propagation method on root architecture and other traits of young citrus rootstock plants. HortScience 52, 1569–1576.

Albrecht, U., Bowman, K.D., 2019. Reciprocal influences of rootstock and scion citrus cultivars challenged with *Ca. Liberibacter asiaticus*. Sci. Hortic. 254, 133–142.

Al-Jaleel, A., Zekri, M., Hammam, Y., 2005. Yield, fruit quality, and tree health of 'Allen Eureka' lemon on seven rootstocks in Saudi Arabia. Sci. Hortic. 105, 457–465.

Anderson, C.M., Castle, W.S., Moore, G.A., 1991. Isozymic identification of zygotic seedlings in Swingle citrumelo *Citrus paradisi* × *Poncirus trifoliata* nursery and field populations. J. Am. Soc. Hortic. Sci. 116, 322–326.

Barry, G.H., Castle, W.S., Davies, F.S., 2004. Soluble solids accumulation in 'Valencia' sweet orange as related to rootstock selection and fruit size. J. Am. Soc. Hortic. Sci. 129, 594–598.

Ben Yahmed, J., Costantino, G., Amiel, P., Talon, M., Ollitrault, P., Morillon, R., Luro, F., 2016. Diversity in the trifoliate orange taxon reveals two main genetic groups marked by specific morphological traits and water deficit tolerance properties. J. Agric. Sci. 154 (3), 495–514.

Bhusal, R.C., Mizutani, F., Rutto, K.L., 2002. Selection of rootstocks for flooding and drought tolerance in *Citrus* species. Pak. J. Biol. Sci. 5 (5), 509–512.

Bitters, W.P., 1986. Citrus Rootstocks: Their Characters and Reactions. Unpublished manuscript. Available 7/5/2018 at http://www.citrusvariety.ucr.edu/links/documents/Bitters.pdf.

Boava, L.P., Cristofani-Yaly, M., Mafra, V.S., Kubo, K., Kishi, L.T., Takita, M.A., Ribeiro-Alves, M., Machado, M.A., 2011. Global gene expression of Poncirus trifoliata, *Citrus* sunki and their hybrids under infection of Phytophthora parasitica. BMC Genomics 12, 39. https://doi.org/10.1186/1471-2164-12-39.

Bové, J.M., 2006. Huanglongbing: a destructive, newly-emerging, century-old disease of citrus. J. Plant Pathol. 88, 7–37.

Bowman, K.D., 1995. Micropropagation of smooth flat Seville and Yuma citrus rootstocks. Proc. Fla. State Hort. Soc. 107, 15–18.

Bowman, K.D., 2001. Notice to Fruit Growers and Nurserymen Relative to the Naming and Release of the US-812 Citrus Rootstock. U.S. Department of Agriculture, Beltsville, MD.

Bowman, K.D., 2007. Raising rootstocks. Florida Grower 100 (4), 24.

Bowman, K.D., 2007a. Notice to Fruit Growers and Nurserymen Relative to the Naming and Release of the US-802 Citrus Rootstock. U.S. Department of Agriculture, Beltsville, MD.

Bowman, K.D., 2007b. Notice to Fruit Growers and Nurserymen Relative to the Naming and Release of the US-897 Citrus Rootstock. U.S. Department of Agriculture, Beltsville, MD.

Bowman, K.D., Albrecht, U., 2015. Comparison of gene expression changes in susceptible, tolerant, and resistant hosts in response to infection with *Citrus* tristeza virus and huanglongbing. J. Cit. Pathol. . iocv_journalcitruspathology_30450.

Bowman, K.D., Albrecht, U., 2017. Efficient propagation of citrus rootstocks by stem cuttings. Sci. Hortic. 225, 681–688.

Bowman, K.D., Garnsey, S.M., 2001. A comparison of five sour orange rootstocks and their response to *Citrus* tristeza virus. Proc. Fla. State Hort. Soc. 114, 73–77.

Bowman, K.D., McCollum, G.M., 2010. Notice to Fruit Growers and Nurserymen Relative to the Naming and Release of the US-942 Citrus Rootstock. U.S. Department of Agriculture, Beltsville, MD.

Bowman, K.D., McCollum, G., 2015. Five new citrus rootstocks with improved tolerance to huanglongbing. HortScience 50, 1731–1734.

Bowman, K.D., McCollum, G.M., 2018a. Notice to Fruit Growers and Nurserymen Relative to the Naming and Release of the US SuperSour 1 Citrus Rootstock. U.S. Department of Agriculture, Beltsville, MD.

Bowman, K.D., McCollum, G.M., 2018b. Notice to Fruit Growers and Nurserymen Relative to the Naming and Release of the US SuperSour 2 Citrus Rootstock. U.S. Department of Agriculture, Beltsville, MD.

Bowman, K.D., McCollum, G.M., 2018c. Notice to Fruit Growers and Nurserymen Relative to the Naming and Release of the US SuperSour 3 Citrus Rootstock. U.S. Department of Agriculture, Beltsville, MD.

Bowman, K.D., Rouse, R.E., 2006. US-812 Citrus rootstock. HortScience 41 (3), 832–836.

Bowman, K.D., Gmitter Jr., F.G., Hu, X., 1995. Relationships of seed size and shape, with polyembryony and the zygotic or nucellar origin of *Citrus* spp. seedlings. HortScience 30 (6), 1279–1282.

Bowman, K.D., Shapiro, J.P., Lapointe, S.L., 2001. Sources of resistance to Diaprepes weevil in subfamily Aurantiodeae, Rutaceae. HortScience 36 (2), 332–336.

Bowman, K.D., Albano, J.P., Graham, J.H., 2002. Greenhouse testing of rootstocks for resistance to Phytophthora species in flatwoods soil. Proc. Fla. State Horticult. Soc. 115, 10–13.

Bowman, K.D., Graham, J.H., Adair Jr., R.C., 2003. Young tree growth in a flatwoods rootstock trial with Diaprepes weevil and Phytophthora diseases. Proc. Fla. State Horticult. Soc. 116, 249–251.

Bowman, K.D., Albrecht, U., Graham, J.H., Bright, D.B., 2007. Detection of Phytophthora nicotianae and P. palmivora in citrus roots using PCR-RFLP in comparison with other methods. Eur. J. Plant Pathol. 119 (2), 143–158.

Bowman, K.D., Faulkner, L., Kesinger, M., 2016a. New citrus rootstocks released by USDA 2001-2010: field performance and nursery characteristics. HortScience 51, 1208–1214.

Bowman, K.D., McCollum, G., Albrecht, U., 2016b. Performance of 'Valencia' orange (*Citrus sinensis* [L.] Osbeck) on 17 rootstocks in a trial severely affected by huanglongbing. Sci. Hortic. 201, 355–361.

Cameron, J.W., Soost, R.K., 1986. C35 and C32: citrange rootstocks for citrus. HortScience 21, 157–158.

Carpenter, J.B., Furr, J.R., 1962. Evaluation of tolerance to root rot caused by Phytophthora parasitica in seedlings of citrus and related genera. Phytopathology 52, 1277–1285.

Castle, W.S., 1987. Citrus rootstocks. In: Rom, R.C., Carlson, R.F. (Eds.), Rootstocks for Fruit Crops. John Wiley and Sons, New York, NY, pp. 361–399.

Castle, W.S., 2010. A career perspective on citrus rootstocks, their development and commercialization. HortScience 45, 11–15.

Castle, W.S., Manthey, J.A., 1998. Screening citrus rootstocks for iron-deficiency tolerance. Fruits 53, 375–381.

Castle, W.S., Tucker, D.P.H., Krezdorn, A.H., Youtsey, C.O., 1993. Rootstocks for Florida Citrus, second ed. University of Florida, Gainesville.

Castle, W.S., Schnell, R.J., Crane, J.H., Grosser, J.W., Gmitter, F.G., Ayala, S.T., Bowman, K.D., 2004. Evaluation of new citrus rootstocks for 'Tahiti' lime production in southern Florida. Proc. Fla. State Horticult. Soc. 117, 174–181.

Castle, W.S., Baldwin, J.C., Muraro, R.P., 2005. 'Hamlin' orange trees on flying dragon trifoliate orange, Changsha mandarin, or Koethen sweet orange × Rubidoux trifoliateorange citrange rootstock at three in-row spacings in a Flatwoods site. Proc. Fla. State Hort. Soc. 120, 92–96.

Castle, W.S., Nunnallee, J., Manthey, J.A., 2009. Screening citrus rootstocks and related selections in soil and solution culture for tolerance to low-iron stress. HortScience 44, 638–645.

Castle, W.S., Bowman, K.D., Baldwin, J.C., Grosser, J.W., Gmitter Jr., F.G., 2011. Rootstocks affect tree growth, yield, and juice quality of 'Marsh' grapefruit. HortScience 46, 841–848.

Castle, W.S., Bowman, K.D., Grosser, J.W., Futch, S.H., Graham, J.H., 2015. Florida Citrus Rootstock Selection Guide, third ed. Extension Digital Information Source (EDIS). SP248/HS1260, University of Florida, Gainesville, FL. http://www.crec.ifas.ufl.edu/extension/citrus_rootstock/templates/guide/explore.html.

Curk, F., Ollitrault, F., Garcia-Lor, A., Luro, F., Navarro, L., Ollitrault, P., 2016. Phylogenetic origin of limes and lemons revealed by cytoplasmic and nuclear markers. Ann. Bot. 117 (4), 565–583.

Deng, Z., Huang, S., Xiao, S., Gmitter, F.G., 1997. Development and characterization of scar markers linked to the *Citrus* tristeza virus resistance gene from Poncirus trifoliata. Genome 40, 697–704.

Deng, Z., Huang, S., Ling, P., Yu, C., Tao, Q., Chen, C., Wendell, M., Zhang, H.B., Gmitter, F.G., 2001. Fine genetic mapping and BAC contig development for the *Citrus* tristeza virus resistance gene locus in Poncirus trifoliata (Raf.). Mol. Gen. Genet. 265, 739–747.

Derrick, K.S., Timmer, L.W., 2000. *Citrus* blight and other diseases of recalcitrant etiology. Annu. Rev. Phytopathol. 38, 181–205.

Duncan, L.W., Cohn, E., 2005. Nematode parasites of citrus. In: Luc, M., Sikora, R.A., Bridge, J. (Eds.), Plant Parasitic Nematodes in Subtropical and Tropical Agriculture. second ed.. CAB International.

Duncan, L.W., Inserra, R.N., O'Bannon, J.H., El-Morshedy, M.M., 1994. Reproduction of a Florida population of Tylenchulus semipenetrans on resistant citrus rootstocks. Plant Dis. 78, 1067–1071.

Dutra de Souza, J., de Andrade Silva, E.M., Coelho Filho, M.A., Morillon, R., Bonatto, D., Micheli, F., et al., 2017. Different adaptation strategies of two citrus scion/rootstock combinations in response to drought stress. PLoS ONE 12 (5), e0177993.

Ebel, R.C., Nesbitt, M., Dozier, W.A., Dane, F., 2008. Freeze risk and protection measures of Satsuma mandarins grown in the southeastern United States. HortScience 43, 287–289.

Eissenstat, D.M., 1991. On the relationship between specific root length and the rate of root proliferation: a field study using citrus rootstocks. New Phytol. 118, 63–68.

Eissenstat, D.M., Achor, D.S., 1999. Anatomical characteristics of roots of citrus rootstocks that vary in specific root length. New Phytol. 141, 309–321.

Fadel, A.L., Stuchi, E.S., Couto, H.T.Z., Ramos, Y.C., Mourão Filho, F.A.A., 2018. Trifoliate hybrids as alternative rootstocks for 'Valencia' sweet orange under rainfed conditions. Sci. Hortic. 235, 397–406.

Federici, C.T., Kupper, R.S., Roose, M.L., 2009. 'Bitters', 'Carpenter' and 'Furr' Trifoliate Hybrids: Three New Citrus Rootstocks. Informational handout from Mikeal Roose. Available 10 September 2018 at https://crec.ifas.ufl.edu/extension/citrus_rootstock/Rootstock_Literature/2009%20Federici,%20Kupper,%20Roose.%20'Bitters',%20'Carpenter'%20and%20'Furr'%20Trifoliate%20Hybrids%20-%20Three%20New%20Citrus%20Rootstocks.pdf.

Ferrarezi, R.S., Wright, A.L., Boman, B.J., Schumann, A.W., Gmitter, F.G., Grosser, J.W., 2017. Protected fresh grapefruit cultivation systems: antipsyllid screen effects on plant growth and leaf transpiration, vapor pressure deficit, and nutrition. Hort. Technol. 27, 666–674.

Folimonova, S.Y., Robertson, C.J., Garnsey, S.M., Gowda, S., Dawson, W.O., 2009. Examination of the responses of different genotypes of citrus to huanglongbing (citrus greening) under different conditions. Phytopathology 99, 1346–1354.

Forner, J.B., Forner-Giner, M.A., Alcaide, A., 2003. Forner-Alcaide 5 and Forner-Alcaide 13: two new citrus rootstocks released in Spain. HortScience 38, 629–630.

Forner-Giner, M.A., Ancillo, G., 2013. Breeding salinity tolerance in citrus using rootstocks. In: Salt Stress in Plants: Signalling, Omics and Adaptations. Springer, New York, NY, ISBN: 9781461461081pp. 355–376.

Forner-Giner, M.A., Alcaide, A., Primo-Millo, E., Forner, J.B., 2003. Performance of 'Navelina' orange on 14 rootstocks in Northern Valencia (Spain). Sci. Hortic. 98 (3), 223–232.

Forner-Giner, M.A., Primo-Millo, E., Forner, J.B., 2009. Performance of Forner-Alcaide 5 and Forner-Alcaide 13, hybrids of Cleopatra mandarin × Poncirus trifoliata, as salinity-tolerant citrus rootstocks. J. Am. Pomol. Soc. 63, 72–80.

Forner-Giner, M.A., Llosá, M.J., Carrasco, J.L., Perez-Amador, M.A., Navarro, L., Ancillo, G., 2010. Differential gene expression analysis provides new insights into the molecular basis of iron deficiency stress response in the citrus rootstock Poncirus trifoliata (L.) Raf. J. Exp. Bot. 61 (2), 483–490.

Forner-Giner, M.A., Legaz, F., Primo-Millo, E., Forner, J.B., 2011. Nutritional responses of citrus rootstocks to salinity: performance of the new hybrids, Forner-Alcaide 5 and Forner-Alcaide 13. J. Plant Nutr. 34, 1–16. https://doi.org/10.1080/01904167.2011.585202.

Forner-Giner, M.A., Rodriguez-Gamir, J., Martínez-Alcántara, B., Quiñones, A., Iglesias, D.J., Primo-Millo, E., Forner, J.B., 2014. Performance of navel orange trees grafted onto two new dwarfing rootstocks (Forner-Alcaide 517 and Forner-Alcaide 418). Sci. Hortic. 179, 376–387.

Fu, L.N., Chai, L.J., Ding, D.K., Pan, Z.Y., Peng, S.A., 2016. A novel citrus rootstock tolerant to iron deficiency in calcareous soil. J. Am. Soc. Hortic. Sci. 141, 112–118.

Gallasch, P.T., Dalton, G.S., 1989. Selecting salt-tolerant citrus rootstocks. Aust. J. Agric. Res. 40, 137–144.

Garcıa-Sanchez, F., Syvertsen, J.P., Gimeno, V., Botıa, P., Perez-Perez, J.G., 2007. Responses to flooding and drought stress by two citrus rootstock seedlings with different water-use efficiency. Physiol. Plant. 130, 532–542.

Garnsey, S.M., Barrett, H.C., Hutchison, D.J., 1987. Identification of citrus tristeza virus resistance in citrus relatives and its potential applications. Phytophylactica 19, 187–191.

Garnsey, S.M., Castle, W.S., Tucker, D.P.H., Rouse, R.E., Wutscher, H.K., Kesinger, M.C., 2001. Budunion incompatibilities and associate declines observed in Florida among trees on Swingle citrumelo and other trifoliate orange-related rootstocks. Proc. Fla. State Hort. Soc. 114, 121–127.

Girardi, E.A., Cerqueira, T.S., Cantuarias-Avilés, T.E., Rodrigues da Silva, S., Stuchi, E.S., 2017. Sunki mandarin and Swingle citrumelo as rootstocks for rain-fed cultivation of late-season sweet orange selections in northern São Paulo state, Brazil. Bragantia, Campinas 76 (4), 501–511.

Gmitter Jr., F.G., Xiao, S.Y., Huang, S., Hu, X.L., Garnsey, S.M., Deng, Z., 1996. A localized linkage map of the citrus tristeza virus resistance gene region. Theor. Appl. Genet. 92, 688–695.

Graham, J.H., 1995. Root regeneration and tolerance of citrus rootstocks to root rot caused by Phytophthora nicotianae. Phytopathology 85, 111–117.

Graham,J.,Feichtenberger,E.,2015.*Citrus* phytophthoradiseases:managementchallengesandsuccesses.J.Cit.Pathol..iocv_journalcitruspathology_27203.

Graham, J.H., Bright, D.B., McCoy, C.W., 2003. Phytophthora-Diaprepes weevil complex: Phytophthora spp. relationships with citrus rootstocks. Plant Dis. 87, 85–90.

Graham, J.H., Bowman, K.D., Bright, D.B., Adair Jr., R.C., 2007. Screening citrus rootstock genotypes for tolerance to the Phytophthora-Diaprepes complex under field conditions. Proc. Fla. State Horticult. Soc. 120, 97–103.

Grosser, J.W., Graham, J.H., McCoy, C.W., Hoyte, A., Rubio, H.W., Bright, D.B., Chandler, J.L., 2003. Development of "Tetrazyg" rootstocks tolerant of the Diaprepes/Phytophthora complex under greenhouse conditions. Proc. Fla. State Hort. Soc. 116, 262–267.

Grosser, J.W., Medina-Urrutia, V., Ananthakrishnan, G., Serrano, P., 2004. Building a replacement sour orange rootstock: somatic hybridization of selected mandarin plus pummelo combinations. J. Amer. Soc. Hort. Sci. 129, 530–534.

Grosser, J.W., Graham, J.H., Hoyte, A., Rubio, H.W., Bright, D.B., Gmitter, J., Chen, C.X., Gmitter Jr., F.G., 2007. Continued development of rootstocks tolerant of the Phytophthora-Diaprepes complex via greenhouse screening. Proc. Fla. State Hort. Soc. 120, 103–109.

Grosser, J.W., Chandler, J.L., Ling, P., Barthe, G.A., 2011. New somatic hybrid rootstock candidates for tree-size control and high juice quality. Proc. Fla. State Hort. Soc. 124, 131–135.

Grosser, J.W., Barthe, G.A., Castle, B., Gmitter, F.G., Lee, O., 2015. The development of improved tetraploid citrus rootstocks to facilitate advanced production systems and sustainable citriculture in Florida. Acta Hortic. 1065, 319–327.

Guo, W.W., Deng, X.X., 2001. Wide somatic hybrids of *Citrus* with its related genera and their potential in genetic improvement. Euphytica 118, 175–183.

Guo, W.-W., Cheng, Y.-J., Deng, X.-X., 2002. Intergeneric somatic hybrids of *Citrus reticulata* with *Poncirus trifoliata* for complementary rootstock improvement and their identification by AFLP analysis. Plant Cell Rep. 20, 829–834.

Hancevic, K., Cerni, S., Nolasco, G., Radic, T., Rosin, J., Gatin, Z., Skoric, D., 2013. *Citrus* Winsonii: biological response to infection with different citrus tristeza virus genotypes. J. Plant Pathol. 95, 615–618.

Hodgson, R.W., 1967. Horticultural varieties of citrus. In: Reuther, W., Webber, H.J., Batchelor, L.D. (Eds.), The Citrus Industry. University of California, Berkeley, CA, pp. 431–459.

Hutchison, D.J., 1974. Swingle Citrumelo—a promising rootstock hybrid. Proc. Fla. State Horticult. Soc. 87, 89–91.

Hutton, R.J., Broadbent, P., Bevington, K.B., 2000. Viroid dwarfing for high density citrus plantings. In: Janick, J. (Ed.), Horticultural Reviews. vol. 24. John Wiley & Sons, Inc., Oxford, pp. 277–317.

Kawase, K., Iwagaki, I., Takahara, T., Ono, S., Hirose, K., 1987. Rootstock studies for citrus varieties in Japan. Jpn. Agric. Res. Q. 20 (4), 253–259.

Kepiro, I.L., Roose, M.L., 2007. Nucellar embryony. In: Khan, I.A. (Ed.), *Citrus* Genetics, Breeding and Biotechnology. CABI Publishing, London, pp. 141–149.

Lacey, K., 2017. Citrus Rootstocks for Western Australia. Government of Western Australia. Available 7/5/2018 at https://www.agric.wa.gov.au/citrus/citrus-rootstocks-western-australia.

Lapointe, S.L., Bowman, K.D., 2002. Is there meaningful plant resistance to Diaprepes abbreviatus (Coleoptera: Curculionidae) in citrus rootstock germplasm? J. Econ. Entomol. 95 (5), 1059–1065.

Lapointe, S.L., Shapiro, J.P., Bowman, K.D., 1999. Identification of sources of plant resistance to Diaprepes abbreviatus (Coleoptera: Curculionidae) by three bioassays. J. Econ. Entomol. 92 (4), 999–1004.

Lea-Cox, J.D., Syvertsen, J.P., 1993. Salinity reduces water use and nitrate-N-use efficiency of citrus. Ann. Bot. 72, 47–54.

Lee, A.T.C., Joubert, J., van Vuuren, S.P., 2009. Rootstock choice. In: Integrated Production Guidelines for Export *Citrus*. vol. 1. Citrus Research International. (Chapter 6).

Levy, Y., Syvertsen, J., 2004. Irrigation water quality and salinity effects in citrus trees. Hortic. Rev. 30, 37–82.

Licciardello, C., Torrisi, B., Allegra, M., Sciacca, F., Roccuzzo, G., Intrigliolo, F., Recupero, G.R., Tononi, P., Delledonne, M., Muccilli, V., 2013. A transcriptomic analysis of sensitive and tolerant citrus rootstocks under natural iron deficiency conditions. J. Amer. Soc. Hort. Sci. 138, 487–498.

Liu, Y.-Z., Deng, X.-X., 2007. *Citrus* breeding and genetics in China. Asian Australas. J. Plant Sci. Biotechnol. 1, 23–28.

Liu, J.J., Chen, K.L., Hu, Q., Yang, M., Zhou, Q.M., Li, H.W., et al., 2008. Preliminary study on Ziyang Xiangcheng (*Citrus junos* Sieb ex Tanaka), a special local citrus germplasm. Southwest China J. Agric. Sci. 21, 1658–1660. (in Chinese with English abstract).

Llosá, M.J., Bermejo, A., Cano, A., Quiñones, A., Forner-Giner, M.A., 2009. The citrus rootstocks Cleopatra mandarin, Poncirus trifoliata, Forner-Alcaide 5 and Forner-Alcaide 13 vary in susceptibility to iron deficiency chlorosis. J. Am. Pomol. Soc. 63, 160–167.

Louzada, E.S., del Rio, H.S., Setamou, M., Watson, J.W., Swietlik, D.M., 2008. Evaluation of citrus rootstocks for the high pH, calcareous soils of South Texas. Euphytica 164, 13–18.

Maas, E.V., 1993. Salinity and citriculture. Tree Physiol. 12, 195–216.

Mademba-Sy, F., Lemerre-Desprez, Z., Lebegin, S., 2012. Use of flying dragon trifoliate orange as dwarfing rootstock for citrus under tropical climatic conditions. HortScience 47, 11–17.

Manthey, J.A., McCoy, D.L., Crowley, D.E., 1994. Stimulation of rhizosphere iron reduction and uptake in response to iron deficiency in citrus rootstocks. Plant Physiol. Biochem. 32, 211–215.

Martinez-Cuenca, M.-R., Primo-Capella, A., Quiñones, A., Bermejo, A., Forner-Giner, M.A., 2017. Rootstock influence on iron uptake responses in *Citrus* leaves and their regulation under the Fe paradox effect. Renaut J, ed.. Peer J. 5, e3553. https://doi.org/10.7717/peerj.3553.

McCarty, C.D., Bitters, W.P., Cole, D.A., 1974. Comparisons between Troyer and Carrizo citrange. Citrograph 59, 294–310.

McCollum, T.G., Bowman, K.D., 2017. Rootstock affects fruit quality among 'ray ruby' grapefruit trees grown in the Indian River district of Florida. HortScience 52 (4), 541–546.

Moore, G.A., Castle, W.S., 1988. Morphological and isozymic analysis of open-pollinated citrus rootstock populations. J. Hered. 79, 59–63.

Morgan, K.T., Rouse, R.E., Ebel, R.C., 2016. Foliar applications of essential nutrients on growth and yield of 'Valencia' sweet orange infected with Huanglongbing. HortScience 51 (12), 1482–1493.

Raga, V., Bernet, G.P., Carbonell, E.A., Asins, M.J., 2014. Inheritance of rootstock effects and their association with salt tolerance candidate genes in a progeny derived from 'Volkamer' lemon. J. Amer. Soc. Hort. Sci. 139 (5), 518–528.

Ream, C.L., Furr, J.R., 1976. Salt tolerance of some *Citrus* species, relatives and hybrids tested as rootstocks. J. Am. Soc. Hortic. Sci. 101, 265–267.

Reforgiato Recupero, G., Caruso, A., Russo, G., 1992. The flying dragon trifoliate orange and BA-300 citrange: effects on scion performance. Proc. Int. Soc. Citricult. 1, 286–290.

Reforgiato Recupero, G., Russo, G., Recupero, S., Zurru, R., Deidda, B., Mulas, M., 2009. Horticultural evaluation of new *Citrus latipes* hybrids as rootstocks for citrus. HortScience 44, 595–598.

Rodríguez-Gamir, J., Primo-Millo, E., Forner, J.B., Forner-Giner, M.A., 2010. Citrus rootstock responses to water stress. Sci. Hortic. 126, 95–102.

Romero, P., Navarro, J.M., Perez-Perez, J., Garcia-Sanchez, F., Gomez-Gomez, A., Porras, I., Martinez, V., Botia, P., 2006. Deficit irrigation and rootstock: their effects on water relations, vegetative development, yield, fruit quality and mineral nutrition of Clemenules mandarin. Tree Physiol. 26, 1537–1548.

Roose, M.L., 1986. Dwarfing rootstocks for citrus. Proc. Intl. Soc. Citrus Nurserymen 1–8.

Roose, M.L., 2007. Citrus rootstock breeding and evaluation. In: Citrus Research Board 2007 Annual Report. Available 10 September 2018 at http://www.citrusresearch.org/wp-content/uploads/2007-Michael-L-Roose-Rootstock-Breeding-and-Evaluation.pdf.

Roose, M.L., 2014. Rootstocks. The Regents of the University of California95–105. (Chapter 5). Available 7/5/2018 at http://www.crec.ifas.ufl.edu/extension/citrus_rootstock/Rootstock_Literature/Univ.%20CA%20CitrusProdManual%20Rootstock%20Chapter%202014.pdf.

Ruiz, D., Martinez, V., Cerdá, A., 1999. Demarcating specific ion (NaCl, Cl, Na) and osmotic effects in the response of two rootstock to salinity. Sci. Hortic. 80, 213–224.

Santana-Vieira, D.D.S., Freschi, L., Almeida, L.A., Moraes, D.H., Neves, D.M., Santos, L.M., et al., 2016. Survival strategies of citrus rootstocks subjected to drought. Sci. Rep. 6, 38775. https://doi.org/10.1038/srep38775.

Saunt, J., 2000. *Citrus* Varieties of the World: An Illustrated Guide, second ed. Sinclair International Limited.

Savage, E.M., Gardner, F.E., 1965. The Troyer and Carrizo citranges. Calif. Citrograph 50 (3), 112–116.

Simón-Grao, S., Nieves, M., Martínez-Nicolás, J.J., Cámara-Zapata, J.M., Alfosea-Simón, M., García-Sánchez, F., 2018. Response of three citrus genotypes used as rootstocks grown under boron excess conditions. Response of three citrus genotypes used as rootstocks grown under boron excess conditions. Ecotoxicol. Environ. Saf. 159, 10–19.

Smith, M., 2017. CT13004 Queensland *Citrus* Improvement Scheme: Finding Better Rootstocks for Australia. Presentation. Available 10 September 2018 at https://www.citrusaustralia.com.au/wp-content/uploads/Smith-Malcolm-Imperial-Rootstock.pdf.

Stansly, P.A., Arevalo, H.A., Qureshi, J.A., Jones, M.M., Hendricks, K., Roberts, P.D., Roka, F.M., 2014. Vector control and foliar nutrition to maintain economic sustainability of bearing citrus in Florida groves affected by huanglongbing. Pest Mgt. Sci. 70, 415–426.

Stover, E., Castle, W.S., 2002. Citrus rootstock usage, characteristics, and selection in the Florida Indian River region. Hort. Technol. 12, 143–147.

Stuchi, E.S., Donadio, L.C., Sempionato, O.R., 2003. Performance of 'Tahiti lime' on Poncirus trifoliata var. monstrosa flying dragon in four densities. Fruits 58, 13–17.

Swingle, W.T., Reece, P.C., 1967. The botany of *Citrus* and its wild relatives. In: Reuther, W., Webber, H.J., Batchelor, L.D. (Eds.), The Citrus Industry. second ed.. vol. I. University of California, Riverside, pp. 190–430.

Sykes, S.R., 2011a. Chloride and sodium excluding capacities of citrus rootstock germplasm introduced to Australia from the People's Republic of China. Sci. Hortic. 128, 443–449.

Sykes, S.R., 2011b. Characterisation of citrus rootstock germplasm introduced as seeds to Australia from the People's Republic of China. Sci. Hortic. 127, 298–304.

Tan, F., Tu, H., Liang, W., Long, J., Wu, X., Zhang, H., Guo, W., 2015. Comparative metabolic and transcriptional analysis of a doubled diploid and its diploid citrus rootstock (C. junos cv. Ziyang Xiangcheng) suggests its potential value for stress resistance improvement. BMC Plant Biol. 15, 450–454.

Timmer, L.W., Garnsey, S.M., Graham, J.H., 2000. Compendium of *Citrus* Diseases, second ed. APS Press, St. Paul, MN.

Verdejo-Lucas, S., Kaplan, D.T., 2002. The citrus nematode: Tylenchulus semipenetrans. In: Starr, J.L., Cook, R., Bridge, J. (Eds.), Plant Resistance to Parasitic Nematodes. CAB International, Wallingford, UK, pp. 207–219.

Verdejo-Lucas, S., Sorribas, F.J., Forner, J.B., Alcaide, A., 1997. Screening hybrid citrus rootstocks for resistance to Tylenchulus semipenetrans Cobb. HortScience 32, 1116–1119.

Verdejo-Lucas, S., Sorribas, F.J., Forner, J.B., Alcaide, A., 2000. Resistance of hybrid citrus rootstocks to a Mediterranean biotype of Tylenchulus semipenetrans Cobb. HortScience 35, 269–273.

Vernière, C., Perrier, X., Dubois, C., Dubois, A., Botella, L., Chabrier, C., Bové, J.M., Duran Vila, N., 2004. *Citrus* viroids: symptom expression and effect on vegetative growth and yield of clementine trees grafted on trifoliate orange. Plant Dis. 88, 1189–1197.

Vidalakis, G., Pagliaccia, D., Bash, J.A., Afunian, M., Semancik, J.S., 2011. *Citrus* dwarfing viroid: effects on tree size and scion performance specific to Poncirus trifoliata rootstock for high-density planting. Ann. Appl. Biol. 158, 204–217.

Vu, J.C.V., Yelenosky, G., 1991. Photosynthetic responses of citrus trees to soil flooding. Physiol. Plant. 81, 7–14.

Wallace, J.M., 1968. Tristeza and seedling yellows. In: Childs, J.F.L. (Ed.), Indexing Procedures for 15 Virus Diseases of *Citrus* Trees. Agric. Handbk. vol. 333. U.S. Dept. Agric., pp. 20–27.

Wang, N., Trivedi, P., 2013. *Citrus* huanglongbing: a newly relevant disease presents unprecedented challenges. Phytopathology 103, 652–665.

Weathers, L.G., Calavan, E.C., Wallace, J.M., Christiansen, D.W., 1955. A budunion and rootstock disorder of Troyer citrange with Eureka lemon tops. Plant Dis. Rptr. 39, 665–669.

Webber, H.J., Reuther, W., Lawton, H.W., 1967. History and development of the citrus industry. In: Reuther, W., Webber, H.J., Batchelor, L.D. (Eds.), The Citrus Industry. University of California, Berkeley, CA, pp. 1–39.

Widmer, T.L., Graham, J.H., Mitchell, D.J., 1998. Histological comparison of fibrous root infection of disease tolerant and susceptible citrus hosts by Phytophthora nicotianae and P. palmivora. Phytopathology 88, 389–395.

Wu, G.A., Prochnik, S., Jenkins, J., Salse, J., Hellsten, U., Murat, F., et al., 2014. Sequencing of diverse mandarin, pummelo and orange genomes reveals complex history of admixture during citrus domestication. Nat. Biotechnol. 32 (7), 656–662.

Wu, G.A., Terol, J., Ibanez, V., Lopez-Garcia, A., Estela, P.-R., Carles, B., et al., 2018. Genomics of the origin, evolution and domestication of citrus. Nature 544, 311–316. https://doi.org/10.1038/nature25447.
Wutscher, H.K., 1979. Citrus rootstocks. Hort. Rev. (Amer. Soc. Hort. Sci.) 1, 237–269.
Wutscher, H.K., 1988. Rootstock effects on fruit quality. In: Ferguson, J.J., Wardowski, W.F. (Eds.), Factors Affecting Fruit Quality. University Florida Citrus Short Course Proceedings. pp. 24–34. Available online 7/5/2018 at https://irrec.ifas.ufl.edu/flcitrus/pdfs/short_course_and_workshop/factors_fruit_quality/Wutscher-Rootstock_Effects.pdf.
Wutscher, H.K., Bistline, F.W., 1988. Performance of 'Hamlin' orange on 30 citrus rootstocks in southern Florida. J. Am. Soc. Hortic. Sci. 113, 493–497.
Wutscher, H.K., Bowman, K.D., 1999. Performance of 'Valencia' orange on 21 rootstocks. HortScience 34 (4), 622–624.
Wutscher, H.K., Hill, L.L., 1995. Performance of Hamlin orange on 16 rootstocks in east Central Florida. HortScience 30, 41–43.
Yelenosky, G., 1985. Cold hardiness in *Citrus*. Hortic. Rev. 7, 201–238.
Yelenosky, G., 1996. An overview of Florida citrus freeze survival. Proc. Fla. State Hort. Soc. 109, 118–123.
Young, R.H., Albrigo, L.G., Cohen, M., Castle, W.S., 1984. Rates of blight incidence in trees on Carrizo citrange and other rootstocks. Citrus Ind. 65, 20–30.
Zambrosi, F.C.B., Mattos, D., Syvertsen, J.P., 2011. Plant growth, leaf photosynthesis, and nutrient-use efficiency of citrus rootstocks decrease with phosphite supply. J. Plant Nutr. Soil Sci. 174, 487–495.
Zandalinas, S.I., Rivero, R.M., Martínez, V., Gómez-Cadenas, A., Arbona, V., 2016. Tolerance of citrus plants to the combination of high temperatures and drought is associated to the increase in transpiration modulated by a reduction in abscisic acid levels. BMC Plant Biol. 16, 105.
Zhou, G.F., Peng, S.A., Liu, Y.Z., Wei, Q.J., Han, J., Islam, M.Z., 2014. The physiological and nutritional responses of seven different citrus rootstock seedlings to boron deficiency. Trees 28, 295–307.

Further reading

Grosser, J.W., Ollitrault, P., Olivares-Fuster, O., 2000. Somatic hybridization in citrus: an effective tool to facilitate variety improvement. In Vitro Cell Dev. Biol. Plant 36, 434–449.
Yelenosky, G., 1973. Tolerance of trifoliate orange selections and hybrids to freezes and flooding. Proc. Fla. State Hort. Soc. 86, 99–104.

Chapter 7

Traditional breeding

Marco Caruso[a], Malcolm W. Smith[b], Yann Froelicher[c], Giuseppe Russo[a], Frederick G. Gmitter Jr.[d]

[a]*CREA Research Centre for Olive, Citrus and Tree Fruit, Acireale, Italy,* [b]*Department of Agriculture and Fisheries, Bundaberg Research Station, Bundaberg, QLD, Australia,* [c]*Unité Mixte de Recherche Amélioration Génétique et Adaptation des Plantes (UMR Agap), Centre de Coopération Internationale en Recherche Agronomique pour le Développement (CIRAD), San Giuliano, France,* [d]*Citrus Research and Education Center (CREC), Institute of Food and Agricultural Sciences (IFAS), University of Florida, Lake Alfred, FL, United States*

7.1 Introduction

The release of cultivars, improved in terms of organoleptic features as well as resistance to pests and diseases represents a major objective for world citriculture. The challenge for breeders and geneticists is the generation of new genotypes that can fulfill market needs, show tolerance or resistance to biotic and abiotic stresses, and that ultimately guarantee higher income to growers.

Many biological and technical constraints limit the possibility of generating improved cultivars. First, many citrus cultivars are clonal selections that resulted from somatic mutations that accumulated over years, decades, or centuries in different growing areas, from a single ancestral genotype. Therefore, there are limited possibilities to improve these cultivars for specific targeted traits via conventional approaches other than clonal selection. For example, successful efforts at reproducing sweet orange or lemon-like cultivars through hybridization have been limited so far.

The long juvenile phase of citrus seedlings is another constraint to genetic progress. Theoretically, the phases of selection and first evaluation of promising potential new cultivars should last on average 15 years; however, practical examples demonstrate that usually the time from hybrid seed to commercial release is 25–30 years. Some of the cultivars that have gained market importance in the last decade, such as Orri (Spiegel-Roy and Vardi, 1992; Vardi et al., 2008) and Nadorcott (Nadori, 2004), are the result of programs that started in the late 1970s and 1980s. Nucellar polyembryony is another impediment to conventional breeding approaches, because it can limit the use of elite cultivars as the seed parent. Some citrus types are highly polyembryonic; zygotic embryo development is hampered by the presence of competitive nucellar embryos, and thus a very limited number of hybrids are generated. Moreover, the identification of hybrids is difficult and time consuming especially in the absence of reliable morphological markers.

Different degrees of male and/or female sterility and self-incompatibility of many elite cultivars may also hamper genetic improvement. While such sterility might represent a useful trait for the generation of seedless cultivars (see Section 7.2.3), poor pollen, and/or seed development in cultivars such as Tahiti lime, Satsuma mandarin, and Navel orange makes them challenging parents for conventional breeding.

Most citrus cultivars are highly heterozygous, and breeding programs are required to select for a large number of horticultural traits. The level of heterozygosity is particularly high in the derived species. A recent genome sequencing project revealed that around 50% of the coding regions of the sweet orange genome contains at least one allelic variant (Xu et al., 2013). The same authors estimated 3.6 SNPs per kb in the whole sweet orange genome. Moreover, Wu et al. (2014), estimating heterozygosity distribution and the number of heterozygous sites per kb in true and derived species, showed peaks of over 20 heterozygous sites per kb in sweet orange and an average of 5–17 sites per kb in mandarins, depending on the level of interspecific admixture. These levels are comparable with those observed in diploid grapevine (8.9 SNPs per kb; Velasco et al., 2007), and higher than in apple (4.4 SNPs per kb; Velasco et al., 2010) and peach (2.47 SNPs per kb on average analyzing 11 peach genomes; M. Cirilli and L. Rossini, personal communication). The high heterozygosity results in a strong segregation leading to remarkable variation and new allelic combinations. This problem is further exacerbated by the large number of traits that must be considered when selecting hybrids. Citrus-breeding programs commonly have selection criteria for more than 20 traits, most of which are under polygenic control and have not been fixed in breeding populations. Hence, the probability of obtaining a new hybrid that satisfies the necessary standards for all traits is extremely

The Genus Citrus. https://doi.org/10.1016/B978-0-12-812163-4.00007-3

low. Consequently, large numbers of individual hybrids need to be evaluated to increase the likelihood of identifying truly improved new candidate cultivars.

The difficulty of introgressing favorable genes from relatives (see Section 7.2.3.2) into commercially acceptable cultivars increases the length of time for breeding programs to successfully achieve objectives. Citrus relatives are a known source of resistance to biotic stresses such as citrus tristeza virus (CTV), Asiatic citrus canker, and Huanglongbing (HLB) and many intergeneric crosses have been performed to obtain resistant hybrids. This has been relatively simple and successful for rootstock breeding, resulting in intergeneric hybrids that are now some of the most important in world citriculture. However, it is much more problematic and time consuming for scion cultivars, because many cycles of backcrossing are needed to eliminate unwanted fruit flavors and other undesirable traits inherited from relatives.

Despite these difficulties and the long times needed, citrus breeders around the world have been able to select valuable new cultivars by means of conventional techniques, namely hybridization and clonal selection. Many of these varieties are still cultivated decades after their initial release.

Citrus-breeding programs based on hybridization and clonal selection started in the late 1800s in Florida with the pioneering work of Swingle and Webber (Traub and Robinson, 1937; Cooper et al., 1962). Preliminary hybridization work occurred as early as 1909 in Japan, with similar efforts in places including Spain, Italy, Israel, and Australia during the early 1930s (Traub and Robinson, 1937). More extensive breeding programs developed at the Citrus Experiment Station in California from 1914 onward (Kender, 2003), from 1937 in Japan (Omura and Shimada, 2016) and Italy in 1940 (Carrante and Bottari, 1951). In fact, the selection of unique phenotypes that arise spontaneously either as seedlings or vegetative mutations has been the primary method for citrus scion genetic improvement since antiquity. In the last three decades there has been a proliferation of new breeding programs throughout the world. More recently, there has been an increased investment of private funds into public citrus-breeding programs, as well as an increasing number of private programs.

After more than a century, many breeding programs around the world are still based on classical approaches and on the phenotypic evaluation of large numbers of progeny and clonal selections, in the field and/or in the greenhouse. These empirical observations during the evaluation phase suffer from a strong influence of climatic conditions where the trees are planted and by the pathogen pressure at the site of selection. Many important fruit quality traits are notoriously difficult to phenotype, necessitating multiple observations across multiple seasons and testing sites.

Recently, important advances related to the identification of horticulturally important genes have been made. These developments include the identification of the Ruby gene controlling anthocyanin pigmentation, along with mapping and marker development for major genes controlling diseases such as *Alternaria* and CTV as well as the important rootstock trait of apomixis. While this is still a very limited number of molecular markers linked to specific traits compared to what has been developed for some other crops, these markers are already making breeding programs more efficient.

Despite the enormous efforts and the relatively low efficiency, traditional breeding methods are so far the only ones that have allowed product innovation in citriculture. New breeding techniques, such as genome editing and cisgenesis, will potentially improve the efficiency of breeding programs, and may be more acceptable to consumers and legislators than previous GMOs. However, these new techniques have yet to be optimized and are still too recent to represent an alternative to classical methods in the short term. Furthermore, the efficiency of these techniques is dependent on knowledge about the genetic control of horticulturally important traits, which is still relatively scarce in citrus compared to other major crops.

7.2 Scion breeding

The objectives of scion breeding vary depending on the species, the characteristics of different growing areas, and the requirements of the particular markets being targeted (Table 7.1). Individual breeding programs differ in the number, relative importance, and selection intensity applied to these various traits, although most programs select for at least 20 traits. Some traits like productivity, flavor, aroma, juiciness, and resistance to common biotic and abiotic stress and postharvest problems are prerequisite in all fresh fruit breeding programs; programs aimed at improving cultivars for processing focus mainly on productivity, juice quality, and extended seasons of maturity.

7.2.1 Somatic mutations and chimeras

Many citrus varieties cultivated worldwide derive from somatic mutations from a single ancestral phenotype. In cultivar types such as sweet orange, lemon, grapefruit, and Clementine and Satsuma mandarins, nearly all phenotypic variability is the result of mutations discovered among trees growing in commercial orchards. These cultivar types accumulated mutations during decades or centuries of cultivation worldwide. Diversity within sweet orange germplasm probably represents the most evident example of how somatic mutations caused intraspecific diversification into hundreds of different cultivated

TABLE 7.1 Traits that have been included in citrus scion breeding programs and are known to segregate in hybrid populations.

Fruit quality traits				
Visual	**Taste/smell/medical**	**Postharvest**	**Pest and disease traits**	**Orchard/business management traits**
Fruit shape	Acid	Albedo adherence to flesh	Alternaria	Freeze tolerance
Fruit size	Anthocyanin	Albedo breakdown	Black spot	On-tree fruit storage
Flesh color	Bitterness	Albedo stringiness	Canker	Maturity time
Juice color	Brix	Chilling injury	CTV	Regular cropping
Navel presence	Chewiness	Essential oil content	Fruit fly	Seed embryony
Navel size	Furanocoumarins	Firmness	Gall wasp	Sunburn
Open core	Juice content	Fruit color uniformity	Greasy spot	Thorns
Puffiness	Lycopene	Fruit size uniformity	HLB	Vigor
Skin color	Off-flavors	Granulation	Mal secco	Yield fruit/ha
Skin markings	Rind oil content	Oleocellosis	Rust mite	Yield solids/ha
Skin sheen	Rind oil smell	Peelability	Scab	
Skin texture	Seediness	Peel strength	Witches' broom	
Skin thickness	Sugar:acid balance	Storage life		
Stem-end neck	Taste			
Wind rub				

clones differing for traits such as fruit size, ripening period, peel and pulp color, presence of seeds, acidity, and pigmentation. Most of the mutations that occur in the field are degenerative (Fig. 7.1) leading to corrugation, variegation, and dry or seedy fruits, but in some cases, they can improve specific characteristics, so they are isolated and propagated by growers and breeders.

Somatic mutations may be evident throughout the whole plant, or in variable portions of the shoot apical meristem (SAM), leading to chimerism. Chimeras have been shown to contribute greatly to phenotypic differences in vegetatively propagated plants. Somatic mutations may occur in all cells of one or two histogenic layers of the SAM (periclinal chimeras), in layer portions (mericlinal chimeras) or traversing all three layers in a specific sector (sectorial chimeras). Somatic mutations can remain as chimeras for centuries or can be unstable. They can be propagated in the whole plant and become solid through cellular rearrangements caused by the invasion of the mutated layer into the others (Pelsy, 2010). In citrus, apical meristems are composed of three layers: L1 gives rise to the epidermal cell layer and is also involved in the formation of juice sacs; L2 produces gametes, subepidermal tissues, and the nucellus; and L3 produces vascular bundles. As in other clonally propagated species, there is much evidence of chimerism in citrus (Frost and Krug, 1942; Cameron and Frost, 1968; Bowman et al., 1991; Caruso et al., 2016).

Among the woody, vegetatively propagated crops, the effect of somatic mutations on phenotypic change has been particularly studied in grapevine by genetics (Hocquigny et al., 2004; Pelsy, 2010) and bioinformatics tools at the whole genome level (Marroni et al., 2017; Carbonell-Bejerano et al., 2017). In citrus, recent findings are shedding light on the molecular basis of somatic mutations leading to horticulturally important phenotypic changes. For example, it is known that all blood orange selections differ from the common sweet oranges because of the presence of a transposable element upstream of a regulatory gene regulating anthocyanin biosynthesis (Butelli et al., 2012). Moreover, resequencing projects are revealing different kinds of mutations leading to cultivar diversification. Terol et al. (2015) revealed chromosomal rearrangements, deletions, and chimerism in two early ripening Clementine mutants, while Scaglione et al. (2014) described the presence of single-nucleotide variants, mobile element insertions and deletions, and chimeric deletions among 10 natural sweet orange mutants. Those tools will be helpful in the identification of other causative mutations underlying horticulturally important traits that have not yet been fully described. They can also be used to generate new markers to assist breeding programs.

FIG. 7.1 Examples of mutations occurring in citrus orchards: (A and B) a Clementine mutated branch producing seedy fruits in solid block; (C) chimeric Tarocco clone producing variegated leaves; and (D) Navel chimeric plant producing greenish and orange fruits.

Graft chimeras have also played a role in cultivar improvement and diversification and can be detected with traditional molecular tools. An efficient method of identify graft chimeras is to genotype different plant tissues derived from different histogenic layers. Aprile et al. (2011) characterized Faris, a lemon genotype producing acid and low-acid fruits on the same tree, by means of microsatellite markers and found that the L1 derived from Millsweet lime, while the L2 derived from an acid lemon. Zhang et al. (2007) identified the navel mutant Zaohong that accumulates β-cryptoxanthin in the pulp. Its characterization demonstrated that Zaohong is actually a periclinal chimera consisting of L1 derived from Satsuma mandarin and L2/L3 from Robertson navel orange.

Cytological chimeras (cytochimeras) are also reported in citrus. They are chimeras that differ in ploidy level and are reported to occur naturally (Frost and Krug, 1942) or after treatment with colchicine or oryzalin (Barrett, 1974; Aleza et al., 2009). They have been used as new tetraploid parents in triploid breeding programs (see Section 7.2.3.3).

7.2.2 Nucellar selections

This method uses the capability of polyembryonic varieties to generate embryos and embryogenic calli from somatic tissues of the seed both in vivo and in vitro. Embryos develop from the nucellar tissues surrounding the embryo sac and can be recovered from developed seeds and undeveloped ovules (Frost and Soost, 1968; Esen and Soost, 1977; Kepiro and Roose, 2007; Wang et al., 2017). The capability of many citrus species and varieties to produce plants identical to the mother plant by seeds has been known for centuries (Cook, 1907). Until the mid-1800s, before the spread of *Phytophthora* root rot, many sweet orange orchards in the Mediterranean, as well as elsewhere in the world, consisted of nongrafted plants propagated by seed (Savastano, 1915; Navarro, 2015).

In recent years, nucellar selections have been developed with the principal purpose of obtaining virus-free varieties (Bruno, 1962; Navarro and Juarez, 1977; Starrantino and Russo, 1980). Nucellar embryogenesis also facilitated germplasm exchange among countries, avoiding or reducing risks of spreading pathogens while maintaining most of the characteristics of the mother plant. Shoot tip grafting has largely replaced nucellar regeneration as a method of sanitation, since nucellar plants often carry many undesirable traits such as thorniness and reversion to a juvenile state (Navarro et al., 1975; Gmitter and Moore, 1986).

Nucellar selection can be considered as a method of genetic improvement and has led to the release of valuable cultivars. In fact, nucellar selections often differ from the mother plant for few or many characteristics, depending on the original genotype from which they originated. Specifically, this method does not create completely true-to-type plants since the nucellar tissues originate only from the L2 histogenic layer. As a result, if a specific phenotype of a mother plant is determined by a somatic mutation occurred in L1, that characteristic will be lost in the nucellar plant (Cameron et al., 1964). Nucellar selections have been generated in many countries, such as China (Liu and Deng, 2007), Japan (Nishiura, 1964), and Italy (Caruso et al., 2016) with the specific purpose of obtaining improved varieties. As indicated above, embryos can be formed through in vitro cultures initiated from undeveloped ovules. In the United States, this phenomenon has been exploited to identify superior somaclonal variants following careful field observations of thousands of Hamlin and Valencia sweet orange trees regenerated from in vitro cultures (Grosser et al., 2015). Interesting and useful variation was found for season of maturity, seed numbers, color, flavor, and possibly disease tolerance. New cultivars have been released from this program and are beginning to be utilized by the Florida citrus industry. The first of these was Valquarius, a Valencia-derived cultivar that matures in the mid-season in Florida usually at least 6 weeks earlier than standard Valencia. This was followed by N7-3, a later maturing and seedless Valencia, with more attractive fresh fruit appearance. A Hamlin-derived cultivar, N13-32, was selected and released because it exceeds the USDA minimum juice color score of 36, which ordinary Hamlin fruit never achieve under Florida conditions. OLL-8 and OLL-4 also have notable deeper juice color as well as much improved sensory attributes; field observations also seem to indicate an enhanced tolerance of HLB. More recently, Florida EV1 and Florida EV2 have become available and are Valencia quality oranges that mature in early November. Since 2015 nearly 1 million trees of these cultivars have been propagated and planted by Florida growers and commercial demand continues to increase.

More than 30 year of breeding using nucellar selection at CREA, Italy, has produced new sweet orange varieties with improved characteristics compared to their progenitor mother clones. These successful examples include several Tarocco clones, such as Scirè D2062, Meli C8158, and Lempso C5787. This breeding research has also shown that it is difficult or perhaps even impossible to develop improved varieties from some mother clones because of the loss of peculiar and valuable characters in the nucellar seedlings (Table 7.2). For example, with the highly pigmented Tarocco Rosso, none of the 43 nucellar selections evaluated showed any anthocyanin pigmentation in the peel, and so the only virus-free clone released to growers was the one obtained by shoot tip grafting. Similarly, in the case of Cara Cara navel orange, none of the four nucellar selections obtained at CREA retained lycopene pigmentation. The chimeric nature of this variety has been confirmed in other breeding programs, with plants producing nonpink fruits or variegated leaves (http://www.citrusvariety.ucr.edu/citrus/caracara.html). By contrast, high variability was observed among Moro nucellar selections in terms of anthocyanin pigmentation. Within a population of 28 nucellar selections, 3 completely lost anthocyanin pigmentation and some others showed less flesh color, while 3 showed increased anthocyanin compared with the Moro mother clone and are now under evaluation in different environments. In another case, nucellar embryogenesis from a degenerative mutant with brown and

TABLE 7.2 Examples of nucellar selections of blood and navel oranges obtained at CREA, Sicily.

Sweet orange variety	Number of nucellar lines evaluated in the field	Number of degenerative mutations	Nucellar clones released to the growers	Notes
Cara Cara	4	4	0	All clones have lost lycopene pigmentation
Moro	28	5	3	The degenerative mutations include nonpigmented and low-pigmented clones
Tarocco Rosso	43	43	0	All nucellar clones have lost peel pigmentation, some have lost internal pigmentation
Tarocco Lempso	12	0	1	
Tarocco Meli	13	0	1	
Tarocco Messina	10	2	1	
Tarocco Scirè	11	0	2	
Tarocco Fondaconovo	10	10	0	All clones contain lower anthocyanin in peel and pulp
Tarocco TDV	10	0	1	

FIG. 7.2 Comparison between the commercial Tarocco TDV nucellar selection (on the *left*) and the old line showing smaller fruit size and the degeneration of the juice vesicles.

dry juice vesicles resulted in nondegenerative clones, including one named Tarocco TDV (Reforgiato Recupero and Russo, 1977). This improved clone has now become one of the most widely grown Tarocco selections due to its earliness in maturation and outstanding pulp pigmentation (Fig. 7.2).

All the differences in terms of pulp-and-peel coloration between nucellar selections and their progenitor mother clones are similar to those observed on graft periclinal chimeras (Zhou et al., 2002; Zhang et al., 2007) in which pulp-and-peel coloration develop autonomously without layer interaction. Therefore, it is likely that, in the above-cited cases, the mutations characterizing the old lines occurred in the L1 and were lost in plants derived from nucellar embryos. However, it is not yet known with certainty if the phenotypic differences that occur between mother plants and nucellar selections are caused solely by the somatic mutations that accumulated in the mother plant and are not retained by nucellar clones, or if there might be additional mutations occurring during the in vitro culture process necessary to grow plants from undeveloped ovules.

7.2.3 Hybridization

Obtaining new commercially valuable cultivars from hybridization is a major challenge for citrus breeders. Due to the long juvenile phase, this method has generally relied on the use of existing elite commercial cultivars to generate hybrid populations. Decades-long breeding programs may also take advantage of their past selections to generate subsequent generations of hybrids and begin to fix traits of interest within breeding populations.

Breeding by hybridization may benefit from recent advances in citrus phylogeny (Wu et al., 2018). The assessment of diversity among and within species, as well as the elucidation of many controversial issues regarding the phylogeny of the true and derived species by genome sequencing is certainly helpful to plan new breeding programs starting from ancestral species. On the other hand, these studies have also revealed the complex ancestry of important cultivated types, such as sweet orange, confirming the difficulty and the limited success that has been achieved in creating new selections of some citrus types via hybridization. By contrast, hybridization is one of the most used breeding methods to obtain new mandarin varieties. After more than 100 years of systematic breeding, there are many examples of successful mandarin varieties diffused worldwide obtained through controlled crosses (see Chapter 5). The selection of seedlings from uncontrolled hybridization events, such as in the case of Clementine, Nadorcott, Ellendale, Imperial, and Murcott has also played a role in the generation of successful hybrids.

To conduct an efficient breeding program, a large number of progenies should be generated. It is not easy to estimate the most appropriate number of hybrids that must be evaluated to find something promising. It depends on the objectives of the breeding program, the quality of available parents, and on what screening can be performed by marker-assisted selection (MAS) or greenhouse phenotyping before the evaluation in the field. Experience suggests that screening of thousands or even tens of thousands of hybrid progenies is needed to select superior cultivars, although there are examples of successful cultivars being selected from very small populations.

7.2.3.1 Diploid breeding

Crosses between diploid individuals have several advantages compared to other strategies. First, they allow the generation of numerous progenies compared to triploids generated by 2x×2x and 2x×4x strategies (see Section 7.2.3.3); so, a higher selection pressure for many quality traits can be applied. In addition, the strategy does not need facilities for embryo rescue. Moreover, new interesting allelic combinations found in F_1 hybrids might be used for further hybridizations.

Diploid breeding programs, especially if involving seedy parents, usually lead to the production of seedy individuals. Seedlessness is a prerequisite for any new variety to be commercially successful in many markets around the world, and different strategies have been used to avoid the presence of seeds in diploid cultivars (Vardi et al., 2008). Some strategies exploit natural sterility that can be conferred by specific parents to part of the progenies. As a general rule, seedless or low-seeded parents frequently produce seedless or low-seeded hybrids (Yamamoto et al., 1995).

Seedlessness can be achieved through gametophytic self-incompatibility. Some cultivars such as Clementine, Nadorcott, Fortune, and Nova are practically seedless if grown in solid blocks due to their self-incompatibility coupled to parthenocarpy (Distefano et al., 2009). Despite the fact that many genes associated to the trait have been identified in several mutants, the genetic control of gametophytic incompatibility is still not well understood, and no markers linked to the trait are currently available. Moreover, in breeding trials, self-incompatible hybrids cannot be easily recognized due to cross-pollination. Thus, the absence of molecular markers and difficulties of preventing cross-pollination in breeding trials containing multiple genotypes have made it impractical to exploit gametophytic self-incompatibility in large-scale breeding programs.

A more practical possibility to obtain seedless varieties is using parents that confer seedlessness due to aberrations occurring after fertilization. An example is represented by seedless Kishu (Mukaku Kishu), a mutant of seedy *Citrus kinokuni*. In this cultivar, and in approximately half of its progenies when crossed to seedy mandarin accessions, seedlessness is caused by arrested seed development at an early stage (Yamasaki et al., 2007, 2009). The variety has been widely used in different programs in Japan and United States, and a few hybrids have been released, such as Bingo (USPP27778P3) and UFGlow (USPP27581P2). Chavez and Chaparro (2011) reported the identification of RAPD markers linked to seedlessness in the progenies of GS (a Robinson mandarin open pollinated seedling)×Mukaku Kishu; however, these markers were found to be population specific: they did not always work in other progenies derived from different seed parents (unpublished data, F.G. Gmitter).

Genetic male sterility is another mechanism that can be exploited to produce seedless genotypes. The inheritance of this trait has been reported to be polygenic but is still poorly understood. It is known that Satsuma selections, as well as Satsuma hybrids like Kiyomi and Harehime, confer undeveloped anthers and reduced pollen production to most of their progenies resulting in seedless fruit when grown in solid blocks (Nakano et al., 2001; Goto et al., 2016). Yamamoto et al. (1997) provided evidence that aborted anthers resulting from the use of Satsuma hybrids and Encore as seed parents were caused by an underlying nuclear gene-cytoplasmic interaction. As with self-incompatibility, it is somewhat difficult to identify male sterility in large breeding trials because their potential for seedlessness is obscured by cross-pollination. This also makes them less desirable for mixed commercial plantings where pollen from neighboring cultivars can cause seediness.

If seedless hybrids cannot be obtained through the exploitation of self-incompatibility, male and/or female sterility, or embryo disruption, and seedy but potentially valuable cultivars are generated, a commonly used strategy to induce sterility is the use of mutagenic agents (see Section 7.2.4).

7.2.3.2 Interspecific crosses with wild relatives to introgress resistance genes or other useful traits

One of the major applications of hybridization in traditional breeding approaches is the introgression of useful traits from relatives, especially resistance to pathogens. The potential value of crossing edible citrus varieties with inedible citrus relatives was recognized by pioneering breeders more than a century ago when Walter Swingle tried to produce cold-tolerant varieties. He used the extremely cold-tolerant relative *Poncirus trifoliata* (syn. *Citrus trifoliata*) and readily produced large numbers of fertile hybrids, many of which had superior cold tolerance than conventional citrus. Despite their improved cold tolerance, none of these hybrids was edible and subsequent breeders have struggled to resolve this problem. However, Swingle's efforts were not in vain because many of his hybrids, and more importantly his strategy of using a wild relative in breeding, have had a profound impact on world citriculture as rootstocks. Many useful and unique genes controlling important traits such as nematode, CTV, and *Phytophthora* tolerance are now present in commercial citrus production thanks to breeding programs that include *Poncirus* as a parent.

Efforts to hybridize *Citrus* spp. with relatively distant genera were made afterward, so that it is now known that many compatible combinations with *Citrus* are possible (Barrett, 1977; Iwamasa et al., 1988; Medina-Filho et al., 1998;

FIG. 7.3 Wide intergeneric hybrids between *Citrus wakonai* and *Atalantia ceylanica* (A); *Citropsis gabunensis* (B, C); *Hesperethusa crenulata* (D, E); and *Severinia buxifolia* (F, G) produced using conventional breeding techniques at Bundaberg Research Station, Australia. (A) Five vigorous hybrids ready for vegetative propagation and evaluation as rootstocks; (B, C) fruit of *Citrus wakonai* (*left*), two intergeneric hybrids (*center*) and *C. gabunensis* (*right*); (D) epigeous seed germination inherited from the *Hesperethusa* parent; (E) budded onto Troyer rootstock and superficially resembling a *Poncirus* hybrid; (F) fruiting field-grown hybrid on its own roots; and (G) intensely pigmented new growth under aphid attack.

Smith et al., 2013; Siebert Wooldridge, 2016). Unfortunately, none of these relatives have proven as commercially useful as *Poncirus*, and many have created difficulties from a practical breeding point of view. For example, although hybrids have been successfully generated from crosses between *Citrus* and distant genera including *Atalantia*, *Citropsis*, *Hesperethusa*, and *Severinia* (Fig. 7.3), these often fail to flower, and when they do have invariably been pollen sterile and produced seedless fruit.

Closer crosses such as those between *Microcitrus* and *Poncirus* or between *Citrus* and *Fortunella* sometimes develop a genetic condition called late-acting lethality in which well-established plants collapse and die even if they have been grafted onto a conventional rootstock. A distant relative can also have hidden faults that do not become obvious until they are used as parents. For example, the desert-adapted *Citrus glauca* (syn. *Eremocitrus glauca*) might be expected to contain useful genes for cold and drought tolerance, but in rootstock breeding its progeny has proven highly susceptible to *Phytophthora* and sensitive to herbicides. Lack of knowledge of the useful and detrimental traits of different citrus relatives has limited their use in breeding and partly explains why they have attracted only sparse and intermittent attention from breeders. For example, the failure of *Microcitrus* and its hybrids as rootstocks is now known to be caused by their extreme sensitivity to

CTV, and this knowledge has led to a breeding strategy in which bridging species are used to introgress CTV resistance from *Poncirus*. In the Australian breeding program, some advanced *Citrus australasica* (syn. *Microcitrus australasica*) hybrids containing CTV resistance genes from *Poncirus* are now showing considerable promise as rootstocks, and have *Phytophthora* resistance far superior to even *Poncirus*.

To obtain new scion cultivars by means of interspecific (i.e., with nonedible *Citrus* species) or intergeneric crosses, breeders must perform backcrosses to eliminate unwanted flavors and traits. Citrus-breeding programs with such ambitious goals must be planned on a decade-long basis. The USDA breeding program in Florida has been pursuing the development of edible hybrids that incorporate genetic elements from *P. trifoliata*, following on from Swingle's first efforts more than 100 years ago. Barrett (1990) released US 119 intergeneric hybrid from a cross of a CTV-resistant citrumelo with Succory sweet orange. This hybrid combines cold hardiness and CTV resistance from *Poncirus* with a degree of improved fruit quality and has been used as a breeding parent for subsequent generations of backcrossing. More recently, a new mandarin hybrid with 1/16 pedigree of *P. trifoliata* named US SunDragon (Stover et al., 2018) was released in the United States. This complex hybrid contains only a slight hint of the off-flavor typical of *Poncirus* hybrids, and more importantly appears to have a high degree of tolerance of HLB.

Recently, a potential new source of resistance to HLB was found in *Eremocitrus* and *Microcitrus* spp. (Ramadugu et al., 2016). These relatives are able to resist or recover from *C*Las infection; and therefore, they may represent a weapon to obtain long-term resistance in new cultivars. As a consequence of this discovery, *Eremocitrus* and *Microcitrus* are now being used extensively in many breeding programs around the world. To apply conventional breeding strategies more rapidly, breeders might take advantage of recently developed biotechnological approaches. For instance, a viral vector described by Velazquez et al. (2016) could dramatically decrease the length of such introgression strategies by shortening the juvenile period of each generation. Furthermore, if the genes associated with disease resistance can be identified then it may be possible to use cisgenesis or genome editing to avoid the need for hybridization.

7.2.3.3 Triploid breeding

The major interest in triploids is to generate sterility. In parthenocarpic plants, seedless fruits are obtained in most cases. Seedless fruit is one of the important characteristics for the fresh market, but fruit with reduced numbers of seeds can also be useful for the juice industry. Few natural triploids exist in *Citrus* germplasm (Curk et al., 2016). The most popular is without doubt the Tahiti lime, also called Bearss or Persian. The triploid Tahiti lime (*Citrus latifolia*) is thought to be a hybrid between a haploid ovule of *Citrus limon* and a diploid gamete of *Citrus aurantifolia* (Curk et al., 2016; Rouiss et al., 2018). Diploid gametes may have two possible origins, unreduced gametes from diploids and tetraploid meiosis. In the first case, the occurrence of spontaneous triploid seedlings from diploids was reported many years ago. Lapin (1937) mentioned about 4% triploidy in hybrid seedlings of *C. limon* crossed with eight other diploids. Later, a correlation was established between size of seeds and triploid embryos in monoembryonic mandarins. The percentage of triploids is higher in the small seeds (<5 mm) (Esen and Soost, 1973; Geraci et al., 1975; Ollitrault et al., 2007; Aleza et al., 2010). In the second case, tetraploids can be one parent of the triploids. Citrus trees are mostly diploid, but a few tetraploids exist in the germplasm like Giant Key lime or *Fortunella hindsii*. Most of them come from spontaneous tetraploidization events by chromosome doubling of nucellar cells (Aleza et al., 2011; Curk et al., 2016). These polyploidization events have been exploited by different breeding programs. Several methods have been developed by breeders to obtain triploids (Starrantino and Reforgiato Recupero, 1981; Williams and Roose, 2004; Ollitrault et al., 2007; Vardi et al., 2008; Aleza et al., 2010; Grosser and Gmitter, 2011; Navarro et al., 2015). The first dedicated triploid breeding programs began in the late 1970s, and some triploid cultivars were released in the last two decades. Some of those have not been successful, while others are performing well in specific environments. New promising triploids have been released in the last few years.

Triploid production capacity in 2x × 2x crosses is widely exploited in Spain, Italy, France, Israel, Morocco, and China, particularly in mandarin breeding programs (Geraci et al., 1975; Ollitrault et al., 2008; Aleza et al., 2010; Cuenca et al., 2011). The identification of triploid-producing monoembryonic female parents, like Fortune and Imperial mandarin has increased the efficiency of triploid breeding and many thousands have now been produced around the world. Second division restitution is the predominant mechanism involved in the origin of unreduced female gametes in different Clementine clones and mandarins (Luro et al., 2004; Cuenca et al., 2011, 2015). Many triploids obtained by this method have been released, like Winola (Spiegel-Roy and Vardi, 1992), Safor (IVIA, Spain; Cuenca et al., 2010), and Florina (CIRAD, France). For lemon, 1%, 5%, 5.8%, and 8.6% of triploid progenies were recovered from 2x × 2x sexual hybridizations using Lisbon, Eureka, Verna, and Fino lemons as the female parents, respectively (Geraci et al., 1975; Pérez-Tornero et al., 2012). Rouiss et al. (2017a) observed a similar frequency of 2n gamete production for Eureka Frost (4.9%) and Fino (8.3%) and they found that 88% of the hybrids arose from second division restitution, 7% from first-division restitution or premeiotic doubling, and 5% from postmeiotic genome doubling. In contrast, pummelo and citron produce very few triploids in diploid crosses.

The 2x×4x strategy have been used in many breeding programs but there are several factors that limits its efficiency. Formerly, the low number of tetraploids in germplasm collections was a limitation. However, the contribution of biotechnologies, such as flow cytometry and somatic hybridization, has made it possible to select many tetraploids in seedlings of polyembryonic plants (doubled diploid) (Aleza et al., 2011), after treatment of buds with colchicine (Aleza et al., 2009) or somatic hybrids produced by protoplast fusion (Grosser et al., 2000; Ollitrault et al., 2007); currently, there exists a rather substantial collection of tetraploid breeding parents globally. The second limit is these interploidy crosses generally yield mostly aborted seeds, and embryo rescue and in vitro culture are required to generate progenies. In addition, in some parental combinations the percentage of triploids is often equivalent to the percentage of tetraploids. Indeed, such crosses enable the encounter between a diploid-restituted gamete from the diploid female parent and a diploid gamete coming from the tetraploid pollen parent. Some triploid hybrids have been released, like Mandared or Mandalate coming from Nules Clementine (2x)×Tarocco sweet orange (4x) and Fortune mandarin (2x)×Avana mandarin (4x), respectively (Reforgiato Recupero et al., 2005). A pummelo-grapefruit hybrid cultivar, 914, was developed in Florida from an embryo rescued from undeveloped seed of Acidless pummelo crossed with pollen from a tetraploid, red-fleshed grapefruit that originated from a gigas-sectored fruit chimera (USPP26177P3; Bowman et al., 1991). 914 fruits are seedless and contain very low levels of the furanocoumarin compounds implicated in the grapefruit-drug interaction (Cancalon and Gmitter, 2013; Chen et al., 2015). The first triploid hybrid created using a somatic hybrid as pollen parent, C4-15-19 was recently released in Florida (USPP26086P3).

The 4x×2x strategy is feasible when tetraploid monoembryonic seed parents are available to breeders. Tetraploids have generally been obtained by the treatment of meristems with colchicine as in the case of tetraploid Clementine (Aleza et al., 2009). The other possibility is to select monoembryonic tetraploid hybrids from crosses between 4x and 2x. First-division restitution is the main mechanism of diploid pollen formation in mandarin hybrids (Rouiss et al., 2017b). The 4x×2x method provides the highest efficiency for triploid hybrid production in comparison with the others, because normally developed seeds result and there is no need for embryo rescue (Aleza et al., 2012). Shasta Gold, Tahoe Gold, and Yosemite Gold triploid mandarins were released by the Riverside breeding program in California (Williams and Roose, 2004), from 4x×2x hybridizations.

7.2.4 Mutation breeding

Mutation breeding techniques have made an important contribution to modern citrus production, with many commercial cultivars derived in part from the use of these methods. Indeed, citrus has benefited more than any other tree crop from the use of mutation breeding and within the last few decades new grapefruit, lemon, mandarin, and orange cultivars have been produced by both public and private breeding programs in different parts of the world.

Mutation breeding is predicated on the high frequency of natural mutations that occur in cultivated citrus, and the fact that many of these mutations have proven extremely beneficial to the development of world citriculture. Breeding using induced mutation is seen as a way of expanding and hastening this process. A range of mutagenic agents have been used for citrus breeding including chemical mutagens, UV light, high-energy beams, X-rays, fast neutrons, thermal neutrons, and gamma rays, although it is gamma rays that has become the standard mutagen and is the agent responsible for most of the commercially successful mutation-derived cultivars. Treated source material has included flowers and fruitlets, immature fruit, zygotic and nucellar seeds and seedlings, mature budwood, whole trees, and callus cultures. Treatment of mature budwood is now used almost exclusively in mutation breeding primarily because it enables well-characterized selections or cultivars to be phenotyped within just a few years for changes in a few traits, most notably seediness. Indeed, the technique has been employed for seed number reduction in citrus more than any other crop (Spiegel-Roy, 1990) and it is one of the few examples of an important commercial trait that can be reliably improved in citrus breeding in a relatively short time.

Mutagenic treatment of citrus germplasm was first reported in the 1930s (Haskins and Moore, 1935; Moore and Haskins, 1932) although it was not until the late 1950s that the technique was employed as a breeding tool (Hensz, 1960) resulting in the first mutation-induced commercial release (Star Ruby grapefruit) in 1970 (Hensz, 1977) in the United States. It has subsequently been employed by breeding programs in many regions of the world (Table 7.3).

Although originally employed as a way of improving existing commercial cultivars, mutation breeding has increasingly been used in a two-stage breeding process, originally envisaged by Vardi et al. (1996), in which high-quality diploid selections from conventional crosses are subjected to mutation treatment to create new seedless cultivars. The cultivar Moncalina (Pardo et al., 2012) is an example from this two-stage breeding process and many more are anticipated in the coming decades as breeding programs avoid releasing seedy diploids.

TABLE 7.3 Citrus mutation breeding programs in different regions of the world, including the year of the beginning of the program and examples of the released cultivars.

Country	Beginning of the program	Reference	Targeted citrus group	Released cultivars
USA, Texas	1959	Hensz (1977)	Grapefruit, orange	Star Ruby
China	1963	Deng (2000)	Orange	Zhongyu No 7
Israel	1970	Spiegel-Roy et al. (1972)	Mandarin, orange	Orri, Mor
USA, Florida	1970	Hearn (1984), McCollum and Hearn (2011)	Orange, grapefruit, mandarin	Early Pride
Italy	1974	Russo et al. (1981), Starrantino et al. (1988b)	Mandarin, lemon	Femminello Siracusano 2Kr
South Africa	1988	Plooy et al. (1992)	Mandarin, orange	Nova SL
Australia	1991	Smith (2006)	Mandarin	IrM1, IrM2
USA, California	1995	Roose and Williams (2007)	Mandarin	Tango
Turkey	2000	Gulsen et al. (2007)	Lemon	Gulsen, Eylul
Spain	2000	Bermejo et al. (2012)	Mandarin	Murta, Moncalina
Thailand	2000	Somsri et al. (2008)	Pummelo, mandarin	
Spain	2002	López-García et al. (2015)	Clementine	Nero, Neufina
USA, Florida	2007	Gasic et al. (2016)	Mandarin	Seedless-Snack
Pakistan	Unknown	Khalil et al. (2011)	Mandarin	Low-seeded Kinnow

Many researchers have examined the effect of varying doses (exposure), aiming to maximize the frequency of mutations without destroying viability. Consequently, LD_{50}'s have been established for seed and budwood of different citrus types and cultivars, although there is considerable discrepancy between studies. For example, Talon et al. (2004) established an LD_{50} of 78 Gy for the gamma-ray treatment of Nules Clementine, whereas Russo et al. (1981) working with Monreal Clementine had growth in only 40% of budsticks treated at 60 Gy and complete death at 80 Gy. Such discrepancies may be the result of factors such as moisture content (Hearn, 1984), which can have a dramatic impact on cellular damage from radiation (Van Harten, 1998). Pummelo is generally considered the most-sensitive citrus type and lemons the most tolerant. In practice, most breeding programs that use gamma rays have settled on a dose somewhere in the range of 30–80 Gy based on local experience with the genotypes important to their program. Vardi et al. (1996) argue that it is sufficient to use doses somewhat below the LD_{50} and present data showing that the frequency of identified mutation at 13 dose levels within the range of 27.5–65 Gy is no better at higher doses than it was at lower doses. This seemingly conflicts with the findings of Russo et al. (1981), who obtained variants from treatment at 20, 40, and 60 Gy, but reduced seed numbers were only observed with irradiation at 40 and 60 Gy. Similarly, Hearn (1984) also suggests that the most efficient dosage for seedlessness was higher than the LD_{50}. Gulsen et al. (2007) found that 50 Gy caused small modifications while 70 and 90 Gy caused large alterations when lemon budwood was subjected to irradiation. In a related study, Uzun et al. (2010) showed that high doses (70 Gy) in lemon breeding resulted in selections with larger fruit size whereas lower doses tended to decrease the fruit size of selected mutations. Furthermore, some recently released cultivars from mutation breeding programs have been generated at quite high doses. It would seem that more research is warranted in order to confirm whether the increased lethality of higher treatment dosage is compensated for by an increased frequency of commercially useful mutations.

Budwood irradiation causes single-cell events in multicellular bud apices, and so a vegetative shoot that develops from a single-treated bud can contain new buds that developed from different mutation events (as well as nonmutated cells). Fear of selecting unstable mericlinal chimeras has stimulated some breeders (Russo et al., 1981; Plooy et al., 1992; Bermejo et al., 2011) to employ a nursery subpropagation system in which buds that shoot away following irradiation are only allowed to develop into a shoot with 5–9 buds, each of which is then rebudded onto a new rootstock (Lapins, 1983). This strategy was also performed to obtain different low-seeded varieties from the Israeli breeding program (Goldenberg et al., 2014), choosing the 7th to the 12th buds for subpropagation. Thus, the first vegetative generation after mutation

(V_1M_1 or MV1) is quickly converted into the second vegetative after mutation (V_2M_1 or MV2) prior to field planting and phenotyping for traits that may have been altered by the mutation treatment. In this way, each tree from the V_2M_1 should be genetically uniform and can be phenotyped as a single unit. By contrast, other breeding programs (Smith, 2006; Roose and Williams, 2007) choose to plant the original V_1M_1 trees in the field and phenotype individual limbs, knowing that different limbs may have developed from different mutated cells. With this system, care must be taken to ensure limbs that develop from the first few buds are not pruned during field maintenance, and phenotyping is slower in order to ensure mutant tree sectors are not overlooked. Both propagation systems have been successfully employed in the development of new commercial citrus cultivars and there is no evidence that either system promotes the inadvertent selection of unstable periclinal chimeras. Indeed, Hearn (1986) found that plants without subpropagation gave the highest frequency of stable mutants. Consequently, the relative merits of each propagation system need to be aligned with the resources of individual breeding programs. In practice, there is little evidence of genetic instability in mutation-derived citrus cultivars, and most of the variability that has been attributed to genetic instability (Starrantino et al., 1988b; Qi et al., 2010) can be explained by environmental effects on phenotyping. It is not uncommon to select a putative low-seeded limb, based on a limited number of fruit available for testing, only to discover that daughter trees derived from this limb have significantly higher seed counts. The original limb selection was not unstable and has not reverted (back-sporting (Van Harten, 1998)) to higher seed number, but instead it is a case of incorrect phenotyping during the original selection process. For example, the data of Qi et al. (2010) from daughter trees of 14 selections, show that selections that averaged <1 seed per fruit for the whole 8-year assessment period never exceeded 1.5 seeds/fruit in any particular season. For these reasons, the system of V_2M_1 is effective not so much because it reduces the chance of selecting unstable chimeras but because it increases the accuracy of phenotyping during the breeding process. Field experience suggests that mutated selections need to be tested as multiple daughter trees for at least three seasons to ensure they have been accurately phenotyped. The relative performance of different selections varies little between geographical locations (Qi et al., 2010) and so it is more efficient to conduct most of the initial testing of selections at a single location.

Nearly all mutation breeding work aimed at inducing seedlessness has been conducted with citrus, reflecting the commercial importance of this trait and the relative ease with which it can be manipulated. The progenitors of cultivated citrus were intrinsically fertile, producing abundant flowers highly attractive to pollinating insects, copious quantities of pollen, and very large numbers of seed per fruit. It is not surprising then that mutagenesis, with its well-established detrimental effects on fertility, has been used so successfully in a crop where seedlessness is commercially desirable. However, the extent to which seed number is reduced following irradiation can vary enormously within the same genotype treated under identical conditions, and breeding programs invariably develop a number of putative low-seeded selections from each progenitor genotype. These low-seeded selections are then subject to broader testing before the best cultivar can be identified and commercialized. It has proven very easy to obtain statistically significant reductions in seed number, even with quite small mutation breeding populations. However, achieving complete seedlessness, or even very low seed numbers has proven far more difficult, except perhaps in highly parthenocarpic lemons (Spiegel-Roy et al., 1985) and mandarins (Williams and Roose, 2010), pointing to the importance of having strong parthenocarpic traits within breeding populations. Some successes with mutation breeding (Smith, 2006) have resulted in cultivars that are still too seedy for many markets, particularly as consumers increasingly demand truly seedless fruit. More research is needed on techniques to increase the efficiency with which completely seedless accessions can be generated, and mutation programs that combine commercial objectives with genomic analysis (Talon et al., 2004) may help in this regard.

Breeders employing mutation techniques have often observed significant changes in fruit acidity and/or maturity time (Talon et al., 2004; Hensz, 1977; Smith, 2006; Hearn, 1986) even within accessions primarily selected for lower seed number. These effects have been thoroughly studied in one of the Spanish mutation breeding programs (Bermejo et al., 2011, 2012, 2015) and show that significant changes in important traits like acidity and sugar composition can be induced via budwood irradiation. Flavonoid levels were also highly variable between the different mutated selections suggesting an opportunity to improve the nutritional value of citrus using mutation breeding. Excessively high acidity is a common problem in many breeding programs and mutation breeding solely for the purpose of acid-level modification may prove useful. Problems of efficiently identifying mutations (Spiegel-Roy, 1990) in a trait like acidity have probably restricted progress, although the use of NIR (Ruslan et al., 2012) and other noninvasive in-field technology has not been fully exploited.

Aside from seedlessness and acidity, other important traits have been altered, or have the potential to be improved, using mutation breeding techniques. For example, there are reports of tree size reduction (Spiegel-Roy and Vardi, 1989), improved fruit color (Smith, 2006), increased fruit size (Talon et al., 2004), and increased disease tolerance, particularly to mal secco (Starrantino et al., 1988a; Gulsen et al., 2007). However, in the case of the Italian breeding program the reported increased tolerance of an irradiated genotype (Starrantino et al., 1988a) was not confirmed after extensive field trials. In other crops, attempts to improve disease tolerance through mutation breeding have generally been disappointing relative to the amount of research investment, in part because mutations predominantly go from dominant to recessive whereas most resistances are dominant (Van Harten, 1998). This is not the case for some important diseases of citrus, and more mutation breeding for disease resistance may be anticipated particularly as better screening techniques are developed.

Mutation breeding will continue to be an important seed-reduction technique by which citrus breeders can improve the commercial value of germplasm. It will also prove useful in revealing genetic variability masked by dominant genes at heterozygous loci and remains to be fully exploited in the improvement of traits as diverse as acidity, color, and disease resistance.

7.3 Rootstock breeding

7.3.1 Objectives

Plant adaptability to specific soil and climatic conditions, water quality and availability, nutrient use efficiency, and biotic stresses mostly depends on rootstock choice. The generation and release of an improved rootstock requires different and usually longer protocols compared to scions. Rootstock breeding programs primarily take into account all the aspects related to plant resilience, but, at the same time, a good rootstock should confer high productivity and high fruit quality to the grafted scions. Objectives vary depending on the major threats and limitations of the different growing areas. Selection is generally performed for vigor, resistance to root rots, CTV and nematodes, and tolerance to biotic stresses such as salinity, high-lime soil, drought, and cold. A first evaluation is made directly in the field or in a greenhouse for specific traits. Later, each promising rootstock is evaluated in combination with different varieties to verify its effects on fruit quality (sugars, acidity, granulation, fruit size, fruit color), yield and yield efficiency, and possible graft incompatibility, that can be visible several years after grafting (Bitters, 1986). A comprehensive list of traits controlled or influenced by the rootstock was published by Castle (2010). The need for the evaluation of multiple characters limits the possibility of releasing new rootstocks in a short time frame.

7.3.2 Conventional methods to generate new rootstocks

Hybridization played a major role in the development of many citrus rootstocks. With the exception of a few true and derived species such as trifoliate orange, sour orange, rough lemon, Yuzu, Rangpur lime, Volkamer lemon, and *Citrus macrophylla*, many rootstocks that are now showing the best performances worldwide are derived from systematic breeding programs based on controlled crosses.

Many citrus species and sexually compatible relatives have been used to develop rootstocks based on their tolerance to biotic and abiotic stresses and on their ability to confer good qualitative features to the scion. *P. trifoliata* and its hybrids are essential parents in many breeding programs, since they confer tolerance or resistance to CTV, *Phytophthora*, nematodes, cold, and high fruit quality (Castle, 1987). Moreover, all F_1 hybrids show the dominant trifoliate character so they can be easily recognized by their morphology, while segregation of trifoliate character using *Poncirus* hybrids as parents is more complex (Cameron and Soost, 1980; Caruso et al., 2014). Other parents that have been commonly used in breeding programs are Cleopatra mandarin for its tolerance to salinity (Ream and Furr, 1976), and Sunki mandarin for its tolerance to blight (Cristofani et al., 1999).

Most of the programs around the world are based on diploid hybridization, but polyploidy also plays a role in the development of improved genotypes. Autopolyploid rootstocks have been shown to improve tolerance to salt stress (Saleh et al., 2008) and drought (Allario et al., 2013), and to reduce tree size (Lee et al., 1990). Moreover, allotetraploid parents obtained via protoplast fusion have been used to obtain new progenies, called tetrazygs, through conventional hybridization (Grosser and Gmitter, 2011).

7.3.3 Propagation

A high degree of polyembryony has traditionally been considered an essential characteristic of a good rootstock, because it favors true-to-type propagation from seeds. However, nucellar polyembryony, particularly useful for nurserymen, has been an obstacle for rootstock breeders since it hampered the production of large segregating hybrid populations. Typically, highly polyembryonic hybrids are obtained from highly polyembryonic parents, but with this strategy a small number of hybrid offspring can be recovered.

An alternative strategy relies on the use monoembryonic seed parents. This allows the generation of large hybrid families, but on the other hand a variable number of monoembryonic or scarcely polyembryonic individuals will be discarded in the absence of an efficient method for clonal propagation. The efficiency of this strategy also seems to depend on the proportion of nucellar seedlings that result when using specific pollen parents. At the CREA, Sicily, breeders hybridized a monoembryonic accession of *Citrus latipes* with three pollen parents, each of which possessed different degrees of polyembryony as seed parents, namely sour orange, trifoliate orange, and Volkamer lemon (Reforgiato Recupero and Russo, 1992). Around 500 offsprings were phenotyped for 2–5 years for their seed embryo number, and only a very small percentage of them had high enough polyembryony for them to be included in rootstock trials (Reforgiato Recupero et al., 2009). The majority of hybrids (especially those derived from sour orange and Volkamer lemon pollen parents) were discarded for their low polyembryony.

Even when rootstocks are considered highly polyembryonic and can be propagated by seeds, accurate phenotyping and/or genotyping should be done to avoid the presence of off-types that can increase variability and lead to a wrong estimation of rootstock performance (Xiang and Roose, 1988; Hussain et al., 2011). The recent development of molecular markers linked to polyembryony (Wang et al., 2017) should certainly help breeding programs; however, the proportion of zygotic seedlings produced by nucellar hybrids will still need to be verified.

Techniques for the vegetative propagation of citrus rootstocks have been available for many decades (Johnston et al., 1959; Fucik and Hensz, 1969; Barlass and Skene, 1982; Ferguson et al., 1985), although this has had little impact on industry preference for rootstocks that come true-to-type from seed (Roose, 1996). More recently, clonal propagation by stem cuttings (Bowman and McCollum, 2015; Bowman and Albrecht, 2017) and in vitro propagation (Albrecht et al., 2017) have been optimized and are being performed on a large scale by research institutions and private companies. This certainly enables the use of valuable rootstocks that are monoembryony or have low polyembryonic capacity, as well as overcoming problems caused by a lack of sufficient quantities of seeds (which is currently occurring for Forner Alcaide 5 in Europe and different US-series rootstocks in Florida).

Since the widespread use of vegetative propagation for rootstock propagation is relatively recent, there is limited information regarding the performance of rootstocks obtained by cuttings or tissue culture compared to nucellar seedlings in long-term field trials. Sykes (2010) compared the performance of micropropagated and seedling rootstocks for four different rootstock cultivars and could find no difference. He carefully examined tree growth and fruit quality on 10-year-old trees for four consecutive seasons and presents a convincing case that vegetatively propagated rootstocks perform in the same way as seedlings. Analysis of root architecture and growth revealed a more developed root system in rooted cuttings and micropropagated plants than in nucellar seedlings (Albrecht et al., 2017) but whether this translates into field performance differences has yet to be determined.

7.3.4 Phenotyping methods for diseases and abiotic stress resistance before field trials

The most efficient programs prescreen plants in the greenhouse to quickly eliminate unwanted material. A range of protocols have been used to test new hybrids for *Phytophthora* resistance (Furr and Carpenter, 1961; Alvarez et al., 2008). To test susceptibility to citrus CTV, new hybrids can be graft inoculated with different strains and then tested for virus replication. These methods use different grafting protocols (Garnsey et al., 1987) and verify virus replication and movement using ELISA or direct tissue blot immunoassay (Gmitter et al., 1996; Smith et al., 2016). Well-established methods have been optimized to test resistance to nematodes (Verdejo-Lucas et al., 1997; Ling et al., 2000). Regarding biotic stresses, protocols have been released to test seedlings and grafted plants for their tolerance to water stress (Rodríguez-Gamir et al., 2010), salinity (Sykes, 1985; Forner-Giner et al., 2011), and iron chlorosis (Pestana et al., 2005; Castle et al., 2009; Martinez-Cuenca et al., 2016). However, while there are existing and well-established methods to screen for many of the main biotic and abiotic conditions, these methods are costly and time consuming, especially if done using multiple replicates of each individual hybrid. Therefore, the ability of citrus rootstock breeders to screen large segregating populations of hybrids for all important traits prior to field trials remains more of a goal than a reality. Rootstock breeders are forced to prioritize and balance the importance of any given trait against the expense of screening for it.

For some diseases, there are no quick or simple screening techniques available. For example, HLB symptom expression requires several months or even years to be seen; so, performance in the field under natural pressure is required. A brute force approach to screening rootstock hybrid progeny has been employed at the University of Florida and has been called The Gauntlet (J.W. Grosser, personal communication). Seedlings are collected annually from crosses made and then germinated in a heavy calcareous soil with high pH values and inoculated with two species of *Phytophthora* common in Florida. Only healthy seedlings that survive move to the next stage. The seedlings are top grafted with a budstick, not individual buds, that came from an HLB-affected tree; the rootstock seedling top is removed and rooted under mist to retain the original hybrid. Trees that produce two healthy flushes after grafting with few or no HLB symptoms move to the next phase: Asian citrus psyllid (ACP) inoculation in growth rooms or greenhouses harboring *C*Las infected ACP. Individual selections displaying few or no symptoms are then moved to a field site where ACP and HLB pressure is very high, ACP control is very minimal, and trees are allowed to grow for further observations for several years. This approach has identified several very promising candidates that have been repropagated from the original rooted cutting for advanced and replicated field trials.

7.3.5 Rootstock trials

Prescreening for biotic and abiotic stress might be helpful to eliminate unwanted material. However, only field trials provide a reliable estimation of the true value of new rootstocks. Castle (2010) suggested that a trial can give a reliable estimation of a rootstocks performance 7–8 years after planting, while Reforgiato Recupero et al. (2009) reported that, in the case of iron chlorosis, symptoms were visible within the first 10 years after planting. Similarly, Smith et al. (2004) examined rootstock effects on cumulative yield over 23 consecutive seasons and concluded that there was only marginal value from assessments beyond the 10th season of cropping.

Traditional rootstock trials often involve a large number of replicates of multiple tree plots and are evaluated for many years, thus requiring significant financial and labor inputs and a sustained commitment. Trials usually include one or more standards as reference. Cumulative production and qualitative features are evaluated for at least 3–4 years, although data from more years are preferred to have a better estimate of a rootstock potential. Adequate replication and proper statistical design are essential to the identification of superior rootstocks particularly for traits that can be influenced by erratic and nonuniform spatial effects such as soil variation and pest and disease pressure. However, traditional approaches to field design can sometimes cause genetic diversity to be sacrificed in order to satisfy replication requirements. An alternative approach to field trials is to utilize new statistical techniques that account for spatial effects and pedigree structure. These new statistical approaches have been used to great effect in other fields of agriculture and resulted in improved breeding efficiency. Their application in citrus rootstock field trials has the potential to generate genetic information and identify the best parents and families rather than simply compare a particular set of different rootstocks. More efficient methods of conducting field testing and increasing the capacity to include greater genetic variability requires continued attention, particularly when exploring new genetic combinations.

Field trials are also essential to verify the possible tolerance of different specific rootstocks or graft combinations to HLB (Albrecht and Bowman, 2012; Castle et al., 2015; Stover et al., 2016; see description of The Gauntlet, above). HLB is a disease unlike almost any other in citrus, in terms of severity and lack of options to minimize its impact. It requires a different approach to the traditional conservative strategy of multiyear, multiscion, and multilocation rootstock trials. Both the USDA and UF breeding programs in Florida have released several new rootstocks for the industry, based almost exclusively on their apparent tolerance of HLB. Tolerance is not described in terms of pathology, but rather as an economic concept. Slower rates of infection and symptom expression, and the ability of the rootstock to retain a healthier and more consistently productive canopy over time have been the criteria driving the release of many of these new rootstocks, without the luxury of longer evaluation periods. This unproven methodology has resulted in new and speculative rootstock releases, and time will tell whether they are adopted by growers and assist in combating the impacts of HLB. For many growers they represent the only option until more robust tolerant or resistant types become available.

7.4 Perspectives

Traditional breeding still relies on the generation of large hybrid families and on phenotyping, which is commonly the bottleneck that slows down breeding programs and exponentially increases their costs. In the last few years, major advances have been made to improve the efficiency of breeding programs, due to the optimization of phenotyping protocols to speed up early selection and more frequent use of MAS. Compared to other major fruit crops, markers linked to major traits are still limited. However, there has been significant progress made over the last decade enabled MAS for traits such as polyembryony, *Alternaria* brown spot resistance and anthocyanin pigmentation, and several QTLs associated with key traits have been identified (Yu et al., 2016; Imai et al., 2017; Curtolo et al., 2017). Recently, the application of GWAS and genomic selection in citrus populations revealed the potential benefits of these strategies to understand the genetic control of other major traits, such as fruit weight, and peel and flesh color (Minamikawa et al., 2017). Stronger collaboration with molecular biologists and geneticists, to identify the genes controlling other major traits, is required to make breeding programs more efficient.

References

Albrecht, U., Bowman, K.D., 2012. Tolerance of trifoliate citrus rootstock hybrids to *Candidatus* Liberibacter asiaticus. Sci. Hortic. 147, 71–80.

Albrecht, U., Bordas, M., Lamb, B., Meyering, B., Bowman, K.D., 2017. Influence of propagation method on root architecture and other traits of young citrus rootstock plants. HortScience 52, 1569–1576.

Aleza, P., Juárez, J., Ollitrault, P., Navarro, L., 2009. Production of tetraploid plants of non-apomictic citrus genotypes. Plant Cell Rep. 28, 1837–1846.

Aleza, P., Juárez, J., Cuenca, J., Ollitrault, P., Navarro, L., 2010. Recovery of citrus triploid hybrids by embryo rescue and flow cytometry from 2x × 2x sexual hybridisation and its application to extensive breeding programs. Plant Cell Rep. 29 (9), 1023–1034.

Aleza, P., Froelicher, Y., Schwarz, S., Agustí, M., Hernández, M., Juárez, J., Luro, F., Morillon, R., Navarro, L., Ollitrault, P., 2011. Tetraploidization events by chromosome doubling of nucellar cells are frequent in apomictic citrus and are dependent on genotype and environment. Ann. Bot. 108 (1), 37–50.

Aleza, P., Juárez, J., Hernández, M., Ollitrault, P., Navarro, L., 2012. Implementation of extensive citrus triploid breeding programs based on 4x × 2x sexual hybridizations. Tree Genet. Genomes 8 (6), 1293–1306.

Allario, T., Brumos, J., Colmenero-Flores, J.M., Iglesias, D.J., Pina, J.A., Navarro, L., Talon, M., Ollitrault, P., Morillon, R., 2013. Tetraploid Rangpur lime rootstock increases drought tolerance via enhanced constitutive root abscisic acid production. Plant Cell Environ. 36, 856–868.

Alvarez, L.A., Vicent, A., De la Roca, E., Bascón, J., Abad Campos, P., Armengol, J., García-Jiménez, J., 2008. Branch cankers on citrus trees in Spain caused by *Phytophthora citrophthora*. Plant Pathol. 57, 84–91.

Aprile, A., Federici, C., Close, T., De Bellis, L., Cattivelli, L., Roose, M., 2011. High and low acid lemons: origin and transcriptome comparisons. Acta Hortic. 892, 37–42.

Barlass, M., Skene, K.G.M., 1982. In vitro plantlet formation from *Citrus* species and hybrids. Sci. Hortic. 17 (4), 333–341.

Barrett, H.C., 1974. Colchicine-induced polyploidy in *Citrus*. Bot. Gaz. 135, 29–41.

Barrett, H.C., 1977. Intergeneric hybridization of *Citrus* and other genera in citrus cultivar improvement. Proc. Int. Soc. Citricult. 586–589.

Barrett, H.C., 1990. US119, an intergeneric hybrid citrus scion breeding line. HortScience 25 (12), 1670–1671.

Bermejo, A., Pardo, J., Cano, A., 2011. Influence of gamma irradiation on seedless citrus production: pollen germination and fruit quality. Food Nutr. Sci. 2, 169–180.

Bermejo, A., Pardo, J., Cano, A., 2012. Murcott seedless: influence of gamma irradiation on citrus production and fruit quality. Span. J. Agric. Res. 10, 768–777.

Bermejo, A., Pardo, J., Zaragoza, S., 2015. Influence of gamma irradiation on seedless citrus production: pollen germination and fruit quality. Acta Hortic. 1065, 229–237.

Bitters, W.P., 1986. Citrus Rootstocks: Their Characters and Reactions. UC Riverside Science Library. http://www.citrusvariety.ucr.edu/links/documents/Bitters.pdf.

Bowman, K.D., Albrecht, U., 2017. Efficient propagation of citrus rootstocks by stem cuttings. Sci. Hortic. 225, 681–688.

Bowman, K.D., McCollum, G., 2015. Five new citrus rootstocks with improved tolerance to huanglongbing. HortScience 50, 1731–1734.

Bowman, K.D., Gmitter, F.G., Moore, G.A., Rouseff, R.L., 1991. Citrus-fruit sector chimeras as a genetic resource for cultivar improvement. J. Am. Soc. Hortic. Sci. 116 (5), 888–893.

Bruno, A., 1962. A vegetative method for producing virus-free nucellar citrus budwood. Nature 195, 1121–1122.

Butelli, E., Licciardello, C., Zhang, Y., Liu, J., Mackay, S., Bailey, P., et al., 2012. Retrotransposons control fruit-specific, cold-dependent accumulation of anthocyanins in blood oranges. Plant Cell 24, 1242–1255.

Cameron, J.W., Frost, H.B., 1968. Genetics, breeding, and nucellar embryony. In: Reuther, W., Batchelor, L.D., Webber, H.J. (Eds.), The Citrus Industry. vol. 2. University of California Press, Berkeley, pp. 325–370.

Cameron, J.W., Soost, R.K., 1980. Leaf types of F_1 hybrids and back-crosses involving unifoliate *Citrus* and trifoliate *Poncirus*. J. Am. Soc. Hortic. Sci. 105, 517–519.

Cameron, J.W., Soost, R.K., Olson, E.O., 1964. Chimeric basis for color in pink and red grape fruit. J. Hered. 55, 23–28.

Cancalon, P.F., Gmitter Jr., F.G., 2013. New grapefruit and pummelo cultivars with very low furanocoumarin contents are good candidates to provide a solution to the drug interaction problem. Fruit Process. 3, 126–129.

Carbonell-Bejerano, P., Royo, C., Torres-Pérez, R., Grimplet, J., Fernandez, L., Franco-Zorrilla, J.M., et al., 2017. Catastrophic unbalanced genome rearrangements cause somatic loss of berry color in grapevine. Plant Physiol. 175 (2), 786–802.

Carrante, V., Bottari, V., 1951. Miglioramento genetico del limone e ricerca di varietà resistenti al "mal secco". Ann. Sper. Agrar. 6, 323–346. [in Italian].

Caruso, M., Distefano, G., Pietro Paolo, D., La Malfa, S., Russo, G., Gentile, A., Reforgiato Recupero, G., 2014. High resolution melting analysis for early identification of citrus hybrids: a reliable tool to overcome the limitations of morphological markers and assist rootstock breeding. Sci. Hortic. 180, 199–206.

Caruso, M., Ferlito, F., Licciardello, C., Allegra, M., Strano, M.C., Silvestro, D., et al., 2016. Pomological diversity of the Italian blood orange germplasm. Sci. Hortic. 213, 331–339.

Castle, W.S., 1987. Citrus rootstocks. In: Rom, R.C., Carlson, R.F. (Eds.), Rootstocks for Fruit Crops. Wiley, New York, pp. 361–399.

Castle, W.S., 2010. A career perspective on citrus rootstocks, their development, and commercialization. HortScience 45, 11–15.

Castle, W.S., Nunnallee, J., Manthey, J.A., 2009. Screening citrus rootstocks and related selections in soil and solution culture for tolerance to low-iron stress. HortScience 44, 638–645.

Castle, B., Grosser, J., Bowman, K., Stover, E., 2015. An HLB-tolerant citrus rootstock: what exactly does that mean? Citrus Ind. 16–19.

Chavez, D.J., Chaparro, J.X., 2011. Identification of markers linked to seedlessness in *Citrus kinokuni* hort. ex Tanaka and its progeny using bulked segregant analysis. HortScience 46, 693–697.
Chen, C., Yu, Q., Wei, X., Cancalon, P.F., Gmitter Jr., F.G., 2015. Identification of genes associated with low furanocoumarin content in grapefruit. Genome 10, 537–545.
Cook, M.T., 1907. Notes on polyembryony. Torreya 7, 113–117.
Cooper, W.C., Reece, P.C., Furr, J.R., 1962. *Citrus* breeding in Florida: past, present and future. Proc. Fla. State Hortic. Soc. 75, 5–13.
Cristofani, M., Machado, M.A., Grattapaglia, D., 1999. Genetic linkage maps of *Citrus sunki* Hort. ex. Tan. and *Poncirus trifoliata* (L.) Raf. and mapping of citrus tristeza virus resistance gene. Euphytica 109, 25–32.
Cuenca, J., Aleza, P., Juárez, J., Pina, J.A., Navarro, L., 2010. 'Safor' mandarin: a new citrus mid-late triploid hybrid. HortScience 45 (6), 977–980.
Cuenca, J., Froelicher, Y., Aleza, P., Juárez, J., Navarro, L., Ollitrault, P., 2011. Multilocus half-tetrad analysis and centromere mapping in citrus: evidence of SDR mechanism for 2n megagametophyte production and partial chiasma interference in mandarin cv 'Fortune'. Heredity 107, 462–470.
Cuenca, J., Aleza, P., Juárez, J., García-Lor, A., Froelicher, Y., Navarro, L., Ollitrault, P., 2015. Maximum-likelihood method identifies meiotic restitution mechanism from heterozygosity transmission of centromeric loci: application in citrus. Sci. Rep. 5, 9897.
Curk, F., Ollitrault, F., Garcia-Lor, A., Luro, F., Navarro, L., Ollitrault, P., 2016. Phylogenetic origin of limes and lemons revealed by cytoplasmic and nuclear markers. Ann. Bot. 117 (4), 565–583.
Curtolo, M., Cristofani-Yaly, M., Gazaffi, R., Takita, M.A., Figueira, A., Machado, M.A., 2017. QTL mapping for fruit quality in *Citrus* using DArTseq markers. BMC Genomics 18, 289–305.
Deng, X.X., 2000. *Citrus* cultivars released during the past 10 years in China. Proc. Int. Soc. Citricult. 1, 37–38.
Distefano, G., Las Casas, G., La Malfa, S., Gentile, A., Tribulato, E., 2009. Pollen tube behavior in different mandarin hybrids. J. Am. Soc. Hortic. Sci. 134, 583–588.
Esen, A., Soost, R.K., 1973. Precocious development and germination of spontaneous triploid seeds in *Citrus*. J. Hered. 64 (3), 147–154.
Esen, A., Soost, R.K., 1977. Adventive embryogenesis in *Citrus* and its relation to pollination and fertilization. Am. J. Bot. 64, 607–614.
Ferguson, J., Young, M., Halvorson, J., 1985. The propagation of citrus rootstocks by stem cuttings. Proc. Fla. State Hortic. Soc. 98, 39–42.
Forner-Giner, M.A., Legaz, F., Primo-Millo, E., Forner, J., 2011. Nutritional responses of Citrus rootstocks to salinity: performance of the new hybrids Forner-Alcaide 5 and Forner-Alcaide 13. J. Plant Nutr. 34, 1437–1452.
Frost, H.B., Krug, C.A., 1942. Diploid-tetraploid periclinal chimeras as bud variants in citrus. Genetics 27, 619–634.
Frost, H.B., Soost, R.K., 1968. Seed reproduction: development of gametes and embryos. In: Reuther, W., Batchelor, L.D., Webber, H.B. (Eds.), The Citrus Industry. vol. 2. University of California, Riverside, CA, USA, pp. 290–324.
Fucik, J.E., Hensz, R.A., 1969. Rooting of unbudded and budded citrus cuttings. J. Rio Grande Valley Hortic. Soc. 23, 10–17.
Furr, J.R., Carpenter, J.B., 1961. Program for breeding citrus rootstocks tolerant to Phytophthora foot rot. Proc. Fla. State Hortic. Soc. 74, 18–23.
Garnsey, S.M., Barrett, H.C., Hutchison, D.J., 1987. Identification of citrus tristeza virus resistance in citrus relatives and its potential applications. Phytophylactica 19, 187–191.
Gasic, K., Preece, J.E., Karp, D., 2016. Register of new fruit and nut cultivars list 48. HortScience 51 (6), 620.
Geraci, G., Esen, A., Soost, R.K., 1975. Triploid progenies 2x X 2x crosses of *Citrus* cultivars. J. Hered. 66, 177–178.
Gmitter Jr., F.G., Moore, G.A., 1986. Plant regeneration from undeveloped ovules and embryogenic calli of *Citrus*: embryo production, germination and plant survival. Plant Cell Tissue Organ Cult. 6, 139–147.
Gmitter Jr., F.G., Xiao, S.Y., Huang, S., Hu, X.L., Garnsey, S.M., Deng, Z., 1996. A localized linkage map of the citrus tristeza virus resistance gene region. Theor. Appl. Genet. 92, 688–695.
Goldenberg, L., Yaniv, Y., Porat, R., Carmi, N., 2014. Effect of gamma irradiation mutagenesis for induction of seedlessness, on the quality of mandarin fruit. Food Nutr. Sci. 5, 943–952.
Goto, S., Yoshioka, T., Ohta, S., Kita, M., Hamada, H., Shimizu, T., 2016. Segregation and heritability of male sterility in populations derived from progeny of Satsuma mandarin. PLoS One 11, e0162408.
Grosser, J.W., Gmitter, F.G., 2011. Protoplast fusion for production of tetraploids and triploids: applications for scion and rootstock breeding in citrus. Plant Cell Tissue Organ Cult. 104 (3), 343–357.
Grosser, J.W., Ollitrault, P., Olivares-Fuster, O., 2000. Somatic hybridization in citrus: an effective tool to facilitate variety improvement. In Vitro Cell. Dev. Biol. Plant 36 (6), 434–449.
Grosser, J., Gmitter, F.G., Castle, W.S., 2015. New sweet orange cultivars. Citrus Ind. 96 (5), 10–18.
Gulsen, O., Uzun, A., Pala, H., Canihos, E., Kafa, G., 2007. Development of seedless and mal secco tolerant mutant lemons through budwood irradiation. Sci. Hortic. 112, 184–190.
Haskins, C.P., Moore, C.N., 1935. Growth modifications in citrus seedlings grown from X-rayed seed. Plant Physiol. 10, 179–185.
Hearn, C.J., 1984. Development of seedless orange and grapefruit cultivars through seed irradiation. J. Am. Soc. Hortic. Sci. 109, 270–273.
Hearn, C.J., 1986. Development of seedless grapefruit cultivars through budwood irradiation. Soc. Hortic. Sci. 111, 304–306.
Hensz, R.A., 1960. Effects of X-rays and thermal neutrons on citrus propagating material. J. Rio Grande Valley Hortic. Soc. 14, 21–25.
Hensz, R.A., 1977. Mutation breeding and the development of the 'Star Ruby' grapefruit. Proc. Int. Soc. Citricult. 2, 582–585.
Hocquigny, S., Pelsy, F., Dumas, V., Kindt, S., He´loir, M.C., Merdinoglu, D., 2004. Diversification within grapevine cultivars goes through chimeric states. Genome 47, 579–589.
Hussain, S., Curk, F., Ollitrault, P., Morillon, R., Luro, F., 2011. Facultative apomixis and chromosome doubling are sources of heterogeneity in citrus rootstock trials: impact on clementine production and breeding selection. Sci. Hortic. 130, 815–819.

Imai, A., Yoshioka, T., Hayashi, T., 2017. Quantitative trait locus (QTL) analysis of fruit-quality traits for mandarin breeding in Japan. Tree Genet. Genomes 13, 79.

Iwamasa, M., Nito, N., Ling, J.T., 1988. Intra and intergeneric hybridization in the orange subfamily, Aurantioideae. Proc. Int. Soc. Citricult. 123–130.

Johnston, J.C., Opitz, K.W., Frolich, E.F., 1959. *Citrus* propagation. Calif. Agr. Expt. Sta. Cir. 475, 1–22.

Kender, W.J., 2003. *Citrus*. HortScience 38, 1043–1047.

Kepiro, J.L., Roose, M.L., 2007. Nucellar embryony. In: Khan, I.A. (Ed.), *Citrus* Genetics, Breeding and Biotechnology. CAB International, Wallingford, UK, pp. 141–150.

Khalil, S.A., Sattar, A., Zamir, R., 2011. Development of sparse-seeded mutant Kinnow (*Citrus reticulata* Blanco) through budwood irradiation. Afr. J. Biotechnol. 10, 14562–14565.

Lapin, W.K., 1937. Investigation on polyploidy in *Citrus*. U.S.S.R. All-Union Sci. Res. Inst. Humid Sub-tropic Works 1 (4), 1–68.

Lapins, K.O., 1983. Mutation breeding. In: Moore, J.N., Janick, J. (Eds.), Methods in Fruit Breeding. Purdue University Press, West Lafayette, pp. 74–99.

Lee, L.S., Gillespie, D., Shaw, R., 1990. Prospects for using *citrus* tetraploids for rootstocks. In: Proceedings of the 3rd World Congress of the International Society of Citrus Nurserymen, Australia.

Ling, P., Duncan, L.W., Deng, Z., Dunn, D., Xu, X., Huang, S., Gmitter, F.G., 2000. Inheritance of citrus nematode resistance and its linkage with molecular markers. Theor. Appl. Genet. 101, 1010–1017.

Liu, Y.Z., Deng, X.X., 2007. *Citrus* breeding and genetics in China. Asian Australas. J. Plant Sci. Biotechnol. 1, 23–28.

López-García, A., Terol, J., Tadeo, F.R., Herrero-Ortega, A., Ibañez, V., Talón, M., 2015. Three new cultivars of clementine: 'Clemenverd', 'Nero' and 'Neufina'. Acta Hortic. 1065, 239–243.

Luro, F., Maddy, F., Jacquemond, C., Froelicher, Y., Morillon, R., Rist, D., Ollitrault, P., 2004. Identification and evaluation of diplogyny in clementine (*Citrus clementina*) for use in breeding. Acta Hortic. 663 (2), 841–848.

Marroni, F., Scaglione, D., Pinosio, S., Policriti, A., Miculan, M., Di Gaspero, G., et al., 2017. Reduction of heterozygosity (ROH) as a method to detect mosaic structural variation. Plant Biotechnol. J. 7, 791–793.

Martinez-Cuenca, M.R., Quinones, A., Forner-Giner, M.A., 2016. Screening of 'King' mandarin (*Citrus nobilis* Lour) x *Poncirus trifoliata* ((L.) Raf.) hybrids as citrus rootstocks tolerant to iron chlorosis. Sci. Hortic. 198, 61–69.

McCollum, T.G., Hearn, C.J., 2011. 'US Early Pride', a very low-seeded, early-maturing mandarin hybrid. HortScience 46, 1695–1697.

Medina-Filho, H.P., Bordognon, R., Ballvé, R., 1998. Sunkifolias and Buxisunkis: sexually obtained reciprocal hybrids of *Citrus sunki* x *Severinia buxifolia*. Genet. Mol. Biol. 21, 1–9.

Minamikawa, M.F., Nonaka, K., Kaminuma, E., Kajiya-Kanegae, H., Onogi, A., Goto, S., Yoshioka, T., et al., 2017. Genome-wide association study and genomic prediction in citrus: potential of genomics-assisted breeding for fruit quality traits. Sci. Rep. 7 (1), 4721–4734.

Moore, C.N., Haskins, C.P., 1932. Note on premature flowering in grapefruit from X-rayed seeds. Science 76, 167–168.

Nadori, E., 2004. Nadorcott mandarin: a promising new variety. Proc. Int. Soc. Citricult. 1, 356–359.

Nakano, M., Nesumi, H., Yoshioka, T., Yoshida, T., 2001. Segregation of plant with undeveloped anthers among hybrids derived from the seed parent, 'Kiyomi' (*Citrus unshiu* × *C. sinensis*). J. Jpn. Soc. Hortic. Sci. 70, 539–545.

Navarro, L., 2015. The Spanish citrus industry. Acta Hortic. 1065, 41–48.

Navarro, L., Juarez, J., 1977. Tissue culture techniques used in Spain to recover virus-free citrus plants. Acta Hortic. 78, 425–435.

Navarro, L., Roistacher, C.N., Murashige, T., 1975. Improvement of shoot-tip grafting in vitro for virus-free citrus. J. Am. Soc. Hortic. Sci. 100, 471–479.

Navarro, L., Aleza, P., Cuenca, J., Juárez, J., Pina, J.A., Ortega, C., Navarro, A., Ortega, V., 2015. The mandarin triploid breeding program in Spain. Acta Hortic. 1065, 389–396.

Nishiura, M., 1964. *Citrus* breeding and bud selection in Japan. Proc. Fla. State Hortic. Soc. 77, 79–83.

Ollitrault, P., Luro, F., Yamamoto, M., 2007. Seedlessness and ploidy manipulations. In: Khan, I.A. (Ed.), *Citrus* Genetics, Breeding and Biotechnology. CAB International, Wallingford, UK, pp. 197–218.

Ollitrault, P., Dambier, D., Luro, F., Froelicher, Y., 2008. Ploidy manipulation for breeding seedless triploid citrus. Plant Breed. Rev. 30, 323–352.

Omura, M., Shimada, T., 2016. *Citrus* breeding, genetics and genomics in Japan. Breed. Sci. 66, 3–17.

Pardo, J., Bermejo, A., Cano, A., Zaragoza, S., 2012. Development and characteristics of mandarin cv. Moncalina. Levante Agricola 51, 8–14.

Pelsy, F., 2010. Molecular and cellular mechanisms of diversity within grapevine varieties. Heredity 104, 331–340.

Pérez-Tornero, O., Córdoba, F., Moreno, M., Yuste, L., Porras, I., 2012. Classic methods and biotechnical tools in lemon breeding, preliminary results. Acta Hortic. 928, 259–263.

Pestana, M., De Varennes, A., Abadia, J., Faria, E.A., 2005. Differential tolerance to iron deficiency of citrus rootstocks grown in nutrient solution. Sci. Hortic. 104, 25–36.

Plooy, C.P.D., Steenekamp, P.J., Vincent, A.P., Human, C.F., Koekemoer, P.J.J., 1992. Mutation breeding strategy for citrus in South Africa. Proc. Int. Soc. Citricult. 1, 110–112.

Qi, J., Liu, Z., Huang, L., Chen, H., Liu, J., Dai, Z., 2010. Effect of gamma ray treatment on seedless mutation in budwood of 'Jincheng' orange [*Citrus sinensis* (L.) Osbeck]. Proc. Int. Soc. Citricult. 1, 146–148.

Ramadugu, C., Keremane, M.L., Halbert, S.E., Duan, Y., Roose, M.L., Stover, E., et al., 2016. Long-term field evaluation reveals Huanglongbing resistance in citrus relatives. Plant Dis. 100 (9), 1858–1869.

Ream, C.L., Furr, J.R., 1976. Salt tolerance of some citrus species, relatives and hybrids tested as rootstocks. J. Am. Soc. Hortic. Sci. 101, 265–267.

Reforgiato Recupero, G., Russo, F., 1977. Un caso di degenerazione delle vescichette del succo dell'arancio 'Tarocco' [*Citrus sinensis* (L.) Osbeck]. *Ann. Ist. Sper. Agrumic. Acireale* 9–10, 3–8. [in Italian].

Reforgiato Recupero, G., Russo, M.P., 1992. Percents of nucellar embryony in some hybrid populations of *Citrus latipes*. Proc. Int. Soc. Citricult. 1, 130–134.

Reforgiato Recupero, G., Russo, G., Recupero, S., 2005. New promising *Citrus* triploid hybrids selected from crosses between monoembryonic diploid female and tetraploid male parents. HortScience 40 (3), 516–520.

Reforgiato Recupero, G., Russo, G., Recupero, S., Zurru, R., Deidda, B., Mulas, M., 2009. Horticultural evaluation of new *Citrus latipes* hybrids as rootstocks for citrus. HortScience 44, 595–598.

Rodríguez-Gamir, J., Primo-Millo, E., Forner, J.B., Forner-Giner, M.A., 2010. Citrus rootstock responses to water stress. Sci. Hortic. 126, 95–102.

Roose, M.L., 1996. Rootstock breeding at the University of California, Riverside. Proc. Int. Soc. Citricult. 2, 1254.

Roose, M.L., Williams, T.E., 2007. Mutation breeding. In: Khan, I.A. (Ed.), *Citrus* Genetics, Breeding and Biotechnology. CAB International, Wallingford, UK, pp. 345–352.

Rouiss, H., Cuenca, J., Navarro, L., Ollitrault, P., Aleza, P., 2017a. Unreduced megagametophyte production in lemon occurs via three meiotic mechanisms, predominantly second-division restitution. Front. Plant Sci. 8, 1211.

Rouiss, H., Cuenca, J., Navarro, L., Ollitrault, P., Aleza, P., 2017b. Tetraploid citrus progenies arising from FDR and SDR unreduced pollen in 4x X 2x hybridizations. Tree Genet. Genomes 13 (1), 10.

Rouiss, H., Bakry, F., Froelicher, Y., Navarro, L., Aleza, P., Ollitrault, P., 2018. Origin of *C. latifolia* and *C. aurantiifolia* triploid limes; the preferential disomic inheritance of doubled-diploid 'Mexican' lime is consistent with an interploid hybridization hypothesis. Ann. Bot. 121 (3), 571–585.

Ruslan, R., Ehsani, R., Lee, W.S., 2012. Quantification of total soluble solids and titratable acidity for citrus maturity using portable Vis-NIR spectroradiometer. Appl. Eng. Agric. 28, 735–743.

Russo, F., Donini, B., Starrantino, A., 1981. Mutagenesis applied for citrus improvement. Proc. Int. Soc. Citricult. 1, 68–70.

Saleh, B., Allario, T., Dambier, D., Ollitrault, P., Morillon, R., 2008. Tetraploid citrus rootstocks are more tolerant to salt stress than diploid. C. R. Biol. 331, 703–710.

Savastano, L., 1915. Agrumi siciliani. Monografia per Giuseppe Inzenga, in Annali della R. Stazione sperimentale di agrumicoltura e frutticoltura di Acireale, III, Tip. Orario delle ferrovie. 1–42. [in Italian].

Scaglione, D., Reforgiato Recupero, G., Cattonaro, F., Licciardello, C., Caruso, M., Catara, A., et al., 2014. Nucleotide and structural variation in citrus genomes: resequencing and de novo approaches. In: International Plant and Animal Genome Conference XXII. W170.

Siebert Wooldridge, T.J., 2016. Cross-Compatibility, Graft-Compatibility, and Phylogenetic Relationships in the Aurantioideae: New Data From the Balsamocitrinae. MSc Thesis, University of California, Riverside. https://escholarship.org/uc/item/1904r6x3.

Smith, M.W., 2006. Variety 'IrM2'. Plant Var. J. 19, 260–264.

Smith, M.W., Shaw, R.G., Chapman, J.C., Owen-Turner, J., Lee, L.S., McRae, K.B., et al., 2004. Long-term performance of 'Ellendale' mandarin on seven commercial rootstocks in sub-tropical Australia. Sci. Hortic. 102 (1), 75–89.

Smith, M.W., Gultzow, D.L., Newman, T.K., 2013. First fruiting intergeneric hybrids between *Citrus* and *Citropsis*. J. Am. Soc. Hortic. Sci. 138 (1), 57–63.

Smith, M.W., Newman, T.K., Gultzow, D.L., Parfitt, S.C., Barkley, P., 2016. *Citrus tristeza virus* replication and movement in seedling trees of 71 rootstock genotypes. Citrus R&T 37 (2), 156–164.

Somsri, S., Putivoranat, M., Kanhom, P., Thayamanont, P., Meecharoen, S., Jompook, P., 2008. Improvement of tropical and subtropical fruit trees: tangerine, pummelo in Thailand through induced mutation and biotechnology. Acta Hortic. 787, 127–140.

Spiegel-Roy, P., 1990. Economic and agricultural impact of mutation breeding in fruit trees. Mutat. Breed. Rev. 5, 1–26.

Spiegel-Roy, P., Vardi, A., 1989. Induced mutation in citrus. In: Iyama, S., Takeda, G. (Eds.), Proceedings of the 6th International Congress of SABRAO. SABRAO, Tsukuba, Japan, pp. 773–776.

Spiegel-Roy, P., Vardi, A., 1992. 'Shani,' 'Orah' and 'Winola': three new selections from our breeding program. Proc. Int. Soc. Citricult. 1, 72–73.

Spiegel-Roy, P., Teich, A.H., Kochba, J., 1972. Gamma irradiation and pollen cultivar influence on polyembryony of "Satsuma" (*Citrus unshiu* Marc.). Radiat. Bot. 12, 365–367.

Spiegel-Roy, P., Vardi, A., Elhanati, A., 1985. Seedless induced mutant in lemon (*Citrus limon*). Mutat. Breed. Newsl. 26, 1–2.

Starrantino, A., Reforgiato Recupero, G., 1981. *Citrus* hybrids obtained in vitro from 2x females and 4x male. Proc. Int. Soc. Citricult. 1, 31–32.

Starrantino, A., Russo, F., 1980. Seedlings from undeveloped ovules of ripe fruits of polyembryonic citrus cultivars. HortScience 15, 296–297.

Starrantino, A., Mannino, P., Russo, F., 1988a. The genetic stability of three seedless clonal selections obtained by gamma rays of the seedy 'Monreal' clementine. Proc. Int. Soc. Citricult. 1, 175–182.

Starrantino, A., Russo, F., Donini, B., Spina, P., 1988b. Lemon mutants obtained by gamma irradiation of the nucellus cultured in vitro. Proc. Int. Soc. Citricult. 1, 231–235.

Stover, E., Inch, S., Richardson, M., Hall, D.G., 2016. Conventional citrus of some scion/rootstock combinations show field tolerance under severe Huanglongbing disease pressure. HortScience 51, 127–132.

Stover, E., Chaparro, J.X., Driggers, R.E., McCollum, T.G., 2018. Release of 'US SunDragon' HLB-Tolerant Sweet Orange-Like Hybrid. USDA/ARS.

Sykes, S.R., 1985. Effects of seedling age and size on chloride accumulation by juvenile citrus seedlings treated with sodium chloride under glasshouse conditions. Aust. J. Exp. Agric. 25 (4), 943–953.

Sykes, S.R., 2010. The effect of nucellar seedling and micropropagated clonal rootstocks on the performance of Imperial and Sunset mandarins. Proc. Int. Soc. Citricult. 1, 44–49.

Talon, M., Iglesias, D.J., Soler, G., Tadeo, F.R., Cercos, M., Zaragoza, S., 2004. *Citrus* mutants. New tools for the characterization of the citrus genome. Proc. Int. Soc. Citricult. 1, 57–60.

Terol, J., Ibanez, V., Carbonell, J., Alonso, R., Estornell, L.H., Licciardello, C., et al., 2015. Involvement of a citrus meiotic recombination TTC-repeat motif in the formation of gross deletions generated by ionizing radiation and MULE activation. BMC Genomics 16, 69.

Traub, H.P., Robinson, T.R., 1937. Improvement of sub-tropical fruit crops: citrus. In: Yearbook of the United States Department of Agriculture, pp. 749–826.

Uzun, A., Gulsen, O., Kafa, G., Seday, U., 2010. Effect of budwood irradiation on fruit weight of 'Kutdiken' lemon. Proc. Int. Soc. Citricult. 1, 98–100.

Van Harten, A.M., 1998. Mutation Breeding: Theory and Practical Applications. Cambridge University Press, Cambridge.

Vardi, A., Elhanati, A., Frydman-Shani, A., Neumann, H., 1996. Strategies and consideration in mandarin improvement programmes. Proc. Int. Soc. Citricult. 1, 109–112.

Vardi, A., Levin, I., Carmi, N., 2008. Induction of seedlessness in citrus: from classical techniques to emerging biotechnological approaches. J. Am. Soc. Hortic. Sci. 133 (1), 117–126.

Velasco, R., Zharkikh, A., Troggio, M., Cartwright, D.A., Cestaro, A., Pruss, D., et al., 2007. A high quality draft consensus sequence of the genome of a heterozygous grapevine variety. PLoS One 2 (12), e1326.

Velasco, R., Zharkikh, A., Affourtit, J., Dhingra, A., Cestaro, A., Kalyanaraman, A., et al., 2010. The genome of the domesticated apple (Malus × domestica). Nat. Genet. 42, 833–841.

Velazquez, K., Aguero, J., Vives, M.C., Aleza, P., Pina, J.A., Moreno, P., et al., 2016. Precocious flowering of juvenile citrus induced by a viral vector based on citrus leaf blotch virus: a new tool for genetics and breeding. Plant Biotechnol. J. 14 (10), 1976–1985.

Verdejo-Lucas, S., Sorribas, F.J., Forner, J.B., Alcaide, A., 1997. Screening of hybrid citrus rootstocks for resistance to *Tylenchulus semipenetrans* Cobb. HortScience 32 (6), 1116–1119.

Wang, X., Xu, Y., Zhang, S., Cao, L., Huang, Y., Cheng, J., et al., 2017. Genomic analyses of primitive, wild and cultivated citrus provide insights into asexual reproduction. Nat. Genet. 49, 765–772.

Williams, T.E., Roose, M.L., 2004. 'TDE2' Mandarin hybrid (Shasta Gold Mandarin), 'TDE3' Mandarin hybrid (Tahoe Gold Mandarin) and 'TDE4' Mandarin hybrid (Yosemite Gold Mandarin): three new mid and late-season triploid seedless mandarin hybrids from California. Proc. Int. Soc. Citricult. 1, 394–398.

Williams, T.E., Roose, M.L., 2010. 'Tango'—a new, very low-seeded, late-season irradiated selection of 'W. Murcott' mandarin from the University of California Riverside. Proc. Int. Soc. Citricult. 1, 202. (Abstract).

Wu, G.A., Prochnik, S., Jenkins, J., Salse, J., Hellsten, U., Murat, F., et al., 2014. Sequencing of diverse mandarin, pummelo and orange genomes reveals complex history of admixture during citrus domestication. Nat. Biotechnol. 32 (7), 656–662.

Wu, G.A., Terol, J., Ibanez, V., López-García, A., Pérez-Román, E., Borredá, C., et al., 2018. Genomics of the origin and evolution of *Citrus*. Nature 554, 311–316.

Xiang, C., Roose, M.L., 1988. Frequency and characteristic of nucellar and zygotic seedlings in 12 citrus rootstocks. Sci. Hortic. 37, 47–59.

Xu, Q., Chen, L.-L., Ruan, X., Chen, D., Zhu, A., Chen, C., et al., 2013. The draft genome of sweet orange (*Citrus sinensis*). Nat. Genet. 45, 59–66.

Yamamoto, M., Matsumoto, R., Yamada, Y., 1995. Relationship between sterility and seedlessness in *Citrus*. J. Jpn. Soc. Hortic. Sci. 64, 23–29.

Yamamoto, M., Matsumoto, R., Okudai, N., Yamada, Y., 1997. Aborted anthers of *Citrus* result from gene-cytoplasmic male sterility. Sci. Hortic. 70, 9–14.

Yamasaki, A., Kitajima, A., Ohara, N., Tanaka, M., Hasegawa, K., 2007. Histological study of expression of seedlessness in *Citrus kinokuni* 'Mukaku Kishu' and its progenies. J. Am. Soc. Hortic. Sci. 132, 869–875.

Yamasaki, A., Kitajima, A., Ohara, N., Tanaka, M., Hasegawa, K., 2009. Characteristics of arrested seeds in Mukaku Kishu-type seedless citrus. J. Jpn. Soc. Hortic. Sci. 78, 61–67.

Yu, Y., Chen, C., Gmitter, F.G., 2016. QTL mapping of mandarin (*Citrus reticulata*) fruit characters using high-throughput SNP markers. Tree Genet. Genomes 12, 77.

Zhang, M., Deng, X.X., Qin, C.P., Chen, C.L., Zhang, H.Y., Liu, Q., et al., 2007. Characterization of a new natural periclinal navel-satsuma chimera of citrus: "Zaohong" navel orange. J. Am. Soc. Hortic. Sci. 132 (3), 374–380.

Zhou, J.M., Hirata, Y., Nou, I.S., Shiotani, H., Ito, T., 2002. Interactions between different genotypic tissues in citrus graft chimeras. Euphytica 126 (3), 355–364.

Chapter 8

Genomic breeding

Tokurou Shimizu
Division of Citrus Research, Institute of Fruit Tree and Tea Science, National Agriculture and Food Research Organization, Shizuoka, Japan

8.1 Introduction

In contrast to mutation breeding that improves a particular trait of an existing variety with little effort, crossbreeding is an efficient approach for developing entirely new varieties. To date, genome-sequencing studies have revealed approximately 29,000 genes in the genus *Citrus* (Xu et al., 2013; Wu et al., 2014b; Wang et al., 2017; Shimizu et al., 2017), and each gene may have different alleles. The purpose of citrus breeding is to identify the best combination of alleles for individual genes to generate a trait of interest. However, breeding for better genetic combinations involved much trial and error, and have a low success rate. Existing elite varieties are regarded as having accumulated the desired alleles; using these varieties could increase the probability of selecting for even better varieties. Nevertheless, crossbreeding requires continued efforts and high costs to select one promising seedling because of the restrictions imposed by the tree size and their long juvenile period (Ollitrault and Navarro, 2012). In particular, both the available area for breeding and the number of seedlings to be grafted in an orchard primarily determine the scale of crossbreeding.

The aim of marker-assisted selection (MAS) is to select candidate seedlings using DNA markers that are associated with the trait of interest (Roose, 2007; Ollitrault and Navarro, 2012). Discarding undesirable seedlings during the juvenile phase before first blooming reduces the cost of breeding by increasing the probability of selecting for promising offspring in an orchard (Fig. 8.1) (Roose, 2007; Ollitrault and Navarro, 2012). The success of this strategy strictly depends on the availability of selectable DNA markers sufficient to predict a particular trait. Initial attempts with MAS in citrus breeding started in the 1980s, using isozymes (Torres et al., 1985). By the time DNA-marker technologies became available, various types of DNA markers had been used in earlier studies (Roose, 2007). Subsequent efforts detected valuable quantitative trait loci (QTL) for citrus tristeza virus (CTV) resistance, stress resistance, dwarfness, fruit acidity, and polyembryony (Chen et al., 2008; Gmitter et al., 2007a,b; Machado et al., 2011; Talon and Gmitter, 2008; Roose and Close, 2008). In proportion to the increase in genome-sequence information, valuable codominant DNA markers have been developed and then replaced dominant markers (Table 8.1) (Talon and Gmitter, 2008; Gmitter et al., 2012; Roose, 2007). These DNA markers have higher transferability for divergent varieties and have improved the accuracy of linkage and QTL analysis.

Linkage analysis of a trait of interest still requires long-term phenotypic evaluation of a breeding population (Fig. 8.2). Furthermore, current QTL analysis is adjusted for a simple trait that is regulated by a few genes. A promising strain to be selected by crossbreeding should be prominent for all traits, that is, not only for traits such as Brix, acidity, fruit weight, and coloring, but also for aroma, productivity, maturation period, and resistance to pests. Due to the usual citrus breeding strategy of selecting a promising offspring following a single hybridization, improving the selection efficiency is to selecting more promising progenies with less cost and effort per generation per acre (Fig. 8.1). In this context, expanding the MAS for diverse breeding populations and many complex traits is fundamentally required to improve the breeding efficiency.

Genomic breeding, or genomics-assisted breeding, is an extended approach of existing MAS. Genomic breeding is intended to predict multiple complex polygenic traits in a breeding population by integrating high-throughput genotyping technology, whole-genome sequence information, a large amount of genotype data for a wide array of varieties, and statistical genetics-analysis methods. In particular, two key approaches, association mapping (AM) and genomic selection (GS), which enable breeders to avoid the constraints of conventional linkage mapping and QTL analysis have been introduced (van Nocker and Gardiner, 2014; Iwata et al., 2016; Bhat et al., 2016; Hall et al., 2010; Oraguzie et al., 2007b). The aim of AM is to develop high-precision DNA markers available for particular trait by detecting linkage disequilibrium (LD) within wide and diverse varieties (Oraguzie et al., 2007b; Hall et al., 2010; Khan and Korban, 2012). Another approach, GS, has been developed to select for a quantitative trait that is regulated by many minor genes with small genetic gains and predict

The Genus Citrus. https://doi.org/10.1016/B978-0-12-812163-4.00008-5

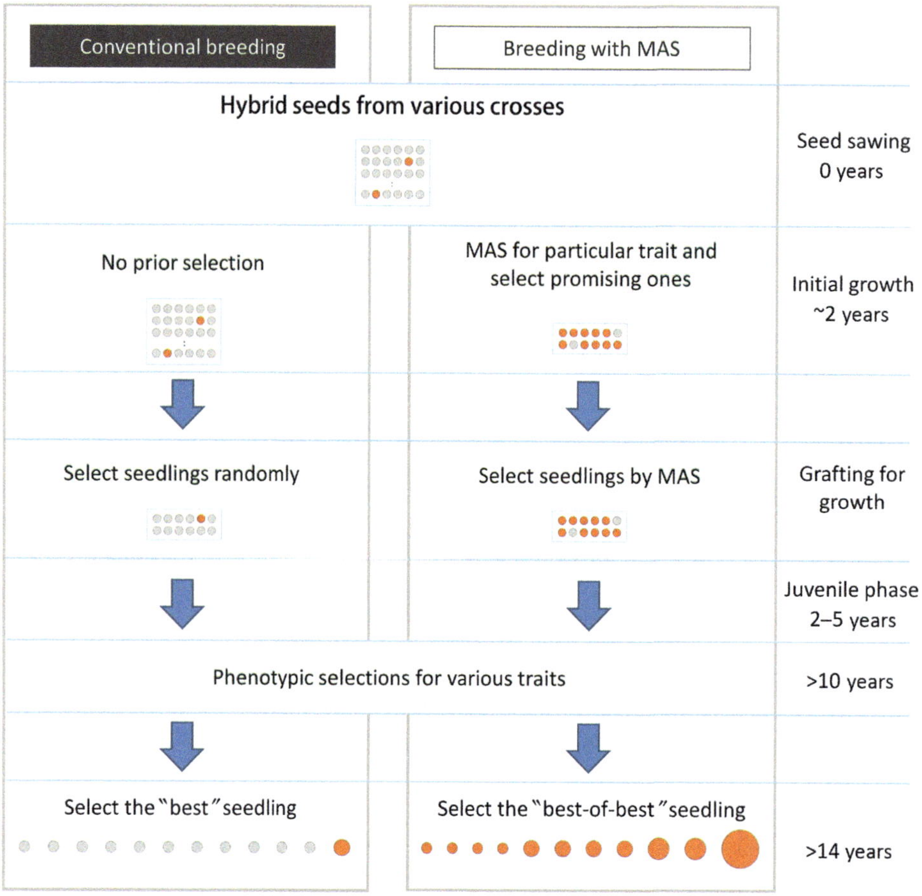

FIG. 8.1 How MAS can improve the efficiency of candidate selection. Left side: typical workflow of conventional citrus crossbreeding without MAS. Right side: a workflow of citrus crossbreeding that is integrating MAS. *Circles* in box represent probable promising seedlings. Years in the right represent minimum periods required for those steps.

TABLE 8.1 Summary of studies performed to develop and evaluate transferrable codominant markers in citrus.

References	Marker type	Numbers	Sources of marker design	Detection
Kijas et al. (1995)	SSR	2	Microsatellite-enriched genomic library of a cross of (*Citrus* × *limonia*) × *P. trifoliata*	PAGE
Kijas et al. (1997)	SSR	14	Microsatellite-enriched genomic library of a cross of (*Citrus* × *limonia*) × *P. trifoliata*	FA
Deng et al. (2000)	CAPS	9	Divergent regions in NBS-LRR genes	PAGE
Novelli et al., (2000)	SSR	18	Microsatellite-enriched genomic library of *C. sinensis*	PAGE
Oliveira et al. (2000)	SSR	4	Microsatellite-enriched genomic library of *C. sinensis*	AGE
Cristofani et al. (2000)	SSR	11	Microsatellite-enriched genomic library of *C. sinensis*	AGE
Omura et al. (2000)	CAPS, SSR	194	Citrus EST from cDNA of fruit and seed	AGE
Scarano et al. (2003)	SSR	20	Commercial products	AGE
Ruiz and Asins, (2003)	SSR	27	Public genome sequence, genomic library of *P. trifoliata*	FA
Ahmad et al. (2003)	SSR	26	Genomic library and microsatellite-enriched genomic library of *C. sinensis*	FA

TABLE 8.1 Summary of studies performed to develop and evaluate transferrable codominant markers in citrus—cont'd

References	Marker type	Numbers	Sources of marker design	Detection
Novelli et al. (2004)	SNP, indel	10	*C. sinensis* genomic library	SQ
Close et al. (2006)	SNP	5,023	Citrus EST and UniGenes databases	GeneChip
Golein et al. (2006)	SSR	7	*C. limon* genomic library	PAGE
Barkley et al. (2006)	SSR	11	Microsatellite-enriched genomic library of *C. maxima*	FA
Jiang et al. (2006)	SSR	12	Public citrus EST database	AGE
Novelli et al. (2006)	SSR	4	Microsatellite-enriched genomic library of *C. sinensis*	PAGE
Chen et al. (2006)	SSR	87	Public citrus ESTs database	FA
Palmieri et al. (2007)	SSR	68	EST database from 7 citrus varieties	AGE
Chen et al. (2008)	SSR	74	Public citrus EST database	FA
Froelicher et al. (2008)	SSR	43	Microsatellite-enriched genomic library of *C. reticulata*	PAGE
Hong et al. (2008)	SSR	78	Public citrus ESTs database	PAGE
Luro et al. (2008)	SSR	41	EST sequence of *C. clementina* 'Nules'	PAGE
Ollitrault et al. (2010)	SSR	79	*C. clementina* BAC end sequences	PAGE
Xu et al. (2010)	CAPS	10	BAC clones linked with *Tyr1*	AGE
Jiang et al. (2010)	CAPS	14	Public citrus EST database	AGE
Aleza et al. (2011)	SSR	8	N/A	PAGE
Shimizu and Yano (2011)	SSR, SNP	6	Public citrus EST database	FA
İncesu et al. (2011)	SSR	14	N/A	FA
Ohta et al. (2011)	CAPS, SNP	46	*P. trifoliata* genomic sequence	AGE, TaqMan
Cuenca et al. (2011)	SSR	13	Microsatellite-enriched genomic library of *C. reticulata*	FA
Kamiri et al. (2011)	SSR	7	Microsatellite-enriched genomic library of *C. reticulata*	PAGE
Cristofani-Yaly et al. (2011)	SSR	22	Microsatellite-enriched genomic library of *C. sinensis*	AGE
García-Lor et al. (2012)	Indel	12	Public citrus EST database for 10 genes	FA
Ma et al. (2012)	SSR	173	Citrus BAC end sequences	PAGE
Ollitrault et al. (2012b)	SNP	1,457	*C. clementina* BAC end sequences	GA
Fujii et al. (2012)	SNP	384	Public citrus EST database	GA
Ollitrault et al. (2012a)	SNP, SSR	83	Citrus EST and genomic sequence resources	FA, GA
Garcia-Lor et al. (2013a)	SNP	24	18 genes from 44 citrus accessions	KASPar
Garcia-Lor et al. (2013b)	Indel	12	18 genes from 44 citrus accessions	FA
Chai et al. (2013)	SSR	48	*C. maxima* EST sequences	PAGE
Cuenca et al. (2013)	SSR, SNP	6	*C. clementina* reference genome sequence	FA, KASPar
Chen and Gmitter Jr, (2013)	SNP	81	Public citrus EST database	SNaPshot
Liu et al. (2013)	SSR	95	EST, cDNA, and genomic sequences of *C. clementina*	AG
Hou et al. (2014)	SSR	15	MADS-box flanking sequences of citrus	PAGE
Wu et al. (2014a)	SNP	25	EST and genome sequences of 3 citrus varieties	HRM

(Continued)

TABLE 8.1 Summary of studies performed to develop and evaluate transferrable codominant markers in citrus—cont'd

References	Marker type	Numbers	Sources of marker design	Detection
Shimada et al. (2014)	CAPS	593	Public citrus EST database	AGE
Chen et al. (2015)	SSR	10	Genomic library of *C. japonica*	FA
Curk et al. (2015)	SNP	73	454-amplicon sequences of four basic citrus taxa	KASPar
Biswas et al. (2015)	SSR	61	Public *P. trifoliata* EST database	PAGE
Ramadugu et al. (2015)	SSR	18	*C. clementina* reference genome sequence	PAGE
Shimizu et al. (2016a)	SNP	768	NGS reads of 15 citrus accessions	GA
Shimizu et al. (2016b)	SSR, indel	214	Public citrus EST, cDNA, and genomic databases	FA
Yu et al. (2016)	SNP	1,536	*C. clementina* reference genome sequence	GA
Minamikawa et al. (2017)	SNP	1,536	NGS reads of 15 citrus accessions	GA
Eck (2017)	SNP	1.4 M, 58 K	NGS reads of 41 citrus accessions	Axiom

AGE: Agarose gel electrophoresis, FA: Fragment analysis, GA: GoldenGate assay, KASPar: Kompetitive allele-specific PCR (Smith and Maughan, 2015), N/A: not available, NGS: next-generation sequencing, PAGE: polyacrylamide gel electrophoresis, SQ: sequencing.

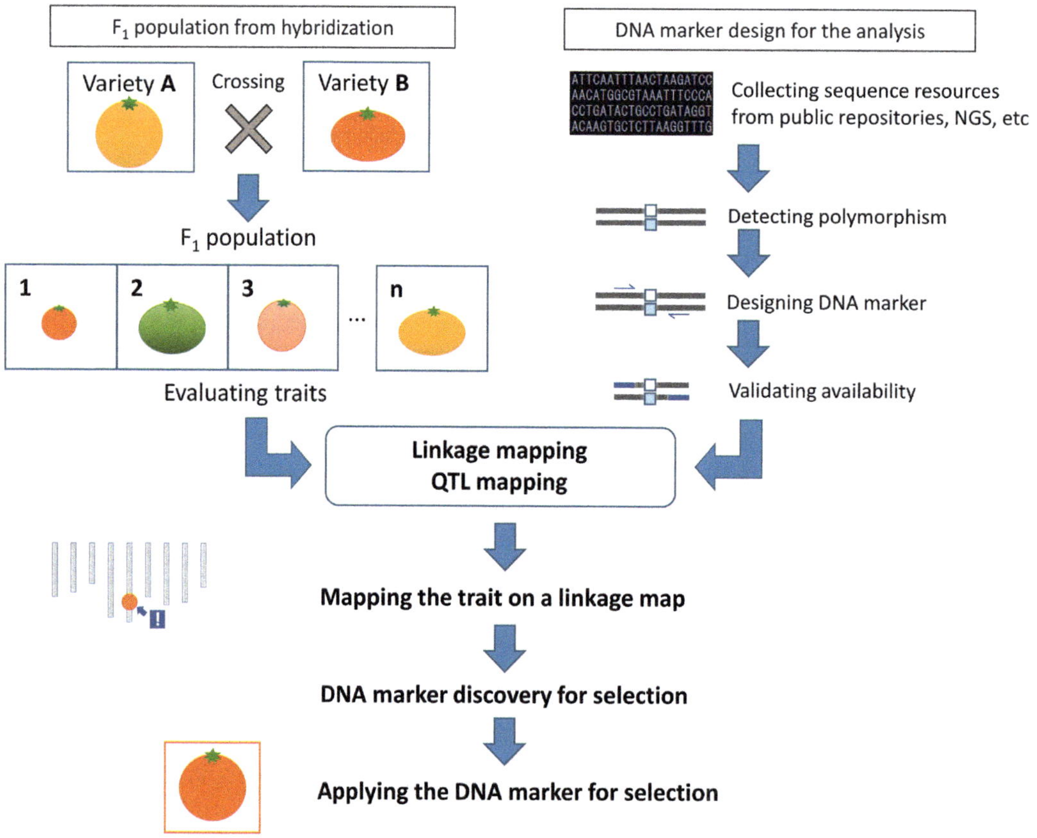

FIG. 8.2 Typical workflow of DNA-marker development for MAS in citrus.

the performance of a trait from whole-genome genotypes obtained from genome-wide dense markers (Meuwissen et al., 2001; van Nocker and Gardiner, 2014). Although AM and GS are both costly when used in a breeding program, the dramatic drop in analysis costs for high-throughput sequencing technologies could enable the introduction of these approaches in conventional breeding.

8.2 DNA markers

8.2.1 Types of DNA markers

Any nucleotide-sequence differences in a locus that are observed between different varieties are referred to as "polymorphisms," which represent the alleles of the varieties being studied. Many types of DNA-marker and their analysis methods have been developed and utilized to detect polymorphisms (Roose, 2007; Batley, 2015). Dominant-type DNA markers were used in early studies since no requirement for prior genome sequence was needed to design the marker (Roose, 2007). These are multi-locus marker systems that help to identify polymorphisms from many bands, but are difficult to transfer to different varieties. Another type of DNA markers, codominant markers, can be used to distinguish allele homozygosity and heterozygosity at a particular locus and have been widely used for precision breeding purposes. Simple sequence repeats (SSRs), indel, single-nucleotide polymorphisms (SNPs), and restriction fragment length polymorphisms (RFLPs) markers are typical codominant markers, but they detect different types of polymorphisms (Roose, 2007). SSR markers are used to determine the repeat numbers of simple sequence motifs, and indel markers detect length differences between varieties caused by the insertion or deletion of a short nucleotide sequence within a region. SNP markers detect nucleotide changes at a particular position in the genome (Henry, 2008; Slate et al., 2009). RFLP markers detect polymorphisms at restriction endonuclease digestion sites. RFLP markers have been used widely in many studies; however, this marker system requires a significant amount of DNA sample and substantial experimental efforts (Semagn et al., 2006). Cleaved amplified polymorphic sequence (CAPS; also known as PCR-RFLP) is an alternative to replace RFLP markers by amplifying a DNA sample by PCR and reduce the quantity of sample required (Semagn et al., 2006; Roose, 2007). Current citrus breeding involves the use of a codominant marker system to ensure high transferability among a wide set of varieties and high throughput (Batley, 2015). Ollitrault et al. reported differences in the availability of DNA markers for citrus-genome analysis in detail (Ollitrault et al., 2015). Codominant DNA markers require a genome sequence for their design; however, associating them with genomic information will expand their availability by connecting different linkage maps among different varieties (Ollitrault et al., 2012b,c, 2015; Luro et al., 2008; Garcia-Lor et al., 2013a). Associating DNA markers with genomic information also enables researchers to evaluate the correlation between genotypes and phenotypes. The first codominant marker was reported by Kijas in 1995 (Kijas et al., 1995), and 55 studies have since then reported the development of codominant markers (Table 8.1). Among those codominant DNA markers, SNP markers have been used in high-throughput genotyping studies (Henry, 2008).

8.2.2 SSR and indel markers

SSR motifs have been identified in the citrus genome and used for DNA-marker design (Chen et al., 2006; Liu et al., 2013; Hou et al., 2014; García-Lor et al., 2012; Ma et al., 2012; Hong et al., 2008; Jiang et al., 2006). These repeat regions are susceptible to mutation, and SSR markers exhibit a high discriminatory power for determining the presence of two or more alleles. In initial attempts by Kijas et al. (1995, 1997), SSR markers were developed from microsatellite-enriched genomic libraries because of the lack of sufficient genome-sequence information for citrus plants. Because preparing an enriched genomic library is a complex and time-consuming process (Kijas et al., 1994), researchers designed SSR markers from genomic sequences as they became available. Among the polymorphisms found in the genome, those found in translated regions tend to be maintained more stably in varieties than those of untranslated regions and to have higher transferability. Thus, nucleotide sequences deposited in the EST and UniGene databases, as well as cloned genes have been used to design SSR markers with higher transferability (Luro et al., 2008). Nucleotide sequences from nongenic regions have been used to design dense, genome-wide DNA markers, but showed less transferability to different varieties (Ollitrault et al., 2010). Indel markers reflect the length polymorphisms induced by insertion or deletion occurring at a particular position within the genome. Ollitrault et al. reported more stable maintenance of indel markers in diverse citrus varieties than with SSR markers (Ollitrault et al., 2015). SSR and indel polymorphisms are observed as size variation of several base pairs. A fragment-analysis method that uses DNA sequencing and separates fluorescently labeled DNA fragments under denaturing conditions can be used to estimate the fragment length accurately (Mason, 2015). Automated fragment-analysis systems can expedite the processing of many samples in a short duration. Although fragment analysis involves the use of expensive fluorescent labeled custom oligo primers, a simple multiplexed post-labeling method achieves low-cost analysis (Shimizu and Yano, 2011).

8.2.3 SNP markers

SNP are ubiquitously found in the genome and show low mutation rates over generations; therefore, SNPs have been used as stable DNA markers (Henry, 2008; Garvin et al., 2010). SNPs involve just two alleles and low discriminative power. On the other hand, studying only two alleles is advantageous because it simplifies the statistical analysis. In initial studies, SNPs were mined by comparing nucleotide sequences in several varieties for specific genes followed by conversion to CAPS markers (Deng et al., 2000; Omura et al., 2000). When many genome-sequence resources became available, SNPs were mined using comprehensive approaches (Table 8.1). Various types of analysis methods for genotyping SNP markers have been used in citrus studies. Those methods, for example, can detect from one to millions of polymorphisms by sequencing, single-nucleotide extension, TaqMan assays, allele-specific PCR analysis (Smith and Maughan, 2015), high-resolution melting-point analysis (HRM) (Simko, 2016), GeneChip, and beadarray systems (Table 8.1; refer also to books by Henry, 2008; Batley, 2015 for detailed descriptions of these methods). Among those methods, high-throughput genotyping is an indispensable tool for achieving genomic breeding. High-throughput genotyping platforms with large SNP markers associated with genome sequences enables dense genetic-map construction, linkage mapping and QTL analysis, AM, or genome-wide association study (GWAS) (Khan and Korban, 2012; Oraguzie et al., 2007b; Zhu et al., 2008) and GS (van Nocker and Gardiner, 2014; Iwata et al., 2016; Bhat et al., 2016). Although large-scale genotyping is costly, the unit price per marker per sample is more affordable than those of small-scale systems. The first high-throughput genotyping platform, a GeneChip for analyzing 5023 SNPs, was developed from EST and UniGene database sequences generated by the University of California (UC) at Riverside in 2006 (Close et al., 2006). This GeneChip was designed not only for genotyping assays, but also for genome-wide expression assays. Other types of high-throughput genotyping arrays that applied the GoldenGate assay system were developed by groups in France (Ollitrault et al., 2012b), Japan (Fujii et al., 2012; Shimizu et al., 2016a; Minamikawa et al., 2017), and the United States (Yu et al., 2016). These groups designed SNP markers from EST sequences, partial genome sequences, and whole-genome sequences, and evaluated their performance. High-throughput and accurate genotyping performances were benefits of the GoldenGate assay; however, Illumina terminated the distribution of the customized GoldenGate assay in 2015, unfortunately. In 2017, Eck in Mike Roose's laboratory at UC Riverside reported the development of two new high-throughput genotyping arrays built on the Axiom system (Eck, 2017). These new arrays contain 1.4 M or 58 K SNP markers designed from the whole-genome sequences of 41 citrus accessions (Eck, 2017), and their applications are anticipated to promote genomic breeding in citrus.

8.2.4 RFLP and CAPS markers

RFLP markers are codominant DNA markers that were used in early studies. Liou first constructed a linkage map of their BC_1 population using RFLP markers and isozymes for genotyping in 1990 (Liou, 1990). The use of RFLP markers for linkage-map construction and QTL analysis continued until 2001 (Table 8.1). Cloning of a genomic DNA fragment from a randomly amplified polymorphic DNA (RAPD) marker or from a cloned gene is mandatory for RFLP marker development (Semagn et al., 2006). CAPS (PCR-RFLP) markers were introduced as a substitute for RFLP markers in 2000 since hybridization-based RFLP markers require considerable amounts of DNA sample, and long and complex operations to detect the polymorphisms (Semagn et al., 2006; Roose, 2007). The nucleotide sequences of isolated genes or ESTs of citrus were often utilized to design the CAPS markers (Deng et al., 2000; Omura et al., 2000; Jiang et al., 2010; Ohta et al., 2011; Shimada et al., 2014). CAPS markers detect polymorphisms at restriction endonuclease recognition sites by digesting an amplified DNA fragment with a restriction endonuclease. Size differences observed as polymorphisms by the CAPS markers are long enough to be detected by agarose gel electrophoresis and have been widely used in many studies. Although CAPS markers are biallelic DNA markers, they can be extended to detect polymorphisms at different loci by changing the restriction endonuclease used for digestion. In contrast, it is difficult to automate CAPS marker analysis because of the complicated process used for digestion after PCR amplification. Fujii et al. (2012) reported the conversion of CAPS markers to SNP markers by mining SNPs within the amplified fragment.

8.2.5 NGS-based high-throughput genotyping

The recent rapid progress of the next-generation sequencing (NGS; also known as high-throughput sequencing) technologies has expanded the availability for genome-wide genotyping without the need for designing DNA markers beforehand. In response to the sharp decrease in the cost of NGS analysis, many high-throughput genotyping applications have been developed that are intended to be used in breeding. Several types of methods, including Diversity Arrays Technology sequencing (DArTSeq) (Sansaloni et al., 2011), restriction-site-associated DNA sequencing (RADSeq) (Davey and Blaxter, 2010;

TABLE 8.2 Summary of studies for NGS-based high-throughput genotyping in citrus.

Technology	References	Marker used	Number of markers
DArTSeq	Gois et al. (2016)	DArT markers	5,287
	Curtolo et al. (2017)	DArT markers	661
	Sagawa et al. (2017)	DArT markers	37,260
RADseq	Penjor et al. (2014)	SNPs	42,613
	Guo et al. (2015)	SNPs	1,543
	Penjor et al. (2016)	SNPs	1,600
Genotyping by sequencing (GBS)	Imai et al. (2017)	SNPs	442
	Oueslati et al. (2017)	SNPs, indels	9,955 SNPs 1,178 indels

Davey et al., 2011), and genotyping by sequencing (GBS) (Elshire et al., 2011) have been developed that differ in their protocol and data-analysis procedures. These methods reduce the complexity of the genome by digesting genomic DNA with a restriction endonuclease and are then used to identify polymorphic loci in many samples based on their sequences. DArTSeq is used to mine for polymorphic loci and SNP from NGS sequences, which are then reported as the presence or the absence of DArTScores (Sansaloni et al., 2011). Both RADSeq and GBS can mine NGS reads, and report SNP and indel loci. Many variations of these methods have been developed, and their selectivity within the genome and the accuracy of the obtained genotypes were reported (Mastretta-Yanes et al., 2015; Heslot et al., 2013).

In 2014, Penjor et al. first reported the use of RADSeq in the discovery of polymorphic loci for phylogenetic analysis (Penjor et al., 2014) (Table 8.2). The authors reported a follow-up study in 2016, as well (Penjor et al., 2016). Meanwhile, Guo et al. detected 1543 SNPs by RADSeq analysis of pummelos and constructed a dense linkage map (Guo et al., 2015). A series of DArTSeq genotyping studies for GWAS (Gois et al., 2016), QTL mapping (Curtolo et al., 2017), and genetic-diversity analysis (Sagawa et al., 2017) were reported from research groups in Brazil. In addition, a research group in Japan reported QTL analysis using GBS analysis (Imai et al., 2017), and another group in France used GBS analysis for genome-wide diversity analysis (Oueslati et al., 2017). Most of those studies detected 10,000 or more polymorphic loci (Table 8.2). Unlike conventional DNA-marker analysis, the loci detected are not conserved in each NGS-based genotyping method. Consequently, NGS-based high-throughput genotyping is not suitable for genotyping at particular loci. However, the genotype of particular loci can be estimated from the association of the NGS reads to the reference genome sequence when sufficient NGS data are obtained. Obtaining genotype data from those of NGS reads without the reference genome (Catchen et al., 2011) is another benefit of this technology. These data from NGS analysis include type-I and type-II errors, which must be considered during sequence analysis (Mastretta-Yanes et al., 2015). Whereas NGS-based genotyping analysis has several constraints and requires a careful validation process, it offers an obvious advantage in the ability to obtain sufficient genotype data in short duration by eliminating the extensive efforts required to develop DNA markers. These NGS-based approaches are anticipated to be used in breeding populations for GS (Bhat et al., 2016).

8.3 Linkage mapping analysis toward MAS

8.3.1 Linkage-map construction using transferrable DNA markers

Linkage mapping is a basic technique used for discovering DNA markers for MAS (Fig. 8.2) (Gmitter et al., 2007a,b; Talon and Gmitter, 2008; Machado et al., 2011; Roose, 2007). Connecting several linkage maps constructed from different populations using transferrable codominant markers enables the transfer of a DNA marker from a certain population to different population and expands the availability of the DNA marker for MAS. Transferring an existing DNA marker is advantageous for citrus varieties because it requires an extended period for developing the map and DNA-marker discovery for MAS (Fig. 8.2). Any linkage map constructed using a codominant marker is valid for mapping a recessive trait. In citrus varieties, breeding populations have been used for linkage mapping because of the difficulty in preparing the population for test crosses. Selecting a promising seedling from a breeding population obtained from the single hybridization is a typical procedure used in citrus breeding (Figs. 8.1 and 8.2). Due to the higher heterozygosity of parents and progenies, pseudo-testcross analysis (Grattapaglia and Sederoff, 1994; Grattapaglia et al., 1995) has been used to construct linkage maps in citrus varieties. The pseudo-testcross analysis is used to construct an integrated map from two linkage maps for each parent that were constructed using DNA-marker genotypes with BC_1 segregation. Next, these two linkage maps are combined with common DNA markers, after which an integrated map is obtained. Cristofani et al. first applied pseudo-testcross analysis to construct a linkage map using RAPD genotypes (Cristofani et al., 1999). The first linkage maps constructed with codominant markers were reported by Omura et al. (2000) and Cristofani et al. (2000). Dominant-type DNA markers, RAPDs, amplified fragment length polymorphisms (AFLPs), inter-simple sequence repeats (ISSRs) (Semagn et al., 2006), and sequence-related amplified polymorphisms (SRAPs) (Li and Quiros, 2001), were also used in early studies with codominant DNA markers to construct linkage maps (Table 8.3). According to the published papers, linkage analysis has been conducted with approximately 37 citrus breeding populations, and 47 linkage maps containing five integrated maps have been constructed to date with codominant markers (Table 8.3).

8.3.2 DNA-marker development for monogenic traits

Bulked-segregation analysis (BSA) (Michelmore et al., 1991) has been successfully used in many studies for the rapid identification of candidate DNA markers associated with a monogenic trait. BSA offers an advantage for the rapid identification of candidate markers from a population that is either too small or too large to conventional linkage-mapping strategies. Both dominant and codominant DNA markers are available for BSA, which has expanded the utility of BSA in a wide range of studies. On the contrary, the use of BSA is limited for simple monogenic traits with high heritability that show clear and stable segregation; applying BSA for analyzing a low heritability trait or a quantitative trait is difficult. The distance in a linkage map between the trait and candidate marker suggests that such associations should be examined separately. Many BSA studies have been reported for the identification of linked DNA markers to CTV resistance (Gmitter et al., 1996; Cristofani et al., 1999; Fang et al., 1998), dwarfing (Cheng and Roose, 1995), citrus nematode resistance (Ling et al., 2000), freeze tolerance (Weber et al., 2003), seedlessness (JinPing et al., 2009), polyembryony/nucellar embryony (García et al., 1999; Nakano et al., 2008), *Alternaria* brown spot (ABS) resistance (Cuenca et al., 2013), male fertility (Chae et al., 2011), and citrus canker (Choi et al., 2007). Closely linked DNA markers for CTV resistance (Rai, 2006; Ohta et al., 2011), ABS resistance (Cuenca et al., 2016), and polyembryony (Wang et al., 2017) had been developed by performing linkage-map analysis or whole-genome sequence analysis.

8.3.3 Polyembryony

Polyembryony (apomixis or nucellar embryony) is an attractive trait of citrus varieties that enables asexual seed propagation (Kepiro and Roose, 2007). Many citrus varieties show polyembryony (Frost and Soost, 1968; Ueno et al., 1967), and this trait has also been used for selecting mutants (Nishiura, 1967). Detection of QTL for polyembryony was first reported by Garcia et al., who developed dominant DNA markers associated with loci responsible for polyembryony (García et al., 1999). Subsequent studies performed to develop DNA markers associated with polyembryony-linked loci were reported by Nakano et al. using a CAPS marker (Nakano et al., 2008), and by Kepiro and Roose with an AFLP marker (Kepiro and Roose, 2009). In a later study, Nakano et al. isolated BAC clones containing a locus associated with polyembryony and revealed the nucleotide sequence for 70 candidate genes (Nakano et al., 2012). In 2017, Wang et al. further evaluated the polyembryony locus in detail by evaluating genotypic traits associated with 100 accessions of primitive, wild, and cultivated citrus using high-throughput sequencing technology (Wang et al., 2017). Based on their study, Wang et al. identified the gene responsible for polyembryony, *CitRWP*, as a homolog of an *Arabidopsis RKD*-family gene (Wang et al., 2017). The identified gene and DNA markers developed will be useful for selecting polyembryonic or monoembryonic seedlings by MAS.

8.3.4 Fruit traits

Fruit traits are key factors that influence the commercial value of fruits. Various fruit traits are needed to be considered, but many scientists have been interested in the accumulation of carotenoids and have isolated genes involved in their metabolism (Alquézar et al., 2008; Zhang et al., 2012; Ikoma et al., 2016). Sugiyama et al. reported the correlation between the expression levels of zeaxanthin epoxidase (*ZEP*) genes involved in carotenoid biosynthesis and alleles for those genes (Sugiyama et al., 2010). They further detected QTLs that influence the expression levels of four genes involved in

TABLE 8.3 Summary of studies performed to construct linkage maps using transferable codominant markers and to detect QTLs.

References	Type(s) of marker	Populations	Pr[a]	Mapped varieties	NM[b]	LG[c]	LN[d]	QTL or locus
Liou (1990)	RFLP, isozymes	LB 1–21[e] × *C. clementina* 'Clementine'	65	Integrated map	35	8	314	–
Jarrell et al. (1992)	RFLP, isozymes	Sacaton (*C. paradisi* × *P. trifoliata*) × Troyer (*C. sinensis* × *P. trifoliata*)	60	Integrated map	38	10	1,700	–
Durham et al. (1992)	RFLP, isozymes	*C. maxima* 'Thong Dee' × USDA 17–40[f]	65	Integrated map	52	11	533	–
		LB 1–21[e] × *C. clementina* 'Clementine'	65	Integrated map	32	8	314	–
Cai et al. (1994)	RAPD, RFLP, isozymes	*C. maxima* 'Thong Dee' × USDA 17–40[f]	60	Integrated map	160	9	1,192	Cold tolerance, and salt tolerance
Cheng and Roose, (1995)	RFLP, RAPD, isozymes	*P. trifoliata* 'Flying Dragon' (open pollinated)	41	'Flying Dragon'	3	1	51	Dwarfing
Luro et al. (1996)	RFLP, RAPD, ISSR, isozymes	FAO 30573 (*C. reshni* × *P. trifoliata*) × *C. maxima*	52	FAO 30573 *C. maxima*	95 34	12 7	1,503 600	– –
De Simone et al. (1998)	RFLP, RAPD, isozymes	*C. latipes* × *C. aurantium*	50	*C. latipes* *C. aurantium*	92 247	12 20	600 1,000	– –
Fang et al. (1998)	RFLP, RAPD	10 populations of *P. trifoliata* progenies	554	*P. trifoliata* F1 *P. trifoliata* selfed 'Sacaton' (*P. trifoliata* × *C. paradisi*) × 'Troyer' (*C. sinensis* × *P. trifoliata*)	6 6 6	1 1 1	3.9 3.5 12.7	CTV resistance CTV resistance CTV resistance
Omura et al. (2000)	CAPS	Kiyomi (*C. unshiu* × *C. sinensis*) × *C. unshiu* 'Miyagawa Wase'	125	Consensus map	134	9	801.7	–
		C. unshiu 'Jutaro' × *C. sinensis* 'Trovita'	53	Consensus map	135	8	655.2	–
Cristofani et al. (2000)	SSR, RAPD, AFLP	*C. sunki* × *P. trifoliata*	80	*C. sunki* *P. trifoliata*	91 86	11 11	N/A N/A	– –
Ruiz and Asins, (2003)	SSR, IRAP, RAPD, RFLP, Isozymes, SCAR	*C. aurantium* 'Afin Verna' × *P. trifoliata* 'Flying Dragon' *C. volkameriana* × *P. trifoliata* 'Rubidoux' *P. trifoliata* 'Flying Dragon' (selfed)	66 80 57	'Flying Dragon' 'Afin Verna' 'Rubidoux' *C. volkameriana* 'Flying Dragon' (selfed)	40 104 43 79 43	5 15 8 10 10	275.2 441.5 341.9 460.1 269.7	– – – – –
Weber et al. (2003)	CAPS,RAPD, SCAR, STS	*C. maxima* 'DPI6–4' × *P. trifoliata* 'Rubidoux'	99	*C. maxima* 'DPI6–4' 'Rubidoux'	117 100	9 9	704 638	 Freeze tolerance
Chen et al. (2008)	SSR	*C. sinensis* 'Sanford,' 'Fiwicke,' 'Ruby' × *P. trifoliata* 'Argentina'	97	*C. sinensis* 'Argentina' Consensus map	113 45 123	11 8 13	525.6 320.4 514.5	– – –
Hong et al. (2008)	SSR	*C. sinensis* 'Li 2' × Wan 2 (*C. unshiu* × *C. sinensis*)	68	Consensus map	173	9	804	–

(Continued)

TABLE 8.3 Summary of studies performed to construct linkage maps using transferable codominant markers and to detect QTLs—cont'd

References	Type(s) of marker	Populations	Pr[a]	Mapped varieties	NM[b]	LG[c]	LN[d]	QTL or locus
Gulsen et al. (2010)	SSR, SRAP, ISSR, POPG, RGA, RAPD	*C. clementina* 'Clementine' × Orlando (*C. paradisi* × *C. tangerina*)	164	'Clementine' Orlando	215 189	9 9	760 740	Alternaria brown spot resistance
Gulsen et al. (2011)								Early fruit setting
Sugiyama et al. (2011) Sugiyama et al. (2014)	CAPS, SNP	Okitsu-46 (Sweet spring[g] × *C. sinensis*) × Nou-5 (Lee[h] × *C. kinokuni*)	93	Okitsu-46 Nou-5	345 254	9 12	660 642	Carotenoid content eQTL for 9 carotenoid biosynthesis genes
Ollitrault et al. (2012a)	SSR, SNP, indel	*C. maxima* 'Chandler' × *C. clementina* 'Nules' *C. clementina* 'Nules' × *C. maxima* 'Pink' *C. sinensis* × *P. trifoliata*	156 140 147	*C. clementina* F *C. clementina* M *C. clementina* F+M 'Chandler' 'Pink' *C. sinensis*	783 869 961 151 81 569	9 9 9 9 9 9	923.54 1,164.26 1,084.07 828.62 633.9 669.61	– – – – – –
Hong et al. (2015)	SSR	*C. sinensis* 'Li 2' × Wan 2 (*C. unshiu* × *C. sinensis*)	68	Consensus map	452	9	936.1	Freeze tolerance
Guo et al. (2015)	RAD, SSR	*C. maxima* 'Guanxi' × *C. maxima* 'Pingshan'	124	'Guanxi' 'Pingshan' Integrated consensus map	511 500 1,563	9 9 9	602.07 641.01 976.58	– – –
Yu et al. (2016)	SNP	Fortune[i] × Murcott[j]	116	Fortune[i] Murcott[j]	189 106	9 9	681.07 395.25	48 QTLs for 15 fruit traits
Yu et al. (2017)								206 candidate QTLs for 94 volatile compounds in juice
Imai et al. (2017)	GBS	Harehime[k] × *C. reticulata*	110	Harehime[k] *C. reticulata*	442 332	9 9	635.8 892.9	9 QTLs for 4 fruit traits
Curtolo et al. (2017)	DArTSeq	Murcott[j] × *C. sinensis* 'Pera'	278	Consensus map	661	13	2,774.29	19 QTLs for 12 fruit traits

[a] *Pr: Number of progenies used for the analysis.*
[b] *NM: Number of markers used for the linkage-map construction.*
[c] *LG: Number of the linkage groups.*
[d] *LN: Whole lengths of the constructed map (cM).*
[e] *LB-21:C. clementina 'Clementine' × C. paradisi 'Duncan.'*
[f] *USDA 17–40:C. maxima 'Thong Dee' × P. trifoliata 'Pomeroy.'*
[g] *Sweet spring: C. unshiu × C. Hassaku.*
[h] *Lee: C. clementina × Orlando (C. paradisi × C. tangerina).*
[i] *Fortune: C. clementina × Orlando (C. paradisi × C. tangerina) (Shimizu et al., 2016b).*
[j] *Murcott: A possible selection of King mandarin (C. nobilis) (Shimizu et al., 2016b).*
[k] *Harehime: (Kiyomi (C. unshiu × C. sinensis) x Osceola (C. clementina × Orlando (C. paradisi × C. tangerina))) × C. unshiu.*

carotenoid biosynthesis, namely zeaxanthin epoxidase (*ZEP*), beta-ring hydroxylase (*HYb*), phytoene synthase (*PSY*), and phytoene desaturase (*PDS*), based on expression-QTL (eQTL) analysis (Sugiyama et al., 2014). These findings will contribute to the MAS of carotenoid-enriched fruit in the future.

Though the detection of fruit-related QTLs using dominant markers was reported (Şahin-Çevik and Moore, 2012), studies for QTL analysis of various fruit traits using codominant marker have been long awaited. Since 2016, several groups reported QTL-based detection of various fruit traits (Yu et al., 2016; Gois et al., 2016; Minamikawa et al., 2017; Curtolo et al., 2017; Imai et al., 2017). An attempt to detect QTLs of fruit traits was reported by Yu et al. (2016). They detected 48 possible QTLs for 15 fruit traits by QTL analysis using 116 progenies from an F_1 population derived from Fortune × Murcott cross (Yu et al., 2016), and 10 of the detected QTLs were confirmed over several years. They detected QTLs for flavedo color that showed large LOD scores, and two genes involved in carotenoid biosynthesis (*CCD4*/NCED4 and *HPD*/*PDF1*) were proposed to contribute to coloring (Yu et al., 2016). Those QTLs associated with coloring showed high scores and could be valuable for MAS. Curtolo et al. (2017) evaluated 14 fruit traits in 278 randomly selected full-sib progenies obtained from Murcott × sweet orange 'Pera' and detected 19 QTLs for 12 traits. Large effects were observed with the QTLs for the number of segments per fruit, ease of peeling, technology index, and fruit per box. The authors revealed that these three QTLs were derived from Murcott, and proposed associations of 16 candidate genes for these QTLs by comparing the DNA markers to the reference sweet orange genome sequence (Curtolo et al., 2017). Another QTL-detection study was reported by Imai et al. (2017); they detected nine QTLs for four fruit traits (fruit weight, sugar content, peel puffing, and water rot) with 110 progenies of the F1 population of Harehime × ponkan and associated them to their parent varieties. These studies employed large-scale genotyping methods for QTL analysis, and the authors proposed the effectiveness of those methods for QTL analysis. However, they detected those QTLs in one breeding population, and their effects in other populations or varieties are unknown. Continuous evaluations of their effect in other populations are expected to assess the wide applicability of these QTLs and associated DNA markers or candidate genes.

8.3.5 Aroma

Aroma is another key factor that influences the preferences of consumers. Citrus is known as a rich source of many aromatic compounds (Miyazaki et al., 2011; Zhang et al., 2017), and correlations between the content of each aromatic compound and the characteristic flavor of individual varieties were reported (Miyazaki et al., 2011, 2012). However, any DNA marker that would be useful for selecting seedlings having distinct flavors has not been made publicly available yet. Recent studies focusing on the detection and characterization of each compound have been performed, and two noteworthy studies were published in 2017. An attempt by Yu et al. identified 148 volatile compounds within 116 progenies from an F_1 population derived from Fortune × Murcott cross (Yu et al., 2017). According to map-based QTL analysis, they detected 206 QTLs for 94 volatile compounds and confirmed 25 QTL to be consistent over more than two harvest times. Several QTLs corresponded to multiple compounds and four candidate genes (*GPS1*, *TSS3*, *TPS4*, and *TPS14*) for some of those QTLs were identified (Yu et al., 2017). Another study led by Alquézar et al. identified 55 terpene-synthase genes (TPSs) in the sweet orange genome, which were then mapped (Alquézar et al., 2017). Expression analysis of these genes in *E. coli* confirmed the function of seven sesquiterpene-synthase genes (Alquézar et al., 2017). Those two studies were conducted with several varieties. However, the QTLs and genes identified should provide valuable starting points toward the MAS of aromas in the future.

8.3.6 Disease, stress resistance, and other traits

Developing a variety that is resistant to viral, fungal, or bacterial infection is expected to increase the yield and fruit quality, and decrease the cost of spraying agrichemicals. Moreover, physiological damages caused by low-temperature exposure often reduce production and fruit quality. Consequently, many efforts to develop MAS for breeding new scions with resistance to disease and stress have been conducted (Gmitter et al., 2007a,b; Machado et al., 2011). Because CTV causes severe symptoms in citrus, Gmitter et al. performed MAS to introduce CTV resistance from *Poncirus trifoliata* by distant hybridization (Gmitter et al., 1996). Deng et al. reported the association of the NBS-LRR-class resistance gene with CTV and nematode resistance (*Tyr*1) (Deng et al., 2000). Then, they developed CAPS markers associated with *Tyr*1 and CTV resistance (Xu et al., 2010). Fine mapping of the locus that conferred CTV resistance in the *P. trifoliata* implicated a 282-kb genomic region (Yang et al., 2003). The following study by Rai (2006) placed the loci within a 121-kb genomic region, and the authors suggested candidate *R* genes. In 2012, Asins et al. demonstrated that the genomic region associated with CTV resistance is conserved among *C. grandis*, *C. aurantium*, and *P. trifoliata* by comparing their linkage maps with common DNA markers (Asins et al., 2012). In addition, Ohta et al. developed DNA markers closely linked to CTV resistance (Ohta et al., 2011) and showed their availability for MAS in distant hybridization (Ohta et al., 2015).

Cuenca et al. showed that the ABS resistance locus was recessive and located it within a 3.3-Mb genomic region near the centromere of chromosome III based on association analysis of linkage maps and genome sequences, using DNA markers (Cuenca et al., 2013). Then, they conducted a large-scale, map-based cloning study of the ABS resistance region to localize the locus within a 366-kb genomic region (Cuenca et al., 2016). Furthermore, they identified a candidate gene from the gene annotation of that region and developed a DNA marker valid for MAS of ABS resistance (Cuenca et al., 2016).

In another study, Hong et al. (2015) mapped QTL for freeze tolerance to the linkage map of the sweet orange 'Li 2' × Wan 2. To date, QTLs for citrus canker (Choi et al., 2007), citrus leprosis (Bastianel et al., 2009), *Phytophthora* gummosis (Siviero et al., 2006) and citrus leafminer (Bernet et al., 2005) susceptibility have been reported by linkage-mapping analysis. Tozlu et al. reported the detection of QTL for salinity and Na+ and Cl− accumulation (Tozlu et al., 1999a,b). Other stress-related QTLs associated with cold-acclimation-responsive loci (Cai et al., 1994) and freeze tolerance (Weber et al., 2003) were reported. Also, Siviero et al. (2003) reported the QTL detection of rooting stem cutting. Despite the progress made in detecting important QTLs, DNA markers used in these studies were of the dominant type. Further evaluation of these QTLs by converting dominant markers to codominant markers and transferring the associating DNA markers to other varieties are anticipated. However, valid DNA markers for MAS of citrus canker, citrus black spot, and Huanglongbing (HLB) disease have not been reported yet, and continuous challenges are demanded.

8.4 MAS for complex traits

8.4.1 QTL analysis of complex traits

As mentioned in previous sections, many QTLs have been identified for monogenic traits, and valuable DNA markers to select for CTV and ABS resistance, and polyembryony had been developed. In contrast, many fruit traits are complex and polygenic, with large divergence occurring among citrus varieties. For example, Combrink et al. reported differences in the distribution of fruit size and shape (Combrink et al., 2013) and rind color (Combrink et al., 2015) within and between half-sib families of Kiyomi tangor. Conventional linkage analysis is useful for dissecting simple inherited traits, but applying this strategy for polygenic traits faces many constraints (Iwata et al., 2016). QTL analysis only detects variations observed between parents, and it is difficult to detect QTLs in other varieties. Segregation of polygenic traits depends on the parental genotypes; thus, the effects of QTLs for those polygenic traits should be examined for each population.

Two alternative analysis methods, MetaQTL and FlexQTL, for identifying common QTLs in diverse varieties have been developed. MetaQTL (Veyrieras et al., 2007) integrates multiple results of QTL studies to estimate the actual QTL among them, based on the Akaike criterion (Veyrieras et al., 2007; Costa, 2015). FlexQTL is a pedigree-based QTL-estimation method for full-sib families (Bink et al., 2014). Applications of both MetaQTL and FlexQTL for identifying fruit traits were demonstrated in apples (Bink et al., 2014; Costa, 2015). However, using these methods, great efforts must be made to develop several hybrid populations and phenotype their progenies; consequently, no study using these methods has been reported in citrus thus far. Two new approaches, AM and GS, have been successfully applied in many studies to overcome the constraints of conventional strategies for selection.

8.4.2 Linkage disequilibrium (LD) analysis

Recent statistical-genetics studies demonstrated the potential of whole-genome LD analysis for detecting QTL and identifying candidate genes (Sorkheh et al., 2008; Gupta et al., 2005; Oraguzie et al., 2007a). According to Mackay and Powell, LD is defined as a nonrandom association of alleles at separate loci located on the same chromosome (Mackay and Powell, 2007). Strong associations indicate a close genetic linkage between these loci, but the genetic structure of a population and a selective bias for the persistence of particular haplotypes also influence LD (Sorkheh et al., 2008; Mackay and Powell, 2007; Oraguzie et al., 2007a). A large distribution of LD values was observed within plants in a previous study (Oraguzie et al., 2007a). A large LD value represents a strong association between two loci on the same chromosome, which helps to detect QTL with fewer DNA markers (Rafalski, 2002). In citrus, very frequent and significant LD values were observed in an F_1 population obtained from 'Chandler' × 'Clementine' (Ollitrault et al., 2012b). Similarly, Minamikawa et al. also observed similar significant LD values ($r^2 > 0.2$) within the citrus genome, and LD in a breeding population was lower than those of the parental population (Minamikawa et al., 2017). The estimated recombination frequency in citrus was as small as 0–2 times per generation per chromosome (Ohta et al., 2015), in agreement with the large LD values observed. These studies indicated that increasing the population size would not proportionally increase the accuracy of linkage analysis. Furthermore, the strong association caused by reduced recombination disturbs the development of closely linked DNA markers and the resolution of linkage and QTL analyses (Rafalski, 2002).

8.4.3 Association mapping (AM) analysis

AM analysis, which is also known as LD mapping, can be used to detect QTLs by evaluating the LD observed within a group of varieties (Balding, 2006; Rikkerink et al., 2007; Sorkheh et al., 2008; Zhu et al., 2008). AM analysis is based on the assumption that the varieties studied randomly inherited part of the ancestral genome. A QTL in an ancestral genome will be LD with the haplotype of neighboring SNP markers, and the extent of their association mostly depends on their genetic distance and the number of generations separating them from their ancestors. As a result, higher resolution can be achieved by AM versus conventional QTL analysis that uses a hybrid population obtained from a single hybridization (Rafalski, 2002; Zhu et al., 2008; Oraguzie et al., 2007a). A benefit of AM is that it does not require a hybrid population to detect QTLs. GWAS are performed to map candidate QTLs to the genome by evaluating LD using a large SNP marker set that covers the whole-genome region. Conventional QTL analysis is only used to evaluate phenotypic differences and polymorphisms occurring between the parents. In contrast, AM considers all phenotypic and genetic variations observed in the varieties studied. Therefore, the detected QTLs and associated DNA markers are anticipated to have wide availability for diverse varieties. Eliminating the time-consuming process of producing an F_1 hybrid population is another advantage of AM. The number of studies performed using AM are growing rapidly for many crops and fruit trees (Khan and Korban, 2012).

In citrus, Nakano et al. (2012) identified a genomic region involved in polyembryony by combining BSA and AM approach. Likewise, Cuenca et al. identified a genomic region of chromosome III that was involved in ABS resistance in a similar manner (Cuenca et al., 2013). GWAS data had shown that various factors, such as the recombination rate, mating system, genetic isolation, population subdivision, selection, or population size are known to influence the LD in a population (Oraguzie et al., 2007b; Rafalski and Morgante, 2004; Mackay and Powell, 2007; Iwata et al., 2016). Among those factors, the population structure is a major factor that causes false association (Balding, 2006), and an exact understanding of the structure is highly recommended to improve the accuracy of GWAS (Balding, 2006; Rikkerink et al., 2007). Minamikawa et al. evaluated the associations among whole-genome SNP genotypes with 17 fruit-trait data obtained from 787 parental varieties and hybrid progenies over 6 years (Minamikawa et al., 2017). They detected SNPs associated with 11 fruit traits (fruit weight, fruit hardness, the pericarp color, peeling, the flesh color, flesh hardness, juiciness, firmness of the locule membrane, the seed number, bitterness, and acidity) by integrating the population structure of those plant samples into the GWAS model (Fig. 8.3). The SNPs detected showed significant scores, not only for the continuous data, but also for the categorical data. By referring to the gene annotations of those detected SNPs, two genes that showed significant associations with pericarp and flesh color were identified as phytoene desaturase 1 (*pds1*) and carotenoid cleavage dioxygenase 4 (*ccd4*). These selected genes and their locations in the genome coincided with the genes that were identified by QTL analysis by Yu et al. (2016) and demonstrated that the power of detection by GWAS was comparable to that of QTL analysis.

8.4.4 Genomic selection (GS)

Meuwissen first proposed the concept of GS for the prediction of total genetic value in 2001 (Meuwissen et al., 2001). GS predicts the performance of a trait of interest from a prediction model that integrates both phenotypic data and whole-genome SNP-marker genotype data of a training population. Unlike AM and GWAS, GS does not predict an association of a particular locus or SNP with a trait. In contrast, GS considers the effects of whole-genome genotypes of a target population for the prediction and that enables to select a polygenic trait even though each locus has a faint genetic gain. GS has enabled significant achievements in Holstein breeding (García-Ruiz et al., 2016). The first study of GS in apples was reported by Kumar et al. (2012).

In citrus, Gois et al. (2016) examined the utility of GS using a hybrid population obtained from Murcott × sweet orange 'Pera'. They evaluated the prediction accuracy for 10 fruit traits by GS with 5287 DArTSeq marker scores. The scores obtained by the random regression-best linear unbiased predictor-based prediction model were high and ranged from 0.59 to 0.91. They also demonstrated that the proportion of genetic variance explained by the DArTSeq markers increased in proportion to the number of markers. Meuwissen et al. used a single family to build a prediction model in their first study (Meuwissen et al., 2001), and Gois et al. also evaluated a single family to build their prediction model (Gois et al., 2016). Single-family-based predictions can be extended to integrate data from multiple families. In this extended prediction approach, a prediction model was built from phenotypic and genotype data obtained from multiple families and existing varieties. Considering the large LD values in citrus, integrating multiple families for GS was anticipated to improve the accuracy of prediction by decreasing the LD. Minamikawa et al. assessed the prediction accuracy for 15 fruit traits by building prediction models from phenotype and genotype data of 111 existing varieties and 35 full-sib families (Minamikawa et al., 2017). Prediction models employ multivariate regression analysis, and various methods have been proposed so far. Minamikawa et al. compared the prediction performance of 12 regression equations by incorporating phenotypic data for

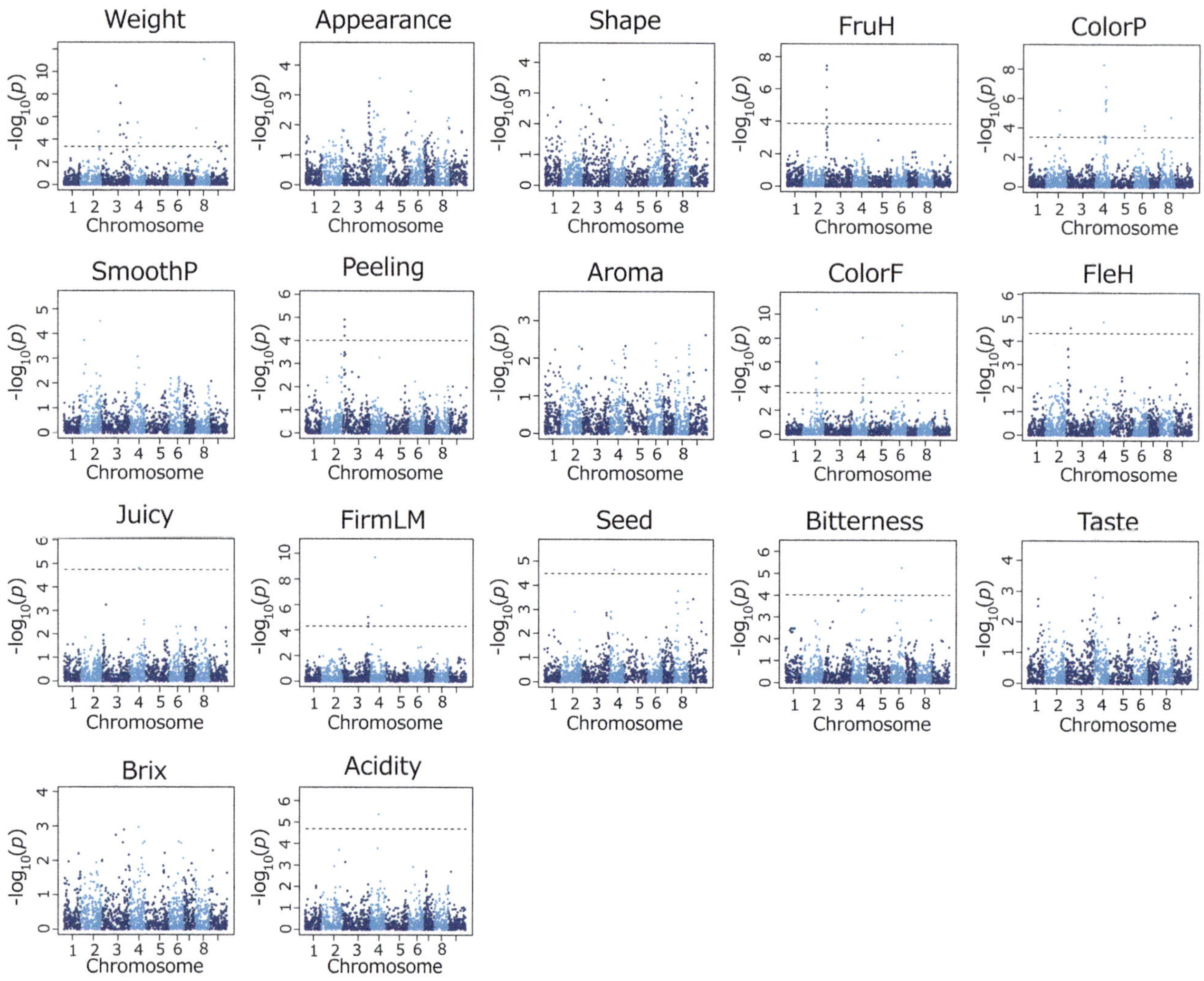

FIG. 8.3 Manhattan plots for 17 fruit-quality traits in combined parental and breeding populations by GWAS. A mixed-linear model employed four principal components of the population structure as covariates. *Dashed lines* indicate a false-discovery rate of 0.05. *(Figure courtesy of Dr. M. Minamikawa. Refer Minamikawa, M.F. et al., 2017. Genome-wide association study and genomic prediction in citrus: potential of genomics-assisted breeding for fruit quality traits. Sci. Rep., 7(1), p. 4721. Available at: http://www.nature.com/articles/s41598-017-05100-x for each trait.)*

15 fruit traits obtained over 6 years. The prediction accuracies varied from 0.2 to 0.9 according to the traits, but no evident differences for those prediction models were confirmed among the most traits. In contrast, they demonstrated a proportional increase in the prediction accuracy by increasing the number of individuals used for modeling. The traits with lower prediction scores were often the categorical data, which suggested that their measuring accuracy for those traits was not high enough for analysis. However, their results demonstrated the possibility of improving the accuracy by integrating sufficient data, even at low precision. This is one of the features of using GS that is difficult to attain in map-based analyses. Furthermore, integrating multiple family data has the added benefit of avoiding strong bias with a particular family. Another interesting feature observed in multiple-family-based GS is the future possibility of expanding this technique as a versatile tool for MAS in citrus. Minamikawa et al. (2017) used a mandarin-based population to construct their prediction models, and an insufficient number of pummelo and their progenies were used in their study. According to the rationale of GS and genetic background of the population used for the modeling, the prediction accuracies obtained from the constructed prediction model were supposed to be high in mandarin progenies, but low in pummelo progenies. However, the models predicted the fruit weights of pummelos at higher accuracies than expected (Fig. 8.4). This result seems to conflict with the rationale of GS. On the other hand, recent studies confirmed the hypothesis that existing citrus varieties emerged from the natural hybridization of ancestral citrus varieties in three taxa: mandarin, pummelo, and citron. Based on these studies, a region of the genome from ancestral pummelos integrated into significant numbers of existing mandarin varieties (Oueslati et al., 2017; Shimizu et al., 2016b; Curk et al., 2014, 2015; Wu et al., 2014b). The admixed pummelo genome would contribute

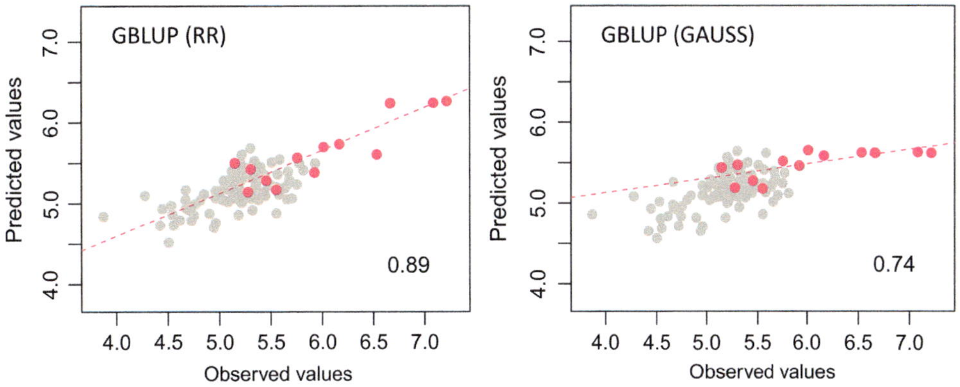

FIG. 8.4 Prediction for fruit weight distribution tails by genomic selection. All varieties except those in the large-fruit group, which are indicated by pink data points, were used as the training set for the prediction model; the large-fruit group was used as the test set. Gray points indicate the results of leave-one-out cross-validation. Predicted and observed values indicate log-transformed weights. Pearson's correlation coefficient (r) between the predicted and observed values for the pink data points are shown on the plots. *GAUSS*: Nonlinear Gaussian kernel regression, *RR*: linear ridge kernel regression. *(Figure courtesy of Dr. M. Minamikawa).*

to predicting the fruit weight of pummelos from mandarin-based prediction models. Single-family-based GS considers the correlation of phenotypic and genotypic variations observed in a single family, which suggests a limit in the utility of this approach. In contrast, multiple-family-based GS modeling is used to consider the correlation between phenotype and genotype data observed within diverse citrus varieties and their progenies. Similar to AM, multiple-family-based GS would be advantageous for citrus breeding by avoiding the need to prepare large breeding populations for individual purposes.

8.5 Future trends

The rapid growth of citrus-genome resources opens the door for developing genome-wide, precision-DNA markers and has enabled the development of DNA markers for selecting disease resistance. Whereas considerable numbers of candidate QTLs associated with fruit traits have been detected, and genes involved in fruit colors have been identified, efforts for developing a method to MAS of various fruit traits is still ongoing. The substantial effort required for developing hybrid populations is an obstacle for high-precision linkage analysis and QTL mapping in citrus. The use of AM and GS has enabled high-resolution study without a hybrid population. However, the detection power of AM and the prediction accuracy of GS are still insufficient for certain traits. Further improvements in their accuracy are required and may be achieved by increasing the amount and quality of data, optimizing the prediction models, or by integrating epistasis effects into the model. In addition to the growth of genome information, several high-throughput genotyping methods have become available at a lower cost. In contrast, progress in phenotyping has not kept up with the demand for high-throughput analysis. Synchronous progress in high-throughput phenotyping is key to the success of applying AM and GS to citrus breeding (Agostino, 2017). Although the subject was not covered in this review, automated phenotyping based on image analysis and unmanned aerial vehicles (Haris et al., 2016) will increase the throughput of phenotyping, which could then help improve the accuracy over a short duration. In addition, simulation studies for predicting the performance of progenies for each combination of varieties (Iwata et al., 2013) are essential for eliminating undesirable combinations.

References

Agostino, N.D., 2017. NGS-based genotyping, high-throughput phenotyping and genome-wide association studies laid the foundations for next-generation breeding in horticultural crops. Diversity 9 (3), 38. Available at: http://www.mdpi.com/1424-2818/9/3/38.

Ahmad, R., Struss, D., Southwick, S.M., 2003. Development and characterization of microsatellite markers in *Citrus*. J. Am. Soc. Hortic. Sci. 128 (4), 584–590. Available at: http://www.ashs.org.

Aleza, P., et al., 2011. Tetraploidization events by chromosome doubling of nucellar cells are frequent in apomictic citrus and are dependent on genotype and environment. Ann. Bot. 108 (1), 37–50. Available at: https://academic.oup.com/aob/article/108/1/37/99538.

Alquézar, B., Rodrigo, M.J., Zacarías, L., 2008. Carotenoid biosynthesis and their regulation in Citrus fruits. Tree For. Sci. Biotechnol. 2 (1), 23–35. Available at: http://www.globalsciencebooks.info/Online/GSBOnline/images/0812/TFSB_2(SI1)/TFSB_2(SI1)23-35o.pdf.

Alquézar, B., et al., 2017. Genomic analysis of terpene synthase family and functional characterization of seven sesquiterpene synthases from *Citrus sinensis*. Front. Plant Sci. 8, 1481. Available at: http://journal.frontiersin.org/article/10.3389/fpls.2017.01481/full.

Asins, M.J., et al., 2012. The position of the major QTL for *Citrus tristeza virus* resistance is conserved among *Citrus grandis, C. aurantium* and *Poncirus trifoliata*. Mol. Breed. 29 (3), 575–587. Available at: https://doi.org/10.1007/s11032-011-9574-x.

Balding, D.J., 2006. A tutorial on statistical methods for population association studies. Nat. Rev. Genet. 7 (10), 781–791. Available at: https://www.nature.com/articles/nrg1916.

Barkley, N.A., et al., 2006. Assessing genetic diversity and population structure in a citrus germplasm collection utilizing simple sequence repeat markers (SSRs). Theor. Appl. Genet. 112 (8), 1519–1531. Available at: https://doi.org/10.1007/s00122-006-0255-9.

Bastianel, M., et al., 2009. Quantitative trait loci analysis of citrus leprosis resistance in an interspecific backcross family of (*Citrus reticulata* Blanco × *C. sinensis* L. Osbeck) × *C. sinensis* L. Osb. Euphytica 169 (1), 101–111. Available at: https://link.springer.com/article/10.1007/s10681-009-9950-3.

Batley, J. (Ed.), 2015. Plant genotyping—methods and protocols. Springer New York, New York, NY. Available at: http://link.springer.com/10.1007/978-1-4939-1966-6.

Bernet, G.P., et al., 2005. Genetic analysis of citrus leafminer susceptibility. Theor. Appl. Genet. 110 (8), 1393–1400. Available at: https://link.springer.com/article/10.1007%2Fs00122-005-1943-6.

Bhat, J.A., et al., 2016. Genomic selection in the era of next generation sequencing for complex traits in plant breeding. Front. Genet. 7, 1–11. Available at: https://doi.org/10.3389/fgene.2016.00221.

Bink, M.C.A.M., et al., 2014. Bayesian QTL analyses using pedigreed families of an outcrossing species, with application to fruit firmness in apple. Theor. Appl. Genet. 127 (5), 1073–1090. Available at: https://link.springer.com/article/10.1007/s00122-014-2281-3.

Biswas, M.K., et al., 2015. Novel polymorphic EST-based microsatellite marker isolation and characterization from *Poncirus trifoliata* (Rutaceae). Front. Agric. Sci. Eng. 2 (1), 60. Available at: http://engineering.cae.cn/fase/EN/10.15302/J-FASE-2015048.

Cai, Q., Guy, C.L., Moore, G.A., 1994. Extension of the linkage map in *Citrus* using random amplified polymorphic DNA (RAPD) markers and RFLP mapping of cold-acclimation-responsive loci. Theor. Appl. Genet. 89 (5), 606–614. Available at: https://link.springer.com/content/pdf/10.1007/BF00222455.pdf.

Catchen, J.M., et al., 2011. *Stacks*: Building and genotyping loci *de novo* from short-read sequences. G3 (Bethesda) 1 (3), 171–182. Available at: http://g3journal.org/lookup/doi/10.1534/g3.111.000240.

Chae, C.-W., et al., 2011. Development of a SCAR marker linked to male fertility traits in "Jinkyool" (*Citrus sunki*). J. Life Sci. 21 (12), 1659–1665. Available at: http://koreascience.or.kr/journal/view.jsp?kj=SMGHBM&py=2011&vnc=v21n12&sp=1659.

Chai, L., et al., 2013. Transferability, polymorphism and effectiveness for genetic mapping of the pummelo (*Citrus grandis Osbeck*) EST-SSR markers. Sci. Hortic. 155, 85–91. Available at: https://doi.org/10.1016/j.scienta.2013.02.024.

Chen, C., Gmitter Jr., F.G., 2013. Mining of haplotype-based expressed sequence tag single nucleotide polymorphisms in citrus. BMC Genomics 14 (1), 746. Available at: http://bmcgenomics.biomedcentral.com/articles/10.1186/1471-2164-14-746.

Chen, C., et al., 2006. Mining and characterizing microsatellites from citrus ESTs. Theor. Appl. Genet. 112 (7), 1248–1257. Available at: http://link.springer.com/10.1007/s00122-006-0226-1.

Chen, C., et al., 2008. EST-SSR genetic maps for *Citrus sinensis* and *Poncirus trifoliata*. Tree Genet. Genomes 4 (1), 1–10. Available at: http://link.springer.com/10.1007/s11295-007-0083-3.

Chen, Y.-Z., et al., 2015. Isolation and characterization of ten microsatellite loci for wild *Citrus japonica* (Rutaceae). J. Genet. 94 (S1), 41–43. Available at: https://doi.org/10.1007/s12041-014-0368-x.

Cheng, F., Roose, M., 1995. Origin and inheritance of dwarfing by the citrus rootstock *Poncirus trifoliata* "Flying Dragon". J. Am. Soc. Hortic. Sci. 120 (2), 286–291. Available at: http://journal.ashspublications.org/content/120/2/286.

Choi, Y.A., et al., 2007. Identification of AFLP markers associated with *Citrus* canker resistance within *Citrus*. Acta Hortic. 738, 201–206. Available at: http://www.actahort.org/books/738/738_18.htm.

Close, T.J., et al., 2006. A GeneChip® for citrus. In: Plant and Animal Genome XIV. San Diego, CA, USA, p. 82. Available at: http://www.intl-pag.org/14/PDF/0114_citrus_close.PDF.

Combrink, N.K., Labuschagne, M.T., Bijzet, Z., 2013. Variation of fruit size and shape in Kiyomi tangor families. Sci. Hortic. 162, 357–364. Available at: https://doi.org/10.1016/j.scienta.2013.08.010.

Combrink, N.K., et al., 2015. Genotypic variation of rind colour in citrus tangor Kiyomi families. Acta Hortic. 1065, 439–448. Available at: https://doi.org/10.17660/ActaHortic.2015.1065.54.

Costa, F., 2015. MetaQTL analysis provides a compendium of genomic loci controlling fruit quality traits in apple. Tree Genet. Genomes 11 (1), 1–11. Available at: https://doi.org/10.1007/s11295-014-0819-9.

Cristofani, M., Machado, M.A., Grattapaglia, D., 1999. Genetic linkage maps of *Citrus sunki* Hort. ex. Tan. and *Poncirus trifoliata* (L.) Raf. and mapping of citrus tristeza virus resistance gene. Euphytica 109 (1), 25–32. Available at: https://doi.org/10.1023/A:1003637116745.

Cristofani, M., et al., 2000. Construction of linkage maps of *Poncirus trifoliata* and *Citrus sunki* based on microsatellite markers. In: Davies, F.S. (Ed.), Proceedings of International Society of Citriculture. Orland, FL. USA, pp. 175–178. Available at: http://internationalsocietyofcitriculture.org/proceedingscontent/2000_florida.html.

Cristofani-Yaly, M., et al., 2011. Transferability and level of heterozygosity of microsatellite markers in *Citrus* species. Plant Mol. Biol. Report. 29 (2), 418–423. Available at: https://doi.org/10.1007/s11105-010-0241-x.

Cuenca, J., et al., 2011. Multilocus half-tetrad analysis and centromere mapping in citrus: evidence of SDR mechanism for 2*n* megagametophyte production and partial chiasma interference in mandarin cv "Fortune.". Heredity 107 (5), 462–470. Available at: http://www.nature.com/doifinder/10.1038/hdy.2011.33.

Cuenca, J., et al., 2013. Genetically based location from triploid populations and gene ontology of a 3.3-Mb genome region linked to Alternaria brown spot resistance in citrus reveal clusters of resistance genes. PLoS One 8 (10), 1–18. Available at: http://journals.plos.org/plosone/article?id=10.1371/journal.pone.0076755.

Cuenca, J., et al., 2016. Fine mapping for identification of citrus Alternaria brown spot candidate resistance genes and development of new SNP markers for marker-assisted selection. Front. Plant Sci. 7, 1–13. Available at: http://journal.frontiersin.org/article/10.3389/fpls.2016.01948/full.

Curk, F., et al., 2014. Next generation haplotyping to decipher nuclear genomic interspecific admixture in *Citrus* species: analysis of chromosome 2. BMC Genet. 15 (1), 152. Available at: http://www.biomedcentral.com/1471-2156/15/152.

Curk, F., et al., 2015. Nuclear species-diagnostic SNP markers mined from 454 amplicon sequencing reveal admixture genomic structure of modern *Citrus* varieties. PLoS One 10 (5), 1–25. Available at: http://journals.plos.org/plosone/article?id=10.1371/journal.pone.0125628.

Curtolo, M., et al., 2017. QTL mapping for fruit quality in *Citrus* using DArTseq markers. BMC Genet. 18 (1), 289. Available at: https://doi.org/10.1186/s12864-017-3629-2.

Davey, J.W., Blaxter, M.L., 2010. RADSeq: Next-generation population genetics. Brief. Funct. Genomics 9 (5–6), 416–423. Available at: https://doi.org/10.1093/bfgp/elq031.

Davey, J.W., et al., 2011. Genome-wide genetic marker discovery and genotyping using next-generation sequencing. Nat. Rev. Genet. 12 (7), 499–510. Available at: http://www.nature.com/nrg/journal/v12/n7/abs/nrg3012.html.

De Simone, M., et al., 1998. Construction of genetic maps for *Citrus aurantium* and *C. latipes* based on AFLP, RAPD and RFLP markers. Fruits 53 (6), 383–390.

Deng, Z., et al., 2000. Cloning and characterization of NBS-LRR class resistance-gene candidate sequences in citrus. Theor. Appl. Genet. 101, 814–822. Available at: https://doi.org/10.1007/s001220051548.

Durham, R.E., et al., 1992. Linkage of restriction fragment length polymorphisms and isozymes in *Citrus*. Theor. Appl. Genet. 84 (1–2), 39–48. Available at: https://doi.org/10.1007/BF00223979.

Eck, Y., 2017. Application of high throughput genotyping SNP array for citrus. In: Plant and Animal Genome XXV. San Diego, CA, p. W148. Available at: https://pag.confex.com/pag/xxv/webprogram/Paper25696.html.

Elshire, R.J., et al., 2011. A robust, simple genotyping-by-sequencing (GBS) approach for high diversity species. PLoS One 6 (5), e19379. Available at: https://doi.org/10.1371/journal.pone.0019379.

Fang, D., Federici, C., Roose, M., 1998. A high-resolution linkage map of the citrus tristeza virus resistance gene region in *Poncirus trifoliata* (L.) Raf. Genetics 150, 883–890. Available at: http://www.genetics.org/content/150/2/883.

Froelicher, Y., et al., 2008. Characterization of microsatellite markers in mandarin orange (*Citrus reticulata* Blanco). Mol. Ecol. Resour. 8 (1), 119–122. Available at: https://doi.org/10.1111/j.1471-8286.2007.01893.x.

Frost, H.B., Soost, R.K., 1968. Seed reproduction: development of gametes and embryos. In: Reuther, W., Batchelor, L.D., Webber, H.J. (Eds.), The Citrus Industry. vol. II. University of California Press, Berkeley, California, pp. 290–324.

Fujii, H., et al., 2012. High-throughput genotyping in citrus accessions using an SNP genotyping array. Tree Genet. Genomes 9 (1), 145–153. https://doi.org/10.1007/s11295-012-0542-3.

García, R., et al., 1999. Genetic analysis of apomixis in *Citrus* and *Poncirus* by molecular markers. Theor. Appl. Genet. 99, 511–518. Available at: https://doi.org/10.1007/s001220051264.

García-Lor, A., et al., 2012. Comparative use of InDel and SSR markers in deciphering the interspecific structure of cultivated citrus genetic diversity: a perspective for genetic association studies. Mol. Genet. Genomics 287 (1), 77–94. Available at: https://doi.org/10.1007/s00438-011-0658-4.

Garcia-Lor, A., Ancillo, G., et al., 2013a. *Citrus* (Rutaceae) SNP markers based on competitive allele-specific PCR; transferability across the Aurantioideae subfamily. Appl. Plant Sci. 1 (4), 1200406. Available at: https://doi.org/10.3732/apps.1200406.

Garcia-Lor, A., Curk, F., et al., 2013b. A nuclear phylogenetic analysis: SNPs, indels and SSRs deliver new insights into the relationships in the "true citrus fruit trees" group (Citrinae, Rutaceae) and the origin of cultivated species. Ann. Bot. 111 (1), 1–19. Available at: https://doi.org/10.1093/aob/mcs227.

García-Ruiz, A., et al., 2016. Changes in genetic selection differentials and generation intervals in US Holstein dairy cattle as a result of genomic selection. Proc. Natl. Acad. Sci. 113 (28), E3995–E4004. Available at: https://doi.org/10.1073/pnas.1519061113.

Garvin, M.R., Saitoh, K., Gharrett, A.J., 2010. Application of single nucleotide polymorphisms to non-model species: a technical review. Mol. Ecol. Resour. 10 (6), 915–934. Available at: https://doi.org/10.1111/j.1755-0998.2010.02891.x/full.

Gmitter, F.G., et al., 1996. A localized linkage map of the citrus tristeza virus resistance gene region. Theor. Appl. Genet. 92 (6), 688–695. Available at: https://doi.org/10.1007/BF00226090.

Gmitter, F.G., Deng, Z., Chen, C., 2007a. Cloning and characterization of disease resistance genes. In: Kahn, I.A. (Ed.), *Citrus* Genetics, Breeding and Biotechnology. CABI, Wallingford, pp. 287–305. Available at: http://www.cabi.org/cabebooks/ebook/20083096352.

Gmitter, F.G., et al., 2007b. Citrus Fruits. In: Fruits and Nuts. Springer Berlin Heidelberg, Berlin, Heidelberg, pp. 265–279. Available at: https://doi.org/10.1007/978-3-540-34533-6_14.

Gmitter, F.G., et al., 2012. *Citrus* genomics. Tree Genet. Genomes 8 (3), 611–626. Available at: https://doi.org/10.1007/s11295-012-0499-2.

Gois, I.B., et al., 2016. Genome wide selection in citrus breeding. Genet. Mol. Res. 15 (4), 1–14. Available at: https://doi.org/10.4238/gmr15048863.

Golein, B., et al., 2006. Development and characterization of new microsatellite loci from lemon (*Citrus limon*). Int. J. Agric. Biol. 8, 172–174. Available at: http://www.fspublishers.org/ijab/past-issues/IJABVOL_8_NO_2/9.pdf.

Grattapaglia, D., Sederoff, R., 1994. Genetic linkage maps of *Eucalyptus grandis* and *Eucalyptus urophylla* using a pseudo-testcross: mapping strategy and RAPD markers. Genetics 137 (4), 1121–1137. Available at: http://www.genetics.org/content/137/4/1121.

Grattapaglia, D., Bertolucci, F.L., Sederoff, R.R., 1995. Genetic mapping of QTLs controlling vegetative propagation in *Eucalyptus grandis* and *E. urophylla* using a pseudo-testcross strategy and RAPD markers. Theor. Appl. Genet. 90 (7–8), 933–947. Available at: https://doi.org/10.1007/BF00222906.

Gulsen, O., et al., 2010. A new citrus linkage map based on SRAP, SSR, ISSR, POGP, RGA and RAPD markers. Euphytica 173 (2), 265–277. Available at: https://doi.org/10.1007/s10681-010-0146-7.

Gulsen, O., et al., 2011. QTL analysis and regression model for estimating fruit setting in young *Citrus* trees based on molecular markers. Sci. Hortic. 130 (2), 418–424. Available at: https://doi.org/10.1016/j.scienta.2011.07.010.

Guo, F., et al., 2015. Construction of a SNP-based high-density genetic map for pummelo using RAD sequencing. Tree Genet. Genomes 11 (1), 2. Available at: https://doi.org/10.1007/s11295-014-0831-0.

Gupta, P.K., Rustgi, S., Kulwal, P.L., 2005. Linkage disequilibrium and association studies in higher plants: Present status and future prospects. Plant Mol. Biol. 57 (4), 461–485. Available at: https://doi.org/10.1007/s11103-005-0257-z.

Hall, D., Tegström, C., Ingvarsson, P.K., 2010. Using association mapping to dissect the genetic basis of complex traits in plants. Brief. Funct. Genomic. Proteomic. 9 (2), 157–165. Available at: https://doi.org/10.1093/bfgp/elp048.

Haris, M., et al., 2016. Construction of a high-resolution digital map to support citrus breeding using an autonomous multicopter. Acta Hortic. 1135, 73–84. Available at: http://www.actahort.org/books/1135/1135_9.htm.

Henry, R.J. (Ed.), 2008. Plant Genotyping II: SNP Technology. CABI, Wallingford. Available at: http://www.cabi.org/cabebooks/ebook/20083134963.

Heslot, N., et al., 2013. Impact of marker ascertainment bias on genomic selection accuracy and estimates of genetic diversity. PLoS One 8 (9), e74612. Available at: https://doi.org/10.1371/journal.pone.0074612.

Hong, Q., et al., 2008. A consensus map constructed with SSR and EST-SSR markers using progeny population from sweet orange tangor. In: Deng, X., Xu, J., Lin, S., Guan, R. (Eds.), Proceedings of 11th International Society of Citriculture. 1. China Agricultural Press, Wuhan, China, pp. 284–288.

Hong, Q., et al., 2015. QTL mapping of citrus freeze tolerance. Acta Hortic. 1065, 467–474. Available at: http://www.actahort.org/books/1065/1065_57.htm.

Hou, X.J., et al., 2014. Genome-wide identification, classification, expression profiling, and SSR marker development of the MADS-box gene family in *Citrus*. Plant Mol. Biol. Report. 32 (1), 28–41. Available at: https://doi.org/10.1007/s11105-013-0597-9.

Ikoma, Y., Matsumoto, H., Kato, M., 2016. Diversity in the carotenoid profiles and the expression of genes related to carotenoid accumulation among citrus genotypes. Breed. Sci. 66 (1), 139–147. Available at: https://www.jstage.jst.go.jp/article/jsbbs/66/1/66_139/_article.

Imai, A., Yoshioka, T., Hayashi, T., 2017. Quantitative trait locus (QTL) analysis of fruit-quality traits for mandarin breeding in Japan. Tree Genet. Genomes 13 (4), 79. Available: https://doi.org/10.1007/s11295-017-1162-8.

İncesu, M., et al., 2011. Molecular diversification and preliminary evaluations of some Satsuma selections'performance under mediterranean conditions. Afr. J. Biotechnol. 10 (21), 4347–4357. Available at: http://www.academicjournals.org/journal/AJB/article-abstract/147E21B32996.

Iwata, H., et al., 2013. Genomic prediction of trait segregation in a progeny population: a case study of Japanese pear (*Pyrus pyrifolia*). BMC Genet. 14 (1), 81. Available at: https://doi.org/10.1186/1471-2156-14-80.

Iwata, H., et al., 2016. Genomics-assisted breeding in fruit trees. Breed. Sci. 66 (1), 100–115. Available at: https://doi.org/10.1270/jsbbs.66.100.

Jarrell, D.C., et al., 1992. A genetic map of citrus based on the segregation of isozymes and RFLPs in an intergeneric cross. Theor. Appl. Genet. 84 (1–2), 49–56. Available at: https://doi.org/10.1007/BF00223980.

Jiang, D., Guang-Yan, Z., Hong, Q.-B., 2006. Analysis of microsatellites in *Citrus* unigenes. Acta Genet. Sin. 33 (4), 345–353. Available at: http://linkinghub.elsevier.com/retrieve/pii/S0379417206600607.

Jiang, D., et al., 2010. The mining of citrus EST-SNP and its application in cultivar discrimination. Agric. Sci. China 9 (2), 179–190. Available at: http://linkinghub.elsevier.com/retrieve/pii/S1671292709600821.

JinPing, X., et al., 2009. Identification of AFLP fragments linked to seedlessness in Ponkan mandarin (*Citrus reticulata* Blanco) and conversion to SCAR markers. Sci. Hortic. 121 (4), 505–510. Available at: https://doi.org/10.1016/j.scienta.2009.03.006.

Kamiri, M., et al., 2011. Evidence for non-disomic inheritance in a *Citrus* interspecific tetraploid somatic hybrid between *C. reticulata* and *C. limon* using SSR markers and cytogenetic analysis. Plant Cell Rep. 30 (8), 1415–1425. Available at: https://doi.org/10.1007/s00299-011-1050-x.

Kepiro, J.L., Roose, M.L., 2007. Nucellar embryony. In: Khan, I.A. (Ed.), *Citrus* Genetics, Breeding and Biotechnology. CABI, CAB international, Nosworthy way Wallingford, pp. 141–149. Available at: https://doi.org/10.1079/9780851990194.0000.

Kepiro, J.L., Roose, M.L., 2009. AFLP markers closely linked to a major gene essential for nucellar embryony (apomixis) in *Citrus maxima* × *Poncirus trifoliata*. Tree Genet. Genomes 6 (1), 1–11. Available at: https://doi.org/10.1007/s11295-009-0223-z.

Khan, M.A., Korban, S.S., 2012. Association mapping in forest trees and fruit crops. J. Exp. Bot. 63 (11), 4045–4060. Available at: https://doi.org/10.1093/jxb/ers105.

Kijas, J.M., et al., 1994. Enrichment of microsatellites from the citrus genome using biotinylated oligonucleotide sequences bound to streptavidin-coated magnetic particles. Biotechniques 16 (4), 656–662. Available at: http://www.ncbi.nlm.nih.gov/pubmed/8024786.

Kijas, J.M., Fowler, J.C., Thomas, M.R., 1995. An evaluation of sequence tagged microsatellite site markers for genetic analysis within *Citrus* and related species. Genome 38 (2), 349–355. Available at: http://www.nrcresearchpress.com/doi/pdf/10.1139/g95-045.

Kijas, J.M.H., et al., 1997. Integration of trinucleotide microsatellites into a linkage map of *Citrus*. Theor. Appl. Genet. 94 (5), 701–706. Available at: https://doi.org/10.1007/s001220050468.

Kumar, S., et al., 2012. Genomic selection for fruit quality traits in apple (*Malus×domestica* Borkh.). PLoS One 7 (5), e36674. Available at: https://doi.org/10.1371/journal.pone.0036674.

Li, G., Quiros, C.F., 2001. Sequence-related amplified polymorphism (SRAP), a new marker system based on a simple PCR reaction: Its application to mapping and gene tagging in *Brassica*. Theor. Appl. Genet. 103 (2–3), 455–461. Available at: https://doi.org/10.1007/s001220100570.

Ling, P., et al., 2000. Inheritance of citrus nematode resistance and its linkage with molecular markers. Theor. Appl. Genet. 100 (7), 1010–1017. Available at: https://doi.org/10.1007/s001220051382.

Liou, P.-C., 1990. A Molecular Study of the *Citrus* Genome through Restriction Fragment Length Polymorphism and Isozyme Mapping. The University of Florida. Available at: https://archive.org/details/molecularstudyof00liou.

Liu, S.R., et al., 2013. Development and characterization of genomic and expressed SSRs in citrus by genome-wide analysis. PLoS ONE 8 (10), e75149. Available at: https://doi.org/10.1371/journal.pone.0075149.

Luro, F., et al., 1996. *Citrus* genome mapping with molecular markers:two maps obtained by segregation analysis of progeny of one intergeneric cross. In: International Citrus Congress. Sun City, South Africa, volume 21996. pp. 862–866. Available at: http://agritrop.cirad.fr/390226/.

Luro, F.L., et al., 2008. Transferability of the EST-SSRs developed on Nules clementine (*Citrus clementina* Hort ex tan) to other *Citrus* species and their effectiveness for genetic mapping. BMC Genomics 9, 287. Available at: https://doi.org/10.1186/1471-2164-9-287.

Ma, X., et al., 2012. Development of SSR markers from citrus BAC end sequences and their integration into linkage map. Acta Botan. Boreali-Occiden. Sin. 32 (6), 1112–1117. Available at: http://www.xbzwxb.com/ch/reader/view_abstract.aspx?file_no=20120607&flag=1.

Machado, M.A., Cristofani-Yaly, M., Bastianel, M., 2011. Breeding, genetic and genomic of citrus for disease resistance. Rev. Bras. Frutic. 33 (1), 158–172. Available at: https://doi.org/10.1590/S0100-29452011000500019.

Mackay, I., Powell, W., 2007. Methods for linkage disequilibrium mapping in crops. Trends Plant Sci. 12 (2), 57–63. Available at: https://www.cell.com/trends/plant-science/pdf/S1360-1385(06)00315-3.pdf.

Mason, A.S., 2015. SSR genotyping. In: Batley, J. (Ed.), Plant Genotyping. Humana Press, New York, NY, pp. 77–89. Available at: https://doi.org/10.1007/978-1-4939-1966-6_6.

Mastretta-Yanes, A., et al., 2015. Restriction site-associated DNA sequencing, genotyping error estimation and *de novo* assembly optimization for population genetic inference. Mol. Ecol. Resour. 15 (1), 28–41. Available at: https://doi.org/10.1111/1755-0998.12291.

Meuwissen, T.H.E., Hayes, B.J., Goddard, M.E., 2001. Prediction of total genetic value using genome-wide dense marker maps. Genetics 157 (4), 1819–1829. Available at: http://www.genetics.org/content/157/4/1819.

Michelmore, R.W., Paran, I., Kesseli, R.V., 1991. Identification of markers linked to disease-resistance genes by bulked segregant analysis: a rapid method to detect markers in specific genomic regions by using segregating populations. Proc. Natl. Acad. Sci. U. S. A. 88 (21), 9828–9832. Available at: https://doi.org/10.1073/pnas.88.21.9828.

Minamikawa, M.F., et al., 2017. Genome-wide association study and genomic prediction in citrus: potential of genomics-assisted breeding for fruit quality traits. Sci. Rep. 7 (1), 4721. Available at: http://www.nature.com/articles/s41598-017-05100-x.

Miyazaki, T., et al., 2011. Distribution of aroma volatile compounds in tangerine hybrids and proposed inheritance. J. Sci. Food Agric. 91 (3), 449–460. Available at: https://doi.org/10.1002/jsfa.4205/abstract.

Miyazaki, T., et al., 2012. Aroma characterization of tangerine hybrids by gas-chromatography-olfactometry and sensory evaluation. J. Sci. Food Agric. 92 (4), 727–735. Available at: https://doi.org/10.1002/jsfa.4663/epdf.

Nakano, M., et al., 2008. Mapping and haplotyping of the flanking region of the polyembryony locus in *Citrus unshiu* Marcow. J. Jpn. Soc. Hortic. Sci. 77 (2), 109–114. Available at: http://joi.jlc.jst.go.jp/JST.JSTAGE/jjshs1/77.109.

Nakano, M., et al., 2012. Characterization of genomic sequence showing strong association with polyembryony among diverse *Citrus* species and cultivars, and its synteny with *Vitis* and *Populus*. Plant Sci. 183, 131–142. Available at: http://www.sciencedirect.com/science/article/pii/S0168945211002238.

Nishiura, M., 1967. *Citrus* breeding through nucellar seedling selection. JARQ: Jpn Agric. Res. Q. 2 (1), 15–19. Available at: http://www.jircas.affrc.go.jp/english/publication/jarq/02-1/02-1index.html.

Novelli, V.M., Cristofani, M., Machado, M.A., 2000. Evaluation of microsatellite markers in cultivars of sweet orange (*Citrus sinensis* [L.] Osbeck). Acta Hortic. 535, 47–50. Available at: http://www.actahort.org/books/535/535_4.htm.

Novelli, V.M., Takita, M.A., Machado, M.A., 2004. Identification and analysis of single nucleotide polymorphisms (SNPs) in citrus. Euphytica 138 (3), 227–237. Available at: https://doi.org/10.102/B:EUPH.0000047086.47988.82.

Novelli, V.M., et al., 2006. Development and characterization of polymorphic microsatellite markers for the sweet orange (*Citrus sinensis* L. Osbeck). Genet. Mol. Biol. 29 (1), 90–96. Available at: http://www.scielo.br/scielo.php?script=sci_arttext&pid=S1415-47572006000100018.

Ohta, S., et al., 2011. PCR primers for marker assisted backcrossing to introduce a CTV resistance gene from *Poncirus trifoliata* (L.) Raf. Into *Citrus*. J. Jpn. Soc. Hortic. Sci. 80 (3), 295–307. Available at: https://doi.org/10.2503/jjshs1.80.295.

Ohta, S., et al., 2015. Construction of genetic linkage map and graphical genotyping of pseudo-backcrossed F2 (BC'2) progeny to introduce a CTV resistance from *Poncirus trifoliata* (L.) Raf. Into *Citrus* by introgression breeding. Tree Genet. Genomes 11 (1), 797. Available at: https://doi.org/10.1007/s11295-014-0797-y.

Oliveira, A.C., et al., 2000. Identification of sexual and nucellar citrus seedlings through single and duplex PCR of simple sequence repeat loci. In: Davies, F.S. (Ed.), Proceedings of International Society of Citriculture, Orland, FL, USA, pp. 138–141. Available at: http://internationalsocietyofcitriculture.org/proceedingscontent/2000_florida.html.

Ollitrault, P., Navarro, L., 2012. *Citrus*. In: Badenes, M.L., Byrne, D.H. (Eds.), Fruit Breeding. Boston, MA, Springer US, pp. 623–662. Available at: https://doi.org/10.1007/978-1-4419-0763-9_16.

Ollitrault, F., et al., 2010. Development of SSR markers from *Citrus clementina* (Rutaceae) BAC end sequences and interspecific transferability in *Citrus*. Am. J. Bot. 97 (11), e124–e129. Available at: https://doi.org/10.3732/ajb.1000280.

Ollitrault, P., Terol, J., Chen, C., et al., 2012a. A reference genetic map of *C. clementina* hort. Ex tan.; citrus evolution inferences from comparative mapping. BMC Genomics 13 (1), 593. Available at: https://doi.org/10.1186/1471-2164-13-593.

Ollitrault, P., Terol, J., Garcia-Lor, A., et al., 2012b. SNP mining in *C. clementina* BAC end sequences; transferability in the *Citrus* genus (Rutaceae), phylogenetic inferences and perspectives for genetic mapping. BMC Genomics 13 (1), 13. Available at: https://doi.org/10.1186/1471-2164-13-13.

Ollitrault, F., et al., 2012c. Development of indel markers from *Citrus clementina* (Rutaceae) BAC-end sequences and interspecific transferability in *Citrus*. Am. J. Bot. 99 (7), 268–273. Available at: https://doi.org/10.3732/ajb.1100569.

Ollitrault, P., et al., 2015. Comparative values of SSRs, SNPs and InDels for citrus genetic diversity analysis. Acta Hortic. 1065, 457–466. Available at: http://www.actahort.org/books/1065/1065_56.htm.

Omura, M., et al., 2000. EST mapping of citrus. In: Davies, F.S. (Ed.), Proceedings of International Society of Citriculture. Orland, FL. USA, p. 71–74.

Oraguzie, N., et al., 2007a. Linkage disequilibrium. In: Oraguzie, N.C., et al. (Eds.), Association Mapping in Plants. Springer-Verlag New York, pp. 11–39. Available at: https://doi.org/10.1007/978-0-387-36011-9_2.

Oraguzie, N.C., et al., 2007b. In: Oraguzie, N.C., et al. (Eds.), Association Mapping in Plants. Springer New York, New York, NY. Available at: https://doi.org/10.1007/978-0-387-36011-9.

Oueslati, A., et al., 2017. Genotyping by sequencing reveals the interspecific *C. maxima*/*C. reticulata* admixture along the genomes of modern citrus varieties of mandarins, tangors, tangelos, orangelos and grapefruits. PLoS One 12 (10), e0185618. Available at: https://doi.org/10.1371/journal.pone.0185618.

Palmieri, D.A., et al., 2007. Frequency and distribution of microsatellites from ESTs of citrus. Genet. Mol. Biol. 30 (3 suppl), 1009–1018. Available at: http://www.scielo.br/scielo.php?script=sci_arttext&pid=S1415-47572007000500029&lng=en&tlng=en.

Penjor, T., et al., 2014. Characterization of limes (*Citrus aurantifolia*) grown in Bhutan and Indonesia using high-throughput sequencing. Sci. Rep. 4, 1–9. Available at: https://www.nature.com/articles/srep04853.

Penjor, T., et al., 2016. RAD-Seq analysis of typical and minor *Citrus* accessions, including Bhutanese varieties. Breed. Sci. 66 (5), 797–807. Available at: https://www.jstage.jst.go.jp/article/jsbbs/66/5/66_16059/_article.

Rafalski, A., 2002. Applications of single nucleotide polymorphisms in crop genetics. Curr. Opin. Plant Biol. 5 (2), 94–100. Available at: https://doi.org/10.1016/S1369-5266(02)00240-6.

Rafalski, A., Morgante, M., 2004. Corn and humans: recombination and linkage disequilibrium in two genomes of similar size. Trends Genet. 20 (2), 103–111. Available at: https://doi.org/10.1016/j.tig.2003.12.002.

Rai, M., 2006. Refinement of the *Citrus tristeza virus* resistance gene (*Ctv*) positional map in *Poncirus trifoliata* and generation of transgenic grapefruit (*Citrus paradisi*) plant lines with candidate resistance genes in this region. Plant Mol. Biol. 61 (3), 399–414. Available at: https://doi.org/10.1007/s11103-006-0018-7.

Ramadugu, C., et al., 2015. Genetic analysis of citron (*Citrus medica* L.) using simple sequence repeats and single nucleotide polymorphisms. Sci. Hortic. 195, 124–137. Available at: https://doi.org/10.1016/j.scienta.2015.09.004.

Rikkerink, E.H.A., Oraguzie, N.C., Gardiner, S.E., 2007. Prospects of association mapping in perennial horticultural crops. In: Oraguzie, N.C., et al. (Eds.), Association Mapping in Plants. Springer New York, New York, NY, pp. 249–269. Available at: https://doi.org/10.1007/978-0-387-36011-9_11.

Roose, M.L., 2007. Mapping and marker-assisted selection. In: Kahn, I.A. (Ed.), *Citrus* Genetics, Breeding and Biotechnology. CABI, Wallingford, pp. 275–285. Available at: http://www.cabi.org/cabebooks/ebook/20083096351.

Roose, M.L., Close, T.J., 2008. Genomics of citrus, a major fruit crop of tropical and subtropical regions. In: Moore, P.H., Ming, R. (Eds.), Genomics of Tropical Crop Plants. New York, NY, Springer New York, pp. 187–202. Available at: https://doi.org/10.1007/978-0-387-71219-2_8.

Ruiz, C., Asins, M., 2003. Comparison between *Poncirus* and *Citrus* genetic linkage maps. Theor. Appl. Genet. 106 (5), 826–836. Available at: https://doi.org/10.1007/s00122-002-1095-x.

Sagawa, C.H.D., et al., 2017. Assessing genetic diversity of *Citrus* by DArT_seq™ genotyping. Plant Biosyst. 1–6. Available at: https://doi.org/10.1080/11263504.2017.1341438.

Şahin-Çevik, M., Moore, G.A., 2012. Quantitative trait loci analysis of morphological traits in *Citrus*. Plant Biotechnol. Rep. 6 (1), 47–57. Available at: https://doi.org/10.1007/s11816-011-0194-z.

Sansaloni, C., et al., 2011. Diversity arrays technology (DArT) and next-generation sequencing combined: Genome-wide, high throughput, highly informative genotyping for molecular breeding of *Eucalyptus*. BMC Proc. 5 (Suppl 7), P54. Available at: https://doi.org/10.1186/1753-6561-5-S7-P54.

Scarano, M.T., et al., 2003. Flow cytometry, SSR and modified AFLP markers for the identification of zygotic plantlets in backcrosses between "Femminello" lemon cybrids (2n and 4n) and a diploid clone of "Femminello" lemon (*Citrus limon* L. Burm. F.) tolerant to *mal secco*. Plant Sci. 164 (6), 1009–1017. Available at: https://doi.org/10.1016/S0168-9452(03)00088-8.

Semagn, K., Bjornstad, A., Ndjiondjop, M.N., 2006. An overview of molecular marker methods for plants. Afr. J. Biotechnol. 5 (25), 2540–2568. Available at: https://academicjournals.org/journal/AJB/article-abstract/B798A239082.

Shimada, T., et al., 2014. Construction of a citrus framework genetic map anchored by 708 gene-based markers. Tree Genet. Genomes 10 (4), 1001–1013. Available at: https://doi.org/10.1007/s11295-014-0738-9.

Shimizu, T., Yano, K., 2011. A post-labeling method for multiplexed and multicolored genotyping analysis of SSR, indel and SNP markers in single tube with bar-coded split tag (BStag). BMC. Res. Notes 4 (1), 161. Available at: http://www.biomedcentral.com/1756-0500/4/161.

Shimizu, T., Kaminuma, E., et al., 2016a. A genomic approach to selecting robust and versatile SNP sets from next-generation sequencing data for genome-wide association study in citrus cultivars. Acta Hortic. 1135, 23–32. Available at: http://www.actahort.org/books/1135/1135_4.htm.

Shimizu, T., Kitajima, A., et al., 2016b. Hybrid origins of citrus varieties inferred from DNA marker analysis of nuclear and organelle genomes. PLoS One 11 (11), e0166969. Available at: https://doi.org/10.1371/journal.pone.0166969.

Shimizu, T., et al., 2017. Draft sequencing of the heterozygous diploid genome of Satsuma (*Citrus unshiu* Marc.) using a hybrid assembly approach. Front. Genet. 8, 180. Available at: https://doi.org/10.3389/fgene.2017.00180/full.

Simko, I., 2016. High-resolution DNA melting analysis in plant research. Trends Plant Sci. 21 (6), 528–537. Available at: https://doi.org/10.1016/j.tplants.2016.01.004.

Siviero, A., Cristofani, M., Machado, M.A., 2003. QTL mapping associated with rooting stem cuttings from *Citrus sunki* vs. *Poncirus trifoliata* hybrids. Crop Breed. Appl. Biotechnol. 3 (1), 83–88. Available at: http://cbab.sbmp.org.br/arquivos/1101.

Siviero, A., et al., 2006. Identification of QTLs associated with citrus resistance to *Phytophthora* gummosis. J. Appl. Genet. 47 (1), 23–28. Available at: https://doi.org/10.1007/BF03194595.

Slate, J., et al., 2009. Gene mapping in the wild with SNPs: guidelines and future directions. Genetica 136 (1), 97–107. Available at: https://doi.org/10.1007/s10709-008-9317-z.

Smith, S.M., Maughan, P.J., 2015. SNP genotyping using KASPar assays. In: Batley, J. (Ed.), Plant Genotyping. Humana Press, New York, NY, pp. 243–256. Available at: https://doi.org/10.1007/978-1-4939-1966-6_18.

Sorkheh, K., et al., 2008. Linkage disequilibrium, genetic association mapping and gene localization in crop plants. Genet. Mol. Biol. 31 (4), 805–814. Available at: http://www.scielo.br/scielo.php?pid=S1415-47572008000500001&script=sci_arttext&tlng=es.

Sugiyama, A., et al., 2010. Structure and expression levels of alleles of *Citrus* zeaxanthin epoxidase genes. J. Jpn. Soc. Hortic. Sci. 79 (3), 263–274. Available at: http://joi.jlc.jst.go.jp/JST.JSTAGE/jjshs1/79.263.

Sugiyama, A., et al., 2011. Quantitative trait loci (QTL) analysis of carotenoid content in *Citrus* fruit. J. Jpn. Soc. Hortic. Sci. 80 (2), 136–144. Available at: https://www.jstage.jst.go.jp/article/jjshs1/80/2/80_2_136/_article.

Sugiyama, A., et al., 2014. Expression quantitative trait loci analysis of carotenoid metabolism-related genes in citrus. J. Jpn. Soc. Hortic. Sci. 83 (1), 32–43. Available at: http://jlc.jst.go.jp/DN/JST.JSTAGE/jjshs1/CH-054.

Talon, M., Gmitter, F.G., 2008. *Citrus* genomics. Int. J. Plant Genomics 2008, 1–17. Available at: https://www.hindawi.com/journals/ijpg/2008/528361/.

Torres, A.M., et al., 1985. Segregation distortion and linkage of *Citrus* and *Poncirus* isozyme genes. J. Hered. 76 (4), 289–294. Available at: https://doi.org/10.1093/oxfordjournals.jhered.a110094.

Tozlu, I., Guy, C.L., Moore, G.A., 1999a. QTL analysis of morphological traits in an intergeneric BC1 progeny of *Citrus* and *Poncirus* under saline and non-saline environments. Genome 42 (4), 692–705. Available at: https://doi.org/10.1139/g99-035.

Tozlu, I., Guy, C.L., Moore, G.A., 1999b. QTL analysis of Na+ and Cl- accumulation related traits in an intergeneric BC1 progeny of *Citrus* and *Poncirus* under saline and nonsaline environments. Genome 42, 692–705. Available at: https://doi.org/10.1139/g99-003.

Ueno, I., Iwamasa, M., Nishiura, M., 1967. Embryo number of various varieties of *Citrus* and it's relatives. Bull. Hortic. Res. Sta. Jpn, Ser. B 7, 11–21. Available at: http://iss.ndl.go.jp/books/R000000004-I8118695-00.

van Nocker, S., Gardiner, S.E., 2014. Breeding better cultivars, faster: applications of new technologies for the rapid deployment of superior horticultural tree crops. Hortic. Res. 1 (March), 14022. Available at: http://www.nature.com/articles/hortres201422.

Veyrieras, J.-B., Goffinet, B., Charcosset, A., 2007. MetaQTL: A package of new computational methods for the meta-analysis of QTL mapping experiments. BMC Bioinf. 8, 49. Available at: https://doi.org/10.1186/1471-2105-8-49.

Wang, X., et al., 2017. Genomic analyses of primitive, wild and cultivated citrus provide insights into asexual reproduction. Nat. Genet. 49 (5), 765–772. Available at: http://www.nature.com/doifinder/10.1038/ng.3839.

Weber, C.A., et al., 2003. Mapping freeze tolerance quantitative trait loci in a *Citrus grandis* x *Poncirus trifoliata* F-1 pseudo-testcross using molecular markers. J. Am. Soc. Hortic Sci. 128 (4), 508–514. Available at: https://doi.org/10.21273/JASHS.128.4.0508.

Wu, B., et al., 2014a. Identification of pummelo cultivars by using a panel of 25 selected SNPs and 12 DNA segments. PLoS One 9 (4), e94506. Available at: https://doi.org/10.1371/journal.pone.0094506.

Wu, G.A., et al., 2014b. Sequencing of diverse mandarin, pummelo and orange genomes reveals complex history of admixture during citrus domestication. Nat. Biotechnol. 32, 656–662. Available at: https://doi.org/10.1038/nbt.2906.

Xu, X., et al., 2010. Developing specific markers and improving genetic mapping for a major locus *Tyr1* of citrus nematode resistance. Biosci. Methods 1 (1), 1–8. Available at: http://bio.sophiapublisher.com/detial_art.php?article_id=62&journal_id=11&type=bm.

Xu, Q., et al., 2013. The draft genome of sweet orange (*Citrus sinensis*). Nat. Genet. 45 (1), 59–66. Available at: https://doi.org/10.1038/ng.2472.

Yang, Z., et al., 2003. Sequence analysis of a 282-kilobase region surrounding the citrus Tristeza virus resistance gene (*Ctv*) locus in *Poncirus trifoliata* L. Raf. Plant Physiol. 131 (2), 482–492. Available at: https://doi.org/10.1104/pp.011262.

Yu, Y., Chen, C., Gmitter, F.G., 2016. QTL mapping of mandarin (*Citrus reticulata*) fruit characters using high-throughput SNP markers. Tree Genet. Genomes 12 (4), 77. Available at: https://doi.org/10.1007/s11295-016-1034-7.

Yu, Y., et al., 2017. Identification of QTLs controlling aroma volatiles using a "fortune" x "Murcott" (*Citrus reticulata*) population. BMC Genomics 18 (1), 646. Available at: http://bmcgenomics.biomedcentral.com/articles/10.1186/s12864-017-4043-5.

Zhang, L., et al., 2012. Regulation of carotenoid accumulation and the expression of carotenoid metabolic genes in citrus juice sacs *in vitro*. J. Exp. Bot. 63 (2), 871–886. Available at: https://doi.org/10.1093/jxb/err318.

Zhang, H., et al., 2017. Comprehensive comparative analysis of volatile compounds in citrus fruits of different species. Food Chem. 230, 316–326. Available at: https://doi.org/10.1016/j.foodchem.2017.03.040.

Zhu, C., et al., 2008. Status and prospects of association mapping in plants. Plant Genome 1 (1), 5–20. Available at: https://www.crops.org/publications/tpg/abstracts/1/1/5.

Chapter 9

Citrus biotechnology

Maria Antonietta Germanà[b], Pablo Aleza[c], Jude W. Grosser[a], Manjul Dutt[a], Nian Wang[d], Jose Cuenca[c], Prabhjot Kaur[a]

[a]Citrus Research and Education Center, Department of Horticultural Sciences, University of Florida, Lake Alfred, FL, United States, [b]Department of Agricultural, Food and Forest Sciences (SAAF), University of Palermo, Palermo, Italy, [c]Centro de Citricultura y Producción Vegetal, Instituto Valenciano de Investigaciones Agrarias (IVIA), Valencia, Spain, [d]Citrus Research and Education Center, Department of Microbiology, University of Florida, Lake Alfred, FL, United States

9.1 Introduction

Agricultural crops that can better withstand the changing climatic and pathogen landscape have been produced through natural selection throughout the millennia and, in recent years, through the process of human-assisted plant breeding and selection. However, a lack of genetic diversity in many commercially cultivated crops (due to monoculture) has made them more vulnerable to biotic and abiotic stresses (Esquinas-Alcazar, 2005; Keneni et al., 2012). Citrus, belonging to the Rutaceae family, is one of the most important commercial woody fruit crops in the tropical and subtropical areas of the world. The origin of citrus is traced back to parts of tropical and subtropical Southeast Asia (Wu et al., 2018). In addition to fresh fruit and juice, essential oils and pectin are also produced from citrus (Fisher and Phillips, 2008; Davies and Albrigo, 1994). Citrus products play an important role in human diets as citrus fruits are a rich source of vitamin C and also contain other important macro- and micronutrients (Ting, 1980). When consumed as a fresh fruit, citrus is a good source of dietary fiber (Marin et al., 2007). Moreover, citrus fruits are rich in antioxidants (Liu et al., 2012) and citrus flavonoid compounds have anticancer and antiinflammatory properties (Benavente-Garcia and Castillo, 2008; see Chapter 24 for a detailed description on this subject). In citrus, despite its wide genetic diversity, large-scale production is restricted to a few cultivars. In the Mediterranean region, citrus production is restricted to Clementine, blood and other orange, and lemon cultivars (Demirkeser et al., 2009). Similarly in Japan and Korea, Satsuma mandarins account for the majority in acreage (Choi, 2004; Hong et al., 2007). In Florida, the cultivars Hamlin and Valencia predominate (Omar et al., 2007; Alva et al., 2003). Citrus has a high cell-to-plant regeneration capacity via both somatic embryogenesis and organogenesis, making it amenable to numerous biotechnology applications. Several elite genetically improved citrus rootstock and scion cultivars have been developed in recent years. All these cultivars have been derived from natural or induced mutations, somaclonal variation, or traditional breeding (Talon and Gmitter, 2008; Peña and Navarro, 2000; Deng, 2005). The worldwide citrus industry requires new varieties to help overcome barriers to production and to help create new marketing opportunities. The complex reproductive biology of citrus, including production of nucellar seedlings, long juvenile periods, and the complex taxonomic relationships among cultivar groups, is one of the reasons for the low-level impact of conventional breeding in citrus genetic improvement (Talon and Gmitter, 2008). Biotechnological interventions have been necessary to aid in the rapid germplasm enhancement of existing cultivars (Peña and Navarro, 2000) and the development of new ones. This chapter will detail the most relevant and important biotechnology applications and resulting progress toward citrus production and improved cultivar development. Topics covered in the chapter include: micropropagation (shoot multiplication and rooting); haploid production (gametic embryogenesis); somaclonal variation and somatic cybridization; transformation; and CRISPR gene editing. Applications of these biotechnologies are already making significant contributions to improving fruit quality for the fresh market and processing as well as solving industry threatening disease problems, especially HLB and citrus canker. Emerging biotechnologies continue to expand the toolbox available for citrus improvement, making it a model system for woody fruit tree crops.

9.2 Micropropagation

Citrus trees are usually asexually propagated by grafting scions onto nucellar seedlings used as rootstocks or by cuttings. Most species of this genus are characterized by a high degree of polyembryony, consisting in the production, in a single

The Genus Citrus. https://doi.org/10.1016/B978-0-12-812163-4.00009-7

seed, of 1–40 adventive embryos by the nucellus (Furusato et al., 1957) so that two or more embryos develop. Nucellar embryos are similar to the mother plant, while zygotic embryos tend to be weak and variable, and very often do not survive because they have to compete for nutrients and space with nucellar embryos (Frost et al., 1968). Due to polyembryony, open-pollinated seed germination is a common procedure used in citrus for large-scale nursery propagation of rootstocks (Wutscher, 1979).

Tissue culture, one of the most interesting areas of plant biotechnology, consists of the ability to establish and maintain plant tissues (callus, cells, protoplasts, etc.) and plant organs (embryos, shoots, roots, flower parts) in a nutritive growth medium, in a culture vessel, in aseptic conditions, and under controlled environmental conditions. Micropropagation is a valuable alternative to traditional propagation, especially when shoot regeneration is achieved by in vitro direct organogenesis. With the aim of increasing the number and the quality of cultivars and rootstocks available to growers, it provides to the nursery activity many practical advantages, such as speeding up and improving the vegetative propagation and the production of true-to-type and virus-free plants; the mass propagation of clones; and plant production in a short time in a limited space and throughout the year (Barlass and Skene, 1982). Micropropagation could appear to be of no commercial importance for *Citrus*, just because most species are characterized by a high degree of polyembryony. However, there are many other advantages for which this technique is very important, such as when cultivars are monoembryonic, seedless, or producing few seeds, or when newly introduced citrus cultivars require many years to produce sufficient seeds. Moreover, other applications of in vitro propagation are the possibility to perform genetic transformation (Cervera et al., 2008) and mutagenesis, and to provide a rapid system to multiply the progeny obtained by breeding programs, providing a valuable system for germplasm conservation.

In citrus, higher multiplication rates are obtained using juvenile explants (Pérez-Molphe-Balch and Ochoa-Alejo, 1997), but plants regenerated from these tissues usually exhibit strong undesirable juvenile characteristics and are slow to come into bearing. Above all, to introduce genetic variation by genetic transformation or mutagenesis, it is fundamental to have available efficient protocols starting from adult tissues. Only a few protocols for the micropropagation of mature citrus plants have been developed because of the high level of contamination (Drew, 1988), the low morphogenetic potential (Bonga, 1982), and the poor rooting ability of the regenerated shoots (George and Sherrington, 1984). For instance, shoots were regenerated by Altman and Goren (1974) from lateral buds collected from a mature, sweet orange plant and plantlets were regenerated by Barlass and Skene (1982) from nodal segments of mature plants of Cleopatra mandarin, Carrizo citrange, Rangpur lime, and sweet orange. Shoot tips from moderately mature Assam emon were used as explants by Singh et al. (1994) to produce shoots while Kotsias and Roussos (2001) employed nodal segments of Interdonato lemon seedlings to investigate the effect of different plant growth regulators. Moreover, the influence of the basal medium and different plant growth regulators on the micropropagation of nodal explants from mature trees of different lemon cultivars was investigated by Pérez-Tornero et al. (2010).

In vitro regeneration has been obtained from several types of citrus through different in vitro techniques and routes of regeneration, such as somatic embryogenesis and organogenesis. Direct (without the callus production) or indirect morphogenesis has been observed in different citrus tissues. Particularly, in vitro regeneration has been described in several citrus species using as explants different sources such as internodal segments, epicotyl segments (Fig. 9.1), hypocotyl segments with attached or not cotyledon fragments (Pérez-Molphe-Balch and Ochoa-Alejo, 1997; Edriss and Burger, 1984; Maggon and Deo, 1995; Ghorbel et al., 1998; Bordón et al., 2000; Moreira-Dias et al., 2000; Germanà et al., 2008), roots (Burger and Hackett, 1986; Sim et al., 1989; Sauton et al., 1982), leaf sections (Sim et al., 1989; Chaturvedi and Mitra, 1974; Gill et al., 1994), and reproductive organs (Germanà et al., 1994; Perez et al., 1998; Germanà, 2003, 2005; Cardoso et al., 2016). Also a transverse thin cell layer (tTCL) excised from stem internodes was used as an explant (Van Le et al., 1999). Several studies have shown that many factors affect the response to in vitro culture: the genotype (Ghorbel et al., 1998; Bordón et al., 2000; Perez et al., 1998; Vardi et al., 1982); the explant type, age, cut, and orientation of explants; the age and the physiological conditions of explant donors (Burger and Hackett, 1986; Sim et al., 1989; Gill et al., 1994; Perez et al., 1998; Garcia-Luis et al., 1999; Moreira-Dias et al., 2001); the culture medium composition (Maggon and Deo, 1995; Burger and Hackett, 1986; Sim et al., 1989; Sauton et al., 1982; Goh et al., 1995; Carimi et al., 1998); and the incubation conditions (Pérez-Molphe-Balch and Ochoa-Alejo, 1997; Bordón et al., 2000; Perez et al., 1998; Garcia-Luis et al., 1999; Duran-Vila et al., 1992; Cabasson et al., 1997).

9.3 Organogenesis and rooting

Several internal or external factors and their interactions affect the in vitro organogenic response of citrus rootstocks and scions to different culture media and, unfortunately, protocols are genotype- and season-dependent. Particularly because the regeneration potential of the explant results are correlated with the physiological conditions of the donor plants, dependent on different levels of endogenous metabolites and plant growth regulators (also involved in the regeneration process), it is very difficult to standardize a protocol when the starting explant is different from year to year. Similarly, cultural conditions and, particularly, the exposure to a dark period seem to be beneficial. This is possibly because they modify the proportion of endogenous hormones, such as cytokinins and auxins, interacting with growth regulators added to the culture medium (Almeida et al., 2003; Gu and Zhang, 2005). The regeneration response of the explants has to be evaluated considering the percentage of explants producing shoots, the average number of shoots per explant, and the shoot-forming capacity. When the starting explant is a seedling epicotyl segment, there are conflicting results in the literature concerning the influence of its distance from the cotyledonary node and the influence of its different position on the medium (horizontal versus vertical)

FIG. 9.1 Adventitious shoots produced by epicotyl segments of Carrizo citrange (*Citrus sinensis* × *Poncirius trifoliata*) cultured in the horizontal position *(up)* and regrowth of encapsulated microcutting of Carrizo citrange *(down)*.

and of the cut in different ways (longitudinally, transversally, and obliquely) on the morphogenic response (Yu et al., 2002; Costa et al., 2004; Peña et al., 2004a; Duan et al., 2007; Germanà et al., 2011a).

It is very important that organogenesis is completed directly, without callus formation, because plants produced by direct organogenesis show greater genetic stability than those produced from callus (Lee and Phillips, 1988; Karam and Al-Majathoub, 2000). Actually, somaclonal variation can avoid limiting the adventitious instead axillary shoot multiplication, reducing the number of subcultures (even if there is the economic interest to increase them), and monitoring the clonal fidelity by molecular markers. RAPD analysis carried out after in vitro multiplication of a seedless lemon cv. Kaghzi Kalan using nodal explants showed that the regeneration system maintained high genetic fidelity and for micropropagation at the commercial level, at least, two cycles of multiple shooting can be performed (Goswami et al., 2013), as previously reported in rough lemon by Savita et al. (2012) and Khawale and Singh (2005).

Evaluating different aspects of the morphogenic process in Carrizo citrange, Germanà et al. (2011a) observed that the best combination to achieve a good response of direct organogenesiswas that of 1 cm long epicotyl segments cultured in the horizontal position (Fig. 9.1). They also confirmed that benzyladenine (BA) is a very effective cytokinin for citrus organogenesis, as previously reported by several authors (Moreira-Dias et al., 2000; Germanà et al., 2008; Garcia-Luis et al., 1999; Hassanein and Azooz, 2003). Meta-Topolin (mT) turned out to be a good alternative to BA for in vitro sprouting of C35 and Troyer citrange rootstocks, providing a comparable multiplication rate at a lower concentration (Chiancone et al., 2016, 2017), as already reported by Niedz and Evens (Niedz and Evens, 2011). In epicotyls of orange, mT exhibited more positive effects over a wider concentration range than BA, that produced an increasingly negative effect as its concentration increased. Direct and indirect in vitro regeneration was reported in the rough lemon by Ali and Mirza (Ali and Mirza, 2006). Maximum shoot regeneration responses of 70% in callus and of 83% during direct shoot regeneration were observed with the same medium, while rooting was induced in up to 70% of these shoots. Low rooting efficiency has been reported as a major problem for in vitro production of citrus plants (Ali and Mirza, 2006; Duran-Vila et al., 1989). To develop a protocol for callus induction and regeneration in citrus rootstocks, shoot tips and nodal segments were used as explants from *Citrus depressa*, Troyer citrange, rough lemon, and Cleopatra mandarin (Sharma et al., 2009), but the results were not good enough.

A study on ex vitro mycorrhization of vitro-derived plantlets of Carrizo citrange was reported by Germanà et al. (2011b). The symbiosis induced a better growth of the rooting apparatus, although results confirmed the low mycorrhizal-dependence of citrange, also in in vitro derived plantlets (Casales et al., 2015).

9.4 Gametic embryogenesis

Gametic embryogenesis is one of the biotechnological tools that can enhance the efficacy of traditional breeding programs. This process, allowing production in a single step of haploids and double haploids and, therefore, of homozygous lines at all loci, can increase the efficiency of perennial crop breeding programs. Haploids (Hs) are sporophytes with the gametophytic chromosome number (n instead of $2n$), and doubled haploids (DHs) are haploids that underwent, in a spontaneous or stimulated way, chromosome doubling (Yahata et al., 2015). Conventional methods, involving several generations of selfing, are not utilizable in *Citrus* to generate homozygous lines because of the long period of the generation cycle, the high heterozygosity of the genomes, and often, the self-incompatibility (Germanà, 2006, 2007, 2009, 2011a, b; Seguí-Simarro, 2010). Gametic embryogenesis permits the induced immature gametes to deviate from the normal gametophytic developmental pathway toward the sporophytic one. Haploids are advantageous for breeding purposes (Germanà et al., 2013).

Most cultivated forms of *Citrus* and related genera (*Poncirus*, *Fortunella*, etc.) are diploid with a monoploid number of chromosomes ($x=9$) (Frost, 1925), but also triploid and tetraploid forms of *Citrus* exist. Several methods are available to obtain haploidization in *Citrus*. Gametic embryogenesis is termed gynogenesis when initiated from the female gamete and microspore embryogenesis (sometimes referred to as androgenesis) when it originates from the microspore, that is, the immature male gamete. Currently, microspore embryogenesis is the favorite technique because of its better productivity; however, gynogenesis has been employed with species that do not respond to other approaches (Forster et al., 2007).

Gynogenesis can be obtained via in situ parthenogenesis (i.e., the production of an embryo from an egg cell without the participation of the male gamete), induced by pollination with irradiated pollen (followed by in vitro embryo culture), or by in situ or in vitro parthenogenesis induced by pollen from a triploid plant (followed by in vitro embryo culture). Irradiated pollen or pollen grains from a triploid plant are able to germinate but do not result in fertilization, while stimulating the development of haploid embryoids from ovules (Germanà, 2012). For instance, three haploid plants were obtained from in vivo crosses of two monoembryonic diploids (Clementine and Lee) × a triploid hybrid of Kawano natsudaidai (Oiyama and Kobayashi, 1993). In situ parthenogenesis induced by irradiated pollen and in vitro embryo rescue culture allowed the production of nine haploid plantlets, which did not survive, and two embryogenic callus lines in Clementine SRA 63 after in situ pollination with irradiated pollen of Meyer lemon (Ollitrault et al., 1996). However, Froelicher et al. (2007) obtained haploid plants of Fortune mandarin and Ellendale tangor through pollination with gamma ray-irradiated pollen also of Meyer lemon, followed by embryo rescue. After pollination of Clemenules Clementine with Fortune mandarin irradiated pollen and embryo rescue, haploid plants were also obtained by Aleza et al. (2009). It has been also reported that the in vitro stigmatic pollination technique, that is, the application of pollen of Oroblanco, a triploid grapefruit, to the apical part of the stigma of an excised gynoecium implanted in solid culture medium, allowed haploid plantlet regeneration in Clemenules Clementine (Germanà and Chiancone, 2001).

Microspore embryogenesis can, instead, be achieved through anther and isolated microspore culture. In many crops, anther culture is usually the technique chosen for haploid and doubled haploid production because of its ease. Since the experiments of Guha and Maheshwari (Guha and Maheshwari, 1964), showing that through the in vitro culture of immature anthers of *Datura innoxia*, the gametophytic development could be reprogrammed to produce sporophytes carrying a haploid chromosome number, anther culture (Fig. 9.2) become the method of choice for microspore embryogenesis in most crops. In citrus, Haploid plantlets obtained through microspore embryogenesis were recovered from *Poncirus trifoliata* (Hidaka et al., 1979) and *C. madurensis* (Chen, 1980); one doubled haploid plantlet was also generated from the hybrid No. 14 of *C. ichangensis* × *C. reticulata* (Deng et al., 1992); haploid plantlets and highly embryogenic haploid calli were reported also from Clementine (Germanà et al., 1994, 2000; Germanà and Chiancone, 2003); haploid, but albino embryos from Mapo tangelo (Germanà and Reforgiato Recupero, 1997); haploid and diploid calli, embryos, and leafy structures but no green plants from lemon (Germanà et al., 1992); and haploid embryos of *Clausena excavata* (Froelicher and Ollitrault, 2000) were also obtained.

Regarding sweet orange anther culture, short-lived homozygous plantlets have been obtained from the anthers of Rhode Red Valencia (Cao et al., 2011) while Wang et al. (2015) described homozygous lines derived from Early Gold and Rohde Red Valencia. Cardoso et al. (2014) obtained homozygous callus from a hybrid between *C. clementina* × *C. sinensis* Hamlin. Cardoso et al. (2016) also reported homozygous and triploid regenerant production (microspore-derived embryos) from Hernandina and Corsica Clementines, whereas all regenerants from sweet oranges, three cultivars of Tarocco and Moro, produced heterozygous and diploid plants similar to the parental genotypes (somatic embryos). Anther-derived calli can be nonmorphogenic or highly embryogenic, and they can conserve their embryogenic potential for a long time. The morphogenic calli in citrus appear friable and white. Occasionally, two diverse lobes of an anther can generate calli. The embryogenic calli can yield clumps of embryos. Isozyme analyses (Germanà et al., 1994, 2000; Ollitrault et al., 1996; Germanà and Reforgiato Recupero, 1997), RAPD markers, and microsatellites (Cardoso et al., 2016; Germanà and Chiancone, 2003; Germanà et al., 2005; Cao et al., 2011; Wang et al., 2015) have been employed to characterize regenerants assessing the gametic origin of calluses and plantlets. Flow cytometry (Bohanec, 2003) is used to determine the ploidy of the regenerants. It has been also reported that Clementine anther culture produces tri-haploids indicating that this technique could be a reliable method of obtaining new triploid genotypes (Germanà et al., 2005).

The studies on isolated microspore culture in citrus are not abundant and almost all are recent. This technique is increasingly the favorite method to produce DHs (double haploids) in many crops, although it requires better equipment and

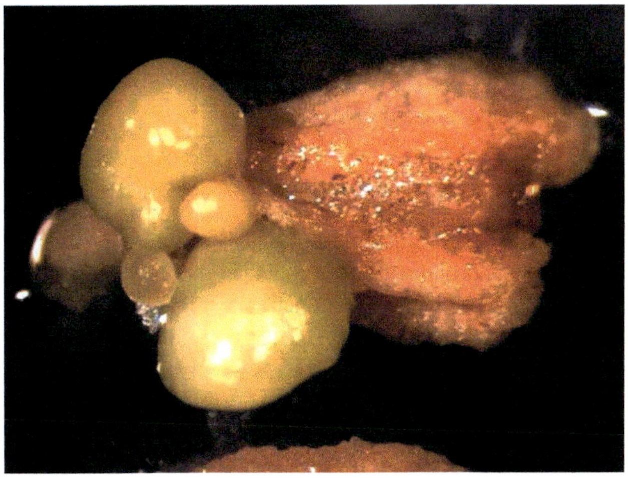

FIG. 9.2 In vitro anther culture of *Citrus clementina* cv. Monreal, with microspore-derived embryos.

more skill (Germanà, 2011a, b; Ferrie and Caswell, 2011). In initial investigations on isolated microspore culture in several citrus species it was reported that isolated citrus microspores produced multinucleated structures, pseudobulbils, and small proembryos, which failed to develop any further (Germanà et al., 1996). The first successful report in citrus microspore embryogenesis via isolated microspore culture was observed in Monreal Rosso and Clemenules Clementines (Chiancone et al., 2015b). Microspore embryogenesis through isolated microspore culture was also induced, using meta-Topolin instead of benzyladenine or zeatin, in the culture medium (Chiancone et al., 2015b). After 11 months of isolated microspore culture, microspore reprogramming and sporophytic development were observed and the presence of multinucleated calli and microspore-derived embryos was evident. In addition, homozygous early embryo regeneration was also obtained in the Tardivo di Ciaculli mandarin, a cultivar very recalcitrant to microspore embryogenesis (Cimò et al., 2016). While the application of mT (replacing BA) in the culture media was not significant, a thermal shock treatment at 35°C for 30 days had a critical role in the induction of microspore embryogenesis and for triggering pollen totipotency, proving to be essential for embryo production (Cimò et al., 2016). Other successful studies on isolated microspore culture were carried out in Moro sweet orange (Germanà et al., unpublished).

A number of endogenous and exogenous factors and their interactions affect the embryogenic response of immature gametes in culture (Smýkal, 2000; Wang et al., 2000). Among them, the genotype, physiological status, and growth conditions of donor plants; the pollen developmental stage; the pretreatment applied to anthers or to floral buds; the culture media composition (macro- and microelements, carbon source, plant growth regulators, etc.); and the conditions of incubation are all factors that greatly influence the explant response to the in vitro culture (Fig. 9.2) (Germanà, 2011a, b). Unfortunately, different species and even cultivars within a species, frequently require different conditions and supplies for embryogenic development and regeneration via microspore embryogenesis (Germanà, 2003, 2011b, 1997; Cardoso et al., 2016, 2014; Wang et al., 2015; Germanà et al., 1996; Chiancone et al., 2015b, 2006).

9.5 Somaclonal variation

The term somaclonal variation is defined as genetic variation that is present in plants regenerated from tissue cultures, either uncovered or induced by a tissue culture process (Larkin and Scowcroft, 1981). The first reports of somaclonal variation were primarily from solanaceous or cereal crops, affecting a wide range of traits including plant height, growth habit, flower, fruit and leaf morphology, juvenility, maturity date, disease resistance, yield, and various biochemical characteristics. Sweet orange and other polyembryonic citrus cultivars are well suited for studies of somaclonal variation due to their biology and good performance in tissue culture. The sweet orange is polyembryonic and forms adventive somatic embryos from nucellar tissue, including embryogenic callus and suspension cultures (Kochba et al., 1972). The sweet orange was also shown early on to be amenable to protoplast regeneration via somatic embryogenesis (Vardi et al., 1982) and to plant regeneration via organogenesis through adventitious shoot induction (Grinblat, 1972), which is a key to its efficient *Agrobacterium*-mediated transformation (Moore et al., 1992).

The UF/CREC citrus improvement program has produced and evaluated > 2000 sweet orange somaclones regenerated from Hamlin, Valencia, Vernia, and a noncommercial selection designated as OLL (Orie Lee Late) for fresh market and processing potential (Grosser et al., 1997). Cultivars released for commercial production from this work are provided in Table 9.1. Improved attributes include earlier maturity, seedlessness, and improved fruit/juice quality, all achieved while maintaining cultivar integrity. Somaclonal variation has proven to be a valuable yet underexploited source of genetic variation for citrus improvement, especially in cultivars that are difficult to improve by conventional breeding. Somaclonal variation could also be exploited in monoembryonic cultivars such as Clementines, depending on the ability of each selection to respond efficiently on shoot induction (organogenesis) media.

9.6 Allotetraploids via somatic hybridization

Production of allotetraploid somatic hybrids via protoplast fusion is now a routine technique in citrus tissue culture labs, as hundreds of allotetraploid hybrids have been produced that combine complementary diploid parents (Grosser and Gmitter, 1990; Grosser et al., 2000, 2010; Omar et al., 2016). The general model for producing allotetraploid somatic hybrids is the chemically induced (PEG) (Grosser and Gmitter, 1990) or electrofusion of protoplasts isolated from an embryogenic callus or suspension culture of one parent (thus far restricted to polyembryonic genotypes), with leaf-derived protoplasts of the second complementary parent (Hidaka and Omura, 1992). Several such somatic hybrids have been flowering and fruiting for years, and their use in interploid crosses has resulted in the production of thousands of new triploid hybrids. As more triploid hybrid fruits from interploid hybridizations involving allotetraploid somatic hybrids are being evaluated, it is now possible to generate general performance characteristics of individual somatic hybrids as breeding parents.

TABLE 9.1 University of Florida commercial sweet orange cultivars developed via somaclonal variation.

Cultivar	Parental variety	Tissue culture source	Primary advantage
Hamlin N13-32	Hamlin	Protoplasts/SE[a]	Enhanced juice color
EV-1 (B7-70)	Valencia	Nucellar seedling/ORG[a]	Very early maturity
EV-2 (SF14W-65)	Valencia	Protoplasts/SE	Very early maturity
SF14W-62 (Valquarius)	Valencia	Protoplasts/SE	Early maturity
B9-65	Valencia	Nucellar seedling/ORG	Yield/fruit quality
N7-3	Valencia	Protoplasts/SE	Seedlessness
OLL-4	Orie Lee Late	Callus/SE	Yield/fruit quality
OLL-8	Orie Lee Late	Callus/SE	Yield/fruit quality

See FFSP–Florida Foundation Seed Producers website http://www.ffsp.net/varieties/citrus/ for more details
[a] *SE*, somatic embryogenesis; *ORG*, organogenesis.

The initial assessments of somatic hybrid performance as breeding parents for mandarin improvement in the UF/CREC citrus improvement program are as follows:

Nova + Succari: Extremely fertile, progeny highly variable, difficult to recover zipperskinned (easy peeling) hybrids, but does transmit good flavor, high Brix, and high ratios at good frequency. Some hybrids exhibit granulated fruit in the first years of fruiting.

Nova + Osceola: Good fertility, transmits small fruit size, low percentage of zipperskinned hybrids, transmits excellent rich flavor and high Brix, excellent transmission of deep external/internal fruit color not requiring cold induction. Many hybrids exhibit granulation in the first few years of fruiting. Most hybrids are not thorny.

Succari + Murcott: Good fertility, transmits large fruit size, rich flavor, high Brix, high ratios, excellent internal color, low percentage of zipperskin progeny.

Page + [Clementine × Satsuma]: Good fertility, transmits good fruit size, and best parent so far for transmission of zipperskin trait. Transmits good flavor and color. Hybrids can be thorny.

Succari + Ponkan: (highly HLB susceptible), Transmits good Ponkan flavor and good fruit size. Some hybrids easy to peel (Note that a Hamlin + Ponkan somatic hybrid was lost to HLB before any crosses could be made). Some granulation observed in early flowering of some hybrids.

Murcott + LB8-9 (Clementine × Minneola): Good fertility, transmits good fruit size, excellent internal color, high Brix, and rich flavor. Difficult to recover zipperskinned hybrids.

Additional allotetraploid parents used to produce triploid mandarin hybrids now in the field, many beginning to flower and fruit, are: Murcott + LB8-8; Hamlin + Dancy; Page + Ortanique; Rhode Red Valencia + Dancy; Succari + Dancy; Valencia + Page; Valencia + Minneola; Valencia + [Robinson × Temple]; Valencia + Murcott; Murcott + Dancy; W. Murcott + UF-03B; Page + Furr; Murcott + Osceola; and Murcott + Ortanique. Monoembryonic females used in interploid crosses include Monreal Clementine, Nules Clementine, FallGlo, Lee, Temple, LB8-9 and LB8-8 (Clementine × Minneola), Fortune, Ambersweet, G-96 (edible trifoliate hybrid); and many other noncommercial selections made by Dr. Fred Gmitter. Allotetraploid somatic hybrids are also being used in interploid crosses for grapefruit/pummelo and acid-fruit (lemons/limes) improvement. Two allotetraploid somatic hybrids produced from the fusion of Murcott with seedlings of Chandler pummelo are showing excellent tolerance of both HLB and canker, and are being used extensively in UF/CREC interploid crosses. Allotetraploid somatic hybrids combining lemon with the Lakeland limequat are exhibiting exceptional fragrance and bright yellow fruit. These hybrids are monoembryonic and can be used successfully as females in interploid crosses, eliminating the need for embryo rescue. Triploid hybrids recovered from lemon + limquat × lemon are generally precocious, and one exceptional lime-like hybrid has already been recovered and is being considered for commercial release. As more information is generated on the performance of various allotetraploid somatic hybrids as breeding parents, the efficiency of their creation and utilization will continue to improve, thereby enhancing citrus improvement programs around the world.

The parental heterozygosity restitution (PHR) of tetraploid parents (either allotetraploid somatic hybrids or Doubled Diploids, DD) depends on the preferential chromosome pairing and consequently the ratio between disomic and tetrasomic segregation. In extreme allotetraploids, there are two sets of homeologous chromosomes, each consisting of two homologous chromosomes. Disomic inheritance occurs if a chromosome pairs exclusively with its homolog; this often emerges as fixed heterozygosity in genetic analyses, with 100% of the interspecific heterozygosity transmitted by each gamete. In extreme autotetraploids, the presence of four homologous chromosomes results in equal opportunities to pair at meiosis, leading to multivalent formation and tetrasomic inheritance with PHR rates that vary between 55% and 66%, depending on the double reduction rate (DR). Aleza et al. (2015) demonstrated that the diploid gamete population produced by the DD Clementine displayed predominant tetrasomic segregation (PHR average around 65%) with PHR values relatively constant across a chromosome with genotypic combinations that are closer to the Clementine. Recently, Rouiss et al. (2017a) demonstrated that disomic segregation predominated in the DD Mexican lime, limiting effective interspecific recombination and the diversity of the diploid gamete population. Nevertheless, Kamiri et al. (2011) observed in allotetraploid somatic hybrids between the mandarin and lemon a tetrasomic and inheritance intermediate between disomic and tetrasomic segregation models. For these reasons, it is very important to increase the knowledge of the inheritance of tetraploid parents to predict the progenies in breeding programs based on interploid hybridizations; this also has important implications for linkage mapping and population genetics.

9.7 Somatic cybridization

A second outcome or byproduct from the widely used citrus somatic hybridization model, where protoplasts isolated from embryogenic callus/cell suspension cultures are fused with the protoplasts derived from nonembryogenic leaf mesophyll cells of the second parent, is the production of cybrid plants (reviewed by Guo et al., 2004a). Somatic hybrids contain a hybrid tetraploid nucleus, whereas cybrids retain the diploid nucleus from the nonembryogenic parent (Grosser et al., 2000; Guo et al., 2013). Citrus cybrid plants regenerated from protoplast fusion events almost always possess the mitochondrial DNA (mtDNA) of the embryogenic parent, indicating a critical role of the mitochondrial genome in the regeneration of cybrids (Moreira-Dias et al., 2000; Yamamoto et al., 1995; Moriguchi et al., 1996; Saito et al., 1993; Cabasson et al., 2001). Chloroplast regeneration in citrus cybrid plants is generally random from either parent, with mixtures not observed. Cybridization in citrus provides another unique source of genetic variation that can be exploited for cultivar development by the creation of novel alloplasmic combinations of the nuclear and organelle genomes that are present in fusion parents.

The first citrus fruit trait targeted for improvement via cybridization was the reduction of seed content in otherwise commercial fresh fruit candidates. This has been partially successful as seed count averages have been significantly reduced, but no completely seedless selections have been released for commercialization so far (Guo et al., 2013, 2004b). Dancy tangerine cybrids putatively containing the Satsuma cytoplasm are showing improved tolerance to *Alternaria*, a significant problem with traditional Dancy cultivation (JW Grosser, unpublished data). Cybrids of the Ruby Red and Duncan grapefruits containing the Dancy cytoplasm showed a significantly extended harvest period with minimal granulation and suppressed seed germination (Satpute et al., 2015). The mechanism for this improved phenotype has not yet been determined. Cybrids of Flame, White Marsh, and Ruby Red somaclone N11-11 containing Meiwa kumquat cytoplasm (Omar et al., 2017) were shown to have significantly increased tolerance to citrus canker in greenhouse assays (Murata, 2018). Data showed that the elevated canker tolerance observed in these grapefruit cybrids was due to the kumquat chloroplasts, rather than the kumquat mitochondria as originally hypothesized. Overall, there is an increasing body of information showing that cybridization in citrus has the potential to improve fruit characteristics and disease resistance, and possibly other important traits. Evolving molecular genetics techniques should improve our limited understanding of cytoplasmic/nuclear genetic interactions in citrus.

9.8 Molecular marker development for Alternaria brown spot disease

Alternaria brown spot (ABS) is an important disease caused by the tangerine pathotype of the fungus *Alternaria alternata* (Fr.) Keissl. It affects many citrus cultivars that show necrotic lesions on fruits and young leaves, ultimately resulting in defoliation and fruit drop (Akimitsu et al., 2003). Fruits affected by ABS are unacceptable for fresh-fruit marketing. The tangerine pathotype of *A. alternata* is a necrotrophic pathogen, and carries a gene cluster (*ACTT*) located in a small chromosome responsible for ACT-toxin biosynthesis (Ajiro et al., 2010). This host-specific toxin is released during the germination of conidia, rapidly affecting the plasma membrane integrity of susceptible host cells (Kohmoto et al., 1993). Currently, ABS control is primarily based on the programmed application of 4-10 fungicide sprays per year, depending on the climate of the region and the susceptibility of the cultivar, although damage reduction is not always satisfactory. Moreover, the systematic application of fungicides for ABS control over years may create environmental problems and public health concerns (Vicent et al., 2009). These constraints are forcing growers to replace susceptible cultivars (Cuenca et al., 2013) with resistant ones. Thus, genetic resistance remains as the best option for disease control (Peres and Timmer, 2006; Vicent et al., 2007).

Despite the susceptibility of many cultivars, some of them are currently used as parents in citrus breeding programs, both at the diploid and triploid levels (Grosser et al., 2010; McCollum, 2007; Aleza et al., 2010). This is the case of the nonapomictic cultivar Fortune, widely used as the female parent in triploid breeding programs in Spain (Navarro et al., 2015), France, and Italy (Reforgiato Recupero et al., 2005; Froelicher et al., 2012). Moreover, susceptible cultivars such as Murcott and Ponkan are widely used as male parents in many diploid breeding programs carried out in Japan and Brazil (Xiao et al., 2009; Pacheco et al., 2012). Also, this is the case for Dancy, Minneola, Nova, Fairchild, Fremont, Page, Orlando, Pixie, and Daisy cultivars that are also used as male parents in diploid and triploid breeding programs or for inducing mutations (Navarro et al., 2015).

Genetic mapping of agronomical traits is a key tool in breeding programs, especially in tree crops with long generation times, as in citrus. The usefulness of genetic mapping has been widely demonstrated for marker-assisted selection (MAS), candidate gene identification, and map-based cloning (Dirlewanger et al., 2004). Bulked segregant analysis (BSA) has been used to map many agronomical traits in several crops, including resistance genes to fungal pathogens. Using BSA in diploid progenies, it was proposed that the inheritance of ABS resistance in citrus is controlled by a single recessive allele (Dalkilic et al., 2005).

As stated, diploid progenies are predominantly used for genetic mapping of traits. Genetic mapping and marker-trait association in polyploids are complicated by the diversity of the meiotic process (and therefore, recombination mechanisms) involved as well as the distribution of markers into multiple dosage classes. Despite these limitations, genetic maps based on segregating molecular markers have been generated for a wide range of polyploids (Xu et al., 1995; Reinisch et al., 1994). Similarly, triploid progenies arising from unreduced gametes can be used for genetic mapping. In sexual polyploidization, two factors affect the transmission of parental heterozygosity to the offspring: the mechanism of $2n$ gamete formation (i.e., FDR or SDR) and the genetic distance from the locus of interest to the centromere. Therefore, Half-Tetrad Analysis (HTA) based on $2n$ gametes is an efficient means of genetic mapping (Mendiburu and Peloquin, 1979; Tavoletti et al., 1996). In 2x × 2x citrus crosses, the $2n$ gamete is mainly transmitted by the female parent and SDR has been reported for $2n$ female gametes in all analyzed mandarins and lemons (Luro et al., 2004; Cuenca et al., 2011, 2015; Rouiss et al., 2017b).

Combining HTA based on $2n$ gametes with the BSA-genome scan approach, Cuenca et al. (2013) identified a genomic region containing the ABS resistance locus. The recessive inheritance of the ABS resistance was confirmed in triploid populations through trait segregation in crosses between resistant and heterozygous ABS-susceptible parents or between two heterozygous ABS-susceptible ones. Based on this segregation of 368 SDR $2n$ gametes, they first located the ABSr locus at 10 cM from a centromere. Then, a BSA over 93 triploid hybrids from a Fortune × Willowleaf population was performed. They identified a genomic region containing numerous markers with a high probability (> 99%) of association with phenotype variation, located on chromosome III. This region includes 25 significant SNP markers within an interval of 13.1 cM. The genomic region between these two markers contains around 15 Mb. Next, new SSR and SNP markers were developed to perform a linkage genetic mapping on the identified genomic region delimiting a 3.3-Mb genome region in the physical map. More recently, Cuenca et al. (2016) performed fine mapping on this region using a 268-diploid mapping population arising from a heterozygous-susceptible × resistant hybridization. Fine mapping showed that the *ABSr* locus is located in a region of 1.1 cM between the markers SNP05/SNP06/SNP07/AT21 (at 0.7 cM) and SNP08 (at 0.4 cM), limiting the chromosome region to the positions 25.496.094 and 25.862.085 within chromosome 3. The closest marker, SNP08, was then selected and the identified region contained eight genes harboring NBS-LRR repeats, which should be considered candidate genes for ABS resistance (Fig. 9.3).

In citrus, a few characteristics of agronomic interest have been linked to molecular markers, such as RAPD markers linked to dwarfing (Cheng and Roose, 1995) or fruit acidity (Fang et al., 1997), SSR markers linked to CTV resistance from *P. trifoliata* (Gmitter et al., 1996; Bernet et al., 2004), AFLP markers linked to polyembryony (Wang et al., 2017), and the dominant PCR assay for the anthocyanin content of pulp from blood orange due to a transposable element insertion in the 5′ end of the *Ruby* gene (Butelli et al., 2012). Other characteristics of interest have been tagged to QTLs, such as salinity tolerance (Ben-Hayyim and Moore, 2007) and nematode resistance (Ling et al., 2000). The new developed SNP08 marker, mapped at 0.4 cM from the ABS resistance gene, greatly improves the selection of resistant genotypes in early development stages and avoids growing and evaluating susceptible genotypes. Flanking markers at 0.7 cM from the *ABSr* locus were also identified on the other side of the gene, and coupling the SNP08 with one of these markers, the probability to select erroneously resistant genotypes would be 0.0028%, that is, less than three false resistant hybrids per 100,000 genotypes analyzed. Moreover, the transferability of the SNP08 between different citrus species or interspecific hybrids was confirmed by comparing the results observed for some varieties with segregation progenies obtained from previous works. Therefore, this marker appears to be phased with the ABS resistance gene in the analyzed germplasm and it could be very useful not only for discriminating between resistant and susceptible cultivars, but also for inferring the allele configuration of the ABS resistance gene. It is therefore a valuable tool for selecting susceptible heterozygous genotypes that can be used as parents in

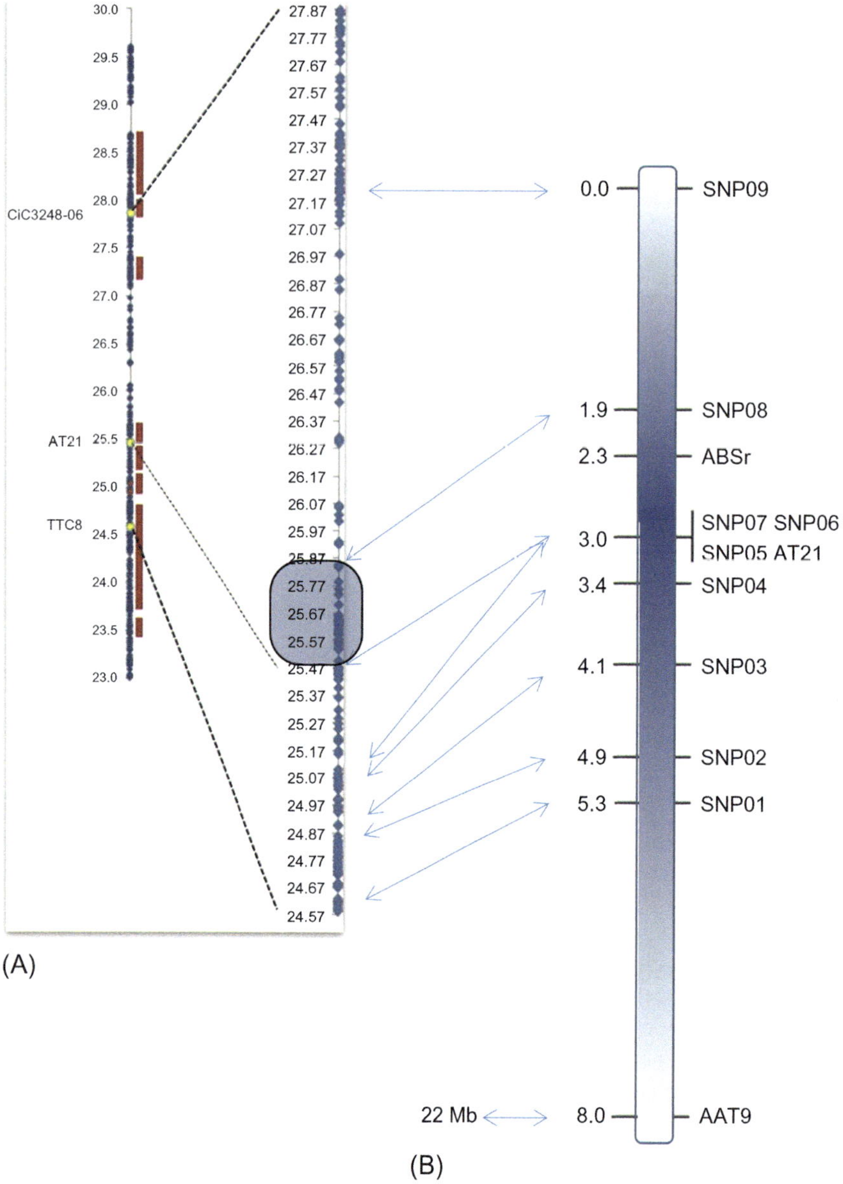

FIG. 9.3 Physical (A) and genetic (B) maps of the region on chromosome III linked to the resistance to Alternaria Brown Spot disease. ABSr indicates the recessive resistance locus flanked by markers at 0.4 and 0.7 cM, comprising a region of 366 kb.

breeding programs and conversely to discard homozygous susceptible varieties. The fact that susceptible yet heterozygous genotypes can be used as parents allows the exploitation of more genetic diversity in citrus breeding programs.

The SNP08 marker was also tested for association in 40 mandarin genotypes (20 resistant and 20 susceptible) used as breeding parents. Results showed a perfect association between the SNP08 and the response to *Alternaria* infections. The SNP08 is currently used for the assisted selection of ABS-resistant genotypes in the citrus breeding programs carried out at IVIA and CIRAD. Since its development, it has been successfully applied to select 2187 resistant hybrids from 4517 total hybrids arising from 10 different parental combinations. This analysis prevented growing >2000 susceptible genotypes that were removed from the breeding program at a very early stage of development, therefore saving a considerable amount of time, personnel, and money resources.

9.9 Reducing juvenility via viral vectors

Sexual hybrids display a long juvenile period that often goes on for >5–6 years. This characteristic is one of the most important constraints for citrus breeding programs. Thus, the long juvenile phase of citrus plants involves a long time period for field evaluation to select a new variety, hampers the recovery of segregation progenies by back crossing, and makes it very difficult to address genetic studies related to fruit characteristics.

Viral vectors have been used in plants for both expression of foreign genes and suppression of endogenous genes by virus-induced gene silencing (VIGS) in the infected plant (Senthil-Kumar and Mysore, 2011). To induce early flowering, different viral vectors expressing the FLOWERING LOCUS T (FT) protein have been used in several plant species such as soybean, cotton, apple, pear, and citrus (Yamagishi and Yoshikawa, 2011; McGarry and Ayre, 2012; Yamagishi et al., 2016; Velázquez et al., 2016). In citrus, Velázquez et al. (2016) developed a viral vector based on the citrus leaf blotch virus (CLBV) genome encoding the FT protein from *Arabidopsis thaliana* and its orthologs in citrus (*clbvINpr-AtFT* or *clbvINpr-CiFT* vectors, respectively). CLBV is an excellent virus to be used as a viral vector because it does not produce symptoms in most commercial cultivars (Galipienso et al., 2000), it is not restricted to phloem (Agüero et al., 2013), and it is not transmitted by vectors. Therefore, it could be used for field experiments. Citrus juvenile plants inoculated by grafting with these viral vectors start to produce flowers four months postinoculation and continue producing flowers over a period of at least five years, depending on the genotype and time of the year. Similar citrus plants that are not inoculated do not flower, but remain vegetatively growing under the same conditions. The reproductive biology of inoculated plants is not modified in comparison with adult plants, producing flowers, pollen, and fruits with the same appearance, viability, and characteristics of the noninoculated plants. In fact, the flowers and pollen of inoculated juvenile plants were used to recover new hybrids that were analyzed with several molecular markers. Genetic analysis confirmed that all plants were hybrids between the parents used.

The advantages of this biotechnological tool in comparison with other strategies such as genetic transformation are: (i) the infection of inoculated plants is fast, about one month, (ii) the viral vector can be easily graft inoculated onto hundreds of different genotypes in a very short period, and (iii) inoculated plants flowered four months postinoculation and these successfully produced fruits with normal seeds. Therefore, this technique allows obtaining seeds for the next-generation progeny within one year without any genome modification in the plant. In the framework of our triploid breeding program carried out at IVIA (Navarro et al., 2015), the viral vector based on the CLBV genome encoding the FT protein has been routinely incorporated. Tetraploid parents were produced by sexual hybridization and screened by molecular markers for some desired traits, such as resistance to the *Alternaria alternata* fungus (Cuenca et al., 2013, 2016) or accumulation of anthocyanins in fruits (Butelli et al., 2012). Then, selected hybrids were inoculated with the viral vector to induce precocious flowering and used in sexual hybridizations between diploids and tetraploids. Until now, >60 crosses have been performed and >1500 hybrids have been recovered from juvenile plants inoculated with the viral vector. This result allows us to conclude that this biotechnological tool has enormous practical value to accelerate genetic studies and breeding programs based on sexual hybridizations. Moreover, the viral vector is not integrated in the genome of the inoculated plant and is not transmitted either by insect vectors (Vives et al., 2008) or by pollen. Although a low seed transmission rate has been detected in a few genotypes, the virus is easily detected in the infected plants by RT-qPCR and could be easily eliminated by shoot tip grafting in vitro (Navarro et al., 1975) to recover any desired genotype vector free.

9.10 Genetic transformation of citrus

Research on the genetic modification of citrus resulting in the incorporation of foreign DNA into the genome has made significant progress since Kobayashi and Uchimiya (Kobayashi and Uchimiya, 1989) proposed direct DNA incorporation into citrus protoplasts to produce transgenic citrus plants. Similarly, biolistics also utilizes direct DNA delivery into the plant cell (Tomes et al., 1995). Both these methods utilize purified plasmid DNA as a vector for DNA insertion into the genome. In addition to direct DNA uptake, *Agrobacterium* has been utilized as a vehicle for DNA delivery and has become the most widely utilized method of citrus transformation. Both juvenile and mature citrus tissues can be transformed using *Agrobacterium*.

9.11 Direct DNA incorporation into citrus

9.11.1 Protoplast transformation

Protoplasts are plant cells with their cell walls mechanically or enzymatically removed (Cocking, 1972; Eeckhaut et al., 2013). Fungal enzymes such as cellulase, macerozyme, and pectinase are used for the digestion of cell walls (Cocking, 1972; Ruesink, 1971). Protoplasts are totipotent cells and can regenerate into fertile plants under optimum tissue culture conditions (Takebe et al., 1971). Uptake of DNA into protoplasts can be stimulated either with polyethylene glycol (PEG) treatment that promotes agglutination of protoplasts and induces the uptake of DNA via endocytosis or via electroporation that triggers destabilization of the plasma membrane and subsequent DNA uptake (Davey et al., 1989; Hansen and Wright, 1999; Joersbo and Brunstedt, 1996; Kao and Michayluk, 1974; Shillito et al., 1985). The protoplast transformation system has been successfully utilized in monocots such as rice, wheat, and maize that are recalcitrant to *Agrobacterium*-mediated transformation (Lörz et al., 1985; Rhodes et al., 1988; Toriyama et al., 1988; Shimamoto et al., 1989; Zhang and Wu, 1988). In citrus, embryogenic suspension cell cultures have been utilized to obtain totipotent protoplasts (Spiegel-Roy and Kochba, 1980; Vardi et al., 1990). The PEG-mediated protoplast transformation method is the most common and well-established method in protoplast transformation of citrus (Omar et al., 2007; Fleming et al., 2000). Protoplasts and DNA are physically mixed and incubated with 40% PEG 8000 to stimulate the movement of DNA into protoplasts, where it can be transiently expressed or stably integrated into the genome (Omar et al., 2007). The transformation efficiency remains typically low with this method (Guo et al., 2005). The low transformation rate might be due to various reasons including the enzymatic degradation of DNA, the stage of the suspension subculturing, the cell cycle, the culture medium, PEG, and environmental factors, which influence the ability of protoplasts to express their totipotency and develop into fertile plants (Davey et al., 2005). To improve transgenic plant selection, a visual selection system utilizing EGFP (enhanced GFP) was developed to facilitate the selection of transformed cells, tissues, or plants (Fleming et al., 2000). Prior to this, PEG-mediated protoplast transformation protocols utilized antibiotic resistance genes for the identification of transformed cells (Kobayashi and Uchimiya, 1989; Vardi et al., 1990). The EGFP-based marker was subsequently utilized for selecting transgenic Valencia plants transformed with the *Xa21* putative canker tolerance gene (Guo and Grosser, 2004) or a fruit quality gene (Guo et al., 2005). Transgenic Hamlin plants were also produced with Xa21 (Omar et al., 2007) and citrus tristeza virus RNAi constructs (Olivares-Fuster et al., 2003), using EGFP as a selectable marker.

The use of electroporation in citrus genetic transformation has been restricted to a few studies with varied success. Hidaka and Omura (Hidaka and Omura, 1993) used this technique for transformation of Ohta Ponkan mandarin. The use of electroporation was also reported for protoplast transformation of Hamlin sweet orange utilizing a plasmid construct encoding an endoplasmic reticulum-targeted EGFP that resulted in more stable and brighter expression of EGFP as compared to nontargeted cytoplasmic EGFP (Niedz et al., 2003). The same plasmid construct (pARS108) was used to illustrate the comparison between targeted and nontargeted cytoplasmic GFP, but with a PEG-mediated transformation system (Omar and Grosser, 2008).

9.11.2 Particle bombardment/biolistics

The particle bombardment technique is an emerging field of plant transformation that has the potential to transform both embryogenic and meristematic tissues (Altpeter et al., 2005; Barampuram and Zhang, 2011). However, this technique is in the developmental process and has had a low rate of success in citrus. It was utilized for transformation of citrus embryogenic suspension cells and epicotyl tissues (Bespalhok Filho et al., 2003; Yao et al., 1996), but neither of those studies reported any stable transgenic plants. Recently, biolistic transformation of Carrizo citrange epicotyl explants has resulted in the successful regeneration of transgenic plants (Wu et al., 2016). The success rate of this method is limited by the integration of multiple copies of the desired gene and the unnecessary DNA that can potentially silence the gene of interest in the plant genome (Barampuram and Zhang, 2011).

9.11.3 *Agrobacterium*-mediated transformation of citrus

This method of citrus transformation has gained popularity due to the higher transformation efficiencies obtained in most cultivars when utilizing *Agrobacterium* (Yu et al., 2002; Peña et al., 1995). The first report of *Agrobacterium* transformation of citrus was published for the transformation of embryogenic suspension cells of Washington navel, Trovita sweet orange, Ohta Ponkan, and Kara mandarin cultivars utilizing two strains of *A. tumefaciens* containing neomycin phosphotransferase (nptII) and hygromycin phosphotransferase (hptII) antibiotic selectable marker genes (Hidaka et al., 1990). *Agrobacterium*-mediated transformation was subsequently utilized for the transformation of the Carrizo citrange and trifoliate orange by using intermodal and epicotyl stem segments as explants, respectively (Moore et al., 1992; Kaneyoshi et al., 1994; Peña et al., 1995, 1997). Since then, numerous reports have been published utilizing the *Agrobacterium* method for transformation of different citrus species (Ghorbel et al., 2000; Gutierrez et al., 1997; Febres et al., 2003; Yang et al., 2000; Cervera et al., 1998; Dominguez et al., 2000). nptII has been the most frequently used marker gene cotransformed with the desired genes into explants (Peña et al., 2004b) while GUS has been the most widely used reporter gene for the selection of transformants (Moore et al., 1992; Peña et al., 1997; Gutierrez et al., 1997). Of late, the green fluorescent protein (GFP) and its enhanced derivatives (EGFP) derived from *Aequorea victoria* have been utilized as reporter genes for the selection of transformants (Mankin and Thompson, 2001; Wang, 2006; Ghorbel et al., 1999). Despite significant progress in the development of different methodologies using *Agrobacterium* for incorporation of transgenes into citrus and the ease of culture in vitro, the genus remains recalcitrant to genetic transformation (Peña et al., 2004b). There could be several factors that influence the relatively low transformation efficiencies as reported in the literature. Some of these factors could include the fact that citrus is not a natural host for *Agrobacterium* coupled with the production of high levels of escapes during selection and problems of rooting of regenerated transgenic shoots (Ghorbel et al., 1999). This together with the time taken for growth and development of transgenic shoots make transformation challenging (Moore et al., 1992).

Most genetic transformation protocols have focused on the optimization of conditions that result in producing the maximum number of transgenic lines (Dutt et al., 2011). Factors such as conditions required for *Agrobacterium* infection, conditions during cocultivation, age of explant used, and an efficient plant regeneration system following a successful infection are important parameters affecting the genetic transformation process (Febres et al., 2011). It has been shown that proper cultural conditions as well as phytohormone treatments enhance host cell susceptibility to *Agrobacterium* (Sangwan et al., 1992). Several factors such as the effect of different growth regulators, culture media, and antibiotics were also tested to improve transformation efficiency (Oliveira et al., 2010).

Agrobacterium can infect different types of juvenile citrus explants such as stem and leaf segments (Khan et al., 2009, 2012) or epicotyl pieces (Bond and Roose, 1998). Embryogenic suspension cells derived from calluses can also be used as explants to transform the seedless sweet orange and the difficult-to-transform mandarin and lemon cultivars (Dutt and Grosser, 2010). Mature tissue transformation has also been utilized but with lower efficiency. The major advantage of using mature tissue as an explant is that transgenic plants start flowering early, within 12–14 months (Cervera et al., 1998; Kobayashi et al., 2003). Genetic transformation of mature citrus tissues was initially proposed by Leandro Peña's group at the Instituto Valenciano de Investigaciones Agrarias (IVIA), Valencia, Spain, by using mature explants of the Pineapple sweet orange (Cervera et al., 1998, 1999). Mature tissue transformation was then performed in sweet orange cultivars by utilizing internodal stem segments, which also exhibited early flowering (Almeida et al., 2003). Mature citrus tissues are recalcitrant to *Agrobacterium* infection and the transformation efficiency is several orders of magnitude lower than that obtained from juvenile citrus tissues (Almeida et al., 2003; He et al., 2011). There is a need to optimize the plant regeneration medium because different cultivars behave differently when placed under similar regeneration conditions (Rodríguez et al., 2008). The addition of $2\,mg\,L^{-1}$ $AgNO_3$ to the tissue culture medium helped minimize the leaf drop in regenerating shoots (Marutani-Hert et al., 2011). The addition of nonionic surfactants aided the growth and regeneration from internodal stem segments of the sweet orange (Curtis and Mirkov, 2012). Utilizing certified clean bacteria- and virus-free materials (Wu et al., 2015), the selection of vigorous first flushes for in vitro culture (Cervera et al., 2008), improving cellular regeneration using bioreactors (Acanda et al., 2017), the addition of high concentrations of cytokinins (Fávero et al., 2012), and incubating *Agrobacterium*-treated stem explants in the dark (Marutani-Hert et al., 2012) has helped improve the transformation efficiency and subsequent plant regeneration.

Alternate protocols for the selection of transgenic lines have also been developed due to public opposition to the use of antibiotic-resistant genes in the *Agrobacterium* method (Ballester et al., 2008; Dutt et al., 2018). Marker-free Pineapple sweet orange plants were developed by utilizing the multi autotransformation (MAT) vector system that combines the isopentenyl transferase (ipt) gene for positive selection with a site-specific recombination and DNA removal system with 65% efficiency (Ballester et al., 2007). This system was, however, not suitable for the Carrizo citrange, suggesting the genotype specificity of this method.

9.12 CRISPR gene editing

The CRISPR/Cas system is an adaptive immune system of prokaryotes against invading phages and plasmids by cleaving the foreign DNA in a sequence-dependent manner (Jinek et al., 2012; Barrangou et al., 2007). About 84% of archaea and 48% of bacteria genomes contain CRISPR-Cas systems (Marraffini, 2013). The potential of the CRISPR/Cas9 system used for gene editing was realized with the pioneering work of Jinek and colleagues (Jinek et al., 2012; Gasiunas et al., 2012). Since then, the CRISPR/Cas system has been widely used for genome editing of different organisms, including eukaryotes and prokaryotes.

Cas9/sgRNA technology has been successfully used for genome editing of citrus (Jia and Wang, 2014a, b; Jia et al., 2016a). The first genome editing of citrus is a proof of concept by applying Cas9/sgRNA technology on the genome editing. In that study, Jia and Wang (Jia and Wang, 2014a) transiently expressed Cas9 and the *CsPDS*-targeting sgRNA in sweet orange leaf tissues using *Xanthomonas citri* (Xcc)-mediated agroinfiltration. Previously, it has been difficult to implement agroinfiltration-mediated transient expression in citrus leaves (Febres et al., 2011). Xcc pretreatment can enhance transient protein expression in citrus leaves, thus increasing the efficacy of the Cas9 protein.

CRISPR technology has been used to generate genome-modified plants via transgenic approaches (Jia et al., 2016a, b; Peng et al., 2017; LeBlanc et al., 2017; Zhang et al., 2017). Specifically, Cas9/sgRNA has been utilized to modify the PthA4 effector binding elements (EBEs) in the promoter as well as the coding region of the *CsLOB1* (*Citrus sinensis* Lateral Organ Boundaries) gene (Jia et al., 2016a, b). *CsLOB1* is a susceptibility gene for citrus canker, a severe disease caused by *X. citri*, and is induced by the pathogenicity factor PthA4, which binds to the EBEPthA4-*CsLOBP* to induce *CsLOB1* gene expression (Hu et al., 2014). Genome editing of the disease susceptibility gene *CsLOB1* in citrus has lead to the development of canker-resistant plants (Jia et al., 2016b, c).

Multiple approaches have been used to improve the efficacy of citrus genome editing via the CRISPR technology as well as broaden its application. Zhang et al. (2017) significantly increased the efficacy of citrus genome editing by adopting the *A. thaliana* YAO promoter to drive the expression of Cas9 with sgRNA under the control of the U6 promoter. Interestingly, LeBlanc et al. (2017) also reported that heating stress increased the targeted mutations by CRISPR/Cas9.

The aforementioned genome editing of citrus has been performed using the Cas9 from the *Streptococcus pyogenes*, which works in accord with the 5′NGG PAM sequences (Jinek et al., 2012). *Staphylococcus aureus*-derived SaCas9/sgRNA, whose preferred PAM sequence is 5′ NNGRRT (Ran et al., 2015), has also been shown to be able to modify the citrus genome via transient or transgenic expression in citrus (Jia et al., 2017). It is important to note that no off-target mutagenesis has been observed for either SpCas9- or SaCas9-induced mutations in citrus.

Genetically modified plants generated via transgenic expression of Cas9/sgRNA contain foreign DNA sequences, requiring a rigorous deregulation process before commercialization (Wolt et al., 2016). Additionally, constant expression of Cas9/sgRNA in transgenic plants may lead to the accumulation of off-target effects. Cas9/sgRNA in genetically edited crop plants can be removed by backcrossing them to wild-type plants. However, the approach for citrus is laborious and time consuming, particularly considering the long juvenile period for citrus. Backcrossing of citrus will lead to a loss of traits of the parental cultivars. Transient expression of either Cas9-sgRNA ribonucleoproteins, Cas9/sgRNA DNA, or RNA has been used successfully to generate nontransgenic genome-modified plants (Zhang et al., 2016; Liang et al., 2017; Svitashev et al., 2016; Woo et al., 2015). Even though nontransgenic genome editing of citrus has not been reported, it is likely that such an approach will eventually be available for citrus.

9.13 Concluding remarks

Citrus improvement is a continuum and requires a sustained long-term effort for success. Emerging biotechnologies are continuing to provide the research community with new tools that can increase the speed and efficiency of the process. New gene-editing tools have created a lot of excitement in the citrus improvement research area; however, additional critical research is needed to bring this technology to fruition, especially regarding identifying the correct genes to edit, and developing an efficient gene editing system(s) that will not leave a GMO signature. It is important that citrus researchers thoroughly understand all the biotechnology tools available, as some less glamorous tools that may be more appropriate for a specific research objective may be overlooked. For example, very few programs are exploiting somaclonal variation, which has remarkably led to the release of eight new sweet orange cultivars from the UF/CREC program. Also, micropagation via shoot multiplication and rooting has become a critical technique for the mass propagation of new improved rootstocks by tissue culture companies. Due to a limited or nonexistent seed supply for most new disease-resistant/tolerant rootstocks, micropropagation will be necessary to produce the millions of trees needed for replant efforts following HLB devastation. The challenges to maintaining a profitable and sustainabale worldwide citrus industry have never been greater, and applications of the emerging biotechnologies will play a critical role for future success.

References

Acanda, Y., Canton, M., Wu, H., Zale, J., 2017. Kanamycin selection in temporary immersion bioreactors allows visual selection of transgenic citrus shoots. Plant Cell Tissue Organ Cult. 129 (2), 351–357.

Agüero, J., Vives, M.C., Velázquez, K., Ruiz-Ruiz, S., Juárez, J., Navarro, L., et al., 2013. *Citrus leaf blotch virus* invades meristematic regions in *Nicotiana benthamiana* and citrus. Mol. Plant Pathol. 14 (6), 610–616.

Ajiro, N., Miyamoto, Y., Masunaka, A., Tsuge, T., Yamamoto, M., Ohtani, K., et al., 2010. Role of the host-selective ACT-toxin synthesis gene ACTTS2 encoding an Enoyl-reductase in pathogenicity of the tangerine pathotype of alternaria alternata. Phytopathology 100 (2), 120–126.

Akimitsu, K., Peever, T.L., Timmer, L., 2003. Molecular, ecological and evolutionary approaches to understanding alternaria diseases of citrus. Mol. Plant Pathol. 4 (6), 435–446.

Aleza, P., Juárez, J., Hernández, M., Pina, J.A., Ollitrault, P., Navarro, L., 2009. Recovery and characterization of a *Citrus clementina* Hort. ex Tan. Clemenules haploid plant selected to establish the reference whole citrus genome sequence. BMC Plant Biol. 9 (1), 110.

Aleza, P., Juárez, J., Cuenca, J., Ollitrault, P., Navarro, L., 2010. Recovery of citrus triploid hybrids by embryo rescue and flow cytometry from 2x×2x sexual hybridisation and its application to extensive breeding programs. Plant Cell Rep. 29 (9), 1023–1034.

Aleza, P., Cuenca, J., Hernández, M., Juárez, J., Navarro, L., Ollitrault, P., 2015. Genetic mapping of centromeres in the nine *Citrus clementina* chromosomes using half-tetrad analysis and recombination patterns in unreduced and haploid gametes. BMC Plant Biol. 15 (1), 80.

Ali, S., Mirza, B., 2006. Micropropagation of rough lemon (*Citrus jambhiri* Lush.): effect of explant type and hormone concentration. Acta Bot. Croat. 65 (2), 137–146.

Almeida, W.A.B., Mourão Filho, F.A.A., Pino, L.E., Boscariol, R.L., Rodriguez, A.P.M., Mendes, B.M.J., 2003. Genetic transformation and plant recovery from mature tissues of *Citrus sinensis* L. Osbeck. Plant Sci. 164 (2), 203–211.

Altman, A., Goren, R., 1974. Growth and dormancy cycles in *Citrus* bud cultures and their hormonal control. Physiol. Plant. 30 (3), 240–245.

Altpeter, F., Baisakh, N., Beachy, R., Bock, R., Capell, T., Christou, P., et al., 2005. Particle bombardment and the genetic enhancement of crops: myths and realities. Mol. Breed. 15 (3), 305–327.

Alva, A.K., Paramasivam, S., Graham, W.D., Wheaton, T.A., 2003. Best nitrogen and irrigation management practices for citrus production in sandy soils. Water Air Soil Pollut. 143 (1), 139–154.

Ballester, A., Cervera, M., Peña, L., 2007. Efficient production of transgenic citrus plants using isopentenyl transferase positive selection and removal of the marker gene by site-specific recombination. Plant Cell Rep. 26 (1), 39–45.

Ballester, A., Cervera, M., Peña, L., 2008. Evaluation of selection strategies alternative to nptII in genetic transformation of citrus. Plant Cell Rep. 27 (6), 1005–1015.

Barampuram, S., Zhang, Z.J., 2011. Recent advances in plant transformation. In: Birchler, J.A. (Ed.), Plant Chromosome Engineering: Methods and Protocols. Humana Press, Totowa, NJ, pp. 1–35.

Barlass, M., Skene, K.G.M., 1982. In vitro plantlet formation from citrus species and hybrids. Sci. Hortic. 17 (4), 333–341.

Barrangou, R., Fremaux, C., Deveau, H., Richards, M., Boyaval, P., Moineau, S., et al., 2007. CRISPR provides acquired resistance against viruses in prokaryotes. Science 315 (5819), 1709–1712.

Benavente-Garcia, O., Castillo, J., 2008. Update on uses and properties of *Citrus* flavonoids: new findings in anticancer, cardiovascular, and anti-inflammatory activity. J. Agric. Food Chem. 56 (15), 6185–6205.

Ben-Hayyim, G., Moore, G.A., 2007. Recent advances in breeding citrus for drought and saline stress tolerance. In: Jenks, M.A., Hasegawa, P.M., Jain, S.M. (Eds.), Advances in Molecular Breeding toward Drought and Salt Tolerant Crops. Springer, Dordrecht, pp. 627–642.

Bernet, G.P., Bretó, M.P., Asins, M.J., 2004. Expressed sequence enrichment for candidate gene analysis of citrus tristeza virus resistance. Theor. Appl. Genet. 108 (4), 592–602.

Bespalhok Filho, J.C., Kobayashi, A.K., Pereira, L.F., Galvão, R.M., Vieira, L.G., 2003. Transient gene expression of beta-glucuronidase in citrus thin epicotyl transversal sections using particle bombardment. Braz. Arch. Biol. Technol. 46 (1), 1–6.

Bohanec, B., 2003. Ploidy determination using flow cytometry. In: Maluszynski, M., Kasha, K.J., Forster, B.P., Szarejko, I. (Eds.), Doubled Haploid Production in Crop Plants: A Manual. Springer, Dordrecht, pp. 397–403.

Bond, J.E., Roose, M.L., 1998. Agrobacterium-mediated transformation of the commercially important citrus cultivar Washington navel orange. Plant Cell Rep. 18, 229.

Bonga, J.M., 1982. Vegetative propagation in relation to juvenility, maturity, and rejuvenation. In: Bonga, J.M., Durzan, D.J. (Eds.), Tissue Culture in Forestry. Springer, Dordrecht, pp. 387–412.

Bordón, Y., Guardiola, J.L., García-Luis, A., 2000. Genotype affects the morphogenic response in vitro of epicotyl segments of Citrus rootstocks. Ann. Bot. 86 (1), 159–166.

Burger, D.W., Hackett, W.P., 1986. Gradients of adventitious bud formation on excised epicotyl and root sections of citrus. Plant Sci. 43 (3), 229–232.

Butelli, E., Licciardello, C., Zhang, Y., Liu, J., Mackay, S., Bailey, P., et al., 2012. Retrotransposons control fruit-specific, cold-dependent accumulation of anthocyanins in blood oranges. Plant Cell 24 (3), 1242–1255.

Cabasson, C., Alvard, D., Dambier, D., Ollitrault, P., Teisson, C., 1997. Improvement of *Citrus* somatic embryo development by temporary immersion. Plant Cell Tissue Organ Cult. 50 (1), 33–37.

Cabasson, C., Luro, F., Ollitrault, P., Grosser, J., 2001. Nonrandom inheritance of mitochondrial genomes in citrus hybrids produced by protoplast fusion. Plant Cell Rep. 20 (7), 604–609.

Cao, H., Biswas, M.K., Lü, Y., Amar, M.H., Tong, Z., Xu, Q., et al., 2011. Doubled haploid callus lines of Valencia sweet orange recovered from anther culture. Plant Cell Tissue Organ Cult. 104 (3), 415–423.

Cardoso, J.C., Martinelli, A.P., Germanà, M.A., Latado, R.R., 2014. In vitro anther culture of sweet orange (*Citrus sinensis* L. Osbeck) genotypes and of a *C. clementina* × *C. sinensis* Hamlin hybrid. Plant Cell Tissue Organ Cult. 117 (3), 455–464.

Cardoso, J.C., Abdelgalel, A.M., Chiancone, B., Latado, R.R., Lain, O., Testolin, R., et al., 2016. Gametic and somatic embryogenesis through in vitro anther culture of different *Citrus* genotypes. Plant Biosyst. 150 (2), 304–312.

Carimi, F., Tortorici, M.C., De Pasquale, F., Crescimanno, F.G., 1998. Somatic embryogenesis and plant regeneration from undeveloped ovules and stigma/style explants of sweet orange navel group [*Citrus sinensis* (L.) Osb.]. Plant Cell Tissue Organ Cult. 54 (3), 183–189.

Casales, F.G., Chiancone, B., Germanà, M.A., Micheli, M. (Eds.), 2015. Study on conversion of encapsulated vitro-derived propagules of c35 citrange [*C. sinensis* (L.) Osb. × *P. trifoliata* (L.) Raf.]. International Society for Horticultural Science (ISHS), Leuven, Belgium.

Cervera, M., Juarez, J., Navarro, A., Pina, J.A., Duran-Vila, N., Navarro, L., et al., 1998. Genetic transformation and regeneration of mature tissues of woody fruit plants bypassing the juvenile stage. Transgenic Res. 7 (1), 51–59.

Cervera, M., Juárez, J., Navarro, A., Pina, J.A., Duran-Vila, N., Navarro, L., et al., 1999. Bypassing juvenile stage in genetic transformation of *Citrus* plants. In: Altman, A., Ziv, M., Izhar, S. (Eds.), Plant Biotechnology and in Vitro Biology in the 21st Century: Proceedings of the IXth International Congress of the International Association of Plant Tissue Culture and Biotechnology Jerusalem, Israel, 14–19 June 1998. Springer, Dordrecht, pp. 115–118.

Cervera, M., Navarro, A., Navarro, L., Peña, L., 2008. Production of transgenic adult plants from Clementine mandarin by enhancing cell competence for transformation and regeneration. Tree Physiol. 28 (1), 55–66.

Chaturvedi, H., Mitra, G., 1974. Clonal propagation of *Citrus* from somatic callus cultures. HortScience 118–120.
Chen, Z., 1980. The induction of *Citrus* pollen plants in artificial media. Acta Genet. Sin. 7, 189–191.
Cheng, F.S., Roose, M.L., 1995. Origin and inheritance of dwarfing by the citrus rootstock *Poncirus trifoliata* flying dragon. J. Am. Soc. Hortic. Sci. 120 (2), 286–291.
Chiancone, B., Tassoni, A., Bagni, N., Germanà, M.A., 2006. Effect of polyamines on in vitro anther culture of *Citrus clementina* Hort. ex Tan. Plant Cell Tissue Organ Cult. 87 (2), 145–153.
Chiancone, B., Casales, F.G., Mondello, V., Torta, L., Burruano, S., Germanà, M.A. (Eds.), 2015a. Ex vitro mycorrhization of vitro-derived plantlets of Carrizo citrange [*C. sinensis* (L.) Osb. × *P. trifoliata* (L.) Raf.]. International Society for Horticultural Science (ISHS), Leuven, Belgium.
Chiancone, B., Karasawa, M.M.G., Gianguzzi, V., Abdelgalel, A.M., Bárány, I., Testillano, P.S., et al., 2015b. Early embryo achievement through isolated microspore culture in *Citrus clementina* Hort. ex Tan., cvs. Monreal Rosso and Nules. Front. Plant Sci. 6, 413.
Chiancone, B., Martorana, L., Casales, F., Germanà, M.A. (Eds.), 2016. The Effects of Benzyladenine and Meta-Topolin on In Vitro Sprouting and Regrowth After Encapsulation of C35 Citrange [*C. sinensis* (L.) Osb. × *Poncirus trifoliata* (L.) Raf.], *Microcuttings*. International Society for Horticultural Science (ISHS), Leuven, Belgium.
Chiancone, B., Martorana, L., Gianguzzi, V., Germanà, M.A. (Eds.), 2017. The Effects of Meta-Topolin and Benzyladenine on In Vitro Organogenesis From Epicotyl Cuttings of Troyer Citrange (*C. sinensis* (L.) Osbeck × *Poncirus trifoliata* (L.) Raf.). International Society for Horticultural Science (ISHS), Leuven, Belgium.
Choi, H.-S., 2004. Volatile constituents of Satsuma mandarins growing in Korea. Flavour Fragr. J. 19 (5), 406–412.
Cimò G, Casamento D, Marinoni DT, Botta R, Germanà MA, Gametic embryogenesis through isolated microspore culture in Mandarin (*Citrus reticulata* Blanco), Mandarino Tardivo Di Ciaculli: effect of meta-Topolin and temperature treatments. Citrus Res. Technol. 37 (2), 2016, 113–122.
Cocking, E.C., 1972. Plant cell protoplasts-isolation and development. Annu. Rev. Plant Physiol. 23 (1), 29–50.
Costa, M.G.C., Alves, V.S., Lani, E.R.G., Mosquim, P.R., Carvalho, C.R., Otoni, W.C., 2004. Morphogenic gradients of adventitious bud and shoot regeneration in epicotyl explants of *Citrus*. Sci. Hortic. 100 (1), 63–74.
Cuenca, J., Froelicher, Y., Aleza, P., Juárez, J., Navarro, L., Ollitrault, P., 2011. Multilocus half-tetrad analysis and centromere mapping in citrus: evidence of SDR mechanism for 2n megagametophyte production and partial chiasma interference in mandarin cv. fortune. Heredity 107 (5), 462.
Cuenca, J., Aleza, P., Vicent, A., Brunel, D., Ollitrault, P., Navarro, L., 2013. Genetically based location from triploid populations and gene ontology of a 3.3-Mb genome region linked to alternaria brown spot resistance in citrus reveal clusters of resistance genes. PLoS One 8 (10), e76755.
Cuenca, J., Aleza, P., Juárez, J., García-Lor, A., Froelicher, Y., Navarro, L., et al., 2015. Maximum-likelihood method identifies meiotic restitution mechanism from heterozygosity transmission of centromeric loci: application in citrus. Sci. Rep. 5, 9897.
Cuenca, J., Aleza, P., Garcia-Lor, A., Ollitrault, P., Navarro, L., 2016. Fine mapping for identification of citrus Alternaria brown spot candidate resistance genes and development of new SNP markers for marker-assisted selection. Front. Plant Sci. 7, 1948.
Curtis, I.S., Mirkov, T.E., 2012. Influence of surfactants on growth and regeneration from mature internodal stem segments of sweet orange (*Citrus sinensis*) cv. Hamlin. Plant Cell Tissue Organ Cult. 108 (2), 345–352.
Dalkilic, Z., Timmer, L., Gmitter, F.G., 2005. Linkage of an Alternaria disease resistance gene in mandarin hybrids with RAPD fragments. J. Am. Soc. Hortic. Sci. 130 (2), 191–195.
Davey, M.R., Rech, E.L., Mulligan, B.J., 1989. Direct DNA transfer to plant cells. Plant Mol. Biol. 13 (3), 273–285.
Davey, M.R., Anthony, P., Power, J.B., Lowe, K.C., 2005. Plant protoplasts: status and biotechnological perspectives. Biotechnol. Adv. 23 (2), 131–171.
Davies, F.S., Albrigo, L.G., 1994. *Citrus*. Cab International.
Demirkeser, T., Kaplankıran, M., Toplu, C., Yıldız, E., 2009. Yield and fruit quality performance of Nova and Robinson mandarins on three rootstocks in Eastern Mediterranean. Afr. J. Agric. Res. 4 (4), 262–268.
Deng, X., 2005. Advances in worldwide citrus breeding. Acta Hort. Sinica 32 (6), 1140–1146.
Deng, X., Deng, Z., Xiao, S., Zhang, W. (Eds.), 1992. Pollen derived plantlets from anther culture of Ichang papeda hybrids No. 14 and trifoliate orange. Proc. Int. Soc. Citricult. 1, 190–192.
Dirlewanger, E., Graziano, E., Joobeur, T., Garriga-Calderé, F., Cosson, P., Howad, W., et al., 2004. Comparative mapping and marker-assisted selection in rosaceae fruit crops. Proc. Natl. Acad. Sci. U. S. A. 101 (26), 9891–9896.
Dominguez, A., Guerri, J., Cambra, M., Navarro, L., Moreno, P., Peña, L., 2000. Efficient production of transgenic citrus plants expressing the coat protein gene of citrus tristeza virus. Plant Cell Rep. 19, 427.
Drew, R., 1988. Rapid clonal propagation of papaya in vitro from mature field-grow trees. HortScience 23 (3), 609–611.
Duan, Y.-X., Liu, X., Fan, J., Li, D.-L., Wu, R.-C., Guo, W.-W., 2007. Multiple shoot induction from seedling epicotyls and transgenic citrus plant regeneration containing the green fluorescent protein gene. Bot. Stud. 48 (2), 165–171.
Duran-Vila, N., Ortega, V., Navarro, L., 1989. Morphogenesis and tissue cultures of three citrus species. Plant Cell Tissue Organ Cult. 16 (2), 123–133.
Duran-Vila, N., Gogorcena, Y., Ortega, V., Ortiz, J., Navarro, L., 1992. Morphogenesis and tissue culture of sweet orange (*Citrus sinensis* (L.) Osb.): effect of temperature and photosynthetic radiation. Plant Cell Tissue Organ Cult. 29 (1), 11–18.
Dutt, M., Grosser, J.W., 2010. An embryogenic suspension cell culture system for *Agrobacterium*-mediated transformation of citrus. Plant Cell Rep. 29 (11), 1251–1260.
Dutt, M., Vasconcellos, M., Grosser, J.W., 2011. Effects of antioxidants on *Agrobacterium*-mediated transformation and accelerated production of transgenic plants of Mexican lime (*Citrus aurantifolia* Swingle). Plant Cell Tissue Organ Cult. 107 (1), 79.
Dutt, M., Zambon, F.T., Erpen, L., Soriano, L., Grosser, J., 2018. Embryo-specific expression of a visual reporter gene as a selection system for citrus transformation. PLoS One 13 (1), e0190413.
Edriss, M.H., Burger, D.W., 1984. In vitro propagation of Troyer citrange from epicotyl segments. Sci. Hortic. 23 (2), 159–162.

Eeckhaut, T., Lakshmanan, P.S., Deryckere, D., Van Bockstaele, E., Van Huylenbroeck, J., 2013. Progress in plant protoplast research. Planta 238 (6), 991–1003.

Esquinas-Alcazar, J., 2005. Protecting crop genetic diversity for food security: political, ethical and technical challenges. Nat. Rev. Genet. 6, 946.

Fang, D.Q., Federici, C.T., Roose, M.L., 1997. Development of molecular markers linked to a gene controlling fruit acidity in citrus. Genome 40 (6), 841–849.

Fávero, P., de FAVM, F., LCL, S., BMJ, M., 2012. Genetic transformation of three sweet orange cultivars from explants of adult plants. Acta Physiol. Plant. 34 (2), 471–477.

Febres, V., Niblett, C., Lee, R., Moore, G., 2003. Characterization of grapefruit plants (*Citrus paradisi* Macf.) transformed with citrus tristeza closterovirus genes. Plant Cell Rep. 21 (5), 421–428.

Febres, V., Fisher, L., Khalaf, A., Moore, G.A., 2011. *Citrus* transformation: challenges and prospects. In: Genetic Transformation. Intech.

Ferrie, A.M.R., Caswell, K.L., 2011. Isolated microspore culture techniques and recent progress for haploid and doubled haploid plant production. Plant Cell Tissue Organ Cult. 104 (3), 301–309.

Fisher, K., Phillips, C., 2008. Potential antimicrobial uses of essential oils in food: is citrus the answer? Trends Food Sci. Technol. 19 (3), 156–164.

Fleming, G.H., Olivares-Fuster, O., Del Bosco, S.F., Grosser, J.W., 2000. An alternative method for the genetic transformation of sweet orange. In Vitro Cell. Dev. Biol. Plant 36, 450.

Forster, B.P., Heberle-Bors, E., Kasha, K.J., Touraev, A., 2007. The resurgence of haploids in higher plants. Trends Plant Sci. 12 (8), 368–375.

Froelicher, Y., Ollitrault, P. (Eds.), 2000. Effects of the Hormonal Balance on Clausena Excavata Androgenesis. International Society for Horticultural Science (ISHS), Leuven, Belgium.

Froelicher, Y., Bassene, J.-B., Jedidi-Neji, E., Dambier, D., Morillon, R., Bernardini, G., et al., 2007. Induced parthenogenesis in mandarin for haploid production: induction procedures and genetic analysis of plantlets. Plant Cell Rep. 26 (7), 937.

Froelicher, Y., Bouffin, J., Dambier, D., Ollitrault, P. (Eds.), 2012. Triploid Seedless Mandarin Breeding in France: S02O02. ISC.

Frost, H.B., 1925. The chromosomes of *Citrus*. J. Wash. Acad. Sci. 15 (1), 1–3.

Frost, H., Soost, R., Reuther, W., Batchelor, L., Webber, H., 1968. The Citrus Industry. vol. 2.

Furusato, K., Ohta, Y., Ishibashi, K., 1957. Studies on polyembryony in *Citrus*. Seiken Ziho 8, 40–48.

Galipienso, L., Navarro, L., Ballester-Olmos, J., Pina, J., Moreno, P., Guerri, J., 2000. Host range and symptomatology of a graft-transmissible pathogen causing bud union crease of citrus on trifoliate rootstocks. Plant Pathol. 49 (2), 308–314.

Garcia-Luis, A., Bordon, Y., Moreira-Dias, J.M., Molina, R.V., Guardiola, J.L., 1999. Explant orientation and polarity determine the morphogenic response of epicotyl segments of troyer citrange. Ann. Bot. 84 (6), 715–723.

Gasiunas, G., Barrangou, R., Horvath, P., Siksnys, V., 2012. Cas9-crRNA ribonucleoprotein complex mediates specific DNA cleavage for adaptive immunity in bacteria. Proc. Natl. Acad. Sci. U. S. A. 109 (39), E2579–E2586.

George, E.F., Sherrington, P.D., 1984. Plant Propagation by Tissue Culture. Exegetics Ltd.

Germanà, M.A., 1997. Haploidy in citrus. In: Jain, S.M., Sopory, S.K., Veilleux, R.E. (Eds.), In Vitro Haploid Production in Higher Plants: Vol. 5—Oil, Ornamental and Miscellaneous Plants. Springer, Dordrecht, pp. 195–217.

Germanà, M.A., 2003. Somatic embryogenesis and plant regeneration from anther culture of *Citrus aurantium* and C-reticulata. Biologia 58 (4), 843–850.

Germanà, M.A., 2005. Protocol of somatic embryogenesis from *Citrus* spp. anther culture. In: Jain, S.M., Gupta, P.K. (Eds.), Protocol for Somatic Embryogenesis in Woody Plants. Springer, Dordrecht, pp. 191–207.

Germanà, M.A., 2006. Doubled haploid production in fruit crops. Plant Cell Tissue Organ Cult. 86 (2), 131.

Germanà, M.A., 2007. 7 Haploidy. Citrus Genet. Breed. Biotechnol. 167.

Germanà, M.A., 2009. Haploids and doubled haploids in fruit trees. In: Touraev, A., Forster, B.P., Jain, S.M. (Eds.), Advances in Haploid Production in Higher Plants. Springer, Dordrecht, pp. 241–263.

Germanà, M.A., 2011a. Gametic embryogenesis and haploid technology as valuable support to plant breeding. Plant Cell Rep. 30 (5), 839–857.

Germanà, M.A., 2011b. Anther culture for haploid and doubled haploid production. Plant Cell Tissue Organ Cult. 104 (3), 283–300.

Germanà, M.A., 2012. Use of irradiated pollen to induce parthenogenesis and haploid production in fruit crops. In: Plant Mutation Breeding and Biotechnology. CABI, FAO, Oxfordshire.

Germanà, M.A., Chiancone, B., 2001. Gynogenetic haploids of citrus after in vitro pollination with triploid pollen grains. Plant Cell Tissue Organ Cult. 66 (1), 59–66.

Germanà, M.A., Chiancone, B., 2003. Improvement of the anther culture protocol in *Citrus clementina* Hort. ex Tan. Plant Cell Rep. 22 (3), 181–187.

Germanà, M.A., Reforgiato Recupero, G., 1997. Haploid embryos regeneration from anther culture of Mapo tangelo (*Citrus deliciosa* × *C. paradisi*). Adv. Hortic. Sci. 11 (3), 147–152.

Germanà, M.A., Crescimanno, F.G., De Pasquale, F., Yu Ying, W. (Eds.), 1992. Androgenesis in 5 Cultivars of Ctrus limon L. Burm. f. International Society for Horticultural Science (ISHS), Leuven, Belgium.

Germanà, M.A., Ying Wang, Y., Barbagallo, M.G., Iannolino, G., Crescimanno, F.G., 1994. Recovery of haploid and diploid plantlets from anther culture of *Citrus clementina* Hort. ex Tan. and *Citrus reticulata* Blanco. J. Hort. Sci. 69 (3), 473–480.

Germanà, M.A., Scarano, M., Crescimanno, F. (Eds.), 1996. First results on isolated microspore culture of citrus. Proc. Int. Soc. Citricult. 2, 882–885.

Germanà, M.A., Crescimanno, F.G., Reforgiato Recupero, G., Russo, M.P. (Eds.), 2000. *Preliminary Characterization of Several Doubled Haploids of Citrus clementina* cv. Nules. International Society for Horticultural Science (ISHS), Leuven, Belgium.

Germanà, M.A., Chiancone, B., Lain, O., Testolin, R., 2005. Anther culture in *Citrus clementina*: a way to regenerate tri-haploids. Aust. J. Agric. Res. 56 (8), 839–845.

Germanà, M.A., Macaluso, L., Patricolo, G., Chiancone, B., 2008. Morphogenic response in vitro of epicotyl segments of *Citrus macrophylla*. Plant Biosyst. 142 (3), 661–664.

Germanà, M.A., Lambardi, M., Ozudogru, E.A. (Eds.), 2011a. Preliminary Results on Desiccation and pvs2-Vitrification of carrizo Citrange Synthetic Seeds as Pre-Treatments to Cryopreservation. International Society for Horticultural Science (ISHS), Leuven, Belgium.

Germanà, M.A., Micheli, M., Chiancone, B., Macaluso, L., Standardi, A., 2011b. Organogenesis and encapsulation of in vitro-derived propagules of Carrizo citrange [*Citrus sinensis* (L.) Osb.×*Poncirus trifoliata* (L.) Raf.]. Plant Cell Tissue Organ Cult. 106 (2), 299–307.

Germanà, M.A., Aleza, P., Carrera, E., Chen, C., Chiancone, B., Costantino, G., et al., 2013. Cytological and molecular characterization of three gametoclones of *Citrus clementina*. BMC Plant Biol. 13 (1), 129.

Ghorbel, R., Navarro, L., Duran-Vila, N., 1998. Morphogenesis and regeneration of whole plants of grapefruit (*Citrus paradisi*), sour orange (*C. aurantium*) and alemow (*C. macrophylla*). J. Hortic. Sci. Biotechnol. 73 (3), 323–327.

Ghorbel, R., Juárez, J., Navarro, L., Peña, L., 1999. Green fluorescent protein as a screenable marker to increase the efficiency of generating transgenic woody fruit plants. Theor. Appl. Genet. 99, 350.

Ghorbel, R., Domínguez, A., Navarro, L., Peña, L., 2000. High efficiency genetic transformation of sour orange (*Citrus aurantium*) and production of transgenic trees containing the coat protein gene of citrus tristeza virus. Tree Physiol. 20 (17), 1183–1189.

Gill, M.I.S., Singh, Z., Dhillon, B.S., Gosal, S.S., 1994. Somatic embryogenesis and plantlet regeneration on calluses derived from seedling explants of Kinnow mandarin (*Citrus nobilis* Lour × *Citrus deliciosa* Tenora). J. Hort. Sci. 69 (2), 231–236.

Gmitter, F.G., Xiao, S.Y., Huang, S., Hu, X.L., Garnsey, S.M., Deng, Z., 1996. A localized linkage map of the citrus tristeza virus resistance gene region. Theor. Appl. Genet. 92 (6), 688–695.

Goh, C.J., Sim, G.E., Morales, C.L., Loh, C.S., 1995. Plantlet regeneration through different morphogenic pathways in pommelo tissue culture. Plant Cell Tissue Organ Cult. 43 (3), 301–303.

Goswami, K., Sharma, R., Singh, P.K., Singh, G., 2013. Micropropagation of seedless lemon (*Citrus limon* L. cv. Kaghzi Kalan) and assessment of genetic fidelity of micropropagated plants using RAPD markers. Physiol. Mol. Biol. Plants 19 (1), 137–145.

Grinblat, U., 1972. Differentiation of *Citrus* stem *in vitro*. J. Am. Soc. Hort. Sci. J. 97, 599–603.

Grosser, J.W., Gmitter, F.G., 1990. Protoplast fusion and citrus improvement. Plant Breed. Rev. 8, 339–374.

Grosser, J., Gmitter, F., Chandler, J., 1997. Development of improved sweet orange cultivars using tissue culture methods. Proc. Fla. State Hort. Soc. 110, 13–16.

Grosser, J.W., Ollitrault, P., Olivares-Fuster, O., 2000. Somatic hybridization in citrus: an effective tool to facilitate variety improvement. In Vitro Cell. Dev. Biol. Plant 36 (6), 434–449.

Grosser, J.W., Calovic, M., Louzada, E.S., 2010. Protoplast fusion technology—somatic hybridization and cybridization. In: Plant Cell Culture: Essential Methods. John Wiley and Sons, New York, USA, pp. 175–198.

Gu, X.F., Zhang, J.R., 2005. An efficient adventitious shoot regeneration system for zhanhua winter jujube *(Zizyphus jujuba* Mill.) using leaf explants. Plant Cell Rep. 23 (12), 775–779.

Guha, S., Maheshwari, S.C., 1964. In vitro production of embryos from anthers of datura. Nature 204, 497.

Guo, W.W., Grosser, J.W. (Eds.), 2004. Transfer of a Potential Canker Resistance Gene into *Citrus* Protoplasts Using GFP as the Selectable Marker. International Society for Horticultural Science (ISHS), Leuven, Belgium.

Guo, W.W., Cai, X.D., Grosser, J.W., 2004a. Somatic cell cybrids and hybrids in plant improvement. In: Daniell, H., Chase, C. (Eds.), Molecular Biology and Biotechnology of Plant Organelles: Chloroplasts and Mitochondria. Springer, Dordrecht, pp. 635–659.

Guo, W.W., Prasad, D., Cheng, Y.J., Serrano, P., Deng, X.X., Grosser, J.W., 2004b. Targeted cybridization in citrus: transfer of satsuma cytoplasm to seedy cultivars for potential seedlessness. Plant Cell Rep. 22 (10), 752–758.

Guo, W., Duan, Y., Olivares-Fuster, O., Wu, Z., Arias, C.R., Burns, J.K., et al., 2005. Protoplast transformation and regeneration of transgenic Valencia sweet orange plants containing a juice quality-related pectin methylesterase gene. Plant Cell Rep. 24 (8), 482–486.

Guo, W.-W., Xiao, S.-X., Deng, X.-X., 2013. Somatic cybrid production via protoplast fusion for citrus improvement. Sci. Hortic. 163, 20–26.

Gutierrez, E., Luth, D., Moore, G.A., 1997. Factors affecting *Agrobacterium*-mediated transformation in *Citrus* and production of sour orange (*Citrus aurantium* L.) plants expressing the coat protein gene of citrus tristeza virus. Plant Cell Rep. 16, 745.

Hansen, G., Wright, M.S., 1999. Recent advances in the transformation of plants. Trends Plant Sci. 4 (6), 226–231.

Hassanein, A.M., Azooz, M.M., 2003. Propagation of citrus reticulata via in vitro seed germination and shoot cuttings. Biol. Plant. 47 (2), 173–177.

He, Y., Chen, S., Peng, A., Zou, X., Xu, L., Lei, T., et al., 2011. Production and evaluation of transgenic sweet orange (*Citrus sinensis* Osbeck) containing bivalent antibacterial peptide genes (Shiva A and Cecropin B) via a novel *Agrobacterium*-mediated transformation of mature axillary buds. Sci. Hortic. 128 (2), 99–107.

Hidaka, T., Omura, M., 1992. Regeneration of somatic hybrid plants obtained by electrical fusion between satsuma mandarin (*Citrus unshiu*) and rough lemon (*C. jambhiri*) or yuzu (*C. junos*). Jpn. J. Breed. 42 (1), 79–89.

Hidaka, T., Omura, M., 1993. Transformation of *Citrus* protoplasts by electroporation. J. Jpn. Soc. Hort. Sci. 62, 371.

Hidaka, T., Yamada, Y., Shichijo, T., 1979. In vitro differentiation of haploid plants by anther culture in *Poncirus trifoliata* (L.) Raf. Jpn. J. Breed. 29 (3), 248–254.

Hidaka, T., Omura, M., Ugaki, M., Tomiyama, M., Kato, A., Ohshima, M., et al., 1990. *Agrobacterium*-mediated transformation and regeneration of *Citrus* spp. from suspension cells. Jpn. J. Breed. 40, 199.

Hong, S.-I., Lee, H.-H., Kim, D., 2007. Effects of hot water treatment on the storage stability of *Satsuma mandarin* as a postharvest decay control. Postharvest Biol. Technol. 43 (2), 271–279.

Hu, Y., Zhang, J., Jia, H., Sosso, D., Li, T., Frommer, W.B., et al., 2014. Lateral organ boundaries 1 is a disease susceptibility gene for citrus bacterial canker disease. Proc. Natl. Acad. Sci. U. S. A. 111 (4), E521–E529.

Jia, H., Wang, N., 2014a. Targeted genome editing of sweet orange using Cas9/sgRNA. PLoS One 9 (4), e93806.

Jia, H., Wang, N., 2014b. Xcc-facilitated agroinfiltration of citrus leaves: a tool for rapid functional analysis of transgenes in citrus leaves. Plant Cell Rep. 33 (12), 1993–2001.

Jia, H., Orbovic, V., Jones, J.B., Wang, N., 2016a. Modification of the PthA4 effector binding elements in type I Cs LOB 1 promoter using Cas9/sg RNA to produce transgenic Duncan grapefruit alleviating XccΔpthA4: dCs LOB 1.3 infection. Plant Biotechnol. J. 14 (5), 1291–1301.

Jia, H., Zhang, Y., Orbović, V., Xu, J., White, F.F., Jones, J.B., et al., 2016b. Genome editing of the disease susceptibility gene CsLOB1 in citrus confers resistance to citrus canker. Plant Biotechnol. J. 15 (7), 817–823.

Jia, H., Orbovic, V., Jones, J.B., Wang, N., 2016c. Modification of the PthA4 effector binding elements in type I CsLOB1 promoter using Cas9/sgRNA to produce transgenic Duncan grapefruit alleviating XccΔpthA4:dCsLOB1.3 infection. Plant Biotechnol. J. 14 (5), 1291–1301.

Jia, H., Xu, J., Orbović, V., Zhang, Y., Wang, N., 2017. Editing citrus genome via SaCas9/sgRNA system. Front. Plant Sci. 8, 2135.

Jinek, M., Chylinski, K., Fonfara, I., Hauer, M., Doudna, J.A., Charpentier, E., 2012. A programmable dual-RNA-guided DNA endonuclease in adaptive bacterial immunity. Science 337 (6096), 816–821.

Joersbo, M., Brunstedt, J., 1996. Electroporation and transgenic plant production. In: Lynch, P.T., Davey, M.R. (Eds.), Electrical Manipulation of Cells. Springer, Boston, MA, pp. 201–222.

Kamiri, M., Stift, M., Srairi, I., Costantino, G., Moussadik, A.E., Hmyene, A., et al., 2011. Evidence for nondisomic inheritance in a citrus interspecific tetraploid somatic hybrid between *C. reticulata* and *C. limon* using SSR markers and cytogenetic analysis. Plant Cell Rep. 30 (8), 1415–1425.

Kaneyoshi, J., Kobayashi, S., Nakamura, Y., Shigemoto, N., Doi, Y., 1994. A simple and efficient gene transfer system of trifoliate orange (*Poncirus trifoliata* Raf.). Plant Cell Rep. 13, 541.

Kao, K.N., Michayluk, M.R., 1974. A method for high-frequency intergeneric fusion of plant protoplasts. Planta 115 (4), 355–367.

Karam, N.S., Al-Majathoub, M., 2000. Direct shoot regeneration and microtuberization in wild *Cyclamen persicum* Mill. Using seedling tissue. Sci. Hortic. 86 (3), 235–246.

Keneni, G., Bekele, E., Imtiaz, M., Dagne, K., 2012. Genetic vulnerability of modern crop cultivars: causes, mechanism and remedies. Int. J. Plant Res. 2 (3), 69–79.

Khan, E.U., Fu, X.-Z., Wang, J., Fan, Q.-J., Huang, X.-S., Zhang, G.-N., et al., 2009. Regeneration and characterization of plants derived from leaf in vitro culture of two sweet orange (*Citrus sinensis* (L.) Osbeck) cultivars. Sci. Hortic. 120 (1), 70–76.

Khan, E.U., Fu, X.-Z., Liu, J.-H., 2012. *Agrobacterium*-mediated genetic transformation and regeneration of transgenic plants using leaf segments as explants in Valencia sweet orange. Plant Cell Tissue Organ Cult. 109 (2), 383–390.

Khawale, R.N., Singh, S.K., 2005. In vitro adventitive embryony in citrus: a technique for citrus germplasm exchange. Curr. Sci. 1309–1311.

Kobayashi, S., Uchimiya, H., 1989. Expression and integration of a foreign gene in orange (*Citrus sinensis* Osb.) protoplasts by direct DNA transfer. Jpn. J. Genet. 64 (2), 91–97.

Kobayashi, A.K., Bespalhok, J.C., Pereira, L.F.P., Vieira, L.G.E., 2003. Plant regeneration of sweet orange (*Citrus sinensis*) from thin sections of mature stem segments. Plant Cell Tissue Organ Cult. 74 (1), 99–102.

Kochba, J., Spiegel-Roy, P., Safran, H., 1972. Adventive plants from ovules and nucelli in *Citrus*. Planta 106 (3), 237–245.

Kohmoto, K., Itoh, Y., Shimomura, N., Kondoh, Y., Otani, H., Kodama, M., et al., 1993. Isolation and biological activities of two host-specific toxins from the tangerine pathotype of alternaria alternata. Phytopathology 83 (5), 495–502.

Kotsias, D., Roussos, P.A., 2001. An investigation on the effect of different plant growth regulating compounds in in vitro shoot tip and node culture of lemon seedlings. Sci. Hortic. 89 (2), 115–128.

Larkin, P.J., Scowcroft, W.R., 1981. Somaclonal variation—a novel source of variability from cell cultures for plant improvement. Theor. Appl. Genet. 60 (4), 197–214.

LeBlanc, C., Zhang, F., Mendez, J., Lozano, Y., Chatpar, K., Irish, V., et al., 2017. Increased efficiency of targeted mutagenesis by CRISPR/Cas9 in plants using heat stress. Plant J. 93 (2), 377–386.

Lee, M., Phillips, R.L., 1988. The chromosomal basis of somaclonal variation. Annu. Rev. Plant Physiol. Plant Mol. Biol. 39 (1), 413–437.

Liang, Z., Chen, K., Li, T., Zhang, Y., Wang, Y., Zhao, Q., et al., 2017. Efficient DNA-free genome editing of bread wheat using CRISPR/Cas9 ribonucleoprotein complexes. Nat. Commun. 8, 14261.

Ling, P., Duncan, L.W., Deng, Z., Dunn, D., Hu, X., Huang, S., et al., 2000. Inheritance of citrus nematode resistance and its linkage with molecular markers. Theor. Appl. Genet. 100 (7), 1010–1017.

Liu, Y., Heying, E., Tanumihardjo, S.A., 2012. History, global distribution, and nutritional importance of citrus fruits. Compr. Rev. Food Sci. Food Saf. 11 (6), 530–545.

Lörz, H., Baker, B., Schell, J., 1985. Gene transfer to cereal cells mediated by protoplast transformation. Mol. Gen. Genet. MGG 199 (2), 178–182.

Luro, F., Maddy, F., Jacquemond, C., Froelicher, Y., Morillon, R., Rist, D., et al. (Eds.), 2004. Identification and Evaluation of Diplogyny in Clementine (*Citrus clementina*) for Use in Breeding. International Society for Horticultural Science (ISHS), Leuven, Belgium.

Maggon, R., Deo, S.B., 1995. Promotion of adventitious bud regeneration by ABA in combination with BAP in epicotyl and hypocotyl explants of sweet orange (*Citrus sinensis* L. Osbeck). Sci. Hortic. 63 (1), 123–128.

Mankin, L.S., Thompson, W.F., 2001. New green fluorescent protein genes for plant transformation: intron-containing, ER-localized, and soluble-modified. Plant Mol. Biol. Report. 19 (1), 13–26.

Marin, F.R., Soler-Rivas, C., Benavente-García, O., Castillo, J., Pérez-Alvarez, J.A., 2007. By-products from different citrus processes as a source of customized functional fibres. Food Chem. 100 (2), 736–741.

Marraffini, L.A., 2013. CRISPR-Cas immunity against phages: its effects on the evolution and survival of bacterial pathogens. PLoS Pathog. 9 (12), e1003765.

Marutani-Hert, M., Evens, T.J., McCollum, G.T., Niedz, R.P., 2011. Bud emergence and shoot growth from mature citrus nodal stem segments. Plant Cell Tissue Organ Cult. 106 (1), 81–91.

Marutani-Hert, M., Bowman, K.D., McCollum, G.T., Mirkov, T.E., Evens, T.J., Niedz, R.P., 2012. A dark incubation period is important for *Agrobacterium*-mediated transformation of mature internode explants of sweet orange, grapefruit, citron, and a citrange rootstock. PLoS One 7 (10), e47426.

McCollum, T.G. (Ed.), 2007. Update of the USDA, ARS citrus scion improvement project. Proc. Fla. State Hort. Soc.120. pp. 285–287.

McGarry, R.C., Ayre, B.G., 2012. Geminivirus-mediated delivery of Florigen promotes determinate growth in aerial organs and uncouples flowering from photoperiod in cotton. PLoS One 7 (5), e36746.

Mendiburu, A.O., Peloquin, S.J., 1979. Gene-centromere mapping by 4x-2x matings in potatoes. Theor. Appl. Genet. 54 (4), 177–180.

Moore, G.A., Jacono, C.C., Neidigh, J.L., Lawrence, S.D., Cline, K., 1992. *Agrobacterium*-mediated transformation of *Citrus* stem segments and regeneration of transgenic plants. Plant Cell Rep. 11 (5), 238–242.

Moreira-Dias, J.M., Molina, R.V., BordÓN, Y., Guardiola, J.L., GarcÍA-Luis, A., 2000. Direct and indirect shoot organogenic pathways in epicotyl cuttings of Troyer Citrange differ in hormone requirements and in their response to light. Ann. Bot. 85 (1), 103–110.

Moreira-Dias, J.M., Molina, R.V., Guardiola, J.L., García-Luis, A., 2001. Daylength and photon flux density influence the growth regulator effects on morphogenesis in epicotyl segments of troyer citrange. Sci. Hortic. 87 (4), 275–290.

Moriguchi, T., Hidaka, T., Omura, M., Motomura, T., Akihama, T., 1996. Genotype and parental combination influence efficiency of cybrid induction in citrus by electrofusion. HortScience 31 (2), 275–278.

Murata M., 2018. Study of Grapefruit Cybrids With Improved *Citrus* Canker Resistance. PhD Dissertation, University of Florida, Gainesville, FL.

Navarro, L., Roistacher, C., Murashige, T., 1975. Improvement of shoot-tip graftingin vitro for virus-free citrus. J. Am. Soc. Hortic. Sci. 100 (5), 471–479.

Navarro, L., Aleza, P., Cuenca, J., Juárez, J., Pina, J.A., Ortega, C., et al., 2015. The mandarin triploid breeding program in Spain. Acta Hortic. 1065 (1), 389–396.

Niedz, R.P., Evens, T.J., 2011. The effects of benzyladenine and meta-Topolin on in vitro shoot regeneration of sweet orange. ARPN J. Agric. Biol. Sci. 6, 64–73.

Niedz, R.P., McKendree, W.L., Shatters, R.G., 2003. Electroporation of embryogenic protoplasts of sweet orange (*Citrus sinensis* (L.) osbeck) and regeneration of transformed plants. In Vitro Cell. Dev. Biol. Plant 39 (6), 586–594.

Oiyama, I., Kobayashi, S., 1993. Haploids obtained from diploid × triploid crosses of citrus. J. Jpn. Soc. Hort. Sci. 62 (1), 89–93.

Olivares-Fuster, O., Fleming, G.H., Albiach-Marti, M.R., Gowda, S., Dawson, W.O., Grosser, J.W., 2003. *Citrus tristeza virus* (CTV) resistance in transgenic citrus based on virus challenge of protoplasts. In Vitro Cell. Dev. Biol. Plant 39 (6), 567–572.

Oliveira, M.L.P., Costa, M.G.C., CVD, S., Otoni, W.C., 2010. Growth regulators, culture media and antibiotics in the in vitro shoot regeneration from mature tissue of citrus cultivars. Pesq. Agrop. Brasileira 45, 654–660.

Ollitrault P, Allent V, Luro F, 1996. Production of Haploid Plants and Embryogenic Calli of Clementine (# *Citrus* Reticulum# Blanco) After# In Situ# Parthenogenesis Induced by Irradiated Pollen, Published in AGRIS (FAO) 2019.

Omar, A.A., Grosser, J.W., 2008. Comparison of endoplasmic reticulum targeted and nontargeted cytoplasmic GFP as a selectable marker in citrus protoplast transformation. Plant Sci. 174 (2), 131–139.

Omar, A.A., Song, W.Y., Grosser, J.W., 2007. Introduction of Xa21, a Xanthomonas-resistance gene from rice, into Hamlin sweet orange [*Citrus sinensis* (L.) Osbeck] using protoplast-GFP co-transformation or single plasmid transformation. J. Hortic. Sci. Biotechnol. 82 (6), 914–923.

Omar, A.A., Dutt, M., Gmitter, F.G., Grosser, J.W., 2016. Somatic embryogenesis: still a relevant technique in citrus improvement. In: Germanà, M.A., Lambardi, M. (Eds.), In Vitro Embryogenesis in Higher Plants. Springer, New York, NY, pp. 289–327.

Omar, A.A., Murata, M., Yu, Q., Gmitter, F.G., Chase, C.D., Graham, J.H., et al., 2017. Production of three new grapefruit cybrids with potential for improved citrus canker resistance. In Vitro Cell. Dev. Biol. Plant 53 (3), 256–269.

Pacheco, C.A., Martelli, I.B., Polydoro, D.A., Schinor, E.H., Pio, R.M., Kupper, K.C., et al., 2012. Resistance and susceptibility of mandarins and their hybrids to Alternaria alternata. Sci. Agric. 69, 386–392.

Peña, L., Navarro, L., 2000. Transgenic citrus. In: YPS, B. (Ed.), Transgenic Trees. Springer, Berlin, Heidelberg, pp. 39–54.

Peña, L., Cervera, M., Juárez, J., Ortega, C., Pina, J., Durán-Vila, N., et al., 1995. High efficiency *Agrobacterium*-mediated transformation and regeneration of citrus. Plant Sci. 104 (2), 183–191.

Peña, L., Cervera, M., Juarez, J., Navarro, A., Pina, J.A., Duran-Vila, N., et al., 1995. *Agrobacterium*-mediated transformation of sweet orange and regeneration of transgenic plants. Plant Cell Rep. 14, 616.

Peña, L., Cervera, M., Juarez, J., Navarro, A., Pina, J.A., Navarro, L., 1997. Genetic transformation of lime (*Citrus aurantifolia* swing): factors affecting transformation and regeneration. Plant Cell Rep. 16, 731.

Peña, L., Perez, R.M., Cervera, M., JuÁRez, J.A., Navarro, L., 2004a. Early events in *Agrobacterium*-mediated genetic transformation of citrus explants. Ann. Bot. 94 (1), 67–74.

Peña, L., Cervera, M., Fagoaga, C., Pérez, R., Romero, J., Juárez, J., et al., 2004b. *Agrobacterium*-mediated transformation of citrus. In: Curtis, I.S. (Ed.), Transgenic Crops of the World: Essential Protocols. Springer, Dordrecht, pp. 145–156.

Peng, A., Chen, S., Lei, T., Xu, L., He, Y., Wu, L., et al., 2017. Engineering canker-resistant plants through CRISPR/Cas9-targeted editing of the susceptibility gene CsLOB1 promoter in citrus. Plant Biotechnol. J.

Peres, N.A., Timmer, L.W., 2006. Evaluation of the alter-rater model for spray timing for control of Alternaria brown spot on Murcott tangor in Brazil. Crop Prot. 25 (5), 454–460.

Perez, R.M., Galiana, A.M., Navarro, L., Duran-Vila, N., 1998. Embryogenesis in vitro of several *Citrus* species and cultivars. J. Hortic. Sci. Biotechnol. 73 (6), 419–429.

Pérez-Molphe-Balch, E., Ochoa-Alejo, N., 1997. In vitro plant regeneration of Mexican lime and mandarin by direct organogenesis. HortScience 32 (5), 931–934.

Pérez-Tornero, O., Tallón, C.I., Porras, I., 2010. An efficient protocol for micropropagation of lemon (*Citrus limon*) from mature nodal segments. Plant Cell Tissue Organ Cult. 100 (3), 263–271.

Ran, F.A., Cong, L., Yan, W.X., Scott, D.A., Gootenberg, J.S., Kriz, A.J., et al., 2015. In vivo genome editing using *Staphylococcus aureus* Cas9. Nature 520, 186.

Reforgiato Recupero, G., Russo, G., Recupero, S., 2005. New promising *Citrus* triploid hybrids selected from crosses between monoembryonic diploid female and tetraploid male parents. HortScience 40 (3), 516–520.

Reinisch, A.J., Dong, J.M., Brubaker, C.L., Stelly, D.M., Wendel, J.F., Paterson, A.H., 1994. A detailed RFLP map of cotton, *Gossypium hirsutum* x *Gossypium barbadense*: chromosome organization and evolution in a disomic polyploid genome. Genetics 138 (3), 829–847.

Rhodes, C., Pierce, D., Mettler, I., Mascarenhas, D., Detmer, J., 1988. Genetically transformed maize plants from protoplasts. Science 240 (4849), 204–207.

Rodríguez, A., Cervera, M., Peris, J.E., Peña, L., 2008. The same treatment for transgenic shoot regeneration elicits the opposite effect in mature explants from two closely related sweet orange (*Citrus sinensis* (L.) Osb.) genotypes. Plant Cell Tissue Organ Cult. 93 (1), 97–106.

Rouiss, H., Bakry, F., Froelicher, Y., Navarro, L., Aleza, P., Ollitrault, P., 2017a. Origin of C. latifolia and C. aurantiifolia triploid limes: the preferential disomic inheritance of doubled-diploid Mexican lime is consistent with an interploid hybridization hypothesis. Ann. Bot. 121 (3), 571–585.

Rouiss, H., Cuenca, J., Navarro, L., Ollitrault, P., Aleza, P., 2017b. Tetraploid citrus progenies arising from FDR and SDR unreduced pollen in 4x × 2x hybridizations. Tree Genet. Genomes 13 (1), 10.

Ruesink, A.W., 1971. [16] Protoplasts of plant cells. In: Methods in Enzymology. vol. 23. Academic Press, pp. 197–209.

Saito, W., Ohgawara, T., Shimizu, J., Ishii, S., Kobayashi, S., 1993. *Citrus* cybrid regeneration following cell fusion between nucellar cells and mesophyll cells. Plant Sci. 88 (2), 195–201.

Sangwan, R.S., Bourgeois, Y., Brown, S., Vasseur, G., Sangwan-Norreel, B., 1992. Characterization of competent cells and early events of *Agrobacterium*-mediated genetic transformation in *Arabidopsis thaliana*. Planta 188 (3), 439–456.

Satpute, A.D., Chen, C., Gmitter, F.G., Ling, P., Yu, Q., Grosser, M.R., et al., 2015. Cybridization of grapefruit with Dancy mandarin leads to improved fruit characteristics. J. Am. Soc. Hortic. Sci. 140 (5), 427–435.

Sauton, A., Mouras, A., Lutz, A., 1982. Plant regeneration from citrus root meristems. J. Hort. Sci. 57 (2), 227–231.

Savita, Bhagat, A., Pati, P.K., Virk, G.S., Nagpal, A., 2012. An efficient micropropagation protocol for *Citrus jambhiri* Lush and assessment of clonal fidelity employing anatomical studies and RAPD markers. In Vitro Cell. Dev. Biol. Plant 48 (5), 512–520.

Seguí-Simarro, J.M., 2010. Androgenesis revisited. Bot. Rev. 76 (3), 377–404.

Senthil-Kumar, M., Mysore, K.S., 2011. New dimensions for VIGS in plant functional genomics. Trends Plant Sci. 16 (12), 656–665.

Sharma, S., Prakash, A., Ajinath, T., 2009. In vitro propagation of citrus rootstocks. Not. Bot. Horti. Agrobot. Cluj Napoca 37 (1), 84.

Shillito, R.D., Saul, M.W., Paszkowski, J., Müller, M., Potrykus, I., 1985. High efficiency direct gene transfer to plants. Bio/Technology 3, 1099.

Shimamoto, K., Terada, R., Izawa, T., Fujimoto, H., 1989. Fertile transgenic rice plants regenerated from transformed protoplasts. Nature 338, 274.

Sim, G.E., Goh, C.J., Loh, C.S., 1989. Micropropagation of citrus mitis blanco-multiple bud formation from shoot and root explants in the presence of 6-benzylaminopurine. Plant Sci. 59 (2), 203–210.

Singh, S., Ray, B.K., Bhattacharyya, S., Deka, P.C., 1994. In vitro propagation of *Citrus reticulata* Blanco and *Citrus limon* Burm.f. HortScience 29 (3), 214–216.

Smýkal, P., 2000. Pollen embryogenesis—the stress mediated switch from gametophytic to sporophytic development. Current status and future prospects. Biol. Plant. 43 (4), 481–489.

Spiegel-Roy, P., Kochba, J. (Eds.), 1980. Embryogenesis in *Citrus* Tissue Cultures. Springer, Berlin, Heidelberg.

Svitashev, S., Schwartz, C., Lenderts, B., Young, J.K., Mark Cigan, A., 2016. Genome editing in maize directed by CRISPR-Cas9 ribonucleoprotein complexes. Nat. Commun. 7, 13274.

Takebe, I., Labib, G., Melchers, G., 1971. Regeneration of whole plants from isolated mesophyll protoplasts of tobacco. Naturwissenschaften 58 (6), 318–320.

Talon, M., Gmitter, F.G., 2008. *Citrus* genomics. Int. J. Plant Genomics 2008, 17.

Tavoletti, S., Bingham, E.T., Yandell, B.S., Veronesi, F., Osborn, T.C., 1996. Half tetrad analysis in alfalfa using multiple restriction fragment length polymorphism markers. Proc. Natl. Acad. Sci. 93 (20), 10918–10922.

TG, M.C. (Ed.), 2007. Update of the USDA, ARS citrus scion improvement project. Proc. Fla. State Hort. Soc. 120, 285–287.

Ting, S.V., 1980. Nutrients and nutrition of citrus fruits. In: *Citrus* Nutrition and Quality. ACS Symposium Series vol. 143. American Chemical Society, pp. 3–24.

Tomes, D.T., Ross, M.C., Songstad, D.D., 1995. Direct DNA transfer into intact plant cells via microprojectile bombardment. In: Gamborg, O.L., Phillips, G.C. (Eds.), Plant Cell, Tissue and Organ Culture: Fundamental Methods. Springer, Berlin, Heidelberg, pp. 197–213.

Toriyama, K., Arimoto, Y., Uchimiya, H., Hinata, K., 1988. Transgenic rice plants after direct gene transfer into protoplasts. Bio/Technology 6, 1072.

Van Le, B., Thanh Ha, N., Anh Hong, L.T., Trân Thanh Vân, K., 1999. High frequency shoot regeneration from trifoliate orange (*Poncirus trifoliata* L. Raf.) using the thin cell layer method. C. R. Acad. Sci. III 322 (12), 1105–1111.

Vardi, A., Spiegel-Roy, P., Galun, E., 1982. Plant regeneration from *Citrus* protoplasts: variability in methodological requirements among cultivars and species. Theor. Appl. Genet. 62 (2), 171–176.

Vardi, A., Bleichman, S., Aviv, D., 1990. Genetic transformation of *Citrus* protoplasts and regeneration of transgenic plants. Plant Sci. 69, 199.

Velázquez, K., Agüero, J., Vives, M.C., Aleza, P., Pina, J.A., Moreno, P., et al., 2016. Precocious flowering of juvenile citrus induced by a viral vector based on citrus leaf blotch virus: a new tool for genetics and breeding. Plant Biotechnol. J. 14 (10), 1976–1985.

Vicent, A., Armengol, J., García-Jiménez, J., 2007. Rain fastness and persistence of fungicides for control of Alternaria Brown spot of *Citrus*. Plant Dis. 91 (4), 393–399.

Vicent, A., Armengol, J., García-Jiménez, J., 2009. Protectant activity of reduced concentration copper sprays against alternaria brown spot on fortune mandarin fruit in Spain. Crop Prot. 28 (1), 1–6.

Vives, M., Moreno, P., Navarro, L., Guerri, J., Rao, G., Myrta, A., et al., 2008. *Citrus leaf blotch virus*. In: Characterization, Diagnosis & Management of Plant Viruses. Vol. 2: Horticultural Crops. CABI, pp. 55–67.

Wang, K., 2006. *Agrobacterium* Protocols. Springer.

Wang, M., van Bergen, S., Van Duijn, B., 2000. Insights into a key developmental switch and its importance for efficient plant breeding. Plant Physiol. 124 (2), 523–530.

Wang, S.-M., Lan, H., Cao, H.-B., Xu, Q., Chen, C.-L., Deng, X.-X., et al., 2015. Recovery and characterization of homozygous lines from two sweet orange cultivars via anther culture. Plant Cell Tissue Organ Cult. 123 (3), 633–644.

Wang, X., Xu, Y., Zhang, S., Cao, L., Huang, Y., Cheng, J., et al., 2017. Genomic analyses of primitive, wild and cultivated citrus provide insights into asexual reproduction. Nat. Genet. 49, 765.

Wolt, J.D., Wang, K., Yang, B., 2016. The regulatory status of genome-edited crops. Plant Biotechnol. J. 14 (2), 510–518.

Woo, J.W., Kim, J., Kwon, S.I., Corvalán, C., Cho, S.W., Kim, H., et al., 2015. DNA-free genome editing in plants with preassembled CRISPR-Cas9 ribonucleoproteins. Nat. Biotechnol. 33 (11), 1162–1164.

Wu, G.A., Terol, J., Ibanez, V., Lopez-Garcia, A., Estela, P.-R., Carles, B., et al., 2018. Genomics of the origin, evolution and domestication of citrus. Nature 544, 311–316. https://doi.org/10.1038/nature25447.

Wu, H., Acanda, Y., Shankar, A., Peeples, M., Hubbard, C., Zale, J., 2015. Genetic transformation of commercially important mature citrus scions. Crop Sci. 55 (6), 2786–2797.

Wu, H., Acanda, Y., Jia, H., Wang, N., Zale, J., 2016. Biolistic transformation of Carrizo citrange (*Citrus sinensis* Osb. ×*Poncirus trifoliata* L. Raf.). Plant Cell Rep. 35 (9), 1955–1962.

Wutscher, H.K., 1979. Citrus rootstocks. Hortic. Rev. 1, 237–269.

Xiao, J.P., Chen, L.G., Xie, M., Ye, W.Q., 2009. Identification of AFLP fragments linked to seedlessness in Ponkan mandarin (*Citrus reticulata* Blanco) and conversion to SCAR markers. Sci. Hortic. 121 (4), 505–510.

Xu, W.W., Sleper, D.A., Chao, S., 1995. Genome mapping of polyploid tall fescue (*Festuca arundinacea* Schreb.) with RFLP markers. Theor. Appl. Genet. 91 (6), 947–955.

Yahata, M., Nukaya, T., Sudo, M., Ohta, T., Yasuda, K., Inagaki, H., et al., 2015. Morphological characteristics of a doubled haploid line from Banpeiyu pummelo [*Citrus maxima* (Burm.) Merr.] and its reproductive function. Hortic. J. 84 (1), 30–36.

Yamagishi, N., Yoshikawa, N., 2011. Expression of FLOWERING LOCUS T from *Arabidopsis thaliana* induces precocious flowering in soybean irrespective of maturity group and stem growth habit. Planta 233 (3), 561–568.

Yamagishi, N., Li, C., Yoshikawa, N., 2016. Promotion of flowering by apple latent spherical virus vector and virus elimination at high temperature allow accelerated breeding of apple and pear. Front. Plant Sci. 7, 171.

Yamamoto, M., Matsumoto, R., Yamada, Y., 1995. Relationship between sterility and seedlessness in citrus. J. Jpn. Soc. Hort. Sci. 64 (1), 23–29.

Yang, Z.N., Ingelbrecht, I.L., Louzada, E., Skaria, M., Mirkov, T.E., 2000. *Agrobacterium*-mediated transformation of the commercially important grapefruit cultivar rio red (*Citrus paradisi* Macf.). Plant Cell Rep. 19 (12), 1203–1211.

Yao, J.-L., Wu, J.-H., Gleave, A.P., Morris, B.A.M., 1996. Transformation of citrus embryogenic cells using particle bombardment and production of transgenic embryos. Plant Sci. 113 (2), 175–183.

Yu, C., Huang, S., Chen, C., Deng, Z., Ling, P., Gmitter, F.G., 2002. Factors affecting *Agrobacterium*-mediated transformation and regeneration of sweet orange and citrange. Plant Cell Tissue Organ Cult. 71 (2), 147–155.

Zhang, W., Wu, R., 1988. Efficient regeneration of transgenic plants from rice protoplasts and correctly regulated expression of the foreign gene in the plants. Theor. Appl. Genet. 76 (6), 835–840.

Zhang, Y., Liang, Z., Zong, Y., Wang, Y., Liu, J., Chen, K., et al., 2016. Efficient and transgene-free genome editing in wheat through transient expression of CRISPR/Cas9 DNA or RNA. Nat. Commun. 7, 12617.

Zhang, F., LeBlanc, C., Irish, V.F., Jacob, Y., 2017. Rapid and efficient CRISPR/Cas9 gene editing in *Citrus* using the YAO promoter. Plant Cell Rep. 36, 1883–1887.

Chapter 10

Vegetative growth

Eduardo Primo-Millo[a], Manuel Agustí[b]
[a]Centro de Citricultura y Producción Vegetal, Instituto Valenciano de Investigaciones agrarias (IVIA), Valencia, Spain, [b]Mediterranean Agroforestry Institute, Polytechnic University of Valencia, Valencia, Spain

10.1 Seed germination

Seed germination includes a series of physiological processes, which begin with the imbibition of water and culminate with the emergence of the seedling through the seed coats.

The absorption of water by the seed triggers a sequence of metabolic changes, which include the activation of the respiratory process, the protein synthesis, and the mobilization of the reserves. Subsequently, cell division and elongation in the radicle causes the rupture of the seed coats where the emergence of the seedling will occur.

Therefore, it is necessary to provide some external circumstances for germination such as:

- a moist substrate,
- enough oxygen to allow aerobic respiration,
- a suitable temperature to activate the embryo's metabolism and the development of the seedling.

Nevertheless, in contrast to other species, *Citrus* seeds do not require environmental stimuli, such as low temperatures or certain lighting conditions, to break dormancy.

During germination, the following events occur:

- Imbibition of water
- Increase of respiration
- Protein synthesis
- Mobilization of reserves
- Seedling development

10.1.1 Imbibition of water

Water absorption by the seed is the first and indispensable step for germination.

In a moist soil or substrate, the hydration capacity of the seed at the beginning of germination depends largely on its water potential (ψ_h), that is, that of the cotyledon cells. Water potential has three components:

$$\psi_h = (-)\psi_o + (-)\psi_m + \psi_t$$

where

ψ_o = osmotic potential.

ψ_m = matric potential.

ψ_t = turgor potential.

The values of the first two components are negative while that of the third is positive. In the dry seed, the sum of the three terms is negative, mainly due to the matric component. Nevertheless, the water is attached to the soil or substrate where the seed germinates with a certain force and, therefore, the gradient between the water potential of the seed and that of the substrate is a factor on which largely depends the intensity of the water flow to the seed.

At the beginning of imbibition, the difference between the water potentials of the seed and that of a well humid soil or substrate is usually very high, producing a strong water flow to the seed and causing a volume increase. However, as the humidity of the seed increases, its water potential increases (becomes less negative), while that of the surrounding

The Genus Citrus. https://doi.org/10.1016/B978-0-12-812163-4.00010-3

environment usually decreases. This means that both potentials tend to equalize and, consequently, water flow to the seed, that is, imbibition, decreases. This flow (F_i) can be expressed according to the following function:

$$F_i = \frac{\psi_e - \psi_i}{c_e + c_i} \cdot f_{sc}$$

where

ψ_e = water potential of the soil or substrate.
ψ_i = water potential of the seed.
c_e = hydraulic conductivity of the soil or substrate.
c_i = hydraulic conductivity of the seed.
f_{sc} = factor dependent on the surface of contact between soil or substrate particles and the seed.

When seeds germinate in optimal conditions, seeds show a water absorption system in three phases:

- The first one is of rapid absorption and occurs mainly as a consequence of the matric forces, due to the hydration capacity of the macromolecules of the walls and the interior of the cells of the seed.
- The second phase constitutes a much slower period of water absorption, as a consequence of the initial increase of the water potential in the seed cells and the resistance to the water imbibition due to the seed coats.
- The third phase, in which the absorption of water is reactivated, is associated with the emergence of the radicle, which causes the rupture of the coats and facilitates the water flow to the embryos. This flow is associated with a negative increase of the osmotic potential of the cotyledons produced by the solutes generated by the hydrolysis of the reserves. Another component of this phase is the absorption of water by the developing seedling.

The duration of each one of these phases depends on certain inherent properties of the seeds, the content in hydratable compounds, and the permeability of the coats to water and oxygen among them. These phases are also affected by environmental conditions, such as the relative humidity, the characteristics and composition of the substrate, temperature, etc. Obviously, the removal of the seed coats accelerates the process of water imbibition.

Another interesting aspect is the relationship of these phases with the metabolism of the seed. The first phase is independent of the metabolic activity; so, it is produced in a similar way in viable and nonviable seeds, without being affected by temperature. In the second phase, hydration activates the metabolism prior to germination. The third phase occurs only in viable seeds and is associated with a strong metabolic activity that includes the mobilization of reserves and the start of seedling growth. Therefore, this last phase is activated by temperature.

10.1.2 Breathing

The seeds require that, during germination, the medium be sufficiently aerated, so that the availability of oxygen must be guaranteed.

The "dry" seed shows a little respiratory activity, which is stimulated when imbibition of water begins. From then, respiration of the seeds can be divided into four phases:

- Phase I: It is characterized by rapid increase in respiration, which generally occurs 12 h after the beginning of imbibition. This increase in respiratory activity is proportional to the increase in the hydration of the seed tissues and it is probably due, in part, to the activation of the enzymes of the respiratory pathways, which occurs with the entry of water into the cells. Since the content of sucrose or other soluble sugars is very low in citrus seeds, the substrates used in this phase are possibly keto acids such as α-ketoglutarate and pyruvate.
- Phase II: In this phase, respiration stabilizes between 12 and 24 h after the beginning of the imbibition. In this phase, as in the previous one, probably, the coats of the seeds, which still remain intact, limit the entrance of oxygen, producing a certain degree of anaerobic respiration, with the activation of the enzyme alcohol dehydrogenase, to form ethanol from acetaldehyde. The elimination of the testa can shorten or cancel this phase. However, this stationary respiratory phase may also reflect the time necessary for the development of a more efficient mitochondrial respiratory system, replacing the one that operated when the germination starts.
- Phase III: There is a second increase in respiratory activity, which is associated with a greater availability of oxygen as a result of the rupture of the testa produced by the emergence of the radicle. Another factor that may contribute to the increase in respiration may be the activity of newly formed mitochondria in the dividing cells of the embryonic axis. Obviously, aerobic respiration predominates in this phase.
- Phase IV: In this last period, after the disintegration of the tissues of the seed (coats, cotyledons, etc.), only the own respiration of the developing seedling remains active.

10.1.3 Protein synthesis

At the beginning of seed germination, an activation of the protein synthesis takes place. It gives rise, among other proteins, to the biosynthesis of hydrolytic enzymes that produce the mobilization of the reserves. Thus, protein synthesis, which is practically nil in dry seeds, begins when the cells are hydrated enough.

The dry cotyledons store important quantities of mRNA, whose function could be the synthesis of enzymes in the first stage of germination. The mRNA accumulated in the dry seed is the result of the transcription that takes place during the maturation of the seed. It is thought that this mRNA is activated during hydration, directing the synthesis of proteins with important specific functions during the initial period of imbibition. However, it is likely that the function of this mRNA will be restricted to the needs of a short period. Subsequently, during the first days of germination, there is a strong increase in mRNA in embryos. This means that, regardless of whether they contain stored mRNA, at the beginning of germination a new transcription is initiated by which new mRNAs are synthesized. In addition, in this initial phase, numerous ribosomes appear in cotyledon cells (García-Agustín and Primo-Millo, 1989), which form polysomes with RNA. All this indicates an active synthesis of the proteins that are required in the early stages of germination.

The protein synthesis focuses mainly on two functions: the development of the whole enzymatic and metabolic system for the mobilization of the reserves contained in the cotyledons, and the development and metabolism of the embryonic axis.

10.1.4 Mobilization of the reserves contained in cotyledons

10.1.4.1 Reserve compounds

The seed contains food reserves to maintain the growth and development of the seedling until it is able to feed itself. In the case of *Citrus*, these reserves are stored in the cotyledons.

The compounds accumulated in these organs are mostly lipids, which account for >50% of their dry weight, and also proteins, which account for about 20% (García-Agustín and Primo-Millo, 1989). The concentration of carbohydrates is very low—starch represents only 0.04% of dry weight. The cotyledons also contain mineral elements, forming part of organic and inorganic compounds, providing essential nutrients for the seedling, before the radicle is able to absorb them by itself from the external solution. Table 10.1 shows, as an example, the composition of cotyledons of *Citrus limon* seeds.

TABLE 10.1 Composition of cotyledons of *C. limon*.

Organic compounds	% DW
Lipids	51.7
Proteins	20.8
Free amino acids	2.3
Starch	0.04
Soluble sugars	0.08
Mineral elements	**% DW**
Nitrogen	4.8
Phosphorus	0.5
Potassium	1.1
Magnesium	0.2
Calcium	0.3
Minor elements	**ppm**
Iron	50.6
Zinc	27.2
Manganese	5.8
Copper	7.7

Lipids constitute a group of chemically heterogeneous substances that are soluble in organic solvents such as petroleum ether, hexane, or chloroform. The reserve lipids of the seeds are esters of glycerol and long-chain monocarboxylic acids that form triglycerides. Among them, it can be found saturated or unsaturated fatty acids. Saturated fatty acids have an even number of carbon atoms, between 4 and 24. The most abundant of them is palmitic acid ($n=14$). However, the predominant fatty acids are unsaturated ones, especially oleic ($n=18{:}1$) and linoleic ($n=18{:}2$), which account for about 60% of the total of the oil of the seed.

The main reserve proteins are globulins (soluble in saline solutions), which constitute approximately 26% of total proteins, and in a smaller proportion albumins (soluble in water) with about 8% (García-Agustín and Primo-Millo, 1989). Within the former, the main component (called citrine) is dissociated into two subunits with molecular weights close to 22 and 33 kDa (García-Agustín and Primo-Millo, 1990; Koltunow et al., 1996). This last polypeptide accounts for about 28% of the total globulins.

The globulins contain relatively high proportions of glutamine, asparagine, arginine, and lysine, which represent >50% of their weight, and show a low content of cysteine and methionine. In the fraction of albumins a component of, approximately, 12 kDa stands out, which represents about 30% of the said fraction. In this fraction, the major amino acids are glutamine, lysine, and arginine (García-Agustín and Primo-Millo, 1990).

The cotyledons of the latent seed present a relatively low content in total free amino acids (2.3% with respect to the dry weight), the majority being proline, glutamic acid, arginine, and asparagine (García-Agustín and Primo-Millo, 1990).

10.1.4.2 *Ultrastructural location of reserves*

The reserves stored in the cotyledons are mainly forming intracellular bodies that contain lipids and proteins.

The ultrastructural study of cotyledon cells (García-Agustín and Primo-Millo, 1989) shows that they contain abundant spheroidal or ovoid bodies, which constitute vesicles where reserve triglycerides are deposited. These bodies have a size between 0.6 and 1.6 μm in diameter and occupy a large part of the interior space of the cell. Although it has not been demonstrated that a membrane is found on the surface of these bodies, it is likely that they are limited by a membranous subunit, in which certain proteins (oleosins) would be embedded in a monolayer of peripheral phospholipids.

The reserve proteins are also located in well-defined bodies of oval or circular section of 1.0 and 2.25 μm in diameter. These protein bodies are surrounded by a unitary lipid-protein membrane and inclusions with crystalloid or globoid forms are frequently observed in them. The former are due to singular proteins and the latter have been described as sites of deposition of organic salts.

In the dormant seeds, no starch grains are seen in the cells of the cotyledons. Their cytoplasm is confined to a narrow band next to the cell wall, the rest of the volume being full of lipid and protein bodies. In this state, there are hardly any organelles in the cytoplasm except scattered ribosomes and some small vesicles of the endoplasmic reticulum (Plate 10.1A).

10.1.4.3 *Transformation of reserve compounds*

The mobilization of reserves requires a previous process of hydrolysis to release compounds of lower molecular weight, which can be used during the initial growth of the seedling. In many cases, the products of hydrolysis undergo a series of metabolic transformations before being transported to the embryonic axis in development.

10.1.4.3.1 Metabolization of cotyledon lipids during germination

Three types of organelles are involved in the mobilization and metabolism of the lipid reserves of the seeds: the oil vesicles (lipid bodies), the glyoxysomes, and the mitochondria. The process occurs in four phases:

I. Lipolysis to yield fatty acids and glycerol. It is produced in the lipid bodies by the action of lipases.
II. Oxidation of fatty acids to acetyl CoA and subsequent formation of succinate in glyoxysomes.
III. Conversion of succinate to oxaloacetate in the mitochondria.
IV. Formation of sucrose from oxaloacetate in the cytoplasm.

In cotyledons of citrus seeds, lipid degradation begins immediately after imbibition, although it accelerates during the second and third week of germination, when a strong lipolysis occurs, and when about 58% of these compounds are metabolized. After this period, the process slows down (Fig. 10.1A).

The initial hydrolysis of the lipids is due to the activity of the lipases (Fig. 10.1B). Two different lipases are found in citrus cotyledons, one with an optimal activity at pH 5 (acid lipase) and another one with an optimum pH between 7.5 and 8 (alkaline lipase). Acid lipase activity has been detected mainly in the supernatant fat layer obtained from crude extracts of cotyledons as in other seeds; it is assumed that it can be located in the membrane of the lipid vesicles. The alkaline lipase is associated with the membrane of glyoxysomes.

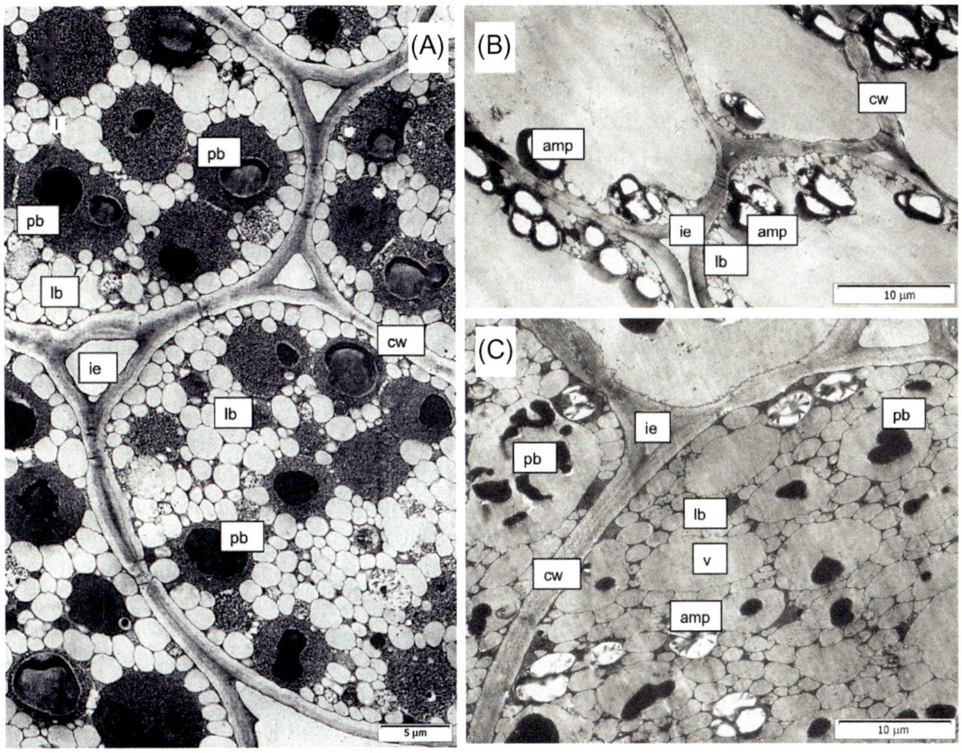

PLATE 10.1 (A) Ultrastructure of the cells from the reserve tissue of cotyledons before the onset of seed germination. Cell wall (cw), intercellular space (ie), protein body (pb), lipid body (lb). Cotyledon cells at 1 (B) and 3 (C) weeks after the beginning of the seed germination process. Cell wall (cw), intercellular space (ie), protein body (pb), lipid body (lb), vacuole (v), and amyloplast (amp).

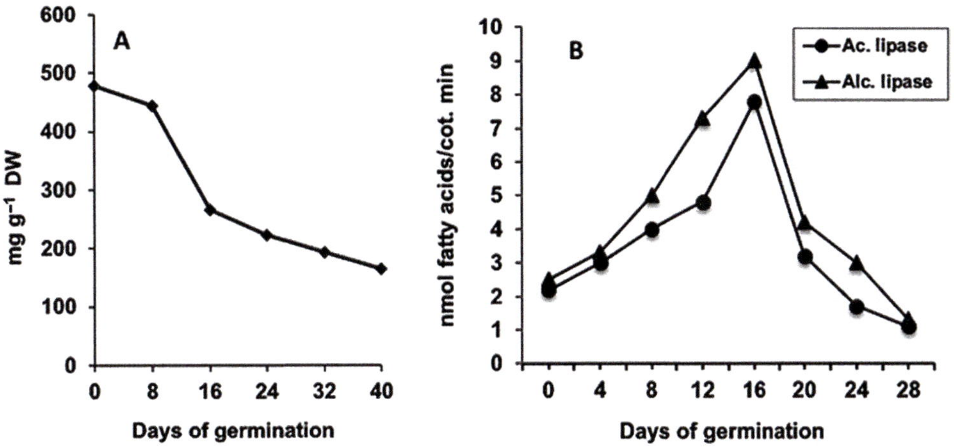

FIG. 10.1 Changes in the concentration of lipids in the cotyledons (A), and time course of lipase activity (acid, Ac. and alkaline, Alc.) during seed germination (B).

The activity of both enzymes shows a pattern of parallel evolution during germination, so that in both cases it is low in dry cotyledons and increases from the beginning of imbibition until reaching a maximum at 15–20 days of germination. The development of this activity correlates inversely with the decrease in lipid content (García-Agustín et al., 1992).

Regarding the control of the mobilization of triglycerides in the cotyledons of citrus fruits, it is observed that lipolysis is markedly superior in cotyledons attached to the embryonic axis than in those separated from it. This indicates that the embryonic axis partially controls the mobilization rate of the reserve triglycerides. This process could be regulated by isocitrate lyase activity, which is a key enzyme in the glyoxylate cycle, and whose activity increases markedly in the cotyledons during germination. The elimination of the embryonic axis partially inhibits its activity in the cotyledons, which supports its function in the regulation of lipolysis (García-Agustín et al., 1992).

10.1.4.3.2 Mobilization of cotyledon proteins during germination

The mobilization of protein reserves of cotyledons begins during the first week from the onset of imbibition, and before the initial growth of the radicle is detected. From then, the protein content of these organs decreases, especially in the period between 8 and 20 days, when their concentration is reduced by half, approximately (Fig. 10.2). In this period, the main component of the globulin reserve (33 kDa) practically disappears. Later on, the degradation of proteins is much slower (García-Agustín and Primo-Millo, 1989).

The hydrolysis of the reserve proteins is catalyzed by different types of proteolytic enzymes that decompose them into peptides of lower molecular weight and amino acids. These enzymes can be classified into two groups: endopeptidases and exopeptidases. The endopeptidases break the internal bonds of the polypeptide chains, giving rise to smaller peptides. The exopeptidases break the polypeptide chain sequentially by the terminal amino acid, thus giving rise to free amino acids. This last group of enzymes is further subdivided into aminopeptidases, which break the polypeptide chain by the free amino terminus, and carboxypeptidases, which do so at the carboxyl terminus. From the imbibition, there is a marked increase in the activity of enzymes belonging to these groups.

In parallel, during the course of germination, the hydrolysis of the reserve proteins produces an increase in the concentration of free amino acids in the cotyledons (Fig. 10.2). A small proportion of these amino acids could be reused for protein synthesis or to provide energy by oxidation of their carbon skeleton, after their deamination. The ammonium produced in the deamination is fixed in the glutamic and aspartic acids to produce glutamine and asparagine, preventing the accumulation of this ion at toxic levels. However, most of the free amino acids are translocated via phloem to the growing seedling. Obviously, the highest rate of accumulation of amino acids in the cotyledons coincides with the highest protein hydrolysis, which occurs during the first 20 days of germination.

The displacement of the amino acids from the cotyledons toward the axis begins around 8–10 days from the start of germination, just when the radicle begins to develop. This process can continue for several weeks, until the disintegration of cotyledons occurs, and is accompanied by an increase in the content of free amino acids and proteins in the seedling. The amino acid that reaches a higher concentration in the cotyledon during proteolysis is asparagine, which amounts to about 60% of the total free amino acids at 40 days of germination. This increase is also reflected in the tissues of the seedling where asparagine constitutes >60% of the total free amino acids. All this clearly indicates that it is asparagine through which the hydrolyzate of the reserve proteins is mainly transported.

When cotyledons are separated from the embryonic axis, the decrease in protein content is more pronounced than in intact cotyledons, producing in the former an accumulation of N α-amino. This last effect is due not only to a higher speed in the degradation of the proteins in the separated cotyledons, but also to the impossibility for the free amino acids to be translocated to the seedling (García-Agustín et al., 1991).

In separate cotyledons, the activity of the exopeptidases (carboxy and amino) increases strongly with respect to that detected in bound cotyledons. This effect has been related to a possible acceleration of the senescence process in the isolated cotyledons (García-Agustín et al., 1991) and discards the possibility that the mobilization of the protein reserves is stimulated by the embryonic axis.

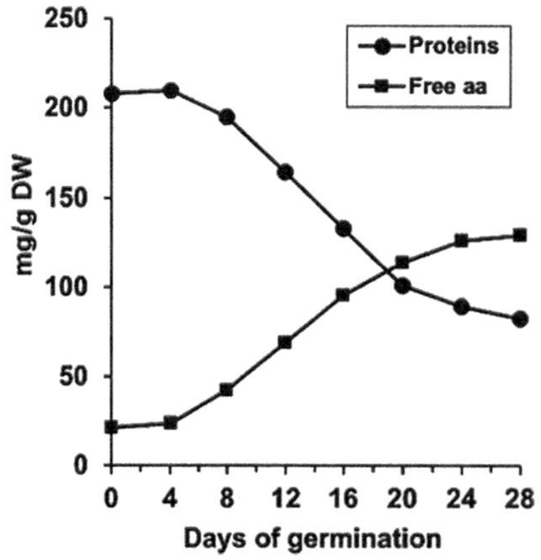

FIG. 10.2 Changes in the concentration of proteins and free amino acids in citrus cotyledons during seed germination.

10.1.4.3.3 Changes in the hydrocarbon fractions of cotyledons during germination

The concentration of total soluble sugars increases in the cotyledons during the first 2 weeks of germination. During this period, the concentration of sucrose shows a rapid initial increase, which is followed by glucose and fructose although in a lower extent. After reaching their maximum values, the concentrations of the soluble sugars decrease, probably as a consequence of their translocation to the embryonic axis in development. The starch content also increases strongly in the first 2 weeks of germination and then remains constant until at least 40 days (Fig. 10.3). The excess carbon produced in the metabolism of lipids is used for the synthesis of soluble sugars and starch (García-Agustín and Primo-Millo, 1989).

10.1.4.4 Ultrastructural changes in the cotyledons during germination

The progressive digestion of the reserve lipids contained in the vesicles causes a gradual decrease in their size until exhaustion. Six days after imbibition, some vacuoles are observed in the volume occupied by the lipid bodies consumed. As germination progresses, the vacuoles become larger and fuse, and 20 days after the imbibition a large vacuole is formed in a central position, the remaining lipid bodies being then located in the vicinity of the cell wall. After the total hydrolysis of the triglyceride reserves, the vacuole almost completely occupies the cytoplasm (Plate 10.1B).

Degradation of protein bodies begins very soon after imbibition occurs. After 6 days of germination, in some cells the disappearance of at least 50% of these bodies is already observed. After 14 days of germination, >75% of the protein bodies have disappeared and at 20 days only very few of them are detectable. The digestion of the protein bodies usually begins in the inner cells of the cotyledon and gradually extends toward the periphery. As the protein bodies disappear in the cells, new vacuoles are formed from the resulting vesicles (Plate 10.1C).

As germination progresses, grains of starch begin to appear, increasing in size and number over time. The maximum density of starch grains is detected between the second and third week of germination, although later on they tend to decrease as the cells are vacuolated. In advanced stages of germination (40 days from imbibition), the cells of the outer layers of the cotyledons may present chloroplasts with starch grains and a system of thylakoids.

Other subcellular organelles appear better defined after imbibition of water. Thus, from the 6 days of germination the following changes are observed:

- Larger endoplasmic reticulum cisterns in the cytoplasm.
- Increase the number of ribosomes, many of which appear grouped (possibly forming polysomes).
- Mitochondria develop structurally, presenting numerous crests.
- Glyoxysomes closely match oil vesicles.

All these changes are maintained in the cells of the cotyledons until the senescence, when almost all cells are vacuolated (approximately after 40 days of germination) (García-Agustín and Primo-Millo, 1989).

10.1.5 Seedling development

When the seed is buried in a moist soil or substrate, it quickly absorbs water and the cotyledons soften. Subsequently, if the temperature is adequate, the radicle begins to grow and its apex emerges by breaking the micropylar end of the testa.

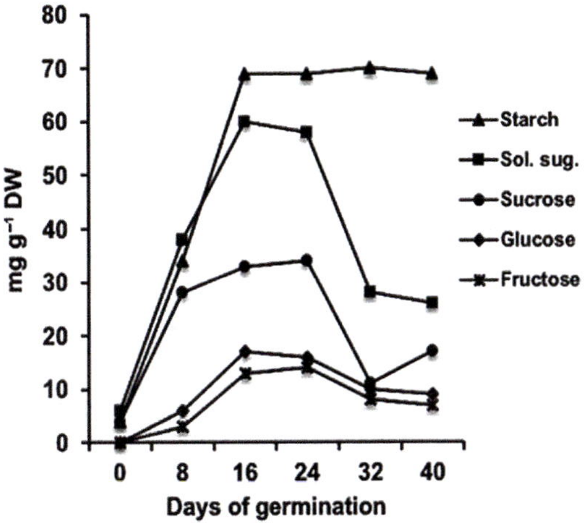

FIG. 10.3 Changes in the concentration of soluble sugars and starch in the cotyledons of citrus during seed germination.

PLATE 10.2 Early stages of seedling development (A). Embryos extracted from a monoembryonic seed (B), and from a polyembryonic seed (C). Morphological factors associated with juvenility: fragment of a shoot from a juvenile plant (D) with strong spinosity, and from an aged plant (E) with no thorns.

Sometimes, this coat can offer resistance to the root apex emergence in the area of the micropyle, and it bends. On the other hand, it is frequent, especially in polyembryonic seeds, that the position of the embryo forces the radicle to twist toward the soil. Afterward, the radicle continues to grow rapidly downward, forming a relatively thick primary root. Meanwhile, the plumule, which appears in a rudimentary state in the dormant seed, lengthens its first internode (epicotyl) and begins to develop its first two true leaves (Plate 10.2A). Cotyledons are usually hypogeous, that is, they remain in the soil during germination. When the root reaches a considerable development, the plumule emerges by tearing the testa, subsequently growing upward. It is usual that its apex remains bent while pushing through the ground, but, as it emerges, it straightens up quickly. Then, the first two true leaves, which remain yellow below the surface of the soil, take on green color as they emerge and expand occupying opposite positions on both sides of the stem. These leaves are different in shape from the leaves generated later.

The secondary roots begin to appear when the primary root has reached a size close to 8–10cm and the first leaves are close to reaching their final size. Then, it could be considered that the germination has been completed. The pattern of increase in dry weight of the seedling during the beginning of its development and that of the loss of the cotyledons during this period is shown in Fig. 10.4.

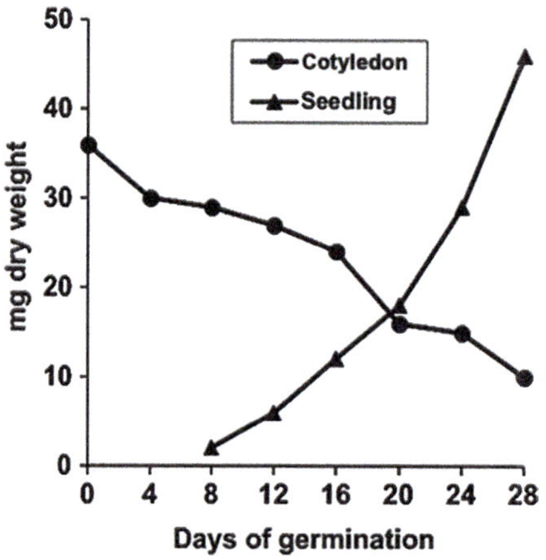

FIG. 10.4 Evolution of the weight of the cotyledons and seedling during seed germination.

10.1.6 Polyembryony

Citrus fruit is distinguished from other fruits because the seeds of many of their species and varieties are polyembryonic. This characteristic allows the cells of the nucellus to generate somatic embryos (Plate 10.2B and C).

Not all the embryos germinate in the polyembryonic seeds, since the smallest ones although swell and, in some cases, emit a small root, they cannot continue the development. In general, only the largest embryos (>5 mm) are able to evolve completely. Nevertheless, despite the development of several embryos, most of seeds only produce a seedling, sometimes two or three.

The proportion of zygotic and nucellar embryos that germinate is variable, depending on the variety and how they develop. However, it seems that the smaller the number of embryos per seed, the greater the probability that hybrid seedlings will develop and vice versa.

Under a reproductive point of view, polyembryony is of great importance, since nucellar embryos have a somatic origin and, therefore, their seedlings have the same genetic constitution as that of the tree from which the seeds come, except if a mutation takes place. The asexual propagation, made possible by nucellar embryony, has had a great impact on the evolution, improvement, and cultivation of *Citrus* species, since it guarantees a broad homogeneity in offspring. Breeding by polyembryonic seeds is widely used in the rootstocks propagation, but it is not suitable for varieties for three reasons: (a) the plants from embryos have juvenile characteristics, for which they take a long time to produce; (b) most commercial varieties do not produce seeds; and (c) in any case, it is necessary to graft a variety onto a rootstock for a better resistance of the soil adverse conditions.

The sexual embryos give rise to a wide variability in the offspring, and, therefore, the seeds of monoembryonic varieties, or with low polyembryony, should not be used in commercial propagation.

10.1.7 Juvenile characters

Plants arising directly from seeds, either sexually (zygotic embryo) or asexually (nucellar embryos), show a series of typical traits, called juvenile characters, during the first years of life.

Among them, the most notable is, possibly, the presence of large thorns on the trunk, branches, and twigs (Plate 10.2D and E), which are stronger in shoots with greater development, especially in vigorous genotypes.

Trees that come directly from a seed, as well as those propagated from them by means of vegetative propagation, tend to reduce the number and size of thorns in new shoots as they age. This effect is especially pronounced in the most external shoots, where they sometimes disappear, while the trunk and the lower parts of the main branches conserve the ability to produce thorny shoots for many years. Therefore, the aged varieties, in which a long time has elapsed since their last reproduction by seed, usually have few thorns, small and weak.

When buds of a juvenile plant are grafted, shoots also present large thorns that persist for a long time in the successive flushes; however, when the grafted bud comes from an aged tree, the thorns usually reappear in the first flush, but they diminish rapidly in the following ones.

The number of thorns and their size varies with the species and variety, as well as with the vigor and position of the shoot. According to the origin of the buds, outbreaks with or without thorns, the trees from grafted buds will also be thorny or not; the selection of buds in this aspect has allowed, after successive grafts, the generation of varieties or lines without thorns.

Within the same variety, the juvenile trees are more vigorous than the old ones and, likewise, the growth tends to be more erect in the former than in the latter.

Juvenility is also characterized by the inability of the trees to flower and, in the case of producing some flowers, to fruit set. The poor flowering associated with *Citrus* juvenility has been positively correlated with the expression of a homolog of the *TFL1* gene (*CsTFL*), which produces the delay of flowering (Pillitteri et al., 2004a); in contrast, this characteristic has also been related to the lack of activity of the homologs of the inducing genes of flowering *LEAFY* (*CsLFY*) and *APETALA1* (*CsAP1*) (Pillitteri et al., 2004b). In this sense, the constitutive expression of the *LEAFY* and *APETALA1* genes reduces the nonreproductive period of *Citrus* (Peña et al., 2001).

Fruits from juvenile trees are larger, and their peel thicker and rougher, than those of the aged ones. Besides, they produce fewer seeds, if any, than adult trees.

Obviously, juvenile characters are lost overtime, although in some species persist for a long time. For example, Satsuma mandarin loses them in about 5–7 years, while some sweet orange cultivars take between 15 and 20 years.

10.2 Dormancy and vegetative activity

Citrus, like other woody species, do not develop continuously, but throughout the year, alternate periods of growth with others of vegetative inactivity. These changes are due to seasonal climatic variations.

When the weather is unfavorable, the plants enter a dormant period; that is, they reduce their vegetative activity, as a defense mechanism in the face of circumstances that could put their survival at risk. When the environmental conditions return to be favorable, normal development is recovered. This depends mainly on the temperature and soil humidity, so that when the daily average values of the first parameter are lower than 12°C or the availability of water is insufficient to cover the needs of the plant, the tree stops the development.

In the Mediterranean area, under irrigation conditions, the plants enter into dormancy as temperatures drop, which usually happens from mid-autumn. In late winter or early spring, when temperature rises again, buds sprout, and the plant recovers the vegetative activity. Therefore, weather conditions can advance or delay the bud to enter and/or exit from dormancy. During the rest period, vegetative organs, especially roots and branches, are more resistant to low temperatures, being able to withstand values below −10°C for many hours without damage. The most tender twigs and leaves, especially those of the last flushes, can freeze at somewhat higher temperatures than those referred, increasing their sensitivity when the plant is active. Mature fruit is very sensitive to frost, which can cause serious damage to the crop, regardless of the state of the plant.

In tropical areas, where the temperature remains high throughout the year, *Citrus* stop the vegetative activity due to lack of water during the dry season, unless irrigated.

The intensity of the dormancy varies from one species to another and is manifested in the lower or greater ease to recover the activity after a short warm period during the winter. For example, *Poncirus trifoliata* has a deep dormancy, while lemon, lime, and citron have a mild one. Orange, mandarin, and grapefruit are in an intermediate position.

A deep dormancy for a long time usually results in abundant flowering, sometimes to the detriment of spring vegetative growth.

Regarding root activity, during dormancy, soil heating promotes root development by increasing the absorption of water and mineral nutrients that the plant needs to start the vegetative development.

10.3 Development of the canopy

10.3.1 Stem growth

Plants from seed usually have a single trunk with an approximately circular section. The main stem, when it is very young, is green, which soon changes to grayish brown or brown, due to the formation of suber in the bark. This presents a smooth surface, although with small longitudinal fissures.

When reaching a certain height, usually 0.75 and 1 m, the trunk emits lateral branches, which will constitute the canopy skeleton. From these arise secondary branches, which, in turn, give rise to the tertiary ones, and so on. The shoots that sprout from branches of higher rank do so forming angles whose amplitude depends, in a large extent, on the species or variety. Thus, when the angle of insertion of the shoots in the trunk or branches tends to be acute, the tree will have an erect appearance, whereas if it is more open, the tree will be wider. Other characters that contribute to the general appearance of the trees are the density of the branches and twigs or the relative development of the inner and outer branches. Height, distribution, density, and orientation of the branches in the trunk can be modified by pruning; however, usually the tree tends to maintain its characteristic appearance.

The vertical branches have a round section, while that of the horizontal and inclined branches may appear somewhat crushed on the sides.

The stems of very young shoots are tender and have an angular shape, due to the thickened area that appears under the point of insertion of the leaves. This thickening makes the cross section of the young stems triangular. However, with secondary growth, the section of stems becomes rounded. The young stem is pale green in color, which changes to dark green after two or three subsequent flushes. This new color is maintained for some time, until it turns grayish brown or brown color with growth.

When the tree becomes adult, the base of the trunk may have a grooved appearance, because it thickens above the insertion zone of the larger lateral roots.

The tissues and primary organs of the stem come from the apical meristem, which is a primary meristem from the embryo. This meristem is located at the upper end of the shoot and has the shape of a dome; on its sides, the leaf primordia that cover the meristem protrude, forming the caulinar apex (Plate 10.3A). Apical meristems are formed by small isodiametric cells that maintain the embryonic characteristics. These organs contain cells called initial or mother cells that are continuously divided during the period of vegetative activity. After each division, one of the descendant cells does not differentiate and maintains the ability to divide indefinitely, which characterizes the initial cells. The other cell, called derivative, undergoes later divisions and, as it separates from the apex, grows, and differentiates according to a specific pattern, becoming part of the corresponding tissue. Consequently, the former remains as an initial cell which continuously

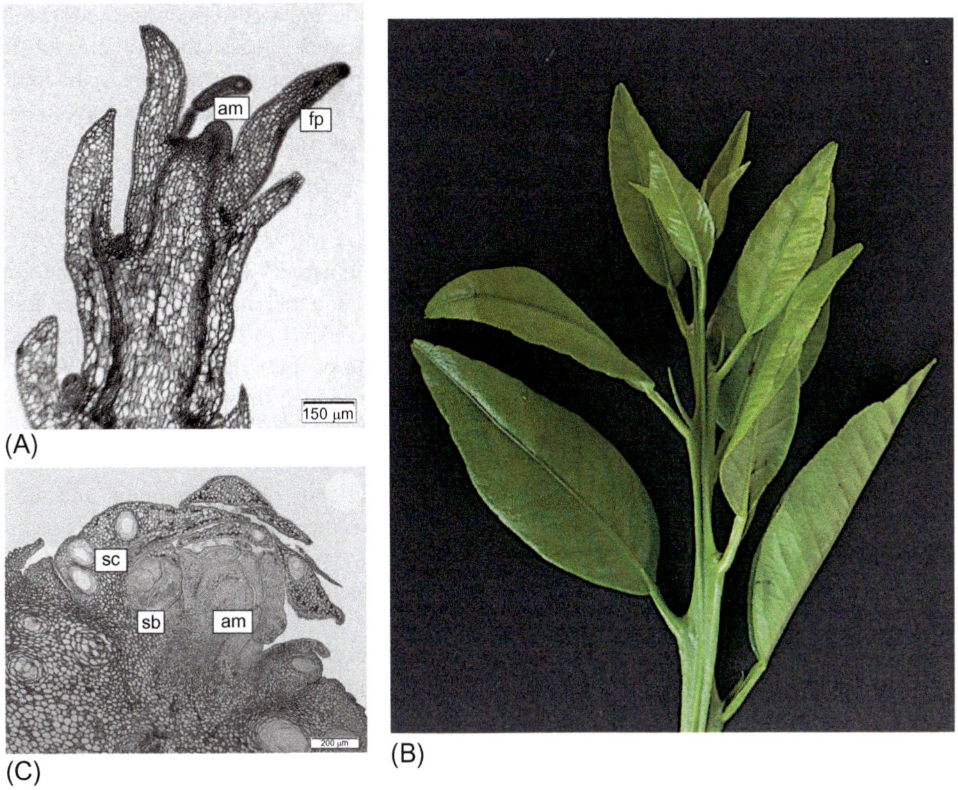

PLATE 10.3 Shoot development. (A) Longitudinal section of an actively growing stem tip. Apical meristem (am) and foliar primordia (fp). (B) Longitudinal section of a bud. Filotaxis is appreciated in the disposition of leaves. (C) Development of a shoot from the active growth of its apex, which forms leaf primordia separated by internodal stem segments, as well as buds and thorns in the axils of the leaves; scales (sc), secondary bud (sb).

regenerates the meristem, that is, is perpetuated indefinitely. The initial cells usually divide slowly, while the derivative ones do so quickly and, before their division ceases, their differentiation can already be appreciated as specific cells of determinate tissues.

The apical meristem consists of three superimposed cellular layers, which are distinguished by their planes of division. The outer one (L1) is formed by a single stratum of cells, which are divided according to an anticlinal plane (perpendicular to the surface). The second layer (L2) is formed by several strata of cells, which are divided anticlinally in the central zone and periclinally (according to a plane parallel to the surface) in the lateral zones. The cells of the innermost layer (L3) are divided in all directions. Each layer has its own initial cells.

The cells of the outermost layer (L1), when dividing anticlinally, cause the tissue to grow superficially forming the protodermis, whose cells differentiate to give rise to the epidermis. The cells of the L3 layer are divided in different directions, to produce a growth in volume inside the plant. From this layer, the pith and vascular tissue are derived through intermediate meristematic zones. In addition, between the provascular cylinder and the protodermis, another intermediate meristematic zone appears, coming from the L2 layer, whose cells are larger and, when differentiated, produce the parenchyma of the cortex and the leaves (Schneider, 1968).

The apical meristem presents seasonal activity, which is normally more intense in spring, decreasing during the summer and early autumn, entering in dormancy at the end of this season.

This meristem produces, at regular intervals, leaf primordia, which will later become leaves, as well as axillary buds and thorns, which are inserted into the nodes (Plate 10.3B). Subsequently, the activity of the intercalary meristems, which are located at the base of the internodes, causes a growth in length.

The vegetative buds, which are found in the axilla of the leaves, generate lateral twigs. These buds contain an apical meristem with leaf primordia, covered by several scales, and next to it there may be a thorn. In the axilla of the scales, some buds can be also formed and, therefore, in the axilla of each leaf there are usually several buds (Plate 10.3C).

The growth of the apical bud inhibits lateral bud burst, up to a certain distance. This phenomenon is known as apical dominance. Removal of the apical bud usually results in the growth of one or more lateral shoots. The IAA synthesized in the apical meristem and transported basipetally seems to be responsible for the inhibition of sprouting of the axillary buds.

The maintenance of axillary bud inhibition by applying IAA locally to a decapitated shoot reinforces the idea. In addition, the application of an auxin transport inhibitor (TIBA) under the caulinar apex releases the axillary buds from inhibition.

Cytokinins may also be involved in this process. In fact, the direct application of a cytokinin to the axillary buds stimulates their growth, counteracting the inhibitory effect of the shoot apex. Auxins cause the shoot apex to be a sink for cytokinins from the root, an effect that may be involved in the apical dominance.

10.3.2 Secondary stem growth

The vascular tissues are renewed, and their extension is increased by the activity of the vascular cambium, which is the cause of the growth in thickness of the stems. The procambium remains in a meristematic state, after completing the primary growth (Plate 10.4A) and becomes the cambium. This meristem consists of two types of cells: the spindle-shaped initials, elongated and sharp, generate the longitudinal or vertical system of the xylem and the phloem; the initial radial cells, almost isodiametric and relatively small, originate those of the rays, which form the transversal or horizontal system of the xylem and the phloem. In the stem, the cambium takes the form of a continuous cylinder that extends longitudinally throughout. When a trunk branches, its cambium continues with that of the branches derived.

The vascular tissues originate from periclinal divisions of the initial cells of the cambium, and the continuous cell divisions, give rise to the radial rays. As the xylem increases in thickness, the cambium also extends its circumference by longitudinal anticlinal divisions.

The cambium remains indefinitely in the same relative position, between the xylem and the phloem, producing secondary phloem outside and secondary xylem inside (Plate 10.4B).

The functional phloem is located closest to the cambium, since it is developed during the last period of vegetative activity. In this tissue, the cells derived from the fusiform initials of the cambium differentiate into sieve tubes, companion cells, parenchyma cells, and sclerenchyma fibers.

It is estimated that the phloem can maintain its functionality for about 2 years since it was generated from the cambium. The first symptom of degeneration and loss of functionality of the sieve tubes is the appearance of callose deposits in the sieve plates, which are very marked in the cell ring around the outer margin of the functional phloem.

The nonfunctional phloem is found abundantly in trunks and branches of a certain age, occupying the outer part of the tissue. This phloem consists of collapsed sieve tubes with their corresponding dead companion cells also attached, living parenchyma cells and bundles of fibers.

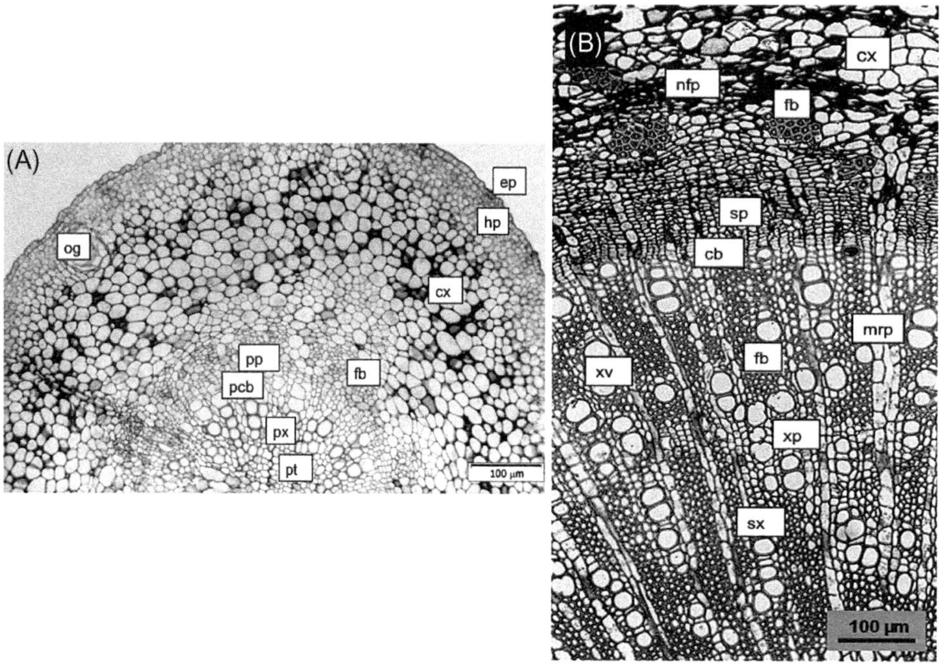

PLATE 10.4 (A) Cross section of primary stem tissues. Epidermis (ep), hypodermis (hp), oil gland (og), cortex (cx), protophloem fibers (fb), metaphloem (pp), procambium (pcb), metaxylem (px), pith (pt). (B) Cross section of secondary stem tissues. Cortex (cx), nonfunctional phloem (nfp), fibers (fb), secondary phloem (sp), cambium (cb), xylem vessel (xv), medullary ray parenchyma (mrp), xylem parenchyma (xp), and secondary xylem (sx).

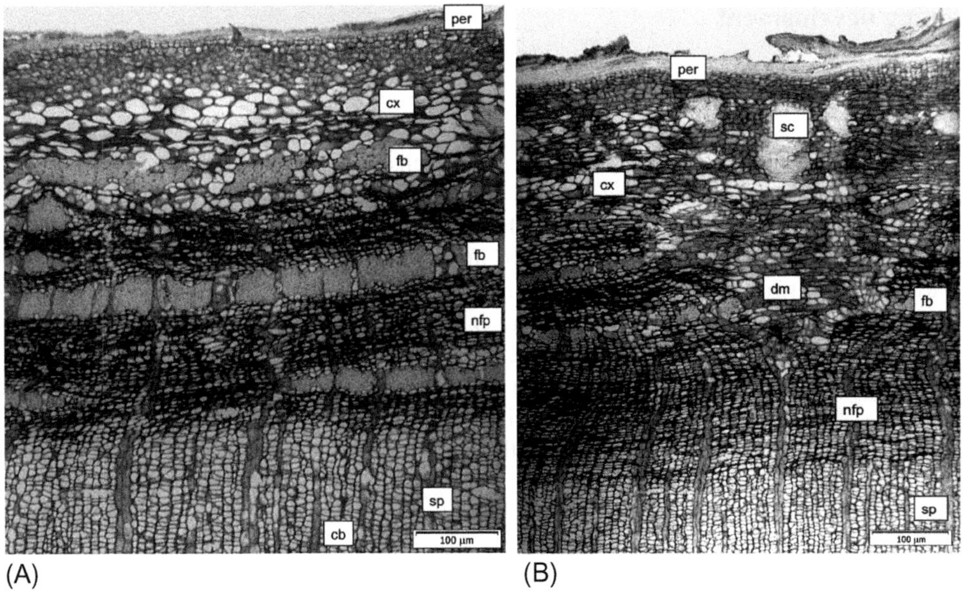

PLATE 10.5 (A) Section of stem bark with secondary growth. Peridermis (per), cortex (cx), fibers (fb), nonfunctional phloem (nfp), secondary phloem (sp), and cambium (cb). (B) Section of stem bark with lateral growth mediated by radially arranged meristematic cells (called dilation meristems). Peridermis (per), sclereids (sc), cortex (cx), dilation meristem (dm), fibers (fb), nonfunctional phloem (nfp), and secondary phloem (sp).

The formation of fibers usually occurs when cambial activity stops, which normally takes place at the entry of dormancy; however, some sporadic circumstances stopping vegetative activity can also promote it. The fibers are located forming separate groups within the phloem tissue, which, viewed in cross section, are arranged in concentric circles. In longitudinal section, the groups of fibers form vertical bands parallel to the sieve tubes.

The xylem generated during secondary growth is composed of vessels, parenchyma cells, medullar rays, and fibers. The vessels are sheathed by living parenchyma cells. When, at the end of autumn, the growth is ending because of the cold, the cambium forms a narrow band of small parenchyma cells together with some narrow vessels, which make it possible to distinguish among annual growth rings. However, as a consequence of certain climatic adversities, the growth can stop during the period of vegetative activity, giving rise to false rings. Over time, the woody tissue loses stored water and nutrients, while its cells accumulate other organic substances that permeate the walls of the cells or infiltrate inside them, causing the older xylem to have a darker color.

When the secondary growth occurs, the older xylem and the pith are covered by the new conductive tissue, while the older phloem is pushed outside, undergoing a compression that causes its obliteration, except in the fiber bands (Plate 10.5A). Accordingly, the cortex and the epidermis must adapt to the increase in width forced by the growth in thickness of internal tissues. Some cells of the inner cortical parenchyma die crushed, whereas in the rest of this tissue they are stretched tangentially or divided anticlinally. In the old stems, the dilatation of cortex is produced by meristematic formations—called dilatation meristems—which are arranged radially, extending from the radii of the secondary phloem to the phellogen. Parenchymal and sclereid cells are differentiated from the cells derived from the dilatation meristems (Plate 10.5B).

When the secondary development of the young stems begins, the cuticle and the walls of the epidermis rupture in some areas of its surface, giving rise to the appearance of vertical cracks. Beneath them the phellogen begins to form. This tissue function is to replace the epidermis as a protective tissue, when it is destroyed. This meristem comes from the dedifferentiation of subepidermal cortical cells. From the phellogen, the phellem (cork bark), a corky tissue constituted by cells with suberized walls, is produced toward the outside; in the opposite direction, that is, toward the interior of the plant, the phelloderm is generated; it is formed by a few layers of parenchyma cells.

The activity of the phellogen and that of the dilation meristems are usually associated, with which the new periderm appears in vertical bands facing the dilation meristems, so that both meristems are arranged perpendicularly (Plate 10.5B).

The central part of the stems is occupied by a solid cylinder of parenchymal cells, which constitute the pith. These cells develop thick cell walls that, over time, are lignified, with which the pith exerts as a supporting tissue that gives consistency to the stem.

10.3.3 Sprouting development

As mentioned above, in temperate zones, the vegetative development of citrus is a discontinuous process, which is restarted every year, after a period of winter rest. At the end of winter or at the beginning of spring, when the ambient temperature increases, the activity begins from the meristems, producing large amounts of growth regulators (auxins, gibberellins, cytokinins, etc.), which induce cell division and elongation. As the cells generated in the meristem are displaced toward the basal part, the differentiation occurs.

The appearance and development of new sproutings throughout the year occurs in well-defined periods separated by others of inactivity, so that when a flush finishes its growth, after a while, the next one begins (Fig. 10.5A). The number of annual flushes varies between two and four, being generally three in well-cultivated trees.

The first flush is the most abundant in terms of the number of sprouted buds, and they can be both vegetative and floral buds. This flush usually begins in late winter and continues to grow during the first half of spring. The second flush takes place in early summer, just after the physiological fruitlet abscission, and the last one in the autumn, giving rise, in both cases, only to vegetative shoots. Generally, vegetative spring shoots are more numerous and smaller in size than those of summer or fall that, also, are fewer (Fig. 10.5B and C). As the tree ages, spring flush is mainly constituted by inflorescences with and without leaves.

Bud burst varies from 1 year to other for a given tree, depending on the weather conditions. In subtropical climates, temperature is the main factor that determines bud sprouting. The area of cultivation and variety also affect the periods of vegetative growth. Shoot development follows a sigmoid curve, being gradual at the beginning, increasing in speed afterward, to slow down at the end (Fig. 10.5D).

The intensity of a flush usually determines the following, in a complementary way. So, if the spring is very abundant, summer is more scarce and vice versa. The same happens with summer and autumn, although the latter has larger leaves.

In young trees, the sprouting periods are not so defined, but rather they occur continuously.

Under tropical conditions of high temperature and high relative humidity, shoot growth takes place continuously, since the separation between flushes is not appreciated. The lemon, lime, and citron retain their tropical character, even in the warm areas, producing new shoots continuously during the period of vegetative activity.

Shoot growth can be interrupted by the abscission of the caulinar apex, which, before being released, acquires a pale-yellow color.

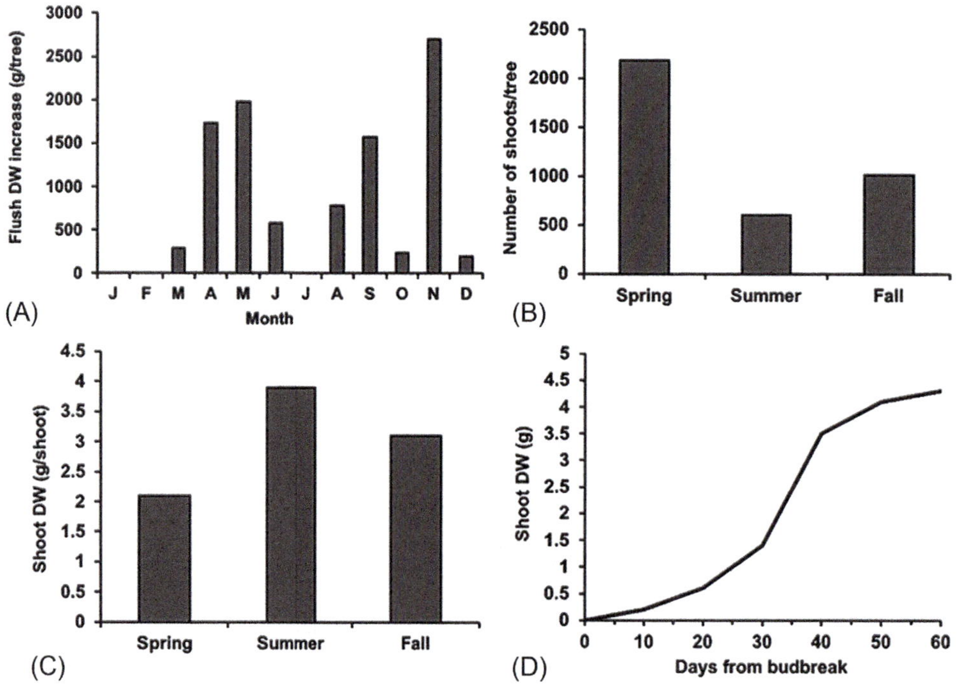

FIG. 10.5 Dry weight increase of the vegetative shoots throughout the year (A). Vegetative shoots of spring, summer, and autumn flushes: number per tree (B), individual dry weight (C), and growth evolution (D).

10.3.4 Factors affecting vegetative development

Several factors, both endogenous and exogenous factors, affect vegetative development in *Citrus*:

(1) Shoot size and distribution are strongly affected by temperature. In subtropical areas, the spring flush produces a large number of twigs with short internodes. In this period, the average temperature oscillates between 12°C and 20°C. In contrast, the summer flush produces fewer twigs but with longer internodes. The temperature in this period varies between 25°C and 35°C on average. In tropical regions, with high average temperature throughout the year, vegetative development does not stop, unless the water is limited because of the alternation of a wet and dry season. In this case, a large number of shoots are produced, the length of the internodes being determined by the plant water status.

(2) Carbohydrates availability is a crucial factor for new shoots development. Therefore, for this process to occur normally, it is necessary that the leaves maintain a high rate of assimilation of CO_2. The maximum of this parameter is reached with luminous intensities between 700 and 1000 $\mu mol\, m^{-2} s^{-1}$ (Iglesias et al., 2002). Lower illumination, as usually occurs inside the canopy, gives rise to weaker twigs.

(3) Endogenous hormonal levels (especially those of auxin and gibberellin) regulate shoot growth. This is evidenced in Carrizo citrange plants upregulating *CcGA20ox1* gene expression that encodes the enzyme GA20 oxidase in the biosynthesis of gibberellins, since the higher GA_1 content the larger the shoots. In these transgenic lines, cell division, more than cell elongation, is upregulated (Fagoaga et al., 2007).

(4) The amount of biomass produced by the different flushes is inversely related with flowering intensity, in the case of spring flush, and with crop load, in that of summer and autumn. In general, fruit set and development reduce vegetative development (Martínez-Fuentes et al., 2010; Monselise and Goldschmidt, 1982; Verreynne and Lovatt, 2009), affecting the three flushes, both in shoot number and shoot development (Martínez-Alcántara et al., 2015).

(5) Water stress reduces shoot development. This is because a moderate water deficit increases abscisic acid (ABA) level in the leaf, causing stomata closure. As a consequence, a decline in net CO_2 uptake occurs and, therefore, a reduction of shoot growth (Forner-Giner et al., 2011). For high water-deficit values, a loss of turgor takes place firstly and then the wilting of the leaves, giving rise to a decrease of global photosynthesis and practically null shoot growth.

(6) Poor nitrogen fertilization reduces shoot development. On the contrary, an excessive supply of nitrogen tends to increase its vigor (Chapman, 1968).

(7) Mineral deficiencies reduce vegetative development, especially that of the affected shoots (Chapman, 1968).

(8) Trees with a broad and deep root system tend to maintain vigorous and large shoots.

(9) Loose, deep, and well-aerated soils favor root development and, thus, vegetative growth.

(10) When the canopy/root ratio is reduced, as a consequence of a cultural practice (pruning, overgrafting, etc.) or due to a meteorological accident (frost, etc.), shoot vigor increases.

(11) On the contrary, certain pests or diseases affecting the roots (nematodes, *Phytophthora*, virosis, etc.) produce vegetative weakness.

(12) Some rootstocks (*Citrus macrophylla*, rough lemon, Carrizo citrange, etc.) induce a strong vigor in the grafted variety, while others (Cleopatra mandarin, sour orange, etc.) reduce it. It has been related to the hydraulic conductivity of the roots, which is a factor strongly involved in the movement of water through the soil system. The aforementioned vigorous rootstocks have a greater hydraulic conductivity than the less vigorous ones (Rodriguez-Gamir et al., 2010; Syvertsen, 1981), which present lower values of stomatal conductance and transpiration rate (Rodriguez-Gamir et al., 2010; Syvertsen and Graham, 1985). These factors are related to the width and density of the xylem vessels, whose values tend to be higher in vigorous rootstocks (Rodriguez-Gamir et al., 2010; Vasconcelos and Castle, 1994). An extreme case of reducing the vigor of the variety is represented by the dwarfing rootstocks, whose characteristics are described later.

10.4 Leaf development

The apical meristem located at the end of the vegetative shoot generates leaf primordia, which are arranged in close succession, forming a spiral around it, so that the youngest are closer to the apex. The newly formed primordia appear superimposed and curved on the apical meristem, but as they develop, they acquire a clear erect position.

Auxin plays a fundamental role both in the initiation of the primordia and in the determination of their position at the caulinar apex. In this way, in certain points of the peripheral zone of the apical meristem a greater concentration of IAA is found which precedes the initiation of a new primordium. This effect can be explained by the transport of the auxin from there to the preexisting primordia, which reduces the concentration of auxin in the cells that surround them, inhibiting the emergence of new neighboring primordia. On the contrary, the concentration of the IAA used to be higher in some points far from the primordia, where the new ones are generated. In this way, auxin induces the formation of the primordia and determines their position.

The surface of the leaf primordia is covered by a layer of meristematic cells (protodermis), which is a continuity of the L1 layer and where, when mature, epidermis will be formed.

The layers of cells below the protodermis are derived from the L2 layer of the apical meristem and are differentiated into procambium and inner meristem. The vascular tissues of the leaf are differentiated from the first one, and the mesophyll from the second one (Plate 10.6A).

The development of leaf primordia gives rise to the leaf blade and the petiole. The wings of the latter and the two halves of the blade develop from marginal meristematic zones found on the sides of the primordium, while its inner cylindrical part forms the central vein. A constriction appears in the area where the petiole meets the leaf blade, which is associated with the formation of an abscission layer. The glands of essential oils begin to differentiate early from the inner meristem that forms the mesophyll.

The axillary buds have their origin in a source of initial cells that are activated in the interior angle of the axils of the emergent leaf primordia.

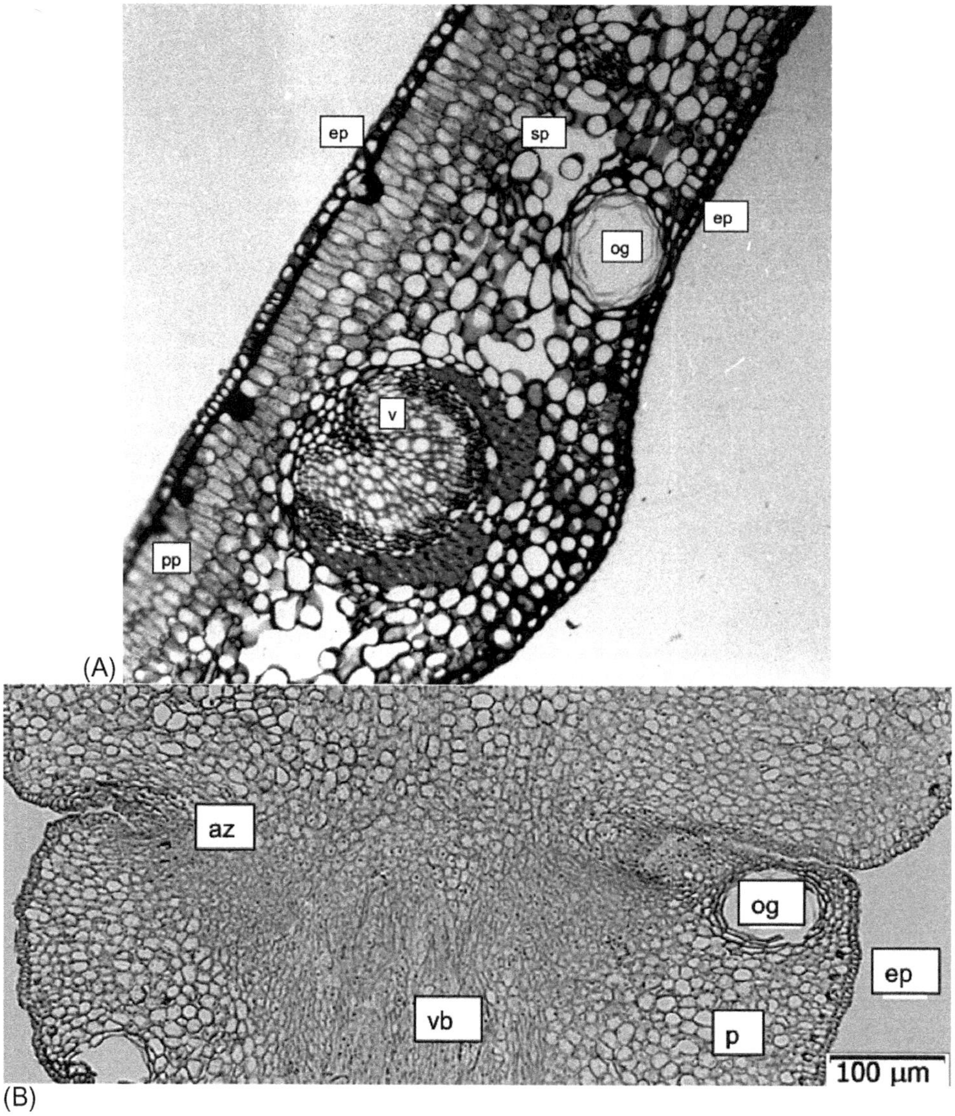

PLATE 10.6 (A) Section of a developing leaf. Epidermis (ep), palisade parenchyma (pp), spongy parenchyma (sp), oil gland (og), and vein (v). (B) Abscission zone between the lamina and the petiole. Abscission zone (az), vascular bundle (vb), parenchyma (p), oil gland (og), and epidermis (ep).

The leaves expand during the period of elongation of the stem in which they are inserted. In an early stage of development, they are thin and light green; nevertheless, after about 2 months, when they reach their final size, take on a dark-green color and become thicker and more consistent. The dry weight of a young leaf is 30%, approximately, and becomes 45% when mature (1 year old). This effect is due to the deposition of large amounts of materials in the cell walls, as well as the accumulation of starch and other reserve polysaccharides.

In *Citrus*, leaf is formed by a single blade with a relatively short petiole that joins it to the stem. According to the species, the blade adopts, more or less, an elliptical, ovate, or lanceolate form. Besides, there is a wide variability in leaf shape and size, depending on several factors, especially the vigor of the shoot where they are located. Leaf color is dark green at the upper surface and pale green at the lower surface.

In most species, the petiole is winged; sour orange and grapefruit have large wings, whereas in sweet orange they are narrow, and lemon lacks wings. Finally, *P. trifoliata* and its hybrids (citranges, citrumelo, etc.) have trifoliate leaves (three leaflets).

The leaves of *Citrus* are perennial and, therefore, do not lose the leaves during the winter rest period. The life of a leaf depends on its location on the tree: those located in fruitful branches live around 15 months, whereas those located in a vegetative shoot usually last >2 years. As leaves are generated with successive flushes, leaves of different ages coexist on the tree, abscissing along the whole year. Nevertheless, leaf abscission is usually more intense during flowering (April-May), followed by autumn fall. *P. trifoliata*, on the other hand, is deciduous, so that in this species leaf abscission takes place in autumn. The leaves are arranged on the stem following a predetermined geometric scheme, called phyllotaxy; according to it, the insertion of each leaf forms an angle with the immediately inferior one, with which they present a spiral arrangement around the stem, in such a way that the shade produced by the new leaves interfere as little as possible with those that are below so as not to affect the photosynthesis (Plate 10.3B). The phyllotaxy of the sweet orange is 3/8, which means that the 8th leaf is superimposed with the 1st one after the spiral formed by the leaves gives three turns to the stem. This spiral can turn to the right or to the left.

Below the base of each petiole appears a prominent elevation in the stem that extends several centimeters along the same. Within it is the leaf trace, which connects the conductive tissue of the shoot with that of the leaf.

10.4.1 Leaf abscission

Leaf abscission is the process by which the plant discards the leaves that have aged or suffered damage, through mechanisms that cause its fall. In this way, the plant renews the leaves by replacing the older ones. In this way, the tree adapts its structure to the changes due to its own development and maintains, permanently, its productive capacity. On the other hand, the exposure area is lowered and the harmful effects exerted on the plant by some types of stress, such as drought, wind, or the attack of certain pathogens, are reduced. The detachment of the leaves is also a way to eliminate part of toxic ions (Cl^-, B), which accumulate in leaves when they are at excessive concentrations in the soil solution. As the process of leaf senescence advances, and before its abscission occurs, a significant proportion of the most mobile mineral elements found therein (N, P, K, etc.) are exported to the new organs, where they recycle.

10.4.1.1 Abscission zones

The leaves are detached by the petiole, where two typical abscission zones are distinguished: one between the blade and the petiole, another one between the petiole and the stem.

The former is activated only when the leaf enters senescence upon completion of its life cycle. It can be activated by ethylene and, also, by adverse environmental conditions. Morphologically, the blade-petiole abscission zone presents a strangulation or groove where the cuticle is very thin. In the internal part of the constriction are between 10 and 15 layers or strata of cortical cells that form a cylinder that surrounds the vascular bundle. These cells, which essentially constitute the abscission zone, are morphologically distinguished from those of the upper and lower layers because they are isodiametric and smaller (Plate 10.6B). As regards the ultrastructural characteristics, the former has dense cytoplasm, interconnected with those of the neighboring cells by means of numerous branched plasmodesms. Inside of them, there are abundant grains of starch, whereas the adjacent cortical cells practically lack them. The walls of the cells of the abscission zone are cellulosic and do not have a lignified secondary wall. In this area, no phloem or xylem fibers or sclereids are formed. The sieve tubes and vessels crossing this zone do so in a zigzag trajectory (Huberman et al., 1983).

10.4.1.2 The abscission process

The activation of the leaf abscission process requires a series of events prior to fall.

Under normal conditions, aging of leaves is the main natural trigger of abscission. Leaf senescence is characterized by the loss of net assimilation capacity of CO_2 and by protein degradation. The amino acids released from their hydrolysis constitute the way in which the N stored in the old leaves is exported to the new organs. Although the citrus leaves age slowly (many of them live 2 or more years), their functionality decreases over time until abscission. Besides, certain stresses—both abiotic and biotic ones—such as water deficit, salinity, toxic levels of certain elements (B, Fe, etc.), nutritional deficiencies, mechanical damage, or the incidence of certain pests, can cause defoliation, which, in many cases, constitute a protection mechanism to prevent plant death.

Before abscission, leaves must undergo a series of internal changes, mainly hormonal in nature, which act as signals that trigger the metabolic reactions that lead to the rupture of the abscission layer.

A model of hormonal control of abscission has been proposed, in which ethylene and indole-3-acetic acid (IAA) play a fundamental role. Ethylene promotes abscission and, to this end, it must bind to specific receptors (Sisler et al., 1985). Conversely, IAA inhibits abscission by making the tissue insensitive to ethylene (Jaffe and Goren, 1979).

The IAA synthesized in the leaf blade moves in a polar direction toward the abscission zone. Thus, it remains unchanged while the flow of IAA from the blade to the petiole is sufficient; however, in the aged leaves, the auxin gradient is less marked, so that its transport rate toward the petiole is less intense. In this way, the level of IAA in the abscission zone decreases while its sensitivity to ethylene increases. This means that the cells of this zone acquire the capacity to respond to low concentrations of ethylene—both endogenous and exogenous—which is the trigger of the abscission process. Therefore, treatments that maintain a high level of auxins in the leaf delay abscission. Ethylene inhibits the transport of auxins by the central nerve of the leaves (Riov and Goren, 1979); this effect can play an important role in the promotion of abscission by ethylene, since it favors the reduction of the endogenous level of IAA in the abscission zone. The inhibition of the IAA translocation by ethylene has been related to the increase in the formation of its conjugates with organic molecules, mainly glucose (IAGlu) to the detriment of the conjugate with the amino acid asparagine (IAAasp) (Sagee et al., 1989). Ethylene also alters the metabolic pathway of IAA, increasing its catabolism. Thus, indole-3-carboxylic acid is formed, which rapidly binds to glucose to yield the ICCGlu glycoside (Sagee et al., 1990). Both, the conjugation and the catabolism of IAA, reduce the amount of the free form of this hormone that can be transferred to the abscission zone.

ABA applied at a relatively high concentrations to leaf explants, induces ethylene production, through the increase of the synthesis of ACC (1-aminocyclopropane-1-carboxylic acid) and the increase of the enzymes ACC synthase and ACC oxidase activity. The ethylene produced increases the activity of the hydrolytic enzymes of the cell wall, which ultimately cause abscission (Sagee et al., 1980).

In *Citrus*, leaf abscission caused by water stress is produced by a characteristic process. Thus, when the trees suffer a prolonged drought, the leaves may dry but do not detach from the branches. However, if after a sufficiently intense water deficit, rain or irrigation occurs, abscission of the leaves occurs immediately. The explanation of this phenomenon is based on the fact that the water deficit promotes the synthesis of ABA that, in turn, promotes and modulates 1-aminocyclopropane-1-carboxylic acid (ACC) biosynthesis in the roots. When the rehydration of the plant occurs and the flow of sap is restored via xylem, ACC is transported to the stem, where it is oxidized to produce ethylene, which activates leaf abscission (Tudela and Primo-Millo, 1992; Gómez-Cadenas et al., 1996).

In salinized plants, leaf abscission is also mediated by the endogenous production of ethylene, in this case, however, induced by a high concentration of Cl^- (Gómez-Cadenas et al., 1998).

Although the abscission zone is differentiated in an early state of the leaf development, its function, which is to propitiate the detachment, remains repressed until a signal, which can be induced by different causes, both internal and external, starts the separation process. The effector mechanism is the hydrolysis of the cell walls of a few layers of cells that are within the abscission zone, also called the separation zone. These cells are considered as "target" cells that respond to stimuli (such as ethylene) by synthesizing and secreting hydrolytic enzymes that degrade the cellulose skeleton of cell walls and the middle lamella. The main enzymes involved in this process are cellulases and polygalacturonases, while no relationship has been found with the activity pectin methylesterase (Ratner et al., 1969; Huberman and Goren, 1979; Rasmussen, 1973; Riov, 1974). Ethylene accelerates abscission by promoting cellulase and polygalacturonase activities in the abscission zone (Huberman et al., 1975; Rasmussen, 1975). This hormone would act by activating the expression of genes that act on the cell walls of the abscission layer (Agustí et al., 2008, 2009, 2012; Estornell et al., 2013). On the other hand, auxin inhibits the increase of the previous activities in the abscission zone and delays the fall (Rasmussen, 1973).

10.4.1.3 Anatomical and ultrastructural changes in the abscission layer

The first histological change in the abscission zone is the formation of the separation zone that affects only some of the cell layers (from 1 to 5) close to the strangulation.

In the cytoplasm of these cells, as the process of abscission progresses, abundant lysosomes appear as the starch grains decrease. Afterward, before the separation takes place, numerous vesicles from the endoplasmic reticulum and the Golgi apparatus appear, binding to the plasmalemma, probably to excrete the hydrolytic enzymes and digest the cell wall.

In the separation zone, the cells of one of the layers increase in volume (up to about three times) and the middle lamella dissolves, whereupon the fracture starts. The growing cells layer is that closest to the stem of the plant, whereas the adjacent layer is closest to the leaf blade.

Subsequently, the cell wall acquires a gelatinous appearance, indicating that degradation is taking place. Toward the end of the separation process, the primary wall dissolves, while the plasmalemma remains intact and, hence, the protoplast is free.

The fracture of the petiole through the separation zone is produced by wind or another mechanical factor that also breaks the vascular bundle, which usually suffers little changes during the whole process. With the excision of the petiole, the process of abscission culminates when the leaf blade is detached from the plant. Subsequently, the wound heals through the deposition of lignin and suberin, as well as gummy substances that clog the discovered ends of the vessels.

It is worthy of consideration that ethylene promotes the deposition of callose in the phloem of the midrib, which has been related to the reduction of the transport of auxins and others assimilated toward the zone of abscission, and is one of the causes of activation of this process (Jaffe and Goren, 1988).

10.5 Formation of the root system

10.5.1 Development of the primary root

The primary root develops from the radicular meristem of the embryo, which forms a taproot that penetrates vertically into the soil. As the main root lengthens, secondary lateral roots (pioneer roots) appear, which tend to develop horizontally at shallow depths. The youngest ones are those closest to the apical meristem of the primary root, and the older ones are closest to the root neck.

Frequently, the apical meristem of the main root can be damaged, and it stops growing, its functions being assumed by the lateral roots. These, in turn, branch out to constitute the root system. The roots that derived directly from the primary root are called secondary roots or the first order, those coming out from them are tertiary or the second order, and so on. Root diameter decreases as the root branches. Primary and lateral roots show a secondary growth.

A fibrous root system forms from the primary root, in the young trees, and from the lateral ones, in adult ones. These groups of roots that originate from ramifications of a main fibrous root, become thinner as they separate, and present a high degree of mortality at the end of each developmental cycle. Fibrous roots are responsible to absorb water and mineral elements from the soil.

During the vegetative period, root growth is intensified in stages in which most apices are active simultaneously. A period of time must elapse from the end of a stage of growth until the next one begins. This kind of growth is manifested initially by the emergence of a large number of fibrous roots that, afterward, also increase in length.

The growing young roots are whitish in color, but as they age, they take on a light-brown color. Usually, fibrous dark-brown roots are those dead.

The root grows and develops from the apex, from which four parts can be distinguished: the calyptra or root cap, the apical meristem, the elongation zone, and the maturation zone (Schneider, 1968). As a whole, the apex has a length of about 2 mm from the root end, although its upper limit is diffuse, as happens between the different parts, with the exception of the root cap (Plate 10.7A).

10.5.1.1 The calyptra or root cap

It is a cone-shaped mass of tissue that covers the end and sides of the apical meristem. It is considered as a structure that protects this meristem and facilitates the penetration of the root in the soil during its growth. The mucilaginous formations in the walls of the outermost cells of the calyptra possibly reduce the friction between the end of the growing root and the earth. The calyptra is constituted by living cells of parenchymatic nature, with a strong mechanical consistency to be able to separate hard particles from the ground. As the initial cells produce new cells, the older ones are progressively displaced toward the end. When the root is lengthened, these cells are compressed against the ground, which can break or detach tissue. When the cells of the root cap differentiate, they acquire the ability to perceive the stimuli of gravity (geotropism).

10.5.1.2 The apical meristem of the root

It is located above the root cap and its function is the production of the cells that form the root. This meristem does not form lateral roots.

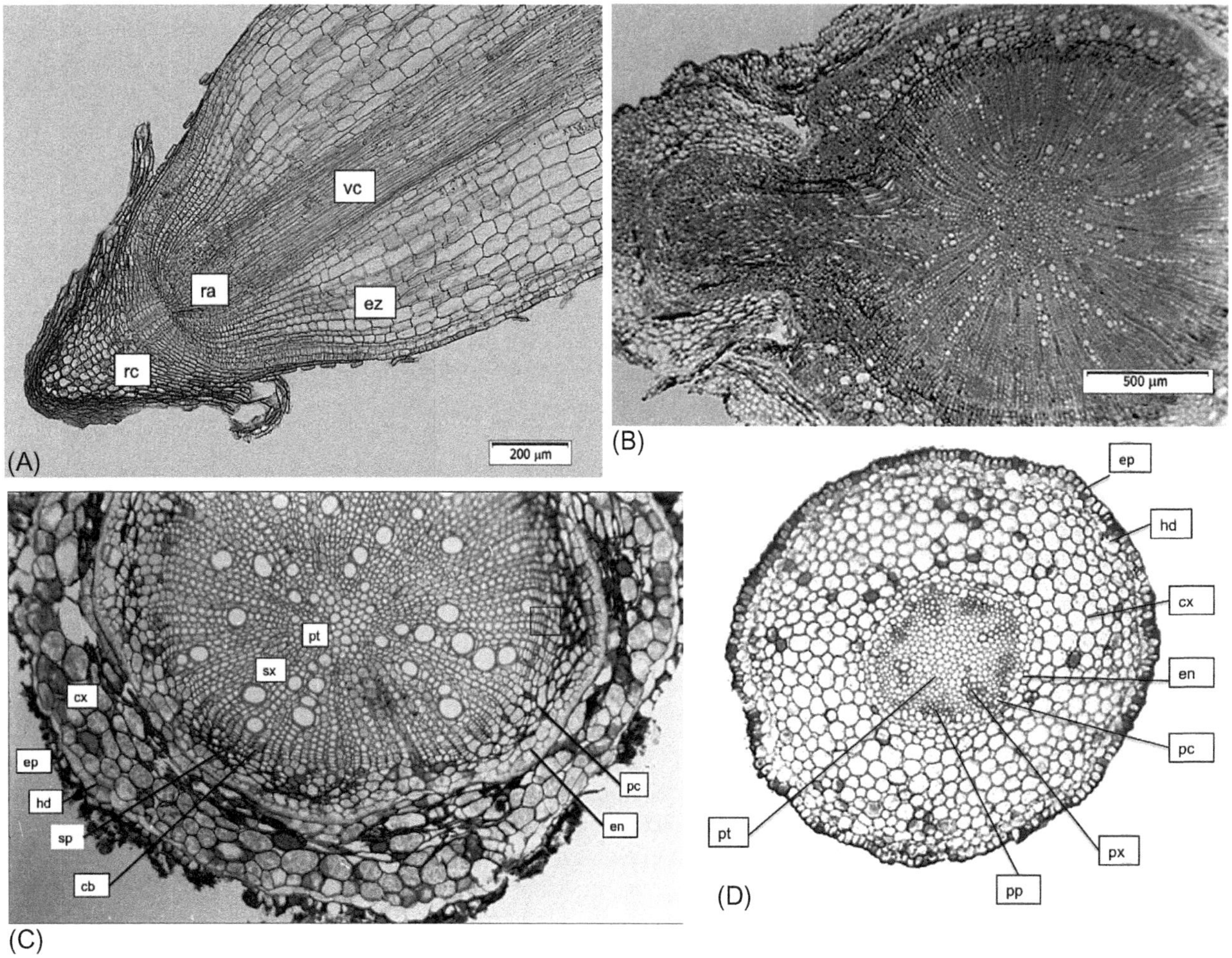

PLATE 10.7 (A) Longitudinal section of the root tip. Root cap (rc), root apical meristem (ra), elongation zone (ez), vascular cylinder (vc). (B) Emergency of a lateral root generated in the pericycle of another of higher order. (C) Cross section of secondary root tissues. Epidermis (ep), hypodermis (hd), cortex (cx), endodermis (en), pericycle (pc), secondary phloem (sp), cambium (cb), secondary xylem (sx), and pith (pt). (D) Cross section of primary tissues from a fibrous root. Epidermis (ep), hypodermis (hd), cortex (cx), endodermis (en), pericycle (pc), primary phloem (pp), primary xylem (px), and pith (pt).

The apical meristem of the root is structured around a small group of cells, whose rate of division is very low. This nucleus, which is located in the center of the root apex, is called the quiescent center (CQ). The initial cells surround the CQ and produce longitudinal rows of derived cells.

The group of initial cells below the CQ produces columns of cells outward through periclinal divisions. These extend axially and mature to form the cells of the calyptra, which are parenchymal in nature and do not undergo subsequent divisions. However, the cells located on both sides of the initial cells undergo oblique divisions to form the curved parts of the apex and subsequently divide anticlinally forming a region of intense meristematic activity, called protodermis. The cells of this tissue continue to be divided anticlinally, with which it grows superficially and, as it matures from the apex to the basal zone, they generate the epidermis.

Around the CQ is a ring of cells that divide periclinally to establish the layers that will give rise to the cortex and the endodermis. Subsequently, the cells generated maintain a high meristematic activity and by anticlinal divisions monostratified columns of cells are derived, which subsequently elongate axially and differentiate into cortical cells. The outermost cell layer originates the hypodermis and the innermost one the endodermis.

The initial cells that are immediately above the CQ divide anticlinally, forming columns of cells that generate the central cylinder.

The external cell layer originates the pericycle and within this the provascular tissues are differentiated.

The CQ represses the differentiation of the initial cells, since if it is eliminated or inactivated (by laser or irradiation) the initial cells differentiate rapidly and the meristem collapses.

During dormancy, only the apical part is perpetuated as a meristematic tissue; the nearby tissues mature and form those of the primary root.

10.5.1.3 Elongation zone

The transition from the apical meristem to the mature primary body of the root occurs through an area located above the active cell division, where the cells lengthen rapidly. Although some cells may continue to divide as they lengthen in this area, their division rate decreases to nullity as they separate from the meristem. In this zone, the epidermal cells immediately derived from the protodermis increase considerably in length, while their width tends to decrease. It is common in this state to form root hairs. Simultaneously, the hypodermic cells adjacent to the previous ones also suffer a strong elongation that may be higher than that of the epidermal cells. The cortical cells also grow axially, although their differentiation is slower than that of the previous ones.

Ethylene induces the radial expansion of the root tips, changing the polarity of the growth, which produces shorter and wider cortical cells. The endogenous GAs counteract and modulate this effect by promoting the elongation of the cells (Tadeo et al., 1997).

10.5.1.4 Maturation zone

Once the division and elongation of the cells ceased, they reach the maturation zone, developing characteristics of specialization. The differentiation can be started much earlier, but it cannot be completed until the cells reach the mature state.

10.5.2 Lateral root development

The lateral roots develop at a certain distance from the apical meristem of a higher root and originate from the pericycle (Plate 10.7B). During the initiation of a lateral root, a group of cells of the pericycle undergoes periclinal and anticlinal divisions, giving rise to the formation of a protuberance, which is the primordium of a lateral root. When growing, the incipient root crosses the cortex of the main one, but before it emerges to the surface, the apical meristem and the calyptra appear already differentiated. In this phase, the endodermis divides anticlinally and forms a layer of cells on the surface of the primordium; shortly before the new root comes out the tissue derived from the endodermis disintegrates.

The vascular systems of the main and lateral roots are not totally independent. When the second one begins in the pericycle of the first one, the space that separates both conductive tissues is small and there are intermediate cells derived from the pericycle which differentiate into conductive cells, communicating both vascular systems.

Auxins are involved in the formation of lateral roots and are specifically required for the initial asymmetric divisions that give rise to the root primordium. Thus, the initiation of the lateral roots is preceded by an increase of auxin content. In addition, the localized application of synthetic auxins, such as indolebutyric acid or naphthaleneacetic acid, favors rooting (Ferguson et al., 1985; Moore, 1986; Primo-Millo, 1976; Primo-Millo and Harada, 1976).

The development of lateral roots is partially mediated by the transcription factor *NAC1*, whose expression is induced by auxins.

The formation of lateral root primordia can be controlled by the apical root meristem. In fact, detaching it stimulates lateral roots initiation.

10.5.3 Secondary root growth

The primary and lateral roots show secondary growth, which is responsible for increase in thickness (Plate 10.7C). The main fibrous roots show a low secondary growth rate, that is practically null in those branched from them (Plate 10.7D).

In the former, the vascular cambium initially presents on the inner edge of the phloem cords, where it generates some secondary elements. At the same time, the cells of the pericycle, which are on the outside of the poles of the protoxylem, divide, so that the internal cells derived from them join the cambium bands located on the inner side of the phloem cords, forming a complete cylinder of cambium. Seen in cross section, the circular contour of this cylinder is due to the fact that the secondary xylem is generated on the inside edge of the phloem cords before that on the outside of the primary xylem.

When the cambium begins the activity, the secondary vascular tissues form a ring that completely envelops the primary xylem. Sieve tubes of the primary phloem are crushed, and some cells of the axial phloem parenchyma differentiate in fibers. The latter also appear in the secondary phloem. The cambium that originates in the pericycle, outside the poles of the xylem, forms the rays (Schneider, 1968).

Secondary growth puts pressure on the epidermal and cortical tissues of the root, the intensity depending on its magnitude. In the fibrous roots, which do not grow in thickness, these tissues remain unchanged throughout the life of the root. When the secondary growth of the fibrous roots is moderate, part of the cortical cells adapts to the same stretching, while others collapse. Thus, the cortical parenchyma narrows while the vascular cylinder widens. It is common, in these cases, to produce longitudinal cracks in the surface of the root, which affect the epidermis, hypodermis, and outer layers of the cortex. Beneath these cracks, a periderm is formed from the cells of the hypodermis or from the outermost part of the cortex, which can extend laterally at a short distance. Nevertheless, this type of root usually retains most of its cortex (Schneider, 1968).

In the roots that present a great growth in thickness, a periderm is generated from the pericycle; the process is as follows: the periclinal divisions of the cells of the pericycle form several layers of cells that precede the formation of the peridermis. The phellogen originates from the outermost cells of the proliferated pericycle. This phellogen produces cork tissue toward the outside and phellodermis toward the interior. When this process finishes, the cortex, together with the endodermis, emerges pushed by the secondary growth. Frequently, subsurface periderm is formed firstly, and later, a deeper one derived from the pericycle.

10.5.4 Root distribution

The distribution of the root system of adult trees varies considerably with the soil, the rootstock, and the irrigation system. However, under appropriate conditions for its expansion, the root follows a general pattern of distribution. In the most superficial layer of the soil, at a depth of 30–40 cm, there is >70% of the root mass, formed by a network of lateral roots that support a wide network of fibrous roots. Longitudinally, in a radius of about 50 cm around the trunk, the thickest roots predominate, while from this to a radius of approximately 2 m, there are mainly medium and small lateral roots, together with most of the fibrous ones. At a greater depth there is another layer of roots, formed from lateral roots that grow downward, more or less vertically. From them, several groups of fibrous roots that develop below 40 cm in depth are derived (Castle, 1978).

The structure of the root system described represents an adaptive strategy of the tree, according to which the upper fibrous roots rapidly absorb water and nutrients applied to the topsoil; the lower layer acts under water stress conditions by absorbing water from the deeper layers of the soil, as well as leached nutrients (as is the case of nitrate).

The deeper root layer is quantitatively much less important than the surface layer and, in some circumstances, it is very poorly developed. Such is the case of shallow or clayey soils, where practically all roots are superficial.

The introduction of drip irrigation, along with the techniques of fertirrigation, caused great changes in the structure of the root system. However, the citrus roots adapt to these new conditions quickly, concentrating the active fibrous roots around the drippers. The highest density of these roots is found in a volume of soil of 10–30 cm radius from the dripper and at a depth of 10–30 cm.

10.5.5 Factors affecting root development

(1) Under appropriated aeration and humidity conditions, root growth depends, to a great extent, on the soil temperature. Growth increases linearly with soil temperatures between 18°C and 30°C, and from this temperature tends to stabilize. Therefore, root growth begins in spring when soil temperature exceeds 13°C, although the growth rate is quite limited below 18°C and also above 36°C

(2) Root growth is very sensitive to soil moisture deficit, so that, when irrigation stops and the matric potential of the soil falls below—0.05 MPa, growth also stops. When rehydrating the soil, the root growth recovers. However, if it was very dry, it should take about 48 h before new rootlets emerge. Therefore, under lack of watering conditions, arrests and reinitiations of root growth usually occur (Bevington and Castle, 1985).

(3) The development of the root system is greater in sandy soils than in clay ones. Best aeration, highest heating capacity, and least penetration resistance of the first one explain it.

(4) Root growth competes with vegetative growth, and thus the former is reduced when a vegetative flush takes place. Accordingly, a period of intense root development occurs after a period of vegetative growth. These alternating growth cycles indicate the existence of the competition between roots and shoots, mentioned above, for available nutrients. Importantly, after a strong pruning, the branches sprout intensely to the detriment of the root growth, which indicates that in this competitive process the vegetative development dominates over that of the roots (Syvertsen, 1994).

(5) Fruiting strongly affects the development of the root system. Thus, in the trees with a heavy crop load root growth is very scarce, probably because the depletion of carbohydrates that occurs under such a condition (Goldschmidt and Golomb, 1982).

10.6 Trees of reduced size

Some varieties of the Satsuma group (Wase type), such as Clausellina, Okitsu, and others, present a series of common characteristics such as their small size, rapid entry into production, and early ripening of their fruits. Their scarce vegetative development coexists with a high productive efficiency (i.e., the relation between the weight of the harvest and the volume of the canopy in kg/m^3), with which, for a high density of plantation (>1000 trees/ha), good yields can be achieved after 3–4 years of planting. In addition, plants of small size are of easy access, allow certain cultural practices (pruning, thinning, pest treatments, harvesting, etc.) to be carried out with greater comfort, which results in a reduction of costs (Plate 10.8).

Through the use of dwarfing rootstocks, plants of small size of any variety can be obtained, offering the aforementioned advantages (high productive efficiency, rapid entry into production, ease to manage and reduction of cultivation, and harvesting expenses). These kinds of trees should be planted at a high density, ranging between 800 and 1200 plants/ha. The best-known dwarfing rootstock in citrus is Flying Dragon (FD), which is a mutant of *P. trifoliata* (Bitters et al., 1979; Cheng and Roose, 1995; McCarty and Cole, 1982). This rootstock (*P. trifoliata*) has important advantages, since it is tolerant to Tristeza virus, and resistant to the fungus *Phytophthora* spp., and to the nematode *Tylenchulus semipenetrans*. It is also tolerant to cold, a characteristic that transmits to the grafted variety, which, in addition, enhanced fruit quality. The main drawback of this rootstock is its sensitivity to iron chlorosis when grown in alkaline or calcareous soils and, therefore, it is only indicated for acid soils. To solve this problem, other rootstocks with dwarf capacity have been obtained through experimental hybridizations. Some of them, like the FD, reduce tree volume by up to 75%, and their productive efficiency is greater than that of a normal vigor rootstock, especially during the first years of cultivation (Bitters et al., 1979; Forner-Giner et al., 2014; McCarty and Cole, 1982; Martínez-Alcántara et al., 2013).

The dwarfing effect of these rootstocks has been related to the water relations of the trees grafted onto them. Thus, when climatic conditions demand a high transpiration rate, the leaf water potential is lower in the variety grafted on a dwarfing rootstock than on a normal vigor rootstock. This effect is related to the lower hydraulic conductance of the dwarfing rootstocks both in the roots and in the graft area. Since the hydraulic conductance of the root and stem determines the ability to transport water, under high transpiration conditions this parameter limits the water flow to the canopy, causing a decrease in water potential, which, in turn, induces the closure of the stomata to prevent cavitation in the vessels of the xylem and damage to the hydraulic system. There are two reasons to explain the effect of this whole process on the reduction of growth of the varieties induced by the dwarfing rootstocks: (1) diurnal variations in the water potential of the aerial part influence the elongation of the shoots, and (2) the decrease in gas exchange as a consequence of the lower stomatal conductance, which, in turn, is regulated by the water state of the leaf; this effect would cause a reduction in the net assimilation of CO_2 that would have adverse effects on growth.

The hydraulic conductance of the root, which appears as a first determinant of the dwarfing action of these rootstocks, is strongly related to the width of the transversal section of the xylem vessels (distribution of diameters) or with their density (number of vessels per unit area of xylem tissue).

Another factor to take into account is the marked decrease in the hydraulic conductance in the graft area, which may be due to vascular deformations as a consequence of defective joints. This anomaly not only contributes to the alteration

PLATE 10.8 Tree of reduced size (<1.5 m height) and high productivity.

of water relationships, but increases the resistance to sucrose transport, causing a reduction of its translocation from the leaves to the roots, and decreasing the concentration of soluble sugars and starch below the graft area. This deficiency in carbohydrates content is the way how dwarfing rootstocks reduce root development and, thus, tree vigor. For the same reason, the availability of nutrients in the canopy is increased, allowing for a greater translocation of carbohydrates to the fruit. This hydrocarbon resources distribution favor the sink capacity of the fruit and explains the high productive efficiency of these kind of rootstocks, despite the lower net assimilation of CO_2 of varieties grafted onto them (Forner-Giner et al., 2014; Martínez-Alcántara et al., 2013).

10.7 Control of vegetative development through the use of growth retardants

The development of vegetative shoots is associated to high levels of endogenous gibberellins in the apex (Goldschmidt and Monselise, 1970). This is the reason why the application of some substances inhibiting the biosynthesis of gibberellins has been tried to reduce the elongation of shoot internodes. The purpose of this is to control the excessive vegetative growth and get trees with less development, which facilitate some cultivation practices (pruning, treatments, collection, etc.), without losing productivity per unit area, which can be achieved by increasing the density of plantation.

AMO-1618 (4-hydroxyl-5-isopropyl-2-methylphenyltrimethylammonium-piperidine carboxylate chloride), CCC (2-chloroethyltrimethylammonium chloride), and SADH (2,2-dimethylhydrazide of succinic acid) are some of the main growth retardants acting in this way. Another inhibitor of gibberellin biosynthesis that has been used in this purpose is PP333 or paclobutrazol [1-(4-chlorophenyl)-4,4-dimethyl-2-(1*H*-1,2,4-triazol-1-yl)pentan-3-ol]. It inhibits kaurene oxidase, thereby blocking the conversion of *ent*-kaurene to *ent*-kaurenoic acid and, thereby, inhibiting gibberellin biosynthesis from an initial metabolic step. Besides, some IAA transport inhibitors, such as BTOA (benzothiazole 2-oxyacetate) and TIBA (triiodobenzoic acid) (Harty and Van Staden, 1988), have also been tested.

These growth retardants are applied at a concentration between 500 and 3000 mg/L, and even higher, in one or several times, by foliar spray or wetting the roots. However, they have very erratic results, since in most cases the treatment does not affect the growth. Furthermore, when some inhibitory effect occurs, it is weak and short in time, since the plants recover normal development in $<$45 days (Delgado et al., 1986a,b; Lenz and Karnatz, 1975; Monselise, 1979; Monselise et al., 1966; Monselise and Halevy, 1962; Phillips, 1972; Yelenosky, 1973).

Nevertheless, paclobutrazol shows a greater efficiency than other retardants, in terms of the inhibition of shoot length. Thus, its foliar sprays at a concentration of 500–3000 mg/L produces a significant reduction of internode length and, thus, a decrease of tree size as a whole (Aron et al., 1985). On the other hand, the applications to the soil (2.5–10 g/tree) also reduce internode length and canopy volume of orange and mandarin trees (Delgado et al., 1986a,b). In some cases, supply of paclobutrazol to the roots is more effective than a foliar spray (Bausher and Yelenoski, 1986).

However, nowadays, these treatments are not used in normal agricultural practice, mainly due to their secondary effects, their limited effectiveness, and because some of them are not authorized in most countries.

References

Agustí, J., Merelo, P., Cercós, M., Tadeo, F.R., Talón, M., 2008. J. Exp. Bot. 59, 2717–2733.
Agustí, J., Merelo, P., Cercós, M., Tadeo, F.R., Talón, M., 2009. BMC Plant Biol. 9, 127.
Agustí, J., Gimeno, J., Merelo, P., Serrano, R., Cercós, M., Conesa, A., Talón, M., Tadeo, F.R., 2012. J. Exp. Bot. 63, 6079–6091.
Aron, Y., Monselise, S.P., Goren, R., Costo, J., 1985. HortScience 20, 96–98.
Bausher, M.G., Yelenoski, G., 1986. HortScience 21, 141–143.
Bevington, K.B., Castle, W.S., 1985. J. Am. Soc. Hortic. Sci. 110, 840–845.
Bitters, W.P., Cole, D.A., McCarty, C.D., 1979. Citrograph 64, 54–56.
Castle, W.S., 1978. Proc. Int. Soc. Citricult. 1, 62–69.
Chapman, H.D., 1968. In: Reuther, W., Batchelor, L.D., Webber, H.J. (Eds.), The Citrus Industry. vol. 2. University of California Press, pp. 127–289.
Cheng, F.S., Roose, M.L., 1995. J. Am. Soc. Hortic. Sci. 120, 286–291.
Delgado, R., Casamayor, R., Rodriguez, J.L., Cruz, P., Fajardo, R., 1986a. Acta Hortic. (179), 537–544.
Delgado, R., Casamayor, R., Rodriguez, J.L., Cruz, P., Fajardo, R., 1986b. Acta Hortic. (179), 545–548.
Estornell, L.H., Agustí, J., Merelo, P., Talón, M., Tadeo, F.R., 2013. Plant Sci. 199–200, 48–60.
Fagoaga, C., Tadeo, F.R., Iglesias, D.J., Huerta, L., Lliso, I., Vidal, A.M., Talón, M., Navarro, L., García-Martínez, J.L., Peña, L., 2007. J. Exp. Bot. 59, 1407–1420.
Ferguson, J., Young, M., Halvorson, J., 1985. Proc. Fla. State Hortic. Soc. 98, 39–42.
Forner-Giner, M.A., Rodriguez-Gamir, J., Primo-Millo, E., Iglesias, D.J., 2011. J. Plant Growth Regul. 30, 353–366.

Forner-Giner, M.A., Rodriguez-Gamir, J., Martínez-Alcántara, B., Quiñones, A., Iglesias, D.J., Primo-Millo, E., Forner, J., 2014. Sci. Hortic. 179, 376–387.

García-Agustín, P., Primo-Millo, E., 1989. J. Exp. Bot. 40, 383–390.

García-Agustín, P., Primo-Millo, E., 1990. Sci. Hortic. 43, 69–81.

García-Agustín, P., Benaches-Gastaldo, M.J., Primo-Millo, E., 1991. J. Sci. Food Agric. 56, 435–443.

García-Agustín, P., Benaches-Gastaldo, M.J., Primo-Millo, E., 1992. J. Plant Physiol. 140, 1–7.

Goldschmidt, E.E., Golomb, A., 1982. J. Am. Soc. Hortic. Sci. 107, 206–208.

Goldschmidt, E.E., Monselise, S.P., 1970. In: Carr, D.J. (Ed.), Plant Growth Substances. Springer Verlag, pp. 758–766.

Gómez-Cadenas, A., Tadeo, F.R., Talón, M., Primo-Millo, E., 1996. Plant Physiol. 112, 401–408.

Gómez-Cadenas, A., Tadeo, F.R., Primo-Millo, E., Talón, M., 1998. Physiol. Plant. 103, 475–484.

Harty, A.R., Van Staden, J., 1988. Isr. J. Bot. 37, 155–164.

Huberman, M., Goren, R., 1979. Physiol. Plant. 45, 189–196.

Huberman, M., Goren, R., Birk, Y., 1975. Plant Physiol. 55, 941–945.

Huberman, M., Goren, R., Zamski, A., 1983. Physiol. Plant. 59, 445–454.

Iglesias, D.J., Lliso, I., Tadeo, F.R., Talón, M., 2002. Physiol. Plant. 116, 563–572.

Jaffe, M.J., Goren, R., 1979. Bot. Gaz. 140, 373–383.

Jaffe, M.J., Goren, R., 1988. Citriculture 1, 407–415.

Koltunow, A.M., Hidaka, T., Robinson, S.P., 1996. Plant Physiol. 110, 599–609.

Lenz, F., Karnatz, A., 1975. Acta Hortic. 49, 147–155.

Martínez-Alcántara, B., Rodriguez-Gamir, J., Martínez-Cuenca, M.R., Iglesias, D.J., Primo-Millo, E., Forner-Giner, M.A., 2013. Trees 27, 629–638.

Martínez-Alcántara, B., Iglesias, D.J., Reig, C., Mesejo, C., Agustí, M., Primo-Millo, E., 2015. J. Plant Physiol. 176, 108–117.

Martínez-Fuentes, A., Mesejo, C., Reig, C., Agustí, M., 2010. J. Sci. Food Agric. 90, 1936–1943.

McCarty, C.D., Cole, D.A., 1982. Citrograph 67, 71–72.

Monselise, S.P., 1979. Sci. Hortic. 11, 151–162.

Monselise, S.P., Goldschmidt, E.E., 1982. Hortic. Rev. 4, 128–173.

Monselise, S.P., Halevy, A.H., 1962. Am. J. Bot. 49, 405–412.

Monselise, S.P., Goren, R., Halevy, A.H., 1966. Proc. Am. Soc. Hortic. Sci. 89, 195–200.

Moore, G.A., 1986. HortScience 21, 300–301.

Peña, L., Martín-Trillo, M., Juárez, J., Pina, J.A., Navarro, L., Martínez-Zapater, J.M., 2001. Nature 19, 263–267.

Phillips, R.L., 1972. HortScience 7, 252–254.

Pillitteri, L.J., Lovatt, C.J., Walling, L.L., 2004a. Plant Physiol. 135, 1540–1551.

Pillitteri, L.J., Lovatt, C.J., Walling, L.L., 2004b. J. Am. Soc. Hortic. Sci. 129, 846–856.

Primo-Millo, E., 1976. Rev. Agroquim. Tecnol. Aliment. 16, 546–556.

Primo-Millo, E., Harada, H., 1976. An. Inst. Nal. Invest. Agr. 6, 9–26.

Rasmussen, G.K., 1973. Plant Physiol. 51, 626–628.

Rasmussen, G.K., 1975. Plant Physiol. 56, 765–767.

Ratner, A., Goren, R., Monselise, S.P., 1969. Plant Physiol. 44, 1717–1723.

Riov, J., 1974. Plant Physiol. 53, 312–316.

Riov, J., Goren, R., 1979. Plant Cell Environ. 2, 83–89.

Rodriguez-Gamir, J., Intrigliolo, D.S., Primo-Millo, E., Forner-Giner, M.A., 2010. Physiol. Plant. 139, 159–169.

Sagee, O., Goren, R., Riov, J., 1980. Plant Physiol. 66, 750–753.

Sagee, O., Goren, R., Riov, J., 1989. In: Clijsters, H., De Proft, M., Marcelle, R., van Poucke, M. (Eds.), Biochemical and Physiological Aspects of Ethylene Production in Lower and Higher Plants. Kluver Academic Publishers, pp. 157–163.

Sagee, O., Riov, J., Goren, R., 1990. Plant Physiol. 92, 54–60.

Schneider, H., 1968. In: Reuther, W., Batchelor, L.D., Webber, H.J. (Eds.), The Citrus Industry. vol. 2. University of California Press, pp. 1–85.

Sisler, E.C., Goren, R., Huberman, M., 1985. Physiol. Plant. 63, 114–120.

Syvertsen, J.P., 1981. J. Am. Soc. Hortic. Sci. 106, 378–381.

Syvertsen, J.P., 1994. Tree Physiol. 14, 497–508.

Syvertsen, J.P., Graham, J.H., 1985. J. Am. Soc. Hortic. Sci. 110, 865–869.

Tadeo, F.R., Gómez-Cadenas, A., Ben-Cheikh, W., Primo-Millo, E., Talón, M., 1997. Planta 202, 370–378.

Tudela, D., Primo-Millo, E., 1992. Plant Physiol. 100, 131–137.

Vasconcelos, L.A.B.C., Castle, W.S., 1994. J. Am. Soc. Hortic. Sci. 119, 185–194.

Verreynne, J.S., Lovatt, C.J., 2009. J. Am. Soc. Hortic. Sci. 134, 299–307.

Yelenosky, G., 1973. HortScience 8, 102–103.

Chapter 11

Flowering and fruit set

Manuel Agustí[a], Eduardo Primo-Millo[b]
[a]*Mediterranean Agroforestry Institute, Polytechnic University of Valencia, Valencia, Spain,* [b]*Centro de Citricultura y Producción Vegetal, Instituto Valenciano de Investigaciones agrarias (IVIA), Valencia, Spain*

11.1 The process of flowering

Citrus trees tend to bloom profusely, to the point that an adult tree can produce >200,000 flowers annually. In areas of temperate climate, the main flowering occurs at the beginning of spring, coinciding with the first vegetative sprouting. Summer and autumn sprouting do not develop flowers, and only exceptionally can form a small number. In tropical climates, *Citrus* trees bloom continuously throughout the year, although the greatest number of flowers occurs also in spring.

The process of flowering begins with the floral bud induction, followed by the differentiation of primordia and, finally, by the development of the floral organs. Each of these three stages is regulated differently and independently.

The floral bud induction begins with a stimulus received by the leaf that allows the transition from vegetative to floral buds. In *Citrus* trees grown in subtropical areas, the process occurs in the second half of autumn (November-December, in the Mediterranean area), when the plants are exposed to low temperatures and short photoperiods, whereas in those grown in the tropical climate, flowering is induced by water stress after a period of drought. Therefore, environmental conditions, mainly low temperatures and water deficit, regulate floral bud induction in citrus.

In subtropical climates, floral bud differentiation occurs in the second half of winter (February in the Mediterranean area), when the buds release from dormancy and the primordia begin to develop.

Flower development can be subdivided into different phenological stages according to the evolution, both external and internal, of its organs. When the flower bud is very small, before the petals are visible, sepals bend around the other parts of the flower. Subsequently, the flower buds pass through four characteristic phases: in the floral stage A, the petals begin to be visible, although the sepals cover them almost completely. The apical ends of the anthers are located at the stigma level. Subsequently, the corolla grows and pushes the sepals outward, emerging over them. In the floral stage B, the petals reach almost half of their final size and the apical ends of the sepals are halfway between the base and the apex of the petals. At this stage, the stigma emerges from the anthers. In floral stage C, the petals have almost reached their definitive size, although they still overlap each other, forming a completely closed cover. The filaments of the stamens have already lengthened considerably, until reaching the level of stigma. In floral stage D, the petals are close to opening and, indeed, they can do so by applying light pressure over them. In this stage, some dehiscent anthers can already be found, although the number depends on the variety. The next floral stage corresponds to the fully open flower (anthesis), when their parts have already matured. The whole process can last from February, when the flower buds can already be easily recognized, until about 2 months later (April), in full bloom, when most of the flowers have opened.

The flowering of *Citrus* has been studied at the molecular level, having obtained the isolation and characterization of homologous genes of the floral bud induction, *FLOWERING LOCUS T* (*CiFT*; Endo et al., 2005; Nishikawa et al., 2007), and differentiation, *TERMINAL FLOWER*, *LEAFY*, and *APETALA1* (*CsTFL*, *CsLFY*, and *CsAP1*, respectively; Pillitieri et al., 2004a,b), and also *SUPRESSOR OF OVEREXPRESSION OF CONSTANS1* (*CsSL1*), *APETALA3* (*CitMADS8*), and *WUSCHEL* (*CsWUS*) (Tan and Swain, 2007). A homolog of the *FLOWERING LOCUS C* (*FLC-like*), repressor of the *FT* activity, has also been related with the process (Michaels and Amasino, 1999; Michaels et al., 2005). These genes, whose expression is modulated by different endogenous and exogenous factors, interact in a complex way, regulating different phases of induction, differentiation, and floral development (Dornelas et al., 2007a,b). *CiFT* seems to play a crucial role in the flowering of *Citrus* (Endo et al., 2005; Nishikawa et al., 2007).

The Genus Citrus. https://doi.org/10.1016/B978-0-12-812163-4.00011-5

11.2 Type of inflorescences

Citrus trees never develop all their buds into flowers. Thus, some of them do not differentiate, maintaining vegetative growth and contributing to the polycarpic growth habit, as shown in other species.

The floral buds are mainly produced on small branches formed the previous year, but may also appear sporadically on older branches.

Citrus species produce leafy and leafless flowered shoots, the number of both flowers and leaves varying among them. Thus, there are single-flowered and multiflowered leafy and leafless shoots, and vegetative shoots (VS) (Fig. 11.1), the number and proportion of each type of shoots depends on the species and varieties and is affected by climatic factors and the status of the plant.

- Multiflowered leafless shoot (RF)
- Multiflowered leafy shoot (BM)
- Single-flowered leafless shoot (FA)
- Single-terminal flowered leafy shoot (BC)

As a rule, VS and BM are proportionally the most abundant shoots. In general, very heavy blooms, typical of weak or old trees, have a higher proportion of RF and FA to the detriment of VS. On the contrary, when the bloom is scarce the latter predominate.

11.3 Control of flowering

11.3.1 Environmental control

11.3.1.1 Temperature

As mentioned above, floral bud induction in citrus grown in subtropical areas coincides with the end of autumn, when the plants are exposed to low temperatures and short photoperiods. Under natural conditions, cold usually coincides with short

FIG. 11.1 Type of shoots in *Citrus* according to the number of leaves and flowers.

days, so it is difficult to distinguish the specific effects of each of these factors on flowering. By means of experiments carried out under controlled environmental conditions, it has been shown that the exposure of plants to a short-day period is not sufficient to induce flowering. However, short-day period favorably predisposes the tree for floral bud induction due to low temperatures (Lenz, 1969). In any case, in plants subjected to both long-day and short-day conditions, floral bud induction requires low temperatures, whose threshold value is below 20°C, the inductive effect being dependent on the time of exposure (Moss, 1969; Southwick and Davenport, 1986; García-Luis et al., 1992). It has been proposed that a temperature of ≈15°C for 1.5 months is needed to induce *Citrus* trees to flower (Inoue, 1990). On the contrary, plants that grow at high temperatures produce VS and do not form inflorescences, regardless of the photoperiod (Moss, 1969, 1977; Southwick and Davenport, 1986).

Therefore, low temperature in the late fall or early winter is the main factor responsible for floral bud induction and dormancy (Moss, 1969; Altman and Goren, 1978; Valiente and Albrigo, 2004; Southwick and Davenport, 1986; García-Luis et al., 1992). Because of this dependence, flowering intensity undergoes significant variations from one year to another, since lower temperatures favor flowering while high temperatures reduce it (Moss, 1977). Similarly, the increase in *CiFT* gene expression during the period of floral bud induction seems to be strongly dependent on the temperature since the activity of their homologs have been increased by an artificial cold treatment (Nishikawa et al., 2007).

11.3.1.2 Water stress

In tropical climates, citrus trees bloom continuously throughout the year, although the largest number of flowers are produced in spring (Monselise, 1985; Spiegel-Roy and Goldschmidt, 1996). Under these conditions, water stress seems to be the main inducer of flowering (Southwick and Davenport, 1986). This is supported by the fact that, after a period of drought, more or less prolonged, if a rain occurs, the plants bloom (Cassin et al., 1969).

Water deficit increases the accumulation of transcripts of the *CsFT* gene, while reduces those of *CsSL1*, *CsAP1*, and *CsLFY* genes. When the water deficit is removed by irrigation, the expression of *CsFT* returns to the levels prior to the treatment and the activities of *CsSL1*, *CsAP1*, and *CsLFY* increase. Therefore, drought induces flowering by overregulation of the *CsFT* gene (Chica and Albrigo, 2013).

The development of flowers until anthesis requires a minimum temperature threshold value close to 10°C (Lovatt et al., 1984). The moderate temperatures below 20°C during the development of flower delay anthesis, whereas temperatures above 25°C advance it. Therefore, given the great influence of temperature on the development of flowers, the period of full bloom varies considerably from year to year (Moss, 1976).

11.3.2 Other factors affecting flowering

Flowering intensity depends on several factors such as the cultivar, the age and condition of the plant, cultural practices, and the crop load and harvest time.

11.3.2.1 Citrus cultivar

Some varieties, such as the Navel sweet orange group, tend to bloom more abundantly than others like Satsuma, whose flowering is usually more moderate and delayed than the former. The lemon, lime, and citron can bloom any time, within the period of activity (Nir et al., 1972).

11.3.2.2 State and age of plants

Weak or aging trees bloom much more intensely, suggesting that their reproductive behavior constitutes a survival strategy (Bustan and Goldschmidt, 1998).

11.3.2.3 Carbohydrate content

It has been proposed that carbohydrate may be a determinant in the process of flowering, although the evidence in this regard is only circumstantial (Goldschmidt et al., 1985). Thus, in alternate bearing trees, the concentration of carbohydrates, mainly starch, in leaves, branches, and roots, is much lower in ON trees than in OFF ones (Jones et al., 1970, 1975; Goldschmidt and Golomb, 1982). The fact that the former develops a very low number of flowers, unlike the latter, suggests a possible function of carbohydrates in floral induction.

On the other hand, the removal of a strip of bark from the branches or a simple cut of the bark tissue (girdling or ringing, respectively) in summer or autumn increases flowering. This technique causes a transient accumulation of carbohydrates in the upper part of the girdling by interrupting the phloem flow, suggesting a relationship between carbohydrate concentration and the ability of the branch to flower (Schaffer et al., 1985; Goldschmidt et al., 1985; García-Luis et al., 1995; Li et al., 2003). However, the above effects do not show a causal carbohydrates and flowering relationship. In fact, the promotion of flowering by low temperatures is independent of the carbohydrate level (Goldschmidt et al., 1985), and early fruit thinning, which strongly increases flowering with respect to late harvest, hardly has any effect on carbohydrate content (García-Luis et al., 1988; Martínez-Fuentes et al., 2010).

Therefore, carbohydrates do not seem to have a specific function in the regulation of flowering, although this process requires a threshold of energy that would be supplied by these sugars.

11.3.2.4 Nutritional status

In some cases, mineral element deficiencies produce a weakened VS, which is usually accompanied by abundant flowering. Particularly, trees with low leaf levels of nitrogen show copious blooms and a low vegetative vigor. On the contrary, the excess of nitrogen gives rise to a strong vegetative development and a reduction of flowering. The promotion of flowering as a consequence of nutritional deficiencies seems to be a nonspecific response associated with the vegetative weakness of the tree.

11.3.2.5 Source of nitrogen

Nitrogen metabolism has been indirectly related to the induction of flowering, on the basis that low temperatures or water stress increase the N-ammonium level in the leaves. Furthermore, foliar applications of urea in winter or after moderate water stress increased the number of flowers per tree under experimental conditions and it was related with a raise of N-ammonium content (Lovatt et al., 1988; Albrigo, 1999). It has been proposed that the accumulation of ammonium increases the biosynthesis of polyamines, which could play a role in floral bud formation (Lovatt et al., 1988). Nevertheless, this mechanism of action has not been confirmed.

11.3.2.6 The influence of fruit on flowering

In citrus, fruit reduces flowering when they remain on the tree up to the floral bud inductive period (November-December) (Moss, 1971; Lewis et al., 1964; Goldschmidt and Golomb, 1982; García-Luis et al., 1986; Martínez-Fuentes et al., 2010). So, in the Mediterranean basin, when fruits are harvested before November, they hardly influence the next flowering. But when harvest is carried out later, there is a partial inhibition of flowering, which increases as the fruit remains on the tree. In such circumstances, the number of fruits also strongly influences this repressive process, so that the greater it is, the lower the number of flowers produced in the spring of the next cycle. Therefore, in general terms, a heavy crop harvested late has a strong depressive effect on flowering the following spring, whereas a light crop early harvested positively affects flower production (Fig. 11.2).

The effect of crop load on flowering varies from one cultivar to another: in the very sensitive ones a moderate crop load can reduce the flowering up to practically nil; in some others, however, a heavy crop load hardly affects the flowering.

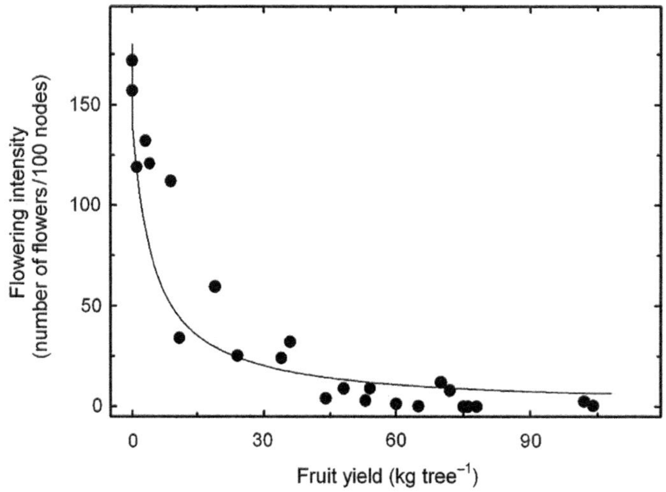

FIG. 11.2 Fruit load and flowering intensity relationship in Moncada mandarin. The nonlinear regression is significant at $P \leq .05$. *(From Muñoz-Fambuena N, Mesejo C, Gonzalez-Más MC, Primo-Millo E, Agustí M, Iglesias DJ. 2011. Ann. Bot. 108: 511–519.)*

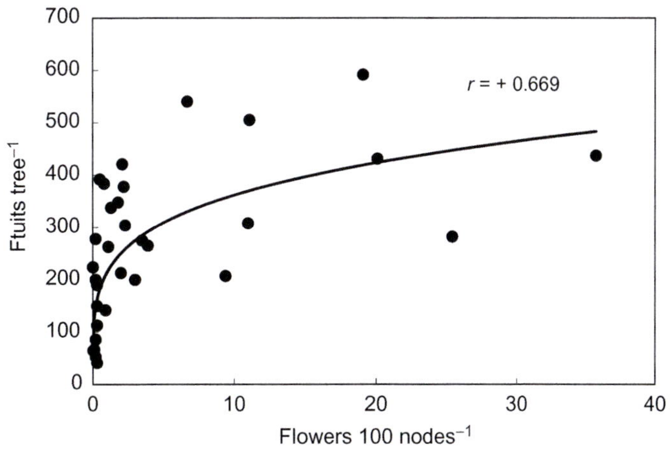

FIG. 11.3 Flowering intensity and yield relationship in Navelina sweet orange.

The effect of fruits and date of harvest inhibiting the potential of buds to differentiate into flowers has been suggested as the cause of alternate bearing (Moss, 1971). After a heavy fruit yield season (ON season) there is a lack of flowers (Fig. 11.3) and, consequently, a lack of fruit yield (OFF season) that, in turn, allows the next flowering and fruit yield, and so on. Alternate bearing mainly affects late ripening citrus varieties, producing enormous annual economic losses (Schaffer et al., 1985). This effect can be counteracted by advancing the harvest date (Goldschmidt and Golomb, 1982; García-Luis et al., 1986; Martínez-Fuentes et al., 2010).

Besides, in the ON year, trees produce many fruits of small size, commercially unappreciated. On the contrary, in the OFF year, trees produce few and large fruits of thick and rough skins, also unappreciated by the markets.

In *Citrus*, it has been shown that fruit inhibits flowering by repressing *FLOWERING LOCUS T* (*FT*) gene expression in leaves (Muñoz-Fambuena et al., 2011). Indeed, the expression of the *CiFT* gene progressively increases during the floral bud inductive period (November) in the leaves of the OFF trees, whereas in those of ON trees the increase does not occur and gene activity remains at a very low level (Fig. 11.4). It agrees with the effect of fruit upregulating the activity of the repressors *CsTFL1*, *CsTFL2*, and *FLC-like* genes.

A similar effect was found when comparing the expression of the *CiFT* gene in the buds of the OFF and ON trees both in November, at the time of floral bud induction (Muñoz-Fambuena et al., 2012a). The fruit also inhibits the expression of the genes of floral differentiation in buds, *CsAP1* and *CsLFY* genes.

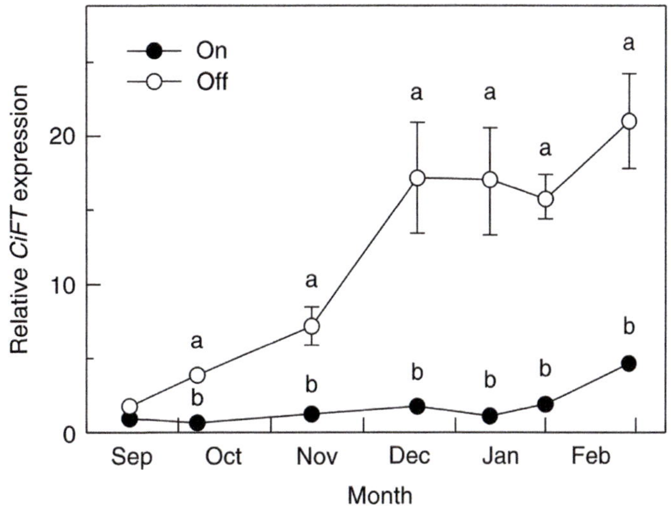

FIG. 11.4 Time course of *CiFT* expression in Moncada mandarin leaves from September to February for fully loaded (ON) or without fruit (OFF) trees. Different letters indicate significant differences ($P \leq .05$) for each sampling date. *(From Muñoz-Fambuena N, Mesejo C, Gonzalez-Más MC, Primo-Millo E, Agustí M, Iglesias DJ. 2011. Ann. Bot. 108: 511–519.)*

11.3.3 Control of flowering

11.3.3.1 The application of gibberellic acid

Gibberellic acid (GA_3) applied at the floral bud inductive period significantly reduced the blossom (Monselise and Halevy, 1964). Treatments applied in mid-November at 25 mg L^{-1} concentration have been shown effective to reduce flowering by 50%, approximately. Treatment also reduced the number of sprouted buds and changed the shoot-type distribution, increasing the number of VS and reducing that of the reproductive ones. The treatment has been shown to be effective in the sweet orange group (Moss, 1970), Clementine (Deidda and Agabio, 1977), Satsuma mandarin (Iwahori and Oohata, 1981; Guardiola et al., 1982; García-Luis et al., 1986), and Tahití lime (Davenport, 1983).

Accordingly, the opposite effect, that is, to promote flowering, has been tried by applying some inhibitors of the gibberellin (GA) biosynthesis, such as triiodobenzoic acid (TIBA), 2-chloroethyl trimethyl ammonium chloride (Cycocel; CCC), benzothiazole-2-oxyacetate (BTOA), and *N*-dimethylamino succinamic acid (SADH). These products have been applied one or several times, as foliar sprays, at a concentration of 500 mg L^{-1} of TIBA, 500–3000 mg L^{-1} of CCC, 25–50 mg L^{-1} de BTOA, and 500–2500 mg L^{-1} de SADH (Monselise et al., 1966; Nir et al., 1972; Monselise and Goren, 1969; Iwahori, 1978; Davenport, 1983). Although in some cases flowering has been promoted, results have been erratic and, consequently, these substances are not used in citriculture nowadays. Similarly, paclobutrazol (PBZ), which is a powerful inhibitor of the GA biosynthesis, has also been used with this purpose (Iwahori and Tominaga, 1986; Delgado et al., 1986a,b; Harty and Van Staden, 1988; Ogata et al., 1996). This compound can be supplied either to the soil, dissolved in the irrigation water (2.5–10 g $tree^{-1}$), or as foliar sprays (up to 1000 mg L^{-1}). Depending on the concentration applied, PBZ increases bud sprouting and flowering, increasing the number of flowering leafless shoots and reducing the vegetative ones and, consequently, reducing vegetative growth and fruit set (Greenberg et al., 1993).

The effect of GA_3 reducing flowering intensity is mediated by a downregulation of *CiFT* gene expression, whereas PBZ upregulates it (Muñoz-Fambuena et al., 2012b). But the effect of PBZ is mediated by the number of fruits remaining on the tree at treatment date; thus, the fruit counteract the effect of PBZ and no increase in the number of flowers is obtained in an ON year (Martínez-Fuentes et al., 2013).

In lemon, the application of GA_3 also inhibits the flowering induced by drought (Nir et al., 1972). In general, the application of GA_3 during the vegetative growth allows to control the out-of-season blooms (Monselise et al., 1981).

It is important to note that GA_3 applied when fruits have not yet completed the process of coloration will cause a delay in color change, which may have commercial repercussions.

11.3.3.2 Water deficit

A prolonged period without irrigation, which is not accompanied by rains, can induce flowering out of season (Nir et al., 1972). The degree of induction is proportional to the intensity and duration of water stress (Southwick and Davenport, 1986). This effect was used in lemon to extend the period of production and commercialization by inducing extemporaneous bloom (Davis and Albrigo, 1994). In practice, the suppression of watering in the middle of summer, for about 2 months, causes flowering toward the end of summer or early autumn. The resulting fruits can be harvested at the beginning of next summer.

11.3.3.3 Girdling or ringing branches

Girdling promotes flowering in citrus (Furr and Amstrong, 1956; Cohen, 1981; Monselise et al., 1983; Goldschmidt et al., 1985; Erner, 1988; Agustí et al., 1992). Best results, in terms of increasing flowering, are achieved when girdling (or ringing) is performed in a period between midsummer and the end of autumn. Its effect is through a stimulus on autumn bud sprouting rather than a direct effect on floral bud induction (Agustí et al., 1992).

This technique is applied to the trees with excessive vegetative vigor and low ability to flower, in order to reverse this trend and improve their productivity. It can be also used to advance the production of young trees, usually 5-to-6-year-old trees, although it must be taken into account that girdling can delay tree development. It has also been reported that girdling promotes leafless inflorescences (flower clusters) and inhibits vegetative growth, which can be an inconvenience for adult trees with moderate or scarce vigor (Erner, 1988).

11.4 Pollination and fertilization

Pollination consists in the transport of the pollen grain from the anther to the stigma, where it is deposited. This process can be carried out with pollen grains from the same plant, or from another plant; in the first case the process is called self-pollination, in the second one cross pollination, unless the pollen comes from another plant of the same clone, in which case it is still a process of self-pollination.

Cross pollination in citrus fruits is mainly produced by bees, which carry the pollen from the flowers of a certain species or variety to those of another one, whereas the wind has little importance in pollinating these species. This is because citrus pollen is heavy, viscous, and adherent and, therefore, it is difficult to be displaced by wind, but easy to be transported by insects. The bees go to the flowers attracted by the aroma and by the abundant nectar they secrete. When it lands on the flower, the pollen remains adhered to its abdomen and can be transported to other flowers. Cross-pollination under natural conditions takes place mainly in this way.

In some flowers, the stigma is considerably above the anthers due to the short length of their filaments with respect to the height of the pistil. In this way, it is difficult for the pollen to deposit on the stigma of its own flower, avoiding self-pollination and favoring cross pollination.

Self-pollination can also occur through the direct contact of the anthers with the stigma or by the displacement of the pollen at a very short distance, in both cases within the same flower. Protandry, by which the pollen matures before the anthesis, favors this type of pollination, since the anthers open and begin to shed pollen while they are still pressed against the stigma in the closed flowers. This allows pollination before opening the flower, and encourages self-pollination.

The sticky liquid secreted by the stigma not only retains pollen grains, but also provides the right conditions for germination. Generally, the stigma is receptive from 1 to 3 days before the anthesis until 6–8 days after it. The effective pollination period (EPP), measured as the capacity to set seeds after pollination, is 8–9 days for *Citrus clementina* and *Citrus sinensis*, and 2–3 days for *Citrus unshiu* (Mesejo et al., 2007).

An important characteristic that affects some varieties in relation to pollination is *self-incompatibility*. This phenomenon consists of the impediment of the pollen tube development in the pistil itself when the pollen is self-incompatible. Therefore, the ovules cannot be fertilized and, thus, the embryo cannot develop by self-pollination, even if the gametes are viable. Normally, the pollen grain germinates upon contact with the exudate of the incompatible stigma; the growth of the pollen tube begins, but it cannot progress beyond the stigmatic papillae, and it is normal that under such circumstances an irregular deposit of callose is formed. Therefore, self-incompatible varieties can only be cross pollinated, which is essential for fruit setting for nonparthenocarpic varieties.

The pollen grain absorbs water from the stigmatic liquid and develops the pollen tube, which penetrates the stigma and descends through a stylar canal until it reaches the locules. However, the pollen tubes can also penetrate through the intercellular spaces of the cortical cells of the style. This path requires constant changes in the direction of pollen tube growth, which causes an abundant deposition of callose. Once it reaches the locular cavity, the pollen tube is directed toward the ovule, where it penetrates through the micropyle. Afterwards, it crosses the nucellus until reaching the embryonic sac, in which it is introduced and disorganizes one or both synergids. The pollen tube carries near its end the two spermatic nuclei, one of which fuses with the oosphere, giving rise to the zygote and, thus, complete the fertilization. The two polar nuclei fuse with the second spermatic nucleus, originating the triploid nucleus of the endosperm (Frost and Soost, 1968).

The time that elapses between pollination and fertilization is variable, since pollen tube development can be affected by environmental factors. However, it is normal to occur between 2 and 8 days.

11.5 Fruit set

During the week after anthesis, the petal fall, and about 2 weeks later the style with the stigma also abscise, marking the transition from the ovary to the developing fruitlet.

In general, a citrus tree produces a number of flowers higher than that of fruits that it is able to maintain and to develop fully. From petal fall to the end of June or beginning of July (Northern Hemisphere) a massive abscission of ovaries and developing fruitlets occurs, and the more intense the flowering, the greater the number of abscised fruitlets (Erickson and Brannaman, 1960). During this period, some small fruits grow faster, keeping their dark green and bright color, while others grow more slowly or stop to grow, acquiring a dull yellowish color. The latter fall from the tree after a short time, while the former mostly continue their development and reach maturity. This process is called "fruit set" and, in citrus, it can be defined as the transition from the stage of slow development of the ovary to the rapid growth of fruitlet. Fruit set is a highly regulated process to face a competitive situation among the reproductive organs and even between these and VS that are in active development during this period. This competence is important for the tree, because, through certain signals, it determines the number of fruits that can pass to the rapid development phase. Hence, fruit set is the main factor limiting fruit production in *Citrus*.

Usually, the number of fruits that reaches maturity does not exceed the 10% of flowers, and it is normal for the percentage of fruit set to be between 1% and 5%. In the case of excessive blooms, the percentage of fruit set may drop to values close to 0.1%.

Characteristically, abscission of the developing organs presents two peaks, corresponding to two periods of intense abscission, separated by a time interval in which it is less pronounced. The two periods of abscission usually appear overlapped, to a greater or lesser extent, although normally the maximums stand out clearly.

These two peaks correspond to two characteristic stages of development. The first occurs mostly during the 3–4 weeks after petal fall and mainly affects ovaries and fruitlets in an early stage of development (between 5 and 10 mm in diameter), although some complete flowers can also be detached. The second peak of abscission (known as "June drop" in the Northern Hemisphere) begins about 2 weeks after the first one, extending from the end of May to the end of June. In this period, small developing fruitlets abscises until they reach a certain size (3–4 cm in diameter for the oranges and 2–3 cm for the mandarins), from which the risk of abscission is much lower. In general, most of the fruits that exceed this second abscission period remain in the tree until harvest.

The pattern of fruitlet abscission may vary depending on the variety, climate, and intensity of flowering. Regarding the variety, fruit set ability is usually greater in those that produce seeds than in the asperms, and within the latter there are great differences in terms of their parthenocarpic ability. The climatic conditions can modify the abscission process, moving the peaks over time, and making them appear more or less overlapping. Regarding flowering intensity, the greater the number of flowers, the more pronounced the fruit drop peaks, as mentioned above.

The final cause of fruitlet drop is the formation of an abscission layer in which there is a weakening of the tissues. During the first drop period the abscission layer appears where the peduncle joins the stem (ZA: abscission zone A). During the second one, when the fruitlets have reached a more advanced stage of development, they are detached from the calyx disk (ZC: abscission zone C).

11.6 Parthenocarpy

As a general rule, in seedy citrus varieties, pollination is required at anthesis for a successful ovule fertilization as a previous step to the development of seeds. Normally, cell division and, consequently, ovary growth slow down during the anthesis, until the pollination and fertilization occurs. Afterwards, the development of the ovary reactivates to form the fruit and seeds. In these varieties, the lack of pollination causes the fruit set to fail, leading to the fall or premature abscission of the ovary.

On the contrary, many varieties of *Citrus* species can naturally set fruits without seeds. This phenomenon, called parthenocarpy, is defined as the ability to produce fruits without fertilization of the ovules and, therefore, without seeds development. Parthenocarpy depends on a genetic factor that favors the maintenance of a relatively high hormone level in the ovary during the anthesis and immediately after it, regardless of pollination and fertilization. The hormonal nature of the stimulus that induces fruit set is confirmed by the fact that parthenocarpy can be artificially induced by the application of certain phytohormones.

The most frequent cases of parthenocarpy in citrus fruits are found in:

(a) Self-incompatible varieties that are unable to produce seeds by self-pollination and which, in the absence of cross pollination, can develop fruits without seeds (facultative parthenocarpy). This is the case, for example, of some hybrid varieties of mandarin, such as Nova, Fortune, Nadorcott, Moncada, etc., or some highly parthenocarpic Clementines, such as Marisol. However, some other self-incompatible varieties with low parthenocarpic ability show low productivity, unless they are pollinated with compatible pollen, setting and developing a large proportion of seedy fruits. This is the case of some Clementine cultivars originated by bud mutations.

(b) Varieties with female gametic sterility, which are unable to produce seeds. This kind of sterility appears absolutely in the triploid varieties. It also occurs in other cultivars, such as Navel sweet orange, Satsuma mandarins, and Marsh seedless grapefruit, where a generalized degeneration of the embryo sacs occurs. In a few cases, some of them may develop seeds in a very small number.

Parthenocarpy is of great importance in cultivated varieties of citrus cultivars for fresh consumption, since the market requires seedless fruits.

11.7 Endogenous regulation of fruit set

The regulation of fruit set is an important physiological process, since it determines the final number of fruits that reach maturity and, therefore, to a large extent, crop yield. Hormonal content of the ovary, as well as the capacity of nutrient supply (photoassimilates) to the developing fruitlets, is of great importance in the process.

11.7.1 Influence of hormone levels

Pollination and the subsequent fertilization of the ovules activate the hormonal signals required for the transition from the ovary to the developing fruit.

The study of the molecular basis of the fruit setting indicates that there are several phytohormones regulating the process [GAs, auxins, cytokinins, abscisic acid (ABA), and ethylene], whose endogenous levels are regulated by the activity of the genes that encode enzymes of their biosynthesis and catabolism. All of them usually present peaks of activity shortly after anthesis, which indicate that they are part of the hormonal stimulus that regulates fruit set.

11.7.1.1 Gibberellins

In the ovaries and developing fruitlets, GAs belong to two different biosynthetic pathways: the 13-hydroxylation and the nonhydroxylation routes, the former being the one that exerts a major function in fruiting (Turnbull, 1989; Talón et al., 1990b; Poling, 1991). In the 13-hydroxylation pathway, GA_{53} is metabolized in several stages to give GA_{19}, which is converted to GA_{20} by the enzyme GA20-oxidase, and subsequently in GA_1 by the activity of GA3-oxidase. Gibberellin GA_1 and its precursor GA_{20} are catabolized by GA2-oxidase, producing, respectively, GA_8 and GA_{29}, which are considered inactive forms (Yamaguchi, 2008; Hedden and Thomas, 2012).

GA20-oxidase activity mainly regulates the flow of the GA_1 biosynthesis pathway. In citrus, two homologous genes have been found, which encode enzymes of this type (Huerta et al., 2009).

GA_1 is considered to be the biologically active GA, which seems to play a major role in set and early development of citrus fruit (Talón et al., 1990b, 1992). Accordingly, it has been shown that in seedless (Salustiana) and seedy (Blanca) sweet orange cultivars, a GA_1 peak appears in the ovary at anthesis. Besides, in a seedy sweet orange cultivar (Pineapple) it has also found that pollination induces the synthesis of GA_1 in the ovaries, while, when pollination is prevented by emasculating flowers, the level of this hormone reduces in the ovaries and fruitlet abscission increases. This effect, however, can be counteracted by applying GA_3 that increases fruit set of the nonpollinated ovaries. Moreover, the application of PBZ (an inhibitor of the biosynthesis of GAs) to the pollinated ovaries reduces GA content and increases abscission (Ben-Cheikh et al., 1997).

In this cultivar (Pineapple), the greatest synthesis of GAs takes place in the ovules, after pollination, coinciding with the maximum expression of the genes GA20ox2, GA3ox1, and GA2ox1. In the seedless parthenocarpic variety Washington Navel the level of GAs in both the ovules and the walls of the ovary is lower than in the Pineapple and their synthesis seems to be regulated constitutively. The higher level of GA_1 detected in both the fertilized ovules and the walls of the ovary of the seedy variety with respect to the sterile one is associated with a lower rate of early abscission of ovaries and fruits during the postflowering period. Later on, during the June drop, the concentration of GA_1 in the pericarp of the developing fruitlets remains relatively high in both varieties, reaching similar levels, although it decreases with respect to that reached by the ovary wall after the anthesis. In seedless varieties, the ability to maintain the concentration of GA_1 in the pericarp of the fruit during the June drop is associated with its ability to overcome the abscission and, therefore, with the parthenocarpic ability (Bermejo et al., 2016).

When comparing two varieties of seedless mandarins with different parthenocarpic ability, it was found that Satsuma (highly parthenocarpic mandarin) contains higher concentrations of the GAs of the 13-hydroxylation pathway (except for the inactive GA_8) than Clementine cv. Oroval (with low natural parthenocarpic ability) (Talón et al., 1992). At petal fall, the GA_1 level in the Satsuma ovaries was five times higher than that of Clementine. In the latter, GA_3 applied locally to the ovaries significantly improves set, but not in the former. In contrast, PBZ increases natural abscission in Satsuma but not in Clementine (Talón et al., 1992). Recently, it has been shown that GAs trigger and maintain cell division in the ovary walls causing parthenocarpic fruit set (Mesejo et al., 2017). The autonomous activation of GA synthesis in Satsuma ovary preceded cell division and *CYCA1.1* upregulation (a G2-stage cell cycle regulator) at anthesis setting a high proportion of fruits, whereas Clementine lacked this mechanism increasing GA biosynthesis and *CYCA1.1* upregulation, failing in fruit set. In situ hybridization experiments revealed a tissue-specific expression of GA20ox2 only in the dividing tissues of the pericarp. Furthermore, *CYCA1.1* expression correlated with endogenous GA_1 content and with GA_3 treatment, which stimulated cell division and ovary growth, mostly in *C. clementina*. Instead, PBZ negated cell division and reduced fruit set. Results suggest that in parthenocarpic citrus the specific GA synthesis in the ovary walls at anthesis triggers cell division and, thus, the necessary ovary growth rate to set fruit (Mesejo et al., 2017).

In the Clementine cultivars Marisol and Clemenules, whose fruits, in the absence of cross pollination do not develop seeds and show high and low parthenocarpic ability, respectively, GA_1 levels in the ovaries were higher in the first one than in the second (Mesejo et al., 2013). All these evidences indicate that the endogenous content of GAs in the ovaries—especially that of GA_1—is the main factor controlling parthenocarpy and that some varieties—such as the Oroval and Clemenules Clementines—do not reach enough levels of GAs in the ovary to set fruits parthenocarpically.

In the seedy Murcott mandarin, emasculation to prevent pollination drastically reduces GA_1 level in the ovary and fruit abscise. Pollination/fertilization increases the expression of the GA20ox2 and GA3ox1 genes, which encode for the main enzymes that regulate GA_1 biosynthesis through the 13-hydroxylation pathway (Bermejo et al., 2015). These evidences confirm that in the seedy varieties, pollination/fertilization induces the synthesis of GA_1 in the ovules, where their concentration strongly increases, as in the wall of the ovary, although in smaller extension. This hormone stimulates the initial growth of the ovary and prevents early abscission. In seedless varieties, GA_1 which is constitutively generated in the ovary exerts the same functions, so that the parthenocarpic ability of these varieties depends, to a large extent, on the level of this hormone in the ovary. On the other hand, both in seedy and seedless varieties, maintenance of the GA_1 concentration in the pericarp of the developing fruitlet allows to control, in a large extent, late abscission (June drop). Accordingly, as in other physiological processes, GA_1 is the active GA controlling fruit set and its endogenous level is a limiting factor in the process.

11.7.1.2 Auxins

Indole-3-acetic acid (IAA) concentration in the ovary increases strongly short after anthesis, and reaches a maximum shortly after petal fall. This pattern is given in both seedy and seedless fruitlets, although there are somewhat higher concentrations in the latter (Ali-Dinar et al., 1988; García-Papi and García-Martinez, 1984). However, no differences have been found between IAA concentration in the ovaries of seedless varieties with different parthenocarpic abilities, such as Satsuma and Clementine cv. Oroval (Talón et al., 1990a; Mesejo et al., 2013). Accordingly, under natural conditions, the content of IAA does not constitute a limiting factor for fruit set and early fruit development, as demonstrated by the lack of response to exogenous application of auxin.

However, when pollination is prevented experimentally in seedy varieties, IAA concentration in the ovary remains at a very low level (Bermejo et al., 2015). It indicates that in seedy fruitlets the strong increase in IAA biosynthesis, which normally occurs just after anthesis, depends on pollination/fertilization. In fact, the lack of pollination reduces the content of IAA in the ovule and in the pericarp, at the same time that the concentration of GA_{20} and GA_1 in both parts of the ovary also decreases. In contrast, the exogenous application of IAA to the nonpollinated ovaries restores their content in GA_{20} and GA_1. This effect paralleled the expression of *GA20ox2* and *GA3ox1* genes, which encode for the key enzymes in the biosynthesis of GA_1, and occurs only in the ovules, and not in the pericarp, which seems to be more insensitive to IAA. This indicates that, probably, IAA induces the biosynthesis of GAs in the ovules, from where they are transported to the pericarp. It has been confirmed that the application of TIBA (an inhibitor of IAA translocation) to the pollinated ovaries decreases IAA, GA_{20}, and GA_1 concentration in all parts of the ovary. This effect indicates that when the translocation of IAA from the cells of the ovule, where its biosynthesis takes place, to the rest of the ovary is restricted, the biosynthesis of GAs is decoupled. In addition, the activity of GA2ox enzyme (responsible for the catabolism of GA_{20} and GA_1) that increases in nonpollinated ovaries, is also inhibited by exogenous IAA.

Therefore, in seedy fruit varieties (nonparthenocarpic varieties), fruit set depends on successful pollination and fertilization. These processes increase GA levels by upregulating the genes encoding the key enzymes of its biosynthesis (such as *GA20ox2* and *GA3ox1*), and downregulating those related with their catabolism (such as *GA2ox1*), these effects being induced by AIA (Bermejo et al., 2018). Consequently, the increase of the endogenous level of AIA in the ovary after pollination promotes the biosynthesis of GA_1, whose signals trigger the program of fruit development.

From petal fall, IAA concentration declines rapidly with ovary development, so that at the beginning of June drop its content in the fruitlets is very low. This decrease coincides with an increase in the conjugated IAA and with the auxin oxidase activity, responsible for its catabolism.

11.7.1.3 Abscisic acid

In seedy fruit varieties and also in seedless ones with a high parthenocarpic ability, GA_1 increases in the ovary shortly after anthesis, and as a consequence, subsequently, the level of ABA remains low. On the contrary, in the absence of pollination, that is, in the self-incompatible varieties with low parthenocarpic ability (which is the case of many Clementine cultivars) ovary barely has an increase in GA_1 content shortly after anthesis, and after a few days there is a transient increase in ABA, followed by a large fruitlet abscission.

The inability of some Clementine cultivars to set fruits parthenocarpically is not only associated with a low concentration of GA_1 in the ovaries but also with the poor capacity to inactivate the ABA through its conjugation (Talón et al., 1990a, 1992; Mesejo et al., 2013). The exogenous application of GA_3 to low parthenocarpic ability varieties (e.g., Clementine cv. Oroval) markedly reduces ABA content in the ovaries just after anthesis, thus reducing fruitlet abscission. In contrast, the inhibition of the synthesis of GAs with PBZ in the ovaries of parthenocarpic varieties (e.g., Satsuma) increases both the level of ABA and abscission (Zacarías et al., 1995). Treatment with fluridone, an inhibitor of carotenoid biosynthesis and thus indirectly of ABA, delays the abscission of developing fruitlets (Zacarías et al., 1995).

In nonpollinated ovaries of seedy varieties—as in the case of the Murcott mandarin—the ABA content increases shortly after anthesis as a consequence of the low level of GA_1 and causes total abscission of the reproductive organs. On the contrary, pollination/fertilization inhibits the increase of ABA in the ovaries and prevents their abscission (Bermejo et al., 2015).

A high concentration of ABA in the ovaries favors the synthesis of 1-aminociclopropane-carboxilic acid (ACC) that results in ethylene production, which acts as a strong effector of abscission (Gómez-Cadenas et al., 2000).

All these evidences indicate an antagonism between the GA_1 and ABA content, whose balances determine, to a large extent, the abscission of the developing ovaries and fruitlets.

11.7.1.4 Cytokinins

The most abundant cytokinins in the citrus ovaries are trans-zeatin (tZ) and 2-isopentenyl adenine (2-iP) (Hernandez-Miñana et al., 1989; Bermejo et al., 2015), although the riboside of zeatin and isopentenyl adenosine have also been identified in flower buds (Stewart and Barthe, 1984) and ovaries (Hernandez-Miñana et al., 1989). The activity of cytokinins in the reproductive organs of citrus strongly increases at anthesis, reaching a maximum toward petal fall (Hernandez-Miñana et al., 1989). Afterwards, it drastically decreases as fruitlet develops (Saidha et al., 1985). The time course of cytokinin activity in the ovaries is similar in seedy and seedless varieties, although it can differ quantitatively from one variety to another (Hernandez-Miñana and Primo-Millo, 1990; Bermejo et al., 2015). In the seedy variety Murcott, tZ and 2-iP contents are similar in the pollinated and nonpollinated ovaries, indicating that their biosynthesis is constitutive and is independent of pollination or fertilization (Bermejo et al., 2015).

11.7.2 Competition for photoassimilates

As mentioned above, citrus trees produce an abundant flowering that, after petal fall, gives rise to a large number of developing ovaries, which, nevertheless, cannot develop in full to become mature fruits. This is because its carbon consumption would exceed the possibilities of the tree. Therefore, the net assimilation capacity of CO_2 by the leaves as a whole, especially in critical periods of high demand for carbon compounds, is considered a limiting factor for fruiting. In this case, the photosynthetic ability of the tree and, therefore, the available photoassimilates constitute a determinant factor responsible for the percentage of flowers that set per tree. The idea that the tree only retains those fruits that it is capable of maintaining, and that those that are in worse conditions to receive the available nutrients are detached from the tree, is very suggestive and has been assumed practically in a general way. According to this, citrus would have an endogenous self-regulating mechanism that adjusts the production of fruits to its capacity of synthesis of metabolites.

The competition established during the fruit set period means that the ovaries or fruitlets do not maintain an appropriate growth rate, because the flow of nutrients that reaches them is insufficient. In fact, maximum competition corresponds to the peaks of abscission of the reproductive organs.

At the beginning of the first fruitlet abscission period (postflowering), there are many ovaries in the tree with a high rate of cell division and a maximum respiratory activity per unit of tissue weight (Bain, 1958; Blanke and Bower, 1991). The scarce external surface of the ovaries implies a low CO_2 absorption ability, insufficient to cover their carbohydrate needs. Therefore, during this period of fruitlet development, its survival depends mainly on the contribution of photoassimilates translocated from the mature leaves more than a young spring ones, which has a low export ability of carbohydrates since they have not yet completed the development. In this sense, there are evidence that until the fall of styles, the ovaries and the developing VS can compete for photoassimilates partitioning from old leaves, many of which—the oldest—have a very low ability to assimilate CO_2. The process is aggravated by the demand for carbohydrates from other organs of the tree, especially the roots.

On the other hand, ovaries of closed flowers have a high content of starch, which possibly constitutes a carbohydrate reserve. This starch hydrolyses rapidly at anthesis, so that the ovary has enough sugars to cover the needs of this stage of intense cell division. There is also at this developmental stage a peak in the concentration of soluble sugars (glucose, fructose, and sucrose) that decays to low levels in the following 2 weeks. This rise can be explained both by the rapid hydrolysis of starch and by the high hormonal levels that the ovaries present shortly after anthesis, which would enhance its sink effect increasing the sucrose translocation. All this together indicates that when ovary development starts, it consumes their own reserves of carbohydrates and subsequently a strong competition for photoassimilates is established (Moss et al., 1972), whose deficit results in the abscission of a large percentage of ovaries and fruitlets that lasts for a month after petals fall.

During the second wave of abscission (June drop), the remaining fruitlets on the tree has undergone a strong reduction with respect to the initial number of ovaries. However, these fruits enter a period of faster development, and individually, consume more carbohydrates than the ovaries. Maximum carbon consumption and, therefore, competition for carbohydrates by developing fruitlets take place in June (Martínez-Alcántara et al., 2015). As a consequence, and for high flowering intensity trees, there is a decrease in starch content of mature leaves during the period of fruit set that reaches the minimum toward the end of June (Guardiola et al., 1984; Sanz et al., 1987; Nebauer et al., 2014; Martínez-Alcántara et al., 2015). Partition of reserves stored in the leaves, to meet the demand of the developing fruit, indicates that the consumption of these compounds during the June drop exceeds the supply capacity by photosynthesis. On the other hand, the young leaves of spring flush undergo a change from sink to source of photoassimilates, since the beginning of June, as shown by the efflux of ^{14}C-labeled compounds (Kriedemann, 1969) and the metabolic changes that take place (Schaffer et al., 1987). Therefore, from the beginning of the June drop, new leaves are already developed enough to be able to provide photoassimilates to other organs (Kriedemann, 1970). This has a special importance in the young leaves of the leafy inflorescences that are responsible to feed their fruits (Moss et al., 1972; Powell and Krezdorn, 1977). Presumably, this is the reason why flowers of leafy inflorescences are more prone to set fruit than those located in leafless inflorescences.

The second peak of abscission, therefore, represents the fine adjustment made by the tree to couple the number of fruits that will eventually reach the ripening to its potential to assimilate CO_2, considering that the availability of photoassimilates is the main limiting factor of fruit set in this period.

There is much evidence that indirectly show that carbohydrates are strongly related to fruit set:

(a) Defoliation causes ovay and fruitlet abscission (Mehouachi et al., 1995, 2000), while in plants with a low flower/leaf ratio there is a higher percent of fruit set. In Clausellina Satsuma mandarin, total defoliation at anthesis or at the beginning of June drop, causes 100% abscission of the developing fruits, and the level of fruitlet abscission in partially defoliated trees ranged between that of the control and that of the totally defoliated ones (Mehouachi et al., 1995, 2000).
(b) The proximity of young leaves improves the ability to set fruit (Moss et al., 1972; Powell and Krezdorn, 1977; Primo-Millo et al., 1977a). Ovaries of leafy inflorescences set in a greater proportion than those of leafless inflorescences. The highest percentage of fruit set corresponds to the terminal-flowered leafy shoots (see Fig. 11.1), indicating the ovary dependence on nearby young leaves to complete its development.
(c) The application of inhibitors of photosynthesis during the fruit set period causes a massive fruitlet abscission (Moss, 1972; Mesejo et al., 2012).
(d) The reduction of photosynthesis by shading the trees considerably reduces the rate of fruit set and yield (Syvertsen et al., 2003).
(e) Supplementing sucrose to the tree by continuously injecting a solution into the trunk, starting 30 days before anthesis, significantly increases fruit set (Iglesias et al., 2003).
(f) Girdling or ringing trunks and branches, at petals fall, markedly increases fruit set. This technique causes interruption of the phloem (Wallerstein et al., 1978) and, therefore, the accumulation of carbohydrates in the tree organs above the girdle (Fishler et al., 1983; Schaffer et al., 1985; Goldschmidt et al., 1985; Ogaki et al., 1963; Nebauer et al., 2011). Hence, its beneficial effects on fruit set are due to the increase in the availability of photoassimilates by developing ovaries and fruitlets.
(g) Deficiencies of some mineral elements (Mg, Fe, etc.) markedly reduce photosynthesis and negatively affect fruit set (Chapman, 1968; Embleton et al., 1973).

All these aspects indicate that fruit set is strongly related to the availability of carbohydrates (Goldschmidt, 1999).

11.7.3 Interactions between hormones and photoassimilates

Endogenous hormones and carbohydrates can interact during fruit setting at least in two ways:

(1) The exogenous application of GA_3 or cytokinins individually to ovaries at petal fall, considerably increases their probability to fruit set (Hield et al., 1958; Krezdorn and Cohen, 1962; Krezdorn, 1969; Moss, 1972; García-Martinez and García-Papi, 1979; Primo-Millo et al., 1977b).
The use of ^{14}C-labeled carbohydrates has shown that the application of hormones greatly induces their translocation to the ovaries, which is essential for their setting (Kriedemann, 1968; Powell and Krezdorn, 1977; Kadoya and Tanaka, 1972). This phenomenon might explain, in part, the fact that fruit set requires relatively high levels of GA_1 in the ovaries, as occurs in seedy varieties that have been adequately pollinated and fertilized, or in those of highly parthenocarpic ability.

(2) Sugar shortage to the ovaries and fruitlets during fruit set induces ethylene production (Gómez-Cadenas et al., 2000; Iglesias et al., 2006), and, therefore, fruitlet abscission. This hormone acts both in the abscission zone A (ZAA) and C (ZAC). It is supported by the effect of defoliation reducing sucrose concentration in developing fruits and increasing abscission rates that correlate with ABA and ACC content. The sequence of events in the process is as follows: the concentration of ABA increases with the reduction of sucrose supply from the leaves, inducing the synthesis of ACC and ethylene production, which is the final effector of abscission. Under normal conditions, the competition for photoassimilates implies that many ovaries and fruitlets receive an insufficient supply of carbohydrates, which induces the hormonal response that causes their massive abscission. In this process sucrose, ABA, ACC, and ethylene are the main components of the self-regulating mechanism that adjusts the fruiting process to the possibilities of the tree to feed the fruits.

In summary, evidence indicate that the ovaries require high contents of endogenous GA_1 (or exogenous GA_3) at petal fall, whose function is to stimulate their growth. On the other hand, enough supply of photoassimilates, mainly sucrose, is also necessary to support growth. The failure of any of these two assumptions induces fruitlet abscission.

11.8 Factors affecting fruit set

11.8.1 Temperature

High temperatures (>38°C) during the set period, especially if accompanied by dry winds, markedly increase fruitlet abscission. These climatic conditions occur frequently in June causing, sometimes, a fruitlet drop so important that it may reduce crop load considerably. This effect is due to an increase of transpiration producing a loss of water. Low relative humidity contributes to aggravate the problem. As a consequence, there is a transient decrease of leaf water potential that results in a temporary closure of the stomata. The reduction of stomatal conductance leads to a reduction in gas exchange parameters, with the consequent loss of net CO_2 uptake in a period in which developing fruit presents a high demand for photoassimilates. This effect of high temperature is much more pronounced for a low soil moisture, although it can also occur in well-watered soils.

11.8.2 Irrigation

A water deficit during the fruit set period increases fruitlet abscission, and reduces crop load. This effect is most intense during the June drop, when the fruitlets are in a relatively advanced stage of development. The roots detect the lack of water in the soil and respond by biosynthesizing ABA, which is transported to the canopy via the xylem. When this hormone reaches the leaves, it causes stomata closure, reducing the loss of water vapor and net CO_2 uptake, giving rise to the same problems described above for high temperatures. ABA transported to the canopy can also reach the fruitlet and induce the synthesis of ACC, as a step prior to ethylene production, which originates abscission.

The maintenance of soil moisture above a certain threshold value by irrigating appropriately, that is, frequency and water amount, favors fruit setting.

11.8.3 Nitrogen fertilization

Shortly after anthesis, the ovary starts an active cellular division that starts the fruit development stage. During this period, developing ovaries and fruitlets need, in addition to carbohydrates, nitrogen for protein biosynthesis.

Nitrogen deficiency gives rise to an intense fruitlet abscission that leads to a notable decrease in yield. On the contrary, high foliar N levels produce high rates of fruit set and abundant yield (Embleton et al., 1973).

Generally, during ovary and fruitlet abscission N concentration in old leaves reaches a minimum that is restituted at the end of the spring. This temporary deficiency is attributed to the fact that old leaves export their reserve of N to attend to the high consumption of this element by the vegetative and reproductive development (Legaz et al., 1995). The preponderant role of the reserve N in the early stages of fruiting is due to the low absorption of N at the end of winter and the beginning of spring, as a consequence of the scarce development of new roots and low soil temperature. However, at the end of spring, the absorption of N by the roots undergoes a strong increase that compensates for the previous seasonal N deficiency (Legaz and Primo-Millo, 1981). During the first half of spring, the N absorbed is concentrated mainly in the young leaves and to a lesser extent in the ovaries, establishing a competition for N between both organs.

Young leaves appear as an essential organ in the redistribution of N, since initially they receive the greater part of the N absorbed by the roots and later on they export it to the developing fruitlet (Legaz et al., 1982). Accordingly, during the postflowering period, when the development of the ovaries accelerates, a deficit of N may occur, leading the fruitlet to fall.

11.8.4 Mineral deficiencies

Generally, most mineral deficiencies reduce fruit set, decoupling some step of the photosynthetic process, and as a consequence reduces the crop yield (Chapman, 1968; Embleton et al., 1973). Treatments to correct these deficiencies achieve a maximum efficiency when applied at petals fall.

11.8.5 Flowering intensity

As mentioned above, the percentage of flowers that set fruit (expressed as number of fruits that reach maturity × 100/number of flowers) decreases with increasing flowering. Although there is a wide range of flowering densities for which the final yield is independent of the number of flowers, when flowering is scarce the percentage of fruit set is usually higher, and the opposite, for larger number of flowers fruit set decreases (Fig. 11.5). Furthermore, for excessive number of flowers—typical of old or weakened trees—an excessive nutrient consumption by the developing reproductive organs takes place, causing their massive abscission and, consequently, a reduction of crop load.

At high flowering intensities, reproductive organs (buds, flowers, ovaries, fruitlets) advance abscission, whereas at low flowering intensities most of the reproductive organs fall as developing fruitlets (June drop).

11.8.6 Position of the flower in the tree

The position of the flower in the tree considerably conditions its probability of setting fruit, possibly due to the capacity to access to the nutrients. In this sense, the probability of fruit set depends, in large extent, on the type of inflorescence and, hence, on the number of young leaves supporting ovary development (Moss et al., 1972; Powell and Krezdorn, 1977; Primo-Millo et al., 1977a). Accordingly, ovaries located in a single-flowered leafy shoot (BC) (see Fig. 11.1) set in a greatest proportion, followed by those located in multiflowered leafy shoots (BM); the ovaries from multiflowered leafless shoots (RF) fall almost entirely.

The ovaries from inflorescences located on shoots with horizontal growth set in a higher proportion than those from inflorescences located on vigorous shoots with vertical growth. This effect seems to be explained by the translocation in phloem sap, slower in horizontal than in vertical growing shoots, which favors fruiting of the former to the detriment of vegetative vigor.

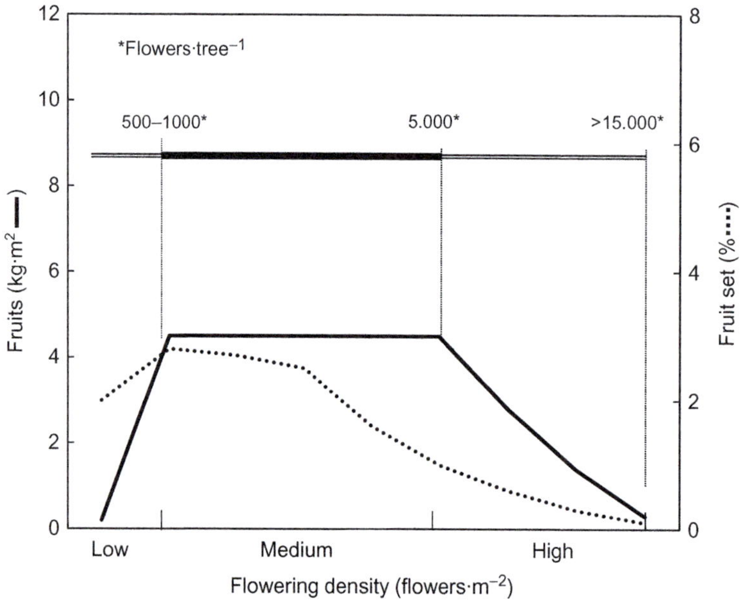

FIG. 11.5 Schematic representation of the relationship between flowering density, fruit set, and crop load in sweet orange. Cultivars must be included in a position in accordance with their tendency to flower.

11.9 Improvement of fruit set: Cultural practices

The application of GA_3 and girdling (ringing) branches are the two main techniques to improve fruit set in citrus.

11.9.1 The application of GA_3

In many species and varieties of citrus, GA_3 improves fruit set when applied individually to the ovaries. Furthermore, when the treatment is carried out to small branches, ovaries located on them also set in a greater proportion (Krezdorn and Cohen, 1962; Krezdorn, 1969; Moss, 1972). However, the application to the whole tree, just after anthesis, does not increase fruit set in the same proportion being much lower, which indicates that the selective treatment limited to small parts of the tree canopy causes their ovaries to be more competitive than others and thus, set in a higher proportion. Accordingly, the application of GA_3 to the whole tree is ineffective in most varieties (Hield et al., 1965; Coggins and Hield, 1968).

In spite of it, GA_3 can be used successfully to improve fruit set in some of the self-incompatible varieties, that is, those with a low parthenocarpic ability, such as some Clementine cultivars. In these varieties, crop load (considered as the total number of fruits harvested per tree) is usually increased with foliar sprays of GA_3 (Soost and Burnett, 1961; Coggins et al., 1966; Del Rivero et al., 1969). The most appropriate date of treatment is shortly after petal fall (García-Martinez and García-Papi, 1979), although in practice, due to the desynchronization of the flowering process, it is recommended when approximately 80% of flowers have lost the petals. Appropriated concentration varies between 5 and 25 $mg\,L^{-1}$ (Damigella et al., 1970; Cuñat et al., 1974; Blondel, 1977).

The application of GA_3 to the entire tree retards abscission of reproductive structures and increases the percentage that reaches the fruitlet stage. This effect, however, becomes significant only for low flower densities but not when the trees flower heavily (Agustí et al., 1982) (Fig. 11.6). This indicates that the interactions described above between carbohydrates and hormones not only affect the action of the endogenous ones, but also that of those applied exogenously.

Bearing in mind that the greater the number of fruits set, the smaller the final size, GA_3 concentration must be carefully established. Therefore, a high concentration can be suitable for varieties with fruits of relatively large size, whereas for small fruit size varieties lower concentrations must be recommended.

11.9.2 Girdling or ringing

Girdling consists of a cut through the phloem, in most cases involving the removal of a strip of bark from the circumference of the trunk or scaffold branches (Goren et al., 2003). Applied after petal fall, girdling reduces the abscission of developing fruitlets and, consequently, increases the number of fruitlets that finally reaches maturity (Damigella et al., 1970; Blondel, 1977; Monselise et al., 1972).

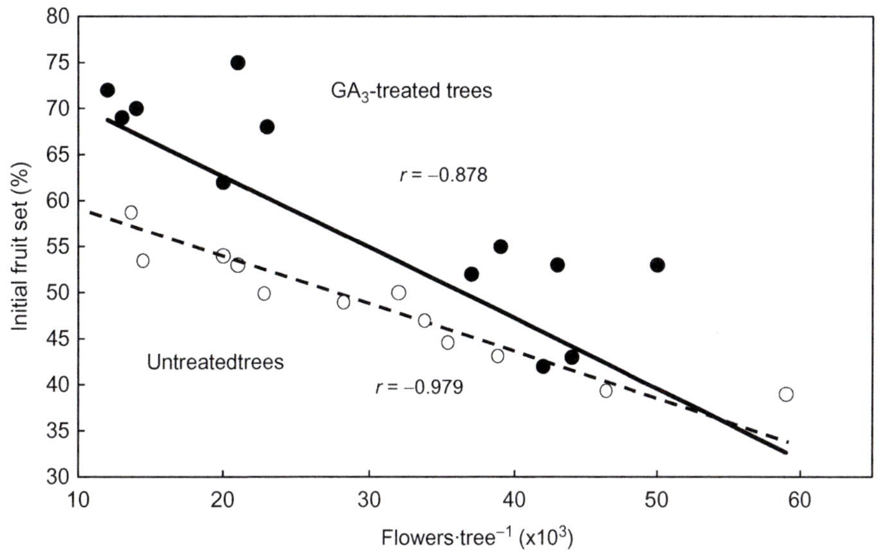

FIG. 11.6 The influence of flowering intensity on the effect of GA3 (10 $mg\,L^{-1}$) sprays to increase the initial fruit set (% of flowers reaching the fruitlet stage) in the Navelate sweet orange. The slopes of the regression lines differ significantly ($P \leq .01$). *(From Agustí M, García-Marí F, Guardiola JL. 1982. Sci. Hortic. 17: 343–352.)*

As stated above, the basis of this practice is the accumulation of assimilates above the girdle by slowing their transport from the leaves (Wallerstein et al., 1974, 1978; Goldschmidt and Golomb, 1982; Rivas et al., 2007). Concomitantly, the concentration of assimilates in the roots of the ringed trees decreases, this effect being more marked when girdling is performed during the period of vegetative activity (Schneider, 1954). Goren et al. (2003) reviewed the physiological and horticultural aspects of girdling.

It is generally accepted that the improvement in fruit set induced by the ringing is due to the greater availability of carbohydrates for the fruits. However, it has been suggested that girdling may alter the hormonal balance of the affected organs, so that promotes the development of the fruitlets or inhibits their abscission. This second hypothesis must still be confirmed (Goren et al., 1971; Wallerstein et al., 1973).

More consistent seems the idea that girdling delays fruitlet abscission and increases quantum yield efficiency of photosystem II of light-adapted leaves (Φ_{PSII}), altering leaf carbohydrate concentrations and photosynthetic performance of various shoot types. Nevertheless, only fruitlets born on leafy shoots increase the final fruit set (Rivas et al., 2007). The delay in fruitlet abscission and the increase in Φ_{PSII} in leafy flowering shoots are the mechanism underlying the progress of fruit set after girdling (Rivas et al., 2007).

11.9.3 Treatment with GA_3 combined with girdling

In some cases of unproductive plantations, it is common to combine foliar spraying with GA_3 and girdling. Girdling acts by accumulating carbohydrates above the girdle, and GA_3 stimulates the sink effect of treated ovaries. Therefore, both treatments complement each other, resulting in the potentiation of their individual effects, in their positive aspect, which is the improvement of the fructification, and in the negative one, which is the reduction of the final fruit size.

To obtain their optimum effectiveness, girdling should be performed within 15–25 days after the application of GA_3 (see Section 11.9.1). As an example, Clementine cv. Fina, a cultivar of seedless fruit and with a very low parthenocarpic ability, increases the crop load higher than 100% with respect to the untreated control as a response to both treatments (Blondel, 1977; Damigella et al., 1970).

11.9.4 Other practices that favor fruit set

11.9.4.1 Flowering control

In old, weakened, or badly cultivated plantations, which generally flower profusely, control of flowering can be beneficial for fruit setting. The inhibition of flowering is carried out by applying GA_3 at a dose of 10–25 $mg\,L^{-1}$, in the middle of autumn (November) coinciding with the period of floral bud induction (Monselise and Halevy, 1964). Treatment reduces the total number of flowers, decreases the proportion of leafless inflorescences, and increases that of the leafy ones, all together increases the setting capability. The application of GA_3 to inhibit flowering is not recommended to the trees with heavy crop, especially if they are late-harvest varieties, since the combined effect can reduce flowering drastically and, thus, the production of the following year.

11.9.4.2 Irrigation and fertilization

Adequate irrigation programming, to maintain an efficient water content in the soil, a balanced fertilization, and the control of the mineral deficiencies are of prime importance to achieve an optimal fruiting. Since irrigation and fertilization are two basic techniques for production, their inadequate practice limits fruit set.

11.9.4.3 Pruning

Pruning performed toward the end of winter has, generally, a positive effect on fruit setting since:

(1) Pruning eliminates numerous weak branches that produce abundant leafless inflorescences. Removing them implies that the number of flowers and, to some extent, the subsequent competition among the reproductive organs is reduced.
(2) Pruning increases the vigor of the subsequent spring flush and, thus, a greater number of leafy inflorescences appears, with a higher ability to fruit set.

11.9.4.4 Arching of branches

As mentioned above, branches with a tendency to grow upward usually produce erect VS fruiting scarcely. This effect is especially important in young trees, which have not yet reached full development, and in excessive vigorous-adult trees, and can be solved by arching down the branches. It is carried out by tying them to other thicker branches or to the trunk, or by tying them to spikes driven into the ground.

This technique drives the phloem sap toward the lateral sides of the canopy, thus moving more slowly along the branches. In this way, excessive vegetative vigor of the tree is largely corrected and, above all, fruiting of arching branches is strongly improved.

11.10 Fruit development

Throughout the fruiting process, the ovary grows and develops into a mature fruit (see Chapter 12 for a wider description). Fruit development begins shortly after anthesis and lasts between 6 and 15 months, depending on the species and variety. In the Mediterranean area, the precocious varieties mature at the beginning of October, whereas in the late ones this process can be extended until the middle of the following spring.

The pattern of growth and development of citrus fruit, measured in size (diameter, length, or volume) or weight, follows a sigmoid curve that can be divided into three stages (Fig. 11.7) (Bain, 1958).

Stage I corresponds to an initial period of moderate growth from the anthesis to the end of the physiological fruitlet abscission (June drop). This stage lasts over 2 months, in which cell division predominates and fruitlet increases their tissues in cell number.

The increase in fruitlet size is mainly due to the growth of the ovary wall tissues (pericarp) that develop the peel of the fruit. At the end of this stage, peel reaches its maximum thickness.

Stage II covers from the end of the June drop until shortly before fruit color change, and corresponds to a period of rapid fruitlet growth. It ranges from 3 to 5 months for precocious or late varieties, respectively, and ends when the fruit is close to reaching its final size.

In this stage, fruit increases in size by elongating the cells rather than by cell division. It is characterized by the expansion of the locules that form the segments. Simultaneously, cells of the external tissues of the vesicles elongate, while those inside them increase considerably in size. In the latter, the vacuoles develop remarkably, and their cytoplasm is confined to a narrow strip around the cell wall. As this stage progresses, spaces of schizogenic and lysogenic origin appear between the walls of the neighboring cells.

As a consequence, the juice vesicles grow to reach their maximum size, filling the segments and developing a structure of mature tissue. When filled with juice, vesicles exert a strong pressure over the peel that undergoes a reduction in thickness. To adapt to the growth, cells of the epidermis of the peel continue to divide anticlinally until the fruit matures.

Phase III corresponds to the period of fruit ripening, the changes associated with this process take place following a nonclimacteric model. This stage is characterized by a reduced fruit growth rate, due to the remaining cell elongation.

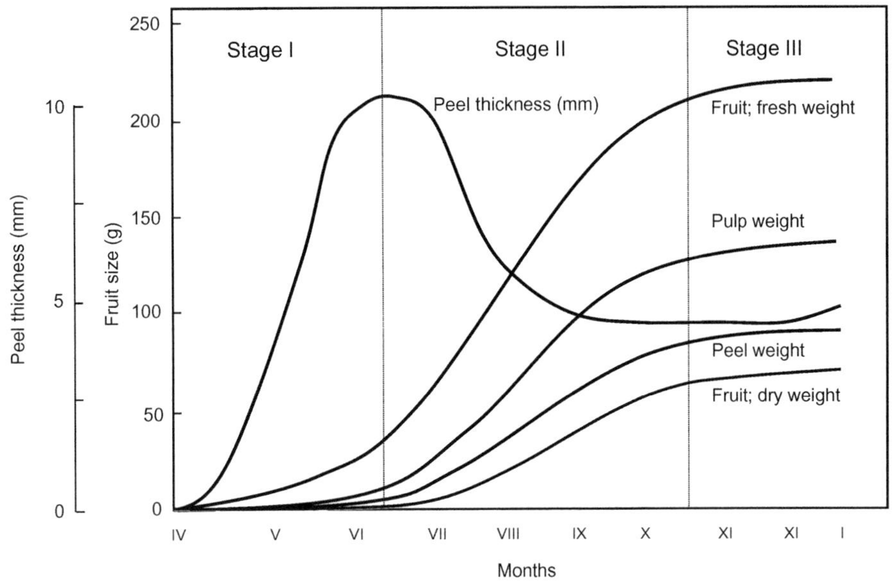

FIG. 11.7 Schematic representation of tissue development of a typical sweet orange fruit. Stages I, II, and III indicate the period of cell division, cell enlargement, and senescence, respectively.

11.11 Factors affecting fruit development

11.11.1 Endogenous factors

11.11.1.1 Genetic potential of the fruit

Genetics determines the differences in size between fruits of different species and varieties. Nevertheless, there is a remarkable variability among them and even among fruits for a given tree, increasing or decreasing to some extent.

11.11.1.2 Hormone levels

As mentioned, the hormonal content of the ovary at anthesis (GAs, auxins, cytokinins, etc.) remarkably increases. Since these hormones participate in the induction of cell division, their levels can determine the number of cells formed in the ovary wall, conditioning the potential for fruit development.

11.11.2 Tree age

In general, young trees produce large fruits with a low percentage of juice and a coarse peel. This may be due to the high vigor of the young trees that reduces fruit set (see Section 11.8.6), giving rise to a fruit much larger than the characteristic one of the variety. When the tree reaches the adult stage and full production, the size and fruit characteristics tend to normalize. On the contrary, old trees tend to produce small-sized fruits, partly because they generally bloom profusely, establishing a strong competition among the reproductive organs that reduces the fruit development and the final size.

11.11.3 Fruit position on the tree

Fruit size depends, to a great extent, on the type of inflorescence, thus the leafy shoots stimulate fruit growth. Differences in size of fruit from different types of inflorescences are stablished shortly after fruit set and increase with time for the leafy ones until ripening. This effect is due to the role of the young leaves as a source of carbohydrates. However, leaves act as sources of photoassimilates when complete their expansion, since during their development they act as a sink and compete with fruitlets at the early stages of development (Moss et al., 1972; Guardiola, 1981, 1992; Agustí et al., 1999).

11.11.4 Foliar area

Fruit weight is also affected by the total leaf area and the number of fruits per tree. Thus, the larger the leaf area per tree the larger the fruit size (Fishler et al., 1983). This confirms that a deficient flow of carbohydrates from the leaves to the fruit restricts their development and, on the contrary, the fruit growth is encouraged when the sap flow is restored.

11.11.5 Seed number

The presence of seeds increases the fruit size, in a dependent manner. This is because the ovaries containing fertilized ovules have higher levels of hormones (Bermejo et al., 2015).

11.11.6 Flowering intensity

It has been proposed that the number of flowers per tree influences fruit development and its final size (Agustí et al., 1999). At anthesis, the ovaries from heavy flowering trees are smaller in size than those from scarcely flowering trees (Guardiola et al., 1984; Guardiola, 1992). Subsequently, the latter grow faster and eventually reach a moderately higher size (Agustí et al., 1999).

This effect indicates that, at least in part, fruit size is established at the early stage of development, although later on some factors, as the competition among developing fruits, also affect fruit growth.

11.11.7 Competition among developing fruits

In citrus, the final fruit size shows an inverse relationship with the number of fruits that reaches maturity (Goldschmidt and Monselise, 1977; Guardiola, 1987; Agustí et al., 1999). Both variables are related to a curvilinear (potential) function (Fig. 11.8), thus for a low number of fruits per tree there is a very steep slope. A short difference between the number of

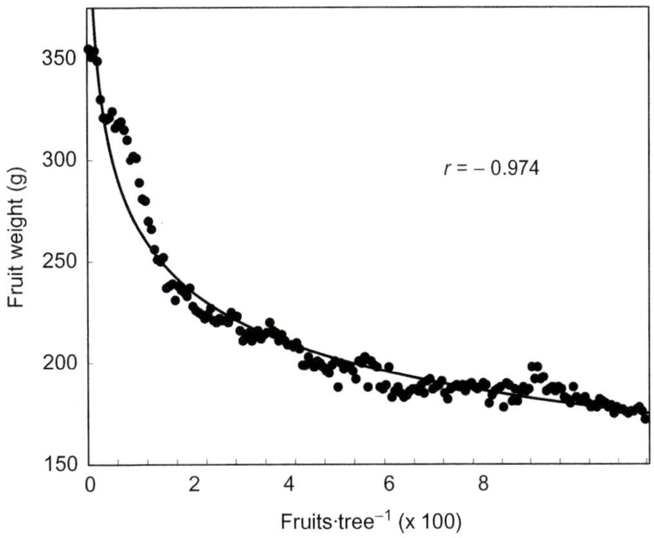

FIG. 11.8 Number of fruits per tree and fruit size relationship in the Navelina sweet orange ($P \leq .01$).

fruits per tree changes markedly fruit size, and on the contrary for a large number of fruits per tree, the fruit size scarcely changes even for large differences in the number of fruits per tree (Guardiola, 1987).

This phenomenon is attributed to the competition between developing ovaries from the flowering onwards. Consequently, the greater the number of developing organs, the stronger the competition for photosynthates and mineral elements, and since the restriction of the nutrients supply limits fruit growth, competition among developing organs may limit their final size.

11.11.8 Environmental factors

11.11.8.1 Temperature

Fruit growth undergoes notable changes throughout the development largely due to temperature (Agustí et al., 1999). Low temperatures in spring delay flower development and have a marked depressive effect on fruitlet growth. On the contrary, the relatively high temperatures during the end of spring and summer, together with a strong luminosity, favor fruit growth by promoting the photosynthetic activity and, thus, increasing carbohydrates supply to the fruit. However, high temperatures in summer, together with a low relative humidity, produce a high rate of evapotranspiration which, in turn, causes stomata closure along the day, reduces the net assimilation rate of CO_2, and decreases the carbohydrate supply to the fruit reducing its growth rate.

Low temperatures during late autumn and winter adversely affect the final phase of fruit growth, which is strongly depressed at temperatures below 5°C.

11.11.8.2 Relative humidity

The effects of low relative humidities, associated with high temperatures, on fruit growth rate have already been discussed in the preceding section.

Water deficits in summer, due to lack of irrigation and/or rain, can cause an irreversible delay in fruit growth, thus reducing its final size (Erickson and Richards, 1955).

In addition, it has been widely observed that rain in summer, and especially in autumn, improves fruit size, irrespective of the irrigation.

11.11.9 Cultural practices

11.11.9.1 Irrigation

Soil moisture has a transcendental role in fruit growth, reducing the fruit size when frequency and dose of water supply are not adequate. The sensitivity of the fruit to water stress is so marked that fruit growth has been proposed as an index of the tree water needs.

11.11.9.2 Fertilization

Potassium content in leaves correlates positively with fruit size and, therefore, an adequate potassium fertilization is necessary to maintain a high rate of fruit growth.

The deficiency of some mineral elements, such as iron and zinc, also reduces the final fruit size and, thus, the correction of these deficiencies is an essential requirement to obtain fruits of normal size and quality. But, once the foliar adequate concentration of these microelements has been reached, subsequent applications do not have any favorable effect (Embleton et al., 1973).

Nevertheless, opposite effects have also been observed. For instance, a strong nitrogen fertilization tends to increase the number of fruits that set and, consequently, reduces its final size (Embleton et al., 1973).

11.11.9.3 Treatments for setting

As mentioned above (see Section 11.9), in some varieties, under certain circumstances, GA_3 applied as a foliar spray, girdling branches or both treatments in combination, are used to improve fruit set.

However, this increase in the number of fruits, in turn, causes competition among them and causes a reduction in the size of fruits that reach maturity (Guardiola, 1981, 1987). Indeed, the greater the response to treatments, the lower the size of the final fruit.

When the application of GA_3 is combined with girdling, the effect on fruit set is strongly enhanced, both in terms of increasing fruiting and reducing fruit size.

11.12 Techniques to improve fruit size

In most mandarin-producing countries, small fruit size is a limiting factor of its quality. Consumers are ready to pay premium prices for larger fruit and, therefore, an increase in fruit size greatly increases growers' returns. Furthermore, increased fruit size markedly enhances frequency distribution of fruit according to size, and there is a significant shift of the crop to larger sizes and a reduction of the discarded fruit.

To overcome this problem, several techniques are used. Among them, pruning (Gilfillan, 1995), girdling (Cohen, 1984a; Agustí et al., 1990), fruit thinning (Gallasch, 1988), and the application of synthetic auxins (Agustí et al., 1994a,b) have shown a positive effect.

11.12.1 Pruning

The regulation of fruit size has always been considered among the main benefits of pruning. It has been suggested that this effect is due to the reduction of competition among developing fruits, either flowers or fruits, or due to their removal, since they are located in weak branches and cannot reach an optimum size.

11.12.2 Manual thinning

Removing the fruits by hand can improve the final size of those remaining on the tree. This technique can be performed in two ways.

11.12.2.1 Manual thinning at random

This manual thinning partially reduces the competition among fruits to the benefit of those remaining on the tree. Experiments carried out in this regard have shown that the optimum time is just after the end of the physiological fruitlet drop, that is, at the end of June or at the beginning of July (HN). Advancing has no sense since fruitlets fall spontaneously; and delaying thinning reduces its effectiveness. On the other hand, to achieve a significant increase in size more than half of the fruits must be removed (Zaragoza et al., 1990), which may imply an unacceptable loss of yield.

11.12.2.2 Selective manual thinning

This technique consists of removal of developing fruits that show a delay in their growth or have visible disorders. It has a double objective: (1) removal of the smallest fruits, which increases the average size of those remaining on the tree but without altering their ability to grow and (2) improve fruit quality by discarding fruits that show any biotical or abiotical disorder. This kind of thinning is usually delayed until the second half of summer, when fruit disorders are clearly visible.

The selective hand thinning has advantages compared to that at random. For example, it does not have to be as intense as it is at random and, therefore, the reduction in crop yield is usually acceptable in economical terms, and the removal of part of the waste fruit in the field improves the harvest value and reduces the packinghouse expenses.

Therefore, the selective hand thinning scarcely affects the competition among developing fruits since (1) fruit growth is not encouraged and (2) only a few fruits are removed. Consequently, what is improved by hand thinning is mainly the outlook appearance of the crop, which seems to be of better quality.

The main drawback of hand thinning is its high cost; therefore, it is usually used in some varieties of mandarins that reach high prices and, in this way, increase growers' returns.

11.12.3 Chemical thinning

Chemical thinning causes a reduction of the reproductive organs, either by inhibiting flowering or by promoting the abscission of developing fruits. As the main advantage, this technique highly reduces labor costs associated with manual thinning, although it has the disadvantages of being less precise and selective.

11.12.3.1 Inhibition of flowering

Inhibition of flowering can be performed by applying GA_3 during the autumn rest period (late November-early December in the NH) coinciding with that of the floral bud inductive period (Monselise and Halevy, 1964). The reduction in the number of flowers caused by treatment increases the ovary size by decreasing the competition among flowers. In addition, treatment gives rise to a redistribution of spring flush, increasing the proportion of flowers located in leafy shoots. It is thought that the greater initial development of the ovary, together with the supplementary supply of photoassimilates by the young leaves, promote fruit growth induced by the treatment. In most cases, flowering is very high and its partial inhibition leaves enough flowers so that the yield is not affected. Nevertheless, since the reduction of flowering increases the fruit set in some cultivars with low parthenocarpic ability (see Section 11.7.2), competition among developing fruits turns into similar in treated and untreated trees, and the overall effects of the treatment on fruit size ends up being limited. On the contrary, in low flowering trees, the yield can decrease to a level lower than the untreated trees due to lack of flowers, reducing markedly the competition among developing fruits and increasing their size notably.

11.12.3.2 Fruitlet abscission

Some products such as naphthalene-acetic acid (100–400 mg L^{-1}) or 2-chloroethyl phosphonic acid (200–300 mg L^{-1}) cause the abscission of fruitlets when applied at an early stage of development (Hield et al., 1962; Ortolá et al., 1991; Monselise, 1979; Galiani et al., 1975; Agustí et al., 1999). It is widely accepted that the best time to perform this chemical thinning is during the final period of the physiological fruit drop.

As in hand thinning, >50% of developing fruits must be removed chemically to diminish competition among them and to increase fruit size significantly. Accordingly, chemical thinners can reduce the crop markedly, although it is mitigated by the increase of the individual fruit weight. It should be noted that this kind of substances affect the smallest fruitlets to a greater degree, establishing a certain selectivity and detaching the fruits that finally give rise to the smaller ones. Hence, although these compounds do not directly stimulate the growth of the fruits that persist in the tree, their average size increases with the detachment of the smaller ones, without diminishing the effect due to the lower competition among the developing fruits, as mentioned above. It is important to note that this effect is lost when the physiological fruitlet abscission ends and the fruit starts the linear stage of development (stage II).

Multiple factors affect the response to these treatments, such as the state of the tree and the climatic conditions at treatment, the cultivar, volume applied, etc. Therefore, the results are usually variable, and even erratic, for the same concentration, and this is the reason why thinner substances are practically in disuse in citrus.

11.12.3.3 Application of synthetic auxins

Some synthetic auxins, applied as a foliar spray during the physiological fruitlet drop, directly stimulate fruit growth in citrus. The effect depends to a great extent on the date of treatment. As a general rule, it is accepted that it should be done toward the end of June drop. In practice, the average fruit diameter should range between 15 and 20 mm for mandarins, and between 25 and 30 mm for oranges, so that a delay in application leads to a reduction in effectiveness.

When the auxin applied has also a thinning effect, the advance of treatment can cause the fall of a significant number of fruits, which can affect the final yield. However, it can be used to improve the response since a minimum and selective thinning effect affecting the smaller fruits can be obtained by adjusting the date of treatment. A combination of both effects, that is, increasing of fruit growth and thinning of the smaller fruits, makes this technique the most effective one to improve fruit size in citrus. It is, as well, easy and economical to apply.

The concentration applied depends on the auxin, and the volume of solution should range between 2500 and 3000 L ha^{-1} (6 L $tree^{-1}$, approx.) when applied by a handgun. A nonionic wetting agent must be added to the solution. Higher concentrations or volumes of solution than recommended can cause excessive thinning, as well as some deformation of the fruits, in general of little importance. Treatments may also affect the development of summer sprouting, in which some rolled leaves could appear. Finally, treatments should not be carried out when the ground is dry or under adverse climatic conditions, especially when the temperature is high or with hot and dry winds. In such cases, the treatment can cause an extra increase of developing fruitlet abscission, reducing the yield markedly.

Among the auxin compounds, 2,4-DP and 3,5,6-TPA are the ones more commonly used in oranges and tangerines (Agustí et al., 1998).

2-4-DP (the butylglycol ester of 2,4-dichlorophenoxy-propionic acid): this is applied at a concentration of 50 mg L^{-1}. This synthetic auxin can increase fruit diameter to between 5% and 10% without hardly producing thinning (Agustí et al., 1994a, 1998). Recently, it has been developed as a new formulation (the 2-ethylhexyl ester), which produces similar effects reducing the concentration by half (25 mg L^{-1}).

3,5,6-TPA (3,5,6-trichloro-2-pyridoxylacetic acid): this synthetic auxin applied at concentrations of 10–20 mg L^{-1} can increase the diameter of the fruits to between 5% and 15%. The response to this substance is very dependent on the species, being more pronounced in mandarins than in orange trees. Its thinning effect is higher than that of 2,4-DP but in most cases, it is compensated by the increase in the final fruit weight and the yield does not decrease (Agustí et al., 1993b, 1995, 1998).

Phenothiol (S-ethyl 4-chloro-o-tolyloxythioacetic acid): this is applied at a concentration of 20 mg L^{-1} under the same conditions as the previous ones, this also increases the diameter of the fruit, but to a lower extent, 5% approximately. Its effect on fruit growth is scarce, but it can be improved by taking advantage of its thinning action, which is optimum at a concentration of 30 mg L^{-1} (Agustí et al., 1993a, 1998; Aznar et al., 1995). This auxin, however, has fallen practically into disuse nowadays to enhance fruit weight.

These substances have replaced some others used in the past, most of them being now out of registration. Among them, the most important were the esters of 2,4,5-TP (2,4,5-trichloro-phenoxy-propionic acid), 2,4-D (2,4-dichlorophenoxy-acetic acid), and 2,4, 5-T (2,4,5-trichlorophenoxy-acetic acid), applied at a concentration of between 5 and 20 mg L^{-1} (Monselise, 1979; Guardiola, 1981; Guardiola et al., 1988; Guardiola and Lázaro, 1987; Stewart et al., 1951, 1952).

Frequency distribution of fruit according to commercial caliber showed that there was a significant shift of the crop to larger sizes with the application of these synthetic auxins.

The distribution of the commercial calibers of the fruits treated with synthetic auxins shows a displacement of these toward those of greater value. The response is general for all mandarin and sweet orange varieties, and gives rise to a remarkable increase of fruits of high commercial quality and a considerable reduction of the discarded ones. Normally, the average increases in fruit diameter achieved with these treatments range between 3 and 6 mm.

It has been suggested that these increased size of fruits have some negative effects on the internal quality of the fruit, and they have been attributed to the auxins. Among them, skin thickness and pulp content seem to be increased. However, these effects are not directly due to their action, but are as a consequence of the fruit's own growth (Agustí et al., 1998).

11.12.4 Girdling or ringing

Ringing branches can be also used to improve the fruit size (Fishler et al., 1983; Cohen, 1984a,b; Agustí et al., 1989, 1990, 1998, 1999).

As in the case of the synthetic auxins, the best response to girdling to increase fruit size is obtained when performed immediately after the physiological fruitlet drop. An increase of between 2 and 4 mm in diameter is generally obtained in mandarin and sweet orange varieties, which gives rise to a better distribution of commercial sizes. And also, a delay in girdling diminishes its effectiveness, although certain effect is maintained up to the end of August (NH). As suggested, the beneficial effect of girdling can be due to the fact that it leads to the accumulation of carbohydrates in branches above the girdle, which increases their availability by the fruit.

It is important to note that an advancement of the treatment has contrary effects. Thus, if it is performed during the physiological fruitlet abscission fruit set increases (see Section 11.9.2) and, consequently, fruit size diminishes.

11.13 Seed development

Seeds come from the fertilized ovules, after a complex process of development, which affects all its parts.

One of the first changes observed in fertilized ovules is the enlargement of the nucellus, at the same time as the endosperm development starts.

After fertilization, the zygote does not begin to divide immediately, but takes between 2 weeks and 2 months. Afterwards, the first divisions occur transversely, while the cells closest to the micropillar end of the embryo sac give rise to a thin suspensor. Concomitantly, those at the free end of the row are divided successively to form a mass of cells, which, in turn, multiply and differentiate throughout the embryo. When this one completes the growth, it will pass successively through globular, heart, and torpedo stages, until the cotyledon development ends.

At the time of fertilization, the antipodes and the synergid cells, which still remain in the embryo sac, degenerate. However, the triploid nucleus of the endosperm, formed by the fusion of the two polar nuclei and the second spermatic nucleus from the pollen tube, is divided to produce a large number of free nuclei that remain scattered in a thin layer of cytoplasm found in the vicinity of the wall of the embryo sac. Meanwhile, this grows destroying and compressing many cells of the nucellus. Afterwards, in the cytoplasm, dividing walls are formed around the multiple nuclei, and the cells thus formed continue to divide to form the endosperm.

As the seed develops, the endosperm expands, surpassing in volume the nucellus, which progressively becomes thinner as a consequence of the lengthening of the seed. Possibly, the nucellus acts as a tissue that feeds the endosperm, before being consumed. In addition, the nucellus exerts a function of absorption and translocation of nutrients from the chalazal tissue, which, in turn, receives them directly from the vascular bundle.

Many *Citrus* varieties, called polyembryonic, can generate embryos from somatic cells of the nucellus. These somatic embryos, called nucellar embryos, develop in the embryo sac together with the sexual embryo. The process of formation is as follows: after fertilization, and shortly before or after the zygote undergoes the first cell division, larger cells appear in the first or second layer of cells of the inner part of the nucellus. These cells, which are distinguished by their large nucleus, are found mostly near the micropillar end of the embryo sac. Some of these cells begin to divide producing masses of cells that invade the endosperm, where they will form nucellar embryos. At the incipient developmental state, these embryos can be distinguished from the sexual because of their irregular shape and because they lack suspensors. The time of appearance of the nucellar embryos with respect to the sexual one is variable. The existence of several embryos in the development phase within a single-embryo sac establishes a competition between them, which depends on several factors, such as the position of the embryo and the time of its appearance. The gametic embryo is frequently removed because it is located in the micropapillary end of the embryo sac, and this location is less favorable to receive the nutrients transported by the vascular bundle, which enter the chalazal region, than that of the nucellar located below. On the other hand, due to the genetic differences between both types of embryo, the vigor of the gametic may be different from that of the nucellar ones, which may affect its survival. The number of nucellar embryos differs from one variety to another and even between seeds of the same variety. Normally, polyembryonic seeds usually contain between 2 and 10 nucellar embryos, although, sometimes, they can be more.

Therefore, the development of the seed is constituted by the embryo or embryos—that occupy the interior of the endosperm, a thin wall with the nucellus in decomposition, and the two teguments.

References

Agustí, M., García-Marí, F., Guardiola, J.L., 1982. Sci. Hortic. 17, 343–352.

Agustí, M., Almela, V., Pons, J., 1989. Agrícola. Vergel. 90, 321–323.

Agustí, M., Almela, V., Aznar, M., 1990. Acta Hortic. 6, 101–106.

Agustí, M., Almela, V., Pons, J., 1992. J. Hortic. Sci. 67, 203–210.

Agustí, M., Almela, V., Juan, M., Aznar, M., Zaragoza, S., Primo-Millo, E., 1993a. Levante Agrícola 323, 117–126.

Agustí, M., Almela, V., Juan, M., Primo-Millo, E., Trénor, I., Zaragoza, S., 1993b. J. Hortic. Sci. 69, 219–223.

Agustí, M., Almela, V., Aznar, M., El-Otmani, M., Pons, J., 1994a. HortScience 29, 279–281.

Agustí, M., Almela, V., Juan, M., Primo-Millo, E., Trenor, I., Zaragoza, S., 1994b. J. Hortic. Sci. 69, 219–223.

Agustí, M., El-Otmani, M., Aznar, M., Juan, M., Almela, V., 1995. J. Hortic. Sci. 70, 955–962.

Agustí, M., Almela, V., Zaragoza, S., Juan, M., Trénor, I., Alonso, E., Primo-Millo, E., 1998. Cuaderno de Tecnología Agraria n° 3. Generalitat Valenciana, Conselleria de Agricultura, Pesca y Alimentación, Valencia, Spain.

Agustí, M., Almela, V., Juan, M., 1999. Levante Agrícola. 347, 134–152.

Albrigo, L.G., 1999. Proc. Fla. State Hortic. Soc. 112, 1–4.

Ali-Dinar, H.M., Krezdorn, A.H., Wheaton, T.A., 1988. Acta Hortic. (218)159–175.

Altman, A., Goren, R., 1978. J. Am. Soc. Hortic. Sci. 103, 120–123.

Aznar, M., Almela, V., Juan, M., El-Otmani, M., Agustí, M., 1995. J. Hortic. Sci. 70, 617–621.

Bain, J.M., 1958. Aust. J. Bot. 6, 1–24.

Ben-Cheikh, W., Perez-Botella, J., Tadeo, F.R., Talón, M., Primo-Millo, E., 1997. Plant Physiol. 114, 557–564.

Bermejo, A., Primo-Millo, E., Agustí, M., Mesejo, C., Reig, C., Iglesias, D.J., 2015. J. Plant Growth Regul. 34, 584–594.

Bermejo, A., Martínez-Alcántara, B., Martínez-Cuenca, M.R., Yuste, R., Mesejo, C., Reig, C., Agustí, M., Primo-Millo, E., Iglesias, D.J., 2016. J. Plant Growth Regul. 35, 1036–1048.

Bermejo, A., Granero, B., Mesejo, C., Reig, C., Agustí, M., Primo-Millo, E., Iglesias, D.J., 2018. J. Plant Growth Regul. 37, 491–501.
Blanke, M.M., Bower, J.P., 1991. Trees 5, 239–243.
Blondel, L., 1977. Proc. Int. Soc. Citric. 2, 375–383.
Bustan, A., Goldschmidt, E.E., 1998. Plant Cell Environ. 21, 217–224.
Cassin, J., Bourdeaut, B., Gougue, F., Furin, V., Gaillard, J.P., 1969. Proc. of the 1st International Citrus Symposium 1. pp. 315–323.
Chapman, H.D., 1968. In: Reuther, W., Batchelor, L.D., Webber, H.J. (Eds.), The Citrus Industry. vol. II. University of California Press, pp. 127–289.
Chica, E.J., Albrigo, L.G., 2013. J. Am. Soc. Hortic. Sci. 138, 88–94.
Coggins, C.W., Hield, H.Z., 1968. In: Reuther, W., Batchelor, L.D., Webber, H.J. (Eds.), The Citrus Industry. Vol 2. University of California Press, pp. 371–389.
Coggins, C.W., Hield, H.Z., Burns, R.M., Eaks, L.L., Lewis, L.N., 1966. Calif. Agric. 20, 12–13.
Cohen, A., 1981. Proc. Int. Soc. Citric. 1, 196–199.
Cohen, A., 1984a. J. Hortic. Sci. 59, 119–125.
Cohen, A., 1984b. J. Hortic. Sci. 59, 567–573.
Cuñat, P., García, J.L., Carot, V., Picó, F., Vayá, J.L., 1974. Revista de Agroquímica y Tecnología de Alimentos 14, 106–112.
Damigella, P., Tribulato, E., Continella, G., 1970. Tecnologia Agraria 5, 508–525.
Davenport, T.L., 1983. HortScience 18, 947–949.
Davis, F.S., Albrigo, L.G., 1994. *Citrus*. CAB International, pp. 134–135.
Deidda, P., Agabio, M., 1977. Proc. Int. Soc. Citric. 2, 688–692.
Del Rivero, J.M., Veyrat, P., Gomez de Barreda, D., 1969. Proceedings of the First International Citrus Symposium. Vol. 3. pp. 1121–1124.
Delgado, R., Casamayor, R., Rodriguez, J.L., Cruz, P., Fajardo, R., 1986a. Acta Hortic. 179, 537–544.
Delgado, R., Casamayor, R., Rodriguez, J.L., Cruz, P., Fajardo, R., 1986b. Acta Hortic. 179, 545–548.
Dornelas, M.C., Camargo, R.L.B., Berger, I.J., Takita, M.A., 2007a. Genet. Mol. Biol. 30, 761–768.
Dornelas, M.C., Camargo, R.L.B., Figueiredo, I., Takita, M.A., 2007b. Genet. Mol. Biol. 30, 769–779.
Embleton, T.W., Jones, W.W., Labanauskas, C.K., Reuther, W., 1973. In: Reuther, W. (Ed.), The Citrus Industry. vol. 3. University of California Press, pp. 183–210.
Endo, T., Shimada, T., Fujii, H., Kobayashi, Y., Araki, T., Omura, M., 2005. Transgenic Res. 14, 703–712.
Erickson, L., Brannaman, B.L., 1960. J. Am. Soc. Hortic. Sci. 75, 222–229.
Erickson, L.C., Richards, S.J., 1955. Proc. Am. Soc. Horticult. Sci. 65, 109–112.
Erner, Y., 1988. Israel. J. Bot. 37, 173–180.
Fishler, M., Goldschmidt, E.E., Monselise, S.P., 1983. J. Am. Soc. Hortic. Sci. 108, 218–221.
Frost, H.B., Soost, R.K., 1968. In: Reuther, W., Webber, H.J., Barchelor, L.D. (Eds.), The Citrus Industry. vol. 2. University of California Press, pp. 290–324.
Furr, J.R., Amstrong, W.W., 1956. Proc. Am. Soc. Horticult. Sci. 67, 176–182.
Galiani, J.L., Monselise, S.P., Goren, R., 1975. HortScience 101, 68–69.
Gallasch, P.T., 1988. Proc. Int. Soc. Citric. 1, 395–405.
García-Luis, A., Almela, V., Monerri, C., Agustí, M., Guardiola, J.L., 1986. Physiol. Plant. 66, 515–520.
García-Luis, A., Fornés, F., Sanz, A., Guardiola, J.L., 1988. Isr. J. Bot. 37, 189–201.
García-Luis, A., Kanduser, M., Santamarina, P., Guardiola, J.L., 1992. Physiol. Plant. 86, 648–682.
García-Luis, A., Fornés, F., Guardiola, J.L., 1995. J. Am. Soc. Hortic. Sci. 120, 222–227.
García-Martinez, J.L., García-Papi, M.A., 1979. Sci. Hortic. 10, 285–293.
García-Papi, M.A., García-Martinez, J.L., 1984. Sci. Hortic. 22, 265–274.
Gilfillan, I.M., 1995. Citrus J. 5, 27–28.
Goldschmidt, E.E., 1999. HortScience 34, 1020–1024.
Goldschmidt, E.E., Golomb, A., 1982. J. Am. Soc. Hortic. Sci. 107, 206–208.
Goldschmidt, E.E., Monselise, S.P., 1977. Proc. Int. Soc. Citric. 2, 668–672.
Goldschmidt, E.E., Aschkenazi, N., Herzano, Y., Shaffer, A.A., Monselise, S.P., 1985. Sci. Hortic. 26, 159–166.
Gómez-Cadenas, A., Mehouachi, J., Tadeo, F.R., Primo-Millo, E., Talón, M., 2000. Planta 210, 636–643.
Goren, R., Goldschmidt, E.E., Monselise, S.P., 1971. J. Hortic. Sci. 46, 443–451.
Goren, R., Huberman, M., Goldschmitd, E.E., 2003. Hortic. Rev. 30, 1–36.
Greenberg, J., Goldschmidt, E.E., Goren, R., 1993. Acta Hortic. 339, 58–61.
Guardiola, J.L., 1981. Levante Agricola 230, 67–78.
Guardiola, J.L., 1987. Proc. Int. Soc. Citric. 1, 381–394.
Guardiola, J.L., 1992. Levante Agrícola, 4° trimestre. pp. 229–242.
Guardiola, J.L., Lázaro, E., 1987. Sci. Hortic. 31, 119–130.
Guardiola, J.L., Monerri, C., Agustí, M., 1982. Physiol. Plant. 62, 297–302.
Guardiola, J.L., García-Marí, F., Agustí, M., 1984. Physiol. Plant. 62, 297–302.
Guardiola, J.L., Almela, V., Barrés, M.T., 1988. Sci. Hortic. 34, pp. 229–237.
Harty, A.R., Van Staden, J., 1988. Isr. J. Bot. 37, 155–164.
Hedden, P., Thomas, S.G., 2012. Biochem. J. 444, 11–25.
Hernandez-Miñana, F.M., Primo-Millo, E., 1990. J. Hortic. Sci. 65, 596–601.

Hernandez-Miñana, F.M., Primo-Millo, E., Primo-Millo, J., 1989. J. Exp. Bot. 40, 1127–1134.
Hield, H.Z., Coggins, C.W., Garber, M.J., 1958. Calif. Agric. 12, 9–11.
Hield, H.Z., Burns, R.M., Coggins, C.W., 1962. Proc. Am. Soc. Horticult. Sci. 81, 218–222.
Hield, H.Z., Coggins, C.W., Garber, M.J., 1965. Hilgardia 36, 297–311.
Huerta, L., García-Lor, A., García-Martínez, J.L., 2009. Tree Physiol. 29, 569–577.
Iglesias, D.J., Tadeo, F.R., Primo-Millo, E., Talón, M., 2003. Tree Physiol. 23, 199–204.
Iglesias, D.J., Tadeo, F.R., Primo-Millo, E., Talón, M., 2006. Trees 20, 348–355.
Inoue, H., 1990. J. Jpn. Soc. Hortic. Sci. 58, 919–926.
Iwahori, S., 1978. Proc. Int. Soc. Citric. 1, 263–270.
Iwahori, S., Oohata, J.T., 1981. Proc. Int. Soc. Citric. 1, 247–249.
Iwahori, S., Tominaga, S., 1986. Sci. Hortic. 28, 347–353.
Jones, W.W., Embleton, T.W., Steinacker, M.L., Cree, C.B., 1970. J. Am. Soc. Hortic. Sci. 95, 380–381.
Jones, W.W., Embleton, T.W., Coggins, C.W., 1975. Hortic. Sci. 10, 514.
Kadoya, K., Tanaka, H., 1972. J. Jpn. Soc. Hortic. Sci. 41, 23–28.
Krezdorn, A.H., 1969. Proceedings of the First International Citrus Symposium. vol. 3. pp. 1113–1119.
Krezdorn, A.H., Cohen, M., 1962. Proc. Fla. State Hortic. Soc. 75, 53–60.
Kriedemann, P.E., 1968. Aust. J. Biol. Sci. 21, 569–571.
Kriedemann, P.E., 1969. J. Hortic. Sci. 44, 273–279.
Kriedemann, P.E., 1970. Aust. J. Agric. Res. 21, 623–632.
Legaz, F., Primo-Millo, E., 1981. Proc. Int. Soc. Citric. 2, 575–582.
Legaz, F., Primo-Millo, E., Primo-Yúfera, E., Gil, C., Rubio, L., 1982. Plant Soil 66, 339–351.
Legaz, F., Serna, M.D., Primo-Millo, E., 1995. Plant Soil 173, 205–210.
Lenz, F., 1969. Proceedings of the 1st International Citrus Symposium. vol. 1. pp. 333–338.
Lewis, L.N., Coggins, C.W., Hield, H.Z., 1964. Proc. Am. Soc. Horticult. Sci. 84, 147–151.
Li, C.Y., Weiss, D., Goldschmidt, E.E., 2003. Ann. Bot. 92, 137–143.
Lovatt, C.J., Streeter, S.M., Minter, T.C., O'Connell, N.V., Flaherty, D.L., Freeman, M.V., Goodall, P.B., 1984. Proc. Int. Soc. Citric. 1, 186–190.
Lovatt, C.J., Zheng, Y., Kaje, K.D., 1988. Isr. J. Bot. 37, 181–188.
Martínez-Alcántara, B., Iglesias, D.J., Reig, C., Mesejo, C., Agustí, M., Primo-Millo, E., 2015. J. Plant Physiol. 176, 108–117.
Martínez-Fuentes, A., Mesejo, C., Reig, C., Agustí, M., 2010. J. Sci. Food Agric. 90, 1936–1943.
Martínez-Fuentes, A., Mesejo, C., Muñoz-Fambuena, N., Reig, C., Gonzalez-Más, M.C., Iglesias, D.J., Primo-Millo, E., Agustí, M., 2013. Sci. Hortic. 151, 122–127.
Mehouachi, J., Serna, D., Zaragoza, S., Agustí, M., Talón, M., Primo Millo, E., 1995. Plant Sci. 107, 189–197.
Mehouachi, J., Iglesias, D.J., Tadeo, F.R., Agustí, M., Primo-Millo, E., Talón, M., 2000. J. Hortic. Sci. Biotechnol. 75, 79–85.
Mesejo, C., Martínez-Fuentes, A., Reig, C., Agustí, M., 2007. Plant Sci. 173, 223–230.
Mesejo, C., Rosito, S., Reig, C., Martínez-Fuentes, A., Agustí, M., 2012. J. Plant Growth Regul. 31, 186–194.
Mesejo, C., Yuste, R., Martinez-Fuentes, A., Reig, C., Iglesias, D.J., Primo-Millo, E., Agustí, M., 2013. Physiol. Plant. 148, 87–96.
Mesejo, C., Yuste, R., Reig, C., Martinez-Fuentes, A., Iglesias, D.J., Muñoz-Fambuena, N., Bermejo, A., Germaná, M.A., Primo-Millo, E., Agustí, M., 2017. Plant Sci. 247, 13–24.
Michaels, S.D., Amasino, R.M., 1999. Plant Cell 11, 949–956.
Michaels, S.D., Himelblau, E., Kim, S.Y., Schomburg, F.M., Amasino, R.M., 2005. Plant Physiol. 137, 149–156.
Monselise, S.P., 1979. Sci. Hortic. 11, 151–162.
Monselise, S.P., 1985. In: Halevy, A.H. (Ed.), CRC Handbook of Flowering. vol. 2. CRC Press, pp. 275–294.
Monselise, S.P., Goren, R., 1969. Proceedings of the 1st International Citrus Symposium. 3. pp. 1105–1112.
Monselise, S.P., Halevy, A.H., 1964. Proc. Am. Soc. Horticult. Sci. 84, 141–146.
Monselise, S.P., Goren, R., Halevy, A.H., 1966. Proc. Am. Soc. Horticult. Sci. 89, 195–200.
Monselise, S.P., Goren, R., Wallerstein, I., 1972. HortScience 7, 514–515.
Monselise, S.P., Brosh, P., Costo, J., 1981. HortScience . 16, 786.
Monselise, S.P., Goldschmidt, E.E., Golomb, A., Rolf, F., 1983. J. Am. Soc. Hortic. Sci. 108, 373–376.
Moss, G.I., 1969. J. Hortic. Sci. 44, 311–320.
Moss, G.L., 1970. Aust. J. Agric. Res. 21, 233–242.
Moss, G.I., 1971. J. Hortic. Sci. 46, 177–184.
Moss, G.I., 1972. Aust. J. Exp. Agric. Anim. Hus. 12, 96–102.
Moss, G.I., 1976. Aust. J. Hortic. Res. 27, 399–407.
Moss, G.I., 1977. Proc. Int. Soc. Citric. 2, 215–222.
Moss, G.I., Steer, B.T., Kriedemann, P.E., 1972. Physiol. Plant. 27, 432–438.
Muñoz-Fambuena, N., Mesejo, C., Gonzalez-Más, M.C., Primo-Millo, E., Agustí, M., Iglesias, D.J., 2011. Ann. Bot. 108, 511–519.
Muñoz-Fambuena, N., Mesejo, C., Gonzalez-Más, M.C., Iglesias, D.J., Primo-Millo, E., Agustí, M., 2012a. J. Plant Growth Regul. 31, 529–536.
Muñoz-Fambuena, N., Mesejo, C., Gonzalez-Más, M.C., Primo-Millo, E., Agustí, M., Iglesias, D.J., 2012b. Ann. Bot. 110, 1109–1118.
Nebauer, S.G., Renau-Morata, B., Guardiola, J.L., Molina, R.V., 2011. Tree Physiol. 31, 169–177.

Nebauer, S.G., Renau-Morata, B., Lluch, Y., Baroja-Fernandez, E., Pozueta-Romero, J., Molina, R.V., 2014. Plant Physiol. Biochem. 80, 105–113.
Nir, I., Goren, R., Leshem, B., 1972. J. Am. Soc. Hortic. Sci. 97, 774–778.
Nishikawa, F., Endo, T., Shimada, T., Fujii, H., Shimizu, T., Omura, M., Ikoma, Y., 2007. J. Exp. Bot. 58, 3915–3927.
Ogaki, C., Fujita, K., Ito, H., 1963. J. Jpn. Soc. Hortic. Sci. 32, 157–167.
Ogata, T., Hasukawa, H., Shiozaki, S., Horiuchi, S., Kawase, K., Iwagaki, L., Okuda, H., 1996. J. Jpn. Soc. Hortic. Sci. 65, 245–253.
Ortolá, A.G., Monerri, C., Guardiola, J.L., 1991. Sci. Hortic. 47, 15–20.
Pillitieri, L.J., Lovatt, C.J., Walling, L.L., 2004a. Plant Physiol. 135, 1540–1551.
Pillitieri, L.J., Lovatt, C.J., Walling, L.L., 2004b. J. Am. Soc. Hortic. Sci. 129, 846–856.
Poling, S.M., 1991. J. Agric. Food Chem. 39, 677–680.
Powell, A.A., Krezdorn, A.H., 1977. J. Am. Soc. Hortic. Sci. 102, 709–714.
Primo-Millo, E., Ibañez, R., Martín, B., Legaz, F., 1977a. Ann. Int. Nal. Invest. Agr. 7, 163–182.
Primo-Millo, E., Ibañez, R., Navarro, A., 1977b. Rev. Agroquim. Tecnol. Aliment. 17, 388–390.
Rivas, F., Gravina, A., Agustí, M., 2007. Tree Physiol. 27, 527–535.
Saidha, T., Goldschmidt, E.E., Monselise, S.P., 1985. Sci. Hortic. 26, 35–41.
Sanz, A., Monerri, C., Gonzalez-Ferrer, J., Guardiola, J.L., 1987. Physiol. Plant. 69, 93–98.
Schaffer, A.A., Goldschmidt, E.E., Goren, R., Galili, D., 1985. J. Am. Soc. Hortic. Sci. 110, 574–578.
Schaffer, A.A., Sagee, O., Goldschmidt, E.E., Goren, R., 1987. Physiol. Plant. 69, 151–155.
Schneider, H., 1954. Hilgardia 22, 593–691.
Soost, R.K., Burnett, R.H., 1961. Proc. Am. Soc. Horticult. Sci. 77, 194–201.
Southwick, S.M., Davenport, T.L., 1986. Plant Physiol. 81, 26–29.
Spiegel-Roy, P., Goldschmidt, E.E., 1996. Biology of *Citrus*. Cambridge University Press.
Stewart, I., Barthe, G.A., 1984. Proc. Int. Soc. Citric. 1, 289–290.
Stewart, W.S., Klotz, L.J., Hield, H.Z., 1951. Hilgardia 21, 161–193.
Stewart, W.S., Hield, H.Z., Brannaman, B.L., 1952. Hilgardia 21, 301–329.
Syvertsen, J.P., Goñi, C., Otero, A., 2003. Tree Physiol. 23, 899–906.
Talón, M., Zacarías, L., Primo-Millo, E., 1990a. Physiol. Plant. 79, 400–406.
Talón, M., Hedden, P., Primo-Millo, 1990b. J. Plant Growth Regul. 9, 201–206.
Talón, M., Zacarías, L., Primo-Millo, E., 1992. Plant Physiol. 99, 1575–1581.
Tan, F.C., Swain, S.M., 2007. Physiol. Plant. 131, 481–495.
Turnbull, G.C.N., 1989. J. Plant Growth Regul. 8, 273–282.
Valiente, J.I., Albrigo, L.G., 2004. J. Am. Soc. Hortic. Sci. 129, 158–164.
Wallerstein, I., Goren, R., Monselise, S.P., 1973. J. Hortic. Sci. 48, 75–82.
Wallerstein, I., Goren, R., Monselise, S.P., 1974. Can. J. Bot. 52, 935–937.
Wallerstein, I., Goren, R., Monselise, S.P., 1978. J. Hortic. Sci. 53, 203–208.
Yamaguchi, S., 2008. Annu. Rev. Plant Biol. 59, 225–251.
Zacarías, L., Talón, M., Ben-Cheikh, W., Lafuente, M.T., Primo-Millo, E., 1995. Physiol. Plant. 95, 613–619.
Zaragoza, S., Trénor, I., Alonso, E., 1990. Levante Agrícola, 3° trimetre. pp. 156–160.

Further reading

Davenport, T.L., 1990. Hortic. Rev. 12, 349–408.
Huberman, M., Goren, R., 1979. Physiol. Plant. 45, 189–196.

Chapter 12

Fruit growth and development

Francisco R. Tadeo[a], Javier Terol[a], María J. Rodrigo[b], Concetta Licciardello[c], Avi Sadka[d]

[a]*Centro de Genómica, Instituto Valenciano de Investigaciones Agrarias (IVIA), Valencia, Spain,* [b]*Instituto de Agroquímica y Tecnología de Alimentos (IATA)—Consejo Superior de Investigaciones Científicas (CSIC), Valencia, Spain,* [c]*CREA Research Centre for Olive, Citrus and Tree Fruit, Acireale, Italy,* [d]*Department of Fruit Tree Sciences, Agricultural Research Organization, The Volcani Center, Rishon Le'zion, Israel*

12.1 The long, complex, and intriguing journey from set fruitlets to ripe fruit

Once citrus flowers have lost their reproductive attributes and the ovaries have become fruitlets already set (see Chapter 11), a long, complex, and intriguing journey begins until they turn into mature fruits ready to be harvested and transported either to fruit processing plants to produce juices and essential oils or to packinghouses to be prepared for fresh fruit market.

Citrus trees are broadly adapted to growing and fruiting at a moderately wide range of environmental conditions from tropical to subtropical climates. The only limitation for citrus growth is related to its inability to survive frost and freezing temperatures. Citrus trees bloom and produce commercial fruit in a broad belt around the world that stretches from over 40°N latitude to almost 40°S latitude. Within this zone, there are geoclimatic zones in both the Northern and the Southern Hemispheres that mark off the most suitable areas for commercial cultivation for each of the four horticultural citrus groups: the oranges, the mandarins, the pummelos, and grapefruits, and the common acid citrus including citrons, lemons, and limes (Hodgson, 1967). The orange group has expanded in a wide range of cultivars adapted to many climatic conditions. Therefore, oranges are widely cultivated and represent the highest production of all commercial citrus species. The mandarin group comprises several species as well as interspecific hybrids. Satsuma mandarin is well adapted to cold subtropical regions, although it is cultivated mainly in Japan, its place of origin. Common mandarins are cultivated in a wide range of climatic regions from dry to humid subtropical or tropical conditions. However, Clementine mandarin is not well adapted to these climatic conditions and is widely cultivated throughout much of the regions of the Mediterranean Basin and certain locations in China. Lemons are sensitive to low temperatures and its distribution is limited to semiarid and arid subtropical regions. In addition, lemons are susceptible to fungal diseases and for this reason are not well adapted to humid subtropical or tropical regions. In contrast, the distribution of grapefruits and limes is limited to the tropics and warm, humid subtropical regions. Many other factors, besides the geographical and climatic ones, also have profound effects on citrus growing and fruiting. Thus, soil type and moisture, scion/rootstock combination, cultural practices (irrigation, fertilization, growth regulators, pruning, fruit thinning, pest and disease control, etc.), and even the tree ages have a great influence on the commercial cultivation of citrus and also on the growth and development of the fruit (an extensive review on this subject can be found in Refs. (Erickson, 1968; Davies and Albrigo, 1994; Spiegel-Roy and Goldschmidt, 1996; Grierson, 2006; Ladaniya, 2008; Tadeo et al., 2008)).

In a pioneering study conducted in 1958 on the growth and development of Valencia orange (Bain, 1958), Dr. Joan M. Bain defined three distinct but overlapping periods characterized by cell division (stage I), cell enlargement (stage II), and cell maturation (stage III). The whole duration of the process of fruit growth and development depends on the climatic conditions as well as the citrus species and the cultivar. As an example, the time between blooming and harvesting of Valencia orange lasts from 6 to 7 months in the low tropics, to 14–16 months in Mediterranean-type climates (Davies and Albrigo, 1994). Stage I starts when the flower opens (anthesis) and the first steps in fruit development occur once pollination and fertilization are completed (see Chapter 11). This period is characterized by slow fruit growth rates but high cell division. The fruit rind or peel undergoes a high increase in thickness during this period and the differentiation and growth of pulp vesicles begin. Stage II constitutes a fast period of growth, where the fruit increases in size mostly by cell enlargement and water accumulation in pulp vesicles. Through these two stages, developing fruits switch basic metabolism from "utilization" to "storage" sinks (Tadeo et al., 2008). In the final period, stage III, fruit growth is mainly arrested and fruit metabolism shifts to integrate a number of biochemical and physiological changes that eventually render an edible organ. During this stage, citrus fruits undergo a nonclimacteric ripening process characterized by the absence of a large peak in

The Genus Citrus. https://doi.org/10.1016/B978-0-12-812163-4.00012-7

the respiration rate and a burst in the release of ethylene in their tissues (Erickson, 1968). Therefore, it is through stages II and III when the fruit gradually reaches the final physical (size, shape, color, and texture) and chemical (flavor and aroma) properties and the nutritional value (amounts of minerals, antioxidants, and vitamins) typical of each citrus species. All in all, these are specific quality attributes that are dependent on the regulation of the physiological and metabolic processes determining citrus fruit growth and development. These factors have a strong economic relevance since they mark off the fruit harvesting season and are also related to the consumer preferences and choices. Additionally, these factors can also eventually limit the success of the citrus industry by all means.

The agronomic and physiological aspects of citrus fruit growth and development have already been extensively revised and debated in the citrus literature (see Refs. listed above). In this chapter, special attention is paid to the knowledge accumulated so far that arises both from the transcriptomic and proteomic analyses and the metabolite profiling of citrus fruits in order to provide a biochemical and molecular view of how the growth and development of citrus fruits take place.

12.2 The fruit of citrus is a modified berry called hesperidium

The fruit of citrus is botanically considered a hesperidium, a specific kind of modified berry with a leathery rind and divided internally into segments filled with unusual structures that arise from the adaxial surface of the ovary wall, grow into the locular cavities, and finally develop into elongated sacs which enclose thin-walled cells that contain a watery juice. These pulp vesicles are only found in members of the subtribe *Citrinae*, belonging to the tribe *Citreae*, subfamily *Aurantioideae*, and family *Rutaceae*, and no close homologies are known among any of the higher plants (Swingle and Reece, 1968).

Citrus fruits develop from a polycarpellary, syncarpous, superior, and multilocular ovary with axile placentation (Schneider, 1968; Fahn, 1974). It is considered that the fruits develop after fertilization, although in a number of commercial citrus varieties they develop without the formation of seeds, a phenomenon called parthenocarpy (see Chapter 11). The pericarp of a citrus fruit is divided into three regions: the exocarp, the mesocarp, and the endocarp, which are, respectively, the outer, the middle, and the inner layers of the fruit wall (see Fig. 12.1A). The division of the pericarp into these layers is only for convenience in order to facilitate the anatomical description of mature fruits because those layers do not represent separate tissues from the point of view of their origin (Schneider, 1968; Fahn, 1974). Citrus fruit also consists of a variable number of fused carpels forming locules or segments that are surrounded by a rather tough membrane, the endocarp proper, known as the segment membrane or wall. External to the segment walls and extending radially between adjacent segments are the septa. The number of segments in citrus fruits ranged from as low as four in some kumquats to as high as 17 in some pummelos (Tisserat et al., 1990). Fruit segments are joined to a central axis composed of several vascular bundles surrounded by a spongy tissue. The size and shape of citrus fruits is highly variable (see Section 12.3) with several distinct external and internal tissue layers and structures that are visible and easily distinguishable when cut longitudinally or transversely (see Fig. 12.1A). The structure and the basic anatomy of the fruit of citrus has been described and discussed in some detail in Refs. (Grierson, 2006; Schneider, 1968).

12.2.1 The fruit rind or peel provides an interface of the fruit with the external environment

The outer tissue layers of a citrus fruit are the exocarp or flavedo and the mesocarp or albedo, collectively identifying the fruit rind or peel (Fig. 12.1B). The outermost and colored part of the fruit rind is the flavedo, whereas the albedo is the innermost and colorless (white) or sometimes tinted (as in red grapefruits and blood oranges) part of the fruit rind. There is a wide diversity in terms of the thickness and consistency of the citrus fruit rind. In the case of some pummelos and citrons, the fruit rind is extremely thick and it is very thin in the Palestine sweet lime and also in the Mexican acid lime, in some small-fruited mandarins such as Sunki, and in the calamondin. Regarding the consistency of the fruit rind, the albedo of the pummelo is soft and spongy and it is hard and with a dense fleshiness in the citron (Hodgson, 1967). As a curiosity, the albedo of citrons is commonly used raw in salads or carpaccio, sweetened (frozen as a crunchy sorbet), candied and eaten directly or in pastries, in liqueur, or as a dried zest or confit in the southern Mediterranean countries (Brigand and Nahon, 2016).

The flavedo consists of a single layer of epidermal cells and the underlying layers of hypodermal and subepidermal cells (see Fig. 12.1B and C). The epidermal cells have a cell wall that is thicker on the outer periclinal (tangential) face. The continuity of the epidermis is interrupted by specialized cells, the guard cells, defining the aperture of a stoma. The number of stomata present in the fruit surface is set at anthesis (flower open) and remains constant during fruit growth and development. As an example, about 300 stomata mm^{-2} are present on the surface of young developing fruits of Satsuma mandarins (Hiratsuka et al., 2012). Obviously, the stomatal frequency is reduced as the fruit surface enlarges and the stomata remain functional until the end of the stage II of fruit growth and development when a high percentage of them collapse (Hiratsuka et al., 2012). Beneath the epidermis, both cell size and cell wall thickness increase as the cell layer is deeper in the exocarp.

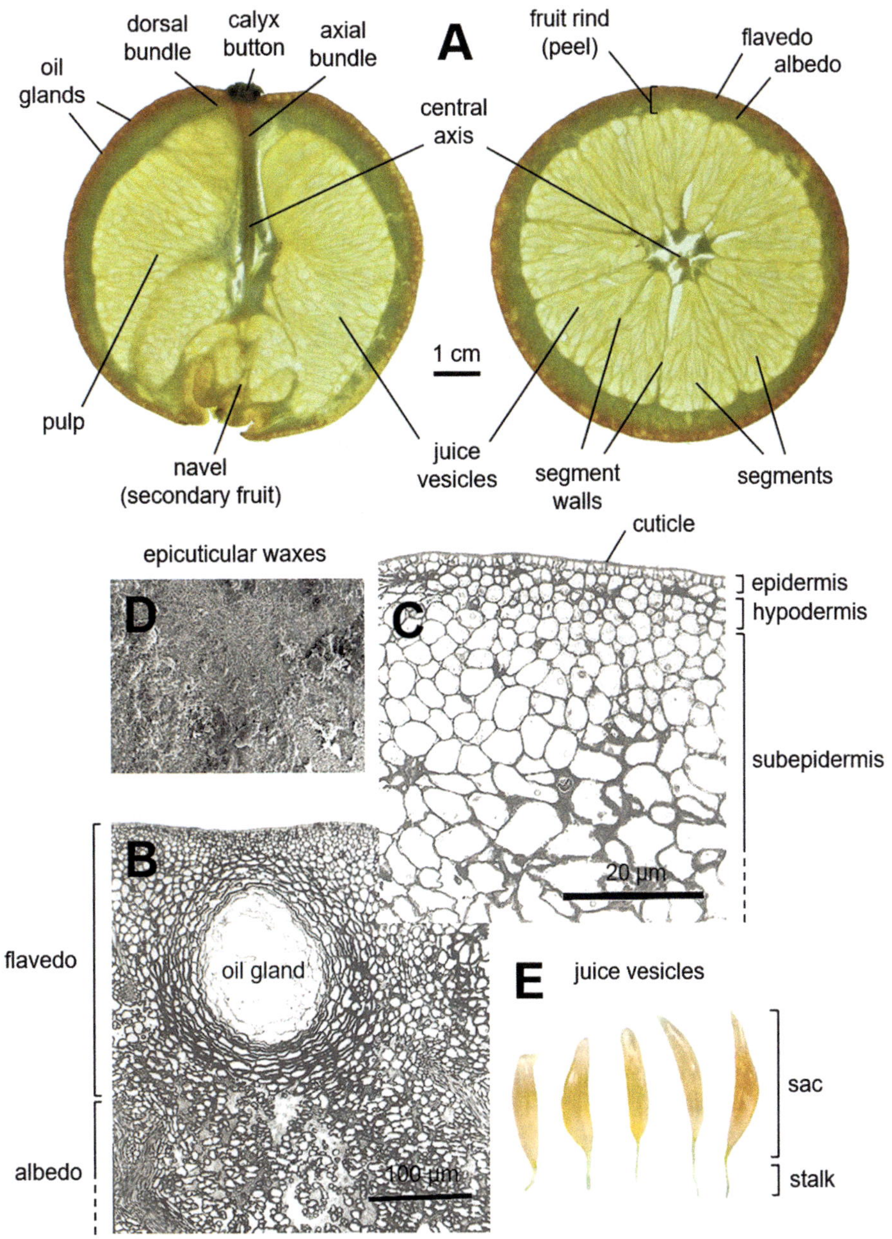

FIG. 12.1 Mature citrus fruit. (A) Longitudinal and equatorial cross sections of a mature citrus fruit. (B) and (C) Light micrographs of longitudinal sections through the fruit rind of a mature citrus fruit showing the exocarp or flavedo and the outer part of the mesocarp or albedo (B) and all the three cell areas of the exocarp or flavedo (C). (D) Scanning electron micrograph of the epicuticular wax deposits on the surface of a mature citrus fruit. (E) Separate juice vesicles showing vesicle bodies or sacs and stalks.

Unlike the hypodermal and subepidermal cells, the cells of the epidermis divide during fruit growth in a strictly anticlinal manner (i.e., perpendicular to the surface). The hypodermal and subepidermal cells fill up the areas between the essential oil glands positioned in the flavedo and blend with the albedo cells near the bottom of the glands (see Fig. 12.1B). The flavedo is a parenchymatous tissue with a normal complement of cell organelles including plastids (Schneider, 1968; Thomson and Platt-Aloia, 1976; Tadeo and Primo-Millo, 1988). Chloroamyloplasts are found in the cytoplasm of both epidermal and hypodermal cells at the onset of fruit growth and are progressively transformed into chloroplasts with a functional photosynthetic apparatus (Tadeo and Primo-Millo, 1988). In fact, the flavedo of young citrus fruits is photosynthetically active although it accounts for only a small part of the requirement for fruit development (Hiratsuka et al., 2017). Along the fruit maturation period (stage III), fruit rind color changes progressively from green to orange as the chloroplasts are converted into chromoplasts that produce and store large amounts of carotenoids (see Section 12.5).

A hydrophobic layer, the cuticle, is coating the epidermal surface of citrus fruits (see Fig. 12.1B). The cuticle is divided into a cutin-rich domain with embedded polysaccharides and an overlying layer enriched in waxes named, respectively, the cuticle layer and the cuticle proper. The cutin matrix of the cuticle proper is impregnated with, and covered by the so-called, respectively, intracuticular and epicuticular waxes. Cutin is a polyester matrix composed of hydroxy and epoxy fatty acid monomers and cuticular waxes consist of long-chain aliphatic compounds [such as very long-chain fatty acids (VLCFAs)], alkanes, alcohols, ketones, esters, diverse cyclic compounds as well as sterols. The amount of cutin in fruit cuticles has been associated with the protection of fruit tissues against pathogenic invasion and the preservation of the biomechanical properties of the cuticle, while the content and composition of cuticular waxes with the permeability of the cuticle to water (Saladié et al., 2007; Isaacson et al., 2009). Epicuticular waxes accumulate in the surface of citrus fruit as a film covered with crystals and platelets of irregular shapes (see Fig. 12.1D). Cutin and waxes are synthesized in the epidermal cells of citrus fruit, which is where genes associated with the biosynthesis of VLCFAs, cutin monomers and waxes, and with secretion are specifically expressed (Matas et al., 2010). The biosynthesis of cutin and waxes occurs in plastids and endoplasmic reticulum and the epidermal cells are rich in these cellular compartments (Thomson and Platt-Aloia, 1976; Tadeo and Primo-Millo, 1988). Cuticle thickness varies widely during the fruit development, from very thin at the onset of the early period of fruit development until reaching its maximum thickness during the next period (Wang et al., 2016). Cuticle thickness remains steady thereafter during fruit maturation. Cuticular wax load, composition, and structure also vary among citrus species and with the fruit position within the canopy (Wang et al., 2014). It is interesting to mention that the proportion of certain wax components is in accordance with the crystal structure of epicuticular waxes. Thus, citrus fruits with high amounts of aldehydes and alkanes show large platelets on the cuticular surface (Wang et al., 2014; Liu et al., 2012). High levels of aldehydes and alkanes in cuticular waxes have been associated with a higher expression of certain genes related with wax biosynthesis such as two aldehyde decarboxylases (*CER1-1* and *CER1-6*), a ketoacyl-CoA synthase (*KCS6*), and *CER3* involved in the biosynthesis of very long-chain alkanes (Wang et al., 2014). Additionally, hormones [gibberellins (GAs), auxins, ethylene, and abscisic acid (ABA)] and environmental conditions may regulate cuticular wax formation as well (Wang et al., 2016; El-Otmani and Coggins, 1985).

The albedo (mesocarp) is the innermost layer of the fruit rind, located between the bottom of the essential oil glands and the pulp segments (see Fig. 12.1B). The mesocarp undergoes an intense meristematic activity during the early period of fruit development (Tadeo and Primo-Millo, 1988). As the fruit enlarges, the inner mesocarp cells increase in size, with a large central vacuole limiting the cytoplasm to a narrow strip attached to the cell wall. Additionally, the intercellular spaces between the inner mesocarp cells enlarge considerably and the cell walls also increase in thickness. Thus, the mesocarp occupies a considerable portion of the fruit volume at the end of stage I of fruit growth and development (Bain, 1958; Schneider, 1968). During the middle period of fruit development (stage II), cell divisions cease in the mesocarp, protuberances develop on cell walls, and schizogenous air spaces begin to form (Schneider, 1968). Thus, the mesocarp is slowly transformed in the spongy tissue with large air spaces and also with scattered vascular traces typical of the citrus hesperidia.

12.2.2 The fleshy pulp is composed of segments, which contain the juice vesicles

The fleshy and edible portion of the citrus mature fruit is the pulp, composed of segments enclosed by a segment wall (see Fig. 12.1A). Filling the segments, there are juice vesicles that consist of a juice vesicle body or sac and a nonvascular stalk (see Fig. 12.1E) and may be also seeds. Juice vesicles grow in length continuously and rapidly during the first two stages of fruit development (Nii and Coombe, 1988; Burns et al., 1992). The elongation of juice vesicles ceases toward the end of stage II of fruit development, which coincides with an increase in the content of soluble solids and a decrease in that of organic acids (Nii and Coombe, 1988). Thus, the juice that vacuoles of vesicle body cells contain is rich in organic acids, sugars, amino acids, and salts.

Juice vesicle primordia initiate by divisions of epidermal and subepidermal cells on the locular side of the endocarp around anthesis (Nii and Coombe, 1988; Burns et al., 1992). The initial development of juice vesicles occurs by cell division of a subterminal meristem and results in the formation of a main juice vesicle body subtended by highly vacuolate stalk cells connected to the endocarp. No vascular connections are found between vascular traces in the endocarp and developing juice vesicles. During this period of juice vesicle growth by cell division, lasting 4–10 weeks in Satsuma mandarin and grapefruit (Nii and Coombe, 1988, 1990; Burns et al., 1992), a defined epidermal layer covered by a thin cuticle encircles the juice vesicle body. Beneath the epidermis, several layers of isodiametric hypodermal cells and a central cavity that consists of rounded cells with dense cytoplasm that accumulate fibrillar material are developed (Burns et al., 1992). Epidermal, hypodermal, and central cavity cells begin to enlarge later during about 12 weeks resulting in highly vacuolated cells. As cells in the central cavity greatly enlarge, central cells begin to lyse and collapse adding fibrillar material, lipid, and cell wall remnants to the central cavity (Burns et al., 1992). During growth of juice vesicles by cell division and cell enlargement, the stalk of juice vesicles shows a central core of elongated cells interconnected by abundant plasmodesmata but without traces of vascular structures (Nii and Coombe, 1988, 1990).

12.2.3 Vascular system of the citrus fruit

The developing fruit requires water and carbon (photoassimilate, sucrose) to maintain cell division and cell expansion and both enter the fruit via the phloem stream (Huang et al., 1992). The vascular system of a citrus fruit forms a broadly branched network of main vascular bundles and subsidiary traces which, for a long time, is known to be confined to the fruit rind and the central axis surrounding the fleshy pulp (Schneider, 1968). The base (stem end) of the fruit is attached to the terminal dome of a cup-like calyx (calyx button) linked to a stem-like peduncle. The vascular cylinder of the peduncle separates into discrete bundles forming three concentric circles as they enter the calyx button (Grierson, 2006; Schneider, 1968). The three vascular circles pass through the fruit abscission zone, located at the interface between the calyx button and the albedo, emerging as the axial, dorsal, and septal bundles (see Fig. 12.2). The axial bundles enter the base of the fruit toward the central axis and branch off near the middle of the fruit giving rise to the ovular bundles. Above the ovules (seeds), the axial bundles diverge into the septa to become the marginal bundles that merge with the dorsal and septal bundles at the apex (stylar end) of the fruit, closing the vascular circuit at this location. The dorsal and the septal bundles go through the inner portion of the albedo opposite, respectively, the segments and the septa in the pericarp. Numerous branches of the dorsal bundles scatter through the pericarp and, along with branches of the septal bundles, go over and contact the entire surface of the septa. There is no vascular continuity at the septa and segment walls (Nii and Coombe, 1988; Koch, 1984). The stalks of juice vesicles that are attached to the inner surface of the segment walls contain no vascular elements (Nii and Coombe, 1988, 1990), so a symplastic and/or an apoplastic route for solute transport into juice vesicles may be operative at this location (see Section 12.4).

12.3 Citrus fruit dimensions are genetically determined but are influenced by environmental and cultural practices

Citrus fruits show a broad range of sizes and shapes with, in addition, special characters at their basal (stem) and apical (stylar) ends. Using diameter as a measure of fruit size, citrus fruits range from very small (about 3 cm in diameter) such as calamondins, kumquats, or some acidic limes, to very large (about 30 cm in diameter) such as citrons or pummelos. Between these limits, there is a wide range of fruit sizes covering the four citrus horticultural groups (Hodgson, 1967). Thanks to the analysis of the citrus genome (see Chapters 2 and 4) it is known that members of some of these horticultural groups are crosses between pure species, with different degrees of admixture (Curk et al., 2016; Wu et al., 2014b, 2018). Ancient mandarins and pummelos are the ancestors of modern mandarins, oranges, and grapefruits and there is a strong degree of correlation between the proportion of pummelo admixture content and their fruit sizes (Wu et al., 2018). Thus, the size of the citrus fruits and possibly also their shapes are reflected in its genomic composition.

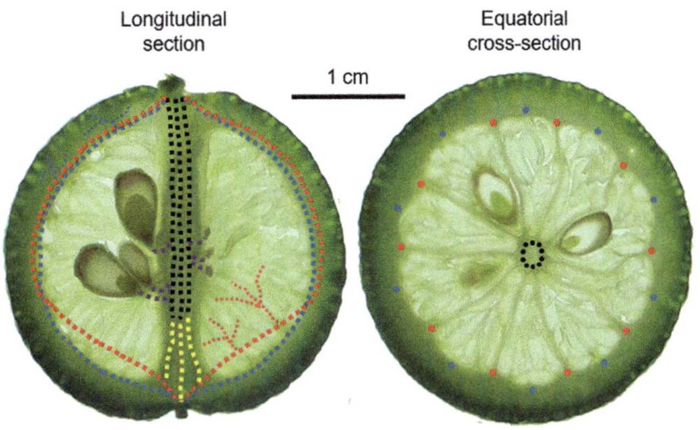

FIG. 12.2 Types and arrangement of vascular bundles in a citrus fruit. Vascular bundles are color coded as follows: axial (*black*), ovular (*purple*), dorsal (*blue*), septal (*red*), and marginal (*yellow*). The dorsal and the septal bundles go through the albedo opposite, respectively, the segments and the septa in the pericarp. Numerous branches of dorsal bundles scatter through the pericarp and, along with branches of septal bundles, go over the septa between segment walls.

FIG. 12.3 Diversity of sizes and shapes of citrus fruits. (A) Fruits of Nagami kumquat (*Fortunella margarita*) are very small, oval in shape, and with a smooth rind surface. (B) Fruits of orangequat (*Citrus unshiu* × *Fortunella crassifolia*) are small, obovate in shape, and necked. (C) Fruits of the small-fruited mandarin *Citrus depressa* (formerly *C. pectinifera*) are oblate in shape and highly depressed at both basal and apical ends. (D) Fruits of Cleopatra mandarin (*Citrus reshni*) are small, oblate in shape, and highly depressed at the apex and with a rough rind surface. (E) Fruits of bergamot (*Citrus bergamia*) are medium-large, round, obovate in shape and usually have a persistent style. (F) Fruits of rough lemon (*Citrus jambhiri*) are medium-sized, elliptic-oblong in shape with basal collar or neck and a rough or bumpy rind surface. (G) Fruits of Moroccan limetta (*Citrus limetta*) are medium-sized, subglobose in shape and with rounded base and apex strongly flattened with broad and deep areolar furrow surrounding a prominent nipple. (H) Fruits of Verna lemon (*Citrus limon*) are medium-sized, elliptical in shape and with a prominent collar or neck at the base and a well-developed nipple at the apex. (I) Fruits of citrangequat (*Fortunella margarita* × [*Citrus sinensis* × *Poncirus trifoliata*]) are medium-sized, pear-shaped (pyriform), and with a medium-rough and bumpy rind surface. (J) Fruits of Indian Khatta lime (*Citrus karna*) are large, oval in shape, and with a broad and prominent nipple at the stylar end and a rind surface ranging from bumpy to (K) warty. (L) Fruits of Washington Navel orange (*Citrus sinensis*) are medium-large, round in shape, and with a protruding navel at the apex. (M) Fruits of Etrog citron (*Citrus medica*) are medium-large, ellipsoid in shape, and with a prominent apical nipple. (N) Fruits of Diamante citron (*Citrus medica*) are large, ellipsoid in shape, and with a low collar or neck at the base and nipple at the apex and a ribbed rind surface. (O) Fruits of Nam Roi pummelo (*Citrus maxima*) are very large and short-pyriform in shape. Fruit images are not to an exact scale. (All pictures were taken at IVIA—Citrus Germplasm Bank.)

The shapes that citrus fruits exhibit are also very diverse and are distinguished by special characters expressed at their basal and apical ends (see Fig. 12.3). In general, oranges are round or globose, mandarins oblate, grapefruits subglobose, lemons elliptical, limes oval, citrons mostly cylindrical, and pummelos oblate and pyriform in some cases (Hodgson, 1967). The base of citrus fruits may be evenly rounded, flattened, moderately or deeply depressed, and also low or highly necked or collared. At the apex, some fruits may include the areola, a small circular area or ring surrounding a nipple (mammilla), or a navel, a pit where the fruit encloses a small secondary fruit. Additionally, the surface of the fruit rind may be smooth or rough with wrinkled, warty, bumpy, or ribbed appearance. The smoothness or roughness of the rind surface may be determined by the size and position of the essential oil glands embedded in the exocarp or flavedo.

All the above-described horticultural characteristics of citrus fruits show a reasonable degree of constancy within a variety in each horticultural group although are influenced to a certain extent by the environmental and the cultural practices applied to the crop on the field (for a review on those issues, see Refs. (Hodgson, 1967; Davies and Albrigo, 1994; Spiegel-Roy and Goldschmidt, 1996); see also Chapter 11). As a summary, only point out that the principal factors that may influence both the size and shape of citrus fruit are the temperature regime and the atmospheric humidity during the fruit developing period, the choice of the rootstock on which the variety grows, the number of set fruits (crop load) and its position in the tree, the application of plant growth regulators, the hand-pruning works to remove interior branches, and the irrigation and the nutritional management to which the plantation is subjected.

The final dimensions of the fruit of citrus may be regulated during multiple stages both throughout the floral development, during the initiation and development of the floral meristem and the development of the ovary, and obviously during fruit development.

The first stage of regulation of the final citrus fruit dimensions is likely to occur in the floral meristems since its size may influence the number of cells that are specified to form the carpel primordia. The second stage of regulation of fruit dimensions is likely controlled by the organization within the floral meristem. Descriptive studies at the anatomical and structural level on organogenesis in the citrus floral meristem have been carried out on Navel orange (Lord and Eckard, 1987) and Satsuma mandarin (Garcia-Luis et al., 2001). The citrus herperidium develops from a syncarpous gynoecium where multiple carpels are fused into a single structure. Since each of the fused carpels gives rise to a segment in the mature fruit, the organization of the floral meristem may regulate the size of the citrus fruit. In fact, a positive relationship exists between the number of fruit segments in cultivars of various citrus species and the size (diameter) of their fruits (Tisserat et al., 1990). However, the outliers of this study [Mexican lime (*Citrus aurantifolia*) and two papedas, kalpa (*C. webberii*) and kaffir lime (*C. hystrix*)] suggest that other regulatory stages outside the floral meristem may be acting to attain the final citrus fruit dimensions. The third stage of regulation may be associated with the specification of the axes of growth by cell division and cell enlargement of the gynoecium: proximal-distal (linked with ovary and fruit length), mediolateral (linked with ovary and fruit width), and abaxial-adaxial (linked with the thickness of each ovary and fruit tissue). The rate and duration of the cell divisions within the developing ovary may impact their final dimensions and, in addition, the differential rates and duration of cell division within distinct tissues in the developing ovary may lead to different fruit shapes. The ovary size prior to anthesis is bigger in leafy than in leafless inflorescences and this difference is evidenced in the final size of fruits (Spiegel-Roy and Goldschmidt, 1996). These differences in final fruit size probably reflect the differences in cell division during ovary development.

Flower anthesis and the occurrence of pollination and fertilization mark the beginning of citrus fruit growth and development. The lack of fertilization leads to a decline in the rate of fruit growth in nonparthenocarpic citrus cultivars and fruit abortion (see Chapter 11). In the case of successful pollination and fertilization or in parthenocarpic cultivars, cell proliferation is then reactivated in the pericarp of the fruit. The mesocarp undergoes intense meristematic activity along the early period of fruit development (Tadeo and Primo-Millo, 1988) in which the fruit rind reaches its maximum thickness. Juice vesicle primordia initiate by cell divisions from epidermal and subepidermal cells of the endocarp and continue growing by cell division into the locules and compactly filling them (Schneider, 1968; Nii and Coombe, 1988; Burns et al., 1992). This is the fourth period of regulation of final citrus fruit dimensions because the activity of cell divisions establishes the number of cells that will make up the mature fruit.

The final stage in the regulation of the final citrus fruit dimensions is the period of cell expansion, the stage II of fruit growth and development (Bain, 1958). Fruit growth during this period occurs by enlargement of juice vesicles and the fruit rind becomes thin as the fruit expands. Climatic conditions, temperature and rainfall regime, seem to influence fruit growth during stage II (Davies and Albrigo, 1994). During stage III, the period of fruit maturation and ripening, fruit growth continues although there is a residual increase in size.

The identification of genes that may participate in the process of citrus fruit growth started with studies of cDNAs resulting from small-scale expressed sequence tag (EST) sequencing projects and now, custom cDNA citrus microarrays (Shimada et al., 2005; Cercos et al., 2006; Goudeau et al., 2008; Liu et al., 2009; Bernardi et al., 2010; Zhang et al., 2017) and RNA-Seq (Yu et al., 2012; Wang et al., 2017a; Liu et al., 2018), are being used to dissect the global transcriptome in different tissues of the growing fruit.

All the transcriptomic analyses carried out so far reveal the occurrence of a massive genetic reprogramming in the growing fruit basically involving changes in the expression of genes associated with carbohydrate metabolism and organic acid and amino acid metabolism. The increase in fruit size is accompanied by a steady uptake and accumulation of water and photoassimilates in the pulp (sugars supplied by nearby leaves via photosynthesis). In addition, there is also an accumulation of organic acids (mostly citric acid) reaching a maximum in stage II and then decline as the fruit matures (see the following section). Therefore, the fact that the bulk of genes regulated during active fruit growth is associated with these metabolic processes is not a huge surprise. Analyzing the data on carbohydrate metabolism with a focus on genes associated with cell wall modification and remodeling, the deposition of new cell wall and the extension of primary walls must be enormously active processes during stages I and II of fruit development. Accordingly, the expression of cellulose synthases, pectin-methylesterases, pectin-acetylesterases, xyloglucan endotransglycosylases, and expansins is upregulated (Cercos et al., 2006; Goudeau et al., 2008; Yu et al., 2012; Wang et al., 2017a). Additionally, it has been pointed out that several genes related to hormonal regulation (auxin and GA), chromatin remodeling, and also transcriptional regulators (MADS-box, growth-regulating factors, ring zinc finger, MYB [R2R3-myb-like], and GRAS [a family consisting of GIBBERELLIN-ACID INSENSITIVE (GAI), REPRESSOR of GA1 (RGA), and SCARECROW (SCR) transcription factors]) may also have a regulatory role in the active increase in size and weight in citrus fruit during the process of fruit growth and development (Liu et al., 2018; Endo et al., 2016; Araújo et al., 2013).

12.4 Metabolism and accumulation of carbohydrate and organic acids, determinants of fruit flavor quality

12.4.1 The physiology and practical aspects of carbohydrate and organic acid accumulation

Pulp acidity of citrus fruit [total acidity (TA)] is correlated with the citric acid concentration, which begin to accumulate during the second phase of fruit development (Hussain et al., 2017). The accumulation reaches a peak when the fruit volume is about 50% of its final value, and then, acid declines gradually as the fruit matures. In most varieties, the major increase in sugar content occurs during the acid decline stage, and toward fruit maturation. During maturation, sugars account for 70%–80% of the total soluble solids (TSS, degrees BRIX). In many cultivars, sucrose, glucose, and fructose accumulate in a ratio of 2:1:1. Sucrose, which is the major photoassimilate found in the phloem of citrus, provides the carbon source for cellular primary metabolism. Indeed, most of the primary and secondary metabolism pathways are present in the juice sac cell; genes encoding the enzymes involved in these pathways have been cloned from the pulp tissue, and/or identified in proteomic analyses.

Practically, the TSS to TA ratio determines whether citrus fruit can be marketed. The ratio will also determine a range of physiological parameters that are important for fruit quality. For instance, undesirable changes in sugar metabolism are associated with the development of some physiological disorders of citrus fruit, such as granulation or sclerification, as well as mandarin puffing (Chen et al., 2005; Ibáñez et al., 2014). Fruit acidity level is a major problem for the development of new citrus cultivars by hybridization and selection because many crosses produce progeny with higher acid levels than either parent. On the other hand, with some easy peeler varieties the problem is often low acid content at maturity. Moreover, reduced acid content is associated with an increase in secondary metabolites (alcohols and aldehydes) which further reduce fruit taste on the tree and during storage, resulting in taste disorders (Porat et al., 2002). Thus, acid and sugar homeostasis in the fruit is critical for the management of existing cultivars, the development of new cultivars, the improvement of preharvest and postharvest strategies, and the control of fruit disorders. Based on high divergence in the sequence of aconitase, catalyzing the first step of citrate catabolism (see below), it was recently suggested that citric acid reduction was a key trait in mandarin domestication (Wang et al., 2018).

12.4.2 The genetic basis of BRIX and TA

BRIX and TA in citrus are likely to be controlled by multiple quantitative trait loci (QTLs), but segregating populations for these traits are relatively limited. Population analyses of acidless pummelo cultivars have suggested that the acidless trait is controlled by a single recessive gene, *acitric*, with three random amplified polymorphic DNA (RAPD) markers showing linkage to the gene (Fang et al., 1997). From crossing of Harehime and Yoshida Ponkan, one QTL controlling TSS was identified, while Fortune and Chandler cross resulted in the identification of seven QTLs controlling TSS and six QTLs controlling TA (Asins et al., 2015; Imai et al., 2017). Citrate is targeted into amino acids through the γ-aminobutyrate (GABA) shunt (see below); indeed, physiological analyses of hybrid lines of two pummelo cultivars with contrasting acidity suggested that the GABA pathway limits citrate degradation (Sheng et al., 2017). While narrow segregating populations hinder map-based cloning of citrus quality traits, rich germplasm collections might facilitate gene cloning by genome-wide association studies (GWAS). A recent work in this direction allowed the identification of single nucleotide polymorphisms (SNPs) associated with 11 traits, including TA and BRIX (Minamikawa et al., 2017). Additionally, another GWAS suggested that mandarin domestication involved the introgression of pummelo genes into a region of chromosome 8 containing 12 genes, including one encoding for the mitochondrial isocitrate dehydrogenase, playing a role in citrate accumulation (Wu et al., 2018).

12.4.3 The biochemistry of carbohydrate and organic acid accumulation

12.4.3.1 Sugar transport, metabolism, and accumulation

Accumulation of sucrose, the major translocated sugar, is detected even in young fruitlets, indicating that its translocation rate is higher than its utilization. Sugar export is described in early works by K. Koch, and these studies still provide the most comprehensive view of the process (Goldschmidt and Koch, 1996; Hiratsuka et al., 2017; Nolte and Koch, 1993a,b; Koch, 1984). The juice sacs are disconnected from the vascular bundles present in the mesocarp (albedo), suggesting apoplastic downloading (Hiratsuka et al., 2017). Therefore, large areas of the fruit display postphloem, nonvascular assimilate transport. Sucrose hydrolysis by invertases and/or sucrose synthase (SuSy) is part of the process. While SuSy is active only in the cytosol, invertases are localized in the cytosol, vacuole, and cell wall (Katz et al., 2007). During stage II, SuSy activity is high in the transport system, including companion cells of the phloem and vascular strand, but relatively low in the sink cells, juice sacs, which display relatively high soluble acid invertase activity. This suggests that SuSy is active in sucrose downloading, while soluble invertase is active in movement in the juice sac (Nolte and Koch, 1993b; Komatsu et al., 1999). Regardless of their cellular localization, all invertases show a reduction in their activities in association with an increase in the transcript level of invertase inhibitor and sucrose accumulation toward fruit maturation (Katz et al., 2011). Three invertase genes showed reduced transcript levels toward maturation (Katz et al., 2011). The citrus genome contains six SuSy genes, with two of them showing relatively high expression in the juice sacs (Komatsu et al., 1999; Islam et al., 2014). Unutilized sucrose is stored in the vacuole, where it is hydrolyzed either by vacuolar invertases or by acid hydrolysis when the vacuolar pH is greatly reduced (Echeverria et al., 1997). Sucrose accumulation does not result only from its nonutilization, but also from active synthesis via the induced activities of sucrose phosphate synthase (SPS) and sucrose phosphatase (SPP); their expression and activities are induced during the later stages of fruit development (Komatsu et al., 1999; Katz et al., 2011).

12.4.3.2 Synthesis, accumulation, and catabolism of citric acid

The synthesis of citric acid occurs as the entry step in the incorporation of the two-carbon acetyl-CoA together with the four-carbon oxaloacetate to yield the six-carbon citrate; a reaction catalyzed by citrate synthase (CS) in the mitochondria, as part of the tricarboxylic acid (TCA) cycle (Fig. 12.4). Oxaloacetate is synthesized by β-carboxylation of phosphoenopyruvate (PEP), a reaction catalyzed by the enzyme phosphoenopyruvate carboxylase (PEPC). It is usually concluded that citrate biosynthesis does not limit citrate accumulation (Sadka et al., 2001; Guo et al., 2016). However, the regulation of the isomerization of citrate to isocitrate by the mitochondrial aconitase plays a major role in citrate accumulation (Sadka et al., 2000a); its activity is correlated with changes in citrate level (Guo et al., 2016; Sadka et al., 2000a). These results provide a plausible scenario for acid accumulation where the reduction in the mitochondrial aconitase generates a local increase in the citric acid level, and the additional acid is removed from the mitochondria to the cytosol and stored in the vacuole (Fig. 12.4). High levels of an inhibitor of the mitochondrial aconitase, citramalate, are present early in fruit development, in association with citrate increase (Degu et al., 2011). Toward fruit maturation, citrate is removed from the vacuole via CsCit1, a H^+/citrate symporter (Shimada et al., 2006), and it is catabolized by a cytosolic aconitase, which is induced in the pulp toward maturation (Sadka et al., 2000a). The citrus genome contains three genes of aconitase, *CitAco1*, *CitAco2*, and *CitAco3* (Terol et al., 2010), depending on the studied system, the mRNAs levels of all three genes or only two of them usually coincide with changes in acid levels (Terol et al., 2010; Chen et al., 2013). Transient overexpression of *CitAco3* resulted in reduced citrate content (Li et al., 2017). Using the yeast-one-hybrid system, four transcription factors (TFs), *bHLH35*, *NAC7*, *bHLH113*, and *TYR*, were reported to interact with the promoter region of *CitAco1* and *CitAco3* (Lu et al., 2016). Moreover, two TFs, *NAC62* and *WRKY1*, interact synergistically to activate *CitAco3* (Li et al., 2017). Regardless, it is not clear yet which gene encodes for which isozyme, mitochondrial, or cytosolic (Degu et al., 2011). Moreover, as in other systems, it seems that major regulation of the enzyme occurs at the posttranscriptional and posttranslational levels (Degu et al., 2011). The cytosolic form of aconitase, but not the mitochondrial one, shows reduction under iron limitation associated with a remarkable induction in citric acid content (Shlizerman et al., 2007). As mentioned above, in most citrus varieties citrate is greatly reduced during the second half of fruit development and toward maturation. Isocitric acid, the product of citrate isomerization, is converted to 2-oxoglutarate (α-ketoglutarate, 2OG) by isocitrate dehydrogenase (Fig. 12.4). The mRNA levels and the activity of the cytosolic form of this enzyme are greatly induced toward fruit maturity (Sadka et al., 2000b), suggesting the cytosol as the site for citric acid catabolism. 2OG is a key compound for glutamate biosynthesis (Cercos et al., 2006), and it has been suggested that some of the citrate is targeted for amino acid biosynthesis generally induced during the second half of fruit development (Sadka et al., 2000b). Glutamate is further metabolized either into glutamine and/or into GABA. Although glutamine is induced by aconitase inhibition and citrate elevation, considerable amount of data pointed out for the importance of the GABA-shunt pathway in citrate consumption (Cercos et al., 2006; Sheng et al., 2017; Katz et al., 2007; Degu et al., 2011; Lin et al., 2015). GABA is synthesized through the protonation of glutamate by glutamate dehydrogenase (GAD). Analysis of juice sacs metabolites has detected significant amounts of GABA and glutamate (Katz et al., 2007), and glutamate and GABA biosynthetic enzymes have been identified in citrus fruit proteomic studies (Katz et al., 2007). Both glutamate and GABA are elevated during acid decline and fruit maturation (Katz et al., 2011). Two genes of GAD were identified in the citrus genome, *CsGAD1* and *CsGAD2*; the mRNA of *CsGAD1* is induced in association with citrate catabolism along with an induced activity of GAD (Liu et al., 2014).

Another path of citrate catabolism is through ATP-citrate lyase (ACL), which catalyzes citrate hydrolysis into acetyl-CoA and oxaloacetate (Fig. 12.4). In plants, this cytosolic enzyme mediates the primary step for acetyl-CoA metabolism feeding into many pathways including fatty acid metabolism, flavonoids, mevalonate, isoprenoid metabolism, etc. (Fatland et al., 2005). Genes encoding for ACL show induction in their mRNA and protein levels toward fruit maturation, in association with acid decline (Cercos et al., 2006; Katz et al., 2007, 2011).

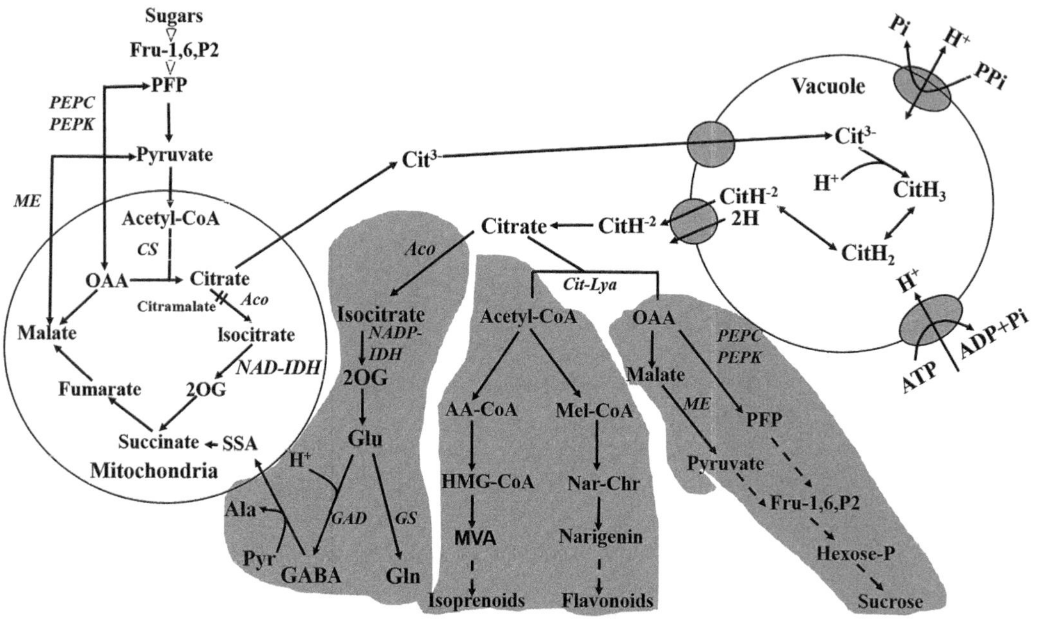

FIG. 12.4 Schematic representation of citric acid metabolism and vacuolar pH homeostasis in the juice sac cell. Citrate is synthesized in the mitochondria as part of the TCA cycle. A metabolic inhibition of aconitase by citramalate creates a local increase in citrate level, and the additional acid is removed to the vacuole, where it is protonated. Vacuolar pH is determined by the activities of the H^+-ATPase and PPase. During the acid decline stage, citrate is removed from the vacuole by H^+/citrate symporter. Citrate could be catabolized into amino acids, acetyl-CoA metabolism, or gluconeogenesis (shaded pathways).

In addition to acetyl-CoA metabolism and amino acid metabolism, the inverse relationships detected between sugar accumulation and acid decline during fruit maturation raise the possibility that some citrate is converted into sugars via gluconeogenesis (Fig. 12.4).

12.4.3.3 Citrate and proton homeostasis in the vacuole, and their transport

The accumulation of citric acid into the vacuole is accompanied by a large influx of protons that is mediated mainly by the vacuolar-type H^+-ATPase (Muller et al., 1997). This influx of protons reduces the vacuolar pH and provides a driving force for additional citrate uptake into the vacuole where citric acid acts buffering the vacuolar acidic pH. Differential quantitative mass spectroscopy of the citrus proteome and expression analyses during fruit development shows that many of the H^+-ATPase subunits are induced from early stage II to late stage II (Katz et al., 2010; Li et al., 2016). The uptake of citrate anion ($^{-3}$Cit) into the plant vacuole is thought to be mainly mediated by the operation of anion channels (Ratajczak et al., 2003), and is protonated to a level dependent on the vacuolar pH (Shimada et al., 2006). Another tonoplastic enzyme which plays a role in proton homeostasis is type 1H^+-pyrophosphatase (PPase), involved in the generation of proton gradients across *endo*-membranes using energy release form inorganic pyrophosphate (Fig. 12.4). Transcript levels of PPase are found to be lower in less acidic fruits, while compared to more acidic ones (Guo et al., 2016). In addition to the above two enzymes, the involvement of other enzymes, plasma membrane (P-type) H^+-ATPase and pH-related gens (pH-like), in citrus fruit pH regulation was also suggested (Aprile et al., 2011; Li et al., 2015; Shi et al., 2015). Citrate retrieval from the vacuole can be mediated by CsCit1, a H^+/citrate symporter (Shimada et al., 2006).

In summary, citrate accumulation and vacuolar pH are two independent processes; however, they are tightly linked by two mechanisms. The first one is the bioenergetics mechanism, which determines the co-accumulation or co-release of citrate with influx of efflux of protons, respectively. The second mechanism is through the operation of the GABA shunt which involves protonation of glutamate during acid decline stage of fruit development.

12.4.4 Contribution of Omics techniques to the understanding of sugar and acid metabolism and accumulation

The use of transcriptomic, proteomic and metabolomic techniques, and their combinations allows a broader view of changes in cellular processes during fruit development. Considering the ease of transcriptomic analysis, especially by RNA deep sequencing, most of the published work use this technology (Cercos et al., 2006; Yu et al., 2012; Lin et al., 2015; Wu et al., 2014a; Lu et al., 2016). In fact, one of the first transcriptomic analysis performed during fruit development allowed the identification of the GABA-shunt pathway and citrate lyase as major routes for citrate catabolism, and these pathways were also identified by proteomic analyses (Cercos et al., 2006; Katz et al., 2007). One advantage of proteomics over transcriptomic is the ability to fractionate the cell. Indeed, most of the enzymes of organic acid metabolism were detected in mitochondrial preparations of juice sac cells, while some were detected in cytosolic preparations, along with most of the enzymes of sugar metabolism (Katz et al., 2007; Muccilli et al., 2009; Zheng et al., 2014). Another understanding resulted from the combined proteomic, metabolomic, and transcriptomic analyses is that organic acid and amino acid accumulation shift toward sugar synthesis during the later stages of fruit development (Katz et al., 2011). Induced gluconeogenesis is supported by reduction in citrate, malate, and oxaloacetate associate with induced activities of PEP carboxykinase and NAD (Nicotinamide Adenine Dinucleotide)-dependent malic enzyme, converting oxaloacetate into PEP and malate into pyruvate, respectively, toward fruit maturation.

12.5 Color change during fruit development and ripening

Internal and external coloration of citrus fruits is a key feature for commercial quality and consumer acceptance. Pigmentation of citrus fruit is highly diverse among species and two main types of isoprenoid-derived pigments are responsible for citrus fruit coloration: chlorophylls (Chls) which impart the green color to the peel of all immature citrus fruits, and carotenoids (Car) providing yellow, orange, and red tints to the peel and pulp of mature fruits. Among the genus *Citrus*, blood oranges are a relevant exception since the purple pigmentation of the fruit is due to the anthocyanins, a group of flavonoid-derived pigments.

12.5.1 Biochemical, molecular, and structural changes related to chlorophylls and carotenoids

In most of the citrus fruit species, the development of the characteristic pigmentation during ripening is not coordinated in the flavedo and pulp. In the flavedo of immature and green mature fruits (stages I and II), the predominant pigments are Chls (Chl-a and -b) being the Chl, the major component (Alós et al., 2008). I The Chls concentrations in immature flavedo are in the range between 300 and 500 $\mu g\, g^{-1}$ FW, while in the pulp much lower contents (<5 $\mu g\, g^{-1}$ FW) are usually detected (Gross, 1987; Jacob-Wilk et al., 1999; Yin et al., 2016). The first step in Chls degradation is the conversion of Chlb to Chla by the action of the Chlb reductase, which is encoded by the non-yellow coloring (*NYC*) locus in citrus (Yin et al., 2016). Previous to the breaker stage, the Chls concentration gradually decreases with a predominant degradation of Chla, and the levels at the ripening stage are nearly undetectable (Alós et al., 2008). Green pale coloration of immature pulp changes progressively and usually earlier than color breaker in flavedo. This is caused by a significant reduced content of Chls and anticipated degradation of Chls and synthesis of orange-colored carotenoids in the pulp compared to flavedo during phase II (Zeng et al., 2015). Some Chl derivatives have been identified in the green flavedo of diverse species of citrus such as OH-chlorophyll *a*, chlorophyllide *a*, pheophorbide *a*, pyropheophorbide *a*, and pheophytin (Alós et al., 2008). In flavedo Chls degradation is highly controlled by the enzyme chlorophyllase which catalyzes the hydrolysis of the ester bond of Chls to chlorophyllide and phytol, the first step in the Chls catabolism (Jacob-Wilk et al., 1999; Trebitsh et al., 1993). Subsequently, magnesium is removed from the chlorophyllide to render pheophorbide a that is converted to the red Chl catabolite (RCC) by a reductase (RCCR) and then to pFCCs by the pheophorbide a oxygenase (PaO) (Hörtensteiner and Kräutler, 2011). The central role of chlorophyllase-regulating Chls catabolism during ripening is supported by a reduced gene expression level in citrus late-ripening mutants at the color break (Wang et al., 2017a; Distefano et al., 2009). However, during natural citrus degreening, the chlorophyllase gene expression is low and constitutive without a significant increase (Jacob-Wilk et al., 1999) and a posttranslational processing of the enzyme seems to be the central regulatory mechanism for its in vivo activity (Azoulay-Shemer et al., 2008, 2011). The sweet orange mutant *nan* ("navel negra," black navel) shows a brown-colored flavedo, retention of Chls, and no accumulation of dephytylated Chl at the ripening stage, associated with a reduced expression of *SGR* (*STAY-GREEN*) genes without alteration of the main enzymatic steps in Chl catabolism (Alós et al., 2008). Thus, the alteration in *nan* mutant may involve an upstream regulatory step different from those in previously described stay-green mutants. Regarding the molecular regulators of Chls degradation process in citrus fruits, a MYB-related TF has been associated with delayed Chls loss in two Clementine mandarin mutants (Rios et al., 2010). Several transcriptional regulators from the ethylene response factors family, *CitAP2/ERF*, have been negatively related with the Chls content (Xie et al., 2014) and, more specifically, transcriptional changes in *ERF13*, closely correlated with sweet orange flavedo degreening during the development or in response to ethylene (Yin et al., 2016).

In the last decade, the characterization of carotenoid biosynthetic pathway in citrus and the correlation with Cars profile has been an active field of research and the corresponding enzymes were deeply characterized (reviewed by Rodrigo et al. (2013a); see also Chapter 24). A schematic representation of the structural pathway of carotenoid biosynthesis as has been proposed in citrus fruits is shown in Fig. 12.5. The basic C5-carotenoid precursor isopentenyl diphosphate (IPP) is synthesized through the 2-*C*-methyl-D-erythritol 4-phosphate (MEP) pathway. The first step catalyzed by phytoene synthase (PSY) is the condensation of two molecules of geranylgeranyl diphosphate (C20, GGPP) to form colorless 15-*Z*-phytoene

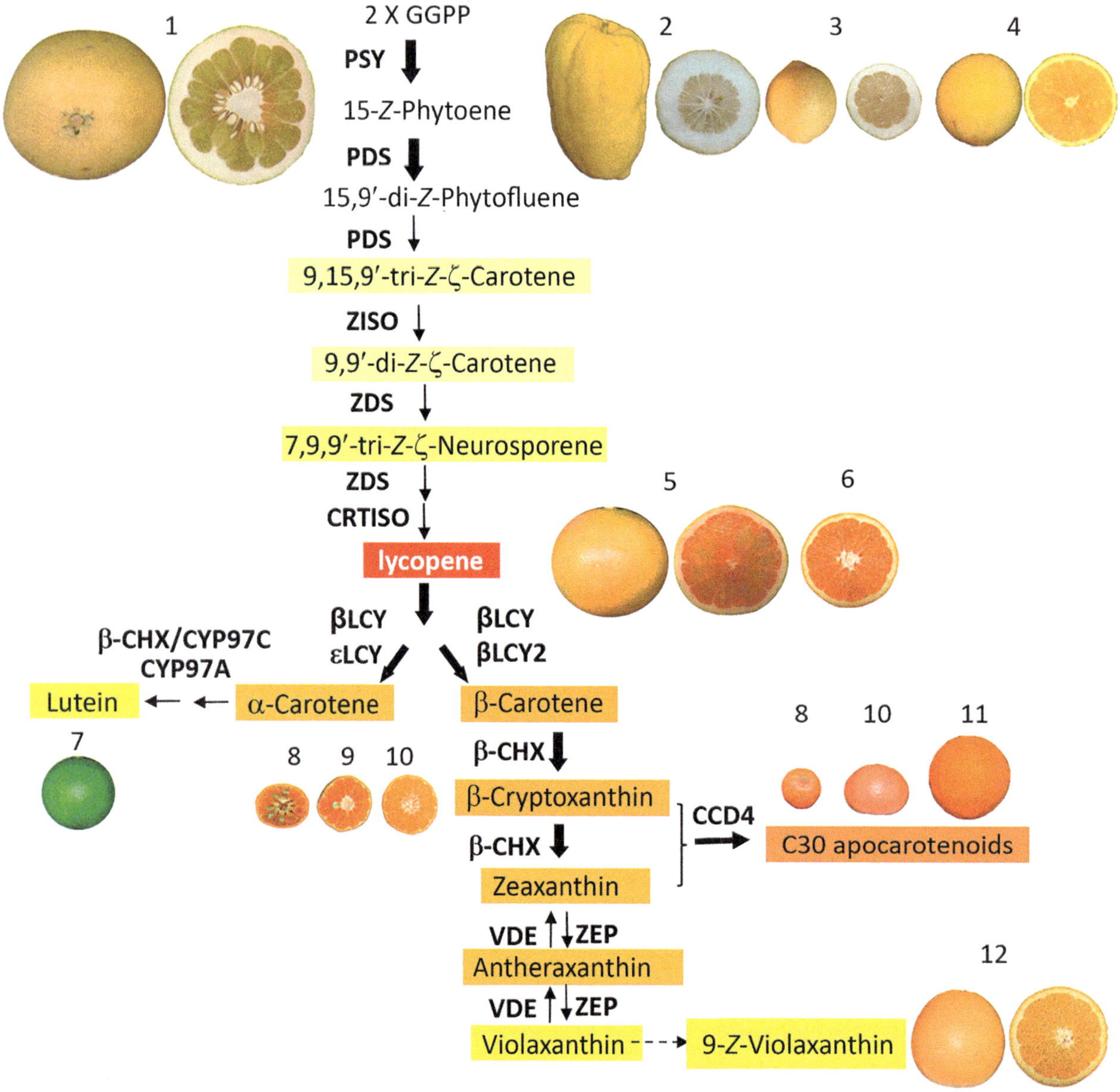

FIG. 12.5 Schematic representation of the carotenoid pathway in the genus *Citrus* showing the diversity in fruit pigmentation. Selected species and varieties are positioned in the pathway at the level of the predominant carotenoid responsible for its coloration. *Bold arrows* indicate the main control points in the pathway determining specific carotenoid accumulation and fruit pigmentation. *Dashed arrow* indicates an unknown enzymatic step. Citrus fruit codes are: (1) Gil pummelo (*C. maxima*), (2) Diamante citron (*C. medica*), (3) Lisbon lemon (*C. limon*), (4) Pinalate sweet orange (*C. sinensis*) (Rodrigo et al., 2003), (5) Star Ruby grapefruit (*C. paradisi*), (6) CaraCara sweet orange (*C. sinensis*), (7) Flavedo of immature *Citrus* sp., (8) Sunki mandarin (*C. reticulata*), (9) Nova mandarin [*C. clementina* × (*C. paradisi* × *C. reticulata*)], (10) Clemenules Clementine (*C. clementina*), (11) Navelina sweet orange (*C. sinensis*), and (12) Navelate sweet orange (*C. sinensis*). Enzyme codes: *PSY*, phytoene synthase; *PDS*, phytoene desaturase; *ZISO*, ζ-carotene isomerase; *ZDS*, ζ-carotene desaturase; *CRTISO*, lycopene isomerase; *ε-LCY*, lycopene ε-cyclase; CYP97A and CYP97C, heme-containing cytochrome P450 hydroxylases; *β-LCY1*, lycopene β-cyclase1; *β-LCY2*, chromoplast-specific lycopene β-cyclase2; *β-CHX*, nonheme b-carotene hydroxylase; *ZEP*, zeaxanthin epoxidase; *VDE*, violaxanthin de-epoxidase; *CCD4*, carotenoid cleavage dioxygenase type 4.

(C40 isoprenoid). Phytoene desaturase (PDS) and ζ-carotene desaturase (ZDS) transform 15-*Z*-phytoene to all-*E*-lycopene via phytofluene, ζ-carotene, and neurosporene. Through the desaturation steps, the activity of two additional isomerases is also required: ζ-carotene isomerase (Z-ISO) and the prolycopene isomerase (CRTISO). Cyclization of lycopene is a key branching point to obtain the α-carotene, with ε- and β-ionone rings, and the β-carotene, with two β-rings. Two types of cyclases participate in these reactions: lycopene β-cyclase (β-LCY) and lycopene ε-cyclase (ε-LCY). A second β-LCY (CYCB or β-LCY2) has been described in citrus which is expressed only in chromoplastic tissues (Alquézar et al., 2009). α-Carotene is converted to lutein by sequential hydroxylation of β-ring by nonheme di-iron carotene hydroxylase (α-CHX/CitHYb) and the α-ring by the CitCYP97A, a heme-containing cytochrome P450-type hydroxylase (Ma et al., 2016). A second β-ring cytochrome P450-type hydroxylase, CitCYP97C, has been recently reported and is likely involved in lutein biosynthesis in citrus fruits (Ma et al., 2016). β-Carotene is converted to zeaxanthin via β-cryptoxanthin by two-step hydroxylation, catalyzed by β-CHX. Zeaxanthin is converted to violaxanthin via antheraxanthin by zeaxanthin epoxidase (ZEP). Violaxanthin can be de-epoxidated into zeaxanthin by violaxanthin de-epoxidase (VDE) in the xanthophyll cycle. The conversion of violaxanthin to neoxanthin, the last carotenoid in the pathway, is catalyzed by the neoxanthin synthase, although this gene has not yet been identified in citrus. The 9-*Z* isomers of violaxanthin and neoxanthin are cleavaged by the 9-*Z*-epoxycarotenoid dioxygenases (NCED) (Rodrigo et al., 2006; Kato et al., 2006), which is the first step for the synthesis of the phytohormone ABA.

The dynamic changes in plastid ultrastructure in citrus fruit tissues during ripening are tightly linked to Chls and Car content and composition. In the green flavedo, the Chls and Car are synthetized and accumulated in the chloroplasts and the suborganellar organization closely resembles those in vegetative green tissues (Gross, 1987; Lado et al., 2015a). Likewise, the Car profile in flavedo of immature and mature green citrus fruits is similar to that found in other chloroplastic tissues, containing the β,ε-xanthophyll lutein as the prevalent Car (Fig. 12.5), and minor proportions of α- and β-carotene, and others β,β-xanthophylls (Rodrigo et al., 2004, 2013a). At immature stages, the relative low level of *PSY*, *PDS*, and *ZDS* expression, involved in the synthesis of linear carotenes, explains the reduced content of early upstream carotenes in green fruits (Rodrigo et al., 2004; Kato et al., 2004). *PSY* has a crucial role regulating the flux of the pathway and three different isoforms have been described; however, only *PSY1* seems to be relevant for Car biosynthesis in fruits (Wei et al., 2017). In immature flavedo, the expression level of *ε-LCY* is high directing the flux of pathway through the β,ε-branch and leading to lutein accumulation (Fig. 12.5). At the onset of color break, in the transition from chloroplast to chromoplast, the level of Car in the flavedo of pigmented citrus species is minimum coincident with the thylakoidal membrane disaggregation (Lado et al., 2015a). In mandarins, hybrids, and sweet oranges, the content and diversity in Car noticeably rises after breaker stage, reaching the maximum level at full mature (Rodrigo et al., 2013a; Matsumoto et al., 2007). This is associated with the formation of novel plastid structures for Car biosynthesis and accumulation such as membranous systems and an increase in number and size of plastoglobules, the predominant structure for Car accumulation citrus chromoplasts (Lado et al., 2015a; Cao et al., 2012). At the molecular level, the expression of *ε-LCY* substantially decreases, while *β-LCY* slightly increases and the chromoplast-specific *β-LCY2* is highly induced, shifting the flux of the pathway into the β,β-branch (Alquézar et al., 2009; Zhang et al., 2012a). The expression of genes involved in upstream carotenes and in the last steps of β,β-xanthophylls is also stimulated in a coordinated manner, resulting in the massive accumulation of xanthophylls (Ikoma et al., 2016).

The Car profile in the pulp of immature green citrus (phase I) has been studied little since the amount of Cars is almost negligible (Gross, 1987; Kato et al., 2004; Alquezar et al., 2013; Lu et al., 2017). One of the most relevant differences between flavedo and pulp is the differentiation process of chromoplasts since in immature citrus pulp the predominant plastid type are amyloplasts (Zeng et al., 2015). At full mature stage, the chromoplasts of juice vesicles are globular and tubular with an increase in the number of plastoglobules and development of membranous structures (Zeng et al., 2015; Lado et al., 2015a; Lu et al., 2017). For most of the citrus species, the Car total content is between 5 and 50 times lower in the juice vesicles compared to flavedo and accumulation in both tissues is not coordinated during ripening (Matsumoto et al., 2007).

In oranges and mandarins, the main carotenoids are the β,β-xanthophylls, being the 9-*Z*-isomer of violaxanthin the carotenoid prevalent in oranges, while mandarins and their hybrids are characterized by an elevated proportion of β-cryptoxanthin, a provitamin A carotenoid (Fig. 12.5) (Kato et al., 2004; Matsumoto et al., 2007; Rodrigo et al., 2013b). Other β,β-xanthophylls and colorless carotenes are also present in variable proportions in the flavedo and pulp (Matsumoto et al., 2007). The preferential accumulation of β-cryptoxanthin in mandarins has been an important focus of investigation. It has been suggested that in mandarins the higher expression ratio between upstream biosynthetic genes (*PSY*, *PDS*, *ZDS*, *β-LCY*, and *β-LCY2*) and downstream genes (*β-CHX* and *ZEP*) is responsible for the elevated β-cryptoxanthin content (Kato et al., 2004). The transcriptional balance between upstream and downstream genes is an important regulatory mechanism in citrus fruit carotenoid composition but not the only one (Wei et al., 2014).

The flavedo of oranges and mandarins accumulates citrus-specific C30 apocarotenoids, mainly β-citraurin and 8-β-apocarotenal, contributing significantly to their characteristic deep orange-red color (Fig. 12.5) (Rodrigo et al., 2013b). The gene responsible for C30-apo synthesis is a novel type 4-carotenoid cleavage dioxygenase (*CCD4*), which cleaves β-cryptoxanthin and zeaxanthin asymmetrically at the 7′,8′double bond. The transcriptional level of this CCD4 enzyme is differentially modulated among the citrus species and highly correlates the external color of citrus fruits (Rodrigo et al., 2013b).

One interesting group of Car-pigmented citrus fruits is those which accumulate lycopene (Fig. 12.5) such as red and pink grapefruits and pummelos, a few spontaneous sweet orange mutants such as CaraCara and Hong Anliu (recently reviewed in Refs. (Ikoma et al., 2016; Chen et al., 2015)). In the lycopene-accumulating citrus varieties, the lycopene increases as the fruit ripens and is usually accompanied by a significant increment in phytoene and phytofluene, small to moderate of β-carotene, and a reduction in the downstream β,β-xanthophylls (Xu et al., 2006; Alquezar et al., 2008; Alquezar et al., 2013). Flavedo and juice vesicles containing lycopene develop an additional type of chromoplasts with needle-like lycopene crystals surrounded by membranes (Zeng et al., 2015; Lado et al., 2015a; Lu et al., 2017). In red grapefruits, accumulation of lycopene and β-carotene is highly controlled by transcriptional regulation of the chromoplast-specific *β-LCY2* and *β-CHX*, with reduced expression levels in red grapefruit compared with sweet orange or white grapefruit (Alquézar et al., 2009; Alquezar et al., 2013). Therefore, an impaired lycopene cyclization seems to be the primary cause for lycopene accumulation in red grapefruits. In addition, at least two alleles of *β-LCY2* gene have been identified in citrus (a and b) with different lycopene cyclase activity and differentially expressed among the citrus species (Alquézar et al., 2009; Zhang et al., 2012a). A deep analysis of Hong Anliu red orange using Omics approaches suggests alterations in the cyclase activity in this mutant (Xu et al., 2009, 2010). In CaraCara, the elevated content of lycopene in the pulp is only partially explained by differential transcriptional regulation of some genes of the MEP pathway and *LCY*s, and further research is needed to elucidate the molecular basis of this mutant (Lu et al., 2017; Alquezar et al., 2008).

In yellow-colored citrus, such as lemon, citron, white grapefruits, or pummelo (Fig. 12.5), the changes in Car profile through fruit development and ripening have been less investigated (Zeng et al., 2015; Alquézar et al., 2009; Kato et al., 2004; Matsumoto et al., 2007; Alquezar et al., 2013). In some yellow species or varieties, an increase in total Car content is observed due to an increase in the colorless carotenes phytoene and phytofluene (Matsumoto et al., 2007). In lemon, a high expression of *PSY* and reduced *PDS* compared to mandarins and oranges has been postulated as the mechanism responsible for the elevated content of upstream carotenes and low downstream β,β-xanthophylls (Kato et al., 2004). In white grapefruits, the reduced Cars concentration has been associated with a deficient desaturation of phytofluene, low *PSY* activity, and altered mechanisms of carotenoid sequestration or increased carotenoid catabolism (Lado et al., 2015a; Alquezar et al., 2013). Interestingly, plastids from white grapefruit tissues showed a lack of well-defined structures, with low electron-dense lipid droplets and achlorophyllous membranes, which is consistent with low Cars content (Lado et al., 2015a). Pinalate is a spontaneous mutant derived from Navel orange, but in contrast with other yellow citrus, the flavedo and the pulp contains extremely high levels of upstream carotenes suggesting a metabolic blockage of the pathway at the ζ-carotene desaturation (Fig. 12.5) (Rodrigo et al., 2003).

In summary, the content and composition in citrus fruit tissues seems to be predominantly controlled at the transcriptional level, but additional regulatory levels have been described such as the allelic diversity in carotenoid biosynthetic genes (Wei et al., 2017; Garcia-Lor et al., 2013) or posttranscriptional mechanisms such as specific carotenoid deposition structures (Lado et al., 2015a; Wei et al., 2017; Ikoma et al., 2016; Kato, 2012). It is also accepted that most of the carotenogenic genes are expressed in a coordinated manner during citrus fruit ripening but little is known about the putative transcriptional factors involved in this process. Recently, a R2R3-MYB transcriptional factor (MYB68) has been shown to regulate the flux of the pathway in the flavedo of mandarins by directly repressing the expression of the β-carotene hydroxylase 2 (*CrBCH2*) (Zhu et al., 2017).

12.5.2 Environmental, nutritional, and hormonal cues affecting chlorophylls and carotenoids

Changes in pigments during citrus fruit ripening are influenced by environmental conditions and endogenous factors such as nutrient status and hormones. Among the environmental conditions, light and temperature are the most critical factors. In subtropical areas, citrus degreening is triggered by rather low temperatures during the night (between 8°C and 15°C) and moderate during the day (20°C) accompanied by a reduction in day length (Alquézar et al., 2009). However, the moderate-low field temperature has a differential effect on individual carotenoids and, therefore, in citrus pigmentation. Moderate-low temperatures stimulate accumulation of β,β-xanthophylls and C30 apocarotenoids, responsible for the orange coloration of citrus fruits, and the expression of key carotenogenic genes (Barry and van Wyk, 2006; Carmona et al., 2012). By contrast, in tropical areas with high temperatures (above 25°C) Chls degradation is delayed and the characteristic increase of β,β-xanthophylls and C30 apocarotenoids does not occur, resulting in yellowish fruits with green peel shades (Stewart and Wheaton, 1972). The accumulation of lycopene in red grapefruits is stimulated by high temperatures (30°C) during the day (Meredith and Young, 1971), most likely because βLCY activity is inhibited at this condition. In the pulp, the induction of Cars accumulation during ripening seems to be temperature independent and not coordinated with flavedo. Nowadays, the primary mechanisms by which citrus fruit tissues are sensing environmental temperatures are largely unknown and deserve further investigation.

Light is a key issue regulating citrus pigmentation. It is known that the peel of oranges and mandarins well exposed to light during the whole development and the ripening period develops more intense coloration that those grown inside the canopy (Cronje et al., 2013). However, the response to light is different depending on the fruit tissue and the carotenoid complement (Lado et al., 2015b). A remarkable example is the flavedo of red grapefruits since light has the opposite effect to that observed for oranges and mandarins. Light avoidance results in mature grapefruits with an intense red pigmentation (Lado et al., 2015b). At the molecular level, the expression level of the carotenoid biosynthetic genes was lower in the flavedo of fruits exposed to dark conditions, as in other citrus species, supporting the stimulatory effect of light on the transcription of carotenogenic genes (Lado et al., 2015b). It has been also shown that the *PSY* promoter from sweet orange contains a light-response motif that may interact with light-related TFs modulating the level of expression (Zeng et al., 2013), but the identification of the factors has not been yet reported. The effect of light on citrus fruit pulp pigmentation has been evaluated in an in vitro culture of juice vesicles. The effects of blue and red lights on Car accumulation were cultivar and species dependent but, in general, blue light treatment was effective inducing Car content and pigmentation, accompanied by an enhanced transcription of key carotenoid biosynthetic genes (Zhang et al., 2015).

In citrus fruit, the onset of degreening has been positively related to ethylene, ABA, and carbohydrates, and negatively related to GAs and nitrogen levels. Among the nutritional factors, sugars and nitrogen are considered the most important affecting citrus fruit pigmentation during on-tree ripening. It is known that there is an inverse relationship between nitrogen and sugars since degreening during ripening is induced under low nitrogen content and high sugar concentration (Huff, 1983). These nutritional factors may interact with environmental conditions as low field temperatures reduce nitrogen tree uptake and concentration in the fruit, triggering the chloroplast to chromoplast transition in the flavedo (Agusti, 1999). This process can be reverted and citrus flavedo can regreens in spring when the field temperature increases, favoring higher nitrogen uptake and reducing the sugar content. In Clementine mandarin, high nitrogen levels in the flavedo delay color changes maintaining the chloroplastic Cars profile, retaining Chls, and hindering upregulation of carotenoid biosynthetic genes (Alós et al., 2006). Contrary to nitrogen, in vivo and in vitro experiments showed that sugar accumulation promotes Chl degradation and biosynthesis of colored carotenoids (Huff, 1983; Iglesias et al., 2001; Gambetta et al., 2012). However, the effect of sugar may be dependent on the fruit developmental stage (Poiroux-Gonord et al., 2012).

The plant hormones ethylene, ABA, and GAs are involved in Chls and Cars metabolism in the flavedo of citrus playing a crucial role in fruit pigmentation. Exogenous ethylene treatment to detached green fruits induces expression of Chlase and key carotenogenic genes which correlates with Chls degradation and carotenoid accumulation (Jacob-Wilk et al., 1999; Rodrigo and Zacarias, 2007). Moreover, a Clementine mutant with a slow rate of natural coloration (Distefano et al., 2009) has decreased sensitivity to the Chls and Cars ethylene-induced gene expression compared to parental cultivar (Alós et al., 2014). Overall data suggest that ethylene produced during citrus fruit maturation, although at minute levels, or the changes in its sensitivity, may be necessary to sustain external pigmentation. The effect of active GAs on citrus flavedo coloration has been extensively studied and displays the opposite effects to those promoted by exogenous ethylene. Elevated GAs in the fruit or exogenously applied may prevent the chloroplast to chromoplast transition and the molecular mechanisms associated with Chls degradation (Jacob-Wilk et al., 1999; Gambetta et al., 2012) and accumulation of colored carotenoids (Alós et al., 2006; Fujii et al., 2008) supporting that citrus color development requires a decrease in active GA levels in the flavedo. The content of ABA, a carotenoid-derived phytohormone, increases in flavedo as the fruit ripens and parallels color development but does not seem to trigger the process (Gambetta et al., 2012; Lafuente et al., 1997).

The effect of nutrients and hormones on the development of citrus pulp pigmentation during ripening has been studied little. An in vitro culture system was used to modulate the sucrose, ABA, and GA content in the juice vesicle from lemon, mandarin, and sweet orange, and the effect on Cars and their metabolism gene expression profile was evaluated (Zhang et al., 2012b). The results showed an enhancement of *PSY* expression in sucrose-treated juice vesicles which contributed to an increment in the Car contents in the three citrus cultivars supporting a positive correlation between sugar concentration and Cars accumulation. However, the application of ABA led to decreases in total carotenoid content in juice vesicles. Regarding GA, the treatment also reduced total and individual Cars contents in all three species, but more markedly in mandarins and sweet oranges compared to lemon that may affect the development of color in pigmented varieties (Zhang et al., 2012b).

12.5.3 General aspects of blood oranges during fruit development and ripening

Based on the fruit morphological characteristics, chemical constituents, and also for convenience, sweet oranges are separated into four groups: the common or round oranges, the Navel oranges, the pigmented (blood) oranges, and the acidless oranges (Davies and Albrigo, 1994). During the last 15 years, the interest on blood oranges increased due to the nutraceutical and antioxidant power linked to their characteristic anthocyanin accumulation. In 1790, Loureiro for the first time describes a citrus fruit pigmented both inside and outside. In fact, red-purple pigmentation typically characterizes the flesh and peel, even though differently in terms of percentage composition of individual components (Fabroni et al., 2016) and in relation to the color distribution in the fruit. Anthocyanin pigmentation, characterizing exclusively blood oranges and their hybrids, increases during the ripening but not in a similar manner in both fruit tissues, neither uniformly nor constantly, in all pigmented genotypes (Fig. 12.6A). The red and purple color in blood oranges is hugely and extremely variable, influenced not only by genotype but also by climatic conditions.

FIG. 12.6 Variability in the purple pigmentation (A) differently in the flesh and rind of sweet orange: (1) Conzamustu, (2) Moro non pigmentato, (3) Moro nucellare 58-8D-I, (4) Tarocco Vitale, (5) Tarocco Ippolito, and (6) Tarocco Lempso and (B) Sunred (hybrid of Oroval Clementine × Moro orange) in November (on the *left*) and February (on the *right*).

In citrus, such as in other plants, it seems that there is a close relationship among pH, acidity content, and anthocyanin pigmentation, whose molecular mechanisms are currently being investigated in depth. Some evidences support the hypothesis about the existence of a different genetic basis among acidic, low acidic, and acidless character among pigmented varieties (Licciardello et al., 2016). In fact, anthocyanins are colored water-soluble pigments that once synthesized in the cytosolic surface of the endoplasmic reticulum accumulate in the vacuoles. The red and purple color of pigmented citrus fruit tissues requires low vacuolar pH to develop.

12.5.3.1 Environmental factors controlling anthocyanin accumulation in fruit

Blood oranges are typically cultivated in the Mediterranean area, although their cultivation has spread to other citrus-growing areas (China, Australia, and the United States). In Italy, the cultivation of blood oranges is mainly located in the east part of Sicily. In this area, climatic conditions meet the weather needs of typical blood varieties, which for their extreme plasticity, are able to adapt and produce balanced ratio sugar/acidity and anthocyanin pigmentation. The unstable and irregular pigmentation in the flesh and in the peel prevent the possibility to promote the consumption of blood oranges merely considering the constant purple color.

Variability in anthocyanin content is influenced not only by the genotype but also by climatic conditions. Environmental requirements, specifically temperatures and light, in addition to pedoclimatic circumstances, represent the fundamental elements for purple coloration. The anthocyanin pigmentation is cold dependent. The induction of anthocyanin pigmentation is notoriously known to be influenced by thermal clines (Fig. 12.6B), although it is hard to associate the differences in thermal clines within cultivar pigmentation. The climatic differences, which occurred during ripening, likely affect anthocyanin accumulation, but not in the same way for all varieties. The Sicilian germplasm collection counts about a hundred among early (January-February), medium (March-April), and late (May-June) Tarocco clones. This indicates that the genetic basis regulating pigmentation and its relation with the external stimuli are rather complex and highly variable among and within varietal groups.

Blood oranges display different degrees of anthocyanin pigmentation in the pulp and peel, varying from a few red vesicles to a deep red and purple color. The concept according to which blood oranges require low temperatures to activate anthocyanin production is widely demonstrated, not only under natural conditions, but also during postharvest cold storage at temperatures ranging from 4°C to 9°C (Lo Piero et al., 2005).

Rootstock is known to influence citrus fruit maturation (Reforgiato Recupero et al., 2009). Swingle citrumelo seems to induce less pigmentation compared to sour orange, Carrizo citrange, and trifoliate orange. The effect of quantity of pigmentation differs according to scion-rootstock interaction, and, as a consequence, it is strictly dependent on the ripening period of the pigmented variety and on the specific climatic conditions of the year (Caruso et al., 2016). The effect of hormones on the retardation of anthocyanin pigmentation has not yet been reported.

12.5.3.2 Potential targets to control purple pigmentation of blood oranges

Anthocyanins are synthetized in the cytosol via the flavonoid pathway and, subsequently, are transported and fixed in highly pigmented bodies located inside the vacuole (for a review, see Ref. (Lo Piero, 2015)). Two groups of genes are required for anthocyanin biosynthesis: the structural genes, encoding the enzymes directly implicated in all of the metabolic reactions, and the regulatory genes encoding the TFs controlling the expression of genes encoding such enzymes. Genes acting earlier in the flavonoid pathway are usually encoded by larger gene families, because they are involved in diverse metabolic functions requiring a specific control of gene expression (Lo Piero, 2015). On the contrary, enzymes directly involved in the later steps are encoded by single active genes (Licciardello et al., 2008). Knowledge on the genes coding for enzymes responsible for anthocyanins biosynthesis (phenylalanine ammonia-lyase, flavanone 3 hydroxylase, chalcone synthase, dihydroflavonol reductase, anthocyanidin synthase, UDP-glucose: flavonoid 3-*O*-glucosyltransferase, glutathione S-transferase) has been widely elucidated both in common (Valencia or Biondo) and pigmented (Moro and Tarocco) varieties. It has become clear that there is a parallelism between the increase in the expression of transcripts and proteins and the accumulation of anthocyanins during the ripening of citrus fruits (Bernardi et al., 2010; Wang et al., 2017a; Muccilli et al., 2009; Butelli et al., 2008, 2012).

The expression of the structural genes required for anthocyanin biosynthesis, modification, and transport is governed by a regulatory complex, known as the MYB-bHLH-WD repeat (MBW) transcriptional complex, composed of MYB (R2R3 conserved domain), MYC (basic helix-loop-helix), and WD-repeat TFs (Butelli et al., 2008). Two R2R3 MYB TFs arranged in tandem in the *Citrus* genome, called *Ruby1* and *Ruby2*, have been shown to have a regulatory function on anthocyanin biosynthesis (Butelli et al., 2012, 2017; Huang et al., 2018). Mutations (deletions or insertions) and insertions of transposable elements in both the coding and the promoter regions of these genes are able to explain anthocyanin pigmentation in all citrus species and varieties. Interestingly, blood oranges can accumulate anthocyanin due to the insertion of a Copia-like long terminal repeat (LTR) transposable element promoting the expression of *Ruby1* by cold temperature (Butelli et al., 2012). On the other hand, a MYC TF, called *Noemi*, has also been identified in citrus acting together with *Ruby1* on organ pigmentation and also as a regulator of fruit acidity (Butelli et al., 2019). An additional citrus MYC TF bearing high homology to *Anthocyanin1* from petunia, called *CitAn1*, has also been identified as a regulator of fruit acidity (Strazzer et al., 2019). *CitAn1* regulates the expression of two vacuolar P-ATPase proton pumps, *CitPH1* and *CitPH5*, involved in generating the pH gradient through the tonoplast to drive massive transport of citrate into the vacuole of juice vesicles. As in the case of citrus MYB TFs, mutations and transposon insertions in *Noemi* and *CitAn1* can explain many of the variations in pH and acidity of anthocyanin-pigmented citrus fruits. However, additional genetic, biochemical, and molecular studies are still needed to fully understand how anthocyanin accumulation and fruit acidity are regulated in the genus *Citrus*.

12.6 Preharvest drop impacts on and determines, respectively, citrus fruit production and harvesting time

Citrus fruits begin to be harvested at horticultural maturity when the TSS/TA ratio (maturity index) is appropriate to be either processed or marketed fresh. In any case, they may be held on the tree (what is called on-tree storage) for rather long periods of time as long as both the external and internal quality and the attachment force to the calyx are retained. Whereas common and late-season Navel oranges, lemons, and most of mandarins hold fairly or especially well on the tree maintaining good quality for several months, other citrus fruits such as early and mid-season Navel oranges, limes, pummelos and most of grapefruits hold poorly on the tree after maturity (Hodgson, 1967). Therefore, certain citrus fruits must be harvested promptly in order to prevent production losses. This natural decline in the attachment force of the fruit not only prevents on-tree storage shortening the harvesting season but also hinders the orderly fruit marketing. Mature fruit drop or preharvest abscission has a serious economic influence especially in those citrus-producing areas dedicated to the fresh fruit market such as, for example, the Mediterranean Basin.

The shedding or abscission of mature citrus fruit is an active, coordinated, and highly regulated process of cell separation (for a review, see Refs. (Tadeo et al., 2008; Burns, 2002)). The separation events that make fruit abscission possible occur in a group of functionally specialized cells, the calyx abscission zone (AZ-C), located at the interface between the calyx button and the pericarp and the nectary or floral disk (Merelo et al., 2017). This location assumes that the axial, dorsal, and septal vascular bundles diverge from the calyx button toward the albedo and the central axis of the fruit, passing through the core of the fruit AZ. In general, fruit abscission is regulated by hormones, mainly auxins and ethylene (for a recent review, see Ref. (Estornell et al., 2013)). The continuous transport of auxins from their sites of synthesis (i.e., young but fully developed leaves) to the fruit seems to be important to prevent a decline in the attachment force of the mature fruit (Yuan et al., 2003). In addition, spraying the tree canopy with synthetic auxins is the sole treatment that have been shown to be effective in delaying preharvest abscission (as an example, see Ref. (Stewart and Klotz, 1947)). However, the attachment force of fruits to the calyx slowly but continuously decreases during fruit ripening in certain citrus species, while auxin levels [indole acetic acid (IAA)] increase in fruit AZs (Yuan et al., 2003; Yuan et al., 2001). Citrus leaves, besides actively synthesizing auxins, produce and export other metabolites to the fruits. In fact, photoassimilates produced by mature leaves are key factors in both the incidence of fruitlet abscission during the physiological drop (see Chapter 11) and in citrus fruit development (Koch, 1984; Goldschmidt and Koch, 1996). On the other hand, preharvest abscission occurs in parallel to overmaturity of fruits and matches with the occurrence of various physiological disorders. Thus, factors other than hormonal fluctuations, such as those associated with fruit senescence or metabolites entering the fruit, may be sensed by AZ-C cells and trigger the decline in the attachment force to the calyx and preharvest abscission.

The citrus fruit AZ-C is constituted by 10–15 cell layers distributed in two clusters of histologically different cell types (Merelo et al., 2017). In addition to a cell cluster integrated by cells containing amyloplasts, the starch-rich cell area previously identified by Wilson and Hendershott (1968) at the distal side of the AZ-C, adjacent to the fruit rind, there is another cell cluster composed of recently divided cells stacked on each other following a regular pattern. This cell cluster, the divided cells area, is located at the proximal side of the AZ-C, adjacent to the calyx button. The biochemical and molecular processes leading to cell separation within the AZ-C are restricted to the starch-rich cell area, whereas cell expansion occurs at the divided cell area. The transcriptional studies performed on abscission-activated citrus fruit AZs show that sugar and polysaccharide metabolism is highly regulated during abscission (Merelo et al., 2017; Cheng et al., 2015). A strong activation of cell wall metabolism occurs in the AZ-C during abscission through both degradation and biosynthesis. Genes encoding enzymes that hydrolyze the middle lamella and the cell wall (pectin-methyl and acetyl-esterases, endo-1,4-β-glucanases, polygalacturonases, pectate lyases, or xyloglucan endotransglycosylases) are responsible for the cell wall disassembly observed in the starch-rich cell area leading to fruit shedding. Other cell wall-related genes involved in the biosynthesis of new cell wall components (cellulose, hemicellulose, and pectin) are likely related to cell expansion in the divided cell area (cellulose synthase, UDP-glucose-4-epimerse, UDP-glucose dehydrogenase, mannan synthase, or rhamnose synthase). In addition, there are also genes involved in the synthesis and deposition of lignin at the distal side of the AZ-C (phenylalanine ammonia-lyases, 4-coumarate:CoA ligases, cinnamoyl-CoA reductases, cinnamyl alcohol dehydrogenases, or Casparian strip membrane domain proteins) likely having a role in generating a tension in the fracture plane of AZ-C to facilitate cell wall breakage and abscission. On the other hand, abscission also regulates the expression of genes involved in signaling (calmodulin, EF-hand calcium-binding proteins, protein kinases, or receptor-like protein kinases) and transcription regulation (ethylene-response factors, auxin-response factors, Aux/IAA, MADS-box, NAC, or MYB TFs) that may be involved in the control of cell separation at the fruit AZ (Merelo et al., 2017; Cheng et al., 2015).

12.7 Transcriptome evolution during ripening: A next-generation view

High-throughput RNA-Seq technique has been used in citrus to analyze transcriptome evolution during ripening, focusing on comparative studies between wild sweet orange varieties and several mutants: Hong Anliu orange displays a red flesh phenotype (Yu et al., 2012), while mutant orange Fengwan (Wu et al., 2014a) and Jincheng (Zhang et al., 2014) produce late-ripening fruits. The thousands of genes differentially expressed during ripening, 19,440 in Anliu and 18,829 in the mutant Hong Anliu (Yu et al., 2012), show the complexity of citrus fruit ripening, a highly coordinated, genetically programmed process that involves many different regulatory pathways. A large number of the differentially expressed genes are involved in cell wall biosynthesis, carbohydrate and citric acid metabolism, carotenoid metabolism, and the response to stress that would be regulated by ABA, sucrose, jasmonic acid, and ethylene by interacting with each other (Yu et al., 2012; Zhang et al., 2014).

Comparison between the mutants and the wild-type sweet orange shows the large number of genes governing each phenotype: >89% of the 18,829 genes showed differential expression during fruit development and ripening in Hong Anliu (Yu et al., 2012), while 13,412 unigenes were identified as differentially expressed unigenes (DEGs) in Jincheng mutant, accounting for 30.20% of the total (Zhang et al., 2014). In Fengwan mutant, with a more restrictive approach, a total of 628 genes were differentially expressed (Wu et al., 2014a).

The red-fleshed mutant Hong Anliu displayed differential expression in genes involved in the major biological processes occurring in the mitochondria (TCA cycle, coupling electron transfer, and oxidative phosphorylation), as well as in carotenoid biosynthesis, suggesting possible cross talk between electron transfer and carotenoid accumulation in sweet oranges (Yu et al., 2012).

The comparison of the late-ripening mutants, Fengwan and Jincheng, allowed the identification of potential regulators and metabolism pathways involved in fruit ripening. Functional categorization of the differentially expressed genes showed that cell wall biosynthesis, carbohydrate and citric acid metabolism, carotenoid metabolism, and the response to stress were the most differentially regulated processes occurring during fruit development and ripening. Overall, it was shown that ABA, sucrose, and jasmonic acid all took part in orange fruit ripening process and might interact with each other. These findings provide an insight into the intricate process of sweet orange fruit ripening (Yu et al., 2012; Zhang et al., 2014).

Genome-wide identification of the TFs involved in citrus fruit ripening of the late-ripening mutant Fengwan sweet orange (MT) and its wild-type counterpart (WT) identified a total of 159 TFs, with 52 of them being differentially expressed. Several important TFs that correlated to ABA, citric acid, fructose, glucose, and sucrose were identified, such as *RD26*, *NTT*, *GATA7*, and *MYB21/62/77*. The analysis of TF family distribution indicated that the TFs of bHLH, C2H2, Dof, ERF, and MYB families might play significant roles during citrus fruit ripening, particularly those of C2H2 and Dof families, which were among the top three families identified in both WT and MT. Most of these TFs were involved in "response to gibberellin," "response to salicylic acid," and "response to ethylene processes," enforcing the pivotal roles of this regulation pathways during fruit ripening (Wu et al., 2016a).

Regulatory mechanisms of citrus fruit ripening also happen at the posttranscriptional and epigenetic levels, involving the regulation by mi-RNAs as well as changes in methylome. In all, 107 known and 21 novel miRNAs, as well as 225 target genes were identified in Fengwan sweet orange. Three of them, csi-miR156k, csi-miR159, and csi-miR166d suppressed specific TFs (*GAMYBs*, *SPLs*, and *ATHBs*) that are supposed to be important regulators involved in citrus fruit development and ripening (Wu et al., 2016b). The expression profiles of several methylation-related genes like *CsMET1*, *CsCMT1*, *CsCMT2*, *CsDRM1*, and *CsDMEs* showed higher expression in the peel than in the flesh, suggesting that methylation levels are not static during fruit development and ripening of citrus (Xu et al., 2015).

12.8 Future perspective: Basic knowledge and advanced techniques should result in improved products

A large body of knowledge exists regarding the physiological, biochemical, and molecular processes associated with citrus fruit growth and development. Understanding how the environment and cultural practices modify these processes has objectively helped producing quality citrus fruits. Nevertheless, a wide set of target genes on which citrus breeding programs are based on is not yet available. The use of transcriptomics, proteomics, and metabolomics may still yield new and interesting knowledge about citrus fruit growth and development through the characterization of mutant genotypes in fruit size and shape, color, flavor or maturation, and harvesting time. Additionally, the availability of high-quality reference genomes of citrus species and wild relatives (Wang et al., 2017b, 2018; Xu et al., 2013; Wu et al., 2014b) and the use of high-throughput sequencing technologies to sequence many citrus varieties and segregating populations (Wu et al., 2018; Wang et al., 2017b, 2018; Minamikawa et al., 2017; Shimizu et al., 2017) may greatly facilitate genomic-based approaches to identify fragments of the citrus genome-containing genes involved in fruit quality traits. At this point, it is important to note that the understanding of the genetic variants that contribute to phenotypic variability in fruit quality traits may be advantageous to select parental genotypes to be used in citrus breeding programs. The identification and validation of genes that support mutations and QTLs associated with fruit quality may be addressed using different molecular technologies such as the already available virus-induced gene silencing (VIGS) or the promising CRISPR gene-editing tools that are being developed now for citrus. Progress in basic understanding of citrus biology and in processes related to fruit growth and development is expected to result in trees which can cope better with changing environmental conditions and in improved products to the citrus industry.

Author contribution

FRT and JT were in charge of writing Sections 12.1–12.3, 12.6, and 12.7. AS was in charge of writing Section 4. MJR and CL were in charge of writing Section 12.5. FRT and AS wrote Section 8. FRT coordinated the assembly of all chapter sections. All authors read and approved the content of the chapter.

References

Agusti, M., 1999. Preharvest factors affecting postharvest quality of citrus fruit. In: Advances in postharvest diseases and disorders control of citrus fruit Trivandrum, India. Research Singpost, pp. 1–34.

Alós, E., Cercos, M., Rodrigo, M.J., Zacarias, L., Talon, M., 2006. Regulation of color break in citrus fruits. Changes in pigment profiling and gene expression induced by gibberellins and nitrate, two ripening retardants. J. Agric. Food Chem. 54, 4888–4895.

Alós, E., Roca, M., Iglesias, D.J., Minguez-Mosquera, M.I., Damasceno, C.M., Thannhauser, T.W., et al., 2008. An evaluation of the basis and consequences of a stay-green mutation in the *navel negra* citrus mutant using transcriptomic and proteomic profiling and metabolite analysis. Plant Physiol. 147, 1300–1315.

Alós, E., Distefano, G., Rodrigo, M.J., Gentile, A., Zacarías, L., 2014. Altered sensitivity to ethylene in 'Tardivo', a late-ripening mutant of Clementine mandarin. Physiol. Plant. 151, 507–521.

Alquezar, B., Rodrigo, M.J., Zacarías, L., 2008. Regulation of carotenoid biosynthesis during fruit maturation in the red-fleshed orange mutant Cara Cara. Phytochemistry 69, 1997–2007.

Alquézar, B., Zacarías, L., Rodrigo, M.J., 2009. Molecular and functional characterization of a novel chromoplast-specific lycopene beta-cyclase from *Citrus* and its relation to lycopene accumulation. J. Exp. Bot. 60, 1783–1797.

Alquezar, B., Rodrigo, M.J., Lado, J., Zacarías, L., 2013. A comparative physiological and transcriptional study of carotenoid biosynthesis in white and red grapefruit (*Citrus paradisi* Macf). Tree Genet. Genomes 9, 1257–1269.

Aprile, A., Federici, C., Close, T.J., De Bellis, L., Cattivelli, L., Roose, M.L., 2011. Expression of the H+-ATPase AHA10 proton pump is associated with citric acid accumulation in lemon juice sac cells. Funct. Integr. Genomics 11, 551–563.

Araújo, P., Cesarino, I., Carmello-Guerreiro, S., Dornelas, M.C., 2013. *Citrus sinensis* L. Osbeck orthologs of *FRUITFULL* and *SHATTERPROOF* are differentially expressed during fruit development. Plant Growth Regul. 70, 1–13.

Asins, M.J., Raga, V., Bernet, G.P., Carbonell, E.A., 2015. Genetic analysis of reproductive, vegetative and fruit quality traits to improve *Citrus* varieties. Tree Genet. Genomes 11, 117.

Azoulay-Shemer, T., Harpaz-Saad, S., Belausov, E., Lovat, N., Krokhin, O., Spicer, V., et al., 2008. *Citrus* chlorophyllase dynamics at ethylene-induced fruit color-break: a study of chlorophyllase expression, posttranslational processing kinetics, and in situ intracellular localization. Plant Physiol. 148, 108–118.

Azoulay-Shemer, T., Harpaz-Saad, S., Cohen-Peer, R., Mett, A., Spicer, V., Lovat, N., et al., 2011. Dual N- and C-terminal processing of citrus chlorophyllase precursor within the plastid membranes leads to the mature enzyme. Plant Cell Physiol. 52, 70–83.

Bain, J.M., 1958. Morphological, anatomical, and physiological changes in the developing fruit of the Valencia orange, *Citrus sinensis* (L) Osbeck. Aust. J. Bot. 6 (1), 1–23.

Barry, G.H., van Wyk, A.A., 2006. Low-temperature cold shock may induce rind colour development of 'Nules Clementine' mandarin (*Citrus reticulata* Blanco) fruit. Postharvest Biol. Technol. 40, 82–88.

Bernardi, J., Licciardello, C., Russo, M.P., Luisa Chiusano, M., Carletti, G., Reforgiato Recupero, G., et al., 2010. Use of a custom array to study differentially expressed genes during blood orange (*Citrus sinensis* L. Osbeck) ripening. J. Plant Physiol. 167, 301–310.

Brigand, J.-P., Nahon, P., 2016. Gastronomy and the citron tree (*Citrus medica* L.). Int. J. Gastronomy and Food Sci. 3, 12–16.

Burns, J.K., 2002. Using molecular biology tools to identify abscission materials for citrus. HortScience 37, 459–464.

Burns, J.K., Achor, D.S., Echeverria, E., 1992. Ultrastructural studies on the ontogeny of grapefruit juice vesicles (*Citrus paradisi* Macf. CV star ruby). Int. J. Plant Sci. 153, 14–25.

Butelli, E., Titta, L., Giorgio, M., Mock, H., Matros, A., Peterek, S., et al., 2008. Enrichment of tomato fruit with health-promoting anthocyanins by expression of select transcription factors. Nat. Biotechnol. 26, 1301–1308.

Butelli, E., Licciardello, C., Zhang, Y., Liu, J., Mackay, S., Bailey, P., et al., 2012. Retrotransposons control fruit-specific, cold-dependent accumulation of anthocyanins in blood oranges. Plant Cell 24, 1242–1255.

Butelli, E., Garcia-Lor, A., Licciardello, C., Casas, G.L., Hill, L., Recupero, G.R., et al., 2017. Changes in anthocyanin production during domestication of *Citrus*. Plant Physiol. 173, 2225–2242.

Butelli, E., Licciardello, C., Ramadugu, C., Durand-Hulak, M., Celant, A., Reforgiato Recupero, G., et al., 2019. *Noemi* controls production of flavonoid pigments and fruit acidity and illustrates the domestication routes of modern citrus varieties. Curr. Biol. 29, 1–7.

Cao, H., Zhang, J., Xu, J., Ye, J., Yun, Z., Xu, Q., et al., 2012. Comprehending crystalline β-carotene accumulation by comparing engineered cell models and the natural carotenoid-rich system of citrus. J. Exp. Bot. 63, 4403–4417.

Carmona, L., Zacarías, L., Rodrigo, M.J., 2012. Stimulation of coloration and carotenoid biosynthesis during postharvest storage of 'Navelina' orange fruit at 12°C. Postharvest Biol. Technol. 74, 108–117.

Caruso, M., Ferlito, F., Licciardello, C., Allegra, M., Strano, M.C., Di Silvestro, S., et al., 2016. Pomological diversity of the Italian blood orange germplasm. Sci. Hortic. 213, 331–339.

Cercos, M., Soler, G., Iglesias, D., Gadea, J., Forment, J., Talon, M., 2006. Global analysis of gene expression during development and ripening of citrus fruit flesh. A proposed mechanism for citric acid utilization. Plant Mol. Biol. 62, 513–527.

Chen, K., Xu, C., Li, F., Zhang, S., 2005. Postharvest granulation of 'Huyou' (*Citrus changshanensis*) fruit in response to calcium. Isr. J. Plant Sci. 53, 35–40.

Chen, M., Xie, X., Lin, Q., Chen, J., Grierson, D., Yin, X., et al., 2013. Differential expression of organic acid degradation-related genes during fruit development of navel oranges (*Citrus sinensis*) in two habitats. Plant Mol. Biol. Report. 31, 1131–1140.

Chen, C., Piero, A.R.L., Gmitter, F., 2015. Pigments in citrus. In: Chen, C. (Ed.), Pigments in Fruits and Vegetables. Springer, New York, pp. 165–187.

Cheng, C., Zhang, L., Yang, X., Zhong, G., 2015. Profiling gene expression in citrus fruit calyx abscission zone (AZ-C) treated with ethylene. Mol. Genet. Genomics 290, 1991–2006.

Cronje, P.J.R., Barry, G.H., Huysamer, M., 2013. Canopy position affects pigment expression and accumulation of flavedo carbohydrates of 'Nules Clementine' mandarin fruit, thereby affecting rind condition. J. Am. Soc. Hortic. Sci. 138, 217–224.

Curk, F., Ollitrault, F., Garcia-Lor, A., Luro, F., Navarro, L., Ollitrault, P., 2016. Phylogenetic origin of limes and lemons revealed by cytoplasmic and nuclear markers. Ann. Bot. 117, 565–583.

Davies, F.S., Albrigo, L.G., 1994. *Citrus*. Oxon, Cab International.

Degu, A., Hatew, B., Nunes-Nesi, A., Shlizerman, L., Zur, N., Katz, E., et al., 2011. Inhibition of aconitase in citrus fruit callus results in a metabolic shift towards amino acid biosynthesis. Planta 234, 501–513.

Distefano, G., Las Casas, G., Caruso, M., Todaro, A., Rapisarda, P., La Malfa, S., et al., 2009. Physiological and molecular analysis of the maturation process in fruits of Clementine mandarin and one of its late-ripening mutants. J. Agric. Food Chem. 57, 7974–7982.

Echeverria, E., Gonzalez, P.C., Brune, A., 1997. Characterization of proton and sugar transport at the tonoplast of sweet lime (*Citrus limmetioides*) juice cells. Physiol. Plant. 101, 291–300.

El-Otmani, M., Coggins, C.W., 1985. Fruit development and growth regulator effects on normal alkanes of "Washington" navel orange fruit epicuticular wax. J. Agric. Food Chem. 33, 656–663.

Endo, T., Fujii, H., Sugiyama, A., Nakano, M., Nakajima, N., Ikoma, Y., et al., 2016. Overexpression of a citrus basic helix-loop-helix transcription factor (*CubHLH1*), which is homologous to Arabidopsis activation-tagged bri1 suppressor 1 interacting factor genes, modulates carotenoid metabolism in transgenic tomato. Plant Sci. 243, 35–48.

Erickson, L.C., 1968. The general physiology of citrus. In: Reuther, W., Webber, H.J., Batchelor, L.D. (Eds.), The Citrus Industry. vol. 2. University of California Press, Berkeley, pp. 86–126.

Estornell, L.H., Agustí, J., Merelo, P., Talón, M., Tadeo, F.R., 2013. Elucidating mechanisms underlying organ abscission. Plant Sci. 199-200, 48–60.

Fabroni, S., Ballistreri, G., Amenta, M., Rapisarda, P., 2016. Anthocyanins in different *Citrus* species: an UHPLC-PDA-ESI/MSn-assisted qualitative and quantitative investigation. J. Sci. Food Agric. 96, 4797–4808.

Fahn, A., 1974. Plant Anatomy, second revised ed. Pergamon Press, Oxford.

Fang, D.Q., Federici, C.T., Roose, M.L., 1997. Development of molecular markers linked to a gene controlling fruit acidity in citrus. Genome 40, 841–849.

Fatland, B.L., Nikolau, B.J., Wurtele, E.S., 2005. Reverse genetic characterization of cytosolic acetyl-CoA generation by ATP-citrate lyase in *Arabidopsis*. Plant Cell 17, 182–203.

Fujii, H., Shimada, T., Sugiyama, A., Endo, T., Nishikawa, F., Nakano, M., et al., 2008. Profiling gibberellin (GA3)-responsive genes in mature mandarin fruit using a citrus 22K oligoarray. Sci. Hortic. 116, 291–298.

Gambetta, G., Martínez-Fuentes, A., Bentancur, O., Mesejo, C., Reig, C., Gravina, A., et al., 2012. Hormonal and nutritional changes in the flavedo regulating rind color development in sweet orange [*Citrus sinensis* (L.) Osb.]. J. Plant Growth Regul. 31, 273–282.

Garcia-Lor, A., Curk, F., Snoussi-Trifa, H., Morillon, R., Ancillo, G., Luro, F., et al., 2013. A nuclear phylogenetic analysis: SNPs, indels and SSRs deliver new insights into the relationships in the 'true citrus fruit trees' group (*Citrinae*, *Rutaceae*) and the origin of cultivated species. Ann. Bot. 111, 1–19.

Garcia-Luis, A., Duarte, A.M.M., Kanduser, M., Guardiola, J.L., 2001. The anatomy of the fruit in relation to the propensity of citrus species to split. Sci. Hortic. 87 (1), 33–52.

Goldschmidt, E.E., Koch, K.E., 1996. *Citrus*. In: Zamski, E., Schaffer, A.A. (Eds.), Photoassimilate Distribution in Plants and Crops: Source-Sink Relations. Marcel Dekker, New York, pp. 797–823.

Goudeau, D., Uratsu, S.L., Inoue, K., daSilva, F.G., Leslie, A., Cook, D., et al., 2008. Tuning the orchestra: selective gene regulation and orange fruit quality. Plant Sci. 174, 310–320.

Grierson, W., 2006. Anatomy and physiology. In: Wardowski, W.F., Miller, W.M., Hall, D.J., Grierson, W. (Eds.), Fresh Citrus Fruit. second ed.. Florida Science Source Incorporated, Longboat Key, pp. 1–122.

Gross, J., 1987. Pigments in Fruits. Academic Press, London, Orlando.

Guo, L.X., Shi, C.Y., Liu, X., Ning, D.Y., Jing, L.F., Yang, H., et al., 2016. Citrate accumulation-related gene expression and/or enzyme activity analysis combined with metabolomics provide a novel insight for an orange mutant. Sci. Rep. 6, 29343.

Hiratsuka, S., Yokoyama, Y., Nishimura, H., Miyazaki, T., Nada, K., 2012. Fruit photosynthesis and phosphoenolpyruvate carboxylase activity as affected by lightproof fruit bagging in Satsuma mandarin. J. Am. Soc. Hortic. Sci. 137, 215–220.

Hiratsuka, S., Nakayama, S., Tamura, S., Nada, K., 2017. Translocation and accumulation of fruit-fixed photosynthate in Satsuma mandarin. Plant Growth Regul. 81, 277–282.

Hodgson, R.W., 1967. Horticultural varieties of citrus. In: Reuther, W., Webber, H.J., Batchelor, L.D. (Eds.), The Citrus Industry, Revised 2nd. vol. 1. University of California Press, Berkeley, pp. 431–588.

Hörtensteiner, S., Kräutler, B., 2011. Chlorophyll breakdown in higher plants. Biochim. Biophys. Acta 1807, 977–988.

Huang, T.-B., Darnell, R.L., Koch, K.E., 1992. Water and carbon budgets of developing citrus fruit. J. Am. Soc. Hortic. Sci. 117, 287–293.

Huang, D., Wang, X., Tang, Z., Yuan, Y., Xu, Y., He, J., et al., 2018. Subfunctionalization of the *Ruby2–Ruby1* gene cluster during the domestication of citrus. Nat. Plants 4, 930–941.

Huff, A., 1983. Nutritional control of regreening and degreening in citrus peel segments. Plant Physiol. 73, 243–249.

Hussain, S.B., Shi, C.Y., Guo, L.X., Kamran, H.M., Sadka, A., Liu, Y.Z., 2017. Recent advances in the regulation of citric acid metabolism in citrus fruit. Crit. Rev. Plant Sci. 36, 241–256.

Ibáñez, A.M., Martinelli, F., Reagan, R.L., Uratsu, S.L., Vo, A., Tinoco, M.A., et al., 2014. Transcriptome and metabolome analysis of Citrus fruit to elucidate puffing disorder. Plant Sci. 217, 87–98.

Iglesias, D.J., Tadeo, F.R., Legaz, F., Primo-Millo, E., Talon, M., 2001. In vivo sucrose stimulation of colour change in citrus fruit epicarps: interactions between nutritional and hormonal signals. Physiol. Plant. 112, 244–250.

Ikoma, Y., Matsumoto, H., Kato, M., 2016. Diversity in the carotenoid profiles and the expression of genes related to carotenoid accumulation among citrus genotypes. Breed. Sci. 66, 139–147.

Imai, A., Yoshioka, T., Hayashi, T., 2017. Quantitative trait locus (QTL) analysis of fruit-quality traits for mandarin breeding in Japan. Tree Genet. Genomes 13, 79.

Isaacson, T., Kosma, D.K., Matas, A.J., Buda, G.J., He, Y., Yu, B., et al., 2009. Cutin deficiency in the tomato fruit cuticle consistently affects resistance to microbial infection and biomechanical properties, but not transpirational water loss. Plant J. 60, 363–377.

Islam, M.Z., Hu, X.M., Jin, L.F., Liu, Y.Z., Peng, S.A., 2014. Genome-wide identification and expression profile analysis of citrus sucrose synthase genes: investigation of possible roles in the regulation of sugar accumulation. PLoS One 9, e113623.

Jacob-Wilk, D., Holland, D., Goldschmidt, E.E., Riov, J., Eyal, Y., 1999. Chlorophyll breakdown by chlorophyllase: isolation and functional expression of the *Chlase1* gene from ethylene-treated Citrus fruit and its regulation during development. Plant J. 20, 653–661.

Kato, M., 2012. Mechanism of carotenoid accumulation in Citrus fruit. J. Jpn. Soc. Hortic. Sci. 81, 219–233.

Kato, M., Ikoma, Y., Matsumoto, H., Sugiura, M., Hyodo, H., Yano, M., 2004. Accumulation of carotenoids and expression of carotenoid biosynthetic genes during maturation in citrus fruit. Plant Physiol. 134, 824–837.

Kato, M., Matsumoto, H., Ikoma, Y., Okuda, H., Yano, M., 2006. The role of carotenoid cleavage dioxygenases in the regulation of carotenoid profiles during maturation in citrus fruit. J. Exp. Bot. 57, 2153–2164.

Katz, E., Fon, M., Lee, Y.J., Phinney, B.S., Sadka, A., Blumwald, E., 2007. The citrus fruit proteome: insights into citrus fruit metabolism. Planta 226, 989–1005.

Katz, E., Fon, M., Eigenheer, R.A., Phinney, B.S., Fass, J.N., Lin, D., et al., 2010. A label-free differential quantitative mass spectrometry method for the characterization and identification of protein changes during citrus fruit development. Proteome Sci. 8, 68.

Katz, E., Boo, K.H., Kim, H.Y., Eigenheer, R.A., Phinney, B.S., Shulaev, V., et al., 2011. Label-free shotgun proteomics and metabolite analysis reveal a significant metabolic shift during citrus fruit development. J. Exp. Bot. 62, 5367–5384.

Koch, K.E., 1984. The path of photosynthate translocation into citrus fruit. Plant Cell Environ. 7, 647–653.

Komatsu, A., Takanokura, Y., Moriguchi, T., Omura, M., Akihama, T., 1999. Differential expression of three sucrose-phosphate synthase isoforms during sucrose accumulation in citrus fruits (*Citrus unshiu* Marc.). Plant Sci. 140, 169–178.

Ladaniya, M.S., 2008. Citrus Fruit—Biology, Technology and Evaluation. Elsevier Incorporated, London.

Lado, J., Zacarías, L., Gurrea, A., Page, A., Stead, A., Rodrigo, M.J., 2015a. Exploring the diversity in Citrus fruit colouration to decipher the relationship between plastid ultrastructure and carotenoid composition. Planta 242, 645–661.

Lado, J., Cronje, P., Alquézar, B., Page, A., Manzi, M., Gómez-Cadenas, A., et al., 2015b. Fruit shading enhances peel color, carotenes accumulation and chromoplast differentiation in red grapefruit. Physiol. Plant. 154, 469–484.

Lafuente, M.T., Martínez-Téllez, M.A., Zacarías, L., 1997. Abscisic acid in the response of 'fortune' mandarins to chilling. Effect of maturity and high-temperature conditioning. J. Sci. Food Agric. 73, 494–502.

Li, S., Liu, X., Xie, X., Sun, C., Grierson, D., Yin, X., et al., 2015. *CrMYB73*, a PH-like gene, contributes to citric acid accumulation in citrus fruit. Sci. Hortic. 197, 212–217.

Li, S.J., Yin, X.R., Xie, X.L., Allan, A.C., Ge, H., Shen, S.L., et al., 2016. The *Citrus* transcription factor, *CitERF13*, regulates citric acid accumulation via a protein-protein interaction with the vacuolar proton pump, CitVHA-c4. Sci. Rep. 6, 20151.

Li, S., Yin, X., Wang, W., Liu, X., Zhang, B., Chen, K., 2017. *Citrus CitNAC62* cooperates with *CitWRKY1* to participate in citric acid degradation via up-regulation of *CitAco3*. J. Exp. Bot. 68, 3419–3426.

Licciardello, C., Russo, M.P., Vale', G., Reforgiato Recupero, G., 2008. Identification of differentially expressed genes in the flesh of blood and common oranges. Tree Genet. Genomes 4, 315–331.

Licciardello, C., Las Casas, G., Caruso, M., Caruso, P., Russo, M.P., Pietro, P., et al., 2016. The evolution of citrate metabolism in acidic and acidless citrus genotypes during fruit development and ripening. Acta Hortic. 1135, 53–60.

Lin, Q., Wang, C., Dong, W., Jiang, Q., Wang, D., Li, S., et al., 2015. Transcriptome and metabolome analyses of sugar and organic acid metabolism in Ponkan (*Citrus reticulata*) fruit during fruit maturation. Gene 554, 64–74.

Liu, Q., Zhu, A., Chai, L., Zhou, W., Yu, K., Ding, J., et al., 2009. Transcriptome analysis of a spontaneous mutant in sweet orange [*Citrus sinensis* (L.) Osbeck] during fruit development. J. Exp. Bot. 60, 801–813.

Liu, D., Zeng, Q., Ji, Q., Liu, C., Liu, S., Liu, Y., 2012. A comparison of the ultrastructure and composition of fruits' cuticular wax from the wild-type 'Newhall' navel orange (*Citrus sinensis* [L.] Osbeck cv. Newhall) and its glossy mutant. Plant Cell Rep. 31, 2239–2246.

Liu, X., Hu, X., Jin, L., Shi, C., Liu, Y., Peng, S., 2014. Identification and transcript analysis of two glutamate decarboxylase genes, *CsGAD1* and *CsGAD2*, reveal the strong relationship between *CsGAD1* and citrate utilization in citrus fruit. Mol. Biol. Rep. 41, 6253–6262.

Liu, T., Li, Y., Zhou, J., Hu, C., Zhang, J., 2018. Genome-wide genetic variation and comparison of fruit-associated traits between kumquat (*Citrus japonica*) and Clementine mandarin (*Citrus clementina*). Plant Mol. Biol. 96, 493–507.

Lo Piero, A.R., 2015. The state of the art in biosynthesis of Anthocyanins and its regulation in pigmented sweet oranges [(*Citrus sinensis*) L. Osbeck]. J. Agric. Food Chem. 63, 4031–4041.

Lo Piero, A.R., Puglisi, I., Rapisarda, P., Petrone, G., 2005. Anthocyanins accumulation and related gene expression in red orange fruit induced by low temperature storage. J. Agric. Food Chem. 53, 9083–9088.

Lord, E.M., Eckard, K.J., 1987. Shoot development in *Citrus sinensis* L. (Washington navel orange). II. Alteration of developmental fate of flowering shoots after GA3 treatment. Bot. Gaz. 148, 17–22.

Lu, X.P., Cao, X.J., Li, F.F., Li, J., Xiong, J., Long, G.Y., et al., 2016. Comparative transcriptome analysis reveals a global insight into molecular processes regulating citrate accumulation in sweet orange (*Citrus sinensis*). Physiol. Plant. 158, 463–482.

Lu, P., Wang, C., Yin, T., Zhong, S., Grierson, D., Chen, K., et al., 2017. Cytological and molecular characterization of carotenoid accumulation in normal and high-lycopene mutant oranges. Sci. Rep. 7, 761.

Ma, G., Zhang, L., Yungyuen, W., Tsukamoto, I., Iijima, N., Oikawa, M., et al., 2016. Expression and functional analysis of citrus carotene hydroxylases: unravelling the xanthophyll biosynthesis in citrus fruits. BMC Plant Biol. 16, 148.

Matas, A.J., Agusti, J., Tadeo, F.R., Talon, M., Rose, J.K., 2010. Tissue-specific transcriptome profiling of the citrus fruit epidermis and subepidermis using laser capture microdissection. J. Exp. Bot. 61, 3321–3330.

Matsumoto, H., Ikoma, Y., Kato, M., Kuniga, T., Nakajima, N., Yoshida, T., 2007. Quantification of carotenoids in citrus fruit by LC-MS and comparison of patterns of seasonal changes for carotenoids among citrus varieties. J. Agric. Food Chem. 55, 2356–2368.

Meredith, F.I., Young, R.H., 1971. Changes in lycopene and carotene content of Redblush grapefruit exposed to high temperatures. HortScience 6, 233–234.

Merelo, P., Agustí, J., Arbona, V., Costa, M.L., Estornell, L.H., Gómez-Cadenas, A., et al., 2017. Cell wall remodeling in abscission zone cells during ethylene-promoted fruit abscission in citrus. Front. Plant Sci. 8, 126.

Minamikawa, M.F., Nonaka, K., Kaminuma, E., Kajiya-Kanegae, H., Onogi, A., Goto, S., et al., 2017. Genome-wide association study and genomic prediction in citrus: potential of genomics-assisted breeding for fruit quality traits. Sci. Rep. 7, 13.

Muccilli, V., Licciardello, C., Fontanini, D., Russo, M.P., Cunsolo, V., Saletti, R., et al., 2009. Proteome analysis of *Citrus sinensis* L. (Osbeck) flesh at ripening time. J. Proteomics 73, 134–152.

Muller, M.L., Irkens-Kiesecker, U., Kramer, D., Taiz, L., 1997. Purification and reconstitution of the vacuolar H+-ATPases from lemon fruits and epicotyls. J. Biol. Chem. 272, 12762–12770.

Nii, N., Coombe, B.G., 1988. Anatomical aspects of juice sacs of Satsuma mandarin in relation to translocation. J. Jpn. Soc. Hortic. Sci. 56, 375–381.

Nii, N., Coombe, B.G., 1990. Ultrastructural changes in the primordia of juice sacs of Satsuma mandarin fruits. J. Jpn. Soc. Hortic. Sci. 59, 35–41.

Nolte, K.D., Koch, K.E., 1993a. Carbohydrate supply affects the level and localization of the sh1 sucrose synthase protein. Plant Physiol. 102, 43.

Nolte, K.D., Koch, K.E., 1993b. Companion-cell specific localization of sucrose synthase in zones of phloem loading and unloading. Plant Physiol. 101, 899–905.

Poiroux-Gonord, F., Fanciullino, A., Bert, L., Urban, L., 2012. Effect of fruit load on maturity and carotenoid content of clementine (*Citrus clementina* Hort. Ex tan.) fruits. J. Sci. Food Agric. 92, 2076–2083.

Porat, R., Daus, A., Weiss, B., Cohen, L., Droby, S., 2002. Effects of combining hot water, sodium bicarbonate and biocontrol on postharvest decay of citrus fruit. J. Hortic Sci. Biotech. 77, 441–445.

Ratajczak, R., Lüttge, U., Gonzalez, P., Etxeberria, E., 2003. Malate and malate-channel antibodies inhibit electrogenic and ATP-dependent citrate transport across the tonoplast of citrus juice cells. J. Plant Physiol. 160, 1313–1317.
Reforgiato Recupero, G., Russo, G., Recupero, S., Zurru, R., Deidda, B., Mulas, M., 2009. Horticultural evaluation of new *Citrus latipes* hybrids as rootstocks for citrus. HortScience 44, 595–598.
Rios, G., Naranjo, M.A., Rodrigo, M.J., Alós, E., Zacarias, L., Cercos, M., et al., 2010. Identification of a GCC transcription factor responding to fruit colour change events in citrus through the transcriptomic analyses of two mutants. BMC Plant Biol. 10, 276.
Rodrigo, M.J., Zacarias, L., 2007. Effect of postharvest ethylene treatment on carotenoid accumulation and the expression of carotenoid biosynthetic genes in the flavedo of orange (*Citrus sinensis* L. Osbeck) fruit. Postharvest Biol. Technol. 43, 14–22.
Rodrigo, M., Marcos, J.F., Alférez, F., Mallent, M.D., Zacarías, L., 2003. Characterization of Pinalate, a novel *Citrus sinensis* mutant with a fruit-specific alteration that results in yellow pigmentation and decreased ABA content. J. Exp. Bot. 54, 727–738.
Rodrigo, M., Marcos, J.F., Zacarías, L., 2004. Biochemical and molecular analysis of carotenoid biosynthesis in flavedo of orange (*Citrus sinensis* L.) during fruit development and maturation. J. Agric. Food Chem. 52, 6724–6731.
Rodrigo, M., Alquezar, B., Zacarías, L., 2006. Cloning and characterization of two 9-cis-epoxycarotenoid dioxygenase genes, differentially regulated during fruit maturation and under stress conditions, from orange (*Citrus sinensis* L. Osbeck). J. Exp. Bot. 57, 633–643.
Rodrigo, M.J., Alquézar, B., Alós, E., Lado, J., Zacarías, L., 2013a. Biochemical bases and molecular regulation of pigmentation in the peel of *Citrus* fruit. Sci. Hortic. 163, 46–62.
Rodrigo, M.J., Alquézar, B., Alós, E., Medina, V., Carmona, L., Bruno, M., et al., 2013b. A novel carotenoid cleavage activity involved in the biosynthesis of Citrus fruit-specific apocarotenoid pigments. J. Exp. Bot. 64, 4461–4478.
Sadka, A., Dahan, E., Cohen, L., Marsh, K.B., 2000a. Aconitase activity and expression during the development of lemon fruit. Physiol. Plant. 108, 255–262.
Sadka, A., Dahan, E., Or, E., Cohen, L., 2000b. NADP(+)-isocitrate dehydrogenase gene expression and isozyme activity during citrus fruit development. Plant Sci. 158, 173–181.
Sadka, A., Dahan, E., Or, E., Roose, M.L., Marsh, K.B., Cohen, L., 2001. Comparative analysis of mitochondrial citrate synthase gene structure, transcript level and enzymatic activity in acidless and acid-containing *Citrus* varieties. Aust. J. Plant Physiol. 28, 383–390.
Saladié, M., Matas, A.J., Isaacson, T., Jenks, M.A., Goodwin, S.M., Niklas, K.J., et al., 2007. A reevaluation of the key factors that influence tomato fruit softening and integrity. Plant Physiol. 144, 1012–1028.
Schneider, H., 1968. The anatomy of citrus. In: Reuther, W., Webber, H.J., Batchelor, L.D. (Eds.), The Citrus Industry. vol. 2. University of California Press, Berkeley, pp. 1–23.
Sheng, L., Shen, D., Yang, W., Zhang, M., Zeng, Y., Xu, J., et al., 2017. GABA pathway rate-limit citrate degradation in postharvest citrus fruit evidence from HB pumelo (*Citrus grandis*) × Fairchild (*Citrus reticulata*) hybrid population. J. Agric. Food Chem. 65 (8), 1669–1676.
Shi, C., Song, R., Hu, X., Liu, X., Jin, L., Liu, Y., 2015. *Citrus* PH5-like H(+)-ATPase genes: identification and transcript analysis to investigate their possible relationship with citrate accumulation in fruits. Front. Plant Sci. 6, 135.
Shimada, T., Fuiii, H., Endo, T., Yazaki, J., Kishimoto, N., Shimbo, K., et al., 2005. Toward comprehensive expression profiling by microarray analysis in citrus: monitoring the expression profiles of 2213 genes during fruit development. Plant Sci. 168, 1383–1385.
Shimada, T., Nakano, R., Shulaev, V., Sadka, A., Blumwald, E., 2006. Vacuolar citrate/H^+ symporter of citrus juice cells. Planta 224, 472–480.
Shimizu, T., Tanizawa, Y., Mochizuki, T., Nagasaki, H., Yoshioka, T., Toyoda, A., et al., 2017. Draft sequencing of the heterozygous diploid genome of Satsuma (*Citrus unshiu* Marc.) using a hybrid assembly approach. Front. Plant Sci. 8, 180.
Shlizerman, L., Marsh, K., Blumwald, E., Sadka, A., 2007. Iron-shortage-induced increase in citric acid content and reduction of cytosolic aconitase activity in Citrus fruit vesicles and calli. Physiol. Plant. 131, 72–79.
Spiegel-Roy, P., Goldschmidt, E.E., 1996. Biology of *Citrus*. Cambridge University Press, Cambridge.
Stewart, W.S., Klotz, L.J., 1947. Some effects of 2,4-dichlorophenoxyacetic acid on fruit-drop and morphology of oranges. Bot. Gaz. 109, 150–162.
Stewart, I., Wheaton, T.A., 1972. Carotenoids in citrus. Their accumulation induced by ethylene. J. Agric. Food Chem. 20, 448–449.
Strazzer, P., Spelt, C.E., Li, S., Bliek, M., Federici, C.T., Roose, M.L., et al., 2019. Hyperacidification of Citrus fruits by a vacuolar proton pumping P-ATPase complex. Nat. Commun. 10, . 744.
Swingle, W.T., Reece, P.C., 1968. The botany of *Citrus* and its wild relatives. In: Reuther, W., Webber, H.J., Batchelor, L.D. (Eds.), The Citrus Industry. vol. 1. University of California Press, Berkeley, pp. 190–430.
Tadeo, F.R., Primo-Millo, E., 1988. Ultrastructural changes in the developing ovaries of citrus. In: Goren, R., Mendel, K. (Eds.), Sixth International Citrus Congress. Balaban Publishers, Philadelphia/Rehovot, Israel, pp. 417–430.
Tadeo, F.R., Cercós, M., Colmenero-Flores, J.M., Iglesias, D.J., Naranjo, M.A., Ríos, G., et al., 2008. Molecular physiology of development and quality of citrus. Adv. Bot. Res. 47, 147–223.
Terol, J., Soler, G., Talon, M., Cercos, M., 2010. The aconitate hydratase family from *Citrus*. BMC Plant Biol. 10, 222.
Thomson, W.W., Platt-Aloia, K., 1976. Ultrastructure of the epidermis of developing, ripening and senescing navel organs. Hilgardia 44, 61–82.
Tisserat, B., Galletta, P., Jones, D., 1990. Carpel polymorphism in citrus fruit. Bot. Gaz. 151, 54–63.
Trebitsh, T., Goldschmidt, E.E., Riov, J., 1993. Ethylene induces de novo synthesis of chlorophyllase, a chlorophyll degrading enzyme, in Citrus fruit peel. Proc. Natl. Acad. Sci. U. S. A. 90, 9441–9445.
Wang, J., Hao, H., Liu, R., Ma, Q., Xu, J., Chen, F., et al., 2014. Comparative analysis of surface wax in mature fruits between Satsuma mandarin (*Citrus unshiu*) and 'Newhall' navel orange (*Citrus sinensis*) from the perspective of crystal morphology, chemical composition and key gene expression. Food Chem. 153, 177–185.

Wang, J., Sun, L., Xie, L., He, Y., Luo, T., Sheng, L., et al., 2016. Regulation of cuticle formation during fruit development and ripening in 'Newhall' navel orange (*Citrus sinensis* Osbeck) revealed by transcriptomic and metabolomic profiling. Plant Sci. 243, 131–144.

Wang, J., Liu, J., Chen, K., Li, H., He, J., Guan, B., et al., 2017a. Comparative transcriptome and proteome profiling of two *Citrus sinensis* cultivars during fruit development and ripening. BMC Genomics 18, 984.

Wang, X., Xu, Y., Zhang, S., Cao, L., Huang, Y., Cheng, J., et al., 2017b. Genomic analyses of primitive, wild and cultivated citrus provide insights into asexual reproduction. Nat. Genet. 49, 765–772.

Wang, L., He, F., Huang, Y., He, J., Yang, S., Zeng, J., et al., 2018. Genome of wild mandarin and domestication history of mandarin. Mol. Plant https://doi.org/10.1016/j.molp.2018.06.001.

Wei, X., Chen, C., Yu, Q., Gady, A., Yu, Y., Liang, G., et al., 2014. Novel expression patterns of carotenoid pathway-related genes in citrus leaves and maturing fruits. Tree Genet. Genomes 10, 439–448.

Wei, X., Hu, H., Tong, H., Gmitter, F.G., 2017. Profiles of gene family members related to carotenoid accumulation in citrus genus. J. Plant Biol. 60, 1–10.

Wilson, W.C., Hendershott, C., 1968. Anatomical and histochemical studies of abscission of oranges. Am. Soc. Horticult. Sci. 92, 203–210.

Wu, J., Xu, Z., Zhang, Y., Chai, L., Yi, H., Deng, X., 2014a. An integrative analysis of the transcriptome and proteome of the pulp of a spontaneous late-ripening sweet orange mutant and its wild type improves our understanding of fruit ripening in citrus. J. Exp. Bot. 65, 1651–1671.

Wu, G.A., Prochnik, S., Jenkins, J., Salse, J., Hellsten, U., Murat, F., et al., 2014b. Sequencing of diverse mandarin, pummelo and orange genomes reveals complex history of admixture during citrus domestication revealed by genome analysis. Nat. Biotechnol. 32, 56–662.

Wu, J., Fu, L., Yi, H., 2016a. Genome-wide identification of the transcription factors involved in citrus fruit ripening from the transcriptomes of a late-ripening sweet orange mutant and its wild type. PLoS One 11, e0154330.

Wu, J., Zheng, S., Feng, G., Yi, H., 2016b. Comparative analysis of miRNAs and their target transcripts between a spontaneous late-ripening sweet orange mutant and its wild-type using small RNA and degradome sequencing. Front. Plant Sci. 7, 1416.

Wu, G.A., Terol, J., Ibanez, V., López-García, A., Pérez-Román, E., Borredá, C., et al., 2018. Genomics of the origin and evolution of *Citrus*. Nature 554, 311–316.

Xie, X.L., Shen, S.L., Yin, X.R., Xu, Q., Sun, C., Grierson, D., et al., 2014. Isolation, classification and transcription profiles of the AP2/ERF transcription factor superfamily in citrus. Mol. Biol. Rep. 41, 4261–4271.

Xu, C., Fraser, P.D., Wang, W., Bramley, P.M., 2006. Differences in the carotenoid content of ordinary citrus and lycopene-accumulating mutants. J. Agric. Food Chem. 54, 5474–5481.

Xu, Q., Yu, K., Zhu, A., Ye, J., Liu, Q., Zhang, J., et al., 2009. Comparative transcripts profiling reveals new insight into orange (*Citrus sinensis*) red-flesh mutant. BMC Genomics 15, 1–15.

Xu, Q., Liu, Y., Zhu, A., Wu, X., Ye, J., Yu, K., et al., 2010. Discovery and comparative profiling of microRNAs in a sweet orange red-flesh mutant and its wild type. BMC Genomics 11, 246.

Xu, Q., Chen, L.-L., Ruan, X., Chen, D., Zhu, A., Chen, C., et al., 2013. The draft genome of sweet orange (*Citrus sinensis*). Nat. Genet. 45, 59–66.

Xu, J., Xu, H., Xu, Q., Deng, X., 2015. Characterization of DNA methylation variations during fruit development and ripening of sweet orange. Plant Mol. Biol. Report. 33, 1–11.

Yin, X., Xie, X., Xia, X., Yu, J., Ferguson, I.B., Giovannoni, J.J., et al., 2016. Involvement of an ethylene response factor in chlorophyll degradation during citrus fruit degreening. Plant J. 86, 403–412.

Yu, K., Xu, Q., Da, X., Guo, F., Ding, Y., Deng, X., 2012. Transcriptome changes during fruit development and ripening of sweet orange (*Citrus sinensis*). BMC Genomics 13 (1), 10.

Yuan, R., Hartmond, U., Kender, W.J., 2001. Physiological factors affecting response of mature 'Valencia' orange fruit to CMN-pyrazole. II. Endogenous concentrations of indole-3-acetic acid, abscisic acid, and ethylene. J. Am. Soc. Hortic. Sci. 126, 420–426.

Yuan, R., Kender, W.J., Burns, J.K., 2003. Young fruit and auxin transport inhibitors affect the response of mature 'Valencia' oranges to abscission materials via changing endogenous plant hormones. J. Am. Soc. Hortic. Sci. 128, 302–308.

Zeng, W., Huang, M., Wang, X., Ampomah-Dwamena, C., Xu, Q., Deng, X., 2013. Identification and functional characterization of the promoter of a phytoene synthase from sweet orange (*Citrus sinensis* Osbeck). Plant Mol. Biol. Report. 31, 64–74.

Zeng, Y., Du, J., Wang, L., Pan, Z., Xu, Q., Xiao, S., et al., 2015. A comprehensive analysis of chromoplast differentiation reveals complex protein changes associated with plastoglobule biogenesis and remodeling of protein systems in sweet orange flesh. Plant Physiol. 168, 1648–1665.

Zhang, L., Ma, G., Shirai, Y., Kato, M., Yamawaki, K., Ikoma, Y., et al., 2012a. Expression and functional analysis of two lycopene β-cyclases from citrus fruits. Planta 236, 1315–1325.

Zhang, L., Ma, G., Kato, M., Yamawaki, K., Takagi, T., Kiriiwa, Y., et al., 2012b. Regulation of carotenoid accumulation and the expression of carotenoid metabolic genes in citrus juice sacs in vitro. J. Exp. Bot. 63, 871–886.

Zhang, Y., Wang, X., Wu, J., Chen, S., Chen, H., Chai, L., et al., 2014. Comparative transcriptome analyses between a spontaneous late-ripening sweet orange mutant and its wild type suggest the functions of ABA, sucrose and JA during citrus fruit ripening. PLoS One 9, e116056.

Zhang, L., Ma, G., Yamawaki, K., Ikoma, Y., Matsumoto, H., Yoshioka, T., et al., 2015. Effect of blue LED light intensity on carotenoid accumulation in citrus juice sacs. J. Plant Physiol. 188, 58–63.

Zhang, S., Shi, Q., Albrecht, U., Shatters Jr., R.G., Stange, R., Mccollum, G., et al., 2017. Comparative transcriptome analysis during early fruit development between three seedy citrus genotypes and their seedless mutants. Hortic Res. 4, 17041.

Zheng, B., Fang, Y., Pan, Z., Sun, L., Deng, X., Grosser, J.W., et al., 2014. iTRAQ-based quantitative proteomics analysis revealed alterations of carbohydrate metabolism pathways and mitochondrial proteins in a male sterile cybrid pummelo. J. Proteome Res. 13, 2998–3015.

Zhu, F., Luo, T., Liu, C., Wang, Y., Yang, H., Yang, W., et al., 2017. An R2R3-MYB transcription factor represses the transformation of α- and β-branch carotenoids by negatively regulating expression of *CrBCH2* and *CrNCED5* in flavedo of *Citrus reticulate*. New Phytol. 216, 178–192.

Chapter 13

Citrus in changing environments

Christopher Vincent[a], Raphaël Morillon[b], Vicent Arbona[c], Aurelio Gómez-Cadenas[c]
[a]*University of Florida Citrus Research and Education Center, Lake Alfred, FL, United States,* [b]*Equipe "Amélioration des Plantes à Multiplication Végétative", UMR AGAP, Département BIOS, CIRAD, Guadeloupe, France,* [c]*Ecophysiology and Biotechnology, Department of Agricultural and Environmental Sciences, Universitat Jaume I, Castelló de la Plana, Spain*

13.1 Limitations to geographical expansion of citrus

Across the earth, natural landscapes have experienced profound and rapid changes in arable land use throughout history and especially during the second half of the 20th century. These changes have arisen from developments in technology and management driven by socioeconomic and political forces (Rounsevell et al., 2003). Geographical localization of a particular crop species is constrained by abiotic and resulting biotic factors that impair their vegetative and reproductive cycles making them biologically or economically unviable in certain world regions (Korres et al., 2016). These processes are not static but shift with socioeconomic and environmental changes.

Citrus is an important fruit tree crop group worldwide, whose origins can be traced back to Ancient Asia from where they were exported to Southern Europe as medicinal and ornamental plant species (e.g., citron [*C. medica* L.] and sour orange [*C. aurantium* L.]). From there, citrus were brought to the new world including North America and Brazil, current important producers. Citrus production areas are located in the subtropical regions including Southern Europe, subtropical North and South America, Southern Asia, South Africa, and Australia. In general, cultivated citrus species adapt well to various climates, although they are limited by low-temperature constraints to the tropical and subtropical regions. Under subtropical climatic conditions, citrus sprout at least twice a year: spring flush that generates flowers and subsequently fruits and one or more summer-fall flushes that generate branches on which spring flushes will develop (Verreynne and Lovatt, 2009).

All cultivated species, except for the citrus-related species trifoliate orange (*Poncirus trifoliata* L. Raf.), are evergreen perennial trees and, therefore, their foliage is subjected to large seasonal environmental variations throughout the year. Indeed, the complexity of the responses of citrus to environmental challenges is related to the number of interactions of field-grown trees: soil temperature conditioning root metabolism, variations in light exposure of leaves and air temperature, along with other abiotic constraints (Ribeiro and Machado, 2007). Interestingly, high temperatures do not seem to be a limiting parameter to citrus growth (Santos et al., 2017; Zandalinas et al., 2016). Indeed, different *Citrus* species tolerate air temperatures of up to 40°C with no apparent damage to photosystems or major alterations in metabolism (Pereira et al., 2017; Zandalinas et al., 2016, 2017b). Nevertheless, when several abiotic stresses occur simultaneously, the resulting situation is far more deleterious than any of the factors acting individually (Zandalinas et al., 2017a). This is the case of high temperatures occurring simultaneously with water scarcity, causing severe effects on carbon assimilation and inducing oxidative damage (Zandalinas et al., 2016). Such combinations of factors affect current geographic distribution, and as the global climate changes, environmental conditions will affect distribution as well as productivity and sustainability of citrus plantings.

13.2 Predicted climate in citrus growing regions

Climate change is expected to result in challenges to citrus production related to high temperatures, drought, and soil salinity, with problems in limited regions associated with flooding, and cold temperatures. Although long-term climate predictions still bear a great deal of imprecision, especially in the prediction of weather, general trends in probability of precipitation and mean and extreme temperatures are well developed. Because climate types among continents are broadly similar both in current conditions and in predicted climate change, we will discuss the predicted climatic conditions according to Mediterranean, semiarid, and humid subtropical climates. Except where otherwise indicated, information in this segment is derived from the Intergovernmental Panel on Climate Change (IPCC) Fifth Assessment Report (2014).

The Genus Citrus. https://doi.org/10.1016/B978-0-12-812163-4.00013-9

13.2.1 Mediterranean climates

Citrus is produced commercially in Mediterranean climates in North and South America, the actual Mediterranean, and South Africa. Despite the variety of locations in which these climate types exists, Mediterranean climates in all continents are expected to undergo similar shifts. In the Mediterranean Basin where citrus are grown, global warming is expected to induce before the end of the century an increase in average air temperatures of between 2.2°C and 5.1°C and an increase in drought periods related to a high frequency of dry days during which the temperature would exceed 30°C (Giannakopoulos et al., 2009). Thus, major environmental impacts on citrus will be related to the change of the water cycle due to a rise in evaporation and a decrease in rainfall. A minimum decrease in rainfall of −4% in the Mediterranean region is expected (Parry et al., 2007). Lobell et al. (2006) studied yield trends in California and concluded that the area suitable for cultivation of sweet orange [*C. sinensis* (L.) Osbeck] would decline with climate change, which could be mitigated by human activities such as irrigation infrastructure, thus agreeing with the conclusions of Pereira et al. (2017) regarding sweet orange growth and yield: irrigation availability will be the primary determinant of productivity in Mediterranean climate zones.

Increases in irrigation in response to these temperature and precipitation patterns are expected to impact soil water availability and quality. In semiarid areas, the groundwater extraction for irrigation led to the lowering of groundwater and thus to a concomitant increase in salt content (Cortina and Mora, 2003). The consequence of a long-term use of salinized water for irrigation is permanent soil degradation because of dispersion of soil aggregates (Ondrasek et al., 2011). Also, saline intrusion problems in many coastal areas of Italy, Spain, Greece, and North Africa are additional threats to irrigation quality (Aru, 1996). Saltwater intrusion is a challenge in coastal areas of other climate types, as well.

13.2.2 Humid subtropics

In the humid subtropical citrus-producing regions of Brazil, China, and Florida there is greater variability among predictions. Relative to the Mediterranean climates, increases in mean temperatures are expected to be minimal, with some subregions expected to remain at the same mean temperatures. Additionally, annual rainfall is expected to remain the same or increase, with increases associated with more frequent high-volume rainfall events. Florida, in particular, is expected to differ from Brazil and China, in that mean temperature increases are expected to be minimal. Additionally, in Florida, the probability of advective freeze events resulting from the incursion of polar fronts is a historic challenge. Although mean temperatures have increased in Florida, there is little evidence that the risk of freeze events has changed. Rogers and Rohli (1991) determined that freeze events in Florida are associated with relative locations of high- and low-pressure zones during polar weather events in the Northern Great Plains region. Thus, mean temperature increases may not result in decreased risk of freezes in the region. Modeling efforts concluded that citrus in Florida would expand northward and generally maintain yields as negative effects of temperature on growth is mitigated by increasing [CO_2] (Tubiello and Ewert, 2002). Liu et al. (2010) did not find a strong trend associated with agricultural water availability and agricultural yields in China. This is in accordance with IPCC predictions of moderate increases in both temperature and rainfall, a trend similar to the citrus-producing region of Brazil. Thus, if both water demand and availability increase, water availability may have little effect on yields, though the cost of production could increase.

13.2.3 Semiarid regions

The primary semiarid citrus production region covers North India and Pakistan. This region is expected to experience 2.5–5.5°C increases by the end of the century, with moderate increases in rainfall, almost exclusively from increased monsoon precipitation (Kumar et al., 2006; Sivakumar et al., 2005). Unirrigated citrus production in this region is likely to see significant reductions in growth and yield, as mean summer temperatures already exceed optimal temperatures for most citrus varieties (Kumar et al., 2006).

13.2.4 Overall climate trends in citrus-producing regions

The type and degree of effects vary greatly by region and climate type. However, several general trends are evident: (1) Mediterranean climates will likely undergo more frequent and lengthy droughts. (2) Most coastal and arid to semiarid regions will face challenges related to salinity. (3) Most regions will have less frequent freeze events, with the exception of Florida. (4) Nearly all citrus-producing regions will see increased average temperatures, the effects of which will primarily depend on water availability and quality.

13.3 Citrus responses to climate-influenced environmental factors

Future climate-influenced trends in citrus production will depend on citrus physiological and growth cycle responses to climate effects on environmental factors. *Citrus* as a genus displays ecophysiological responses different from many model species, especially annual agronomic species. Most of these unique responses reflect either its nature as a woody perennial, its subtropical origin, or its evergreen growth habit.

13.4 Soil moisture

13.4.1 Optimal water requirements in citrus

Water availability is the most important factor in determining plant performance and, ultimately, survival (Verslues et al., 2006). In this respect, both water scarcity and excess are considered important abiotic stressors, restricting colonization of different world regions by sessile plant organisms. It is difficult to narrowly define the optimal water requirements for a given plant species since it is determined by different components: (1) whole plant transpiration, (2) soil texture, (3) ambient humidity, (4) water quality, and (5) plant characteristics.

Whole plant transpiration constitutes an important force driving water and solute intake into the roots and translocation to the aerial organs; moreover, it is an essential mechanism to regulate leaf surface temperature thus protecting photosystems from thermal degradation under high temperatures (Zandalinas et al., 2016). Transpiration intensity depends on endogenous regulation of stomatal opening that is tightly controlled by plant hormones such as abscisic acid (ABA) and local signals such as reactive oxygen species (ROS) and Ca^{2+} waves (Morillon and Chrispeels, 2001; Munemasa et al., 2013).

Optimal soil textures for citrus cultivation are those predominantly containing a high proportion of sand and less clay with a comparatively heavier subsurface (Srivastava and Singh, 2009). In water-limited regions, soil texture has an impact on water availability. This influences soil water movement and the energy state of the water in the soil column, hence determining soil wetness values and the availability of water to plants (Fernandez-Illescas et al., 2001). Indeed, the abundance of fine-textured particles (clay) in the root-colonized layer of soils is associated with lower hydraulic conductivity and water availability. The amount of clay in soil correlates negatively with the amount of feeder roots in all soil layers (Srivastava and Singh, 2009).

The ambient relative humidity also determines transpiration intensity and the rate at which leaves dehydrate. The vapor pressure deficit (VPD) between the air and the leaf, defined as the simple difference between the saturation vapor pressure and the actual vapor pressure of the air, determines water use efficiency. High VPDs induce unbalance between transpired water and assimilated carbon. High evaporative demands also act as powerful stimuli inducing stomatal closure (Lobell and Gourdji, 2012).

In citrus, the use of reclaimed water with high salinity values reduces gas exchange parameters and fruit quality (Romero-Trigueros et al., 2017) forcing growers to implement specific water management practices (Chapter 20).

Plant characteristics such as the particular scion-rootstock combination, canopy size, and crop load influence transpiration traits, and determine the specific water requirements (Santana-Vieira et al., 2016; Souza et al., 2017). Despite these variations, for practical reasons, it is necessary to estimate the amount of water required by citrus orchards to maintain an adequate production while water is not unnecessarily spent. Therefore, the estimation of water requirements needs to consider all the above-mentioned aspects.

Throughout the year, citrus go through several phenological stages starting from spring bud break around February and March in the Northern Hemisphere, a second flush or set of flushes during summer and fall, and a quiescent winter period, when air temperatures decrease (Fig. 13.1A). Water limitations during spring growth had no effect on shoot extension, but during summer or early fall, it reduced or even arrested shoot growth (Ginestar and Castel, 1996). Moreover, fruits are prone to water stress at different developmental stages negatively affecting fruit set, inducing abnormal fruit development and abscission resulting in reduced fruit yield. Working with mature (7–10 years-old) Valencia orange trees grafted onto trifoliate orange rootstocks, differences in the total ET were found between drip-irrigated and rainfed trees (767 mm vs 620 mm, respectively). The average seasonal potential evapotranspiration rates (ETc) increased from 1.3 mm day^{-1} in winter to 2.8 mm day^{-1} in spring, to 3.0 mm day^{-1} in summer, before the decline in fall. These values are extrapolable to those found in different citrus-producing areas such as Florida, Valencia (Spain), and semiarid regions in Israel. The potential water use for mature orange trees in a subtropical region is equivalent to 80 L $tree^{-1}$ day^{-1}, although this value varies among varieties, seasons, and producing regions (García Petillo and Castel, 2007; Marin and Angelocci, 2011; Rana et al., 2005; Villalobos et al., 2009). When VPD is kept constant, photosynthetic responses of young, mature leaves to soil moisture

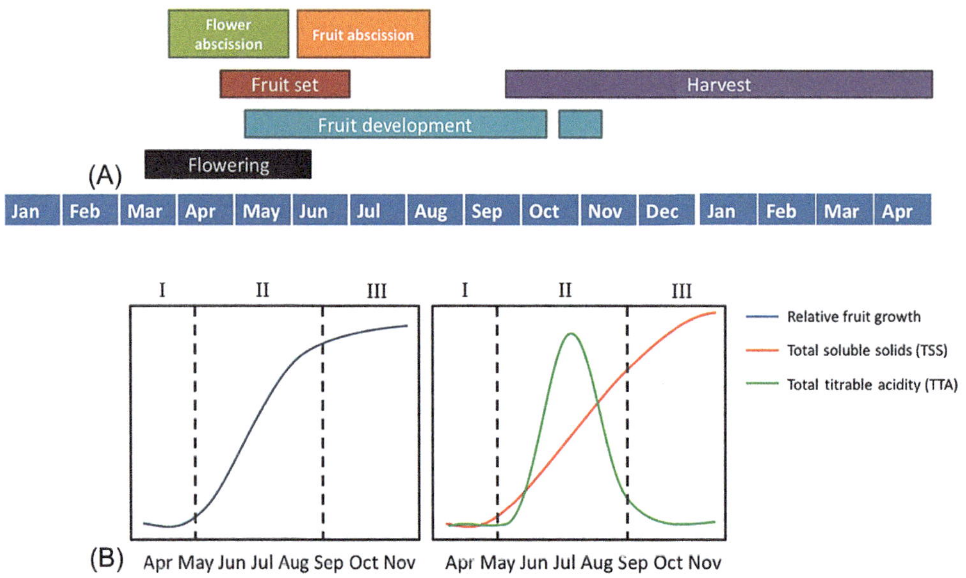

FIG. 13.1 Seasonal fruit growth and development stages of typical sweet orange. Months are according to seasons in subtropical Northern Hemisphere.

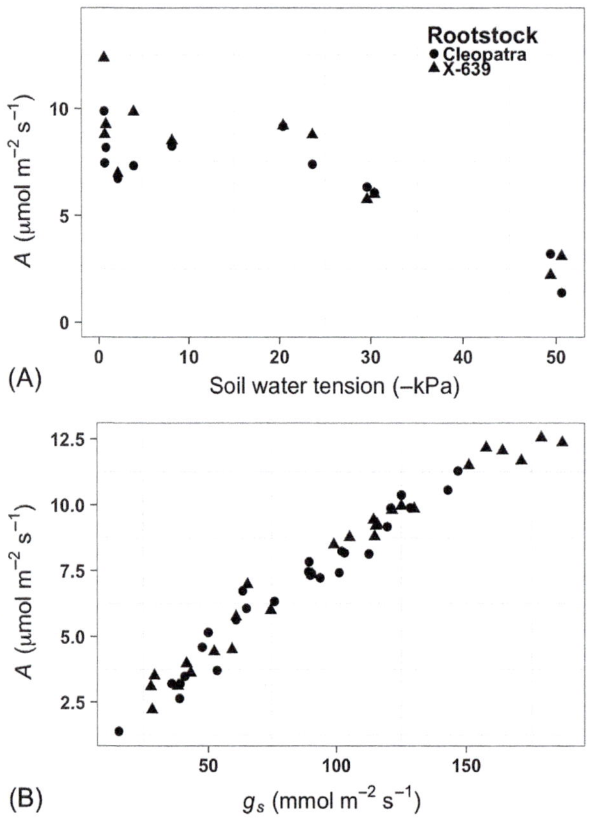

FIG. 13.2 (A) Relationship between net CO_2 assimilation (A) and soil water tension and (B) relationship between A and stomatal conductance (g_s) of Hamlin sweet orange on X-639 or Cleopatra rootstock over a progressive soil drying period.

indicate moderate sensitivity to water deficit, with moderate reduction of photosynthesis at −30 kPa and severe reduction by −50 kPa (Fig. 13.2). Due to variations among species, varieties, rootstock-scion combinations, and their interactions with environmental characteristics, the accurate estimation of citrus water requirements is still challenging; hence, it is preferable to apply the above expected water requirements in order to avoid stomatal limitations that might also influence growth and reproductive parameters.

13.4.2 Flooding

In many citrus-producing regions such as the Mediterranean coast, Florida, Brazil, or China intense rains are common and often flood agricultural landscapes (Arbona et al., 2009; Rodríguez-Gamir et al., 2011). Waterlogging causes a quick depletion of soil O_2 among other disturbances to soil chemistry such as changes in redox potential, alteration in the oxidation state of different soil constituents, and changes in the soil microbiota (Sauter, 2013; Schreiber et al., 2011).

Cultivated *Citrus* species do not possess any specific adaptation to soil flooding; nevertheless, there are differences in the ability to tolerate prolonged flooding among cultivars (Arbona et al., 2009; Arbona and Gómez-Cadenas, 2008; García-Sánchez et al., 2007). Other cultivated species, such as tomato (Vidoz et al., 2016) or soybean (Shimamura et al., 2014), undergo developmental modifications aimed at withstanding the flooded period: adventitious root formation, shoot/stem elongation, and aerenchyma formation. Waterlogging-tolerant citrus rootstocks such as Carrizo citrange (*C. sinensis* × *P. trifoliata*) or citrumelo CPB-4475 (*C. paradisi* × *P. trifoliata*) showed enhanced gas exchange parameters that remained at control values during most of the stress period (Arbona et al., 2009). Moreover, tolerant genotypes also showed an efficient induction of the antioxidant defenses, whereas in sensitive Cleopatra mandarin, soil flooding depressed antioxidant enzyme activities resulting in increased oxidative damage (Arbona et al., 2008; Hossain et al., 2009).

13.4.3 Management of soil flooding in citrus orchards

In many regions, soil waterlogging is a seasonal event linked to heavy rains occurring after the summer season (Arbona et al., 2008; Jackson et al., 2008). The incidence and severity of this adverse condition cannot be predicted based on climatic models although it is expected to rise due to global warming (Lobell and Gourdji, 2012). If a certain nonreturn threshold is not surpassed and plants are not subjected to waterlogging for long periods, they can recover almost fully (Hossain et al., 2009). Nevertheless, cultivars vary in their ability to withstand and subsequently recover from the stress period (Arbona et al., 2008, 2009). Because soil or water management mitigation practices are difficult to implement, variety selection should be a primary concern of growers in flood-prone locations.

Soils with poor drainage are not adequate for citrus plantation as the excess water supply is not properly evacuated. In addition, grove areas where the water table is located near the root zone could also induce soil flooding symptoms. Whenever possible, these types of soils and cultivation areas should be avoided for citrus planting. However, when this is not possible, ensuring optimal drainage by installing a system of ditches, beds, or tiling should be implemented including water managing techniques aimed at reducing watering to avoid excess water from affecting citrus roots (Zekri et al., 2015). In addition, tolerant citrus rootstock cultivars such as Carrizo citrange or citrumelo CPB 4475 should be chosen as stress tolerance of the rootstock is transferred to the grafted scion (Hardy et al., 2012; Moya et al., 2002). Moreover, under field conditions, prolonged soil waterlogging along with warm conditions favors the propagation of several diseases such as Phytophthora root rot or *Alternaria alternata* brown spot. Therefore, protective treatments with fungicides should be performed along with organic mulching that helps to manage fungal diseases. Nutritional imbalances caused by waterlogging can be addressed by application of foliar fertilizers (Smith, 2011).

13.4.4 Water deficit—Effects on crop productivity

Limitation of water supply affects citrus growth and development and severely impairs the reproductive processes leading to reductions in yield. In most Mediterranean citrus-producing regions, the drought period occurs during summer coincident with the phase II of citrus fruit growth (Fig. 13.1B). Inadequate water supply during this stage leads to reductions in fruit size and quality and also results in important yield losses due to fruit abscission (Romero et al., 2006).

Water stress causes stomatal closure aimed to minimize water loss by transpiration. At the same time, CO_2 intake is also reduced impairing photosynthesis and subsequently affecting electron transport between photosystems leading to an increase in the incidence of oxidative stress (Pinheiro and Chaves, 2011). As the water availability in the soil decreases, the ability of plants to maintain water potential is reduced. To cope with this, plants undergo osmoregulation by accumulating osmotically active substances such as inorganic salts and compatible osmolytes such as the amino acid proline and nonmetabolic carbohydrates such as raffinose (Gimeno et al., 2009). By accumulating osmotically active substances, plants reduce their endogenous osmotic potential to maintain water uptake and supply to plant tissues. Although all cultivated *Citrus* species are drought sensitive, there are some cultivars with certain ability to tolerate water stress such as Rangpur lime (*C. limonia*) following an avoidance strategy and maintaining growth, whereas other *Citrus* species such as mandarin (*C. reticulata*) exhibit a tolerance strategy focused on plant survival (Allario et al., 2012; Santana-Vieira et al., 2016). These tolerance strategies are associated with physiological, hormonal, and morphological changes such as variations in cell wall elasticity, enhanced osmolyte and hormone accumulation, and improved stomatal control contributing to survival under low soil water potentials (Gonçalves et al., 2016).

One of the primary responses of plants to limitations in water supply is stomatal closure to prevent water loss through transpiration followed by an increase in root-to-shoot ratio. Nevertheless, under prolonged water deficit periods, the ability of plants to balance water uptake and transpiration through the stomatal aperture, reducing leaf water potential is not sufficient to maintain function. Therefore, they need other strategies to ensure survival (Santana-Vieira et al., 2016). These strategies can be summarized as: (1) induction of physiological and developmental mechanisms aimed at maintaining growth and development during the drought period, known as dehydration avoidance and (2) induction of osmolyte accumulation and protective responses (e.g., antioxidant enzyme activity) oriented to ensure survival (Santana-Vieira et al., 2016), known as dehydration tolerance. Under prolonged stress periods, the tolerance strategy seems to be more effective (Allario et al., 2012; Santana-Vieira et al., 2016; Souza et al., 2017). In this respect, when comparing the physiological response of sweet orange scions grafted onto diploid or tetraploid Rangpur lime seedlings, the latter showed enhanced drought tolerance resulting from dehydration tolerance responses such as increased hormone and solute levels and lower stomatal conductance (g_s; Allario et al., 2012).

13.4.5 Irrigation to mitigate water deficit

Despite the reported stomatal sensitivity of *Citrus* species to water limitations, it is possible to grow citrus under water deficit maintaining productivity and fruit quality (Navarro et al., 2015; Pérez-Pérez et al., 2008). In oranges, deficit irrigation (DI) had an impact on yield, a decrease of 20% of control water supply (60% of Class A pan evaporation) had only 5% penalty in yield (reflected in fruit size, not number; Castel and Buj, 1990). In many citrus-producing areas with semiarid climate with dry hot summers and high evaporative demand, scarcity in water resources is an important limiting factor. A common approach is to use strategies collectively known as DI and these strategies involve (1) reduction of water supply throughout the year based on potential seasonal ETc or (2) implementation of moderate to severe reductions in water supply during the less stress-sensitive periods of plant development. The relative success of the application of these techniques relies on the rootstock cultivar used. For example, fruit yield and water use efficiency are less affected in trees grafted onto Cleopatra mandarin than in those onto Carrizo citrange (Pérez-Pérez et al., 2008; Romero et al., 2006).

Another group of water-saving techniques known as "partial rootzone drying" or PRD can be readily applied to drip-irrigated orchards. This strategy involves the irrigation of a part of the root system while the other part is kept dry and after a certain period of time, the irrigation is switched (Consoli et al., 2014). Besides being a water-saving strategy, the application of PRD had positive effects on water use efficiency of citrus orchards compared to full irrigation (Consoli et al., 2017). Increase of water use efficiency in orchards subjected to different DI strategies is thought to operate through a reduction in vegetative growth, altering the balance between vegetative and reproductive growth (Consoli et al., 2014).

In citrus fruit production, not only yield is an important parameter but also quality. The effect of limiting water supply throughout the vegetative and reproductive stages of citrus or during different stages of fruit growth and development has been the focus of several studies. DI strategies efficiently increased total soluble solids (TSS), total titratable acidity (TTA) (Navarro et al., 2015; Pérez-Pérez et al., 2008), and flavonoid accumulation (Grilo et al., 2017), whereas fruit physiopathies such as albedo breakdown were reduced (Carr, 2012; Grilo et al., 2017).

13.5 Soil salinity

Due to the use of saline water for irrigation, the increase of soil salinity is one of the major abiotic constraints for the citrus industry, especially where irrigation is required (Grattan et al., 2015). In the Gaza Strip, Ashour and Al-Najar (2012) showed that an increase of temperatures by +1°C or +2°C caused an increase of the annual average evapotranspiration leading to an increase of irrigation requirements by 3.28% and 6.68%, respectively, and similar trends were found by Fares et al. (2017). In the Souss region of Tunisia, groundwater depletion and repeated droughts forced farmers to uproot high-value citrus orchards (Shah, 2014). To tackle these problems, treated wastewater was tested for citrus irrigation (Grattan et al., 2015). This trend of salinity due to groundwater depletion has occurred in numerous Mediterranean horticultural regions (Libutti and Monteleone, 2017). Likewise, a variety of coastal citrus-producing regions including the Mediterranean and Florida have seen more frequent problems of irrigation water salinity due to saltwater intrusion into superficial aquifers. In the case of low-lying areas such as Florida, rising sea levels exacerbate saltwater intrusion, and in some areas, groundwater extraction also contributes (Langevin and Zygnerski, 2013).

Citrus trees are classified as "salt sensitive." Therefore salinity reduces growth and fruit productivity by inducing plant physiological disorders (Syvertsen and Garcia-Sanchez, 2014). Salinity can strongly impact citrus trees and cause leaf drop, twig dieback, reduce plant growth, and reduce fruit production (Levy and Syvertsen, 2004; Maas, 1993). However, mild salinity can induce a slight increase in TSS, sometimes accompanied by a similar increase in TTA (Boman, 2000). Thus, in the orchard, precise irrigation fertilization management is required to limit the impact of salt toxicity.

Salinity is a complex stress that affects plants through ionic and osmotic factors occurring both in the soil and in the canopy, leading to the limitation of transpiration by closing the stomata, accumulation of ions, and nutritional imbalances (Hussain et al., 2012; Forner-Giner et al., 2011). Large differences in salt stress tolerance have been found among *Citrus* species and related genera (Hussain et al., 2012, 2015) with sensitive types exhibiting decreased g_s, CO_2 assimilation rates (A), and leaf osmotic potential. One important mechanism through which salinity affects citrus plants is through foliar ionic accumulation, thus practices and varietal selection that reduce this accumulation will reduce salt stress.

13.5.1 Salinity effect on vegetative and reproductive growth

Salt stress importantly alters seed germination. Although citrus varieties are usually propagated by budding, most of the rootstock multiplication is based on seeds (Barlass and Skene, 1986). Therefore, salinity could affect seed emergence and early establishment of seedlings (Zekri, 1993). However, the length of primary roots seems to overcome salinity after emergence by increasing proportionally with the increase in salt concentration (Kaushal et al., 2013). It has been hypothesized that salt stress reduces the levels of seed endogenous gibberellins while increasing the concentration of ABA, the main hormones that antagonistically control germination (Gómez-Cadenas et al., 2015). More recent data indicate that salinity induces solubilization and mobilization of a specific protein, citrin, in citrus seeds, and that this protein could be detrimental for seedling establishment because it causes premature depletion of citrin-like storage compounds that are normally useful during seedling establishment under nonstressed conditions (Ziogas et al., 2017).

Salt stress also reduces plant growth. Hence, when salt concentration increases beyond the tolerance limits, plant height, stem diameter, leaf area, and fresh and dry weight decrease. Leaf chlorosis, necrosis, and eventually abscission are also observed (Gómez-Cadenas et al., 2003). High mortality is registered when salt concentrations increase over $3\,dS\,m^{-1}$ (Simpson et al., 2015). Salinity also promotes leaf succulence (Cerdá et al., 1977). Microscopic examination showed that the increase in leaf thickness was due to the development of larger cells in the spongy mesophyll (Zekri and Parsons, 1990a,b).

Root growth is often less sensitive to salinity than shoot growth, so the root:shoot ratio increases at high salinity conditions (Rewald et al., 2012). The distribution of roots is also affected by soil salinity and increased salinity results in a reduction of root development in the layer below 0.3 m and increased roots in the upper layers (Alva and Syvertsen, 1991).

13.5.2 Salinity effects on mineral nutrition

Although the effects of high salinity on plant nutrition are dependent on the rootstock used, in general, NaCl addition to the growth media increased N, P, and K^+ levels in citrus cells and reduced Ca and Mg concentration (Banuls et al., 1991; Nieves et al., 1990; Zekri and Parsons, 1990a,b). Unlike most other crops, in citrus, Cl^- is the main ion that causes damage (Bañuls and Primo-Millo, 1995). In terms of chloride transport, several works indicate that some rootstocks are much more efficient than others in restricting the uptake and/or transport of Cl^- to shoots (Hussain et al., 2012, 2015; López-Climent et al., 2008; Moya et al., 2002; see Chapter 14). It has also been described that rootstocks with high ability for chloride exclusion are unable to restrict Na^+ uptake to the leaves, suggesting that separate mechanisms exist for the uptake and/or transport of Cl^- and Na^+ in citrus (Walker et al., 1983). Behboudian et al., (1986) revealed that the ability of rootstocks to exclude Na^+ from the xylem in the proximal root and basal stem and to sequester it in both the wood and the bark of these regions was related to a drop in K^+ levels in these tissues, suggesting Na–K exchange. The "fine root turnover" phenomenon in which continuous root formation apparently used by the plants to remove the excess of ions in this tissue has been well established in trifoliate orange (Tozlu et al., 2000). Applications of Ca and N can be beneficial for citrus under stress.

13.5.3 Salinity effects on citrus physiology

High salinity, A, in citrus plants having nonstomatal factors has a marked influence in this reduction. It has been reported that, under salt stress, g_s decreases concomitantly with A, maintaining, therefore, values of substomatal CO_2 concentration invariable or even increased in stressed plants (García-Sánchez et al., 2007; Perez-Perez et al., 2007). These results are supported by the damages in leaf anatomy or changes in chlorophyll fluorescence, which indicate a possible impairment in the electron transport chain in salinized citrus plants (López-Climent et al., 2008).

Despite the relative tolerance of various cultivars, salinity induces severe alterations to the photosynthetic metabolism that could end up with the leakage of electrons and the overproduction of ROS (Arbona et al., 2003). Under these situations, a progressive reduction of photosynthesis occurs and, therefore, photorespiration is activated in peroxisomes. The oxidative reactions that take place in this organelle render large amounts of H_2O_2 that can cross membranes and affect any surrounding structures or molecules. Excess ROS produced under stress react with many different molecules, yielding several oxidative by-products such as lipid hydroperoxides that initiate chain reaction cascades with other lipid molecules. Lipid hydroperoxides also produce another subproduct named malondialdehyde widely used as a marker for oxidative damage (Arbona et al., 2003, 2008; Hossain et al., 2009). It has also been shown that the production of ROS under nonstressful conditions is involved in the control of cell differentiation and organogenesis in citrus (Faltin et al., 2010). Proteins also constitute a target for ROS: amino acids in active sites can be oxidized and inhibit protein function (Garg and Manchanda, 2009). Citrus leaves must activate antioxidant defense mechanisms to cope with these additional stresses.

Evidence also suggests that, in parallel to oxidative stress, production of reactive nitrogen species and particularly nitric oxide (NO) occurs during exposure of citrus to salinity, leading to nitrosative stress (Tanou et al., 2009a,b). Preexposure of roots to H_2O_2 or NO could ensure plant survival under salt stress and supports the notion that interaction between H_2O_2 and NO production may be locally and systematically sensed in citrus plants (Tanou et al., 2012).

13.5.4 Mitigation of salinity effects

Reduction in water use under saline soil conditions will limit the translocation of toxic saline ions into the leaves (Syvertsen and Garcia-Sanchez, 2014; Moya et al., 1999). Similarly, genome duplication was shown to improve tolerance to saline toxicity, caused by lower Cl^- accumulation associated with the reduced transpiration rate of tetraploid genotypes (Ruiz et al., 2016b). In this sense, Carrizo citrange tetraploid seedlings appeared to be more tolerant to salinization related to morphological and histological traits of roots leading to reduced hydraulic conductance and transpiration rate (Ruiz et al., 2016a). A schematic representation with different scenarios for salt tolerance of rootstock-scion combinations is shown in Fig. 13.3. A section regarding the best genotypes for breeding is presented in Chapter 14. Under well-watered saline conditions, osmotic adjustment often increases leaf turgor above that of nonsalinized leaves (Garcia-Sanchez et al., 2006). Therefore, producers in

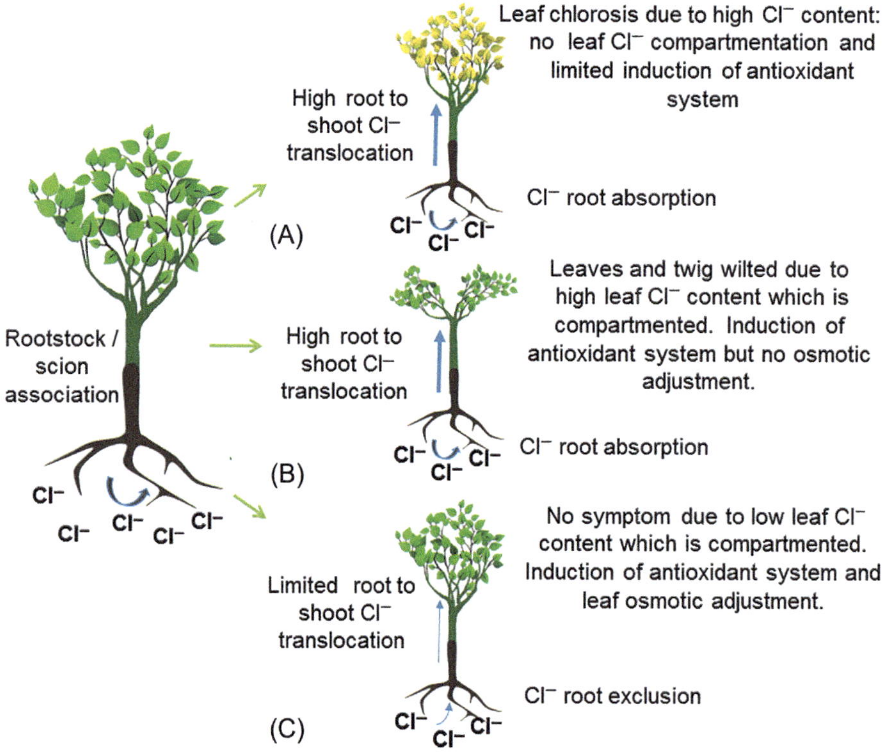

FIG. 13.3 Different scenarios of tolerance for rootstock/scion associations subjected to salt stress. (A) If the rootstock is not Cl^- excluder, Cl^- are actively translocated from root to shoot in the transpiration stream. In the case of a scion variety which is not able to compartmentalize Cl^- and if the antioxidant system is not triggered, chlorosis in the leaf is quickly observed (B) with the same rootstock than in (A), if the scion variety is able to compartmentalize Cl^- and to induce antioxidant systems but without osmotic adjustment, fast wilting may occur but chlorosis will be delayed. (C) If the rootstock is Cl^- excluder and if the variety is able to limit its transpiration stream by closing the stomata, a limited amount of Cl^- may be translocated. Also, if the scion variety compartmentalizes Cl^-, induces antioxidant systems and osmotic adjustment, higher tolerance to salt stress may be expected.

salinity-prone regions should choose rootstock and scion varieties that minimize water use. Additionally, dry season irrigation practices should rely on water-conserving practices such as PRD (this chapter and Chapter 20). Additionally, low salinity water sources for irrigation should be prioritized, since water above $1.5\,dS\,m^{-1}$ will likely lead to decreased growth and survival (Simpson et al., 2015).

13.6 Air moisture

Among the numerous morphological and physiological features that contribute to citrus water conservation nature, stomatal sensitivity to air moisture is prominent. As an isohydric genus, high VPD induces stomatal closure (Tardieu and Simonneau, 1998). Even under sufficient soil moisture, citrus responds to increasing VPD with reduced g_s (Khairi and Hall, 1976a). Included in this VPD response is the midday depression of photosynthesis, which occurs in citrus under most conditions conducive to photosynthesis, but is more pronounced when VPD is higher (Jifon and Syvertsen, 2001, 2003). This process had long been observed on a leaf level for short time scales but was recently reported for field-scale transpiration (Er-Raki et al., 2009; Maestre-Valero et al., 2017; Villalobos et al., 2009). Maestre-Valero et al. (2017) reported VPD was a primary driver of crop coefficient (K_c; crop water use relative to evaporation) water demand on a monthly basis (Fig. 13.4).

Although the general tendency of stomatal VPD sensitivity is ubiquitous across the genus, different species have been found to vary both in the degree of response. For example, under humid conditions lemon (*C. limon* [L.] Osbeck) maintained *A* under temperatures of up to 35°C, while sweet orange *A* declined after 30°C (Kriedemann, 1968). However, in the same study under dry conditions, *A* declined above 20°C in both species, with a steeper decline in *A* of lemon than that of sweet orange. Citrus are capable of acclimating to a range of suboptimal air conditions. Khairi and Hall (1976a) found that both sweet orange and grapefruit (*C. paradisi* Macf.) increased photosynthetic rates over 3–5 days under constant daytime conditions, except under extremes, such as temperatures above 30°C and VPD above −2.5 kPa.

The strong stomatal sensitivity appears to be a mechanism of protection of xylem water status. Nortes et al. (2010) found that evapotranspiration of citrus groves had a maximum above which increased evaporative demand did not increase evapotranspiration. Nortes et al. attributed this effect to limitations of whole plant conductance, in which case stomatal sensitivity would protect the xylem and leaf water status from failure that would otherwise result when plant to atmosphere flux outstripped internal water conductance capacity. This hypothesis is consistent with a xylem vulnerability study which found approximately 40% loss in xylem conductivity between −1 and −2 MPa xylem pressure, and no subsequent loss between −2 and −4 MPa (Poggi et al., 2007). Additionally, the same study found a steep decline in g_s between −0.3 and −0.6 MPa xylem tension and no additional decline after −1.2 MPa.

Thus, increasingly dry air conditions during warm growing seasons are likely to have moderate effects on the relative water demand of citrus crops, mitigated by water conservation stomatal responses. However, the same stomatal responses

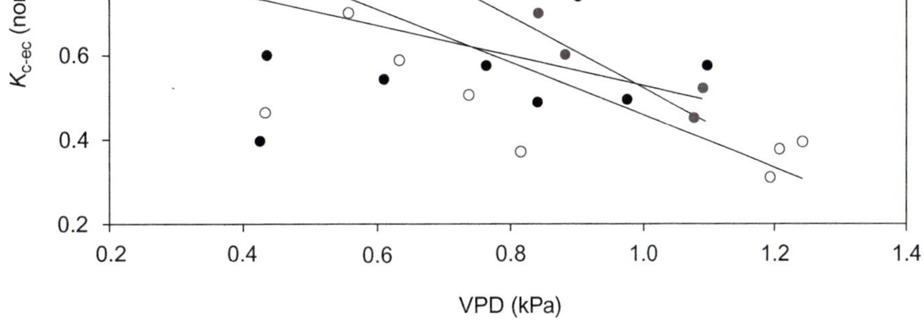

FIG. 13.4 Daily crop coefficients $K_{c\text{-}ec}$ determined by the eddy covariance system vs vapor pressure deficit (VPD; kPa) in 2008, 2009, and 2010. Each data point is a monthly average value.

likely limit photosynthesis. Mesophyll conductance is highly correlated with photosynthesis across varieties (Romero-Aranda et al., 1997). Additionally, stomatal sensitivity to VPD negatively impacts photosynthesis, as g_s and A were highly correlated over a gradient of water deficits in Hamlin sweet orange (Fig. 13.2B).

One important characteristic of citrus stomatal responses to VPD and to temperature is acclimation. Although relatively little attention has been paid in climate change related studies, *Citrus* species generally increase g_s and A over 3–5 days as a given set of conditions persists. For example, sweet orange reduced g_s from 183 to 93 mmol m^{-2} s^{-1} when the air was dried from 2.5 to 4.1 kPa at 38°C (Khairi and Hall, 1976b). However, by the third day under these hot and dry conditions, g_s had recovered to 121 mmol m^{-2} s^{-1}. Similar results were reported for grapefruit, and lemon, as well as other noncultivated relatives (Khairi and Hall, 1976a,b).

13.7 Temperature

13.7.1 Heat

Drought and heat are often cooccurring stresses that drastically affect global agricultural systems. Consequently, higher temperatures associated with high evaporative losses coupled with increasing water demand will result in decreasing water availability for agriculture (Hatfield and Prueger, 2015), and some citrus-producing regions are expected to undergo higher temperatures in the coming decades, exposing producers to heat impacts on water loss and on production and quality (Fares et al., 2017; Khairi and Hall, 1976a). Heat alone, apart from soil water deficit, affects citrus plants through increases in VPD, oxidative stress, and other effects on vegetative and reproductive growth and development.

13.7.2 Mechanisms involved in heat tolerance

Impacts on plant physiology and mechanisms of heat tolerance in plants were reviewed by Bita and Gerats (2013). High temperature will affect photosynthesis, respiration, water relations, and membrane stability, leading to the production of ROS. Heat tolerance mechanisms include maintenance of membrane stability, scavenging of ROS, production of antioxidants and compatible solutes to favor osmotic adjustment, and induction of MAPK and CDPK signaling cascades, transcriptional activation as well as chaperone expression to aid in protein refolding.

Temperature stresses on annual and perennial crops have an impact on all phases of plant growth and development (Hatfield and Prueger, 2015). The optimum temperature range in citrus was estimated to be 22–34°C (Cole and McCloud, 1985). If temperatures exceed 35°C the optimum range during citrus fruit development, fruit drop, and a reduction in TSS, TTA, and fruit size occur (Hutton and Landsberg, 2000). Thus, temperature increases are predicted to lead to high loss of crop productivity.

Investigation of heat stress in citron has shown a large decrease of photosynthetic rate and carboxylation efficiency which was associated with chloroplast ultrastructure alteration, oxidative damage resulting from ROS accumulation, and activation of enzymes for their detoxification (Guo et al., 2006). Higher temperature is associated with enhanced sunlight irradiance that may induce photooxidative stress. Sun-acclimated leaves and shade-acclimated leaves returned to natural sunlight irradiance in *C. medica* L., *C. reticulata* Blanco, *C. maxima*, and *Fortunella japonica* showed that kumquat and pummelo have greater tolerance to photooxidative stress than citron and mandarin (Santini et al., 2012).

Carrizo citrange and Cleopatra mandarin were recently shown to present different ability to tolerate drought and heat applied alone or in combination (Zandalinas et al., 2016). Physiological responses (gas exchange parameters and chlorophyll fluorescence) in Cleopatra mandarin evidenced a higher susceptibility of that cultivar. In that study, Carrizo was found to be more tolerant to high temperatures by retaining higher photosynthetic activity and transpiration favoring leaf cooling than Cleopatra mandarin. Metabolite changes related to the accumulation of photoprotective antioxidant secondary metabolites were also observed in Cleopatra mandarin that was interpreted by authors as an activation of energy metabolism along with metabolic pathways, oriented to mitigate the damaging effects of stress (Zandalinas et al., 2017a). Carrizo was able to cope with oxidative stress by enzymatic means preventing the accumulation of antioxidant metabolites. Investigation of the antioxidant defense system in both cultivars confronted with drought and heat stress showed that Carrizo citrange has a much more efficient and coordinated ROS detoxification system than Cleopatra mandarin (Zandalinas et al., 2017a).

13.7.3 Management of high temperature and breeding heat-tolerant varieties

Irrigation management associated with painting tree trunks white may prevent cracking and splitting of the tender new bark especially for young trees or heavily pruned trees over the winter.

Important breeding programs were developed in citrus to make seedless varieties. One of the most advanced strategies consisted in developing triploid varieties (Ollitrault et al., 2008). Stomata size increased with the level of ploidy while their density decreased. High stomatal density is associated with high g_s and transpiration rates. Also, small stomata lead to greater stomatal resistance in water stress. Because small stomata can open and close more rapidly, it is expected that diploid citrus are more adapted to water deficit. Tetraploid citrus seedlings were shown to present higher leaf water content than diploids (Allario et al., 2011) with more efficient antioxidant systems compared to their diploid counterparts (Wang et al., 2013). Yet, investigation of the impact of polyploidy on plant transpiration and water stress tolerance showed that tetraploid citrus seedlings are more tolerant to water deficit than the respective diploid (Oliveira et al., 2017). Thus, further investigations are required to determine whether triploids present better adaption to heat and drought.

13.7.4 Chilling and freezing

Since citrus are grown in tropical and subtropical areas, chilling and freezing events affect only specific areas where frost events are infrequent and often mild. Predictions regarding climate change call for fewer extreme frost events in most areas and an increase in warm extreme temperature events (Kharin et al., 2007). However, air temperatures and precipitation increases are predicted that may lead to episodic frost events in early fall or late spring when plant cold acclimation is low (Rigby and Porporato, 2008). In citrus, advanced flowering date due to global warming citrus may thus lead to frost risks. Cold (below 13°C) will impair plant growth and development and will reduce productivity (Ribeiro and Machado, 2007). Damage will result when citrus are exposed to temperatures below −2.2°C (Yelenosky, 1985). However, relatively cold winters are advantageous in that they induce uniform flowering and the development of good fruit color (Levy and Syvertsen, 2004).

13.7.5 Management of freezing temperature in citrus orchards

Freeze damage on citrus trees may occur when water freezes into the plant tissue. Then, leaves may dry out and fall. In fruit, pitting can occur and they may drop if they are badly damaged. Periods of acclimation favor cold tolerance in citrus (Davies and Albrigo, 1994). To minimize crop losses from freezes, sprinkler irrigation through microsprinklers is effective with moderate freezes depending on wind speeds.

13.7.6 Mechanisms involved in chilling and freezing tolerance

Extensive investigations associating functional genomics, genetics, physiology, and biochemistry have been carried out to decipher plant responses to cold. Because of its specific deciduous character, trifoliate orange is considered a major source for cold tolerance improvement in citrus. In that species, early cold-induced genes that may favor cold acclimation were identified (Peng et al., 2012). Overexpression of trifoliate orange genes that code for basic helix-loop-helix transcription factors (PtrbHLH) is shown to promote tolerance by increasing the expression of peroxidase genes favoring, in turn, ROS removal (Huang et al., 2013). More recently Huang et al. (2015) showed that under cold conditions ICE1, which is an inducer of C-repeat-binding factor (CBF) gene modulates polyamine levels through interactions with arginine decarboxylase, leading in turn to lower H_2O_2 and O_2^- contents, and higher activity of antioxidant enzymes, such as superoxide dismutase and catalase. In lemon, knocking out PtrBAM1, a gene coding a β-amylase in trifoliate orange, was shown to reduce starch degradation and decrease the contents of maltose and soluble sugars leading in turn to more limited cold adaption (Peng et al., 2014). Cold-responsive trifoliate orange miRNAs through high-throughput sequencing were identified (Zhang et al., 2014). More recently, transgenic lemon plants overexpressing ptr-MIR396b, a cold-responsive miRNA from trifoliate orange which guides the cleavage of 1-aminocyclopropane-1-carboxylic acid oxidase (ACO), are shown to display enhanced cold tolerance. The reduction of ACO transcript levels, thereby led to a repression of ethylene synthesis simultaneously promoting polyamine synthesis, leading to enhanced ROS scavenging (Zhang et al., 2016). Proteomic and metabolomic profiling of Valencia orange fruit after natural frost exposure performed by Perotti et al. (2015) suggested that redistribution of photoassimilates in the frost-exposed plants occurred, associated with increased sugar levels.

Rootstock-scion combinations allow the improvement of tolerance to different biotic and abiotic stresses. However, little is known regarding the mechanisms that allow the rootstock to improve scion cold tolerance. Numerous rootstock hybrids using trifoliate orange as a parent were produced to promote cold tolerance in citrus cultivars. However, the impact was found to be very limited. Controlled salinity stress under greenhouse conditions was shown to enhance cold-induced freeze tolerance in citrus seedlings by reducing physiological activity and growth (Syvertsen and Yelenosky, 1988). Recently, the use of tetraploid Carrizo citrange as rootstock was shown to promote chilling tolerance in Clementine via the antioxidant system (Oustric et al., 2017).

13.7.7 Breeding cold-tolerant varieties

In citrus, trifoliate orange has been widely investigated since it has deciduous leaves and undergoes winter dormancy. Trifoliate orange cultivars are very resistant to chilling and freezing events since they can withstand temperatures of −26°C when cold-acclimated (Yelenosky, 1985). However, when trifoliate orange and trifoliate orange hybrids such as Carrizo citrange are grafted with a scion, leaves are no longer deciduous, making rootstock-scion associations much more sensitive to cold. Fruit of trifoliate orange are inedible, making them useless for breeding varieties with better cold tolerance. Mandarins are usually considered the most cold-tolerant varieties (Yelenosky, 1985), evidenced by low downregulation of photosynthetic parameters and low accumulation of oxidized compounds (Santini et al., 2013). If pummelos are considered to be vulnerable to negative temperatures by Yelenosky (1985), Santini et al. (2013) ranked them as tolerant, probably because the process of cold acclimation in both experiments was not the same. In the last study, kumquat was ranked as intermediate tolerant and citron as the most sensitive because it presented a marked decrease in photosynthetic performance and the greatest accumulation of oxidative compounds (Santini et al., 2013). In order to select germplasm better adapted to cold for further citrus breeding, Inch et al. (2014) surveyed citrus germplasm in the field for freeze damage as an indication of cold tolerance. They determined the following ranking in the citrus genus for tolerance to cold: *C. maxima* > *C. reticulata* = *C. sinensis* = *C. aurantium* > *C. medica* L.

13.8 Increased carbon dioxide

Generally, increased ambient carbon dioxide concentrations ([CO_2]) are expected to benefit citrus more than other crop species. Although elevated CO_2 increases the supply of substrate for carbon fixation, plant acclimation responses generally lead to diminishing returns to additional CO_2 in terms of total plant growth (Idso and Kimball, 1992). The widely accepted physiological explanation for this process is that as CO_2 becomes less limiting, plants downregulate photosynthesis in favor of conservation of water and nitrogen nutrition (Ainsworth and Bush, 2011). However, results of studies performed at elevated [CO_2] indicate that although the general process of stomatal acclimation occurs in citrus, many varieties are capable of increasing growth linearly with [CO_2] up to at least 720 ppm, though this relationship varies widely within the genus depending on genotype and conditions.

13.8.1 Effects of CO_2 on overall growth and photosynthesis

Generally, citrus and related species have been found to be more responsive to elevated [CO_2] than other species. A long-term study found that sour orange at elevated [CO_2] (ambient + 300 ppm) from 1987 to 2004 grew approximately 70% more than a treatment held at ambient concentrations (Kimball et al., 2007). Root:shoot ratio was unaffected by [CO_2] in this study, while allocation to increased numbers of fruits contributed to increased growth. Potted studies of doubling [CO_2] for a wider variety of citrus and *Citrus* × *Poncirus* found relative growth effects ranging from −12% to 121% averaging a 54% increase (Koch et al., 1986; Syvertsen et al., 2000). Consistent with Kimball et al. (2007), several studies have reported very little attenuation of growth effects over time in response to elevated [CO_2] (Idso and Kimball, 1992; Kimball et al., 2007; Martin et al., 1995). This general response included increased A across a [CO_2] gradient, with minimal response of respiration (Idso and Kimball, 1992). Although Koch et al. (1986) did find a downregulation of RuBP carboxylase, their study was in potted plants in growth chambers while Idso and Kimball (1992) used plants grown in soil outdoors. Vu et al. (2002) found that elevated [CO_2] increased A while acclimation reduced Rubisco concentration and activity, leading to increased water use efficiency and decreased N cost of photosynthesis in plants grown under elevated [CO_2]. Thus, although citrus acclimates to increased [CO_2] by attenuating g_s and carbon fixing enzymes, many genotypes retain the capacity for long-term increases in growth.

13.8.2 Effects on partitioning and growth habit

Although Kimball et al. (2007) found no long-term change in root:shoot, several changes in biomass allocation have been found. The numerous studies that found changes in root:shoot ratios may be artifacts of the frequent use of potted plants for such studies (Koch et al., 1986; Syvertsen et al., 2000). However, results of Kimball et al. may be specific to *C. aurantium*, as comparisons of multiple species have found widely varying responses. Syvertsen et al. (2000) likewise found greater growth benefits of elevated CO_2 concentration in genotypes with higher root:shoot ratios. Downton et al. (1987) found increased yield in Valencia sweet orange in response to elevated [CO_2], which they attributed to reductions in fruit drop. This hypothesis is further supported by findings of elevated starch and soluble sugar concentrations in plants subjected to elevated [CO_2] (Vu et al., 2002). Fruit drop has been linked to low carbohydrate supply, thus elevated carbohydrate levels are expected to reduce early fruit abscission.

In addition to differences in allocation, morphological and growth habit alterations have been found in response to elevated CO_2 concentration. In addition to a greater number of leaves, specific leaf weight (SLW, mg DW cm^{-2}) increased under elevated [CO_2] (García-Sánchez and Syvertsen, 2006) though not all studies found differences (Martin et al., 1995). Martin et al. (1995) found that elevated [CO_2] led to reduced branching but increased shoot extension and leaf production in Eureka lemon, and argued that this was advantageous to total plant growth because it reduced intracanopy shading. Therefore, increasing [CO_2] is expected to increase both growth and yield in citrus, resulting from increased carbohydrate allocation to fruits and changes in canopy architecture facilitating increased light penetration.

13.8.3 Interactions with other abiotic stresses

In general, elevated [CO_2] leads to altered responses to other abiotic factors. Several growth-reducing factors can be mitigated by increased [CO_2]. Vu et al. (2002) found that high temperatures (6°C above ambient) reduced water use efficiency of *A*, but elevated [CO_2] (720 ppm) restored water use efficiency. This is likely because increased ambient [CO_2] leads to increased diffusion into the mesophyll even when g_s is low. Similar results were found with [CO_2] enrichment mitigation of negative effects of salinity on growth (García-Sánchez and Syvertsen, 2006). In this case, elevated [CO_2] did not reduce the toxicity of salts or mineral imbalances that resulted. However, it did make growth more water- and nutrient-efficient, resulting in increased growth despite salinity stress. Therefore, although increasing [CO_2] is expected to contribute to other environmental stresses, such as increased frequency of high temperatures, these stresses will be mitigated by the unusually high benefit of increased [CO_2] to citrus. A growth model of sweet orange predicted that within the range of increases in [CO_2] expected by the IPCC, growth in response to elevated [CO_2] would overcome the negative effects due to supraoptimal temperatures (Pereira et al., 2017). Thus, under irrigated conditions, increased temperatures and [CO_2] are likely to have no effect and even increase growth and yield in citrus.

13.8.4 Maximizing benefits of increased [CO_2]

The wide variation of results of growth gains among different species and cultivars should lead agricultural researchers to consider the potential mechanisms by which some take advantage of increased [CO_2]. Although they considered only two cultivars, Koch et al. (1986) attributed the differences between Carrizo citrange and Swingle citrumelo to the differences in allocation of photoassimilates. They hypothesized that increased relative allocation to sinks allowed Swingle to drive growth more effectively in response to increased CO_2 supply than Carrizo. Although no study has set out to expressly test this hypothesis in citrus, the results of several studies comparing genotypic interactions with ambient or elevated [CO_2] align closely, with increased root allocation (the presumptive sink in vegetative plants) leading to greater increases in growth in response to elevated [CO_2] (Fig. 13.5). This hypothesis has been borne out by similar comparisons in other species, such as soybean (*Glycine* max L.; Ainsworth and Bush, 2011). Thus, germplasm development considering the allocation patterns is likely to reap the increasing benefit as global ambient [CO_2] continues to rise.

13.9 Conclusions

Globally, climate change presents one important opportunity and four major challenges. The opportunity is the unique capacity of some citrus species or cultivars to maintain higher growth rates in response to elevated [CO_2]. The major challenges include (1) high temperatures, (2) flooding, (3) access to irrigation, and (4) salinity.

Irrigation is the primary practice to mitigate the increased hydraulic demand due to increasing temperatures and increased frequency of drought. Stable growth and yields in nearly all climatic regions will depend on access to quality irrigation water. However, increased irrigation may also contribute to increased salinity, especially in the Mediterranean and semiarid regions. Thus, the future of citrus production will be greatly improved by further development of water conservation irrigation practices and attention in planting planning of access to quality irrigation water.

Other abiotic challenges must be addressed by combining site selection with germplasm development. Selecting for traits of ROS detoxification, osmoprotectants, and increased whole plant and leaf conductance will likely lead to more

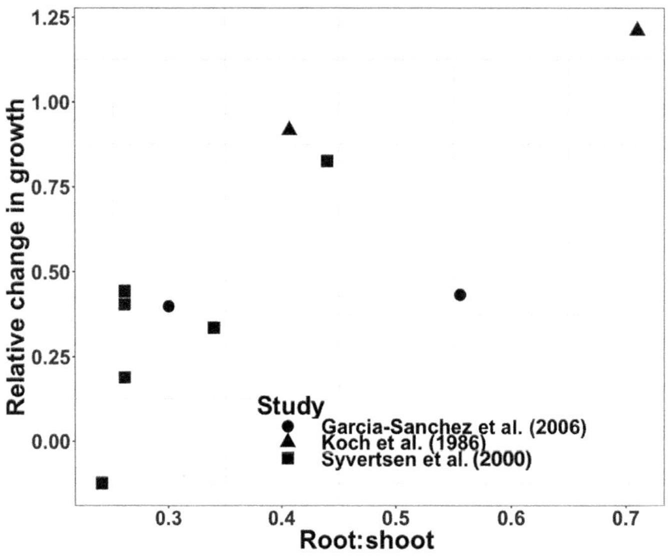

FIG. 13.5 Relationship in three experiments between mean varietal root:shoot allocation and a relative increase in total dry weight due to doubling of ambient CO_2 concentration.

heat-tolerant cultivars. Likewise, for flood-prone areas existing flood-tolerant varieties can be selected, and germplasm development should focus on rootstock traits of antioxidant expression and scion traits associated with maintenance of leaf conductance under anoxic soil conditions.

There is some evidence that genotypic variation within the *Citrus* genus could lead to very wide variations in the relative benefits of increased [CO_2]. Thus, germplasm development could improve global resilience by selection for the greatest growth responses to elevated [CO_2]. The studies reviewed in this chapter indicate that these responses, under irrigated conditions, are likely to mitigate the negative effects of heat and salinity. Although it has not been clearly elucidated, there may be a relationship between growth allocation patterns and plant capacity to increase growth in response to elevated [CO_2], thus germplasm development may benefit from selection for increased allocation to sinks as a proxy trait associated with increased responsiveness. Selecting for such capacity and the various tolerance traits examined in this chapter is likely to improve production resilience to detrimental abiotic conditions.

Acknowledgments

The authors acknowledge funds from Spanish Ministerio de Economía y Competitividad (MINECO) and Universitat Jaume I through grants AGL2016-76574-R and UJI-B2016-24, respectively, to V.A. and A.G.-C. and from the "Cavalbio" FEDER project to R.M.

References

Ainsworth, E.A., Bush, D.R., 2011. Carbohydrate export from the leaf: a highly regulated process and target to enhance photosynthesis and productivity. Plant Physiol. 155, 64–69. https://doi.org/10.1104/pp.110.167684.

Allario, T., Brumos, J., Colmenero-Flores, J.M., Tadeo, F., Froelicher, Y., Talon, M., Navarro, L., Ollitrault, P., Morillon, R., 2011. Large changes in anatomy and physiology between diploid Rangpur lime (*Citrus limonia*) and its autotetraploid are not associated with large changes in leaf gene expression. J. Exp. Bot. 62, 2507–2519.

Allario, T., Brumos, J., Colmenero-Flores, J.M., Iglesias, D.J., Pina, J.A., Navarro, L., Talon, M., Ollitrault, P., Morillon, R., 2012. Tetraploid Rangpur lime rootstock increases drought tolerance via enhanced constitutive root abscisic acid production. Plant Cell Environ. 36, 856–868. https://doi.org/10.1111/pce.12021.

Alva, A.K., Syvertsen, J.P., 1991. Irrigation water salinity affects soil nutrient distribution, root density, and leaf nutrient levels of citrus under drip fertigation. J. Plant Nutr. 14 (7), 715–727.

Arbona, V., Gómez-Cadenas, A., 2008. Hormonal modulation of citrus responses to flooding. J. Plant Growth Regul. 27, 241–250. https://doi.org/10.1007/s00344-008-9051-x.

Arbona, V., Flors, V., Jacas, J., García-Agustín, P., Gómez-Cadenas, A., 2003. Enzymatic and non-enzymatic antioxidant responses of Carrizo citrange, a salt-sensitive citrus rootstock, to different levels of salinity. Plant Cell Physiol. 44 (4), 388–394.

Arbona, V., Hossain, Z., López-Climent, M.F., Pérez-Clemente, R.M., Gómez-Cadenas, A., 2008. Antioxidant enzymatic activity is linked to waterlogging stress tolerance in citrus. Physiol. Plant. 132, 452–466. https://doi.org/10.1111/j.1399-3054.2007.01029.x.

Arbona, V., López-Climent, M.F., Pérez-Clemente, R.M., Gómez-Cadenas, A., 2009. Maintenance of a high photosynthetic performance is linked to flooding tolerance in citrus. Environ. Exp. Bot. 66, 135–142. https://doi.org/10.1016/j.envexpbot.2008.12.011.

Aru, A., 1996. The Rio Santa Lucia site: an integrated study of desertification. In: Brandt, C.J., Thornes, J.B. (Eds.), Mediterranean Desertification and Land Use. John Wiley and Sons, Chichester, pp. 189.

Ashour, E.K., Al-Najar, H., 2012. The impact of climate change and soil salinity in irrigation water demand in the Gaza strip. J. Earth Sci. Clim. Chang. 3, 120.

Bañuls, J., Primo-Millo, E., 1995. Effects of salinity on some citrus scion-rootstock combinations. Ann. Bot. 76, 97–102.

Banuls, J., Legaz, F., Primo-Millo, E., 1991. Salinity-calcium interactions on growth and ionic concentration of citrus plants. Plant Soil 133 (1), 39–46.

Barlass, M., Skene, K.G.M., 1986. Fruit trees. 6. *Citrus* (*Citrus* species). In: Biotechnology in Agriculture and Forestry. Springer-Verlag, Berlin.

Behboudian, M.H., Törökfalvy, E., Walker, R.R., 1986. Effects of salinity on ionic content, water relations and gas exchange parameters in some citrus scion—rootstock combinations. Scientia Horticult. 28 (1–2), 105–116.

Bita, C.E., Gerats, T., 2013. Plant tolerance to high temperature in a changing environment: scientific fundamentals and production of heat stress-tolerant crops. Front. Plant Sci. 4. https://doi.org/10.3389/fpls.2013.00273.

Boman, B., 2000. Effect of saline irrigation water on citrus production. Univ. of Florida, Indian River Research and Education Center, Ft. Pierce, FL.

Carr, M.K.V., 2012. The water relations and irrigation requirements of citrus (*Citrus* spp.): a review. Exp. Agric. 48 (3), 347–377.

Castel, J.R., Buj, A., 1990. Response of Salustiana oranges to high frequency deficit irrigation. Irrig. Sci. 11, 121–127. https://doi.org/10.1007/BF00188448.

Cerdá, A., Caro, M., Fernández, F.G., Guillen, M.G., 1977. Foliar contents of sodium and chloride on citrus rootstocks irrigated with saline waters. In: International Conference on Managing Saline Water for Irrigation, Planning for the Future, Lubbock, TX, USA, 1976.

Cole, P., McCloud, P., 1985. Salinity and climatic effects on the yields of citrus. Aust. J. Exp. Agric. 24 (3), 711–717.

Consoli, S., Stagno, F., Roccuzzo, G., Cirelli, G.L., Intrigliolo, F., 2014. Sustainable management of limited water resources in a young orange orchard. Agric. Water Manag. 132, 60–68. https://doi.org/10.1016/j.agwat.2013.10.006.

Consoli, S., Stagno, F., Vanella, D., Boaga, J., Cassiani, G., Roccuzzo, G., 2017. Partial root-zone drying irrigation in orange orchards: effects on water use and crop production characteristics. Eur. J. Agron. 82, 190–202. https://doi.org/10.1016/j.eja.2016.11.001.

Cortina, L.M., Mora, N.H., 2003. The role of groundwater in Spain's water policy. Water Int. 28, 313–320.

Davies, F.S., Albrigo, L.G., 1994. *Citrus*. In: Davies, F.S., Albrigo, L.G. (Eds.), Crop Production Science in Horticulture Series. CAB International, Wallingford, UK, pp. 134.

Downton, W.J.S., Grant, W.J.R., Loveys, B.R., 1987. Carbon dioxide enrichment increases yield of Valencia orange. Funct. Plant Biol. 14, 493–501. https://doi.org/10.1071/pp9870493.

Er-Raki, S., Chehbouni, A., Guemouria, N., Ezzahar, J., Khabba, S., Boulet, G., Hanich, L., 2009. *Citrus* orchard evapotranspiration: comparison between eddy covariance measurements and the FAO-56 approach estimates. Plant Biosyst. 143, 201–208.

Faltin, Z., Holland, D., Velcheva, M., Tsapovetsky, M., Roeckel-Drevet, P., Handa, A.K., et al., 2010. Glutathione peroxidase regulation of reactive oxygen species level is crucial for in vitro plant differentiation. Plant Cell Physiol. 51 (7), 1151–1162.

Fares, A., Bayabil, H.K., Zekri, M., Mattos, D., Awal, R., 2017. Potential climate change impacts on citrus water requirement across major producing areas in the world. J. Water Clim. Chang. 8 (4), 576–592. https://doi.org/10.2166/wcc.2017.182.

Fernandez-Illescas, C.P., Porporato, A., Laio, F., 2001. The ecohydrological role of soil texture in a water-limited ecosystem. Water Resour. Res. 37, 2863–2872.

Forner-Giner, M., Legaz, F., Primo-Millo, E., Forner, J., 2011. Nutritional responses of citrus rootstocks to salinity: performance of the new hybrids Forner-Alcaide 5 and Forner-Alcaide 13. J. Plant Nutr. 34, 1437–1452.

García Petillo, M., Castel, J.R., 2007. Water balance and crop coefficient estimation of a citrus orchard in Uruguay. Span. J. Agric. Res. 5, 232–243.

García-Sánchez, F., Syvertsen, J.P., 2006. Salinity tolerance of Cleopatra mandarin and Carrizo citrange citrus rootstock seedlings is affected by CO_2 enrichment during growth. J. Am. Soc. Hortic. Sci. 131, 24–31.

Garcia-Sanchez, F., Syvertsen, J.P., Martinez, V., Melgar, J.C., 2006. Salinity tolerance of 'Valencia' orange trees on rootstocks with contrasting salt tolerance is not improved by moderate shade. J. Expt. Bot. 57, 697.

García-Sánchez, F., Syvertsen, J.P., Gimeno, V., Botía, P., Perez-Perez, J.G., 2007. Responses to flooding and drought stress by two citrus rootstock seedlings with different water-use efficiency. Physiol. Plant. 130, 532–542. https://doi.org/10.1111/j.1399-3054.2007.00925.x.

Garg, N., Manchanda, G., 2009. ROS generation in plants: boon or bane? Plant Biosyst. 143 (1), 81–96.

Giannakopoulos, C., Le Sager, P., Bindi, M., Moriondo, M., Kostopoulou, E., Goodess, C.M., 2009. Climatic changes and associated impacts in the Mediterranean resulting from a 2-°C global warming. Glob. Planet. Chang. 68, 209–224.

Gimeno, J., Gadea, J., Forment, J., Pérez-Valle, J., Santiago, J., Martínez-Godoy, M.A., Yenush, L., Bellés, J.M., Brumós, J., Colmenero-Flores, J.M., Talón, M., Serrano, R., 2009. Shared and novel molecular responses of mandarin to drought. Plant Mol. Biol. 70, 403–420. https://doi.org/10.1007/s11103-009-9481-2.

Ginestar, C., Castel, J.R., 1996. Responses of young clementine citrus trees to water stress during different phenological periods. J. Hortic. Sci. 71, 551–559. https://doi.org/10.1080/14620316.1996.11515435.

Gómez-Cadenas, A., et al., 2003. Physiological and molecular responses of citrus to salinity. Recent Res. Dev. Plant Mol. Biol. 1, 281–298.

Gómez-Cadenas, A., et al., 2015. Abscisic acid: a versatile phytohormone in plant signaling and beyond. Curr. Protein Peptide Sci. 16 (5), 413–434.

Gonçalves, L.P., Alves, T., Martins, C.P.S., de Sousa, A.O., dos Santos, I.C., Pirovani, C.P., Almeida, A.F., Coelho Filho, M.A., da Silva Gesteira, A., dos Santos Soares Filho, W., Girardi, E.A., Costa, M.G.C., 2016. Rootstock-induced physiological and biochemical mechanisms of drought tolerance in sweet orange. Physiol. Plant. 38, 174. https://doi.org/10.1007/s11738-016-2198-3.

Grattan, S., Díaz, F.J., Pedrero, F., Vivaldi, G.A., 2015. Assessing the suitability of saline wastewaters for irrigation of citrus spp.: emphasis on boron and specific-ion interactions. Agric. Water Manag. 157, 48–58.

Grilo, F.S., Di Stefano, V., Lo Bianco, R., 2017. Deficit irrigation and maturation stage influence quality and flavonoid composition of 'Valencia' orange fruit. J. Sci. Food Agric. 97, 1904–1909. https://doi.org/10.1002/jsfa.7993.

Guo, Y., Zhou, H., Zhang, L., 2006. Photosynthetic characteristics and protective mechanisms against photooxidation during high temperature stress in two citrus species. Sci. Hortic. 108, 260–267.

Hardy, S., Barkley, P., Creek, A., Donovan, N., 2012. Impacts and management of flooding and waterlogging in citrus orchards. Primefact 1189, 1–9.

Hatfield, J.L., Prueger, J.H., 2015. Temperature extremes: effect on plant growth and development. Weather Clim. Extrem. 10, 4–10.

Hossain, Z., López-Climent, M.F., Arbona, V., Pérez-Clemente, R.M., Gómez-Cadenas, A., 2009. Modulation of the antioxidant system in *Citrus* under waterlogging and subsequent drainage. J. Plant Physiol. 166, 1391–1404. https://doi.org/10.1016/j.jplph.2009.02.012.

Huang, X.S., Wang, W., Zhang, Q., Liu, J.H., 2013. A basic helix-loop-helix transcription factor, PtrbHLH, of *Poncirus trifoliata* confers cold tolerance and modulates peroxidase-mediated scavenging of hydrogen peroxide. Plant Physiol. 162, 1178–1194.

Huang, X.S., Zhang, Q., Zhu, D., Fu, X., Wang, M., Zhang, Q., Moriguchi, T., Liu, J.H., 2015. ICE1 of *Poncirus trifoliata* functions in cold tolerance by modulating polyamine levels through interacting with arginine decarboxylase. J. Exp. Bot. 66, 3259–3274.

Hussain, S., Luro, F., Costantino, G., Ollitrault, P., Morillon, R., 2012. Physiological analysis of salt stress behaviour of citrus species and genera: low chloride accumulation as an indicator of salt tolerance. S. Afr. J. Bot. 81, 103–112.

Hussain, S., Morillon, R., Anjum, M.A., Ollitrault, P., Costantino, G., Luro, F., 2015. Genetic diversity revealed by physiological behavior of citrus genotypes subjected to salt stress. Acta Physiol. Plant. 37, 1740.

Hutton, R.J., Landsberg, J.J., 2000. Temperature sums experienced before harvest partially determine the post-maturation juicing quality of oranges grown in the Murrumbidgee Irrigation Areas (MIA) of New South Wales. J. Sci. Food Agric. 80, 275–283.

Idso, S.B., Kimball, B.A., 1992. Effects of atmospheric CO_2 enrichment on photosynthesis, respiration, and growth of sour orange trees. Plant Physiol. 99, 341–343.

Inch, S., Stover, E., Driggers, R., Lee, R.F., 2014. Freeze response of *Citrus* and *Citrus*-related genotypes in a Florida field planting. HortScience 49, 1010–1016.

IPCC, 2014. Climate change 2014: synthesis report. In: Pachauri, R.K., Meyer, L.A. (Eds.), Contribution of Working Groups I, II and III to the Fifth Assessment Report of the Intergovernmental Panel on Climate Change [Core Writing Team]. IPCC, Geneva, pp. 151.

Jackson, M.B., Ishizawa, K., Ito, O., 2008. Evolution and mechanisms of plant tolerance to flooding stress. Ann. Bot. 103, 137–142. https://doi.org/10.1093/aob/mcn242.

Jifon, J.L., Syvertsen, J.P., 2001. Effects of moderate shade on citrus leaf gas exchange, fruit yield, and quality. Proc. Fla. State Hortic. Soc. 114, 177–181.

Jifon, J.L., Syvertsen, J.P., 2003. Moderate shade can increase net gas exchange and reduce photoinhibition in citrus leaves. Tree Physiol. 23, 119–127. https://doi.org/10.1093/treephys/23.2.119.

Kaushal, M., et al., 2013. Effect of salinity on survival and growth performance of in vitro grown rough lemon (*Citrus jambhiri* Lush.) seeds. Indian J. Biotechnol. 12, 284–286.

Khairi, M., Hall, A.E., 1976a. Temperature and humidity effects on net photosynthesis and transpiration of citrus. Physiol. Plant. 36, 29–34.

Khairi, M., Hall, A.E., 1976b. Comparative studies of net photosynthesis and transpiration of some *Citrus* species and relatives. Physiol. Plant. 36, 35–39.

Kharin, V.V., Zwiers, F.W., Zhang, X., Hegerl, G.C., 2007. Changes in temperature and precipitation extremes in the IPCC ensemble of global coupled model simulations. J. Clim. 20, 1419–1444.

Kimball, B.A., Idso, S.B., Johnson, S., Rillig, M.C., 2007. Seventeen years of carbon dioxide enrichment of sour orange trees: final results. Glob. Chang. Biol. 13, 2171–2183. https://doi.org/10.1111/j.1365-2486.2007.01430.x.

Koch, K.E., Jones, P.H., Avigne, W.T., Allen, L.H., 1986. Growth, dry matter partitioning, and diurnal activities of RuBP carboxylase in citrus seedlings maintained at two levels of CO_2. Physiol. Plant. 67, 477–484. https://doi.org/10.1111/j.1399-3054.1986.tb05766.x.

Korres, N.E., Norsworthy, J.K., Tehranchian, P., Gitsopoulos, T.K., Loka, D.A., Oosterhuis, D.M., Gealy, D.R., Moss, S.R., Burgos, N.R., Miller, M.R., Palhano, M., 2016. Cultivars to face climate change effects on crops and weeds: a review. Agron. Sustain. Dev. 36, 1–22. https://doi.org/10.1007/s13593-016-0350-5.

Kriedemann, P.E., 1968. Some photosynthetic characteristics of citrus leaves. Aust. J. Biol. Sci. 21, 895–906. https://doi.org/10.1071/bi9680895.

Kumar, K.R., Sahai, A.K., Kumar, K.K., Patwardhan, S.K., Mishra, P.K., Revadekar, J.V., Kamala, K., Pant, G.B., 2006. High-resolution climate change scenarios for India for the 21st century. Curr. Sci. 90, 334–345.

Langevin, C.D., Zygnerski, M., 2013. Effect of sea-level rise on salt water intrusion near a coastal well field in southeastern Florida. Groundwater 51, 781–803.

Levy, Y., Syvertsen, J.P., 2004. Irrigation water quality and salinity effects in citrus trees. Hortic. Rev. 30, 37–82.

Libutti, A., Monteleone, M., 2017. Soil vs. groundwater: the quality dilemma. Managing nitrogen leaching and salinity control under irrigated agriculture in Mediterranean conditions. Agric. Water Manag. 186, 40–50.

Liu, C., Liu, H., Fang, J., Li, J., Tan, K., Zhou, L., Ciais, P., Friedlingstein, P., Piao, S., Peng, S., Zhang, T., Huang, Y., Ding, Y., Yu, Y., Ma, Y., Shen, Z., 2010. The impacts of climate change on water resources and agriculture in China. Nature 467, 43. https://doi.org/10.1038/nature09364.

Lobell, D.B., Gourdji, S.M., 2012. The influence of climate change on global crop productivity. Plant Physiol. 160, 1686–1697. https://doi.org/10.1104/pp.112.208298.

Lobell, D.B., Field, C.B., Cahill, K.N., Bonfils, C., 2006. Impacts of future climate change on California perennial crop yields: model projections with climate and crop uncertainties. Agric. For. Meteorol. 141, 208–218. https://doi.org/10.1016/j.agrformet.2006.10.006.

López-Climent, M.F., Arbona, V., Pérez-Clemente, R.M., Gómez-Cadenas, A., 2008. Relationship between salt tolerance and photosynthetic machinery performance in citrus. Environ. Exp. Bot. 62 (2), 176–184.

Maas, E.V., 1993. Salinity and citriculture. Tree Physiol. 12, 195–216.

Maestre-Valero, J.F., Testi, L., Jiménez-Bello, M.A., Castel, J.R., Intrigliolo, D.S., 2017. Evapotranspiration and carbon exchange in a citrus orchard using eddy covariance. Irrig. Sci. 35, 397–408. https://doi.org/10.1007/s00271-017-0548-6.

Marin, F.R., Angelocci, L.R., 2011. Irrigation requirements and transpiration coupling to the atmosphere of a citrus orchard in Southern Brazil. Agric. Water Manag. 98, 1091–1096. https://doi.org/10.1016/j.agwat.2011.02.002.

Martin, C.A., Stutz, J.C., Kimball, B.A., Idso, S.B., Akey, D.H., 1995. Growth and topological changes of *Citrus limon* (L.) Burm. f. ‘Eureka’ in response to high temperatures and elevated atmospheric carbon dioxide. J. Am. Soc. Hortic. Sci. 120, 1025–1031.

Morillon, R., Chrispeels, M.J., 2001. The role of ABA and the transpiration stream in the regulation of the osmotic water permeability of leaf cells. Proc. Natl. Acad. Sci. U. S. A. 98, 14138–14143. https://doi.org/10.1073/pnas.231471998.

Moya, J.L., Primo-Millo, E., Talon, M., 1999. Morphological factors determining salt tolerance in citrus seedlings: the shoot to root ratio modulates passive root uptake of chloride ions and their accumulation in leaves. Plant Cell Environ. 22, 1425–1433.

Moya, J.L., Tadeo, F.R., Gómez-Cadenas, A., Primo-Millo, E., Talón, M., 2002. Transmissible salt tolerance traits identified through reciprocal grafts between sensitive Carrizo and tolerant Cleopatra citrus genotypes. J. Plant Physiol. 159, 991–998.

Munemasa, S., Muroyama, D., Nagahashi, H., Nakamura, Y., Mori, I.C., Murata, Y., 2013. Regulation of reactive oxygen species-mediated abscisic acid signaling in guard cells and drought tolerance by glutathione. Front. Plant Sci. 4, 472. https://doi.org/10.3389/fpls.2013.00472.

Navarro, J.M., Botía, P., Pérez-Pérez, J.G., 2015. Influence of deficit irrigation timing on the fruit quality of grapefruit (*Citrus paradisi* Mac.). Food Chem. 175, 329–336. https://doi.org/10.1016/j.foodchem.2014.11.152.

Nieves, M., Martinez, V., Cerda, A., Guillen, M.G., 1990. Yield and mineral composition of ‘Verna’ lemon trees as affected by salinity and rootstock combination. J. Hortic. Sci. 65 (3), 359–366.

Nortes, P.A., Baille, A., González-Real, M.M., Ruiz-Salleres, I., Verhoef, A., Martin-Gorriz, B., Egea, G., 2010. Effects of high temperature and vapour pressure deficit on net ecosystem exchange and energy balance of an irrigated orange orchard in a semi-arid climate (Southern Spain). In: XXVIII International Horticultural Congress on Science and Horticulture for People (IHC2010): International Symposium, 922, pp. 149–156.

Oliveira, T.M., Yahmed, J.B., Dutra, J., Maserti, B.E., Talon, M., Navarro, L., Ollitraut, P., da Silva Gesteira, A., Morillon, R., 2017. Better tolerance to water deficit in doubled diploid Carrizo citrange compared to diploid seedlings is associated with more limited water consumption. Acta Physiol. Plant. 39, 204.

Ollitrault, P., Dambier, D., Luro, F., Froelicher, Y., 2008. Ploidy manipulation for breeding seedless triploid citrus. Plant Breed. Rev. 30, 323–354.

Ondrasek, G., Rengel, Z., Veres, S., 2011. Soil salinisation and salt stress in crop production. In: Shanker, A., Venkateswarlu, B. (Eds.), Abiotic Stress in Plants—Mechanisms and Adaptations. InTech.

Oustric, J., Morillon, R., Luro, F., Herbette, S., Lourkisti, R., Giannettini, J., Berti, L., Santini, J., 2017. Tetraploid Carrizo citrange rootstock (*Citrus sinensis* Osb.×*Poncirus trifoliata* L. Raf.) enhances natural chilling stress tolerance of common clementine (*Citrus clementina* hort. ex Tan). J. Plant Physiol. 214, 108–115.

Parry, M.L., Canziani, O.F., Palutikof, J.P., van der Linden, P.J., Hanson, C.E., 2007. Climate change 2007: impacts, adaptation and vulnerability. In: Contribution of Working Group II to the Fourth Assessment Report of the Intergovernmental Panel on Climate Change. IPCC.

Peng, T., Zhu, X.F., Fan, Q.J., Sun, P.P., Liu, J.H., 2012. Identification and characterization of low temperature stress responsive genes in *Poncirus trifoliata* by suppression subtractive hybridization. Gene 492, 220–228.

Peng, T., Zhu, X., Duan, N., Liu, J.H., 2014. PtrBAM1, a beta-amylase-coding gene of *Poncirus trifoliata*, is a CBF regulon member with function in cold tolerance by modulating soluble sugar levels. Plant Cell Environ. 37, 2754–2767.

Pereira, F.F.S., Sánchez-Román, R.M., González, A.M.G.O., 2017. Simulation model of the growth of sweet orange (*Citrus sinensis* L. Osbeck) cv. Natal in response to climate change. Clim. Chang. 143, 101–113. https://doi.org/10.1007/s10584-017-1986-0.

Perez-Perez, J.G., Syvertsen, J.P., Botia, P., Garcia-Sanchez, F., 2007. Leaf water relations and net gas exchange responses of salinized Carrizo citrange seedlings during drought stress and recovery. Ann. Bot. 100, 335–345.

Pérez-Pérez, J.G., Romero, P., Navarro, J.M., Botía, P., 2008. Response of sweet orange cv “Lane late” to deficit-irrigation strategy in two rootstocks. II: flowering, fruit growth, yield and fruit quality. Irrig. Sci. 26, 519–529. https://doi.org/10.1007/s00271-008-0113-4.

Perotti, V.E., Moreno, A.S., Tripodi, K.E., Meier, G., Bello, F., Cocco, M., Vazquez, D., Anderson, C., Podesta, F.E., 2015. Proteomic and metabolomic profiling of Valencia orange fruit after natural frost exposure. Physiol. Plant. 153, 337–354.

Pinheiro, C., Chaves, M.M., 2011. Photosynthesis and drought: can we make metabolic connections from available data? J. Exp. Bot. 62, 869–882. https://doi.org/10.1093/jxb/erq340.

Poggi, I., Polidori, J.J., Gandoin, J.M., Paolacci, V., Battini, M., Albertini, M., Améglio, T., Cochard, H., 2007. Stomatal regulation and xylem cavitation in Clementine (*Citrus clementina* Hort) under drought conditions. J. Hortic. Sci. Biotechnol. 82, 845–848. https://doi.org/10.1080/14620316.2007.11512316.

Rana, G., Katerji, N., De Lorenzi, F., 2005. Measurement and modelling of evapotranspiration of irrigated citrus orchard under Mediterranean conditions. Agric. For. Meteorol. 128, 199–209. https://doi.org/10.1016/j.agrformet.2004.11.001.

Rewald, B., Raveh, E., Gendler, T., Ephrath, J.E., Rachmilevitch, S., 2012. Phenotypic plasticity and water flux rates of Citrus root orders under salinity. J. Exp. Bot. 63 (7), 2717–2727.

Ribeiro, R.V., Machado, E.C., 2007. Some aspects of citrus ecophysiology in subtropical climates: re-visiting photosynthesis under natural conditions. Braz. J. Plant Physiol. 19, 393–411. https://doi.org/10.1590/S1677-04202007000400009.

Rigby, J.R., Porporato, A., 2008. Spring frost risk in a changing climate. Geophys. Res. Lett. 35, 12703.

Rodríguez-Gamir, J., Ancillo, G., González-Mas, M.C., Primo-Millo, E., Iglesias, D.J., Forner-Giner, M.A., 2011. Root signalling and modulation of stomatal closure in flooded citrus seedlings. Plant Physiol. Biochem. 49, 636–645. https://doi.org/10.1016/j.plaphy.2011.03.003.

Rogers, J.C., Rohli, R.V., 1991. Florida citrus freezes and polar anticyclones in the Great Plains. J. Clim. 4, 1103–1113.

Romero, P., Navarro, J.M., Pérez-Pérez, J., García-Sánchez, F., Gómez-Gómez, A., Porras, I., Martinez, V., Botía, P., 2006. Deficit irrigation and rootstock: their effects on water relations, vegetative development, yield, fruit quality and mineral nutrition of Clemenules mandarin. Tree Physiol. 26, 1537–1548.

Romero-Aranda, R., Bondada, B.R., Syvertsen, J.P., Grosser, J.W., 1997. Leaf characteristics and net gas exchange of diploid and autotetraploid citrus. Ann. Bot. 79, 153–160.

Romero-Trigueros, C., Nortes, P.A., Alarcón, J.J., Hunink, J.E., Parra, M., Contreras, S., Droogers, P., Nicolás, E., 2017. Effects of saline reclaimed waters and deficit irrigation on citrus physiology assessed by UAV remote sensing. Agric. Water Manag. 183, 60–69. https://doi.org/10.1016/j.agwat.2016.09.014.

Rounsevell, M.D., Annetts, J., Audsley, E., Mayr, T., Reginster, I., 2003. Modelling the spatial distribution of agricultural land use at the regional scale. Agric. Ecosyst. Environ. 95, 465–479. https://doi.org/10.1016/S0167-8809(02)00217-7.

Ruiz, M., Quinones, A., Martinez-Cuenca, M.R., Aleza, P., Morillon, R., Navarro, L., Primo-Millo, E., Martinez-Alcantara, B., 2016a. Tetraploidy enhances the ability to exclude chloride from leaves in Carrizo citrange seedlings. J. Plant Physiol. 205, 1–10.

Ruiz, M., Quiñones, A., Martinez-Alcantara, B., Aleza, P., Morillon, R., Navarro, L., Primo-Millo, E., Martinez-Cuenca, M., 2016b. Effects of salinity on diploid (2x) and doubled diploid (4x) *Citrus macrophylla* genotypes. Sci. Hortic. 207, 33–40.

Santana-Vieira, D.D.S., Freschi, L., da Hora Almeida, L.A., de Moraes, D.H.S., Neves, D.M., dos Santos, L.M., Bertolde, F.Z., dos Santos Soares Filho, W., Coelho Filho, M.A., da Silva Gesteira, A., 2016. Survival strategies of citrus rootstocks subjected to drought. Sci. Rep. 6, 38775. https://doi.org/10.1038/srep38775.

Santini, J., Giannettini, J., Herbette, S., Pailly, O., Ollitrault, P., Luro, L., Berti, L., 2012. Physiological and biochemical response to photooxidative stress of the fundamental *Citrus* species. Sci. Hortic. 147, 126.

Santini, J., Giannettini, J., Pailly, O., Herbette, S., Ollitrault, P., Berti, L., Luro, F., 2013. Comparison of photosynthesis and antioxidant performance of several *Citrus* and *Fortunella* species (Rutaceae) under natural chilling stress. Trees 27, 71–83.

Santos, J.A., Costa, R., Fraga, H., 2017. Climate change impacts on thermal growing conditions of main fruit species in Portugal. Clim. Chang. 140, 273–286. https://doi.org/10.1007/s10584-016-1835-6.

Sauter, M., 2013. Root responses to flooding. Curr. Opin. Plant Biol. 16, 282–286. https://doi.org/10.1016/j.pbi.2013.03.013.

Schreiber, C.M., Zeng, B., Temperton, V.M., Rascher, U., Kazda, M., Schurr, U., Höltkemeier, A., Kuhn, A.J., 2011. Dynamics of organic acid occurrence under flooding stress in the rhizosphere of three plant species from the water fluctuation zone of the Three Gorges Reservoir, P.R. China. Plant Soil 344, 111–129. https://doi.org/10.1007/s11104-011-0732-z.

Shah, T., 2014. Groundwater governance and irrigated agriculture. In: TEC Background Papers 69.

Shimamura, S., Yoshioka, T., Yamamoto, R., Hiraga, S., Nakamura, T., Shimada, S., Komatsu, S., 2014. Role of abscisic acid in flood-induced secondary aerenchyma formation in soybean (*Glycine max*) hypocotyls. Plant Prod. Sci. 17, 131–137. https://doi.org/10.1626/pps.17.131.

Simpson, C.R., Nelson, S.D., Melgar, J.C., Jifon, J., King, S.R., Schuster, G., Volder, A., 2014. Growth response of grafted and ungrafted citrus trees to saline irrigation. Scientia Horticult. 169, 199–205.

Simpson, C.R., Nelson, S.D., Melgar, J.C., Jifon, J., Schuster, G., Volder, A., 2015. Effects of salinity on physiological parameters of grafted and ungrafted citrus trees. Scientia Horticul. 197, 483–489.

Sivakumar, M.V.K., Das, H.P., Brunini, O., 2005. Impacts of present and future climate variability and change on agriculture and forestry in the arid and semi-arid tropics. Clim. Chang. 70, 31–72. https://doi.org/10.1007/s10584-005-5937-9.

Smith, M., 2011. Managing citrus orchards affected by wet weather. In: Technical Note. Department of Employment, Economic Development, and Innovation, Queensland Government, Australia.

Souza, J.D., Mario, E., Silva, D.A., Anto, M., Bonatto, D., Micheli, F., Silva, A., 2017. Different adaptation strategies of two citrus scion/rootstock combinations in response to drought stress. PLoS One, 1–23. https://doi.org/10.1371/journal.pone.0177993.

Srivastava, A.K., Singh, S., 2009. *Citrus* decline: soil fertility and plant nutrition. J. Plant Nutr. 32, 197–245. https://doi.org/10.1080/01904160802592706.

Syvertsen, J.P., Garcia-Sanchez, F., 2014. Multiple abiotic stresses occurring with salinity stress in citrus. Environ. Exp. Bot. 103, 128–137. Response to Abiotic stresses of plants of Mediterranean-type ecosystems.

Syvertsen, J.P., Yelenosky, G., 1988. Salinity can enhance freeze tolerance of citrus rootstock seedlings by modifying growth, water relations, and mineral nutrition. J. Am. Soc. Hortic. Sci. 113, 889.

Syvertsen, J.P., Lee, L.S., Grosser, J.W., 2000. Limitations on growth and net gas exchange of diploid and tetraploid citrus rootstock cultivars grown at elevated CO_2. J. Am. Soc. Hortic. Sci. 125, 228–234.

Tanou, G., Filippou, P., Belghazi, M., Job, D., Diamantidis, G., Fotopoulos, V., Molassiotis, A., 2012. Oxidative and nitrosative-based signaling and associated post-translational modifications orchestrate the acclimation of citrus plants to salinity stress. Plant J. 72 (4), 585–599.

Tanou, G., Job, C.F., Rajjou, L.F., Arc, E.F., Belghazi, M.F., Diamantidis, G.F., Molassiotis, A.F., Job, D., 2009a. Proteomics reveals the overlapping roles of hydrogen peroxide and nitric oxide in the acclimation of citrus plants to salinity. Plant J. 60, 795–804.

Tanou, G., Molassiotis, A., Diamantidis, G., 2009b. Hydrogen peroxide- and nitric oxide-induced systemic antioxidant prime-like activity under NaCl-stress and stress-free conditions in citrus plants. J. Plant Physiol. 166, 1904–1913.

Tardieu, F., Simonneau, T., 1998. Variability among species of stomatal control under fluctuating soil water status and evaporative demand: modelling isohydric and anisohydric behaviours. J. Exp. Bot. 49, 419–432.

Tozlu, I., Moore, G.A., Guy, C.L., 2000. Regulation of growth and differential tissue dry mass accumulation by *Citrus grandis*, *Poncirus trifoliata* and their F1 under salinized and non-salinized environments. Funct. Plant Biol. 27 (1), 27–33.

Tubiello, F.N., Ewert, F., 2002. Simulating the effects of elevated CO_2 on crops: approaches and applications for climate change. Eur. J. Agron. 18 (1–2), 57–74.

Verreynne, J.S., Lovatt, C.J., 2009. The effect of crop load on budbreak influences return bloom in alternate bearing 'Pixie' mandarin. J. Am. Soc. Hortic. Sci. 134, 299–307.

Verslues, P.E., Agarwal, M., Katiyar-Agarwal, S., Zhu, J., Zhu, J.-K., 2006. Methods and concepts in quantifying resistance to drought, salt and freezing, abiotic stresses that affect plant water status. Plant J. 45, 523–539. https://doi.org/10.1111/j.1365-313X.2005.02593.x.

Vidoz, M.L., Mignolli, F., Aispuru, H.T., Mroginski, L.A., 2016. Rapid formation of adventitious roots and partial ethylene sensitivity result in faster adaptation to flooding in the aerial roots (aer) mutant of tomato. Sci. Hortic. 201, 130–139. https://doi.org/10.1016/j.scienta.2016.01.032.

Villalobos, F.J., Testi, L., Moreno-Perez, M.F., 2009. Evaporation and canopy conductance of citrus orchards. Agric. Water Manag. 96, 565–573. https://doi.org/10.1016/j.agwat.2008.09.016.

Vu, J.C.V., Newman, Y.C., Allen Jr, L.H., Gallo-Meagher, M., Zhang, M.-Q., 2002. Photosynthetic acclimation of young sweet orange trees to elevated growth CO_2 and temperature. J. Plant Physiol. 159 (2), 147–157. https://doi.org/10.1078/0176-1617-00689.

Walker, R.R., Douglas, T.J., 1983. Effect of salinity level on uptake and distribution of chloride, sodium and potassium ions in citrus plants. Aust. J. Agric. Res. 34 (2), 145–153.

Wang, Z., Wang, M., Liu, L., Meng, F., 2013. Physiological and proteomic responses of diploid and tetraploid black locust (*Robinia pseudoacacia* L.) subjected to salt stress. Int. J. Mol. Sci. 14, 20299–20325.

Yelenosky, G., 1985. Cold hardiness in citrus. Hortic. Rev. 7, 201.

Zandalinas, S.I., Rivero, R.M., Martínez, V., Gómez-Cadenas, A., Arbona, V., 2016. Tolerance of citrus plants to the combination of high temperatures and drought is associated to the increase in transpiration modulated by a reduction in abscisic acid levels. BMC Plant Biol. 16, 105. https://doi.org/10.1186/s12870-016-0791-7.

Zandalinas, S.I., Balfagon, D., Arbona, V., Gómez-Cadenas, A., 2017a. Modulation of antioxidant defense system is associated with combined drought and heat stress tolerance in citrus. Front. Plant Sci. 8. https://doi.org/10.3389/fpls.2017.00953.

Zandalinas, S.I., Sales, C., Beltrán, J., Gómez-Cadenas, A., Arbona, V., 2017b. Activation of secondary metabolism in citrus plants is associated to sensitivity to combined drought and high temperatures. Front. Plant Sci. 7, 1–17. https://doi.org/10.3389/fpls.2016.01954.

Zekri, M., 1993. Salinity and calcium effects on emergence, growth and mineral composition of seedlings of eight citrus rootstocks. J. Hortic. Sci. 68 (1), 53–62.

Zekri, M., Parsons, L.R., 1990a. Comparative effects of NaCl and polyethylene glycol on root distribution, growth, and stomatal conductance of sour orange seedlings. Plant Soil 129 (2), 137–143.

Zekri, M., Parsons, L.R., 1990b. Calcium influences growth and leaf mineral concentration of citrus under saline conditions. HortScience 25 (7), 784–786.

Zekri, B.M., Parsons, L., Boman, B., 2015. Flooding injury and importance of drainage. Citrus Ind. 96 (6), 24–26.

Zhang, X., Li, X., Liu, J., 2014. Identification of conserved and novel cold-responsive microRNAs in trifoliate orange (*Poncirus trifoliata* (L.) raf.) using high-throughput sequencing. Plant Mol. Biol. Rep. 32, 328–341.

Zhang, X., Wang, W., Wang, M., Zhang, H., Liu, J., 2016. The miR396b of *Poncirus trifoliata* functions in cold tolerance by regulating ACC oxidase gene expression and modulating ethylene and polyamine homeostasis. Plant Cell Physiol. 57, 1865–1878.

Ziogas, V., et al., 2017. Characterization of β-amino-and γ-amino butyric acid-induced citrus seeds germination under salinity using nanoLC-MS/MS analysis. Plant Cell Rep. 36 (5), 787–789.

Further reading

Díaz-Vivancos, P., Clemente-Moreno, M.J., Rubio, M., Olmos, E., García, J.A., Martínez-Gómez, P., Hernández, J.A., 2008. Alteration in the chloroplastic metabolism leads to ROS accumulation in pea plants in response to plum pox virus. J. Exp. Bot. 59, 2147–2160. https://doi.org/10.1093/jxb/ern082.

Grosser, J.W., Ollitrault, P., Olivares-Fuster, O., 2000. Somatic hybridization in citrus: an effective tool to facilitate variety improvement. In Vitro Cell. Dev. Biol. Plant 36, 434–449.

IPCC, 2007. Climate change 2007: synthesis report. In: Pachauri, R.K., Reisinger, A. (Eds.), Contribution of Working Groups I, II and III to the Fourth Assessment Report of the Intergovernmental Panel on Climate Change. IPCC, Geneva, Switzerland.

Martínez-Alcántara, B., Martínez-Cuenca, M.R., Oliver, A.Q., Iglesias, D.J., Primo-Millo, E., Forner-Giner, M.A., 2015. Comparative expression of candidate genes involved in sodium transport and compartmentation in citrus. Environ. Exp. Bot. 111, 52–62.

Yahmed, J.B., Novillo, P., Garcia-Lor, A., Salvador, A., Mimoun, M.B., Luro, F., Talon, M., Ollitrault, P., Morillon, R., 2015. Salt tolerance traits revealed in mandarins (*Citrus reticulata* Blanco) are mainly related to root-to-shoot Cl− translocation limitation and leaf detoxification processes. Sci. Hortic. 191, 90–100.

Chapter 14

Salinity and water deficit

José M. Colmenero-Flores[a], Vicent Arbona[b], Raphaël Morillon[c], Aurelio Gómez-Cadenas[b]

[a]*Institute of Natural Resources and Agrobiology, Spanish National Research Council (CSIC), Seville, Spain,* [b]*Ecophysiology and Biotechnology, Department of Agricultural and Environmental Sciences, Universitat Jaume I, Castelló de la Plana, Spain,* [c]*Equipe "Amélioration des Plantes à Multiplication Végétative", UMR AGAP, Département BIOS, CIRAD, Guadeloupe, France*

14.1 Introduction

Citrus is a subtropical crop and, therefore, is poorly tolerant to water deficit and salt stress conditions. Its cultivation has been extended to different climatic regions between 40° north-south latitudes, where citrus production is particularly high in arid and semiarid regions. In major citrus-producing regions such as the Mediterranean Basin, Southern California (USA), South Africa, and Australia, where summer weather is usually dry and hot, and available water is of low quality due to excessive concentrations of soluble salts, water deficit and salinity become limiting to citrus productivity (revised in Maas, 1993; Storey and Walker, 1999; Gómez-Cadenas et al., 2003). Furthermore, climate change in citrus-producing regions is expected to amplify the effects of drought and salinity stress (see Chapter 13).

Physiological perturbations caused by salt stress and water deficit in plants have much in common. Both reduce the ability of plants to take up water, and this quickly causes reductions in growth rate, along with a suite of metabolic changes. If excessive amounts of salt enter the plant, salt-specific effects have later an impact on growth when saline ions eventually rise to toxic levels in the older transpiring leaves, causing premature senescence, and reducing the photosynthetic leaf area of the plant to a level that cannot sustain growth (Munns and Tester, 2008). Therefore, molecular responses that are associated with the acclimatization to water deficit and salt stress are interrelated, thus hindering the dissection of signaling processes and the identification of major determinants of stress resistance (Pardo, 2010). To better understand the responses of citrus plants to water deficit and salinity at the level of the organism and cell, we will refer to stress avoidance as the first response to minimize water loss or excessive accumulation of ions from sensitive plant tissues; and, in the longer term, if the stress becomes more severe, stress tolerance mechanisms will protect cellular structures and integrity of biomolecules. Whereas avoidance strategies maintain plant growth and productivity, tolerance mechanisms ensure continued growth and survival. We will use the term stress resistance in cases where it is not possible or not desirable to refer to a more specific mechanism (Verslues et al., 2006).

Citrus can be classified as a "salt-sensitive" crop and high concentrations of salts seriously affect its productivity. Toxicity triggered by salinity results from ion accumulation in leaves with the subsequent appearance of symptoms such as chlorosis and necrosis. However, other effects of salinity on plant physiology have been reported. Among them, an initial osmotic effect, metabolic disturbances, and nutrient imbalances are very common. Although both sodium (Na^+) and chloride (Cl^-) ions have been reported to cause ion toxicity in plants, including citrus, Cl^- exclusion from the canopy tissues is a general mechanism of salinity stress avoidance in citrus species (Storey and Walker, 1999).

Drought is the most threatening abiotic stress that affects agriculture by limiting water supply to the plant aerial part and constitutes an important determinant of citrus yield and quality in regions like China or Brazil (de Souza et al., 2017). In citrus, water deficit reduces growth and metabolism, leading to a reduction in fruit yield and quality. Drought also increases costs of juice extraction and reduces peel thickness making citrus fruit more vulnerable to damage during handling and shipping (Carr, 2012).

Rootstocks improve the resistance of citrus crops to abiotic factors, increase the efficiency of natural (water and soil) resources utilization, and decrease the use of chemical inputs, representing an essential tool for food security (Romero et al., 2006; Garcia-Sanchez et al., 2007; Rodriguez-Gamir et al., 2010; Albacete et al., 2015). Root traits based on root and xylem anatomy, water relations, hormonal and biochemical profiles, or distinct ion transport efficiencies can be transferred to the grafted scion and clearly contribute to plant productivity under drought and salt stress conditions. Therefore, the use of adequate rootstocks is essential to cultivate citrus under these stressing factors, and important differences have been reported

The Genus Citrus. https://doi.org/10.1016/B978-0-12-812163-4.00014-0

in the ability of diverse citrus rootstocks to improve stress avoidance and/or tolerance mechanisms. The ability of plants to cope with water deficit relies on physiological traits that lead to changes in stomatal conductance, plant water relations, and cell tolerance acquisition (Santana-Vieira et al., 2016). Moreover, the ability of citrus plants to cope with salt stress mainly relies on the rootstock capacity to restrict excessive Na^+ and Cl^- accumulation in the shoot, a property that can, in turn, be associated to other physiological attributes like vigor, water use, and shoot/root ratio (Moya et al., 2003).

Harmful effects of water and saline stress on plant tissues have been discussed in the previous chapter of this book. In this chapter, stress components, citrus plant responses to water deficit and salinity, resistance mechanisms including available information on various citrus genotypes, as well as agronomic and biotechnological approaches to improve crop stress resistance will be addressed.

14.2 Salinity

Citrus is very often grown in areas with low precipitation regimes and generally requires supplemental irrigation systems, which are particularly prone to salinization. In many citrus-producing regions, such as the Mediterranean Basin, summer weather is usually dry and hot and the available water is of low quality due to excessive concentrations of soluble salts and, therefore, salinity becomes a major environmental constraint (Maas, 1993; Storey and Walker, 1999; Syvertsen and García-Sánchez, 2014).

The general term "salinity" refers to the presence of the dissolved cations sodium (Na^+), calcium, boron, magnesium, and potassium, and the anions chloride (Cl^-), sulfate, and carbonate in soil solution in a concentration high enough to reduce crop growth (Ferguson and Grattan, 2005). However, in most of the studies, it has been observed that salinity stress is caused by sodium chloride (NaCl) accumulation. About a 13% decrease of citrus yield occurs for each $1\,dS\,m^{-1}$ increase in salinity over $1.3\,dS\,m^{-1}$ in the saturated soil extract, the values being over $3\,dS\,m^{-1}$, critical for citrus production (García-Sánchez et al., 2003; Simpson et al., 2014). Not only vegetative and reproductive development, but also fruit set, growth, maturation, rind color, internal quality, and yield are parameters affected by high salinity (revised in Chapter 13).

14.2.1 Salinity components

Detrimental effects of salinity in plant physiology are a consequence of osmotic effects, causing rapid and often transient changes in plant water relations, and specific ion toxicities, which occur later and more gradually (Munns and Tester, 2008). Harmful effects of the osmotic and ion-toxicity components are difficult to differentiate and the relative contribution of each factor probably depends on salt stress severity. Under mild NaCl conditions, the dominant component seems to be osmotic given that citrus fruit yield reductions occur without excessive tissue accumulations of Cl^- or Na^+ and without any apparent toxicity symptoms (Maas, 1993). Under more severe salinity stress, producing excessive accumulation of Cl^- or Na^+ in leaves, the ion-toxicity contribution is more relevant due to its pernicious effect on the photosynthetic machinery and growth.

14.2.1.1 Osmotic component

Osmotic effects are caused by the total concentration of osmolytes in the soil solution, mainly produced by a combination of natural soil salts, irrigation water quality, and fertilization. Because of the difficulty of quantifying the specific contribution of osmotic tolerance, the relative importance of this parameter in salt stress resistance remains unknown for most species (Munns and Tester, 2008). As salinity in the soil solution increases, soil water becomes less available to the plant. Contrary to halophytes, glycophyte plants, like citrus, are supposedly inefficient in compartmentalizing saline ions into intracellular organelles, so that osmotic adjustment is incomplete and the plants respond as if they were water-stressed (Ferguson and Grattan, 2005). However, it has been shown recently that also in glycophytes, cell osmotic adjustment is a primary function to be carried out under salt stress conditions even at the expense of an effective exclusion of toxic ions (Alvarez-Aragon and Rodriguez-Navarro, 2017). This means that NaCl-stressed plants promote root Na^+ and Cl^- uptake and tissue accumulation to ensure efficient osmotic adjustment (Li et al., 2017b), even if these processes lead to long-term toxicity. In citrus plants, the observation that the osmotic pressure produced by salinity is less harmful than that produced by direct water deficit caused by polyethylene glycol (Zekri and Parsons, 1990b) indicates that, to some extent, citrus plants use saline ions to adjust osmotically and avoid water deficit. This is particularly true in leaves of citrus plants subjected to salt stress, which maintain better water status in comparison to nonstressed plants (Gómez-Cadenas et al., 1998; Pérez-Pérez et al., 2007). However, this effective mechanism to avoid cell water deficit has deleterious effects on plant physiology because there is an increase in Na^+ and Cl^- concentration in leaf tissue. Plants growing in salinized media also reduce internal water deficits through the synthesis of organic solutes for osmotic adjustment and cell protection (Section 14.4.1).

14.2.1.2 Ion-toxicity component

Under salt stress conditions, ion toxicity is commonly associated with Na^+ accumulation in most plant species. However, the damage caused by salinity in citrus trees has been also associated with excessive accumulation of other ions like Cl^- and sulfate in the scion leaves. Delimiting the toxicity degree of these ions is a difficult task since it varies widely depending on the type of salts (accompanying ions) and the rootstock/scion combinations used (reviewed in Maas, 1993; Storey and Walker, 1999). For instance, the concentration of calcium (Ca^{2+}) in the soil solution is an important factor controlling the severity of specific ion toxicities since it ameliorates some of the effects of salinity in citrus (Section 14.5.2).

Probably the most relevant factor to consider is the capacity of ion exclusion from shoot tissues, mostly regulated by the rootstock. For example, the ability of the Trifoliate orange rootstock to restrict Na^+ accumulation in the leaves and its low Cl^- exclusion capacity (Walker, 1986) determine that Cl^- is the prevalent ion toxicity in citrus varieties grafted on this rootstock or its hybrids. Due to the greater presence of Cl^- salts in salinized soils or the generalized low Cl^- exclusion capacity of citrus genotypes, Cl^- is commonly the ion that accumulates at higher concentrations in the leaves of citrus plants subjected to salt stress. Although photosynthesis and transpiration are also affected by Na^+ buildup, Cl^- is considered the major toxic component of salt stress in citrus (Bañuls and Primo-Millo, 1995; Romero-Aranda et al., 1998). For instance, both NaCl and KCl reduce citrus growth markedly, whereas $NaNO_3$ has little effect (Bañuls et al., 1997). Consequently, ion-toxicity indicators like growth reduction, leaf injury, and abscission are clearly associated with leaf Cl^-, rather than Na^+, accumulation.

This does not imply that Cl^- is more metabolically toxic than Na^+ for citrus plants. Rather, this and other woody perennial crops like stone fruits and grapevines are better at excluding Na^+ from leaf blades, which is efficiently accumulated into the woody tissue of roots and the basal stem, presumably through xylem retrieval (Li et al., 2017a; Walker, 1986). Thus Cl^-, which continues passing to the leaves, becomes the more significant toxic component of the saline solution (Munns and Tester, 2008). Therefore, the physiological basis for citrus resistance to salt stress is mostly related to the ability to restrict Cl^- transport from root to shoot, a mechanism whose efficiency is particularly dependent on rootstock performance (Brumós et al., 2010).

14.2.2 Salinity avoidance mechanisms

Genotypes that are able to restrict the entry of Cl^- and Na^+ ions into the shoot or compartmentalize them into intracellular organelles will be more resistant to salt stress. On the one side, salt avoidance mechanisms prevent the excessive accumulation of Cl^- and Na^+ ions in shoot organs. On the other side, compartmentalization of Cl^- and Na^+ ions in leaf cell vacuoles represents a tolerance mechanism that allows the plant to survive osmotic and ionic damage in sensitive (e.g., photosynthetic) plant tissues (Section 14.4). Resistance to salinity in glycophytes such as citrus plants is largely based on avoidance (Cl^- and Na^+ exclusion) mechanisms, rather than on the effective vacuolar compartmentalization of these ions in leaf cells.

14.2.2.1 Differences between genotypes and breeding programs

For years, farmers and breeders have searched for the best combinations of rootstock/scion to cultivate citrus in soils affected by salinity. According to salt stress resistance, a rough classification of citrus rootstocks includes Cleopatra and Sunki mandarins as good salt resistant; sour orange and Volkameriana lemon as medium salt tolerant; and trifoliate orange and its hybrids as salt sensitive (Cooper, 1961). It is a well-known phenomenon that citrus genotypes have a pronounced influence on the amount of Cl^- and/or Na^+ accumulation in the foliage of grafted and ungrafted trees. Although the effect of salinity on growth and yield is mostly determined by the rootstock, the scion can also influence (Storey and Walker, 1999). Shoot Cl^- exclusion is mainly regulated by the rootstock, probably because it mostly depends on anion transporters present in the root, as explained below. However, Na^+ exclusion also relies importantly on the scion, most likely because relevant genes for Na^+ exclusion in plants function at both the root and shoot levels (Pardo, 2010).

Few correlations exist in the ability of different citrus genotypes to restrict sodium vs chloride ions (Maas, 1993). The range in leaf ion accumulation among genotypes, obtained when the most efficient excluder is compared with the least efficient one, is more than 10-fold for Cl^- and about 4- to 6-fold for Na^+ (Storey and Walker, 1999). This wide variation for ion exclusion and evidence showing that Na^+ and Cl^- exclusion mechanisms are heritable traits (Sykes, 1992), stimulated breeding programs aimed at producing new salt-tolerant rootstocks (Forner-Giner et al., 2009; Raga et al., 2014, 2016). The exclusion of both Cl^- and Na^+ ions was observed as a continuous character, with progenies that segregate widely for their ability to restrict the accumulation of these ions in shoot tissues that perform as well as, and sometimes better than, the parent varieties (Sykes, 1992; Forner-Giner et al., 2009). Maas (1993) ranked the best ion excluder genotypes, identifying sour orange as an excellent Na^+ excluder, having also good Cl^- exclusion capacity. Its desirable agronomic characteristics made it a reasonably good rootstock to cope with salinity problems, but trees grafted on this genotype are susceptible to tristeza virus. Unfortunately, tristeza-tolerant rootstocks such as trifoliate orange and its hybrids (citranges and Swingle citrumelo) are sensitive to salinity. Although most of the breeding effort aimed at obtaining dwarfing and disease tolerance, new rootstock genotypes have been recently evaluated for salinity tolerance (Forner-Giner et al., 2009, 2011; Sykes, 2011), pointing to Cleopatra mandarin (*Citrus reshni*) as an excellent parental genotype in breeding programs to obtain salt-resistant citrus hybrids (Forner-Giner et al., 2009; Raga et al., 2016). Some of them, for example, Forner-Alcaide 5 (FA-5), combine good agronomic properties with high salinity resistance. Salt stress tolerance within the genus *Citrus* seems to be characteristic of mandarins (*C. reticulata*) and pummelos (*C. maxima*), while all secondary genotypes that presented good resistance to salt stress shared mandarin or pummelo as a female parent (Hussain et al., 2015). Combava (*C. hystrix*), which belongs to the subgenus Papeda, was also qualified as resistant.

Obtaining new rootstocks resistant to salinity through biotechnological approaches requires a deeper understanding of the physiological and molecular mechanisms that regulate Na^+ and Cl^- homeostasis in plants.

14.2.2.2 Sodium exclusion

The concerted activity of the Na^+ transporters HKT1 and SOS1 regulates Na^+ and K^+ homeostasis in plants under salt stress (Wang et al., 2014). SOS1 facilitates Na^+ exclusion by extruding the cation from root tip epidermal cells (Shi et al., 2002). In xylem parenchyma cells, SOS1 loads Na^+ into the xylem sap, indirectly facilitating effective osmotic adjustment of shoot tissues through vacuolar compartmentalization, whereas HKT1 mediates the reverse flux, unloading Na^+ from xylem vessels to prevent Na^+ overaccumulation in photosynthetic organs (Berthomieu et al., 2003). Coordination of both HKT1 and SOS1 Na^+ transporters allows adequate partition of the toxic cation between organs. Sodium retrieved from xylem sap through HKT1 may be transported into the phloem for shoot-to-root transport. However, Na^+ phloem reallocation represents a small fraction (10% or less) of the amount imported in the transpiration stream via the xylem (Pardo, 2010). Therefore, in woody perennial plants like citrus, the excess of Na^+ retrieved by HKT1 from the root and the stem xylem sap may eventually represent an important portion of the Na^+ immobilized in woody tissues. A correlation has been established between the expression of *SOS1* and *HKT1* genes and the Na^+ exclusion capacity of two genotypes, Cleopatra mandarin and trifoliate orange (Martinez-Alcantara et al., 2015).

14.2.2.3 Chloride nutrition

Effective Cl^- exclusion from the shoot has been broadly recognized as the most relevant mechanism avoiding salt stress in citrus plants. To properly understand how citrus regulates Cl^- homeostasis, we must pay attention to the functions that this ion plays in the biology of plants. In recent years, there has been a paradigm shift in this regard since chloride has gone from being considered a harmful ion, accidentally absorbed by plants through passive mechanisms or nitrate transporters, to be considered a preferential macronutrient whose transport is finely regulated by plants (Tadeo et al., 2008; Brumós, 2009; Franco-Navarro et al., 2016, 2019; Cubero-Font et al., 2016; Wege et al., 2017).

In addition to being an essential micronutrient, Cl^- has been recently defined as a beneficial macronutrient for higher plants (Franco-Navarro et al., 2016). When supplied at concentrations in excess of those needed to satisfy the micronutrient requirements, but insufficient to cause toxicity (1–5 mM Cl^-), Cl^- treatments resulted in leaf accumulation levels which are typical of the content of a macronutrient in citrus (Brumós et al., 2010), tobacco (Franco-Navarro et al., 2016), and many other glycophyte species (Colmenero-Flores, unpublished results). Besides a preferential osmoticum, macronutrient levels of Cl^- play specific roles in regulating water relations, resulting in higher water-use efficiency and higher resistance to water deficit in tobacco, tomato, and citrus plants (Brumós, 2009; Rosales et al., 2012; Franco-Navarro et al., 2016, 2019). In tobacco plants, Cl^- specifically increases the mesophyll diffusion conductance to CO_2, stimulates nitrate-use efficiency and higher leaf cell size, and leads to a moderate increase of dry biomass, mainly due to higher shoot expansion (Franco-Navarro et al., 2016, 2019). The role of Cl^- in osmotic regulation and cell elongation has been assessed in *Arabidopsis* and rice mutant lines with lack of function of the cation-chloride cotransporter (*CCC*) gene (Colmenero-Flores et al., 2007; Chen et al., 2016). Chloride has been also proposed to play important roles in the regulation of photosynthetic electron transport and photoprotective mechanisms in chloroplasts since adequate Cl^- homeostasis, regulated by the thylakoid anion channels AtCLCe, AtVCCN1, and AtVCCN2, is required to adjust photosynthesis to fluctuating light and environmental conditions (Herdean et al., 2016a,b; Duan et al., 2016). Although it is an abundant and ubiquitous element in nature, the need to apply chloride fertilization in certain soils and for certain crops is increasingly accepted (Schwenke et al., 2015). An adequate Cl^- nutrition is therefore required to optimize the efficiency in the use of water, nitrogen, and carbon/energy of crops.

14.2.2.3.1 Cl⁻ homeostasis in citrus

If higher plants manage Cl^- homeostasis as a nutrient rather than a toxic ion, it is expected that the differences between rootstocks that confer resistance to salinity lie in plasma membrane transport mechanisms that regulate Cl^- uptake and its long-distance transport. Under nonsaline treatments (e.g., 4.5 mM Cl^-), Cl^- is actively taken up and accumulated into leaf tissues of citrus plants to levels that exceed the critical requirement content by one order of magnitude in Cl^--excluder rootstocks and by two orders in Cl^--includer (salt-sensitive) rootstocks through symplastically regulated transport mechanisms (Brumós et al., 2010). Both includer and excluder rootstocks modulate Cl^- influx according to the availability of the nutrient since uptake capacity was induced by Cl^- starvation but inhibited after Cl^- resupply. The strong inhibition of net Cl^- uptake that occurs after application of low Cl^- (4.5 mM) to Cl^--starved rootstocks resembles the reduction observed after salt stress application (Fernández-Ballester et al., 2003; Moya et al., 2003) and suggests a homeostatic regulatory mechanism induced after high Cl^- influx into the plant root symplast (Fig. 14.1). The most relevant difference between the Cl^- includer Carrizo citrange and the excluder Cleopatra mandarin is the rate of root xylem Cl^- translocation (Fig. 14.1). These results indicate that (i) similar transport mechanisms operate when Cl^- is managed as a nutrient or in response to a salt stress imposition and (ii) the key rate-limiting regulatory step modulating Cl^- accumulation in the shoot of citrus plants is the loading of Cl^- from the root stelar symplast to the xylem apoplast (Brumós et al., 2010).

14.2.2.4 Molecular mechanisms regulating xylem Cl⁻ translocation

Because Cl^- is a negatively charged molecule, Cl^- influx is very likely governed by high- and low-affinity Cl^-/H^+ symport transporters under low and moderate external Cl^- concentrations (Felle, 1994), while Cl^- ions are released passively from the cell through anion-efflux channels (Roberts, 2006) following its electrochemical potential gradient (Fig. 14.1A, B, D, E). Anion channels from epidermal cells can also mediate Cl^- influx under salt stress conditions (Fig. 14.1C, F) (Skerrett and Tyerman, 1994). In the root vasculature, Cl^- release to the xylem sap from cells adjacent to xylem vessels is mostly expected to rely on anion channels (Fig. 14.1).

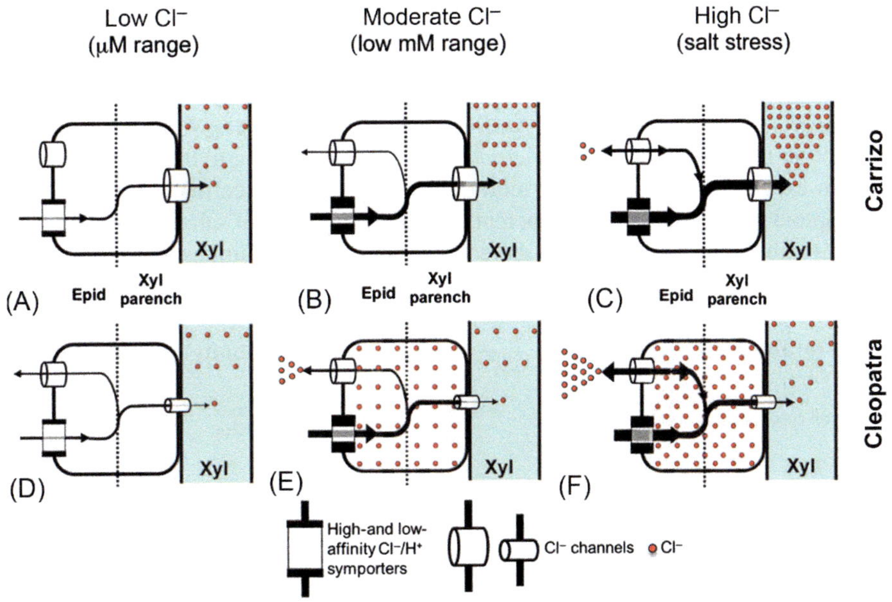

FIG. 14.1 Proposed model for the symplastic regulation of Cl^- homeostasis in the root of the includer Carrizo citrange and the excluder Cleopatra mandarin rootstocks. For clarity, Cl^- transport mechanisms of endosomal membranes are not shown. Under low external Cl^- concentrations (A, D), high-affinity Cl^-/H^+ symporter(s) govern(s) external Cl^- uptake. Following the addition of Cl^- at the low millimolar range a high influx is generated (B, E) and the plant readjusts Cl^- uptake to lower values through modification of the abundance or kinetic properties of influx transporters and activation of plasma membrane anion-efflux channels from epidermal cells. Anion channels from epidermal cells can also mediate Cl^- influx under salt stress (C, F). The main difference between the symplastic regulation of the includer and excluder rootstocks is related to the rate of Cl^- translocation to the xylem. Cl^- xylem translocation is clearly restricted in the salt-tolerant rootstock Cleopatra mandarin, but not in the sensitive Citrange carrizo. Lower Cl^- translocation in the salt-tolerant rootstock Cleopatra mandarin gives rise to higher Cl^- accumulation in distal roots and probably to higher Cl^- release to the rhizosphere mediated by Cl^- channels from epidermal cells. *Epid*, epidermal cells; *Xyl*, xylem vessel; *Xyl parench*, xylem parenchyma cells. *Modified from Brumós, J., Talón, M., Bouhlal, R.Y.M., Colmenero-Flores, J.M., 2010. Cl- homeostasis in includer and excluder citrus rootstocks: transport mechanisms and identification of candidate genes. Plant Cell Environ. 33, 2012–2027.*

The slow-type anion channel SLAH3 is a member of the SLAC/SLAH family with a high selectivity for NO_3^- over Cl^- (Geiger et al., 2011). In the root, SLAH3 localizes to pericycle cells facing xylem vessels (Cubero-Font et al., 2016). SLAH1, another member of the same family, colocalizes with SLAH3 in the xylem-pole pericycle. SLAH1 functions like a molecular switch that regulates xylem Cl^- translocation according to environmental cues (Cubero-Font et al., 2016). The *AtSLAH1* gene is highly expressed in the absence of stress, and its product interacts with SLAH3 forming a SLAH1/SLAH3 heteromer that determines a significant increase in Cl^- conductance of the SLAH3 channel under optimal growing conditions (Fig. 14.2). Under salt stress conditions, the expression of the *AtSLAH1* gene is strongly repressed in an ABA-dependent manner, significantly reducing the Cl^- conductance of SLAH3 and therefore diminishing xylem Cl^- translocation (Cubero-Font et al., 2016; Qiu et al., 2017). The stelar-localized NPF2.4 plasma membrane protein has also been recently identified as a transporter that facilitates root-to-shoot transfer of Cl^- (Li et al., 2016). Mutation in citrus rootstocks of SLAH1 and NPF2.4 transporters through CRISPR/Cas9 technology could lead to salt-resistant citrus crops.

14.2.2.5 Morphophysiological factors and ploidy level

Shoot ion accumulation has been linked also to physiological and morphological factors like plant growth, transpiration, and water use (Moya et al., 2002, 2003; Syvertsen et al., 2000). Because rapidly growing trees always use more water than slower-growing trees, leaves on rapidly growing trees would be exposed to relatively more Cl^- in the transpiration stream than leaves on slower-growing trees. Thus, some of the mechanisms underlying salt stress avoidance in citrus rootstocks have been attributed to their low growth vigor, low water use, and/or a high shoot/root ratio. It should be expected, however, that the greater growth of vigorous genotypes is proportional to the amount of water and ions consumed so that the concentration of chloride should remain relatively constant. In grapevine, another Cl^--sensitive crop, root apoplastic transport and water relations cannot account for the differences in Cl^- transport of two grapevine rootstocks differing in salt tolerance (Abbaspour et al., 2014). In citrus, however, salt-tolerant rootstocks tend to produce trees with slower growth or that use less water than trees on salt-sensitive rootstocks (Lopez-Climent et al., 2008).

The importance of ploidy level on salinity resistance in citrus has been explored recently (Dambier et al., 2011). Tetraploid citrus plants are thicker and more succulent than the corresponding diploid genotypes. In comparison to the diploid parent, its double diploid or autotetraploid genotype shows differential gene expression (Allario et al., 2011) and improved salinity avoidance. As previously stressed, some of the proposed mechanisms accounting for a greater Cl^- exclusion capacity are morphological and histological traits of roots leading to lower hydraulic conductance and transpiration rate (Ruiz et al., 2016). Roots of autotetraploid genotypes are thicker, less branched, with a lower number of roots and with higher deposition of suberin in cell walls of the exodermis cells in sections close to the root apex. Allotetraploid somatic and zygotic hybrids also exclude Na^+ and Cl^- ions more efficiently than their corresponding diploid parents (Mouhaya et al., 2010). While tetraploid genotypes represent a potentially valuable source of salt-resistant citrus rootstocks, it will be important to determine whether salt tolerance is conserved under real-field conditions as some works have indicated that tetraploids are more sensitive to salinity that their diploid relatives under extreme stress conditions.

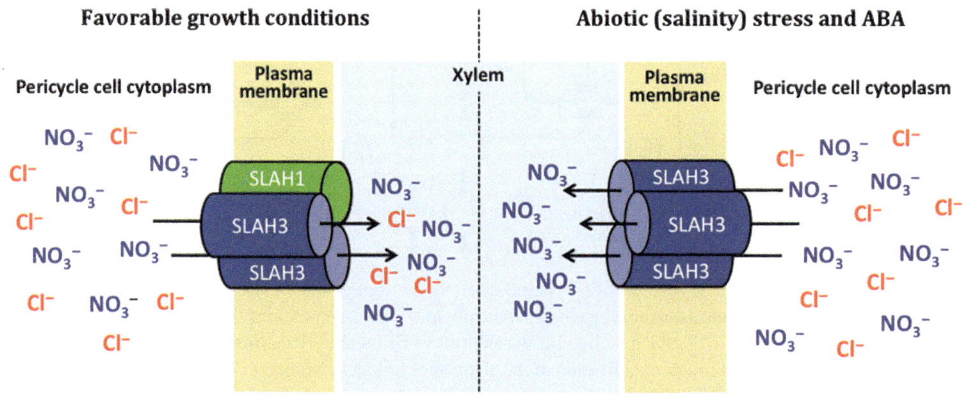

FIG. 14.2 Regulation of root xylem Cl^- translocation by the anion channels SLAH1 and SLAH3. Under favorable growth conditions (*left*), high transcriptional activity of both *SLAH1* and *SLAH3* genes in the xylem-pole pericycle determines the formation of heteromeric SLAH1/SLAH3 complexes. The SLAH1 channel does not transport anions itself but modifies the kinetic properties of SLAH3, which increases its chloride conductance by seven times and mediates xylem translocation of both chloride (Cl^-) and nitrate (NO_3^-) anions. However, under salt stress (*right*), gene expression of *AtSLAH1* is strongly reduced by abscisic acid (ABA)-dependent regulatory pathway, favoring the formation of the SLAH3/SLAH3 homomers that significantly reduce Cl^- conductance, decreasing xylem Cl^- translocation and shoot accumulation (Cubero-Font et al., 2016).

14.2.3 Citrus responses to salinity

14.2.3.1 Global gene expression

Functional genomics and physiological studies have been conducted in which the response of a salt-sensitive rootstock (Carrizo citrange) was compared with that of a resistant, Cl^--excluder, rootstock (Cleopatra mandarin) in a wide range of time, between hours and weeks (Brumós et al., 2009; Brumós, 2009). In the short term, the root of the resistant genotype shows a faster accumulation of abscisic acid (ABA), followed by a strong increase in Cl^- accumulation (Brumós, 2009), as a consequence of reduced xylem Cl^- translocation (Fig. 14.1). These responses, not verified in the sensitive rootstock, correlate with a greater number of differentially expressed genes involved in signaling, chromatin silencing, ABA and osmotic-stress responses, and jasmonic acid (JA) biosynthesis in the resistant rootstock. After 3 weeks, differential gene expression in the root becomes even greater as the resistant genotype specifically intensifies the induction of genes involved in stress responses, catabolism of carbohydrates and amino acids, and biosynthesis of secondary metabolites.

In leaves subjected to salt stress for several weeks, Cl^- toxicity, rather than Na^+ toxicity or the concomitant osmotic perturbation, was the primary factor involved in the molecular responses of both rootstocks to salinity (Brumós et al., 2009). Similar to the response in the root, the resistant rootstock strongly induced genes involved in stress acclimatization mechanisms, while repressed genes are involved in central metabolic processes such as photosynthesis and carbon utilization. These features agreed with important reductions in stomatal conductance, net photosynthetic rate, Fv/Fm, and Φ_{PSII} in the resistant rootstock. These molecular responses and acclimatization mechanisms occur in the resistant rootstock before the leaves reach harmful levels of chloride, suggesting that they are a consequence of early signaling mechanism elicited from the root.

These results are in line with two different strategies identified in citrus in response to salt stress. Whereas some genotypes like the salt-sensitive Carrizo keep a high photosynthetic activity and induce dissipative mechanisms at the antennae (which contribute to the reduction of damage to the light-harvesting complexes), others, like the salt-resistant Cleopatra mandarin show a rapid induction of defensive mechanisms and reduction of stomatal conductance and photosynthesis (Lopez-Climent et al., 2008). Therefore, it is proposed that salt stress avoidance in the resistant genotype is based on early ABA biosynthesis and signaling responses elicited in the root that determines higher ability to exclude Cl^-, induction of stronger stress responses and defensive mechanisms in roots and shoot including the stimulation of secondary metabolism, reduction of transpiration, and inhibition of primary metabolism.

In citrus trees acclimatized to moderate salinity (NaCl 30 mM) after continuous treatment for 2 years, photosynthetically active young leaves maintain nontoxic Cl^- concentrations, but exhibit a much stronger response to salinity (10 times greater number of differentially expressed genes) than photosynthetically active mature leaves that accumulate higher Cl^- concentration (Perez-Perez et al., 2015). It is proposed that long-term acclimation to salt stress in citrus requires: (i) differential compartmentalization of Cl^- according to leaf age; (ii) a release of the repression of genes involved in primary metabolism, enabling the plant to resume photosynthetic activity; (iii) to adjust to a number of physiological, metabolic, and cellular processes to the new homeostatic conditions related to the cell wall structure, defense mechanisms, lipid metabolism, endomembrane system activity, and antioxidant activity.

More recently, Xie et al. (2017) have shown similar results in citrus root comparing salt and dehydration treatments. Differentially expressed genes, including 76 miRNAs were functionally associated with carbohydrate metabolism, hormone signal transduction, reactive-oxygen species (ROS) system, and phenylalanine metabolism. About 20% of the differentially expressed genes were common to the salt stress and the water deficit responses, indicating similar molecular responses underlying dehydration and salinity resistance. An important number of genes coding for transcription factors (NACs, MYBs, CBFs, ERFs, WRKYs, ZFPs, CAMTAs, and bZIPs) was involved in salt stress and evidence has shown that citrus roots deal with salinity mainly by regulating transcription factors, which integrate carbohydrate metabolism, ROS system, and hormone signaling pathway into a complex network.

14.2.3.2 Hormone regulation

One of the most apparent effects of salt stress on citrus is the activation of leaf abscission that affects mainly the old leaves that accumulate chloride and abscise (Gómez-Cadenas et al., 2003). The analysis of endogenous levels of plant hormones such as ABA, ethylene, and its direct precursor, 1-aminocyclopropane-1-carboxylic acid (ACC), revealed a biphasic response that paralleled the chloride accumulation in salt-stressed plants. Therefore, changes of ABA and ACC levels showed an initial transient increase after the onset of the treatment followed by a gradual and continuous accumulation. It has been hypothesized that the initial transitory hormonal rises are induced during the osmotic adjustment that takes place shortly after salinization, whereas progressive chloride buildup determines subsequent hormonal accumulations. Interestingly, during the second phase, chloride levels correlated with abscission rates in leaves (Gómez-Cadenas et al., 1998). Polyamines have also been proposed as signal molecules of salt stress in citrus plants (Tanou et al., 2014).

14.3 Water deficit

Among the various abiotic stresses, drought is the major factor limiting crop productivity worldwide. In regions with Mediterranean climates, the vegetative growth season is initiated by a rise in air temperature, hot dry summers lead to annual drought-stress conditions, and years of low rainfall occur recurrently. In tropical regions, where the citrus culture frequently occurs in dryland (e.g., Brazil and China), shoot growth takes place in distinct flushes initiated by water availability (Carr, 2012) and water deficit can severely limit citrus crop productivity.

Rapid stomatal closure is induced after water deficit perception so that the primary physiological parameter affected is photosynthesis, which will reduce sharply the carbon dioxide (CO_2) assimilation capacity of the tree. More detailed descriptions on the deleterious effects of water deficit on citrus biology and the use of water management practices have been addressed in Chapters 13 and 20, respectively. We will focus here on the water deficit response mechanisms, the physiological and molecular bases of water deficit acclimatization, and the use of efficient rootstocks and varieties improving water use efficiency.

14.3.1 Resistance mechanisms and differences among genotypes

Due to the existence of other breeding priorities (e.g., dwarfism, pests, and disease resistance), and because it is a complex quantitative trait controlled by many genes, drought resistance has not been a central objective in citrus-breeding programs. Since the use of rootstocks is a rapid tool in tailoring plants to better adapt to suboptimal growing conditions (Kumar et al., 2017), the performance of different rootstocks and rootstock/scion combinations under water deficit has been a subject of study. Thus, different genotypes have been compared resulting in the following order of drought resistance: mandarins (*C. reticulata* spp., including Sunki, Cleopatra, Cleopatra hybrids, *C. erythrosa*), Rangpur lime, Rough lemon, sour orange, and *Citrus macrophylla* classified as good resistant; lemon, trifoliate orange, citrange hybrids (*Citrus sinensis* × *Poncirus trifoliata*), and *C. chuana* classified as medium resistant; and sweet orange, *C. verrucosa*, and grapefruit classified as poor resistant (Rodriguez-Gamir et al., 2010; Ma and Fan, 2014; Zaher-Ara et al., 2016), with citrange hybrids and grapefruit classified as intermediate or poor resistant depending on the study.

Whole-plant water deficit avoidance defines the mechanisms that allow the plant to maintain tissue water potential (Ψ_w) and water content close to the unstressed level by increasing water uptake or limiting water loss (Levitt, 1972). Regulation of stomatal closure is a critical avoidance mechanism that operates in the very short term (explained below). In the longer term, elasticity in leaf rolling, the increase of the root/shoot ratio by developing a deeper and thicker root system, leaf biomass reduction, higher cuticular resistance, and regulation of root water conductivity can be of greatest importance for crop plants (Rodriguez-Gamir et al., 2010; Gowda et al., 2011; Hu and Xiong, 2014). Since stomatal closure is considered the most important mechanism to regulate gas exchange and water loss, genotypes presenting lower transpiration rate, associated with higher ability to close stomata rapidly, such as Cleopatra mandarin, are considered to perform better under water deficit conditions (López-Climent et al., 2008). When grafted with Valencia orange, both rootstocks Cleopatra mandarin and FA-5 (Cleopatra mandarin × *P. trifoliata* hybrid) consume less water from the soil and use it more efficiently than *P. trifoliata* (Rodriguez-Gamir et al., 2010). Rootstocks play a major role in drought avoidance by controlling and adjusting the water supply to meet the transpiration demand of the shoots. Since water deficit reduces root hydraulic conductance (Rodriguez-Gamir et al., 2011), the downregulation of genes encoding water channels or aquaporins (CsTIPS, CsXIPs) has been proposed as a mechanism to minimize water loss under water deprivation (Martins et al., 2015).

The high resistance to water deficit of Rangpur lime, or citrus plants grafted on this rootstock, has been associated with enhanced root growth; maintenance of the total carbohydrate pool and its remobilization to the root under water deficit (Pedroso et al., 2014); increased root hydraulic conductivity (Medina et al., 1998); and increased cell wall elasticity and osmotic adjustment that contributes to the maintenance of turgor, photosynthesis, and growth over a wider range of soil-water potentials (Goncalves et al., 2016; Santana-Vieira et al., 2016).

Under conditions of prolonged or severe water deficit, when the plant is unable to regulate water loss and a decrease in tissue Ψ_w occurs, the plant must adopt strategies to tolerate dehydration. Under these conditions, Sunki mandarin performs better than Rangpur lime (Santana-Vieira et al., 2016). Similar to the salt-stress resistance strategy of Cleopatra mandarin (Section 14.2.2), Sunki mandarin reduces plant production accompanied by robust mechanisms of cell protection. This strategy is associated with an increase in the rigidity of cell walls, which determines faster loss of turgor, possibly leading to earlier water-stress sensing and responses (Goncalves et al., 2016). Accordingly, this rootstock induces greater concentrations of ABA, salicylic acid (SA), and soluble carbohydrates, such as trehalose and raffinose (Santana-Vieira et al., 2016). SA has been implicated as one of the regulatory signals involved in promoting the activity of antioxidant enzymes (Aimar et al., 2011). Interestingly dehydration tolerance traits are manifested also in the scion grafted on Sunki mandarin rootstocks (Santana-Vieira et al., 2016).

The hybrid FA-5 follows an intermediate, although apparently very efficient, strategy. This hybrid is as productive as its parental *P. trifoliata* in terms of high photosynthetic activity (*A*). When confronted with water deficit, FA-5, or plants grafted on FA-5, are able to maintain higher *A* and transpiration values in comparison to both parental genotypes. In addition, FA-5 plants consume less water than those of the drought-sensitive *P. trifoliata* during the first days of the dry period. They also develop a more efficient osmotic adjustment than its tolerant parental Cleopatra mandarin during the period of maximum stress, being able to combine optimally the mechanisms of water deficit avoidance and tolerance (Rodriguez-Gamir et al., 2010).

Therefore, interesting genotypes for using as appropriate progenitors in breeding programs aiming at obtaining drought resistant rootstocks are: Rangpur lime, providing drought avoidance traits; mandarins like Cleopatra and Sunki, providing drought tolerance traits; and specific genotypes of the genus *Poncirus* that are more resistant to water deficit (Ben Yahmed et al., 2016), providing dwarfing as well as tolerance to Citrus Tristeza virus and cold.

14.3.1.1 Ploidy level

Citrus genome duplication limits growth and development (Guerra et al., 2014), associated with greater epidermal and stomatal cell size, lower cell frequency, reduced transpiration rate (Mouhaya et al., 2010; Allario et al., 2011) and, therefore, improved water-use efficiency. These traits can be transferred to the grafted scion without apparently impairing their quality (Hussain et al., 2012). Important changes in citrus polyploid genotypes (3× or 4×) include thicker and greener leaves, which are associated with higher leaf water content (Allario et al., 2011; Padoan et al., 2013). Lower CO_2 assimilation rate per unit leaf area measured in tetraploids is probably related to a reduced stomatal conductance, increased leaf thickness, and larger mesophyll cell volume, limiting, in turn, CO_2 diffusion conductance to the site of carboxylation in chloroplasts (Romero-Aranda et al., 1997).

When subjected to water deficit, autotetraploid Carrizo citrange seedlings reduce water consumption and avoid water deficit more efficiently than the diploid parental (Oliveira et al., 2017). Tetraploidization also induces large changes in leaf primary and secondary metabolite accumulation in citrus seedlings that were proposed to be beneficial when the plant is subjected to stress (Tan et al., 2015). Autotetraploid plants of Rangpur lime present constitutively higher ABA contents than the correspondent diploids, which correlate with constitutive overexpression of drought-responsive genes; limited shoot gas exchange and growth under control conditions; and greater resistance to water deficit when used as rootstock of sweet orange scion (Allario et al., 2013). In-silico interactomic analyses suggest that the proteome of the double diploid Rangpur lime is better suited to tolerate cellular dehydration (de Souza et al., 2017). Genes involved in abiotic stress tolerance like antioxidant enzymes and heat shock proteins (HSPs) showed higher constitutive levels in tetraploid plants of Cleopatra mandarin when compared with the diploid ones (Podda et al., 2013). Long-term investigations are required to confirm the interest of polyploid genotypes as a source of stress-resistant rootstocks and its influence on fruit quality and yield under both optimal and suboptimal field conditions.

14.3.2 Citrus responses to water deficit

14.3.2.1 Perception and hormonal regulation

Regulation of water deficit response in plants is mainly orchestrated by ABA that can be locally or systemically produced (Gonzalez-Guzman et al., 2014). ABA accumulation in plant tissues is one of the first responses observed once water deficit is perceived. Resistance to water deficit has been significantly and positively correlated to the content of ABA in several citrus genotypes (Neves et al., 2013; Oliveira et al., 2015; Santana-Vieira et al., 2016). A relationship between ABA accumulation in response to drought and the expression of genes of the carotenoid cleavage dioxygenase family has been reported in both vegetative and reproductive tissues (Agustí et al., 2007). However, ABA accumulation in water-stressed citrus roots does not rely on either carotenoid content in this organ (Manzi et al., 2016) or on the expression of several key genes involved in carotenoid and ABA biosynthetic pathways in the root (Manzi et al., 2017). Interestingly, leaf dehydration is needed to induce ABA accumulation in roots of citrus plants, indicating that changes in the leaf evapotranspiration demand, rather than root dehydration, trigger the water stress response at the whole plant level (Manzi et al., 2017).

In leaves, the fast ABA-mediated stomatal closure operates through binding of the sesquiterpene plant hormone to PYR/PYL/RCAR receptors (Geiger et al., 2011). In the absence of ABA, PP2C phosphatases (ABI1 and ABI2) inhibit autophosphorylation of the Ca^{2+}-independent and Ca^{2+}-dependent kinases SnRK2.6/OST1 and CPKs, respectively, rendering those kinases inactive. Upon perception of ABA by RCAR/PYR/PYL receptors in guard cells, ABI1 and ABI2 phosphatases are bound to the ABA-receptor complex and are inactivated, relieving the inhibition of SnRK2.6/OST1 and CPK. The kinases, in turn, phosphorylate and activate the anion channels SLAC1 and SLAH3 (the same anion channel involved in root-to-shoot xylem translocation, Fig. 14.2), depolarizing the guard cell membrane potential and triggering the cascade of events that lead to the loss of turgor pressure in the guard cells and to stomatal closure. As a result of structural studies performed with *C. sinensis* (sweet orange) and *Solanum lycopersicum* (tomato) ABA receptors, the role of PP2C as a bona fide coreceptor required to increase ABA-binding affinity has been proposed (Moreno-Alvero et al., 2017).

ABA also regulates signaling pathways leading to transcriptional regulation of multiple genes (the ABA slow response), which includes dehydration tolerance mechanisms, as will be explained below. In citrus leaves, dehydration has a minor effect on the transcriptional regulation of the genes encoding for the CsRCAR/PYR/PYL ABA receptors and the CsSnRK2 activating kinases, whereas genes encoding for the clade-A PP2C phosphatases CsABI1, CsAHG1, and CsAHG3, are induced by water deficit (Romero et al., 2012). Since PP2CA phosphatases are the negative regulators of the ABA signal transduction pathway, transcriptional induction of these genes may reveal a negative feedback regulation of the ABA response or its importance as a coreceptor as previously indicated.

Another well-known response of citrus to water deprivation is abscission of aerial organs including leaves, stems, and fruits. It involves ethylene as the main plant hormone regulating fruit and leaf shedding (Estornell et al., 2013; Iglesias et al., 2007; Merelo et al., 2017), although previous ABA accumulation is required for abscission to occur (Gómez-Cadenas et al., 1996) (Fig. 14.3). Recently, another hormonal player in the regulation of water stress in citrus has come on stage: the oxylipin JA. As a result of the relative accumulation kinetics of JA and ABA in citrus subjected to different abiotic stress conditions, it was proposed that JA could act upstream of the ABA signal (Fig. 14.3), thereby constituting an additional mediator between stress perception and the regulation of the physiological response (Arbona and Gomez-Cadenas, 2008; de Ollas et al., 2013). Using JA and ABA biosynthesis inhibitors alternatively, it was shown that an early and transient JA accumulation was required to induce a subsequent ABA accumulation. However, this response was only observed in roots whereas in leaves, no clear conclusion could be drawn (de Ollas et al., 2013). Using *Arabidopsis thaliana* accessions impaired in ABA or JA biosynthesis or perception, the regulatory role of jasmonates upstream of ABA was confirmed and further expanded allowing the identification of JA-Ile as the signaling molecule mediating stress perception and ABA accumulation (Fig. 14.3). Moreover, the tissue specificity of this response was also confirmed (de Ollas et al., 2015).

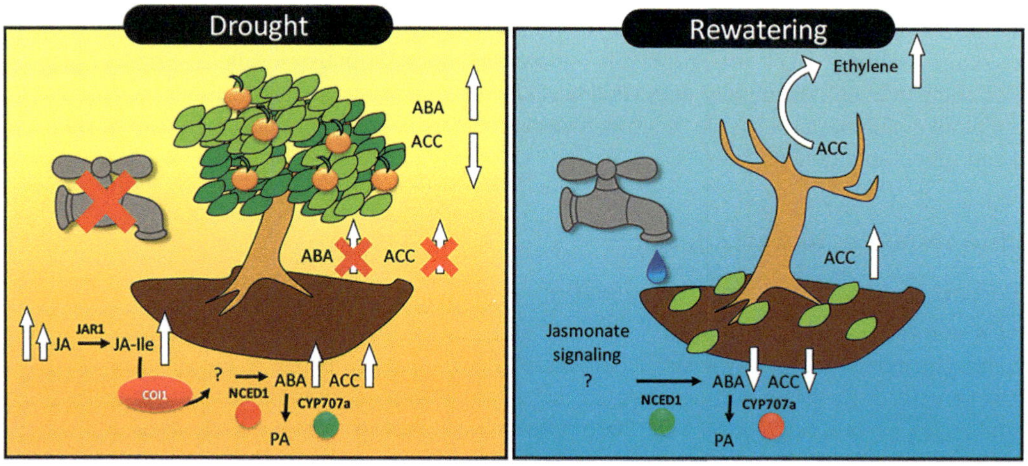

FIG. 14.3 Hormonal regulation of drought-induced leaf abscission in citrus. Under drought conditions (*left panel*), JA is transiently accumulated in citrus roots and subsequently converted to JA-Ile by the jasmonic acid-amido synthetase JAR1, which induces *NCED1* overexpression in a *COI1*-dependent manner (de Ollas et al., 2013, 2015). On the other hand, *CYP707a*, encoding an ABA 8′-hydroxylase is repressed, leading to the accumulation of ABA (Arbona et al., 2017). Along with ABA, 1-aminocyclopropane, 1-carboxylic acid (ACC) is also accumulated (Gómez-Cadenas et al., 1996). When the watering conditions are restored (*right panel*), *NCED1* is repressed and *CYP707a* is induced, leading to the accumulation of phaseic acid (PA) and the reduction of ABA (Arbona et al., 2017). Restoration of xylem flow transports the ACC accumulated in the root to the aerial tissues, where it is oxidized to ethylene triggering leaf and fruit abscission (Gómez-Cadenas et al., 1996).

14.3.2.2 Global responses

Functional genomic studies have revealed that the expression of a wide range of genes is regulated in response to water deficit in citrus plants. The transcriptome of Clemenules mandarin grafted onto Cleopatra mandarin subjected to water deficit shows induction of proteins involved in lysine, proline and raffinose catabolism, hydrogen peroxide reduction, vacuolar malate transport, RCI2 proteolipids, and cell-protecting proteins (including hydrophilins, osmotins, and HSPs). In addition, novel drought-responsive genes were identified in this study suggesting that water-deficit resistance in citrus may benefit from inhibition of proteolysis, activation of zeaxanthin and linolenoyl synthesis, reinforcement of ribosomal structure, and downregulation of ethylene response (Gimeno et al., 2009a). The gene encoding for the NAC transcription factor *CsNAC1* is strongly induced by water deficit in the leaves of Cleopatra mandarin and Rangpur lime (Oliveira et al., 2011). In addition, proteomic studies in different genotypes reveal that acclimatization to water deficit involves downregulation of primary metabolism, activation of ROS, detoxification, synthesis of cell-protective proteins (see Section 14.4), activation of DNA repair, and increased abundance of proteins involved in transport (Oliveira et al., 2015). Metabolomic studies in the roots of citrus rootstocks reinforce the importance of both ABA and JA phytohormones and suggest the involvement of proline and phenylpropanoids, as well as their derivatives, in response of citrus to water deficit (Argamasilla et al., 2014). In addition, higher concentrations of proline, IAA, and phenylpropanoids, together with the higher number of putative metabolites found in the roots of nonstressed Cleopatra plants, could be an effective physiological mechanism for preventing or delaying drought-derived damage in the stress-resistant rootstock. It has been also proposed that recurrent water deficit produces epigenetic alterations that improve drought resistance in citrus plants (Neves et al., 2017).

14.4 Resistance and tolerance mechanisms common to water deficit and salinity

In most cases, the plant's first response to drought or soil salinity is to avoid stress, preventing the reduction of Ψ_w or toxic ion accumulation in sensitive leaf tissues. For example, short-term stomatal closure reduces water loss and root-to-shoot Cl^- and Na^+ translocation, maintaining the plant close to the unstressed levels. In the case of mild stress or stress of a limited duration, avoidance mechanisms by themselves can be sufficient to maintain plant performance (Verslues et al., 2006). If the stress becomes more severe and the plant is no longer able to maintain adequate water or ion homeostasis, stress-tolerance mechanisms become important for plant survival or efficient recovery from the stressful situation.

Tolerance mechanisms to water deficit include: (i) cell dehydration avoidance mechanisms (osmotic adjustment and cell wall hardening responses); (ii) cell dehydration tolerance mechanisms characterized by the accumulation of osmoprotectants, antioxidants, and ROS scavengers, as well as the biosynthesis of cell-protecting proteins, including HSPs and hydrophilins.

14.4.1 Osmotic adjustment and synthesis of compatible osmolytes

Osmotic adjustment in plants is preferentially achieved through vacuolar compartmentalization of inorganic ions, a process requiring energy consumption 10 times lower than the synthesis of organic solutes (Raven, 1985). Under water deficit conditions caused by salt stress, citrus respond by absorbing Na^+ and Cl^- and compartmentalizing the saline ions in the vacuole for efficient detoxification, osmotic adjustment, and turgor maintenance (as explained in Section 14.2.1). Under a water deficit caused by drought stress, osmotic adjustment is more difficult to achieve due to the lower availability of inorganic ions. Even so, in response to water deficit or osmotic shock, plant cells activate mechanisms of fast turgor adjustment based on the activation of the plasma membrane H^+-ATPase for specific stimulation of K^+, Cl^-, and Na^+ uptake (Shabala and Lew, 2002) and vacuolar compartmentalization of saline ions.

To equilibrate the Ψ_w of sensitive cell compartments like the cytoplasm and organelles with that of the vacuole, both K^+ and the synthesis of compatible osmolytes, such as proline, are required. Besides osmoregulation, compatible solutes have been proposed to play other important roles for acquisition of water deficit and salt stress tolerance (Bohnert and Sheveleva, 1998; Shabala and Shabala, 2011): (i) stabilization of cell macromolecules like membranes and protein complexes; (ii) adjustment of metabolic pathways to stressful conditions; and (iii) ROS scavenging.

Higher drought tolerance in some citrus rootstocks has been related to the accumulation of compatible solutes, such as carbohydrates and osmoprotectant amino acids such as proline (Rodriguez-Gamir et al., 2010; Pedrosa et al., 2015; Zandalinas et al., 2016). The role of nonstructural carbohydrates as active osmolytes has been studied in many plant species with contrasting results depending on the genotype, the severity and duration of the stress, and the tissue studied. In leaves and roots of citrus, salt stress causes progressive depletion of starch, but the participation of high abundant soluble carbohydrates such as glucose, hexoses, sucrose, and fructose in stress tolerance has not been clearly established (Arbona et al.,

2005b; Anjum, 2008; Roussos et al., 2013; Tsabarducas et al., 2015; Santana-Vieira et al., 2016). However, the sucrose-hydrolyzing enzyme alkaline/neutral invertase (A/N-INV) is induced by different abiotic stresses and ABA in citrus, and overexpression of the *P. trifoliata* gene *PtrA/NINV* enhances resistance to salinity and water deficit (Dahro et al., 2016). Higher trehalose and raffinose accumulation correlate with higher dehydration tolerance of the resistant rootstock Sunki mandarin (Santana-Vieira et al., 2016), although to levels that are not compatible with an osmoregulatory role, between 500 and 1000 times lower than that of sucrose. It is well known that metabolic intermediaries of trehalose at very low concentrations play an important signaling role in the regulation of carbohydrate homeostasis according to developmental and environmental conditions, playing important roles in drought tolerance (Paul et al., 2008).

14.4.2 Antioxidant defense

A secondary effect of dehydration and salt stress is the increase of ROS, ubiquitous and highly reactive agents that oxidize lipids, proteins, and nucleic acids, acting as an important source of stress for cells. To alleviate ROS toxicity, plants respond by overproducing antioxidant enzymes, including superoxide dismutase, catalase, and peroxidase, and metabolites such as ascorbate, and glutathione. It is widely accepted that antioxidant defenses are required for abiotic stress tolerance and, therefore, the ROS-response correlates positively with the degree of resistance of citrus plants (Balal et al., 2012; Seday et al., 2014). But the opposite correlation has also been observed, possibly when stress avoidance mechanisms prevent significant oxidative damage in photosynthetic tissues (López-Climent et al., 2008). For instance, when Cleopatra mandarin and Carrizo citrange are subjected to identical salinity conditions, only the sensitive genotype, Carrizo, increased antioxidant defenses whereas Cleopatra, the resistant genotype, did not (Arbona et al., 2004).

14.4.3 Synthesis of protective proteins

Other important mechanisms of plant cell tolerance induced in vegetative tissues subjected to water deficit or ion toxicity consist of the accumulation of proteins that play different roles in cell protection: hydrophilins and HSPs. Both constitute a diversity of protein families highly associated with the acquisition of desiccation tolerance in plants and other organisms. Most of them are developmentally regulated during embryo maturation and are strongly accumulated at the onset of seed desiccation. In vegetative tissues, the content of hydrophilins and HSPs is usually undetectable or very small, but in response to osmotic stress, a strong ABA-dependent transcriptional activation is induced, which results in an important accumulation of these proteins (Colmenero-Flores et al., 1997). Hydrophilins, including late embryogenesis abundant (LEA) proteins, enable the cells to survive protoplasmic water depletion (Garay-Arroyo et al., 2000; Reyes et al., 2005). A comprehensive study of hydrophilins (LEA protein families) has been performed in citrus (Pedrosa et al., 2015). Under conditions of cellular dehydration HSPs, and particularly small-type HSP of plants, function as molecular chaperones by preventing stress-induced aggregation of proteins and facilitating the subsequent reactivation. Numerous studies have corroborated the participation of hydrophilins and HSP protective proteins in the response of citrus plants to water deficit and salinity stress (Brumós et al., 2009; Gimeno et al., 2009a; Podda et al., 2013; Pedrosa et al., 2015; Xiao et al., 2017).

14.5 Agronomic and biotechnological approaches to improve crop stress resistance

14.5.1 Agronomic and palliative practices

14.5.1.1 Agronomic palliatives

Different palliative treatments have been assayed to increase yield under salt stress and water deficit conditions. First, judicious manipulations of irrigation and fertilizer applications are considered essential to cope with this problem in a commercial orchard (Chapter 15). Other horticultural practices such as the use of orange varieties as interstocks (Gimeno et al., 2009b), shading (Simon-Grao et al., 2016), treatments with persistent analogues of ABA (Arbona et al., 2006), the use of hydrogels (Arbona et al., 2005a), and exogenous application of SA (Khoshbakht and Asgharei, 2015), and polyamines (Tanou et al., 2014) also improved citrus performance under drought and/or salt stress conditions.

Different works have dealt with the effect of increasing calcium or nitrates in the nutritional solution as palliative treatments of salt stress. In the presence of an adequate concentration of calcium (Ca^{2+}), plants exclude Na^+ more efficiently and withstand the effects of relatively high NaCl concentration (Zekri and Parsons, 1990a; Bañuls et al., 1991; Romero-Aranda et al., 1998). This is probably a consequence of the Ca^{2+}-dependent inhibition of nonselective cation channels, which constitute a major pathway for passive Na^+ entry into the root under salt stress (Demidchik and Maathuis, 2007).

It has been repeatedly demonstrated that nitrate and other nitrogen-derived compounds like urea or ammonium have marked beneficial effects on the growth response of citrus (Iglesias et al., 2004; Gimeno et al., 2009c; Fernández-Crespo et al., 2012). Addition of nitrate seems to have two separate effects that probably improve the performance of citrus seedlings under saline conditions. First, nitrate supplementation stimulated photosynthesis and growth and reduced leaf abscission. Second, the nitrogen-induced increase in leaf biomass results in chloride dilution, the critical parameter for salt damage (Iglesias et al., 2004; Khoshbakht et al., 2014). In addition, foliar potassium nitrate application improves the tolerance of citrus seedlings to drought conditions (Gimeno et al., 2014).

14.5.1.2 Mycorrhizas

Inoculation of citrus plants with arbuscular mycorrhizal fungi has also been demonstrated as an effective way to increase resistance to salt stress and water deficit in certain rootstocks by favoring plant performance, possibly due to improved water and nutrient uptake (especially P nutrition); effective spatial configuration of root system; elevated concentration of tetramine spermine; improving osmotic adjustment through organic and inorganic osmolytes; scavenging ROS through antioxidant enzymes and antioxidants; soil structural improvements, etc. (Wu et al., 2013; Navarro et al., 2014; Huang et al., 2014; Zhang et al., 2017).

14.5.1.3 Priming

Chemical priming has been proposed as a promising field in plant stress physiology and crop stress management. Plants are apparently capable of inducing stress "memory," or "stress imprinting" following a first stress exposure that leads to acclimation to a later biotic or abiotic stress. Through priming (also known as hardening), plants are able to induce responses to a range of stresses, providing low-cost protection in relatively high stress-pressure conditions (Borges et al., 2014). ROS in the form of hydrogen peroxide (H_2O_2) and reactive nitrogen species in the form of nitric oxide (NO) induce priming toward salinity and drought in citrus plants (Molassiotis et al., 2016; Tanou et al., 2009). These chemical agents, including also sodium nitroprusside, sodium hydrosulfide, melatonin, and polyamines can potentially confer enhanced tolerance in the field against multiple abiotic stresses.

14.5.2 Molecular approaches

Many factors related to the plant response to water and salinity stresses have been identified, and many of the genes responsible for such traits have already been shown effective for engineering stress tolerance in model plants (Pardo, 2010). In citrus plants, different works have shown the effectiveness of this technology to the improvement of drought and salinity resistance (Gong and Liu, 2013). It is expected that the manipulation of signaling factors produces superior resistance because they control a broad range of downstream events. Thus, overexpression of the ABA-signaling AtCBF3/DREB1A transcription factor results in higher salt tolerance of transgenic *C. macrophylla* plants (Alvarez-Gerding et al., 2015). Overexpression of the transcription factors WRKY70 in lemon confer enhanced tolerance to drought stresses (Gong et al., 2015). Although not performed in citrus, it is interesting the observation that *MhNCED3*, the rate-limiting enzyme in the ABA biosynthesis pathway in *Malus hupehensis*, enhances plant tolerance to Cl^- stress by reducing Cl^- accumulation (Zhang et al., 2015). Finally, overexpression of the betaine aldehyde dehydrogenase gene involved in the biosynthesis of the compatible osmolyte glycinebetaine enhances salt tolerance in transgenic *P. trifoliata* plants (Fu et al., 2011). To improve transgenic plant performance, effective expression systems, including cell type-specific and stress-inducible promoters are required to fine-tune the citrus plant response to the specific circumstances of the crop (e.g., phenological stage), or the environmental stress condition (e.g., extent, intensity, etc.).

Since easy editing of genes has been recently demonstrated in citrus (Jia and Wang, 2014; Weeks et al., 2016; Jia et al., 2017), new biotechnological advances regarding improved resistance of citrus plants to water deficit and salinity are expected in the near future.

14.6 Concluding remarks

Relevant factors related to plant response to water deficit and salinity stress have been identified in citrus plants. Under salinity stress, salt ions contribute to osmotic adjustment significantly, so that leaf ion toxicity becomes the main problem to be solved. Citrus has some capacity to exclude sodium, but chloride, a plant nutrient with specific symplastic transport mechanisms, is less efficiently excluded. Therefore, resistance to salt stress is mainly associated with the rootstock ability to exclude chloride. Physiological and molecular approaches, recently supported on omics technologies, have identified different stress resistance mechanisms in various citrus genotypes. Greater resistance has been associated with induction of earlier and stronger stress responses; stress avoidance mechanisms consisting on efficient ion exclusion mechanisms and stringent control of transpiration; rapid inhibition of photosynthesis and primary metabolism; acquisition of cell tolerance including efficient osmotic adjustment, adequate control of oxidative damage, enhancement of secondary metabolism, and biosynthesis of protective molecules. In the stress-resistant rootstocks commonly used today, these responses lead to rapid growth inhibition. Breeding programs should identify new rootstocks that effectively combine stress avoidance and tolerance mechanisms to optimize both plant production under adverse environmental conditions and efficient crop recovery after stress. Molecular mechanisms that have already been shown effective for engineering stress resistance in model and crop plants under controlled conditions could be transferred to field-grown crops in the near future taking advantage of technologies with enormous potential such as targeted genome editing.

References

Abbaspour, N., Kaiser, B., Tyerman, S., 2014. Root apoplastic transport and water relations cannot account for differences in Cl- transport and Cl-/NO3 (-) interactions of two grapevine rootstocks differing in salt tolerance. Acta Physiol. Plant. 36, 687–698.

Agustí, J., Zapater, M., Iglesias, D.J., Cercûs, M., Tadeo, F.R., Talûn, M., 2007. Differential expression of putative 9-cis-epoxycarotenoid dioxygenases and abscisic acid accumulation in water stressed vegetative and reproductive tissues of citrus. Plant Sci. 172, 85–94.

Aimar, D., Calafat, M., Andrade, A.M., Carassay, L., Abdala, G., Molas, M.L., 2011. Drought tolerance and stress hormones: from model organisms to forage crops. Plants Environ. 272, 137–164.

Albacete, A., Martinez-Andujar, C., Martinez-Perez, A., Thompson, A.J., Dodd, I.C., Perez-Alfocea, F., 2015. Unravelling rootstockxscion interactions to improve food security. J. Exp. Bot. 66, 2211–2226.

Allario, T., Brumos, J., Manuel Colmenero-Flores, J., Tadeo, F., Froelicher, Y., Talon, M., Navarro, L., Ollitrault, P., Morillon, R., 2011. Large changes in anatomy and physiology between diploid Rangpur lime *(Citrus limonia)* and its autotetraploid are not associated with large changes in leaf gene expression. J. Exp. Bot. 62, 2507–2519.

Allario, T., Brumos, J., Colmenero-Flores, J.M., Iglesias, D.J., Pina, J.A., Navarro, L., Talon, M., Ollitrault, P., Morillon, R., 2013. Tetraploid Rangpur lime rootstock increases drought tolerance via enhanced constitutive root abscisic acid production. Plant Cell Environ. 36, 856–868.

Alvarez-Aragon, R., Rodriguez-Navarro, A., 2017. Nitrate-dependent shoot sodium accumulation and osmotic functions of sodium in Arabidopsis under saline conditions. Plant J. 91, 208–219.

Alvarez-Gerding, X., Espinoza, C., Inostroza-Blancheteau, C., Arce-Johnson, P., 2015. Molecular and physiological changes in response to salt stress in *Citrus macrophylla* W plants overexpressing Arabidopsis CBF3/DREB1A. Plant Physiol. Biochem. 92, 71–80.

Anjum, M.A., 2008. Effect of NaCl concentrations in irrigation water on growth and polyamine metabolism in two citrus rootstocks with different levels of salinity tolerance. Acta Physiol. Plant. 30, 43–52.

Arbona, V., Gomez-Cadenas, A., 2008. Hormonal modulation of citrus responses to flooding. J. Plant Growth Regul. 27, 241–250.

Arbona, V., Jacas, J., Gómez-Cadenas, A., 2004. Response of the antioxidant machinery of two citrus rootstocks to salt stress. Proc. Int. Soc. Citricult. 2, 644–648.

Arbona, V., Iglesias, D.J., Jacas, J., Primo-Millo, E., Talon, M., Gómez-Cadenas, A., 2005a. Hydrogel substrate amendment alleviates drought effects on young citrus plants. Plant Soil 270, 73–82.

Arbona, V., Marco, A.J., Iglesias, D.J., Lopez-Climent, M.F., Talon, M., Gomez-Cadenas, A., 2005b. Carbohydrate depletion in roots and leaves of salt-stressed potted *Citrus clementina* L. Plant Growth Regul. 46, 153–160.

Arbona, V., López-Climent, M.F., Mahouachi, J., Pérez-Clemente, R.M., Abrams, S.R., Gómez-Cadenas, A., 2006. Use of persistent analogs of abscisic acid as palliatives against salt-stress induced damage in citrus plants. J. Plant Growth Regul. 25, 1–9.

Arbona, V., Zandalinas, S.I., Manzi, M., Gonzalez-Guzman, M., Rodriguez, P.L., Gomez-Cadenas, A., 2017. Depletion of abscisic acid levels in roots of flooded Carrizo citrange (*Poncirus trifoliata* L. Raf. x *Citrus sinensis* L. Osb.) plants is a stress-specific response associated to the differential expression of PYR/PYL/RCAR receptors. Plant Mol. Biol. 93, 623–640.

Argamasilla, R., Gomez-Cadenas, A., Arbona, V., 2014. Metabolic and regulatory responses in citrus rootstocks in response to adverse environmental conditions. J. Plant Growth Regul. 33, 169–180.

Balal, R.M., Khan, M.M., Shahid, M.A., Mattson, N.S., Abbas, T., Ashfaq, M., Garcia-Sanchez, F., Ghazanfer, U., Gimeno, V., Iqbal, Z., 2012. Comparative studies on the physiobiochemical, enzymatic, and ionic modifications in salt-tolerant and salt-sensitive citrus rootstocks under NaCl stress. J. Am. Soc. Hortic. Sci. 137, 86–95.

Bañuls, J., Primo-Millo, E., 1995. Effects of salinity on some *Citrus* scion-rootstock combinations. Ann. Bot. 76, 97–102.

Bañuls, J., Legaz, F., Primo-Millo, E., 1991. Salinity-calcium interactions on growth and ionic concentration of *Citrus* plants. Plant Soil 133, 39–46.

Bañuls, J., Serna, M.D., Legaz, F., Talon, M., Primo-Millo, E., 1997. Growth and gas exchange parameters of *Citrus* plants stressed with different salts. J. Plant Physiol. 150, 194–199.

Ben Yahmed, J., Costantino, G., Amiel, P., Talon, M., Ollitrault, P., Morillon, R., Luro, F., 2016. Diversity in the trifoliate orange taxon reveals two main genetic groups marked by specific morphological traits and water deficit tolerance properties. J. Agric. Sci. 154, 495–514.

Berthomieu, P., Conejero, G., Nublat, A., Brackenbury, W.J., Lambert, C., Savio, C., Uozumi, N., Oiki, S., Yamada, K., Cellier, F., Gosti, F., Simonneau, T., Essah, P.A., Tester, M., Very, A.A., Sentenac, H., Casse, F., 2003. Functional analysis of AtHKT1 in Arabidopsis shows that Na+ recirculation by the phloem is crucial for salt tolerance. EMBO J. 22, 2004–2014.

Bohnert, H.J., Sheveleva, E., 1998. Plant stress adaptations—making metabolism move. Curr. Opin. Plant Biol. 1, 267–274.

Borges, A.A., Jimenez-Arias, D., Exposito-Rodriguez, M., Sandalio, L.M., Perez, J.A., 2014. Priming crops against biotic and abiotic stresses: MSB as a tool for studying mechanisms. Front. Plant Sci. 5, 642. https://doi.org/10.3389/fpls.2014.00642.

Brumós, J., 2009. Aproximación genómico-funcional, molecular y fisiológica en el estudio de la salinidad y la homeostasis de cloruro en los cítricos. PhD thesis, Universidad Politécnica de Valencia.

Brumós, J., Colmenero-Flores, J.M., Conesa, A., Izquierdo, P., Sánchez, G., Iglesias, D.J., López-Climent, M.F., Gómez-Cadenas, A., Talón, M., 2009. Membrane transporters and carbon metabolism implicated in chloride homeostasis differentiate salt stress responses in tolerant and sensitive Citrus rootstocks. Funct. Integr. Genomics 9, 293–309.

Brumós, J., Talón, M., Bouhlal, R.Y.M., Colmenero-Flores, J.M., 2010. Cl- homeostasis in includer and excluder citrus rootstocks: transport mechanisms and identification of candidate genes. Plant Cell Environ. 33, 2012–2027.

Carr, M.K.V., 2012. The water relations and irrigation requirements of citrus (*Citrus* spp.): a review. Exp. Agric. 48, 347–377.

Chen, Z.C., Yamaji, N., Fujii-Kashino, M., Ma, J.F., 2016. A cation-chloride cotransporter gene is required for cell elongation and osmoregulation in rice. Plant Physiol. 171, 494–507.

Colmenero-Flores, J.M., Campos, F., Garciarrubio, A., Covarrubias, A.A., 1997. Characterization of Phaseolus vulgaris cDNA clones responsive to water deficit: identification of a novel late embryogenesis abundant-like protein. Plant Mol. Biol. 35, 393–405.

Colmenero-Flores, J.M., Martinez, G., Gamba, G., Vazquez, N., Iglesias, D.J., Brumos, J., Talon, M., 2007. Identification and functional characterization of cation-chloride cotransporters in plants. Plant J. 50, 278–292.

Cooper, W.C., 1961. Toxicity and accumulation of salts in *Citrus* trees on various rootstocks in Texas. Proc. Fla. State Hortic. Soc. 74, 95–104.

Cubero-Font, P., Maierhofer, T., Jaslan, J., Rosales, M.A., Espartero, J., Díaz-Rueda, P., Müller, H.M., Hürter, A.-L., Al-Rasheid, K.A., Marten, I., Hedrich, R., Colmenero-Flores, J.M., Geiger, D., 2016. Silent S-type anion channel subunit SLAH1 gates SLAH3 open for chloride root-to-shoot translocation. Curr. Biol. 26, 2213–2220.

Dahro, B., Wang, F., Peng, T., Liu, J.-H., 2016. PtrA/NINV, an alkaline/neutral invertase gene of Poncirus trifoliata, confers enhanced tolerance to multiple abiotic stresses by modulating ROS levels and maintaining photosynthetic efficiency. BMC Plant Biol. 16, 76.

Dambier, D., Benyahia, H., Pensabene-Bellavia, G., KaÇar, Y.A., Froelicher, Y., Belfalah, Z., Lhou, B., Handaji, N., Printz, B., Morillon, R., Yesiloglu, T., Navarro, L., Ollitrault, P., 2011. Somatic hybridization for citrus rootstock breeding: an effective tool to solve some important issues of the Mediterranean citrus industry. Plant Cell Rep. 30, 883–900.

Demidchik, V., Maathuis, F.J.M., 2007. Physiological roles of nonselective cation channels in plants: from salt stress to signalling and development. New Phytol. 175, 387–404.

de Ollas, C., Hernando, B., Arbona, V., Gomez-Cadenas, A., 2013. Jasmonic acid transient accumulation is needed for abscisic acid increase in citrus roots under drought stress conditions. Physiol. Plant. 147, 296–306.

de Ollas, C., Arbona, V., Gomez-Cadenas, A., 2015. Jasmonoyl isoleucine accumulation is needed for abscisic acid build-up in roots of Arabidopsis under water stress conditions. Plant Cell Environ. 38, 2157–2170.

de Souza, J.D., Silva, E.M.D., Coelho, M.A., Morillon, R., Bonatto, D., Micheli, F., Gesteira, A.D., 2017. Different adaptation strategies of two citrus scion/rootstock combinations in response to drought stress. PLoS One 12, e0177993. https://doi.org/10.1371/journal.pone.0177993.

Duan, Z., Kong, F., Zhang, L., Li, W., Zhang, J., Peng, L., 2016. A bestrophin-like protein modulates the proton motive force across the thylakoid membrane in Arabidopsis. J. Integr. Plant Biol. 58, 848–858.

Estornell, L.H., Agusti, J., Merelo, P., Talon, M., Tadeo, F.R., 2013. Elucidating mechanisms underlying organ abscission. Plant Sci. 199, 48–60.

Felle, H.H., 1994. The H+/Cl- symporter in root-hair cells of Sinapis alba (an electrophysiological study using ion-selective microelectrodes). Plant Physiol. 106, 1131–1136.

Ferguson, L., Grattan, S.R., 2005. How salinity damages citrus: osmotic effects and specific ion toxicities. HortTechnology 15, 95–99.

Fernández-Ballester, G., García-Sánchez, F., Cerdá, A., Martínez, V., 2003. Tolerance of citrus rootstock seedlings to saline stress based on their ability to regulate ion uptake and transport. Tree Physiol. 23, 265–271.

Fernández-Crespo, E., Camañes, G., García-Agustín, P., 2012. Ammonium enhances resistance to salinity stress in citrus plants. J. Plant Physiol. 169, 1183–1191.

Forner-Giner, M.A., Primo-Millo, E., Forner, J.B., 2009. Performance of Forner-Alcaide 5 and Forner-Alcaide 13, hybrids of Cleopatra mandarin x Poncirus trifoliata, as salinity-tolerant citrus rootstocks. J. Am. Pomol. Soc. 63, 72–80.

Forner-Giner, M.A., Legaz, F., Primo-Millo, E., Forner, J., 2011. Nutritional responses of citrus rootstocks to salinity: performance of the new hybrids forner-alcaide 5 and forner-alcaide 13. J. Plant Nutr. 34, 1435–1452.

Franco-Navarro, J.D., Brumos, J., Rosales, M.A., Cubero-Font, P., Talon, M., Colmenero-Flores, J.M., 2016. Chloride regulates leaf cell size and water relations in tobacco plants. J. Exp. Bot. 67, 873–891.

Franco-Navarro, J.D., Rosales, M.A., Álvarez, R., Cubero-Font, P., Calvo, P., Díaz-Espejo, A., Colmenero-Flores, J.M., 2019. Chloride as macronutrient increases water use efficiency by anatomically-driven reduced stomatal conductance and increased mesophyll diffusion to CO2. Plant J. 99, 815–831. https://doi.org/10.1111/tpj.14423.

Fu, X.Z., Khan, E.U., Hu, S.S., Fan, Q.J., Liu, J.H., 2011. Overexpression of the betaine aldehyde dehydrogenase gene from Atriplex hortensis enhances salt tolerance in the transgenic trifoliate orange (Poncirus trifoliata L. Raf.). Environ. Exp. Bot. 74, 106–113.

Garay-Arroyo, A., Colmenero-Flores, J.M., Garciarrubio, A., Covarrubias, A.A., 2000. Highly hydrophilic proteins in prokaryotes and eukaryotes are common during conditions of water deficit. J. Biol. Chem. 275, 5668–5674.

García-Sánchez, F., Carvajal, M., Cerdá, A., Martínez, V., 2003. Response of 'Star Ruby' grapefruit on two rootstocks to NaCl salinity. J. Hortic. Sci. Biotechnol. 78, 859–865.

Garcia-Sanchez, F., Syvertsen, J.P., Gimeno, V., Botia, P., Perez-Perez, J.G., 2007. Responses to flooding and drought stress by two citrus rootstock seedlings with different water-use efficiency. Physiol. Plant. 130, 532–542.

Geiger, D., Maierhofer, T., Al-Rasheid, K.A., Scherzer, S., Mumm, P., Liese, A., Ache, P., Wellmann, C., Marten, I., Grill, E., Romeis, T., Hedrich, R., 2011. Stomatal closure by fast abscisic acid signaling is mediated by the guard cell anion channel SLAH3 and the receptor RCAR1. Sci. Signal. 4, ra32.

Gimeno, J., Gadea, J., Forment, J., Perez-Valle, J., Santiago, J., Martinez-Godoy, M.A., Yenush, L., Belles, J.M., Brumos, J., Colmenero-Flores, J.M., Talon, M., Serrano, R., 2009a. Shared and novel molecular responses of mandarin to drought. Plant Mol. Biol. 70, 403–420.

Gimeno, V., Syvertsen, J.P., Nieves, M., Simón, I., Martinez, V., Garcia-Sanchez, F., 2009b. Orange varieties as interstocks increase the salt tolerance of lemon trees. J. Hortic. Sci. Biotechnol. 84, 625–631.

Gimeno, V., Syvertsen, J.P., Nieves, M., Simón, I., Martínez, V., García-Sánchez, F., 2009c. Additional nitrogen fertilization affects salt tolerance of lemon trees on different rootstocks. Sci. Hortic. 121, 298–305.

Gimeno, V., Diaz-Lopez, L., Simon-Grao, S., Martinez, V., Martinez-Nicolas, J.J., Garcia-Sanchez, F., 2014. Foliar potassium nitrate application improves the tolerance of *Citrus macrophylla* L. seedlings to drought conditions. Plant Physiol. Biochem. 83, 308–315.

Gómez-Cadenas, A., Tadeo, F.R., Talón, M., Primo-Millo, E., 1996. Leaf abscission induced by ethylene in water stressed intact seedlings of (*Citrus reshni* Hort. ex Tan.) requires previous abscisic acid accumulation in roots. Plant Physiol. 112, 401–408.

Gómez-Cadenas, A., Tadeo, F.R., Primo-Millo, E., Talon, M., 1998. Involvement of abscisic acid and ethylene in the responses of citrus seedlings to salt shock. Physiol. Plant. 103, 475–484.

Gómez-Cadenas, A., Iglesias, D.J., Arbona, V., Colmenero-Flores, J.M., Primo-Millo, E., Talón, M., 2003. Physiological and molecular responses of citrus to salinity. Recent Res. Dev. Plant Mol. Biol. 1, 281–298.

Goncalves, L.P., Alves, T.F.O., Martins, C.P.S., De Sousa, A.O., Dos Santos, I.C., Pirovani, C.P., Almeida, A.A.F., Coelho, M.A., Gesteira, A.S., Soares, W.D., Girardi, E.A., Costa, M.G.C., 2016. Rootstock-induced physiological and biochemical mechanisms of drought tolerance in sweet orange. Acta Physiol. Plant. 38, 174. https://doi.org/10.1007/s11738-016-2198-3.

Gong, X.-Q., Liu, J.-H., 2013. Genetic transformation and genes for resistance to abiotic and biotic stresses in *Citrus* and its related genera. Plant Cell Tissue Org. Cult. 113, 137–147.

Gong, X., Zhang, J., Hu, J., Wang, W., Wu, H., Zhang, Q., Liu, J.-H., 2015. FcWRKY70, a WRKY protein of Fortunella crassifolia, functions in drought tolerance and modulates putrescine synthesis by regulating arginine decarboxylase gene. Plant Cell Environ. 38, 2248–2262.

Gonzalez-Guzman, M., Rodriguez, L., Lorenzo-Orts, L., Pons, C., Sarrion-Perdigones, A., Fernandez, M.A., Peirats-Llobet, M., Forment, J., Moreno-Alvero, M., Cutler, S.R., Albert, A., Granell, A., Rodriguez, P.L., 2014. Tomato PYR/PYL/RCAR abscisic acid receptors show high expression in root, differential sensitivity to the abscisic acid agonist quinabactin, and the capability to enhance plant drought resistance. J. Exp. Bot. 65, 4451–4464.

Gowda, V.R.P., Henry, A., Yamauchi, A., Shashidhar, H.E., Serraj, R., 2011. Root biology and genetic improvement for drought avoidance in rice. Field Crop Res. 122, 1–13.

Guerra, D., Schifino Wittmann, M.T., Schwarz, S.F., Dutra de Souza, P.V., Gonzatto, M.P., Weiler, R.L., 2014. Comparison between diploid and tetraploid citrus rootstocks: morphological characterization and growth evaluation. Bragantia 73, 1–7.

Herdean, A., Nziengui, H., Zsiros, O., Solymosi, K., Garab, G., Lundin, B., Spetea, C., 2016a. The Arabidopsis thylakoid chloride channel AtCLCe functions in chloride homeostasis and regulation of photosynthetic electron transport. Front. Plant Sci. 7, 115. https://doi.org/10.3389/fpls.2016.00115.

Herdean, A., Teardo, E., Nilsson, A.K., Pfeil, B.E., Johansson, O.N., Unnep, R., Nagy, G., Zsiros, O., Dana, S., Solymosi, K., Garab, G., Szabo, I., Spetea, C., Lundin, B., 2016b. A voltage-dependent chloride channel fine-tunes photosynthesis in plants. Nat. Commun. 7, 11654. https://doi.org/10.1038/ncomms11654.

Hu, H., Xiong, L., 2014. Genetic engineering and breeding of drought-resistant crops. In: Merchant, S.S. (Ed.), Annu. Rev. Plant Biol. 65, 715–741.

Huang, Y.-M., Srivastava, A.K., Zou, Y.-N., Ni, Q.-D., Han, Y., Wu, Q.-S., 2014. Mycorrhizal-induced calmodulin mediated changes in antioxidant enzymes and growth response of drought-stressed trifoliate orange. Front. Microbiol. 5, 682.

Hussain, S., Curk, F., Dhuique-Mayer, C., Urban, L., Ollitrault, P., Luro, F., Morillon, R., 2012. Autotetraploid trifoliate orange (Poncirus trifoliata) rootstocks do not impact clementine quality but reduce fruit yields and highly modify rootstock/scion physiology. Sci. Hortic. 134, 100–107.

Hussain, S., Morillon, R., Anjum, M.A., Ollitrault, P., Costantino, G., Luro, F., 2015. Genetic diversity revealed by physiological behavior of citrus genotypes subjected to salt stress. Acta Physiol. Plant. 37, 1740. https://doi.org/10.1007/s11738-014-1740-4.

Iglesias, D.J., Levy, Y., Gómez-Cadenas, A., Tadeo, F.R., Primo-Millo, E., Talon, M., 2004. Nitrate improves growth in salt-stressed citrus seedlings through effects on photosynthetic activity and chloride accumulation. Tree Physiol. 24, 1027–1034.

Iglesias, D.J., Cercos, M., Colmenero-Flores, J.M., Naranjo, M.A., Rios, G., Carrera, E., Ruiz-Rivero, O., Lliso, I., Morillon, R., Tadeo, F.R., Talon, M., 2007. Physiology of citrus fruiting. Braz. J. Plant Physiol. 19, 333–362.

Jia, H., Wang, N., 2014. Targeted genome editing of sweet orange using Cas9/sgRNA. PLoS One 9, e93806. https://doi.org/10.1371/journal.pone.0093806.

Jia, H., Xu, J., Orbovic, V., Zhang, Y., Wang, N., 2017. Editing citrus genome via SaCas9/sgRNA system. Front. Plant Sci. 8, 2135.

Khoshbakht, D., Asgharei, M.R., 2015. Influence of foliar-applied salicylic acid on growth, gas-exchange characteristics, and chlorophyll fluorescence in citrus under saline conditions. Photosynthetica 53, 410–418.

Khoshbakht, D., Ghorbani, A., Baninasab, B., Naseri, L.A., Mirzaei, M., 2014. Effects of supplementary potassium nitrate on growth and gas-exchange characteristics of salt-stressed citrus seedlings. Photosynthetica 52, 589–596.

Kumar, P., Rouphael, Y., Cardarelli, M., Colla, G., 2017. Vegetable grafting as a tool to improve drought resistance and water use efficiency. Front. Plant Sci. 8, 1130. https://doi.org/10.3389/fpls.2017.01130.

Levitt, J., 1972. Responses of Plants to Environmental Stresses. Academic Press, New York.

Li, B., Byrt, C., Qiu, J., Baumann, U., Hrmova, M., Evrard, A., Johnson, A.A.T., Birnbaum, K.D., Mayo, G.M., Jha, D., Henderson, S.W., Tester, M., Gilliham, M., Roy, S.J., 2016. Identification of a stelar-localized transport protein that facilitates root-to-shoot transfer of chloride in Arabidopsis. Plant Physiol. 170, 1014–1029.

Li, B., Tester, M., Gilliham, M., 2017a. Chloride on the move. Trends Plant Sci. 22, 236–248. https://doi.org/10.1016/j.tplants.2016.12.004.

Li, Z., Peng, D.D., Zhang, X.Q., Peng, Y., Chen, M., Ma, X., Huang, L.K., Yan, Y.H., 2017b. Na+ induces the tolerance to water stress in white clover associated with osmotic adjustment and aquaporins-mediated water transport and balance in root and leaf. Environ. Exp. Bot. 144, 11–24.

Lopez-Climent, M.F., Arbona, V., Perez-Clemente, R.M., Gomez-Cadenas, A., 2008. Relationship between salt tolerance and photosynthetic machinery performance in citrus. Environ. Exp. Bot. 62, 176–184.

Ma, W.-T., Fan, W.-G., 2014. Relationship between drought resistance and endogenous hormone content in different citrus species. Ying Yong Sheng Tai Xue Bao 25, 147–154.

Maas, E.V., 1993. Salinity and citriculture. Tree Physiol. 12, 195–216.

Manzi, M., Lado, J., Jesus Rodrigo, M., Arbona, V., Gomez-Cadenas, A., 2016. ABA accumulation in water-stressed *Citrus* roots does not rely on carotenoid content in this organ. Plant Sci. 252, 151–161.

Manzi, M., Pitarch-Bielsa, M., Arbona, V., Gomez-Cadenas, A., 2017. Leaf dehydration is needed to induce abscisic acid accumulation in roots of citrus plants. Environ. Exp. Bot. 139, 116–126.

Martinez-Alcantara, B., Martinez-Cuenca, M.R., Quinones, A., Iglesias, D.J., Primo-Millo, E., Forner-Giner, M.A., 2015. Comparative expression of candidate genes involved in sodium transport and compartmentation in citrus. Environ. Exp. Bot. 111, 52–62.

Martins, C.D.S., Pedrosa, A.M., Du, D.L., Goncalves, L.P., Yu, Q., Gmitter, F.G., Costa, M.G.C., 2015. Genome-wide characterization and expression analysis of major intrinsic proteins during abiotic and biotic stresses in sweet orange (*Citrus sinensis* L. Osb.). PLoS One 10, e0138786. https://doi.org/10.1371/journal.pone.0138786. eCollection 2015.

Medina, C.L., Machado, E.C., Pinto, J.M., 1998. Photosynthesis of 'Valencia' orange tree grafted on four rootstocks and submitted to water deficit. Bragantia 57, 1–14.

Merelo, P., Agusti, J., Arbona, V., Costa, M.L., Estornell, L.H., Gomez-Cadenas, A., Coimbra, S., Gomez, M.D., Perez-Amador, M.A., Domingo, C., Talon, M., Tadeo, F.R., 2017. Cell wall remodeling in abscission zone cells during ethylene-promoted fruit abscission in citrus. Front. Plant Sci. 8, 126. https://doi.org/10.3389/fpls.2017.00126. eCollection 2017.

Molassiotis, A., Job, D., Ziogas, V., Tanou, G., 2016. *Citrus* plants: a model system for unlocking the secrets of NO and ROS-inspired priming against salinity and drought. Front. Plant Sci. 7, 229. https://doi.org/10.3389/fpls.2016.00229. eCollection 2016.

Moreno-Alvero, M., Yunta, C., Gonzalez-Guzman, M., Lozano-Juste, J., Luis Benavente, J., Arbona, V., Menendez, M., Martinez-Ripoll, M., Infantes, L., Gomez-Cadenas, A., Rodriguez, P.L., Albert, A., 2017. Structure of ligand-bound intermediates of crop ABA receptors highlights PP2C as necessary ABA co-receptor. Mol. Plant 10, 1250–1253.

Mouhaya, W., Allario, T., Brumós, J., Andrés, F., Talon, M., Ollitrault, P., Morillon, R., 2010. Sensitivity to high salinity in tetraploid citrus seedlings increases with water availability and correlates with expression of candidate genes. Funct. Plant Biol. 37, 674–685.

Moya, J.L., Tadeo, F.R., Gómez-Cadenas, A., Primo-Millo, E., Talón, M., 2002. Transmissible salt tolerance traits identified through reciprocal grafts between sensitive Carrizo and tolerant Cleopatra citrus genotypes. J. Plant Physiol. 159, 991–998.

Moya, J.L., Gómez-Cadenas, A., Primo-Millo, E., Talón, M., 2003. Chloride absorption in salt-sensitive Carrizo citrange and salt-tolerant Cleopatra mandarin citrus rootstocks is linked to water use. J. Exp. Bot. 54, 825–833.

Munns, R., Tester, M., 2008. Mechanisms of salinity tolerance. Annu. Rev. Plant Biol. 59, 651–681.

Navarro, J.M., Pérez-Tornero, O., Morte, A., 2014. Alleviation of salt stress in citrus seedlings inoculated with arbuscular mycorrhizal fungi depends on the rootstock salt tolerance. J. Plant Physiol. 171, 76–85.

Neves, D.M., Coelho Filho, M.A., Bellete, B.S., Da Silva, M.F.G.F., Souza, D.T., Soares Filho, W.D.S., Costa, M.G.C., Gesteira, A.S., 2013. Comparative study of putative 9-cis-epoxycarotenoid dioxygenase and abscisic acid accumulation in the responses of Sunki mandarin and Rangpur lime to water deficit. Mol. Biol. Rep. 40, 5339–5349.

Neves, D.M., Almeida, L.A.D., Santana-Vieira, D.D.S., Freschi, L., Ferreira, C.F., Soares, W.D.S., Costa, M.G.C., Micheli, F., Coelho, M.A., Gesteira, A.D., 2017. Recurrent water deficit causes epigenetic and hormonal changes in citrus plants. Sci. Rep. 7, 13684.

Oliveira, T.M., Cidade, L.C., Gesteira, A.S., Coelho Filho, M.A., Soares Filho, W.S., Costa, M.G.C., 2011. Analysis of the NAC transcription factor gene family in citrus reveals a novel member involved in multiple abiotic stress responses. Tree Genet. Genomes 7, 1123. https://doi.org/10.1007/s11295-011-0400-8.

Oliveira, T.M., Da Silva, F.R., Bonatto, D., Neves, D.M., Morillon, R., Maserti, B.E., Filho, M.A.C., Costa, M.G., Pirovani, C.P., Gesteira, A.S., 2015. Comparative study of the protein profiles of Sunki mandarin and Rangpur lime plants in response to water deficit. BMC Plant Biol. 15, 69.

Oliveira, T.M., Ben Yahmed, J., Dutra, J., Maserti, B.E., Talon, M., Navarro, L., Ollitraut, P., Gesteira, A.D.S., Morillon, R., 2017. Better tolerance to water deficit in doubled diploid 'Carrizo citrange' compared to diploid seedlings is associated with more limited water consumption. Acta Physiol. Plant. 39, 204.

Padoan, D., Mossad, A., Chiancone, B., Germana, M.A., Khan, P.S.S.V., 2013. Ploidy levels in *Citrus* clementine affects leaf morphology, stomatal density and water content. Theor. Exp. Plant Physiol. 25, 283–290.

Pardo, J.M., 2010. Biotechnology of water and salinity stress tolerance. Curr. Opin. Biotechnol. 21, 185–196.

Paul, M.J., Primavesi, L.F., Jhurreea, D., Zhang, Y., 2008. Trehalose metabolism and signaling. Annu. Rev. Plant Biol. 59, 417–441.

Pedrosa, A.M., Santos Martins, C.D.P., Goncalves, L.P., Cardoso Costa, M.G., 2015. Late embryogenesis abundant (LEA) constitutes a large and diverse family of proteins involved in development and abiotic stress responses in sweet orange (*Citrus sinensis* L. Osb.). PLoS One 10, e0145785. https://doi.org/10.1371/journal.pone.0145785.

Pedroso, F.K.J.V., Prudente, D.A., Bueno, A.C.R., Machado, E.C., Ribeiro, R.V., 2014. Drought tolerance in citrus trees is enhanced by rootstock-dependent changes in root growth and carbohydrate availability. Environ. Exp. Bot. 101, 26–35.

Pérez-Pérez, J.G., Syvertsen, J.P., Botía, P., García-Sánchez, F., 2007. Leaf water relations and net gas exchange responses of salinized Carrizo citrange seedlings during drought stress and recovery. Ann. Bot. 100, 335–345.

Perez-Perez, J.G., Gomez-Gomez, A.G., Botia, P., Brumos, J., Talon, M., Colmenero-Flores, J.M., 2015. Transcriptional profile analysis of young and mature leaves of *Citrus* trees acclimated to salinity. In: Sabatermunoz, B., Moreno, P., Pena, L., Navarro, L. (Eds.), XII International Citrus Congress—International Society of Citriculture.

Podda, A., Checcucci, G., Mouhaya, W., Centeno, D., Rofidal, V., Del Carratore, R., Luro, F., Morillon, R., Ollitrault, P., Maserti, B.E., 2013. Salt-stress induced changes in the leaf proteome of diploid and tetraploid mandarins with contrasting Na+ and Cl- accumulation behaviour. J. Plant Physiol. 170, 1101–1112.

Qiu, J., Henderson, S.W., Tester, M., Roy, S.J., Gilliham, M., 2017. SLAH1, a homologue of the slow type anion channel SLAC1, modulates shoot Cl- accumulation and salt tolerance in Arabidopsis thaliana. J. Exp. Bot. 68, 4495–4505.

Raga, V., Bernet, G.P., Carbonell, E.A., Asins, M.J., 2014. Inheritance of rootstock effects and their association with salt tolerance candidate genes in a progeny derived from 'Volkamer' lemon. J. Am. Soc. Hortic. Sci. 139, 518–528.

Raga, V., Intrigliolo, D.S., Bernet, G.P., Carbonell, E.A., Asins, M.J., 2016. Genetic analysis of salt tolerance in a progeny derived from the citrus rootstocks Cleopatra mandarin and trifoliate orange. Tree Genet. Genomes 12, 34. https://doi.org/10.1007/s11295-016-0991-1.

Raven, J.A., 1985. Regulation of pH and generation of osmolarity in vascular plants—a cost-benefit analysis in relation to efficiency of use of energy, nitrogen and water. New Phytol. 101, 25–77.

Reyes, J.L., Rodrigo, M.J., Colmenero-Flores, J.M., Gil, J.V., Garay-Arroyo, A., Campos, F., Salamini, F., Bartels, D., Covarrubias, A.A., 2005. Hydrophilins from distant organisms can protect enzymatic activities from water limitation effects in vitro. Plant Cell Environ. 28, 709–718.

Roberts, S.K., 2006. Plasma membrane anion channels in higher plants and their putative functions in roots. New Phytol. 169, 647–666.

Rodriguez-Gamir, J., Primo-Millo, E., Forner, J.B., Forner-Giner, M.A., 2010. Citrus rootstock responses to water stress. Sci. Hortic. 126, 95–102.

Rodriguez-Gamir, J., Ancillo, G., Aparicio, F., Bordas, M., Primo-Millo, E., Forner-Giner, M.A., 2011. Water-deficit tolerance in citrus is mediated by the down regulation of PIP gene expression in the roots. Plant Soil 347, 91–104.

Romero, P., Navarro, J.M., Perez-Perez, J.G., Garcia-Sanchez, F., Gomez-Gomez, A., Porras, I., Martinez, V., Botia, P., 2006. Deficit irrigation and rootstock: their effects on water relations, vegetative development, yield, fruit quality and mineral nutrition of Clemenules mandarin. Tree Physiol. 26, 1537–1548.

Romero, P., Lafuente, M.T., Rodrigo, M.J., 2012. The *Citrus* ABA signalosome: identification and transcriptional regulation during sweet orange fruit ripening and leaf dehydration. J. Exp. Bot. 63, 4931–4945.

Romero-Aranda, R., Bondada, B.R., Syvertsen, J.P., Grosser, J.W., 1997. Leaf characteristics and net gas exchange of diploid and autotetraploid citrus. Ann. Bot. 79, 153–160.

Romero-Aranda, R., Moya, J.L., Tadeo, F.R., Legaz, F., Primo-Millo, E., Talon, M., 1998. Physiological and anatomical disturbances induced by chloride salts in sensitive and tolerant citrus: beneficial and detrimental effects of cations. Plant Cell Environ. 21, 1243–1253.

Rosales, M.A., Vázquez-Rodríguez, A., Franco-Navarro, J.D., Cubero-Font, P., Colmenero-Flores, J.M., 2012. Chloride nutrition improves water use efficiency and drought tolerance in tomato plants. In: Bonilla, I., Hernández, L.E., Lucena, J.J. (Eds.), La nutrición mineral de las plantas como base de una agricultura sostenible. Universidad Autónoma de Madrid, Madrid.

Roussos, P.A., Gasparatos, D., Kyriakou, C., Tsichli, K., Tsantili, E., Haidouti, C., 2013. Growth, nutrient status, and biochemical changes of sour orange plants subjected to sodium chloride stress. Commun. Soil Sci. Plant Anal. 44, 805–816.

Ruiz, M., Quiñones, A., Martínez-Cuenca, M.R., Aleza, P., Morillon, R., Navarro, L., Primo-Millo, E., Martínez-Alcántara, B., 2016. Tetraploidy enhances the ability to exclude chloride from leaves in Carrizo citrange seedlings. J. Plant Physiol. 205, 1–10.

Santana-Vieira, D.D.S., Freschi, L., Almeida, L.A.D., De Moraes, D.H.S., Neves, D.M., Dos Santos, L.M., Bertolde, F.Z., Soares, W.D., Coelho, M.A., Gesteira, A.D., 2016. Survival strategies of citrus rootstocks subjected to drought. Sci. Rep. 6, 38775.

Schwenke, G.D., Simpfendorfer, S.R., Collard, B.C.Y., 2015. Confirmation of chloride deficiency as the cause of leaf spotting in durum wheat grown in the Australian northern grains region. Crop Pasture Sci. 66, 122–134.

Seday, U., Gulsen, O., Uzun, A., Toprak, G., 2014. Response of citrus rootstocks to different salinity levels for morphological and antioxidative enzyme activites. J. Anim. Plant Sci. 24, 512–520.

Shabala, S.N., Lew, R.R., 2002. Turgor regulation in osmotically stressed Arabidopsis epidermal root cells. Direct support for the role of inorganic ion uptake as revealed by concurrent flux and cell turgor measurements. Plant Physiol. 129, 290–299. https://doi.org/10.1104/pp.020005.

Shabala, S., Shabala, L., 2011. Ion transport and osmotic adjustment in plants and bacteria. Biomol. Concepts 2, 407–419.

Shi, H., Quintero, F.J., Pardo, J.M., Zhu, J.-K., 2002. The putative plasma membrane Na+/H+ antiporter SOS1 controls long-distance Na+ transport in plants. Plant Cell 14, 465–477.

Simon-Grao, S., Simon, I., Lidon, V., Conesa, A., Manera, J., Brotons, J.M., Martinez-Nicolas, J.J., Garcia-Sanchez, F., 2016. Effects of shade screens and mulching on the color change of fruits from "Fino 49" lemon trees irrigated with water of different salinity or irrigation regimes. Sci. Hortic. 209, 316–322.

Simpson, C.R., Nelson, S.D., Melgar, J.C., Jifon, J., King, S.R., Schuster, G., Volder, A., 2014. Growth response of grafted and ungrafted citrus trees to saline irrigation. Sci. Hortic. 169, 199–205.

Skerrett, M., Tyerman, S.D., 1994. A channel that allows inwardly directed fluxes of anions in protoplasts derived from wheat roots. Planta 192, 295–305.

Storey, R., Walker, R.R., 1999. *Citrus* and salinity. Sci. Hortic. 78, 39–81.

Sykes, S.R., 1992. The inheritance of salt exclusion in woody perennial fruit species. Plant Soil 146, 123–129.

Sykes, S.R., 2011. Chloride and sodium excluding capacities of citrus rootstock germplasm introduced to Australia from the People's Republic of China. Sci. Hortic. 128, 443–449.

Syvertsen, J.P., García-Sánchez, F., 2014. Multiple abiotic stresses occurring with salinity stress in citrus. Environ. Exp. Bot. 103, 128–137.

Syvertsen, J.P., Lee, L.S., Grosser, J.W., 2000. Limitations on growth and net gas exchange of diploid and tetraploid Citrus rootstock cultivars grown at elevated CO2. J. Am. Soc. Hortic. Sci. 125, 228–234.

Tadeo, F.R., Cercos, M., Colmenero-Flores, J.M., Iglesias, D.J., Naranjo, M.A., Rios, G., Carrera, E., Ruiz-Rivero, O., Lliso, I., Morillon, R., Ollitrault, P., Talon, M., 2008. Molecular physiology of development and quality of citrus. In: Kader, J.C., Delseny, M. (Eds.), Botanical Research: Incorporating Advances in Plant Pathology. vol. 47. Academic Press, pp. 147–223. https://doi.org/10.1016/s0065-2296(08)00004-9.

Tan, F.-Q., Tu, H., Liang, W.-J., Long, J.-M., Wu, X.-M., Zhang, H.-Y., Guo, W.-W., 2015. Comparative metabolic and transcriptional analysis of a doubled diploid and its diploid citrus rootstock (C. junos cv. Ziyang xiangcheng) suggests its potential value for stress resistance improvement. BMC Plant Biol. 15, 89.

Tanou, G., Molassiotis, A., Diamantidis, G., 2009. Hydrogen peroxide- and nitric oxide-induced systemic antioxidant prime-like activity under NaCl-stress and stress-free conditions in citrus plants. J. Plant Physiol. 166, 1904–1913.

Tanou, G., Ziogas, V., Belghazi, M., Christou, A., Filippou, P., Job, D., Fotopoulos, V., Molassiotis, A., 2014. Polyamines reprogram oxidative and nitrosative status and the proteome of citrus plants exposed to salinity stress. Plant Cell Environ. 37, 864–885.

Tsabarducas, V., Chatzistathis, T., Therios, I., Koukourikou-Petridou, M., Tananaki, C., 2015. Differential tolerance of 3 self-rooted *Citrus* limon cultivars to NaCl stress. Plant Physiol. Biochem. 97, 196–206.

Verslues, P.E., Agarwal, M., Katiyar-Agarwal, S., Zhu, J., Zhu, J.-K., 2006. Methods and concepts in quantifying resistance to drought, salt and freezing, abiotic stresses that affect plant water status. Plant J. 45, 523–539. https://doi.org/10.1111/j.1365-313X.2005.02593.x.

Walker, R.R., 1986. Sodium exclusion and potassium-sodium selectivity in salt treated trifoliate orange (Poncirus trifoliata) and Cleopatra mandarin (*Citrus reticulata*) plants. Aust. J. Plant Physiol. 1 (13), 293–303.

Wang, Q., Guan, C., Wang, S.M., 2014. Coordination of AtHKT1;1 and AtSOS1 facilitates Na+ and K+ homeostasis in Arabidopsis thaliana under salt stress. J. Plant Biol. 57, 282–290.

Weeks, D.P., Spalding, M.H., Yang, B., 2016. Use of designer nucleases for targeted gene and genome editing in plants. Plant Biotechnol. J. 14, 483–495.

Wege, S., Gilliham, M., Henderson, S.W., 2017. Chloride: not simply a 'cheap osmoticum', but a beneficial plant macronutrient. J. Exp. Bot. 68, 3057–3069.

Wu, Q.S., Srivastava, A.K., Zou, Y.N., 2013. AMF-induced tolerance to drought stress in citrus: a review. Sci. Hortic. 164, 77–87.

Xiao, J.P., Zhang, L.L., Zhang, H.Q., Miao, L.X., 2017. Identification of genes involved in the responses of tangor (C-reticulata x C-sinensis) to drought stress. Biomed. Res. Int. 2017, 8068725. https://doi.org/10.1155/2017/8068725.

Xie, R., Zhang, J., Ma, Y., Pan, X., Dong, C., Pang, S., He, S., Deng, L., Yi, S., Zheng, Y., Lv, Q., 2017. Combined analysis of mRNA and miRNA identifies dehydration and salinity responsive key molecular players in citrus roots. Sci. Rep. 7, 42094.

Zaher-Ara, T., Boroomand, N., Sadat-Hosseini, M., 2016. Physiological and morphological response to drought stress in seedlings of ten citrus. Trees 30, 985–993.

Zandalinas, S.I., Rivero, R.M., Martinez, V., Gomez-Cadenas, A., Arbona, V., 2016. Tolerance of citrus plants to the combination of high temperatures and drought is associated to the increase in transpiration modulated by a reduction in abscisic acid levels. BMC Plant Biol. 16, 105. https://doi.org/10.1186/s12870-016-0791-7.

Zekri, M., Parsons, L.R., 1990a. Calcium influences growth and leaf mineral concentration of citrus under saline conditions. HortScience 25, 784–786.

Zekri, M., Parsons, L.R., 1990b. Comparative effects of NaCl and polyethylene-glycol on root distribution, growth, and stomatal conductance of sour orange seedlings. Plant Soil 129, 137–143.

Zhang, W.-W., Yang, H.-Q., You, S.-Z., Fan, S.-L., Ran, K., 2015. MhNCED3, a gene encoding 9-cis-epoxycarotenoid dioxygenase in Malus hupehensis Rehd., enhances plant tolerance to Cl- stress by reducing Cl- accumulation. Plant Physiol. Biochem. 89, 85–91.

Zhang, Y.C., Wang, P., Wu, Q.H., Zou, Y.N., Bao, Q., Wu, Q.S., 2017. Arbuscular mycorrhizas improve plant growth and soil structure in trifoliate orange under salt stress. Arch. Agron. Soil Sci. 63, 491–500.

Chapter 15

Soil and nutrition interactions

Dirceu Mattos, Jr.[a], Davie M. Kadyampakeni[b], Ana Quiñones Oliver[c], Rodrigo Marcelli Boaretto[a], Kelly T. Morgan[d], Jose Antonio Quaggio[e]

[a]*Sylvio Moreira Citrus Research Center, Instituto Agronômico (IAC), Cordeirópolis, Brazil,* [b]*Citrus Research and Education Center, University of Florida, Lake Alfred, FL, United States,* [c]*Centro de Desarrollo de Agricultura Sostenible, Instituto Valenciano de Investigaciones Agrarias (IVIA), Valencia, Spain,* [d]*Southwest Florida Research & Education Center, University of Florida, Immokalee, FL, United States,* [e]*Center of Soils and Environmental Resources, Instituto Agronômico (IAC), Campinas, Brazil*

15.1 Introduction

Citrus trees, despite their mesophytic characteristics, are cultivated in many regions of the world under different edaphic and climatic conditions between the latitudes of 40° N and 40° S (Agustí, 2003), where their production surpasses that of other fruit groups. Globally, the main producers are China and Brazil (29.6 and 19.0×10^6 tons, respectively), followed by the United States (9.4×10^6 Mg). Mediterranean countries produce approximately 20% of the global harvest (24.2×10^6 Mg) (FAO, 2015).

The potential of citrus trees to produce fruits is determined by the genetic value of the canopy and the rootstock variety (Spiegel-Roy and Goldschmidt, 1996; Davies and Albrigo, 1998). In addition to the interest in growing different varieties, either for the fresh fruit market or juice processing, the use of rootstocks has become fundamental to overcome the juvenility of plants, regulate tree vigor, increase nutrient uptake efficiency and fruit quality, and induce resistance/tolerance of trees to biotic and abiotic stresses.

High citrus yields with efficient fruit production require proper soil management in fruit orchards. To guarantee the sustainability of citriculture, it is therefore crucial to obtain information on soil attributes, nutrient availability, plant development, nutritional status, and fertilization practices (Kadyampakeni et al., 2015; Mattos Jr. et al., 2007a,b; Morgan and Kadyampakeni, 2012).

In general, shallow soils, as well as clayey soils with poor drainage and high compaction, impede root growth and limit tree vigor. Similarly, sandy soils with low water retention capacity as well as highly acidic or alkaline soils also negatively impact tree vigor and fruit yield (Morgan et al., 2006a). However, as long as adequate irrigation and nutrient supply are provided, citrus trees demonstrate great adaptability to different environments and sustain high productivity. Adequate irrigation schedules are particularly important for soils with low water-holding capacities (Morgan et al., 2006b; Kadyampakeni and Morgan, 2017).

This chapter aligns the global contributions on the nutrient management of citrus, considering soil and plant interactions as supported by the diagnostic of soil fertility and plant nutritional status. Based on this, fertilization programs of orchards are established to guarantee improved productivity and fruit quality.

15.2 Soils of major citrus-producing regions in the world

Soils around the world vary in color, mineralogy, texture, and structure, which influence their chemical, physical, and biological characteristics. The intrinsic interactions of such characteristics make up the environment of the root system and regulate water and nutrient availability, thereby affecting citrus tree performance.

Understanding the soil type, based on its parent material and development, allows one to predict soil behavior, limitations, and qualities. Moreover, establishing citrus orchards based on the capacity of the land allows growers to adopt soil management strategies that prevent soil degradation as well as to run economically viable businesses.

Citrus are produced commercially throughout the tropical and subtropical regions from the northern to the southern Americas (Argentina, Brazil, Mexico, and the USA), in Asia (China, India, Japan, and South Korea) and Australia as well in the Mediterranean regions (Morocco, Israel, Italy, Spain, and Turkey), and in Africa (South Africa), where soils vary greatly. Information on soil units is available from the World Soil Classification (IUSS Working Group WRB, 2015), with reference to pedogenetic processes, although different countries use modified systems, such as the USDA Soil Taxonomy (Soil Survey Staff, 2014). Here, we describe a collection of soils under citrus cultivation based on their intrinsic properties and relationships with orchard performance.

The Genus Citrus. https://doi.org/10.1016/B978-0-12-812163-4.00015-2

15.2.1 Soils of tropical and subtropical regions

Tropical and subtropical regions are characterized by hot summers and mild to cool winters; frosts are rare and according to the period during which there is sufficient water for plant growth, soil moisture regimes are ustic to udic (Soil Survey Staff, 2014).

The main citrus-producing areas in the tropics are found on Oxisols and Ultisols, followed by Entisols and Alfisols and, to a minor extent, Inceptisols, which likely present a lateritic, cemented continuous soil layer in the subsurface, impairing root growth. In some temperate areas such as Argentina, Mollisols, high in organic matter, predominate.

The soils found in São Paulo and Triângulo Mineiro, where most citrus are grown in Brazil, are highly heterogeneous, limiting sustainable use and management. They generally develop from either basalt or sedimentary rocks and are sandy (<10% clay) to clay-type soils (50%–75% clay), such as Oxisols, Entisols, Alfisols, and Ultisols (Tables 15.1 and 15.2). Most of these soils are highly weathered and inherently infertile, with low levels of organic matter and potassium (K), calcium (Ca), and magnesium (Mg), and a low cation exchange capacity (CEC; pH-dependent charge); the phosphorus (P) fixing capacity is generally high, with low micronutrient availability. These soils are usually acidic, with high levels of toxic aluminum (Al), and present a low water-holding capacity when the texture is sandy; when clayey, they are susceptible to compaction.

In Florida (USA), a humid subtropical region, citrus is irrigated and grown in well-drained sands low in mineral nutrients and organic matter (Entisols), in fine-textured lowland soils with a sandy surface, often developed on calcareous marine deposits (Alfisols), or in sandy, acidic, coarse-textured, and poorly drained soils (Spodosols) (Alva and Tucker, 1999) (Tables 15.3 and 15.4). Entisols are sandy mineral soils low in organic matter, natural fertility, and water-holding capacity (Weil and Brady, 2016). They have weak or no diagnostic subsurface layers and are well drained to excessively well-drained (Obreza and Collins, 2008). Spodosols are sandy mineral soils low in organic matter and natural fertility in the surface layer; they contain an acidic subsurface restrictive layer composed of Al and iron (Fe) "cemented" together with organic matter (Obreza

TABLE 15.1 Physical and chemical properties of typical root zones[a] in common soil orders found in citrus groves in the São Paulo state, Brazil.

	Texture					
Soil orders[b]	Sand ($g\,kg^{-1}$)	Silt ($g\,kg^{-1}$)	Clay ($g\,kg^{-1}$)	Organic matter ($g\,dm^{-3}$)	pH	Cation exchange capacity ($mmol_c\,dm^{-3}$)
Ultisol	663–810	34–70	120–300	6–13	5.2–5.7	40–66
Oxisol	666–787	40–64	160–270	5–11	4.1–4.2	24–54
Entisol	848–867	13–18	120–135	8–10	4.3–4.9	32–49

[a] *Up to 90-cm soil depth.*
[b] *According to the USDA Soil Taxonomy (Soil Survey Staff, 2014).*

Adapted from Corá, J.E., Silva, G.O., Martins Filho, M.V., 2005. Manejo do solo sob citros. In: Mattos Jr., D., De Negri, J.D., Pio, R.M., Pompeu Jr., J. (Ed.), Citros. Instituto Agronômico e Fundag, Campinas, pp. 347–368.

TABLE 15.2 Physical and chemical properties of the subsurface diagnostic layers of typical common soil orders found in citrus groves in the São Paulo state, Brazil.

		Texture					
Soil orders	Layer[a]	Sand ($g\,kg^{-1}$)	Silt ($g\,kg^{-1}$)	Clay ($g\,kg^{-1}$)	Organic matter ($g\,dm^{-3}$)	pH	Cation exchange capacity ($mmol_c\,dm^{-3}$)
Ultisol	Bt	611–681	37–49	270–345	4–8	5.5–5.9	32–45
Oxisol	Bw	691–705	39–40	255–270	4–5	5.1–5.7	23–24
Entisol	C	840–867	13–17	120–145	6–10	4.2–4.9	28–35

[a] *Diagnostic subsurface horizons according to the USDA Soil Taxonomy (Soil Survey Staff, 2014).*

Adapted from Corá, J.E., Silva, G.O., Martins Filho, M.V., 2005. Manejo do solo sob citros. In: Mattos Jr., D., De Negri, J.D., Pio, R.M., Pompeu Jr., J. (Ed.), Citros. Instituto Agronômico e Fundag, Campinas, pp. 347–368.

TABLE 15.3 Physical and chemical properties of typical root zones[a] in common soil series found in citrus groves in Florida.

Soil orders[b]	Soil texture: Sand ($g\,kg^{-1}$)	Silt ($g\,kg^{-1}$)	Clay ($g\,kg^{-1}$)	Organic matter ($g\,dm^{-3}$)	Water-holding capacity: $cm\,m^{-1}$	cm in the root zone	pH	Cation exchange capacity ($mmol_c\,dm^{-3}$)
Entisols	970–985	50–125	75–125	5–10	2.5–6.6	1.5–5.3	3.6–7.3	20–40
Alfisols	850–965	20–60	15–90	5–3	2.5–10.7	1.3–4.6	4.5–8.4	20–180
Spodosols	960–985	10–35	05–10	10–30	2.5–6.6	1.3–3.8	3.6–7.3	20–60

[a] *Top 90cm of soil for central Ridge Entisols and top 45cm of soil for flatwoods Alfisols, Spodosols, and Entisols.*
[b] *According to the USDA Soil Taxonomy (Soil Survey Staff, 2014).*

Adapted from Obreza, T.A., Collins, M.E., 2008. Common soils used for citrus production in Florida. SL 193. Florida Coop. Ext. Serv., Gainesville.

and Collins, 2008). Alfisols are sandy mineral soils low in organic matter in the surface layer, but higher in relative natural fertility compared with Spodosols (Obreza and Collins, 2008). Alfisols contain a subsurface layer of loamy material (a mixture of mostly clay and sand with little silt) that has a relatively high water-holding capacity (Weil and Brady, 2016).

China also has important citrus-growing areas in tropical and subtropical regions, where the physical diversity is characterized by lowlands, flatlands, foothills, and coasts, encompassing red to oxidic and alluvial soils, including some lateritic soils with varying fertility, depending on the parent materials from which they developed. The red to oxidic soils, classified as Oxisols, Ultisols, and some Alfisols in the US taxonomy, dominated by varying charge minerals, are acidic, with low CEC, nutrient availability, and organic matter (He et al., 2004).

Interestingly, although they are not commonly cultivated with citrus, the Andisols found on the Jeju Island, South Korea, are derived from volcanic ashes. Those are brown- to dark-colored soils, clay to silty textured, with high Al concentrations, associated with the occurrence of poorly crystalline Al and Fe minerals, low in nutrients and high in organic matter (Chon et al., 2008). Andisols present predominantly positive surface charges at soil pH, which, along with the high water content, requires adequate nutrient management practices for high citrus yields.

15.2.2 Soils of the Mediterranean and similar regions

The Mediterranean climate can be found in different regions around the world, although it is most distinct in the Mediterranean Sea Basin. Citrus is cultivated in more than half of the Mediterranean-climate regions. Similar regions are California and the Western Cape Province of South Africa (Verheye and de la Rosa, 2005). In the Mediterranean Basin, the main citrus-producing countries are Spain, Egypt, Italy, Turkey, and Morocco. Although the soils largely differ, in California (USA),

TABLE 15.4 Physical and chemical properties of the subsurface diagnostic layers of typical Alfisols and Spodosols found in flatwoods citrus groves[a] in Florida.

Soil orders	Layer[b]	Soil texture: Sand ($g\,kg^{-1}$)	Silt ($g\,kg^{-1}$)	Clay ($g\,kg^{-1}$)	Organic matter ($g\,dm^{-3}$)	Water-holding capacity ($cm\,m^{-1}$)	pH	Cation exchange capacity ($mmol_c\,dm^{-3}$)
Alfisols	Bt	770–810	35–70	130–195	10–12	10.0–20.0	5.1–8.4	40–260
Spodosols	Bh	905–950	15–50	15–50	10–45	10.0–25.0	3.3–5.3	50–250

[a] *These layers may reside in an undisturbed state beneath the root zone, or they may be partially excavated and mixed into the root zone soil during the bedding process.*
[b] *Diagnostic subsurface horizons according to the USDA Soil Taxonomy (Soil Survey Staff, 2014).*

Adapted from Obreza, T.A., Collins, M.E., 2008. Common soils used for citrus production in Florida. SL 193. Florida Coop. Ext. Serv., Gainesville.

TABLE 15.5 Physical and chemical properties for common soil orders in Spain (Duiker et al., 2001).

	Soil texture					
Soil orders[a]	Sand ($g\,kg^{-1}$)	Silt ($g\,kg^{-1}$)	Clay ($g\,kg^{-1}$)	Organic matter ($g\,dm^{-3}$)	pH	Cation exchange capacity ($mmol_c\,dm^{-3}$)
Alfisol	502	230	268	26	6.4	17.8
Inceptisol	345	272	383	26	7.8	30.3
Vertisol	93	370	537	31	7.7	46.5

[a] *According to the USDA Soil Taxonomy (Soil Survey Staff, 2014).*

they are generally more fertile than in the other Mediterranean-climate regions such as South Africa and the Mediterranean Basin, where the soils are largely alkaline.

The most common soil pattern in Mediterranean areas, expressed in terms of temperature and soil moisture regimes, is xeric, where most of the rainfall occurs in winter, followed by an important dry period during summer. Mean annual soil temperature ranges between 15°C and 22°C (Soil Survey Staff, 2014). The CEC and base saturation (V, %) are generally high, except for some small areas where the lithology is dominated by acid minerals (Table 15.5). Since the CEC is almost saturated, most of these soils are rich in nutrients such as P and K, usually supplied from mineral weathering, with high illitic clay contents and adequate levels of Ca and Mg. However, Fe, copper (Cu), manganese (Mn), zinc (Zn), and boron (B) are frequently deficient in calcareous soils. Additionally, in irrigated land, salinization may become an important problem due to the low quality of irrigation water and/or water-saving systems, in combination with the specific climatic characteristics of the Mediterranean.

In some parts of California (USA), moderate soil weathering and clay illuviation to the deeper B horizons are the main processes in the building up of Alfisol soils. In these soils, the hematite-induced reddening of the clays is due to summer dehydration of free Fe oxyhydroxides. In other types of soils, such as Xerochrept (Calcisols), the main soil-forming processes are carbonate dissolution and reprecipitation, which build up calcic horizons in semiarid regions. Another minor soil type of citrus cultivation regions is Vertisol, which mainly occurs in lowlands where deep layers of swelling/cracking clays have sedimented.

Other soils used for citrus production are shallow soils on nearly bare slopes, mostly as a result of erosion after deforestation. They can be classified as Entisols or Inceptisols, depending on the degree of development after erosion (Spain, Italy, Morocco, Egypt, South Africa, and California).

15.3 The role of mineral nutrients in citrus production

About 95% of plant dry matter is formed by carbon, hydrogen, and oxygen fixed through photosynthesis. The mineral nutrients, taken up mostly by plant roots and, to a lesser extent, by leaves, represent the other 5%. Mineral nutrient disorders cause abnormal development of plants and, when severe, result in plant decay and significant yield loss. Adequate nutrient supply positively affects growth, plant morphology and anatomy, meeting the requirements to produce large amounts of high-quality fruits.

Based on the quantities required for plant growth (from fractions of a milligram to grams per kilogram of dry matter), mineral nutrients are classified as (i) macronutrients, such as nitrogen (N), P, K, Ca, Mg, and sulfur (S) and (ii) micronutrients, such as Fe, Cu, Mn, Zn, B, chloride (Cl), molybdenum (Mo), and nickel (Ni). These elements are absorbed by plants as NO_3^-, NH_4^+, NO_x, NH_3, $H_2PO_4^-$, K^+, Ca^{+2}, Mg^{+2}, SO_4^{-2}, SO_2, H_2S, Fe^{+2}, Cu^{+2}, Mn^{+2}, Zn^{+2}, H_3BO_3, Cl^-, MoO_4^{-2}, and Ni^{+2}. Other elements might also be taken up, but they do not meet the essential criteria to be classified as nutrients (Marschner, 2012). Cobalt (Co), selenium (Se), silicon (Si), and sodium (Na) are beneficial elements. High concentrations of an element in plant tissues may affect the metabolic pathways and plant cell functions, causing toxicity.

15.3.1 Nitrogen

Citrus growth and production are largely regulated by N supply (Alva and Paramasivam, 1998; Cantarella et al., 2003; Quaggio et al., 2014), which affects photosynthetic rates and carbohydrate synthesis (Kato, 1986), specific leaf weight (Syvertsen and Smith Jr., 1984), and biomass production and allocation in different plant organs (Lea-Cox et al., 2001;

Morgan et al., 2006a; Dovis et al., 2016). Significant amounts of N are exported with fruit harvest (1.2–1.9 kg per ton of fresh fruits) (Paramasivam et al., 2000; Mattos Jr. et al., 2003a; Morgan et al., 2006a; Kadyampakeni et al., 2016).

Research has focused on the sustainability of citriculture, supported by the high N requirements, either native in the soil or added by fertilizers, by increasing the efficiency of the nutrient acquisition by roots and/or biomass production (Morgan et al., 2007; Hirel et al., 2011; Santa-María et al., 2015). A simple increase in N-fertilizer applications can cause nutritional imbalances in plants, resulting in nutrient losses (Syvertsen and Smith, 1996; De Paz and Ramos, 2004). In this context, the efficient use of nitrogenous fertilizers has become a first-order concern in modern citrus production due to nitrate (NO_3^-) contamination of groundwater and surface waters, increased production costs, and crop profitability (Ramos et al., 2002; Quiñones et al., 2005; Kadyampakeni et al., 2014a). Nowadays, fertilization guidelines are addressed to match crop demand with nutrient supply, taking into account soil conditions, irrigation, fertilizer timing, organic management, and other factors (Quiñones et al., 2008; Morgan et al., 2009a; Quaggio et al., 2010; Martínez-Alcántara et al., 2016).

Studies, mostly using isotope labeling (^{15}N), have characterized the fate of N in the soil and plant system, considering seasonal uptake, storage in plant parts, and redistribution to developing organs such as leaves and fruits (Legaz et al., 1995; Mattos Jr. et al., 2003b; Roccuzzo et al., 2017). Reduced N fertilization may not immediately affect fruit production. However, when N doses are lower than recommended, trees may undergo a gradual reduction of leaves and growth, which will consequently lead to losses in fruit production in the later years.

Nitrogen uptake and assimilation are greatest during spring and summer when water availability and temperature favor metabolism and plant growth; new plant parts represent the strongest sinks for N (Mattos Jr. et al., 2003b; Morgan et al., 2007). Salinity, associated with irrigation water, also plays a role in N uptake, decreasing chlorophyll content, root hydraulic conductivity, plant transpiration, N acquisition, and growth. On the other hand, NO_3^- could compete with chloride uptake and/or reduce the concentration in roots through the reallocation of chloride to new leaves, thus enhancing salt stress acclimation of citrus (Iglesias et al., 2004).

The horticultural characteristics of citrus rootstocks make N use efficiency of trees a complex process, with more vigorous varieties having higher N requirements than slow-growing ones (Syvertsen et al., 1993; Syvertsen and Smith, 1995). Similarly, the canopy of the variety regulates N use efficiency. For instance, the production of oranges (Quaggio et al., 1998), mandarins, and Murcott tangor (Mattos Jr. et al., 2004) is nonsignificant at leaf N contents above 28 $g\,kg^{-1}$. In the case of lemons and acid limes, the adequate N content in the leaves is approximately 22 $g\,kg^{-1}$ (Quaggio et al., 2002; Mattos Jr. et al., 2003c). Increased CO_2 assimilation per unit of N in leaves and preferential allocation of carbon to the canopy in relation to the roots explain the greater vigor of lemon compared with orange trees (Dovis et al., 2016). Combining both parameters, whole-plant carbon assimilation is high. However, other characteristics of these species still need to be explained in view of partitioning leaf N among structural components of plant biomass, N assimilation per unit of chlorophyll, or intrinsic CO_2 stomatal conductance.

More recently, studies have focused on the molecular, regulatory, and functional characterization of proteins, identified as high (HATS) and low (LATS) affinity N transporters across cell membranes. Although there is a significant amount of information available, our knowledge of citrus trees in this regard is still incomplete. Nitrate influx into root cells follows a multiphase transport system, either constitutive or inducible, depending primarily on the N demand, but also on the external NO_3^- concentration (Cerezo et al., 2007). An ammonium (NH_4^+) transporting system, mediated by the *CitAMT1* gene, has been identified, whose expression is upregulated or downregulated, depending on the NH_4^+ concentration in the root substrate (Camañes et al., 2007). These authors concluded that citrus exhibit a greater capacity to take up NH_4^+ compared with NO_3^-.

Despite the recent advances in the investigation of the molecular mechanisms of N uptake, considering the role of individual genes, a comprehensive understanding at the level of system biology has yet to be achieved. For instance, a long-term field study of N fertigation of citrus trees has demonstrated that the highest fruit yield was attained with an $NO_3^-:NH_4^+$ ratio of 8:1 in the soil solution (Quaggio et al., 2014). Noteworthy, this ratio correlated with that found in the sap extract of trees that favored Ca uptake by roots. The Ca uptake of plants is limited to the maintenance of the equilibrium of charge (cation-anion balance) required for greater uptake of other nutrients (Hinsinger et al., 2003), which is highlighted in the case of citrus trees due to the high demand of this nutrient (Mattos Jr. et al., 2003a).

15.3.2 Phosphorus

Despite the structural role of P in DNA and proteins as well as in phosphorylating and dephosphorylating reactions in plant cells, required for normal growth and fruit production, P supply in tropical orchards has not received much attention in the past decades. This is mainly because early decisions for nutrient management were based on the information that citrus trees did not respond to P. In many regions, citrus are cultivated in soils developed from P-rich sediments (Jackson et al.,

1995), whereas tropical soils are generally P-deficient. Therefore, response curves between soil P reserve and fruit yield were fitted, which described the contribution of P supply to plant growth and fruit yield (Quaggio et al., 1998, 2006a).

Rootstock varieties play an important role in P acquisition from the soil solution (Quaggio et al., 2004; Mattos Jr. et al., 2006; Obreza et al., 2008a), and the regulation of P distribution and remobilization in plant parts is characterized by the activity of the acid phosphatase in roots and leaves (Zambrosi et al., 2012a). Trees on Rangpur lime and Swingle citrumelo rootstocks more efficiently take up and use P compared with trees on mandarin-type rootstocks such as Sunki and Cleopatra (Zambrosi et al., 2012b, 2013).

The use of phosphite (Phi; PO_3^-), a reduced form of P, is common in fruit cultivation because of the positive effects on the control of *Phytophthora* root rot and the impacts on flowering and fruit quality. However, there is some misunderstanding on the contribution of Phi formulations, since PO_3^- cannot be oxidized in plants and used as a nutrient (P-PO_4^-) (Zambrosi et al., 2011). In fact, P-deficient trees sprayed with PO_3^- formulations exhibit impaired leaf tissue organization and growth (Zambrosi et al., 2017). Furthermore, Phi can only act as an inducer of biotic and abiotic stress tolerance in citrus orchards when the plants receive sufficient amounts of P.

15.3.3 Potassium

Potassium is required for the maintenance of the osmotic potential of the cells, regulating stomatal opening and closure as well as cell elongation. Carbohydrate transport, enzyme activation, stabilization of cell pH, and other processes in citrus are also related to K nutrition.

Potassium is removed by citrus fruit harvest at similar amounts as N (1.2–1.9 kg of K per ton of fresh fruits). Its supply increases fruit size and peel thickness, positively affecting fresh fruit quality, but negatively impacting juice production (Alva et al., 2006a,b; Quaggio et al., 2006a). Optimizing size and internal quality (expressed by the concentrations of soluble solids and citric acid) is achieved by balancing N and K fertilization since increased doses of N tend to increase the number of fruits in the plant, albeit at the cost of fruit size. Excess K can lead to yield losses due to nutritional imbalances, marked by decreases in Ca and Mg contents (Mattos Jr. et al., 2004). Potassium depletion can result in thin fruit rind which is prone to splitting, thereby reducing fruit quality and yield (Morgan et al., 2005).

Rootstock responses to K supply differ, with Swingle citrumelo exhibiting increased fruit yield with increased K rates compared with Rangpur lime and Cleopatra mandarin (Mattos Jr. et al., 2006). Despite the differences in growth and yield of trees grafted onto those varieties, studies on the interaction of B and K supply show that K uptake is facilitated by B because of an increased hyperpolarization of the plasma membrane (Schon et al., 1990). Since Swingle is also responsive to B (Mesquita et al., 2016), physiological and molecular characteristics, more than simply differences in root growth, are likely to regulate K transport in Swingle.

Caballero et al. (2013) have isolated and characterized genes encoding putative K transporters in citrus rootstocks with high homology with the *Arabidopsis* AtKUP3. However, when comparing Carrizo citrange and Cleopatra mandarin, no specific response to high-affinity K uptake was established between the tested varieties.

15.3.4 Calcium, magnesium, and sulfur

Citrus trees, unlike most cultivated crops, show higher leaf tissue concentrations of Ca compared with those of other nutrients (Table 15.6). Furthermore, trees are sensitive to soil acidity, in which excess Al and Ca deficiency are relevant, reducing plant growth by the damaging roots (Quaggio et al., 1992a).

Calcium in plants is mostly found as pectates that form the cell wall, playing a role in cell division and elongation (White and Broadley, 2003). This nutrient regulates membrane permeability and stomatal opening and closing as a secondary messenger, also associated with calmodulins that are important for cell repair after abiotic and biotic stresses (Marschner, 2012).

In citrus trees, Ca deficiency (leaf Ca = 21.0 g kg^{-1}) reduces concentrations of NO_3^- as well as chlorophyll and soluble protein levels and the activities of nitrate reductase and RuBisco compared with well-nourished trees (leaf Ca = 52.0 g kg^{-1}; Lavon et al., 1999). The uptake of Ca by citrus is likely dependent on the uptake of N-(NH_4^+:NO_3^-), whose ratio affects the maintenance of the balance of charges in the plant (van Beusichem et al., 1988; Hinsinger et al., 2003). The charge balance estimate based on the sum of ions absorbed by trees, expressed in terms of equivalents: $\sum(\text{cations}) = (NH_4^+ + Ca^{+2} + Mg^{+2} + K^+ + Na^+)$ and $\sum(\text{anions}) = (NO_3^- + SO_4^- + H_2PO_4^- + Cl^-)$ (Kirkby and Knight, 1977; Haynes and Goh, 1978), tends to be in equilibrium. Moreover, the Ca and N interdependence in citrus is supported by the positive correlation of fruit yield and N use efficiency with the NO_3^-:Ca^{+2} ratio in the sap extract and the total Ca: N ratio in the leaves (Quaggio et al., 2014).

Because of the structural role Ca plays in the cell wall, it likely affects the resistance of plants to diseases by increasing anatomical characteristics that favor the formation of a physical barrier imposed by the host on pathogens. Such anatomical characteristics

TABLE 15.6 Optimum leaf nutrient concentrations in citrus trees.

Nutrient	Brazil[a,b]	China[c,d]	India[c,e]	South Africa[a,f]	Spain[c,g]	USA[c,h]
			$g\,kg^{-1}$			
N	25–30[i]	24–26	20–24	20–27	28–30	25–27
P	1.2–1.6	1.3–1.5	0.9–1.7	1.1–1.6	1.3–1.6	1.2–1.6
K	12–16	–	11–18	7–15	7.1–10	12–17
Ca	35–50	34–48	17–30	35–55	30–50	30–49
Mg	3.5–5.0	2.0–2.9	3.2–6.9	3.0–5.5	2.5–4.5	3.0–4.9
S	2.0–3.0	2.3–2.5	–	–	2.0–3.0	–
			$mg\,kg^{-1}$			
B	75–150	17–19	–	50–150	31–100	36–100
Cu	10–20	3.7–10	4–16	5–16	6–14	5–16
Fe	50–150	4046	53–137	80–300	61–100	60–120
Mn	35–70	14–23	42–87	30–150	26–60	25–100
Zn	50–75	23–30	12–29	20–70	26–70	25–100
Mo	0.5–2.0	–	–	–	0.1–3.0	0.1–2.0

–, not available.
[a] *Based on 4- to 6-month-old spring flush leaves from fruiting terminals with fruit 2–4 cm in diameter.*
[b] *Adapted from Quaggio et al. (2010).*
[c] *Based on 4- to 6-month-old spring flush leaves.*
[d] *Adapted from Srivastava and Singh (2004).*
[e] *Adapted from Srivastava (2013).*
[f] *Adapted from Du Plessis and Koen (1996).*
[g] *Adapted from Quiñones et al. (2012).*
[h] *Adapted from Obreza et al. (2008b).*
[i] *For lemons and acid limes, optimum range = 20–24 g kg^{-1}.*

are related to the development of thicker cell walls and medium lamellae of leaves and flower petals, attained with higher Ca concentrations (Petená et al., 2016). In addition, Ca also affects the formation and arrangement of vascular vessels (Petená, 2016).

Magnesium is essential for chlorophyll and an important cofactor of several enzymes. It also plays a prominent role in the phloem loading by facilitating the transport of carbohydrates from the leaf mesophyll to companion cells and vascular tissue. Thus, Mg promotes a better distribution of sugars in the plant, thereby increasing the root system and fruit quality (Quaggio et al., 1992b; Morton et al., 2008). Magnesium also plays a role in the reduction of reactive oxygen species (ROS) as a defense of the plant to biotic and abiotic stresses (Sewelam et al., 2016).

Hunger signs of Mg in citrus have first been identified by Averna-Saccá (1912) in citrus orchards in Brazil. Leaf symptoms of Mg deficiency are widespread in orchards on acidic soils, where the availability of this nutrient in the CEC is low. Magnesium deficiency in citrus is also induced by high contents of K and Ca in soils as a result of unbalanced fertilization (Mattos Jr. et al., 2004; Quaggio et al., 1992b, 2014).

Unlike other macronutrients, sulfur is not a key element for nutrient management in citrus orchards (Mattos Jr. et al., 2007a,b; Quaggio et al., 2011). Furthermore, an adequate S supply of citrus trees can be achieved with the application of nitrogen fertilizers, phosphates, and micronutrients as well as with pesticides containing sulfur.

15.3.5 Micronutrients

Citrus trees require an adequate supply of micronutrients for high productivity. Most nutritional disorders in citrus cultivation are associated specifically with deficiencies in B, Fe, Mn, and Zn (Zekri and Obreza, 2015). Recently, Cu deficiencies appeared to be more relevant and associated with excessive plant vigor due to high N supplies (Mattos Jr. et al., 2010; Hippler et al., 2018a).

On the other hand, B and Cu toxicities are increasingly found under specific growing conditions. This is the case in arid and semiarid regions where B accumulates in the root soil layer or because of the uneven application of B fertilizers (Papadakis et al., 2003; Bell and Dell, 2008), but also in areas where Cu accumulates in the soil due to Cu deposition as a result of foliar application of pesticides (Fan et al., 2011; Hippler et al., 2017a). Copper-induced stress damages to plants is minimized with the supply of N or Ca by enhancing the antioxidant system, while maintaining photosynthesis (Hippler et al., 2018b).

In general, Cu, Fe, and Mn deficiencies impair the photosynthesis of citrus trees, whereas B and Zn play a role in meristematic growth. Several enzyme systems are also regulated by micronutrients, such as the antioxidant system that comprises the first line of defense against damages induced by ROS. However, few studies have correlated the use of micronutrients with the production of citrus plants (Quaggio et al., 2003), and little is known about the effects of those on the flowering of citrus. However, these elements might play an important role in the increase of flower and fruit quality (Pestana et al., 2001a, 2005a,b).

In addition, Fe is essential in plant-sustaining processes (Abadía, 1992). Despite being relatively abundant in a large range of cultivated soils, Fe deficiency is a common nutritional disorder that reduces the yield and quality of fruit trees in the Mediterranean Basin, where 20%–50% of fruit orchards suffer from this problem (Martínez-Cuenca et al., 2017). Frequently, Fe chlorosis disorders on leaves appear due to impaired Fe absorption by roots and translocation to leaves, generated by a short Fe supply due to high levels of bicarbonate or microelements or because of the flooding conditions in soils, as well as in high pH substrates used in nursery cultivation (Martínez-Cuenca et al., 2015; Ferrarezi et al., 2007).

Iron deficiency/tolerance is determined by the rootstock, which exhibits differences in susceptibility (Hamzé et al., 1986; Pestana et al., 2005b; Fu et al., 2017). For example, Cleopatra mandarin is highly tolerant to chlorosis (Castle and Manthey, 1998). On the contrary, *Poncirus trifoliata*, Carrizo citrange, and Swingle citrumelo are susceptible to Fe chlorosis and require expensive Fe-chelate treatments to improve their horticultural performance. Forner-Alcaide 5 (*Citrus reshni* × *P. trifoliata*), obtained through traditional hybridization (Forner et al., 2003), exhibits greater tolerance to chlorosis than Carrizo (Llosá et al., 2009).

Tissue lignification and bud-take of nursery trees depend on adequate Cu supply and vary with rootstock varieties, of which Sunki mandarin is more responsive to Cu and the plant N nutritional status than Rangpur lime (Mattos Jr. et al., 2010). Moreover, structural and adequate ultrastructural arrangements in tree leaves and roots and, consequently, water and gas exchange are achieved with optimum B supply (Mesquita et al., 2016; Wu et al., 2017). Trees grafted on Swingle rootstocks sustain better growth under low B availability in the root medium as compared with trees on other rootstocks, but still respond positively to increased B levels by combining higher B absorption and root growth as well as showing a better organization of xylem vessels (Boaretto et al., 2008; Mesquita et al., 2016). Rootstocks more susceptible to B deficiency exhibit increased expression levels of chlorophyll decomposition-related genes, glucose and lignin synthesis-related genes, and a reduced expression of carotenoid-related genes (Liu et al., 2017).

Boron is also an important effector of H^+-ATPase activity of the plasma membrane of root cells (Pollard et al., 1977; Cara et al., 2002), which, in specific concentrations, modulates the activity of the enzyme promoting the plasma cytoplasmic hyperpolarization and the subsequent K^+ uptake as a response to the maintenance of a cellular electrochemical equilibrium (Schon et al., 1990). This is likely in line with the improved response of trees on Swingle citrumelo to K fertilization (Ferreira et al., 2014).

More recently, the importance of Mo in nutrient management for citrus production has been revisited (Hippler et al., 2017b). Molybdenum application via foliar spraying increases the assimilation efficiency and use of NO_3^- as a result of increased nitrate reductase activity in leaves and roots. Plants with a high N supply show higher Mo uptake than those with a low N supply, which likely supports the need of Mo for improved N use efficiency in citrus orchards.

15.4 Monitoring soil fertility and plant nutritional status

Monitoring soil fertility and plant nutrition status allows the establishment of optimal nutrient management in citrus groves, based on the 4R nutrient stewardship principles, which seek a balanced supply of nutrients. These principles consist of considering the characteristics of fertilizers to deliver plant-available forms (right source), assessment and decisions based on soil nutrient supply and plant demand (right rate), dynamics of plant uptake and soil-plant-atmosphere interactions, which play a role in nutrient conservation (right time), and improved crop use in the field by placing fertilizers in the root zone (right place).

15.4.1 Soil analysis

The analysis of the soil is important for predicting the reaction of citrus plants to fertilization with P and K. Responses to fertilization are expected when the soil is low or very low in the specific nutrient, while no response is expected when the soil tests show optimum or high to very high nutrient levels (Obreza and Morgan, 2008; Kadyampakeni et al., 2014d). It is also important to know the pH of the soil, because, for most soils, nutrients become available for citrus between pH 5.5 and 7.0, but are largely unavailable or reach toxicity levels at pH values below 5.0 or above 7.0, respectively (Havlin et al., 2004).

Considering the characteristics of soils and the production environment, research programs with annual and perennial crops have developed procedures for soil laboratory analysis. For instance, in the 1980s, the Instituto Agronômico (IAC), Brazil, proposed the use of ion exchange resins to simultaneously extract plant-available P, Ca, Mg, and K. This technique proved to be more suited to the characteristics of tropical soils compared to traditional methods such as the Mehlich method (or double acid) or the use of neutral ammonium acetate or ammonium acetate (pH = 4.8) (Mehlich, 1984; van Raij et al., 1986; Hanlon et al., 1995). Resin extraction is also suited for calcareous and saline soils (van Raij et al., 2001).

Laboratory efforts toward soil analysis have to be supported by efficient soil sampling in relation to the area under citrus cultivation. Earlier research could not demonstrate consistent correlations between soil and leaf analysis for citrus, based on long-term field studies (Reitz and Koo, 1959; Koo, 1962; Hanlon et al., 1995), which discouraged the use of soil testing for fertilizer recommendation for citrus in the past.

However, when soil samples are taken from portions where fertilization might affect soil conditions and root growth, significant correlations are obtained. This is illustrated when samples are taken along the planting row or from both the planting row and the proximal mid-row (Quaggio et al., 1992a,b). Soil acidification and K buildup, resulting from fertilizer application along the planting row, are high and decrease Mg availability in the soil, limiting correlations for interpreting laboratory results. In contrast, taking subsamples from both the planting row and the proximal mid-row, where more roots are found and optimum nutrient equilibrium is observed in the soil CEC, significant correlations between soil and leaf nutrients are observed (Fig. 15.1).

Similarly, orchards under fertigated management will require soil sampling to be carried out within the soil wet bulb, provided by driplines or microsprinklers. Fig. 15.2 illustrates the distribution of soil-P concentration in a citrus grove when three different fertigation systems are used.

In citrus orchards, soil sampling is generally carried out in areas homogeneous in color and texture, orchard management, tree age, canopy, and rootstock combination and productivity. Soil samples should be collected at a depth of 0–20 cm to determine the need for liming and fertilization and at 20–40 cm to diagnose potential chemical barriers for root development, such as Ca deficiency with or without excess Al.

Calibration curves for P and K soil analyses, using data from a network of long-term NPK factorial experiments, established critical levels for optimum citrus production with soil exchangeable-K = 2.0 mmol$_c$ dm^{-3} and P_{resin} = 20 mg dm^{-3} (Cantarella et al., 1992; Quaggio et al., 1998). Based on this, interpretation of classes for soil analysis results (from very low to high levels) and tree responses to fertilizer application in the orchards (Fig. 15.3) have been established. Calibration curves have also been established for Mg and B (Quaggio et al., 1992b, 2003).

Guidelines for the interpretation of soil macronutrient and micronutrient fertility specific to citrus has to be provided for citrus-growing regions on a global level, considering specific soil characteristics and laboratory methods to assess nutrient availability.

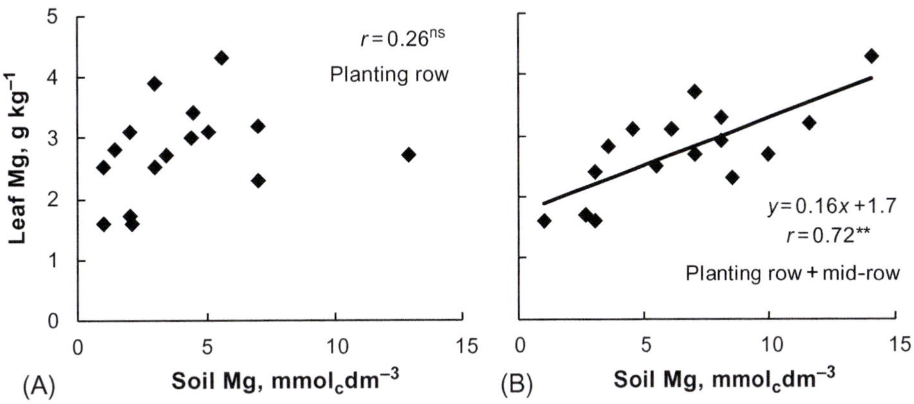

FIG. 15.1 Correlation between soil-Mg and leaf-Mg for two different soil sampling sites: (A) planting row and (B) planting row + mid-row. *(Adapted from Quaggio, J.A., Teófilo Sobrinho, J., Dechen, A.R., 1992a. Response to liming of 'Valencia' orange tree on Rangpur lime: effects of soil acidity on plant growth and yield. Proc. Int. Soc. Citricult. 2, 628–632; Quaggio, J.A., Teófilo Sobrinho, J., Dechen, A.R., 1992b. Magnesium influences on fruit yield and quality of 'Valencia' sweet orange on Rangpur lime. Proc. Int. Soc. Citricult. 2, 633–637.)*

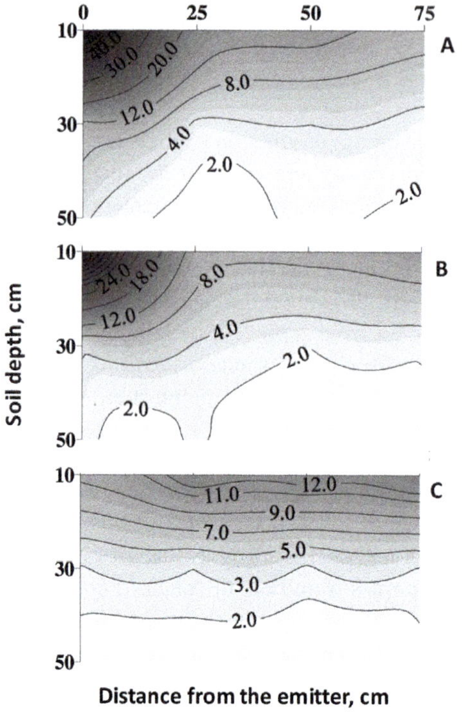

FIG. 15.2 Phosphorus concentration isolines (resin extract; mg dm^{-3}) in the soil profile and distance from the point of water emission by drip systems with one dripline (A), two driplines (B), and a microsprinkler (C). *(Adapted from Laurindo, V.T., Silva, G.O., Pavani, L.C., Quaggio, J.A., 2010. Padrão de distribuição de K$^+$, Ca^{++}, Mg^{++} e P no solo de um pomar de citros em função da fertirrigação. Eng. Agríc. 30 (5), 909–921.)*

15.4.2 Plant analysis

Citrus trees store significant amounts of nutrients in the biomass, which are partly redistributed to developing organs such as leaves and fruits (Mattos Jr. et al., 2003b). For this reason, leaf analysis is a useful tool in combination with soil chemical analysis to assess the nutritional balance of plants. Moreover, in the case of N, where methods of soil analysis lack consistency in diagnosis (Cantarella et al., 1994; Mattos Jr. et al., 1995), citrus leaf N analysis is used for evaluating its availability (Quaggio et al., 1998).

The use of leaf nutrient concentrations as criteria for assessing the nutritional status of plants has first been suggested by Lagatu and Maume (1934), based on the relationship between growth rate and nutrient content in the dry matter.

According to Bould (1983), leaf analysis is based on four assumptions: (i) the leaf is the most important site of plant metabolism, (ii) changes in nutrient supply are reflected in the composition of the index tissue such as the leaf or its petiole,

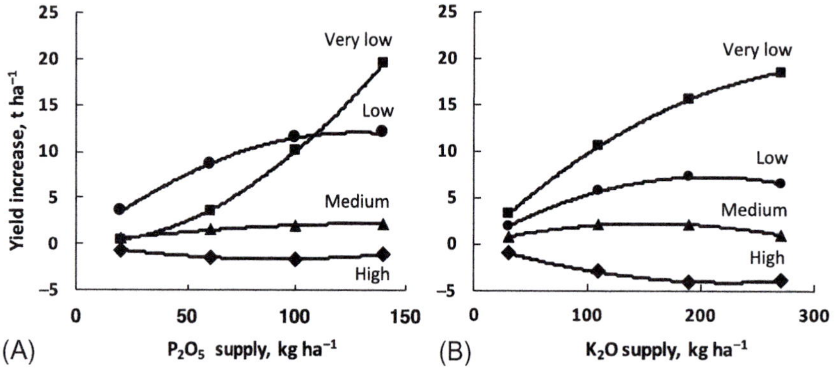

FIG. 15.3 Yield response curves for P and K fertilization according to soil classes of nutrient availability in the soil (P-resin: very low <6, low = 6–12, medium = 13–30, and high >30 mg dm^{-3}; exch-K: very low <0.8, low = 0.8–1.5, medium = 1.6–3.0, and high >3.0 mmol$_c$ dm^{-3}). *(Adapted from Quaggio, J.A., Cantarella, H., van Raij, B., 1998. Phosphorus and potassium soil test and nitrogen leaf analysis as a basis for citrus fertilization. Nut. Cycl. Agroecosyst. 52, 67–74.)*

(iii) changes are more pronounced at certain developmental stages, and (iv) the concentration of the nutrients in the leaf at a specific growth stage is related to crop performance. Therefore, within certain limits, increases in nutrient concentrations are related to increased growth or productivity.

Concentrations of different nutrients are more stable in 6–8-month-old leaves, and for this reason, this moment is the most adequate period for leaf sampling (Smith, 1966; Hanlon et al., 1995; Legaz et al., 1995; Quaggio et al., 1998; Carranca, 1999; Menino, 2012). It is of great importance not only that the leaves are of the same age but also that they have developed in the same way. The presence of flowers or small growing fruits in a branch significantly diminishes leaf nutrient concentrations, as the nutrients are translocated to those organs (Legaz et al., 1995; Davies and Albrigo, 1998). Thereby, differences in the concentrations of various nutrients in these samples are expected due to the sink effects (Khan et al., 2000).

Nutrient concentration in non-fruiting shoot leaves from spring flushes are used to indicate the nutritional status of trees in the USA, Australia, and Spain; whereas nutrient concentrations of fruiting shoot leaves are used in Brazil and South Africa. The interpretation of the results of this analysis is based on ranges of concentrations established in studies for particular leaf types, performed in long-term experiments, as illustrated in Fig. 15.4.

Ideally, the interpretation of the results of leaf and soil analyses for nutrient management recommendations should consider data series obtained in different evaluation years, since year-to-year changes due to environmental conditions and crop size may occur (Guardiola et al., 1973). In fact, these data should be observed associated with the production data series and fruit quality. With the advance of new canopy and rootstock varieties, it will be necessary to revise reference tables for the interpretation of laboratory results (Rozane et al., 2015).

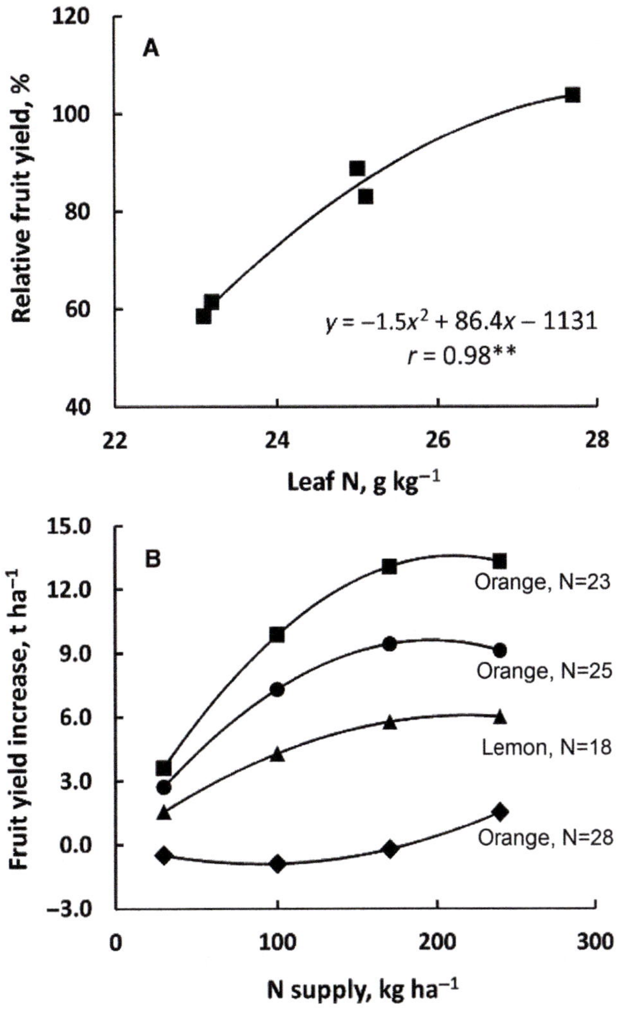

FIG. 15.4 Fruit yield as a function of leaf N content of citrus (A) and response curves either for orange or lemon with different N contents ($N = g\,kg^{-1}$) (B) (Quaggio et al., 1998).

More recently, several authors have defended the analysis of other plant organs, such as flowers (Pestana et al., 2001a) or fruits (Lacertosa et al., 2001), but approved reference values or even a consensus about the preference for one of those organs as the most adequate base for fertilization management is still lacking (Menino, 2012).

Visual symptoms of nutritional disorders are typically identified in the leaves when the problem is acute due to molecular lesions and subcellular changes that lead to cell and tissue deformations. These are useful as an aid in interpreting the diagnosis of the nutritional status of citrus. However, they can easily be confused if deficiency and toxicity result from the disturbed concentration of more than one element in plants. Moreover, when the visual stage is found, tree development and yield are already affected.

15.5 Nutrient management strategies

15.5.1 Acidity and alkalinity

Citrus are sensitive to soil acidity; when grown at pH < 5, root growth is limited due to excess Al^{+3} and H^+ (Smith, 1966). Based on this, several studies have demonstrated pronounced positive responses of orange trees to increasing rates of Ca and acidity neutralization in acidic soils (Anderson and Martin, 1969; Anderson, 1971, 1987; Quaggio et al., 1992a; Silva et al., 2007). Liming also helps to increase microbial activity and nutrient availability and uptake.

Limestone application correlates with soil V (%), tree vigor, growth, and fruit number per tree, as well as with fruit yield. Maximum fruit yield is attained with V = 70%, which corresponds to pH 5.5 (determined in $CaCl_2$ 0.01 mol L^{-1}) (Quaggio et al., 1992a). Soil liming, using dolomitic limestone, is also required to supply trees with Mg (Quaggio et al., 1992b).

Furthermore, the use of agricultural gypsum ($CaSO_4.2H_2O$), although not affecting soil pH, improves the root environment by increasing Ca availability in the deep layers of dystrophic or alic soils, consequently increasing NO_3^- uptake (Carvalho and van Raij, 1997).

Citrus trees do not grow well under alkaline conditions, which are commonly found either in calcareous soils or in those that became alkaline from the repeated application of high-bicarbonate irrigation water (Spiegel-Roy and Goldschmidt, 1996; Llosá et al., 2009). The presence of $CaCO_3$ in the soil solution increases the pH and affects the availability of nutrients. For this reason, lime-induced chlorosis, associated with Fe deficiency, is frequently observed in citrus trees.

Soil and foliar applications, as well as trunk injections, can correct Fe chlorosis. Foliar applications are fast-acting and less expensive than soil-applied treatments (Pestana et al., 2001b, 2003). However, foliar applications are effective only when the trees show symptoms of moderate chlorosis (Rombolà et al., 2000). Soil application of chelates, mainly Fe (III)-EDDHA, is widely used since the nutrient largely remains available to plants (Pestana et al., 2003).

Lowering the soil pH in those soils is achieved with the application of elemental S, sulfuric acid, and others (via irrigation water), thiosulfate-type fertilizers, as well as with ammonium-containing fertilizers. However, care must be taken to avoid root damage.

15.5.2 Fertilization with solubles and solids

The recommendations of fertilization for citrus aim to supply nutrients according to different stages of tree growth: (i) from planting to (ii) nonbearing trees to (iii) adult bearing trees. Nutrient rates are optimized for groups of varieties of oranges, lime and lemons, and mandarins as well as rootstocks. Also, fertilization recommendations consider the quality (internal and external) and destination of the fruit (fresh fruit or industry), mostly by adjusting N and K rates (Quaggio et al., 2010).

Fertilization is conducted to meet the nutrient demands of trees in early spring, when vegetative and reproductive flushes of growth occur, and extends to the beginning of autumn, assuring adequate nutrient reserves and nutrient balance in the plant biomass to facilitate normal flowering and fruit set in the following crop year (Bustan and Goldschmidt, 1998).

Soluble dry fertilization in citrus production involves fertilizer application by soil banding or broadcasting. Such fertilizers mostly include ammonium sulfate, urea, ammonium nitrate, potassium nitrate, calcium nitrate, superphosphates, monoammoniun and diammonium phosphates, monopotassium phosphate, potassium chloride, potassium sulfate, potassium nitrate, and magnesium sulfate. The application of soluble dry fertilizers is the easiest and cheapest way of applying fertilizers in citrus and has been in use for a long time in several parts of the world (Obreza and Rouse, 1993; Obreza and Morgan, 2008; Morgan et al., 2009b). However, problems associated with the use of dry granular fertilizers include root injury and reduced water uptake (Morgan et al., 2009b). This method is also labor intensive compared with other fertilizer application methods.

Soluble dry fertilization is conducted in three to four applications during the crop year to increase fertilization efficiency by reducing nutrient losses in the soil with drainage water, which occurs mainly in sandy soils, and by adjusting the demand for nutrients in different periods of citrus development (from flowering to fruit ripening) (Alva et al., 2006a, 2008). Moreover, the improved use of N fertilizers is important to avoid losses via ammonia volatilization, which is likely to account for >50% of nutrient losses (Cantarella et al., 2003; Mattos Jr. et al., 2003b); volatilized ammonia is not significantly further absorbed by leaves (Boaretto et al., 2012).

Water-soluble P sources should also be used to increase the efficiency of phosphate fertilizers, consequently improving root growth (Zambrosi et al., 2013). No differences in crop yield and quality are expected with the use of K sources (Quaggio et al., 2011) when salinity is not a limiting factor in the citrus orchard.

The use of specialty forms, such as slow-release formulations and urease- and nitrification-inhibiting N fertilizers, is still limited in the citriculture because of the lack of consistent results to support improved efficiency, mostly when used in bearing orchards, as well as because of the higher costs compared with conventional sources (Alva et al., 2006a,b; Chien et al., 2009).

The efficiency of micronutrient applications via the soil, besides B, varies greatly with mineralogy and texture, as well as organic matter. Therefore, the availability of Cu, Mn, and Zn is most likely limited in soils with a more clayey texture (Hippler et al., 2014, 2015a,b).

15.5.3 Fertigation

Fertigation is the practice of applying fertilizer solutions with irrigation water, typically through a microsprinkler or a drip system (Obreza and Boman, 2008). This practice has been developed in Israel to cope with the arid climate and aims to conserve reclaimed water; it has the potential to reduce nutrient leaching and subsequent groundwater contamination.

Willis et al. (1991) showed that a fertigation frequency of only 5 times a year resulted in higher soil ammonium and nitrate levels in the top 15 cm within 1 week after fertilization compared with fertigating 10 or 30 times a year. Alva et al. (1998) also found that an increased fertigation frequency substantially minimized nitrate-N movement under microsprinkler-irrigated mature orange trees. Studies by Kadyampakeni et al. (2014b,c, 2015, 2016) and Morgan et al. (2012) further supported the development of open hydroponic systems for constant applications of nutrients in irrigation water on sandy soils with low nutrient retention (Morgan and Kadyampakeni, 2012). Kusakabe et al. (2006) noted that the N rate, rather than the fertigation frequency, affects concentrations of residual nitrate. Thus, a careful consideration of nutrient concentrations in fertigation programs is also required to contain nutrient levels within acceptable contaminant limits. The use of frequent fertigation, combined with improved irrigation scheduling, results in improved fertilizer uptake efficiency, enhanced water use efficiency, increased residence time of nutrients in the root zone, and reduced potential for groundwater pollution (Obreza et al., 1999, 2008b; Quiñones et al., 2005, 2007). Compared with conventional ground application, fertigation can produce similar or better tree growth as well as yield and fruit quality with less amounts of fertilizer (Schumann et al., 2003; Obreza et al., 2008b; Morgan et al., 2009b; Kadyampakeni et al., 2014b,c, 2016).

Water, fertilizer placement, and application frequency are managed more efficiently compared with a dry fertilization program (Obreza et al., 2008b; Kadyampakeni et al., 2014a). Productivity gains with irrigation and fertigation, using the same fertilizer doses, increased nutrient use efficiency by 12% and 25%, respectively, for localized irrigation with solid fertilization and fertigation compared with irrigation and with solid fertilizer (Fig. 15.5; Quaggio et al., 2006b) and a water use

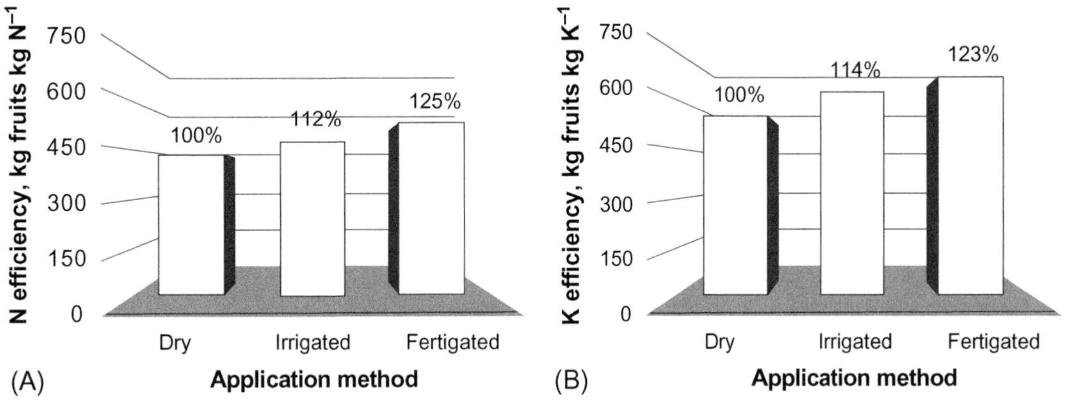

FIG. 15.5 Comparison of nitrogen (A) and potassium (B) use efficiencies in conventional solid fertilization, applied in rain-fed and irrigated treatments in comparison to fertigation. Values represent the means of four harvests. *(Adapted from Quaggio, J.A., Mattos Jr. D., Souza, T.R., Krontal, Y., 2006. Dinâmica de íons no solo e ganhos de eficiência fertilizante devido à irrigação e fertirrigação na citricultura. In: Reunião Brasileira de Fertilidade do Solo e Nutrição de Plantas, 27. Proceedings... Embrapa Agropecuária Oeste, Dourados; CD-ROM.)*

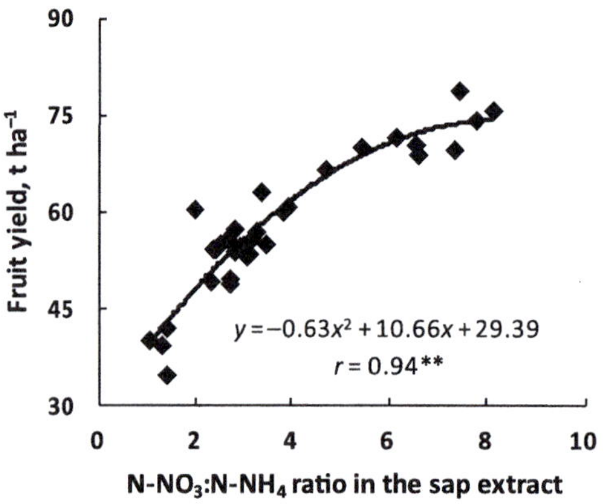

FIG. 15.6 Productivity of Valencia orange as a function of the NO_3^-:NH_4^+ ratio measured in the sap extract of trees (values represent the means of four crops) (Quaggio et al., 2014).

efficiency of 15% (Quiñones et al., 2007). For this reason, it is recommended to reduce N and K doses by 20% when nutrients are applied through fertigation in relation to the nutrient levels recommended for solid fertilization without irrigation.

Yield increases with fertigation also result from an optimal ionic balance achieved in the plants, that is, improved absorption of cations and anions, as illustrated by the increase in the ratio of NO_3^-:NH_4^+, measured in the sap extract of the trees (Fig. 15.6) that correspond to the average concentrations of NO_3^- and NH_4^+ found in the soil solution (Quaggio et al., 2006b). This increased ratio in the sap extract indicates the highest supply of N in the form of NO_3^-, which enhances the absorption of cations such as Ca, Mg, and K and minimizes acidification of the wet bulb. Therefore, to increase nutrient use efficiency and to ensure high fruit yields, it is essential to use at least 60% of N fertilizer doses using nitrogen sources in the form of NO_3^-, avoiding the exclusive use of N in the amidic or ammoniacal forms.

Analysis of tree sap extracts during the growing season demonstrates that N and Ca concentrations vary characteristically during the phenological stages, with two peaks of demand during full bloom (spring) and floral differentiation (late summer to early fall), when the plants differentiate vegetative organs into reproductive organs to produce flowers and fruits in the following harvest (Souza et al., 2012). Such information is valuable to better schedule nutrient application via fertigation in citrus orchards.

The application of micronutrients via fertigation is possible, especially for B (Boaretto et al., 2011), even though Cu, Fe, Mn, and Zn will depend on the source type (i.e., soluble or chelated) and the soil thermodynamics in the soil-root wetting zone (Zekri and Koo, 1992).

Fertigation costs are likely lower than those of dry or foliar application methods since fertilization is incorporated into the normal irrigation schedule (Obreza et al., 2008b). However, if not carefully managed, especially in the rainy season, fertigation could result in groundwater pollution (Recous et al., 1992; Quiñones et al., 2007). The duration of irrigation during fertigation as well as the flush time to clean the lines should be relatively short to avoid excess water drainage, which can contribute to the leaching of water and nutrients below the root zone (Alva et al., 2006b).

Moreover, checking the compatibility between fertilizers for fertigation is important to avoid precipitation and clogging of irrigation lines. These problems also arise associated with poor water quality, either by excess Fe or carbonate contents or by microbial growth.

15.5.4 Foliar fertilization

Foliar fertilization for citrus trees is an important practice to supplement the nutritional program, especially in terms of micronutrients such as B, Cu, Mn, Mo, and Zn. Foliar application is, however, not intended to replace a soil fertilization program. Conversely, foliar spraying can be used successfully to supplement nutrition, providing specific nutrients during the critical stages of tree growth, flowering, fruit set, and fruit development (Obreza et al., 2008b; Lovatt, 2013).

A foliar nutrition program should ideally supplement soil fertilizer applications, particularly when the supply of nutrients via the root system is not sufficient to meet the demands of the tree because of limitations in the root system, periods of development of the main growth flushes, conditions favorable for nutrient soil adsorption, cold weather, prolonged periods of drought, and limitations in the mobility of the nutrient in the trees (Quaggio et al., 2010).

Despite the thick cuticle layer of citrus leaves, trees respond relatively well to foliar nutrition due to a high number of stomata on the lower leaf surface as well as high amounts of cuticle pores (Smolen, 2012). Foliar applications should target the main growth flushes (spring and summer) when the leaves are two-thirds to nearly fully expanded and have a poorly developed cuticle to facilitate the absorption of nutrients. Also, foliar sprays are recommended during early morning or late evening, when the air temperature is relatively low and humidity is high (Obreza et al., 2008b; Quaggio et al., 2010).

Although foliar spraying appears to be a rapid way to correct nutritional disorders in citrus trees, a well-planned foliar fertilization program, especially for micronutrients, should not be used only in a corrective way, but also as a preventive measure. Boron, Mn, and Zn are poorly redistributed in the phloem of citrus plants (Embleton et al., 1965; Boaretto et al., 2002, 2004). However, due to the large storage in the tree biomass (Mattos Jr. et al., 2003b), even if small amounts of micronutrients are redistributed in the plant, they play a key role in the growth of new shoots.

Despite the numerous benefits of foliar fertilization, it presents some practical limitations: (i) micronutrients have a low mobility in the phloem; thus, continuous sprays to new leaf flushes are required, (ii) low amounts of micronutrients are sprayed to the plants to avoid toxicity caused by saline effects of products, (iii) micronutrient supply is carried out by multiple applications, and this is frequently associated with other cultural practices for controlling pests/diseases and not necessarily with a time when micronutrient demand is the highest, and (iv) growers have also reduced spraying volumes per application in citrus groves with the aim of reducing costs, which worsened the cited problems (Boaretto et al., 2002, 2004; Clapp, 2009; Fernández et al., 2013; Macedo et al., 2017).

The general recommendations for foliar fertilization to citrus are to prepare mixtures of soluble salts in the following concentrations ($mg\,L^{-1}$): Zn (500–1000), Mn (300–700), and B (200–300), as well as $5\,g\,L^{-1}$ of urea as an adjuvant. Lower concentrations are recommended for tree maintenance, while higher concentrations should be used when there are visible symptoms of deficiency. These doses have been established for fertilizers in the sulfate form, but when fertilizers are in the chloride or nitrate forms, concentrations must be reduced to avoid leaf burn. Applications of B should preferably be made to the soil, especially in young orchards (Quaggio et al., 2010).

Furthermore, foliar nutrition with K and Ca has been proposed to enhance the quality and postharvest stability of citrus fruits. Mostafa and Saleh (2006) reported that spraying potassium nitrate, combined with girdling, had a positive effect on leaf mineral content, fruit set, and yield as number or fruit weight of citrus. Such treatments increased the fruit diameter, soluble solid content (SS), and the SS: acid ratio. Moreover, foliar applications of Ca reduce skin discoloration in citrus fruits (Crisosto et al., 1997).

15.5.5 Organic fertilization

The use of organic fertilizers, in addition to mineral fertilization or certified organic production systems, represents a very specific market niche in citriculture, dependent on the regulation of sanitary agencies (Abreu Jr. et al., 2005) as well as the quality and availability of raw materials and waste (Canali et al., 2012; Martínez-Alcántara et al., 2016), and the profitability, which is considerably associated with the value of the commercialized fruit. These fertilizers are primarily N sources, and many of them can benefit soil quality.

15.5.6 Fertilization and stress alleviation

The supply of N, Ca, and Mg, as well as some micronutrients, affects the structure and integrity of plant tissues, biochemical (carbohydrate metabolism and oxidative stress) and physiological responses, and biomass partitioning, potentially resulting in increased tolerance and/or resistance to stress and, subsequently, pests and diseases (Dordas, 2008; Underwood, 2012; Sewelam et al., 2016) in citrus orchards. This has received the attention of citrus growers ever since phytosanitary problems have affected citrus farming and increased fruit production costs, decreasing the sustainability of the sector.

In this context, some diseases caused by fungi (postbloom fruit drop and black spot) can be mitigated by the supply of nutrients such as Ca, which improves the integrity of tissues and increases the structuring of conducting vessels in plants. Diseases caused by bacteria (such as citrus canker) may be diminished (both in severity and incidence) by the natural barrier promoted by Ca. Also, in terms of bacterial infections, the destructive effects of huanglongbing (HLB) can probably be mitigated by the improved activity of the antioxidant enzymatic system, which is regulated by Cu, Fe, Mn, and Zn, as well as by the balanced metabolism and partition of carbohydrates in the plant, achieved through, for example, adequate Mg supply. Foliar applications using amounts higher than those recommended (specifically regarding Mn and Zn) increased growth and yield of trees affected by HLB in Florida (Morgan et al., 2016).

A more recent trend in fruit crops is the use of foliar or soil applications of biostimulants, such as mixtures of polypeptides, oligopeptides, amino acids, a combination of mineral elements, plant growth-promoting bacteria, fungi, and other priming molecules to improve fruit yield/size, internal and external quality, as well as plant tolerance to other environmental stresses such as high temperatures (Brown and Saa, 2015; Tanou et al., 2017).

However, the great diversity of components and plant responses under varying environmental conditions still interfere with the establishment of precise practical recommendations.

15.6 Concluding remarks

The available information for the management of soils and nutrients in citriculture is comprehensive and demonstrates the scientific progress achieved by studies carried out in the last decades, which elucidated the role of mineral nutrients in citrus production and quality. It is also based on our capacity to interpret diagnoses of nutrient availability in soils and on our understanding of plant responses to nutrient supply under different management systems. In spite of these advances, we still need more detailed information on the factors that lead to increased absorption efficiency and mineral nutrient use by citrus, considering the different combinations of canopy and rootstock varieties and based on biochemical and molecular processes. Other aspects that appear to be top priorities in modern citriculture are associated with the beneficial role of mineral nutrients in plant tolerance to biotic and abiotic stresses in several production niches, where the occurrence of phytosanitary problems, besides issues related to climatic changes, limit citrus production on a global level.

References

Abadía, J., 1992. Leaf responses to Fe deficiency: a review. J. Plant Nutr. 15 (10), 1699–1713.

Abreu Jr., C.H., Oliveira, F.C., Berton, R., Silva, F.C., 2005. Uso de resíduos orgânicos no pomar. In: Mattos Jr., D., De Negri, J.D., Pio, R.M., Pompeu Jr., J., (Eds.), Citros. vol. 29. Instituto Agronômico e Fundag, Campinas, pp. 871–896.

Agustí, M., 2003. Citricultura. Mundi-Prensa, Madrid, p. 422.

Alva, A.K., Paramasivam, S., 1998. Nitrogen management for high yield and quality of citrus in sandy soils. Soil Sci. Soc. Am. J. 62, 1335–1342.

Alva, A.K., Tucker, D.P.H., 1999. Soils and citrus nutrition. In: Timmer, L.W., Duncan, L.W. (Eds.), *Citrus* Health Management. APS Press, St. Paul, MN, pp. 59–71.

Alva, A.K., Paramasivam, S., Graham, W.D., 1998. Impact of nitrogen management practices on nutritional status and yield of Valencia orange trees and groundwater nitrate. J. Environ. Qual. 27, 904–910.

Alva, A.K., Mattos Jr., D., Paramasivam, S., Patil, B., Dou, H., Sajwan, K., 2006a. Potassium management for optimizing citrus production and quality. Int. J. Fruit Sci. 6 (1), 3–43.

Alva, A.K., Paramasivam, S., Obreza, T.A., Schumann, A.W., 2006b. Nitrogen best management practice for citrus trees I. Fruit yield, quality, and leaf nutritional status. Sci. Hortic. 107, 233–244.

Alva, A.K., Mattos Jr., D., Quaggio, J.A., 2008. Advances in nitrogen fertigation of citrus. J. Crop Improv. 22, 121–146.

Anderson, C.A., 1971. Effects of soil pH and calcium on yields and fruits quality of young 'Valencia' oranges. Proc. Fla. State Horticult. Soc. 84, 4–11.

Anderson, C.A., 1987. Fruits yields, trees size and mineral nutrition relationships in 'Valencia' orange trees as affected by liming. J. Plant Nutr. 10, 1907–1916.

Anderson, C.A., Martin, F.G., 1969. Effects of soil pH and calcium on the growth and mineral uptake of young citrus trees. Proc. Fla. State Horticult. Soc. 82, 7–12.

Averna-Saccá, R.A., 1912. Chlorose da laranjeira e de outras plantas nas terraferruginosas. Bol. Agric. 13, 129–150.

Bell, R.W., Dell, B., 2008. Micronutrients for Sustainable Food, Feed, Fiber and Bioenergy Production. International Fertilizer Industry Association, Paris, p. 175.

Boaretto, A.E., Boaretto, R.M., Muraoka, T., Nascimento Filho, V.F., Tiritan, C.S., Mourão Filho, F.A.A., 2002. Foliar micronutrient application effects on citrus fruit yield, soil and leaf concentrations and ^{65}Zn mobilization within the plant. Acta Hortic. 594, 203–209.

Boaretto, A.E., Boaretto, R.M., Contin, T.L.M., Muraoka, T., 2004. É móvel ou imóvel o boro em laranjeiras? Laranja 25, 195–208.

Boaretto, R.M., Quaggio, J.A., Mourão Filho, F.A.A., Gine, M.A., Boaretto, A.E., 2008. Absorption and mobility of boron in young citrus plants. Commun. Soil Sci. Plant Anal. 39, 2501–2514.

Boaretto, R.M., Quaggio, J.A., Mattos Jr., D., Muraoka, T., Boaretto, A.E., 2011. Boron uptake and distribution in field grown citrus trees. J. Plant Nutr. 34, 839–849.

Boaretto, R.M., Mattos Jr., D., Quaggio, J.A., Cantarella, H., Trivelin, P.C.O., 2012. Absorption of $^{15}NH_3$ volatilized from urea by *Citrus* trees. Plant Soil 357, 1–6.

Bould, C., 1983. Diagnosis of Mineral Disorder in Plants—Principles. H.M. Stationery Office, New York, p. 170.

Brown, P., Saa, S., 2015. Biostimulants in agriculture. Front. Plant Sci. 6, https://doi.org/10.3389/fpls.2015.00671.

Bustan, A., Goldschmidt, E.E., 1998. Estimating the cost of flowering in a grapefruit tree. Plant Cell Environ. 21, 217–224.

Caballero, F., García-Sánchez, F., Gimeno, V., Syvertsen, J.P., Martínez, V., Rubio, F., 2013. High-affinity potassium uptake in seedlings of two citrus rootstocks Carrizo citrange (*Citrus sinensis* [L.] Osb. × *Poncirus trifoliata* [L.] Raf.) and Cleopatra mandarin (*Citrus reshni* hort. Ex Tanaka). Aust. J. Crop. Sci. 7 (5), 538–542.

Canali, S., Rocuzzo, G., Tittarelli, F., Intrigliolo, F., 2012. Organic citrus: soil fertility and plant nutrition management. In: Srivastava, A.K. (Ed.), Advances in *Citrus* Nutrition. Springer Science, Dordrecht, pp. 353–368.

Cantarella, H., Quaggio, J.A., Bataglia, O.C., van Raij, B., 1992. Response of citrus to NPK fertilization in a network of field trials in São Paulo State, Brazil. Proc. Int. Soc. Citricult. 2, 607–612.

Cantarella, H., Mattos Jr., D., van Raij, B., 1994. Lime effect on soil N availability indexes as measured by plant uptake. Commun. Soil Sci. Plant Anal. 25 (3), 989–1006.

Cantarella, H., Mattos Jr., D., Quaggio, J.A., Rigolin, A.T., 2003. Fruit yield of Valencia sweet orange fertilized with different N sources and the loss of applied N. Nutr. Cycl. Agroecosyst. 67, 1–9.

Cara, F.A., Sánchez, E., Ruiz, J.M., Romero, L., 2002. Is phenol oxidation responsible for the short-term effects of boron deficiency on plasma-membrane permeability and function in squash roots? Plant Physiol. Biochem. 40, 853–858.

Carranca, C., 1999. Diagnóstico do estado nutritivo dos citrinos. Métodos de recomendação da fertilização. Agron. Lusit. 46, 1–9.

Carvalho, M., van Raij, B., 1997. Calcium sulphate, phosphogypsum and calcium carbonate in the amelioration of acid subsoils for root growth. Plant Soil 192 (37), 37–48.

Castle, W.S., Manthey, J.A., 1998. Screening citrus rootstocks for iron-deficiency tolerance. Fruits 53, 375–381.

Cerezo, M., Camañes, G., Flors, V., Primo-Millo, E., García-Agustín, P., 2007. Regulation of nitrate transport in citrus rootstocks depending on nitrogen availability. Plant Signal. Behav. 2 (5), 337–342.

Chien, S.H., Prochnow, L.I., Cantarella, H., 2009. Recent developments of fertilizer production and use to improve nutrient efficiency and minimize environmental impacts. Adv. Agron. 102, 267–322.

Chon, C., Ahn, J., Kim, K., Park, K., 2008. Soil properties affecting the reductive capacity of volcanic ash soils in Korea. In: American Geophysical Union, Fall Meeting 2008. abstract #V11B-2025.

Clapp, J.G., 2009. Let's be careful when defining salt index. Fluid J. 17 (4).

Crisosto, C.H., Johnson, R.S., De Jong, T., 1997. Orchard factors affecting post-harvest stone fruit quality. Hort. Sci. 32 (5), 820–823.

Davies, F.S., Albrigo, L.G., 1998. *Citrus*. CAB International, Wallingford, p. 244.

De Paz, J.M., Ramos, C., 2004. Simulation of nitrate leaching for different nitrogen fertilization rates in a region of Valencia (Spain) using a GIS-GLEAMS system. Agric. Ecosyst. Environ. 103 (1), 59–73.

Dordas, C., 2008. Role of nutrients in controlling plant diseases in sustainable agriculture. A review. Agron. Sustain. Dev. 29, 1–14.

Dovis, V.L., Martinatti, J., Boaretto, R.M., Machado, E.C., Erismann, N.M., Mattos Jr., D., 2016. Nitrogen use efficiency in lemon and orange trees: a growth and resources allocation analysis. Citrus Res. Technol. 37 (2), 165–172.

Du Plessis, S.F., Koen, T.J., 1996. South Africa. Proc. Int. Soc. Citricult. 1, 551–552.

Duiker, S.W., Flanagan, D.C., Lal, R., 2001. Erodibility and infiltration characteristics of five major soils of southwest Spain. Catena 45, 103–121.

Embleton, T.W., Wallihan, E.F., Goodall, G.E., 1965. Effectiveness of soil vs. foliar applied zinc and of foliar applied manganese on California lemons. Proc. Am. Soc. Horticult. Sci. 86, 253–259.

Fan, J., He, Z., Ma, L.Q., Stoffella, P.J., 2011. Accumulation and availability of copper in citrus grove soils as affected by fungicide application. J. Soils Sediments 11, 639–648.

FAO, 2015. Citrus Fruits. Annual Statistics. Food and Agriculture Organization. viewed November 2017, http://www.fao.org/faostat/en/#data/QC.

Fernández, V., Sotiropoulos, T., Brown, P.H., 2013. Foliar Fertilisation: Principles and Practices. International Fertilizer Industry Association (IFA), Paris, p. 140.

Ferrarezi, R.S., Bataglia, O.C., Furlani, P.R., Schammass, E.A., 2007. Iron sources for citrus rootstock development grown on pine bark/vermiculite mixed substrate. Sci. Agric. 64 (5), 520–531.

Ferreira, G.A., Hippler, F.W.R., Boaretto, R.M., Quaggio, J.A., Façanha, A.R., Mattos Jr., D., 2014. Absorption of boron and potassium and the relationship with the H^+-ATPase activity of citrus rootstocks. In: Abstracts… Simpósio Luso-Espanhol de Nutrição Mineral das Plantas, XV. NutriPLANTA 2014. Universidade de Lisboa, Lisbon.

Forner, J.B., Forner-Giner, M.A., Alcaide, A., 2003. Forner-Alcaide 5 and Forner-Alcaide 13: two new citrus rootstocks released in Spain. Hort. Sci. 38, 629–630.

Fu, L., Zhu, Q., Sun, Y., Du, W., Pan, Z., Peng, S., 2017. Physiological and transcriptional changes of three citrus rootstock seedlings under iron deficiency. Front. Plant Sci. 8, 1104. https://doi.org/10.3389/fpls.2017.01104.

Guardiola, J.L., Vivas, S., Gonzálues-Sicilia, E., 1973. The effectiveness of leaf nitrogen content as an index to nitrogen nutrition. Proc. Int. Soc. Citricult. 1, 151–157.

Hamzé, M., Ryan, J., Zaabout, M., 1986. Screening of citrus rootstocks for lime-induced chlorosis tolerance. J. Plant Nutr. 9 (3), 459–469.

Hanlon, E.A., Obreza, T.A., Alva, A.K., 1995. Tissue and soil analysis. In: Tucker, D.P.H., Alva, A.K., Jackson, L.K., Wheaton, T.A. (Eds.), Nutrition of Florida *Citrus* Trees. University of Florida (SP 169), Gainesville, pp. 13–20.

Havlin, J.L., Beaton, J.D., Tisdale, S.L., Nelson, W.L., 2004. Soil Fertility and Fertilizers: An Introduction to Nutrient Management, seventh ed. Pearson Education, Inc., Upper Saddle River, NJ, p. 528.

Haynes, R.J., Goh, K.M., 1978. Ammonium and nitrate nutrition of plants. Biol. Rev. 53, 465–510.

He, Z., Zhang, M., Wilson, M.J., 2004. Distribution and classification of red soils in China. In: Wilson, M.J., He, Z., Yang, X. (Eds.), The Red Soils of China. Springer, Dordrecht, pp. 29–33.

Hinsinger, P., Plassard, C., Tang, C., Jaillard, B., 2003. Origins of root-mediated pH changes in the rhizosphere and their responses to environmental constraints: a review. Plant Soil 248, 43–59.

Hippler, F.W.R., Reis, I.M.S., Boaretto, R.M., Quaggio, J.A., Mattos Jr., D., 2014. Características adsortivas de solos e o suprimento de zinco e manganês para os citros. Citrus Res. Technol. 35, 73–83.

Hippler, F.W.R., Boaretto, R.M., Quaggio, J.A., Azevedo, R.A., Mattos Jr., D., 2015a. Towards soil management with Zn and Mn: estimates of fertilisation efficacy of *Citrus* trees. Ann. Appl. Biol. 166, 484–495.

Hippler, F.W.R., Boaretto, R.M., Quaggio, J.A., Boaretto, A.E., Abreu-Junior, C.H., Mattos Jr., D., 2015b. Uptake and distribution of soil applied zinc by citrus trees-addressing fertilizer use efficiency with ^{68}Zn labeling. PLoS ONE 10, e0116903.

Hippler, F.W.R., Boaretto, R.M., Quaggio, J.A., Mattos Jr., D., 2017a. Copper in *Citrus* production: required but avoided. Citrus Res. Technol. 38 (1). https://doi.org/10.4322/crt.ICC067.

Hippler, F.W.R., Boaretto, R.M., Verónica, L.D., Gomes, G.O.F., Quaggio, J.A., Quiñones, A., Mattos Jr., D., 2017b. Revisiting nutrient management for *Citrus* production: to what extent does molybdenum affect nitrogen assimilation of trees? Sci. Hortic. 225 (18), 462–470.
Hippler, F.W.R., Boaretto, R.M., Teixeira, L.A.J., Quaggio, J.A., Mattos Jr., D., 2018a. Copper supply and fruit yield of young *Citrus* trees: fertiliser sources and application methods. Bragantia 77.
Hippler, F.W.R., Boaretto, R.M., Docis, V.L., Quaggio, J.A., Azevedo, R.A.A., Mattos Jr., D., 2018b. Oxidative stress induced by Cu nutritional disorders in *Citrus* depends on nitrogen and calcium availability. Sci. Rep. 8. https://doi.org/10.1038/s41598-018-19735-x.
Hirel, B., Tétu, T., Lea, P.J., Dubois, F., 2011. Improving nitrogen use efficiency in crops for sustainable agriculture. Sustainability 3, 1452–1485.
Iglesias, D.J., Levy, Y., Gómez-Cadenas, A., Tadeo, F.R., Primo-Millo, E., Talón, M., 2004. Nitrate improves growth in salt-stressed citrus seedlings through effects on photosynthetic activity and chloride accumulation. Tree Physiol. 24 (9), 1027–1034.
IUSS Working Group WRB, 2015. World Reference Base for Soil Resources 2014, Update 2015. International Soil Classification System for Naming Soils and Creating Legends for Soil Maps. World Soil Resources Reports No. 106, FAO, Rome.
Jackson, L.K., Alva, A.K., Tucker, D.P.H., Calvert, D.V., 1995. Factors to consider in developing a nutrition program. In: Tucker, D.P.H., Alva, A.K., Jackson, L.K., Wheaton, T.A. (Eds.), Nutrition of Florida *Citrus* Trees. University of Florida (SP 169), Gainesville, pp. 3–12.
Kadyampakeni, D.M., Morgan, K.T., 2017. Irrigation scheduling and soil moisture dynamics influence water uptake by Huanglongbing affected trees. Sci. Hortic. 224, 272–279.
Kadyampakeni, D.M., Morgan, K.T., Schumann, A.W., Nkedi-Kizza, P., Mahmoud, K., 2014a. Ammonium and nitrate distribution in the soil using drip and microsprinkler irrigation for citrus production. Soil Sci. Soc. Am. J. 78, 645–654.
Kadyampakeni, D.M., Morgan, K.T., Schumann, A.W., Nkedi-Kizza, P., 2014b. Effect of irrigation pattern and timing on root density of young citrus trees infected with Huanglongbing disease. Hort. Technol. 24 (2), 209–221.
Kadyampakeni, D.M., Morgan, K.T., Schumann, A.W., Nkedi-Kizza, P., Obreza, T.A., 2014c. Water use in drip and microsprinkler irrigated citrus trees as a function of tree size, soil characteristics and Huanglongbing infection. Soil Sci. Soc. Am. J. 78 (4), 1351–1361.
Kadyampakeni, D.M., Morgan, K.T., Schumann, A.W., Nkedi-Kizza, P., Mahmoud, K., 2014d. Phosphorus and potassium distribution and adsorption on two Florida sandy soils. Soil Sci. Soc. Am. J. 78, 325–334.
Kadyampakeni, D.M., Morgan, K.T., Nkedi-Kizza, P., Kasozi, G.N., 2015. Nutrient management options for Florida citrus: a review of NPK application and analytical methods. J. Plant Nutr. 38 (4), 568–583.
Kadyampakeni, D.M., Morgan, K.T., Schumann, A.W., 2016. Biomass, nutrient accumulation and tree size relationships for drip-and microsprinkler-irrigated orange trees. J. Plant Nutr. 39 (5), 589–599.
Kato, T., 1986. Nitrogen metabolism and utilization in citrus. Hortic. Rev. 8, 181–216.
Khan, I.A., Srikandakumar, A., Embleton, T.W., 2000. Interrelationship of nutrient status of orange leaves from fruiting and non-fruiting terminals and their conversion factors. Proc. Int. Soc. Citricult, 433–434.
Kirkby, E.A., Knight, A.H., 1977. Influence of the level of nitrate nutrition on ion uptake and assimilation, organic acid accumulation, and cation-anion balance in whole tomato plants. Plant Physiol. 60, 349–353.
Koo, R.C.J., 1962. The use of leaf, fruit, and soil analysis in estimating potassium status of orange trees. Proc. Fla. State Horticult. Soc. 75, 67–72.
Kusakabe, A., White, S.A., Walworth, J.L., Wright, G.C., Thompson, T.L., 2006. Response of microsprinkler-irrigated navel oranges to fertigated nitrogen rate and frequency. Soil Sci. Soc. Am. J. 70, 1623–1628.
Lacertosa, G., Giuffrida, A., Mennone, C., et al., 2001. Stato nutrizionale e qualità dei frutti in agrumeti del metapontino. In: XVII Convegno SICA, pp. 387–394.
Lagatu, H., Maume, L., 1934. Recherches sur le diagnostic foliaire. Ann. École Nat. Agric. Montpellier 22, 257–306.
Lavon, R., Salomon, R., Goldschmidt, E.E., 1999. Effect of potassium, magnesium, and calcium deficiencies on nitrogen constituents and chloroplast components in *Citrus* leaves. J. Am. Soc. Hortic. Sci. 124, 158–162.
Lea-Cox, J.D., Syvertsen, J.P., Graetz, D.A., 2001. Springtime ^{15}Nitrogen uptake, partitioning, and leaching losses from young bearing *Citrus* trees of differing nitrogen status. J. Am. Soc. Hortic. Sci. 126, 242–251.
Legaz, F., Serna, M.D., Primo-Millo, E., 1995. Mobilization of the reserve N in citrus. Plant Soil 173, 205–210.
Liu, X., Zhang, J.-W., Guo, L.-X., Liu, Y.-Z., Jin, L.-F., Hussain, S.B., Du, W., Deng, Z., Peng, S.-A., 2017. Transcriptome changes associated with boron deficiency in leaves of two citrus scion-rootstock combinations. Front. Plant Sci. 8. https://doi.org/10.3389/fpls.2017.00317.
Llosá, M.J., Bermejo, A., Cano, A., Quiñones, A., Forner-Giner, M.A., 2009. The citrus rootstocks Cleopatra mandarin, *Poncirus trifoliata*, Forner-Alcaide 5 and Forner-Alcaide 13 vary in susceptibility to iron deficiency chlorosis. J. Am. Pomol. Soc. 63 (4), 160–167.
Lovatt, C.J., 2013. Properly timing foliar-applied fertilizers increases efficacy: a review and update on timing foliar nutrient applications to citrus and avocado. Hort. Technol. 23 (50), 536–541.
Macedo, L.O., Boaretto, R.M., Jacobassi, R.C., Carr, N.F., Quaggio, J.A., Mattos Jr, D., 2017. Use of sparingly soluble micronutrients sources for citrus production. Citrus Res. Technol. 38 (1). https://doi.org/10.4322/crt.ICC099.
Marschner, P., 2012. Mineral Nutrition of Higher Plants, third ed. Elsevier, London, p. 643.
Martínez-Alcántara, B., Martínez-Cuenca, M.R., Bermejo, A., Legaz, F., Quiñones, A., 2016. Liquid organic fertilizers for sustainable agriculture: nutrient uptake of organic versus mineral fertilizers in citrus trees. PLoS ONE 11 (10). https://doi.org/10.1371/journal.pone.0161619.
Martínez-Cuenca, M.R., Quiñones, A., Primo-Millo, E., Forner-Giner, M.A., 2015. Flooding impairs Fe uptake and distribution in citrus due to the strong down-regulation of genes involved in strategy responses to Fe deficiency in roots. PLoS ONE 10 (4). https://doi.org/10.1371/journal.pone.0123644.
Martínez-Cuenca, M.R., Primo-Capella, A., Quiñones, A., Bermejo, A., Forner-Giner, M.A., 2017. Rootstock influence on iron uptake responses in *Citrus* leaves and their regulation under the Fe paradox effect. Peer J. 5. https://doi.org/10.7717/peerj.3553.

Mattos Jr., D., Cantarella, H., van Raij, B., 1995. Manuseio e conservação de amostras de solo para preservação do N inorgânico. Rev. Bras. Ciênc. Solo 19, 423–431.

Mattos Jr., D., Quaggio, J.A., Cantarella, H., Alva, A.K., 2003a. Nutrient content of biomass components of Hamlin sweet orange trees. Sci. Agric. 60, 155–160.

Mattos Jr., D., Graetz, D.A., Alva, A.K., 2003b. Biomass distribution and ^{15}N-nitrogen partitioning in citrus trees on a sandy Entisol. Soil Sci. Soc. Am. J. 67, 555–563.

Mattos Jr., D., Quaggio, J.A., Cantarella, H., 2003c. Manejo da Fertilidade do solo. In: Mattos Jr., D., De Negri, D., Figueiredo, J.O. (Eds.), Lima ácida Thaiti. Campinas, Instituto Agronômico, pp. 67–80.

Mattos Jr., D., Quaggio, J.A., Cantarella, H., Carvalho, S.A., 2004. Modelos de resposta do tangor Murcott à fertilização com N, P e K. Rev. Bras. Frutic. 26, 164–167.

Mattos Jr., D., Quaggio, J.A., Cantarella, H., Alva, A.K., 2006. Response of young citrus trees on selected rootstocks to nitrogen, phosphorus, and potassium fertilization. J. Plant Nutr. 29, 1371–1385.

Mattos Jr., D., Quaggio, J.A., Cantarella, H., 2007a. *Citrus*. In: Johnston, E.A. (Ed.), Fertilizing for High Yield Quality Tropical Fruits. Switzerland, International Potash Institute, pp. 70–88.

Mattos Jr., D., Quaggio, J.A., Cantarella, H., Boaretto, A.E., 2007b. Nitrogênio e enxofre na cultura dos citros. In: Yamada, T., Stipp e Abdalla, S.R., Vitti, G.C. (Eds.), Nitrogênio e enxôfre na agricultura brasileira. International Plant Nutrition Institute, Piracicaba, pp. 413–443.

Mattos Jr., D., Ramos, U.M., Quaggio, J.A., Furlani, P.R., 2010. Nitrogênio e cobre na produção de mudas de citros em diferentes porta-enxertos. Bragantia 69, 135–147.

Mehlich, A., 1984. Mehlich 3 soil test extractant: a modification of Mehlich 2 extractant. Commun. Soil Sci. Plant Anal. 14, 1409–1416.

Menino, M., 2012. Leaf analysis in citrus: interpretation tools. In: Srivastava, A.K. (Ed.), Advances in *Citrus* Nutrition. Springer Science, Dordrecht, pp. 59–79.

Mesquita, G.L., Zambrosi, F.C.B., Tanaka, F.A.O., Boaretto, R.M., Quaggio, J.A., Ribeiro, R.V., Mattos Jr., D., 2016. Anatomical and physiological responses of citrus trees to varying boron availability are dependent on rootstock. Front. Plant Sci. 7. https://doi.org/10.3389/fpls.2016.00224.

Morgan, K.T., Kadyampakeni, D., 2012. Open field hydroponics: concept and application in citrus. In: Srivastava, A.K. (Ed.), Advances in *Citrus* Nutrition. Springer Science, Dordrecht, pp. 271–281.

Morgan, K.T., Rouse, R.E., Roka, F.M., Futch, S.H., Zekri, M., 2005. Relationship among rootstock, mineral content, and peel thickness of Hamlin orange. Proc. Fla. State Horticult. Soc. 118, 19–21.

Morgan, K.T., Scholberg, J.M.S., Obreza, T.A., Wheaton, T.A., 2006a. Size, biomass, and nitrogen relationships with sweet orange tree growth. J. Am. Soc. Hortic. Sci. 131 (1), 149–156.

Morgan, K.T., Obreza, T.A., Scholberg, J.M.S., Parsons, L.R., Wheaton, T.A., 2006b. *Citrus* water uptake dynamics on a sandy Florida Entisol. Soil Sci. Soc. Am. J. 70 (1), 90–97.

Morgan, K.T., Obreza, T.A., Scholberg, J.M.S., 2007. Characterizing orange tree root distribution in space and time. J. Am. Soc. Hortic. Sci. 132 (2), 262–269.

Morgan, K.T., Schumann, A.W., Castle, W.S., Stover, E.W., Kadyampakeni, D., Spyke, P., Roka, F.M., Muraro, R., Morris, R.A., 2009a. *Citrus* production systems to survive greening: horticultural practices. Proc. Fla. State Horticult. Soc. 122, 114–121.

Morgan, K.T., Wheaton, T.A., Castle, W.S., Parsons, L.R., 2009b. Response of young and maturing citrus tree grown on a sandy soil to irrigation scheduling, nitrogen fertilizer rate, and nitrogen application method. Hort. Sci. 44, 145–150.

Morgan, K.T., Scholberg, J.M.S., Obreza, T.A., Wheaton, T.A., 2012. Seasonal nitrogen storage dynamics and budgets of mature citrus trees on a sandy Entisol. J. Plant Nutr. 35 (13), 2009–2023.

Morgan, K.T., Rouse, R.E., Ebel, R.C., 2016. Foliar applications of essential nutrients on growth and yield of 'Valencia' sweet orange with huanglongbing. Hort. Sci. 51, 1482–1493.

Morton, A.R., Trolove, S.N., Kerckhoffs, L.H.J., 2008. Magnesium deficiency in citrus grown in the Gisborne district of New Zealand. N. Z. J. Crop. Hortic. Sci. 36 (3), 199–213.

Mostafa, E.A.M., Saleh, M.M.S., 2006. Response of Balady mandarin trees to girdling and potassium sprays under sandy soil conditions. Res. J. Agric. Biol. Sci. 2 (3), 137–141.

Obreza, T.A., Boman, B.J., 2008. Fertilizer sources and formulations. In: Obreza, T.A., Morgan, K.T. (Eds.), Nutrition of Florida *Citrus* Trees. Soil and Water Science Department, Institute of Food and Agricultural Sciences, University of Florida, SL253, Gainesville, pp. 38–42.

Obreza, T.A., Collins, M.E., 2008. Common Soils Used for *Citrus* Production in Florida. SL 193. Florida Coop. Ext. Serv., Gainesville.

Obreza, T.A., Morgan, K.T., 2008. Nutrition of Florida *Citrus* Trees. SL253. University of Florida, IFAS Extension, Gainesville.

Obreza, T.A., Rouse, R.E., 1993. Fertilizer effects on early growth and yield of 'Hamlin' orange trees. Hort. Sci. 28, 111–114.

Obreza, T.A., Alva, A.K., Hanlon, E.A., Rouse, R.E., 1999. *Citrus* Grove Leaf Tissue and Soil Testing: Sampling, Analysis and Interpretation. SL115. University of Florida Cooperative and Extension Service, Gainesville.

Obreza, T.A., Rouse, R.E., Morgan, K.T., 2008a. Managing phosphorus for citrus yield and fruit quality in developing orchards. Hort. Sci. 43, 2162–2166.

Obreza, T.A., Boman, B.J., Zekri, M., Futch, S., 2008b. In: Obreza, T.A., Morgan, K.T. (Eds.), Nutrition of Florida *Citrus* Trees. Soil and Water Science Department, Institute of Food and Agricultural Sciences, University of Florida, SL253, Gainesville, pp. 43–47.

Papadakis, I.E., Dimassi, K.N., Therios, I.N., 2003. Response of two citrus genotypes to six boron concentrations: concentration and distribution of nutrients, total absorption, and nutrient use efficiency. Aust. J. Agric. Res. 54, 571–580.

Paramasivam, S., Alva, A.K., Hostler, K.H., Easterwood, G.W., Southwell, J.S., 2000. Fruit nutrient accumulation of four orange varieties during fruit development. J. Plant Nutr. 23, 313–327.

Pestana, M., Correia, P.J., Varennes, A., Abadía, J., Faria, E.A., 2001a. The use of floral analysis to diagnose the nutritional status of orange trees. J. Plant Nutr. 24, 1913–1923.

Pestana, M., David, M., De Varennes, A., Abadía, J., Araújo, E., Faria, E.A., 2001b. Responses of Newhall orange trees to iron deficiency in hydroponics: effects on leaf chlorophyll, photosynthetic efficiency, and root ferric chelate reductase activity. J. Plant Nutr. 24, 1609–1620.
Pestana, M., Varennes, A., Araújo, E., Faria, E.A., 2003. Diagnosis and correction of iron chlorosis in fruit trees: a review. Food Agric. Environ. 1, 46–51.
Pestana, M., Beja, P., Correia, P.J., Verennes, A., Faria, E.A., 2005a. Relationships between nutrient composition of flowers and fruit quality in orange trees grown in calcareous soil. Tree Physiol. 25, 761–767.
Pestana, M., de Varennes, A., Abadía, J., Araújo Faria, E., 2005b. Differential tolerance to iron deficiency of citrus rootstocks grown in nutrient solution. Sci. Hortic. 104, 25–36.
Petená, G., 2016. Estudos anatômicos e ultraestruturais dos *Citrus*: efeitos de fontes de nitrogênio em pomar fertirrigado. Instituto Agronômico. Dissertação (Mestrado em Agricultura Tropical e Subtropical).
Petená, G., Tanaka, F.A.O., Mesquita, G.L., Boaretto, R.M., Zambrosi, F.C.B., Quaggio, J.A., Mattos Jr., D., 2016. Scanning electron microscopy of leaf and petal cuts of citrus trees fertigated with two nitrogen sources. Citrus Res. Technol. 37 (2). https://doi.org/10.4322/crt.ICC119.
Pollard, A.S., Parr, A.J., Loughman, B.C., 1977. Boron in relation to membrane function in higher plants. J. Exp. Bot. 28, 831–834.
Quaggio, J.A., Teófilo Sobrinho, J., Dechen, A.R., 1992a. Response to liming of 'Valencia' orange tree on Rangpur lime: effects of soil acidity on plant growth and yield. Proc. Int. Soc. Citricult. 2, 628–632.
Quaggio, J.A., Teófilo Sobrinho, J., Dechen, A.R., 1992b. Magnesium influences on fruit yield and quality of 'Valencia' sweet orange on Rangpur lime. Proc. Int. Soc. Citricult. 2, 633–637.
Quaggio, J.A., Cantarella, H., van Raij, B., 1998. Phosphorus and potassium soil test and nitrogen leaf analysis as a basis for citrus fertilization. Nutr. Cycl. Agroecosyst. 52, 67–74.
Quaggio, J.A., Mattos Jr., D., Cantarella, H., Almeida, E.L.E., Cardoso, S.A.B., 2002. Lemon yield and fruit quality affected by NPK fertilization. Sci. Hortic. 96, 151–162.
Quaggio, J.A., Mattos Jr., D., Cantarella, H., Tank Jr., A., 2003. Fertilização com boro e zinco no solo em complementação à aplicação via foliar em laranjeira Pêra. Pesq. Agrop. Brasileira 38 (5), 627–634.
Quaggio, J.A., Mattos Jr., D., Cantarella, H., Stuchi, E.S., Sempionato, O.R., 2004. Sweet orange trees grafted on selected rootstocks fertilized with nitrogen, phosphorus and potassium. Pesq. Agrop. Brasileira 39, 55–60.
Quaggio, J.A., Mattos Jr., D., Souza, T.R., Krontal, Y., 2006a. Dinâmica de íons no solo e ganhos de eficiência fertilizante devido à irrigação e fertirrigação na citricultura. In: Reunião Brasileira de Fertilidade do Solo e Nutrição de Plantas, 27. Proceedings…. Embrapa Agropecuária Oeste, Dourados. CD-ROM.
Quaggio, J.A., Mattos Jr., D., Cantarella, H., 2006b. Fruit yield and quality of sweet oranges affected by nitrogen, phosphorus and potassium fertilization in tropical soils. Fruits 61 (5), 293–302.
Quaggio, J.A., Mattos Jr., D., Boaretto, R.M., 2010. Citros. In: Prochnow, L.I., Casarin, V., Stipp, S.R. (Eds.), Boas práticas para uso eficiente de fertilizantes. Piracicaba, International Plant Nutrition, pp. 371–409.
Quaggio, J.A., Mattos Jr., D., Boaretto, R.M., 2011. Sources and rates of potassium for sweet orange production. Sci. Agric. 68, 369–375.
Quaggio, J.A., Souza, T.R., Zambrosi, F.C.B., Boaretto, R.M., Mattos Jr., D., 2014. Nitrogen-fertilizer forms affect the nitrogen-use efficiency in fertigated citrus groves. J. Plant Nutr. Soil Sci. 177, 404–411.
Quiñones, A., Bañuls, J., Primo-Millo, E., Legaz, F., 2005. Recovery of the ^{15}N-labelled fertilizer in citrus trees in relation with timing of application and irrigation system. Plant Soil 268, 367–376.
Quiñones, A., Martínez-Alcántara, B., Legaz, F., 2007. Influence of irrigation system and fertilization management on seasonal distribution of N in the soil profile and on N-uptake by citrus trees. Agric. Ecosyst. Environ. 122, 399–409.
Quiñones, A., Martínez-Alcántara, B., Legaz, F., 2008. Optimization of N fertilization management in citrus trees: ^{15}N as a tool in NUE improvement studies. In: Elsworth, L.R., Paley, W.O. (Eds.), Fertilizers: Properties, Applications, and Effects. Nova Science Publishers, Inc., Hauppauge, pp. 181–206.
Quiñones, A., Martínez-Alcántara, B., Primo-Millo, E., Legaz, F., 2012. Fertigation: Concept and Application in *Citrus*. In: Srivastava, A.K. (Ed.), Advances in *Citrus* Nutrition. Springer Science, Dordrecht, pp. 281–301.
Ramos, C., Agut, A., Lidon, A.L., 2002. Nitrate leaching in important crops of the Valencian community region (Spain). Environ. Pollut. 118 (2), 215–223.
Recous, S., Machet, J.M., Mary, B., 1992. The partitioning of fertilizer-N between soil and crop: comparison of ammonium and nitrate applications. Plant Soil 144, 101–111.
Reitz, H.J., Koo, R.C., 1959. Effect of nitrogen and potassium fertilization on yield and fruit quality of Valencia orange in calcareous soil. Proc. Fla. State Horticult. Soc. 72, 12–16.
Roccuzzo, G., Scandellari, F., Allegra, M., Torrisi, B., Stagno, F., Tanja, T., Zanotelli, D., Gioacchini, P., Millard, P., Tagliavini, M., 2017. Seasonal dynamics of root uptake and spring remobilisation of nitrogen in field grown orange trees. Sci. Hortic. 229 (19), 223–230.
Rombolà, A.D., Brüggemann, W., Tagliavini, M., Marangoni, B., Moog, P.R., 2000. Iron source affects Fe reduction and re-greening of kiwifruit (*Actinidia deliciosa*) leaves. J. Plant Nutr. 23, 1751–1765.
Rozane, D.E., Mattos Jr., D., Parent, S.-E., Natale, W., Parent, L.E., 2015. Meta-analysis in the selection of groups in varieties of citrus. Commun. Soil Sci. Plant Anal. 46 (15). https://doi.org/10.1080/00103624.2015.1069307.
Santa-María, G.E., Moriconi, J.I., Oliferuk, S., 2015. Internal efficiency of nutrient utilization: What is it and how to measure it during vegetative plant growth? J. Exp. Bot. 28. https://doi.org/10.1093/jxb/erv162.
Schon, M.K., Novacky, A., Blevins, D.G., 1990. Boron induces hyperpolarization of sunflower root cell membranes and increases membrane permeability to K^+. Plant Physiol. 93, 566–571.
Schumann, A.W., Fares, A., Alva, A.K., Paramasivam, S., 2003. Response of 'Hamlin' orange to fertilizer source, annual rate, and irrigated area. Proc. Fla. State Horticult. Soc. 116, 256–260.

Sewelam, N., Kazan, K., Schenk, P.M., 2016. Global plant stress signaling: reactive oxygen species at the cross-road. Front. Plant Sci. 7, 187.
Silva, M.A.C., Natale, W., Prado, R.M., Corrêa, M.C.M., Stuchi, E.S., Andrioli, I., 2007. Aplicação superficial de calcário em pomar de laranjeira pêra em produção. Rev. Bras. Frutic. 29 (3), 606–612.
Smith, P.F., 1966. Leaf analysis of citrus. In: Childers, N.F. (Ed.), Nutrition of Fruit Crops. Somerset Press, New Jersey, pp. 208–228.
Smolen, S., 2012. Foliar nutrition: current state of knowledge and opportunities. In: Srivastava, A.K. (Ed.), Advances in *Citrus* Nutrition. Springer Science, Dordrecht, pp. 41–58.
Soil Survey Staff, 2014. Keys to Soil Taxonomy, 12th ed. USDA-Natural Resources Conservation Service, Washington, DC.
Souza, T.R., Bôas, R.L.V., Quaggio, J.A., Salomão, L.C., Foratto, L.C., 2012. Dinâmica de nutrientes na solução do solo em pomar fertirrigado de citros. Pesq. Agrop. Brasileira 47 (6), 846–854.
Spiegel-Roy, P., Goldschmidt, E.E., 1996. Biology of *Citrus*. Cambridge University Press, Cambridge, p. 230.
Srivastava, A.K., 2013. Nutrient deficiency symptomology in citrus: an effective diagnostic tool or just an aid for post-mortem analysis. Agric. Adv. 2 (6), 177–194.
Srivastava, A.K., Singh, S., 2004. Leaf and soil nutrient guides in citrus—a review. Agric. Rev. 25 (4), 235–251.
Syvertsen, J.P., Smith, M.L., 1995. Nitrogen leaching, N uptake efficiency and water use from citrus trees fertilized at three n rates. Proc. Fla. State Horticult. Soc. 108, 151–155.
Syvertsen, J.P., Smith, M.L., 1996. Nitrogen uptake efficiency and leaching losses from lysimeter-grown *Citrus* trees fertilized at three nitrogen rates. J. Am. Horticult. Sci. Alexandria 121, 57–62.
Syvertsen, J.P., Smith Jr., M.L., 1984. Light acclimation in *Citrus* leaves. I. Changes in physical characteristics, chlorophyll, and nitrogen content. J. Amer. Soc. Hortic. Soc. 109, 807–812.
Syvertsen, J.P., Smith, M.L., Boman, B.J., 1993. Tree growth, mineral nutrition and nutrient leaching losses from soil of salinized citrus. Agric. Ecosyst. Environ. 45, 319–334.
Tanou, G., Ziogas, V., Molassiotis, A., 2017. Foliar nutrition, biostimulants and prime-like dynamics in fruit tree physiology: new insights on an old topic. Front. Plant Sci. 8. https://doi.org/10.3389/fpls.2017.00075.
Underwood, W., 2012. The plant cell wall: a dynamic barrier against pathogen invasion. Front. Plant Sci., 3, 1–6.
van Beusichem, M.L., Kirkby, E.A., Baas, R., 1988. Influence of nitrate and ammonium nutrition on the uptake, assimilation, and distribution of nutrients in *Ricinus communis*. Plant Physiol. 86, 914–921.
van Raij, B., Quaggio, J.A., Silva, N.M., 1986. Extraction of phosphorus, potassium, calcium, and magnesium from soils by an ion-exchange resin procedure. Commun. Soil Sci. Plant Anal. 17, 547–566.
van Raij, B., Andrade, J.C., Cantarella, H., Quaggio, J.A., 2001. Análise química para avaliação da fertilidade de solos tropicais. Campinas, Instituto Agronômico, p. 285.
Verheye, W., de la Rosa, D., 2005. Mediterranean Soils in Land Use and Land Cover, From Encyclopedia of Life Support Systems (EOLSS), Developed Under the Auspices of the UNESCO. Eolss Publishers, Oxford, UK. http://www.eolss.net. Retrieved December 21, 2005.
Weil, R.R., Brady, N.C., 2016. The Nature and Properties of Soils. Pearson, Columbus, p. 1104.
White, P.J., Broadley, M.R., 2003. Calcium in plants. Ann. Bot. 92, 487–511.
Willis, L.E., Davies, F.S., Graetz, D.A., 1991. Fertigation and growth of young 'Hamlin' orange trees in Florida. Hort. Sci. 26, 106–109.
Wu, X., Riaz, M., Yan, L., Du, C., Liu, Y., Jiang, C., 2017. Boron deficiency in trifoliate orange induces changes in pectin composition and architecture of components in root cell walls. Front. Plant Sci. 8. https://doi.org/10.3389/fpls.2017.01882.
Zambrosi, F.C.B., Mattos Jr., D., Syvertsen, J.P., 2011. Plant growth, leaf photosynthesis, and nutrient-use efficiency of citrus rootstocks decrease with phosphite supply. J. Plant Nutr. Soil Sci. 174, 487–495.
Zambrosi, F.C.B., Mattos Jr., D., Furlani, P.R., Quaggio, J.A., Boaretto, R.M., 2012a. Eficiência de absorção e utilização de fósforo em porta-enxertos cítricos. Rev. Bras. Ciênc. Solo 36, 485–496.
Zambrosi, F.C.B., Mattos Jr., D., Boaretto, R.M., Quaggio, J.A., Muraoka, T., Syvertsen, J.P., 2012b. Contribution of phosphorus absorption and remobilization for citrus growth. Plant Soil 355, 353–362.
Zambrosi, F.C.B., Mattos Jr., D., Quaggio, J.A., Cantarella, H., Boaretto, R.M., 2013. Phosphorus uptake by young citrus trees in low-P soil depend on rootstock varieties and nutrient management. Commun. Soil Sci. Plant Anal. 44, 2107–2117.
Zambrosi, F.C.B., Mesquita, G.L., Menino, G., Tanaka, F.A.O., Mattos Jr., D., Quaggio, J.A., 2017. Anatomical and ultrastructural damage to citrus leaves from phosphite spray depends on phosphorus supply to roots. Plant Soil 415, 1–13.
Zekri, M., Koo, R.C.J., 1992. Application of micronutrients to citrus trees through microirrigation systems. J. Plant Nutr. 15 (11), 2517–2529.
Zekri, M., Obreza, T.A., 2015. Micronutrient Deficiencies in *Citrus*: Iron, Zinc, and Manganese. University of Florida Cooperative Extension Service, Institute of Food and Agricultural Sciences, EDIS.

Further reading

Corá, J.E., Silva, G.O., Martins Filho, M.V., 2005. Manejo do solo sob citros. In: Mattos Jr., D., De Negri, J.D., Pio, R.M., Pompeu Jr., J., (Eds.), Citros. Instituto Agronômico e Fundag, Campinas, pp. 347–368.
Laurindo, V.T., Silva, G.O., Pavani, L.C., Quaggio, J.A., 2010. Padrão de distribuição de K^+, Ca^{++}, Mg^{++} e P no solo de um pomar de citros em função da fertirrigação. Eng. Agríc. 30 (5), 909–921.
Quaggio, J.A., Cantarella, H., Mattos Jr., D., 1996. Soil testing and leaf analysis in Brazil—recent developments. Proc. Int. Soc. Citricult. 2, 1269–1275.
Yaalon, D.H., 1997. Soils in the Mediterranean region: what makes them different? Catena 28 (3–4), 157–169.

Chapter 16

Citrus pests in a global world

Alberto Urbaneja[a], Tim G. Grout[b], Santin Gravena[c], Fengnian Wu[d], Yijing Cen[d], Philip A. Stansly[e]

[a]*Centro de Protección Vegetal y Biotecnología, Instituto Valenciano de Investigaciones Agrarias (IVIA), Valencia, Spain,* [b]*Citrus Research International, Nelspruit, South Africa,* [c]*Paulista State University (UNESP, Retired) and Member of GCONCI-Consultant Citrus Group, Jaboticabal, Brazil,* [d]*South China Agricultural University, Guangzhou, China,* [e]*University of Florida—IFAS, Southwest Florida Research & Education Center, Immokalee, FL, United States*

16.1 Introduction

The agro-ecological citrus system is characterized by richness and abundance of phytophagous species as well as their natural enemies. However, only about 20 of these pests can be considered key, in part due to effective biological control. As early as the fourth century, Chinese citrus growers were known to distribute arboreal nests of the weaver ant, *Oecophylla smaragdina* (F.) (Hymenoptera: Formicidae) to manage pests, such as citrus stink bugs [*Rhynchocoris humeralis* (Thunberg) (Hemiptera: Pentatomidae)], leaf beetles [*Podagricomela nigricollis* Chen (Coleoptera: Chrysomelidae)], weevils [*Hypomeces squamosus* (Fabricius) (Coleoptera: Curculionidae)], and larvae of citrus butterflies and fruit borers (*Anomala cupripes* Hope (Coleoptera: Scarabaeidae)] (Yang, 2002). This practice continued into the 20th century in some parts of Southeast Asia, Burma, and China (Li, 1982). Another early well-known example is the first successful introduction to California of an exotic natural enemy, *Rodolia* (=*Novius*) *cardinalis* (Mulsant) (Coleoptera: Coccinellidae) from Australia to control the cottony cushion scale *Icerya purchasi* (Maskell) (Hemiptera: Monophlebidae) a pest of Asian origin (Fig. 16.1). *R. cardinalis* has now been introduced worldwide with the result that *I. purchasi* is rarely considered more than an occasional pest.

Citrus cultivation has been exported throughout the world's tropics and subtropics from origins in Southeast Asia, unfortunately along with many of its pests but not their natural enemies. This process continues at an accelerated pace in concert with increasing international commerce and travel. Consequently, many citrus pests are now worldwide in distribution and citrus industries have suffered from these and vectored diseases, notably huanglongbing (HLB) and citrus tristeza virus transmitted by psyllids and aphids, respectively. Two other vector and disease combinations: citrus variegated chlorosis (CVC) and citrus leprosis are transmitted by certain Cicadellidae and *Brevipalpus* (Acari: Tenuipalpidae) mites, respectively, and threaten to spread from their origins in South America.

Ever-increasing citrus pest diversity plus evolving pest control methods have resulted in continuously changing management models. The earliest was based on classical biological control, focusing on the introduction and establishment of exotic natural enemies from Southeast Asia to new citrus-growing regions. The next stage following WWII was marked by the growing use of synthetic pesticides capable of controlling principal pests but often without consideration for auxiliary fauna. Secondary pest outbreaks, pesticide resistance, and growing public awareness of health and environmental problems associated with pesticides led to the third stage of integration centered on biological control and selective chemistry. However, the appearance of the insect- and mite-transmitted diseases mentioned above, particularly HLB, has given way to a most recent stage of citrus pest management characterized by a return to intensive pesticide use for vector control.

While globalization and increased international traffic of raw materials have facilitated the distribution of many citrus pests throughout the world, the absence of some of these in specific areas and the influence of different climates and management options vary among zones. In this chapter, citrus pests and their management are detailed for five major citrus-growing areas of the world, Asia, the Mediterranean basin, North America, South America, and Africa. In each of these zones, pests have been classified based on three levels of control: (1) *key pests* for which chemical or biorational intervention is usually necessary, (2) *occasional pests* requiring occasional intervention, and (3) *secondary pests* generally controlled by natural enemies unless upset by the overuse of pesticides.

The Genus Citrus. https://doi.org/10.1016/B978-0-12-812163-4.00016-4

FIG. 16.1 Adults of the predator *Rodolia cardinalis* preying upon nymphs of the cottony cushion scale *Icerya purchasi*. *Photography courtesy of Jose Catalan, IVIA.*

16.2 Citrus pest management in Asia

Citrus is originally from Asia and produced in most Asian countries. The main citrus producers include China, India, Indonesia, Iran, Japan, Korea, Malaysia, Pakistan, Syria, Thailand, and Vietnam, together with approximately 38% of the total area planted to citrus worldwide. China is the world's greatest citrus producer with an estimated 2.5 million ha producing 36.6 million tons in 2015. Another large citrus producer is India, with close to a million hectares and production of over 7 million tons (Santoshkumar et al., 2013). Iran and Pakistan also produce over 2 million tons of citrus, while Indonesia, Syria, Vietnam, Korea, Japan, and Thailand produce over 0.5 million tons per year (Qi and Qi, 2016). Most Asian citrus fruit is oriented to fresh consumption. The requirement of high exterior quality necessitates frequent insecticide and acaricide use for citrus pest management.

Major citrus pests vary by country with over 800 pest species recorded from China. However, only a few of these are widely distributed and have consistently caused economic loss during the past 40 years. Similarly, 250 species of insects and mites have been reported infesting different species of citrus in India (Santoshkumar et al., 2013). Most infest young flush, hampering citrus growth and fruit development.

Inappropriate pest control of pests adversely affects citrus quantity and quality including size, brix, and appearance. Management tactics include physical, chemical, biological, and cultural approaches. Chemical control is often preferred because it is easy to use and provides quick results. However, overuse of pesticides disrupts natural biological systems, induces pest resistance, and eventually leads to a resurgence of pest populations. Increasing resistance to common insecticides such as organophosphates, pyrethroids, and neonicotinoids has led to more frequent applications. The overuse of pesticides has also raised serious environmental and human health concerns. Thus, alternative control options, such as physical, cultural, and biological measures, are always needed and have been evolving, particularly during the past few decades. Among these control measures, biological control has been favored by both scientists and citrus growers.

Biological control of citrus pests has a long history in Asia long before the presence of chemical insecticides. As mentioned above, the use of the golden weaver ant (*O. smaragdina*) for pest control was first recorded in *Plants of the Southern Regions* (Nanfang Caomu Zhuang, or 南方草木状) written in AD 304 by Ji Han in China). Given the diversity and abundance of natural enemies in Asian citrus orchards, natural control would be effective if not disrupted by chemical insecticides (Chen, 1985). Predacious mites have also been mass reared and released for citrus mite pests control in Asia (Huang, 2011). *Amblyseius* species (Phytoseiidae) such as *Amblyseius nicholsi* Ehara et Lee, and *Amblyseius deleoni* (Muma & Denmark) (Acarina: Phytoseiidae) are the most abundant predacious mites in citrus orchards in China. These species can be reared on pollen and were successfully used for citrus red mite control during the 1980s (Huang, 1989). However, production was limited by the availability of quality pollen. *Amblyseius cucumeris* (Oudemans) and *Amblyseius barkeri* (Hughes) (Acarina: Phytoseiidae) have been successfully cultured with factitious prey since the early 2000s, and are effectively used in the organic culture of citrus in China (Zhang et al., 2003; Pei et al., 2014).

Cultural control also has an important role in citrus pest management. The use of organic fertilizers is reported to reduce populations of many pest species, such as mites, scales, aphids, and whiteflies (Pang and Dong, 2012). Citrus interplants with a common weed, *Ageratum conyzoides* Linn. (Asteraceae), reportedly provide refuge and pollen for predatory mites

FIG. 16.2 Adult of the Asian citrus psyllid, *Diaphorina citri*.

and is used for citrus red mite control (Huang, 2011). Citrus interplanted with guava [*Psidium guajava* Linn. (Myrtaceae)] is used to reduce the Asian citrus psyllid (ACP) *Diaphorina citri* Kuwayama (Hemiptera: Liviidae) populations in Viet Nam (Gottwald et al., 2007) (Fig. 16.2). Windbreaks were found to reduce ACP population in citrus orchards (South China Agricultural College, 1981).

Mineral oils are widely used for citrus integrated pest management (IPM) in Asia. As early as the 1920s, oil-emulsifiable formulations were produced and sold for spider mite control (Ota, 2013). Petroleum spray oils were developed and promoted since the 1980s in China for control of mites, scales, whiteflies, and aphids (Cen et al., 1999). However, these oils were mainly used for winter sprays when citrus trees are in dormancy because of the high risk of phytotoxicity. Two research projects, "Citrus IPM in China," and "Citrus IPM in China and Southeast Asia" funded by Australian Centre for International Agricultural Research (ACIAR) were carried out from 1993 to 2000 in China, Vietnam, Thailand, Indonesia, and Malaysia. High-quality horticultural mineral oils characterized by narrow boiling point ranges and low levels of phytotoxic aromatic ring compounds were shown to be safe to crops and to control citrus pests such as ACP, citrus leafminer, and oriental fruit fly, mainly due to behavioral effects (Rae et al., 1996, 1997, 2000; Beattie et al., 2002; Ouyang et al., 2010). Since then, oils have been widely promoted for citrus pest control in Asia.

HLB was first described from south China and has been recognized as a major threat for citrus production in Asia during the last 50 years. Insecticidal control of the ACP and *Cacopsylla citrisuga* (Yang & Li) (Fig. 16.3) is a key measure for HLB management. Neonicotinoid, pyrethroid, organophosphorus, and carbamate insecticides are widely used for dormant (winter) season sprays and also applied two to three times during flushing periods for ACP control. One unwanted result of increased pesticide use has been an upsurge of acaricide sprays against the now resistant citrus red mite *Panonychus citri* (McGregor) (Acari: Tetranychidae) (Chen et al., 1980) (Fig. 16.4), with up to 58 applications a year being recorded, one-third of which were acaricides. Most insecticides used for ACP control disrupt beneficial arthropod populations, and copper compounds commonly applied for canker control suppress entomopathogens of citrus red mite. Thus, the demands of ACP and canker control have come into conflict with citrus red mite management in citrus culture in Asia.

FIG. 16.3 The psyllid *Cacopsylla citrisuga* is mainly present in southwestern China at high altitude and it has been found to vector HLB.

FIG. 16.4 Female and male citrus red mite, *Panonychus citri*.

16.3 Citrus pest management in the Mediterranean basin

The citrus industry in the Mediterranean basin represents approximately 20% of total citrus production worldwide. Commercial citriculture is found in Spain, Italy, Cyprus, Greece and France, Morocco, Egypt, Algeria and Tunisia, Turkey, and Israel. The Mediterranean industry is mainly oriented to fresh consumption requiring high-quality standards, especially in the main producing countries such as Spain, Turkey, and Morocco serving export markets. The Spanish industry provides an example as it exports more than 56% of its production. Consequently, "cosmetic thresholds" have been developed for some of the most important pests (Jacas and Urbaneja, 2010; Jacas et al., 2010).

Another important factor in the Mediterranean panorama is the absence until now of HLB, CVC and citrus canker, obviating a need to control psyllids, sharpshooters (Cicadellidae), or apply inordinate amounts of copper (Vacante and Gerson, 2012; Urbaneja et al., 2017). The remaining pests can be managed with a minimum of pesticide treatments compared to citrus-growing areas where these diseases are present. In addition, restrictions on the availability of active ingredients since January 2014 (EC, 2009), has stimulated the development of pest management programs based on conservation of natural enemies (Zappalà, 2010; Magrama, 2014; Mazih, 2015). These programs place a special emphasis on minimizing pesticide use in concert with consumer demand for high-quality citrus without pesticide residues. This is especially true in Northern Europe, which is the destination for most exported fruit.

As in many other citrus industries, the Mediterranean has been challenged by introductions of exotic pests. The response since the early 20th century has been numerous introductions of exotic natural enemies. Many have established stable populations and become highly valuable members of the natural fauna. The result has been better plant health compared to many other citrus-producing areas in the world.

Secondary pests controlled by natural enemies below economic injury levels (EILs) in the Mediterranean include *I. purchasi* and *Lepidosaphes gloverii* (Packard) (Hemiptera: Diaspididae) (Table 16.1). Occasional pests whose natural control is generally but not always satisfactory include *Aleurothrixus floccosus* (Maskell) (Hemiptera: Aleyrodidae), *Panonychus citri*, *Planococcus citri* Risso (Hemiptera: Pseudococcidae), and *Phyllocnistis citrella* Stainton (Lepidoptera: Gracillariidae). Key pests whose population oscillations that typically exceed EILs include *Ceratitis capitata* (Wiedemann) (Diptera: Tephritidae) (Fig. 16.5), *Aonidiella aurantii* (Maskell) (Hemiptera: Diaspididae), *Tetranychus urticae* Koch (Acari: Tetranychidae), and *Aphis spiraecola* Patch. (Hemiptera: Aphididae) in Clementine mandarins. The key to successful pest control is to maintain key pests below EIL without undoing suppression of natural enemies that would result in outbreaks of secondary and occasional pests. Pesticide timing and choice for efficacy and selectivity is critical to provide adequate control while avoiding disruption of biological control.

There has been a growing interest in recent years to promote and increase ecological infrastructures that enhance and preserve the action of natural enemies. In particular, cover crop management has minimized the use of pesticides of key

TABLE 16.1 Main citrus pests in the principal citrus production areas, Asia, the Mediterranean basin, North America, South America, and Africa in order of decreasing economic importance.

	Asia	Mediterranean Basin	North America	South America	Africa
Key pests	*Panonychus citri* *Diaphorina citri* *Phyllocnistis citrella* *Phyllocoptruta oleivora* *Aphis citricola* *Toxoptera citricidus* *Toxoptera aurantii* *Aphis gossypii* *Papilio xuthus* *Papilio polytes* *Anoplophora chinensis* *Bactrocera dorsalis*	*Ceratitis capitata* *Aonidiella aurantii* *Pezothrips kellyanus* *Delotococcus aberiae*[1] *Aphis spiraecola*[2] *Tetranychus urticae*[2] *Prays citri*[3]	*Diaphorina citri*[1,2] *Phyllocnistis citrella*[1] *Phyllocoptruta oleivora*[1] *Diaprepes abbreviatus*[1] *Aonidiella aurantii*[2,3] *Coccus pseudomagnoliarum*[3] *Scirtothrips citri*[3] *Eutetranychus banksi*[3] *Panonychus citri*[2,3]	*Diaphorina citri* *Brevipalpus* spp.[1] *Phyllocoptruta oleivora* *Polyphagotarsonemus latus*[2] *Gymnandrosoma aurantiana* *Ceratitis capitata* *Anastrepha fraterculus* *Phyllocnistis citrella*[3] *Cicadellinae (Sharpshooters)*[4]	*Ceratitis capitata* *Ceratitis rosa* *Thaumatotibia leucotreta* *Bactrocera dorsalis* *Trioza erytreae* *Aonidiella aurantii* *Scirtothrips aurantii* *Planococcus citri*
Occasional pests	*Cacopsylla citrisuga*[1] *Dialeurodes citri*[2] *Aleurocanthus spiniferus* *Parlatoria pergandii* *Chrysomphalus ficus* *Aonidiella aurantii* *Unaspis yanonensis* *Icerya purchasi* *Planococcus citri* *Scirtothrips citri* *Scirtothrips dorsalis* *Myzus persicae* *Eotetranychus kankitus* *Eutetranychus orientalis*[2] *Aleurocanthus woglumi*[2] *Bactrocera minax* *Chelidonium argentatum* *Nadezhdiella cantori* *Buzura suppressaria benescripta* *Anomala cupripes* *Rhynchocoris humeralis* *Platymycteropsis mandarinus* *Sympiezomias citri* *Hypomeces squamosus*	*Aspidiotus nerii*[3] *Aphis gossypii*[2] *Planococcus citri* *Phyllocnistis citrella* *Aceria sheldoni* *Paraleyrodes minei* *Lepidosaphes beckii* *Parlatoria pergandei* *Aleurothrixus floccosus* *Ceroplastes sinensis* *Coccus hesperidum* *Panonychus citri*	*Aculops pelekassi* *Panonychus citri* *Solenopsis invicta* *Eutetranychus banksii* *Aphis spiraecola* *Aphis gossypii* *Papilio cresphontes* *Toxoptera citricidus* *Unaspis citri* *Anastrepha suspensa*	*Planococcus citri* *Unaspis citri* *Parabemisia myricae* *Aleurothrixus floccosus* *Aleurocanthus woglumi* *Tetranychus mexicanus* *Panonychus citri* *Frankliniella occidentalis*[1] *Scirtothrips citri*[2] *Oxydia apidania* *Toxoptera citricidus* *Aphis spiraecola* *Toxoptera aurantii* *Naupactus* spp.	*Helicoverpa armigera* *Aceria sheldoni* *Lozotaenia capensana* *Panonychus citri* *Aleurothrixus floccosus* *Aphis gossypii* *Empoasca distinguenda* *Eutetranychus orientalis* *Lepidosaphes beckii* *Naupactus godmani* *Papilio* spp. *Phyllocoptruta oleivora* *Phyllocnistis citrella* *Prays citri* *Chrysomphalus aonidum* *Pseudococcus longispinus* *Toxoptera citricidus*

(Continued)

TABLE 16.1 Main citrus pests in the principal citrus production areas, Asia, the Mediterranean basin, North America, South America, and Africa in order of decreasing economic importance—cont'd

	Asia	Mediterranean Basin	North America	South America	Africa
Secondary pests	*Lepidosaphes beckii* *Nipaecoccus vastator* *Cappaea taprobarensis* *Ceroplastes rubens* *Parlatoria zizyphus* *Oraesia emarginata* *Oraesia excavata* *Adris tyrannus* *Adoxophyes cyrtosema* *Homona coffearia* *Bactrocera tsuneonis* *Bactrocera zonata*[2] *Papilio demoleus*[2] *Eudocima fullonia*[2] *Cappaea taprobanensis* *Diaphorina communis*[3]	*Saissetia oleae* *Toxoptera aurantii* *Chrysomphalus dyctiospermi* *Ectomyelois ceratoniae* *Cacoecimorpha pronubana* *Icerya purchasi*	*Saissetia oleae* *Chrysomphalus aonidum* *Parlatoria pergandei* *Lepidosaphes beckii* *Lepidosaphes gloverii* *Planococcus citri*	*Icerya purchasi* *Cornuaspis beckii* *Chrysomphalus aonidum* *Selenaspidus articulatus* *Parlatoria cinerea* *Parlatoria ziziphus* *Orthezia praelonga*	*Icerya purchasi* *Coccus hesperidum* *Ceroplastes brevicauda* *Ceroplastes destructor* *Ectomyelois ceratoniae* *Pulvinaria aethiopica* *Nipaecoccus viridis* *Delottococcus aberiae* *Saissetia oleae* *Anoplolepis custodiens* *Pheidole megacephala*

Key pests: their numbers exceed every season the economic injury level (EIL); occasional pests: EIL are exceeded every 2–4 years, secondary pests: EIL are rarely exceeded. Within each group pests are ordered by importance from top to bottom.
Asia: [1]Mainly present in southeastern China with high elevation, and was found associated with HLB. [2]Key pests in India. [3]Mainly on lemons in Yunnan Province, China.
Mediterranean Basin: [1]Only present in Middle East of Spain. [2]Mainly on Clementine mandarins. [3]Mainly on lemons.
North America: [1]Major pest in Florida. [2]Major pest in Texas. [3]Major pest in California.
South America: [1]Only in Oranges. [2]Mainly in fruits for fresh market. [3]Associated with the citrus canker. [4]*Acrogonia citrina, Oncometopia facialis*, and *Dilobopterus costalimai*.

FIG. 16.5 Female of the Mediterranean fruit fly, *Ceratitis capitata.*

FIG. 16.6 Colony of the green citrus aphid, *Aphis spiraecola.*

pests such as *T. urticae* and *A. spiraecola* in Clementine mandarins (Fig. 16.6). Establishment of a monoculture ground cover of the grass *Festuca arundinacea* Schreb. (Poaceae) has promoted an increase of specific predators of Tetranychidae and hindered the transfer of spider mites between the citrus canopy and the ground cover due to feeding specialization (Aguilar-Fenollosa et al., 2011a,b). Similarly, a sown grass cover enriched with wild forbs serves as host for alternate prey to build populations of aphid predators that later move into the citrus canopy (Gómez-Marco et al., 2016a,b). These positive results have strengthened conservation biological control in the Mediterranean citrus industry and reduced chemical treatments. In response, the Valencian Community (Eastern Spain) has launched an area-wide program against the Mediterranean fruit fly, *C. capitata*, based on sterile insect release (Juan-Blasco et al., 2014). Biorational control strategies are also being employed for control of California red scale (CRS), *A. aurantii*, including mass release of the parasitoid *Aphytis melinus* (Tena et al., 2015) and use of pheromones for mating disruption (Vacas et al., 2010).

Unfortunately, pest management in the Mediterranean is too often disrupted by exotic pests, most recently the mealybug *Delottococcus aberiae* (De Lotto) (Hemiptera: Pseudococcidae) (Fig. 16.7) and the thrips species *Pezothrips kellyanus* (Bagnall) (Thysanoptera: Thripidae) and *Chaetanophothrips orchidii* (Moulton) (Thysanoptera: Thripidae). The urgency to control these pests with insecticides compromises conservation biological control. Detection of HLB or its vectors would undoubtedly change the rules of the game. Prevention is the best way to forestall what could otherwise spell disaster for the Mediterranean fruit industry.

FIG. 16.7 Infested fruit with nymphs of *Delottococcus aberiae*. *Photography courtesy of Alejandro Tena, IVIA.*

16.4 Citrus pest management in North America

Citrus production in the United States is confined primarily to three regions: Florida, Texas, and California including southern Arizona. Mexico, Costa Rica, the Dominican Republic are also significant players in North American process and fresh fruit industries. The Florida citrus industry is largely oriented toward orange juice production (ca. 90%), whereas the other two regions in the United States supply the fresh fruit market with the full gamut of citrus types. Production of oranges in Florida has been second only to Brazil but has plummeted in recent years due to HLB first detected in 2005. Therefore, control of the HLB vector, the ACP *D. citri,* has become the primary goal of insect pest management in the state as well as in Mexico, Central America, and the Caribbean. At the other extreme, ACP is still found only in relatively small areas of California with HLB confined to dooryard citrus. Monitoring and control of ACP there is thus a growing concern, although traditional pest management is still a major goal. The situation in the smaller south Texas industry is intermediate, with HLB well established already.

16.4.1 Florida

First detected in 1998, ACP swept through Florida in record time with speed comparable to the spread of citrus leafminer *P. citrella* 5 years earlier (Fig. 16.8). Movement of both pests was aided by commercial transport of plants throughout the state. HLB or citrus greening disease was detected 7 years later in pomelos in the extreme southeast of the peninsula. Subsequent delimiting surveys discovered positive citrus trees well up the east coast and across the state. Tree removal proved ineffective, and so began a protracted battle with vector control and tree health management as basic tools, gradually aided by a reactivated breeding program.

Management of ACP in the United States is primarily through inscticidal control. This is a year-round endeavor for all commercial citrus in Florida and Texas, whereas control in California and Arizona is primarily in response to ACP presence

FIG. 16.8 Citrus leaf infested with the citrus leafminer, *Phyllocnistis citrella*. *Photography courtesy of Elena LLácer, IVIA.*

FIG. 16.9 Orange fruit damaged by the citrus rust mite, *Phyllocoptruta oleivora*.

as determined by monitoring. Key to the management of ACP in Florida are the so-called "dormant sprays" during winter when bearing trees are generally not flushing and psyllids must, therefore, survive as adults on hardened foliage or alternate hosts (Qureshi and Stansly, 2010). The objective is to reduce populations as much as possible before the spring flush when conditions are most favorable for ACP reproduction and subsequent spread of HLB. Another advantage of dormant sprays is conservation of predatory insects such as ladybeetles and lacewings that are generally absent from the groves due to lack of prey associated with a young flush. For this and reasons of economy and efficacy, broad-spectrum insecticides such as organophosphates and especially pyrethroids are products of choice whereas neonicotinoids are discouraged due to their importance as soil-applied systemic insecticides in young plantings. More selective insecticides are recommended during the growing season, both to address occasional or secondary pests as well as to conserve beneficial arthropods (Qureshi et al., 2014). Such secondary pests include citrus leafminer, *P. citrella*, the citrus rust mite *Phyllocoptruta oleivora* (Ashmead) (Fig. 16.9) and *Aculops pelekassi* (Keifer) (Acari: Eriophyoidea) and the red spider mites *P. citri* and *Eutetranychus banksi* (McGregor) (Acari: Tetranychidae). Leafminer is of special concern due to its role of exacerbating the spread of citrus canker disease by facilitating entry of the causative bacteria, *Xanthomonas axonopodis* pv. *citri* (Xanthomonadales: Xanthomonadaceae). Another pest of concern is the weevil *Diaprepes abbreviatus* L. (Coleoptera: Curculionidae), whose larvae are voracious root feeders exacerbating root loss from HLB and *Phytophthora* spp. (Peronosporales: Peronosporaceae).

Effects of natural enemies on ACP survival in Florida prior to widespread insecticide use were evaluated by Michaud (2002, 2004) and Qureshi and Stansly (2009) using exclusion techniques. Mortality of 90% or greater was attributed to predators such as ladybeetles, lacewings, syrphid flies, and spiders. A later study by Monzo et al. (2014) showed significant effects of insecticides on three classes of beneficial arthropods in response to different regimes and intensities of insecticide use in the field. These studies signaled the largely untapped potential of beneficial arthropods to contribute in a significant way to ACP control if fewer and more selective insecticides are used. Economic thresholds for mature trees with high incidence of HLB based on "tap" sampling (Monzo et al., 2015) have been shown to be effective in reducing insecticide use and optimizing profits (Monzo and Stansly, 2017).

New plantings represent the future of the industry and a special challenge for ACP control due to frequent flushing and susceptibility to HLB. Effective soil-applied systemic insecticides are largely confined to the neonicotinoid (4A) class, although cyantraniliprole (group 28) is effective but expensive. Thus, rotations with foliar-applied products are necessary to forestall selection for resistance. A cultural practice that is gaining acceptance in Florida is the use of metalized reflective polyethylene film as a ground cover in the plant row (Croxton and Stansly, 2014; Stansly et al., 2016). The mulch mirrors the sky, reflecting all visible wavelengths including UV, disrupting insect flight. The effect is reduced ACP infestation and incidence of HLB. This and the advantages of additional light and efficiency of the required drip irrigation, fertigation, and chemigation system results in increased growth rate and early yield.

16.4.2 California and Texas

In contrast to Florida with its cool dry winters and wet hot summers, the climate in California is Mediterranean with cool wet winters and hot dry summers. Likewise, the citrus industry there is also primarily fresh fruit. It is not surprising then, that the pest complex is similar in California and the Mediterranean region. Major pests in California not adequately controlled biologically include CRS *A. aurantii* (Fig. 16.10) and the citrus thrips *Scirtothrips citri* (Moulton) (Thysanoptera: Thripidae). In addition, the citricola scale, *Coccus pseudomagnoliarum* (Kuwana) (Hemiptera: Coccidae) and the citrus red mite, *P. citri*, can also be considered key pests under certain conditions of cultivation (Table 16.1). Therefore, a combination of biological, chemical, and cultural controls is needed for these pests.

FIG. 16.10 Citrus fruit infested by California red scale *Aonidiella aurantii*. *Photography courtesy of Alejandro Tena, IVIA.*

CRS is a key pest in California and Texas, causing significant damage to all aerial parts of the tree and causing the fruit to be culled if infested with more than 10 scales. Pheromone-baited sticky traps are used for monitoring populations, although correlations with female numbers can be problematic, necessitating visual inspection. *A. melinus* DeBach, *A. lingnanensis* Compere (Hymenoptera: Aphelinidae), and *Comperiella bifasciata* (Howard) (Hymenoptera: Encyrtidae) are important parasitoids of CRS in different regions of California. Mass release of *A. melinus* is a viable management tactic in most regions, although generally incompatible with carbamates or organophosphate insecticides. The insect growth regulators buprofezin and pyriproxyfen are effective and compatible with parasitoids but suppress Coccinellidae, especially vedalia beetles needed for control of cottony cushion scale. Spirotetramat is active against all scale stages and compatible with parasitoids but suppresses predacious mites. Oil is least damaging to natural enemies and effective against young scale stages, although this tends to synchronize scale populations, later creating a dearth of susceptible third instar hosts needed by parasitoids.

A similar pest complex occurs in Texas with *A. aurantii*, *P. citri*, and *E. banksi* (McGregor) (Acari; Tetranychidae) as key pests. However, ACP was found in Texas in 2001 and HLB confirmed in commercial citrus in 2012. Thus, ACP control has become the main thrust of citrus pest management in the Lower Rio Grande Valley, employing similar strategies to those used in Florida.

16.5 Citrus pest management in South America

The citrus industry of South America (SA) represents approximately 38% of the total citrus production worldwide and includes oranges, grapefruit, tangerines, and lemons. Citriculture is important to the economies of Brazil, Argentina, Uruguay, Venezuela, and Peru. About 80% of the citrus production in SA is oriented to processing with no reason to create a standard of high external quality fruit.

The major pest management problem is the ACP, *D. citri*, a vector of HLB first detected in Brazil in 2004, and now also present in Argentina and Venezuela. ACP with HLB is capable of infecting and killing large numbers of plants in a relatively short time even at very low density. In response to this threat, the use of insecticides has intensified to 36–40 sprays a year applied at regular intervals in an attempt to wipe out the vector.

Unlike other countries where ACP is a recent invader, the citrus in Brazil has lived with the psyllid for at least 76 years as documented by the entomologist Angelo da Costa Lima (Costa Lima, 1942). Long residence in Brazil probably resulted in a biological balance with predatory arthropods compared to what is seen in the United States, Mexico, and elsewhere in the Americas (Huffaker et al., 1984). Given the rapidity with which HLB appeared in other countries following the establishment of ACP, it is likely that the disease was introduced in recent times, probably in plants or plant parts brought into the country. Curiously, the parasitoid *Tamarixia radiata* (Waterston) (Hymenoptera: Eulophidae) was never found in South America before the detection of HLB (Gravena et al., 1996). Thus, a recent introduction of the parasitoid by the same process seems likely.

Development and application of citrus IPM began in the late 1970s and improved until 2004. Only two key pests were present before 1987, both mites: (1) a complex of *Brevipalpus* spp. (Acari: Tenuipalpidae) that transmit citrus leprosis

FIG. 16.11 Leprosis virus symptoms on fruits and branches transmitted by the tenuipalpid citrus mite complex *Brevipalpus* spp.

virus (Fig. 16.11) and (2) Citrus Rust Mite, *P. oleivora*. Little insecticide was used as only acaricides were needed which controlled these pests with little impact on beneficial insects. The first major change occurred in 1987 with the discovery of CVC transmitted by sharpshooters (Cicadellinae), mainly *Acrogonia citrina* Marucci & Cavichioli, *Oncometopia facialis* (Signoret), and *Dilobopterus costalimai* Young (Hemiptera: Cicadellidae). Insecticides were used to forestall transmission of the bacterial disease caused by *Xylella fastidiosa* (Xanthomonadales: Xanthomonadaceae) inhabiting the xylem. The bacteria caused dieback in the canopy that could be controlled by the timely pruning of symptomatic branches. IPM continued despite this setback, albeit with greater difficulty due to imbalances caused by the multiple sprayings of insecticides, mainly during the formation of new orchards.

The "Pest Sampler" organization was formed in 1977 consisting of about 6000 professionals dedicated to monitoring insect and mite pest populations in Brazilian citriculture. Many short courses for Samplers and Citrus IPM Managers were taught at the Paulista State University (UNESP) until 1993 and for another 12 years by a private company. The IPM program consisted of four subsystems focusing on the key and some secondary pests: (1) conservation and increase of natural enemies, (2) sampling with regard to levels of action and inaction, (3) use of selective insecticides/acaricides, and selective application, and (4) environmental pest management. Thus, most orchards among about 750,000 ha of citrus, during the 1980–1990s, were managed with some level of IPM. An average of only one to two annual sprays were applied until 1987 and five to six sprays from 1987 until 2004 when HLB was first detected in Brazil. But, now the bacterium is almost totally extinct from Brazilian citriculture according to Fundecitrus.

Everything changed for pest management in the citrus industry of SA with detection of HLB. Instead of selective sprays based on EILs as before, applications followed preestablished spraying schedules with up to 36–40 sprays a year using broad-spectrum insecticides. It was assumed that these sprays also controlled the other pests, although secondary pest outbreaks occurred due to unavoidable imbalances and evidence of resistance by ACP and other insects such as citrus rust and citrus leprosis mites (*Brevipalpus* spp.)

Brevipalpus mites are the number 1 key pest after ACP in orange orchards in Brazil because they vector leprosis virus that can cause devastating damage to citrus in São Paulo and Minas Gerais states and has spread as far as southern Mexico. The sample units for this pest are: (1) two inner canopy and two outer canopy fruits of the present season or in their absence of the previous season and (2) 30 cm of two productive external branches with a young fruit at the tip. The recommended procedure is to first focus on the peduncle, then the blossom end of the fruit, and finally the rest of the fruit using a 10× loupe. One adult mite is sufficient to consider the fruit infested. The IPM unit is a block of 2000–3000 plants (or 10 ha) of which 1% is sampled. An acaricide is applied when mites are found on 10% of the fruits and leprosis is present or at 15% in the absence of virus symptoms. The average number of sprays using this system was 1.5/year. Orchards, where acaricides are not applied on the day after reaching EIL, use a lower EIL (3%–5% infestation). Leprosis is not a problem in mandarins, lemons, and limes in Brazil and SA (Gravena, 2005).

Rust mite is key pest number 2 in orchards under IPM in Brazil, SA, as it is in Mexico and Florida. Damage is mechanical and fruits of all ages are monitored by size. Rust mite damage is primarily a concern for the fresh fruit industry, but in extreme cases may cause significant loss of weight and juice. A 10× loupe is recommended with 1 cm^2 of the field of view to estimate the number of mites. The lens is positioned on a side of the fruit where mites will most likely be present. Three external fruits per plant and 1% of the plants in a block are monitored. Five or more mites per cm^2 on 20%–30% of the fruit signals a spray for either fresh or juice; 20 or more mites per cm^2 on 10% of the fruit for fresh; and 30 or more mites/cm^2 on 10% of the fruit for juice. The sampling frequency is 7 days in spring and summer and 10–15 days in autumn and winter for both key mite pests (Gravena, 2005). Two or three applications for rust mite using sulfur, abamectin or another approved acaricide has been an average using this system.

Balanced against the need for practically eliminating ACP to avoid the demise of whole orchards is the considerable risk posed by toxic insecticide residues in fruits and juice as well as catastrophic effects on the citrus agroecosystem. One option might be a combination of management systems: one for ACP and another for the rest of the pest complex. Clearly, the usual IPM program in use since 1970–2004 will no longer function the same, but the system of scheduled sprays could be improved by alternating chemical groups to avoid resistance in ACP and other pests, mainly *Brevipalpus* spp. For example, broad-spectrum sprays for ACP could be replaced by biopesticides or IGRs to mitigate harmful effects to natural enemies. Thus, a new IPM is needed that encompasses ACP until genetically based technologies are available to manage the causal bacterium *Candidatus* Liberibacter asiaticus.

16.6 Citrus pest management in Africa

Citrus is grown only for local or regional consumption in most of Africa south of the Sahara because of warm tropical conditions that do not permit sufficient color break to compete in international markets. However, citrus production from approximately 17°S southward that includes Zimbabwe, South Africa, and Eswatini is largely directed at lucrative export markets (Grout and Moore, 2015).

Distinct approaches to pest management are required for local vs export destinations. Production pests that reduce crop yield are more important than cosmetic pests for fruits destined for local consumption, especially when harvested green. Pesticides are seldom required and input costs are low, opening the door for biological control. In contrast, cosmetic pests are important where the fruit is produced for export or local high-quality consumption, requiring more intensive control measures with costs that must be offset by higher yields. In addition, some pests of minor importance for local markets may be of phytosanitary concern at the export destination, requiring levels of control incompatible with natural enemies. Biorational techniques are thus preferred to avoid disruption of biological control of secondary or occasional pests.

16.6.1 Production pests

The African citrus psyllid *Trioza erytreae* (Del Guercio) (Hemiptera: Triozidae) is a vector of the African citrus greening disease (African HLB) caused by the bacterium *Candidatus* Liberibacter africanus (Khamis et al., 2017) (Fig. 16.12). *T. erytreae* is a key pest in cooler climates in Africa where maximum temperatures seldom reach 35°C. Pest status is low where this psyllid occurs in the absence of greening, for example, the Eastern Cape Province of South Africa (Rwomushana et al., 2017). This is also the case for low altitude, hot parts of East Africa where ACP is reported but not either form of HLB (Roberts et al., 2017). However, vector control is essential in the presence of HLB and whenever new foliage is present or production declines over a period of 4–10 years until the orchard is no longer profitable. For this reason, the larger citrus orchards at low altitude are more viable where HLB is not present.

FIG. 16.12 Adult of the African citrus psyllid, *Trioza erytreae*. *Photography courtesy of Peter R. Stephen, CRI.*

The CRS, *A. aurantii*, is widespread in Africa but is usually not much more than a cosmetic pest due to biological control by *Aphytis africanus* Quednau (Hymenoptera: Aphelinidae) and coccinellid beetles. However, when these are eliminated by disruptive chemicals or dust from nearby roads, high populations can cause fruit drop, twig dieback, and even tree death. This can also happen to a lesser degree with cottony cushion scale *I. purchasi* when *Rodolia* spp. ladybeetles are compromised by sprays.

Internal feeders such as fruit flies and lepidopteran fruit borers cause preharvest fruit drop although their role as phytosanitary pests is often of greater concern. The color break does not occur until shortly before harvest in the tropics, so fruit is less attractive to fruit flies than in the subtropics where the attack begins 2 months before harvest. False codling moth, *Thaumatotibia leucotreta* (Meyrick) (Lepidoptera: Tortricidae), can often be maintained at subeconomically damaging levels for local market fruit by timely destruction of infested fruit while sparing egg and larval parasitoids. However, high populations may cause significant fruit drop from 6 weeks after petal fall to harvest where alternative hosts such as chilies or paprika are grown nearby.

16.6.2 Cosmetic pests

The citrus thrips, *Scirtothrips aurantii* Faure (Thysanoptera: Thripidae), is widespread in Africa and causes different types of fruit damage depending on when feeding occurs. Most severe damage occurs under the calyx or at the bottom of the fruit soon after petal fall. However, late-season populations can build up on new foliage and then move to the fruit, causing scribbling and russeting that resembles mite damage and reduces pack out. The citrus rust mite, *P. oleivora*, can be a serious cosmetic pest in warm, humid climates if not controlled. The fungicide mancozeb used to manage leaf spot of citrus (*Pseudocercospora angolensis* (T. Carvalho & O. Mendes) Crous & U. Braun (Capnodiales: Mycosphaerellaceae)) and citrus black spot (*Phyllosticta citricarpa* (Kiely) (Botryosphaeriales: Phyllostictaceae)) will also control citrus rust mite, although resistance has developed in parts of South Africa (Grout and Stephen, 2002). Lepidopteran pests such as leafrollers (Tortricidae) and the African bollworm, *Helicoverpa armigera* (Hübner) (Lepidoptera: Noctuidae), can cause sporadic cosmetic damage soon after petal fall in the form of deep scars in developing fruitlets. Feeding by the African mealybugs, *Nipaecoccus viridis* (Newstead) (Hemiptera: Pseudococcidae) and *D. aberiae*, causes bumps on citrus fruit but these species are rare compared with the citrus mealybug, *P. citri*. Mealybug feeding can also result in red spots on the rind if the weather turns cold before harvest and sooty mold residues can be a cosmetic problem. However, rind damage from leaf spot of citrus may be worse than any caused by arthropod pests in tropical Africa if rain falls on developing fruitlets (Harling et al., 2010).

16.6.3 Phytosanitary pests

Internal fruit pests pose the greatest phytosanitary threat because of strict tolerance requirements at the export destination and difficult postharvest detection. In addition, pests like Fuller's rose beetle, *Naupactus godmani* (Crotch) (Coleoptera: Curculionidae) are not considered economically damaging to citrus but are phytosanitary pests for certain Asian countries and thus require control (Mo and Stevens, 2013; Morse and Grafton-Cardwell, 2013).

Internal fruit pests are mostly controlled in Southern Africa using biorational techniques. False codling moth *T. leucotreta* is managed by sterile insect release, mating disruption during summer months, as well as attract and kill, and sprays of granulovirus. Nevertheless, importing countries often demand postharvest control, usually in the form of cold treatments (Moore et al., 2017).

Several species of fruit fly (Tephritidae) attack citrus fruits shortly before harvest in Africa. *C. capitata* is the most widespread, occurring in all climatic regions of Africa. *Ceratitis rosa* Karsch (Diptera: Tephritidae) attacks citrus on the eastern side of the continent at lower elevations, but declines in dry weather. The oriental fruit fly, *Bactrocera dorsalis* (Hendel) (Diptera: Tephritidae) (Fig. 16.13), has spread throughout Africa south of the Sahara with the exception of the southern and western parts of South Africa, although citrus is not considered a good host (Theron et al., 2017). Fortunately, most fruit flies can be controlled with sparsely applied protein bait spray droplets containing a toxicant with little disruption of IPM. Bait stations and attract-and-kill approaches are also available. Unfortunately, the sterile insect technique is only feasible in some parts of the Western Cape Province of South Africa where only *C. capitata* is dominant. Some importing countries demand postharvest cold treatments for citrus grown in Africa to control fruit flies but these requirements are less stringent than for false codling moth which is often also present.

FIG. 16.13 Female of the oriental fruit fly, *Bactrocera dorsalis*.

16.7 Epilogue

The citrus industry has always met pest management challenges in the past and will doubtless find effective solutions for the current hurdle of HLB. While the anticipated genetic solution for citrus susceptibility to Liberibacter is still pending, growers in affected areas can still survive thanks to higher prices needed to finance better vector control and enhanced horticultural practices that maintain production levels. Areas still vector free need to step up vigilance to prevent or delay entry of either of the two HLB vectors as long as possible. At the end of the day, continued demand for citrus fruits, juice, and other derived products will provide the impetus necessary to develop appropriate technologies that deal effectively with this and future problems threatening the citrus industry.

References

Aguilar-Fenollosa, E., Ibáñez-Gual, M.V., Pascual-Ruiz, S., Hurtado, M., Jacas, J.A., 2011a. Effect of ground-cover management on spider mites and their phytoseiid natural enemies in clementine mandarin orchards (II): top-down regulation mechanisms. Biol. Control 59, 171–179.

Aguilar-Fenollosa, E., Ibáñez-Gual, M.V., Pascual-Ruiz, S., Hurtado, M., Jacas, J.A., 2011b. Effect of ground-cover management on spider mites and their phytoseiid natural enemies in clementine mandarin orchards (I): bottom-up regulation mechanisms. Biol. Control 59, 158–170.

Beattie, A., Watson, D., Stevens, M., Rae, D., Spooner-Hart, R., 2002. Spray Oils Beyond 2000: Sustainable Pest and Disease Management: Proceedings of a Conference Held From 25 to 29 October 1999 in Sydney, New South Wales, Australia. University of Western Sydney.

Cen, Y.J., Xu, C.B., Tian, M.Y., 1999. Advances in the use of petroleum spray oils in control of pests in citrus. J. South China Agric. Univ. 20, 118–122.

Chen, S.J., 1985. Integrated pest management especially by natural control in citrus orchards. Nat. Enemies Insects 7, 223–231.

Chen, S.J., Chou, F.W., Zhuang, S.G., Chen, H.L., 1980. An investigation on the cause of the rampancy of citrus red mite, *Panonychus citri* McG. and control measures. J. South China Agric. Coll. 1, 101–111.

Costa Lima, A., 1942. Insetos do Brasil (3° Tomo)—Homópteros. Escola Nacional de Agronomia SÉRIE DIDÁTICA N.° 4—1942, p. 101.

Croxton, S.D., Stansly, P.A., 2014. Metalized polyethylene mulch to repel Asian citrus psyllid, slow spread of huanglongbing and improve growth of new citrus plantings. Pest Manag. Sci. 70, 318–323.

EC, 2009. Sustainable use directive. European Parliament and of the Council of 21 October 2009 establishing a framework for community action to achieve the sustainable use of pesticides. Off. J. Eur. Union L309, 71–86.

Gómez-Marco, F., Tena, A., Jacas, J.A., Urbaneja, A., 2016a. Early arrival of predators controls *Aphis spiraecola* (Hemiptera: Aphididae) colonies in citrus clementine. J. Pest. Sci. 89, 69–79.

Gómez-Marco, F., Urbaneja, A., Tena, A., 2016b. A sown grass cover enriched with wild forb plants improves the biological control of aphids in citrus. Basic Appl. Ecol. 17, 210–219.

Gottwald, T.R., Hall, D.G., Beattie, G.A.C., Ichinose, K., Nguyen, M.C., Le, Q.D., Bar-Joseph, M., Lapointe, S., Stover, E., Parker, P.E., McCollum, G., Hilf, M.E., 2007. Investigations of the effect of guava as a possible tool in the control/management of huanglongbing. In: Proceedings of the 17th Conference of the International Organization of Citrus Virologists, Adana, Turkey, October 22–26.

Gravena, S., 2005. Manual prático de manejo ecológico de pragas dos citros. Jaboticabal, S. Gravena. 372 pp.

Gravena, S., Beretta, M.J.G., Paiva, P.E.B., Gallão, R., Yamamoto, P.T., 1996. Seasonal abundance and natural enemies of *Diaphorina citri* (Hemiptera: Psyllidae) in citrus orchards of São Paulo state, Brasil. In: Proceedings. 13th IOVC Conference.

Grout, T.G., Moore, S.D., 2015. *Citrus*. In: Prinsloo, G.L., Uys, M. (Eds.), Insects of Cultivated Plants and Natural Pastures in Southern Africa. Entomological Society of Southern Africa, Hatfield, Pretoria, South Africa, pp. 448–499.

Grout, T.G., Stephen, P.R., 2002. *Citrus* rust mite develops resistance to mancozeb. SA Fruit J. 1, 40–43.

Harling, R., Shamie, I.M.O., Sesay, S.S., Kamara, A.B., Reeder, R., Boa, E., Kelly, P., 2010. First report of *Pseudocercospora angolensis* causing leaf and fruit spot of *Citrus* in Sierra Leone. New Dis. Rep. 22, 1.

Huang, M.D., 1989. Studies on the Integrated Management of *Citrus* Insect Pests. Academic Books and Periodicals Publishing Company.

Huang, M.D., Yang, Y.P., Ouyang, G.C., Xiong, J.J., 2011. Studies and Application of Phytoseiids in China. Chung Tai Educational Press, pp. 42–49.

Huffaker, C.B., Berryman, A.A., Laing, J.E., 1984. Natural control of insect populations. In: Huffaker, C.B., Rabb, R.L. (Eds.), Ecological Entomology. International Center for Integrated and Biological Control. North Caroline State University, pp. 359–398. 844 pp.

Jacas, J., Urbaneja, A., 2010. Biological control in citrus in Spain: from classical to conservation biological control. In: Ciancio, A., Mukerji, K.G. (Eds.), Integrated Management of Arthropod Pests and Insect Borne Diseases. Springer, Dordrecht, pp. 61–72.

Jacas, J.A., Karamaouna, F., Vercher, R., Zappalà, L., 2010. *Citrus* pest management in the northern Mediterranean basin (Spain, Italy and Greece). In: Ciancio, A., Mukerji, K.G., Dordrecht, The Netherlands (Eds.), Integrated Management of Arthropods Pests and Insect Borne Diseases. Springer, NL, pp. 3–27.

Juan-Blasco, M., Sabater-Muñoz, B., Pla, I., Argilés, I., Castañera, P., Jacas, J.A., Ibáñez-Gual, M.V., Urbaneja, A., 2014. Estimating SIT-driven population reduction in Mediterranean fruit fly from sterile mating. Bull. Entomol. Res. 104, 233–242.

Khamis, F.M., Rwomushana, I., Ombura, L.O., Cook, G., Mohamed, S.A., Tanga, C.M., Nderitu, P.W., Borgemeister, C., Sétamou, M., Grout, T.G., Ekesi, S., 2017. DNA barcode reference library for the African *Citrus* Triozid, *Trioza erytreae* (Hemiptera: Triozidae): vector of African *Citrus* greening. J. Econ. Entomol. 110, 2637–2646. https://doi.org/10.1093/jee/tox283.

Li, H.-L., 1982. Nan-fang ts'ao-mu chuang, a fourth century flora of Southeast Asia. The Chinese University Press, Hong Kong.

Magrama, 2014. Guía de Gestión Integrada para el cultivo de los cítricos. http://www.mapama.gob.es/es/agricultura/temas/sanidad-vegetal/GUIACITRICOS_tcm7-348110.pdf. Cited 11 Nov 2017.

Mazih, A., 2015. Status of citrus IPM in the southern Mediterranean basin Morocco, North Africa. Acta Hortic. (1065), 1097–1103.

Michaud, J.P., 2002. Biological control of Asian citrus psyllid (Homoptera: Psyllidae) in Florida. A preliminary report. Entomol. News 113, 216–222.

Michaud, J.P., 2004. Natural mortality of Asian citrus psyllid (Homoptera: Psyllidae) in Central Florida. Biol. Control 29, 260–269.

Mo, J., Stevens, M.M., 2013. Vertical and horizontal movements of Fuller's rose weevil (Coleoptera: Curculionidae) in Australian citrus groves. Entomol. Sci. 16 (1), 26–33.

Monzo, C., Stansly, P.A., 2017. Economic injury levels for Asian citrus psyllid control in process oranges from mature trees with high incidence of huanglongbing. PLoS One 12 (4), e0175333.

Monzo, C., Qureshi, J.A., Stansly, P.A., 2014. Insecticide sprays, natural enemy assemblages and predation on Asian citrus psyllid, Diaphorina citri (Hemiptera: Psyllidae). Bull. Entomol. Res. 104, 576–585. https://doi.org/10.1017/S0007485314000315.

Monzo, C., Arevalo, H.A., Jones, M.M., Vanaclocha, P., Croxton, S.D., Qureshi, J.A., Stansly, P.A., 2015. Sampling methods for detection and monitoring of the Asian citrus psyllid (Hemiptera: Psyllidae). Environ. Entomol. 1–9. https://doi.org/10.1093/ee/nvv032.

Moore, S.D., Kirkman, W., Stephen, P.R., Albertyn, S., Love, C.N., Grout, T.G., Hattingh, V., 2017. Development of an improved postharvest cold treatment for *Thaumatotibia leucotreta* (Meyrick) (Lepidoptera: Tortricidae). Postharvest Biol. Technol. 125, 188–195.

Morse, J.G., Grafton-Cardwell, B., 2013. Bifenthrin trunk sprays as a strategy for Fuller rose beetle (FRB) field control in 2013. Citrograph 26 (28), 30–33 (March/April).

Ota, H., 2013. Historical development of pesticides in Japan. J. Pestic. Sci. 38, 161–166.

Ouyang, G.C., Cen, Y.J., Yu, S.K., Huang, M.D., Liang, G.W., Zhang, B.L., 2010. Assay of repellent components of mineral oils against the oriental fruit fly, *Bactrocera dorsalis* (Diptera: Tephritidae). Acta Entomol. Sin. 53 (6), 655–663.

Pang, S.T., Dong, Y.H., 2012. Effects of soil fertilization on herbivores infestation. Soils 44 (5), 719–726.

Pei, Q., Feng, C.Q., Chen, L., He, C.L., Huang, Z.H., Wang, Y.M., 2014. Control efficacy of *Amblyseius barkeri* on citrus red mite. China Plant Prot. 34 (11), 33–36.

Qi, L., Qi, C.J., 2016. Status quo and development trend of world's citrus industry. Agric. Outlook 12, 46–52.

Qureshi, J.A., Stansly, P.A., 2009. Exclusion techniques reveal significant biotic mortality suffered by Asian citrus psyllid *Diaphorina citri* (Hemiptera: Psyllidae) populations in Florida citrus. Biol. Control 50, 129–136.

Qureshi, J.A., Stansly, P.A., 2010. Dormant season foliar sprays of broad spectrum insecticides: an effective component of integrated management for *Diaphorina citri* (Hemiptera: Psyllidae) in citrus orchards. Crop Prot. 29, 860–866.

Qureshi, J.A., Kostyk, B.C., Stansly, P.A., 2014. Insecticidal suppression of Asian citrus psyllid *Diaphorina citri* (Hemiptera: Liviidae) vector of huanglongbing pathogens. PLoS One https://doi.org/10.1371/journal.pone.0112331.

Rae, D.J., Waston, D.M., Liang, W.G., Tan, B.L., Li, M., Huang, M.D., Ding, Y., Xiong, J.J., Du, D.P., Tang, J., Beattie, G.A.C., 1996. Comparison of petroleum spray oils, abamectin, cartap, and methomyl for control of citrus leafminer (Lepidoptera:Gracillariidae) in Southern China. Hortic. Ent. 89, 493–500.

Rae, D.J., Liang, W.G., Watson, D.M., Beattie, G.A.C., Huang, M.D., 1997. Evaluation of petroleum spray oils for control of the Asian citrus psylla, *Diaphorina citri* (Kuwayama) (Hemiptera: Psyllidae), in China. Int. J. Pest Manag. 43, 71–75.

Rae, D.J., Watson, D.M., Huang, M.D., Cen, Y.J., Wang, B.Z., Beattie, G.A.C., Liang, W.G., Tan, B.L., Liu, D.G., 2000. Efficacy and phytotoxicity of multiple petroleum oil sprays on sweet orange (*Citrus sinensis* (L.)) and pummelo (*C. grandis* (L.)) in southern China. Int. J. Pest Manag. 46, 125–140.

Roberts, R., Cook, G., Grout, T.G., Khamis, F., Rwomushana, I., Nderitu, P.W., Seguni, Z., Materu, C.L., Steyn, C., Pietersen, G., Ekesi, S., le Roux, H.F., 2017. Resolution of the identity of 'Candidatus Liberibacter' species from huanglongbing-affected citrus in East Africa. Plant Dis. 101, 1481–1488.

Rwomushana, I., Khamis, F.M., Grout, T.G., Mohamed, S.A., Sétamou, M., Borgemeister, C., Heya, H.M., Tanga, C.M., Nderitu, P.W., Seguni, Z., Materu, C.L., Ekesi, S., 2017. Detection of *Diaphorina citri* Kuwayama (Hemiptera: Liviidae) in Kenya and potential implication for the spread of Huanglongbing disease in East Africa. Biol. Invasions 19, 2777–2787.

Santoshkumar, C.K., Kumaranag, K.M., Datta, M.B., 2013. 12 Important Insect Pests of *Citrus* and Their Management. http://www.krishisewa.com/articles/disease-management/225-citrus_ip.html. Cited 12 Dec 2017.

South China Agricultural College, 1981. Agricultural Entomology. China Agriculture Press833–834.

Stansly, P., Croxton, S., Sherrod, J., 2016. Big boost in young tree growth and yield from insecticides and metalized mulch. Citrus Ind. 97 (3), 20–22.

Tena, A., Pekas, A., Cano, D., Wäckers, F.L., Urbaneja, A., 2015. Sugar provisioning maximizes the biocontrol service of parasitoids. J. Appl. Ecol. 52, 795–804.

Theron, C.D., Manrakhan, A., Weldon, C.W., 2017. Host use of the oriental fruit fly, *Bactrocera dorsalis* (Hendel) (Diptera: Tephritidae), in South Africa. J. Appl. Entomol. 141, 810–816.

Urbaneja A, Catalán J, Tena A, Jacas JA (2017) Gestión integrada de plagas de cítricos. http://www.gipcitricos.ivia.es. Cited 11 Nov 2017.

Vacante, V., Gerson, U., 2012. Integrated Control of *Citrus* Pests in the Mediterranean Region. Bentham Science Publishers. 287 pp.

Vacas, S., Alfaro, C., Navarro-Llopis, V., Primo, J., 2010. Mating disruption of California red scale, *Aonidiella aurantii* Maskell (Homoptera: Diaspididae), using biodegradable mesoporous pheromone dispensers. Pest Manag. Sci. 66, 745–751.

Yang, P., 2002. Historical perspective of the red tree ant, *Oecophylla smaragdina* and its utilization against citrus insect pests. Chin. J. Biol. Control 18 (1), 28–32.

Zappalà, L., 2010. *Citrus* integrated pest management in Italy. In: Ciancio, A., Mukerji, K.G. (Eds.), Integrated Management of Arthropod Pests and Insect Borne Diseases. Springer, Dordrecht, pp. 73–100.

Zhang, Y.X., Lin, J.Z., Ji, J., Chen, C.P., Lin, Z., 2003. Studies on *Amblyseius cucumeris* Oudemans for controlling the pest mites on citrus trees. Plant Prot. 29 (5), 31–33.

Chapter 17

Diseases caused by fungi and oomycetes

Ozgur Batuman[a], Mark Ritenour[b], Antonio Vicent[c], Hongye Li[d], Jae-Wook Hyun[e], Vittoria Catara[f], Haijie Ma[d,g], Liliana M. Cano[h]

[a]*Department of Plant Pathology, Southwest Florida Research and Education Center, University of Florida, Immokalee, FL, United States,* [b]*Department of Horticultural Science, Indian River Research and Education Center, University of Florida, Fort Pierce, FL, United States,* [c]*Centro de Protección Vegetal y Biotecnología, Instituto Valenciano de Investigaciones Agrarias (IVIA), Valencia, Spain,* [d]*Institute of Biotechnology, Zhejiang University, Hangzhou, People's Republic of China,* [e]*Citrus Research Institute, National Institute of Horticultural and Herbal Science, Jeju, South Korea,* [f]*Department of Agriculture, Food and Environment, University of Catania, Catania, Italy,* [g]*Department of Horticultural Science, Citrus Research and Education Center, University of Florida, Lake Alfred, FL, United States,* [h]*Department of Plant Pathology, Indian River Research and Education Center, University of Florida, Fort Pierce, FL, United States*

17.1 Greasy spot disease of citrus caused by *Zasmidium citri-griseum*

17.1.1 Introduction

Zasmidium citri-griseum (= *Mycosphaerella citri*) belongs to the *Mycosphaerellaceae* family and is the causal agent of the greasy spot (Whiteside, 1970). *Z. citri-griseum* infects citrus inducing leaf spots, premature defoliation, which results in reduced fruit size, fruit drop, and yield loss (Aguilera-Cogley et al., 2017). Rind blotch occurs when the pathogen directly infects the fruit. Greasy spot is widespread in the southeast North America (Florida and Eastern Mexico), Central America, Northern South America, and the islands of the Caribbean (Timmer and Gottwald, 2000). Outside the American continent, Japan, Egypt, and China have also reported greasy spot in citrus to be caused or associated with this same species (Leki, 1986; Haggag, 2012; Huang et al., 2015b). Similar greasy spot-like symptoms have been reported on citrus in other countries, however, this was associated with different *Mycosphaerellaceae* species (Huang et al., 2015b; Aguilera-Cogley et al., 2017).

17.1.2 Disease symptoms

Leaf tissues infected with *Z. citri-griseum* show chlorosis and yellow-brown pustules on the abaxial surface (Fig. 17.1A and B). Fruits infected with *Z. citri-griseum* show a typical rind blemish which significantly reduces the fruit quality for the fresh market (Fig. 17.1C). The pathogen incubation period is long and disease symptoms are only observed after 4–6 months (Mondal and Timmer, 2006). Microscopic observations show that *Z. citri-griseum* exhibits epiphytic growth on the leaf surface (Timmer et al., 2001). Secretion of nutrients or honeydew on the surface of the leaves also maximizes fungus survival (Whiteside, 1974).

17.1.3 Infection process

Z. citri-griseum belong to the *Mycosphaerellaceae* family that are characterized by the formation of pseudothecia that are in contact with the plant tissue. The typically ovoid and bitunicate asci carry ascospores, which also are responsible for the infection (Dewdney and Graham, 2016). Ascospores germinate and form epiphytic mycelia on the leaf surface that penetrate through stomata (natural openings of the abaxial leaf surface). Fungal growth in the leaf tissue is slow and symptoms develop several months later. Decomposing leaf litter in the orchard is the main source of airborne ascospores and primary inoculum (Mondal and Timmer, 2006).

The Genus Citrus. https://doi.org/10.1016/B978-0-12-812163-4.00017-6

FIG. 17.1 Greasy spot disease symptoms in citrus caused by *Zasmidium citri-griseum*. Brown necrotic spots in the abaxial leaf surface in sweet orange (A) and in grapefruit (B). Rind blotch on grapefruit (C).

17.1.4 Pathogenicity and virulence

Zasmidium citri-griseum infects plants from the *Rutaceae* family such as citrus. Most citrus species appear to be susceptible, including the citrus relative *Poncirus*. Greasy spot development is favored in humid growing regions of citrus from the Caribbean to Florida in the US and in some cases are also present in semiarid areas. Greasy spot can affect most commercial citrus and its relatives. Fresh fruit market value of grapefruit (*Citrus paradisi*) is threatened by greasy spot rind blotch since grapefruit is highly susceptible. Besides grapefruit, other susceptible citrus species are sweet orange (*C. sinensis*), lemon (*C. limon*), and citrus hybrids such as tangelos are also considered susceptible based on field observations (Timmer et al., 2001).

17.1.5 Disease management

Copper is effective and still widely used to control greasy spot on leaves as it can directly kill ascospores and prevent infection. Copper can be used together with petroleum oils as additives to prevent penetration of the epiphytic mycelia. Petroleum oil sprays alone are not sufficient to control greasy spot rind blotch. Homogenous spray distribution of copper and oils on the leaf area is essential to protect the tree from future infections due to the epiphytic growth of *Z. citri-griseum* (Timmer et al., 2001). Copper applications can result in phytotoxicity on fruits when the temperature is >35°C and under dry weather conditions (Dewdney and Graham, 2016). Application of strobilurin or triazole fungicides is required to control this fungus in more susceptible citrus such as grapefruit, sweet orange, or in hot weather conditions. Timely spraying maximizes the efficacy of fungicide applications for greasy spot, and it is recommended to apply fungicide preventively prior to the ascospore penetration phase of the leaf surface (Dewdney, 2019).

17.2 Root rot, foot rot, brown rot of fruits, canopy blight, and damping-off diseases of citrus caused by *Phytophthora*

17.2.1 Introduction

Phytophthora species are among the most destructive oomycete plant pathogens known (Erwin and Ribeiro, 1996; Blair et al., 2008). They have a broad host range and are known as soilborne pathogens that mainly spread through the movement of infested soil, water, and infected plants (Goodwin, 1997). In Florida, *Phytophthora* species are the most economically important soilborne disease of citrus causing yield losses of up to 30% (Timmer et al., 2000). They infect all parts of the tree, including roots, stem, branches, twigs, leaves, and fruits. Major diseases caused by *Phytophthora* are root rot, foot rot or gummosis, fruit brown rot, canopy blight, and damping-off of seedlings (Cacciola et al., 2008) (Fig. 17.2). About 10 species of *Phytophthora* have been reported worldwide (Cacciola et al., 2008). *Phytophthora nicotianae, P. citrophthora,* and *P. palmivora* are the most predominant species in citrus (Graham and Menge, 1999; Graham and Timmer, 2006).

17.2.2 Disease symptoms

In general, *Phytophthora* species are favored by wet or flooded conditions (Graham and Feichtenberger, 2015). Foot rot (gummosis) is one of the most serious diseases caused by *Phytophthora* in citrus. Infection can occur at the ground

FIG. 17.2 Foot rot on Carrizo crown and roots (A) and brown rot on sweet orange (B) caused by *Phytophthora*. Strain of *P. nicotianae* cultured on V8 juice agar after 7 days showing typical white filamentous growth (C).

level or up the trunk of the tree (Fig. 17.2A). Symptoms of brown rot of fruit include light brown and leathery fruit rind (Fig. 17.2B). Brown rot is a common preharvest decay of citrus fruit that is favored by rainfall and results in preharvest fruit drop. Canopy blight symptoms include leaf chlorosis, defoliation, twig dieback, and weak growth of flushes and are commonly observed in citrus nurseries (Graham and Feichtenberger, 2015). Damping-off (seedling rot) is a disease of citrus in nurseries that affects the seedlings causing up to 80% mortality just before or after they emerge from the soil.

17.2.3 Infection process

Phytophthora exhibits filamentous growth similar to fungi (Fig. 17.2C), although they are genetically distant and more closely related to brown algae. Citrus tissues are most commonly infected through indirect contact with motile biflagellate zoospores that are released from sporangia, and less commonly from direct germination of sporangia (Klotz and De Wolfe, 1960; Graham and Timmer, 2006). Chlamydospores are resistant structures that allow the pathogen to survive in unfavorable conditions. Oospores are formed after sexual reproduction and act also as organs of preservation. Chlamydospore and oospore germination can occur in the presence of nutrient extracts from citrus (Savita et al., 2012).

17.2.4 Pathogenicity and virulence

Most commercial scion cultivars and rootstock can be infected (Feichtenberger et al., 1992). However, the use of tolerant scion-rootstock combinations is a useful tool toward overcoming root rot caused by *Phytophthora*. Rootstock responses to *Phytophthora* vary according to the species (Graham, 1995).

In Florida, *Phytophthora* species can interact with other plant pathogen microbes and pests (Graham and Feichtenberger, 2015). Diaprepes root weevil (*Diaprepes abbreviatus*) when in complex with *P. palmivora* can increase the severity of root rot damage and result in higher populations of the pathogen in the rhizosphere (Graham, 2000). *P. nicotianae* infecting roots also infected with *Candidatus* Liberibacter asiaticus (*C*Las, associated with HLB) has been reported to favor attraction of zoospores to roots and accelerate *Phytophthora* infection (Graham et al., 2011).

17.2.5 Disease management

It is essential to avoid *Phytophthora* in citrus nurseries by heat-treating seeds (50°C for 10 min) and planting them into sterilized, soilless media (Graham and Timmer, 2006). Adequate soil drainage in the orchards is essential because waterlogging induces buildup of *Phytophthora* populations in the soil, increasing the risk of infection (Graham and Menge, 1999). In established citrus orchards fungicides like copper have been highly effective for foot rot in citrus. Systemic fungicides like fosetyl-Al and metalaxyl/mefenoxam are used for foot root and root rot control in mature orchards. Metalaxyl/mefenoxam are also used in preplant applications to prevent damping-off (Graham and Timmer, 2006). Fungicide recommendations should be based on results of propagule counts in soil by conducting annual sampling. Fungicide applications are recommended to be continued for at least 1 year after initiation. Application methods include soil applications (soil drenching) and foliar applications (spraying the canopy). For control of foot rot, the microsprinkler irrigation system has been more effective. Brown rot of fruit can be managed with foliar applications of contact or systemic fungicides. Although chemical control is available, it can also be challenging as emerging fungicide resistant isolates have been reported (Timmer et al., 1998;

Thind et al., 2009). To avoid fungicide resistance development, it is important to first recognize other potential abiotic or biotic stress causing a decline of the trees before applying these fungicides (Graham and Dewdney, 2015). Chemical control cannot fully eradicate *Phytophthora* from citrus orchards. Another method of control is using *Phytophthora* resistant or tolerant citrus cultivars and rootstocks, for example, some *C. reticulata* (mandarin) and *P. trifoliata* and some of its hybrids with *Citrus*, show resistance or tolerance to *P. nicotianae*. Budding high on the rootstock and planting the trees on ridges avoid contact of the susceptible bark with rain or irrigation water carrying infective propagules. (Graham, 1990; Zhou et al., 1997; Graham and Menge, 1999; Guo and Deng, 2001; Mourã Filho et al., 2008; Boava et al., 2011; Yan et al., 2017; Dalio et al., 2018).

17.3 Melanose disease of citrus caused by *Diaporthe citri*

17.3.1 Introduction

Melanose is a disease of citrus forming corky pustules on twigs, young leaves, and fruits, which is caused by *Diaporthe citri* (= *Phomopsis citri*) (Fig. 17.3) (Fawcett, 1936). Melanose is only a serious problem in tropical and subtropical regions with fruit grown for the fresh market, in areas with high humidity and substantial rainfall during early fruit development (Gopal et al., 2014). Such places include, among others, Florida and eastern Australia. While the fungus may be found in many arid citrus-producing regions, it is usually not a problem. Leaf infections are not particularly important because, while leaf infection results in lesions, they are unlikely to significantly affect tree productivity or fruit internal quality. However, the disease results in superficial blemishes on the rind that can reduce the grade or render the fruit completely unsuitable for the fresh market. Melanose is not a serious problem when fruit is grown for processing (i.e., juice). After initial infection and pustule formation, melanose symptoms do not progress further.

17.3.2 Disease symptoms

The disease usually forms small (~1 mm diameter) reddish-brown to black pustules on fruit rind that may be randomly distributed on the surface ("flyspeck" melanose), or in patterns following locations where free water carried spores along the fruit surface ("tearstain" melanose) (Whiteside, 2000; Gopal et al., 2014). These symptoms have been sometimes confused with rust mite (*Phyllocoptruta oleivora*) damage, but rust mite blemishes are relatively smooth, whereas melanose pustules are raised and rough. Advances in multispectral imaging and computer programming have made distinguishing melanose from other rind blemishes possible using cameras on a grading line, although the technology is not yet commercially available (Zhang et al., 2018). Unfortunately, under certain conditions, copper sprays used to control the disease can themselves result in blemishes that look very similar to melanose and are sometimes called "star" or "stellate" melanose (Zhang et al., 2018). When *D. citri* infection is severe and early in fruit development (i.e., soon after petal fall), the pustules converge and form large patches of scar tissue that cracks as the tissue expands to form "sharkskin" or "mudcake" melanose symptoms (Fig. 17.3A). Lesions on leaves form small, black raised pustules that may be surrounded by a yellow halo early in development, but the chlorotic areas regreen over time (Fig. 17.3B). Leaves can become deformed if the shoot apex is severely infected when leaves are unfolding and, if leaf infection is severe enough, may cause general chlorosis and leaf abscission. Symptoms on twigs are similar to leaf symptoms, with severe infections resulting in shoot dieback.

FIG. 17.3 Melanose disease symptoms on a fruit (A) and leaf (B) of grapefruit caused by *Diaporthe citri*.

17.3.3 Infection process

Diaporthe citri lives, grows, and reproduces primarily on deadwood, especially on twigs that died within the past 2–4 months. Asexually produced conidia are by far the most abundant spore produced and are primarily extruded under wet conditions and spread relatively short distances by rain and other water sources. Sexually formed ascospores can also be produced and are disseminated relatively long distances by wind but are not the main cause of infection. Spores landing on fruit or leaves directly penetrate epidermal cells via germ tubes secreting pectic enzymes, resulting in invasion and death of both epidermal and subepidermal cells up to about six cell layers deep, which become filled with a reddish-brown, gummy substance. Although the affected cells shrink by forming an initial slight depression, periderm forms beneath the dead cells about a week after infection, lifting the pustules and rupturing the cuticle to release the gum that then turns brown and hardens, imparting the characteristic raised and rough pustules.

The fungus does not develop after pustule formation on fruits and leaves and so does not produce spores, thus, fruits and leaves are not a source of inoculum for further infections. The amount of inoculum for new infections is related to the number of twigs within the canopy that died in recent months and are colonized by the fungus. Conditions that increase twig death, such as freezing injury, storm damage, insect pressure, mechanical pruning, etc. often lead to greater inoculum for disease development (Whiteside, 2000; Mondal et al., 2004).

17.3.4 Pathogenicity and virulence

While all citrus species are susceptible to melanose, grapefruit and lemon appear most affected, being more visible on the smooth, light-colored surface than on other varieties, and with larger pustules forming on grapefruit (Mondal et al., 2007; Dewdney, 2018). Leaves and fruits are susceptible to infection only during a limited period of development. Leaves are resistant after they become fully expanded. Fruits are resistant for about 3 months after petal fall, but the time period can vary based on the growing region (Gopal et al., 2014).

17.3.5 Disease management

Management of the disease is usually only needed when the fruit is intended for the fresh market and when the trees become old enough to have deadwood to support inoculum production. Managing trees to remove or avoid deadwood formation within the canopy is one way to minimize the disease. While pruning can be used, labor costs often preclude such practices, especially for large trees. Larger trees may benefit from topping if large amounts of deadwood are present in the canopy.

Fungicides are the main method used to control the disease, including copper- and strobilurin-containing fungicides (Dewdney, 2018). As a practical approach to control infection of deadwood or subsequent sporulation, fungicides would need to be applied continuously to have any effect. To protect fruit, copper fungicides are most commonly used because they are effective and more economical than other fungicides. However, copper can injure fruit if applied during hot weather, thus, other fungicides are often used during these periods (Albrigo et al., 2005).

The timing and frequency of fungicide application are critical considerations, especially in relation to fruit development and rain events that wash residues away. Fruit growth exposes new surfaces that are vulnerable to infection until the next fungicide application. In addition, localized heavy thunderstorms require more frequent fungicide reapplications. The first fungicide application must be applied before new fruit are exposed to rainy conditions. In Florida, this means they often do not occur until mid- to late-April, before the spring rainy season starts. After spraying starts, frequent fungicide applications at lower rates are usually the most effective. Repeated applications every 3 weeks are usually sufficient to control the disease, but a web-based tool is also available, which accounts for fruit growth and local weather events to calculate when the next copper application is needed (Dewdney et al., 2013). Fungicide applications can be ceased after the fruit has grown to a point when they are no longer susceptible (e.g., for grapefruit, about 7 cm diameter).

17.4 Citrus black spot disease caused by *Phyllosticta citricarpa*

17.4.1 Introduction

The citrus black spot disease (CBS) was first mentioned by Benson (1895) in New South Wales, Australia, consisting of fruit blemishes and premature fruit drop. The fungus was subsequently identified as *Phoma citricarpa* (McAlpine, 1899) and Koch's postulates were fulfilled by Lee (1920) and later by Kiely (1948), who also described the teleomorph of the fungus as *Guignardia citricarpa*. Based on current nomenclature, the accepted name of the CBS pathogen is *Phyllosticta citricarpa* (Glienke et al., 2011). Other *Phyllosticta* spp. have been associated with fruit and leaf spots in citrus, but in most cases, pathogenicity tests are lacking.

Sawada (1959) reported the presence of CBS in Taiwan. Lee (1920) indicated the presence of the disease also in mainland China, based on isolations from citrus fruit imported from China into the Philippines. In international databases, CBS was recorded as present in New Zealand, but further molecular studies concluded that the original isolate was indeed misidentified (Everett and Rees-George, 2006). In South Africa, CBS was reported by Doidge (1929) initially as a disease of minor importance in KwaZulu-Natal province. The disease is now widespread in many areas of sub-Saharan Africa. In Brazil, CBS was mentioned by Averna-Saccá (1940) but it was most likely a misdiagnosis, with a confirmed report four decades later. In Tucuman and Entre Rios, Argentina, CBS was first reported in the mid-1980s (Foguet et al., 1985; Garrán, 1996). In the Caribbean and North America, the disease was officially reported in Cuba in 2007 (Hidalgo-Góngora and Pérez-Vicente, 2010) and it was first observed in Florida in 2010 (Schubert et al., 2012). Citrus-growing areas in the Mediterranean basin were free of CBS, but *P. citricarpa* has been reported in Portugal, Malta, and Italy by Guarnaccia et al. (2017), however those findings were not confirmed by a later survey performed on the same sampling sites by the National Plant Protection Organizations of the three countries (EFSA Panel on Plant Health, 2018). Recently, outbreaks of CBS caused by *P. citricarpa* have been reported in India and Tunisia (Das et al., 2018; EPPO, 2019). In general, CBS epidemics are characterized by a long lag phase. The pathogen may be present for many years in an area before symptoms are first observed, and then decades until the disease reaches epidemic proportions, depending on climate and host susceptibility (Kotzé, 1981).

17.4.2 Disease symptoms

The CBS disease causes several types of rind blemishes on the fruit, reducing its marketability for fresh consumption. Lesions on fruit may be rather variable in appearance with symptoms generally being classified into different categories. Different categories of symptoms with different stages of development can be present on the same fruit. Moreover, some CBS symptoms can be easily confused with other diseases, abiotic disorders or pest damage. "Hard spot" is considered the most typical symptom of CBS, consisting of depressed lesions in the rind with a grayish center, dark margin, and yellow-to-green halo depending on fruit maturity (Fig. 17.4A). Hard spot lesions may remain 3–10 mm in diameter or coalesce into large lesions. "Freckle spot" consists of small spots about 1–3 mm in diameter, slightly depressed, mostly reddish and without any halo. Freckle spot lesions can also coalesce to form larger lesions. "False melanose" appears as small raised dark pustules, resembling those of the melanose disease caused by *D. citri*. Symptoms of "Virulent spot" extend into the albedo, forming sunken irregular lesions, variable in color, covering a great part of the fruit surface. Two additional miscellaneous symptoms have also been described, "Lacy Spot" consisting of superficial yellowish lesions, and "cracked spot" associated with the rust mite (IPPC, 2016).

Fruit symptoms usually develop on mature fruit at the time of color change, starting on the side most exposed to sunlight. Infected fruit may develop postharvest symptoms during transport or storage, particularly by increasing temperature (Brodrick and Rabie, 1970). Disease incidence is usually higher in old trees, with poor vigor and particularly in those subjected to drought stress (Kotzé, 1981). Disease incidences from 13% to 98% of fruit affected have been reported in untreated plots (Makowski et al., 2014). In addition to external fruit blemishes, CBS can also induce premature fruit drop and yield losses over 60% have been reported in Brazil (Silva-Junior et al., 2016). On leaves, CBS is usually asymptomatic and *P. citricarpa* develops as quiescent infections. In highly susceptible citrus species, such as lemon (*C. limon*) or stressed

FIG. 17.4 Citrus black spot disease symptoms on fruit (A) and leaf (B) of sweet orange caused by *Phyllosticta citricarpa*. Strain of *P. citricarpa* grown on potato dextrose agar (PDA) (C).

trees, foliar symptoms of CBS may appear as circular light brown spots up to 3 mm in diameter, with dark margins and a yellow halo (Fig. 17.4B). Similar symptoms can also be present on twigs, but 0.5–2 mm in diameter (IPPC, 2016).

17.4.3 Infection process

The species *P. citricarpa* is heterothallic and complementary mating types are needed for sexual reproduction (Tran et al., 2017). Sexual reproduction occurs in the leaf litter, where pseudothecia, asci, and ascospores are produced. Alternate wetting and drying and rising temperatures favor pseudothecia maturation and ascospore production. Conversely, an excessively moist environment induces more rapid leaf litter decay and the formation of pseudothecia is negatively affected (Lee and Huang, 1973). Studies conducted in South Africa indicated that once mature, the release of ascospores is mainly driven by rain (Fourie et al., 2013). Ascospores are airborne and can disseminate relative long distances, though quantitative data on their dispersal range are not available. In Taiwan, airborne ascospores were trapped mainly from May to October, in Brazil from November to March, and in South Africa from October to February (Lee and Huang, 1973; Reis et al., 2006; Fourie et al., 2013). Nevertheless, the spore traps used by these studies could not differentiate *P. citricarpa* from the coexisting endophyte *P. capitalensis*. Therefore, results should be interpreted with caution.

Asexual reproduction of *P. citricarpa* occurs in fruit lesions, twigs, and leaf litter. Conidia are produced in pycnidia and are dispersed by rain splash. Laboratory studies with simulated wind-driven rain indicated that *P. citricarpa* conidia can disseminate at least 8 m and to heights of 75 cm (Perryman et al., 2014). Based on studies conducted in South Africa, ascospores were generally considered the main, if not the only, source of inoculum for CBS epidemics. Conidia were historically deemed to have epidemiologically a minor role, only for fruit-to-fruit infections when mature fruit with lesions remain on the canopy after fruit set. However, studies conducted in Brazil indicated that conidia produced on twigs may be another important source of inoculum (Spósito et al., 2008; Silva-Junior et al., 2016). Epidemics of CBS in Florida, have recently been associated with a clonal population of *P. citricarpa*, with only one mating type, further supporting the importance of asexual reproduction and conidia under certain conditions (Wang et al., 2016; Hendricks et al., 2017).

Ascospores and conidia of *P. citricarpa* germinate and infect under adequate conditions of temperature and moisture. Nevertheless, the specific temperatures and wetness durations required for infection have not been determined. Citrus leaves remain susceptible to infection for up to 8–10 months after their emergence (Truter, 2010). In South Africa, fruit are considered susceptible to infection during 4 months after fruit set (Kotzé, 1981) but longer susceptibility periods up to 6–7 months were reported in Ghana and Brazil (Brentu et al., 2012; Lanza et al., 2018). The pathogen penetrates into the host by forming appressoria and establishes latent infections between the cuticle and epidermis (McOnie, 1967). The latent, asymptomatic period lasts for several months and CBS symptoms are visible only when fruit is fully developed, and ripening starts (i.e., color break). Symptom expression is enhanced by rising temperatures, high light intensity, drought stress, and tree age (Kotzé, 1981).

17.4.4 Pathogenicity and virulence

All commercial citrus species and cultivars are considered to be susceptible to CBS, except for sour orange (*C. aurantium*) and Tahiti lime (*C. latifolia*) (Kotzé, 1981; Baldassari et al., 2008). Apparently, *P. citricarpa* can colonize tissues of these two citrus species but without inducing CBS symptoms (Baldassari et al., 2008; Wickert et al., 2009). Lemon is considered the most susceptible to CBS and in some areas the first disease outbreaks were spotted on this citrus species. Conversely, other regions reported the first disease foci in sweet orange orchards. Late-maturing cultivars of sweet orange, like those of the Valencia group, were considered more susceptible to CBS and indeed they show higher disease severity. However, this may be due to the longer period the fruit of these late-maturing cultivars hang on the canopy, interacting with the environmental and physiological factors leading to symptom expression.

17.4.5 Disease management

Quarantines limiting or prohibiting the movement of citrus material were established as an exclusion measure to avoid the entry of *P. citricarpa* into new areas (Martínez-Minaya et al., 2015). For eradication, Whiteside (1967) proposed drastic pruning, leaving only the framework limbs, together with leaf litter elimination as an alternative to tree removal. However, despite the relatively slow epidemic progression of CBS, successful eradication was not accomplished anywhere. As Kotzé (1981) pointed out, CBS advances slowly but rarely retreats.

Using healthy plant propagation material and cleaning plant debris from incoming vehicles are essential to avoid the introduction of CBS into the orchard. Once established in the orchard, cultural practices are recommended to manage CBS.

Maintaining optimal tree condition by proper fertilization and avoiding drought stress may reduce CBS severity. Irrigation can be also used to synchronize flowering and avoid secondary infections from mature to young fruit. Early harvest, before symptom expression and subsequent fruit drop, may reduce yield losses. Nevertheless, latent infections on fruit can still develop symptoms during transport and storage. Various cultural practices were evaluated to reduce ascospore inoculum in the orchard by leaf litter removal, enhancing leaf decomposition or establishing cover crops (Bellotte et al., 2009, 2013; Van Bruggen et al., 2017). However, poor disease control was obtained in Brazil when implementing some of these practices at large scale (Spósito et al., 2011), probably owing to the concurrence of additional inoculum sources like conidia produced on twig lesions.

Fungicide treatments are generally needed for CBS management in the field. Different combinations of copper compounds, dithiocarbamates, strobilurins, and benzimidazoles have been used (Makowski et al., 2014). In South Africa, the disease was effectively controlled by a single spray of benomyl when applied as a curative treatment after the infection period (Kotzé, 1981). Over time, *P. citricarpa* developed resistance to benomyl (Schutte et al., 2003) and nowadays CBS control is based on preventive spray programs to protect the fruit from infections. The number of sprays per growing season may vary from 2 to 6 depending on the prevailing climate conditions, inoculum availability, cultivar, and fruit susceptibility period (EFSA, 2014b). The destination of the fruit also determines the spray programs, because fruit quality standards are much more stringent for fresh consumption than for industrial processing. Postharvest treatments cannot effectively control latent infections in fruit (Yan et al., 2016).

17.5 Brown spot, leaf spot and black rot diseases of citrus caused by *Alternaria*

17.5.1 Introduction

Alternaria species are generally known as necrotrophyc fungi, and several are pathogenic to citrus and cause significant yield losses. Alternaria diseases of citrus include Alternaria brown spot of tangerine (*C. reticulata*) and their hybrids, Alternaria leaf spot of rough lemon (*C. jambhiri*) and Rangpur lime (*C. limonia*), Alternaria black rot of citrus fruits, and "mancha foliar" disease, caused by *A. limicola*, that occurs only on Mexican lime in Mexico (Palm and Civerolo, 1994). *Alternaria* species causing these four citrus diseases are difficult to distinguish morphologically, thus, pathogenicity tests, toxin assays, and phylogenetic analysis based on sequence data from several genes are usually required to distinguish them.

Symptoms of Alternaria brown spot of tangerine were first reported on "Emperor" mandarin in Australia in 1903 (Cobb, 1903). It appeared in Florida in 1974 and subsequently was identified in other citrus production areas including Israel (Solel, 1991), Cuba (Herrera-Isla, 1992), Turkey (Canihos et al., 1997), South Africa, Colombia (Timmer et al., 2000), Spain (Vicent et al., 2000), Italy (Bella et al., 2001), Brazil, and Argentina (Peres et al., 2003). Recently, this disease was also reported in Southeast Asia and caused significant losses on several important cultivars in China (Wang et al., 2010; Huang et al., 2015a). Alternaria leaf spot of rough lemon was first reported in South Africa in 1929 (Doidge, 1929). This disease affects only rough lemon and Rangpur lime, two important rootstocks in some citrus-growing areas. Thus, Alternaria leaf spot of rough lemon is only significant in nurseries and seed production blocks. Alternaria black rot of fruit was first described as early as 1900s in California (Pierce, 1902; Cobb, 1903). It is a significant postharvest disease for all species of citrus under the appropriate conditions but occurs most commonly on Navel oranges (*C. sinensis*) in the field prior to harvest. Alternaria black rot occurs in all citrus-producing regions but is more prevalent in production areas with Mediterranean climates, characterized by cool and moist winters and hot and dry summers, than in humid areas.

17.5.2 Disease symptoms

Host specificity and pathogenicity of Alternaria brown spot and Alternaria leaf spot pathogens are associated with the production of host-selective toxins (Ohtani et al., 2009). Alternaria brown spot affects young leaves, twigs, and fruits, producing brown-to-black lesions surrounded by a yellow halo (Kohmoto et al., 1979; Meena et al., 2017a) (Fig. 17.5A and B). Leaf lesions are generally circular, extending along the veins as the toxin spreads in vascular tissues, which gives the lesions an eye-spot appearance with a tail (Timmer et al., 2004). On young leaves of very susceptible cultivars, lesions can appear as early as 24 h after infection. Lesions enlarge as leaves mature and can vary in size from 1 to 10 mm and will extend to whole leaves if the conditions, such as moisture, temperature, and susceptibility of cultivars, are appropriate (Fig. 17.5A). The yellow halo is usually more evident on full-size but immature leaves. When this disease is severe, the affected leaves often abscise. Infected young shoots also show brown-to-black sunken lesions, the entire young shoots will dieback, especially if the leaves have fallen (Timmer et al., 2000).

FIG. 17.5 Symptoms of Alternaria brown spot in "Fortune" mandarin leaves (A) and fruit (B). Symptoms of Alternaria black rot in sweet orange fruit (C). Strain of *Alternaria alternata* tangerine pathotype on the PDA (D).

In the case of Alternaria brown spot, young fruits can be infected shortly after petal fall; the lesions are brown to black, sunken without a yellow halo surrounding them. When the fruits grow, larger than to 2–3cm in diameter, the lesions become sunken, turn to gray in the center and are surrounded by a yellow halo (Fig. 17.5B). The infected fruits drop off rapidly, reducing yield. For the very susceptible cultivars, such as 'Hongju' (also known in the western countries as "Dancy"), the fruit is susceptible to Alternaria brown spot during the whole growing season but, in general, fruit is susceptible for 4months after petal fall. Even when the fruit is no longer susceptible, some fruits may fall as the result of earlier infections, especially if lesions are near the fruit peduncle. Alternaria leaf spot of rough lemon affects only rough lemon and Rangpur lime. Symptoms on leaves are very similar to those produced by Alternaria brown spot and do not appear for about 3days after inoculation. Symptoms on fruit are merely small brown specks and are quite reduced relative to the tangerine pathotype (Timmer et al., 2000). Alternaria black rot affects the central columella of the fruit (Fig. 17.5C) and can affect all species of citrus (Brown and McCornack, 1972). External symptoms are often not apparent and, if present, appears as a small brown-to-gray spot on the stylar end of the fruit (Brown and Eckert, 2000). Affected fruit are more brightly colored than normal fruit due to ethylene generated in response to infection.

17.5.3 Infection process

The pathogen of Alternaria brown spot overwinters mainly on diseased fruits, twigs, or mature leaves. In the spring of the following year, the spores produced on the diseased leaves spread through the airflow and then are deposited on susceptible tissues. Rainfall and dew favor the infection of this pathogen. When environmental conditions are conducive, spores germinate and then produce ACT-toxin to kill the host cells for nutrients. The ACT-toxin quickly spreads through the vascular system, accelerates cell necrosis and enlargement of the lesion, and in severe cases, eventually leads to leaf and fruit drop. Spore production on the diseased tissue begins about 10days after the lesions are well developed and continues for about 50days (Dewdney, 2016).

The pathogen of Alternaria black rot overwinters mainly as spores or mycelium on diseased fruits or other substrates. Under suitable temperature and humidity conditions, this fungus produces spores that are spread by airflow on to the receptacle or young fruit, and then those spores remain dormant until the anaphase of fruit growth. During the anaphase of fruit growth or postharvest storage, the hyphae invade from the natural crack, wound or pedicle, causing decay. This disease is greatly affected by environmental factors and the degree of fruit maturity at harvest. The moist conditions facilitate the growth and sporulation of the pathogen on dead citrus tissue. The optimum temperature for growth of the pathogen is around 25°C (Timmer et al., 2000). It is not conducive to mycelial growth when the temperature is lowered to 12–14°C. Poor irrigation and drainage, poor cultivation management, citrus orchards with weak trees, fruits suffering from sunburn, insect caused wounds, or mechanical damage promote the occurrence of Alternaria black rot.

17.5.4 Pathogenicity and virulence

Alternaria brown spot and Alternaria leaf spot pathogens are under taxonomic reassessment (Woudenberg et al., 2015). Alternaria brown spot was described to be caused by *A. alternata* tangerine pathotype on some cultivars of tangerine (*C. reticulata*) and some hybrids of tangerine × grapefruit (*C. reticulata* × *C. paradisi*) or tangerine × sweet orange (*C. reticulata* × *C. sinensis*). This pathogen produces the host-selective ACT-toxin (Kohmoto et al., 1993), which can cause lesions or kill the whole leaf even when it does not colonize the leaf tissue. Young leaves are very susceptible to this pathogen whereas old leaves are resistant. Alternaria leaf spot of rough lemon and Rangpur lime was described to be caused by *A. alternata* rough lemon pathotype (Timmer et al., 2003). This pathogen produces the host-selective ACR-toxin or ACRL-toxin (Gardner et al., 1985; Nakatsuka et al., 1986; Meena et al., 2017b) that kills host cells for nutrients and reproduction on diseased tissue. Alternaria black rot of fruit is caused by *A. alternata* or some other species (Brown and Eckert, 2000) on all citrus cultivars such as lemon, lime, sweet orange, mandarin, and grapefruit (Gardner et al., 1985; Katoh et al., 2006). Natural cracks or wounds are required for penetration of the fungus (Timmer et al., 2003). Some studies have shown that the occurrence of Alternaria black rot of fruit is related to the activity of endopolygalacturonase (Isshiki et al., 2001).

17.5.5 Disease management

Fungicides are the primary means for the control of Alternaria brown spot. Copper compounds are the most economical fungicide to control Alternaria brown spot on citrus (Vicent et al., 2007). Spraying at spring flush (2–3 cm) with a second application 7–14 days later using copper compounds protect young leaves from infection and inhibits sporulation. Further sprays are aimed mainly to protect fruit from infection. The strobilurin fungicides can also be used but with a proper strategy to avoid fungicide resistance as a substitute when the weather is dry and hot to avoid copper damage to fruit (Timmer et al., 2003). Using disease-free planting material, pruning affected tissues, leaf litter management, and improving cultivation measures can be used to reduce disease severity. It is important for susceptible cultivars to be planted rather with a wide-spacing on higher ground with good air ventilation and light penetration so that the canopy may dry more rapidly. Avoiding excessive watering and nitrogen application, and increasing organic fertilizer, phosphorus and potassium applications can reduce disease severity in susceptible citrus cultivars.

Cultural practices such as trimming to improve the ventilation and light penetration, removal of diseased fruits, twigs, and leaves to reduce inoculum sources and avoiding overfertilizing and watering are effective in controlling Alternaria leaf spot (Timmer et al., 2000). For the control of Alternaria black rot, preventing fruit stress, trimming and pruning trees to improve air circulation and light permeability of the orchard, cutting off dead, weak, diseased, or dense branches to prevent fruit abrasions and stylar damage, as well as harvesting fruits at optimum maturity can reduce disease incidence (Timmer et al., 2000, 2003).

17.6 Postbloom fruit drop disease caused by *Colletotrichum*

17.6.1 Introduction

Postbloom fruit drop (PFD) disease of citrus, caused by *Colletotrichum acutatum,* is an economically important fungal disease of citrus occurring in the humid tropical areas of the Americas and sporadically in the humid subtropics (Timmer et al., 1994). In Florida, yield losses due to PFD have been observed in some years when rainfall was high during spring flowering. The pathogen infects all citrus cultivars and causes peach-to-orange-brown necrotic lesions on petals with salmon pink acervuli, necrosis of stigma, and premature fruit drop (Denham and Waller, 1981) and retention of calices called "buttons" (Fig. 17.6). Although PFD-like disease symptoms were observed in various areas of Central and South America in the 1950s and 1960s (Timmer et al., 1994), it was not reported until in the late 1970s from Belize (Fagan, 1979). The disease is now widespread in all humid citrus areas from Florida (USA) in the north to Uruguay in the south. It is endemic in many humid tropical areas such as southern Mexico, Belize, Costa Rica, and many islands in the Caribbean (Timmer and Peres, 2015).

In other areas such as Brazil, it is also a problem, when flowering coincides with a rainy period in which the crop losses can exceed 80% (De Goes et al., 2008). Thus far, the disease appears to be confined in the Americas because it has not been reported from other parts of the world.

Morphological characteristics such as colony morphology, conidial shape, the presence or absence of setae and sclerotia, and appressorium shape and size have been traditionally used for differentiation of species in the genus *Colletotrichum* (Sutton, 1992). Most of the fungi currently included in *C. acutatum* were formerly assigned to *C. gloeosporioides* with the teleomorph being *Glomerella cingulata* (Baxter et al., 1983; Van Der Aa et al., 1990; Sutton, 1992). In fact, the citrus PFD pathogen was also originally identified as *C. gloeosporioides* (Agostini et al., 1992; Timmer et al., 1994), but was later found to be *C. acutatum* (Brown et al., 1996). Nevertheless, the species *C. gloeosporioides* has also been reported recently as a causal agent of PFD (Lima et al. 2011; McGovern et al. 2012). Recent comprehensive studies of *Colletotrichum* species-complexes using multilocus analyses (Damm et al., 2012; Weir et al., 2012), and a new locus like ApMat (Sharma et al., 2015.) have substantially contributed to the better understanding of *Colletotrichum* diversity on citrus. At least 15 *Colletotrichum* species have been reported associated with citrus (Damm et al., 2012; Weir et al., 2012; Sharma et al., 2015). Known *Colletotrichum* diversity on citrus is based on studies with spatially limited sampling, thus, a geographically comprehensive study with proper pathogenic characterization is still lacking (Peres et al., 2008; Hou et al., 2010).

Colonies of *C. acutatum* are usually white initially and later become covered with pink-to-orange conidial aggregates (Zulfiqar et al., 1996). Conidia are primarily produced in acervuli; however, *C. acutatum* is also capable of forming secondary conidia on the surface of living leaves (MacKenzie et al., 2010), and it has recently been reported that secondary conidia can survive for up to 3 months on different weed species commonly found in citrus orchards in Brazil (Frare et al., 2016). Conidia of *C. acutatum* are usually ellipsoid and fusiform at one end and are different than those of *C. gloeosporioides*, which typically are rounded at both ends (Adaskaveg and Förster, 2000).

17.6.2 Disease symptoms

The disease in citrus is usually first observed as a peach-to-orange-colored necrotic lesions on open flowers (Fig. 17.6). Petals on open flowers are the most susceptible to infection, however, unopened flower buds may be infected (Timmer and Peres, 2015). The internal organs of the flower, as well as the pollen, can also be infected (Lin et al., 2001; Marques et al., 2013). In severe cases, whole flower clusters may be attacked causing blight of entire blossoms with shriveled necrotic petals stick firmly to inflorescences. After petal fall, the calyces and floral disks which normally abscise if no fruit is set, remain attached to the twig (Fig. 17.6A). These calyces, commonly called "buttons," persist for the life of the twig. Thus, signs of the disease are usually present year-round even though the infection occurs only during bloom. The buttons are characteristic of the disease and are rarely associated with any other disorder. Leaves surrounding infected flowers are often distorted with twisted laminae and enlarged veins (Timmer and Peres, 2015).

17.6.3 Infection process

The fungus is highly dependent on rainfall, especially frequent rain events when flowers are present in the orchard. Infection usually happens within 24 h of a favorable weather event, and symptoms occur in approximately 4–5 days (Agostini et al., 1992; Feichtenberger and Spósito, 2000b), In petals, penetration occurs intra- and intercellularly and

FIG. 17.6 Disease symptoms of postbloom fruit drop on Valencia sweet orange in the calyx (button) (A) and typical orange lesions on flower buds (B). Strain of *Colletotrichum acutatum* isolated from citrus flowers grown on potato dextrose agar (PDA) (C).

through the stomata (Marques et al., 2013). Copious conidia formed in acervuli on the surface of infected flowers are readily dispersed to healthy flowers with water/rain splash (Timmer et al., 1994). During the season, the disease continues to spread in an orchard as long as flowers are available particularly if trees are under continuous stress (e.g., HLB) and environmental conditions are favorable. Conidia deposited on vegetative surfaces, including weed hosts on the orchard floor, form appressoria, and latent infections that serve as survival structures during periods when flowers are absent. These appressoria will germinate to produce secondary conidia when exposed to petal leachates under moist conditions (Agostini et al., 1992; Timmer et al., 1994; Zulfiqar et al., 1996; MacKenzie et al., 2010; Frare et al., 2016). Thus, the first petals produced in the spring probably stimulate the germination of appressoria and the formation of conidia to initiate a new infection cycle. Moreover, at least in Brazil, weeds may also serve as an inoculum source for infection of citrus flowers during periods of bloom (Frare et al., 2016). This might indeed explain the sudden outbreaks of PFD in an orchard during years that had favorable climatic conditions for the disease, even in areas where the disease had not been previously reported (Timmer et al., 1994).

The lifestyles of *Colletotrichum* species can be broadly categorized as necrotrophic, hemibiotrophic, latent or quiescent and endophytic; of which hemibiotrophic is the most common. The differences in lifestyle depend on the *Colletotrichum* species, the host species, the physiological maturity of the host and environmental conditions. In citrus, *C. acutatum* infection strategies lead to a unique lifestyle; as a necrotroph on flowers and as a biotroph on leaves, thus, reproduces primarily on flower petals and does not readily colonize and form acervuli on other tissues (Agostini et al., 1992; Timmer et al., 1994; Zulfiqar et al., 1996; Peres et al., 2005).

While windblown rain is the main means of disease spread, human activity contributes to long-distance spread through equipment movement, especially if such movement occurs while the canopy is wet. Bees, fruit flies, and other flower-visiting insects are known to carry conidia of *Colletotrichum* (Pena and Duncan, 1989). Honeybees (*Apis mellifera*), as pollinating insects, are usually abundant in citrus orchards during the bloom period. In fact, recent studies have shown that the conidia of *Colletotrichum* were attached to parts of honeybee bodies and were dispersed among citrus plants in the greenhouse by honeybees (Gasparoto et al., 2017). Additionally, pollen grains can also be infected by the pathogen (Marques et al., 2013). Thus, it is also possible that bees carry conidia of *C. acutatum* directly on their legs and that the pathogen may survive in beehives. Nevertheless, this assumptions and roles of both infected pollen and honeybee or other pollinators in the epidemiology of PFD under field conditions should be determined with further studies.

17.6.4 Pathogenicity and virulence

Several *Colletotrichum* species have been shown to exhibit substantial pathogenic variation (Cannon et al., 2012). Differences in pathogenicity of strains from different hosts have been reported in several studies. Some fruit diseases caused by the *C. acutatum*-complex have been shown to be caused by distinct phylogenetic lineages (Peres et al., 2008), and strawberry fruit rot, in particular, was rarely found to be caused by isolates from diverse hosts (MacKenzie et al., 2009). The comparison of isolates of *C. acutatum* from sweet orange, Key lime, strawberry, blueberry, and leatherleaf fern have shown little evidence of cross pathogenicity. Furthermore, sequences of some genes in isolates from the same host were found similar or identical to each other but distinct from those isolates from other hosts (MacKenzie et al., 2010).

The mechanisms involved in young fruit abscission are suggested to be associated with a series of physiological modifications caused by an imbalance of plant hormones and/or growth regulators (Lahey et al., 2004). Until recently, PFD has been attributed exclusively to *C. acutatum*, whereas *C. gloeosporioides* was thought to only cause postharvest anthracnose in citrus fruits and can be found epiphytically on leaves and flowers (Brown et al., 1996; Peres et al., 2005). It was speculated that epiphytic *C. gloeosporioides* isolates might have shifted into being pathogenic in Brazil and those isolates of *C. gloeosporioides* that cause PFD are apparently less aggressive than those of *C. acutatum* (Lima et al., 2011). Consistent with this hypothesis, disease incidence, assessed by blossom blight or by persistent calyces, was found to be lower when plants were inoculated with *C. gloeosporioides* than with *C. acutatum* (Lima et al., 2011).

17.6.5 Disease management

The management of PFD requires an integrated disease strategy consisting of adequate cultural practices and timely application of effective fungicides or biocontrol agents. Effective cultural practices to prevent PFD outbreaks include avoiding stress conditions in citrus trees due to abiotic (i.e., irrigation, fertigation, or chemical) and biotic (i.e., disease) factors. This, in turn, will promote more compact bloom periods and will delay inoculum buildup and prevent serious epidemics in the orchard. As a matter of fact, PFD in Florida in recent years has reemerged with severe outbreaks after many years of minimal incidences in the state and is probably due to increased tree decline and off-season flowering caused by HLB, which is now endemic in the state. In some areas, overhead irrigation was used, which greatly extended the wetting periods and splash dispersed inoculum within trees and to nearby trees. Switching to micro sprinkler or drip irrigation has helped reduce PFD severity (Timmer and Peres, 2015).

Adequate control of PFD is highly dependent on timely applications of effective fungicide sprays in the bloom period. Timing is highly specific to the orchard situation, including overall flowering stage and local weather conditions and can be complicated because of inconsistent and off-season bloom that usually occur in HLB endemic areas (e.g., Florida and Brazil).

There are a number of fungicides shown to be effective on *Colletotrichum* species in citrus. Among those, benomyl and captafol were the most effective fungicides for PFD control in the past (Fagan, 1984; Timmer and Zitko, 1992). Captan, maneb, and other contact fungicides also provided some control of the disease. Carbendazim was also effective against PFD (De Goes et al., 2008; Silva-Junior et al., 2014) and would be a substitute for benomyl, which has been removed from the market. However, the use of the benzimidazole fungicides, like thiabendazole, carbendazim, and thiophanate-methyl, has also been recently restricted (Timmer and Peres, 2015). Folpet, a member of the same fungicide group as captan, has also proven quite effective for control of PFD in Brazil (Feichtenberger and Spósito, 2000a; De Goes et al., 2008). In Florida, Timmer and Zitko (1998) found that strobilurins, such as trifloxystrobin and azoxystrobin, provided some control of the disease, but were not as effective as benomyl. Silva-Junior et al. (2014) found that the mixture of trifloxystrobin + tebuconazole was the most effective in studies in Brazil. Although many products on the market provide some degree of control of PFD, growers have been left with no highly effective materials. Thus, control of PFD has been made even more challenging (Timmer and Peres, 2015).

17.7 Citrus scab diseases caused by *Elsinoë*

17.7.1 Introduction

The causal agents of scab diseases of citrus are *Elsinoë fawcettii* (= *Sphaceloma fawcettii*) and *E. australis* (= *S. australis*). Scab diseases of citrus are distributed in many citrus-producing countries worldwide in humid zones except Mediterranean regions. These pathogens affect the fruits, leaves, and twigs of susceptible species of citrus, including lemons, grapefruits, many tangerines and their hybrids causing external blemishes and reducing the acceptability of fruits for the fresh market (Chung, 2011). Two scab diseases are currently recognized on citrus: citrus scab, caused by *E. fawcettii*, and sweet orange scab, caused by *E. australis*. *Elsinoë* isolates were differentiated into at least six pathotypes by pathogenicity and genetic analysis: Florida broad host range (FBHR), Florida narrow host range (FNHR), Tryon's, "Lemon," "Jingeul," and *E. australis* natsudaidai pathotype (Timmer et al., 1996; Hyun et al., 2009). The teleomorphs of both species of *Elsinoë* were only found in Brazil. The conidia have two types: hyaline conidia and spindle-shaped colored conidia (Timmer et al., 2000). Colony morphology of *E. australis* and *E. fawcettii* cultured on PDA has two types, mucoid and tomentose (Fig. 17.7A–C). Although the same isolate exhibits two colony morphologies, it is not certain whether the colony morphology type is determined under specific conditions. However, most of the *Elsinoë* strains isolated from lesions initially have a tomentose type, and when they are subcultured, they appear to be mainly mucoid type.

17.7.2 Disease symptoms

The scab pathogens induce typical symptoms on adaxial as well as abaxial surfaces of the leaves (Fig. 17.7D). These symptoms begin to appear 3 days after pathogen invasion, with tissues around the infection sites becoming yellowish and a small spot appears in the center. Initially, scab pustules are slightly raised and pink-to-light (Fig. 17.7E) brown and, as the pustules develop, they become warty and cracked and turn yellowish-brown and eventually dark gray (Fig. 17.7F) (Whiteside, 1975; Chung, 2011). Elevation and size of the pustules vary according to disease sensitivity, tissue age at the time of infection, and environmental conditions. Elevated protuberances form on the side of the leaf that has been invaded, and a corresponding indentation forms on the opposite side. In severe cases, the young leaves or young fruits become deformed and even drop from the tree. Symptoms vary depending on the developmental stage of the leaf or fruit (Timmer et al., 2004). Protuberance-type symptoms are induced in early infections and gradually become flatter as the tissue grows larger. However, as the fruit grows larger, the protruding-type or the scab-type lesion turns into a flatter or scurfy-type lesion. Scab symptoms, especially the flatter and scurfy-type lesions that occur on fruits, are easily confused with gray mold (*Botrytis cinerea*). In addition, in resistant cultivars or almost expanded leaves and when environmental conditions are not suitable, lesions may appear as black spots.

FIG. 17.7 Colony morphology of *Elsinoë* showing mucoid (A) and tomentose type (B and C) grown for 20–30 days on plates with potato dextrose agar (PDA). Symptoms of scab showing elevated protuberances in leaves (D) and scab pustules in fruits from *light* (E) to *dark gray* (F).

17.7.3 Infection process

Conidia of the citrus scab pathogens infect on adaxial as well as abaxial surfaces of the leaf. Conidia extend their germ tubes from 1-day postinoculation with globose appressorium-like structures at their tip. Cuticle and mesophyll cells including epidermis in the infected dark region degrade 1–2 day-postinoculation, resulting in cellular necrosis. The adjacent cells of the necrotic region enlarge and are gradually covered leaving a groove-shaped necrotic center and appear as a first visible symptom after 3 days (Paudyal et al., 2017). Hyphae extend through the necrotic epidermis and colonize intra- and intercellularly in mesophyll tissue. Also, lamellated cell-walls are observed as a boundary between infected and noninfected regions in the lesion after typical symptoms were formed. Cross sections through the lesion reveal hyphal colonization in epidermis and mesophyll tissues. In response to the pathogen colonization, host cell walls are lignified, inner cells are rapidly compartmentalized, and a semicircular boundary is formed that separates the infected region from the noninfected region, and finally prevents the intercellular pathogen spread (Paudyal and Hyun, 2015). The pathogen overwinters in lesions on twigs, leaves and remaining fruit, and when the lesions are wet at 15–20°C for 2 days, it initiates production of conidia for infection. The rate of disease progression is fastest at around 25°C and the first symptom appears within 3 days after infection. The disease does not occur above 35°C. Overwintered lesions on leaves are very important as a first inoculum source in the field. Once conidia are formed, pathogens can invade even if the plant surface is wet for about 3 h at the optimum temperature (24–28°C) (Whiteside, 1975). If the disease occurs in the spring flushes, it may develop rapidly depending on weather conditions, and eventually, the fruit symptoms become more severe. The conidia formed in the lesions are disseminated by air mixed in raindrops and spread to distant places. Leaves are most susceptible just after emergence and are resistant to infection by the time they reach half expansion. Fruits are susceptible to infection for 6–8 weeks after petal fall (Timmer et al., 2000). However, if the environmental conditions are conducive, infections may occur even after 3–4 months of petal fall.

17.7.4 Pathogenicity and virulence

The citrus scab pathogen, *E. fawcettii*, can infect commercial species of citrus such as sweet orange (*C. sinensis*), lemon (*C. limon*), grapefruit (*C. paradisi*), many tangerine, Clementine and mandarin lines (*C. reticulata*), Satsuma mandarin (*C. unshiu*), and their hybrids, as well as species used as rootstocks such as sour orange (*C. aurantium*), rough lemon (*C. jambhiri*), and Cleopatra mandarin (*C. reshni*) (Hyun et al., 2009). The sweet orange scab pathogen, *E. australis*, is mainly distributed in South America and usually affects only sweet orange fruits. Scab diseases do not occur in the Mediterranean citrus-growing areas. It is also common to find *E. fawcettii* infecting ornamental citrus and rootstocks like calamondin (*C. madurensis*) and *P. trifoliata*. It is not common to find *E. fawcettii* in Persian limes, kumquats, pummelos, citrons or limes (Chung, 2011).

17.7.5 Disease management

Growing citrus cultivars resistant to scab diseases may be ideal, but there are some practical limitations. Therefore, for the control of scab diseases it is recommended to remove the affected plant tissues to reduce the pathogen inoculum (Timmer et al., 2004). Since new conidia are generated in the next growing season in lesions of twigs, leaves and remaining fruit infected in the previous season, it is necessary to remove these diseased tissues as much as possible at harvest and by pruning. Even when there is a small amount of diseased tissue left, the disease can develop quickly under favorable weather conditions. Thinning and pruning allow exposure to sunlight and enhance air movement, reducing the humidity in the orchard, and thereby minimizing disease development. Overhead irrigation should not be used on susceptible varieties for production of fresh fruit because it greatly increases the amount of infection.

When scab is severe and there is substantial inoculum carryover from the previous season, fungicides should be applied before the spring flush is one-fourth expanded. If the disease develops in the spring flush, fungicide sprays should be applied more than two times after petal fall until the fruit becomes resistant to infection. However, in the case of sweet orange scab, it is not necessary to control it in the spring flush. Strobilurin fungicides are of the most effective, reaching over 98% of disease control. Copper fungicides can be also used, but they are less effective. Benomyl, and sterol-biosynthesis-inhibiting fungicides were also reported effective for scab control, but their use is currently restricted in many countries. Fungicides with the same mode of action should not be used consecutively because resistant isolates can develop. When sprayed on fruits for scab control, fungicides can be chosen also to control black spot, melanose or other diseases simultaneously (Timmer et al., 2004).

17.8 Mal secco disease caused by *Plenodomus tracheiphilus*

17.8.1 Introduction

Mal secco is a vascular disease of lemon (*C. limon*) and other species of the genera *Citrus*, *Fortunella*, *Poncirus*, and *Severinia* and their hybrids, caused by *Plenodomus tracheiphilus* (= *Phoma tracheiphila*). The pathogen enters the tree through wounds in leaves, branches, and roots and colonizes the xylem vessels, causing wilt and leaf drop and ultimately dieback of twigs and branches, hence the Italian name "mal secco" (dry disease) (Cutuli et al., 1984; Perrotta and Graniti, 1988). Two other syndromes were described: "mal fulminante" (sudden death) and "mal nero" (black disease) due to root infections. "Mal secco" is most prevalent and severe on lemon, lime (*C. aurantifolia*), citron (*C. medica*), and bergamot (*Citrus* × *bergamia*). The disease was first described by Petri (1929), but symptoms were already observed in the Greek islands of Chios and Poros (Perrotta and Graniti, 1988). During the years its devastating effects have affected almost all the lemon producing countries of the Mediterranean Basin (EFSA, 2014a).

17.8.2 Disease symptoms

The typical symptoms consist of red discolored strands in the xylem of stems, leaf vein chlorosis, wilt, and leaf drop and ultimately dieback of twigs and branches (Fig. 17.8A). Symptom expression starts as slight discoloration of the primary and secondary veins, leaf and shoot chlorosis, wilting of leaves, leaf drop, and twig and branch dieback (Migheli et al., 2009). The petioles often remain for a period after leaves drop. Newly infected shoots, branches, and trunks show a salmon-pink or reddish-brown color, along with gum production, within the xylem vessels. On the withered twigs, immersed, flask-shaped, or globose pycnidia develop under the epidermis and emerge as black points within lead-gray or ash-gray areas. Frequently, a sprouting of shoots from the base of affected branches and rootstock suckers is observed. When infection starts at the base of a twig or branch the disease progresses more rapidly, whole shoots or branches wilt followed by leaf and fruit drop.

17.8.3 Infection process

Infection occurs by conidia that are produced in pycnidia on withered twigs or by phialoconidia produced by phialides formed on "free" hyphae grown on exposed wood surfaces, wounded plant tissues, and within the xylem elements (Migheli et al., 2009; Nigro et al., 2011). Conidia are usually dispersed by water-splash and wind-blown rain. Conidial germination may occur at temperatures between 5°C and 30°C with an optimum at 25°C. Temperatures above 30°C inhibit mycelial growth but do not kill the pathogen. Warm moist conditions are favorable to the infection and disease development. The pathogen enters susceptible host tissues (leaves, branches, twigs, and roots) by both conidia and mycelium through wounds/injuries. Penetration through stomata has never been demonstrated (Zucker and Catara, 1985). Once in the host, the fungus reaches the lumen of the xylem and then spreads systemically, mostly upward, by the conidia transported in the xylem sap.

FIG. 17.8 Dieback of twigs and branches (A) caused by mal secco disease; (B) *Plenodomus tracheiphilus* colonies arising from infected wood; and (C) colonies with different pigmentation.

The infection period depends on local climatic and seasonal conditions. Any rainy period is conducive for infection provided the inoculum is present. The length of the incubation period varies depending on the climatic conditions, the age of the host and the form of the disease. In the case of "mal secco" it ranges between 2 and 7 months, whereas in “mal nero” it may last for several years. Near saturation humidity and temperatures between 20°C and 25°C are favorable for the production of phialides on hyphae present in wounds and leaf scars. *P. tracheiphilus* can survive as mycelium and conidia within pycnidia on infected leaves, twigs, and branches of susceptible citrus hosts. Infections occurring in lemon seeds do not have any epidemiological role. The survival potential of the pathogen in infected host plant debris lying on the soil varies between 30 days to 1 year depending on the soil type. The number of propagules in the soil decreases during the dry season (Russo et al., 2011).

17.8.4 Pathogenicity and virulence

Symptom distribution and severity are correlated with xylem colonization as in other tracheomycotic diseases (Migheli et al., 2009; Nigro et al., 2011). Conidia could be passively transported by the xylem sap flux reaching rapidly to twigs located far from the inoculation point. *P. tracheiphilus* is able to produce both hydrolytic enzymes and toxins. Toxin transport via-xylem leads to leaf chlorosis. The systemic invasion by the pathogen causes the impairment of water transport in the plant and consequent wilt associated with progressive clogging of the xylem by both gums and mycelium. Later, digestion of the walls of neighboring conduits leads to an easier and faster pathway for fungal diffusion. The lysis of pit membranes provides circumstantial evidence for the production of pectic enzymes by the fungus during the colonization process, which may act as virulence factors.

Culture filtrates of *P. tracheiphilus* induce wilting and chlorosis in lemon and nonhost plant cuttings such as tomato. Two nonselective phytotoxic glycoproteins of 93 and 60 kDa (called Pt60) (generally referred to as malseccin) have been extracted from culture filtrates of *P. tracheiphilus* (Migheli et al., 2009). Both toxins were able to reproduce symptoms of leaf vein chlorosis when injected into citrus leaves. Pt60 toxin has been isolated also from the leaves and the xylem of naturally infected lemon trees showing symptoms of the disease including leaf vein chlorosis, the typical salmon-orange wood discoloration, and leaf drop. One of the most recent studies pointed out the synchronous presence in the fungus of hydrolytic enzymes, toxic compounds, oxidative stress inducers, and membrane transporters. The contemporary differential ability to modulate the lipoperoxidative pathway in the host could play a key role in the interaction between *P. tracheiphilus* and its hosts.

Chromogenic and nonchromogenic variants of the fungus based on the production in the culture of red pigments were described; the nonpigmented isolates showed reduced virulence (Fig. 17.8B and C) (Migheli et al., 2009). ITS sequencing and most DNA fingerprinting techniques failed to show substantial genetic variability among isolates of the fungus, among them f-AFLP was the most discriminant (Nigro et al., 2011; Russo et al., 2011).

17.8.5 Disease management

P. tracheiphilus is of quarantine concern therefore strict regulation is adopted from pathogen-free countries (EPPO, 2019). As the introduction of *P. tracheiphilus* is a serious threat to lemon-growing areas, restrictions on the movement of citrus propagating material are mandatory. Therefore, the certification of plant materials is a critical step in disease control (EFSA, 2014b). Conventional and real-time PCR-based diagnostic methods are available based on SCARS or ITS primers (Nigro et al., 2011). The protocols are amenable for in planta detection as well as for fungus identification. The detection of the fungus in the soil was also achieved.

Where the disease is present, it is important (and part of mandatory protocols) to prune withered shoots, remove suckers, followed by burning of pruned branches to reduce inoculum (Perrotta and Graniti, 1988; Nigro et al., 2011). Pruning is performed in late spring and early summer when symptoms are more easily recognized. If the main trunk is not colonized by the fungus, removal of the diseased branches can limit spread and sometimes cure the tree of the disease. Fungicide applications are recommended in the nursery. In the orchards, applications of copper compounds or other authorized fungicides (e.g., mancozeb) are useful during the rainy season, after hailstorms, frost damage, and pruning. Many efforts have been made for the establishment of lemon cultivars that combine resistance with agronomic and qualitative characteristics. Currently, no citrus cultivars/clones or rootstocks resistant/tolerant to *P. tracheiphilus* infection that have competitive yields and satisfactory bioagronomic features are available for use on a large scale to control the disease in infested citrus-producing areas.

References

Adaskaveg, J.E., Förster, H., 2000. Occurrence and Management of Anthracnose Epidemics Caused by *Colletotrichum* Species on Tree Fruit Crops in California. American Phytopathological Society, St. Paul, MN.

Agostini, J.P., Timmer, L.W., Mitchell, D.J., 1992. Morphological and pathological characteristics of strains of *Colletotrichum gloeosporioides* from citrus. Phytopathology 82, 1377–1382.

Aguilera-Cogley, V.A., Berbegal, M., Catala, S., Brentu, F.C., Armengol, J., Vicent, A., 2017. Characterization of *Mycosphaerellaceae* species associated with citrus greasy spot in Panama and Spain. PLoS One 12, e0189585.

Albrigo, L.G., Beck, H.W., Timmer, L.W., Stover, E.W., 2005. Development and testing of a recommendation system to schedule copper sprays for citrus disease control. J. ASTM Int. 2, 1–12.

Averna-Saccá, R., 1940. Pústulas pretas sobre laranjas doces produzidas pelo *Phoma citricarpa*. Rev. Agric. 15, 668–674.

Baldassari, R.B., Wickert, E., De Goes, A., 2008. Pathogenicity, colony morphology and diversity of isolates of *Guignardia citricarpa* and *G. mangiferae* isolated from *Citrus* spp. Eur. J. Plant Pathol. 120, 103–110.

Baxter, A.P., Van Der Westhuizen, G.C.A., Eicker, A., 1983. Morphology and taxonomy of South African isolates of *Colletotrichum*. S. Afr. J. Bot. 2, 259–289.

Bella, P., Guarina, C., La Rosa, R., Catara, A., 2001. Severe infections of *Alternaria* spp. on a mandarin hybrid. J. Plant Pathol. 83.

Bellotte, J.A.M., Kupper, K.C., Rinaldo, D., Souza, A.D., Pereira, F.D., Goes, A.D., 2009. Acceleration of the decomposition of Sicilian lemon leaves as an auxiliary measure in the control of citrus black spot. Trop. Plant Pathol. 34.

Bellotte, J.A.M., Kupper, K.C., Rinaldo, D., Souza, A.D., Goes, A.D., 2013. The effects of inter-crop cultivation between rows of citrus crop on spreading of *Guignardia citricarpa* ascospores and in the citrus black spot occurrence. Rev. Bras. Frutic. 35, 102–111.

Benson, A.H., 1895. Black spot of the orange. Agric. Gazette NSW 6, 249.

Blair, J.E., Coffey, M.D., Park, S.Y., Geiser, D.M., Kang, S., 2008. A multi-locus phylogeny for *Phytophthora* utilizing markers derived from complete genome sequences. Fungal Genet. Biol. 45, 266–277.

Boava, L.P., Cristofani-Yaly, M., Mafra, V.S., Kubo, K., Kishi, L.T., Takita, M.A., Ribeiro-Alves, M., Machado, M.A., 2011. Global gene expression of *Poncirus trifoliata*, *Citrus sunki* and their hybrids under infection of *Phytophthora parasitica*. BMC Genomics 12, 39.

Brentu, F.C., Oduro, K.A., Offei, S.K., Odamtten, G.T., Vicent, A., Peres, N.A., Timmer, L.W., 2012. Crop loss, aetiology, and epidemiology of citrus black spot in Ghana. Eur. J. Plant Pathol. 133, 657–670.

Brodrick, H.T., Rabie, C.J., 1970. Light and temperature effects on symptom development and sporulation of *Guignardia citricarpa* Kiely, on *Citrus sinensis* (Linn) Osbeck. Phytophylactica 2, 157–163.

Brown, G.E., Eckert, J.W., 2000. Alternaria Rot. APS Press, St. Paul, MN.

Brown, G.E., McCornack, A.A., 1972. Decay caused by *Alternaria citri* in Florida citrus fruit. Plant Dis. Rep. 56, 909–912.

Brown, A.E., Sreenivasaprasad, S., Timmer, L.W., 1996. Molecular characterization of slow-growing orange and key lime anthracnose strains of *Colletotrichum* from citrus as *C. acutatum*. Phytopathology 86, 523–527.

Cacciola, S.O., Di, G.M., Lio, S., 2008. Management of *Citrus* Diseases Caused by *Phytophthora* spp. Springer Science.

Canihos, Y., Erkilic, A., Timmer, L.W., 1997. First report of *Alternaria* brown spot of Minneola tangelo in Turkey. Plant Dis. 81, 1214.

Cannon, P.F., Damm, U., Johnston, P.R., Weir, B.S., 2012. *Colletotrichum*—current status and future directions. Stud. Mycol. 73, 181–213.

Chung, K.R., 2011. *Elsinoë fawcettii* and *Elsinoë australis:* the fungal pathogens causing citrus scab. Mol. Plant Pathol. 12, 123–135.

Cobb, N.A., 1903. Letters on the diseases of plants-*Alternaria* of the citrus tribe. Agric. Gazette NSW 14, 955–986.

Cuturi, G., Laviola, C., Perrotta, G., Salerno, M., Spina, P., 1984. Il mal secco degli agrumi. In: Seminario Internazionale di Studio - Programma di Ricerche Agrimed 1984–88. della Commissione della Comunità Europea, Capo d'Orlando, Italia.

Dalio, R.J.D., Maximo, H.J., Oliveira, T.S., Azevedo, T.M., Felizatti, H.L., Campos, M.A., Machado, M.A., 2018. Molecular basis of *Citrus sunki* susceptibility and *Poncirus trifoliata* resistance upon *Phytophthora parasitica* attack. Mol. Plant Microbe Interact. 31, 386–398.

Damm, U., Cannon, P.F., Woudenberg, J.H., Crous, P.W., 2012. The *Colletotrichum acutatum* species complex. Stud. Mycol. 73, 37–113.

Das, A.K., Nerkar, S., Kumar, A., 2018. First report of *Phyllosticta citricarpa* causing citrus black spot on *Citrus sinensis* and *C. reticulata* in India. Plant Dis. 102 (8), 1661.

De Goes, A., Garrido, R.B.O., Reis, R.F., Baldassari, R.B., Soares, M.A., 2008. Evaluation of fungicide applications to sweet orange at different flowering stages for control of postbloom fruit drop caused by *Colletotrichum acutatum*. Crop Prot. 27, 71–76.

Denham, T.G., Waller, J.M., 1981. Some epidemiological aspects of postbloom fruit drop disease (*Colletotrichum gloeosporioides*) in citrus. Ann. Appl. Biol. 98, 65–77.

Dewdney, M.M., 2016. Alternaria brown spot, Alternaria diseases in citrus SP 043. In: Rogers, M.E., Dewdney, M.M. (Eds.), 2016 Florida Citrus Pest Management Guide. University of Florida, IFAS Extension, Lake Alfred, FL, pp. 81–83.

Dewdney, M.M., 2018. Melanose. In: 2018-2019 Florida Citrus Production Guide. University of Florida, IFAS Extension, Lake Alfred, FL.

Dewdney, M.M., 2019. Greasy spot. In: Plant Pathology Department. University of Florida, IFAS Extension, Florida.

Dewdney, M.M., Graham, J.H., 2016. Foliar Disease Management in 2016. Citrus Industry.

Dewdney, M.M., Fraisse, C.W., Zortea, T., Burrow, J., 2013. A Web-Based Tool for Timing Copper Applications in Florida Citrus. University of Florida, IFAS Extension EDIS, p. 4.

Doidge, E.M., 1929. Some diseases of citrus prevalent in South Africa. S. Afr. J. Sci. 26, 320–325.

EFSA, 2014a. Scientific opinion on the pest categorisation of *Plenodomus tracheiphilus* (Petri) Gruyter, Aveskamp & Verkley [syn. *Phoma tracheiphila* (Petri) L.A. Kantschaveli & Gikashvili]. PLH Panel (EFSA Panel on Plant Health). EFSA J. 12, 3775. 3734 pp.

EFSA, 2014b. Scientific opinion on the risk of *Phyllosticta citricarpa* (*Guignardia citricarpa*) for the EU territory with identification and evaluation of risk reduction options. In: PLH Panel (EFSA Plant Health Panel) EFSA J. 12 (2), 3557. https://doi.org/10.2903/j.efsa.2014.3557.

EFSA Panel on Plant Health, 2018. Evaluation of a paper by Guarnaccia et al. (2017) on the first report of *Phyllosticta citricarpa* in Europe. EFSA J 16 (1), e05114.

EPPO, 2019. EPPO Global Database. Available online.

EPPO (European and Mediterranean Plant Protection Organization), 2019. EPPO Reporting Service No. 07-2019 2019/141 First Report of *Phyllosticta citricarpa* in Tunisia. https://gd.eppo.int/reporting/article-6571.

Erwin, D.C., Ribeiro, O.K., 1996. Phytophthora Diseases Worldwide. APS Press, St Paul, MN.

Everett, K.R., Rees-George, J., 2006. Reclassification of an isolate of *Guignardia citricarpa* from New Zealand as *Guignardia mangiferae* by sequence analysis. Plant Pathol. 55, 194–199.

Fagan, H.J., 1979. Postbloom fruit drop, a new disease of citrus associated with a form of *Colletotrichum gloeosporioides*. Ann. Appl. Biol. 91, 13–20.

Fagan, H.J., 1984. Postbloom fruit drop of citrus in Belize: II. Disease control by ground/aerial spraying. Turrialba 34, 179–184.

Fawcett, H.S., 1936. *Citrus* Diseases and Their Control. McGraw-Hill, New York.

Feichtenberger, E., Spósito, M.B., 2000a. Fungicide treatments for control of postbloom fruit drop of citrus (*Colletotrichum acutatum*) in 'Pera' sweet orange trees. In: Proceedings of the 9th International Society of Citriculture Congress, pp. 145.

Feichtenberger, E., Spósito, M.B., 2000b. Fungicide treatments for control of Postbloom fruit drop of citrus (*Colletotrichum acutatum*) in 'Pera' sweet orange trees. In: Proceedings of the 9th International Society of Citriculture Congress, Orlando, FL.

Feichtenberger, E., Rossetti, V., Pompeu, J.J., TeóFilo Sobrinho, J., Figueiredo, J.O., 1992. Evaluation of tolerance to *Phytophthora* species in scion rootstock combinations of citrus in Brazil—a review. Proc. Int. Soc. Citricult. 7, 854–858.

Foguet, J.L., Vázquez De Ramallo, N.E., Torres Leal, G.J., 1985. Presencia de mancha negra de los citrus en Tucumán. Avan. Agroindustr. 6, 9–10.

Fourie, P.H., Schutte, G.C., Serfontein, S., Swart, S.H., 2013. Modelling the effect of temperature and wetness on *Guignardia* pseudothecium maturation and ascospore release in citrus orchards. Phytopathology 103, 281–292.

Frare, G.F., Couto, H.T.Z., Ciampi-Guillardi, M., Amorim, L., 2016. The causal agent of citrus postbloom fruit drop, *Colletotrichum acutatum*, can survive on weeds. Australas. Plant Pathol. 45, 339–346.

Gardner, J.M., Kono, Y., Tatum, J.H., Suzuki, Y., Takeuchi, S., 1985. Structure of major component of ACRL toxins, host-specific phytotoxic compound produced by *Alternaria citri*. Agric. Biol. Chem. 49, 1235–1238.

Garrán, S.M., 1996. *Citrus* black spot in the northeast of Entre Rios: etiology, epidemiology and control. In: Proceedings of the International Society of Citriculture, Sun City, South Africa.

Gasparoto, M.C.G., Lourenco, S.A., Tanaka, F.A.O., Sposito, M.B., Marchini, L.C., Silva Junior, G.J., Amorim, L., 2017. Honeybees can spread *Colletotrichum acutatum* and *C. gloeosporioides* among citrus plants. Plant Pathol. J. 66, 777–782.

Glienke, C., Pereira, O.L., Stringari, D., Fabris, J., Kava-Cordeiro, V., Galli-Teresawa, L., Cunnington, J., Shivas, R.G., Groenewald, J.Z., Crous, P.W., 2011. Endophytic and pathogenic *Phyllosticta* species, with reference to those associated with citrus black spot. Persoonia 26, 47–56.

Goodwin, S.B., 1997. The population genetics of *Phytophthora*. Phytopathology 87, 462–473.

Gopal, K., Mukunda Lakshmi, L., Sarada, G., Nagalakshmi, T., Gouri Sankar, T., Gopi, V., Ramana, K.T.V., 2014. *Citrus* melanose (*Diaporthe citri* Wolf): a review. Int. J. Curr. Microbiol. App. Sci. 3, 113–124.

Graham, J.H., 1990. Evaluation of tolerance of citrus rootstocks to *Phytophthora* root-rot in chlamydospore-infested soil. Plant Dis. 74, 743–746.

Graham, J.H., 1995. Root regeneration and tolerance of citrus rootstocks to root rot caused by *Phytophthora nicotianae*. Phytopathology 85, 111–117.

Graham, J.H., 2000. Larval feeding injury to citrus roots and its relationship to invasion by soil-borne plant pathogens. In: Futch, S.H. (Ed.), Proceedings of the Diaprepes Short Course. University of Florida, IFAS Cooperative Extension Service, Lake Alfred, FL.

Graham, J.H., Dewdney, M.M., 2015. Florida Citrus Pest Management Guide [Online]. University of Florida, IFAS Extension, Florida. (Accessed).

Graham, J.H., Feichtenberger, E., 2015. *Citrus Phytophthora* diseases: management challenges and successes. J. Cit. Pathol. 2 (1). https://escholarship.org/uc/item/3db485rh.

Graham, J.H., Menge, J.A., 1999. Root Diseases. American Phytopathological Society, St. Paul, MN.

Graham, J.H., Timmer, L.W., 2006. *Phytophthora* Diseases of *Citrus* [Online]. University of Florida, IFAS Extension. (Accessed).

Graham, J.H., Irey, M.S., Taylor, J., 2011. *Phytophthora* damage to roots: a potential contributor to decline of HLB-affected trees. Cit. Ind. 92, 20–23.

Guarnaccia, V., Groenewald, J.Z., Youjoan, L., Glienke, C., Carstens, E., Hattingh, V., Fourie, P.H., Crous, P.W., 2017. First report of *Phyllosticta citricarpa* and description of two new species, *P. paracapitalensis* and *P. paracitricarpa*, from citrus in Europe. Stud. Mycol. 87, 161–185.

Guo, W.W., Deng, X.X., 2001. Wide somatic hybrids of *Citrus* with its related genera and their potential in genetic improvement. Euphytica 118, 175–183.

Haggag, W.M., 2012. First report of *Mycosphaerella citri,* the agent of greasy spot of orange in Egypt. Disease Note 94, .

Hendricks, K.E., Christman, M.C., Roberts, P.D., 2017. Spatial and temporal patterns of commercial citrus trees affected by *Phyllosticta citricarpa* in Florida. Sci. Rep. 7, 1641.

Herrera-Isla, L., 1992. La mancha parda de los cítricos en Cuba. Levante Agrícola 31, 49–50.

Hidalgo-Góngora, E.L., Pérez-Vicente, L., 2010. Diferenciación morfológica, cultural y biológica de *Guignardia citricarpa* y *Guignardia mangiferae* en frutos cítricos de Cuba. Fitosanidad 14, 141–152.

Hou, X.L., Zhou, X.Y., Tan, P., Jiang, Y.L., 2010. The investigation on fungal diseases of citrus in Guizhou. Hubei Agric. Sci. 49, 3052–3054.

Huang, F., Fu, Y., Nie, D., Stewart, J.E., Peever, T.L., Li, H., 2015a. Identification of a novel phylogenetic lineage of *Alternaria alternata* causing citrus brown spot in China. Fungal Biol. 119, 320–330.

Huang, F., Groenewald, J.Z., Zhu, L., Crous, P.W., Li, H., 2015b. Cercosporoid diseases of *Citrus*. Mycologia 107, 1151–1171.

Hyun, J.W., Yi, S.H., MacKenzie, S.J., Timmer, L.W., Kim, K.S., Kang, S.K., Kwon, H.M., Lim, H.C., 2009. Pathotypes and genetic relationship of worldwide collections of *Elsinoë* spp. causing scab diseases of citrus. Phytopathology 99, 721–728.

IPPC, 2016. Diagnostic protocols for regulated pests DP 5: *Phyllosticta citricarpa* (McAlpine) Aa on fruit. In: ISPM 27 (Eds.), I.P.P. Convention. IPPC, FAO, Rome.

Isshiki, A., Akimitsu, K., Yamamoto, M., Yamamoto, H., 2001. Endo-polygalacturonase is essential for citrus black rot caused by *Alternaria citri* but not brown spot caused by *Alternaria alternata*. Mol. Plant Microbe Interact. 14, 749–757.

Katoh, H., Isshiki, A., Masunaka, A., Yamamoto, H., Akimitsu, K., 2006. A virulence-reducing mutation in the postharvest citrus pathogen *Alternaria citri*. Phytopathology 96, 934–940.

Kiely, T.B., 1948. Preliminary studies on *Guignardia citricarpa*, n. sp.: the ascigenous stage of *Phoma citricarpa* McAlp. and its relation to black spot of citrus. Proc. Linnean Soc. NSW 68, 249–292.

Klotz, L.J., De Wolfe, T.A., 1960. The production and use of zoospore suspension of *Phytophthora* spp. for investigations on diseases of citrus. Plant Dis. Rep. 44, 572–573.

Kohmoto, K., Scheffer, R.P., Whiteside, J.O., 1979. Host-selective toxins from *Alternaria citri*. Phytopathology 69, 667–671.

Kohmoto, K., Itoh, Y., Shimomura, N., Kondoh, Y., Otani, H., Kodama, M., Nishi-Mura, S., Nakatsuka, S., 1993. Isolation and biological activities of two host- specific toxins from the tangerine pathotype of *Alternaria alternata*. Phytopathology 83, 495–502.

Kotzé, J.M., 1981. Epidemiology and control of citrus black spot in South-Africa. Plant Dis. 65, 945–950.

Lahey, K.A., Yuan, R., Burns, J.K., Ueng, P.P., Timmer, L.W., Chung, K.R., 2004. Induction of phytohormones and differential gene expression in citrus flowers infected by the fungus *Colletotrichum acutatum*. Mol. Plant Microbe Interact. 17, 1394–1401.

Lanza, F.E., Metzker, T.G., Vinhas, T., Behlau, F., Silva Junior, G.J., 2018. Critical fungicide spray period for citrus black spot control in Sao Paulo state, Brazil. Plant Dis. 102, 334–340.

Lee, H.A., 1920. Black spot of citrus fruits caused by *Phoma citricarpa*. Philipp. J. Sci. 17, 635–641.

Lee, Y.S., Huang, C.S., 1973. Effect of climatic factors on the development and discharge of ascospores of the citrus black spot fungus. J. Taiwan Agric. Res. 22, 135–144.

Leki, H., 1986. The causal fungus of greasy spot in Okinawa distintic of Japan. Ann. Phytopathol. Soc. Jpn. 52, 484–487.

Lima, W.G., Spósito, M.B., Amorim, L., Gonçalves, F.P., Melo De Filho, P.A., 2011. *Colletotrichum gloeosporioides,* a new causal agent of citrus post-bloom fruit drop. Eur. J. Plant Pathol. 131, 157–165.

Lin, Y., Stover, E.W., Sonoda, R., Rosskopf, E., 2001. Stigma and style necrosis is associated with postbloom fruit drop disease in citrus following artificial inoculation. HortScience 36, 1138.

MacKenzie, S.J., Peres, N.A., Barquero, M.P., Arauz, L.F., Timmer, L.W., 2009. Host range and genetic relatedness of *Colletotrichum acutatum* isolates from fruit crops and leatherleaf fern in Florida. Phytopathology 99, 620–631.

MacKenzie, S.J., Peres, N.A., Timmer, L.W., 2010. Colonization of citrus leaves and secondary conidiation response to citrus flower extracts by non-postbloom fruit drop strains of *Colletotrichum acutatum*. Trop. Plant Pathol. 35, 333–342.

Makowski, D., Vicent, A.M.D., Pautasso, M., Stancanelli, G.S., Rafoss, T.T., 2014. Comparison of statistical models in a meta-analysis of fungicide treatments for the control of citrus black spot caused by *Phyllosticta citricarpa*. Eur. J. Plant Pathol. 139, .

Marques, J.P.R., Amorim, L., Spósito, M.B., Marin, D., Appezzato-Da-Glória, B., 2013. Infection of citrus pollen grains by *Colletotrichum acutatum*. Eur. J. Plant Pathol. 136, 35–40.

Martínez-Minaya, J., Conesa, D., López-Quílez, A., Vicent, A., 2015. Climatic distribution of citrus black spot caused by *Phyllosticta citricarpa*, a historical analysis of disease spread in South Africa. Eur. J. Plant Pathol. 143, 69–83.

McAlpine, D., 1899. Fungus Diseases of *Citrus* in Australia and Their Treatment. Agric. Dep, Melbourne, VIC, Australia.

McGovern, R.J., Seijo, T.E., Hendricks, K., Roberts, P.D., 2012. New report of *Colletotrichum gloeosporioides* causing postbloom fruit drop on citrus in Bermuda. Can. J. Plant Pathol. 34, .

McOnie, K.C., 1967. Germination and infection of citrus by ascospores of *Guignardia citricarpa*. Phytopathology 57, 743–746.

Meena, M., Gupta, S.K., Swapnil, P., Zehra, A., Dubey, M.K., Upadhyay, R.S., 2017a. *Alternaria* toxins: potential virulence factors and genes related to pathogenesis. Front. Microbiol. 8, 1451.

Meena, M., Swapnil, P., Upadhyay, R.S., 2017b. Isolation, characterization and toxicological potential of *Alternaria*-mycotoxins (TeA, AOH and AME) in different *Alternaria* species from various regions of India. Sci. Rep. 7, 8777.

Migheli, Q., Cacciola, S.O., Balmas, V., Pane, A., Ezra, D., Di San Lio, G.M., 2009. Mal secco disease caused by *Phoma tracheiphila*: a potential threat to lemon production worldwide. Plant Dis. 93, 852–867.

Mondal, S.N., Timmer, L.W., 2006. Greasy spot, a serious endemic problem for citrus production in the Caribbean Basin. Plant Dis. Rep. 90, .

Mondal, S.N., Agostini, J.P., Zhang, L., Timmer, L.W., 2004. Factors affecting pycnidium production of *Diaporthe citri* on detached ditrus twigs. Plant Dis. 88, 379–382.

Mondal, S.N., Vicent, A., Reis, R.F., Timmer, L.W., 2007. Efficacy of pre- and postinoculation application of fungicides to expanding young citrus leaves for control of melanose, scab, and Alternaria brown spot. Plant Dis. 91, 1600–1606.

Mourã Filho, F.D.A.A., Pio, R., Mendes, B.M.J., De Azevedo, F.A., Schinor, E.H., Entelmann, F.A., Cantuarias-Avilé, T.E., 2008. Evaluation of citrus somatic hybrids for tolerance to *Phytophthora nicotianae* and citrus tristeza virus. Sci. Hortic. (Amsterdam) 115, 301–308.

Nakatsuka, S., Goto, T., Kohmoto, K., Nishimura, S., 1986. Host-Specific Phytotoxins. University of Tokyo Press, Tokyo, Japan.

Nigro, F., Ippolito, A., Salerno, M.G., 2011. Mal secco disease of citrus: a journey through a century of research. J. Plant Pathol. 93, 523–560.

Ohtani, K., Fukumoto, T., Nishimura, S., Miyamoto, Y., Gomi, K., Akimitsu, K., 2009. *Alternaria* pathosystems for study of citrus diseases. Tree For. Sci. Biotechnol. 3, 108–115.

Palm, M.E., Civerolo, E.L., 1994. Isolation, pathogenicity and partial host range of *Alternaria limicola*, causal agent of mancha foliar de los citricos in Mexico. Plant Dis. 78, 879–883.

Paudyal, D.P., Hyun, J.W., 2015. Physical changes in Satsuma mandarin leaf after infection of *Elsinoë fawcettii* causing citrus scab disease. Plant Pathol. J. 31, 421–427.

Paudyal, D., Hyun, J.W., Hwang, R.Y., 2017. Infection and symptom development by citrus scab pathogen *Elsinoë fawcettii* on leaves of Satsuma mandarin. Eur. J. Plant Pathol. 148, 807–816.

Pena, J.E., Duncan, R., 1989. Role of arthropods in the transmission of post bloom fruit drop. Proc. Fla. State Hortic. Soc. 102, 249–251.

Peres, N.A.R., Agostini, J.P., Timmer, L.W., 2003. Outbreaks of Alternaria brown spot of citrus in Brazil and Argentina. Plant Dis. 87, 750.

Peres, N.A., Timmer, L.W., Adaskaveg, J.E., Correll, J.C., 2005. Lifestyles of *Colletotrichum acutatum*. Plant Dis. 89, 784–796.

Peres, N.A., MacKenzie, S.J., Peever, T.L., Timmer, L.W., 2008. Postbloom fruit drop of citrus and key lime anthracnose are caused by distinct phylogenetic lineages of *Colletotrichum acutatum*. Phytopathology 98, 345–352.

Perrotta, G., Graniti, A., 1988. *Phoma tracheiphila* (Petri) Kantschaveli and Gikashvili. Blackwell Scientific Publications, Oxford, United Kingdom.

Perryman, S.A.M., Clark, S.J., West, J.S., 2014. Splash dispersal of *Phyllosticta citricarpa* conidia from infected citrus fruit. Sci. Rep. 4, 6568.

Petri, L., 1929. Sulla posizione sistematica del fungo parassita delle piante di limone affette da "mal secco". Boll. Staz. Patol. Veg. 9, 393–396.

Pierce, N.B., 1902. Black rot of oranges. Bot. Gaz. 33, 234–235.

Reis, R.F., Timmer, L.W., De Goes, A., 2006. Effect of temperature, leaf wetness, and rainfall on the production of *Guignardia citricarpa* ascospores and on black spot severity on sweet orange. Fitopatol. Bras. 31, 29–34.

Russo, M., Grasso, F., Bella, P., Licciardello, G., Catara, A., Catara, V., 2011. Molecular diagnostic tools for the detection and characterization of *Phoma tracheiphila*. Acta Hortic. 892, 207–214.

Savita, G., Virk, S., Avinash, N., 2012. *Citrus* diseases caused by *Phytophthora* species. GERF Bull. Biosci. 3, 18–27.

Sawada, K., 1959. *Phoma citricarpa* (McAlpine). Descriptive Catalogue of Taiwan (Formosan) Fungi. Part XI. Coll. Agric. Nat. Taiwan Univ. vol. 8, 1–268.

Schubert, T.S., Dewdney, M.M., Peres, N.A., Palm, M.E., Jeyaprakash, A., Sutton, B., Mondal, S.N., Wang, N.-Y., Rascoe, J., Picton, D.D., 2012. First report of *Guignardia citricarpa* associated with citrus black spot on sweet orange (*Citrus sinensis)* in North America. Plant Dis. 96, 1225.

Schutte, G.C., Mansfield, R.I., Smith, H., Beeton, K.V., 2003. Application of azoxystrobin for control of benomyl-resistant *Guignardia citricarpa* on 'Valencia' oranges in South Africa. Plant Dis. 87, 784–788.

Sharma, G., Pinnaka, A.K., Shenoy, B.D., 2015. Resolving the *Colletotrichum siamense* species complex using ApMat marker. Fungal Divers. 71, 247–264.

Silva-Junior, G.J., Spósito, M.B., Marin, D.R., Amorim, L., 2014. Efficacy and timing of application of fungicides for control of citrus postbloom fruit drop. Crop Prot. 59, 51–56.

Silva-Junior, G.J., Da Silva Scapin, M., Silva, F.P., Silva, A.R.P., Behlau, F., Ramos, H.H., 2016. Spray volume and fungicide rates for citrus black spot control based on tree canopy volume. Crop Prot. 85, 38–45.

Solel, Z., 1991. Alternaria brown spot on Minneola tangelos in Israel. Plant Pathol. 40, 145–147.

Spósito, M.B., Amorim, L., Bassanezi, R.B., Filho, A.B., Hau, B., 2008. Spatial pattern of black spot incidence within citrus trees related to disease severity and pathogen dispersal. Plant Pathol. 57, 103–108.

Spósito, M.B., Amorim, L., Bassanezi, R.B., Yamamoto, P.T., Felippe, M.R., Czermainski, A.B., 2011. Relative importance of inoculum sources of *Guignardia citricarpa* on the citrus black spot epidemic in Brazil. Crop Prot. 30, 1546–1552.

Sutton, B.C., 1992. The Genus *Glomerella* and its Anamorph *Colletotrichum*. CAB International, Wallingford, UK.

Thind, T.S., Goswami, S., Thind, S.K., Mohan, C., 2009. Resistance in *Phytophthora parasitica* against metalaxyl in citrus orchards. Indian Phytopathol. 62, 536–538.

Timmer, L.W., Gottwald, T.R., 2000. Greasy spot and similar disases. In: Timmer, L.W., Garnsey, S.M., Graham, J.H. (Eds.), Compendium of *Citrus* Diseases. American Phytopathological Society Press, St Paul, MN.

Timmer, L.W., Peres, N.A., 2015. Where have all the flowers gone? Postbloom fruit drop of citrus in the Americas. J. Cit. Pathol. 2 (1). https://escholarship.org/uc/item/13k7c4gf.

Timmer, L.W., Zitko, S.E., 1992. Timing of fungicide applications for control of postbloom fruit drop of citrus in Florida. Plant Dis. 76, 820–823.

Timmer, L.W., Zitko, S.E., 1998. Evaluation of fungicides for control of postbloom fruit drop by *Colletotrichum acutatum*. F&N Tests 53, .

Timmer, L., Agostini, J., Zitko, S., Zulfiqar, M., 1994. Postbloom fruit drop, an increasingly prevalent disease of citrus in Americas. Plant Dis. 78, 329–334.

Timmer, L.W., Priest, M., Broadbent, P., Tan, M.K., 1996. Morphological and pathological characterization of species of *Elsinoë* causing scab disease of citrus. Phytopathology 86, 1032–1038.

Timmer, L.W., Graham, J.H., Zitko, S.E., 1998. Metalaxyl-resistant isolates of *Phytophthora nicotianae*: occurrence, sensitivity, and competitive parasitic ability on citrus. Plant Dis. 82, 254–261.

Timmer, L.W., Garnsey, S.M., Graham, J.M., 2000. Compendium of *Citrus* Diseases. A. Press, St. Paul, MN.

Timmer, L.W., Roberts, P.D., Chung, K.R., 2001. Greay spot. In: Plant Pathology Department. University of Florida, IFAS Extension, Florida.

Timmer, L.W., Peever, T.L., Solel, Z., Akimitsu, K., 2003. Alternaria diseases of citrus—Novel pathosystems. Phytopathol. Mediterr. 42, 99–112.

Timmer, L.W., Mondal, S.N., Peres, N.A.R., Bhatia, A., 2004. Fungal diseases of fruit and foliage of citrus trees. In: Naqvi, S.A.M.H. (Ed.), Diseases of Fruits and Vegetables. Kluwer Academic Publishers, The Netherlands, pp. 191–227.

Tran, N.T., Miles, A.K., Dietzgen, R.G., Dewdney, M.M., Zhang, K., Rollins, J.A., Drenth, A., 2017. Sexual reproduction in the citrus black spot pathogen, *Phyllosticta citricarpa*. Phytopathology 107, 732–739.

Truter, M.P., 2010. Epidemiology of citrus black spot disease in South Africa and its impact on phytosanitary trade restrictions. (PhD Thesis) University of Pretoria.

Van Bruggen, A.H., Sharma, K., Shin, K., 2017. Sugar cane processing residue, bagasse, enhances decomposition of citrus leaves and could contribute to citrus black spot management. Crop Prot. 93, 89–97.

Van Der Aa, H.A., Noordeloos, M.E., De Gruyter, J., 1990. Species concepts in some larger genera of the Coelomycetes. Stud. Mycol. 32, 3–19.

Vicent, A., Armengol, J., Sales, R., García-Jiménez, J., 2000. First report of Alternaria brown spot of citrus in Spain. Plant Dis. 84, 1044.

Vicent, A., Armengol, J., García-Jiménez, J., 2007. Rain fastness and persistence of fungicides for control of Alternaria brown spot of citrus. Plant Dis. 91, 393–399.

Wang, X., Li, Z.A., Tang, K.Z., Zhou, C.Y., 2010. First report of Alternaria brown spot of citrus caused by *Alternaria alternata* in Yunnan province, China. Plant Dis. 94, 375.

Wang, N.Y., Zhang, K., Huguet-Tapia, J.C., Rollins, J.A., Dewdney, M.M., 2016. Mating type and simple sequence repeat markers indicate a clonal population of *Phyllosticta citricarpa* in Florida. Phytopathology 106, 1300–1310.

Weir, B.S., Johnston, P.R., Damm, U., 2012. The *Colletotrichum gloeosporioides* species complex. Stud. Mycol. 73, 115–180.

Whiteside, J.O., 1967. Sources of inoculum of the black spot fungus, *Guignardia citricarpa,* in infected Rhodesian citrus orchards. Rhodesia Zambia Malawi J. Agric. Res. 5, 171–177.

Whiteside, J.O., 1970. Etiology and epidemiology of citrus greasy spot. Phytopathology 60, 1409–1414.

Whiteside, J.O., 1974. Environmental factors affecting infection of citrus leaves by *Mycosphaerella citri*. Phytopathology 64, 115–120.

Whiteside, J.O., 1975. Biological characteristics of *Elsinoë fawettii* pertaining to the epidemiology of sour orange scab. Phytopathology 65, 1170–1177.

Whiteside, J.O., 2000. Melanose. In: Timmer, L.W., Garnsey, S.M., Graham, J.H. (Eds.), Compendium of *Citrus* Diseases. American Phytopathological Society APS Press, St Paul, MN.

Wickert, E., De Goes, A., Lemos, E.G.M., De Souza, A., Da Silveira, E.L., Pereira, F.D., Rinaldo, D., 2009. Relaçoes filogeneticas e diversidade de isolados de *Guignardia* spp. oriundos de diferentes hospedeiros nas regiones ITS1–5,8S-ITS2. Rev. Bras. Frutic. 31, 360–380.

Woudenberg, J.H.C., Seidl, M.F., Groenewald, J.Z., De Vries, M., Stielow, J.B., Thomma, B.P.H.J., Crous, P.W., 2015. Alternaria section Alternaria: species, formae speciales or pathotypes? Stud. Mycol. 82, 1–21.

Yan, J., Dewdney, M.M., Roberts, P.D., Ritenour, M.A., 2016. The effects of postharvest hot water and fungicide treatments on *Guignardia citricarpa* growth and the development of citrus black spot symptoms on 'Valencia' orange fruit. HortScience 51, 1555–1560.

Yan, H., Zhong, Y., Jiang, B., Zhou, B., Wu, B., Zhong, G., 2017. Guanggan *(Citrus reticulata)* shows strong resistance to *Phytophthora nicotianae*. Sci. Hortic. 225, 141–149.

Zhang, Y., Lee, W.S., Li, M., Zheng, L., Ritenour, M.A., 2018. Non-destructive recognition and classification of citrus fruit blemishes based on ant colony optimized spectral information. Postharvest Biol. Technol. 143, 119–128.

Zhou, C.H., Deng, X.X., Grosser, J.W., 1997. On the resistance of citrus somatic hybrids to citrus foot rot. Acta Phytopathol. Sin. 27, 361–364.

Zucker, W.V., Catara, A., 1985. Osservazioni al microscopio elettronico a scansione sulla penetrazione fogliare di *Phoma tracheiphila*. Inf. Fitopatol. 35, 33–35.

Zulfiqar, M., Brlansky, R.H., Timmer, L.W., 1996. Infection of flower and vegetative tissues of citrus by *Colletotrichum acutatum* and *C. gloeosporioides*. Mycologia 88, 121–128.

Chapter 18

Bacterial pathogens of citrus: Citrus canker, citrus variegated chlorosis and Huanglongbing

Dean Gabriel[a], Timothy R. Gottwald[b], Silvio A. Lopes[c], Nelson A. Wulff[c]

[a]Plant Pathology Department, University of Florida, Gainesville, FL, United States, [b]US Department of Agriculture, Agricultural Research Service, Fort Pierce, FL, United States, [c]Research and Development Department, Fundecitrus, Araraquara, Brazil

18.1 Introduction

Commercial citrus groves are invariably planted as dense monocultures comprising a few commercial citrus varieties covering large tracts of land, providing little to no natural variation in resistance to any pest/vector, pathogen, or a combination of both with capacity to exploit it. The majority of bacterial pathogens need no insect vector at all, for example, the xanthomonads causing citrus canker, although disease incidence can be dramatically accelerated by leafminers. These xanthomonads enter through stomata or wounds, actively attacking mesophyll cells from the outside and directly affecting citrus cell gene expression by injection of eukaryotic transcription factors. Nevertheless, as discussed in detail below, there is remarkably little variation in resistance among commercially grown citrus genera, species, and varieties to the Asiatic strains of the disease.

Liberibacters are literally injected by their psyllid vectors directly into living phloem cells. Psyllids free of Liberibacters may be introduced into a citrus growing region free of Huanglongbing (HLB), resulting only in the direct damage caused by the psyllid. Independent of vector introduction, Liberibacters may be introduced by planting infected and asymptomatic citrus into a previously uninfected region, but the pathogen cannot spread without an insect vector with host range on citrus. As discussed in some detail below, the excessively long incubation period required before HLB symptoms become recognizable can prevent or frustrate timely detection of such introductions, even with inspections. Epidemics can result from the intersection of these two independent introduction events, as appears to have occurred relatively recently in Florida with HLB. Of course, both events can be concurrent. Again, however, there is remarkably little variation in resistance among commercially grown citrus varieties to Liberibacter infections.

It is useful to distinguish between parasitism and pathogenicity. A parasite is capable of deriving nutrition from, and reproducing on or within, its host(s), and is not always associated with the disease. For example, a Liberibacter-infected pear tree is nearly asymptomatic (Raddadi et al., 2011). A pathogen causes disease, but not always on all hosts. Pathologists typically pay attention only to pathogens on commercially important crops, but pathogens can have a much wider parasitic host range than expected if they cause no disease or cause the disease on weeds. A recurring theme in this chapter involves the worldwide increase in commerce and in transcontinental dispersal of many pathogens and their vectors. The fact that there is often overlap of parasitic, if not pathogenic, host ranges of multiple bacterial species, and the fact that apparent host range limitations of a given pathogen can be disguised by the host range limitations of its insect vector can result in emerging and sometimes surprisingly severe disease threats. In general terms, if a vector is introduced into a geographic region with an unusually wide host range that may include weed or arboreal crops, it is conceivable that the vector may facilitate exposure of previously isolated bacterial populations to one another, thereby facilitating horizontal gene transfer events, either by cell–cell conjugation or by phage transduction within either the insect vector or plant host. Transcontinental dispersal of pathogens and vectors dramatically increases the opportunities for horizontal gene transfer and allows increased recombination frequencies between pathogenic variants of a species and even across genera.

The Genus Citrus. https://doi.org/10.1016/B978-0-12-812163-4.00018-8

Xylella strains are injected by their various vectors directly into (dead) xylem vessels. In the case of citrus variegated chlorosis (CVC), despite the fact that sharpshooters feeding on both oranges and other citrus varieties are abundantly available, it is therefore perhaps surprising that host range on citrus appears limited to *Citrus sinensis*. Why only sweet orange? One can speculate that since coffee plantations with coffee leaf scorch disease are in close proximity to citrus groves with CVC in Brazil and since both diseases are caused by the same *X. fastidiosa* subspecies pauca (Xfp), the opportunity arose for horizontal gene transfer or recombination of host range factors that allowed emergence of CVC. Recombination is documented between Xf subsp. morus strains pathogenic to mulberry and blueberry in the United States (Almeida and Nunney, 2015), and Xfp strains from coffee (Jacques et al., 2016). The wide variation in *Xylella* pathogenicity and host range illustrates the need for a better understanding of the genetic factors determining both. In contrast with both citrus canker and HLB, there is remarkably strong variation in resistance among commercially grown citrus genera.

In all the three cases, each involving very different pathogen disease cycle and reproductive strategies, the causal bacterial genera exhibited a wide host range, pathogenicity, and genome variations that indicate horizontal gene transfer of pathogenicity factors. Selection in citrus monocultures does the rest. In addition to the three bacterial pathogens most damaging to commercial citrus that are the focus of this chapter, other bacteria causing notable citrus diseases of worldwide importance include *Burkholderia andropogonis* (bacterial brown leaf spot) and a few Mollicute (lacking a cell wall) species such as *Spiroplasma citri* (citrus stubborn disease) and *Phytoplasma* species (witches' broom disease of lime and HLB-like diseases). Currently the worldwide incidence and importance of these bacteria are minor.

18.2 Citrus canker

18.2.1 History and background

Asiatic citrus canker (ACC) is an important bacterial disease of citrus with severe yield, trade, and quarantine implications. The ACC is also referred to as "Canker A" and is by far the most widespread and severe form of the disease. The disease is caused by *Xanthomonas citri* (*ex* Hasse 1915) (Gabriel et al., 1989) subsp. *citri* (Schaad et al., 2006) pathotype A (XccA). Historical synonyms for XccA are *X. citri, Xanthomonas axonopodis* pv. *citri*, and *Xanthomonas campestris* pv. *citri* "strain" A). ACC is believed to have originated in Southeast Asia-India based on herbarium specimens from India and Java in the 1830–1840s. During the 1900s it was found throughout Asia, including Japan, as well as South and Central Africa, Middle East, Australia, New Zealand, Pacific Islands, South America and Florida, USA. Another form of citrus canker, "Cancrosis B," with nearly identical symptoms as ACC (erumpent, hyperplastic lesions on leaves, stems, and fruit are the key diagnostic symptoms) emerged in South America (Carrera, 1933) with host range limited to lemons, Mexican lime, sour orange, and pummelo (Stall and Seymour, 1983). Cancrosis B is also referred to as "False canker" or "Canker B" and is caused by a different and phylogenetically quite distinct species, *Xanthomonas fuscans* subsp. *aurantifolii* (Xfa) (Schaad et al., 2006) pathotype B (XfaB). Cancrosis B is prevalent on lemons in Argentina, Paraguay, and Uruguay, but is less aggressive than ACC and is largely under control (Canteros et al., 2017). Since it is in a separate species, XfaB (Cancrosis B) strains can be easily differentiated by microbiological, serological, genetic, and host range inoculation tests from XccA (ACC) strains.

What do the two species, *X. citri* and *X. fuscans*, have in common to allow their respective subspecies, Xcc and Xfa, to elicit very similar symptoms on citrus? Both encode and inject nearly identical eukaryotic transcriptional activator-like (TAL) protein effectors into citrus cells. Pathogenicity gene *pthA* was the first member of the *avrBs3/pthA* gene family demonstrated to function for pathogenicity and is required for both growth and pathogenicity of XccA strains on citrus (Swarup et al., 1991, 1992). *pthA* is capable of conferring ability to cause canker-like symptoms to strains that cannot otherwise cause canker, such as *X. aurantifolii* pv. *citrumelo*. A very close homolog, *pthB*, is similarly required for pathogenicity of XfaB (Cancrosis B) strain B69 and is found on a self-mobilizing plasmid that can transfer itself between these two species (El-Yacoubi et al., 2007). All canker-causing strains of *X. citri* subsp. citri tested carry *pthA* homologs (Cubero and Graham, 2002; Mavrodieva et al., 2004; Munoz-Bodnar et al., 2017). When *pthA* is transiently expressed in citrus using either *Agrobacterium tumefaciens* or particle bombardment, small canker-like lesions are elicited (Duan et al., 1999). Therefore, *pthA* and its homologs encode the primary causal effectors of the citrus canker disease phenotype.

Not surprisingly, variant pathotypes within both Xcc and Xfa species have emerged that are primarily distinguished on the basis of host range and determined by apparently minor differences in virulence factors. A third citrus canker type, described as Cancrosis C or "Canker C," was discovered in Brazil on Mexican lime by Namekata (1971) with an even more restricted host range than XfaB (Cancrosis B) strains (Stall and Seymour, 1983). Cancrosis C is also caused by Xfa, and strains of this group (XfaC) are primarily limited to Mexican lime, although sour orange can also be a host. XfaC strains are currently distinguished from XfaB strains by pathogenicity tests and from ACC strains by species-level tests.

Variants of *X. citri* subsp. *citri* (Xcc) have also emerged more recently. These also elicit ACC-like lesions but with less aggressive symptoms and with host range largely limited to Mexican lime: XccA* and $XccA^{W}$ (Gordon et al., 2015). XccA* strains were discovered in Oman, Saudi Arabia, Iran, and India with host range limited to Mexican lime (Vernière et al., 1998). The $XccA^{W}$ pathotype is also characterized by host range restricted to Mexican lime and lemon and has to date only been found in Florida and Texas (Munoz-Bodnar et al., 2017). Grapefruit (*Citrus paradisi*), resistant to both XccA* and $XccA^{W}$ strains, may serve to differentiate the two groups, since $XccA^{W}$ strains elicit a strong hypersensitive response (HR) in grapefruit, while XccA* strains do not (Brunings and Gabriel, 2003). A single avirulence gene was found to restrict an $XccA^{W}$ strain from virulence on grapefruit, but this gene was not detected in XccA* strains tested (Rybak et al., 2009). Interestingly, XfaC strains also cause an HR on grapefruit, while XfaB strains do not (Brunings and Gabriel, 2003). Minor genetic differences in one or a few virulence factors may account for the differences from XccA in the restriction of these pathotypes to lime and from each other. The restricted host range variants of Xcc (XccA* and $XccA^{W}$) and Xfa (XfaB and XfaC) are of limited worldwide distribution and economic consequence and therefore further discussion will be limited to ACC, caused by XccA.

18.2.2 Symptoms and epidemiology of ACC

ACC is primarily a leaf and fruit spotting disease characterized by erumpent lesions (cankers) on fruit, foliage, and young stems (Fig. 18.1) of susceptible cultivars of citrus (Schubert et al., 2001; Timmer et al., 1996). Most commercial citrus varieties are moderately to highly susceptible to the disease, although grapefruit is the most susceptible fruit. When the disease is severe, defoliation, dieback, and fruit drop can occur and infected fruit that remain are less valuable or entirely unmarketable (Graham and Gottwald, 1991). ACC is spread by rain splash combined with wind and is greatly exacerbated by tropical storms and hurricanes (Gottwald et al., 2002). ACC epidemics are aggravated by the Asian citrus leaf miner, *Phyllocnistis citrella*, introduced to Florida in 1993 and to Brazil in 1996. Where the leaf miner and citrus canker coincide, canker has increased significantly as a consequence of the insect's feeding activities which create wounds that expose leaf mesophyll tissues to splashed inoculum, thus greatly expanding the probability of infection (Gottwald et al., 1997, 2007).

During seasons when spring and summer rains are combined with wind speeds in excess of 8 m/s damage from the disease can range from nominal to significant (Serizawa et al., 1969; Serizawa and Inoue, 1974). When lesions on leaves are wetted for only 1–2 min, the bacteria ooze to the surface in great numbers and this inoculum release can persist for several hours, continuously providing inoculum for dispersal throughout prolonged storm events (Bock et al., 2010a,b). Distance of spread due to bacterial aerosols is directly related to rain splash and wind speed, from a few cm during gentle rains without wind to several km during tropical storms and hurricanes (Gottwald and Irey, 2007; Bock et al., 2011). For moderate wind-blown rain events, natural windbreaks such as conifers or hardwood trees with interlocking canopies can effectively reduce wind speed and trap windblown inoculum, thereby reducing infection. Citriculturists effectively control ACC in Argentina

FIG. 18.1 Asiatic citrus canker (ACC) lesions on leaves, fruit, and young stems of grapefruit.

by judicious use of such natural windbreaks effectively surrounding plantings 1–5 ha in area (Stall and Seymour, 1983). ACC is highly contagious. Spread is often human mediated by contaminated clothing, vehicles, equipment (e.g., pruners, hedgers, mowers, ladders, harvesters), or animals either in commercial or residential citrus. When the foliage is wetted, the bacteria ooze from the hyperplastic lesions that have broken the foliar surfaces (for a physiological process description, refer Brunings and Gabriel, 2003) and can be transferred when the mechanical device, human, or animal come in contact with the bacterial ooze and inadvertently transfer it to susceptible foliar tissues on adjacent or distant trees.

18.2.3 ACC control through eradication

ACC currently exists in the majority of commercial citrus producing regions of the world and in the Americas has plagued multiple States in the Southeastern US, as well as Argentina, Bolivia, Brazil, Chile, Paraguay, and Uruguay (Braithwaite et al., 2002; Koizumi, 1985; Gottwald et al., 2007). The approach to controlling ACC depends on whether the disease is newly introduced or if it has become endemic. If it is newly introduced, by far the best approach is eradication. ACC infestations have most significant impact on commerce and trade due to the governmental regulatory actions that result in restricted interstate and international transport and sale of fruit originating from known infested areas.

The consequences of ACC introduction to citriculture must be examined through a historical, economic, and sociopolitical lens to fully comprehend its impact. If the disease is recently introduced and in low incidence, eradication has often been attempted to return the area to a disease-free status, thereby avoiding the cost of quarantines and unfavorable trade impediments. Conceptually, eradication is quite simple. Diseased plants must be removed more rapidly than they become diseased. The more rapidly and efficiently diseased individuals are removed, the less chance of continued pathogen spread, and the quicker the disease will be extinguished. In practice, however, eradication is complex. It is not always easy to recognize newly infected plants. A several week-long incubation period between infection and disease expression is quite common. Thus, some infected members of the plant population are "subclinical"; that is, infected but not yet expressing disease. For ACC the subclinical infections can persist for 10–14 days without expressing symptoms and even then, these symptoms are not always easily seen. The average time of detection in residential areas was estimated to be 108 days postinfection providing ample time for polycyclic reproduction and spread (Gottwald et al., 2001a,b).

Additionally, diseased plants are not always recognizably aggregated into convenient or contained groups for easy eradication. The disease is often somewhat considerably diffuse among an otherwise healthy population. The combination of lengthy incubation periods and diffuse patterns of disease necessitates that to achieve eradication, large radii around recognized infections of both diseased and asymptomatic but potentially latently infected plants must be removed or "culled" as well (Gottwald et al., 2001a,b, 2002, 2007; Schubert et al., 2001). For ACC epidemics, the resulting eradication programs have at times exceeded the social tolerance of commercial growers and additionally residential homeowners who resist the removal of their apparently healthy but potentially asymptomatically affected trees in support of the commercial citrus industry.

Several early eradication programs were successful in South Africa, Australia, the Fiji Islands, Mozambique, New Zealand, and the United States (Koizumi, 1985). In Northeast Argentina, ACC was first recognized in 1928 and eradication was attempted in 1976 but subsequently failed in part due to grower's concern and reluctance to remove the large proportions of commercial trees to achieve eradication (Canteros et al., 2017; Hogg, 1985; Rodriguez, 2012). In Australia ACC has been introduced and successfully eradicated three times: 1918, 1984 and a third time in Queensland in 2004 (Gambley et al., 2009). In Brazil ACC was first detected in 1957 and eradication programs were intermittent over the next 60 years eventually leading to establishment of ACC; by 2009 the disease was declared endemic and the eradication program halted (Behlau et al., 2016).

ACC was first detected in Florida in 1910 and it was eradicated by the 1930s after massive removal of infected trees (Schubert et al., 2001). ACC was discovered a second time in Florida in 1986 and was declared eradicated in 1992 but was discovered a third time in multiple residential areas in Dade County in 1995 within a 14 mile^2 area south of the Miami International Airport (Gottwald et al., 2001a,b). This discovery prompted an intensive eradication campaign by a joint state/federal citrus canker eradication program (CCEP). The program was based on an epidemiological study of ACC spread that resulted in an ACC-eradication protocol that required the removal of all "exposed trees" within a 579-m radius of a known infected tree (Gottwald et al., 2001a,b). Over the next 10 years, the canker eradication program cost approximately $1 billion dollars, the largest plant disease eradication program worldwide to date. The program was plagued by legal conflicts between the CCEP and residential homeowners who felt that the regulatory actions being taken protected the citrus industry at too high a cost to residential citrus tree owners. The litigation of numerous injunctions and appeals led to a discontinuous and sporadic eradication program. As the legalities of eradication were debated, the disease continued to spread, and outbreaks in new commercial and residential areas erupted across Florida.

Hurricanes and tropical storms in Florida have been associated with long-range dissemination and increased local incidence of ACC (Gottwald and Irey, 2007; Irey et al., 2006). In 2004, three hurricanes and a major tropical storm crossed the Florida peninsula and exacerbated preexisting ACC infections and caused overwhelming dispersal of the pathogen over the commercial industry. Statistical models developed to predict the spread of ACC due to hurricanes and tropical storms led to numerous discoveries of new ACC outbreaks and more effective and rapid eradication (Irey et al., 2006). However, hurricane Wilma struck Florida in the Fall of 2005 and caused such massive additional spread of the disease that the regulatory agencies, citrus industry, and research scientists alike concluded that the disease had likely become ineradicable, resulting in the termination of the 10-year, $1 billion canker eradication program (Gottwald and Irey, 2007). As a consequence, ACC has greatly increased and is now found throughout the majority of Florida's commercial groves.

18.2.4 Endemic control/mitigation of ACC

Where ACC is endemic, mitigation requires integration of appropriate cultural practices including windbreaks and leafminer control, with frequent applications of copper (copper hydroxide and copper oxychloride) sprays (Leite Jr et al., 1987). Leafminer control via insecticides is particularly important for young trees and certain cultivars that have more frequent and longer duration of growth flush. Copper sprays reduce infection somewhat. Fruit are most susceptible for 90days after petal fall when copper is most efficacious. Grapefruit is highly susceptible to ACC requiring more sprays to provide acceptable control compared to orange or mandarins which are much less susceptible. Antibiotics such as gentamicin, kasugamycin, and streptomycin also have activity for canker control if properly formulated. However, the possibility of eliciting bacterial resistance and human and animal health issues raise concerns about wide scale antibiotic use in agriculture (Graham et al., 2006; Behlau et al., 2012). Recently the control of ACC via the use of nano-formulated zinc oxide was shown to exceed the cuprous oxide and cuprous oxide/zinc oxide bactericides, although deployment has not been wide scale (Graham et al., 2016).

18.2.5 Implications of citrus canker quarantines and trade

The risk of introduction of citrus canker bacteria to new, unaffected citrus producing areas is a major concern for those citrus industries attempting to remain free of citrus canker. If canker is introduced, quarantines and restrictions could ensue that would severely limit or deny access to domestic and international markets. Various regulatory agencies such as USDA (United States Department of Agriculture), APHIS (Animal and Plant Health inspection Service) initiate extensive Pest Risk Assessments and Risk Management Analysis to evaluate the potential for processed fruit as a pathway for ACC introduction. Based on the assessments and analyses, regulatory agencies develop regulations under which fruit may be moved interstate or intercountry from quarantined areas. To address fruit as a pathway, a series of studies were conducted with citrus fruit as a potential pathway for XccA to enter and become established in these areas, including: (1) survival of XccA in disease lesions on fruit and on fruit surface after disinfectant treatment as used by commercial packing houses; (2) survival of XccA in culled fruit that were discarded in cull piles, and (3) dissemination of XccA and disease establishment in simulated rainstorms. Taken as a group, this series of studies on canker bacteria viability in and on fruit surfaces, and bacteria dissemination, demonstrated that harvested and packinghouse-disinfested citrus fruit are extremely unlikely to be a pathway for canker to reach and become established in canker-free areas (Gottwald et al., 2009). These studies have helped to diminish the need of quarantines and maintain open markets and trade.

18.3 Citrus variegated chlorosis

18.3.1 History and background

Xylella fastidiosa (Xf) is a Gram-negative, xylem-limited, insect vectored and often difficult to culture bacterium that causes some of the most serious plant diseases worldwide and in a wide variety of hosts. Its five (to date) described subspecies tend to have strong host range preferences that may overlap, although individual strains appear to have a highly restricted host range. Xf appears to have originated in the Americas, and causes production-limiting diseases including: (1) Xf subsp. *fastidiosa* (Xff), which causes Pierce's disease (PD) of grapevines in the United States; (2) Xf subsp. *pauca* (Xfp), causes CVC in sweet orange and coffee leaf scorch in Brazil (Nunney et al., 2014); (3) Xf subsp. *sandyi* (Xfs) causes leaf scorch of oleander; (4) Xf subsp. *morus* causes leaf scorch of mulberry and blueberry in the United States, and (5) Xf subsp. *multiplex* (Xfm) causes leaf scorch diseases in North America on a wide range of tree species, including almond, peach, oak and more recently associated with quick decline of olive trees in Italy (Almeida and Nunney, 2015).

CVC is transmitted by sharpshooters and has so far been found limited in host range only to sweet orange varieties. The first trees affected by the disease were observed in 1984 in Argentina, where it was named "pecosita" or "falsa mancha grasienta" (Contreras, 1992), and in 1987 in Brazil within the main citrus growing belt of the country in the northern region of the state of São Paulo (Rossetti et al., 1990). Currently, CVC is present in all major Brazilian citrus producing areas. Nevertheless, in the last 30 years of known disease outbreaks in Brazil, despite the wide distribution on many citrus varieties and occurrence of large populations of the species of sharpshooters known to transmit the pathogen (Yamamoto and Gravena, 2000), no other citrus type was found affected by the disease. This indicates that the host range limitation to sweet orange is determined by the Xfp genome.

Xylem colonization of Xfp starts when the insect vector carrying the pathogen feeds on the xylem sap of a healthy tree. Xylem colonization is enhanced by enzymes that degrade the xylem pit membranes, which would otherwise occlude movement of objects as large as Xf (Roper et al., 2006). Xfp colonization ultimately leads to the obstruction of sap flow (Queiróz-Voltan and Paradela, 1999; Garcia et al., 2012), with infected trees expressing symptoms characteristic of the disease first on leaves and later on fruits. Symptoms may be elicited by multiple pathogenicity factors that may not correlate with ability to colonize the xylem (Zhang et al., 2015). CVC-affected leaves show small spots distributed throughout on both the surfaces (Fig. 18.2). As the infection progresses, orange gum appears in the center of the spot present on the lower surface of the leaf. In later infection stages, the yellow spots coalesce and become necrotic. Highly affected leaves usually abscise, leaving behind tufts of small leaves at the tips of affected branches, which commonly exhibit symptoms resembling zinc deficiency. The fruits that appear on the affected branches usually do not drop, as happens in the case of trees affected by HLB. Instead they mature earlier and are smaller (Fig. 18.2) and much harder. The fruit are also richer in sugar but more acidic than the fruit from asymptomatic branches of infected or healthy trees (Laranjeira and Palazzo, 1999). The smaller size causes direct losses since larger numbers of affected fruits are needed to fill a typical commercial box of 40.8 kg (Ayres, 2000). The harder fruits can slow commercial juice production, while the smaller size also makes the fruits less acceptable to consumers in fresh markets.

The disease progress of CVC is faster in younger than older trees, and in trees exposed to seasonal water deficits, as happens more frequently in the northern regions of the state of São Paulo. Under these conditions, tree development is highly affected by the disease, with the decrease in fruit production reaching up to 75% of that in healthy or asymptomatic trees (Ayres, 2000).

18.3.2 Disease management, control, and mitigation

In Brazil, CVC management has relied on the use of preventive measures, particularly eradication of symptomatic citrus, to reduce inoculum sources and the chances of recurrent infections. Although weeds and some arboreal plants may harbor Xfp, the apparently low bacterium titer and short life of weedy hosts greatly limit their importance as sources of CVC (Lopes

FIG. 18.2 Mature leaves and fruits of sweet orange severely affected by citrus variegated chlorosis (CVC). Note the raised, gummy pustules on the abaxial leaf surface in (A). The two green fruits at the top right of (B) are unaffected.

et al., 2003). Correct identification of diseased trees is an important step in this process which, due to constant appearance of new symptomatic trees as a result of constant infections, requires frequent and careful tree inspections. Under certain conditions, the health of diseased trees may be restored by drastic pruning of symptomatic branches. However, for pruning to be successful the tree must be over 2.5 years of age and the symptoms present at the tip of only one secondary branch (Rodas, 1994). Recently, based on the high levels of resistance to Xfp infection observed for the Cravo Rangpur lime rootstock (Garcia et al., 2012), a new mitigation measure was developed (Lopes, 2019). Since the bacteria do not move downwards below the graft union into these rootstocks, new healthy scions can be generated by removing all the infected scion, even from highly affected sweet orange trees, and grafting with healthy budwood onto these rootstocks. Unfortunately, the immunity to CVC exhibited by the rootstocks does not transfer to the scion. However, the new shoots that develop on the rootstock trunk produced new fruits faster and at lower costs than those on the new regularly planted nursery trees.

Since the removal of infected trees is not uniformly followed by all growers, the inoculum pressure from neighboring groves may be high in some areas. Therefore, and as for HLB, aggressive control of the insect vector is a critical component of CVC control. Currently, such control consists in treating younger trees to be used for replants with systemic insecticides applied to the trunk (Yamamoto, 2008). In the state of São Paulo, young trees for planting are produced inside screened nurseries, where they are protected from infected sharpshooters. Clean nursery tree production is a key aspect of CVC control, and has been mandatory in the state of São Paulo since 2003. Systemic insecticide treatments of nursery stock prior to field planting will protect the trees for the next 2–3 months. New applications of systemic and regular spray of contact insecticides must be carried out for the following 2 years after field planting (Yamamoto, 2008), a period when the trees constantly flush with new shoots and become attractive to the vectors. After this period, the effectiveness of systemic insecticide applications is reduced, and the trees must be protected with contact insecticide only, which should be applied more frequently during flushing.

18.3.3 Recent discoveries and advances

In the last three decades, new research and field observations have revealed important aspects of the CVC pathosystem which have contributed to improved cultural practices and strategies to control CVC. The Xfp CVC strain 9a5c was the first plant pathogen to have its genome completely sequenced (Simpson et al., 2000). Functional genomic analyses allowed the development of testable hypotheses that revealed many new aspects of the mechanisms of Xf pathogenicity and bacterial adaptation to the xylem sap. It also revealed important and previously unknown vulnerabilities of these pathogens. For example, while most bacterial pathogens carry multiple redundant copies of TolC, a critical bacterial efflux pump component is needed to efflux toxic phytoanticipins, phytoalexins, and/or antibiotics; Xf strains carry only one copy of TolC and inactivation of this gene is lethal for growth in xylem sap (Reddy et al., 2007). The biology, migration patterns, feeding behavior, and pathogen transmission by the insect vector were also elucidated. New experimental procedures allowed for more precise evaluation of the responses of the distinct citrus varieties and species to pathogen infection (Lopes et al., 2005), and the impact of the environment on disease progress and severity was determined.

During this period of intensive research activity and implementation of integrated pest management strategies, CVC incidence initially continued increasing, reaching 34%–44% of the average 200 million orange trees grown in state of São Paulo from 1999 to 2012. However, the incidence dropped from 37.6% in 2012 to 6.8% in 2015. A further decrease was observed in the 2016 and 2017 surveys, and currently <3.0% of the trees express CVC symptoms in the main citrus belt of Brazil (http://www.fundecitrus.com.br/comunicacao/revistas). The losses attributed to the disease, estimated in the range of $100–120 million per year in the period of high incidence (Bové and Ayres, 2007; Lopes et al., 2004), were in 2017 negligible. Several technical advances help explain the level of control achieved, but other cultural factors were equally important, including: (1) the removal of large areas of abandoned groves, which must have significantly contributed to decrease sources of inoculum; (2) an aggressive inspection and eradication program, and (3) the increased rigor and wider efforts to control the insect vector. Grower attitudes toward stronger and more cooperative integrated control measures were greatly stimulated by the report and fast spread of an even more destructive and incurable insect transmitted citrus disease, as described below.

18.4 Huanglongbing

18.4.1 History and background

HLB, also known as "citrus greening," is caused by uncultured, phloem-restricted alpha-proteobacteria in the genus *Candidatus* Liberibacter. These bacterial plant pathogens cause severe disease symptoms on citrus, but nevertheless are

FIG. 18.3 Classic symptoms of "yellow shoot," "yellow dragon," or Huanglongbing (HLB).

well adapted, circulative, and systemic colonizers of the psyllids that vector them into phloem cells. Three species are currently well described as causing HLB in citrus—*Ca.* L. asiaticus (Las), *Ca.* L. africanus (Laf), and *Ca.* L. americanus (Lam)—and each species appears to have independently evolved in the continent reflecting its name (Beattie et al., 2005; Nelson et al., 2013). HLB is currently the most significant bacterial disease of citrus worldwide. HLB is found in Asia, the Middle East, Africa, and the Americas. It is spread by the psyllid vectors *Diaphorina citri* and *Trioza erytreae*. Other well-described *Ca.* Liberibacter species with different vectors are known to cause serious diseases on solanaceous crops (Liefting et al., 2009; Munyaneza et al., 2009), less serious problems on celery (Teresani et al., 2014), and almost no symptoms at all on pear (Raddadi et al., 2011); these will not be further discussed here.

Reports on the occurrence of HLB as "yellow shoot" disease (Fig. 18.3) date back to 1919 in Southwest Asia, from China to Pakistan, 1928 in Africa, and 2004 in the American continent, although references to maladies similar to HLB are even older. HLB was first called Greening in South Africa, Mottle Leaf in the Philippines, Dieback in India, and Vein Phloem Degeneration in Indonesia; in the Americas, it is most commonly called HLB or citrus greening (Bové, 2006; da Graça, 1991).

The most characteristic symptom of HLB is the appearance of asymmetric, blotchy mottles on the leaves, which is a diagnostic symptom. Before the development of DNA-based diagnostic tests allowing the detection of specific Liberibacters, the occurrence of HLB was confirmed mainly by two tests: a thin layer chromatographic test (Schwarz and van Vuuren, 1970) and a starch test to demonstrate its high abundance in symptomatic samples (Su, 2008). However, both methods lacked sufficient specificity to be used routinely in diagnosis. Although transmission electron microscopy (TEM) was used to identify the presence of walled bacteria in phloem cells, it is time consuming and laborious and does not distinguish bacteria at species level.

In several old descriptions of the occurrence of symptoms resembling HLB, the authors mentioned the presence of psyllids and damage caused by psyllid feeding. It was not until the 1960s that *T. erytreae* in South Africa (McClean and Oberholzer, 1965) and *D. citri* in India (Capoor et al., 1967) were demonstrated to vector the disease. In 1970 bacterial cells were found in phloem sieve tubes (Laflèche and Bové, 1970), characterized as Gram-negative bacteria (Garnier et al., 1984) and later determined to be a species of alpha proteobacteria (Jagoueix et al., 1994). Two species were then proposed: Las for the disease in Asia and Laf for the disease in Africa. In 2004, when HLB was seen for the first time in the Americas (in São Paulo State, Brazil), both Las and a new species, Lam, were proposed as causal agents (Teixeira et al., 2005b).

Koch's postulates have yet to be completed for any Liberibacter, and none of the pathogenic Liberibacters have been cultured, likely because they have lost certain key genes to make them capable of axenic growth (Jain et al., 2017b). The only extant isolates of *Liberibacter crescens* (Lcr) were cultured after isolation from diseased mountain papaya in Puerto Rico (Leonard et al., 2012; Fagen et al., 2014) but none of these could be transmitted back to papaya or any other plant or insect host, despite numerous efforts. Lcr may be nonpathogenic. Although the causal relationship between Liberibacters and HLB is technically associative, strong and accumulating evidence leaves little doubt that Liberibacters are the primary etiological agent of HLB. Much of this conclusion rests upon diagnostic molecular tests, primarily polymerase chain reaction (PCR), which is currently the gold standard for detection and diagnosis. Metagenomics were used to sequence multiple genomes of Las (Duan et al., 2009; Lin et al., 2013; Katoh et al., 2014), Lam (Wulff et al., 2014), and Laf (Lin et al., 2015). Molecular techniques and other tools have been used to help elucidate the insect-Liberibacter as well as the plant-Liberibacter interactions.

18.4.2 Symptoms

Las growth *in planta* is phloem limited (Fig 18.4A), causing perturbed assimilate partitioning, leaf chlorosis, loss of productivity, and eventual decline and death of infected trees after several years. By contrast, the bacterium is systemic and circulative in its psyllid host, causing only localized apoptosis in the midgut tissue and accumulating on the outside of the gut under a biofilm (Fig. 18.4B) but without causing overt disease symptoms (Ghanim et al., 2016). Field diagnoses of HLB affected trees are initially determined visually by trained human inspectors. HLB in Florida was initially discovered by spotting the appearance of "yellow shoot" symptoms (Fig. 18.3). No field deployable device has yet been validated by PCR results as reliably allowing field use for diagnoses of Las infection of symptomatic or asymptomatic trees. However, replicated field studies have indicated that canines can detect Las- and Laf-infected plants with >99% accuracy and preliminary results indicate canines will alert trained human handlers to Lam-infected plants as well (Gottwald et al., 2017). The main symptoms used for field diagnosis is blotchy mottle of the leaves (Fig. 18.5), and also *asymmetric* chlorosis along the axis of the leaves, frequently delimited in color from green to yellow by veins. A similar—but symmetrical—yellowing of the leaves is characteristic of zinc deficiency, and this close similarity can cause some confusion and doubt, even to trained scouts looking for symptomatic trees in areas where HLB was previously unknown.

Within one to multiple years after HLB has been identified in a citrus grove, nearly 100% of the trees typically become infected, and often without apparent symptoms. The severity of the symptoms is correlated with the time that a plant has been infected. After a plant is first infected, a surprisingly lengthy (usually no shorter than 6 months) incubation period is required before any symptoms start to develop (Lee et al., 2015). The first symptoms are a pale or mild leaf mottle, which only slowly progresses to more characteristic leaf mottle symptoms. Symptom progression is usually faster during the fall and winter seasons. Leaf curling, reduced leaf size, and thickening of the midveins or raised corky veins are additional symptoms (Fig. 18.6) often, but not always, found. Twig dieback is also observed but takes longer to develop. Fruits developing from infected branches can be lopsided (Fig. 18.6), harbor aborted seeds, and have discolored columellas (Fig. 18.7). Fruit color inversion is sometimes seen, with yellowing starting from the peduncular end and the stylar end remaining green (these symptoms gave rise to the common name "greening"). A yellow to orange abscission zone is seen in the peduncular scar of the fruits, even in the absence of color inversion (Bové, 2006). Leaf and fruit drop are common characteristics, and infected groves experience very high levels of premature fruit drop, exacerbated by drought conditions. Root systems are

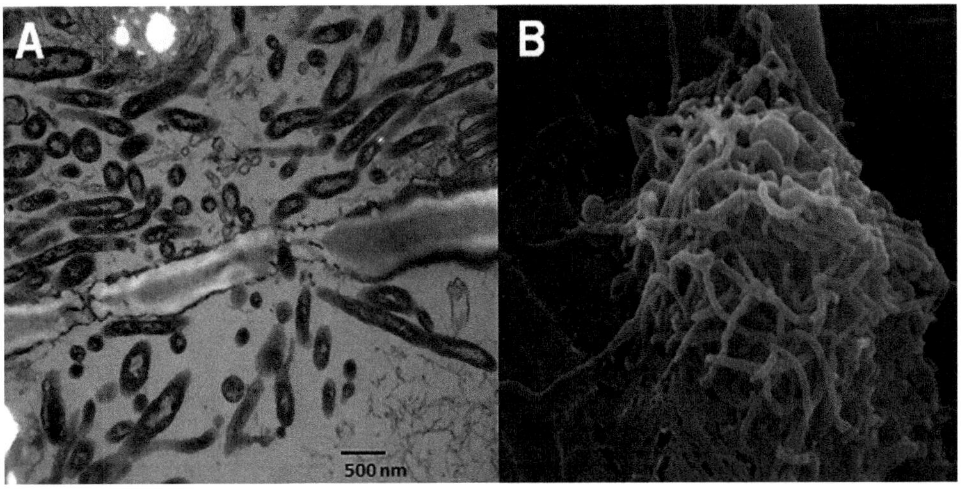

FIG. 18.4 Transmission electron micrographs of (A) *Ca.* Liberibacter americanus (Lam) packed in a sieve tube element and moving through a sieve pore in plant host (periwinkle) phloem cells, and (B) Las in insect host (psyllid), on outer gut surface following bacterial penetration through the gut and growth on the *outer* gut surface, bathed in hemolymph. *(A: Photo courtesy of Dr. Brigitte Batailler, UMR-1332 de Biologie du Fruit et Pathologie, INRA, Université Bordeaux, Villenave d'Ornon, France. Bar = 500 nm; B: Photo courtesy of Prof. Mike Davis, University of Florida.)*

FIG. 18.5 Blotchy mottle symptoms of Huanglongbing (HLB) caused by *Ca.* Liberibacter asiaticus (Las) in sweet orange (A and B), Rangpur lime (C), and Tahiti lime (D).

FIG. 18.6 HLB symptoms of corky veins and misshapen fruit in field grown grapefruit.

FIG. 18.7 HLB symptoms in Valencia fruit (aborted seeds, and misshapen, discolored columellas).

also affected by severely reduced numbers of fibrous roots, and increased susceptibility to *Phytophthora* (Graham et al., 2013). With time, symptoms become severe, branches die and tree decline is evident; skeletonized trees in groves with long-standing HLB becomes all too commonplace (Fig. 18.8).

18.4.3 Etiology and diagnosis

Liberibacters were first found in citrus affected with HLB using TEM (Laflèche and Bové, 1970). Up to the 1990s, TEM was the way to prove the association of phloem-restricted bacteria with HLB. With the advance of molecular techniques and characterization of ribosomal protein and 16SrDNA genes, PCR-based diagnostics and the nature of HLB etiology were

FIG. 18.8 Scenes from commercial citrus Hamlin groves in Florida, 2018 (A, still productive) and 2016 (B, abandoned), about 10 years after HLB was found in the groves.

established. The causal bacteria were named *Ca.* Liberibacter spp.; "*Candidatus*" refers to inability to culture the bacteria and "Liberibacter" refers to the liber or sieve elements in which they were found (Jagoueix et al., 1994). Attempts to raise monoclonal antibodies from the infected tissues failed to yield either Las-specific or universal antibodies, largely due to the inability to grow Las in axenic cultures. However, extensive genome sequencing may now provide adequate targets for antigen production and generation of useful antibodies (Ding et al., 2017).

DNA probes for the detection of nucleic acids from Liberibacters in Southern and dot blots were developed (Villechanoux et al., 1992), but were soon surpassed by PCR-based diagnostic tests. PCR is useful for Liberibacter detection and identification, both in plant and in psyllid samples (Jagoueix et al., 1996; Hocquellet et al., 1999; Teixeira et al., 2005a, 2008a; Li et al., 2006; Roberts and Pietersen, 2017). There are many variations in PCR techniques and primers used, with the most common based on the detection of the 16S and ribosomal protein genes. Real time (semi-quantitative, or qPCR) enjoys widespread use. PCR and qPCR are much less effective in evaluating chemical treatment effects, since DNA from dead Liberibacter cells remains in phloem cells (Trivedi et al., 2009). In plant tissues, small pioneer or fibrous roots and seed coats of developing seeds are the organs with highest Liberibacter cell numbers (Achor et al., 2014). qPCR also enabled the quantification of Las and Lam titers in plants and psyllids. Estimated average highest titers in individual psyllids were 6.3×10^8 for Las and 5.6×10^4 for Lam; and in 1 g of tissue of sweet orange and orange jasmine leaves, respectively, 4×10^7 and 1×10^3 for Las, and 3.2×10^6 and 2×10^4 for Lam (Duan et al., 2009; Lopes et al., 2010; Teixeira et al., 2008a).

Fluorescent in situ hybridization (FISH) indicated that Las is able to colonize the filter chamber, midgut, Malpighian tubules, hemolymph, salivary glands, ovaries, and muscle and fat tissues of *D. citri* (Ammar et al., 2011). Loop-mediated isothermal amplification (LAMP)-based diagnostic tests have the advantage of reduced cost of equipment, but the relative lack of specificity and single-target detection are concerns (Okuda et al., 2005; Li et al., 2007). Conventional PCR can be still very useful for limited sample testing, allow multiple pathogen detection; although it is less sensitive than qPCR, but has the advantage of reduced initial costs. For activity related to sampling and eradication programs, where laboratory diagnoses may guide scouting operations, current molecular techniques are most useful for confirmatory purposes.

The vast majority of publications dealing with diagnosis of HLB have established a consensus that Liberibacters are the most frequently encountered causal agent of HLB. Phytoplasmas can also elicit HLB symptoms, but these agents are of marginal occurrence in relation to Liberibacters (Teixeira et al., 2008b; Chen et al., 2009; Wulff et al., 2015). Outside citrus and Rutaceae, the finding of new *Ca.* Liberibacter species is flourishing. As mentioned above, *Ca.* L. solanacearum was found in 2008 infecting Solanaceae and other crops, such as carrots and celery, causing disease in these hosts.

Ca. L. europaeus was found as an asymptomatic endophyte of pear trees and *Liberibacter crescens* was described in 2012 as the first axenically cultivated member of this genus. Except *L. crescens*, all other Liberibacters—that is, the known pathogenic Liberibacters—are all axenically uncultured to date due at least in part to the loss of the glyoxylase pathway and the presence of phage that are repressed by host factors (Jain et al., 2017a,b; Merfa et al., 2019).

Laf diversity at the subspecies or haplotype level in Africa is noteworthy. Besides having a host range on citrus trees, Laf subsp. capensis (LafC) has been found in cape chestnut (*Calodendrum capense*), while recently, additional rutaceous hosts were described for the following subspecies: Laf subsp. clauseane (LafCl) in *Clausena anisata*, Laf subsp. vepridis (LafV) in *Vepris lanceolata*, Laf subsp. zanthoxyli (LafZ) in *Zanthoxylum capense*, and Laf subsp. tecleae (LafZ) in *Teclea gerrardii* (Roberts et al., 2015, 2017). Additionally, LafCl strains with a wide host range were identified in *Citrus*, with a characteristic pattern of single nucleotide polymorphisms (SNPs) named LafCl ex citrus. Of great importance for diagnostic testing is the fact that the widely used 16S PCR primers used for Las identification (Li et al., 2006) incorrectly identified LafCl ex citrus, and LafCl, LafV, and LafZ were also misidentified as Las or Laf by the same qPCR test. Clearly, additional primer sets are required to distinguish these subspecies. A more general caution is that a great deal of care must be taken when dealing with first occurrences of Liberibacters in unusual places and plant and insect hosts (Roberts et al., 2017).

Currently Las is the prevalent species in HLB samples from Brazil, China, and the United States, where Las is now found in all citrus growing states. In Brazil, only a very limited number of samples harboring groups IX and III phytoplasmas and Lam are now found, where once Lam was the predominant HLB causal agent. One reason for the sharp decrease in Lam incidence in Brazil seems to be its lower titer levels in citrus tissues that reduce the chances of being acquired and transmitted by psyllids (Lopes et al., 2009a). Lam and phytoplasmas would probably show up occasionally in additional places. It should also be noted that since Las and Lam can be dodder transmitted to tobacco and periwinkle, it seems likely that *Ca.* Liberibacter host range may be more limited by the host range of the insect vector responsible for phloem transmission than by any particular pathogenicity or virulence determinants in these pathogens.

18.4.4 Pathogen biology: Phloem restriction in citrus, systemic infection in psyllid, and lack of axenic culture

After the transmission of Las, Lam, and Laf to periwinkle using dodder, the bacteria were found restricted to phloem sieve tubes, as expected. The TEM was also used to characterize uncultured *Ca.* Liberibacters as Gram negative by visualization of the tri-lamellar cell wall envelope (Garnier et al., 1984). Periwinkles were used for the maintenance of multiple isolates *in planta* at INRA in Bordeaux, France for decades by Bové's group (Teixeira et al., 2008a; Wulff et al., 2014). Periwinkles have allowed the maintenance of isolates in a smaller space than citrus and even comparative strain characterizations, such as heat tolerance (Las is exceptionally more heat tolerant than Lam and Laf; Bové et al., 1974; Lopes et al., 2009b), antibody generation (Garnier and Bové, 1993; Ke et al., 1993), and the description of the first Las phage particles (Zhang et al., 2011). Indeed, periwinkle became the source of DNA enriched for genome sequencing of Lam (Wulff et al., 2014). However, transmission from periwinkle to citrus cannot be accomplished because *Diaphorina citri* and *T. erytreae* are restricted in host range to citrus.

Ca. Liberibacter spp. multiply in the sieve tubes of infected plant hosts and there are pockets of sieve cells filled with elongated, electron dense young bacterial cells (Fig. 18.4) while most sieve tubes are free of bacteria. *Ca.* Liberibacters literally squeeze from sieve cell to sieve cell through sieve pores (Su, 2008), apparently without degrading plant cell walls, since there are no cell wall degrading enzymes annotated in any Liberibacter genomes. With maturity, the bacterial cells form less dense spherical bodies of 600–800 nm, eventually plasmolyzing (Su, 2008). As a host response to the spread of the bacterium in the vascular system, an increase in callose deposition occurs and together with bacterial multiplication, blocks the flow of phloem sap between the cells, disrupting the source to sink movement of sugars (Koh et al., 2012). HLB-affected trees are imbalanced in sugars (Fan et al., 2010). Notably, all pathogenic Liberibacters, by stark contrast with *L. crescens*, lack any known mechanism for sucrose import, and appear to rely instead on direct import of ATP from host cells (Jain et al., 2017b).

One consequence of the development of an intracellular lifestyle is the necessity of avoiding (to the greatest extent possible) and suppressing (where avoidance is impossible) the triggering of innate immune responses by any host, plant or insect. This can involve the shedding or downregulation of genes that directly encode such triggers; for example, flagellin, lipopolysaccharides (LPS), or enzymes that synthesize these triggers. Flagellin and LPS are two classic examples of a general group of molecules collectively referred to as "pathogen-associated molecular patterns" (PAMPs), which are recognized by the innate immune systems in eukaryotic hosts of all types (Jones and Dangl, 2006). Although flagella-like appendages have been observed on the surface of *Ca.* L. solanacearum infecting potato (Cicero et al., 2016), flagella have not been observed with any citrus-associated Liberibacter, and the flagellin produced by Las is a PAMP weaker than those of other plant pathogenic bacteria (Zou et al., 2012). Eukaryotic cells typically are highly sensitive to LPS, which serves as a highly protective bacterial outer membrane defensive barrier against plant antimicrobial compounds such as phytoalexins; interestingly, Lam has no LPS genes at all (Wulff et al., 2014). The severe reduction in genome size observed generally with all intracellular pathogenic bacteria is likely driven, in part, by selection pressure to remove genes encoding PAMPs.

Las actively suppresses innate citrus immune responses to infection, helping to explain why HLB symptoms have such a long incubation period following initial introduction by the psyllid vector (a period spanning multiple months to years). The suppression of citrus defense reactions, particularly generation of reactive oxygen species (ROS) and prevention of apoptosis, is accomplished by a phage-encoded peroxidase (Jain et al., 2015) and two chromosomally encoded peroxiredoxins (Jain et al., 2018, 2019). The peroxidase and one peroxiredoxin are both nonclassically secreted directly into host phloem cytoplasm, resulting in: (a) straight detoxification of host-generated H_2O_2, (b) limited catastrophic chain reaction spread of a lipid peroxidation event in the phloem cell membrane, and (c) dampening of the systemic propagation of H_2O_2-mediated defense signaling in the host plant following the initial colonization event. Las peroxiredoxin may represent a critical secreted effector conserved across all pathogenic Liberibacter species (even in the phageless strains) that functions to suppress host symptoms, a tactic used by most biotrophic plant pathogens.

The extent of Las and Lam multiplication and symptom induction in citrus is quite different. Las multiplies to 10-fold higher titers in citrus than Lam, but Lam elicits much more severe symptoms of HLB, despite lower titer (Lopes et al., 2009a). Las can be transmitted by grafting budwood with efficiencies close to 100%, whereas Lam transmission by grafting is usually lower than 50%. There are no comparable data for Laf. Natural transmission by psyllids is influenced by several factors and bacterium multiplication and temperature are key aspects. Las titer in citrus influences *D. citri* transmission; while Las is readily transmitted from citrus to citrus, Lam is less efficiently transmitted. Interestingly and by contrast, Lam titer in *Murraya exotica* (*M. paniculata*) is higher than that of Las (Lopes et al., 2010). *Murraya* spp. is a good host for Lam acquisition and transmission to sweet orange plants, but this is not true for Las (Gasparoto et al., 2010; Damsteegt et al., 2010).

18.4.5 Genomics and taxonomy

Alphaproteobacteria are quite diverse as a class, and often form associations with eukaryotes. Free-living members have genome sizes over 5 Mbp and include the mutualistic endosymbionts in the Rhizobiaceae family: *Rhizobium, Azorhizobium, Bradyrhizobium*, and *Mesorhizobium*, as well as the pathogenic *Agrobacterium* species. Alphaproteobacteria are also well known to adapt an intracellular lifestyle with differing species utilizing independent paths of severe genome reduction. These famously include the mitochondria, as well as the Rickettsiales (including tick-borne diseases in genus *Rickettsia*, and endosymbionts of arthropods in genus *Wolbachia*). Liberibacters form a cluster of closely related alphaproteobacteria in the Rhizobiaceae family with highly reduced genomes as compared with free-living members. Interestingly, they appear much more highly adapted to their insect hosts than plant hosts, spreading inter- and intracellularly across salivary glands, alimentary canal, midgut epithelium and basement membrane, Malpighian tubules, hemolymph, muscle and fat tissues, and ovaries with little or no evidence of pathogenic effects (Ammar et al., 2011).

Only two strains of a single species of *Liberibacter*, *L. crescens* strains BT-0 and BT-1 (NC_019907.1), have been cultured in vitro (Leonard et al., 2012). Both were isolated from the sap of Babaco mountain papaya (*Carica stipulata* × *C. pubescens*) and are not known to be pathogenic, despite numerous efforts in several labs, and they have no known insect host. *L. crescens* has a slightly larger genome size (1.5 Mb) than any of the pathogenic *Ca.* Liberibacters (all about 1.2 Mb). *L. crescens* appears to have diverged from the uncultured, pathogenic *Ca.* Liberibacter spp. early (Nakabachi et al., 2013). The first Liberibacter genome to be completely sequenced was Las strain psy62 (Duan et al., 2009); since then more than nine additional Las genomes have been sequenced. In addition to Las, complete genome sequences were determined for Lam strain São Paulo (Wulff et al., 2014), Laf strain PTSAPSY (Lin et al., 2015), and Lso strain CLso-ZC1 (Lin et al., 2011). All the chromosomes contain three ribosomal operons and genome GC% of 35.4%, while Lam GC% is lowest, at 31.1%.

All pathogenic *Ca.* Liberibacters contain prophage or remnants of prophage not found in *L. crescens* with GC% content of 36% (Zhang et al., 2011). Typically, Las strains have 1–2 complete prophages, but a few strains have only remnants, such as Ishi-1 from Japan (Katoh et al., 2014). Both prophages may be integrated in the chromosome, but one of them typically replicates as an excision plasmid, thus expanding the copy number per cell of the encoded genes. As mentioned above, the peroxidase is encoded on the excision plasmid, can functionally suppress plant defenses and is therefore likely a horizontally acquired lysogenic conversion factor benefitting plant infections (Jain et al., 2015). One of the prophages can be lytic *in planta* and even form particles in citrus (Fu et al., 2015), yet remain as a stable lysogen in psyllids (Zhang et al., 2011). Suppression of Las phage gene expression appears to be critically important in psyllid hosts (Fleites et al., 2014; Jain et al., 2017a). Despite the few horizontally acquired gene additions, comparative genomics and metabolic pathway analyses of the chromosomal genomes have revealed a trend toward reduction or complete absence of multiple biosynthetic pathways, including glycolysis and glyoxylase, metabolic enzymes, LPS biosynthesis, sucrose import and secretion systems in these genomes, helping to explain the failure to obtain sustainable axenic cultures of the pathogenic *Ca.* Liberibacters to date (Wulff et al., 2014; Jain et al., 2017b).

18.4.6 Epidemiology: Latency and incubation

In the absence of a vector, all *Ca.* Liberibacter species remain confined to their plant hosts. Citrus trees infected with various pathogens are not infrequently brought into commercial citrus growing areas by unsuspecting homeowners. If a Las-infected tree is brought into an area without the vector, both the tree and the Las that infects it can remain alive for many years, due to the surprisingly long incubation period between the initial infection and the appearance of even the slightest symptom. The vectors of HLB are also found in disease-free areas and introduction of Las-free insect vectors usually precedes the discovery of HLB by many years. In sharp contrast to the long incubation period for HLB, the latency period of the transmission of Las from citrus to psyllid and back again is measured in weeks, and the multiplication of infected psyllid vectors is explosive (Lee et al., 2015). The importance and surprising nature of the very short latency period for transmission of Las, coupled with the exceptionally long incubation period for HLB cannot be overemphasized. Las epidemics spread very rapidly in completely asymptomatic trees, making early detection critical (Gottwald, 2010). *T. erytreae* is found in the Mediterranean basin, although the bacteria are not, so the disease is not known to be present (Pérez-Otero et al., 2015; Siverio et al., 2017). The absence of both HLB and its vectors, such as in Australia, is a great advantage to the success and sustainability of citriculture.

Psyllid transmission is the only natural mode of dissemination of Las, Lam, and Laf, although spread by infected budwood during citrus propagation may occur locally. Illicit importation of budwood or entire infected plants will spread the disease and perhaps the vector over long distances including internationally. Strict quarantine measures should be applied to restrict the imports of propagative citrus material which originate only from certified regions or from certified plants free of psyllids and Liberibacters.

18.4.7 Control/management

Since there is no cure for HLB, preventive and mitigation actions must be taken, starting with quarantine regulations, implementation of inspections of nursery trees, and production of budwood in insect-proof nurseries. Although systemic or semi-systemic antibiotics have shown promise with trunk injection (e.g., Bové et al., 1980; Hu and Wang, 2016), there is some controversy over the widespread use of even limited types of antibiotics in agriculture. Psyllid monitoring (Miranda et al., 2017) and control (Miranda et al., 2016) by means of insecticide control in commercial areas is a key aspect to avoid the spread of HLB-infection in orchards (Bassanezi et al., 2013). In order to reduce inoculum, scouting for infected trees and removal of symptomatic trees must be carried out. These three strategies have been used in South Africa, Brazil, and Costa Rica (Bové, 2012; Aubert, 1990) and proven to be effective as a regional management strategy (Bassanezi et al., 2013). Due to the very short latency period for Las transmission, the development of presymptomatic Las detection techniques is a research priority. Due to the severe economic impact of HLB in regions and states where HLB infections have recently been detected, deployment of all available early detection resources, such as the use of canines, should be strongly considered.

References

Achor, D., Davis, C.L., Brlansky, R.H., Folimonova, S.Y., 2014. Guidelines for selection of tissues for electron microscopy confirmation of *Candidatus* Liberibacter spp in Huanglongbing-affected citrus. J. Citrus Pathol. 1, 99.

Almeida, R.P.P., Nunney, L., 2015. How do plant diseases caused by *Xylella fastidiosa* emerge? Plant Dis. 99, 1457–1467.

Ammar, E.D., Shatters Jr., R.G., Hall, D.G., 2011. Localization of *Candidatus* Liberibacter asiaticus, associated with citrus Huanglongbing disease, in its psyllid vector using fluorescence in situ hybridization. J. Phytopathol. 159, 726–734.

Aubert, B., 1990. Integrated activities for the control of huanglungbin-greening and its vector *Diaphorina citri* in Asia. In: Aubert, B., Tontyaporn, S., Buangsuwon, D. (Eds.), Rehabilitation of Citrus Industry in the Asia Pacific Region. Proceedings of the Asia Pacific International Conference on Citriculture, Chiang Mai, Thailand. UNDP-FAO, Rome, pp. 133–144.

Ayres, A.J., 2000. Intensidade da clorose variegada dos citros em pomares comerciais de laranja do estado de São Paulo e sul do triângulo mineiro (MSc Thesis). Universidade Estadual Paulista, Faculdade de Ciências Agrárias e Veterinárias, Jaboticabal, Brasil.

Bassanezi, R.B., Montesino, L.H., Gimenes-Fernandes, N., Yamamoto, P.T., Gottwald, T.R., Amorim, L., Bergamin Fo, A.B., 2013. Efficacy of area-wide inoculum reduction and vector control on temporal progress of huanglongbing in young sweet orange plantings. Plant Dis. 97, 789–796.

Beattie, G.A.C., Mabberley, D.J., Holford, P., Broadbent, P., De Barro, P., 2005. Huanglongbing: its possible origins, collaborative research in Southeast Asia, and developing incursion management plans for Australia. In: Second International *Citrus* Canker and Huanglongbing Research Workshop, p. 52.

Behlau, F., Jones, J.B., Myers, M.E., Graham, J.H., 2012. Monitoring for resistant populations of *Xanthomonas citri* subsp. citri and epiphytic bacteria on citrus trees treated with copper or streptomycin using a new semi-selective medium. Eur. J. Plant Pathol. 132, 259–270.

Behlau, F., Fonseca, A.E., Belasque Jr., J., 2016. A comprehensive analysis of the Asiatic citrus canker eradication programme in São Paulo state, Brazil, from 1999 to 2009. Plant Pathol. 65, 1390–1399.

Bock, C.H., Graham, J.H., Gottwald, T.R., Cook, A.Z., Parker, P.E., 2010a. Wind speed effects on the quantity of *Xanthomonas citri* subsp. citri dispersed downwind from canopies of grapefruit tree infected with citrus canker. Plant Dis. 94, 725–736.

Bock, C.H., Graham, J.H., Gottwald, T.R., Cook, A.Z., Parker, P.E., 2010b. Wind speed and wind-associated leaf injury affect severity of citrus canker on Swingle citrumelo. Eur. J. Plant Pathol. 128, 21–38.

Bock, C.H., Parker, P.E., Cook, A.Z., Graham, J.H., Gottwald, T.R., 2011. Infection and decontamination of citrus canker and inoculated the surfaces. Crop Prot. 30, 259–264.

Bové, J.M., 2006. Huanglongbing: a destructive, newly-emerging, century-old disease of citrus. J. Plant Pathol. 88, 7–37.

Bové, J.M., 2012. Huanglongbing and the future of citrus in São Paulo state. Brazil. J. Plant Pathol. 94 (3), 465–467.

Bové, J.M., Ayres, A.J., 2007. Etiology of three recent diseases of citrus in São Paulo state: sudden death, variegated chlorosis and huanglongbing. IUBMB Life 59, 346–354.

Bové, J.M., Calavan, E.C., Capoor, S.P., Cortez, R.E., Schwarz, R.E., 1974. Influence of temperature on symptoms of California stubborn, south African greening, Indian citrus decline and Philippines leaf mottling diseases. In: Conference of the International Organization of Citrus Virologists. Proceedings of 6th Conference IOCV. University of California, Riverside, pp. 12–15.

Bové, J.M., Bonnet, P., Granier, M., Aubert, B., 1980. Penicillin and tetracycline treatments of greening-disease-affected citrus plants in the glasshouse, and the bacterial nature of the prokariote associated with greening. In: Conference of the International Organization of Citrus Virologists. Proceedings of 8th Conference IOCV. University of California, Riverside, pp. 91–102.

Braithwaite, M., Leite, R.P., Smith, J.J., Boa, E., Saddler, G.S., 2002. First report of citrus canker caused by *Xanthomonas campestris* pv. *citri* on *Citrus sinensis* in Bolivia. Plant Pathol. 51, 383.

Brunings, A.M., Gabriel, D.W., 2003. *Xanthomonas citri*: breaking the surface. Mol. Plant Pathol. 4, 141–157.

Canteros, B.I., Gochez, A.M., Moschini, R.C., 2017. Management of citrus canker in Argentina, a success story. Plant Pathol. J. 33, 441–449.

Capoor, S.P., Rao, D.G., Viswanath, S.M., 1967. *Diaphorina citri*: a vector of the greening disease of citrus in India. Indian J. Agric. Sci. 37, 572–576.

Carrera, C., 1933. Informe preliminar sobre una enfermedad nueva comprobada en los citrus de Bella Vista (Corrientes). Bol. Mens. Min. Agric. Nac. Buenos Aires 34, 275–280.

Chen, J., Pu, X., Deng, X., Liu, S., Li, H., Civerolo, E., 2009. A phytoplasma related to *Candidatus* Phytoplasma asteri detected in citrus showing huanglongbing (yellow shoot disease) symptoms in Guangdong, P. R. China. Phytopathology 99 (3), 236–242.

Cicero, J.M., Fisher, T.W., Brown, J.K., 2016. Localization of 'Candidatus Liberibacter solanacearum' and evidence for surface appendages in the potato psyllid vector. Phytopathology 106, 142–154. https://doi.org/10.1094/PHYTO-04-15-0088-R.

Contreras, J.D.V., 1992. Pecosita ou falsa mancha grasienta na Argentina. Laranja e Cia 31, 6.

Cubero, J., Graham, J.H., 2002. Genetic relationship among worldwide strains of *Xanthomonas* causing canker in citrus species and design of new primers for their identification by PCR. Appl. Environ. Microbiol. 68, 1257–1264.

da Graça, J.V., 1991. *Citrus* greening disease. Annu. Rev. Phytopathol. 29, 109–136.

Damsteegt, V.D., Postnikova, E.N., Stone, A.L., Kuhlmann, M., Wilson, C., Sechler, A., Schaad, N.W., Brlansky, R.H., Scheneider, W.L., 2010. *Murraya paniculata* and related species as potential hosts and inoculum reservoirs of '*Candidatus* Liberibacter asiaticus', causal agent of Huanglongbing. Plant Dis. 94, 528–533.

Ding, F., Paul, C., Brlansky, R., Hartung, J.S., 2017. Immune tissue print and immune capture-PCR for diagnosis and detection of *Candidatus* Liberibacter Asiaticus. Sci. Rep. 7, 46467.

Duan, Y.P., Castaneda, A., Zhao, G., Erdos, G., Gabriel, D.W., 1999. Expression of a single, host-specific, bacterial pathogenicity gene in plant cells elicits division, enlargement, and cell death. Mol. Plant Microbe Interact. 12, 556–560.

Duan, Y., Zhou, L., Hall, D.G., Li, W., Doddapaneni, H., Lin, H., Liu, L., Vahling, C.M., Gabriel, D.W., Williams, K.P., Dickerman, A., Sun, Y., Gottwald, T., 2009. Complete genome sequence of citrus huanglongbing bacterium, '*Candidatus* Liberibacter asiaticus' obtained through metagenomics. Mol. Plant Microbe Interact. 22, 1011–1020.

El-Yacoubi, B., Brunings, A.M., Yuan, Q., Shanker, S., Gabriel, D.W., 2007. *In planta* horizontal transfer of a major pathogenicity effector gene. Appl. Environ. Microbiol. 73, 1612–1621.

Fagen, J.R., Leonard, M.T., Coyle, J.F., McCullough, C.M., Davis-Richardson, A.G., Davis, M.J., Triplett, E.W., 2014. *Liberibacter crescens* gen. nov., sp. nov., the first cultured member of the genus *Liberibacter*. Int. J. Syst. Evol. Microbiol. 64 (7), 2461–2466.

Fan, J., Chen, C., Brlansky, R.H., Gmitter Jr., F.G., Li, Z.-G., 2010. Changes in carbohydrate metabolism in *Citrus sinensis* infected with '*Candidatus* Liberibacter asiaticus'. Plant Pathol. 59, 1037–1043.

Fleites, L.A., Jain, M., Zhang, S.J., Gabriel, D.W., 2014. "*Candidatus* Liberibacter asiaticus" prophage late genes may limit host range and culturability. Appl. Environ. Microbiol. 80, 6023–6030.

Fu, S.M., Hartung, J., Zhou, C.Y., Su, H.N., Tan, J., Li, Z.A., 2015. Ultrastructural changes and putative phage particles observed in sweet orange leaves infected with '*Candidatus* Liberibacter asiaticus'. Plant Dis. 99, 320–324.

Gabriel, D.W., Kingsley, M.T., Hunter, J.E., Gottwald, T., 1989. Reinstatement of *Xanthomonas citri* (ex Hasse) and *X. phaseoli* (ex Smith) to species and reclassification of all *X. campestris* pv. citri strains. Int. J. Syst. Bacteriol. 39, 14–22.

Gambley, C.F., Miles, A.K., Ramsden, M., Doogan, V., Thomas, J.E., Parmenter, K., Whittle, P.J.L., 2009. The distribution and spread of citrus canker in Emerald, Australia. Australas. Plant Pathol. 2009 (38), 547–557.

Garcia, A.L., Torres, S.C.Z., Heredia, M., Lopes, S.A., 2012. *Citrus* responses to *Xylella fastidiosa* infection. Plant Dis. 96, 1245–1249.

Garnier, M., Bové, J.M., 1993. *Citrus* greening disease and the greening bacterium. In: Conference of the International Organization of Citrus Virologists. Proceedings of 12th Conference IOCV. University of California, Riverside, pp. 212–219.

Garnier, M., Danel, N., Bové, J.M., 1984. The greening organism is a gram negative bacterium. In: Conference of the International Organization of Citrus Virologists. Proceedings of 9th Conference IOCV. University of California, Riverside, pp. 115–124.

Gasparoto, M.C.G., Bassanezi, R.B., Amorim, L., Montesino, L.H., Lourenço, S.A., Wulff, N.A., Teixeira, D.C., Mariano, A.G., Martins, E.C., Leite, A.P.R., Bergamin Filho, A., 2010. First report of '*Candidatus* Liberibacter americanus' transmission from *Murraya paniculata* (L.) Jack to sweet orange (*Citrus sinensis* Osbeck) by Asian citrus psyllid (*Diaphorina citri* Kuwayama). J. Plant Pathol. 92 (2), 546.

Ghanim, M., Fattah-Hosseini, S., Levy, A., Cilia, M., 2016. Morphological abnormalities and cell death in the Asian citrus spyllid (*Diaphorina citri*) midgut associated with *Candidatus* Liberibacter asiaticus. Sci. Rep. 6, 33418.

Gordon, J.L., Lefeuvre, P., Escalon, A., Barbe, V., Cruveiller, S., Gagnevin, L., Pruvost, O., 2015. Comparative genomics of 43 strains of *Xanthomonas citri* pv. citri reveals the evolutionary events giving rise to pathotypes with different host ranges. BMC Genomics 16, 1098. https://doi.org/10.1186/s12864-015-2310-x.

Gottwald, T.R., 2010. Current epidemiological understanding of citrus Huanglongbing. Annu. Rev. Phytopathol. 48, 119–139.

Gottwald, T.R., Irey, 2007. Post-hurricane analysis of citrus canker II: predictive model estimation of disease spread and area potentially impacted by various eradication protocols following catastrophic weather events. Plant Health Prog. https://doi.org/10.1094/PHP-2007-0405-01-RS April 2007.

Gottwald, T.R., Graham, J.H., Schubert, T.S., 1997. An epidemiological analysis of the spread of citrus canker in urban Miami, Florida, and synergistic interaction with the Asian citrus leafminer. Fruits 52, 383–390.

Gottwald, T.R., Hughes, G., Graham, J.H., Sun, X., Riley, T., 2001a. The citrus canker epidemic in Florida: the scientific basis of regulatory eradication policy for an invasive species. Phytopathology 91, 30–34.

Gottwald, T.R., Sun, X., Riley, T.D., Graham, J.H., Ferrandino, F., Taylor, E.L., 2001b. Geo-referenced, spatiotemporal analysis of the urban citrus canker epidemic in Florida. Phytopathology 92, 361–377.

Gottwald, T.R., Graham, J.H., Schubert, T.S., 2002. *Citrus* canker: the pathogen and its impact. Plant Health Prog. www.plantmanagementnetwork.org/pub/php/review/citruscanker.

Gottwald, T.R., Bassanezi, R.B., Amorim, L., Bergamin-Filho, A., 2007. Spatial pattern analysis of citrus canker infected plantings in São Paulo Brazil and augmentation of infection elicited by the Asian Leafminer. Phytopathology 97, 674–683.

Gottwald, T., Graham, J., Bock, C., Bonn, G., Civerolo, E., Irey, M., Leite, R., McCollum, G., Parker, P., Ramallo, J., Riley, T., Schubert, T., Stein, B., Taylor, E., 2009. The epidemiological significance of post-packinghouse survival of *Xanthomonas citri* subsp. *citri* for dissemination of Asiatic citrus canker via infected fruit. Crop Prot. 29, 508–524.

Gottwald, T., Poole, G., Taylor, E., Hall, D., Hartung, J., Bartels, D., McCollum, D., Hilf, M., Luo, W., Louws, F., Polek, M., Vidalakis, G., Mauk, P., 2017. Canine assisted early detection of HLB. J. Citrus Pathol. 4 (1), 16.

Graham, J.H., Gottwald, T.R., 1991. Research perspectives on eradication of citrus bacterial diseases in Florida. Plant Dis. 75, 1193–1200.

Graham, J.H., Gottwald, T.R., Leite Jr., R.P., 2006. Prospects for control of citrus canker with novel chemical compounds. Proc. Fla. State Hort. Soc. 119, 83–88.

Graham, J.H., Johnson, E.G., Gottwald, T.R., Irey, M.S., 2013. Pre-symptomatic fibrous root decline in citrus trees caused by Huanglongbing and potential interaction with *Phytophthora* spp. Plant Dis. 97, 1195–1199.

Graham, J.H., Johnson, E.G., Myers, M.E., Young, R.P., Das, S., Santra, S., 2016. Potential of nano-formulated zinc oxide for control of citrus canker on grapefruit trees. Plant Dis. 100, 2442–2447.

Hocquellet, A., Toorawa, P., Bové, J.M., Garnier, M., 1999. Detection and identification of the two "*Candidatus* Liberobacter sp." Associated with citrus huanglongbing by PCR amplification of ribosomal protein genes of the beta operon. Mol. Cell. Probes 13, 373–379.

Hogg, D.R., 1985. *Citrus* canker in Argentina: a case history. In: Timmer, L.W. (Ed.), *Citrus* Canker: An International Perspective. Citrus Research & Education Center, Univ. Florida, Lake Alfred, FL, pp. 8–10.

Hu, J., Wang, N., 2016. Evaluation of the spatiotemporal dynamics of oxytetracycline and its control effect against citrus Huanglongbing via trunk injection. Phytopathology 106, 1495–1503.

Irey, M., Gottwald, T.R., Graham, J.H., Riley, T.D., Carlton, G., 2006. Post-hurricane analysis of citrus canker spread and progress towards the development of a predictive model to estimate disease spread due to catastrophic weather events. Plant Health Prog. https://doi.org/10.1094/PHP-2006-0822-01-RS. August 2006 (Online).

Jacques, M.-A., Denancé, N., Legendre, B., Morel, E., Briand, M., Mississipi, S., Durand, K., Olivier, V., Portier, P., Poliakoff, F., Crouzillat, D., 2016. New variants of coffee-infecting *Xylella fastidiosa* issued from homologous recombination. Appl. Environ. Microbiol. 82, 1556–1568.

Jagoueix, S., Bové, J.M., Garnier, M., 1994. The phloem-limited bacterium of greening disease of citrus is a member of the α subdivision of the Proteobacteria. Int. J. Syst. Bacteriol. 44, 379–386.

Jagoueix, S., Bové, J.M., Garnier, M., 1996. PCR detection of the two '*Candidatus*' liberobacter species associated with greening disease of citrus. Mol. Cell. Probes 10, 43–50.

Jain, M., Fleites, L.A., Gabriel, D.W., 2015. Prophage-encoded peroxidase in '*Candidatus* Liberibacter asiaticus' is a secreted effector that suppresses plant defenses. Mol. Plant-Microbe Interact. 28, 1330–1337.

Jain, M., Fleites, L.A., Gabriel, D.W., 2017a. A small *Wolbachia* protein directly represses phage lytic cycle genes in "*Candidatus* Liberibacter asiaticus" within psyllids. mSphere 2, e00171-17. https://doi.org/10.1128/mSphereDirect.00171-17.

Jain, M., Munoz-Bodnar, A., Gabriel, D.W., 2017b. Concomitant loss of the glyoxalase system and glycolysis makes the uncultured pathogen "*Candidatus* Liberibacter asiaticus" an energy scavenger. Appl. Environ. Microbiol. 83, e01670-17. https://doi.org/10.1128/AEM.01670-17.

Jain, M., Munoz-Bodnar, A., Gabriel, D.W., 2019. '*Candidatus* Liberibacter asiaticus' peroxiredoxin (LasBCP) suppresses oxylipin-mediated defense signaling in citrus. J. Plant Physiol 236, 61–65. https://doi.org/10.1016/j.jplph.2019.03.001.

Jain, M., Munoz-Bodnar, A., Zhang, S., Gabriel, D.W., 2018. A secreted '*Candidatus* Liberibacter asiaticus' peroxiredoxin simultaneously suppresses both localized and systemic innate immune responses in planta. Mol. Plant Microbe Interact. 31, 1312–1322.

Jones, J.D., Dangl, J.L., 2006. The plant immune system. Nature 444, 323–329.

Katoh, H., Miyata, S., Inoue, H., Iwanami, T., 2014. Unique features of a japanese '*Candidatus* Liberibacter asiaticus' strain revealed by whole genome sequencing. PLoS One 9 (9), e106109.

Ke, C., Ke, S., Wu, R.J., Yang, H., Hsu, P.T., 1993. Purification and serology of the organism associated with citrus huanglongbing. In: Proceedings of 12th Conference IOCV. University of California, Riverside, pp. 220–223.

Koh, E.J., Zhou, L., Williams, D.S., Park, J., Ding, N., Duan, Y.P., Kang, B.H., 2012. Callose deposition in the phloem plasmodesmata and inhibition of phloem transport in citrus leaves infected with "*Candidatus* Liberibacter asiaticus". Protoplasma 249, 687–697.

Koizumi, M., 1985. *Citrus* canker: the world situation. In: Timmer, L.W. (Ed.), *Citrus* Canker: An International Perspective. University of Florida, Lake Alfred, FL, pp. 2–7.

Laflèche, D., Bové, J.M., 1970. Structures de type mycoplasme dans les feuilles d'orangers atteints de la maladie du greening. C.R. Acad. Sci. Paris 270, 1915–1917.

Laranjeira, F.F., Palazzo, D., 1999. Danos qualitativos à produção de laranja 'Natal' causados pela clorose variegada dos citros. Laranja 20, 77–91.

Lee, J.A., Halbert, S.E., Dawson, W.O., Robertson, C.J., Keesling, J.E., Singer, B.H., 2015. Asymptomatic spread of huanglongbing and implications for disease control. Proc. Natl. Acad. Sci. U. S. A. 112, 7605–7610.

Leite Jr., R.P., Mohan, S.K., Pereira, A.L.G., Campacci, C.A., 1987. Integrated control of citrus canker—effect of genetic resistance and application of bactericides. Fitopatologia-Brasileira 12, 257–263.

Leonard, M.T., Fagen, J.R., Davis-Richardson, A.G., Davis, M.J., Triplett, E.W., 2012. Complete genome sequence of *Liberibacter crescens* BT-1. Stand Genomic Sci. 7, 271–283.

Li, W., Hartung, J.S., Levy, L., 2006. Quantitative real-time PCR for detection and identification of *Candidatus* Liberibacter species associated with citrus huanglongbing. J. Microbiol. Methods 66 (1), 104–115.

Li, W., Hartung, J.S., Levy, L., 2007. Evaluation of DNA amplification methods for improved detection of "*Candidatus* Liberibacter species" associated with citrus huanglongbing. Plant Dis. 91, 51–58.

Liefting, L.W., Weir, B.S., Pennycook, S.R., Clover, G.R.G., 2009. '*Candidatus* Liberibacter solanacearum', associated with plants in the family Solanaceae. Int. J. Syst. Evol. Microbiol. 59, 2274–2276.

Lin, H., Lou, B., Glynn, J.M., Doddapaneni, H., Civerolo, E.L., Chen, C., Vahling, C.M., 2011. The complete genome sequence of '*Candidatus* Liberibacter solanacearum', the bacterium associated with potato zebra chip disease. PLoS One 6, e19135.

Lin, H., Han, C.S., Liu, B., Lou, B., Bai, X., Deng, C., Civerolo, E.L., Gupta, G., 2013. Complete genome sequence of a Chinese strain of "*Candidatus* Liberibacter asiaticus". Genome Announc. 1, e00184-13.

Lin, H., Pietersen, G., Han, C., Read, D.A., Lou, B., Gupta, G., Civerolo, E.L., 2015. Complete genome sequence of "*Candidatus* Liberibacter africanus," a bacterium associated with citrus huanglongbing. Genome Announc. 3, e00733-15.

Lopes, S.A., Marcussi, S., Torres, S.C.Z., Souza, V., Fagan, C., França, S.C., 2003. Weeds as alternative hosts of the citrus, coffee, and plum strains of *Xylella fastidiosa* in Brazil. Plant Dis. 87, 544–549.

Lopes, S.A., Laranjeira, F.F., Amorim, L., Bergamin, F.A., 2004. Clorose variegada: perdas anuais de US$ 100 milhões. Visão Agrícola 1, 20–23.

Lopes, S.A., Teixeira, D.C., Fernandes, N.G., Ayres, A.J., Torres, S.C.Z., Barbosa, J.C., Li, W.B., 2005. An experimental inoculation system to study citrus–*Xylella fastidiosa* interactions. Plant Dis. 89, 250–254.

Lopes, S.A., Bertolini, E., Frare, G.F., Martins, E.C., Wulff, N.A., Teixeira, D.C., Fernandes, N.G., Cambra, M., 2009a. Graft transmission efficiencies and multiplication of "*Candidatus* Liberibacter americanus" and "*Ca.* Liberibacter asiaticus" in citrus plants. Phytopathology 99, 301–306.

Lopes, S.A., Frare, G.F., Bertolini, E., Cambra, M., Fernandes, N.G., Ayres, A.J., Marin, D.R., Bové, J.M., 2009b. Liberibacters associated with citrus huanglongbing in Brazil: '*Candidatus* Liberibacter asiaticus' is heat tolerant, '*Ca.* L. americanus' is heat sensitive. Plant Dis. 93, 257–262.

Lopes, S.A., Frare, G.F., Camargo, L.E.A., Wulff, N.A., Teixeira, D.C., Bassanezi, R.B., Beattie, G.A.C., Ayres, A.J., 2010. Liberibacters associated with orange jasmine in Brazil: Incidence in urban áreas and relatedness to citrus liberibacters. Plant Pathol. 59, 1044–1053.

Lopes, S.A., 2019. A new strategy to control citrus variegated chlorosis disease. Plant Dis. https://doi.org/10.1094/PDIS-02-19-0302-RE.

Mavrodieva, V., Levy, L., Gabriel, D.W., 2004. Improved sampling methods for real-time polymerase chain reaction diagnosis of citrus canker from field samples. Phytopathology 94, 61–68.

McClean, A.P.D., Oberholzer, P.C.J., 1965. *Citrus* psylla, a vector of the greening disease of sweet orange. S. Afr. J. Agric. Sci. 8, 297–298.

Merfa, M.V., Pérez-López, E., Naranjo, E., Jain, M., Gabriel, D.W., De La Fuente, L., 2019. Progress and obstacles in culturing '*Candidatus* Liberibacter asiaticus', the bacterium associated with Huanglongbing. Phytopathology 109, 1092–1101.

Miranda, M.P., Yamamoto, P.T., Garcia, R.B., Lopes, J.P.A., Lopes, J.R.S., 2016. Thiamethoxam and imidacloprid drench applications on sweet orange nursery trees disrupt the feeding and settling behavior of *Diaphorina citri* (Hemiptera: Liviidae). Pest Manag. Sci. 1–9.

Miranda, M.P., Santos, F.L., Bassanezi, R.B., Montesino, L.H., Barbosa, J.C., Sétamou, M., 2017. Monitoring methods for *Diaphorina citri* Kuwayama (Hemiptera: Liviidae) on citrus groves with different insecticide application programmes. J. Appl. Entomol. 142, 1–8.

Munoz-Bodnar, A., Santillana, G., Mavrodieva, V., Liu, Z., Nakhla, M., Gabriel, D.W., 2017. Complete genome sequences of three *Xanthomonas citri* strains from Texas. Genome Announc. 5, e00609–e00617. https://doi.org/10.1128/genomeA.00609-17.

Munyaneza, J.E., Crosslin, J.M., Buchman, J.L., 2009. Seasonal occurrence and abundance of the potato psyllid, *Bactericera cockerelli*, in south Central Washington. Am. J. Pot. Res. 86, 513–518.

Nakabachi, A., Nikoh, N., Oshima, K., Inoue, H., Ohkuma, M., Hongoh, Y., Miyagishima, S.Y., Hattori, M., Fukatsu, T., 2013. Horizontal gene acquisition of Liberibacter plant pathogens from a bacteriome-confined endosymbiont of their psyllid vector. PLoS One 8, e82612.

Namekata, T., 1971. Comparative studies between *Xanthomonas citri* (Hasse) Dow., pathogen of citrus canker and *Xanthomonas citri* (Hasse) Dow N.F. f.sp. aurantifolia, pathogen of "Calego lime canker". In: Boletim de Divulgacao, Escola Superior de Agricultura 'Luiz de Queiroz', Universidade de Sao Paulo. vol. 18. pp. 102–104. 242–244.

Nelson, W.R., Munyaneza, J.E., McCue, K.F., Bové, J.M., 2013. The Pangean origin of *Candidatus* Liberibacter species. J. Plant Pathol. 95 (3), 455–461.

Nunney, L., Ortiz, B., Russell, S.A., Ruiz Sánchez, R., Stouthamer, R., 2014. The complex biogeography of the plant pathogen *Xylella fastidiosa*: genetic evidence of introductions and subspecific introgression in Central America. PLoS One 9, e112463.

Okuda, M., Matsumoto, M., Tanaka, Y., Subandiyah, S., Iwanami, T., 2005. Characterization of the tufB-secE-nusG-rplKAJL-rpoB gene cluster of the citrus greening organism and detection by loop-mediated isothermal amplification. Plant Dis. 89, 705–711.

Pérez-Otero, R., Mansilla, J.P., del Estal, P., 2015. Detección de la psila africana de los cítricos, Trioza erytreae (Del Guercio, 1918) (Hemiptera:Psylloidca: Triozidae), en la península Ibérica. Arquivos Entomolóxicos 13, 119–122.

Queiróz-Voltan, R.B., Paradela, F.O., 1999. Caracterização de estruturas anatômicas de citros infectados com *Xylella fastidiosa*. Laranja 20, 55–76.

Raddadi, N., Gonella, E., Camerota, C., Pizzinat, A., Tedeschi, R., Crotti, E., Mandrioli, M., Bianco, P.A., Daffonchio, D., Alma, A., 2011. '*Candidatus* Liberibacter europaeus' sp. nov. that is associated with and transmitted by the psyllid *Cacopsylla pyri* apparently behaves as an endophyte rather than a pathogen. Environ. Microbiol. 13, 414–426.

Reddy, J.D., Reddy, S.L., Hopkins, D.L., Gabriel, D.W., 2007. TolC is required for pathogenicity of *Xylella fastidiosa* in *Vitis vinifera* grapevines. Mol. Plant Microbe Interact. 20, 403–410.

Roberts, R., Pietersen, G., 2017. A novel subspecies of *Candidatus* Liberibacter africanus found on native *Teclea gerrardii* (family: Rutaceae) from South Africa. Antonie Van Leeuwenhoek 110 (3), 437–444.

Roberts, R., Sttenkamp, E.T., Pietersen, G., 2015. Three novel lineages of *Candidatus* Liberibacter africanus associated with native rutaceous hosts of *Trioza erytreae* in South Africa. Int. J. Syst. Evol. Microbiol. 65, 723–731.

Roberts, R., Cook, G., Grout, T.G., Khamis, F., Rwomushana, I., Nderitu, P.W., Seguni, Z., Materu, C.L., Steyn, C., Pietersen, G., Ekesi, S., Roux, H.F., 2017. Resolution of the identity of *Candidatus* Liberibacter species form huanglongbing-affected citrus in East Africa. Plant Dis. 101, 1481–1488.

Rodas, V., 1994. Convivência com a clorose variegada dos citros. Laranja 15, 129–134.

Rodriguez, D.S., 2012. Origin and Development of *Citrus* in Bella Vista, Corrientes. INTA Books, Bella Vista, Corrientes54.

Roper, M.C., Greve, L.C., Warren, J.G., Labavitch, J.M., Kirkpatrick, B.C., 2006. Xylella fastidiosa requires polygalacturonase for colonization and pathogenicity in *Vitis vinifera* grapevines. Mol. Plant Microbe Interact. 20, 411–419.

Rossetti, V., Garnier, M., Bové, J.M., Beretta, M.J.G., Teixeira, A.R.R., Quaggio, J.A., De Negri, J.D., 1990. Présence de bactéries dans le xylème d'orangers atteints de chlorose variégé, une nouvelle maladie des agrumes au Bresil. C. R. Acad. Sci. Ser. III 30, 345–349.

Rybak, M.A., Minsavage, G.V., Stall, R.E., Jones, J.B., 2009. Identification of *Xanthomonas citri* ssp. *citri* host specificity genes in a heterologous expression host. Mol. Plant Pathol. 2009 (10), 249–262.

Schaad, N.W., Postnikova, E., Lacy, G., Sechler, A., Agarkova, I.V., Stromberg, P.E., Stromberg, V.K., Vidaver, A.M., 2006. Emended classification of xanthomonad pathogens on citrus. Papers in Plant Pathology. Syst. Appl. Microbiol. 29, 690–695.

Schubert, T.S., Rizvi, S.A., Sun, X., Gottwald, T.R., Graham, J.H., Dixon, W.N., 2001. Meeting the challenge of eradicating citrus canker in Florida—again. Plant Dis. 85, 340–356.

Schwarz, R.E., van Vuuren, S.P., 1970. Centrifugal extraction of phenolic markers for indexing citrus greening and avocado sun-blotch diseases. Phytophylactica 2, 65–68.

Serizawa, S., Inoue, K., 1974. Studies on citrus canker, *Xanthomonas citri*. III. The influence of wind on the infection of citrus canker. In: Bull. Shizuoka Prefect. Citrus Exp. Stn. Komagoe Shimizu City, Japan. vol. 11, pp. 54–67.

Serizawa, S., Inoue, K., Goto, M., 1969. Studies on citrus canker I. Dispersal of the citrus canker organism. In: Bull. Schizuoka Pref. Citrus Exp. Stn. vol. 8, pp. 81–85.

Simpson, A.J., Reinach, F.C., Arruda, P., Abreu, F.A., Acencio, M., et al., 2000. The genome sequence of the plant pathogen *Xylella fastidiosa*. The *Xylella fastidiosa* consortium of the organization for nucleotide sequencing and analysis. Nature 406, 151–159.

Siverio, F., Marco-Noales, E., Bertolini, E., Teresani, G.R., Peñalver, J., Mansilla, P., Aguin, O., Pérez-Otero, R., Abelleira, A., Guerra-Garcia, J.A., Hernández, E., Cambra, M., López, M.M., 2017. Survey of huanglongbing associated with *Candidatus* Liberibacter species in Spain: analyses of citrus plants and *Trioza erytreae*. Phytopathol. Mediterr. 56 (1), 98–110.

Stall, R.E., Seymour, C.P., 1983. Canker, a threat to citrus in the Gulf-Coast States. Plant Dis. 67, 581–585.

Su, H.J., 2008. Research and health management of citrus Huanglongbing in Taiwan. In: Proceedings of the International Research Conference on Huanglongbing, Orlando, pp. 57–91.

Swarup, S., Defeyter, R., Brlansky, R.H., Gabriel, D.W., 1991. A pathogenicity locus from *Xanthomonas citri* enables strains from several pathovars of *Xanthomonas campestris* to elicit canker like lesions on citrus. Phytopathology 81, 802–809.

Swarup, S., Yang, Y., Kingsley, M.T., Gabriel, D.W., 1992. A *Xanthomonas citri* pathogenicity gene, *pthA*, pleiotropically encodes gratuitous avirulence on nonhosts. Mol. Plant Microbe Interact. 5, 204–213.

Teixeira, D.C., Danet, J.L., Eveillard, S., Martins, E.C., Jesus Jr., W.C., Yamamoto, P.T., Lopes, S.A., Bassanezi, R.B., Ayres, A.J., Saillard, C., Bové, J.M., 2005a. *Citrus* huanglongbing in São Paulo state, Brazil: PCR detection of the '*Candidatus*' Liberibacter species associated with the disease. Mol. Cell. Probes 19, 173–179.

Teixeira, D.C., Saillard, C., Eveillard, S., Danet, J.L., da Costa, P.I., Ayres, A.J., Bové, J., 2005b. "*Candidatus* Liberibacter americanus", associated with citrus huanglongbing (greening disease) in Sao Paulo State, Brazil. Int. J. Syst. Evol. Microbiol. 55, 1857–1862.

Teixeira, D.C., Saillard, C., Couture, C., Martins, E., Wulff, N.A., Eveillard-Jagoueix, S., Yamamoto, P.T., Ayres, A.J., Bové, J.M., 2008a. Distribution and quantification of *Candidatus* Liberibacter americanus, agent of huanglongbing disease of citrus in Sao Paulo State, Brasil, in leaves of an affected sweet orange tree as determined by PCR. Mol. Cell. Probes 22, 139–150.

Teixeira, D.C., Wulff, N.A., Martins, E.C., Kitajima, E.W., Bassanezi, R., Ayres, A.J., Eveillard, S., Saillard, C., Bové, J.M., 2008b. A phytoplasma closely related to the pigeon pea witches'-broom phytoplasma (16Sr IX) is associated with citrus huanglongbing symptoms in the state of Sao Paulo, Brazil. Phytopathology 98 (9), 977–984.

Teresani, G.R., Bertolini, E., Alfaro-Fernández, A., Martínez, C., Tanaka, F.A.O., Kitajima, E.W., Roselló, M., Sanjuán, S., Ferrándiz, J.C., López, M.M., Cambra, M., Font, M.I., 2014. Association of '*Candidatus* Liberibacter solanacearum' with a vegetative disorder of celery in Spain and development of a real-time PCR method for its detection. Phytopathology 104, 804–811.

Timmer, L.W., Zitko, S.E., Gottwald, T.R., 1996. Population dynamics of *Xanthomonas campestris* pv. *citri* on symptomatic and asymptomatic citrus leaves under various environmental conditions. Proc. Int. Soc. Citricult. 1, 448–451.

Trivedi, P., Sagaram, U.S., Kim, J.-S., Brlansky, R.H., Rogers, M.E., Stelinski, L.L., Oswalt, C., Wang, N., 2009. Quantification of viable *Candidatus* Liberibacter asiaticus in hosts using quantitative PCR with the aid of ethidium monoazide (EMA). Eur. J. Plant Pathol. 124, 553–563.

Vernière, C., Hartung, J.S., Pruvost, O.P., Civerolo, E.L., Alvarez, A.M., Maestri, P., Luisetti, J., 1998. Characterization of phenotypically distinct strains of Xanthomonas *axonopodis* pv. *citri* from Southwest Asia. Eur. J. Plant Pathol. 104, 477. https://doi.org/10.1023/A:1008676508688.

Villechanoux, S., Garnier, M., Renaudin, J., Bove, J.M., 1992. Detection of several strains of the bacterium-like organism of citrus greening disease by DNA probes. Curr. Microbiol. 24, 89–95.

Wulff, N.A., Zhang, S., Setubal, J.C., Almeida, N.F., Martins, E.C., Harakava, R., Kumar, D., Rangel, L.T., Foissac, X., Bove, J., Gabriel, D.W., 2014. The complete genome sequence of *Candidatus* Liberibacter americanus, associated with citrus Huanglongbing. Mol. Plant Microbe Interact. 27, 163–176.

Wulff, N.A., Teixeira, D.C., Martins, E.C., Toloy, R.S., Bianco, L.F., Colletti, D.A.B., Kitajima, E.W., Bové, J.M., 2015. Sunn hemp, a major source-plant of the phytoplasma associated with huanglongbing symptoms of sweet orange in São Paulo State, Brazil. J. Citrus Pathol. 2 (1), 26956.

Yamamoto, P.T., 2008. Controle de insetos vetores de bactérias causadoras em doenças dos citros. In: Manejo integrado de pragas dos citros. Fundo de Defesa da Citricultura (Fundecitrus), Araraquara, Brazil, pp. 237–260.

Yamamoto, P.T., Gravena, S., 2000. Espécies e abundância de cigarrinhas e psilídeos (Homoptera) em pomares cítricos. An. Soc. Entomol. Brasil 29, 169–176.

Zhang, S., Flores-Cruz, Z., Zhou, L., Kang, B.H., Fleites, L., Gooch, M.D., Wulff, N.A., Davis, M.J., Duan, Y., Gabriel, D.W., 2011. *Ca.* Liberibacter asiaticus carries an excision plasmid prophage and a chromosomally integrated prophage that becomes lytic in plant infections. Mol. Plant-Microbe Interact. 24, 458–468.

Zhang, S., Chakrabarty, P.K., Fleites, L.A., Rayside, P.A., Hopkins, D.L., Gabriel, D.W., 2015. Three new Pierce's disease pathogenicity effectors identified using *Xylella fastidiosa* biocontrol strain EB92-1. PLoS One 10 (7), e0133796.

Zou, H., Gowda, S., Zhou, L., Hajeri, S., Chen, G., Duan, Y., 2012. The destructive citrus pathogen, '*Candidatus* Liberibacter asiaticus' encodes a functional flagellin characteristic of a pathogen-associated molecular pattern. PLoS One 7 (9), e46447. https://doi.org/10.1371/journal.pone.004644.

Further reading

Chang, C.J., Garnier, M., Zreik, L., Rossetti, V., Bové, J.M., 1993. *Citrus* variegated chlorosis: cultivation of the causal bacterium and experimental reproduction of the disease. In: Proc. Conf. Int. Org. Citrus Virol. 12th. IOCV, Riverside, CA, pp. 294–300.

Chapter 19

Citrus viruses and viroids

Changyong Zhou[a], John V. da Graça[b], Juliana Freitas-Astúa[c], Georgios Vidalakis[d], Nuria Duran-Vila[e], Irene Lavagi[d]

[a]*Southwest University, Chongqing, People's Republic of China,* [b]*Citrus Center, Texas A&M University-Kingsville, Weslaco, TX, United States,* [c]*Embrapa Cassava and Fruits, Cruz das Almas, Brazil,* [d]*University of California, Riverside, CA, United States,* [e]*Centro de Protección Vegetal y Biotecnología, Valencian Institute of Agrarian Research (IVIA), Valencia, Spain*

19.1 Introduction

Citrus cultivation dates back to at least 4000 years in China (Webber, 1967), and has since spread around the globe to all areas with suitable tropical or subtropical environments. It is therefore not surprising that citrus has acquired a significant number of pathogens, including many virus and virus-like pathogens, along the way. Some may have coevolved with citrus, while others have infected them from alternate hosts as citrus cultivation has increased, as a monoculture. Here we will briefly describe ca. 30 citrus virus and viroid diseases and briefly describe graft-transmissible diseases of unknown etiology, updating a review by Bové et al. (2010). Graft-transmissible diseases associated with prokaryotes are not covered in this chapter.

In 1913, a paper was published by Trabut describing an infectious chlorosis of citrus which was transmissible by grafting. Bové and Garnier (2002) regarded this as the first report of a citrus virus-like disease, but the symptoms do not match with any other known disease, thus its exact identity remains a mystery. The description of graft transmission of psorosis by Fawcett (Fawcett, 1934) in 1934 was the first report of an identifiable graft-transmissible and possible viral citrus disease. A number of other citrus diseases were shown to be graft- or arthropod-transmitted in the subsequent years, including tristeza, exocortis, leprosis, and cachexia-xyloporosis. Consequently, an international conference on these diseases was held in California in 1957. Although at the time none of the etiological agents had been identified and characterized, because of their modes of transmission, it was assumed that they were all virus diseases. The International Organization of Citrus Virologists (IOCV) was formed and the proceedings of the conference were published (Wallace, 1959).

Since then, many of the causal agents of those diseases have been identified, most of which are viruses, while others have been shown to be viroids. The first citrus viroid associated with the exocortis disease was discovered in 1972 (Semancik and Weathers, 1972; Fawcett and Klotz, 1948). The molecular aspects of replication, movement, pathogenesis, and taxonomy of viroids have recently been reviewed (Flores et al., 2017a,b; Pallás and Gómez, 2017; Di Serio et al., 2017). Additional recommended information on viroids can be found in the original publications on the viroids discovery (Diener, 1971), viroid RNA structural domains and evolutionary implications (Keese and Symons, 1985), and the concept of "viroids" as "symbionts akin to viruses" (Altenburg, 1946).

19.2 *Citrus tristeza virus* (*Closterovirus, Closteroviridae*)

Tristeza caused by *Citrus tristeza virus* (CTV) is one of the most devastating citrus diseases, with worldwide distribution (Bar-Joseph et al., 1989). Tristeza ('sadness' in Spanish and Portuguese) was the name given to the outbreak of quick decline and death of over 30 million citrus trees on sour orange rootstock in Argentina (1931) and Brazil (1937) (Wallace, 1978). Other epidemics occurred in California (1939), Florida (1951), Spain (1956), Israel (1970), and Venezuela (1980) (Roistacher, 1993), which led to the rebuilding of many citrus industries using tristeza-tolerant rootstocks. In addition to quick decline, two other syndromes are caused by CTV, stem-pitting (SP) in lime, grapefruit, pummelo, and some sweet

The Genus Citrus. https://doi.org/10.1016/B978-0-12-812163-4.00019-X

FIG. 19.1 Stem pitting and dropped fruits induced by *Citrus tristeza virus* in Star Ruby grapefruit.

orange cultivars (Fig. 19.1), and seedling yellows (SY) in sour orange and lemon used for biological indexing (Bové et al., 2010). In summary, decline-causing isolates of CTV cause trees on sour orange to decline and die, sometimes suddenly with dead leaves remaining, while in others a slow decline ensues. SP results from the death of some phloem elements; affected trees are unthrifty and produce small fruit. SY occurs in inoculated sour orange seedlings, which can recover.

Many CTV reviews have been published previously (Bové et al., 2010; Bar-Joseph et al., 1989; da Graça and van Vuuren, 2010). CTV has a long flexuous particle of 2000 × 11 nm, which contains a 19.3-kb positive sense single-stranded RNA (+ssRNA) genome encapsidated by two coat proteins (CP). The viral genome is organized into 12 open reading frames (ORFs) flanked by two untranslated regions (UTRs). Five ORFs are involved in virion assembly and virus movement, and two in replication. Five ORFs are unique to CTV (Dawson et al., 2013).

It is experimentally graft-transmitted, and naturally transmitted in a semi-persistent manner by several species of aphids; *Toxoptera citricida* is the most efficient, with *Aphis gossypii*, *A. spiraecola*, and *T. aurantii* being far less efficient vectors. Viruliferous aphids are able to acquire CTV within 30 min and transmit it for at least 24 h, infectivity being lost within 48 h after acquisition. The ratio of aphids carrying CTV in the field ranges from 19% to 27%. The same scale increase of CTV incidence in field was found to be 2–4 yr by *T. citricida*, but 8–15 yr by *A. gossypii* (Gottwald et al., 1996, 1997).

Poncirus trifoliata, *Atalantia buxifolia*, *Swinglea glutinosa*, *A. ceylanica*, and *Fortunella crassifolia* are generally resistant to CTV (Mestre and Asins, 1997), although CTV can replicate in protoplasts of some of the above species (Albiach-Marti et al., 2004), and some CTV genotypes can replicate in *P. trifoliata* (Dawson and Mooney, 2000; Harper et al., 2014; Yokomi et al., 2017; Zhou et al., 2017a).

19.3 *Satsuma dwarf virus (Sadwavirus, Secoviridae)*

Satsuma dwarf was noted in the 1930s, and was confirmed to be caused by *Satsuma dwarf virus* (SDV) in Japan (Yamada and Sawamura, 1952). It has since been reported from Turkey, Korea, and China (Azeri, 1973; Moon, 1986; Zhou et al., 1990). Iwanami (2010) reviewed SDV in 2010. In summary, affected Satsuma mandarin trees have shortened internodes and are severely stunted with boat- or spoon-shaped leaves, significant yield losses, and malformed fruits. The disease spreads slowly in orchards in concentric circles, and is suspected to be soil-borne by *Olpidium* or nematode. A symptomless host, China laurestine (*Viburnum odoratissimum*), interplanted as a windbreak can facilitate SDV spread. The virus has a wide host range including many woody and annual species. The SDV virion is an icosahedral particle of diameter 26 nm, which contains a bipartite +ssRNA genome encapsidated by two CPs of 42 and 22 kDa. Three serologically related viruses, *Citrus mosaic virus*, Navel orange infectious mottling virus, and Natsudaidai dwarf virus share over 75% amino acid sequence identity with SDV.

19.4 Citrus leprosis viruses

***Citrus leprosis virus C* and *Citrus leprosis virus C2* (*Cilevirus, Kitaviridae*)**

***Citrus chlorotic spot dichorhavirus*, *Citrus leprosis N dichorhavirus* and *Orchid fleck dichorhavirus* (*Dichorhavirus, Rhabdoviridae*)**

Citrus leprosis (CL) was first observed in Florida, USA, around 1860, and in the south of South America in 1920–30s. It has recently spread throughout South and Central America, to Mexico. CL is an atypical virus disease, characterized by the localized chlorotic and necrotic lesions present on both sides of the leaves, stems, and fruits. Small chlorotic, circular spots can turn into large lesions and become darker and depressed, often with necrotic centers and chlorotic halos. Severe CL causes leaf and fruit drop, decreases plant lifespan, and may lead to the death of young susceptible plants (Bastianel et al., 2010). It is not easily transmissible by grafting, with no systemic spread, and can be caused by at least four viruses belonging to two distinct genera.

Studies on CL revealed two different types of the disease: the cytoplasmic type (CL-C) and the nuclear type (CL-N) (Bastianel et al., 2010). The cileviruses *Citrus leprosis virus C* (CiLV-C) and *Citrus leprosis virus C2* (CiLV-C2) are responsible for CL-C. These viruses have bipartite +ssRNA genomes (RNA1≈8.7 kb and RNA2≈5.0 kb) encoding six ORF. Their enveloped short bacilliform particles of ~110 × 55 nm accumulate in the endoplasmic reticulum and form dense, vacuolated viroplasm-like structures in the cytoplasm of infected cells. CiLV-C has significant intraspecific genome diversity and is the main form responsible for losses from Mexico to Argentina (Ramos-Gonzalez et al., 2016). This virus is probably the causal agent of leprosis reported in South America since early 20th century (Bastianel et al., 2010). CiLV-C2 was reported in Colombia in 2013 and is now prevalent in that country (Roy et al., 2015a). Both viruses cause similar syndromes, infecting primarily sweet orange, but also some mandarins and grapefruits (Fig. 19.2).

CiLV-C and CiLV-C2 are transmitted in a persistent manner by *Brevipalpus yothersi* (Ramos-Gonzalez et al., 2016), but while Roy et al. (2015a) stated that they replicate in their vectors, Tassi et al. (2017) suggested that the CiLV-C-mite interaction is of the persistent circulative type. All active stages of the mite vector can acquire and transmit CiLV-C, which is not transovarially transmitted.

The CL-N form can be caused by three distinct viruses belonging to *Dichorhavirus*, *Rhabdoviridae*, order *Mononegavirales*. The CL-N virus reported infecting several citrus species in Mexico and Colombia was initially named as CL virus nuclear type and citrus necrotic spot virus (Cruz-Jaramillo et al., 2014; Roy et al., 2015b), but it was reclassified as a citrus strain of *Orchid fleck virus* (OFV-citrus) (Afonso et al., 2016). OFV-citrus has a bipartite, -ssRNA genome, and replicates in the nucleus of infected cells. RNA1 of 6.4 kb encodes five proteins, while RNA2 codes only for RNA polymerase. This virus is likely transmitted in a persistent propagative manner by *B. californicus* (García-Escamilla et al., 2017). Another CL-N virus named *Citrus leprosis virus N* (CiLV-N) was recently characterized in sweet oranges in Brazil, and is a species of *Dichorhavirus*. *Citrus chlorotic spot virus C* (CiCSV) is a new dichorhavirus infecting citrus in Northeast Brazil (Chabi-Jesus et al., 2018). It shares 53%–60% of nucleotide sequence identity with OFV-citrus and is transmitted by *B. phoenicis sensu stricto*, probably in a persistent propagative interaction (Ramos-González et al., 2017). The CL-N viruses induce similar localized, nonsystemic symptoms in citrus. However, while the yellow halos are often brighter in leaves, symptoms in fruits and stems are undistinguishable from those caused by the cileviruses (Fig. 19.3).

A putative virus, tentatively named CiLV-N0, was partially sequenced from herbarium samples from Florida, collected in 1948, but it does not occur in nature any longer (Hartung et al., 2015). A CL-N virus reported in Panama has not been yet characterized (Dominguez et al., 2001).

FIG. 19.2 Bright yellow halo caused by *Citrus leprosis virus C* on sweet orange fruits.

FIG. 19.3 Distinct bright yellow halo caused by *Citrus leprosis virus N* on sweet orange leaves.

19.5 *Citrus psorosis virus* (*Ophiovirus, Aspiviridae*)

Citrus psorosis, recorded in Florida as scaly bark in 1896 (Swingle and Webber, 1896), was confirmed a graft-transmissible disease in 1934 (Fawcett, 1934). Its worldwide occurrence led to the earliest eradication programs and virus-free budwood programs (Doidge, 1926; Hiltebrand, 1959). It remains a common and serious problem in Argentina (De Francisco et al., 2015).

Moreno et al. (2015) published an extensive psorosis review. In summary, the virus induces a range of symptoms depending on the strain and the environment. In field trees, the most obvious symptom is the scaling of the bark of trunks and limbs, with or without gum formation and wood discoloration below the bark. While most citrus varieties are susceptible, bark scaling symptoms are most common on sweet orange and grapefruit, and occasionally mandarin. Where the scaling is limited and spreads slowly, the disease is referred to as psorosis A; where the scaling is rampant, it is referred to as psorosis B. Young leaves of the spring flush may develop transient chlorotic flecking, and some shoots die suddenly (shock reaction). Psorosis B is also associated with chlorotic blotches on older leaves, and depressed spots or rings on the fruit. In the greenhouse, inoculated seedlings often show the shock reaction initially, and then develop leaf flecking.

Citrus psorosis virus (CPsV) has a CP of 48 kDa and the virus particles are non-enveloped, naked filamentous nucleocapsids existing as either circular open structures or collapsed pseudo-linear forms. The genome is -ssRNA consisting of three segments (García et al., 2017).

Psorosis is readily transmitted by grafting, and can be experimentally transmitted mechanically to *Gomphrena globosa* and *Chenopodium quinoa*, and back to citrus. There is evidence from field observations that natural spread occurs (Timmer and Garnsey, 1980; Beñatena and Portillo, 1984). Observations in Argentina suggest that an aerial vector such as aphids may be involved (Beñatena and Portillo, 1984), while in Texas evidence of the presence of CPsV RNA associated with the fungus *Olpidium* has been published (Palle et al., 2005). Four other ophioviruses, including *Mirafiori lettuce big-vein virus*, are transmitted by *O. brassicae* (García et al., 2017), but no transmission of CPsV has been confirmed thus far.

19.6 Citrus tatter leaf virus (*Apple stem grooving virus*) (*Capillovirus, Betaflexiviridae*)

Tatter leaf is an example of virus detection prior to disease evidence. The graft-inoculated symptoms of "tatter leaf" in *Citrus excelsa*, and of stunting in Troyer citrange led to suspicion of two virus diseases (Wallace and Drake, 1953, 1968). The two syndromes were experimentally confirmed to be caused by a single agent Citrus tatter leaf virus (CTLV), also causing scion swellings and bud union crease in field Satsuma trees of trifoliate orange in Japan (Roistacher et al., 2000; Miyakawa and Tsuji, 1988). As a latent virus for many citrus species, it induces chlorotic leaf symptoms in Mexican lime,

C. excelsa, trifoliate orange and its hybrids. *C. excelsa* can recover and the virus becomes latent. In trifoliate orange and hybrids, the chlorotic spots are associated with leaf deformities, and the stems have a zigzag growth pattern. The virus can cause economic losses in mandarins on trifoliate orange (Takahara et al., 1988; Zhang et al., 1988).

CTLV has rod-shaped virions of 600–650 × 13 nm, with a CP of 27 kDa and a +ssRNA species of 6.5 kb, and is considered to be an isolate of *Apple stem grooving virus* (ASGV) (Magome et al., 1997). Tataneni et al. (2009) did note, however, that there are differences between CTLV and ASGV in ORF1, variable regions I and II. CTLV is readily transmitted mechanically (Roistacher et al., 1980). Seed transmission has been demonstrated in *Chenopodium* and *Lilium* (Inouye et al., 1979), and there is some evidence of a low rate of seed transmission in lemons (Tanner et al., 2011).

19.7 *Citrus variegation virus/Citrus leaf rugose virus* (*Ilarvirus*, *Bromoviridae*)

Citrus infectious variegation was first recognized as a virus-like disease in 1939 (Fawcett and Klotz, 1939). Because of the association with leaf fleck symptoms, it was considered to be part of the "psorosis-complex." Crinkly leaf, which had been noted in 1936 (Fawcett, 1936), was soon regarded as the same (Fawcett and Bitancourt, 1943). Leaf rugose reported in Florida (Garnsey, 1975) was also found to be caused by a related pathogen. *Citrus variegation virus* (CVV) was the first citrus virus to be mechanically transmitted (Grant and Smith, 1960).

These viruses infect a wide range of citrus hosts, but symptoms are generally mild (Garnsey, 2000). CVV induces more obvious symptoms in lemons, sour orange, grapefruit, and citron; *Citrus leaf rugose virus* (CLRV) induces milder symptoms in the same hosts but causes noticeable leaf flecking in lemons. In addition to the United States, CVV has been reported from several Mediterranean and Latin American countries (Moreira et al., 2011).

All three syndromes are caused by members of genus *Ilarvirus*, of the subtype 2, family *Bromoviridae*. The virions are 26–32 nm in diameter, and the RNA genome is divided. RNA1 (3.4 kb) contains one ORF, RNA 2 is 3 kb in size with one ORF, and RNA3 codes for the CP. CVV is closely related to other ilarviruses than to CLRV. Crinkly leaf virus is considered to be a CVV strain (Ge and Scott, 1994; Scott and Ge, 1995). Lovisolo (Lovisolo, 1993) suggested that ilarviruses may have infected citrus by chance via pollen or thrips, even though there is no evidence of rapid field spread.

19.8 *Citrus leaf blotch virus* (*Citrivirus*, *Betaflexiviridae*)

Roistacher and Blue (Roistacher and Blue, 1968) reported the detection of a graft-transmissible agent from symptomless Cleopatra mandarin which induced chlorotic mottling in 'Dweet' tangor, and it was initially named Dweet mottle virus (DMV). Navarro et al. (1984) reported a new graft-transmissible agent associated with bud union crease of kumquat on Troyer citrange in Corsica (France), which induced vein clearing in several citrus varieties, including 'Dweet' tangor. Stem pitting was found in 'Etrog' citron and flexuous virions of 900 × 14 nm were observed, with a +ssRNA genome of 8.7 kb and a CP of 41 kDa (Galipienso et al., 2001; Vives et al., 2001, 2002). The virus was then named *Citrus leaf blotch virus* (CLBV) and it was proven to have 96%–98% homology with DMV (Vives et al., 2005).

Plants on Troyer citrange inoculated with an infectious clone of CLBV did not induce a bud union crease, indicating that the virus was not the cause of this symptom in the original Cleopatra mandarin tree, even though the bud union crease was transmissible through grafting from this tree (Vives et al., 2008). Thus, no serious field disease of citrus has been associated with CLBV, so far. It has also been detected in sweet cherry trees with mosaic symptoms (Wang et al., 2016), and peony with stunted and gnarling in the United States (Gress et al., 2017). A low rate of seed transmission and invasion of meristematic tissues in citrus has been reported (Guerri et al., 2004; Agüero et al., 2013). In addition to the United States and France, it has been detected in citrus in Japan, Australia, Cuba, and New Zealand (Vives et al., 2005).

19.9 *Citrus vein enation virus* (*Enamovirus*, *Luteoviridae*)

Citrus vein enations on the abaxial leaf surface on rough lemon and Mexican lime were first described in California, USA (Wallace and Drake, 1953). Fraser (Fraser, 1959) described a graft-transmissible disorder called woody gall on rough lemon, and suggested that both galls and vein enation were caused by the same pathogen, which was confirmed by Wallace and Drake (Wallace and Drake, 1961). Virtually all commercial citrus varieties are hosts to this virus, but conspicuous enations occur only on Mexican lime and sour orange, while smaller and fewer enations develop on several other species. Woody galls appear to be limited to rough lemon, Volkamer lemon, and Mexican lime (Klotz et al., 1982), and symptoms are only observed in more temperate climates (Wallace, 1981).

Citrus vein enation virus (CVEV) is transmitted persistently by *Myzus persicae*, *T. citricida*, and *A. gossypii* (Maharaj and da Graça, 1989; Hermoso de Mendoza et al., 1993). Transmission electron microscopy of infected plant and aphid tissue and purified extracts revealed the presence of 30 nm diameter particles (Maharaj and da Graça, 1988), and it was suggested that it may be a luteovirus (da Graça and Maharaj, 1991; Kanrar et al., 2011). Next-generation sequencing (NGS) investigations determined that vein enation is associated with an enamovirus with a +ssRNA genome of 6 kb and five ORFs (Vives et al., 2013).

19.10 *Citrus yellow mosaic virus* (*Badnavirus, Caulimoviridae*)

Citrus yellow mosaic caused by citrus yellow mosaic virus (CYMV) was first reported in India (Ahlawat et al., 1993). It causes serious losses in pummelo and sweet orange with unusual foliar symptoms, which begin as yellow flecks or dots on young leaves, later developing into mottling and bright yellow flecking along veins, sometimes merging to create rings (Ahlawat et al., 1996). CYMV has a DNA genome of 7.6 kb with six putative ORFs, and was most closely related to *Cacao swollen shoot virus* (Huang and Hartung, 2001). CYMV is vectored by the citrus mealybug, *Planococcus citri* (Garnsey et al., 1998).

19.11 *Indian citrus ringspot virus* (*Mandarivirus, Alphaflexivirdae*)

Indian citrus ringspot caused by *Indian citrus ringspot virus* (ICRSV) was described in Kinnow mandarin, causing psorosis-like leaf flecking, but without bark scaling (Ahlawat, 1989). ICRSV is a filamentous rod-shaped virus with a CP of 34 kDa and RNA genome of 7.5 kb with some genetic variation (Rustici et al., 2000, 2002; Hoa and Ahlawat, 2004).

19.12 *Citrus yellow vein clearing virus* (*Mandarivirus, Alphaflexivirdae*)

Citrus yellow vein clearing, initially described in lemon and sour orange with symptoms of vein clearing, leaf crinkling, and water-soaked veins (Fig. 19.4), was first reported from Pakistan in 1988, then in Turkey, China, and Iran (Catara et al., 1988; Önelge, 2003; Zhou and Li, 2010; Bari Hashmian and Aghajanzadeh, 2017). The disease is spreading rapidly in Pakistan and Turkey, and is also threatening the lemon- and lime-producing areas of China and Iran. It is caused by *Citrus yellow vein clearing virus* (CYVCV) that has a virion of 670–700 nm in length (Grimaldi and Catara, 1996) and is closely related to ICRSV, but is serologically distinct and genetically stable while it is latent in many citrus species (Loconsole et al., 2012; Adams et al., 2014; Song et al., 2015; Zhou et al., 2016). CYVCV can be mechanically transmitted and is aphid-borne to beans and lemons (Iftikhar et al., 2010; Önelge et al., 2011; Zhang et al., 2018), and also whitefly-borne to lemons (Zhang et al., 2019).

19.13 *Citrus chlorotic dwarf-associated virus* (*Geminiviridae*)

Citrus chlorotic dwarf (CCD) was first reported from Turkey (Çinar et al., 1994) as a very serious citrus disease (Loconsole et al., 2012). Disease symptoms include leaf crinkling, curling, inverted cupping, reduced size, and deformation typically accompanied by chlorotic patterns and variegation. Shortened internodes result in bushy vegetation and stunted appearance of trees. CCD was found to be transmissible by grafting to several citrus cultivars, with sweet orange showing some tolerance, and vectored by the bayberry whitefly (*Parabemisia myricae*) (Loconsole et al., 2012; Kersting et al., 1996). CCD has been reported in China (Guo et al., 2015) with low incidence (Zhou et al., 2017b). Loconsole et al. (2012) identified and characterized the *Citrus chlorotic dwarf-associated virus* (CCDaV) as a new +ssDNA geminivirus associated with CCD.

FIG. 19.4 Vein clearing, leaf crinkling of Eureka lemon caused by *Citrus yellow vein clearing virus*.

19.14 *Citrus* concave gum-associated virus and *Citrus* virus A (tentative *Coguvirus, Bunyavirales*)

Concave gum was first reported in California (Fawcett and Lee, 1926). Affected trees have broad concavities on the limb or truck, with gum oozing out of any cracks which may appear. The removal of the bark reveals gum. Young leaves of concave gum-infected trees developed mosaic-like symptoms, resembling psorosis, and oak-leaf patterns led to the conclusion that the two diseases were related (Fawcett, 1939). However, cross-protection studies found the psorosis and concave gum did not protect against each other (Roistacher and Calavan, 1965), and after the identification of the CPsV 48 kDa CP, no such protein could be detected in concave gum-infected plants (da Graça et al., 1991). Recently, Navarro et al. (2018a,b) identified two negative-stranded RNA, bipartite, phlebo-like viruses, associated with concave gum. It has been proposed that these two plant infecting viruses are species of a new genus tentatively named *Coguvirus* in the order *Bunyavirales* that are evolutionary linked but distinct to the phlebo-like viruses that primarily infect arthropods (Navarro et al., 2018b). Confirmation of such viruses as causal agents of the concave gum disease is pending.

19.15 Diseases of unknown etiology

While there has been significant progress in identifying several etiological agents of graft-transmissible diseases of citrus, there are still some diseases with unknown causal agents. Citrus blight was first described in Florida (Fawcett, 1936), and subsequently from Brazil and South Africa (Wutscher et al., 1980). Root graft-transmissions (Timmer et al., 1992) indicated possible virus association. Certain strains of CTV and an idaeovirus found in blight-affected roots were suggested as playing a role (Derrick et al., 1988, 2003), while more recently an endogenous citrus pararetrovirus is being investigated as a possible causal agent (Roy et al., 2014).

Two diseases of unknown etiology first discovered in the Mediterranean region, cristacortis and impietratura, induce oak-leaf patterns similar to those induced by concave gum (Roistacher, 1991). The CPsV CP was not detected in any of the affected plants, supporting the suggestion that these are not part of a psorosis group of viruses (da Graça et al., 1991). Viruses with sequence similarities to Citrus concave gum-associated virus and Citrus virus A have recently been detected in cristacortis and impietratura affected trees, but their causal relationship to the diseases remains unknown (Velázquez et al., 2019).

The cause of graft-transmissible abnormal bud unions (Navarro et al., 1984) also remains unidentified, as do the agents of measles in Florida (Lee et al., 1993), yellow ringspot in Spain (Navas-Castillo and Moreno, 1993), Bahia bark scaling and zonate chlorosis in Brazil (Rossetti et al., 1969; Nickel et al., 2007), and concentric ring blotch in South Africa (Dippenaar, 1958).

Citrus yellow vein has been reported from California, and plants coinfected with vein enation have enhanced symptoms (Weathers, 1960). Kanrar et al. (Kanrar et al., 2011) reported both luteovirus- and umbravirus-like sequences in plants with yellow vein symptoms, and since CVEV has since then been identified as an enamovirus, a type of virus known to be closely associated with umbraviruses, yellow vein may be associated with an umbra-like virus (GenBank JX101610), capable of interacting with the vein enation luteovirus in mix infections.

19.16 Citrus sudden death-associated virus (*Marafivirus, Tymoviridae*)

Citrus sudden death (CSD) was first described and is restricted in Southwestern Minas Gerais and Northern São Paulo, Brazil, in early 2000s. CSD is a graft-transmissible disease, has an aerial vector, and has been responsible for the death of 4 million trees, most of them sweet oranges, but also mandarins, tangors, limes, and sweet lime trees (Yamamoto et al., 2011). Visible symptoms start with a general tree decline, a pale green coloration of the leaves, and defoliation. Root death is also observed, but the most typical symptom is the yellow stained layer that appears in the rootstock bark, near the line of the grafting (Bassanezi et al., 2003). After symptoms appear, the plants die rapidly. The disease only affects plants grafted on lemon and few other rootstock species, but not on trifoliate orange, 'Swingle' citrumelo, mandarin, and sour orange. The general CSD symptoms (with the exception of the yellow stain in the rootstock), epidemiology and anatomical pathology are similar to CTV quick decline, thus one or more strains of CTV was initially suggested as associated with the disease (Bassanezi et al., 2003). Maccheroni et al. (2005) proposed that *Citrus sudden death-associated virus* (CSDaV) (*Marafivirus, Tymoviridae*), often present in symptomatic plants, might be the CSD causal agent. Matsumura et al. (2017) reported association of CSD symptoms with a specific CSDaV genotype and also with a citrus endogenous pararetrovirus (CitPRV).

19.17 Miscellaneous viruses

Some other viruses have been detected in citrus. These include *Tobacco necrosis virus* (Yot-Dauthy et al., 1972) which was isolated from citrus affected by cristacortis, concave gum, and psorosis but not shown to be the causal agent, and *Olive latent virus 1* from CCD-affected trees (Martelli et al., 1996). Both are necroviruses (*Tombusviridae*) and appear to be present at low titers, and likely not causing disease. A cucumovirus was detected in tatter leaf-infected plants (Marcó et al., 1991), and unidentified rod-shaped virus particles in a psorosis-infected source (Levy and Gumpf, 1991).

19.18 Citrus exocortis viroid (*Pospiviroid, Pospiviroidae*)

Citrus exocortis viroid (CEVd) was originally identified by Semancik and Weathers in 1972 from the bioamplification host *Gynura aurantiaca* expressing symptoms of stunting, epinasty, and leaf distortion after inoculation with what was considered to be the exocortis virus (Semancik and Weathers, 1972; Weathers and Greer Jr., 1968; Kapur et al., 1974). CEVd was first sequenced in 1982 by two independent groups (Gross et al., 1982; Visvader et al., 1982) as a 371-nucleotide RNA sequence with a predicted rod-like secondary structure of minimal free energy replicating in the nucleus (Duran-Vila and Semancik, 2003; Duran-Vila, 2017). CEVd has been reported to be graft- and mechanically transmitted in most citrus-growing areas around the world, and a large number of sequence variants has been reported from studies that indicate complex interactions between viroid sequences and viroid biology (Hajeri et al., 2011; Gandía et al., 2005; Bernad et al., 2009; Murcia et al., 2011; Flores et al., 2017c; Barbosa et al., 2005a; Kyriakou, 1992). The natural host range of CEVd includes both rutaceous and non-citrus hosts such as grapevine, tomato, eggplant, turnip, carrot, and broad bean (Duran-Vila and Semancik, 2003; Duran-Vila, 2017). Some non-citrus hosts generate larger CEVd molecules by duplications of the right terminal region of the viroid (Semancik et al., 1994; Fadda et al., 2003; Szychowski et al., 2005). Using tomato as an experimental host, Visvader and Symons (Visvader and Symons, 1985, 1986) proposed that CEVd isolates could be classified into severe "Class A" and mild "Class B," which differed by a minimum of 26 nt, mainly affecting two regions in the pathogenicity and variable domains (Keese and Symons, 1985). CEVd isolates classified into these two classes induce distinct symptoms in *Gynura* (Chaffai et al., 2007) but only subtle differences in commercial rootstock/scion combination with similar overall performance of the infected trees (Vernière et al., 2004, 2006). In Etrog citron, CEVd induces a variety of symptoms ranging from severe stunting, leaf epinasty and rugosity, petiole wrinkle and necrosis, midvein necrosis, and browning of the tip of the leaf blade (Duran-Vila and Semancik, 2003; Duran-Vila, 2017; Roistacher et al., 1977; Duran-Vila et al., 1993). In trifoliate orange, CEVd induces the exocortis disease first described by Fawcett and Klotz (Fawcett and Klotz, 1948) as a bark scaling disorder, which was reported as "scaly butt" in Australia (Benton et al., 1949, 1950) and "Rangpur lime disease" in Brazil (Moreira, 1955, 1959), while Koch's postulates were fulfilled by Vernière et al. (Vernière et al., 2004). In Clementine trees grown on trifoliate orange, CEVd was associated with poor appearance of the canopy and severe stunting resulting in significant yield reduction that became more pronounced over time (Vernière et al., 2004). In Washington Navel sweet orange grown on Carrizo citrange, CEVd did not induce the characteristic bark scaling symptoms and caused a slight reduction of tree growth and yield that was not as significant as reported in the case of trees grafted on trifoliate orange (Murcia et al., 2015). CEVd synergistic effects with other citrus viroids have resulted in enhanced bark scaling or reduced tree growth and yield for commercial citrus trees such as Clementine on trifoliate orange (Vernière et al., 2006), lemon on sour orange (Kyriakou et al., 2005), sweet orange, Clementine and tangor on Carrizo citrange (Bani Hashemian et al., 2009, 2010a). On the other hand, CEVd antagonistic reactions with CBCVd and CDVd in coinfected trifoliate orange have been reported (Vernière et al., 2006) (Fig. 19.5).

19.19 Hop stunt viroid (*Hostuviroid, Pospiviroidae*)

Distinct variants of the Hop stunt viroid (HSVd) (295–302 nt) have been identified in citrus both as causal agents of the cachexia-xyloporosis disease (Semancik et al., 1988a,b) and as genetic elements modifying tree performance without inducing any disease (i.e., TsnRNAs) (Semancik et al., 1997). HSVd has been reported to be graft- and mechanically transmitted in most world citrus-growing areas (Flores et al., 2017c; Barbosa et al., 2005a; Hataya et al., 2017). Its natural host range is probably the largest among the citrus viroids and a very large number of sequence variants from different hosts and recombination events have been reported (Duran-Vila and Semancik, 2003; Hataya et al., 2017; Sano et al., 1986; Kofalvi et al., 1997; Roistacher et al., 1973). In Etrog citron, HSVd induces very mild symptoms of necrotic spots in the stem just below the insertion of the leaf petioles and sometimes petiole wrinkling and leaf tip necrosis (Duran-Vila and Semancik, 2003; Duran-Vila et al., 1993). Therefore, the preferred bioindicator for cachexia indexing is Parson's Special mandarin (PSM) grafted on rough lemon. Typical cachexia symptoms of gumming and wood staining develop within

FIG. 19.5 Bark scaling induced by *Citrus exocortis viroid* in Trifoliate orange rootstock with Navel orange scion.

12 months on PSM just above the bud union (Roistacher et al., 1973). The most common symptoms observed in sensitive citrus such as lime, tangelo, mandarin, Clementine, Satsuma, and kumquat are discoloration and gum impregnation of the bark, which could be easily seen by scrapping or removing the outer bark layers, and pegs on the cambial side of the bark, with the corresponding depressions or pits in the wood. Affected trees are stunted and chlorotic, and may decline and die (Duran-Vila and Semancik, 2003). The identification of specific HSVd variants as causal agents of the cachexia disease included a combination of several biological and molecular studies (Semancik et al., 1988a,b; Sano et al., 1986, 1988a,b; Roistacher et al., 1973; Albanese et al., 1988; Diener et al., 1988). The differences between the HSVd variants that cause the cachexia disease (CVd-IIb, 299 nt and -IIc, 296 nt) and those that do not (CVd-IIa, 302 nt) are very subtle, with only six nucleotides found in the variable domain ("cachexia expression motif"), as determinants of cachexia symptom expression (Reanwarakorn and Semancik, 1998, 1999; Palacio-Bielsa et al., 2004). Specific directed mutagenesis confirmed that this motif plays a major role in inducing cachexia symptoms, and that subtle changes within this motif affect symptom severity and may even suppress symptom expression (Serra et al., 2008a). Vernière et al. (Vernière et al., 2004, 2006) confirmed that only the HSVd variants CVd-IIb and -IIc induced cachexia symptoms and reduced fruit production on Clementine trees growing on trifoliate orange, nevertheless they also induced bark-cracking symptoms. On the other hand, the HSVd variant CVd-IIa induced only bark-cracking symptoms on the trifoliate orange and had no significant effect on the size or fruit production of Clementine scion (Vernière et al., 2004, 2006). Semancik et al. (Semancik et al., 1997) reported that CVd-IIa (syn. TsnRNA-IIa) induced a statistically significant reduction in the size of Valencia orange tree on trifoliate orange accompanied by increased yield per tree. Valencia trees with CVd-IIa consistently out-yielded viroid-free trees by c.15%, even though the canopy volume was reduced by 20% (Semancik et al., 1997). The presence of the non-cachexia variant of HSVd, CVd-IIa in the parent Washington navel orange tree, currently over 140-years old, suggests that this RNA species does not have a deleterious effect on tree health and raises the question of whether CVd-IIa actually contributed to the superior performance of this tree that was selected for introduction into the United States (Riverside, California) in 1873 (Hadidi et al., 2017; Semancik, 2003) (Fig. 19.6).

19.20 Citrus bent leaf viroid (*Apscaviroid, Pospiviroidae*)

Citrus bent leaf viroid (CBLVd) was originally identified by Duran-Vila and colleagues after sequential polyacrylamide gel electrophoresis (sPAGE) analysis of symptomatic tissues from the bioamplification and bioindicator host Etrog citron (Duran-Vila and Semancik, 2003; Duran-Vila et al., 1988). Distinct CBLVd variants include CVd-Ia (327 nt), studied for its properties as a dwarfing RNA (Semancik et al., 1997), CVd-Ib (318 nt), transmitted to avocado by heterologous grafting

FIG. 19.6 Gumming and wood staining induced by the citrus cachexia variants of the *Hop stunt viroid* in Parson's Special mandarin on Carrizo citrange rootstock.

(Ashulin et al., 1991), and CVd-I-LSS (327 nt), reported from Japan, Pakistan, and China (Ito et al., 2000; Wu et al., 2014). CBLVd has been reported to be graft- and mechanically transmitted in most world citrus-growing areas (Flores et al., 2017c; Barbosa et al., 2005a). Its natural host range is restricted to the *Rutaceae*. In Etrog citron, CBLVd induces localized midvein necrosis that results in abrupt downward bending of leaves that alternate with flushes of symptomless leaves (Duran-Vila and Semancik, 2003; Duran-Vila et al., 1993). In trifoliate orange, CBLVd has been associated with bark pitting, mild stunting, and canopy volume reduction while it has antagonistic interactions in mixed infection with CDVd (Vernière et al., 2006; Semancik et al., 1997; Roistacher et al., 1993). CBLVd variant CVd-Ia (syn. TsnRNA-Ia) has been reported as the critical element of the TsnRNA mixture required for the production of dwarfed trees on Carrizo citrange rootstock (Vidalakis et al., 2010).

19.21 Citrus dwarfing viroid (*Apscaviroid, Pospiviroidae*)

Citrus dwarfing viroid (CDVd) was originally identified after sPAGE analysis of symptomatic tissues from the bioamplification and bioindicator host Etrog citron (Duran-Vila and Semancik, 2003; Duran-Vila et al., 1988). Distinct CDVd variants include CVd-IIIa (297 nt) and CVd-IIIb (294 nt). CDVd has been reported to be graft- and mechanically transmitted in most world citrus-growing areas with some sequence variability that appears to be conserved within geographic areas (Flores et al., 2017c; Barbosa et al., 2005a; Owens et al., 2000; Murcia et al., 2009; Vidalakis et al., 2011). The natural host range of CDVd is restricted to rutaceous plants. In Etrog citron, CDVd induces general leaf epinasty characterized by petiole ringing and necrosis and general leaf drooping as a result of petiole bending and midvein necrosis (Duran-Vila and Semancik, 2003; Duran-Vila et al., 1993; Murcia et al., 2009). In trifoliate orange, CDVd has been associated with the peculiar symptom of "finger imprint" that is the result of a specific interaction of the viroid, the host, and irrigation (Roistacher et al., 1993; Vidalakis et al., 2004) and it has been reported to have antagonistic interactions in mixed infection with CBLVd (Vernière et al., 2006). In greenhouse studies, CDVd was associated with a decrease in root growth of Carrizo citrange and Rangpur lime seedlings (Owens et al., 2005) and interaction with CTV that increased viroid titer but no symptom expression in Mexican lime (Serra et al., 2014). More importantly, however, CDVd reduces tree growth and yield on trees grown on trifoliate orange (Vernière et al., 2004, 2006). This property is the basis for the use of the specific CDVd variant CVd-IIIb (syn. TsnRNA-IIIb) as a citrus dwarfing agent. CVd-IIIb reduces the size of citrus trees grown on trifoliate orange, increases the yield per canopy volume and yield per land surface unit in high-density plantings, without any negative effects to the fruit quality (Semancik et al., 1997; Semancik, 2003; Vidalakis et al., 2011; Hardy et al., 2007; Hutton et al., 2000).

19.22 Citrus viroid V (*Apscaviroid, Pospiviroidae*)

The discovery of Citrus viroid V (CVd-V) (294 nt) is linked to a study performed to define the response of several *Citrus* species and related genera to viroid infection (Barbosa et al., 2002). *Atalantia citroides* was identified as an unusual viroid host, because when graft-inoculated with a mixture of citrus viroids, it appeared to be immune to infection. However, sPAGE analysis revealed the presence of a viroid-like RNA in *A. citroides* that was absent in the mock inoculated controls. Further biochemical and molecular analyses together with the unusually restricted host range, suggested the existence of a new viroid RNA (Barbosa et al., 2005b; Serra et al., 2009). Even though the original source of CVd-V is uncertain, this viroid was probably present, but overlooked, in field sources containing HSVd or CDVd, which have electrophoretic mobility similar to CVd-V in sPAGE, thus leaving this viroid undetected in indexing programs. Today, CVd-V has been shown to be present in a range of commercial citrus species from several citrus-producing areas around the world (Flores et al., 2017c; Serra et al., 2010; Bani Hashemian et al., 2010b). Host range studies showed that, like other citrus viroids, CVd-V could infect many citrus species. In Etrog citron, CVd-V shows synergistic interactions such as enhanced leaf symptoms and very pronounced dwarfing with the other apscaviroids (CBLVd and CDVd) (Serra et al., 2008b,c). The effects incited by CVd-V on commercial species and cultivars are not known yet.

19.23 Citrus viroid VI (*Apscaviroid, Pospiviroidae*)

Citrus viroid VI (CVd-VI) (330 nt) originally identified with the description of CVd-OS from Japan (Ito et al., 2001) has been recently reported in China (Tessitori, 2017; Cao et al., 2017). The natural host range of CVd-VI includes citrus and Japanese persimmon (*Diospyros kaki*) (Nakaune and Nakano, 2008). In Etrog citron, CVd-VI induces mild petiole necrosis and characteristically very mild leaf bending differing from that induced by other citrus apscaviroids. In trifoliate orange, CVd-VI has been associated with CEVd-like severe bark scaling and stunting of the entire citrus tree grown on trifoliate in complex citrus viroid mixtures that do not contain CEVd (Ito et al., 2002a).

19.24 Citrus viroid VII (tentative *Apscaviroid, Pospiviroidae*)

Citrus viroid VII (CVd-VII) (368 nt) was discovered from a non-symptomatic Lisbon lemon tree in New South Wales, Australia, during routine biological indexing (Chambers et al., 2017). So far, CVd-VII has not been reported in any other country. In Etrog citron, CVd-VII induces epinasty symptoms, leaf bending, and curling associated with midvein necrosis. The effects of CVd-VII on commercial species and cultivars, other than the non-symptomatic lemons (Lisbon and Eureka), are not known yet (Chambers et al., 2017). Molecular, biochemical, and *in silico* analyses revealed that the transmissible and autonomously replicating CVd-VII RNA shared less than 90% similarity with other citrus viroids, and has a predicted rod-like secondary structure conformation with features characteristic of *Apscaviroid*, including the terminal conserved region, a CCR, and two thermodynamically stable hairpins. More information on host range, symptoms and the interaction with other citrus viroids are required before International Committee on Taxonomy of Viruses (ICTV) can formally recognize CVd-VII as a new apscaviroid (Chambers et al., 2017).

19.25 Citrus bark cracking viroid (*Cocadviroid, Pospiviroidae*)

Citrus bark cracking viroid (CBCVd) (284 nt) (Puchta et al., 1991) was originally identified after sPAGE analysis of symptomatic tissues from the bioamplification and bioindicator host Etrog citron (Duran-Vila and Semancik, 2003; Duran-Vila et al., 1988). CBCVd, the smallest of the citrus viroids, is highly conserved and considered as a chimeric-recombinant viroid between CEVd and HSVd, which may explain CBCVd infectivity in citrus and hops (Puchta et al., 1991; Jakse et al., 2015). CBCVd has been reported to be graft- and mechanically transmitted in most world citrus-growing areas (Barbosa et al., 2005a; Lavagi et al., 2017). In Etrog citron, CBCVd induces moderate stunting and random leaf epinasty, typically associated with midvein and petiole necrosis similar to CDVd (Duran-Vila and Semancik, 2003; Duran-Vila et al., 1993). In trifoliate orange, CBCVd induces severe bark cracking and gummy pitting symptoms with no effect on tree size and yield, while both antagonism and synergism have been reported in mixed infections with other citrus viroids (Vernière et al., 2004, 2006). Trees grafted on Carrizo citrange also developed severe bark cracking symptoms but CBCVd also caused a significant reduction in tree size (Murcia et al., 2015) which was associated with a small root system (Bani Hashemian et al., 2009). In spite of the reduction in tree size, CBCVd-infected trees resulted in good yield efficiency making CBCVd a suitable candidate to control the size of trees grafted on Carrizo citrange in high-density plantings, as suggested in the case of CDVd for trees grafted on trifoliate orange (Vidalakis et al., 2011).

19.26 Diagnosis

Prior to the characterization of the citrus viruses and viroids, the only available diagnosis tool for diseases caused by them was biological indexing (Roistacher, 1991) or, in few cases, transmission electron microscopy. As knowledge of the viral pathogens of citrus has accumulated, and new techniques developed, it became possible to rapidly and accurately detect them. Such techniques include sPAGE (Rivera-Bustamante et al., 1986), ELISA (enzyme-linked immunosorbent assay) (Kawai and Nishio, 1990), molecular hybridization (Palacio-Bielsa et al., 1999), and various polymerase chain reaction (PCR) assays (De Francisco et al., 2015; Bennani et al., 2002; Ito et al., 2002b; Hailestones et al., 2002; Locali et al., 2003; Liu et al., 2009; Roy et al., 2005; Bernad and Duran-Vila, 2006; Hilf, 2008; Wang et al., 2009; Vidalakis and Wang, 2013; Cowell et al., 2017; Osman et al., 2017). More recently, probe microarrays, nano-biosensors, and NGS have been used for citrus virus and viroid detection (Jakse et al., 2015; Di Serio et al., 2014; Roodbar Shojaei et al., 2016; Scuderi et al., 2016; Visser et al., 2016; Jooste et al., 2017; Yu et al., 2017).

19.27 Control

Since most citrus viruses and viroids are transmissible through grafting, the most important control measure is the use of propagative materials free of pathogens. Quarantine regulations are restrictive in developed countries and certified budwood programs are now common in major citrus-growing areas (Gergerich et al., 2015; Martin et al., 2016). To produce pathogen-free material, citrus source trees must be derived from therapied plants (e.g., nucellar embryony, thermotherapy, or shoot-tip grafting). Some of the pathogens reviewed in this chapter are mechanically transmitted; therefore, grafting, cutting, and pruning tools must be disinfested preferably with a 1% sodium hypochlorite solution prior to each use (Roistacher, 1991).

Many of the diseases induced by citrus viruses and viroids are rootstock related: CTV, CTLV, CSD, and CEVd are major examples. The use of resistant or tolerant rootstocks has protected the citrus industries in many countries. Where pathogens are endemic and have arthropod vectors, other measures of control are required. For example, in CTV stem pitting-endemic areas such as Brazil, South Africa, and Australia, the use of mild strain cross protection has been essential for citrus production (da Graça and van Vuuren, 2010). Host and isolate specificity makes the selection of CTV protective isolates empirical and time consuming. Efficacy can also be affected by uneven distribution of virus within the preimmunized budwood in different seasons (Bové et al., 2010; Zhou et al., 2002). For leprosis control, chemical control of the mite vectors is recommended (Bastianel et al., 2010).

Breeding for resistance is the preferred approach for long-term control of viral diseases. For CTV, although attempts to identify the regions of resistance gene on trifoliate orange have located a 121-kb region, including 10 genes, the CTV resistance mechanism has not been elucidated and transgenic grapefruit plant lines with individual genes did not show resistance to CTV (Harper et al., 2014; Yokomi et al., 2017; Zhou et al., 2017a).

19.28 Conclusion

Although studies to identify pathogens of graft-transmissible diseases of citrus can be traced back a century, real progress was not made until 1957 when the IOCV was founded. Studies progressed from gathered data on symptomatology, graft-transmissibility, host range, and vectors to electron microscopy (observation of CTV virions in the 1960s) (Kitajima et al., 1964), gel electrophoresis, and the discovery of citrus viroids in the 1970s (Semancik and Weathers, 1972). Serological, molecular, and sequencing diagnostic technologies have led to the identification of ca. 30 citrus viruses and viroids to date, but there are more than 10 other graft-transmissible citrus diseases of unknown etiology. New technologies such as NGS will probably lead to the identification of more citrus pathogens.

Although the basic knowledge of biological, serological, and molecular properties can lead to the identification and characterization of most citrus viral pathogens, the fact that many of them often evolve as diverse populations of distinct yet closely related sequence variants (e.g., quasi species), might cause some confusion between identified genomic sequences and observed symptoms. Furthermore, the bilateral and triangle relationships among pathogens, hosts, and/or vectors are even more complicated. Hence, only new technologies paired with biology and the necessity of Koch's postulates completion will lead to the elucidation of many citrus viral diseases.

Along with globalization, inevitably, movement of citrus trees or budwood involves a risk of accidentally introducing pests of biosecurity concern such as CTV, which has been regulated successfully by quarantine services. However, because of the citrus monoculture in different climatic and geographical areas, "new-encounter" pathogens that have not coevolved with the original citrus species (e.g., leprosis) have caused devastating epidemics (Barkley et al., 2014). The citrus community needs to stay on high alert and have programs in place for the management of citrus viral diseases, such as the use of tolerant rootstocks that managed the 1930s global CTV quick decline outbreaks and the 1960s mild strain cross protection against CTV stem pitting. The development of citrus germplasm therapy technologies of thermotherapy and shoot-tip grafting in the 1970s gave the citrus industry worldwide an exceptional tool for the exclusion of graft-transmissible pathogens (Navarro et al., 1976). Since then the pathogen-free propagation schemes in various countries have been continuously renewed with new technologies, new varieties, and improved pathogen-free clones of older lines. Some countries have mandated the use of healthy propagating material through legislation, which has greatly contributed to the sustainable development of the global citrus industry (Osman et al., 2017; Gergerich et al., 2015). Screening for genotypes resistant or tolerant to citrus virus and viroid diseases is ongoing and traditional breeding methods are being complemented with genetic engineering and genome editing technologies. Although some "engineered" citrus have shown promising results in laboratory experiments, there is still a long way to go before commercial field deployment, and long-term projects for the evaluation of such material are ongoing in several countries.

Acknowledgment

We sincerely thank Principal Scientist Patricia Barkley, retired from the Elizabeth Macarthur Agricultural Institute, NSW DPI, Australia, for her critical review of the virus section of this chapter.

References

Adams, M.J., Lefkowitz, E.J., King, A.M.Q., Carstens, E.B., 2014. Ratification vote on taxonomic proposals to the International Committee on Taxonomy of Viruses. Arch. Virol. 159, 2831–2841.

Afonso, C.L., Amarasinghe, G.K., Bányai, K., Bào, Y., Basler, C.F., Bavari, S., et al., 2016. Taxonomy of the order Mononegavirales: update 2016. Arch. Virol. 161, 2351–2360.

Agüero, J., Vives, M.C., Velázquez, K., Ruiz-Ruiz, S., Juárez, J., Navarro, L., et al., 2013. *Citrus leaf blotch virus* invades meristematic regions in *Nicotiana benthamiana* and citrus. Mol. Plant Pathol. 14, 610–616. https://doi.org/10.1111/mpp.12031.

Ahlawat, Y.S., 1989. Psorosis—a disease of citrus in India. Ind. Phytopathol. 42, 21–25.

Ahlawat, Y.S., Narian, T.K., Sardar, K.K., 1993. Association of a rhabdovirus with yellow vein mosaic, a new disease of citrus. In: Proc 12th Conf IOCV, pp. 455–457.

Ahlawat, Y.S., Varma, A., Lockhart, B.E., Srivastava, M., Chakraborty, N.K., Varma, A., 1996. Association of a badnavirus with citrus mosaic disease in India. Plant Dis. 80, 590–592.

Albanese, G., La Rosa, R., Davino, M., Hammond, R.W., Smith, D.R., Diener, T.O., 1988. A viroid different from *Citrus exocortis viroid* found in commercial *Citrus* in Sicily. In: Proc. 10th Conf IOCV, pp. 165–172.

Albiach-Marti, M.R., Grosser, J.W., Gowda, S., Mawassi, M., Satyanarayana, T., Garnsey, S.M., et al., 2004. *Citrus tristeza virus* replicates and forms infections virions in protoplasts of resistant citrus relatives. Mol. Breed. 14, 117–128.

Altenburg, E., 1946. The "viroid" theory in relation to plasmagenes, viruses, cancer and plastids. Am. Nat. 80 (794), 559–567.

Ashulin, L., Lachman, O., Hadas, R., Bar-Joseph, M., 1991. Nucleotide sequence of a new viroid species, citrus bent leaf viroid (CBLVd) isolated from grapefruit in Israel. Nucleic Acids Res. 19, 4767.

Azeri, T., 1973. First report of satsuma dwarf virus disease on satsuma mandarins in Turkey. Plant Dis. Rep. 57, 149–153.

Bani Hashemian, S.M., Serra, P., Barbosa, C.J., Juárez, J., Aleza, P., Corvera, J.M., et al., 2009. The effect of a field-source mixture of citrus viroids on the performance of 'Nules' clementine and 'Navelina' sweet orange trees grafted on Carrizo citrange. Plant Dis. 93, 699–707.

Bani Hashemian, S.M., Murcia, N., Trenor, I., Duran-Vila, N., 2010a. Low performance of citrus trees grafted on Carrizo citrange is associated with viroid infection. J. Plant Pathol. 92, 511–517.

Bani Hashemian, S.M., Taheri, H., Duran-Vila, N., Serra, P., 2010b. First report of *Citrus* viroid V in Moro Blood sweet orange in Iran. Plant Dis. 94, 129.

Barbosa, C., Pina, J.A., Navarro, L., Duran-Vila, N., 2002. Replication/accumulation and symptom expression of citrus viroids on some species of citrus and related genera. In: Proc. 15th Conf IOCV, pp. 264–271.

Barbosa, C., Serra, P., Pina, J.A., Navarro, L., Darós, J.A., Flores, R., et al., 2005a. Identification and preliminary characterization of a viroid-like RNA in *Atalantia citroides*. In: Proc. 16th Conf IOCV, pp. 264–271.

Barbosa, C.J., Pina, J.A., Pérez-Panadés, J., Bernad, L., Serra, P., Navarro, L., et al., 2005b. Mechanical transmission of citrus viroids. Plant Dis. 89, 749–754.

Bari Hashmian, S.M., Aghajanzadeh, S., 2017. Occurrence of *Citrus yellow vein clearing virus* in citrus species in Iran. J. Plant Pathol. 99 (1), 290.

Bar-Joseph, M., Marcus, R., Lee, R.F., 1989. The continuous challenge of *Citrus tristeza virus* control. Annu. Rev. Phytopathol. 27, 291–316.

Barkley, P., Schubert, T., Schutte, G.C., Godfrey, K., Hattingh, V., Telford, G., et al., 2014. Invasive pathogens in plant biosecurity. Case study: citrus biosecurity. In: The Handbook of Plant Biosecurity. Springer Netherlands, pp. 547–592.

Bassanezi, R.B., Bergamin Filho, A., Amorim, L., Gimenes-Fernandes, N., Gottwald, T.R., Bové, J.M., 2003. Spatial and temporal analyses of citrus sudden death as a tool to generate hypotheses concerning its etiology. Phytopathology 93, 502–512.

Bastianel, M., Novelli, V.M., Kitajima, E.W., Kubo, K.S., Bassanezi, R.B., Machado, M.A., et al., 2010. *Citrus* leprosis: centennial of an unusual mite-virus pathosystem. Plant Dis. 94, 284–292.

Beñatena, H.N., Portillo, M.M., 1984. Natural spread of psorosis in sweet orange seedlings. In: Proc 9th Conf IOCV, pp. 159–164.

Bennani, B., Mendes, C., Zemzani, M., Azeddoug, H., Nolasco, G., 2002. *Citrus variegation virus*: molecular variability of a portion of the RNA3 containing the coat protein gene and design of primers for RT-PCT detection. Eur. J. Plant Pathol. 108, 155–162.

Benton, R.J., Bowman, F.T., Fraser, L., Kebby, R.G., 1949. Selection of citrus budwood to control scaly butt in trifoliata rootstock. Agr. Gaz. NSW 60, 31–34.

Benton, R.J., Bowman, F.T., Fraser, L., Kebby, R.G., 1950. Stunting and scaly butt associated with *Poncirus trifoliata* rootstock. NSW Dept. Agr. Sci. Bull. 70, 1–20.

Bernad, L., Duran-Vila, N., 2006. A novel RT-PCR approach for detection and characterization of citrus viroids. Mol. Cell. Probes 20, 105–113.

Bernad, L., Duran-Vila, N., Elena, S.F., 2009. Effect of citrus hosts on the generation, maintenance and evolutionary fate of genetic variability of citrus exocortis viroid. J. Gen. Virol. 90, 2040–2049.

Bové, J.M., Garnier, M., 2002. A historical case: "*Chloroseinfectieuse des citrus*" (Infectious chlorosis of citrus), first experimentally graft-transmitted disease of citrus. In: Proc 15th Conf IOCV, pp. 368–370.

Bové, J.M., Moreno, P., Duran-Vila, N., da Graça, J.V., 2010. Fifty years of International Organization of Citrus Virologists (IOCV), 1957 to 2007: from graft-transmitted citrus agents to viroids, viruses and endogenous bacteria. In: Proc 17th Conf IOCV, pp. xiii–lxx.

Cao, M.J., Wu, Q., Yang, F.Y., Wang, X.F., Li, R.H., Zhou, C.Y., et al., 2017. Molecular characterization and phylogenetic analysis of citrus viroid VI variants from citrus in China. Eur. J. Plant Pathol. 149 (4), 885–893.

Catara, A.A., Azzaro, A., Moghal, S.M., Khan, D.A., 1988. Virus, viroid and prokaryotic diseases of citrus in Pakistan. In: Proc 6th Conf IOCV, pp. 957–963.

Chabi-Jesus, C., Ramos-González, P.L., Guerra-Peraza, O., Kitajima, E.W., Tassi, A.D., Harakava, R., et al., 2018. Molecular characterization of citrus chlorotic spot virus, a new dichorhavirus associated with citrus leprosis symptoms. Plant Dis. 102 (8), 1588–1598. doi:10.1094/PDIS-09-17-1425-RE.

Chaffai, M., Serra, P., Gandía, M., Hernánadez, C., Duran-Vila, N., 2007. Molecular characterization of CEVd strains that induce different phenotyes in *Gynura aurantiaca*: structure-pathogenicity relationships. Arch. Virol. 152, 1283–1294.

Chambers, G.A., Donovan, N.J., Bodaghi, S., Jelinek, S.M., Vidalakis, G., 2017. A novel citrus viroid found in Australia, tentatively named citrus viroid VII. Arch. Virol. 163 (1), 215–218. https://doi.org/10.1007/s00705-017-3591-y.

Çinar, A., Korkmaz, S., Kersting, U., 1994. Presence of new whitefly-borne citrus disease of possible viral aetiology in Turkey. FAO Plant Prot. Bull. 42, 73–75.

Cowell, S.J., Harper, S.J., Dawson, W.O., 2017. A real-time qPCR assay for the detection of *Citrus tatter leaf virus*. J. Virol. Methods 244, 29–31.

Cruz-Jaramillo, J.L., Ruiz-Medrano, R., Rojas-Morales, L., López-Buenfil, J.A., Morales-Galván, O., Chavarín-Palacio, C., et al., 2014. Characterization of a proposed dichorhavirus associated with the citrus leprosis disease and analysis of the host response. Viruses 6, 2602–2622.

da Graça, J.V., Maharaj, S.B., 1991. *Citrus vein enation virus*, a possible luteovirus. In: Proc 11th Conf IOCV, pp. 391–394.

da Graça, J.V., van Vuuren, S.P., 2010. Managing *Citrus tristeza virus* losses using cross protection. In: Karasev, A.V., Hilf, M.E. (Eds.), *Citrus tristeza virus* Complex and Tristeza Diseases. APR Press MN, pp. 247–260.

da Graça, J.V., Lee, R.F., Moreno, P., Civerolo, E.L., Derrick, K.S., 1991. Comparison of isolates of citrus ringspot, psorosis and other virus-like agents of citrus. Plant Dis. 75, 613–616.

Dawson, T.E., Mooney, P.A., 2000. Evidence for trifoliate resistance breaking isolates of *Citrus tristeza virus* in New Zealand. In: Proc 14th Conf IOCV, pp. 69–76.

Dawson, W.O., Garnsey, S.M., Tatineni, S., Folimonova, S.Y., Harper, S.J., Gowda, S., 2013. *Citrus tristeza virus*-host interactions. Front. Microbiol. 4, 88.

De Francisco, A., Costa, N., Plata, M.I., Garcia, M.L., 2015. Improved detection of *Citrus psorosis virus* and coat-protein derived transgenes in citrus plants: comparison between RT-qPCR and TAS-ELISA. J. Phytopathol. 163, 915–925.

Derrick, K.S., Brlansky, R.H., da Graça, J.V., Lee, R.F., Timmer, L.W., Nguyen, T.K., 1988. Partial characterization of a virus associated with citrus ringspot. Phytopathology 78, 1298–1301.

Derrick, K.S., Beretta, M.J., Barthe, G.A., Kayim, M., 2003. Strains of *Citrus tristeza virus* predominately found in roots with implications for citrus blight and citrus sudden death. Phytopathology 93, S20.

Di Serio, F., Flores, R., Verhoeven, J.T.J., Li, S.F., Pallás, V., Randles, J.W., et al., 2014. Current status of viroid taxonomy. Arch. Virol. 159, 3467–3478.

Di Serio, F., Li, S.F., Pallas, V., Owens, R.A., Randles, J.W., Sano, T., et al., 2017. Viroid taxonomy. In: Hadidi, A., Flores, R., Randles, J.W., Palukaitis, P. (Eds.), Viroids and Satellites. Academic Press, pp. 135–146.

Diener, T.O., 1971. Potato spindle tuber "virus": IV. A replicating, low molecular weight RNA. Virology 45, 411–428.

Diener, T.O., Smith, D.R., Hammond, R.H., Albanese, G., La Rosa, R., Davino, M., 1988. *Citrus* B viroid identified as a strain of hop stunt viroid. Plant Dis. 72, 691–693.

Dippenaar, B.J., 1958. Concentric ring blotch of citrus–its cause and control. S. Afr. J. Agr. Res. 1, 83–106.

Doidge, E., 1926. Scaly bark (psorosis) of citrus trees. J. Dept. Agric. Union S. Afr. 12, 61–67.

Dominguez, F.S., Bernal, A., Childers, C.C., Kitajima, E.W., 2001. First report of citrus leprosis on Panama. Plant Dis. 85, 228.

Duran-Vila, N., 2017. *Citrus* exocortis viroid. In: Hadidi, A., Flores, R., Randles, J.W., Palukaitis, P. (Eds.), Viroids and Satellites. Academic Press, pp. 169–179.

Duran-Vila, N., Semancik, J., 2003. *Citrus* viroids. In: Hadidi, A., Flores, R., Randles, J.W., Semancik, J.S. (Eds.), Viroids. CSIRO Publishing, Collingwood, VIC, pp. 178–194.

Duran-Vila, N., Roistacher, C.N., Rivera-Bustamante, R., Semancik, J.S., 1988. A definition of citrus viroid groups and their relationship to the exocortis disease. J. Gen. Virol. 69, 3069–3080.

Duran-Vila, N., Pina, J., Navarro, L., 1993. Improved indexing of citrus viroids. In: Proc. 12th Conf IOCV, pp. 202–211.

Fadda, Z., Daròs, J.A., Flores, R., Duran-Vila, N., 2003. Identification in eggplant of a variant of citrus exocortis viroid (CEVd) with a 96 nucleotide duplication in the right terminal region of the rod-like secondary structure. Virus Res. 97, 145–149.

Fawcett, H.S., 1934. Is psorosis a virus disease? Phytopathology 24, 659–668.

Fawcett, H.S., 1936. *Citrus* Diseases and Their Control, second ed. McGraw-Hill Book Co, New York and London, p. 656.

Fawcett, H.S., 1939. Psorosis in relation to other virus-like effects on citrus. Phytopathology 29, 6.

Fawcett, H.S., Bitancourt, A.A., 1943. Comparative symptomology of psorosis varieties on citrus in California. Phytopathology 33, 837–864.

Fawcett, H.S., Klotz, L.J., 1939. Infectious variegation of citrus. Phytopathology 29, 670.

Fawcett, H.S., Klotz, L.J., 1948. Exocortis on trifoliate orange. Citrus Leaves 28, 8.

Fawcett, H.S., Lee, H.A., 1926. *Citrus* Diseases and Their Control. McGraw-Hill Book Co, New York and London, p. 582.

Flores, R., Minoia, S., Lopez-Carrasco, A., Delgado, S., Martınez de Alba, A.E., Kalantidis, K., 2017a. Viroid replication. In: Hadidi, A., Flores, R., Randles, J.W., Palukaitis, P. (Eds.), Viroids and Satellites. Academic Press, pp. 71–81.

Flores, R., Di Serio, F., Navarro, B., Owens, R.A., 2017b. Viroid pathogenesis. In: Hadidi, A., Flores, R., Randles, J.W., Palukaitis, P. (Eds.), Viroids and Satellites. Academic Press, pp. 93–103.

Flores, R., Randles, J.W., Palukaitis, P., 2017c. Part VI. Geographical distribution of viroids and viroid diseases. In: Viroids and Satellites. Academic Press, Boston, pp. 459–518.

Fraser, L.R., 1959. Woody gall, a suspected virus disease of rough lemon and other citrus varieties. Proc. Linnean Soc. NSW 84, 332–336.

Galipienso, L., Vives, M.C., Moreno, P., Milne, R.G., Navarro, L., Guerri, J., 2001. Partial characterization of *Citrus leaf blotch virus*, a new virus from Nagami kumquat. Arch. Virol. 146, 357–368.

Gandía, M., Rubio, L., Palacio, A., Duran-Vila, N., 2005. Genetic variation and population structure of a *Citrus exocortis viroid* (CEVd) isolate and the progenies of infectious haplotypes. Arch. Virol. 150, 1945–1957.

García, M.L., Dal Bó, E., da Graça, J.V., Gago-Zachert, S., Hammond, J., Moreno, P., et al., 2017. ICTV taxonomy profile: Ophioviridae. J. Gen. Virol. 98, 1161–1162.

García-Escamilla, P., Duran-Trujillo, Y., Otero-Colina, G., Valdovinos-Ponce, G., Santillán-Galicia, M.T., Ortiz-García, C.F., et al., 2017. Transmission of viruses associated with cytoplasmic and nuclear leprosis symptoms by *Brevipalpus yothersi* and *B. californicus*. Trop. Plant Pathol. https://doi.org/10.1007/s40858-017-0195-8.

Garnsey, S.M., 1975. Purification and properties of citrus leaf rugose virus. Phytopathology 65, 50–57.

Garnsey, S.M., 2000. Ilarviruses Compendium of *Citrus* Diseases, second ed. APS Press, St. Paul, MN, pp. 56–57.

Garnsey, S.M., Behe, C.G., Lockhart, B.E., 1998. Transmission of citrus yellow mosaic badnavirus by the citrus mealybug. Phytopathology 88 (9), S31.

Ge, X., Scott, J.W., 1994. The nucleotide sequence of citrus leaf rugose virus RNA-2. J. Gen. Virol. 75, 2841–2846.

Gergerich, R.C., Welliver, R.A., Gettys, S., Osterbauer, N.K., Kamenidou, S., Martin, R.R., et al., 2015. Safeguarding fruit crops in the age of agricultural globalization. Plant Dis. 99, 176–187.

Gottwald, T.R., Garnsey, S.M., Cambra, M., Moreno, P., Irey, M., Borbón, J., 1996. Differential effects of *Toxoptera citricida* vs. *Aphis gossypii* on temporal increase and spatial patterns of spread of citrus tristeza. In: Proc 13th Conf IOCV, pp. 120–129.

Gottwald, T.R., Garnsey, S.M., Cambra, M., Moreno, P., Irey, M., Borbón, J., 1997. Comparative effects of aphid vector species on increase and spread of *Citrus tristeza virus*. Fruits 52, 397–403.

Grant, T.J., Smith, P.F., 1960. Infectious variegation of citrus found in Florida. Plant Dis. Rep. 44, 426–429.

Gress, J.C., Smith, S., Tzanetakis, I.E., 2017. First report of *Citrus leaf blotch virus* in peony in the U.S.A. Plant Dis. 101, 637.

Grimaldi, V., Catara, A., 1996. Association of a filamentous virus with yellow vein clearing of lemon. In: Proc 13th Conf IOCV, pp. 343–345.

Gross, H.J., Krupp, G., Domdey, H., Raba, M., Jank, P., Lossow, C., et al., 1982. Nucleotide sequence and secondary structure of citrus exocortis and chrysanthemum stunt viroid. Eur. J. Biochem. 121, 249–257.

Guerri, J., Pina, J.A., Vives, M.C., Navarro, L., Moreno, P., 2004. Seed transmission of *Citrus leaf blotch virus*: implications in quarantine and certification programs. Plant Dis. 88, 906.

Guo, J., Lai, X.P., Li, J.X., Yue, J.Q., Zhang, S.Y., Li, Y.Y., et al., 2015. First report on *Citrus chlorotic dwarf associated-virus* on lemon in Dehong Prefecture, Yunnan. China Plant. Dis. 99, 1287.

Hadidi, A., Vidalakis, G., Sano, T., 2017. Economic significance of fruit tree and grapevine viroids. In: Hadidi, A., Flores, R., Randles, J.W., Palukaitis, P. (Eds.), Viroids and Satellites. Academic Press, pp. 15–25.

Hailestones, D., Bryant, K.L., Broadbent, P., Zhou, C., 2002. Detection of *Citrus tatter leaf virus* with reverse transcription—polymerase chain reaction (RT-PCR). Australas. Plant Pathol. 29, 240–248.

Hajeri, S., Ramadugu, C., Manjunath, K., Lee, J.N.R., Vidalakis, G., 2011. *In vivo* generated *Citrus exocortis viroid* progeny variants display a range of phenotypes with altered levels of replication, systemic accumulation and pathogenicity. Virology 417 (2), 400–409.

Hardy, S., Sanderson, G., Barkley, P., Donovan, N., 2007. Dwarfing citrus trees using viroids. NSW Agr. 3, 1–5.

Harper, S.J., Cowell, S.J., Robertson, C.J., Dawson, W.O., 2014. Differential tropism in roots and shoots infected by *Citrus tristeza virus*. Virology 460–461, 91–99.

Hartung, J.S., Roy, A., Fu, S., Shao, J., Schneider, W.L., Brlansky, R.H., 2015. History and diversity of *Citrus leprosis virus* recorded in herbarium specimens. Phytopathology 105, 1277–1284.

Hataya, T., Tsushima, T., Sano, T., 2017. Hop stunt viroid. In: Hadidi, A., Flores, R., Randles, J.W., Palukaitis, P. (Eds.), Viroids and Satellites. Academic Press, pp. 199–210.

Hermoso de Mendoza, A., Pina, J.A., Ballester-Olmos, J.F., Navarro, L., 1993. Persistent transmission of *Citrus vein enation virus* by *Aphis gossypii* and *Myzuspersicae*. In: Proc 12th Conf IOCV, pp. 361–362.

Hilf, M., 2008. An immunocapture RT-PCR procedure using *Apple stem grooving virus* antibodies facilitates analysis of *Citrus tatter leaf virus* from the original Meyer lemon host. Plant Dis. 92, 746–750.

Hiltebrand, W.F., 1959. Certification program for maintenance of virus-free propagation sources of citrus in California. In: Wallace, J.M. (Ed.), *Citrus* Virus Diseases. Univ Cal Div Agr Sci, pp. 229–231.

Hoa, N.V., Ahlawat, Y.S., 2004. Characterization of four isolates of Indian citrus ringspot virus. Ind. Phytopathol. 57, 296–302.

Huang, Q., Hartung, J.S., 2001. Cloning and sequence analysis of an infectious clone of *Citrus yellow mosaic virus* that can infect sweet orange via *Agrobacterium*-mediated inoculation. J. Gen. Virol. 82, 2549–2558.

Hutton, R.J., Broadbent, P., Bevington, K.B., 2000. Viroid dwarfing for high density citrus plantings. Hortic. Rev. 24, 277–317.

Iftikhar, Y., Iqbal, Z., Ahmed, S., Awan, A.R., Saleem, U., Sarwar, G., et al., 2010. Effect of environmental factors on *Yellow vein clearing virus* incidence in lemon. J. Agric. Res. 48, 87–92.

Inouye, N., Maeda, T., Mitsuhata, K., 1979. *Citrus tatter leaf virus* isolated from lily. Ann. Phytopathol. Soc. Japan 45, 712–720.

Ito, T., Ieki, H., Ozaki, K., 2000. A population of variants of a viroid closely related to citrus viroid-I in citrus plants. Arch. Virol. 145, 2105–2114.

Ito, T., Ieki, H., Ozaki, K., Ito, T., 2001. Characterization of a new citrus viroid species tentatively termed *Citrus* viroid OS. Arch. Virol. 146 (5), 975–982.

Ito, T., Ieki, H., Ozaki, K., Iwanami, T., Nakahara, K., Hataya, T., et al., 2002a. Multiple citrus viroids in citrus from Japan and their ability to produce exocortis-like symptoms in citron. Phytopathology 92, 542–547.

Ito, T., Ieki, H., Ozaki, K., 2002b. Simultaneous detection of six citrus viroids and *Apple stem grooving virus* from citrus plants by multiplex reverse transcription polymerase chain reaction. J. Virol. Methods 106, 235–239.

Iwanami, T., 2010. Review: properties and control of *Satsuma dwarf virus*. JARQ 44 (1), 1–6.

Jakse, J., Radisek, S., Pokorn, T., Matousek, J., Javornik, B., 2015. Deep-sequencing revealed citrus bark cracking viroid (CBCVd) as a highly aggressive pathogen on hop. Plant Pathol. 64, 831–842.

Jooste, T.L., Visser, M., Cook, G., Burger, J.T., Maree, H.J., 2017. In silico probe-based detection of citrus viruses in NGS data. Phytopathology 107 (8), 988–993.

Kanrar, S., Afunian, M., Greer, G., Tzanetakis, I., Vidalakis, G., 2011. Viruses associated with yellow vein and vein enation disease of citrus. Phytopathology 101, S87.

Kapur, S.P., Weathers, L.G., Calavan, E.C., 1974. Studies on strains of exocortis virus in citron and *Gynura aurantica*. In: Proc. 6th Conf IOCV, pp. 21–28.

Kawai, A., Nishio, T., 1990. Detection of *Citrus tatterleaf virus* by enzyme-linked immunosorbent assay (ELISA). Ann. Phytopath. Soc. Japan 56, 342–345.

Keese, P., Symons, R.H., 1985. Domains in viroids: evidence of intermolecular RNA rearrangements and their contribution to viroid evolution. Proc. Natl. Acad. Sci. U. S. A. 82, 4582–4586.

Kersting, U., Korkmaz, S., Çinar, A., Ertuğrul, B., Önelge, N., Garnsey, S.M., 1996. *Citrus* chlorotic dwarf, a new whitefly-transmitted disease in the eastern Mediterranean region of Turkey. In: Proc 13th Conf IOCV, pp. 220–225.

Kitajima, E.W., Silva, D.M., Oliveira, A.R., Müller, G.W., Costa, A.S., 1964. Thread-like particles associated with tristeza disease of citrus. Nature 201, 1011–1012.

Klotz, L.J., Calavan, E.C., Weathers, L.G., 1982. Virus and virus-like diseases of citrus. Univ. Cal. Div. Agr. Sci., 75.

Kofalvi, S.A., Marcos, J.F., Canizares, M.C., Pallas, V., Candresse, T., 1997. Hop stunt viroid (HSVd) sequence variants from *Prunus* species: evidence for recombination between HSVd isolates. J. Gen. Virol. 78 (12), 3177–3186.

Kyriakou, A.P., 1992. Incidence in Cyprus of *Citrus exocortis viroid* and its mechanical transmission. Plant Pathol. 41, 20–24.

Kyriakou, A., Ioannou, M., Hadjinicolis, A., Hoffman, R., Antoniou, E., Papayiannis, L., et al., 2005. *Citrus* exocortis and cachexia viroids affect growth, yield and fruit quality of Lapithou lemon on sour orange rootstock in Cyprus. In: Proc. 16th Conf IOCV, pp. 257–263.

Lavagi, I., Matoušek, J., Vidalakis, G., 2017. Other Cocadviroids. In: Hadidi, A., Flores, R., Randles, J.W., Palukaitis, P. (Eds.), Viroids and Satellites. Academic Press, pp. 275–287.

Lee, R.F., Derrick, K.S., Futch, S.H., Tucker, D.P.H., 1993. Graft transmission of citrus measles. In: Proc 12th Conf IOCV 1993, pp. 371–374.

Levy, L., Gumpf, D.J., 1991. Studies on the psorosis disease of citrus and preliminary characterization of a flexuous virus associated with the disease. In: Proc 11th Conf IOCV, pp. 319–336.

Liu, K., Zhou, C., Song, Z., Li, Z., Tang, K., 2009. Detection of *Citrus tatter leaf virus* (CTLV) by real-time PCR. J. Fruit Sci. 26, 748–751 (in Chinese).

Locali, E.C., Freitas-Astua, J., Souza, A.A., Takita, M.A., Astua-Monge, G., Antonioli, R., et al., 2003. Development of a molecular tool for the diagnosis of leprosis, a major threat to the citrus production in the Americas. Plant Dis. 87, 1317–1321.

Loconsole, G., Önelge, N., Potere, O., Giampetruzzi, A., Bozan, O., Satar, S., et al., 2012. Identification and characterization of *Citrus yellow vein clearing virus*, a putative new member of the genus *Mandarivirus*. Phytopathology 102, 1168–1175.

Lovisolo, O., 1993. Agroecology and centers of origin of graft-transmissible disease of citrus. In: Proc 12th Conf IOCV, pp. 406–411.

Maccheroni, W., Alegri, M.C., Greggio, C.C., et al., 2005. Identification and genomic characterization of a new virus (*Tymoviridae* family) associated with citrus sudden death disease. J. Virol. 79, 3028–3037.

Magome, H., Yoshikawa, N., Takahashi, I.T., Miyakawa, T., 1997. Molecular variability of the genomes of capilloviruses from apple, Japanese pear, European pear and citrus trees. Phytopathology 87, 389–396.

Maharaj, S.B., da Graça, J.V., 1988. Observation of virus-like particles associated with citrus vein enation virus-infected citrus and the viruliferous aphid vector *Toxoptera citricidus*. Phytophylactica 20, 357–360.

Maharaj, S.B., da Graça, J.V., 1989. Transmission of citrus vein enation by *Toxoptera citricidus*. Phytophylactica 21, 81–82.

Marcó, G.M., Semancik, J.S., Gumpf, D.J., 1991. Cucumo-like virus isolated from cowpea indicator plants manifesting the *Citrus tatter leaf virus* syndrome. In: Proc 11th Conf IOCV, pp. 352–357.

Martelli, G.P., Yilmaz, M.A., Savino, V., Baloglu, S., Grieco, F., Güldür, M.E., et al., 1996. Properties of a citrus isolate of olive latent virus 1, a new necrovirus. Eur. J. Plant Pathol. 102, 527–536.

Martin, R.R., Constable, F., Tzanetakis, I.E., 2016. Quarantine regulations and the impact of modern detection methods. Annu. Rev. Phytopathol. 4 (54), 189–205.

Matsumura, E.E., Coletta-Filho, H.D., Nouri, S., Falk, B.W., Nerca, L., Oliveira, T.S., et al., 2017. Deep sequencing analysis of RNAs from citrus plants grown in a citrus sudden death-affected area reveals diverse known and putative novel viruses. Viruses 9, 92. https://doi.org/10.3390/v9040092.

Mestre, P.F., Asins, M.J., 1997. Efficient search for new resistant genotypes to the *Citrus tristeza closterovirus* in the orange subfamily Aurantioideae. Theor. Appl. Genet. 95 (1), 1282–1288.

Miyakawa, T., Tsuji, T., 1988. The association of tatterleaf virus with budunion crease of trees on trifoliate orange. In: Proc 10th Conf IOCV, pp. 360–364.

Moon, D.Y., 1986. Viral diseases of horticultural subtropical and tropical crops in Korea. In: Food and Fert Tech. Center for Asia-Pacific Reg Taiwan, Taipei, pp. 12–18.

Moreira, S., 1955. Sintomas de "exocortis" emlimoneirocravo. Bragantia 14, 19–21 (in Portuguese).

Moreira, S., 1959. Rangpur lime disease and its relationship to exocortis. In: Wallace, J.M. (Ed.), *Citrus* Virus Diseases. Univ Cal Div Agr Sci, pp. 135–140.

Moreira, L., Albertazzi, F.J., Garita, L., Ortiz, B., Villalobos, W., 2011. First report of *Citrus variegation virus* in Palestine sweet lime, as coffee shade in Costa Rica. In: Proc 18th Conf IOCV. iocv.org/proceedings/eighteen/Moreira_et_al.pdf.

Moreno, P., Guerri, J., García, M.L., 2015. The psorosis disease of citrus: a pale light at the end of the tunnel. J Citrus Pathol. iocv_journal citrus paythology_28860.

Murcia, N., Bernad, L., Serra, P., Bani Hashemian, S.M., Duran-Vila, N., 2009. Molecular and biological characterization of natural variants of *Citrus dwarfing viroid*. Arch. Virol. 154 (8), 1329–1334.

Murcia, N., Bernad, L., Duran-Vila, N., Serra, P., 2011. Two nucleotide positions in the *Citrus exocortis viroid* RNA associated with symptom expression in Etrog citron but not in experimental herbaceous hosts. Mol. Plant Pathol. 12, 203–208.

Murcia, N., Bani Hashemian, S.M., Serra, P., Pina, J.A., Duran-Vila, N., 2015. *Citrus* viroids: symptom expression and performance of Washington navel sweet orange trees grafted on Carrizo citrange. Plant Dis. 99, 125–136.

Nakaune, R., Nakano, M., 2008. Identification of a new Apscaviroid from Japanese persimmon. Arch. Virol. 153 (5), 969–972.

Navarro, L., Roistacher, C.N., Murashige, T., 1976. Effect of size and source of shoot tips on psorosis-A and exocortis content of navel orange plants obtained by shoot-tip grafting *in vitro*. In: Proc. 7th Conf. IOCV, pp. 194–197.

Navarro, L., Pina, J.A., Ballester-Olmos, J.F., Moreno, P., Cambra, M., 1984. A new graft transmissible disease found in Nagami kumquat. In: Proc 9th Conf IOCV, pp. 234–240.

Navarro, B., Minutolo, M., De Stradis, A., Palmisano, F., Alioto, D., De Sierio, F., 2018a. The first phlebo-like virus infecting plants: a case study on the adaption of negative-stranded RNA viruses to new hosts. Mol. Plant Pathol. 67, 1057–1089.

Navarro, B., Zicca, S., Minutolo, M., Saponari, M., Alioto, D., Di Serio, F., 2018b. A negative-stranded RNA virus infecting citrus trees: the second member of a new genus within the order Bunyavirales. Front. Microbiol. 9 (2340), 1–11.

Navas-Castillo, J., Moreno, P., 1993. Biological diversity of citrus ringspot isolates in Spain. Plant Pathol. 42, 347–357.

Nickel, O., De Jesus, B.C., Santos-Filho, H.P., Passos, O.S., Laranjeira, F.F., 2007. Bahia bark scaling of citrus: a disease of unknown etiology. Pest Technol. 1, 70–75.

Önelge, N., 2003. First report of yellow vein clearing of lemons in Turkey. J. Turk. Phytopathol. 32, 53–55.

Önelge, N., Satar, S., Elibüyük, Ö., Bozan, O., Kamberoğlu, M., 2011. Transmission studies on citrus yellow vein clearing virus. In: Proc 18th Conf IOCV. iocv.org/proceedings/eighteen/Onlege_et_al.pdf.

Osman, F., Dang, T., Bodaghi, S., Vidalakis, G., 2017. One-step multiplex RT-qPCR detects three citrus viroids from different genera in a wide range of hosts. J. Virol. Methods 245, 40–52.

Owens, R.A., Yang, G., Gundersen-Rindal, D., Hammond, R.W., Candresse, T., Bar-Joseph, M., 2000. Both point mutation and RNA recombination contribute to the sequence diversity of citrus viroid III. Virus Genes 20 (3), 243–252.

Owens, R., Thompson, S., Sieburth, P., Hilf, M., 2005. Mutational analysis of *Citrus* viroid III symptom expression. In: Proc. 16th Conf IOCV, pp. 280–290.

Palacio-Bielsa, A., Foissac, X., Duran-Vila, N., 1999. Indexing of citrus viroids by imprint hybridisation. Eur. J. Plant Pathol. 105, 897–903.

Palacio-Bielsa, A., Romero-Durbán, J., Duran-Vila, N., 2004. Characterization of citrus HSVd isolates. Arch. Virol. 149, 537–552.

Pallás, V., Gómez, G., 2017. Viroid movement. In: Hadidi, A., Flores, R., Randles, J.W., Palukaitis, P. (Eds.), Viroids and Satellites. Academic Press, pp. 83–91.

Palle, S.R., Miao, H., Seyran, M., Louzada, E.S., da Graça, J.V., Skaria, M., 2005. Evidence for association of *Citrus psorosis virus* with symptomatic trees and an *Olpidium*-like fungus in Texas. In: Proc 16th Conf IOCV, pp. 423–426.

Puchta, H., Ramm, K., Luckinger, R., Hadas, R., Bar-Joseph, M., Saänger, H.L., 1991. Primary and secondary structure of citrus viroid IV (CVd IV), a new chimeric viroid present in dwarfed grapefruit in Israel. Nucleic Acids Res. 19 (23), 6640.

Ramos-Gonzalez, P.L., Chabi-Jesus, C., Guerra-Peraza, O., Breton, M.C., Arena, G.D., Nunes, M.A., et al., 2016. Phylogenetic and molecular variability studies reveal a new genetic clade of *Citrus leprosis virus C*. Viruses 8, 153.

Ramos-González, P.L., Chabi-Jesus, C., Guerra-Peraza, O., Tassi, A.D., Kitajima, E.W., Harakava, R., et al., 2017. *Citrus leprosis virus N*: a new dichorhavirus causing citrus leprosis disease. Phytopathology 107, 963–976.

Reanwarakorn, K., Semancik, J.S., 1998. Regulation of pathogenicity in hop stunt viroid-related group II. J. Gen. Virol. 79, 3163–3171.

Reanwarakorn, K., Semancik, J.S., 1999. Correlation of hop stunt viroid variants to cachexia and xyloporosis diseases of citrus. Phytopathology 89, 568–574.

Rivera-Bustamante, R.F., Gin, R., Semancik, J.S., 1986. Enhanced resolution of circular and linear molecular forms of viroid and viroid-like RNA by electrophoresis in a discontinuous-pH system. Anal. Biochem. 156, 91–95.

Roistacher, C.N., 1991. Graft Transmissible Diseases of *Citrus*: Handbook for Detection and Diagnosis. FAO, Rome.

Roistacher, C.N., 1993. A review of the citrus tristeza disease with particular reference to cross protection. In: Proc 4th World Cong ISCN, pp. 93–109.

Roistacher, C.N., Blue, R.L., 1968. A psorosis-like virus causing symptoms only on Dweettangor. In: Proc 4th Conf IOCV, pp. 13–18.

Roistacher, C.N., Calavan, E.C., 1965. Cross protection studies with strains of concave gum and psorosis viruses. In: Proc 4th Conf IOCV, pp. 154–161.

Roistacher, C.N., Blue, R.L., Calavan, E.C., 1973. A new test for citrus cachexia. Citrograph 58, 261–262.

Roistacher, C.N., Calavan, E.C., Blue, R.L., Navarro, L., Gonzales, R., 1977. A new more sensitive citron indicator for the detection of mild isolates of citrus exocortis viroid (CEV). Plant Dis. 61, 135–139.

Roistacher, C.N., Nauer, E.M., Wagner, R.L., 1980. Transmissibility of cachexia, Dweet mottle, psorosis, tatterleaf and infectious variegation viruses on knife blades and its prevention. In: Proc 8th Conf IOCV, pp. 425–429.

Roistacher, C.N., Bash, J.A., Semancik, J.S., 1993. Distinct disease symptoms in *Poncirus trifoliata* induced by three citrus viroids from three specific groups. In: Proc. 12th Conf IOCV, pp. 173–179.

Roistacher, C.N., Bash, J., Gumpf, D.J., 2000. Continued attempts over a 22-year period to separate components of the citrus tatter leaf-citrange stunt virus complex. In: Proc 14th Conf IOCV, pp. 179–184.

Roodbar Shojaei, T., Mohd Salleh, M.A., Sijam, K., Abdul Rahim, R., Mohsenifar, A., Safarnejad, R., et al., 2016. Detection of *Citrus tristeza virus* by using fluorescence resonance energytransfer-based biosensor. Spectrochim. Acta A: M BS 169, 216–222.

Rossetti, V., Lasca, C.C., Negretti, S., 1969. New developments regarding leprosis and zonate chlorosis of citrus. In: Proc 1st Intern Citrus Symp 1969. vol. 3, pp. 1453–1456.

Roy, A., Fayad, A., Barthe, G., Brlansky, R.H., 2005. A multiplex polymerase chain reaction method for reliable sensitive and simultaneous detection of multiple viruses in citrus trees. J. Virol. Methods 129, 47–55.

Roy, A., Shao, J., Schneider, W.L., Hartung, J.S., Brlansky, R.H., 2014. Population of endogenous pararetro virus genomes in Carrizo citrange. Genome Announc. 2 (1), e01063-13.

Roy, A., Hartung, J.S., Schneider, W.L., Shao, J., Leon, M.G., Melzer, M.J., et al., 2015a. Role bending: complex relationships between viruses, hosts, and vectors related to citrus leprosis, an emerging disease. Phytopathology 105, 872–884.

Roy, A., Stone, A.L., Shao, J., Otero-Colina, G., Wei, G., Choudhary, N., et al., 2015b. Identification and molecular characterization of nuclear *Citrus leprosis virus*, a member of the proposed *Dichorhavirus* genus infecting multiple citrus species in Mexico. Phytopathology 105, 564–575.

Rustici, G., Accotto, G.P., Noris, E., Masenga, V., Luisoni, E., Milne, R.G., 2000. Indian citrus ringspot: a proposed new species with some affinities to potex-, carla-, fovea- and allexiviruses. Arch. Virol. 145, 1895–1908.

Rustici, G., Milne, R.G., Accotto, G.P., 2002. Nucleotide sequence, genome organization and phylogenetic analysis of *Indian citrus ringspot virus*. Arch. Virol. 147, 2215–2224.

Sano, T., Hataya, T., Sasaki, A., Shikata, E., 1986. Etrog citron is latently infected with hop stunt viroid-like RNA. Proc. Jap. Acad. 62, 325–328.

Sano, T., Hataya, T., Shikata, E., 1988a. Complete nucleotide sequence of a viroids isolated from Etrog citron, a new member of the hop stunt viroid group. Nucleic Acids Res. 16, 347.

Sano, T., Kudo, H., Sugimoto, T., Shikata, E., 1988b. Synthetic oligonucleotide hybridization probes to diagnose hop stunt viroid strains and citrus exocortis viroid. J. Virol. Methods 19, 109–119.

Scott, J.W., Ge, X., 1995. The nucleotide sequence of *Citrus leaf rugose virus* RNA1. J. Gen. Virol. 76, 3233–3238.

Scuderi, G., Lombardo, A., Raspagliesi, D., Russo, M., Catara, A., Licciardello, G., 2016. Development and evaluation of a novel probe microarray for genotyping *Citrus tristeza virus* using an integrated lab-on-chip device. J. Plant Pathol. 98 (1), 25–34.

Semancik, J.S., 2003. Considerations for the introduction of viroids for economic advantage. In: Hadidi, A., Flores, R., Randles, J.W., Semancik, J.S. (Eds.), Viroids. CSIRO Publishing, Collingwood, VIC, pp. 357–362.

Semancik, J.S., Weathers, L.G., 1972. Exocortis virus: an infectious free-nucleic acid plant virus with unusual properties. Virology 46, 456–466.

Semancik, J.S., Roistacher, C.N., Rivera-Bustamante, R., Duran-Vila, N., 1988a. *Citrus* cachexia viroid, a new viroid of citrus: relationship to viroids of the exocortis disease complex. J. Gen. Virol. 69, 3059–3068.

Semancik, J.S., Roistacher, C.N., Duran-Vila, N., 1988b. A new viroid is the causal agent of the citrus cachexia disease. In: Proc. 10th Conf IOCV, pp. 125–135.

Semancik, J.S., Szychowski, J.A., Rakowski, A.G., Symons, R.H., 1994. A stable 463 nucleotide variant of citrus exocortis viroid produced by terminal repeats. J. Gen. Virol. 75, 727–732.

Semancik, J.S., Rakowski, A.G., Bash, J.A., Gumpf, D.J., 1997. Application of selected viroids for dwarfing and enhancement of production of 'Valencia' orange. J. Hort. Sci. 72, 563–570.

Serra, P., Gago, S., Duran-Vila, N., 2008a. A single nucleotide change in hop stunt viroid modulates citrus cachexia symptoms. Virus Res. 138, 130–134.

Serra, P., Barbosa, C.J., Daròs, J.A., Flores, R., Duran-Vila, N., 2008b. *Citrus* viroid V: molecular characterization and synergistic interactions with other members of the genus *Apscaviroid*. Virology 370, 102–112.

Serra, P., Eiras, M., Bani Hashemian, S.M., Murcia, N., Kitajima, E.W., Daròs, J.A., et al., 2008c. *Citrus* viroid V: occurrence, host range, diagnosis and identification of new variants. Phytopathology 98, 1199–1204.

Serra, P., Bani Hashemian, S.M., Pensabene-Bellavia, G., Gago, S., Duran-Vila, N., 2009. An artificial chimeric derivative of *Citrus viroid V* involves the terminal left domain in pathogenicity. Mol. Plant Pathol. 10, 515–552.

Serra, P., Pina, J.A., Duran-Vila, N., 2010. Identification and characterization of a variant of *Citrus* viroid V (CVd-V) in Seminole tangelo. In: Proc 17th Conf IOCV, pp. 150–157.

Serra, P., Bani Hashemian, S.M., Fagoaga, C., Romero, J., Ruiz-Ruiz, S., Gorris, M.T., et al., 2014. Virus-viroid interactions: *Citrus tristeza virus* enhances the accumulation of *Citrus dwarfing viroid* in Mexican lime via virus-encoded silencing suppressors. J. Virol. 88, 1394–1397.

Song, Z., Kurth, E.G., Peremyslov, V.V., Zhou, C.Y., Dolja, V.V., 2015. Molecular characterization of a *Citrus yellow vein clearing virus* strain from China. Arch. Virol. 160, 1811–1813.

Swingle, W.T., Webber, H.J., 1896. The principal diseases of citrus fruits in Florida. USDA Div. Veg. Physiol. Pathol. Bull. 8, 42.

Szychowski, J.A., Vidalakis, G., Semancik, J.S., 2005. Host-directed processing of *Citrus exocortis viroid*. J. Gen. Virol. 86 (2), 473–477.

Takahara, T., Kawase, K., Ono, S., Iwagaki, I., Hirose, K., Yoshinaga, K., 1988. Rootstocks for Ponkan (*Citrus reticulata* Blanco) in relation to tatter leaf virus. Bull. Fruit Tree Res. Sta. D 10, 35–45.

Tanner, J.D., Kunta, M., da Graça, J.V., Skaria, M., Nelson, S.D., 2011. Evidence of seed transmission of citrus tatter leaf virus in citrus. In: Proc Conf 18th IOCV. iocv.org/proceedings/eighteen/Tanner_et_al.pdf.

Tassi, A.D., Garita, L.C., Amorim, L., Novelli, V.M., Freitas-Astúa, J., Childers, C.C., et al., 2017. Virus-vector relationship in the citrus leprosis pathosystem. Exp. Appl. Acarol. 71 (3), 227–241.

Tataneni, S., Afunian, M.R., Hilf, M.E., Gowda, S., Dawson, W.O., Garnsey, S.M., 2009. Molecular characterization of *Citrus tatter leaf virus* historically associated with Meyer lemon trees: complete genome sequence and development of biologically active *in vitro* transcripts. Phytopathology 99, 423–431.

Tessitori, M., 2017. Apscaviroids infecting citrus trees. In: Hadidi, A., Flores, R., Randles, J.W., Palukaitis, P. (Eds.), Viroids and Satellites. Academic Press, pp. 243–249.

Timmer, L.W., Garnsey, S.M., 1980. Natural spread of citrus ringspot virus in Texas and its association with psorosis diseases in Florida and Texas. In: Proc 8th Conf IOCV, pp. 167–173.

Timmer, L.W., Lee, R.F., Brlansky, R.H., Graham, J.H., Albrigo, L.G., Derrick, K.S., Tucker, D.P.H., 1992. The infectious nature of citrus blight. Proc. Florida Hort. Soc. 105, 21.

Velázquez, K., Hervas, A., Pina, J.A., Moreno, P., Guerri, J., Navarro, L., et al., 2019. Complete sequence of three phlebo-like viruses associated to cristacortis, concave gum and impietratura diseases. In: 21st IOCV Conf Abstr., IOCV-02-20.

Vernière, C., Perrier, X., Dubois, C., Dubois, A., Botella, L., Chabrier, C., et al., 2004. *Citrus* viroids: symptom expression and effect on vegetative growth and yield on clementine trees grafted on trifoliate orange. Plant Dis. 88, 1189–1197.

Vernière, C., Perrier, X., Dubois, C., Dubois, A., Botella, L., Chabrier, C., et al., 2006. Interactions between citrus viroids affect symptom expression and field performance of clementine trees grafted on trifoliate orange. Phytopathology 96, 356–368.

Vidalakis, G., Wang, J., 2013. Molecular method for universal detection of citrus viroids. US Patent Publication number 20130115591, 1–20.

Vidalakis, G., Gumpf, D.J., Bash, J.A., Semancik, J.S., 2004. Finger imprint of *Poncirus trifoliata*: a specific interaction of a viroid, a host, and irrigation. Plant Dis. 88 (7), 709–713.

Vidalakis, G., Pagliaccia, D., Bash, J.A., Semancik, J.S., 2010. Effects of mixtures of citrus viroids as transmissible small nuclear RNA on tree dwarfing and commercial scion performance on Carrizo citrange rootstock. Ann. Appl. Biol. 157, 415–423.

Vidalakis, G., Pagliaccia, D., Bash, J.A., Afunian, M., Semancik, J.S., 2011. *Citrus dwarfing viroid*: effects on tree size and scion performance specific to *Poncirus trifoliata* rootstock for high-density planting. Ann. Appl. Biol. 158, 204–217.

Visser, M., Bester, R., Burger, J.T., Maree, H.J., 2016. Next-generation sequencing for virus detection: covering all the bases. Virol. J. 13, 85.

Visvader, J.E., Symons, R.H., 1985. Eleven new sequence variants of citrus exocortis viroid and the correlation of sequences with pathogenicity. Nucleic Acids Res. 13, 2907–2920.

Visvader, J.E., Symons, R.H., 1986. Replication of *in vitro* constructed viroid mutants: location of the pathogenicity modulating domain of citrus exocortis viroid. EMBO J. 5, 2051–2055.

Visvader, J.E., Gould, A., Buening, G., Symons, R.H., 1982. *Citrus exocortis viroid*: nucleotide sequence and secondary structure of an Australian isolate. FEBS Lett. 137, 288–292.

Vives, M.C., Galipienso, L., Navarro, L., Moreno, P., Guerri, J., 2001. The nucleotide sequence and genome organization of *Citrus leaf blotch virus*: candidate type species for a new genus. Virology 287, 225–233.

Vives, M.C., Galipienso, L., Navarro, L., Moreno, P., Guerri, J., 2002. Characterization of two kinds of subgenomic RNAs produced by *Citrus leaf blotch virus*. Virology 295, 328–336.

Vives, M.C., Pina, J.A., Juárez, J., Navarro, L., Moreno, P., Guerri, J., 2005. Dweet mottle disease probably is caused by *Citrus leaf blotch virus*. In: Proc 16th Conf IOCV, pp. 251–256.

Vives, M.C., Martín, S., Ambrós, S., Renovell, A., Navarro, L., Pena, J.A., et al., 2008. Development of a full-genome cDNA clone of *Citrus leaf blotch virus* and infection of citrus plants. Mol. Plant Pathol. 9, 787–797.

Vives, M.C., Velázquez, K., Pina, J.A., Moreno, P., Guerri, J., Navarro, L., 2013. Identification of a new enamovirus associated with citrus vein enation disease by deep sequencing of small RNAs. Phytopathology 103, 1077–1086.

Wallace, J.M., 1959. *Citrus* Virus Diseases. Univ Calif Div Agric Sci.
Wallace, J.M., 1978. Virus and virus-like diseases. In: Reuther W. Webber, HJ, Batchelor LD, editors. The Citrus Industry. Univ. Calif. Div. Agric. Sci. 4, 67–184.
Wallace, J.M., 1981. Vein enation-woody gall. In: Bové, J.M., Rossetti, V.R. (Eds.), Description and Illustration of Virus and Virus-like Diseases of *Citrus*. A Collection of Color Slides. vol. 6. IRFA, Paris.
Wallace, J.M., Drake, R.J., 1953. A virus induced vein enation in citrus. Citrus Leaves 33, 22–24.
Wallace, J.M., Drake, R.J., 1961. Induction of woody galls by wounding citrus infected with vein enation virus. Plant Dis. Rep. 45, 682–683.
Wallace, J.M., Drake, R.J., 1968. Citrange stunt and ringspot, two previously undescribed virus diseases of citrus. In: Proc 4th Conf IOCV, pp. 177–180.
Wang, X.F., Zhou, C.Y., Tang, K.Z., Zhou, Y., Li, Z.A., 2009. A rapid one-step multiplex RT-PCR assay for the simultaneous detection of five citrus viroids in China. Eur. J. Plant Pathol. 124 (1), 175–180.
Wang, J., Zhu, D., Tan, Y., Wei, H., Liu, Q., 2016. First report of *Citrus leaf blotch virus* in sweet cherry. Plant Dis. 100, 1027.
Weathers, L.G., 1960. Yellow-vein disease of citrus and studies of interactions between yellow-vein and other viruses of citrus. Virology 11, 753–764.
Weathers, L.G., Greer Jr., F.C., 1968. Additional herbaceous hosts of the exocortis virus of citrus. Phytopathology 58, 1071.
Webber, H.J., 1967. History and development of the citrus industry. In: Reuther W, Webber, HJ, Batchelor LD, editors. The Citrus Industry. Univ. Calif. Div. Agric. Sci. 1, 1–39.
Wu, Q., Cao, M.J., Su, H.N., Atta, S., Yang, F.Y., Wang, X.F., et al., 2014. Molecular characterization and phylogenetic analysis of *Citrus* viroid I-LSS variants from citrus in Pakistan and China reveals their possible geographic origin. Eur. J. Plant Pathol. 139 (1), 13–17.
Wutscher, H.K., Schwarz, R.E., Campiglia, H.G., Moreira, C.S., Rossetti, V., 1980. Blightlike citrus tree declines in South America and South Africa. Hort. Sci. 15, 588–590.
Yamada, S., Sawamura, K., 1952. Studies on the dwarf disease of satsuma orange, *Citrus unshiu* Marcovitch-preliminary report. Hort. Div. Tokai-Kinki Agr. Exp. Sta. Bull., 67–71.
Yamamoto, P.T., Bassanezi, R.B., Wulff, N.A., Santos, M.A., Sanches, A.L., Toloy, R.S., et al., 2011. *Citrus* sudden death is transmitted by graft-inoculation and natural transmission is prevented by individual insect-proof cages. Plant Dis. 95, 104–112.
Yokomi, R.K., Selvaraj, V., Maheshwari, Y., Saponari, M., Giampetruzzi, A., Chiumenti, M., et al., 2017. Identification and characterization of *Citrus tristeza virus* isolates breaking resistance in trifoliate orange in California. Phytopathology 107 (7), 901–908.
Yot-Dauthy, D., Laflèche, D., Bové, J.M., 1972. *Citrus*, a local lesion host of tobacco necrosis virus. In: Proc 5th Conf IOCV, pp. 212–216.
Yu, Y.Q., Wu, Q., Su, H.N., Wang, X.F., Cao, M.J., Zhou, C.Y., 2017. Small RNA deep sequencing reveals full-length genome of *Citrus yellow vein clearing virus* in Chongqing. China J. Integ. Agr. 16 (2), 503–508.
Zhang, T.M., Liang, X.Y., Roistacher, C.N., 1988. Occurrence and detection of citrus tatter leaf virus (CTLV) in Huangyan, Zhejiang Province, China. Plant Dis. 72, 543–545.
Zhang, Y.H., Liu, C.H., Wang, Q., Wang, Y.L., Zhou, C.Y., Zhou, Y., 2019. Identification of *Dialeurodes citri* as a vector of *Citrus yellow vein clearing virus* in China. Plant Dis. 103, 65–68.
Zhang, Y.H., Wang, Y.L., Wang, Q., Cao, M.J., Zhou, C.Y., Zhou, Y., 2018. Identification of *Aphis spiraecola* as a vector of *Citrus yellow vein clearing virus*. Eur. J. Plant Pathol. 152, 841–844.
Zhou, C.Y., Li, Z.A., 2010. Current occurrence and research status of graft-transmissible citrus diseases in PR China. In: Proc 18th Conf IOCV, pp. 44.
Zhou, C.Y., Zhao, X.Y., Jiang, Y.H., He, X.H., 1990. Identification of satsuma dwarf disease of satsuma mandarin. J. Southwest Agr. Univ. 12 (4), 346–348. (in Chinese).
Zhou, C.Y., Broadbent, P., Hailstones, D.L., Bowyer, J., Connor, R., 2002. Movement and titer of *Citrus tristeza virus* (pre-immunizing isolate PB61) within seedlings and field trees. In: Proc 15th Conf IOCV, pp. 39–47.
Zhou, Y., Chen, H.M., Cao, M.J., Wang, X.F., Jin, X., Liu, K.H., et al., 2016. Occurrence, distribution, and molecular characterization of *Citrus yellow vein clearing virus* in China. Plant Dis. 101, 137–143.
Zhou, Y., Liu, Y.J., Liu, K.H., Yang, F.Y., Zhou, C.Y., 2017a. Distribution and population structure of *Citrus tristeza virus* in *Poncirus trifoliata*. Australas. Plant Pathol. 46, 351–355.
Zhou, Y., Zhang, Y., Liu, Y., Chen, H., Li, T., Zhou, C., 2017b. Distribution and molecular characterization of citrus chlorotic dwarf-associated virus in China. Australas. Plant Pathol. 46 (3), 227–229.

Chapter 20

Horticultural practices

Fernando Alferez
University of Florida, Institute of Food and Agricultural Sciences, Southwest Florida Research and Education Center, Immokalee, FL, United States

Grove management starts with planning of the planting site, and by comprehending every operation until harvesting time in order to achieve the best marketable fruit yield and quality. It includes horticultural practices such as grove establishment, irrigation, nutrition, canopy management, and grove protection. Importantly, fruit quality starts in the grove. In this chapter we are focusing on grove planning with special focus on tree spacing, canopy management, and tree size control, and crop load management from a horticultural point of view.

20.1 Grove planning and tree spacing

Good grove design and management has, among others, the goal of providing a canopy structure able to capture optimum light intensity and minimize competition from neighboring plants. It is well known that for a given cultivar and environment combination, an optimum planting density exists for maximizing yields (Schumann et al., 2012). In planning a citrus grove, attention must be paid to site selection, choosing an adequate variety (scion) and rootstock combination to better adapt to soil characteristics and environmental conditions, and determining the correct tree spacing.

20.1.1 Site selection

According to Futch (1992), several factors determine the suitability of a site for citrus production, among others, well drained and fertile soil and location as it relates to extreme climate exposure, with susceptibility to freeze damage being the most important consideration. In this regard, topography is also a factor to consider, since during radiational freezes cold air settles in the lowest area, resulting in temperatures a few degrees lower than that in slightly higher surrounding areas, whereas during freezes with strong northerly winds, groves on exposed northern slopes and atop small ridges may be subjected to colder temperatures than those occurring on southern slopes (Sauls, 2008).

In general, an ideal citrus grove has good drainage, balanced soil pH, and access to sufficient good quality water for irrigation. The optimum pH of the irrigation water ranges between 6.0 and 6.5. In high-pH soils, a pH-tolerant rootstock variety is recommended (pH sensitivity: Swingle > Carrizo > Sour orange >Cleopatra). The best quality water should not be high in bicarbonates and salt levels (Zekri et al., 2017). Alkalinity caused by carbonate ($CO_3^{=}$), bicarbonate (HCO_3^-), and hydroxyl (OH^-) anions is detrimental to HLB (Huanglongbing)-affected trees as it decreases nutrient uptake and root longevity. In the cases of soils high in bicarbonates, lowering of soil pH through the acidification of irrigation water has proven to be a strategy that releases Ca and Mg from bicarbonate and increases the availability of soil Mn, Zn, and Fe for root uptake (Graham and Morgan, 2015).

20.1.2 Variety and rootstock selection

There are other chapters in this book specific to rootstock and scion selection (see Chapters 5 and 6). However, it is worth mentioning that in order to select the right rootstock, it is important to consider previous and accumulated experience, field and research data, and rootstock-scion compatibility (Castle and Futch, 2015). According to Castle (2010), nowadays breeding citrus rootstocks places emphasis on reducing tree size beyond what is currently possible among commercial rootstocks. In any case, it is desirable for the industry to have different choices to select the adequate rootstock that matches the grove's space needs. Thus, size-controlling rootstocks are important for higher density orchards, and rootstocks that induce scion precocity will allow significant cropping to begin early. Newly produced rootstocks should lead to excellent yields of

The Genus Citrus. https://doi.org/10.1016/B978-0-12-812163-4.00020-6

high-quality fruit and possess other key traits, such as disease tolerance, to be successful. This is especially important under HLB and other biotic and abiotic threats.

20.1.3 Tree spacing

Historically, according to Hutton (1989) worldwide tendencies in land management involve, and are influenced by the following aspects:

(1) Reduced availability of suitable citrus land, with a consequent development of poor soil types which influence tree growth rate and ultimately size and productivity.
(2) Increasing land values and taxes.
(3) Modification of zoning laws affecting the agriculture.
(4) Tree losses from freezes, diseases, and declines.
(5) Increased expenditures for equipment, energy, irrigation systems, and labor.

These factors are increasingly important, as reviewed by Nawaz et al. (2010). To remain profitable, citrus production must use resources more efficiently, and this can be achieved with adequate planning that considers tree spacing as a fundamental characteristic of the grove.

Tree spacing is an important factor determining profitability of a grove. Since sunlight is the source of energy for tree and fruit growth, grove design must address the arrangement of tree canopies to capture irradiance efficiently. Tree rows are typically oriented north to south for maximum sunlight interception. However, row orientation may also depend on the row length and water drainage direction at the site (Zekri et al., 2017). Tree spacing should be based on the vigor of the scion-rootstock combination and the expected life of the grove. The ability of citrus to adapt to a wide range of spacing is well known. In this regard, tree vigor is of fundamental importance in determining tree spacing, density, and pruning regime. Trees of low vigor will never fill their allocated space, preventing generation of maximum economic returns if planted in a wide plantation frame. On the other hand, placing vigorous trees in small spaces will result in poor production and will carry management problems (Wheaton et al., 1995). Tree spacing affects root densities, which may be related to water use. In general, in healthy citrus trees, higher tree densities result in higher root densities (Kaufmann et al., 1972; Castle, 1980).

Tree spacing widely affects the yield and quality of citrus fruit. According to Whitney et al. (1991) citrus fruit yields per unit of land area are related to tree spacing in the early bearing years (Boswell et al., 1975; Phillips, 1974; Wheaton et al., 1986). However, as the trees grow and compete, fruit yields per unit land area become independent of tree spacing, and may even decline at the closer spacings (Boswell et al., 1975). Tree spacing is an increasingly important consideration in citrus grove management due to increasing operating costs and less economic returns to the grower, as well as disease pressure that reduces yield and quality. Generally, closely planted groves provide greater and earlier returns but with a demand for precise management; operations such as spray coverage and harvesting are facilitated at higher density plantings (Phillips, 1978; Zekri, 2000). Higher density plantings require hedging and topping to maintain good light interception. In mature citrus groves that are not hedged regularly, insufficient light and increased shading promote a thin canopy and reduction in fruiting on the lower parts of the tree (Rouse et al., 2006). Factors that affect light interception in a mature grove include width of the drive middle, foliage wall angle (hedging angle), tree height, and row orientation.

With HLB pressure more and more prevalent worldwide, grove management, including site planning and tree spacing, is rapidly evolving toward higher density plantings. The concept of high-density plantings is not new in citrus (see Tucker and Wheaton, 1978, for a comprehensive review), but in the last years the trend toward higher density plantings has increased in search of faster economic returns under HLB pressure. Higher density plantings are an essential component of newer production systems such as CUPS (citrus under protective screen) aimed at excluding the Asian citrus psyllid (*Diaphorina citri*) and therefore HLB. In this case, planting densities may range from 871 to 1361 trees per acre and involve intensive hydroponics with daily or hourly delivery of nutrients by drip fertigation (Schumann et al., 2017).

20.2 Irrigation and water management planning

Irrigation strategies should allow growers to maintain or increase crop production without exhausting water resources. By selecting a proper irrigation scheduling method and application time, water use efficiency could be increased. This is especially important in groves affected by HLB, since water stress can produce a deleterious change in physiological activity of growth and production of citrus trees. It is recommended that for HLB-affected trees, irrigation frequency should be increased and irrigation amounts should be decreased to minimize water stress from drought stress or water shortage, while ensuring optimal water availability in the root zone at all times. It is recommended that growers maintain soil moisture in the root zone (top 3 ft. for Ridge and 18 in. for flatwood soils, in Florida for example) using soil moisture sensors or irrigation apps (Kadyampakeni et al., 2014; Migliaccio et al., 2016).

20.2.1 Water management

In poorly drained soils, control of high water tables and rapid removal of excess surface water from rainfall are essential for citrus production (Chapters 13 and 14). Soils with poor drainage are not adequate for citrus plantation as excess water supply is not properly evacuated. Drainage systems generally include beds, water furrows, lateral ditches, collector ditches, and may include perimeter ditches and discharge pumps. Drainage water from several lateral ditches usually is accumulated in collector ditches and conveyed off-site. This may be achieved by gravity or by discharge pumps. The size of the collector ditches and related pumping facilities depend on factors such as size of the area being served, soils, bed and water furrow design, and slope of ditches (Boman et al., 2018).

20.3 Canopy management and tree size control

Tree canopy and bearing volume are two important factors in fruit production and fruit quality. Generally, citrus trees with larger canopy volumes produce more fruit than smaller canopy trees (Vashisth et al., 2017). Tree size control through canopy management of citrus trees is essential to maintain and improve tree health and increase productivity and fruit quality. In addition, from a managerial perspective, it allows better efficiency while performing regular operations in the grove, since machinery can operate faster and chemical application can better reach targeted areas of the trees.

Canopy management mostly involves pruning operations. Pruning is one of the oldest horticultural practices. It changes the form and growth of a tree. In general, the pruning process has the following characteristics:

(1) Adjusts tree shape and the canopy's ratio of framework to fruit-bearing shell.
(2) Alters the shoot-root ratio.
(3) Changes the carbohydrate status of the tree (Boswell and Cole, 1978a).

When considering a citrus tree canopy, trees typically have 3–5 growth flushes a year, although this depends on growing conditions, i.e., management practices and geographical area. In general, the outer 90 cm of the spherical shaped citrus tree canopy produces most of the fruit. It is estimated that 90% of the solar radiation is absorbed by this outer 90 cm of canopy. When trees are small or in a narrow hedgerow, the total canopy may be considered to be productive (Rouse et al., 2006). However, when trees are large, the inner nonproductive part (scaffold limbs and branches inside the outer 90-cm shell) can be a major component of the total canopy volume. In every flushing period, new flushes add on established flushes, resulting in a drift of the young bearing wood to the outside of the tree canopy (Hardy, 2004). This results in larger trees with increased shading inside the canopy resulting in most of the fruit being carried on the tops and outside of trees. In general, through pruning we are able to remove weak branches and rejuvenate trees, reduce shading while improving the light interception, improve air circulation through the canopy, reduce pest and disease pressure, enhance tree performance, alleviate alternate bearing, increase fruit size and improve fruit quality, and facilitate harvesting.

The type of production will determine the type of pruning to be adopted. There are great differences in the way that tree canopies are managed depending on geographical area (i.e., Mediterranean, milder climates, or subtropical and tropical climates) and type of the fruit produced (fresh fruit or processing fruit varieties). In general, a massive production aimed at processing fruit for juice extraction demands mechanical topping, skirting, and hedging the trees with the objective of inducing lateral secondary branching while facilitating grove operations. In general, hedging down the tree row needs to maintain 8-ft width for passage of equipment, while efficiency of machinery is enhanced with longer tree rows, and turn space is needed at end of rows to accommodate large machines (Rouse and Futch, 2004). This affects also the grove design, as mentioned above. In contrast, high-quality fruit for fresh market demands more specialized manual pruning, with the goal of achieving better fruit coloration and absence of peel blemishes while facilitating manual, careful harvesting by the piece.

20.3.1 Mechanical pruning cuts

This topic has been extensively reviewed elsewhere (Rouse et al., 2006; Zekri, 2015; Vashisth et al., 2017). There exist three types of basic mechanical cuts: hedging, topping, and skirting, which are performed with special machinery and constitute important cultural grove practices. In general, tree response to these depends on several factors including variety, rootstock, tree age, growing conditions, time of pruning, and production practices. Main characteristics and uses of these mechanical pruning cuts are explained below.

20.3.1.1 Hedging

Hedging is cutting back the sides of trees to prevent or alleviate crowding (Fig. 20.1A). Hedging produces numerous cut wood surfaces along the side of the tree canopy from which new sprouts arise and eventually develop into a wall of new foliage (Vashisth et al., 2017). Middles (alleys) between tree rows should be sufficiently wide to accommodate grove equipment and provide adequate light access to the sides of the trees. Middles are usually hedged to a width of 7–8 ft. in Florida, but vary depending on original grove design, scion variety, rootstock, and equipment used in all production practices. Hedging is usually done at an angle, with the boom tilted inward toward the treetops so that the hedged row middles are wider at the bottom than at the top. This angled hedging allows more light to reach the lower skirts of the tree (Vashisth et al., 2017). Furthermore, to be more effective hedging is started before the canopy gets overgrown, and branches targeted are below 0.13″–0.25″ in diameter. Cutting these branches induces new flushing of fruit-bearing shoots (Fig. 20.1B and C).

20.3.1.2 Topping

Topping is the most invigorating form of pruning. For this reason, if done at the wrong time it can result in excessive upright vegetative growth, which causes shading and uses the tree's water and nutrient reserves (Krajewski, 2002; Hardy, 2004). Topping involves cutting off vegetation in the top of the tree, and ideally is performed before trees become excessively tall. No studies have shown an increase in yield due to topping, however, fruit quality is improved and fruit size increases, while management and harvesting costs are reduced (Vashisth et al., 2017). Trees can be flat-topped or roof-topped (topped at angles) (Rouse et al., 2006). Some common topping heights are 10–12 ft. at the shoulder and 13–14 ft. at the peak. As a general rule, topping heights are two times the width of the row middle (Vashisth, 2018).

FIG. 20.1 Mechanical hedging: (A) hedging of a high-density row planting of Valencia trees on the dwarfing rootstock Flying Dragon in southwest Florida in February, (B) massive lateral new flushing in early April after hedging, and (C) details of a pruned lateral branch showing new flushes with bearing capacity.

According to Wheaton (1986, 1997) topping and hedging can be scheduled to reduce crop load during heavy crop years. These operations are usually performed in winter or the early spring before flowering occurs if there is certainty that the coming crop load will be excessive. Alternatively, the decision is made after flowering and fruit set is largely completed. The benefits of reducing crop load are greater the earlier in the season it is done. Thus, it is important to complete the hedging and/or topping operations as soon as possible after crop load is estimated.

20.3.1.3 Skirting

Skirting raises tree skirts. This allows movement of herbicide booms, and mechanical harvesting and other equipment. Fruit and limbs near the ground are often damaged by the passage of such equipment and by herbicide spray and fertilizer contact. Skirting allows uniform distribution of granular fertilizers, good water coverage by microsprinkler irrigation systems, and more uniform herbicide application under tree canopies. Skirting facilitates the inspection and repair of microirrigation systems and reduces the incidence of Phytophthora foot rot and brown rot as it allows good air circulation (Zekri, 2015). In addition, since herbicide booms do not make contact with low hanging foliage and fruit, the chance for spread of canker is reduced (Rouse et al., 2006). Data from a 5-year study showed that average yields were not affected by skirting. In contrast, although fruit from unskirted trees had juice quality equivalent to other fruits on the tree, the fruit had increased blemishes and were smaller. Skirted trees had reduced yield the following year, but rebounded with yields above unskirted trees (Whitney et al., 2003). The current machinery available for mechanical harvesting makes skirting necessary for building a tree that facilitates the operations of this kind of massive harvester.

20.3.2 Manual pruning

Manual pruning involves several practices that comprise all stages of tree development, from young tree formation to maintenance pruning of bearing trees, and rejuvenation pruning of old trees. In the following, these aspects are explained:

20.3.2.1 Young tree formation

The main goal is to identify and preserve branches which will make good future scaffold limbs (those well orientated/angled) and to remove unwanted branches (those poorly orientated/angled) which will have an adverse effect on tree development and production (Taverner et al., 2011). Sometimes, selecting ideal scaffold branches present problems in the grove. For example, regrowth after pruning is unpredictable, and branches, if bent by their own weight, can provoke a shift in dominance. These facts force a careful follow up of the grove after selective formation pruning of young trees (Medina-Urrutia et al., 2002).

20.3.2.2 Young tree training

This type of pruning is performed in the first years (2–6 years, depending on the regions) after the scaffold branches have been established, and is aimed at maintaining the already selected scaffold branches by selectively pruning misplaced new growth such as water sprouts or gourmands. This has to be performed annually and involves no more than 10% of the canopy volume in order to avoid induction of excessive vegetative growth that results in yield reduction (Medina-Urrutia et al., 2002; Zaragoza et al., 2003).

20.3.2.3 Clearing internal canopy

This pruning eliminates vertical branches in the interior of the canopy and is performed mostly in adult trees. These usually vigorous branches do not bear any fruit, and reduce internal air circulation and light penetration (Curtí-Díaz et al., 2000). The removal of this type of branch induces better fruit coloration and reduces peel blemishes caused by wind (wind scars), as well to minimize associated disease problems. It may involve removal of up to 50% of the canopy in some cases.

20.3.2.4 Rejuvenating pruning

In older or poorly managed trees, this type of severe pruning will shape again the tree with newly selected scaffold branches. However, in subtropical areas this type of pruning may delay the new production by 3 years (Kretchman and Jutras, 1962; Kretchman and Krezdorn, 1962; Lewis et al., 1963). This practice is usually performed to remediate trees declining due to age, presence of diseases and pests in the soil, or root loss. However, if scaffold and secondary branches are healthy, this type of pruning does not improve the tree (Boswell and Cole, 1978b).

20.3.3 Effects of pruning on tree physiology

In general, tree response to pruning depends on several factors including variety, rootstock, tree age, growing conditions, timing and type of pruning, and production practices. Pruning affects physiology of citrus trees in several ways, which include changing branch dominance by affecting the number and disposition of buds in the tree canopy (Krajewski and Krajewski, 2011), changing canopy lighting (Krajewski and Pittaway, 2000), and altering air circulation. It also has a clear effect on fruit quality (Krajewski and Krajewski, 2011). According to Tucker et al. (1991), the removal of the terminal bud destroys apical dominance so one or several lateral buds will commence to grow, leading to new branching. Apical dominance explains many of the growth characteristics of trees and their responses to pruning. Severe pruning if performed in late spring or summer may induce only vegetative growth, diminishing fruit production (Boswell and Cole, 1978a; Tucker et al., 1991).

Light penetration is a very important factor for fruit production and quality. In some varieties such as grapefruit, light intensity inside the canopy can be as low as 2% of total sun irradiance if the trees were not pruned (Fischler et al., 1983). According to Krajewski and Pittaway (2000), of all cultural practices, only pruning and training can optimize light distribution and sap flow to fruits. Indeed, in well-managed groves correct pruning increases fruit quality and yield (Bevington et al., 2002). Pruning may affect nitrogen (N) requirements. If N is applied after a severe pruning, there may be excessive vegetative growth and low-quality fruit. In contrast, if pruning is light, N requirements are not affected (Tucker et al., 1991).

Pruning may also have deleterious effects on the tree, affecting fruit production. The inadequate pruning of healthy, mature citrus trees can reduce yields in proportion to the amount of foliage removed, and can delay fruiting of young, nonbearing trees as well (Vashisth et al., 2017). In general, the disadvantages are: (1) reduction in fruit yield when pruning is severe in young nonbearing trees and (2) reduction in stored reserves, including foliage stored carbohydrates and minerals such as Ca and N, and to a lesser extent K, Mg, P, and Na (Calabrese and Pino, 1992). In addition, severe hedging or topping of citrus trees during the winter can reduce cold hardiness because trees with exposed internal scaffold wood and new tender growth are susceptible to cold injury (Zekri, 2015). Other problems associated with these practices when not performed routinely are that long intervals between prunings increase the cost of the operation due to heavy cutting and more brush disposal (Zekri, 2015).

20.3.4 Other considerations

Pruning may affect disease incidence, and at the same time, can be used as a disease management tool in areas where citrus canker, citrus black spot, and HLB, among others, are present. Pruning results in vegetative regrowth that must be carefully protected from all diseases and pests, including vectors transmitting these diseases that feed on new flushes (Krajewski and Krajewski, 2011). Attention to the management of insects that feed on new foliar flush growth has increased as a result of the establishment of invasive plant diseases associated with insects that develop exclusively on these new flushes. Citrus can be monitored to identify peak periods of flush abundance to time insecticide applications for these insects (Hall and Albrigo, 2007). On the other hand, it is recommended to avoid pruning and generating vegetative regrowth at any time when activity of leaf pests such as aphids, thrips, psyllids, and lepidoptera larvae is high. Citrus leaf miner feeding activity aggravates severity of canker (Gottwald et al., 2002) and psyllid feeding activity hastens spread of HLB (Catling, 1969; Gottwald et al., 2007).

20.4 Crop load management

For satisfactory management of groves, a regular production through the seasons is desirable. Alternate bearing is a major problem in citrus production worldwide, especially with mandarin cultivars (Wheaton, 1992). Alternate bearing trees produce a heavy (on) crop followed by a light (off) crop. The impact of alternate bearing on fruit size and quality for fresh market fruit (Chapter 11) is a major factor affecting profitability. Alternate bearing is a common phenomenon in many-seeded citrus cultivars. In these cultivars, high yields are followed by extremely low yields and vice versa. Lack of flowering reduces the following season's yield, and this is the cause of high flower intensity and subsequent yield (Monselise and Goldschmidt, 1982). Also, the cultural practice of holding fruit on the tree to extend the commercial harvest period during an on-crop year exacerbates alternate bearing (Verreyne and Lovatt, 2009).

Carbohydrate availability coupled with hormonal balance in the fruit determine fruit set and development. In alternate bearing sweet orange Salustiana, the accumulation of reserves is inversely related to crop load (Monerri et al., 2011), and changes in carbohydrate reserves during the year reflect variations in supply and demand. Fruiting trees accumulate most fixed carbon in fruits, while no accumulation is observed in roots before harvest. In the no-fruiting trees, however, most

TABLE 20.1 Plant growth regulator general uses.

Variety	Growth regulator	Time of application	Effect
Sweet orange, Temple, grapefruit	2,4-D	November–December	Reduction in preharvest fruit drop
Navel orange	2,4-D	6–8 weeks after bloom	Reduction of June drop
Tangerine and Murcott	NAA	Mid May	Fruit thinning
Seedless grapefruit	GA_3	November–January	Delay of peel senescence process and peel color development in mature-green fruit
Tangelos	GA_3	Full bloom	Improvement of fruit set
Minneola tangelo	GA_3	Early December	Delay of stem-end rind breakdown
Valencia orange	Homobrassinolides	After June drop	Better fruit growth, color development and retention

fixed carbon is transported to roots and utilized in growth processes, and after December, stored as reserves. Reserve carbohydrate accumulation in leaves starts by early December, and the levels in leaves are, until bud sprouting, the same in both the "on" and "off" trees. The heavy flower formation which follows an "off" year causes the rapid mobilization of the stored reserves, which are exhausted at full bloom (reviewed by Nebauer et al., 2014).

Endogenous hormonal levels play a major role in determining fruit set in citrus; this topic has been extensively researched and reviewed elsewhere (Talón et al., 1990, 1992, 1999; Mesejo et al., 2016; see also Chapter 11). In general, exogenous applications of gibberellic acid increase the sink capacity of the fruit (Agustí et al., 1982), and branch girdling enhances carbohydrate availability (Wallerstein et al., 1978; Goldschmidt et al., 1985). The former cannot be adopted in many cases since some varieties are insensitive to gibberellins, whereas the latter can be used always in healthy trees.

Reducing the number of fruit (fruit thinning), manually or with chemical substances, is one of the most common techniques to increase fruit size. Manual thinning is more efficient during physiological drop (Zaragoza et al., 1992). In this case, however, the high number of fruitlets make it economically not viable. Thinning may also reduce alternate bearing (Jones et al., 1977; Monselise et al., 1981) and tree collapse in years "on" (Monselise and Goldschmidt, 1982; Chapman, 1984).

Chemical thinning can be used as well for regulating fruit number and size (El-Otmani et al., 2000). Substances such as naphtalenic acid, ethephon, and ethylclozate if applied at the cell division stage can induce fruitlet abscission, and effectively reduce fruit crop load. These treatments, however, lose efficacy after fruit set, if applied during the cell elongation phase (Agustí et al., 1995). Chemical thinning has more effect on smaller fruitlets that will produce fruit of smaller size (Kamuro and Hirai, 1981). For this reason, eliminating these smaller fruits at a stage when the fruitlets are responsive to the chemicals applied increases average fruit weight (Agustí and Almela, 1984).

Table 20.1 summarizes the applications of plant growth regulators as spray treatments. The table is a general summary of documented effects regardless of geographical/climatic aspects of the citrus-growing region.

References

Agustí, M., Almela, V., 1984. Mejora de la calidad del fruto de la mandarina Satsuma. Bco. De Santander, Madrid, España, ISBN: 84-398-1798-3.

Agustí, M., García-Marí, F., Guardiola, J.L., 1982. Gibberellic acid and fruit set in sweet orange. Sci. Hortic. 17, 257–264.

Agustí, M., El-Otmani, M., Aznar, M., Juan, M., Almela, V., 1995. Effect of 3,5,6-trichloro-2-pyridyloxyacetic acid on 'Clementine' early fruitlet development and on fruit size at maturity. J. Hortic. Sci. 70, 955–962.

Bevington, K.B., Falivene, S., Moulds, G., Krajewski, A.J., 2002. Pruning *Citrus* for Export Fruit Size. CMDG Final Report Project 50, New South Wales Dept. of Agric., Australia.

Boman, B., Wilson, C., Jennings, M., Shukla, S., 2018. Detention/Retention for *Citrus* Stormwater Management. EDIS Circular 1405. Agricultural and Biological Engineering Department, UF/IFAS Extension.

Boswell, S.B., Cole, L.A., 1978a. Pruning citrus in California: a review of origins and history. Citrograph 63 (4), 80–84.

Boswell, S.B., Cole, L.A., 1978b. Pruning citrus in California: a review of origins and history. Citrograph 63 (5), 109–113.

Boswell, S.B., McCarty, C.D., Hench, K.W., Lewis, L.N., 1975. Effect of tree density on the first ten years of growth and production of 'Washington' navel orange trees. J. Amer. Soc. Hort. Sci. 100 (4), 370–373.

Calabrese, F., Pino, M., 1992. Pruning and mineral losses from mandarin orchards in Sicily. Proc. Int. Soc. Citriculture 2, 690–692.

Castle, W.S., 1980. Fibrous root distribution of 'Pineapple' orange trees on rough lemon rootstock at three tree spacings. J. Amer. Soc. Hort. Sci. 105 (3), 478–480.

Castle, W.S., 2010. A career perspective on citrus rootstocks, their development, and commercialization. Hortic. Sci. 45, 11–15.

Castle, W.S., Futch, S.H., 2015. Choose the Right Rootstock. EDIS HS932. Horticultural Sciences Department, UF/IFAS Extension.

Catling, H.D., 1969. The bionomics of the South African citrus psylla, Trioza erytheae (Del Guercio) (Homoptera: Psyllidae). 1. The influence of the flushing rhythm ofcitrus and factors that regulate flushing. J. Ent. Soc. S. Africa 32, 191–208.

Chapman, J.C., 1984. Ethephon as a fruit thinning agent for 'Murcott' mandarins. Sci. Hortic. 24, 135–141.

Curtí-Díaz, S., Loredo–Salazar, X., Díaz-Zorrilla, D., Sandoval-Rincón, J.A., Hernández-Hernández, J., 2000. Tecnología para producir limón Persa. INIFAP. CIRGOC-Campo Experimental Ixtacuaco. Libro técnico # 8. 145 p.

El-Otmani, M., Coggins Jr., C.W., Agustí, M., Lovatt, C.L., 2000. Plant growth regulators in citriculture: world current uses. Crit. Rev. Plant Sci. 19, 395–447.

Fischler, M., Goldschmidt, E.E., Monselise, S.P., 1983. Leaf area and fruit size on girdled grapefruit branches. J. Amer. Soc. Hort. Sci. 108 (2), 218–221.

Futch, S.H., 1992. Site Selection. Fact Sheet HS-141. Horticultural Sciences Department, UF/IFAS Extension.

Goldschmidt, E.E., Aschkenazi, N., Herzano, Y., Schaffer, A., Monselise, S.P., 1985. A role of carbohydrate levels in the control of flowering in citrus. Sci. Hortic. 26, 159–166.

Gottwald, T.R., Graham, J.H., Schubert, T.S., 2002. *Citrus* canker: the pathogen and its impact. Plant Health Prog. https://doi.org/10.1094/PHP-2002-0812-01-RV (Online).

Gottwald, T.R., da Graça, J.V., Bassanezi, R.B., 2007. *Citrus* Huanglongbing: the pathogen and its impact. Plant Health Prog. https://doi.org/10.1094/PHP-2007-0906-01-RV (Online).

Graham, J.H., Morgan, K., 2015. Managing excessive bicarbonates with acidification. Citrus Industry (May 2015).

Hall, D.G., Albrigo, L.G., 2007. Estimating the relative abundance of flush shoots in citrus with implications on monitoring insects associated with flush. Hortic. Sci. 42, 364–368.

Hardy, S., 2004. Growing Lemons in Australia. A Production Manual. NSW Department of Primary Industries, ISBN: 0-7347-1611-7.

Hutton, R.J., 1989. The influence of tree size control and plant density on citrus productivity. Symposium on physiology of productivity of subtropical and tropical tree fruits. Acta Hortic. 175, 12–17.

Jones, W.W., Coggins Jr., C.W., Embleton, T.W., 1977. Growth regulators and alternance bearing. Proc. Int. Soc. Citriculture 2, 657–660.

Kadyampakeni, D.M., Morgan, K.T., Schumann, A.W., Nkedi-Kizza, P., 2014. Effect of irrigation pattern and timing on root density of young citrus trees infected with Huanglongbing disease. Hort. Tech. 24 (2), 209–221.

Kamuro, Y., Hirai, K., 1981. Physiological activity of Ethylchlozate. Fruit thinning and maturity accelerating effects for citrus. Proc. Int. Soc. Citriculture 1, 260–263.

Kaufmann, M.R., Boswell, S.B., Lewis, L.N., 1972. Effect of tree spacing on root distribution of 9-year-old 'Washington' navel orange. J. Amer. Soc. Hort. Sci. 97 (2), 204–206.

Krajewski, A., 2002. Pruning for Improving Quality in *Citrus*. Fruit Doctors.

Krajewski, A.J., Krajewski, A.J., 2011. Canopy management of sweet orange, grapefruit, lemon, lime and mandarin trees in the tropics: principles, practices and commercial experiences. In: Proc. Ist IS on Trop. Hort. Ed.: N. Benkeblia, Acta Hort. 894, ISHS 2011.

Krajewski, A.J., Pittaway, T., 2000. Manipulation of citrus flowering and fruiting by pruning. Proc. Int. Soc. Citriculture 357–360.

Kretchman, D.W., Jutras, P.J., 1962. The influence of pruning on size and quality of Florida grapefruit. Proc. Fla. Sta. Hort. Soc. 35–45.

Kretchman, D.W., Krezdorn, A.H., 1962. Recent developments in pruning citrus. Citrus Industry 43 (1), 6–7. 13–16.

Lewis, L.N., McCarty, C.D., Moore, P.W., 1963. Effects of skeletonizacion on rejuvenation and production of declining orange trees. Am. Soc. Horticult. Sci. 83, 295–299.

Medina-Urrutia, V.M., Robles-González, M.M., Becerra-Rodríguez, S., Orozco-Romero, J., Orozco-Santos, M., Garza-López, J.G., Ovando-Cruz, M.E., Chavez-Contreras, X., Felix-Castro, F.A., 2002. El cultivo del limón Mexicano. INIFAP. Libro Técnico Num. 1 México. 204 p.

Mesejo, C., Yuste, R., Reiga, C., Martínez-Fuentes, A., Iglesias, D.J., Munoz-Fambuena, N., Bermejo, A., Antonietta Germanà, M., Primo-Millo, E., Agustí, M., 2016. Gibberellin reactivates and maintains ovary-wall cell division causingfruit set in parthenocarpic *Citrus* species. Plant Sci. 247, 13–24.

Migliaccio, K.W., Morgan, K.T., Vellidis, G., Zotarelli, L., Fraisse, C., Zurweller, B.A., Andreis, J.H., Crane, J.H., Rowland, D.L., 2016. Smartphone apps for irrigation scheduling. Trans. ASABE 59 (1), 291–301.

Monerri, C., Fortunato-Almeida, A., Molina, R.V., Nebauer, S.G., García-Luis, A., Guardiola, J.L., 2011. Relation of carbohydrate reserves with the forthcoming crop, flower formation and photosynthetic rate, in the alternate bearing 'Salustiana' sweet orange (*Citrus sinensis* L.). Sci. Hortic. 129, 71–78.

Monselise, S.P., Goldschmidt, E.E., 1982. Alternating bearing in fruit trees. Hortic. Rev. 4, 128–173.

Monselise, S.P., Goldschmidt, E.E., Golomb, A., 1981. Alternate bearing in citrus and ways of control. Proc. Int. Soc. Citriculture 1, 232–242.

Nawaz, M.A., Afzal, M., Maqbool, M., Hussain, Z., Ashraf, T., Iftikhar, Y., 2010. The productivity and quality of citrus fruits as influenced by tree spacing. Int. J. Agric. Appl. Sci. 2, 32–35.

Nebauer, S.G., Renau-Morata, B., Lluch, Y., Baroja-Fernandez, E., Pozueta-Romero, J., Molina, R.-V., 2014. Influence of crop load on the expression patterns of starch metabolism genes in alternate-bearing citrus trees. Plant Physiol. Biochem. 80, 105–113.

Phillips, R.L., 1974. Performance of 'Pineapple' orange at three tree spacings. Proc. Fla. State Hort. Soc. 87, 81–84.

Phillips, R.L., 1978. *Citrus* tree spacing and size control. In: Proc. Int. Soc. Citriculture, pp. 319–324.

Rouse, R., Futch, S., 2004. Start Now to Design *Citrus* Groves for Mechanical Harvesting. EDIS HS974. Horticultural Sciences Department, UF/IFAS Extension.

Rouse, R., Parsons, L., Wheaton, A., 2006. Hedging, topping and skirting trees in the citrus canker era. Citrus Industry (December 2006).

Sauls, J.W., 2008. *Citrus* Orchard Establishment. Texas Agri Life Extension Service.

Schumann, A., Hostler, K., Waldo, L., 2012. Advanced citrus production systems-grove designs for higher efficiencies. Citrus Industry (September 2012).

Schumann, A., Singerman, A., Wright, A.L., Ferrarezi, R.S., 2017. *Citrus* under Protective Screen (CUPS) Production Systems. EDIS HS1304. Horticultural Sciences Department, UF/IFAS Extension.

Talón, M., Hedden, P., Primo-Millo, E., 1990. Gibberellins in *Citrus sinensis*: a comparison between seeded and seedless varieties. J. Plant Growth Regul. 9, 201–206.

Talón, M., Zacarías, L., Primo-Millo, E., 1992. Gibberellins and parthenocarpic ability in developing ovaries of seedless mandarins. Plant Physiol. 99, 1575–1581.

Talón, M., Juan, M., Soler, J., Agustí, M., Primo-Millo, E., 1999. Criterios de racionalización de las aplicaciones de ácido giberélico para mejorar el cuajado de los frutos cítricos. Levante Agrícola 347, 128–133.

Taverner, P., Hardiyanto, A., Gunner, L., Shepherd, K., Morey, P., Andri, K.B., Ratule, T., Dewayani, W., Murdolelono, B., 2011. Market Development for *Citrus* From Eastern Indonesia. Australian Center for Agricultural Research, Canberra, Australia, ISBN: 978-1-921738-91-3.

Tucker, D.P.H., Wheaton, T.A., 1978. Trends in higher density plantings. Proc. Fla. State Hort. Soc. 91, 36–40.

Tucker, D.P.H., Wheaton, T.A., Muraro, R.P., 1991. *Citrus* Tree Spacing and Pruning. Florida Cooperative Extension Services. IFAS-University of Florida. SP. 74. 15 p.

Vashisth, T., Zekri, M., Alferez, F., 2017. Canopy Management of *Citrus* Trees. EDIS HS1303. Horticultural Sciences Department, UF/IFAS Extension.

Verreyne, J.S., Lovatt, C.J., 2009. The effect of crop load on budbreak influences return bloom in alternate bearing 'pixie' mandarin. J. Amer. Soc. Hort. Sci. 134 (3), 299–307.

Wallerstein, I., Goren, R., Monselise, S.P., 1978. Rapid and slow translocation of 14C-sucrose and 14C-assimilates in *Citrus* and Phaseolus with special reference to ringing effect. J. Hortic. Sci. 53, 203–208.

Wheaton, T.A., 1986. Alternate bearing. In: *Citrus* Flowering, Fruit Set, and Development. University of Florida Citrus Short Course, pp. 67–72.

Wheaton, T.A., 1992. Alternate bearing of citrus. Proc. Int. Seminar Citriculture 1, 224–228.

Wheaton, T.A., 1997. Alternate bearing of *Citrus* in Florida. In: *Citrus* Flowering and Fruiting Short Course, pp. 87–92.

Wheaton, T.A., Whitney, J.D., Castle, W.S., Tucker, D.P.H., 1986. Tree spacing and rootstock affect growth, yield, fruit quality, and freeze damage of young 'Hamlin' and 'Valencia' orange trees. Proc. Fla. State Hort. Soc. 99, 29–32.

Wheaton, T.A., Whitney, J.D., Castle, W.S., Muraro, R.P., Browning, H., Tucker, D.P.H., 1995. Tree vigor important in citrus tree spacing and topping. Proc. Fla. State Hort. Soc. 108, 63–69.

Whitney, J.D., Elezaby, A., Castle, W.S., Wheaton, T.A., Littell, R.C., 1991. *Citrus* tree spacing effects on soil water use, root density, and fruit yield. Trans. ASAE 34, 129–134.

Whitney, J.D., Wheaton, T.A., Castle, W.S., Tucker, D.P.H., 2003. Tree skirting effects on yield and quality of Valencia oranges. Proc. Fla. State Hort. Soc. 116, 236–239.

Zaragoza, S., Trenor, I., Alonso, E., Primo-Millo, E., Agustí, M., 1992. Treatments to increase the final fruit size on Satsuma 'Clausellina'. Proc. Int. Soc. Citriculture 2, 725–728.

Zaragoza, S., Trenor, I., Medina, A., 2003. Influence of manual pruning on citrus fruit quality and production. Proc. Int. Soc. Citriculture I, 513–514.

Zekri, M., 2000. Evaluation of orange trees budded on several rootstocks and planted at high density on flat woods soil. Proc. Fla. State Hort. Soc. 113, 119–123.

Zekri, M., 2015. Mechanical Pruning of *Citrus* Trees. EDIS HS1267. Horticultural Sciences Department, UF/IFAS Extension.

Zekri, M., Albrecht, U., Vincent, C., Vashisth, T., 2017. Grove Planning and Establishment. EDIS HS1302. Horticultural Sciences Department, UF/IFAS Extension.

Chapter 21

Postharvest technology of citrus fruits

Lorenzo Zacarias[a], Paul J.R. Cronje[b], Lluís Palou[c]

[a]Instituto de Agroquímica y Tecnología de Alimentos, Consejo Superior de Investigaciones Científicas (IATA-CSIC), Valencia, Spain, [b]Citrus Research International, Department of Horticultural Sciences, Stellenbosch University, Stellenbosch, South Africa, [c]Centre de Tecnologia Postcollita, Institut Valencià d'Investigacions Agràries (IVIA), Valencia, Spain

21.1 Introduction

The estimated total production of citrus fruit in 2016 was over 170 million tons, representing the major fruit-tree crop in the world. About half of this total production is destined to the juice processing industry, but the distribution between processing and fresh market is very different among the citrus-producing countries. Thus, in the United States or Brazil around 70% and 90%, respectively, of the production is dedicated to the juice industry, whereas in other countries of the Mediterranean Basin, such as Spain, or South Africa, 80%–85% of the total production is destined for the fresh market either local or international (FAOSTAT, 2018). In this scenario, pre- and postharvest handling and requirements for both industries are absolutely different, and while in the juice industry the external appearance of the fruit is not a basic parameter of quality, for the fresh-fruit market quality and appearance of the rind, the absence of damage or deteriorations and presentation are essential attributes for the market and consumer's acceptance.

Postharvest technology encompasses a number of techniques, processes, and treatments related to handling, processing, storage, transport, etc. of the fruit, aimed to prepare them for market requirements, to extend their commercial life, and to reduce the losses during the whole chain, from harvest to the consumer table. These general objectives are differentially developed in the different citrus-producing countries, according to their needs, market destinations, and other requirements. It is clearly envisaged that postharvest requirements for overseas exportation and international shipments of citrus fruit are absolutely different to that of fruit designed to local markets. Moreover, current globalization and international trades are imposing strict quarantine treatments and prolonged cold storage periods, among other requirements, to reach new long-distance markets. Thus, fruits should be prepared properly and submitted to adequate postharvest treatments in order to withstand prolonged transportation and storage that may exceed the physiological tolerance of the fruits to these extreme conditions.

Modern postharvest biology of citrus fruits also addresses the biological basis of the stress conditions imposed on the fruit during their postharvest life. Understanding the biochemical and molecular mechanisms involved in these processes as well as in the parameters of fruit quality, genotypic differences, or susceptibility to postharvest diseases are new challenges that may improve in the future the postharvest performance of citrus fruits. In this chapter, we summarize and critically review basic aspects and new developments of postharvest biology and technology of citrus fruits. The chapter is organized into three main sections covering the following subjects: (1) physiological disorders and responses to postharvest conditions, (2) fungal diseases and control strategies, and (3) handling, transport and storage technologies.

21.2 Postharvest Physiology

21.2.1 Responses of citrus fruits to postharvest stress conditions

During their postharvest life, citrus fruits are subjected to different abiotic stress conditions that may elicit profound changes in their physiological, biochemical, and molecular responses to cope with these deleterious circumstances. Because of the subtropical origin, fruits of many citrus species and cultivars are sensitive to develop chilling injury (CI) symptoms of

The Genus Citrus. https://doi.org/10.1016/B978-0-12-812163-4.00021-8

the peel that may severely compromise their external quality and marketability. Due to the commercial significance, the responses of chilling-sensitive and chilling-resistant fruits to postharvest cold stress are probably the best studied in the postharvest biology of citrus.

Early work indicated a moderate increase in fruit respiration during storage at low temperature that substantially increased in sensitive fruits few hours after transference to 20°C. By contrast, cold-tolerant tissues usually produce little or no excess of CO_2 after exposure to chilling temperature (Eaks, 1970). Then, activation of the respiratory metabolism and accumulation of ethanol/acetaldehyde as a consequence of the anaerobic respiration appear to be responses of citrus fruit to CI (Schirra, 1992).

A common response of chilling-sensitive citrus fruit to storage at low temperatures is an increase in ethylene production. This upsurge of ethylene during cold storage is of moderated magnitude but paralleled the development of chilling damage in many fruits. Moreover, ethylene production experienced a massive increase after fruit rewarming to shelf life conditions (Schirra, 1992; Lafuente et al., 2001; Zacarıas et al., 2003). These hormonal responses may be indicative of the injury caused by low temperature, but the involvement of ethylene in the regulation of the cold-induced metabolic changes is still unclear. Analysis of the expression of ethylene biosynthetic genes [1-aminocyclopropane-1- carboxylic acid (ACC) synthase (*ACS*) and ACC oxidase (*ACO*)] in grapefruits revealed that *ACS1* and, to a minor extent, *ACS2* and *ACO* followed a damage-related induction, indicating that stimulation of ethylene biosynthesis is a chilling-related response rather than a cold response (Wong et al., 1999; Maul et al., 2008; Lado et al., 2015a,b). Accumulation of ACS and ACO mRNAs during cold storage may explain the rapid upsurge of ethylene production after fruit rewarming once the tissue has reached a temperature near the optimum for enzyme activity (Lado et al., 2015a). However, the changes in ethylene perception and signaling during cold stress are more complex. Different members of the gene families of ethylene receptors followed distinct patterns of expression during cold storage and were differentially affected by ethylene, indicating that ethylene receptor genes are regulated by both factors, cold and chilling, at least in the peel of grapefruits (Lado et al., 2015a, b). These evidences suggest that ethylene appears to be a part of the defense mechanisms of chilling-sensitive citrus fruit to cope with cold stress, and that the naturally low levels of ethylene evolving from the fruit are required to maintain fruit quality during postharvest life (Porat et al., 1999; Lado et al., 2019). This hypothesis is supported by the fact that exogenous ethylene alleviated CI and the ethylene action inhibitor 1-methylcyclopropene (MCP) stimulated not only CI in fruits of different species (Lafuente et al., 2001), but also other postharvest physiological disorders (Figure 21.1M) (Estables-Ortiz et al., 2017) and pathological infections (Marcos et al., 2005). Moreover, it has also been shown that ethylene only induced a small subset (15%) of the genes differentially regulated during cold storage (Lafuente et al., 2017), reinforcing the idea that ethylene is only a factor in the complex metabolic network operating against chilling in citrus fruits. It should also be mentioned that the deleterious effect of 1-MCP to citrus fruits in many postharvest situations is not uniform, but dependent on the species and developmental stages (Lado et al., 2019). Then, even though 1-MCP is effective delaying fruit degreening in fruits of many citrus varieties, its commercial application during the postharvest storage may be detrimental for the external appearance and quality of the fruit (Figure 21.1M) (Estables-Ortiz et al., 2017).

Alteration in cell membrane permeability induced by changes in the saturation of fatty acids has been classically considered the primary effector of CI. These changes implicate an increase in unsaturated-saturated fatty acid ratio and in the double-bond index (Schirra, 1992). Early results indicated that CI altered lipid composition and membrane fluidity in grapefruits (Nordby et al., 1987), but the changes in the unsaturation of lipids were not correlated with the sensitivity/resistance to CI (Mulas et al., 1996). In Fortune mandarins, a hybrid highly sensitive to CI, storage at 2°C increased total fatty acid content and modified their relative composition to a more unsaturated profile (decreasing linolenic acid and galactolipids, and increasing the polar lipids linoleic and oleic acids) (Mulas et al., 1996). In this chilling-sensitive mandarin, it has also been observed an induction in the expression of several lipid desaturases genes and a downregulation of sphingolipids desaturase. Based on these results, it has also been proposed that lipid degradation metabolism affecting membrane fluidity may be more critical for chilling development than lipid unsaturation (Lafuente et al., 2017). Other alterations in lipid metabolism have been postulated to act as signaling molecules in the response to chilling (as arachidonic acid) (Zacarias et al., unpublished). Despite the relevance of lipid metabolism in the response of citrus fruit to low temperature, the mechanisms implicated in the natural resistance or sensitivity to chilling remain still unknown.

A number of experimental evidences implicate oxidative stress and the generation of reactive oxygen species (ROS) in the damage induced by cold stress in citrus fruits (Lado et al., 2016; Sala, 1998). A higher activity of the antioxidant enzymatic system has been demonstrated in the natural tolerance of fruits of different citrus varieties to CI and also in the tolerance induced by heat conditioning (Sala, 1998). In particular, catalase appears to play a key function in the protection against chilling damage (Ghasemnezhad et al., 2008; Lado et al., 2016). By contrast, a higher peroxidase activity has been associated with the development of CI symptoms (Ghasemnezhad et al., 2008). Altogether, oxidative stress appears to be a primary mechanism of the damage induced by exposure of citrus fruit to cold temperatures (Sala and Lafuente, 2000).

FIG. 21.1 Representative pictures of postharvest physiological disorders in citrus fruits: (A) chilling injury in mandarins, (B) chilling injury in grapefruits, (C) chilling injury of oranges—scalding and pitting, (D) nonchilling scalding and pitting of Navel oranges, (E) postharvest chemical phytotoxicity burn in fruit of different species and varieties, (F) stem-end rind breakdown in Valencia oranges, (G) oleocellosis in Navel oranges, (H) peteca in lemons, (I) postharvest heat burns in mandarins, (J) non-chilling peel pitting in Navelate oranges on the tree, and during postharvest in Marsh grapefruit and Fallglo tangerine, (K) rind breakdown of Clementine mandarin, (L) scalding of mandarin, and (M) Clementine mandarin stored at 12°C for 12 days, untreated fruits (*top*) or fruits previously treated with 1000 ppb 1-MCP (*bottom*).

It is worth mentioning that in red-fleshed varieties of oranges (Hong Anliu) or grapefruits (Star Ruby), accumulation of the red pigment lycopene appears to modify the antioxidant status of the cells. Transcriptomic and proteomic profiling in the Hong Anliu orange identified a number of antioxidant enzymes differentially expressed with respect to the wild-type orange (Xu et al., 2009; Pan et al., 2009). Moreover, Lado et al. (2015b, 2016) found that CI symptoms in grapefruits were restricted to the yellow areas of the peel, whereas the red sector accumulating a large amount of lycopene was almost devoid of CI. In these CI-tolerant sectors of the peel, total antioxidant capacity and ascorbic acid content were lower than in the CI-sensitive sector, suggesting that these responses may be related to the chilling-induced damage. Interestingly, the antioxidant enzymatic system was not different between both sectors, but a much higher singlet-oxygen antioxidant capacity (SOAC) was found in the chilling-tolerant red sector. These evidences indicate that singlet oxygen appears to be an injurious ROS inducing chilling symptoms in the peel of citrus fruit and that the presence of lycopene may activate the capacity to scavenge this toxic ROS induced during cold storage in insensitive cultivars. This assumption also implicates that the chromoplast, where lycopene is accumulating, of mature citrus fruits is a key cellular organelle in the signaling and response to CI.

The plant hormone abscisic acid (ABA) has been implicated in the response to different plant tissues to cold stress. However, experimental evidences indicate that ABA does not play a critical role in either fruit response to chilling or the heat-induced chilling tolerance of citrus fruits. This assumption is supported by the postharvest response of fruit of the ABA-deficient orange mutant, Pinalate, during chilling storage and also by the fact that cold and heat followed by cold downregulated the expression of carotenoid biosynthetic genes, including the 9-cisepoxycarotenoid dioxygenase (*NCED*), a key step in the regulation of ABA, and then reduced the concentration of the hormone (Lafuente et al., 2017). However, the involvement of ABA in the response of citrus fruits to altered water relations is more evident. It has been demonstrated that changes in the relative humidity (RH) after fruit detachment and during postharvest management induced rind staining in fruit of different varieties (Alférez et al., 2010) and this response appears to be mediated by changes in ABA. Moreover, ABA also appears to be a signal mediating the activation of several phospholipases induced by dehydration (Romero et al., 2014).

Different studies indicated that carbohydrate metabolism plays a crucial role in the response of citrus fruits to cold stress, acting as protective osmolytes and antioxidants, and probably acting also as signaling molecules in the cross talk with other metabolic pathways. Cold storage also stimulated the expression of invertase genes (converting sucrose into glucose and fructose) and in general repress the glycolysis. Cold also induced the expression of β-amylase (*PtrBAM1*) in *Poncirus trifoliata*, which is regulated by members of the cold-binding factors (CBFs) transcription factors that are involved in cold tolerance (Peng et al., 2014). Prestorage treatments that induce cold tolerance also modulate carbohydrate concentrations. In Ponkan mandarins, starch and sucrose metabolism are downregulated in response to low temperatures (Zhu et al., 2011). Heat conditioning (3 days, 37°C) increases sucrose, but not hexoses, accumulation during the cold storage of Navel oranges (Holland et al., 2005), whereas glucose and fructose content is enhanced in satsuma mandarin (Yun et al., 2013). It appears, therefore, that induction of cold tolerance in citrus fruits implicate dual general mechanisms: inducing protective and stress-related responses and also arresting some chilling-induced mechanisms (Sanchez-Ballesta et al., 2003; Maul et al., 2008; Lafuente et al., 2017).

The involvement of secondary metabolism in the response of citrus fruit to cold is well documented. In particular, the metabolism of phenylpropanoids appears to play a relevant role in chilling tolerance. Phenylalanine ammonia lyase (PAL) activity and expression of the gene are induced by chilling and regulated in an ethylene-dependent and -independent manner (Lafuente et al., 2001; Sanchez-Ballesta et al., 2003). Evidences indicate that PAL activity is a key metabolic step in the response of citrus to chilling. Other enzymes of the secondary metabolism, such as *O*-methyl transferases, are also stimulated in response to cold or to heat treatments, but the general consensus indicate that flavonoids and methylated phenylpropanoids are more likely related to the chilling responses than phenolic compounds (Lafuente et al., 2017).

Large-scale transcriptomic analysis of the changes induced by cold in fruit of different citrus species revealed that chilling provokes a general rearrangement of the metabolism to induce a global set of responses to cope with cold stress. These responses encompass both downregulation of the expression of genes related to cellular metabolism and the activation of processes related to degradation and adaptation to stress conditions (Maul et al., 2008). In blood oranges, it has been also observed that in addition to general transcriptomic changes, cold also increases the synthesis of flavonoids, including anthocyanins, as a response to cold, illustrating the complex interplay of metabolic pathways acting in CI that may differ among fruits of different citrus species (Crifò et al., 2011). Moreover, chilling stimulates the expression of different transcriptions factors, as members the WRKY family, MADS-boxes, ethylene-responsive factors (ERFs), CBFs, or HLH, but their involvement in the chilling-induced damage or in chilling tolerance still needs to be clarified (Lafuente et al., 2017).

21.2.2 Postharvest physiological disorders

Citrus fruits are very prone to develop numerous physiological peel disorders, or peel blemishes, which are manifested by different morphological symptoms that can appear either before or after harvest. The incidence of these symptoms depreciates the external quality of the fruit and is one of the main factors for market rejections and postharvest losses. The peel of citrus fruits contains oil glands full of essential oils that can easily oxidize. Exposure of fruits to stressful conditions that may induce injury and damage can provoke breakdown of the gland cells that may result in the release of the essential oil content. Different environmental pre- and postharvest conditions may cause damage on the fruit surface that are manifested by injuries, darkening of the cells and browning, but the causes of these blemishes may differ and then their identification can be complicated. More than 70 different physiological disorders have been described in fruits of the different citrus species and varieties (Lafuente and Zacarías, 2006; Petracek et al., 2007; Lado et al., 2019). These disorders appear during postharvest as a result of rough handling, inadequate storage, or inappropriate market conditions. In this section, two main classes of postharvest disorders are considered: (a) CI, as a major postharvest disorder by inadequate storage temperature and (b) nonchilling disorders, which include a series of blemishes induced by factors other than chilling temperatures.

21.2.2.1 Chilling injury

Low-temperature storage is the most widely used and efficient system to extend postharvest life of citrus fruit. Moreover, cold exposure below 2°C is also required in the quarantine treatments for export to several markets. However, citrus fruits, like other fruits of subtropical origin, are sensitive to CI when exposed to low nonfreezing temperatures. The critical temperature at which CI develops is variety dependent but can be affected by environmental and growing conditions. It is generally recognized that lemons, grapefruits, and limes are more sensitive to low-temperature storage than oranges and mandarins. Recommendations for optimal temperature and storage period have been previously published (El-Otmani et al., 2011; Lado et al., 2019).

In grapefruit, CI symptoms are manifested as small, depressed light-brown pits that, after long cold storage, are progressively extended forming larger brown collapsed areas (Figure 21.1A) (Platt-Aloia and Thomson, 1976). Chilling-sensitive mandarins and hybrids also developed typical peel-pitting symptoms with brown pits extended over the fruit surface (Figure 21.1B). However, scalding is also observed in certain varieties like Nova mandarin that are aggravated by waxing and processing throughout the packing line (Figure 21.1I). The characteristic darkening of chilling-injured fruit is, at least in part, due to an internal release and oxidation of cell components and particularly the content of oil glands. In fact, in fruits showing chilling symptoms, oil glands remain intact in contrast with fruit showing "oleocellosis" where the glands are broken and their content is released on the fruit surface (Figure 21.1G). An inner darkening of the oil glands without pits or depressions is also a symptom of CI observed in early harvested Clementine mandarins (Figure 21.1K). In lemon fruit, CI is also manifested as an albedo darkening or as browning of the capillary membranes (Petracek et al., 2007). Orange fruits are more resistant to chilling but in sensitive fruits (depending on the varieties, season, or growing conditions), CI is manifested as extended, nondepressed, and superficial browning areas in the flavedo of the fruit, known as superficial scalding (Figure 21.1C), which differs from the typical peel-pitting symptoms of grapefruit (Figure 21.1B). Oranges may also develop typical depressed CI symptoms depending on the year and peel susceptibility (Lafuente and Zacarías, 2006; Lado et al., 2019). Symptoms of CI are usually manifested during cold storage but may become exacerbated after transference of the fruit to ambient temperature (shelf life) or rewarming, whereas fruit with symptoms of CI is less resistant to decay during postharvest storage (Lafuente and Zacarías, 2006).

Ultrastructural changes induced by cold storage in sensitive fruits are similar among the different varieties. In general, these alterations comprise collapse and shriveling of the cells, large empty spaces between the membrane and the cell walls, disruption of the vacuole, and disorganization of chloroplasts and mitochondria. There is also a reduction of starch grains, lipid droplet accumulation, and as the damage progresses the vacuole appears contracted and filled of dark material, extensive cell walls damage, and destruction of the cells (Platt-Aloia and Thomson, 1976; Lado et al., 2015a,b).

The critical threshold temperature for storage of citrus fruits to avoid CI varies greatly among varieties: 10°C–12°C for grapefruit and lemon, 9°C–10°C for mandarin hybrids that are sensitive to CI (i.e., Fortune, Nova, and Nadorcott), 5°C–6°C for mandarin hybrids and tangelos, 3°C–4°C for Clementine and Satsuma mandarins, and 2°C–4°C for oranges (Ladaniya, 2008). It appears that sensitivity to developing CI is genetically related. However, environmental factors may modulate the response of the fruit to CI. In fact, grapefruit is more susceptible to CI early and later in the season, whereas those harvested at mid-season are more resistant (Purvis, 1979). Under Mediterranean conditions, mid-season mandarins are much more susceptible to CI than early- and late-harvested fruit. It appears that fruit postharvest sensitivity to CI is influenced by climatic conditions and the length of exposure to cold temperatures in the field (Lafuente and Zacarías, 2006). Rootstock appears to play an important role also in the incidence of chilling, although whether this effect is related to water relations or to any other factor remains to be determined. Nadorcott mandarin is sensitive to develop superficial peel browning when stored at temperatures below 8°C. Small-sized fruit with thinner peel is more susceptible to the disorder than larger fruits. The rootstock effect on peel thickness was evident, as fruit with thicker rind from trees grafted onto vigorous rootstocks such as *Citrus volkameriana* show lowest blemish incidence when stored at 6°C than fruits from trees grafted on less vigorous rootstocks (El-Otmani et al., 2011). Other postharvest treatments to reduce the incidence of CI during the whole postharvest chain of citrus fruits are considered further in this chapter.

21.2.2.2 Nonchilling disorders

The first step of fruit handling after harvest consist of a number of operations that may negatively impact the quality of the fruit, resulting in different types of blemishes. Environmental conditions at harvest time, delaying the time from harvest to packing, the ripening stage at harvest, ethylene treatment, and other postharvest operations are factors that, to different extents, influence the severity of nonchilling peel disorders. These symptoms are more diverse and may depend on the particular peel nature and morphology among species and varieties. Some of the more relevant nonchilling disorders are mentioned below.

21.2.2.3 Oleocellosis

This disorder is caused by the rupture of oil glands and the release of their content, which produces a phytotoxic effect and necrosis in epidermal cells. Injured fruit shows light yellow rind spots that become sunken and turn dark in color as the damage develops (Figure 21.1G). Damaged areas are more susceptible to decay (Knight et al., 2002). The disorder is usually caused by rough handling of fruit during picking and packing. Conditions that may cause an increase in peel turgor at harvest time such as wet and foggy mornings, or excessive irrigation or rainfall greatly increase incidence of this disorder. Factors such as the pressure required to break the oil gland and the capability of the affected area to tolerate the released oil have a significant effect on the incidence of the disorder (Lafuente and Zacarías, 2006). To control the disorder, fruit should not be harvested in wet conditions leading to a low vapor pressure deficit (VPD), such as after rainfall or harvesting in the afternoon instead of the morning; and the time from harvesting to handling in the packinghouse should be delayed until the following day.

21.2.2.4 Stylar-end breakdown (SEB) blossom-end clearing (BEC)

SEB is a rind blemish that mostly affects Tahiti limes. It is initiated at the blossom end of the fruit and gradually spreads to other fruit areas. In addition, it is characterized by the rupture of juice vesicles and release of their content, producing a water-soaked albedo. Inappropriate handling and dropping the fruit on its stylar end are probably the cause of the disorder. Careful fruit postharvest handling and storing at high humidity may reduce the incidence of the disorder. BEC mainly affects grapefruit and is characterized by a translucent and wet appearance at the blossom end of the fruit. Rough handling is the primary cause of the blemish. Affected fruit shows an open core with the ruptured juice vesicles leaking, giving rise to a wet spot. Fruit with a thinner skin is more prone to the disorder. Symptoms of the disorder can be detected few days after handling and brushing (Echeverria et al., 1999). Higher ambient temperatures during harvest can also result in higher incidence of the disorder.

21.2.2.5 Zebra skin

A disorder that is specific to mandarins and tangerines occurring mainly in the flavedo of early harvested fruit. Satsuma mandarin (*Citrus unshiu*) varieties are especially susceptible to the disorder. Small, light-colored, and thin-skinned fruit are more susceptible to developing zebra skin. The cells in the affected area become dark and necrotic and are mostly located at the center of the fruit segments. Preharvest water stress followed by rain or excessive irrigation causes sudden changes in peel cell turgidity that can cause zebra skin. Severity of the disorder is higher in small fruits with a thin peel and also in fruits from older trees. Degreening tends to increase the development of the disorder. The disorder could lead to increased decay as well as off-flavors in the fruit due to the physical damage. Allowing fruit to reach full color on the tree, delaying harvest after rain, and careful handling and brushing of the fruit in the packing line reduce zebra skin.

21.2.2.6 Peel pitting or staining in oranges, mandarins, and grapefruits

These terms refer to a peel blemish occurring at nonchilling temperatures several days after processing the fruit in the packing line. The disorder is also referred to as rind staining or rind breakdown, depending on the countries, although these terms may be confusing as they have been used to refer to several kinds of postharvest physiological disorders. These disorders have been described during the postharvest of Marsh grapefruit, Navel oranges, Fallglo mandarins and Shamouti oranges, and also in Navelate orange on the tree (Lafuente and Zacarías, 2006; Magwaza et al., 2013; Lado et al., 2019). Symptoms of the disorders in grapefruit and orange initiated as small round depressions randomly distributed over the fruit surface that may progressively develop as clusters of collapsed oil glands that may expand to form scattered areas of collapsed tissue (Figure 21.1D and J). In Florida, the disorder was initially observed in wax-coated grapefruits and it was then suggested to be due to alterations of internal gas concentrations by less permeable shellac-based waxes (Petracek et al., 1995; Petracek et al., 1998a, b). By contrast, modified atmosphere packaging, which also altered fruit internal gas concentration, reduced the incidence of the damage in Shamouti oranges (Porat et al., 2004). The fact that the incidence of the blemish of fruit on the tree (Figure 21.1J) increased after several days of high air RH following a period of low environmental RH indicated that alterations in RH in the orchard may be elicitors of the disorder (Agustí et al., 2001). It can, therefore, be hypothesized that alterations in peel water status are critical factors in the incidence of rind staining. Sudden fluctuations in air RH during storage at non-chilling temperatures, which lead to sudden changes in VPD between fruit and the environment, have been demonstrated to cause rind staining in Navel oranges (Alférez et al., 2003), "Marsh" grapefruit (Alférez and Burns, 2004), and Fallglo mandarin (Alférez et al., 2005). The disorder was induced in non-waxed fruits, indicating that waxing itself was not the direct cause of peel pitting and that a previous peel dehydration was a prerequisite (Alférez and Burns, 2004;

Alférez et al., 2005). It has been proposed that waxing dehydrated fruit impaired normal water exchange and altered turgor potential of both flavedo and albedo, and may create a suction force from the dehydrated albedo that would cause collapse of external epidermal cells (Alférez et al., 2010; Alquézar et al., 2010). The greater the level of dehydration of the fruit, either in the field or after harvest but before storage at high RH, the more prone is the fruit to developing postharvest rind staining. In this scenario, albedo thickness and its ability to adjust to altered water relations is probably an important factor involved in the disorder (Cronje et al., 2017). Moreover, it has been shown a decrease in the water potential of flavedo and albedo of Navelate oranges with fruit maturity, resulting in a restricted ability to adjust water relations, which may explain the higher incidence of rind stating in more mature fruits (Alférez and Zacarías, 2014).

Postharvest rind breakdown in Clementine mandarins is a major problem in several countries. This disorder is particularly important for Clemenules Clementine compared to other varieties and develops after 3–4 weeks of storage (Figure 21.1K). Preharvest factors appear to be responsible for the disorder; in fact, rind breakdown incidence is higher in less-colored fruits growing inside the canopy with lower potassium content (Cronje et al., 2011a, b). In Nadorcott mandarins, it has also been shown that delaying wax application by a previous dehydration period aggravated rind staining and that the disorder is significantly higher in fruit from rough lemon rootstock compared to fruit from Carrizo citrange (Cronje, 2013).

21.2.2.7 Stem-end rind breakdown (SERB)

This blemish is localized in a discrete area around the calyx end of the fruit and characterized by collapse and darkening of the peel around the stem end of the fruit. It is often associated with peel aging and occurs mainly in oranges (Figure 21.1F). Nutritional imbalance and excessive water loss between harvest and waxing appear to be among its causes. Differences in stomata number, wax cuticle thickness, and water potential in the flavedo area surrounding the calyx may explain the particular localization of the disorder (Albrigo, 1972). Fruits with thicker peel or from arid areas are more resistant to the development of SERB than those with thinner peel or from humid environments. Minimizing the time between harvest and waxing, cooling the fruit immediately after harvest, avoiding direct exposure of the fruit to sunlight, and any conditions that would cause fruit desiccation would reduce SERB incidence. Avoiding the use of hard brushes in the packing line and ensuring good coverage of the fruit peel with appropriate wax will also reduce the occurrence of SERB (Petracek et al., 2007).

21.2.2.8 Peteca of lemons

This rind disorder is common in lemon fruit and characterized by deep depressions in the rind and darkening of the oil glands, which result from collapsing of the albedo and oil releasing from flavedo causing damage to the adjacent tissue (Figure 21.1H). The first symptoms develop 3–5 days after harvest and are fully developed within 2 weeks. Incidence of the disorder has been described in several lemon-producing countries around the world. Although the causes of the disorder are not well known, nutritional deficiencies and imbalances have been postulated. Contrary to oleocellosis, peteca in lemons develops in the absence of mechanical damage, being more likely a senescence-related disorder. Rind maturity has been suggested to play a role in the susceptibility to peteca, as the incidence of the disorder is higher in early harvested fruit than in more mature fruits (Cronje, 2015). Cold and wet conditions near to the harvest time aggravate the disorder incidence. Excessive handling in the packinghouse and use of polyethylene-based wax increased the incidence of the disorder. Delaying harvest during humid and foggy days and ethylene application and waxing about 2–3 days after harvest with adequate ventilation to remove CO_2 may decrease the incidence of the disorder (Cronje, 2015). Control strategies are difficult due to the highly erratic incidence of this disorder with large variations between seasons, orchards, and within orchards in one season.

21.3 Postharvest pathology

21.3.1 Main postharvest diseases

Postharvest diseases, also controversially named in the past as pathological disorders, are those postharvest alterations caused by biotic factors. The current tendency is to restrict the name disorders only to those caused by physiological or abiotic factors. Postharvest diseases of citrus fruit are typically caused by filamentous fungi and their common names are based on the symptoms they produce. Some of the most economically important postharvest pathogens infect the fruit through rind wounds or injuries inflicted during harvest, transportation, and postharvest handling. These so-called wound pathogens include *Penicillium digitatum* (Pers.: Fr.) Sacc. and *Penicillium italicum* Wehmer, the cause of green and blue molds, respectively (Figure 21.2A); *Geotrichum citri-aurantii* (Ferraris) E.E. Butler, the cause of sour rot (Figure 21.2E); *Rhizopus stolonifer* (Ehrenb.) Vuill. (syn.: *R. nigricans* Ehrenb.), the cause of Rhizopus rot

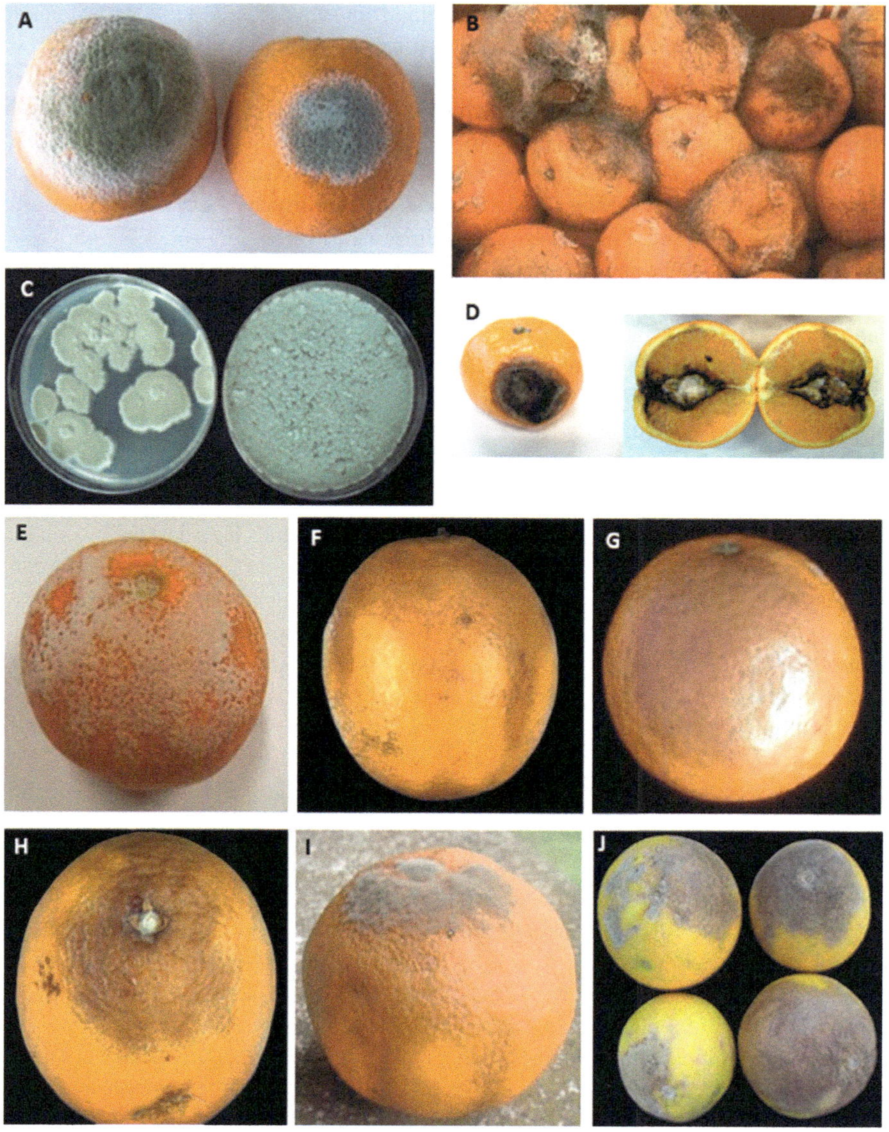

FIG. 21.2 Representative pictures of fungal postharvest diseases of citrus fruits: (A) green and blue molds of orange caused by *Penicillium digitatum* and *Penicillium italicum*, respectively, (B) nest of watery rot caused by *Rhizopus stolonifer* within a mandarin container, (C) colonies of *Penicillium digitatum* (left) and *Penicillium italicum* (right) growing on medium potato dextrose agar (PDA), (D) external (mandarin) and internal (orange) black rot caused by *Alternaria* sp., (E) sour rot of mandarin caused by *Geotrichum citri-aurantii*, (F) stem-end rot of orange caused by *Lasiodiplodia theobromae*, (G) brown rot of orange caused by *Phytophthora* sp., (H) stem-end rot of orange caused by *Phomopsis citri*, (I) gray mold of orange caused by *Botrytis cinerea*, and (J) anthracnose of orange caused by *Colletotrichum gloeosporioides*.

(Figure 21.2B); and *Mucor piriformis* A. Fisch., the cause of Mucor rot. Symptoms of decay caused by the two latter pathogens are similar and these diseases are also commonly known as soft rot or watery rot. Even though they are not very frequent under adequate citrus postharvest handling, these diseases can be especially devastating since nests of fungal mycelia are formed in *Rhizopus*- or *Mucor*-infected fruit stored in piles of bins or boxes in the packinghouse and leakage from decayed tissue, apart from being phytotoxic, carries inoculum that may easily infect adjacent healthy fruit by the action of pectolytic enzymes. The most important difference between these two species is that while *R. stolonifer* does not grow at temperatures lower than 5°C, *M. piriformis* is able to grow, although slowly, at temperatures in the range 0°C–5°C, thus causing decay on cold-stored citrus fruit. Although with considerably less frequency and virulence, Mucor rot can also be caused by the species *M. circinelloides* Tiegh., *M. racemosus F. racemosus* Fresen., *M. hiemalis* Wehmer, and *M. mucedo* L. Other infrequent wound pathogens are *Penicillium ulaiense* H.M. Hsieh, H.J. Su and Tzean, which causes whisker mold and it is difficult to distinguish from *P. italicum*; *Trichoderma viride* Pers. [syn.: *T. lignorum* (Tode) Harz], the cause of Trichoderma rot; *Aspergillus niger* Tiegh, the cause of Aspergillus black rot; and

Cladosporium herbarum (Pers.) Link, the cause of Cladosporium rot (Eckert and Eaks, 1989; Brown and Eckert, 2000; Palou, 2014).

Other important postharvest pathogens, called latent pathogens, infect the fruit in the field during the growing season and remain latent or quiescent until they resume growth after harvest. The principal species in this group include *Lasiodiplodia theobromae* (Pat.) Griffon and Maubl. (syns.: *Diplodia natalensis* Pole-Evans, *Botryodiplodia theobromae* Pat.) and *Phomopsis citri* H.S. Fawc., which cause the diseases commonly known as stem-end rots (Figure 21.2F and H); *Colletotrichum gloeosporioides* (Penz.) Penz. and Sacc., the cause of anthracnose (Figure 21.2J); *Phytophthora* spp., including *P. citrophthora* (R.E. Sm. and E.H. Sm.) Leonian, *P. palmivora* (E.J. Butler) E.J. Butler, *P. syringae* (Kleb.) Kleb. and *P. hibernalis* Carne, the cause of brown rot (Figure 21.2G); *Alternaria* spp., mainly *A. alternata* (Fr.) Keissl., the cause of *Alternaria* black rot (Figure 21.2D); and *Botrytis cinerea* Pers.: Fr., which cause gray mold (Figure 21.2I). Other infrequent latent pathogens are *Sclerotinia sclerotiorum* (Lib.) de Bary, *Fusarium oxysporum* Schltdl., or *Trichothecium roseum* (Pers.) Link (Eckert and Eaks, 1989; Smilanick et al., 2006a).

Because of their high relative importance in terms of commercial losses in most production areas and situations, green and blue molds caused by *Penicillium* spp. are the most economically important postharvest diseases of fresh citrus fruit worldwide. Therefore, decay management programs primarily address the control of these molds. Colonies of both *P. digitatum* and *P. italicum* are plane, heavy sporing, and grow rapidly on potato dextrose agar (PDA) or malt extract agar (MEA), but poorly on Czapek agar. The texture is velutinous, usually with exudate droplets absent or very limited. The colony obverse of *P. digitatum* is olive green and the reverse colorless to cream yellow or pale dull brown. Colonies of *P. italicum* are blue or gray-green colored and often appear granular due to the presence of bundles of conidiophores and conidial heads. The reverse is uncolored or gray to yellow-brown, although it can turn to brownish orange or red brown on some synthetic media (Figure 21.2C). Symptoms of infection of citrus fruit by *P. digitatum* and *P. italicum* start as a water-soaked, soft, circular area surrounding the infected rind wound that, as the fungus grows, expands radially, and develops aerial white mycelium first and colored velutinous masses of spores later. In the case of green mold, the conidial layer is olive green surrounded by a broad band of dense, nonsporulating white mycelium limited by fairly firm decaying peel. In the case of blue mold, the central sporulating area is blue or bluish green surrounded by a very narrow band of nonsporulating white mycelium limited by a broad band of soft, water-soaked peel (Figure 21.2A). At ambient temperatures, *P. digitatum* grows faster than *P. italicum*, and within 10–15 days the entire surface of the fruit is completely covered with spores and begins to shrink. Conversely, *P. italicum* grows better than *P. digitatum* at temperatures below 10°C and blue mold may predominate in fruit subjected to commercial cold storage (Brown and Eckert, 2000; Palou, 2014). On the other hand, *P. digitatum* is the first phytopathogenic *Penicillium* spp. whose complete genome has been entirely sequenced (Marcet-Houben et al., 2012).

The impact of postharvest decay is not only restricted to fruit losses. When rotten fruit is found in export shipments, even though the incidence may be relatively low, wholesale buyers typically reject the load and charge the producer for the transport and handling costs. Furthermore, they can abandon the affected producer and seek for other sources in the market.

21.3.2 Preharvest and postharvest factors affecting disease incidence

Actual losses due to postharvest diseases are quite varied and depend on the area of production, citrus variety, tree age and condition, weather conditions during the growing and harvest season, the extent of physical injury to the fruit during harvest and subsequent operations, the effectiveness of antifungal treatments, and postharvest handling and environment. In general, the incidence of postharvest decay is higher in production areas with abundant summer rainfall, such as Florida, Brazil, or southeastern Asia. The principal diseases in these regions are stem-end rots caused by *L. theobromae* and *P. citri*, which require rain and humid weather for inoculum production and dispersal and subsequent fruit contamination and infection. In production areas with a Mediterranean-type climate, such as Spain and other Mediterranean countries, California, or most citrus areas in South Africa, where summer rainfall is scant, total postharvest decay incidence is lower and the most prevalent causal agents are *Penicillium* spp. and other wound pathogens (Eckert and Eaks, 1989; Smilanick et al., 2006a). Penicillium molds, however, are also very important in humid areas because both *P. digitatum* and *P. italicum* reproduce very rapidly and their spores are ubiquitous in the atmosphere and on fruit surfaces and are readily disseminated by air currents. Therefore, the source of fungal inoculum in citrus groves and packinghouses is practically continuous during the season and the fruit can become contaminated and infected in the grove, the packinghouse, and during distribution and marketing. Furthermore, healthy citrus fruit can become unmarketable when soiled with conidia of these two fungi that are loosened in handling of diseased fruit in the packinghouse (Palou, 2014).

Besides climatic considerations, preharvest orchard operations have a crucial influence on the final incidence of citrus postharvest diseases, especially of those caused by latent pathogens. Cultural practices such as appropriate orchard sanitation (removal of fallen decayed fruit, mummies, infected stems and twigs, as well as dead or senescent prunings and plant materials) are important to reduce inoculum levels in the field, although the high costs involved usually hinder their adoption (Smilanick et al., 2006a). Bird and insect control may contribute to preserve the integrity of the fruit peel by decreasing the infliction of rind punctures and injuries that are entrance gates for both latent and wound pathogens. Field treatments with agrochemicals could also be of use either to reduce pathogenic inoculum levels (fungicides) or protect the fruit (growth regulators, salts, or biocontrol agents) (Youssef et al., 2012).

Fruit harvest and transportation must be careful, gentle, and adequate to minimize the production of rind punctures, wounds, bruises, compression damage, and general mechanical injuries since they are clearly related to the incidence of postharvest decay, especially to that caused by *Penicillium* spp. Stem clipping instead of fruit pulling is mandatory. Unfortunately, the large amounts of fruit to be moved, the high labor costs and, in many cases, the lack or low availability of conscious, skilled and well-trained teams of pickers are factors that may impede the adoption of optimum harvesting practices. Harvest should be avoided after rainfall or when free water is present on the fruit surface.

After harvest, many individual factors and fruit handling operations may influence the incidence of citrus postharvest diseases but, in general, there are three basic aspects that should be taken into consideration when establishing disease management strategies: (i) effective fruit and packinghouse sanitation to reduce atmospheric and superficial pathogenic inoculum levels, (ii) appropriate practices during fruit handling and storage to hinder fungal development and maintain fruit natural disease resistance, and (iii) adoption of suitable postharvest fungicide/antifungal treatments.

Effective cleaning and disinfection of fruit and packinghouse facilities and storage rooms during the citrus processing season is necessary to prevent the buildup of fungal inoculum to high levels. Chlorine (sodium hypochlorite; NaClO) has been for many years the most widely used sanitizer for fruit surface disinfection and prevention of contamination of dip or drench solutions, but it is currently increasingly being substituted in many packinghouses by other compounds such as peracetic acid (PAA), hydrogen peroxide (H_2O_2), or some mixtures. These disinfectants are included in the directory of products for organic production or processing (USDA, 2016) and overcome important problems associated with the use of chlorine and derivatives, such as their deficient performance with high levels of organic matter or potential harmful effects from nitrosamine formation and corrosive activity (Kanetis et al., 2008). Sanitizers, in general, are oxidants that kill free microbial structures, but not those established within rind wounds or developing under the peel of the fruit. Therefore, they are not an alternative to the use of conventional fungicides. Furthermore, they should be applied with caution since their high oxidant activity may injure the fruit rind (Palou et al., 2007a). The mentioned disinfectants and others such as quaternary ammonia, isopropyl alcohol, formaldehyde, or ozone are also used to disinfect previously cleaned packinghouse facilities and equipment, including storage rooms and field containers.

Commercial postharvest fruit handling and storage conditions in the packinghouse can modulate the interaction between the pathogen and the fruit host that may or may not lead to disease development in both latent- and wound-infected fruit. In a particular pathosystem, parameters such as fruit maturity, rind condition, inoculum density, time of infection, or temperature and RH, among others, directly influence the range of complex mechanisms that define the host-pathogen interaction and disease occurrence. The application of exogenous ethylene for degreening of early season citrus fruit that have optimum sensory quality but not the required external rind color is a common practice in packinghouses to meet market requirements and consumers' expectations. While it has been demonstrated that degreening clearly exacerbates the incidence of anthracnose because ethylene stimulates conidia germination and appressoria formation and germination of *C. gloeosporioides* (Brown, 1992), its influence on green and blue molds depends more on the effect of ethylene and degreening environmental conditions on the fruit host than on the pathogen itself. Thus, in Florida and other humid citrus production areas where degreening is performed at temperatures surrounding 30°C and very high RH, the process exerts a curing effect that reduces Penicillium molds by wound lignification. In Spain, California, and other Mediterranean areas where citrus fruits are degreened at 20°C–22°C (near optimum temperatures for the development of *Penicillium* spp.), the effect depends on initial rind color and fruit maturity, and disease severity generally increases on less green (more mature) fruit, presumably through an ethylene senescence effect on the fruit peel (Moscoso-Ramírez and Palou, 2014). Beyond the application of washing, sanitation, and specific antifungal treatments, packingline operations are also important to select, sort, and discard infected and/or decayed fruit. While this has been traditionally performed by hand, research work is in progress to develop and implement vision technologies, e.g., chlorophyll fluorescence imaging, hyperspectral computer vision, or laser-light backscattering imaging (Lorente et al., 2015), able to discriminate, in an accurate and fast manner, citrus fruit with incipient rind infections (caused by *Penicillium* spp. or *G. citri-aurantii*) and/or internal infections (caused by *Alternaria* spp.). Cold storage can be considered as a complementary physical tool for postharvest decay control of fresh citrus fruits. Low temperatures can inhibit or delay fungal growth and, in addition, they considerably reduce the fruit metabolic activity and delay its senescence, thus contributing to the maintenance of fruit resistance to disease. While postharvest pathogens such as *Penicillium* spp., *G. citri-aurantii*, *B. cinerea*, *Alternaria* spp., *Mucor* spp., or *Phytophthora* spp. are able to grow, although slowly, at the commercial cold storage temperatures for most oranges and mandarins (3°C–5°C), other pathogens such as *R. stolonifer*, *L. theobromae*, *P. citri*, *C. gloeosporioides*, or *A. niger* are not (Palou, 2009).

21.3.3 Disease management strategies

As previously mentioned, effective control of green and blue molds is the main goal of any postharvest disease management strategy for fresh citrus fruits. Typically, these and other postharvest diseases have been controlled worldwide for many years by the application after harvest of conventional chemical fungicides. Nevertheless, concerns about environmental pollution and human health risks associated with fungicide residues periodically lead to regulatory reviews and potential restrictions or cancelations on fungicide usage. Likewise, traditional citrus export markets are increasingly demanding products with lower levels of pesticides in order to satisfy the consumer's safety demands. This is especially significant in some EU markets where important trade companies and supermarkets are demanding fruit with residue levels even lower than those established by official regulations. In addition, new high-value markets based on sustainable, organically grown, or "green" agricultural production are currently arising all over the world. Furthermore, the widespread and continuous use of these synthetic compounds has led to a buildup of resistant biotypes of the pathogens in commercial packinghouses that seriously compromise the efficacy of such treatments (Erasmus et al., 2015). There is, therefore, a clear and increasing need to develop and implement control methods alternative to conventional fungicides for the control of postharvest diseases of citrus fruit. If conventional chemicals are not used, the goal is to accomplish satisfactory decay control by adopting "integrated disease management" (IDM) programs (Wisniewski et al., 2016). Since in some citrus-producing countries like Spain, the concept "integrated management" implies fruit production in compliance with particular regulations that still allow the use of postharvest fungicides (textually "only if necessary and under technical supervision"), the alternative control concept that is discussed here could be better named as "nonpolluting integrated disease management" (NPIDM).

21.3.3.1 Control with chemical fungicides

In general, citrus postharvest treatments with synthetic fungicides have a reasonable cost, are easy to apply and, depending on the active ingredient, provide curative, preventive, and antisporulant effects. These chemicals can be applied as aqueous solutions or mixed with waxes or other type of fruit coatings, and common treatment modes are spray, drench, or dip. Although most of the commercial products are single active ingredients, in the market there are also mixtures of two or more different active ingredients. Table 21.1 shows the maximum residue limits (MRLs) of common postharvest fungicides allowed in citrus fruits destined to major export markets.

Imazalil (IMZ) is currently the most used citrus postharvest fungicide worldwide. This imidazole is a sterol 14-α-demethylation inhibitor (DMI) typically applied in water at 500–1000 ppm as a dip, drench, or spray, or in amended waxes at 2000–3000 ppm. It can be applied as IMZ sulfate or IMZ emulsifiable concentrate and it is especially effective against citrus Penicillium molds. Efficacy, particularly against *P. digitatum*, fruit residue load, influence of type of formulation and application mode, and synergy with heat and food additives for mold control have been characterized (Smilanick et al., 2008). The most important problem related to IMZ usage in citrus packinghouses is the proliferation of resistant strains of *P. digitatum* and *P. italicum*, often associated with suboptimal residue loading (Erasmus et al., 2015). Recently, the IMZ resistance level of a broad group of isolates of *P. digitatum* from citrus packinghouses worldwide has been determined (Kellerman et al., 2017).

Thiabendazole (TBZ) is another benzimidazole used for many years in citrus packinghouses that is facing important resistance problems during the past years. Typical modes of application to control Penicillium molds and alleviate CI include water drench, dip, or spray over rotating brushes at concentrations of 500–2000 ppm, and also wax formulations at higher concentrations. TBZ is compatible and synergistic with heat and some antifungal food additives (Schirra et al., 2011). Since *P. italicum* is usually more tolerant than *P. digitatum* to benzimidazole fungicides, blue mold can be prevalent on TBZ-treated citrus fruits (Gutter et al., 1981).

Sodium ortho-phenylphenate (SOPP) is another classical fungicide used for many years to wash, disinfect, and protect citrus fruit. Usual applications are at room temperature in soak tanks or foamer washes at concentrations of 5000–20,000 ppm, with a final rinsing of treated fruit with water. Wax formulations are also available, but they are not as common as water applications (Smilanick et al., 2006a).

Pyrimethanil (PYR) is a relatively new active ingredient classified as a "reduced-risk" fungicide with significant activity against *P. digitatum* and *P. italicum*, particularly against TBZ and IMZ-resistant isolates of these fungi. Common applications are as aqueous solutions at about 500 ppm and in wax over rotating brushes at doses higher than 1000 ppm (Smilanick et al., 2006b). PYR performance is significantly enhanced when applied in combination with heat or some salts classified as

TABLE 21.1 Maximum residue limits (MRL; mg/kg) of postharvest fungicides in citrus fruits destined to major export markets.

Active ingredient	European Union	Switzerland	USA	Canada
Fludioxonil	10.0	10.0	10.0	10.0
Imazalil	5.0[a]	5.0	10.0	5.0
Fosetyl-Al[b]	75.0[a]	75.0[a]	5.0	9.0 OML[c]
Thiophanate methyl[d]	6.0	6.0	NT[e]	10.0[f]
Ortho-phenyl phenol (and Na salt)	5.0[g]	5.0	10.0	10.0
Pyrimethanil	8.0	8.0	10.0 OM[c]	10.0
Prochloraz	10.0	10.0	NT	NT
Propiconazole	9.0 O; 5.0 ML[h]	6.0	8.0	8.0
Tebuconazole	0.9 O; 5.0 ML[c]	NT	1.0 O	1.0 O
Thiabendazole	7.0[i]	5.0	10.0	10.0

[a] *Temporary MRL.*
[b] *Sum of fosetyl and phosphorous acid and salts, expressed as fosetyl.*
[c] *O = Orange; M = Mandarin; L = Lemon.*
[d] *Carbendazim is a potential metabolite with the following MRL: 0.2 O, 0.7 ML in the EU; 0.5 in Switzerland; banned in the USA.*
[e] *NT = No tolerance.*
[f] *Sum of thiophanate methyl and carbendazim residues.*
[g] *MRL of 10.0 mg/kg after 08/08/2018 (Commission Regulation (EU) 2018/78).*
[h] *Use only until March 2020 (Commission Regulation (EU) 2018/1865).*
[i] *MRL of 7.0 mg/kg (previously 5.0 mg/kg) since 21/01/2018 (Commission Regulation (EU) 2017/1164).*

food additives (Smilanick et al., 2006b; Kanetis et al., 2008). It is also used in mixture with IMZ to manage fungal biotypes resistant to this active ingredient. Combined applications are especially suitable because some studies report high risks of development of isolates of *P. digitatum* resistant to PYR in citrus packinghouses (Kanetis et al., 2008).

Fludioxonil (FLU) is a phenylpyrrole derived from the natural antibiotic pyrrolnitrin produced by several *Pseudomonas* spp. It is also classified as a "reduced-risk" fungicide and it is usually applied at doses of 500–1200 ppm for citrus disease control. However, it is less effective than IMZ against green mold due to the lower penetration of the active ingredient into the fruit rind and a limited anti-sporulation activity (Smilanick et al., 2008). FLU shows synergistic activity when applied in combination with TBZ or sodium carbonate. Similar to PYR, managing resistance strategies should be adopted for commercial use of FLU because the risk of resistance development in citrus packinghouses is high (Kanetis et al., 2008).

Other fungicides allowed in the EU for postharvest use on citrus fruit include the benzimidazole thiophanate methyl, the imidazole prochloraz (proposed for cancelation, but extended until 2021), the triazoles tebuconazole (TCZ) and propiconazole (PCZ), and the aluminum salt fosetyl-Al (Table 21.1). While most of these active ingredients are used to control *Penicillium* spp., TCZ and especially PCZ show activity against *G. citri-aurantii*, the cause of sour rot. This is of particular interest after the withdrawal of guazatine in the EU in 2012. PCZ, however, was recently canceled in the EU and can be used only until March 2020. On the other hand, fosetyl-Al is used in both preharvest and postharvest applications against brown rot caused by *Phytophthora* spp. Other active ingredients that have been tested as citrus postharvest fungicides include azoxystrobin, trifloxystrobin, and cyprodinil (Schirra et al., 2011).

21.3.3.2 Nonpolluting integrated disease management

The purpose of NPIDM strategies, based on the knowledge of pathogen biology and epidemiology and the consideration of all preharvest, harvest, and postharvest factors that may influence disease incidence, is to minimize decay losses with no adverse effects on fruit quality by taking cost-effective actions on every one of those factors at the right moment. Besides preharvest, harvest, and other postharvest considerations that have been discussed in the previous section, the basis of successful NPIDM strategies to control citrus postharvest diseases is the commercial adoption of suitable environmentally friendly postharvest antifungal treatments to replace the use of conventional fungicides (Palou, 2009). According to their nature, these alternative methods can be physical, chemical, or biological. Significant advances in the evaluation of these control treatments, either alone or in combination with other treatments of the same or different nature, have been accomplished over the last years and comprehensive reviews have been published (El-Ghaouth et al., 2002; Palou, 2009, Palou et al., 2016; Droby et al., 2016). The main findings are summarized in this section.

Antifungal heat treatments are the most important physical means and increasingly play a key role in NPIDM strategies for citrus postharvest decay control because they leave no residues and are relatively effective, simple, cheap, easy to apply and combine, and synergistic with other physical, chemical, or biological control methods. They include curing, hot water dips (HWDs), and hot water rinsing and brushing (HWRB). Curing of citrus typically employs exposure of fruit for 2–3 days to an air atmosphere heated to temperatures higher than 30°C at RH higher than 90%. In spite of its good efficacy, especially against Penicillium molds, commercial implementation of curing is rare because it can be unpractical, expensive, and risky (it can cause fruit weight loss or rind phytotoxicity if not correctly applied). Brief (2–5 min) HWDs at 45°C–55°C are cheaper and more feasible for heat application than curing, but they are less effective and persistent, and the range of effective yet non-phytotoxic temperatures is very narrow. Due to the synergistic activity of heat, chemical antifungal aqueous solutions are increasingly heated to these non-phytotoxic temperatures in citrus packinghouses. HWRB consists basically in packingline machinery that applies hot water over rotating brushes at high temperature (55°C–65°C) for a very short time (10–30 s). Other physical control means that are under investigation are irradiation and illumination treatments including far-ultraviolet light (UV-C, $100 < \lambda < 280$ nm), blue light ($390 < \lambda < 540$ nm; Ballester and Lafuente, 2017), and ionizing radiations ($\lambda < 100$ nm).

Chemicals alternative to conventional fungicides should be synthetic or natural compounds with known and very low toxicological effects on mammals and impact on the environment. The most important synthetic compounds are inorganic or organic salts, applied to fruit as aqueous solutions, and composite edible coatings formulated with antifungal ingredients. Natural compounds can be of plant or animal origin and include plant extracts and essential oils, antifungal peptides and small proteins, and natural antifungal edible coatings such as chitosan and *Aloe* spp. gels. As substances that will be in contact with fresh fruit, all alternative chemicals should be affirmed as food additives or generally recognized as safe (GRAS) materials by the competent legislation. Furthermore, GRAS compounds are exempt from residue tolerances on all agricultural commodities and are good candidates for use in organic agriculture. In general, these alternative chemicals can be applied to citrus fruit after harvest as aqueous solutions, vapors (fumigants), or coating treatments. General limitations of alternative low-toxicity chemicals include limited effectiveness, lack of preventive activity, and performance highly dependent on the fruit host species and cultivar, maturity stage, and physical condition. In general, these treatments are more effective on oranges than on mandarins, which are naturally more susceptible to disease. Among GRAS salts, sodium carbonate and bicarbonate and potassium sorbate (dip treatments in aqueous solutions at 2%–3% for 60–150 s) are more effective against green and blue molds than other salts and are currently being successfully used for commercial applications in citrus packinghouses worldwide. They have low cost, are synergistic with heat (solutions heated to 45°C–50°C), and suitable for combination with other disease control methods in integrated applications. Other salts with activity against citrus postharvest diseases include sodium benzoate, sodium parabens, and potassium silicate. A wide range of natural antifungal compounds from plants or animals has been evaluated for the control of citrus postharvest diseases, including essential oils and many of their more common volatile compounds (thymol, cinnamaldehyde, citral, eugenol, limonene, etc.), plant extracts such as isothiocyanates, propolis, and those of pomegranate, garlic, *Sanguisorba* sp., *Orobanche* sp., *Acacia* sp., *Whitania* sp., *Parastrephia* sp. etc., and peptides and proteins. Although many of them showed satisfactory antifungal activity in in vitro tests, inefficacy and/or fruit rind phytotoxicity when applied in vivo, together with the production of off-odors and/or off-flavors in treated fruit are general barriers to aqueous or vapor application of volatile oils or plant extracts. Antimicrobial peptides and small proteins, on the other hand, can be naturally obtained or artificially synthesized using plants or microorganisms as biofactories through genetic engineering techniques. Limitations are nonspecific toxicity, low stability and, above all, production and purification costs. Nowadays, a very active research field is the replacement of citrus commercial waxes, which are often a vehicle for the application of synthetic fungicides like IMZ, PYR, TBZ, or SOPP, with natural or synthetic edible coatings with antifungal properties. Chitosan and other derivatives from chitin present in crustacean shells, insect cuticles, and fungal cell walls are the most common natural coatings used for the inhibition of citrus postharvest pathogens. Others are gels from *Aloe vera* and other plant species. While these natural coatings present inherent antifungal activity, synthetic biopolymer-based coatings are formulated with antimicrobial ingredients that, according to their nature, can be synthetic food preservatives or GRAS salts; natural compounds such as essential oils, plant extracts, or peptides; and biocontrol agents. For instance, hydroxypropyl methylcellulose (HPMC)-lipid edible composite films containing GRAS salts such as potassium sorbate, sodium benzoate, or paraben salts showed curative activity and were effective to reduce green and blue molds on oranges and mandarins.

The denomination biological control is usually restricted to the utilization of microbial antagonists (bacteria, yeasts, and even some filamentous fungi) as agents for disease control. In general, these antagonists can be applied as aqueous cell suspensions in postharvest spray, drench, or dip applications; as ingredients of edible fruit coatings; or as biofumigants. Despite the extensive research in the past decades and the identification of many antagonists with biocontrol activity against citrus postharvest pathogens, the commercial use of antagonistic microorganisms remains very limited. In fact, only few biological products are currently registered for postharvest use on citrus fruits, that is, Bio-Save (*Pseudomonas syringae*), Shemer (*Metschnikowia fructicola*), and Pantovital (*Pantoea agglomerans*). The main reasons are their inconsistent performance when applied as stand-alone treatments, excessive specificity and lack of curative activity, disinformation and subsequent limited acceptance by the consumers, limited market and profitability for developing companies, and in some cases (e.g., EU countries), the inappropriate and strict regulatory issues that prevent registration.

Precisely due to their nonpolluting nature and consequent inherent limitations for decay control, alternative physical, chemical, or biological stand-alone treatments cannot presently achieve the performance of conventional fungicides required by the industry. Therefore, intensive research work is devoted to the evaluation of combinations of different treatments. According to Palou et al. (2016), three objectives may be pursued by the integration of two or more treatments: (i) additive and/or synergistic effects to increase the effectiveness and/or the persistence of individual treatments, (ii) complementary effects to combine preventive and curative activities, and (iii) potential commercial implementation of effective treatments that are too unpractical, costly, or risky as single treatments. Heat, biocontrol agents, and GRAS salts and natural compounds are the alternative treatments most utilized in combinations for the control of citrus postharvest diseases.

21.4 Postharvest handling and storage

In all postharvest practices involving the handling of citrus fruit from the orchard to the consumer, certain key factors should be considered as they could impact on fruit quality. The first is related to the non-climacteric nature of the maturation of citrus fruits, with relatively low and constant rate of respiration (4–8 mL $CO_2\,kg^{-1}\,h^{-1}$ at 5°C and 10–15 mL $CO_2\,kg^{-1}\,h^{-1}$ at 20°C) and less ethylene production (0.1 mL $kg^{-1}\,h^{-1}$ at 20°C) (El-Otmani et al., 2011). The second factor is related to the high susceptibility of citrus fruit to CI during cold storage below 4°C for an extended period, which varies not only between citrus species but also cultivars. Lastly, the citrus fruit comprises of two distinct physiological units, that is, rind or peel (pericarp) and the pulp (endocarp), which react separately to environmental conditions and treatments. However, the rind respiration is considerably higher compared with that of the pulp and increases toward harvest, while the pulp remains constant. It should be also kept in mind that optimum maturity and flavor are achieved on the tree and all efforts to retain quality should be made during postharvest handling in order to maintain the full commercial value of the fruit. The non-climacteric nature of the citrus fruit results in the fruit not undergoing dramatic changes during the postharvest period, except the visible increase in rind coloration (Tadeo et al., 2008).

Due to the large variation in citrus types and the numerous existing cultivars and those being developed, it is important not to accept general recommendations as correct under all conditions. Care should be taken to develop detailed postharvest handling protocols in order to maintain fruit quality and ensure full market value. A concept that holds true for all situations is that the impact of postharvest handling is the sum of all the various management actions and that a detrimental postharvest action may be difficult to correct during the remaining postharvest period. In most instances where fruit quality has been compromised in the market, it is the result of a cumulative effect of suboptimal handling processes.

21.4.1 Harvesting and orchard practices

During harvesting in the orchard and subsequent handling of the fruit, including all actions from the orchard to the packinghouse, transport, and cold storage of the fruit, the focus should be on reducing the risk of compromising fruit quality. The fruit quality can be compromised by increased incidences of decay, physiological disorders, mechanical damage, overmaturity, and loss of internal quality. Postharvest handling of citrus fruit in relation to various production areas and industries has been reviewed by El-Otmani et al. (2011) and Arpaia (2014). The harvesting practice can differ drastically according to the production purpose of the orchards: either for processing or for fresh consumption. In those industries focusing on fruit production for processing, various options exist as to mechanical harvesting systems. These technologies are constantly being developed but this chapter is focused on fruits for the fresh market.

The first critical decision in the citrus harvesting process is the correct timing to harvest, within the season as well as during the day, which most often affects the rind quality. The loss of postharvest quality can often be due to the incorrect harvest time, that is, the rind of over-mature fruit being more susceptible to disorders. In addition, too early harvest in the day or when trees and fruit are still wet can cause rind disorders or decay. For example, early season citrus fruit is more susceptible to oleocellosis (Zheng et al., 2010), peteca spot of lemons (Cronje, 2015), and CI (Lafuente and Zacarías, 2006), but could be more resistant against decay. In contrast, later in the season, fruit could be more susceptible to decay and disorders associated with premature senescence such as SERB and rind breakdown (Cronje et al., 2011a, b).

The importance of correct harvesting practices could not be overemphasized as reduced storage potential is often associated with fruit damage during the picking processes. Not only is the incidence of various diseases (Palou, 2014) but also the rate of moisture loss that could lead to disorders such as non-chilling rind pitting (Alférez et al., 2003) and noxan (Ben-Yehoshua et al., 2001). The citrus fruit for fresh consumption cannot be mechanically harvested, due to the high degree of damage on the fruit surface and resultant decay. Two options are used: first, snap picking where the fruit is pulled and twisted at an angle to the twig and second, picking by clipping the stem as close as possible to the calyx with a type of scissors. Due to the injuries associated with long stems on the rind of especially soft citrus most of the citrus industry are now using clippers, even if it results in a higher demand on labor.

After the fruit has been picked and placed in bins or in standard-harvesting boxes, it is important to deliver it in a timely fashion to a packaging house for two reasons: controlling decay and reducing the moisture loss from the rind. It is important to consider the environmental conditions during picking, temporary storage in the orchard and transport to the packinghouse, and employ actions to prevent unnecessary moistures loss (excessive exposure to sunlight, delayed transport, etc.). High temperature and low RH will increase the VPD between the fruit and environment and will lead to moisture loss from the fruit peel, possibly compromising the fruit quality.

21.4.2 Packinghouse practices

The general handling sequence of fresh citrus fruit at the packinghouse level is similar in most countries. The critical aspect to adhere to market demand is traceability, which requires that all fruit received in the market must be able to be traced to a site of production, time of harvest, packinghouse used, and packing date. This information, fixed to each carton, supplied in barcode-type systems enables remote scanning allowing relevant parties to trace fruit back and access all information such as plant protective chemicals used as well as all chemical treatments applied in the packing process.

Even though the flow of the fruit from the orchard through the packinghouse could differ between countries and even production regions, the general principles and systems are similar (Fig. 21.3). On arrival from the orchard, the fruit must be drenched ideally within 24 h to remove field heat, kill fungal spores, and leave fungicide residues on the fruit for protection during degreening. After adequate drying-off, fruits are exposed to degreening if peel coloration is inadequate. Thereafter, fruits are either wet or dry dumped onto the packing line, with wet dumping doubling as a sanitizing action and resulting in

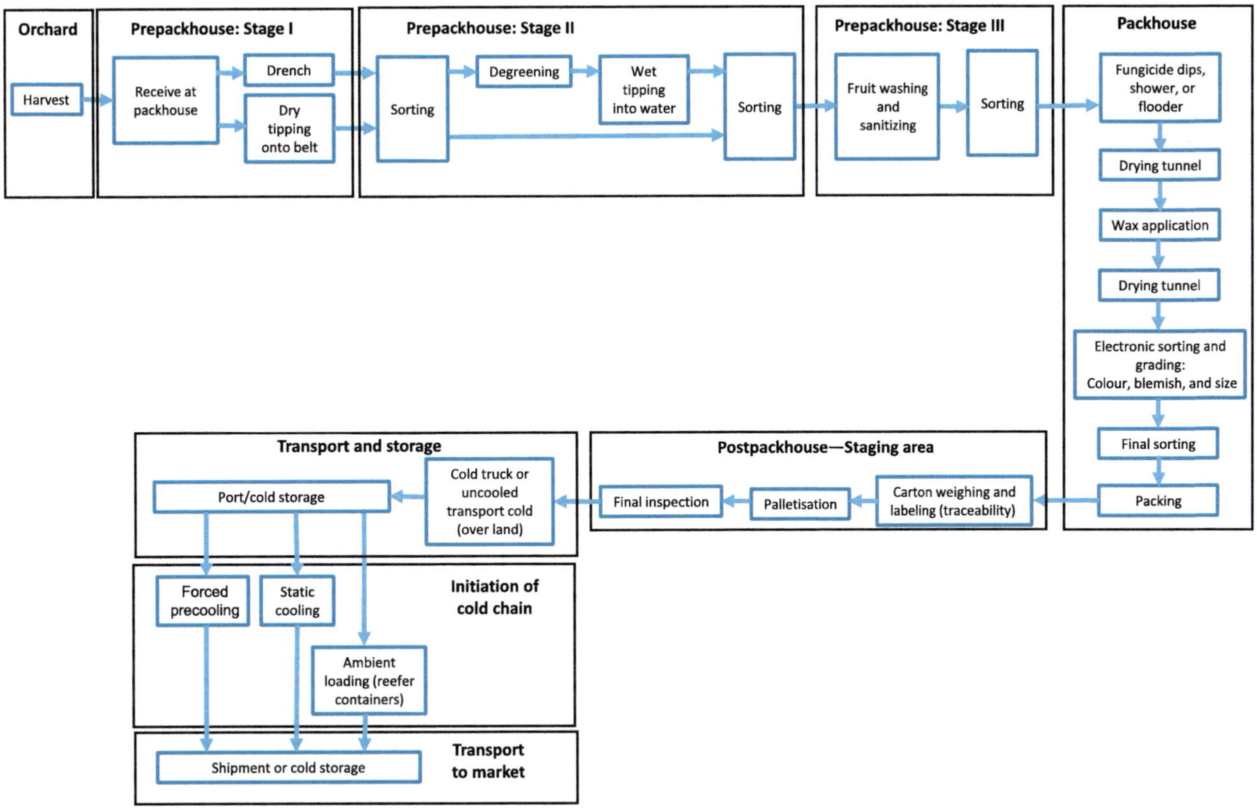

FIG. 21.3 Schematic flow chart of the different steps, procedures, and treatments during the entire postharvest chain of citrus fruit.

fewer injuries. The fruit moves onto a conveyor belt to enable presorting either with mechanical sizers and/or a combination with visual inspections enabling the elimination of substandard fruit. All rejected fruit is removed from the packinghouse via a conveyor system to be collected for processing. After presorting, the fruit is subjected to a fungicide bath (dip treatment or flooder) to facilitate residue loading and thereafter, the fruits are air-dried before waxing. After the wax is applied, the fruits are dried before undergoing computerized classification according to predetermined specifications including size, color development, shape, and blemish incidence. Thereafter, the fruits are packed by hand or with a mechanical packing machine into bags or cartons, labeled with all relevant information to facilitate traceability, that is, cultivar, grade, size, packing date, packingline number, production unit code, and packinghouse code. Packed cartons must conform to a weight specification before being palletized. Each pallet is provided with a unique barcode identification that includes, among others, the destination, date packed, and packaging type, and is linked to the production unit code as presented on each carton, to ensure recording and tracking of each pallet from the packinghouse to the end buyer (Fig. 21.3).

21.4.2.1 Ethylene degreening

Degreening of citrus fruits is a postharvest treatment to induce coloration of the peel without affecting internal fruit quality. This treatment is widely used in many countries to uniform coloration and to advance the commercial and marketing period. Acceptance of citrus fruit by consumers, especially in the Northern Hemisphere and in countries of cold winter, is generally established for the bright and attractive orange color of the peel, whereas the greenish appearance is often associated with unripened fruit. This consideration is subjective as consumers from tropical countries where citrus fruits do not attain the characteristic orange color due to the absence of cold night temperatures are familiar with them and consume mature fruits with acceptable internal quality but of green-yellow peel coloration. It is well known that color changes in citrus fruits require night temperatures around 12°C and also marked differences between day and night temperature (reviewed in Alquezar et al., 2008). Moreover, molecular analysis of carotenoid biosynthesis revealed that at 12°C the expression of key genes of the pathway is stimulated and increased the concentration of the corresponding carotenoids and xanthophylls (Carmona et al., 2012). Citrus fruits growing under subtropical conditions do not develop adequate external color as do fruits growing under Mediterranean conditions or other cooler regions (Porat, 2008).

Ethylene degreening is especially important for early varieties that reach adequate internal quality for consumption, while the peel is still green and is currently applied in countries with important fresh-fruit market, like Spain, Morocco, South Africa, Israel, and the United States. Relevant reviews have outlined the main characteristics of the treatment and optimal conditions for adequate degreening (Porat, 2008).

Commercial degreening of citrus fruits has become a common postharvest practice for more than a century. At that time, fruits were cured at high temperature, in order to heal mechanical injuries, in kerosene stoves that also promoted peel degreening. Denny in 1924 reported for the first time that ethylene was the agent responsible for fruit degreening and since then, extensive technologies have been developed worldwide to ensure adequate degreening and proper coloration of the fruit. The ethylene-releasing agent 2-chloroethylphosphonic acid, also known as Ethephon or Ethrel, was also used in the past or in certain countries where the technology of application of ethylene as a gas is not implemented. Since ethylene is produced by the chemical breakdown of Ethephon by the plant cells, its use in modern citriculture is very limited.

The physiological basis of fruit degreening is well understood and major efforts are concentrated on optimizing technological requirements such as temperature, minimum time to obtain adequate coloration, air circulation, ventilation, etc. Commercial degreening are usually carried out in large rooms of packinghouses equipped with automatic injectors to introduce the required concentration of concentrated ethylene in the room to achieve an ethylene concentration around 2–5 ppm. This system of ethylene application is referred to as intermittent continuous flow of ethylene and is widely used in most countries. Excessive exposition of the fruit to ethylene or election of high ethylene concentration may be detrimental for fruit quality and, in some instances, may originate typical brown spots on the peel surfaces. This effect depends on the cultivar, and the maturation stage. Another interesting phenomenon of degreening with important practical consequences is the so-called "post-ethylene effect." This effect comprises the residual changes in coloration taking place 24 h or more in air after the removal of ethylene. Fruits of different cultivars exhibit variable residual post-ethylene effects but in many cases enable shortening the degreening period and fruit coloration may continue during storage or shipment. Then, a current recommendation to degreen is to shorten exposition to ethylene in order to attain adequate coloration reducing exposition to the gas (Porat, 2008).

Temperature is an important factor determining the efficacy of the degreening process. The range of degreening temperatures may differentially affect chlorophyll degradation and carotenoids accumulation: high temperature (around 30°C) accelerates chlorophyll degradation but inhibits carotenoid accumulation, producing pale fruits. However, in the range of 15°C–20°C, fruit attained good peel coloration but at a lower rate. Under tropical climatic conditions, it is recommended to degreen fruit at 28°C–29 °C, since under warm humid areas many fruit can be affected by a quiescent Lasiodiplodia stem end that will develop more rapidly under ethylene treatment. In general, in countries of continental or Mediterranean climactic conditions, it is recommended to perform the degreening treatments at moderate temperature between 20°C and 23°C and high RH, not <90% (Porat, 2008; Passaro-Carvalho et al., 2012).

Being citrus fruit non-climacteric, exogenous ethylene increases the rate of fruit respiration and then under commercial degreening conditions increases the consumption of oxygen and the release of carbon dioxide inside the degreening chambers. Therefore, adequate ventilation is essential to reduce the build up of carbon dioxide in the ethylene room that should be maintained below 0.15%–0.2%. Degreening rooms are preferred to be large but not very high to facilitate air circulation and ventilation. Modern packinghouses have adopted specific CO_2 sensors to monitor continuously carbon dioxide concentration, and in some instances, ethylene concentration may be also automatically adjusted.

Duration of the degreening period depends obviously on the intensity of green color at harvest (peel chlorophyll content), but should be as short as possible and preferably not exceeding 2–3 days to prevent quality loss. Nowadays, degreening is preferred to be reduced as much as possible and in some countries current legislation restrict degreening to <60 h. Research efforts are addressed to optimize the time of the degreening period as a function of the peel color at harvest for each specific cultivar. There are, however, many problems allowing a standardization of the process, as the rootstock, environmental conditions (fruit from warm climates are more difficult to degreen than those from cold climates), peel thickness, genetic differences between varieties, among others. Under Mediterranean conditions, different lengths of ethylene treatment have been estimated in relation to the color at harvest, in order to obtain a commercial coloration for both oranges and mandarins for export to the European Union, the United States, or Japan (Table 21.2) (Passaro-Carvalho et al., 2012).

It is important that the degreening treatments are applied to fruit with optimum quality requirements of internal maturity and also when the natural coloration has been already initiated. Exposure of immature fruit to ethylene may reduce chlorophyll content and green appearance but proper coloration would not be developed. Rodrigo and Zacarías (2007) have shown that the peel of early harvested Navelate oranges displayed an attenuated expression of carotenoids biosynthetic genes compared to fruit harvested once the natural degreening process was already initiated in the field, indicating that peel should be at a specific ripening stage to be able to respond properly to the ethylene-induced coloration. Other desirable condition for a successful and efficient degreening is a homogeneous coloring of the fruit that will produce uniform coloration and better acceptance in the market. Some countries have adapted the use of color photocells in the packing lines for a previous color calibration and implementation of the required ethylene degreening treatment to reduce disorders.

Degreening with ethylene not only accelerates chlorophyll loss and stimulates coloration of the skin but, depending on the concentration and time of exposure to the gas, and other environmental factors, it may also provoke adverse effects on fruit quality. Increasing respiratory rate is a common feature of ethylene application than in non-climacteric fruits decrease when ethylene is withdrawn. Degreening treatments may produce severe detrimental effects on peel quality, inducing dehydration of peri-peduncular areas around the calyx, blackening and abscission of the calyx, browning, and dehydration of the peel that are commonly manifested during the marketing period, several days after the treatment. In some varieties of grapefruit or oranges, ethylene may also reduce fruit firmness and provoke peel softening, and in some mandarins may increase the content of ethanol, probably related to the build up of carbon dioxide (Passaro-Carvalho et al., 2012).

TABLE 21.2 Degreening treatments for exportation of mandarin and orange fruits to European Union countries and, the USA and Japan.

	European Union		USA or Japan	
Initial color index (IC)[a]	Mandarins	Oranges	Mandarins	Oranges
$-13<IC<-5$	72 h C_2H_4	NA[b]	48–72 h C_2H_4	NA
$-5<IC<+3$	48 h C_2H_4	72 h C_2H_4	48–72 h Air	48–72 h C_2H_4
	72 h Air			
$IC>+3$	24 h C_2H_4	48 h C_2H_4	24 h Air	24 h C_2H_4
	48 h Air	72 h Air		48 h Air

[a] *IC calculated as defined by Jimenez-Cuesta et al. (1981). Higher than 7 are not degreened. Data adapted from Passaro-Carvalho et al. (2012).*
[b] *NA: No application.*

21.4.2.2 Fruit cleaning and sanitation

All practices to clean and sanitize the fruit from the orchard are focused on reducing the risks of decay development from existing fungal infections and production of new infections. Due to the optimum conditions for decay development during degreening, fruit must be treated before this step by a drench and washing action. In general, the fruit is treated within 24h after harvest with a fungicide such as TBZ and IMZ. After degreening, the fruit is waxed or stored for waxing later, depending on the commercial requirement (Smilanick et al., 2006c; Erasmus et al., 2011). In terms of peel physiology and specifically to reduce moisture loss, it is ideal to drench fruit as soon as possible after harvest as the treatment will remove field heat. The loss of moisture and dehydration of the rind during this stage of the postharvest process is known to result in rind breakdown/staining of various cultivars (Alférez et al., 2003, 2005; Cronje et al., 2017). It should be noted that during the washing process, employing a line of brushes and a type of soap or sanitizing agent, could remove some of the natural wax coatings from the fruit surface, thus increasing the water loss from the flavedo (Alférez et al., 2010). Therefore, the timely application of wax after washing is important to reduce the moisture loss.

21.4.2.3 Sorting of citrus fruit according to quality classes

In the global citrus industry sustained commercial success depends on producing and delivering fruit of a consistent quality standard to the fresh markets. The process of sorting fruit according to internationally accepted quality parameters is essential and all efforts should be made to use available technology to consistently deliver fruit of an expected standard to a client. These standards are generally based on the absence of physical injuries, decay and physiological disorders as well as the fruit being of a required shape, size, and color.

In terms of sorting equipment for citrus fruit, significant advances have been made due to the availability of new technology and the increased adoption of nondestructive technology in addition to comprehensive analytical software. These advantages have resulted in a declining reliance on subjective destructive assessment of fruit quality in order to permit fast, accurate, and repeated measurements on one piece of fruit, without causing waste. Sensing and measuring the quality attributes of fruit can now be performed with visible to near-infrared (Vis/NIR) spectroscopy (Vis/NIRS), one of the most advanced technologies with regard to instrumentation, applications, accessories, and chemometric software packages (Magwaza et al., 2012; Ruiz-Altisen et al., 2010; Cubero et al., 2016).

The uses of automated systems to facilitate sorting in real time have different classification functionalities in one packingline. The first traits used in the automated sorting center is appearance, that is, size or dimensions, shape, surface texture, external defects, peel/rind color, and fruit size. Evaluating color and external defects relies mainly on computer vision techniques measuring the reflectance, necessary for color and external defects. Quantification of size shape and surface textures can be evaluated with either reflectance or distance/range data. In terms of assessing citrus fruit quality, a few characteristics are critical in addition to color and size. Even though new technology using hyperspectral imaging has been used and several advancements have been made, the in-line assessment of internal quality in citrus is still a major challenge for research (Magwaza et al., 2012). A noteworthy advance in nondestructive quality detection would be to separate progressive peel problems from nonprogressive appearance-related problems such as wind, hail scaring, oleocellosis, and types of insect's damage. These progressive quality concerns include physiological disorders, that is, peteca, rind pitting and breakdown as well as deleterious fungi causing decay, as *P. digitatum* or *C. gloeosporioides* (Cubero et al., 2016). The importance of identifying progressive quality issues in the packinghouse stems from the high cost associated with the arrival of this type of fruit in the market.

A further major challenge in research on nondestructive sorting in citriculture should increasingly focus on the detection of phytosanitary diseases such as citrus canker, black spot, and insects such as fruit fly larvae. Currently, detection and sorting fruit for a specific organism (disease or insect) still requires human vision; however, new hyperspectral imagining technology could offer new opportunities. Detecting and preventing the export of infected fruit will be an important step to comply with phytosanitary regulations and maintaining the year-round sustainable global supply of citrus to the market.

21.4.2.4 Wax application

The main purpose of a wax application on citrus fruit is to prevent moisture loss and additionally to impart an attractive shiny appearance on the fruit. The third critical aspect is to act as a carrier for fungicides and lastly to reduce the incidence of postharvest physiological disorders associated with moisture loss such as SERB and CI (Hagenmaier, 2002; Petracek et al., 2007). For the wax coating to have these desired effects, it should enable the fruit to continue with all relevant physiological processes, that is, gaseous exchanges during postharvest handling, while reducing moisture loss and preventing decay development. The incorrect wax formulation or application of wax could, therefore, be detrimental to fruit quality, reducing the commercial value.

The water-based waxes used in all citrus industries have specialized formulations containing various types of emulsifiers, drying agents, and solid contents. The solid content of a wax product is important as the effect on fruit quality can be cultivar specific and it may influence peel color and appearance and need to conform market requirements. Furthermore, the incorrect formulation of wax application could lead to off-flavors such as ethanol and acetaldehyde associated with anaerobic respiration (Cohen et al., 1990; Porat et al., 2005).

In general, citrus waxes consist of solids, that is, shellac, wood resin, polyethylene, and carnauba and bases such as ammonia to dry the coating (Hagenmaier, 1993) and are applied after drying of the fungicide application that takes place in the packing line. However, the wax products formulated in the various production areas differ and could also include sugarcane wax, bee wax, and wood resin (colophony). Shellac is a refined product from raw lac, which is secreted by the *Laccifer lacca* insect and is added to formulations to improve the shine of the coating. The natural vegetable wax, carnauba, is processed from the palm leaves of *Copernica cerifera* (Hagenmaier, 1998) and even though it imparts a low level of shine it is recommended for certain markets due to its plant-based origin. Polyethylene, synthesized from ethylene, results in a wax with desirable properties such as high shine and high permeability to gases; but, if applied with a high solid content it could affect fruit quality negatively by increasing off-flavor development. Formulation of a coating is propriety information from commercial companies that constantly reevaluate and develop new products to comply with the market specifications.

Application of wax coatings is done via a variety of nozzle types over a set of rotating brushes, fabricated from horsehair and/or polyethylene (Hagenmaier, 1993). The nozzles apply the coating to the brushes that in turn spread it on the peel at a rate of 1–1.3 L/ton of fruit. Various systems of delivering the desired volume of wax with nozzles to the brushes are used, such as drip or cone shapes. The effective spreading of the coating is not only influenced by the type of hair in a brush but also by how the tufts of hair are arranged in the brush shaft. In terms of the hair type, the wax brush can be made of either horsehair (most expensive) or polyethylene or combinations of these two types. Horsehair is generally regarded as the best in terms of spreading the wax to enable it to be incorporated into a coating. However, due to commercial realities such as cost, combinations of 40%–50% horsehair with polyethylene hairs/strands in brushes are widely used. In addition to the wax type and application method, it is vital to be cognizant of the packing line and the brush speed as this could lead to over- or under-waxing, which may impact negatively on moisture as well as decay control.

21.4.2.5 Carton box design and ventilation

Packaging plays a key role in the citrus postharvest cold chain to ensure the fruit are cooled and protected from the packinghouse to the consumer in an energy-smart, resource-efficient, and cost-effective way. In determining the effectiveness of fresh citrus fruit packaging, an analysis must be performed integrating the simultaneous and quantitative evaluation of product cooling rate, box ventilation, product quality/shelf life, box mechanical strength, and energy consumption to reduce and maintain fruit temperature (Defraeye et al., 2015). The cost of citrus packaging is increasing and can account for 20% of the cost per carton at the farm gate, with production cost being 40%–50% and packinghouse handling 30%. In addition to the cost of the packaging, fruit losses in the cold chain can be substantial due to mechanical injury and packaging failure that can result due to poor design, faulty selection, and use of carton boxes.

The packaging of citrus fruit started in wooden boxes and although being very strong was replaced by more versatile, lighter, and cheaper carton boxes, which allowed printing for market purposes (Vigneault et al., 2007). Carton boxes have a large disadvantage as they only allow limited air circulation during precooling making adequate ventilation critical. Generally, boxes can handle a total opening area (ventilation holes) up to 5% without any significant decrease in the strength of the stack. The vents should be located away from the corners of the box and be designed to avoid blockage by produce and it is recommended to avoid round vents. Because of the many disadvantages of carton boxes, reusable plastic containers are gaining popularity in the agri-food industry. They are more durable than carton boxes and can keep their strength after wetting.

In order to effectively precool citrus fruit and remove heat from the whole pallet to reduce the respiration rate and moisture loss, a high cooling homogeneity must occur through the pallet in a horizontal direction. Thereafter, during shipment of the pallet, cooling and removing of heat from respiration or from the environment occurs in a vertical direction. Therefore, in carton design, linking up adequate ventilation holes in the horizontal and vertical direction is important to reduce and maintain fruit temperature (Defraeye et al., 2015).

21.4.2.6 Moisture loss management

The citrus fruit rind (pericarp), being anatomically isolated from the juice vesicles of the pulp and with leaf-like properties, is the source of moisture loss via transpiration and evaporation (Kaufmann, 1970). In terms of resistance to moisture loss, the citrus fruit has a thin film of epicuticular waxes on the surface of their cutin matrix. Most of the water vapor and other gases are understood to move out of the detached citrus fruit by flowing freely throughout the spongy parenchyma of the mesocarp (albedo) and the central axis, diffusing through the exocarp and the colored portion of the peel (flavedo), and by evaporating on the surface. In this system, the flavedo portion of the peel seems to be the primary site of resistance against moisture loss (Ben-Yehoshua, 1969).

Water vapor and other gases move across the peel by different mechanisms in fruits. Ben-Yehoshua et al. (1985) concluded that stomata of harvested citrus fruits are essentially closed, but that ethylene, O_2, and CO_2 may still diffuse mainly through residual stomatal openings, while water moves preferentially through a liquid aqueous phase in the cuticle. Cuticular transpiration as a mechanism was proposed for fruit, rather than stomatal transpiration. In a study on 'Fortune' mandarin fruit, low temperatures induced a breakdown of the more external cellular strata of the peel which then modified the physiological function of the cuticle and led to a dramatically increased in water permeability (Vercher et al., 1994). The difference in rind anatomy between citrus types could influence the movement of water across the peel as well as the response to the moisture loss. For instance, mandarin fruit has a much thinner albedo compared with that of other citrus fruit and it has been shown that albedo thickness could be important in the incidence of disorders associated with dehydration followed by rehydration (Alférez et al., 2005; Cronje et al., 2017). Moon et al. (2003) observed that the wax density of the rind, and thus also the coverage of stomata, increased with maturity, in both water-stressed and control Okitsu wase Satsuma mandarin. Cohen et al. (1994) reported significant water loss in citrus fruit associated with the development of microscopic cracking of rind tissue when the fruit was moved from 2°C to 20°C. It is, therefore, recommended to reduce the VPD during the period from picking until drenching, as this is a period of high moisture loss. It was reported that fruit exposed to moderate dehydration in the orchard is much prone to develop rind injury after subsequent storage at high RH leading to rehydration of the peel causing water potential difference between the albedo and flavedo (Agustí et al., 2001; Alférez et al., 2003). These alterations in water status in different parts of the peel are thought to be the underlying mechanism of various peel disorders and that the erratic incidence of these physiological rind disorders may be related to the variation in environmental conditions at harvest (Alquézar et al., 2010). During harvesting, when the prevailing climate results in a high VPD, that is, warm days and at low humidity or windy days, moisture loss should be minimized by timely removal of field heat via drenching or by increasing the RH after picking since this could help to maintain firmness and increase resistance to decay during and after storage.

21.4.3 Transport and international shipment procedures

21.4.3.1 Temperature management prior to and during shipments

The cold chain for citrus fruit involves precooling (static or forced air cooling) after packaging, cold storage prior to transport via air, road and cold storage in transit during shipment and subsequent cold storage at wholesalers, retailers, and finally consumers. For most horticultural crops, storage and shipment at temperatures near freezing point is the most widely used technology to reduce the deterioration process as soon as possible after harvest. However, most tropical and subtropical fruit including citrus are not suited to these temperatures. Generally, for all citrus types, temperatures between 4°C and 10°C are commonly used, or 12°C to form lemons, to store or ship the fruit.

In order to reach the desired shipping or storage temperature, citrus fruit commonly undergoes forced-air precooling (FAC), whereby air at low temperature is delivered to the individual fruits inside the carton box to ensure sufficiently rapid cooling. This requires adequate ventilation of the package to ensure cooling homogeneity. During precooling, the half-cooling time and the seven-eighths cooling time (SECT) are important measuring points, as they describe the time to reduce the difference between the fruit and air by half or by seven-eighths, respectively. Precooling and cold storage of fruit have been well documented. However, the transport operations have received less attention especially in the case of citrus shipments. The drastic shift in the past decade toward refrigerated containers "reefers" has almost completely replaced bulk reefer vessels, necessitating a better understanding of this critical link to maintaining the market value of citrus fruit.

In various horticultural industries, research on shipment of fruit in reefer containers has focused on monitoring the temperature and controls of the environment (Defraeye et al., 2015). Holistic experimental evaluation showed the feasibility of using the cooling capacity of the refrigerated container to reduce citrus fruit temperature. By using a standard ambient loading protocol 'Navel' orange fruit reached the SECT in 3 days (Defraeye et al., 2016). During ambient loading, fruit are loaded "warm" into refrigerated containers, therefore, not receiving any precooling and temperatures can range from 15°C to 22°C. The cooling and removal of field heat occur in the reefer container during long-haul transport that results in significant logistical and financial savings and it is being used for Valencia oranges and grapefruit (Defraeye et al., 2015).

A further aspect requiring additional research effort in terms of the cold chain is the increased awareness of the fruit as a potential pathway for phytosanitary pests and diseases. Exporting countries need to comply with a specific temperature range during the export of citrus. This conflict of maintaining a required temperature range, while not compromising the value of the fruit requires detailed handling protocols from multifaceted research projects. It is important not to evaluate a single stage in the cold chain independently but rather as an integrated system where a deviation in one aspect will negatively impact on the later stages and the ability to control fruit temperature without damaging the fruit.

21.4.3.2 Possible negative impact during cold storage: Pre- and postexport

Cold storage and shipment temperature protocols of the different citrus species from various countries and regions have been developed and are being used. The first considerations in these protocols are always the incidence of decay and incorrect cold-chain processes causing breaks and a cooling delay that will increase decay incidence. The second aspect should be the physiological response of the peel. At low temperatures for an extended period, the peel will be negatively affected and CI lesions could develop (Lafuente and Zacarías, 2006) as well as negatively influencing peel color (Van Wyk et al., 2009). It is complex to list specific cold protocols for each situation as it will depend on cultivar, production area/country, and target market. Nevertheless, during development of cold protocols, the citrus type (genetic response), temperature (threshold), and duration of exposure should be carefully considered in order to prevent quality loss. Employing the correct cold protocol for a specific citrus type during storage or shipment will be critical to maintaining quality.

21.4.3.3 Treatments to induce cold resistance to prolonged cold storage

The reduction of CI during the commercial cold storage and shipment of citrus fruit has been a focus in postharvest research in order to develop technology and strategies to reduce the loss of commercial value (Lafuente and Zacarías, 2006). However, no single solution has yet emerged which could be implemented on a commercial scale to reduce the incidence of CI, if fruit is exposed to a too long period below a threshold temperature. Currently, to reduce the impact of postharvest low-temperature stress, a combination of various practices is recommended and included exposing only high-quality fruit to cold temperatures as well as employing packinghouse treatments such as application of the fungicide TBZ and a wax coating.

Certain postharvest fungicide treatments such as TBZ have been shown to reduce the incidence of CI in Toracco orange (Schirra and Mulas, 1995) and Navel orange and grapefruit (Schirra et al., 2000). The efficacy of the fungicides, of which the mechanism is largely unknown, is more pronounced in hot water-dip treatments (40°C–50°C) compared to cold water (<20°C) (Schirra et al., 2000). The second technology and probably the most critical commercial practice that can reduce the development of CI symptoms is the wax application. In addition to preventing moisture loss under nonchilling conditions, it is known that wax reduces water loss in cold-damaged peel tissues, and then preventing the development of chilling symptoms (Wild, 1993). The type of postharvest wax applied could determine the efficacy and for example, Hagenmaier (2000) reported that polyethylene-candelilla wax was better in reducing the incidence of CI than shellac or wood resin wax. Nevertheless, this aspect in postharvest handling of citrus fruit should be further elucidated as new wax formulations are being developed. Hot water or heat treatments during the packaging process of citrus have been shown to be effectively reduce CI (Porat et al., 2000). Ghasemnezhad et al. (2008) reported a reduction in CI in "Fortune" mandarin when it was treated with a temperature of 45°C for 2–5 min or 50°C for 2 min, before storage for 8 weeks at 2 °C. However, in general, temperatures >50 °C for even a short duration can cause peel heat damage seen as a scalding brown symptom (Figure 21.1I). Furthermore, other quality parameters could also be affected leading to a weight decrease, peel coloration, and an increase in the total soluble solids (TSS): acid ratio. Therefore, the practicalities surrounding treatment of large volumes of fruit within a consistent and narrow temperature range for a specific duration have resulted in this technology not being widely commercialized.

The modified atmospheric packaging of fruit via the use of plastic films helps to maintain a higher RH as well as changing the concentrations of CO_2 and O_2. Porat et al. (2004) reported that microfilms such as extended and polyethylene reduced CI by 71% and 35%, respectively, in Shamouti oranges stored at a 2 °C for 6 weeks, possibly due to the higher RH helping to reduce water loss of fruit. Reduction of fruit moisture loss with higher RH is beneficial for export of quality fruit. Although modified atmospheric packaging has proven effective in moisture control, high RH can induce pathogen growth, but using fungicides will prevent the growth of pathogens. Furthermore, anaerobic respiration and development of off-flavors could be stimulated. Porat et al. (2004) indicated that if the concentration inside the packaging is >7%–8% CO_2 in combination with <14%–15% O_2, it will cause anaerobic respiration in citrus, which will lead to the formation of off-flavors that will affect fruit quality. It is accepted that, in general, citrus fruit could not tolerate atmospheres containing either <10% O_2 or >5% CO_2 (Ke and Kader, 1990). Furthermore, differences in peel anatomy and juice vesicle biochemical composition significantly influence the response of different cultivars to low O_2 and high CO_2 conditions (Shi et al., 2005, 2007). It is, therefore, very important to choose the most efficient type of plastic film perforation that will modify atmospheric concentrations of CO_2 and O_2 in the package to such an extent that it will reduce the incidence of CI.

21.4.4 Postharvest quarantine treatments for citrus exports

All countries have the right to maintain strict quarantine measures against various citrus pests and diseases, resulting in detailed protocols being negotiated between the importing and exporting countries. These protocols are based on guidelines from various role players such as the International Plant Protection Convention (IPPC), European and Mediterranean Plant Protection Organization (EPPO), and United States Department of Agriculture-Animal and Plant Health Inspection Service (USDA-APHIS). Various options in terms of postharvest phytosanitary treatments exist and include cold and heat treatments, chemical treatment (i.e., methyl bromide), and gamma irradiation; however, the most widely used commercial option is cold treatments during export. This compulsory low-temperature cold treatment, commonly referred to as a cold-sterilization protocol, is designed to kill insect larvae of listed phytosanitary insects such as the various fruit fly species. However, these treatments lead to an increase in incidences of CI as temperature normally ranges below 2°C for periods of 14–24 days to which citrus fruit are not tolerant. Strict postharvest handling protocols encompassing the factors discussed above need to be adhered to in order to maintain and deliver fruit of high quality in the market. However, even under optimum postharvest handling practices, cold treatments of citrus fruit can lead to a 3%–5% loss of fruit in the market.

Treatment of citrus fruit with methyl bromide is detrimenntal to shelf life and not only reduces fruit quality (external lesions and internal off-flavor) but also is environmentally unsound and used in limited markets. Irradiation is one of the newest phytosanitary measures used to prevent the introduction or spread of regulated pests. A significant advantage of irradiation is that, unlike for cold treatment, the effectiveness against a pest is not only measured in terms of mortality, but also in preventing the successful development of the insect's life stages. Therefore, inspection agencies will be accepting of live (but nonviable) pests, as described by the IPPC approved Guidelines for the Use of Irradiation as a Phytosanitary Measure (ISPM No. 18; IPPC, 2003) (Hallman et al., 2016).

Irradiation not only affects the insect but also the fruit cellular metabolism, and excessive doses would be detrimental. The effect of irradiation on the quality of citrus fruit depends upon various factors such as dose, fruit maturity, and cultivar (Palou et al., 2007b). In general, McDonald et al. (2013) found that the primary effect of irradiation on quality was external damage (pitting and visual damage) of navel oranges treated at 400 Gy and higher and that the damage observed was fairly extensive. To overcome constraints of ionizing radiation and cold treatment as stand-alone disinfestation treatments, a combination treatment consisting of ionizing radiation and cold, both at reduced doses, temperature, and duration has been proposed (Palou et al., 2007b). Hofmeyr et al. (2016) indicated that combination treatments involving ionizing radiation and cold, suppressed larvae of the false codling moth *Thaumatotibia leucotreta* (Meyrick) in citrus fruit and their successive development stages more effectively than the individual treatment. The probit-9 level efficacy of a treatment combination comprising 60 Gy of ionizing radiation followed by cold exposure for 16 days at 2.5 °C was confirmed.

References

Agustí, M., Almela, V., Juan, M., Alférez, F., Tadeo, F.R., Zacarías, L., 2001. Histological and physiological characterization of rind breakdown of Navelate sweet orange. Ann. Bot. 88, 415–422.

Albrigo, L.G., 1972. Distribution of stomata and epicuticular wax on oranges as related to stem-end rind breakdown and water loss. J. Am. Soc. Hortic. Sci. 97, 220–223.

Alférez, F., Agustí, M., Zacarías, L., 2003. Postharvest rind staining in Navel oranges is aggravated by changes in storage relative humidity: effect on respiration, ethylene production and water potential. Postharvest Biol. Technol. 28, 143–152.

Alférez, F., Alquézar, B., Burns, J.K., Zacarías, L., 2010. Variation in water, osmotic and turgor potential in peel of 'Marsh' grapefruit during development of postharvest peel pitting. Postharvest Biol. Technol. 56, 44–49.

Alférez, F., Burns, J., 2004. Postharvest peel pitting at non–chilling temperatures in grapefruit is promoted by changes from low to high relative humidity during storage. Postharvest Biol. Technol. 32, 79–87.

Alférez, F., Zacarías, L., 2014. Influence of fruit maturity in the susceptibility of Navelina oranges to develop postharvest non-chilling peel pitting. Food Sci. Technol. Int. 20, 183–191.

Alférez, F., Zacarías, L., Burns, J.K., 2005. Low relative humidity at harvest and before storage at high humidity influence the severity of postharvest peel pitting in citrus. J. Am. Soc. Hortic. Sci. 130, 225–231.

Alquezar, B., Rodrigo, M.J., Zacarías, L., 2008. Carotenoid biosynthesis and their regulation in citrus fruits. Tree and Forestry Science and Biotechnology 2, 23–35.

Alquézar, B., Mesejo, C., Alférez, F., Agustí, M., Zacarías, L., 2010. Morphological and ultrastructural changes in the peel of 'Navelate' oranges in relation to variations in relative humidity during postharvest storage and development of peel pitting. Postharvest Biol. Technol. 56, 163–170.

Arpaia, M.L., 2014. *Citrus* postharvest handling. In: Ferguson, L., Grafton-Cardwell, E.E. (Eds.), *Citrus* Production Manual. ANR, University of California, St. Davis, CA.

Ballester, A.R., Lafuente, M.T., 2017. LED blue light-induced changes in phenolics and ethylene in citrus fruit: Implication in elicited resistance against *Penicillium digitatum* infection. Food Chem. 218, 575–583.

Ben-Yehoshua, S., 1969. Gas exchange, transpiration and the commercial deterioration in storage of orange fruits. J. Am. Soc. Hortic. Sci. 94, 524–528.

Ben-Yehoshua, S., Burg, S.P., Young, R., 1985. Resistance of citrus fruit to mass transport of water vapor and other gases. Physiol. Plant. 79, 1048–1053.

Ben-Yehoshua, S., Peretz, J., Moran, R., Lavie, B., Kim, J.J., 2001. Reducing the incidence of superficial flavedo necrosis (Noxan) of 'Shamouti' oranges (*Citrus sinensis* Osbeck). Postharvest Biol. Technol. 22, 19–27.

Brown, G.E., 1992. Factors affecting the occurrence of anthracnose on Florida citrus fruit. Proc. Int. Soc. Citricult. 3, 1044–1048.

Brown, G.E., Eckert, J.W., 2000. Postharvest fungal diseases. In: Timmer, L.W., Garnsey, S.M., Graham, J.H. (Eds.), Compendium of *Citrus* Diseases. second ed. APS Press, St. Paul, MN, pp. 37–45.

Carmona, L., Zacarías, L., Rodrigo, M.J., 2012. Stimulation of coloration and carotenoid biosynthesis during postharvest storage of 'Navelina' orange fruit at 12°C. Postharvest Biol. Technol. 74, 108–117.

Cohen, E., Shalom, Y., Rosenberger, I., 1990. Postharvest ethanol buildup and off-flavor in 'Murcott' tangerine fruit. J. Am. Soc. Hortic. Sci. 115, 775–778.

Cohen, E., Shapiro, B., Shalom, Y., Klein, J.D., 1994. Water loss: a non-destructive indicator of enhanced cell membrane permeability of chilling-injured citrus fruit. J. Am. Soc. Hortic. Sci. 119, 983–986.

Crifò, T., Puglisi, I., Petrone, G., Recupero, G.R., Lo Piero, A.R., 2011. Expression analysis in response to low temperature stress in blood oranges: implication of the flavonoid biosynthetic pathway. Gene 476, 1–9.

Cronje, P.J.R., Barry, G.H., Huysamer, M., 2011a. Postharvest rind breakdown of 'Nules Clementine' mandarin is influenced by ethylene application, storage temperature and storage duration. Postharvest Biol. Technol. 60, 192–201.

Cronje, P.J.R., Barry, G.H., Huysamer, M., 2011b. Fruiting position during development of 'Nules Clementine' mandarin affects the concentration of K, Mg and Ca in the flavedo. Sci. Hortic. 130, 829–837.

Cronje, P.J.R., 2013. Postharvest rind disorders of 'Nadorcott' mandarin are affected by rootstock in addition to postharvest treatments. Acta Hortic. (1007), 111–117.

Cronje, P.J.R., 2015. Could ethylene metabolism in lemon fruit influence peteca incidence? Acta Hortic. (1065), 1455–1461.

Cronje, P.J.R., Zacarías, L., Alférez, F., 2017. Differential susceptibility to peel pitting in fruit from citrus cultivars with different albedo thickness. Postharvest Biol. Technol. 123, 77–82.

Cubero, S., Lee, W.S., Aleixos, N., Albert, F., Blasco, J., 2016. Automated systems based on machine vision for inspecting citrus fruit from the field to postharvest: a review. Food Bioprocess Technol. 9, 1623–1639.

Defraeye, T., Cronje, P., Berry, T., Opara, U.L., East, A., Hertog, M., Verboven, P., Nicolai, B., 2015. Towards integrated performance evaluation of future packaging for fresh produce in the cold chain. Trends Food Sci. Tech. 44, 201–225.

Defraeye, T., Nicolai, B., Kirkman, W., Moore, S., van Niekerk, S., Verboven, P., Cronjé, P.-J.R., 2016. Integral performance evaluation of the fresh-produce cold chain: a case study for ambient loading of citrus in refrigerated containers. Postharvest Biol. Technol. 112, 1–13.

Droby, S., Wisniewski, M., Teixidó, N., Spadaro, D., Jijakli, M.H., 2016. The science, development, and commercialization of postharvest biocontrol products. Postharvest Biol. Technol. 122, 22–29.

Eaks, I.L., 1970. Respiratory response, ethylene production, and response to ethylene of citrus fruit during ontogeny. Plant Physiol. 45, 334–338.

Echeverria, E., Burns, J.K., Miller, W.M., 1999. Fruit temperature and maturity affect development of blossom end clearing in grapefruit. Hortic Sci. 34, 1249–1250.

Eckert, J.W., Eaks, I.L., 1989. Postharvest disorders and diseases of citrus fruits. In: Reuter, W., Calavan, E.C., Carman, G.E. (Eds.), The Citrus Industry. University of California Press, Berkeley, CA, pp. 179–260.

El-Ghaouth, A., Wilson, C.L., Wisniewski, M., Droby, S., Smilanick, J.L., Korsten, L., 2002. Biological control of postharvest diseases of citrus fruits. In: Gnanamanickam, S.S. (Ed.), Biological Control of Crop Diseases. M. Dekker, New York, pp. 289–312.

El-Otmani, M., Ait-Oubahou, A., Zacarías, L., 2011. *Citrus* spp.: orange, mandarin, tangerine, clementine, grapefruit, pomelo, lemon and lime. In: Yahia, E.M. (Ed.), Postharvest Biology and Technology of Tropical and Subtropical Fruits. Vol. 2. Woodhead Publishing, pp. 437–514.

Erasmus, A., Lennox, C.L., Jordaan, H., Smilanick, J.L., Lesar, K., Fourie, P.H., 2011. Imazalil residue loading and green mould control in citrus packinghouses. Postharvest Biol. Technol. 62, 193–203.

Erasmus, A., Lennox, C.L., Korsten, L., Lesar, K., Fourie, P.H., 2015. Imazalil resistance in *Penicillium digitatum* and *P. italicum* causing citrus postharvest green and blue mould: impact and options. Postharvest Biol. Technol. 107, 66–76.

Estables-Ortiz, B., Romero, P., Ballester, A.R., Gonzalez-Candelas, L., Lafuente, M.T., 2017. Inhibiting ethylene perception with 1-methylcyclopropene triggers molecular responses aimed to cope with cell toxicity and increased respiration in citrus fruits. Plant Physiol. Biochem. 103, 154–166.

FAOSTAT, 2018. http://www.fao.org/faostat/es/#data/QC.

Ghasemnezhad, M., Marsh, K., Shilton, R., Babalar, M., Woolf, A., 2008. Effect of hot water treatments on chilling injury and heat damage in Satsuma mandarins: antioxidant enzymes and vacuolar ATPase, and pyrophosphatase. Postharvest Biol. Technol. 48, 364–371.

Gutter, Y., Shachnai, A., Schiffmann-Nadel, M., Dinoor, A., 1981. Biological aspects of citrus molds tolerant to benzimidazole fungicides. Phytopathology 71, 482–487.

Hagenmaier, R.D., 1993. Selection of citrus 'wax' coatings on criteria other than short-term gloss. In: Wardowski, W.F. (Ed.), Packinghouse Newsletter. University of Florida, FL, USA.

Hagenmaier, R.D., 1998. Wax microemulsion formulations used as fruit on coatings. Proc. Florida State Hortic. Soc. 111, 251–255.

Hagenmaier, R.D., 2000. Evaluation of a polyethylene-candelilla coating for 'Valencia' oranges. Postharvest Biol. Technol. 19, 147–154.

Hagenmaier, R.D., 2002. The flavor of mandarin hybrids with different coatings. Postharvest Biol. Technol. 24, 79–87.

Hallman, G.J., Hénon, Y.M., Parker, A.G., Blackburn, C.M., 2016. Phytosanitary irradiation: an overview. Florida Entomol. 99 (Special Issue), 2.

Hofmeyr, J.H., Hattingh, V., Hofmeyr, M., Slabbert, J.P., 2016. Post-harvest phytosanitary disinfestation of false codling moth, *Thaumatotibia leucotreta* (Lepidoptera: Tortricidae) in citrus fruit: Validation of an ionizing radiation and cold combination treatment. African Entomol. 24, 217–226.

Holland, N., Menezes, H.C., Lafuente, M.T., 2005. Carbohydrate metabolism as related to high-temperature conditioning and peel disorders occurring during storage of citrus fruit. J. Agric. Food Chem. 53, 8790–8796.

Jimenez-Cuesta, M., Cuquerella, J., Javega, J.M., 1981. Determination of a color index for citrus fruit degreening. Proc. Int. Soc. Citricult. 2, 750–753.

>Kanetis, L., Förster, H., Adaskaveg, J.E., 2008. Optimizing efficacy of new postharvest fungicides and evaluation of sanitizing agents for managing citrus green mold. Plant Dis. 92, 261–269.

Kaufmann, M.R., 1970. Water potential components in growing citrus fruits. Plant Physiol. 46, 145–149.

Ke, D., Kader, A.A., 1990. Tolerance of 'Valencia' oranges to controlled atmospheres, as determined by physiological responses and quality attributes. J. Am. Soc. Hortic. Sci. 115, 779–783.

Kellerman, M., McLeod, A., Beukes, I., Rose, L.J., Erasmus, A., Fourie, P.H., 2017. Classification of imazalil resistance in an international collection of *Penicillium digitatum* isolates. Can. J. Plant Pathol. 39, 133–137.

Knight, T.G., Klieber, A., Sedgley, M., 2002. Structural basic of rind disorder oleocellosis in Washington navel orange (*Citrus sinensis* L. Osbeck). Ann. Bot. 90, 7665–7773.

Ladaniya, M.S., 2008. Citrus Fruit: Biology, Technology and Evaluation. Academic Press, Elsevier, San Diego, CA (558 pp).

Lado, J., Rodrigo, M.J., Zacarías, L., 2015a. Analysis of ethylene biosynthesis and perception during postharvest cold storage of Marsh and star ruby grapefruits. Food Sci. Technol. Inter. 21, 537–546.

Lado, J., Rodrigo, M.J., Cronje, P.J.R., Zacarías, L., 2015b. Involvement of lycopene in the induction of tolerance to chilling injury in grapefruit. Postharvest Biol. Technol. 100, 176–186.

Lado, J., Rodrigo, M.J., López-Climent, M., Gómez-Cadenas, A., Zacarías, L., 2016. Implication of the antioxidant system in chilling injury tolerance in the red peel of grapefruit. Postharvest Biol. Technol. 111, 214–223.

Lado, J., Cronje, P.J.R., Rodrigo, M.J., Zacarias, L., 2019. Postharvest physiological disordes of citrus fruits. In: de Freitas, S.T., Pareek, S. (Eds.), Postharvest Physiological Disorders of Fruits and Vegetables. CRC Press, Boca Raton, FL, pp. 321–342.

Lafuente, M.T., Zacarias, L., Martínez-Téllez, M.A., Sanchez-Ballesta, M.T., Dupille, E., 2001. Phenylalanine ammonia-lyase as related to ethylene in the development of chilling symptoms during cold storage of citrus fruits. J. Afric. Food. Chem. 49, 6020–6025.

Lafuente, M.T., Zacarías, L., 2006. Postharvest physiological disorders in citrus fruit. Stewart Postharvest Rev. 2, 1–9.

Lafuente, M.T., Estables, B., Gonzalez-Candelas, L., 2017. Insights into the molecular events that regulate heat-induced chilling tolerance in citrus fruits. Frontiers Plant Sci. 8, 1113.

Lorente, D., Zude, M., Idler, C., Gómez-Sanchis, J., Blasco, J., 2015. Laser-light backscattering imaging for early decay detection in citrus fruit using both a statistical and a physical model. J. Food Eng. 154, 76–85.

Magwaza, L.S., Opara, U.L., Nieuwoudt, H.H., Cronje, P.J.R., Saeys, W., Nicolaï, B., 2012. NIR spectroscopy applications for internal and external quality analysis of citrus fruit—a review. Food Bioprocess Technol. 5, 425–444.

Magwaza, L.S., Opara, L.U., Cronje, P.J.R., Terry, L.A., Landahl, S., Nicolai, B., 2013. Non-chilling physiological rind disorders in citrus fruit. Hortic. Rev. 41, 131–176.

Marcet-Houben, M., Ballester, A.R., de la Fuente, B., Harries, E., Marcos, J.F., González-Candelas, L., Gabaldón, T., 2012. Genome sequence of the necrotrophic fungus *Penicillium digitatum*, the main postharvest pathogen of citrus. BMC Genomics 13, 646.

Marcos, J.F., Gonzalez-Candelas, L., Zacarías, L., 2005. Involvement of ethylene biosynthesis and perception in the susceptibility of citrus fruits to *Penicillium digitatum* infection and the accumulation of defense-related mRNAs. J. Exp. Bot. 56, 2128–2193.

Maul, P., McCollum, G.T., Popp, M., Guy, C.L., Porat, R., 2008. Transcriptome profiling of grapefruit flavedo following exposure to low temperature and conditioning treatments uncovers principal molecular components involved in chilling tolerance and susceptibility. Plant Cell Environ. 31, 752–768.

McDonald, H., Arpaia, M.L., Caporaso, F., Obenland, D., Were, L., Rakovski, C., Prakash, A., 2013. Effect of gamma irradiation treatment at phytosanitary dose levels on the quality of 'Lane Late' navel oranges. Postharvest Biol. Technol. 91–99.

Moon, D.G., Cho, Y.S., Mizutani, F., Rutto, K.L., Bhusal, R.C., 2003. Wax deposition on the fruit surface of 'Satsuma' mandarin as affected by water stress. Asian J. Plant Sci. 2, 1138–1141.

Moscoso-Ramírez, P.A., Palou, L., 2014. Effect of ethylene degreening on the development of postharvest Penicillium molds and fruit quality of early season citrus fruits. Postharvest Biol. Technol. 91, 1–8.

Mulas, M., Lafuente, M.T., Zacarias, L., 1996. Chilling effects on fatty acid composition of flavedo lipids in stored "Fortune" mandarins. Adv. Hortic. Sci. 2, 85–90.

Nordby, H.E., Purvis, A.C., Yelenosky, G., 1987. Lipids in peel of grapefruit and resistance to chilling injury during cold storage. HortScience 22, 915–917.

Palou, L., Smilanick, J.L., Margosan, D.A., 2007a. Ozone applications for sanitation and control of postharvest diseases of fresh fruits and vegetables. In: Troncoso-Rojas, R., Tiznado-Hernández, M.E., González-León, A. (Eds.), Recent advances in alternative postharvest technologies to control fungal diseases in fruits and vegetables. Transworld Research Network, Trivandrum (Kerala, India), pp. 39–70.

Palou, L., Del Rio, M.A., Marcilla, A., Alonso, M., Jacas-Miret, J.A., 2007b. Combined postharvest X-ray and cold quarantine treatments against the Mediterranean fruit fly in 'Clemenules' mandarins. Spanish J. Agric. Res. 5, 569–578.

Palou, L., 2009. Control of citrus postharvest diseases by physical means. Tree For. Sci. Biotechnol. 3, 127–142.

Palou, L., 2014. *Penicillium digitatum*, *Penicillium italicum* (green mold, blue mold). In: Bautista-Baños, S. (Ed.), Postharvest Decay: Control Strategies. Academic Press, Amsterdam, The Netherlands, pp. 45–102.

Palou, L., Ali, A., Fallik, E., Romanazzi, G., 2016. GRAS, plant- and animal-derived compounds as alternatives to conventional fungicides for the control of postharvest diseases of fresh horticultural produce. Postharvest Biol. Technol. 122, 41–52.

Pan, Z.Y., Liu, Q., Yun, Z., Guan, R., Zeng, W.F., Q, X., Deng, X.X., 2009. Comparative proteomics of a lycopene-accumulating mutant reveals the important role of oxidative stress on carotenogenesis in sweet orange (*Citrus sinensis* L. Osbeck). Proteomics 9, 5455–5470.

Passaro-Carvalho, C.P., Navarro, P., Salvador, A., 2012. Postharvest. In: *Citrus*: growing, postharvest and industrialization. Lasallia Research and Science Series, Bogota, Colombia, pp. 223–285.
Peng, T., Zhu, X., Duan, N., Liu, J.H., 2014. PtrBAM1, a β-amylase-coding gene of Poncirus trifoliata, is a CBF regulon ember with function in cold tolerance by modulating soluble sugar levels. Plant Cell Environ. 37, 2754–2767.
Petracek, P.D., Wardowski, W.F., Brown, G.E., 1995. Pitting of grapefruit that resembles chilling injury. HortScience 30, 1422–1426.
Petracek, P.D., Dou, H., Pao, S., 1998a. The influence of applied waxes on postharvest physiological behavior and pitting of grapefruit. Postharvest Biol. Technol. 14, 99–106.
Petracek, P.D., Montalvo, L., Dou, H., Davis, C., 1998b. Postharvest pitting of 'Fallglo' tangerine. J. Amer. Soc. Hortic. Sci. 123, 130–135.
Petracek, P.D., Kelsey, D.F., Grierson, W., 2007. Physiological peel disorders. In: Wardowski, W.F., Miller, W.M., Hall, D.J., Grierson, W. (Eds.), Fresh Citrus Fruits. second ed. Florida Science Source, Inc., Longboat Key, FL, pp. 397–442.
Platt-Aloia, K.A., Thomson, W.W., 1976. An ultrastructural study of two forms of chilling-induced injury to the rind of grapefruit (*Citrus paradisi*, Macfed). Cryobiology 13, 95–106.
Porat, R., Weiss, B., Cohen, L., Daus, A., Goren, R., Droby, S., 1999. Effect of ethylene and 1-methylcyclopropene on the postharvest qualities of 'Shamouti' oranges. Postharvest Biol. Technol. 15, 155–163.
Porat, R., Pavoncello, D., Peretz, J., Ben-Yehoshua, S., Lurie, S., 2000. Effects of various heat treatments on the induction of cold tolerance and on the postharvest qualities of 'star ruby' grapefruit. Postharvest Biol. Technol. 18, 159–165.
Porat, R., Weiss, B., Cohen, L., Daus, A., Aharoni, N., 2004. Reduction of postharvest rind disorders in citrus fruit by modified atmosphere packaging. Postharvest Biol. Technol. 33, 35–43.
Porat, R., Weiss, B., Cohen, L., Daus, A., Biton, A., 2005. Effects of polyethylene wax content and composition on taste, quality, and emission of off-flavor volatiles in 'Mor' mandarins. Postharvest Biol. Technol. 38, 262–268.
Porat, R.P., 2008. Degreening of citrus fruit. Tree For. Sci. Biotechnol. 2, 71–76.
Purvis, A.C., 1979. Relationship between mid-season resistance to chilling injury and reducing sugar level in grapefruit peel. Hort Sci. 14, 227–229.
Rodrigo, M.J., Zacarías, L., 2007. Effect of postharvest ethylene treatment on carotenoid accumulation and the expression of carotenoid biosynthetic genes in the flavedo of orange (*Citrus sinensis* L. Osbeck) fruit. Postharvest Biol. Technol. 43, 14–22.
Romero, P., Lafuente, M.T., Alferez, F., 2014. A transcriptional approach to unravel the connection between phospholipases A2 and D and ABA signal in citrus under water stress. Plant Physiol. Biochem. 80, 23–32.
Ruiz-Altisent, M., Ruiz-Garcia, L., Moreda, G.P., Hernandez-Sanchez, N., Lu, R., Correa, E.C., Diezma, B., Nicolaï, B., Garcia-Ramos, J., 2010. Sensors for product characterization and quality of specialty crops—a review. Comput. Electron. Agric. 74, 176–194.
Sala, J.M., 1998. Involvement of oxidative stress in chilling injury in cold-stored mandarin fruits. Postharvest Biol. Technol. 13, 255–261.
Sala, J.M., Lafuente, M.T., 2000. Catalase enzyme activity is related to tolerance of mandarin fruits to chilling. Postharvest Biol. Technol. 20, 81–89.
Sanchez-Ballesta, M.T., Lluch, Y., Gosalbes, M.J., Zacarias, L., Granell, A., Lafuente, M.T., 2003. A survey of genes differentially expressed during long-term heat-induced chilling tolerance in citrus fruit. Planta 218, 65–70.
Schirra, M., 1992. Behaviour of 'Star Ruby' grapefruits under chilling and non-chilling storage temperature. Postharvest Biol. Technol. 2, 315–327.
Schirra, M., Mulas, M., 1995. Improving storability of Toracco oranges by postharvest hot-dip fungicide treatment. Postharvest Biol. Technol. 6, 129–138.
Schirra, M., D'hallewin, G., Cabras, P., Azngioni, A., Ben-Yehoshua, S., Lurie, S., 2000. CI and residue uptake in cold-stored 'Star Ruby' grapefruit following thiabendazole and imazalil dip treatments at 20 and 50°C. Postharvest Biol. Technol. 20, 91–98.
Schirra, M., D'Aquino, S., Cabras, P., Angioni, A., 2011. Control of postharvest diseases of fruit by heat and fungicides: Efficacy, residue levels, and residue persistence. A review. J. Agric. Food Chem. 59, 8531–8542.
Shi, J.X., Porat, R., Goren, R., Goldschmidt, E.E., 2005. Physiological responses of 'Murcott' mandarins and 'Star Ruby' grapefruits to anaerobic stress conditions and their relation to fruit taste, quality, and emission of off-flavor volatiles. Postharvest Biol. Technol. 38, 99–105.
Shi, J.X., Porat, R., Goren, R., Goldschmidt, E.E., 2007. Molecular, biochemical and anatomical factors governing ethanol fermentation metabolism and accumulation of off-flavors in mandarins and grapefruit. Postharvest Biol. Technol. 46, 242–251.
Smilanick, J.L., Brown, G.E., Eckert, J.W., 2006a. The biology and control of postharvest diseases. In: Wardowski, W.F., Miller, W.M., Hall, D.J., Grierson, W. (Eds.), Fresh Citrus Fruits. second ed. Florida Science Source, Inc., Longboat Key, FL, pp. 339–396.
Smilanick, J.L., Mansour, M.F., Mlikota-Gabler, F., Goodwine, W.R., 2006b. The effectiveness of pyrimethanil to inhibit germination of *Penicilliium digitatum* and to control citrus green mold after harvest. Postharvest Biol. Technol. 42, 75–85.
Smilanick, J.L., Mansour, M.F., Sorenson, D., 2006c. Pre- and postharvest treatments to control green mold of citrus fruit during ethylene degreening. Plant Dis. 90, 89–96.
Smilanick, J.L., Mansour, M.F., Mlikota-Gabler, F., Sorenson, D., 2008. Control of citrus postharvest green mold and sour rot by potassium sorbate combined with heat and fungicides. Postharvest Biol. Technol. 47, 226–238.
Tadeo, F.R., Cercos, M., Colmenero-Flores, J.M., Iglesias, J.D., Naranjo, M.A., Rios, G., Carrera, E., Ruiz-Rivero, O., Lliso, I., Morillon, R., Ollitrault, P., Talon, M., 2008. Molecular physiology of development and quality of citrus. Adv. Bot. Res. 47, 147–223.
USDA. 2016. Peracetic acid. Handling/processing. USDA National Organic Program, Technical Evaluation Report.
Van Wyk, A.A., Huysamer, M., Barry, G.H., 2009. Extended low-temperature shipping adversely affects rind colour of 'Palmer Navel' sweet orange (*Citrus sinensis* L. Osb.) due to carotenoid degradation but can partially be mitigated by optimising post-shipping holding temperature. Postharvest Biol. Technol. 53, 109–116.
Vercher, R., Tadeo, F.R., Almela, V., Zaragoza, S., Primo-Millo, E., Agustí, M., 1994. Rind structure, epicuticular wax morphology and water permeability of 'Fortune' mandarin fruits affected by peel pitting. Ann. Bot. 74, 619–625.

Vigneault, C., de Castro, L.R., Goyette, B., Markarian, N.R., 2007. Indirect airflow distribution measurement method for horticultural crop package design. Canadian Biosys. Engin. 49, 13–22.

Wild, B.L., 1993. Reduction of chilling injury in grapefruit and orange stored at 1°C by pre-storage hot dip treatments, curing and wax applications. Aust. J. Exp. Agric. 33, 279–285.

Wisniewski, M., Droby, S., Norelli, J., Liu, J., Schena, L., 2016. Alternative management technologies for postharvest disease control: The journey from simplicity to complexity. Postharvest Biol. Technol. 122, 3–10.

Wong, W.S., Ning, W., Xu, P.L., Kung, S.D., Yang, S.F., Li, N., 1999. Identification of two chilling-regulated 1-aminocyclopropane-1-carboxylate synthase genes from citrus (*Citrus sinensis* Osbeck) fruit. Plant Mol. Biol. 41, 587–600.

Xu, Q., Yu, K.Q., Zhu, A.D., Ye, J.L., Liu, Q., Zhang, J.C., Deng, X.X., 2009. Comparative transcripts profiling reveals new insight into molecular processes regulating lycopene accumulation in a sweet orange (*Citrus sinensis*) red-flesh mutant. BMC Genomics 10, 540.

Youssef, K., Ligorio, A., Sanzani, S.M., Nigro, F., Ippolito, A., 2012. Control of storage diseases of citrus by pre- and postharvest application of salts. Postharvest Biol. Technol. 72, 57–63.

Yun, Z., Gao, H., Liu, P., Liu, S., Luo, T., Jin, S., Xu, Q., Xu, J., Cheng, Y., Deng, X., 2013. Comparative proteomic and metabolomic profiling of citrus fruit with enhancement of disease resistance by postharvest heat treatment. BMC Plant Biol. 13, 44.

Zacarıas, L., Lafuente, M.T., Marcos, M., Saladie, M., Dupille, E., 2003. Regulation of ethylene biosynthesis during cold storage of the chilling-sensitive fortune mandarin fruit. In: Vendrell, M., Klee, H., Pech, J.C., Romojaro, F. (Eds.), Biology and Biotechnology of the Plant Hormone Ethylene III. IOS Press, Amsterdam, The Netherlands, pp. 112–117.

Zheng, Y., He, S., Yi, S., Zhou, Z., Maob, S., Zhao, X., Deng, L., 2010. Characteristics and oleocellosis sensitivity of citrus fruits. Sci. Hortic. 123, 312–317.

Zhu, A., Li, W., Ye, J., Sun, X., Ding, Y., Cheng, Y., Deng, X., 2011. Microarray expression profiling of postharvest Ponkan mandarin (*Citrus reticulata*) fruit under cold storage reveals regulatory gene candidates and implications on soluble sugars metabolism. J. Integr. Plant Biol. 53, 358–374.

Chapter 22

Chemistry of citrus flavor

Yu Wang[a], Siyu Wang[a], Simona Fabroni[b], Shi Feng[a], Paolo Rapisarda[b], Russell Rouseff[c]
[a]*Citrus Research and Education Center, Lake Alfred, FL, United States,* [b]*CREA, Research Center for Olive, Citrus and Tree Fruit, Acireale, Italy,* [c]*Citrus Research Institute of China, Chinese Academy of Agriculture Sciences, Southwest University, Chongqing, People's Republic of China*

22.1 Introduction

Citrus aromas are among the most desirable natural flavors, explaining their frequent use in the food, pharmaceutical, and cosmetics industries. The delicate aroma of citrus is due to a specifically proportioned, complex mixture of volatiles. Notably, it is only a small portion of these volatile compounds, typically in low concentrations, that evoke aromatic activities. With the advancement of analytical techniques, aroma-active volatiles responsible for citrus flavor have been comprehensively studied, including their distribution and quantitation, as well as their aroma properties. In this chapter, we have provided information about the flavor of four major citrus fruits, that is, lemons, oranges, mandarins, and grapefruits, each possessing a signature aroma profile. This chapter attempts to comprehensively summarize odorants of each citrus fruit according to their chemical families, as well as their sensory impacts.

The majority of citrus fruits are processed into juice. However, the oil from the outer layer of peel and the condensate from making concentrated juice are also major sources of flavor products from citrus fruit (Berger, 2007). Citrus flavor products can be listed as (i) peel oil, obtained by mechanical rupture of fruit oil glands either prior to, or during juice extraction; (ii) essence oil and aqueous essence (sometimes called aqueous aroma), consisting of volatile juice compounds both formed from the condensate of the evaporation process of citrus juices; (iii) petitgrain oil, a product from steam distillation of citrus leaves and twigs; and (iv) oil of neroli, a floral oil produced from steam distillation of orange blossoms.

22.2 Lemon

Lemon (*Citrus limon*) fruits are marketed for both fresh consumption and for processing; the juice and peel oil are the primary processed products. During the 2016–2017 season 7,631,000 tons of lemons and limes were produced worldwide of which 2,143,000 were industrially processed (United States Department of Agriculture, 2018). Assuming an average yield of about 350 g of essential oil per ton of lemons, about 743 tons of lemon essential oils were produced. The major lemon producers include Mexico, Argentina, and the European Union. Italy is the leading European producer of lemon essential oils producing roughly 39 tons.

Lemon peel oil is more valuable than lemon juice (Berger, 2007). Therefore, there have been numerous investigations into the natural composition of lemon essential oil to determine quality factors, and as a way to detect frauds. Conversely, few studies have been carried out on lemon juice volatiles because of its lower commercial value. Lemon essential oil has a recognizable color that ranges from sharp yellow to green, and a unique fragrance that resembles the fresh fruit. They are used pure or combined in aromatizers and as flavorings in the food and cosmetic industries. Lemon oils are also added to pharmaceutical or para-pharmaceutical preparations and employed in aromatherapy (Arce and Soto, 2008). Due to the commercial relevance of lemon essential oils, many investigations into the composition of volatile and nonvolatile fractions of lemon essential oil have been documented over the past 50 years (Shaw, 1977, 1979; Dugo et al., 2002, 2010; Russo, et al., 2015). The volatile fraction of cold-pressed lemon essential oils represents 96%–98% of the whole oil, while the nonvolatile fraction ranges from 2% to 4%.

Over 70 compounds have been identified in the volatile fraction of cold-pressed lemon essential oils. Typical values and ranges of individual compounds in the volatile fraction of lemon essential oils are shown in Table 22.1 (Dugo et al., 2010). Total values within chemical classes such as hydrocarbons, aldehydes, ketones, alcohols, and esters are also listed.

The Genus Citrus. https://doi.org/10.1016/B978-0-12-812163-4.00022-X

TABLE 22.1 Percentage composition of the volatile fraction of cold-pressed lemon oil: average and range of variation of the individual compounds and their classification in substance classes.

N.		Average values	Min	Max
1	Tricyclene	0.01	0.01	0.01
2	α-Thujene	0.32	0.30	0.35
3	α-Pinene	1.65	1.51	1.97
4	Camphene	0.04	0.04	0.06
5	Sabinene	1.65	1.48	2.37
6	β-Pinene	12.27	10.65	19.31
7	6-Methyl-5-hepten-2-one	tr.	tr.	0.01
8	Myrcene	1.24	1.02	1.34
9	Decane	tr.	tr.	tr.
10	Octanal	0.05	0.04	0.07
11	α-Phellandrene	0.02	0.01	0.03
12	δ-3-Carene	tr.	tr.	tr.
13	α-Terpinene	0.12	0.02	0.16
14	*p*-Cymene	0.32	0.11	1.49
15	Limonene	67.33	60.34	70.81
16	(*Z*)-β-Ocimene	0.06	0.01	0.09
17	(*E*)-β-Ocimene	0.09	0.05	0.13
18	γ-Terpinene	9.03	7.47	9.66
19	Octanol+*cis*-Sabinene hydrate	0.02	0.01	0.06
20	Terpinolene	0.27	0.14	0.31
21	Linalool+*trans*-Sabinene hydrate	0.10	0.06	0.21
22	Nonanal	0.09	0.07	0.14
23	Heptyl acetate	tr.	tr.	0.01
24	*cis*-Limonene oxide	0.01	tr.	0.04
25	*trans*-Limonene oxide	0.01	tr.	0.04
26	Camphor	tr.	tr.	0.01
27	Citronellal	0.07	0.04	0.10
28	Borneol	tr.	tr.	tr.
29	Terpinen-4-ol	0.03	0.01	0.05
30	α-Terpineol	0.12	0.07	0.23
31	Dodecane	tr.	tr.	0.03
32	Decanal	0.03	tr.	0.05
33	Octyl acetate	tr.	tr.	0.01
34	Citronellol	0.01	tr.	0.08
35	Neral	0.63	0.43	0.80
36	Carvone	tr.	tr.	0.01

TABLE 22.1 Percentage composition of the volatile fraction of cold-pressed lemon oil: average and range of variation of the individual compounds and their classification in substance classes—cont'd

N.		Average values	Min	Max
37	Geraniol	tr.	tr.	0.02
38	Piperitone	tr.	tr.	tr.
39	Geranial	1.08	0.71	1.48
40	Perillaldehyde	0.02	0.01	0.04
41	Bornyl acetate	tr.	tr.	0.01
42	Tridecane	tr.	tr.	0.01
43	Undecanal	0.02	tr.	0.04
44	Nonyl acetate	tr.	tr.	0.01
45	Methyl geranoate	tr.	tr.	0.02
46	Citronellyl acetate	0.02	0.01	0.05
47	Neryl acetate	0.37	0.26	0.60
48	Geranyl acetate	0.32	0.18	0.48
49	Tetradec-1-ene	tr.	tr.	0.01
50	Tetradecane	0.01	tr.	0.01
51	*cis*-α-Bergamotene	0.02	0.01	0.02
52	β-Caryophyllene	0.17	0.14	0.25
53	*trans*-α-Bergamotene	0.27	0.21	0.39
54	*cis*-β-Farnesene	0.02	0.02	0.04
55	α-Humulene	0.01	0.01	0.01
56	*cis*-β-Santalene	0.01	tr.	0.01
57	Geranyl propanoate	0.01	tr.	0.01
58	γ-Curcumene	tr.	tr.	tr.
59	*trans*-β-Bergamotene	0.01	0.01	0.02
60	Valencene	0.02	tr.	0.07
61	Bicyclogermacrene	0.04	tr.	0.05
62	*cis*-α-Bisabolene	0.03	0.03	0.05
63	β-Bisabolene	0.44	0.34	0.66
64	*trans*-γ-Bisabolene	tr.	tr.	tr.
65	*trans*-α-Bisabolene	0.01	0.01	0.01
66	*cis*-Sesquisabinene hydrate	tr.	tr.	0.01
67	Germacrene D-4-ol	0.01	0.01	0.06
68	Spathulenol	0.01	tr.	0.04
69	2,3-Dimethyl-3-(4-methyl-3-pentenyl)-2-norbornarol	0.01	tr.	0.02
70	Campherenol	tr.	tr.	0.02
71	α-Bisabolol	0.02	0.01	0.03

Continued

TABLE 22.1 Percentage composition of the volatile fraction of cold-pressed lemon oil: average and range of variation of the individual compounds and their classification in substance classes—cont'd

N.		Average values	Min	Max
72	Nootkatone	tr.	tr.	0.01
	Hydrocarbons	96.68	95.50	97.17
	Aliphatic	0.02	0.01	0.04
	Monoterpene	95.59	94.17	96.14
	Sesquiterpene	1.07	0.86	1.47
	Oxygenated compounds	3.13	2.70	3.89
	Aldehydes	2.01	1.45	2.57
	Aliphatic	0.19	0.15	0.30
	Monoterpene	1.82	1.21	2.39
	Ketones	0.02	0.01	0.09
	Aliphatic	tr.	tr.	0.01
	Monoterpene	0.01	tr.	0.02
	Sesquiterpene	tr.	tr.	0.08
	Alcohols	0.34	0.23	0.69
	Aliphatic	0.02	0.01	0.06
	Monoterpene	0.26	0.16	0.49
	Sesquiterpene	0.06	0.03	0.15
	Esters	0.74	0.51	1.03
	Aliphatic	0.01	tr.	0.02
	Monoterpene	0.74	0.51	1.02
	Oxides	0.01	tr.	0.07

Adapted from Dugo, P., Ragonese, C., Russo, M., Sciarrone, D., Santi, L., Cotroneo, A., Mondello, L., 2010. Sicilian lemon oil: Composition of volatile and oxygen heterocyclic fractions and enantiomeric distribution of volatile components. J. Sep. Sci. 33, 3374–3385.

Quantitatively, limonene, a monoterpene hydrocarbon, is the predominate volatile comprising almost 65% of the total volatile fraction. Other major monoterpene hydrocarbons include β-pinene (~13%) and γ-terpinene (~10%). Other minor terpenes include α-pinene, sabinene, and myrcene and total approximately 2%. Other volatile components are found in much lower amounts. Only, geranial, a monoterpene aldehyde, exceeds 1%. The major sesquiterpene hydrocarbons include β-bisabolene, trans-α-bergamotene, and β-caryophyllene. Among the oxygenated volatiles, aldehydes constitute the most important fraction. Citral is an aroma character impact volatile and is found as two geometric isomers, neral and geranial. The ethyl esters of these two important volatiles, neryl acetate and geranyl acetate, also contribute to the characteristic lemon aroma. Some aliphatic aldehydes such as decanal and nonanal are also important components of lemon aroma. Total esters represent <1% of total volatiles and total alcohols are slightly lower than total esters. Alcohols are mainly represented by α-terpineol and linalool. Nootkatone is the major oxygenated sesquiterpene. A recent study has reported the presence of volatile sulfur compounds in Lisbon lemon peel (Cannon et al., 2015).

22.2.1 Seasonal changes

Lemon is one of the multi-flowering citrus species that typically contain fruits and flowers in diverse stages of development throughout the growing season. The composition of the volatile fraction of lemon essential oil undergoes cyclic variations due to the variable ages of the fruit. Even though the exterior of the harvested fruits might appear to be identical, they will contain fruits from different bloom dates. In Italy, the most important lemon fruit production, period is known commercially as "Primofiore" or winter lemon and occurs between October and April. These fruits are medium sized with a light green to yellow peel. "Bianchetto", or spring lemon, is the fruit collected between April and June. These lemons are large sized with a yellow peel. "Verdello", or summer lemon, is the fruit collected between July and September. These are medium-large-sized fruit with a green peel. During the months of November to April, modest variations in the quantitative composition of lemon essential oils volatiles can be observed. However, some components reach their maximum or minimum levels in the summer harvest period. The summer or "Verdello" lemons are mainly used for the industrial production of cold-pressed lemon essential oils. During the summer harvest period β-pinene, sabinene, nonanal, decanal, esters, and alcohols reach their maximum levels, whereas limonene, neral, geranial, and total aldehydes are found at their lowest levels. Moreover, cold-pressed essential oils obtained during the "Primofiore" harvest season, have some differentiating features such as a low content of esters and a rather high percentage of linalool. The evaluation of the genuineness and quality of lemon essential oils can be more accurately evaluated if the seasonal influence of specific compositional parameters is taken into consideration.

22.2.2 Lemon oil extraction

Lemon essential oils are extracted from the whole fruit or from the peel during citrus juice processing. The resulting oil is typically called cold-pressed oil. Different technologies can be employed, that is, systems that operate on the whole fruit by extracting essential oils first and subsequently the juice, systems that operate on the depulped peel after juice extraction, and systems which simultaneously extract both the juice and essential oils. Each of these systems has positive and negative aspects and are generally chosen based on the quality and quantity of the fruit to be processed as well as the equipment available. The equipment for processing the whole fruit includes technologies designed to extract essential oils and subsequently juice from de-oiled fruits. The machines used for this purpose are characterized by the abrasive action exerted on the flavedo, the outermost layer of the whole fruit. Extracted oils are removed from the intact fruit with water sprays and the comingled mixture separated using centrifuges.

One of the technologies to extract essential oils from peels after juicing is the slow-folding process that originated in Sicily. In this system, peels are folded and compressed in order to break the oil glands causing the oil to spurt out. During this process, the extracted oil is washed off using strong jets of water, and then the water/oil emulsion is separated by centrifugation. These processes should be performed as rapidly as possible. Lengthy contact with juice or with water used to wash the essential oils can deteriorate the resultant oils due to significant losses of flavor-important water-soluble compounds (i.e., oxygenated aldehydes, alcohols, esters, and terpenoids). An alternative industrial practice is to extract the essential oils and juice simultaneously. The most widely used machines to extract lemon oils are the Brown oil extractor and the FMC in-line extractor from Food Machinery Corp. (now JBT FoodTech). Among the industrially extracted lemon essential oils, most appreciated for their aroma, are those extracted by "sfumatura", a true reproduction of the traditional manual technique using sponges. Nonetheless, regardless of the higher quality of essential oils extracted by the "sfumatura" method, processing the whole fruit is better suited for continuous large-scale productions; thus, the method of choice depends solely on the scale of the processing plant. Extraction system influence over the quantitative composition is limited, affecting almost only the alcohol content of the oil, which is present in higher concentration in oils extracted using the simultaneous extractions of the juice and oils compared to oils produced by "sfumatura" equipment. This is possibly due to the higher water:oil ratio used by this latter system, which causes a decrease in water-soluble oxygenated compounds, that is, alcohols, in the oil.

22.2.3 Aroma volatiles

Odor-active chemicals are also present in lemon juice, and they possess the pleasant aroma characteristics of the fresh fruit. Moreover, most of the compounds found in fresh lemon juice are also found in lemon peel oil or in aqueous lemon essence. Although lemon juice is the main commercial product of lemon fruit processing, few studies on its volatile fraction can be found in the literature (Cannon et al., 2015; Moshonas and Shaw, 1972; Shaw and Wilson III, 1982; Komura, 2006; Saura et al., 2012; Zhong et al., 2014; Karabagias, 2017; Cannon and Ho, 2018). In an early study (Moshonas and Shaw, 1972), 27 volatile compounds were identified in lemon juice aqueous essence including 13 alcohols, 4 aldehydes, 4 ketones, 3 hydrocarbons, 2 ethers, and 1 ester. The essence was recovered from the condensation of vapors from the first stage of the juice evaporator. More recently, approximately 30 volatiles from different lemon varieties were identified (Komura, 2006; Allegrone et al., 2006). Moreover, glycosidically bound volatile compounds have also been identified in lemon juice, representing a hidden resource of aroma in fresh fruit (Zhong et al., 2014). Table 22.2 shows the volatile compounds and their relative average amounts (mg/L) identified in hand-squeezed juices of four different Italian lemon productions. Monoterpene hydrocarbons (limonene, α-thujene, α-pinene, camphene, β-pinene, β-myrcene, α-terpinene, β-ocimene, γ-terpinene, and terpinolene) constitute the main volatile group, representing 70%–90% of total lemon juice volatiles. Sesquiterpene hydrocarbons are mainly represented by *β*-caryophyllene, α-bergamotene, α-zingiberene, and *β*-bisabolene, while oxygenated compounds are mainly aldehydes (nonanal, decanal, 9-methyldecanal, undecanal, dodecanal, neral, and geranial), monoterpene alcohols (linalool, fenchol, borneol, 4-terpineol, α-terpineol, nerol, and geraniol), and esters (octyl acetate, nonyl acetate, citronellyl acetate, neryl acetate, and geranyl acetate). In addition, experimental evidence of the presence of volatile sulfides in lemon juice has existed for more than three decades (Shaw and Wilson III, 1982; Allegrone et al., 2006; Cannon and Ho, 2018).

Odor-active chemicals, which characterize the volatile fraction of lemon essential oils and lemon juice, are described in detail in the following sections.

TABLE 22.2 Volatile compounds and their relative average amounts (mg/L) identified in hand-squeezed juices of four different Italian lemon productions.

N.		Verdello Siracusano	Interdonato	Primo Fiore Capo d'Orlando	Femminello Siracusano
1	Ethanol	0.157	0.130	0.001	0.116
2	Acetic acid	0.004	ND	ND	ND
3	α-Thujene	0.083	0.004	0.003	ND
4	α-Pinene	0.285	0.033	0.023	0.007
5	Camphene	0.029	0.003	0.003	ND
6	*β*-Pinene	3.280	0.270	0.170	0.070
7	*β*-Myrcene	0.741	0.116	0.076	0.026
8	α-Terpinene	0.027	0.009	0.091	0.028
9	Limonene	34.597	5.449	3.839	1.421
10	*β*-Ocimene	0.017	0.010	0.008	0.004
11	*γ*-Terpinene	5.731	0.567	0.598	0.252
12	Terpinolene	0.502	0.083	0.081	0.045
13	Nonanal	0.012	0.004	0.017	ND
14	Linalool	0.021	0.027	0.051	0.064
15	Fenchol	0.062	0.005	0.029	0.013
16	Borneol	0.117	0.28	0.071	0.06
17	4-Terpineol	0.372	0.115	0.157	0.098
18	α-Terpineol	0.846	0.242	0.274	0.215
19	Decanal	0.064	0.003	0.005	ND
20	Octyl acetate	0.014	0.002	0.002	ND
21	Nerol	0.024	0.010	0.016	0.021
22	Neral	0.014	0.017	0.030	0.013
23	Geraniol	0.070	0.007	0.020	0.035
24	9-Methyldecanal	0.032	ND	ND	ND
25	Geranial	0.037	0.042	0.080	0.027
26	Undecanal	0.014	ND	0.002	ND

TABLE 22.2 **Volatile compounds and their relative average amounts (mg/L) identified in hand-squeezed juices of four different Italian lemon productions—cont'd**

N.		Verdello Siracusano	Interdonato	Primo Fiore Capo d'Orlando	Femminello Siracusano
27	Nonyl acetate	0.003	ND	ND	ND
28	Citronellyl acetate	0.030	0.045	0.007	0.003
29	Neryl acetate	0.284	0.281	0.011	0.068
30	Geranyl acetate	0.054	0.101	0.005	0.020
31	Dodecanal	0.042	0.014	0.002	ND
32	β-Caryophyllene	0.186	0.028	0.01	0.003
33	α-Bergamotene	0.966	0.206	0.024	0.006
34	α-Zingiberene	0.028	ND	0.003	ND
35	β-Bisabolene	1.414	0.288	0.03	0.008
	Hydrocarbons				
	Monoterpene	45.294	6.544	4.892	1.853
	Sesquiterpene	2.594	0.522	0.067	0.017
	Oxygenated compounds				
	Aldehydes	0.215	0.080	0.136	0.040
	Monoterpene Alcohols	1.512	0.686	0.618	0.506
	Esters	0.385	0.429	0.025	0.091

Adapted from Allegrone, G., Belliardo, F., Cabella, P., 2006. Comparison of volatile concentrations in hand-squeezed juices of four different lemon varieties. J. Agric. Food Chem. 54, 1844–1848.

22.2.4 Aldehydes

Aldehydes are present in citrus essential oils in minor concentrations with respect to other chemical classes. However, they contribute significantly to aroma due to their very low threshold values (Li et al., 2016). Aldehyde components vary widely in the different citrus essential oils, and their qualitative and quantitative compositions depend on the species (Liu et al., 2012). Dugo et al. (2010) reported the average percentage composition and the annual range of variation of volatile components of Italian lemon oil. The authors reported an average composition of 2.57% for total aldehydes, with 2.39% being constituted by monoterpene aldehydes and 0.30% constituted by aliphatic aldehydes. Among other compounds, citral is the main aroma and flavor contributor in lemon oil (Nguyen et al., 2009) and is constituted by the mixture of two monoterpene aldehydes that are double-bond isomers. The *E*- or *trans*- isomer is known as geranial while the *Z*- or *cis*- isomer is known as neral. Their biosynthetic pathway begins with the formation of acyclic compounds such as geraniol (C_{10}), farnesol (C_{15}), geranyl-geraniol (C_{20}), and squalene (C_{30}). Then, from these compounds, other components originate through cyclization, transposition and oxidation reactions, such as the formation of neral and geranial from the oxidation reactions of geraniol. In addition to citral, citronellal, a monoterpene aldehyde, and decanal, dodecanal, nonanal, octanal, and undecanal, all aliphatic aldehydes, have been reported in Italian cold-pressed lemon oil (Table 22.3). Mussinan et al. (1979) studied volatile components of lemon juice. They found that in fresh lemon juice citral is present in a ratio equal to 95% geranial and only 5% neral. This is quite different compared to the same ratio reported for lemon peel oil, which is about 60% geranial to 40% neral (Steven et al., 1977). Allegrone et al. (2006) reported that average aldehyde concentrations in lemon juice are quite different, depending on the lemon cultivar. Indeed, they showed that aldehydes are prevalent in Verdello Siracusano (0.22 mg/L), representing the summer fruit of the cv. Femminello Siracusano, and Primo Fiore Capo d'Orlando (0.14 mg/L) juices, followed by Interdonato (0.080 mg/L) and Femminello Siracusano (0.040 mg/L). Qualitatively, the most represented aldehyde compounds in lemon juice are neral, geranial, nonanal, decanal, undecanal, dodecanal, and 9-methyldecanal.

TABLE 22.3 Percentage average composition of volatile aldehyde fraction of cold-pressed lemon oil.

Aliphatic	
Decanal	0.01–0.08
Dodecanal	Tr
Nonanal	0.04–0.19
Octanal	0.02–0.14
Undecanal	tr-0.05
Monoterpene	
Citronellal	0.04–0.17
Geranial	0.60–2.25
Neral	0.45–1.33

Adapted from Dugo, G., Cotroneo, A., Verzera, A., Bonaccorsi, I., 2002. Composition of the volatile fraction of cold-pressed citrus peel oils. In: Dugo, G., Di Giacomo, A. (Eds.), Citrus, The genus Citrus. CRC Press, Taylor & Francis Group, Boca Raton, FL, pp. 284–317.

Citral is characterized by a typical fresh-like and intense lemon odor and flavor, and its use in the composition of flavor blends for the standardization of different citrus-based juices is a common practice. For that reason, previous studies have been aimed at finding the sensory threshold of this compound in order to find out the relationship between concentration and real sensory impact. Odor (defined as the combination of taste, odor, and other oral sensations) thresholds have been determined in water by Ahmed et al. (1978), who reported values equal to 85.3 and 41.4 ppb, respectively. Further on, Moshonas and Shaw (1994) indicated for this compound an aroma threshold equal to 27.6 ppb while more recently Plotto et al. (2004) showed that orthonasal and retronasal thresholds measured in a reconstituted deodorized orange juice by an untrained sensory panel were significantly higher, with values equal to 1230 and 714 μg/L, respectively. About aliphatic aldehydes, Plotto et al. (2004) confirmed that octanal, decanal, and dodecanal, containing an even number of carbon atoms, exhibited lower orthonasal and retronasal thresholds with respect to aldehydes with an odd number of carbon atoms such as nonanal. Referring to odor and flavor thresholds, Ahmed et al. (1978) reported values equal to 1.41 and 0.52 ppb, 2.53 and 4.25 ppb, 1.97 and 3.02 ppb, 0.53 and 1.07 ppb, for octanal, nonanal, decanal, and dodecanal, respectively. Plotto et al. (2004) have shown values that are >100 times higher in deodorized orange juice than in water. Some indication of the importance of aldehyde components to flavor can be seen by comparing their very low aroma thresholds to their amount in the volatile fraction of lemon essential oils.

Although aldehydes contribute remarkably to the minty, herbaceous, and lemon notes of lemon essential oils due to low odor thresholds, their transformation products have much higher odor thresholds. Therefore, the decrease of aldehyde concentrations has a great effect on the loss of the typical aroma of lemon essential oils. It is well known that lemon oil, as a food flavoring, deteriorates in the presence of acids. Schieberle and Grosh (1988) have found that an aqueous lemon oil/citric acid emulsion, left for 30 days at 37 °C, undergoes dramatic changes which alters its fresh-like characteristics. In particular, they found a significant decrease in the concentration of citral together with the formation of *p*-methylacetophenone, *p*-cresol, fenchyl alcohol, *p*-cymene, and 1-terpinen-4-ol, all compounds that characterize the deterioration of lemon oil in the presence of citric acid. Moreover, further studies have indicated that the concentrations of aldehydes may dramatically decrease during production and storage of citrus essential oils, due to reactions induced by heat, sunlight, and oxygen (Nguyen et al., 2009; Sun et al., 2014).

Although not involving lemon, there is some recent information regarding the chemistry of essential oil degradation. Li et al. (2016) investigated the reaction pathway and inducing factors of aldehyde degradation of pummelo essential oils during storage. They showed that the combined effects of UV irradiation and air exposure are the key factors for aldehyde degradation. In particular, they demonstrated that aliphatic aldehydes were transformed into their organic acids after the co-treatment of UV irradiation and air exposure, as the former provides energy and the latter provides oxygen for the oxidative reactions. Based on the analytical results of this research, putative reactions were suggested showing that neither UV irradiation nor air exposure alone could sustain these reactions in a relatively short time, but only the co-exposure of both factors triggered the reaction (Fig. 22.1A). In the same work, they studied monoterpene aldehydes and demonstrated that *trans–/cis*-citral was transformed into cyclocitral after UV irradiation regardless of air exposure, indicating that UV irradiation is the major factor inducing the generation of cyclocitral. Moreover, citral was converted into neric and geranic acids after the co-treatment of UV irradiation and air exposure, while neither of the two factors alone could induce the reaction (Fig. 22.1B). Thus, reduced freshness in essential oils exposed to sunlight is related to the decrease of neral and geranial (Sun et al., 2014).

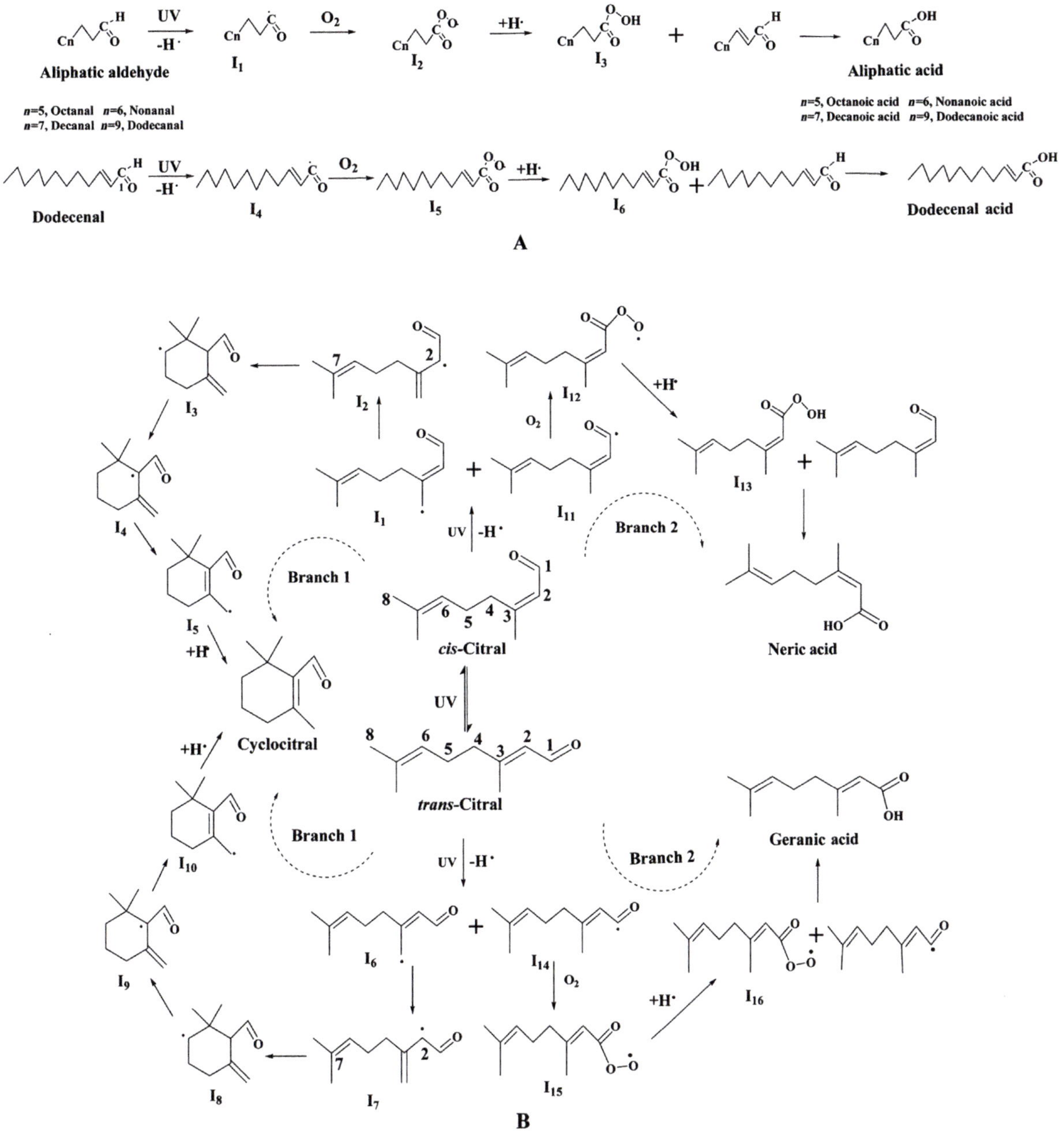

FIG. 22.1 Transformation pathway of aliphatic aldehydes (A) and citral (B) induced by UV irradiation and air exposure (I_n stands for the free radical in the transformation). *(Reprinted with permission from Li, L.J., Hong, P., Chen, F., Sun, H., Yang, Y.F., Yu, X., et al., 2016. Characterization of the aldehydes and their transformations induced by UV irradiation and air exposure of white guanxi honey pummelo (Citrus grandis (L.) Osbeck) essential oil. J. Agric. Food Chem. 64, 5000–5010. Copyright (2016) American Chemical Society.)*

22.2.5 Esters

On average, ester compounds contribute approximately 1% of the volatile fraction of lemon essential oils, with values ranging from 0.51% to 1.03% (Table 22.1). Nevertheless, a slight change in the quantitative composition of ester compounds can significantly alter olfactory perception. Ester compounds of lemon essential oils can be classified into either aliphatics or monoterpenes. The main contributor is monoterpene esters, with an average total of 0.74%, whereas aliphatic esters range from trace amounts to 0.02%. The main aliphatic esters identified in lemon essential oils are heptyl acetate, octyl acetate, nonyl acetate, and bornyl acetate (Dugo et al., 1983, 1999; Dugo, 1986, 1994; Cotroneo et al., 1986, 1988; Trozzi et al., 1993; Verzera et al., 1996). Dugo et al. (1999) also listed decyl acetate, in addition to nonyl- and octyl acetate, as aliphatic ester compounds identified in Italian lemon essential oils. All these aliphatic ester compounds are present in trace amounts. Octyl acetate, with a typical jasmine aroma, is usually used in the formulation of floral perfumes. Odor and flavor thresholds for octyl acetate have been determined in water by Ahmed et al. (1978) who reported values equal to 47 and 210 ppb, respectively. Monoterpene esters are mainly represented by citronellyl acetate, neryl acetate, and geranyl acetate. Neryl- and geranylacetate have been reported present in higher levels compared to citronellyl acetate in Italian lemon essential oils, (Dugo et al., 2011) with concentrations ranging from 0.26% to 0.60% and from 0.18% to 0.48%, respectively, while the average percent of citronellyl acetate is 0.02%. Neryl acetate, characterized fresh rose, is typically used to imitate genuine fresh lemon essential oil. Recently, Gök et al. (2015) evaluated and compared the effectiveness of three different extraction methods, supercritical CO_2 extraction (SFE), cold pressing (CP), and hydrodistillation (HD), on the composition of Kibris lemon essential oils. Among the three extraction methods, that obtained by SFE presented higher amounts of oxygenated compounds, followed by CP and HD oil. The major differences were due to aldehyde amounts (mainly neral and geranial), linalool and geranyl acetate. The total amount of esters determined by SFE, CP, and HD extract was 1.86%, 1.84%, and 1.80%, respectively, with neryl acetate ranging from 1.28% to 1.21%. Notably, these values are significantly higher compared to those observed in Italian lemon genuine CP essential oils (Dugo et al., 2011; Verzera et al., 1999), in a Japanese oil (Sawamura et al., 1999) and in Uruguayan oils (Tirado et al., 1995). Therefore, they are only comparable to those reported for the French Panaché lemon variety (Lota et al., 2002). Ester compounds in lemon juice are mainly represented by neryl acetate, geranyl acetate, citronellyl acetate, octyl acetate, and nonyl acetate (Saura et al., 2012). Their composition may vary depending on the variety. Indeed, as reported in Table 22.2, ester compounds are mainly present in Interdonato (0.43 mg/L) and Verdello Siracusano (0.39 mg/L) juices. Much lower concentration is seen in Femminello Siracusano (0.09 mg/L) and Primo Fiore Capo d'Orlando (0.025 mg/L) juices.

The general biosynthesis of ester species follows esterification reactions that combine acids and alcohols. The interconversion of monoterpenol and monoterpenyl acetate is affected by alcohol acetyltransferase-catalyzed acylation and esterase-catalyzed deacylation reactions. Nevertheless, it has been demonstrated that geranyl acetate esterase (GAE) activity is present almost ubiquitously in plants, independently from the natural availability of its main substrate (geranyl acetate). Thus, the compositional variation of the essential oil cannot be estimated by considering only enzymatic GAE activity (Sangwan et al., 2007). In addition, in the same work it was observed that the GAE activity in plants lacking geranyl acetate can be presumably linked to a generally occurring esterase enzymatic activity with broad substrate specificity, only limited to esters based on a primary hydroxyl function.

22.2.6 Terpenoid hydrocarbon

This class of compounds is quantitatively the most representative (~96%) of the volatile fraction of lemon essential oils. Terpene hydrocarbons are mostly monoterpene (C_{10}) and sesquiterpene (C_{15}) hydrocarbons, which account on average for about the 95.59% and 1.07% of the volatile fraction of lemon essential oil, respectively (Table 22.1). Small but appreciable quantities of aliphatic hydrocarbons are also present (0.02%). Monoterpene hydrocarbons are represented by α-thujene, α-pinene, camphene, sabinene, β-pinene, myrcene, α-phellandrene, δ-3-carene, α-terpinene, *p*-cymene, limonene, (*E*)-β-ocimene, (*Z*)-β-ocimene, γ-terpinene, and terpinolene. As shown in Table 22.1, limonene is the predominant terpene with an average percentage equal to 67.33%, followed by β-pinene and γ-terpinene representing about 12% and 9% of this class of compounds, respectively. α-Pinene, sabinene, and myrcene are present in a percentage that slightly exceeds 1%. Sesquiterpene hydrocarbons are mostly represented by β-bisabolene, *trans*-α-bergamotene, and β-caryophyllene accounting for average percentages equal to about 0.44%, 0.27%, and 0.17%, respectively. Even if terpene hydrocarbons represent the most abundant volatiles of citrus essential oils, they are characterized by poor odor contribution. Odor and flavor thresholds for D-limonene have been determined in water by Ahmed et al. (1978), who reported values equal to 60 and 210 ppb, respectively. Indeed, they scarcely contribute to the typical aroma and flavor of essential oil and their insolubility in aqueous and alcoholic solutions can give rise to problems for their use as combined aromatizers and flavorings in foods and drinks, aside from pharmaceutical and cosmetic products. Moreover, their presence can be detrimental for the quality of essential oil due to their susceptibility to oxidation and subsequent onset of resinification, when exposed to oxygen (Arce and Soto, 2008). Industrial multipurpose application of lemon essential oil involves the preventive removal of terpene hydrocarbons with the subsequent concentration of the more desirable oxygenated components (aldehydes, esters, and ketones). Deterpenation is generally carried out by extraction with conventional solvents (ethanol, methanol, aqueous solutions of ethanol or methanol, ethyl acetate, and acetone) and vacuum distillation. Conventional distillation-based deterpenation methods could be responsible for some loss of quality of citrus essential oils. Therefore, at the industrial-level newer technologies have also been developed such as countercurrent liquid–liquid extraction, membrane separation, adsorption chromatography, and extraction with innovative and environmental-friendly solvents (supercritical fluids and ionic liquids). As for oils, monoterpene hydrocarbons constitute the main volatile group of lemon juice. Allegrone et al. (2006) demonstrated this class of components represents 70%–90% of total volatiles in hand-squeezed juices of four different Italian lemon productions with amounts equal to 45.42, 6.54, 4.89, and 1.81 mg/L for Verdello Siracusano, Interdonato, Primo Fiore Capo d'Orlando, and Femminello Siracusano, respectively. Limonene is reported as the most represented volatile in all juices ranging from 54.1% to 68.8%, followed by γ-terpinene (6.8%–11.4%), β-pinene (2.6–6.5%), terpinolene (1%–1.7%), and β-myrcene (1%–1.5%). Moreover, other minor components (α-thujene, α-pinene, camphene, α-terpinene, and β-ocimene) have also been identified. Sesquiterpene hydrocarbons in lemon juices of the four varieties are reported to be mostly represented by β-bisabolene and α-bergamotene, representing 0.3%–3.4% and 0.2%–1.9% of the total volatiles, followed by β-caryophyllene and α-zingiberene.

The biosynthetic pathway of terpenoid hydrocarbons first leads to the formation of acyclic terpenes such as geraniol (C_{10}), farnesol (C_{15}), geranylgeraniol (C_{20}), and squalene (C_{30}), which subsequently gives rise to all the other terpene compounds through cyclization, transposition, and oxidation reactions. The key intermediate compound is represented by mevalonic acid originating from the aldol condensation of acetyl-co A and acetoacetyl-co A, and subsequent reduction. The phosphorylation of the two alcoholic functions of mevalonic acid produces isopentyl pyrophosphate that undergoes isomerization producing 3,3-dimethylallyl pyrophosphate. Geranyl pyrophosphate originates from this C_5 species, which then produces geraniol, the monoterpene precursor, and farnesyl pyrophosphate, the sesquiterpene precursor (Towers and Stafford, 2012). Many factors can produce changes in the concentration of volatile constituents, such as the growing season, the presence of disease and/or disorders. Recently, Xie et al. (2018) observed significant differences in the volatile profiles and expression of genes involved in terpenoid metabolism in oleocellosis peels compared to healthy ones. In particular, compared to terpene hydrocarbons, they found a decrease in α-pinene, limonene, β-myrcene, and β-caryophyllene in orange and mandarin, for example. Furthermore, it has been widely reported that many volatiles undergo degradation and oxidation during storage, which is caused, for instance, by the transformation of monoterpenes into monoterpene alcohols or monoterpene oxides. Heat and oxygen have been indicated as the major factors affecting the volatile composition of citrus essential oils (Nguyen et al., 2009). For instance, Iwanami et al. (1997) found that lemon essential oil had a dusty flavor after UV irradiation. More recently, Sun et al. (2014) found that reduced freshness of sunlight-exposed pummelo essential oils is related to the degradation and transformation of some aroma-impact volatiles, such as myrcene and limonene, aside from the simultaneous decrease of citral.

22.2.7 Alcohols, ketones, and oxygen heterocycles

Compared to other chemical classes, alcohols are present in minor concentrations in lemon essential oils, contributing to <1% of the volatile fraction, with their average content equal to about 0.34% (Table 22.1). Alcohol components are mainly represented by monoterpene compounds (0.26%), followed by sesquiterpene and aliphatic compounds accounting for approximately 0.06% and 0.02%, respectively. Among monoterpenes, α-terpineol is found in highest concentrations with an average percentage of 0.23%. This volatile may originate from the acid-catalyzed hydration of limonene in an acidic environment. It possesses a characteristic lilac flower fragrance that lends itself to the cosmetic industry. *cis*-Sabinene hydrate and *trans*-sabinene hydrate are present in average percentages of 0.04%, whereas linalool reaches values ranging from 0.05% to 0.18% (Mitiku et al., 2000) Citronellol, geraniol, and terpinen-4-ol are present with average percentages of 0.08%, 0.02%, and 0.05%, respectively. Borneol and octanol are aliphatic alcohols that are present in trace amounts. α-Bisabolol, campherenol, and 2,3-dimethyl-3-(4-methyl-3-pentenyl)-2-norbornanol are the main sesquiterpene alcohol components accounting on average for 0.06% of the total volatile fraction. α-Bisabolol was found in highest concentration (0.03%).

Genetic improvement programs are usually aimed at generating new cultivars with higher resistance to disease and biotic and abiotic stresses, as well as seedless and easy to peel fruits with optimal size and shape and with new organoleptic characteristics. Thus, the interactions of two genomes can produce combined effects on gene expression and phenotypes, and significantly influence the qualitative and quantitative composition of the essential oils of a new citrus hybrid (Fabroni et al., 2012; Ruberto et al., 1994). The essential oil composition of a new hybrid with lemon-like characteristics obtained by crossing a diploid Clementine with a tetraploid lemon have been investigated (Mitiku et al., 2000). The chemical profile of this essential oil clearly showed a predominant contribution from lemon in the hybrid. At the same time in this new hybrid, linalool, an important Clementine essential oil component, was intermediate between that of the parents, producing twice the content, that increased from ~0.20% (in the lemon male parent) to ~0.50%.

Alcohols in lemon juices are mostly represented by monoterpene alcohols such as linalool, borneol, 4-terpineol, α-terpineol, and sesquiterpene alcohols such as nerol and geraniol with a total percentage distribution ranging from 3.00% to 19.29% of the whole volatile fraction, depending on the variety (Allegrone et al., 2006). Saura et al. (2012) recently tested tangential flow filtration for clarifying lemon juice for the food industry. They showed that monoterpene and sesquiterpene alcohols crossed the membrane due to higher water solubility of the alcohols in the juice serum. This differed compared to hydrocarbons, which due to their hydrophobic character, were primarily associated with the colloidal structure of the lemon juice. This higher permeability of the cross-filtration system to the alcohols is basic for its application in the food industry, since these compounds are the main responsible chemicals for the typical lemon aroma of the juices. Odor and flavor thresholds for α-terpineol and linalool have been determined in water by Ahmed et al. (1978), who reported values equal to 280 and 300 ppb, and 5.3 and 3.8 ppb, respectively. Thus, the concentration of these two alcohols, either in lemon essential oils than in juice, exceeds their odor and flavor thresholds.

Ketone group compounds are present in lemon essential oils with average percentages ranging from 0.01% to 0.09%. Aliphatic (6-methyl-5-hepten-2-one), monocyclic monoterpene (carvone and piperitone), bicyclic monoterpene (camphor), and sesquiterpene (nootkatone) ketones are all present in trace amounts, each component with maximum concentrations of 0.01% (Table 22.1). Biogenetically, ketone components are derived from alcohol through oxidation reactions catalyzed by different enzymes, which exert dehydrogenase activity. Due to their relevant contribution to flavor and fragrance of essential oil, recent studies have been aimed at optimizing biocatalytic systems to synthetize ketones, in order to overcome the low yields of their isolation from natural plant resources. For instance, a two-enzyme reaction sequence for the synthesis of nootkatone, known for its grapefruit aroma, has been developed. In the first step, a cytochrome P450 monooxygenase catalyzes the hydroxylation of valencene to nootkatol, which is further oxidized to nootkatone by an alcohol dehydrogenase (Schulz et al., 2015).

22.2.8 Lemon oil sulfur compounds

Volatile sulfur compounds may be partially responsible for the aroma of fresh fruit. The concentration of these compounds in citrus fruits is considerably lower than in most other natural products making them difficult to identify and quantify. Therefore, only a few references can be found regarding their presence in lemon essential oil or juice. The first report (Shaw and Wilson III, 1982) identified the presence of hydrogen sulfide in the headspace of Eureka lemon juice. It was followed 25 years later by the identification and quantitation of dimethyl sulfide in four Italian lemon juices (Allegrone et al., 2006). Primo Fiore Capo d'Orlando lemon juice contained the highest amount of dimethyl sulfide (10.1 μg/L), followed by Verdello Siracusano (7.2 μg/L), Interdonato (6.5 μg/L), and Femminello Siracusano (5.9 μg/L). A recent study reported that different volatile sulfur compounds in lemon essential oils may contribute to the aroma profile of the fresh fruit (Cannon et al., 2015). In this study, novel sulfur-containing volatile compounds from Lisbon lemon oil were identified; among these, two showed odor activity by aroma extract dilution analysis (AEDA): 3-methylbut-2-ene-1-thiol (sulfurous, skunk-like; flavor dilution, representing the strength of the odor, equal to 243) and 3-mercapto-3,7-dimethyl-6-octenyl acetate (sulfurous; flavor dilution equal to 9). It was hypothesized that both may originate through direct addition of hydrogen sulfide to neryl and geranyl acetate or, alternatively; they can be an adduct of L-cysteine or L-glutathione with neryl and geranyl acetate, followed by a specific enzyme to release the thiol.

22.3 Orange and mandarin

Orange (*Citrus sinensis*) juice is the most preferred juice beverage worldwide. Orange and mandarin (*Citrus reticulata*) essential oils are also the most popular of the natural essential oils that are used commercially for food flavoring, toiletry products, cosmetics, and perfumes. The delicate aroma of both orange and mandarin is due to a complex mixture of volatiles. The aim of orange and mandarin flavor and odor research is to determine the aroma characteristics, functionality, and industrial or commercial values of their juices and essential oils.

Gas chromatography-olfactometry (GC-O) identified a number of odor-active volatiles that contribute to the flavor profile of oranges and mandarins. Over 300 volatiles have been reported in orange and mandarin, but only a very small portion of them exert significant odor activities. In addition, the manner in which juice is extracted, fruit maturity, cultivar type as well as processing methods all alter the list of significant odorants. For instance, it is well-documented that processing and subsequent storage reduce orange juice flavor quality due to thermally accelerated odor degradation (Nisperos-Carriedo and Shaw, 1990). Differing from grapefruits and lemons, which contain signature aroma compounds, such as 1-p-methene-8-thiol and citral, respectively, it has been documented that orange and mandarin odors result from a collection of aroma-active volatiles present in low concentrations (Marin et al., 1992; Hinterholzer and Schieberle, 1998; Bazemore et al., 1999).

Aroma compounds in oranges and mandarins are liberated from either rupture of aqueous juice vesicle or oil from the peel. Orange and mandarin aroma compounds consist of a vast array of chemical species. A total of seven different families of compounds have been characterized in orange and mandarin, including alcohols, aldehydes, hydrocarbons, esters, terpenoids, ethers, and acids. In this section, each odor-active chemical family will be discussed in detail.

22.3.1 Aldehydes

Aldehydes play a major role in orange and mandarin flavor (Table 22.4). The concentration of aldehydes is altered significantly after thermal processing due to their high chemical reactivity. Different odor-active aldehydes were identified from saturated aliphatic aldehydes, unsaturated aliphatic aldehydes, terpenic aldehydes, and phenolic aldehydes. Odor-active aldehydes drive from different biosynthetic pathways, including the terpenoid pathway, fatty acid oxidation pathway, and amino acid derivation pathway (Schwab et al., 2008). The terpenic aldehydes and saturated aliphatic aldehydes fractions deliver green notes whereas the unsaturated aliphatic aldehydes contribute green/fatty/metallic odors to the orange juice.

TABLE 22.4 Odor-active aldehydes identified from oranges and mandarins.

Odor-active compound	Chemical structure	Odor quality	Ref.
Acetaldehyde	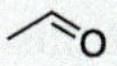	Solvent-like, fruity, fresh	Buettner and Schieberle (2001a), Rega et al. (2003), Mahattanatawee et al. (2005)
Hexanal		Green, grassy, soapy, fruity	Hinterholzer and Schieberle (1998), Tønder et al. (1998), Buettner and Schieberle (2001a), Arena et al. (2006), Rega et al. (2003), Mahattanatawee et al. (2005)
Octanal		Green, citrus-like, geranium, floral	Mahattanatawee et al. (2005), Tønder et al. (1998), Hinterholzer and Schieberle (1998), Rouseff et al. (2001), Zhang et al. (2013), Rega et al. (2003)
Nonanal	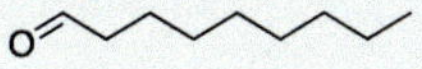	Citrus-like, soapy, floral	Tønder et al. (1998), Hinterholzer and Schieberle (1998), Bazemore et al. (1999), Buettner and Schieberle (2001a), Arena et al. (2006)
Decanal		Green, citrus-like, fatty	Hinterholzer and Schieberle (1998), Rouseff et al. (2001), Rega et al. (2003), Mahattanatawee et al. (2005)
Dodecanal		Soapy	Hinterholzer and Schieberle (1998), Buettner and Schieberle (2001a), Mahattanatawee et al. (2005)
(*Z*)-Hex-3-enal	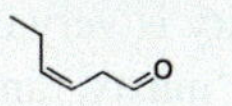	Green, leaf-like, grassy	Rouseff et al. (2001), Bazemore et al. (1999), Buettner and Schieberle (2001a), Arena et al. (2006)
(*Z*)-Non-2-enal	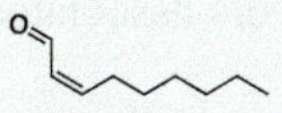	Green, metallic, fatty	Hinterholzer and Schieberle (1998), Buettner and Schieberle (2001a), Mahattanatawee et al. (2005)
(*E*)-Non-2-enal	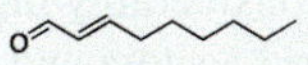	Fatty, tallowy, green	Rouseff et al. (2001), Bazemore et al. (1999), Buettner and Schieberle (2001a)
(*E*,*E*)-2,4-Nonadienal	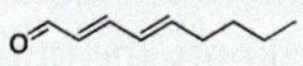	Fatty, waxy, soapy	Hinterholzer and Schieberle (1998), Buettner and Schieberle (2001a), Rouseff et al. (2001), Mahattanatawee et al. (2005)
(*E*,*Z*)-2,6-Nonadienal	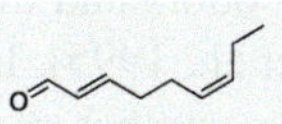	Cucumber-like, green	Rouseff et al. (2001), Hinterholzer and Schieberle (1998), Mahattanatawee et al. (2005)

Continued

TABLE 22.4 Odor-active aldehydes identified from oranges and mandarins—cont'd

Odor-active compound	Chemical structure	Odor quality	Ref.
(*E*,*E*)-2,4-Decadienal		Fatty, waxy, green	Mahattanatawee et al. (2005), Hinterholzer and Schieberle (1998), Buettner and Schieberle (2001a)
Neral	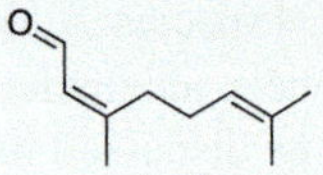	Lemon-like, citrus-like lemongrass, minty	Mahattanatawee et al. (2005), Marin et al. (1992)
Geranial	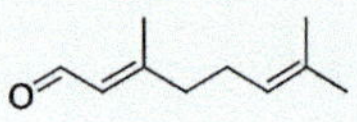	Citrus-like, green, minty	Mahattanatawee et al. (2005), Marin et al. (1992)

Acetaldehyde with a fresh, solvent-like note is the major contributor to the fresh and pungent flavor of freshly squeezed orange and mandarin juice. The concentration of acetaldehyde is significantly reduced after heat treatment that is designed to inactivate enzymatic and microbial activities (Bazemore et al., 1999). The three straight-chain aldehydes originate from the oxidation of fatty acids, octanal, nonanal, and decanal have been identified as important components in orange/mandarin flavor (Maarse, 1991) and deliver a fatty odor. Another saturated aldehyde, hexanal, is a ubiquitous flavor component, formed from both 9- and 13-hydroperoxides of linoleate and other unsaturated aldehydes formed during the oxidation of linoleate (Risch and Ho, 2000). Although hexanal, commonly found with a green note in fresh and stored foods, is not considered an important aroma compound in orange and mandarin flavor.

(Z)-Hex-3-enal, the grassy note in orange and mandarin flavor, is formed via a lipoxygenase-catalyzed reaction of fatty acid hydroperoxides (Hinterholzer and Schieberle, 1998). Due to the instability of this compound, its concentration is lowered significantly during the thermal processing of juice. The grassy odor of (*Z*)-Hex-3-enal does enhance the fresh characteristic of orange and mandarin juice, but the positive relationship ranges in concentration. The addition of (Z)-hex-3-enal to a juice base at 9 μg/mL greatly diminished the hedonic rating (Ahmed et al., 1978).

Terphenic aldehydes (e.g., neral and geranial) that are responsible for the unique scent of lemon oil occur in orange juice at concentrations above their threshold level in water, particularly in processed juice. They are considered key aroma compounds in the flavor of both orange and mandarin (Ahmed et al., 1978).

22.3.2 Esters

Odor-active esters play a critical role in many fruit flavors, including orange and mandarin (Table 22.5). Esters are biosynthesized from Acyl-CoA in the intact fruit, with their concentrations reduced by hydrolase enzymes released as fruit cells are macerated during juicing (Belitz and Grosch, 1999).

Most ester species are ethyl esters, with ethyl butanoate contributing the most to the orange and mandarin flavor. Identified in various orange varieties, including late Hamlin, late Pineapple, mid- and late Valencia, Pera, Navelina, Moro, Tarocco, and Washington navel juices, ester concentrations rise as the fruit matures (Kealey et al., 1979; Arena et al., 2006; Buettner and Schieberle, 2001a; Rega et al., 2003). The concentration of ethyl butanoate and total esters decrease with thermal processing (Nisperos-Carriedo and Shaw, 1990). Notably, ester odor quality is very concentration dependent. For instance, ethyl hexanoate, described as 'fruity' at low concentrations, is considered 'musty, rancid, rotten' at higher concentrations (Baldwin et al., 2014). The alteration of odor quality may cause off-flavors associated with overripe or fermented oranges (Plotto et al., 2008).

In fresh orange and mandarin juice, other ethyl esters are present at levels considerably higher than their aroma threshold in water (Kealey et al., 1979). In addition, the fruity and floral notes of neryl acetate and citronellyl acetate have both been identified in mandarin (Barboni et al., 2009; Tietel et al., 2009, 2010). Numerous studies have contributed to the current understanding of these esters, including an omission study that demonstrated only the omission of ethyl butanoate, and no other ester, can be detected by a sensory panel (Buettner and Schieberle, 2001a).

While studying the correlation of aldehyde structures to both corresponding odors and thresholds, methyl esters have approximately 100 times greater relative thresholds than ethyl esters (Takeoka et al., 1995). In addition, straight-chain esters have much greater thresholds than their branch-chain homologs. Furthermore, another study reported ethyl propanoate had close to 1000 times greater threshold than ethyl 2-methylpropanoate (Plotto et al., 2008).

TABLE 22.5 Odor-active esters identified from oranges and mandarins.

Odor-active compound	Chemical structure	Odor quality	Ref.
Methyl butanoate	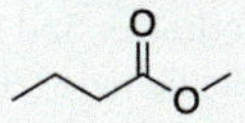	Fruity, strawberry	Rega et al. (2003), Arena et al. (2006)
Ethyl acetate	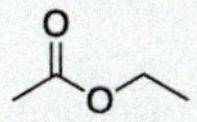	Fruity, solvent-like	Buettner and Schieberle (2001a)
Ethyl butanoate	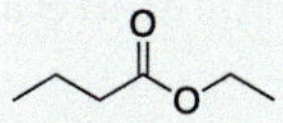	Fruity, pineapple	Marin et al. (1992), Bazemore et al. (1999), Buettner and Schieberle (2001a), Rega et al. (2003), Mahattanatawee et al. (2005), Arena et al. (2006)
Ethyl-2-methylpropanoate	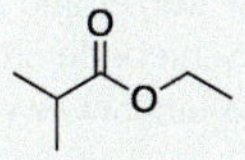	Fruity	Buettner and Schieberle (2001a), Mahattanatawee et al. (2005), Rouseff et al. (2001)
Ethyl-2-methylbutanoate	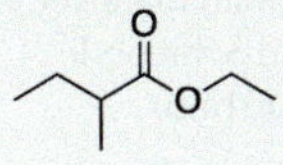	Fruity	Marin et al. (1992), Buettner and Schieberle (2001a), Mahattanatawee et al. (2005), Arena et al. (2006)
Ethyl hexanoate	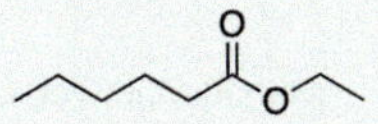	Fruity, orange	Buettner and Schieberle (2001a), Rega et al. (2003), Mahattanatawee et al. (2005), Arena et al. (2006)
Ethyl octanoate	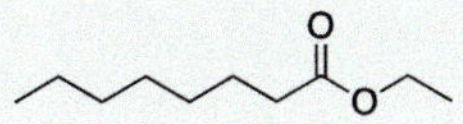	Spicy, floral, fruity	Rega et al. (2003), Arena et al. (2006)

22.3.3 Terpenoid hydrocarbons

Terpenes are the most abundant volatile compounds in orange and mandarin, typically comprising >90% of the total volatiles. However, terpenes are only responsible for very limited odor activity due to high odor thresholds. Terpenoid hydrocarbon structures are vastly diverse since all contain one or more chiral centers, which produce both enantiomers and diastereoisomers with different odor characteristics. The terpenoid compounds listed in Table 22.6 are the major terpenes found in both orange and mandarin juice and peel oil.

Limonene is present in the highest concentration, approximately 400 times that of its odor threshold level in the water, among all the terpenoid flavor compounds. However, due to its low solubility in water, the majority of limonene is adhered to the pulp. Therefore, limonene is found at a lower concentration in freshly squeezed juice compared to processed juice because of the introduction of peel oil by commercial extraction. The role of limonene in the overall flavor profile of orange and mandarin is not clear. Controversial results have shown limonene to be narrowly aroma active, yet a necessary ingredient to mimic orange juice odor (Marin et al., 1992; Ahmed et al., 1978; Buettner and Schieberle, 2001a).

β-Myrcene is another abundant aromatic terpene in orange and mandarin. It is reported as a strong odorant, negatively contributing to the flavor of orange and processed juice (Ahmed et al., 1978). β-Pinene, a bicyclic terpene with a pungent note, is considered to make a positive contribution to orange and mandarin flavor (Ahmed et al., 1978).

22.3.4 Alcohols, ketones, acids, and oxygen heterocycles

A diverse array of volatile alcohols is found both in orange and mandarin juices (Table 22.7), yet there are only six alcohol species that demonstrate aromatic activity. These are formed by oxidation of unsaturated fatty acid producing homologous aldehydes, and their subsequent reduction. Linalool, the most potent alcohol in orange and mandarin juice, possesses a distinctive sweet floral odor. In addition, it is also present at higher concentrations in commercially extracted juice due to the increased levels of peel oil introduced by processing (Bazemore et al., 2003). (*Z*)-3-Hexen-1-ol contributes green and woody notes to freshly squeezed orange juice and is considered an important aliphatic alcohol. It is also a product of linolenic acid oxidation via lipoxygenase and hydroperoxide lyase (Hotchkiss, 1988). The final important odor-active alcohol, 1-Octanol, provides a green herbal note.

The most intense ketone odorant in orange is 1-octen-3-one, which possesses a distinct mushroom odor often described as metallic, and an odor threshold as low as 0.005 μg/L (Peterson and Reineccius, 2002). Nootkatone, a sesquiterpene ketone

TABLE 22.6 Odor-active terpernoid compounds identified oranges and madarins.

Odor-active compound	Chemical structure	Odor quality	Ref.
α-Pinene	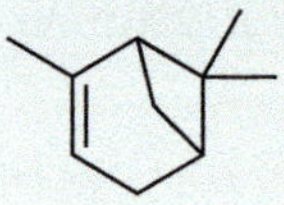	Resin, Pine tree, ethereal	Hinterholzer and Schieberle (1998), Buettner and Schieberle (2001a)
β-Pinene	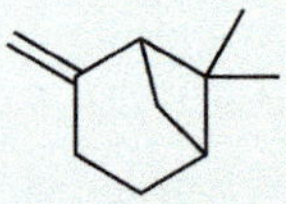	Wood, terpene-like, pungent	Hinterholzer and Schieberle (1998), Rega et al. (2003), Arena et al. (2006)
β-Myrcene	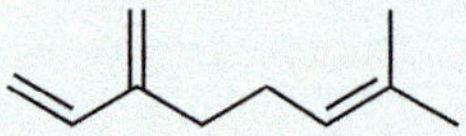	Mossy, geranium, musty	Hinterholzer and Schieberle (1998), Buettner and Schieberle (2001a), Arena et al. (2006), Mahattanatawee et al. (2005)
Limonene	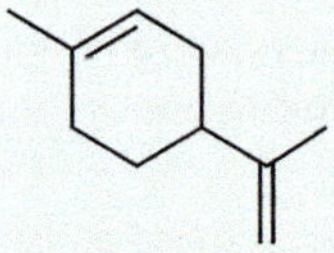	Minty, lemon, citrus-like	Mahattanatawee et al. (2005), Hinterholzer and Schieberle (1998), Buettner and Schieberle (2001a,b)
α-Terpinolene	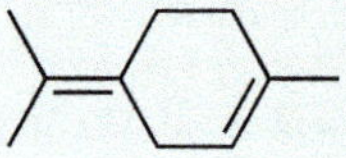	Green, chemical, woody	Mahattanatawee et al. (2005), Arena et al. (2006)

TABLE 22.7 Odor-active alcohols, ketones, acids and oxygen heterocycles identified in oranges and mandarins.

Odor-active compound	Chemical structure	Odor quality	Ref.
Aliphatic alcohols			
1-Hexanol	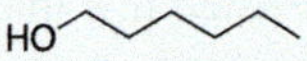	Floral, grass	Arena et al. (2006), Rega et al. (2003)
(*Z*)-3-Hexen-1-ol	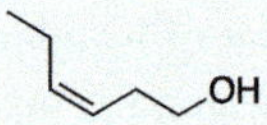	Woody, green	Mahattanatawee et al. (2005), Rega et al. (2003)
1-Octanol	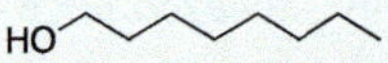	Herbal, green	Rega et al. (2003), Mahattanatawee et al. (2005)
Terpene alcohols			
Linalool	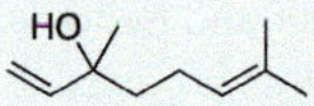	Floral, sweet, fruity	Arena et al. (2006), Buettner and Schieberle (2001a,b), Mahattanatawee et al. (2005), Tønder et al. (1998)
Geraniol	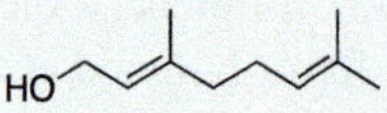	Rose, floral, citrus, geranium	Mahattanatawee et al. (2005), Rega et al. (2003)
Terpinen-4-ol	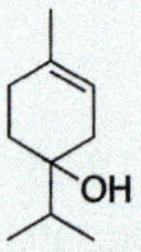	Metallic	Bazemore et al. (1999), Mahattanatawee et al. (2005)
Aliphatic ketones			
1-Octen-3-one	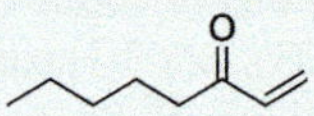	Mushroom	Arena et al. (2006), Hinterholzer and Schieberle (1998), Mahattanatawee et al. (2005), Rega et al. (2003), Tønder et al. (1998)
(*Z*)-Octa-1,5-dien-3-one	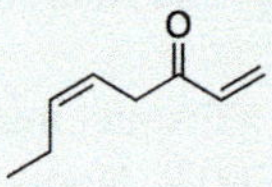	Geranium	Buettner and Schieberle (2001a), Hinterholzer and Schieberle (1998), Mahattanatawee et al. (2005)

TABLE 22.7 Odor-active alcohols, ketones, acids and oxygen heterocycles identified in oranges and mandarins—cont'd

Odor-active compound	Chemical structure	Odor quality	Ref.
Terpene ketones, norisoprenoids			
β-Ionone	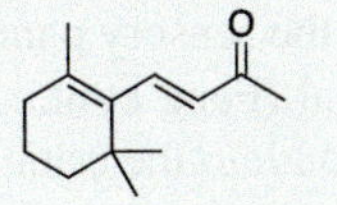	Floral, raspberry	Buettner and Schieberle (2001a), Hinterholzer and Schieberle (1998), Mahattanatawee et al. (2005), Rega et al. (2003)
Nootkatone	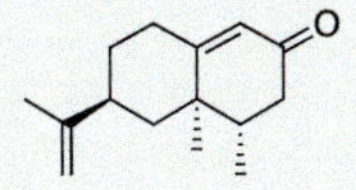	Grapefruit like, green	Mahattanatawee et al. (2005), Hinterholzer and Schieberle (1998)

and characteristic compound of grapefruit, is also found in orange peel oil, but at a much lower concentration (Moshonas and Lund, 1969), and with weak odor intensity (Hinterholzer and Schieberle, 1998; Mahattanatawee et al., 2005).

Biosynthesized from a carotenoid substrate, ionones are widely distributed in nature (Baldwin et al., 2000). They originate from oxidative cleavage of the double bond between carbon 9 and 10 in β-carotene, and in β-cryptoxanthin (Giuffrida et al., 2006). β-Ionone is the most potent norisoprenoid detected in freshly squeezed orange juice and provides sweet floral and raspberry-like odor in orange and mandarin juice (Elmaci and Altug, 2005; Hinterholzer and Schieberle, 1998).

2,5-Dimethyl-4-hydroxy-3(2H)-furanone, also reported in fresh orange juice as an odor-active compound (Buettner and Schieberle, 2001a), is considered a product of the Maillard reaction. Its concentration increases after thermal processing, and some studies report this furanone to contribute an off-flavor (Walsh et al., 1997).

Moreover, volatile acids have also been identified as odor active in fresh orange juice, although making very limited contributions to overall orange flavor.

22.3.5 Sulfur- and nitrogen-containing compounds

Compounds containing sulfur and nitrogen are extremely difficult to detect using current analytical techniques. Nonetheless, they supply an extremely potent aroma to orange flavor. When identifying and investigating sulfur compounds, the challenge stems from proper extraction of compounds found in very low concentrations from complex juice matrices. Sulfur- and nitrogen-containing compounds, such as 3-s-butyl-2-methoxypyrazine and 3-isopropyl-2-methoxypyrazine, have been identified in mandarin (Tietel et al., 2011). These pyrazines, present as metabolic by-products in certain foods and microorganisms, are potent odorants.

The sulfur-containing flavor compounds identified in orange vary by variety. Shaw et al. identified methanethiol and hydrogen sulfide of both Valencia and Navel orange juice (Shaw et al., 1980). Using GC-O and AEDA, methional (cooked potato) and p-menth-1-ene-8-thiol (grapefruit-like), two odor-active sulfur compounds, were reported in Valencia orange (Hinterholzer and Schieberle, 1998). An additional study comparing fresh, pasteurized, and stored juice found methylfuran-3-thiol in Valencia and Shamouti cultivars from Israel (Bezman et al., 2001). 1-Phenylethanethiol was identified in Pontianak orange, a first in a natural product, and described the compound as "sulfurous, resinous" (Fischer et al., 2008). A later omission investigation showed 1-phenylethanethiol is essential to the aroma profile of these fruits (Dharmawan et al., 2008).

22.3.6 Differences between orange and mandarin

The chemical profile of orange and mandarin flavor has been largely investigated over the past few decades. The orange is a complex hybrid derived from the genomic introgression of mandarin and pummelo (*Citrus maxima*). Thus, it is not surprising that orange and mandarin share certain similarities in terms of their aroma profile.

However, orange and mandarin do have distinct flavor differences. For instance, ethyl 2-methylpropanoate, ethyl butanoate, (*S*)-ethyl 2-methylbutanoate, 3a,4,5,7a-tetrahydro-3,6-dimethyl-2(3H)-benzofuranone (wine lactone), (*Z*)-hex-3-enal, α-pinene, (*R*)-limonene, ethyl butanoate, (*S*)-ethyl 2-methylbutanoate, and acetaldehyde are considered the most potent odorants in orange juice based on their high flavor dilution factors and odor-active values (OAVs) (Hinterholzer and Schieberle, 1998; Buettner and Schieberle, 2001a). Whereas in mandarin studies, limonene, linalool, γ-terpinene, β-myrcene, α-pinene, and octanal were demonstrated as key aroma components (Yu et al., 2007), along with α-terpineol, terpinen-4-ol, nonanal, decanal, and carvone (Tietel et al., 2011). The aroma compounds in orange and mandarin greatly overlap, but with substantiated quantitative differences. From a sensory perspective, descriptive attributes are very limited in their ability to help differentiate orange from the mandarin aroma. Recently, a model study revealed that ethyl butanoate, ethyl 2-methylbutanoate, octanal, decanal, and acetaldehyde were key aroma compounds in orange. This means that this combination of compounds could possess a character recognized as orange by the sensory panel. For mandarin, key aroma compounds are linalool, octanal, α-pinene, limonene, and (*E,E*)-2,4-decadienal (Feng et al., 2017). Differences between orange and mandarin propose an interesting perspective for further investigation into the critical genes and enzymes regulating aroma biosynthesis in the citrus genus, as well as their genomic relationships.

22.4 Grapefruit

Grapefruit is a major commercial citrus crop. The world's largest producers of fresh grapefruit include China (these statistics combine grapefruit and pummelo production, with the latter being by far the main product in China, which produces very little true grapefruit), the United States, Mexico, and South Africa (United States Department of Agriculture, 2018). Grapefruit (*Citrus paradisi*) is a hybrid of pummelo and sweet orange. The unique flavor of grapefruit is a combination of a sweet–sour and slightly bitter taste, as well as a pronounced sulfur aroma, all of which differentiate it from the other citrus fruits. Nevertheless, the grapefruit shares a variety of important citrus aroma components with other members of the citrus family. Based on early flavor investigations (Moshonas and Shaw, 1971; Moshonas, 1971; Coleman et al., 1972; Wilson III and Shaw, 1980; Cadwallader and Xu, 1994; Jella et al., 1998), it is not surprising that major grapefruit volatiles (i.e., aldehydes, esters, terpenoid hydrocarbons, alcohols, ketones, acids, and oxygen heterocycles) overlap with those mentioned in Section 22.3. To avoid repetition, in this section, we will focus on the discussion of characteristic grapefruit volatiles. Furthermore, current investigations into grapefruit flavor are also covered in this segment.

The early years (the 1960s) of grapefruit volatile aroma composition investigations identified nootkatone as the first characteristic grapefruit aroma (Beisel et al., 1964). However, further research into nootkatone in grapefruit revealed other compounds, also present in grapefruit, that contributed to its characteristic flavor as well (Stevens et al., 1970; Shaw and Wilson III, 1981). In a series of more recent studies, the aroma profile of grapefruit was discussed in detail (Table 22.8). The most odor-active aroma compounds, with the highest flavor dilution (FD) factors, in fresh grapefruit juice were characterized as ethyl butanoate, p-1-menthene-8-thiol, (*Z*)-3-hexenal, 4,5-epoxy-(*E*)-2-decenal, 4-mercapto-4-methylpentan-2-one, 1-heptene-3-one, and wine lactone (Buettner and Schieberle, 1999). Further quantification and calculation of the OAVs indicated that the fruity esters, ethyl 2-methylpropanoate, ethyl butanoate and (*S*)-ethyl 2-methylbutanoate, the sweet

TABLE 22.8 Potent odorants in hand-squeezed grapefruit juice (Buettner and Schieberle, 2001c, 1999).

Odorant	Odor quality	FD[a] factor	OAV (n)[b]	OAV (rn)[c]
Ethyl butanoate	Fruity	256	70	695
1-Heptene-3-one	Geranium-like	256	13	13
(*Z*)-3-hexenal	Green, leaf-like	256	430	3587
4,5-Epoxy-(*E*)-2-decenal	Metallic	128	26	207
4-Mercapto-4-methylpentane-2-one	Black currant-like	128	–	8000
Wine lactone	Sweet, spicy	128	–	138
p-1-Menthene-8-thiol	Grapefruit-like	128	100	–
Ethyl 2-methylpropanoate	Fruity	64	290	193
(*S*)-Ethyl 2-methylbutanoate	Fruity	32	650	975

[a]*Flavor dilution factor.*
[b]*Odor activity values (n = nasally).*
[c]*Odor activity values (rn = retronasally).*

wine lactone, the grassy note (*Z*)-hex-3-enal, the metallic odor trans-4,5-epoxy-(*E*)-dec-2-enal, and the typical sulfurous, grapefruit-like odor 4-mercapto-4-methylpentan-2-one and 1-p-menthene-8-thiol (Fig. 22.2), were among the most potent odorants of fresh grapefruit juice (Buettner and Schieberle, 2001c). In addition, reconstitution experiments revealed that the omission of grapefruit-like smelling 1-p-menthene-8-thiol in the incomplete aroma mixture led to the perception of a grapefruit-like odor, but with less grapefruit-like impression. On the other hand, the aroma model omitted with the catty, blackcurrant-like 4-mercapto-4-methylpentan-2-one was described as orange-like, while lacking anything grapefruit like. These results imply the typical aroma of hand-squeezed grapefruit-juice is not only due to 1-p-menthene-8-thiol but is also more greatly expressed by 4-mercapto-4-methylpentan-2-one. Due to the importance of sulfurous components characterized in grapefruit, they are discussed in greater detail in the following paragraph.

Sulfur-containing volatiles contribute a fresh and authentic aroma to a variety of subtropical and tropical fruits. Although widely reported in various citrus fruits, sulfurous compounds crucially contribute to the characterization of grapefruit flavor. The identification of 1-p-menthene-8-thiol in canned grapefruit juice in the early 1980s brought attention to the sulfur volatiles in grapefruit (Demole et al., 1982; Janes et al., 1993). This compound provides a powerful, grapefruit-like aroma with a very low taste detection threshold, 0.00002 mg/L for (*R*)-1-p-menthene-8-thiol and 0.00008 mg/L for (*S*)-1-p-menthene-8-thiol. More recently, using high-resolution GC-O and CI-MS, 4-mercapto-4-methylpentan-2-one, a mercaptan, was identified as a key contributor to the intensity of grapefruit-like aroma (Buettner and Schieberle, 1999, 2001c). In plants, sulfur-containing volatiles are synthesized by cleavage reactions of modified amino acids or their precursors (Goff and Klee, 2006). 1-p-Menthene-8-thiol is thought to form from an amino acid (e.g., methionine and cysteine) or hydrogen sulfide and isoprenoid precursors (e.g., limonene and α-pinene) (Janes et al., 1993). Possible pathways for the formation of 4-mercapto-4-methylpentan-2-one are from either the S-cysteine or S-glutathione conjugate or through the precursor of mesityl oxide (Wakabayashi et al., 2004; Roland et al., 2016). Sulfur volatiles make an important contribution to grapefruit flavor. Unfortunately, many challenges present themselves during the identification process of sulfur-containing volatiles in grapefruit. These challenges include the low concentration at which sulfur compounds occur and their potential instability (Cannon and Ho, 2018). Moreover, high concentrations of monoterpenes and sesquiterpenes in grapefruit further hinder identification of sulfur-containing volatiles. To characterize sulfur volatiles in grapefruit, multiple extractions using large volumes of juice are necessary. Additional requirements include aroma enrichment by fractionation on column, and detectors specific to selection and identification of sulfur compounds (e.g., pulsed flame photometric detector). Along with 1-p-menthene-8-thiol and 4-mercapto-4-methylpentan-2-one, other sulfur compounds reported in grapefruit juice include hydrogen sulfide, methyl sulfide, 4-mercapto-4-methyl-2-pentanol, 3-mercapto-1-hexan-1-ol, 3-mercaptohexyl acetate, sulfur dioxide, methanethiol, dimethyl sulfide, carbon disulfide, dimethyl disulfide, 2-methyl thiophene, 3-methyl thiophene, methional, dimethyl trisulfide, 2,8-epithiocis-p-menthane, and methional (Shaw et al., 1980; Lin et al., 2002a,b; Jabalpurwala et al., 2010). However, their sensory role in grapefruit juice aroma requires further investigations.

Rotundone is a sesquiterpenoid that has been identified as an aroma-active compound in Shiraz wine, Shiraz grape, spices, and cypriol oil (Wood et al., 2008; Clery et al., 2016). Most recently, rotundone (Fig. 22.3), responsible for peppery, spicy, and woody odors, was identified in grapefruit using GC-O. The sensory effects of rotundone on model fruit beverages were studied as rotundone was added. Even at subthreshold levels, rotundone significantly affected the overall flavors

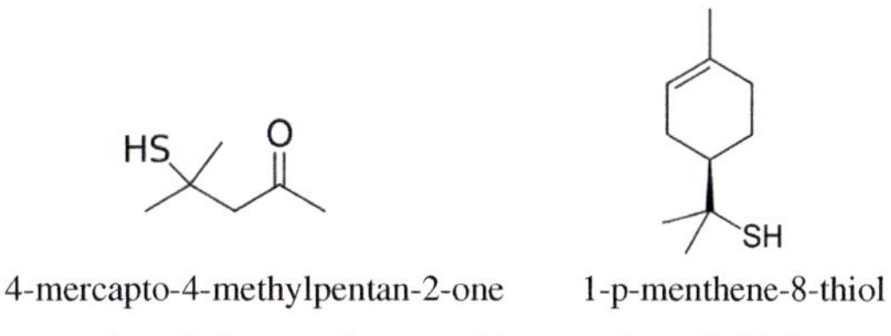

FIG. 22.2 The chemical structures of 4-mercapto-4-methylpentan-2-one and 1-p-menthene-8-thiol.

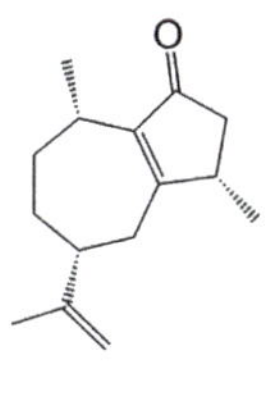

FIG. 22.3 The chemical structure of rotundone.

of the beverages (Nakanishi et al., 2017a). The addition of rotundone to model beverages better approximated the natural fruit flavor. However, it did not directly confer the woody odor. Further investigation of rotundone in grapefruit revealed the highest FD factors in both peel and juice (Nakanishi et al., 2017b). The level of rotundone in grapefruit peel was determined ~2000 ng/kg, while in the grapefruit juice, concentrations of rotundone ranged from 30 to 50 ng/kg. Similarly, in the sensory analyses, rotundone was observed to enhance the perception of various attributes in grapefruit juice without imparting a woody note.

22.5 Conclusion

Citrus aroma compounds have been extensively studied over the past several decades. Citrus juice volatiles are mainly represented by monoterpene hydrocarbons, sesquiterpene hydrocarbons, and oxygenated compounds, mainly represented by aldehydes, monoterpene alcohols, and esters. Volatile compounds are industrially recovered from the condensate from steam distillation/evaporation of juices. This condensate produces two products, essence oil and aqueous essence, the latter sometimes called aqueous aroma, and used as a flavor enhancer in citrus products in the food industry.

Due to their relevance to different fields, citrus essential oils have been widely investigated and several reviews on their qualitative and quantitative composition are present in the literature. Within the volatile fraction of essential oil, the most represented component is limonene, quantitatively followed by β-pinene, γ-terpinene, sabinene, and α-pinene. Among the other components, only two monoterpene aldehydes, neral and geranial, exceed 1%. The content of both aldehydes, defined as "citral," is an important parameter in the commercial transactions of essential oil as these substances determine its olfactory peculiarities.

Citrus flavor is traditionally used in the food, cosmetics, and pharmaceutical industries. In the food industry, these essences are conveniently used in the aromatization of numerous preparations, mostly beverages and confectionery products. In the cosmetic field, since citrus aromas are very popular, they are used in the formulation of original fragrances contributing to the production of perfumes, colognes, and after-shave lotions, and also in the aromatization of toothpastes and deodorants. The pharmaceutical field is another relevant field of application as these essential oils are not only active ingredients but are also used to improve the organoleptic characteristics of some preparations. For centuries, citrus essential oils have also been used in folk medicine. More recently, numerous studies have demonstrated the biological properties of these oils to be an effective anti-inflammatory, antimicrobial, and antioxidant agents (Amorim et al., 2016; Belletti et al., 2004; Lopes Campêlo et al., 2011). Due to their economic and biological relevance in multiple fields, an in-depth understanding of the qualitative and quantitative composition of both volatile and nonvolatile fractions of citrus essential oils is of great interest. Moreover, genetic improvement programs should be encouraged to discover new valuable hybrids containing a high content of oxygenated compound.

References

Ahmed, E.M., Dennison, R.A., Dougherty, R.H., Shaw, P.E., 1978. Flavor and odor thresholds in water of selected orange juice components. J. Agric. Food Chem. 26, 187–191.

Allegrone, G., Belliardo, F., Cabella, P., 2006. Comparison of volatile concentrations in hand-squeezed juices of four different lemon varieties. J. Agric. Food Chem. 54, 1844–1848.

Amorim, J.L., Simas, D.L.R., Pinheiro, M.M.G., Moreno, D.S.A., Alviano, C.S., da Silva, A.J.R., Fernandes, P.D., 2016. Anti-inflammatory properties and chemical characterization of the essential oils of four citrus species. PLoS ONE 11, e0153643.

Arce, A., Soto, A., 2008. *Citrus* essential oils: extraction and deterpenation. Tree For. Sci. Biotechnol. 2, 1–9.

Arena, E., Guarrera, N., Campisi, S., Nicolosi Asmundo, C., 2006. Comparison of odour active compounds detected by gas-chromatography–olfactometry between hand-squeezed juices from different orange varieties. Food Chem. 98, 59–63.

Baldwin, E.A., Scott, J.W., Shewmaker, C.K., Schuch, W., 2000. Flavor trivia and tomato aroma: biochemistry and possible mechanisms for control of important aroma components. Hort. Sci. 35, 1013–1021.

Baldwin, E., Bai, J., Plotto, A., Ritenour, M., 2014. Citrus fruit quality assessment: producer and consumer perspectives. Stewart Postharvest Rev. 10, 1–7.

Barboni, T., Luro, F., Chiaramonti, N., Desjobert, J. M., Muselli, A., Costa, J., 2009. Volatile composition of hybrids *Citrus* juices by headspace solid-phase micro extraction/gas chromatography/mass spectrometry. Food Chem. 116, 382–390.

Bazemore, R., Goodner, K., Rouseff, R., 1999. Volatiles from unpasteurized and excessively heated orange juice analyzed with solid phase microextraction and GC-olfactometry. J. Food Sci. 64, 800–803.

Bazemore, R., Rouseff, R., Naim, M., 2003. Linalool in orange juice: origin and thermal stability. J. Agric. Food Chem. 51, 196–199.

Beisel, C.G., Johnson, V.E., Kustel, P.A., 1964. Personal communication. J. Food Sci. 29, 565.

Belitz, H., Grosch, W., 1999. Food Chemistry. Springer-Verlag.

Belletti, N., Ndagijimana, M., Sisto, C., Guerzoni, M.E., Lanciotti, R., Gardini, F., 2004. Evaluation of the antimicrobial activity of citrus essences on Saccharomyces cerevisiae. J. Agric. Food Chem. 52, 6932–6938.

Berger, R.G., 2007. Flavours and Fragrances: Chemistry, Bioprocessing and Sustainability. Springer Science & Business Media.

Bezman, Y., Rouseff, R.L., Naim, M., 2001. 2-Methyl-3-furanthiol and methional are possible off-flavors in stored orange juice: aroma-similarity, NIF/SNIF GC–O, and GC analyses. J. Agric. Food Chem. 49, 5425–5432.

Buettner, A., Schieberle, P., 1999. Characterization of the most odor-active volatiles in fresh, hand-squeezed juice of grapefruit (*Citrus paradisi* Macfayden). J. Agric. Food Chem. 47, 5189–5193.

Buettner, A., Schieberle, P., 2001a. Application of a Comparative Aroma Extract Dilution Analysis to Monitor Changes in Orange Juice Aroma Compounds During Processing. ACS Publications.

Buettner, A., Schieberle, P., 2001b. Evaluation of aroma differences between hand-squeezed juices from Valencia late and navel oranges by quantitation of key odorants and flavor reconstitution experiments. J. Agric. Food Chem. 49, 2387–2394.

Buettner, A., Schieberle, P., 2001c. Evaluation of key aroma compounds in hand-squeezed grapefruit juice (*Citrus paradisi* Macfayden) by quantitation and flavor reconstitution experiments. J. Agric. Food Chem. 49, 1358–1363.

Cadwallader, K., Xu, Y., 1994. Analysis of volatile components in fresh grapefruit juice by purge and trap/gas chromatography. J. Agric. Food Chem. 42, 782–784.

Cannon, R.J., Ho, C.-T., 2018. Volatile sulfur compounds in tropical fruits. J. Food Drug Anal. 26, 445–468.

Cannon, R.J., Kazimierski, A., Curto, N.L., Li, J., Trinnaman, L., Janczuk, A.J., Agyemang, D., da Costa, N.C., Chen, M.Z., 2015. Identification, synthesis, and characterization of novel sulfur-containing volatile compounds from the in-depth analysis of Lisbon lemon peels (*Citrus limon* L. Burm. f. cv. Lisbon). J. Agric. Food Chem. 63, 1915–1931.

Clery, R.A., Cason, J.R., Zelenay, V., 2016. Constituents of cypriol oil (Cyperus scariosus R. Br.): N-containing molecules and key aroma components. J. Agric. Food Chem. 64, 4566–4573.

Coleman, R.L., Lund, E.D., Shaw, P.E., 1972. Analysis of grapefruit essence and aroma oils. J. Agric. Food Chem. 20, 100–103.

Cotroneo, A., Dugo, G., Licandro, G., Ragonese, C., di Giacomo, G., 1986. On the genuineness of *Citrus* essential oils. Part XII. Characteristics of Sicilian lemon essential oil produced with the FMC extractor. Flav. Frag. J. 1, 125–134.

Cotroneo, A., Verzera, A., Dugo, G., Dugo, G., Licandro, G., 1988. Sulla genuinità delle essenze agrumarie. Nota XXII. La composizione dell'essenza di limone siciliana prodotta industrialmente nell'annata 1984/85. Industria Bevande 17, 209–224.

Demole, E., Enggist, P., Ohloff, G., 1982. 1-p-Menthene-8-thiol: a powerful flavor impact constituent of grapefruit juice (*Citrus parodisi* MACFAYDEN). Helv. Chim. Acta 65, 1785–1794.

Dharmawan, J., Kasapis, S., Sriramula, P., Lear, M.J., Curran, P., 2008. Evaluation of aroma-active compounds in Pontianak orange peel oil (*Citrus nobilis* Lour. Var. microcarpa Hassk.) by gas chromatography—olfactometry, aroma reconstitution, and omission test. J. Agric. Food Chem. 57, 239–244.

Dugo, G., 1986. L'huile essentielle de citron sicilien. Parfums, cosmétiques, arômes, 95–105.

Dugo, G., 1994. The composition of the volatile fraction of the Italian citrus essential oils. Perfum. Flavor 19, 29–51.

Dugo, G., Licandro, G., Cotroneo, A., Dugo, G., 1983. Sulla genuinità delle essenze agrumarie. Nota I. Caratterizzazione di essenze di limone siciliane. Essenze Deriv. Agrum. 53, 173–217.

Dugo, G., Bartle, K.D., Bonaccorsi, I., Catalfamo, M., Cotroneo, A., Dugo, P., Lamonica, G., McNair, H., Mondello, L., Previti, P., 1999. Advanced analytical techniques for the analysis of citrus essential oils. Part 1. Volatile fraction: HRGC/MS analysis. Essenze e derivati agrumari 69, 79–111.

Dugo, G., Cotroneo, A., Verzera, A., Bonaccorsi, I., 2002. Composition of the volatile fraction of cold-pressed citrus peel oils. In: Dugo, G., Di Giacomo, A. (Eds.), *Citrus*. The Genus *Citrus*. CRC Press Taylor & Francis Group, Boca Raton, FL, pp. 284–317.

Dugo, P., Ragonese, C., Russo, M., Sciarrone, D., Santi, L., Cotroneo, A., Mondello, L., 2010. Sicilian lemon oil: composition of volatile and oxygen heterocyclic fractions and enantiomeric distribution of volatile components. J. Sep. Sci. 33, 3374–3385.

Dugo, G., Cotroneo, A., Bonaccorsi, I., Trozzi, A., 2011. Composition of the Volatile Fraction of *Citrus* Peel Oils. Taylor and Francis Group, Boca Raton, FL.

Elmaci, Y., Altug, T., 2005. Flavor characterization of three mandarin cultivars (Satsuma, Bodrum, Clemantine) by using GC/MS and flavor profile analysis techniques. J. Food Qual. 28, 163–170.

Fabroni, S., Ruberto, G., Rapisarda, P., 2012. Essential oil profiles of new *Citrus* hybrids, a tool for genetic citrus improvement. J. Essent. Oil Res. 24, 159–169.

Feng, S., Suh, J.H., Gmitter, F.G., Wang, Y., 2017. Differentiation between the flavors of sweet orange (*Citrus sinensis*) and mandarin (*Citrus reticulata*). J. Agric. Food Chem. 66, 203–211.

Fischer, A., Grab, W., Schieberle, P., 2008. Characterisation of the most odour-active compounds in a peel oil extract from Pontianak oranges (*Citrus nobilis* var. Lour. Microcarpa Hassk.). Eur. Food Res. Technol. 227, 735–744.

Giuffrida, D., Torre, L.L., Manuela, S., Pellicanò, T.M., Dugo, G., 2006. Application of HPLC–APCI–MS with a C-30 reversed phase column for the characterization of carotenoid esters in mandarin essential oil. Flav. Frag. J. 21, 319–323.

Goff, S.A., Klee, H.J., 2006. Plant volatile compounds: sensory cues for health and nutritional value? Science 311, 815–819.

Gök, A., Ismail Kirbaşlar, Ş., Gülay Kirbaşlar, F., 2015. Comparison of lemon oil composition after using different extraction methods. J. Essent. Oil Res. 27, 17–22.

Hinterholzer, A., Schieberle, P., 1998. Identification of the most odour-active volatiles in fresh, hand-extracted juice of Valencia late oranges by odour dilution techniques. Flav. Frag. J. 13, 49–55.

Hotchkiss, J.H., 1988. Food and Packaging Interactions. ACS Publications.

Iwanami, Y., Tateba, H., Kodama, N., Kishino, K., 1997. Changes of lemon flavor components in an aqueous solution during UV irradiation. J. Agric. Food Chem. 45, 463–466.

Jabalpurwala, F., Gurbuz, O., Rouseff, R., 2010. Analysis of grapefruit sulphur volatiles using SPME and pulsed flame photometric detection. Food Chem. 120, 296–303.

Janes, J., Marr, I., Unwin, N., Banthorpe, D., Yusuf, A., 1993. Reaction of monoterpenoids with hydrogen sulphide to form thiols and epi-sulphides of potential organoleptic significance. Flav. Frag. J. 8, 289–294.

Jella, P., Rouseff, R., Goodner, K., Widmer, W., 1998. Determination of key flavor components in methylene chloride extracts from processed grapefruit juice. J. Agric. Food Chem. 46, 242–247.

Karabagias, I., 2017. Volatile compounds of freshly prepared lemon juice from the region of Kalamata. SM Anal. Bioanal. Tech. 2, 1013.

Kealey, K., Kinsella, J., Nagy, S., 1979. Orange juice quality with an emphasis on flavor components. Crit. Rev. Food Sci. Nutr. 11, 1–40.

Komura, H., 2006. Comprehensive two-dimensional gas chromatographic analysis of commercial lemon-flavored beverages. J. Sep. Sci. 29, 2350–2356.

Li, L.J., Hong, P., Chen, F., Sun, H., Yang, Y.F., Yu, X., Huang, G.L., Wu, L.M., Ni, H., 2016. Characterization of the aldehydes and their transformations induced by UV irradiation and air exposure of white Guanxi honey pummelo (*Citrus grandis* (L.) Osbeck) essential oil. J. Agric. Food Chem. 64, 5000–5010.

Lin, J., Jella, P., Rouseff, R., 2002a. Gas chromatography-olfactometry and chemiluminescence characterization of grapefruit juice volatile sulfur compounds. In: Heteroatomic Aroma Compounds. vol. 826, pp. 110–120.

Lin, J., Rouseff, R.L., Barros, S., Naim, M., 2002b. Aroma composition changes in early season grapefruit juice produced from thermal concentration. J. Agric. Food Chem. 50, 813–819.

Liu, C., Cheng, Y., Zhang, H., Deng, X., Chen, F., Xu, J., 2012. Volatile constituents of wild citrus Mangshanyegan (*Citrus nobilis* Lauriro) peel oil. J. Agric. Food Chem. 60, 2617–2628.

Lopes Campêlo, L.M., Moura Gonçalves, F.C., Feitosa, C.M., De Freitas, R.M., 2011. Antioxidant activity of *Citrus* limon essential oil in mouse hippocampus. Pharm. Biol. 49, 709–715.

Lota, M.-L., De Rocca Serra, D., Tomi, F., Jacquemond, C., Casanova, J., 2002. Volatile components of peel and leaf oils of lemon and lime species. J. Agric. Food Chem. 50, 796–805.

Maarse, H., 1991. Volatile Compounds in Foods and Beverages. CRC Press.

Mahattanatawee, K., Rouseff, R., Valim, M.F., Naim, M., 2005. Identification and aroma impact of norisoprenoids in orange juice. J. Agric. Food Chem. 53, 393–397.

Marin, A., Acree, T., Hotchkiss, J., Nagy, S., 1992. Gas chromatography-olfactometry of orange juice to assess the effects of plastic polymers on aroma character. J. Agric. Food Chem. 40, 650–654.

Mitiku, S., Sawamura, M., Itoh, T., Ukeda, H., 2000. Volatile components of peel cold-pressed oils of two cultivars of sweet orange (*Citrus sinensis* (L.) Osbeck) from Ethiopia. Flav. Frag. J. 15, 240–244.

Moshonas, M.G., 1971. Analysis of carbonyl flavor constituents from grapefruit oil. J. Agric. Food Chem. 19, 769–770.

Moshonas, M., Lund, E.D., 1969. Aldehydes, ketones and esters in Valencia orange peel oil. J. Food Sci. 34, 502–503.

Moshonas, M.G., Shaw, P.E., 1971. Analysis of volatile flavor constituents from grapefruit essence. J. Agric. Food Chem. 19, 119–120.

Moshonas, M.G., Shaw, P.E., 1972. Analysis of flavor constituents from lemon and lime essence. J. Agric. Food Chem. 20, 1029–1030.

Moshonas, M.G., Shaw, P.E., 1994. Quantitative determination of 46 volatile constituents in fresh, unpasteurized orange juices using dynamic headspace gas chromatography. J. Agric. Food Chem. 42, 1525–1528.

Mussinan, C., Mookherjee, B., Malcolm, G., 1979. Isolation and identification of the volatile constituents of fresh lemon juice. In: Abstracts of Papers of the American Chemical Society, Amer Chemical Soc 1155 16th ST, NW, Washington, DC, pp. 54.

Nakanishi, A., Fukushima, Y., Miyazawa, N., Yoshikawa, K., Maeda, T., Kurobayashi, Y., 2017a. Identification of Rotundone as a potent odor-active compound of several kinds of fruits. J. Agric. Food Chem. 65, 4464–4471.

Nakanishi, A., Fukushima, Y., Miyazawa, N., Yoshikawa, K., Maeda, T., Kurobayashi, Y., 2017b. Quantitation of Rotundone in grapefruit (*Citrus paradisi*) peel and juice by stable isotope dilution assay. J. Agric. Food Chem. 65, 5026–5033.

Nguyen, H., Campi, E.M., Jackson, W.R., Patti, A.F., 2009. Effect of oxidative deterioration on flavour and aroma components of lemon oil. Food Chem. 112, 388–393.

Nisperos-Carriedo, M.O., Shaw, P.E., 1990. Comparison of volatile flavor components in fresh and processed orange juices. J. Agric. Food Chem. 38, 1048–1052.

Peterson, D., Reineccius, G.A., 2002. Biological Pathways for the Formation of Oxygen-Containing Aroma Compounds. ACS Publications.

Plotto, A., Margaría, C.A., Goodner, K.L., Goodrich, R., Baldwin, E.A., 2004. Odour and flavour thresholds for key aroma components in an orange juice matrix: terpenes and aldehydes. Flav. Frag. J. 19, 491–498.

Plotto, A., Margaría, C.A., Goodner, K.L., Baldwin, E.A., 2008. Odour and flavour thresholds for key aroma components in an orange juice matrix: Esters and miscellaneous compounds. Flav. Frag. J. 23, 398–406.

Rega, B., Fournier, N., Guichard, E., 2003. Solid phase microextraction (SPME) of orange juice flavor: odor representativeness by direct gas chromatography olfactometry (D-GC-O). J. Agric. Food Chem. 51, 7092–7099.

Risch, S.J., Ho, C.-T., 2000. Flavor Chemistry: Industrial and Academic Research. ACS Publications.

Roland, A., Viel, C., Reillon, F., Delpech, S., Boivin, P., Schneider, R., Dagan, L., 2016. First identification and quantification of glutathionylated and cysteinylated precursors of 3-mercaptohexan-1-ol and 4-methyl-4-mercaptopentan-2-one in hops (*Humulus lupulus*). Flav. Frag. J. 31, 455–463.

Rouseff, R., Bazemore, R., Goodner, K., Naim, M., 2001. GC-olfactometry with solid phase microextraction of aroma volatiles from heated and unheated orange juice. Adv. Exp. Med. Biol. 488, 101–112.

Ruberto, G., Biondi, D., Piattelli, M., Rapisarda, P., Starrantino, A., 1994. Essential oil of the new citrus hybrid, *Citrus clementina* x *C. limon*. J. Essent. Oil Res. 6, 1–8.

Russo, M., Bonaccorsi, I., Costa, R., Trozzi, A., Dugo, P., Mondello, L., 2015. Reduced time HPLC analyses for fast quality control of citrus essential oils. J. Essent. Oil Res. 1, 1–9.

Sangwan, N.S., Sharma, P.K., Sangwan, R.S., 2007. Geranyl acetate esterase is commonly present but linalyl acetate esterase occurrence is highly limited in plants. Flav. Frag. J. 22, 173–177.

Saura, D., Marti, N., Valero, M., González, E., Carbonell, A., Laencina, J., 2012. Separation of aromatics compounds during the clarification of lemon juice by cross-flow filtration. Ind. Crop. Prod. 36, 543–548.

Sawamura, M., Poiana, M., Kawamura, A., Itoh, T., Song, H., Ukeda, H., Mincione, B., 1999. Volatile components of peel oils of Italian and Japanese lemon and bergamot Ital. J. Food Sci. 11, 121–130.

Schieberle, P., Grosh, W., 1988. Identification of potent flavor compounds formed in an aqueous lemon oil/citric acid emulsion. J. Agric. Food Chem. 36, 797–800.

Schulz, S., Girhard, M., Gassmeye, S.K., Jäger, V.D., Schwarze, D., Vogel, A., Urlacher, V.B., 2015. Selective enzymatic synthesis of the grapefruit flavor (+)-Nootkatone. ChemCatChem 7, 601–604.

Schwab, W., Davidovich-Rikanati, R., Lewinsohn, E., 2008. Biosynthesis of plant-derived flavor compounds. Plant J. 54, 712–732.

Shaw, P.E., 1977. Essential oils. In: Nagy, S., Shaw, P.E., Veldehuis, M.K. (Eds.), *Citrus* Science and Technology, vol. I. The Avi Publishing Company, Westport, CT, pp. 427–462.

Shaw, P.E., 1979. Review of quantitative analyses of citrus essential oils. J. Agric. Food Chem. 27, 246–257.

Shaw, P.E., Wilson III, C.W., 1981. Importance of nootkatone to the aroma of grapefruit oil and the flavor of grapefruit juice. J. Agric. Food Chem. 29, 677–679.

Shaw, P.E., Wilson III, C.W., 1982. Volatile sulfides in headspace gases of fresh and processed citrus juices. J. Agric. Food Chem. 30, 685–688.

Shaw, P.E., Ammons, J.M., Braman, R.S., 1980. Volatile sulfur compounds in fresh orange and grapefruit juices: Identification, quantitation, and possible importance to juice flavor. J. Agric. Food Chem. 28, 778–781.

Steven, N., Shaw, P.E., Vldhuis, M.K., 1977. *Citrus* Science and Technology, Avi Publishing Company.

Stevens, K., Guadagni, D., Stern, D., 1970. Odour character and threshold values of nootkatone and related compounds. J. Sci. Food Agric. 21, 590–593.

Sun, H., Ni, H., Yang, Y., Wu, L., Cai, H.-N., Xiao, A.-F., Chen, F., 2014. Investigation of sunlight-induced deterioration of aroma of pummelo (*Citrus maxima*) essential oil. J. Agric. Food Chem. 62, 11818–11830.

Takeoka, G.R., Buttery, R.G., Turnbaugh, J.G., Benson, M., 1995. Odor thresholds of various branched esters. LWT- Food Sci. Technol. 28, 153–156.

Tietel, Z., Fallik, E., Porat, R., Lewinsohn, E., Bar, E., 2009. Sensory and aroma profiling of fresh and stored 'Or' mandarins. In: II International Symposium on Citrus Biotechnology, 892, pp. 373–382.

Tietel, Z., Bar, E., Lewinsohn, E., Feldmesser, E., Fallik, E., Porat, R., 2010. Effects of wax coatings and postharvest storage on sensory quality and aroma volatile composition of 'Mor'mandarins. J. Sci. Food Agric. 90, 995–1007.

Tietel, Z., Plotto, A., Fallik, E., Lewinsohn, E., Porat, R., 2011. Taste and aroma of fresh and stored mandarins. J. Sci. Food Agric. 91, 14–23.

Tirado, C.B., Stashenko, E., Combariza, M., Martinez, J., 1995. Comparative study of Colombian citrus oils by high-resolution gas chromatography and gas chromatography-mass spectrometry. J. Chromatogr. A 697, 501–513.

Tønder, D., Petersen, M.A., Poll, L., Olsen, C.E., 1998. Discrimination between freshly made and stored reconstituted orange juice using GC odour profiling and aroma values. Food Chem. 61, 223–229.

Towers, G.N., Stafford, H.A., 2012. Biochemistry of the Mevalonic Acid Pathway to Terpenoids. Springer Science & Business Media.

Trozzi, A., Verzera, A., Del Duce, R., Cotroneo, A., 1993. Sulla genuinita'delle essenze agrumarie. Nota XLIV. Aggiornamento sulla composizione della frazione volatile dell'olio essenziale di limone. Essenze e derivati agrumari 63, 375–394.

Verzera, A., Trozzi, A., Licandro, G., Scuderi, L., 1996. Sulla genuinità delle essenze agrumarie. Nota LVI. La composizione dell'olio essenziale di limone FMC prodotto nella stagione 1995. Essenz. Deriv. Agrum 66, 353–367.

United States Department of Agriculture, 2018. Foreign Agricultural Service, *Citrus*: World Markets and Trade. Available at: http://apps.fas.usda.gov/psdonline/circulars/citrus.pdf.

Verzera, A., Dugo, P., Mondello, L., Trozzi, A., Cotroneo, A., 1999. Extraction technology and lemon oil composition. Ital. J. Food Sci. 11, 361–370.

Wakabayashi, H., Wakabayashi, M., Eisenreich, W., Engel, K.-H., 2004. Stereochemical course of the generation of 3-mercaptohexanal and 3-mercaptohexanol by β-lyase-catalyzed cleavage of cysteine conjugates. J. Agric. Food Chem. 52, 110–116.

Walsh, M., Rouseff, R., Naim, M., 1997. Determination of furaneol and p-vinylguaiacol in orange juice employing differential UV wavelength and fluorescence detection with a unified solid phase extraction. J. Agric. Food Chem. 45, 1320–1324.

Wilson III, C.W., Shaw, P.E., 1980. Glass capillary gas chromatography for quantitative determination of volatile constituents in cold-pressed grapefruit oil. J. Agric. Food Chem. 28, 919–922.

Wood, C., Siebert, T.E., Parker, M., Capone, D.L., Elsey, G.M., Pollnitz, A.P., Eggers, M., Meier, M., Vössing, T., Widder, S., 2008. From wine to pepper: rotundone, an obscure sesquiterpene, is a potent spicy aroma compound. J. Agric. Food Chem. 56, 3738–3744.

Xie, J., Deng, L., Zhou, Y., Yao, S., Zeng, K., 2018. Analysis of changes in volatile constituents and expression of genes involved in terpenoid metabolism in oleocellosis peel. Food Chem. 243, 269–276.

Yu, Q., Xie, B.-J., Zhang, Y., Zhou, H.-Y., Pan, S.-Y., 2007. Study on aroma components in fruit from three different Satsuma mandarin varieties. Agric. Sci. China 6, 1487–1493.

Zhang, J.-X., Guan, S.-H., Yang, M., Feng, R.-H., Wang, Y., Zhang, Y.-B., Chen, X., Chen, X.-H., Bi, K.-S., Guo, D.-A., 2013. Simultaneous determination of 24 constituents in cortex Lycii using high-performance liquid chromatography–triple quadrupole mass spectrometry. J. Pharm. Biomed. Anal. 77, 63–70.

Zhong, S., Ren, J., Chen, D., Pan, S., Wang, K., Yang, S., Fan, G., 2014. Free and bound volatile compounds in juice and peel of eureka lemon. Food Sci. Technol. Res. 20, 167–174.

Further reading

Bonamonte, D., Foti, C., Lionetti, N., Rigano, L., Angelini, G., 2010. Photoallergic contact dermatitis to 8-methoxypsoralen in Ficus carica. Contact Dermatitis 62, 343–348.

Dugo, P., Russo, M., 2011. The oxygen heterocyclic components of *Citrus* essential oils. In: *Citrus* Oils: Composition, Advanced Analytical Techniques, Contaminants, Biological Activity, pp. 405–443.

Dugo, P., Mondello, L., Dugo, L., Stancanelli, R., Dugo, G., 2000. LC-MS for the identification of oxygen heterocyclic compounds in citrus essential oils. J. Pharm. Biomed. Anal. 24, 147–154.

Frérot, E., Decorzant, E., 2004. Quantification of total furocoumarins in citrus oils by HPLC coupled with UV, fluorescence, and mass detection. J. Agric. Food Chem. 52, 6879–6886.

Masson, J., Liberto, E., Beolor, J.-C., Brevard, H., Bicchi, C., Rubiolo, P., 2016. Oxygenated heterocyclic compounds to differentiate *Citrus* spp. essential oils through metabolomic strategies. Food Chem. 206, 223–233.

Russo, M., Torre, G., Carnovale, C., Bonaccorsi, I., Mondello, L., Dugo, P., 2012. A new HPLC method developed for the analysis of oxygen heterocyclic compounds in *Citrus* essential oils. J. Essent. Oil Res. 24, 119–129.

Steck, W., El-Dakhakhny, M., Brown, S.A., 1969. The Role of Marmesin and Columbianetin in the Biosyntheses of Furanocoumarins. National Research Council of Canada Saskatoon (Saskatchewan) Prairie Regional Lab.

Weber, B., Hartmann, B., Stöckigt, D., Schreiber, K., Roloff, M., Bertram, H.-J., Schmidt, C.O., 2006. Liquid chromatography/mass spectrometry and liquid chromatography/nuclear magnetic resonance as complementary analytical techniques for unambiguous identification of polymethoxylated flavones in residues from molecular distillation of orange peel oils (*Citrus sinensis*). J. Agric. Food Chem. 54, 274–278.

Chapter 23

Global economics and marketing of citrus products

Thomas H. Spreen[a,b], Zhifeng Gao[a], Waldir Fernandes, Jr.[c,d], Marisa L. Zansler[b]

[a]*Food and Resource Economics Department, University of Florida, Gainesville, FL, United States,* [b]*Economic and Marketing Research Department, Florida Department of Citrus, Gainesville, FL, United States,* [c]*Economics, Business Administration and Education Department, São Paulo State University, Jaboticabal, Brazil,* [d]*College of Technology, State Center of Technological Education, São Jose do Rio Preto, Brazil*

23.1 Introduction

Citrus is grown throughout the tropical and subtropical regions of the world. It is marketed in a variety of outlets, but an important distinction is the utilization of citrus in fresh versus processed form. Processed citrus products include juice, both single-strength and concentrated, and canned citrus products. Processed citrus also yields marketable citrus by-products, including citrus pulp and peel (generally transformed into animal feed), molasses, citrus oils, D-limonene, and essences.

In this chapter, the discussion is organized around the four major varieties of commercial citrus products: sweet oranges, mandarins, grapefruit (including pummelos), and lemons (including limes). The categories are fresh utilization, processed utilization (juice and canned sections), and by-products from juice processing.

The largest citrus producers include (in order) China, Brazil, and the United States. The three largest producing countries have citrus industries that vary greatly. In China, production mostly occurs on small, mixed crop farms and nearly all production is marketed domestically. Brazil is characterized by very large farms, and a large proportion of its production is sent for processing into juice. Nearly all Brazilian produced orange juice is exported to Western Europe and North America. There are small and medium-sized growers in Brazil producing fresh fruit for the domestic market. In both China and Brazil, the domestic market is informal, with little grading of fruit, little or no government inspection and oversight, and a little collection of data on volume and prices.[a] The United States has a range of small, medium, and large farms. The most important citrus-producing states (in order) are Florida, California, Texas, and Arizona. Florida orange production is mainly sent for processing into juice, while the other states focus on production for the fresh market. There is a great deal of government oversight in terms of fruit grading, market standards, and data collection compared to China and Brazil.

23.2 Sweet oranges

A list of the leading orange producing countries is shown in Table 23.1 for the period between 2000 and 2016, the most recent year with published data. Among the three largest producing countries, Brazil and the United States have experienced substantial production declines in recent years; the US production declined by nearly 50% over the sample period. Huanglongbing (HLB and also known as citrus greening) is the main contributor to the production decline.

Allocation of sweet orange production between fresh and processed utilization by major producing countries is shown in Table 23.2, based on the most recent data available. Brazil and the United States are the two primary suppliers of orange juice to the world market, although, in recent years, Mexico's share of world supply has grown with the decline in the US production. Other suppliers of orange juice include Costa Rica, Belize, China, Australia, South Africa, Spain, and Italy. Of these smaller producing countries, Costa Rica, Belize, and South Africa export all or nearly all of their production while the others supply their respective domestic markets.[b]

a. However, these observations are evolving in Brazil as industry-based organizations have recently released an orange tree census and production estimates; and likewise, in China very recently there has been a substantial increase in e-commerce, bringing with it much higher quality expectations among consumers.

b. In the case of Spain and Italy, "domestic market" is defined as the EU market.

The Genus Citrus. https://doi.org/10.1016/B978-0-12-812163-4.00023-1

TABLE 23.1 Sweet orange production by major growing countries, selected years, 1999–2000 through 2015–16.

	Million MT					
	1999–2000	*2004–05*	*2009–10*	*2011–12*	*2013–14*	*2015–16*
Brazil	16,524	16,606	15,830	20,482	16,850	14,350
China	2,710	4,200	6,500	6,600	7,600	7,000
United States	11,894	8,293	7,478	8,166	6,153	5,371
Mexico	2,881	4,120	4,051	3,666	4,400	3,535
EU-27[a]	n/a	n/a	6,244	6,023	6,712	6,055
Spain[a]	2,710	2,700	n/a	n/a	n/a	n/a
Italy[a]	1,900	1,007	n/a	n/a	n/a	n/a
Egypt	1,730	1,750	2,401	2,350	2,570	2,750
Turkey	1,050	1,280	1,690	1,650	1,700	1,700
South Africa	1,047	1,120	1,459	1,466	1,620	1,560
Greece[a]	950	820	n/a	n/a	n/a	n/a
World total	43,196	45,434	49,151	53,830	51,008	45,763

[a]*Production data for countries of the European Union were aggregated beginning in 2009–10.*
https://www.fas.usda.gov/databases/production-supply-and-distribution-online-psd

TABLE 23.2 Allocation of sweet orange production between fresh and processed utilization by major growing countries, selected years, 2009–10 through 2015–16.

	Thousand MT				
	2007–08	*2009–10*	*2011–12*	*2013–14*	*2015–16*
Fresh utilization[a]					
Brazil	5,059	5,035	7,262	6,038	5,304
China	5,205	6,298	6,380	6,885	6,300
United States	1,942	1,924	2,102	1,720	1,791
Mexico	3,297	3,171	2,836	3,333	2,800
EU-27	4,974	5,030	4,967	5,076	4,955
Egypt	2,709	2,331	2,350	2,485	2,830
Turkey	1,327	1,590	1,550	1,600	1,700
South Africa	1,114	1,179	1,217	1,252	1,133
Argentina	702	686	461	600	530
Costa Rica	104	135	95	84	105
Processed utilization					
Brazil	11,791	10,795	13,220	11,832	9,058
China	245	202	520	715	600
United States	7,199	5,554	6,064	4,420	3,732
Mexico	1,000	880	830	1,200	1,600

TABLE 23.2 Allocation of sweet orange production between fresh and processed utilization by major growing countries, selected years, 2009–10 through 2015–16—cont'd

	Thousand MT				
	2007–08	*2009–10*	*2011–12*	*2013–14*	*2015–16*
EU-27	1,518	1,214	1,056	1,474	1,286
Egypt	50	70		85	100
Turkey	100	100	100	100	100
South Africa	412	280	249	471	142
Argentina	238	84	104	200	270
Costa Rica	165	235	275	136	230
Total production					
Brazil	16,850	15,830	20,482	17,870	14,362
China	5,450	6,500	6,900	7,600	6,900
United States	9,141	7,478	8,166	6,140	5,523
Mexico	4,297	4,051	3,666	4,533	4,400
EU-27	6,492	6,244	6,023	6,550	6,241
Egypt	2,759	2,401	2,350	2,570	2,930
Turkey	1,427	1,690	1,650	1,700	1,800
South Africa	1,526	1,459	1,466	1,723	1,275
Argentina	940	770	565	800	800
Costa Rica	269	370	370	220	335

[a]*Includes both domestic consumption and net exports.*
Source: FAS, USDA.

The market for fresh sweet oranges differs from processed oranges in two distinct ways. First, there are close to 50 countries with commercial fresh market orange production, and production is found on every continent except Antarctica. Supply is much more dispersed across countries with China, Spain, Mexico, and the United States being the largest supplying countries. Second, the proportion of fresh sweet oranges marketed in domestic markets around the world is much higher compared to orange juice. China, Mexico, and the United States are all examples in which a large proportion of fresh produce is sold in the domestic market. Spain ships to other countries in the European Union. The largest exporter of fresh produce, after Spain, is South Africa, which is located in the southern hemisphere and hence does not compete directly with northern hemisphere exporters such as Spain (and to a lesser extent, the United States).

The processed orange market (primarily orange juice) was dominated for years by Florida. The development of frozen concentrated orange juice (FCOJ) occurred soon after the end of World War II. With its ease of shipment and storage, FCOJ spurred a rapid increase in orange juice consumption in the United States during the 1950s and 1960s. Nearly all of this supply originated in Florida. A freeze that struck Florida in 1962 resulted in a reduced crop for that season and future years.

A major Florida processor formed a partnership with a Brazilian entity and FCOJ production was initiated in the state of São Paulo, Brazil. Throughout the remainder of the 1960s and the 1970s, however, Florida remained the primary supplier of FCOJ to the United States and Canada.

In the 1980s, a series of freezes visited Florida with the two most significant events occurring in December 1983 and January 1985. Millions of trees were destroyed, and Florida production was severely impacted. As the industry started to recover, another major freeze struck in December 1989. In the 1989–90 season, Florida orange production was approximately one-half the level realized before the freezes occurred. These events provided a major impetus for the processed orange sector in Brazil to rapidly expand production. Brazil soon became the leading producer and supplier of orange juice to the world market.

In the 1990s, two significant innovations took place, one in São Paulo, Brazil and a second in Florida that changed the world market for orange juice. The first was the development of bulk shipment of FCOJ in tanker ships. Before this innovation, FCOJ was put into lined 55-gal drums before shipment. Drums are bulky and require labor for the loading, off-loading, and eventual transfer of FCOJ to the final destination. Engineers in Brazil developed a system of pumping FCOJ into tanker trucks. FCOJ was hauled to the large port of Santos located near the city of São Paulo. It was pumped from the tanker truck directly to large ships with holds specifically designed for FCOJ. FCOJ at 65°Brix has a relatively high freezing point so that the product could remain frozen for shipment to the United States. Later, Brazilian companies found markets in Europe for orange juice, and bulk shipment helped spur the expansion of the European market by greatly lowering the cost of transport.

The second innovation was the development of aseptic storage of orange juice in the United States. Aseptic storage was pioneered by Tropicana, a large orange juice marketing company based in Florida,[c] and it allows the marketing of not-from-concentrate (NFC) orange juice. NFC orange juice consumption grew rapidly in the United States and Canada in the 1990s and the first decade of the 21st century. It now accounts for >50% of the US orange juice consumption. Recently, the US brands such as Tropicana and Minute Maid have begun a push to market NFC orange juice in Europe, supplied from Brazil, shipped in bulk tanker ships with aseptic storage.

23.2.1 Sweet orange industry organization in Brazil

Brazil is the world's largest sweet orange producer, accounting for about one-quarter of global production. Although it has a conspicuous domestic market of over 200 million people for fresh fruit, processed utilization represents >60% of the total produced in the Brazilian territory.

The processed orange industry of the state of São Paulo dominates the world market for orange juice. Brazil itself produces virtually 50% of the orange juice consumed worldwide. In addition, São Paulo-based processors Citrosuco and Cutrale have processing assets also located in the state of Florida, the Unites States.

In São Paulo, despite the diseases citrus variegated chlorosis (CVC) and more recently, HLB, the combination of abundant land, good soils, and sufficient rainfall to allow reasonable production, little threat of freezes, and lower production and harvesting costs, has allowed the Brazilian citrus industry to become the leader in world sweet orange production.

Even suffering threats from a phytosanitary standpoint, such as: the citrus tristeza virus (CTV) in the 1950s; an outbreak of citrus canker in the 1970s; the beginning of CVC in 1987; and the first case of HLB in 2004, São Paulo has adapted to the new scenarios, relying on its ability to switch to more efficient plant control systems and migrate to new production areas. The 2017–18 sweet orange crop is expected to be 15.3 MMT compared to the record crop of 17.3 MMT in 1996–97.

Despite the entry of Citrovita in 1989 and Cambuhy in 1992, a robust process of concentration and consolidation in the processed orange industry of São Paulo occurred in the 1990s. After that decade, more mergers and acquisitions took place, resulting in a degree of concentration in which, by 2017, three companies [Citrosuco, Cutrale, and Louis Dreyfus (a French company)] are responsible for about 85% of total Brazilian orange juice exports.

Processors in São Paulo have developed a highly efficient system of fruit procurement, processing, and logistics, chiefly in bulk transportation of orange juice through trucks and vessels as well as ownership or partnership of port terminals. Therefore, the citrus industry in São Paulo is geared to exporting orange juice. Strategic alliances have been formed between citrus processors and distributors of dairy products, however, especially for NFC distribution in South America, since the demand for NFC products in Brazil has been increasing. With that, small processors started their operations with local brands in the 2010s, whose consumption of pasteurized orange juice in Brazil has increased significantly, utilizing approximately 10 million boxes (408,000 MT), still a small portion of the crop. Unlike Florida, the processed orange industry in São Paulo, however, is not vertically integrated into retail distribution. On the other hand, processors in São Paulo are considerably vertically integrated into fruit production.

São Paulo citrus growers have two major organizations which work for their benefit. Associtrus (The Brazilian Association of Citrus Growers), founded in 1974, which is the only citrus growers' association recognized by CitrusBR (founded in June 2009 by the largest Brazilian producers and exporters of citrus juices and derivatives: Cutrale, Citrosuco, and Louis Dreyfus), although it has few members and lacks representativeness. Fundecitrus (Fund for Citrus Plant Protection), founded in 1977, is an interprofessional organization jointly funded by processors and growers. Processors and

c. Tropicana was an independent company at the time it introduced NFC orange juice to the US market. It is now owned by PepsiCo.

growers pay a per-box rate based on quantities negotiated between them. Fundecitrus is responsible for coordinating the monitoring, eradicating and avoiding the spreading of pests and diseases, such as citrus canker, CVC, and HLB in the citrus area as well as leading the research program on CVC, blight, the sudden death of citrus disease, and more recently, HLB.

Until 2015, the USDA's Foreign Agriculture Service (FAS) was the primary source of Brazil's citrus production data with semiannual publications available in June and December of each year. In 2015, the São Paulo citrus industry empowered Fundecitrus with gathering and publishing a commercial tree census, disaggregated by variety and location using a methodology similar to information published by the National Agricultural Statistics Service (NASS) of the USDA for Florida citrus. Using the tree census data, in May of 2016, an orange production forecast was released by Fundecitrus for the states of São Paulo and lower Minas Gerais. Throughout the 2016–17 season, the forecast was updated several times. This process was repeated in the 2017–18 season.

23.2.2 Sweet orange industry organization in Florida

The organization of the citrus industry in Florida shares similarities and differences compared to São Paulo. The Florida Department of Citrus (FDOC), established in 1935, is an agency of the state of Florida whose prime objective is to promote the state's citrus industry. Operating as a marketing order, The Florida Citrus Commission (FCC), whose members are appointed by the governor of Florida, is tasked with overseeing and guiding the activities of the FDOC, conducting a variety of industry programs, and regulating the quality standards of citrus grown in Florida. The FDOC proposes the budget to fund programs and research each season, and the FCC has the authority to impose a per-box assessment on certified fresh and processed citrus each season based on the volume of citrus produced by the industry. Revenues from the assessments are used to promote the consumption of Florida citrus products, support economic and marketing research, and regulating quality standards for the industry.

Both the Agricultural Research Service (ARS) of the USDA and the University of Florida conduct research programs related to a wide array of issues related to citrus. These include research on new scions and rootstocks, plant protection, irrigation management, production, and development of new technology. The state of Florida has established the Citrus Research and Development Foundation (CRDF) to administer programs that deal with HLB and other diseases afflicting Florida citrus. Outside Florida, other US universities including Texas A & M University and the University of California—Riverside also conduct research related to citrus production. The California Citrus Research Board provides research money to researchers that focus on issues that affect citrus production in California. In Florida, Florida Citrus Mutual is the largest and most influential grower organization responsible for lobbying activities at both the state and federal level to support citrus growers. There are several other substate organizations including the Indian River Citrus League, Gulf Citrus Growers, and Peace River Valley Citrus Growers that work with Florida Citrus Mutual and the FDOC.

The Florida Citrus Processors Association (FCPA) is the association that represents citrus processors in the state. With the influx of foreign-based companies, this organization has undergone considerable change. Consolidation in the Florida processing industry has also reduced its membership; six large processing companies remain in Florida and two are owned by Brazilian companies. FCPA used to be the entity that collected price information on prices paid by processors for fruit delivered to processing plants. The FDOC now collects and publishes this data.

NASS, USDA located in central Florida publishes data on tree number, fruit production, utilization (fresh vs. processed), and farm-level prices (known as "on-tree" prices) on an annual basis. The first crop forecast released in early October is a much-anticipated event. The crop forecast is updated monthly throughout the harvest season with the final figure published in July of the following year.

23.2.3 Trade of fresh sweet oranges

Fresh orange exports by major exporting countries are shown in Table 23.3. The three largest exporting countries are Egypt, South Africa, and the United States. Published data on world exports of fresh oranges can be misleading because it includes intra-EU trade, and fails to accurately depict the role of the EU as an exporter. Noticeably absent from this list is Brazil and Mexico, both major orange producers. In the case of Brazil, their lack of participation in the export market stems from both high utilization of fruit from São Paulo state by the processing sector and high domestic demand for fresh fruit grown in other states. A similar observation is appropriate for Mexico with a domestic fresh market serving as the main user of fresh oranges (processed utilization is approximately 20% of the crop).

Fresh orange imports by major importing countries are shown in Table 23.4. The five leading importing regions are the EU, Russia, Saudi Arabia, Hong Kong, and China. Again, the EU figure is affected by intra-EU trade, but since EU imports

TABLE 23.3 Fresh orange exports by major importing countries, 2006–07 through 2016–17 seasons (MT).

	Seasons										
Country	*2006/2007*	*2007/2008*	*2008/2009*	*2009/2010*	*2010/2011*	*2011/2012*	*2012/2013*	*2013/2014*	*2014/2015*	*2015/2016*	*2016/2017*
World	3,168,000	3,666,000	3,505,000	3,769,000	3,996,000	3,932,000	3,889,000	4,002,000	4,059,000	4,468,000	4,629,000
Egypt	620,000	850,000	774,000	850,000	1,000,000	900,000	1,000,000	1,100,000	1,200,000	1,464,000	1,520,000
South Africa	934,000	971,000	869,000	1,045,000	942,000	1,088,000	1,162,000	1,144,000	1,160,000	1,064,000	1,120,000
United States	350,000	613,000	493,000	670,000	750,000	695,000	678,000	506,000	522,000	655,000	635,000
Turkey	179,000	155,000	256,000	209,000	339,000	357,000	244,000	349,000	305,000	371,000	395,000
EU	260,000	242,000	236,000	272,000	318,000	279,000	322,000	346,000	297,000	319,000	300,000
Australia	125,000	108,000	134,000	88,000	114,000	133,000	127,000	126,000	156,000	161,000	180,000
Hong Kong	26,000	45,000	53,000	62,000	70,000	67,000	45,000	49,000	74,000	107,000	140,000
Morocco	257,000	296,000	305,000	161,000	175,000	138,000	82,000	111,000	130,000	89,000	120,000
China	79,000	124,000	155,000	158,000	92,000	129,000	83,000	108,000	53,000	74,000	62,000

https://www.fas.usda.gov/databases/production-supply-and-distribution-online-psd

TABLE 23.4 Fresh orange imports by major importing countries, 2006–07 through 2016–17 seasons (MT).

	Seasons										
Country	*2006/2007*	*2007/2008*	*2008/2009*	*2009/2010*	*2010/2011*	*2011/2012*	*2012/2013*	*2013/2014*	*2014/2015*	*2015/2016*	*2016/2017*
World	3,194,000	3,360,000	3,109,000	3,414,000	3,495,000	3,668,000	3,628,000	3,423,000	3,708,000	4,006,000	4,115,000
EU	1,042,000	1,040,000	846,000	959,000	800,000	848,000	883,000	819,000	927,000	972,000	980,000
Russia	500,000	517,000	436,000	478,000	573,000	495,000	512,000	469,000	440,000	473,000	440,000
Saudi Arabia	270,000	280,000	270,000	302,000	312,000	348,000	274,000	274,000	448,000	435,000	435,000
Hong Kong	144,000	177,000	176,000	193,000	200,000	188,000	217,000	230,000	256,000	286,000	320,000
China	36,000	62,000	66,000	80,000	99,000	98,000	88,000	88,000	146,000	220,000	300,000
United Arab Emirates	85,000	106,000	172,000	182,000	167,000	196,000	201,000	220,000	230,000	222,000	220,000
Canada	170,000	214,000	177,000	204,000	211,000	190,000	199,000	183,000	190,000	204,000	195,000
Iraq	10,000	28,000	66,000	70,000	124,000	196,000	169,000	189,000	180,000	189,000	190,000
United States	116,000	77,000	90,000	106,000	102,000	119,000	139,000	143,000	155,000	164,000	185,000
Korea, South	78,000	112,000	71,000	108,000	142,000	173,000	152,000	100,000	111,000	154,000	145,000
Malaysia	91,000	86,000	90,000	83,000	88,000	96,000	104,000	100,000	102,000	101,000	100,000
Japan	90,000	99,000	96,000	104,000	120,000	127,000	113,000	87,000	83,000	100,000	95,000

https://www.fas.usda.gov/databases/production-supply-and-distribution-online-psd

exceed EU exports, the EU is a net importer of fresh oranges. The portion of the fresh oranges imported into Hong Kong is likely destined for consumption in south China markets.

23.3 Grapefruit and pummelos

Although grapefruit (both red and white fleshed) and pummelos are not close substitutes, most international organizations [such as the Food and Agricultural Organization (FAO) of the United Nations] aggregate production statistics for these two scion types together. Grapefruit originated in the Caribbean. It is grown for western markets such as North America and Western Europe, although Japan is a high-valued market for fresh grapefruit. Pummelos originated in China and are consumed primarily in East Asian countries such as China, Thailand, and Vietnam. With immigration from these countries to the West, pummelo consumption in the United States and Western Europe has been increasing.

Grapefruit and pummelo production among major producing countries since 2000 is shown in Table 23.5. Note the steep decline in grapefruit production in the United States over the past 10 years. This is a result of several factors. First, CTV found its way to Florida and since a large proportion of the grapefruit planted in Florida was on sour orange rootstock, which is highly susceptible to the CTV, tree numbers began to decline. Second, once citrus canker was found in Florida in the 1990s, an eradication program was initiated by the State of Florida in conjunction with the Animal and Plant Health Inspection Service (APHIS) of the USDA in an attempt to rid the state of the canker disease. Third, four hurricanes visited Florida in 2004 and 2005. Three of the storms passed through the main grapefruit production area along the east coast of Florida, resulting in tree damage. But more importantly, the storms spread citrus canker throughout the state forcing abandonment of the tree eradication program, Fourth, HLB was first identified in Florida in 2005. Grapefruit is one of the most susceptible of citrus scions to HLB. Consequently, grapefruit productivity has plummeted and very few new plantings of grapefruit are occurring.

A similar story can be told regarding plummeting grapefruit production in Cuba, once the third largest producer behind Florida and Israel. The combination of multiple hurricanes and the presence of HLB have drastically reduced grapefruit production in Cuba (Table 23.5).

Another factor negatively affecting grapefruit is an apparent downward shift in demand, especially by consumers in developed economies such as the United States and Western Europe. Several factors can be identified which have negatively affected grapefruit demand. Among these is the presence in grapefruit of a family of chemicals called furanocoumarins. These compounds interact with statins such as simvastatin (Zocor) and atorvastatin (Lipitor) by increasing the quantity of statins taken up by the bloodstream, creating a potential for negative health consequences; this grapefruit-drug interaction is sometimes referred to as the Grapefruit Juice Effect (GJE). As both fresh and processed grapefruit consumption is concentrated among the older age cohorts of the population, the discovery of the GJE has had a chilling effect on

TABLE 23.5 Production of pummelos and grapefruit by major growing countries, selected years, 1999–2000 through 2015–16.

	Million MT					
	1999–2000	*2004–05*	*2009–10*	*2011–12*	*2013–14*	*2015–16*
China	n/a	1,724	2,900	3,200	3,717	4,300
United States	2,530	914	1,123	1,047	955	735
Mexico	165	310	401	415	423	430
South Africa	200	270	343	305	390	330
Israel	365	247	235	245	236	185
Turkey	120	110	191	230	235	200
Argentina	n/a	170	150	160	n/a	n/a
EU-27	n/a	n/a	95	102	113	97
World total	3,662	3,795	5,288	5,544	6,069	6,277

https://www.fas.usda.gov/databases/production-supply-and-distribution-online-psd

grapefruit demand. Another factor is that fresh grapefruit is a relatively large fruit, which must be either peeled or sectioned requiring preparation time. Increased availability of other fruits such as bananas, berries, and imported stone fruits has provided convenient alternatives for fresh grapefruit during the winter season in the Northern Hemisphere.

Pummelos have experienced a different supply/demand path over recent years compared to grapefruit. Pummelo production has remained relatively stable as Asian growers have developed techniques to manage HLB which is endemic throughout the region. As immigrants from China and other East Asian countries have moved to the United States and Western Europe, they have brought the consumption habits with them including a fondness for pummelo. Although there is no separate documentation of the growth of pummelo consumption in the West, increased availability of pummelos in supermarkets in the United States is one indication of increased demand outside of East Asia.

Exports of grapefruit/pummelos by major exporting countries are shown in Table 23.6. The four leading exporting countries are South Africa, China, Turkey, and the United States. Note that as recently as the 2011–12 season, the United States held the leading position, but the steep decline in grapefruit production Florida, has led to the decline in the United States position as a supplier of grapefruit to the world market. As noted elsewhere, nearly all of the exports from China are pummelos while the other countries shown in Table 23.6 are exporting grapefruit.

Imports of grapefruit/pummelos by major importing countries are shown in Table 23.7. The three leading importers are the EU, Russia, and Japan. In the case of the EU, most of these imports would come from outside the EU primarily South Africa, Turkey, and the United States, and recently there has also been the importation of pummelos to the EU from China. With the decline of the US production, there has also been a marked decline in the importation of grapefruit worldwide.

While organizations that support the production and marketing of grapefruit/pummelos exist in several countries, the Florida Administrative Committee administers a marketing order that affects the marketing of grapefruit from Florida in the early portion of the season. This organization also collects and disseminates prices and volumes of fresh citrus fruits across oranges, grapefruit, and tangerines that are sold fresh in the US market.

23.4 Mandarins/tangerines

World production, consumption, and trade of mandarins/tangerines have increased significantly from the early 2000s to 2016. Production increased from 14.5 MMT in 2001 to 28.5 MMT in 2016, and consumption increased from 13.260 MMT to 26.5 MMT. Total world exports and imports increased at a faster rate from 2001 to 2016, by 180% for export and 169% for import. Mandarins/tangerines used for processing only increased by 25%, from 1.2 MMT to 1.5 MMT. Despite the increase, the proportion of mandarins/tangerines used for processing to the total production decreased from 8.2% to 5.2%, indicating a worldwide trend of increasing consumption of fresh fruit and vegetables (Table 23.8).

23.4.1 Production and consumption by country

China is the world largest producer and consumer of mandarins/tangerines, and it is largely responsible for increased global production and fresh consumption of mandarins/tangerines. From 2001 to 2016, mandarin/tangerine production of China increased from 6.3 million tons to 19.3 million tons. During this same period, its share of the total world production increased from 43.3% to 67.7%. Consumption in China increased at a slightly faster rate, from 5.8 MMT in 2001 to 18.1 MMT in 2016, accounting for 43.6% and 67.4% in 2001 and 2016, respectively, of total world fresh consumption. Other than China, the production and fresh consumption of mandarins/tangerines of other countries have remained relatively stable. EU is the second largest producer and consumer of mandarins/tangerines, followed by Japan (Figs. 23.1 and 23.2). Despite a slight increase in quantity, over the years, both the EU and Japan's share in world total production and fresh consumption declined. EU's and Japan's share of production decreased from 19.8% to 11.3% and 10.6% to 3.5%, respectively, from 2001 to 2016. In the meanwhile, the share of consumption decreased from 20.4% to 11.4% for the EU, and from 10.6% to 3.4% for Japan (USDA, 2017).

23.4.2 Exports and imports by country

World trade of mandarins/tangerines is more diverse (Figs. 23.3 and 23.4). China is not only the largest producer, it is also the world's largest exporter in most years. Chinese exports increased from 0.169 MMT in 2001 to 0.8 MMT in 2011, the highest in its export history, and then decreased to 0.6 MMT in 2016. The export of some small producers of mandarins/tangerines is comparable to that of China because of low domestic demand. For instance, Turkey is the second largest exporter of mandarins/tangerines, followed by Morocco, EU, and South Africa. Other than the EU who experienced a decreasing trend of export starting in 2012, the export of Turkey, Morocco, and South Africa kept increasing from 2001 to 2016.

TABLE 23.6 Fresh grapefruit exports by major exporting countries, 2006–07 through 2016–17 seasons (MT).

Country	Seasons										
	2006/2007	*2007/2008*	*2008/2009*	*2009/2010*	*2010/2011*	*2011/2012*	*2012/2013*	*2013/2014*	*2014/2015*	*2015/2016*	*2016/2017*
World	872,000	851,000	825,000	830,000	808,000	800,000	813,000	825,000	736,000	784,000	756,000
South Africa	215,000	196,000	210,000	187,000	217,000	174,000	242,000	217,000	221,000	203,000	235,000
China	51,000	101,000	102,000	119,000	84,000	118,000	130,000	165,000	124,000	159,000	170,000
Turkey	133,000	133,000	128,000	154,000	153,000	177,000	132,000	177,000	145,000	190,000	126,000
United States	332,000	270,000	247,000	242,000	227,000	209,000	184,000	147,000	141,000	124,000	107,000
Israel	81,000	81,000	85,000	84,000	83,000	78,000	79,000	78,000	61,000	61,000	70,000
Mexico	12,000	14,000	11,000	18,000	17,000	19,000	18,000	14,000	19,000	22,000	22,000
EU	17,000	21,000	21,000	22,000	20,000	18,000	21,000	19,000	15,000	14,000	15,000
Hong Kong	2,000	2,000	4,000	4,000	7,000	7,000	7,000	8,000	10,000	11,000	11,000

https://www.fas.usda.gov/databases/production-supply-and-distribution-online-psd

TABLE 23.7 Fresh grapefruit imports by major importing countries, 2006–07 through 2016–17 seasons (MT).

Country	Seasons										
	2006/2007	*2007/2008*	*2008/2009*	*2009/2010*	*2010/2011*	*2011/2012*	*2012/2013*	*2013/2014*	*2014/2015*	*2015/2016*	*2016/2017*
World	829,000	840,000	789,000	796,000	752,000	731,000	743,000	752,000	670,000	700,000	688,000
EU	399,000	430,000	399,000	389,000	348,000	341,000	337,000	360,000	339,000	364,000	350,000
Russia	74,000	95,000	86,000	112,000	117,000	113,000	141,000	133,000	101,000	117,000	110,000
Japan	221,000	188,000	180,000	168,000	167,000	149,000	134,000	109,000	100,000	82,000	75,000
China	3,000	3,000	6,000	7,000	12,000	13,000	17,000	26,000	31,000	33,000	42,000
Canada	55,000	51,000	48,000	46,000	45,000	44,000	43,000	42,000	40,000	39,000	38,000
United States	22,000	14,000	12,000	12,000	7,000	1,000	13,000	13,000	10,000	16,000	24,000
Ukraine	15,000	19,000	16,000	21,000	23,000	27,000	30,000	27,000	15,000	18,000	18,000
Hong Kong	13,000	12,000	16,000	18,000	18,000	24,000	15,000	16,000	15,000	16,000	16,000

TABLE 23.8 World mandarins/tangerines distribution from 2001 to 2016 (unit: 1000 MT).

Market year	Production	Consumption	Export	Import	For processing	For processing/production
2001	14,468	13,260	871	853	1190	8.23%
2002	14,478	13,122	1009	912	1259	8.70%
2003	15,244	13,816	1090	1014	1352	8.87%
2004	16,669	15,079	1331	1209	1468	8.81%
2005	17,058	15,589	1288	1346	1527	8.95%
2006	17,120	15,201	1523	1501	1897	11.08%
2007	19,031	17,502	1564	1609	1574	8.27%
2008	20,391	18,846	2020	1863	1388	6.81%
2009	22,125	20,638	1975	1941	1453	6.57%
2010	21,898	20,333	2104	2034	1495	6.83%
2011	23,761	22,177	2389	2216	1411	5.94%
2012	24,489	23,057	2165	2117	1384	5.65%
2013	26,474	24,847	2483	2281	1425	5.38%
2014	28,294	26,527	2330	2177	1614	5.70%
2015	28,730	27,186	2316	2181	1409	4.90%
2016	28,527	26,883	2445	2292	1491	5.23%

https://www.fas.usda.gov/databases/production-supply-and-distribution-online-psd

Another country that experienced an increasing trend of export is Israel, which surpassed Argentina in 2014, becoming the world's sixth largest exporter of mandarins/tangerines. Percentagewise, China accounted for 24.1% of world total exports in 2016. The export share of other countries in 2016 was 28.8% for Turkey, 20.9% for Morocco, 9.8% for EU, 8.6% for South Africa, and 4.3% for Israel. From 2001 to 2016, the share of the EU export of mandarins/tangerines in the world's total export decreased from 16.1% to 9.8%, the share of South Africa kept almost unchanged (8.2% vs. 8.6%), while the share of all other four countries increased (USDA, 2017).

Russia is the world's largest importer of mandarins/tangerines. Its import increased fivefold from 150,000 MT in 2001 to 800,000 MT in 2016. Based on the import quantity of 2016, the other two largest importers are the EU and the United States. Canada, Ukraine, Thailand, and Vietnam are in the second-tier countries; Indonesia, Philippines, and Malaysia are in the third-tier countries; and China, Japan, and South Africa are in the fourth-tier countries. All the countries experienced an increase in import from 2001 to 2016. Percentagewise, Russia's share of world total import increased from 18.1% in 2001 to 35.8% in 2016. From 2001 to 2016, the import share changed from 42.0% to 18.8% of EU; from 6.2% to 10.3% for the United States; from 9.7% to 6.5% for Canada; from 4.6% to 6.1% for Ukraine; from 0.1% to 5.2% for Thailand; from 2.1% to 5.5% for Vietnam. Despite the increases in imports, the import share of China, Japan, and South Africa was <1.5% (USDA, 2017).

Changes in the shares of export and import countries imply some significant structural changes in the trade of mandarins/tangerines.[d] In 2001, the top four largest mandarin/tangerine exporters were Turkey, China, South Africa, and Argentina, with a total export value of US$417 million, accounting for 82% of world mandarin/tangerine trade. In 2016, the top four largest mandarins/tangerines exporters were Turkey, China, Morocco, and South Africa, with a total export value of US$1639 million, also accounting for 82% of world mandarin/tangerine trade (USDA, 2017).

In 2001, the key mandarin/tangerine trade partners of China included the Philippines, Malaysia, Indonesia, Russia, Canada, Vietnam, and Singapore (GTA, 2017). These countries import accounted for 24.3%, 24.0%, 12.5%, 9.8%, 8.6%, 7.3%, and 5.2%, respectively, of China's total export. Turkey's key mandarin/tangerine trade partners were Russia (16.6%),

d. EU is not included in the discussion because the discussion is focused on trade between individual countries.

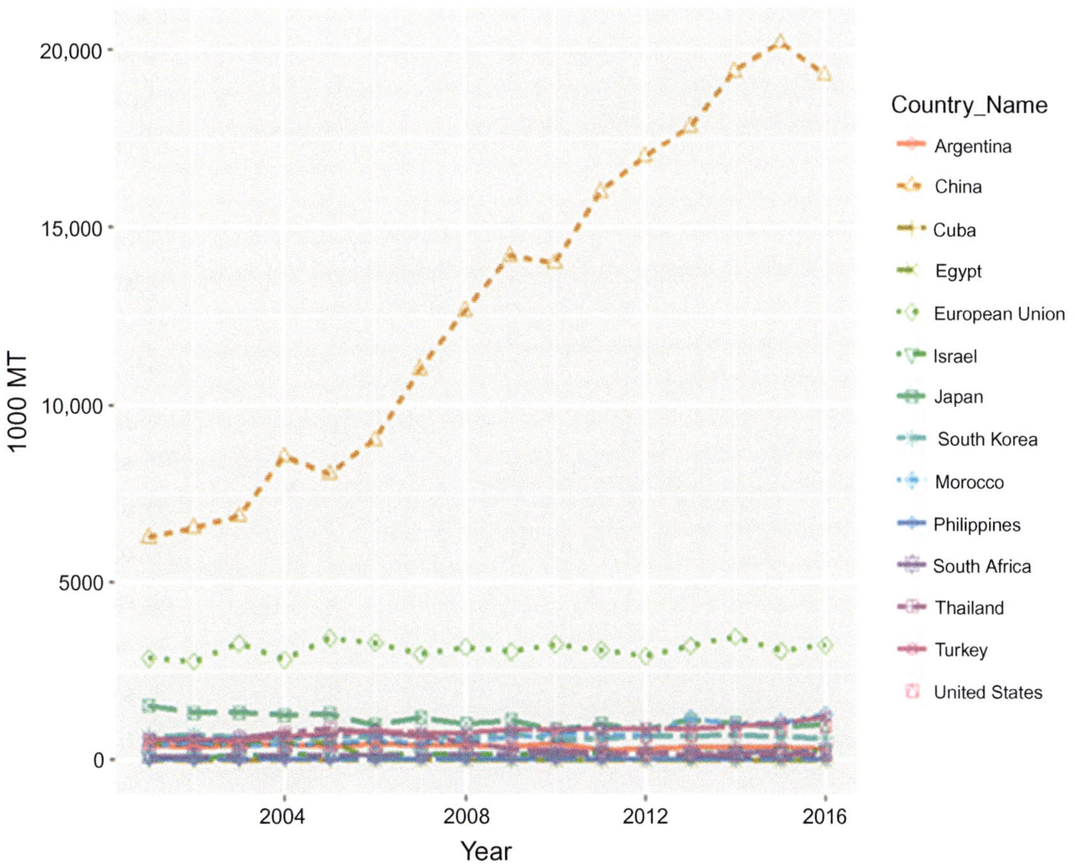

FIG. 23.1 World Mandarin/Tangerine production by country. *https://www.fas.usda.gov/databases/production-supply-and-distribution-online-psd*

Ukraine (15.4%), the United Kingdom (8.8%), Saudi Arabia (7.5%), Australia (7.0%), and Romania (6.9%). For South Africa, the United Kingdom was its dominant mandarin/tangerine importer, accounting for 44.6% of its export. The Netherlands, the United States, Saudi Arabia, and Belgium accounted for 22.3%, 6.1%, 5.7%, 5.6%, respectively, of South Africa's mandarin/tangerine export. The Netherlands, the United Kingdom, Canada, Russia, and the Philippines accounted for 34.2%, 22.3%, 11.0%, 10.8%, and 5.3%, respectively, of Argentina's mandarin/tangerine export.

In 2016, Thailand and Vietnam became the two largest importers of Chinese mandarins/tangerines. The import of Thailand, Vietnam, Malaysia, Russia, Philippines, and Kazakhstan accounted for 20.0%, 18.3%, 14.1%, 13.4%, 6.4%, and 6.0%, respectively, of China's mandarin/tangerine export. In 2016, the importers of Turkey's mandarins/tangerines concentrated to Russia, Iraq, and Ukraine, accounting for 39.5%, 28.3%, and 13.9%, respectively, of Turkey's mandarin/tangerine export. For Morocco, the new major exporter of mandarins/tangerines, Russia, France, Canada, the Netherlands, and the United States accounted for 40.8%, 20.0%, 11.8%, 10.1%, and 6.0%, respectively, of its total export. For South Africa, the United Kingdom, and the Netherlands were still its top two largest mandarin/tangerine importers, accounting for 34.9% and 24.9%, respectively, of its export. Russia and the United States accounted for 7.6% and 5.4%, respectively, of South Africa's export.

23.4.3 Mandarins/tangerines used for processing by country

Most mandarins/tangerines used for processing are for canned mandarins/tangerines. China and the EU dominated the mandarins/tangerines used for processing (Fig. 23.5). Before 2007, the amount of mandarins/tangerines used for processing was comparable between China and the EU. However, in 2008, the EU experienced a significant drop in mandarins/tangerines used for processing, and China surpassed the EU becoming the world largest mandarins/tangerines processor. The third largest mandarins/tangerines processor is the United States, followed by South Korea, Argentina, Japan, and Israel. South Africa, Turkey, and Morocco also process a small account of mandarins/tangerines. In general, the amount of mandarins/tangerines used for processing started keeping at a relatively stable level since 2012 for most countries.

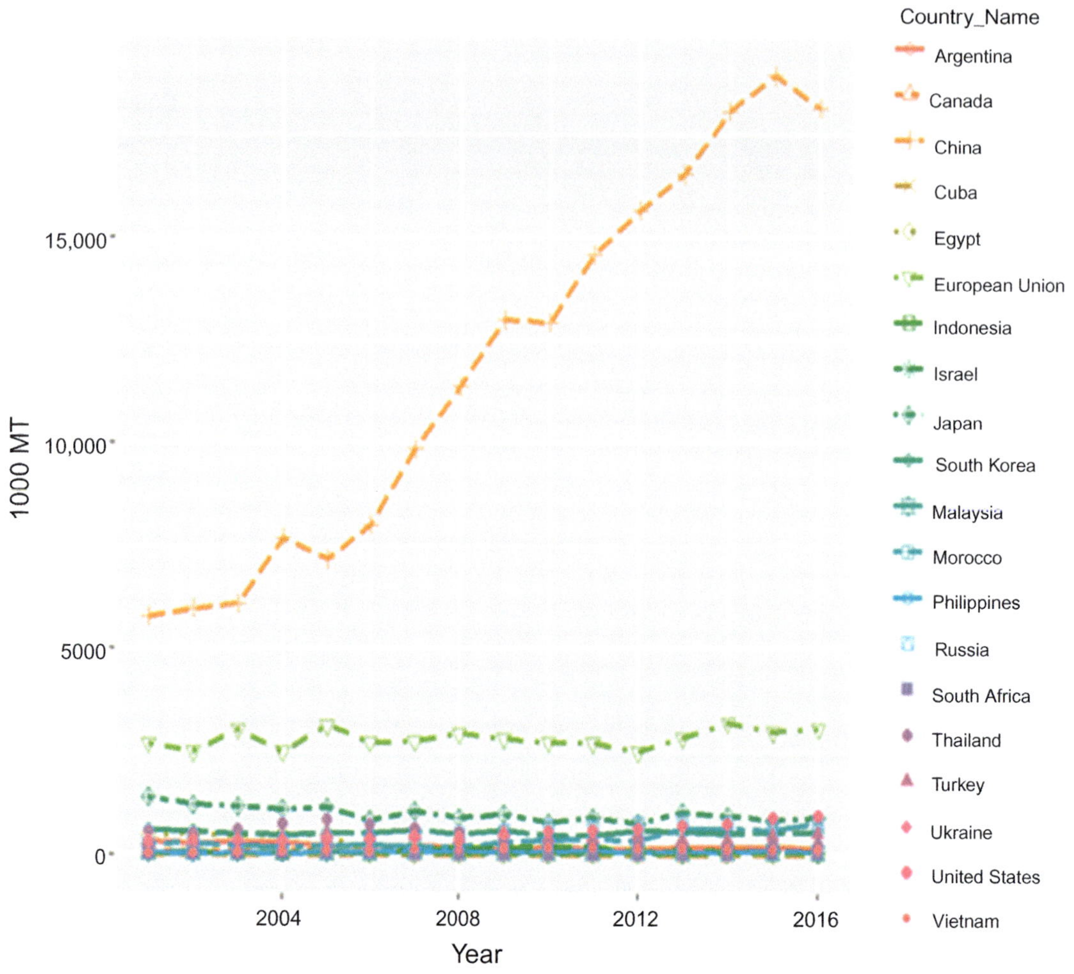

FIG. 23.2 World Mandarin/Tangerine consumption by country. *https://www.fas.usda.gov/databases/production-supply-and-distribution-online-psd*

23.5 Lemons and limes

Lemons and limes are similar to grapefruit and pummelos in that they are imperfect substitutes, but for statistical purposes, their production is aggregated by the FAO and other organizations that report world production statistics. World lemon/lime production in major producing countries is shown in Table 23.9. It is well-known that Mexico, the largest producing country primarily grows Persian (seedless) and Mexican (or Key) (seedy) limes. Other lime producing countries are found primarily in the tropical portion of the Western Hemisphere including the Bahamas, Brazil, and Guatemala even though lime production can be found scattered across the world. The other countries shown in Table 23.9 grow primarily lemons. Argentina is the largest producer of lemons, followed closely by Spain (whose production is now reported with the other EU countries). South Africa has also seen its production grow in recent years while production in Turkey and the United States (California and Arizona) has remained flat. As seen in Table 23.9, there has been a steady increase in world production of lemons and limes.

Lemons and limes differ from other citrus varieties in that they are almost always consumed in association with other foods as a flavor enhancer, garnish, or with alcoholic and nonalcoholic beverages.

Exports of lemons/limes by major exporting countries are shown in Table 23.10. The leading exporting countries are Mexico, Turkey, South Africa, and Argentina. Mexico's exports are nearly exclusively limes while the other countries export lemons. Note the strong growth in exports exhibited by Mexico, Turkey, and South Africa. In the case of Mexico, it has benefited from the demise of the Florida lime industry as a result of citrus canker. Mexico, however, has good markets in other countries besides the United States including Japan and the EU. With stagnating lemon production with the EU, Turkey has become a major supplier to both the EU and Russia. South Africa exports lemon to many destinations including the EU, Saudi Arabia, and Russia.

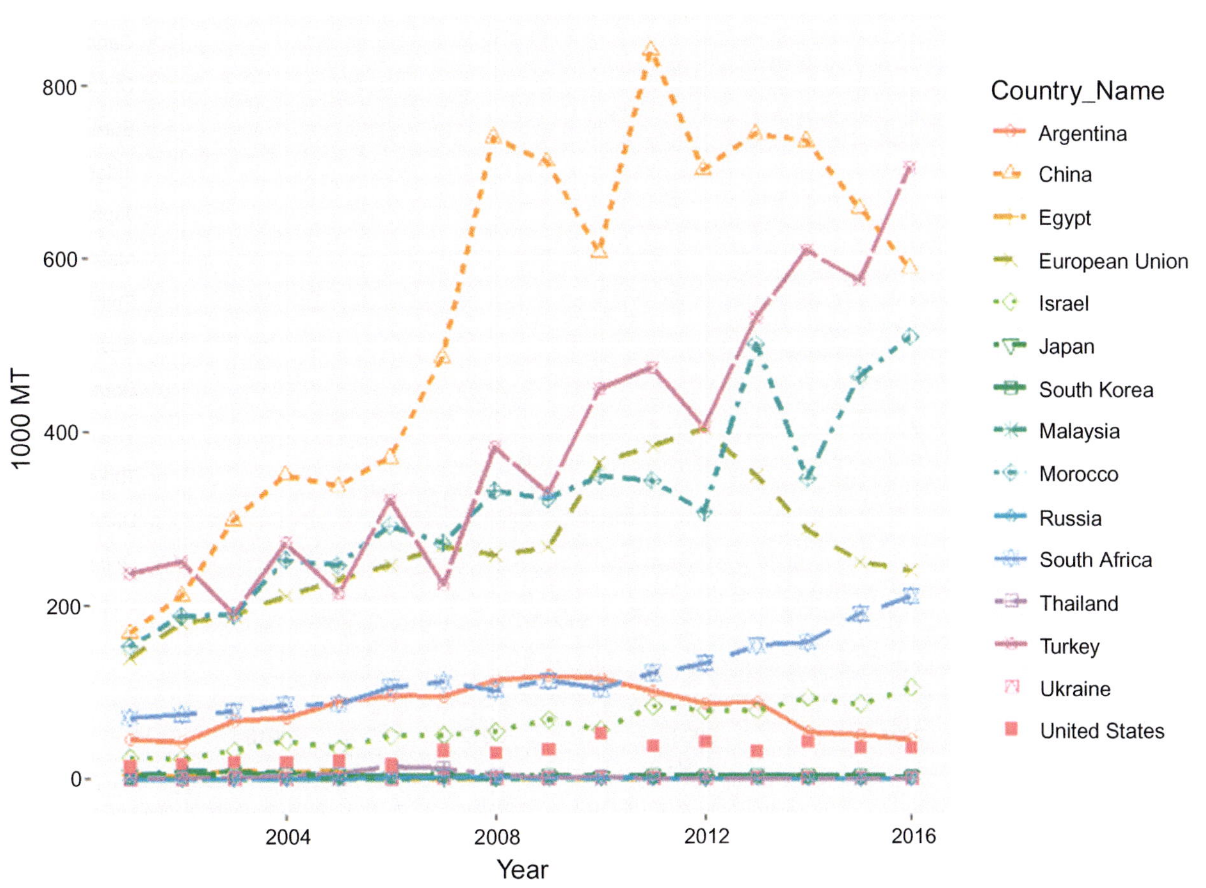

FIG. 23.3 World Mandarin/Tangerine export by country. *https://www.fas.usda.gov/databases/production-supply-and-distribution-online-psd*

Imports of lemons/limes by major importing countries are shown in Table 23.11. The four largest importing countries are the United States, the EU, Russia, and Saudi Arabia. The US imports are mostly limes from Mexico as the US is a net exporter of lemons. As noted elsewhere, the EU, Russia, and Saudi Arabia are sourcing lemons from Turkey, South Africa, and Argentina.

23.6 Price Determination for citrus

Pricing of citrus differs by country and use (fresh vs. processed). In many less developed countries, the fruit may be sold at the orchard or at local markets. In the former case, a broker visits the orchard site and will offer to buy all of the fruit for a fixed price regardless of the volume of fruit harvested. The broker arranges harvest and transport to a regional market or a packinghouse where the fruit is sorted. A portion of the fruit will be deemed suitable for the fresh market, the remaining fruit may be processed or sent to a lower price outlet.

In the latter case, the grower is responsible for harvest and transport to the local market. At that point, the fruit may be sold to households or purchased by brokers who then transport the fruit to larger regional markets. In some countries, ownership of the fruit may change several times before eventually purchased by the final user.

In most developed countries, a more formal market is likely to exist. Smaller growers may market their fruit in a manner similar to that found in less developed countries. Larger growers, however, may have contracts with supermarkets or other larger purchasers. These growers may have their own harvesters or contract with harvest providers to complete harvest and transport to a packinghouse facility. At the packinghouse, the fruit is sorted and graded. Fruit meeting fresh market standards is then sent to retail outlets (after washing and packing into transport containers generally referred to as cartons). Poorer quality fruit is sent to the processing plant or sold in lower value markets. The grower receives proceeds from both high quality and lower quality fruit. His/her final price per unit depends on the proportion of fruit allocated among high and low quality.

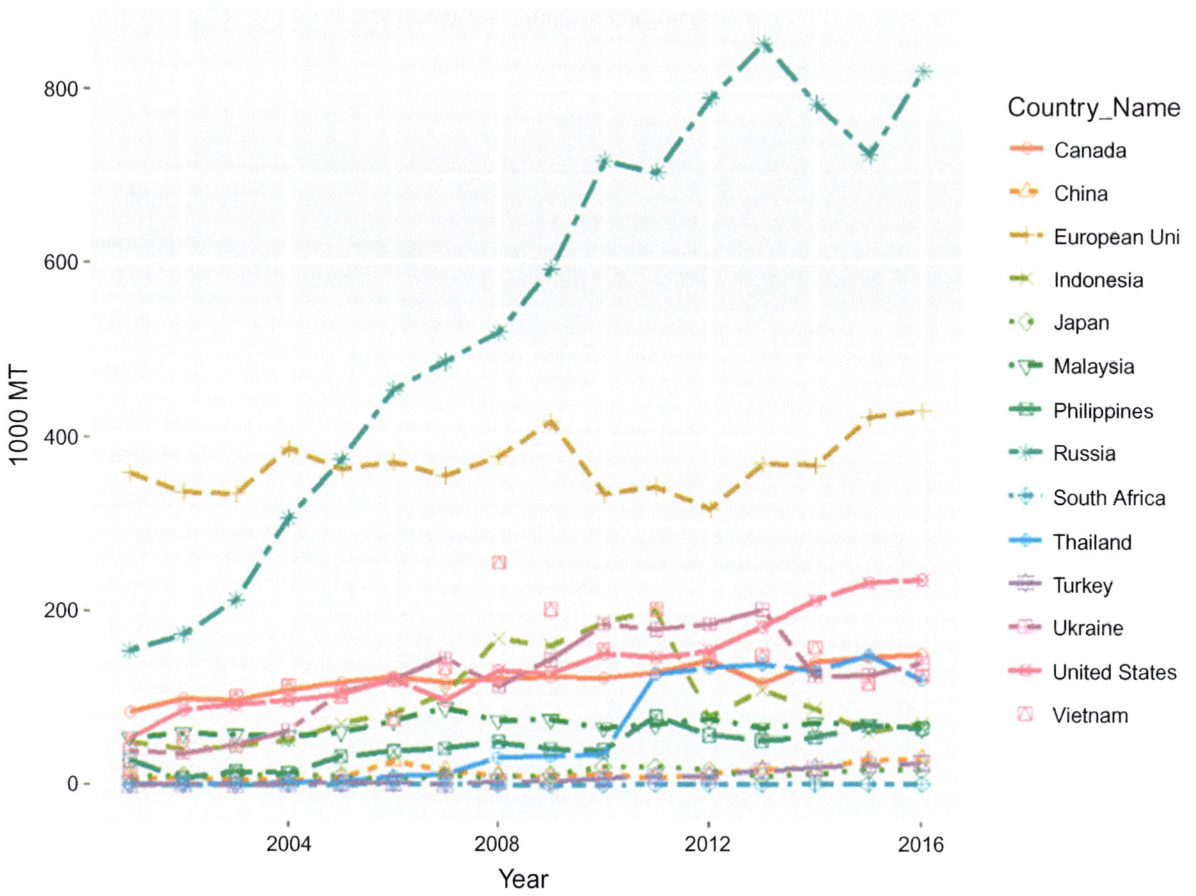

FIG. 23.4 World Mandarin/Tangerine import by country. *https://www.fas.usda.gov/databases/production-supply-and-distribution-online-psd*

FIG. 23.5 Mandarin/Tangerine used for processing by country.

TABLE 23.9 **Production of lemons and limes by major growing countries, selected years, 1999–2000 through 2015–16.**

	Million MT					
	1999–2000	*2004–05*	*2009–10*	*2011–12*	*2013–14*	*2015–16*
Argentina	1,050	1,300	1,000	1,200	780	1,500
Spain[a]	899	900	n/a	n/a	n/a	n/a
United States	797	738	800	771	748	847
Mexico	1,230	1,890	1,850	2,055	2,167	2,270
EU-27[a]	n/a	n/a	1,160	1,264	1,308	1,260
Turkey	500	535	783	750	760	670
South Africa	110	180	216	260	312	345
World total	4,101	4,323	5,951	6,524	6,216	7,001

[a]*Production data for countries of the European Union were aggregated beginning in 2009–10.*
https://www.fas.usda.gov/databases/production-supply-and-distribution-online-psd

Another payment system exists for both fresh and processed fruit, although it is more common in fresh fruit marketing. This system is called "participation." The term derives from the fact that the grower "participates" in the marketing process and is not paid until the fruit is sold, either to a broker or a supermarket. Generally, participation is organized through a packinghouse.[e] The packinghouse organizes the harvest, de-greens the fruit (if necessary), sorts and grades, packs the fruit into cartons (or other containers such as poly bags if desired by the next buyer), and may store the fruit if it is advantageous to do so. Fruit not suitable for the fresh market is sent to a processing plant. This fruit is called "eliminations" since it has been "eliminated" from the fresh marketing channel. Once the fruit is sold, the costs of the services provided by the packinghouse are deducted from the proceeds including harvest and transport to the packinghouse, and the remainder is paid to the grower. A disadvantage of this approach is that the grower must wait until the end of the harvest season for payment. An advantage is that middlemen in the marketing channel are avoided; it is likely the grower will receive a higher price.

A variation of participation arrangements is cooperatives. In the United States, cooperatives are found across a wide range of commodities including grains, dairy, sugar, and many types of fruit including citrus. The two most important cooperatives are Florida's Natural Growers, a processing cooperative located in central Florida, and Sunkist, a fresh fruit cooperative based in California. Florida's Natural Growers is known as a federated cooperative since it is owned by several fresh fruit cooperatives. In its early years, then known as Citrus World, it mainly processed eliminations from the fresh fruit cooperatives. With the advent of FCOJ and later NFC, Florida's Natural Growers became one of the three national brands of orange juice (competing with Tropicana and Minute Maid). An individual grower can belong to one of the fresh fruit cooperatives or belong to an entity that only delivers fruit to the processing plant. A pool is calculated for early and midseason oranges and another pool is computed for late (Valencia) oranges. There are also separate pools for red and white seedless processed grapefruit and for specialty varieties including tangerines.

Sunkist is a nonstock cooperative. Growers contract with packinghouses to pack fruit. Sunkist then oversees marketing, both domestic and international. Sunkist markets navel and Valencia oranges, lemons, and more recently mandarins from 6000 growers located in California and Arizona. Separate pools are computed for each variety of fruit.

The main difference between a cooperative and a private entity that offers fruit purchase through a participation plan is that growers own the physical assets in a cooperative. In the case of Florida's Natural Growers, the fresh fruit cooperatives "own" the processing plant operated by the cooperative.

For processed utilization, especially processed utilization of oranges, growers sell fruit to processing plants. This may be accomplished directly or through an intermediary. In the case of direct sale, in both Brazil and Florida, medium-term contracts extending 3–5 years are widely used. The contract may be a fixed price over the life of the contract. There may also be provisions to allow the price to escalate if supply–demand conditions warrant. This contract is characterized as a floor price with a rise tied to a third-party reporting service. In Brazil, the FCOJ futures contract traded at the New York

e. Although the grower may own the packinghouse individually or jointly with other growers. In the latter case, the growers may form a cooperative to jointly own the packinghouse and coordinate the harvest.

TABLE 23.10 Fresh lemon/lime exports by major exporting countries, 2006–07 through 2016–17 seasons (MT).

	Seasons										
Country	2006/2007	2007/2008	2008/2009	2009/2010	2010/2011	2011/2012	2012/2013	2013/2014	2014/2015	2015/2016	2016/2017
World	1,460,000	1,526,000	1,371,000	1,474,000	1,489,000	1,689,000	1,552,000	1,591,000	1,722,000	1,822,000	1,891,000
Mexico	436,000	508,000	445,000	456,000	432,000	625,000	523,000	519,000	610,000	662,000	680,000
Turkey	328,000	220,000	351,000	434,000	457,000	429,000	369,000	426,000	433,000	434,000	494,000
South Africa	110,000	166,000	130,000	145,000	162,000	165,000	175,000	220,000	246,000	237,000	270,000
Argentina	360,000	400,000	250,000	264,000	255,000	267,000	280,000	150,000	185,000	280,000	220,000
United States	120,000	157,000	93,000	93,000	102,000	95,000	110,000	127,000	114,000	110,000	115,000
EU	92,000	59,000	87,000	67,000	68,000	92,000	77,000	101,000	105,000	68,000	75,000
Hong Kong	5,000	4,000	8,000	6,000	7,000	8,000	7,000	33,000	18,000	17,000	20,000
Morocco	0	0	0	6,000	5,000	7,000	10,000	10,000	7,000	9,000	12,000

https://www.fas.usda.gov/databases/production-supply-and-distribution-online-psd

TABLE 23.11 Fresh lemon/lime imports by major importing countries, 2006 through 2016-17 seasons (MT).

	Seasons										
Country	***2006/2007***	***2007/2008***	***2008/2009***	***2009/2010***	***2010/2011***	***2011/2012***	***2012/2013***	***2013/2014***	***2014/2015***	***2015/2016***	***2016/2017***
World	13,53,000	14,50,000	13,61,000	14,50,000	14,25,000	14,97,000	15,26,000	15,17,000	16,02,000	18,12,000	18,05,000
United States	3,91,000	4,24,000	3,98,000	4,01,000	3,94,000	4,75,000	4,78,000	4,81,000	5,63,000	6,15,000	6,85,000
EU	4,18,000	5,15,000	4,05,000	4,71,000	4,21,000	4,40,000	4,26,000	3,80,000	3,99,000	5,57,000	4,50,000
Russia	2,05,000	1,91,000	2,03,000	2,11,000	2,22,000	2,00,000	2,12,000	2,09,000	2,07,000	1,86,000	2,10,000
Saudi Arabia	79,000	60,000	1,00,000	1,03,000	1,04,000	85,000	88,000	85,000	1,03,000	1,21,000	1,25,000
Canada	58,000	58,000	58,000	65,000	78,000	88,000	1,00,000	99,000	87,000	1,02,000	1,00,000
United Arab Emirates	44,000	51,000	44,000	54,000	55,000	63,000	78,000	85,000	94,000	91,000	90,000
Japan	68,000	62,000	52,000	53,000	57,000	55,000	51,000	51,000	51,000	51,000	52,000
Ukraine	60,000	55,000	58,000	62,000	64,000	62,000	63,000	54,000	44,000	41,000	45,000
Hong Kong	25,000	19,000	27,000	21,000	26,000	25,000	26,000	66,000	48,000	37,000	40,000

https://www.fas.usda.gov/databases/production-supply-and-distribution-online-psd

Board of Trade (NYBOT) is often used as the reference price. In Florida, where a large proportion of the crop is used to produce NFC orange juice, the FCOJ futures contract is not as useful. In its place, the FDOC post-estimate price is often used. A price collected by Florida Citrus Mutual, the largest grower organization in Florida, called the Mutual Mostlys price, may also be used.

Intermediaries are found in the three largest processed orange producers: Brazil, the United States, and Mexico. In Florida, the term "bird dog" is used to describe intermediaries. Bird dogs make contracts with small and medium growers to purchase fruit and then sell in larger lots to processors. They charge a small fee for their service. They may also provide harvesting and hauling activities as a part of their service. In Mexico, which lacks publicly available data on fruit production and prices, these intermediaries are called "coyotes." This term has a negative connotation, but coyotes in Mexico play a role quite similar to bird dogs in Florida.

An important distinction regarding pricing in the processed orange market in the United States compared to Brazil and Mexico is the unit of measurement. In Brazil and Mexico, most fruit purchased from growers is priced based on the weight of the fruit. Brazil has adopted the Florida system of using 90-pound (40.8 kg) boxes as the unit of measurement while Mexico uses metric tons. In Florida, growers are paid based on the juice content of the fruit which is measured in pound solids (or soluble solids). The State of Florida has test houses located at all processing plants which test a sample of each load of fruit delivered to establish its pound solids content. This use of a government entity to serve as a third-party arbiter is unique in the world.

23.7 Trade agreements and citrus

Trade agreements implemented over the past 50 years have affected the patterns of trade among citrus exporters and importers. The largest and most important agreement is the EU. Other agreements include the North American Free Trade Agreement (NAFTA), the Southern Common Market also known as MERCOSUR, the US–Japan Beef and Citrus Agreement, the Caribbean Basin Economic Recovery Act (CBERA) also known as the Caribbean Basin Initiative (CBI), and preferences offered by the EU to select countries including Israel and Egypt. Recently, the United States and South Korea entered into a free trade agreement and that will affect the US exports of fresh and processed citrus to South Korea.

The EU, since its inception, has adopted a stance of supporting domestic agriculture. This posture also extends to citrus that is grown in Spain, Portugal, Italy, and Greece. Growers are provided ample production subsidies that are not found in other major citrus producing countries apart from China. In addition, the EU imposes seasonal tariffs to limit imports from competing countries. The EU also imposes a relatively small tariff on imported orange juice.

The NAFTA was implemented in 1994. It formed a free trade zone among Canada, the United States, and Mexico. Tariffs were phased out for most agricultural products in fresh and processed citrus. This meant that the US orange juice tariff was eliminated on imports from Mexico with total elimination occurring in 2001. Canada has never imposed a tariff on orange juice. Mexico also limited restrictions on the importation of fresh citrus from the United States. Phytosanitary-based restrictions remained in place, however, limiting imports of fresh oranges and tangerines from Mexico because of the Mexican fruit fly.

The Southern Common Market also known as MERCOSUR includes Brazil, Argentina, Uruguay, and Paraguay. It has facilitated trade in citrus products among these countries also phytosanitary concerns limit trade in fresh citrus products.

The US–Japan Beef and Citrus Agreement was signed in 1984. It led to an expansion of Japanese imports of beef and citrus products from the United States. Exports of fresh grapefruit from Florida and fresh oranges from California expanded rapidly. Japan became the largest importer of Florida grapefruit.

The CBERA, more commonly called the CBI was signed in 1983. It reduced tariffs for a wide range of products including most agricultural products among not only Caribbean countries such as Jamaica and the Dominican Republic, but also several Central American countries, including Belize, Costa Rica, Honduras, and the United States. The intention of the act was to expand trade among the small economies of the Caribbean countries and the United States. The US tariff on imported orange juice was eliminated for countries participating in the agreement. Costa Rica, Belize, Honduras, and the Dominican Republic all established plantings of sweet orange trees intended for processing as well as the construction of processing plants.

In March 2012, the US–South Korea Free Trade Agreement went into effect. One impact of the agreement was to loosen restrictions imposed by South Korea on the importation of both fresh and processed citrus products from the United States. The result has been expanded imports of fresh orange and grapefruit as well as increased importation of the US produced orange juice.

Other agreements affect citrus trade. Notably, the EU grants trade preferences to selected exports including Israel (grapefruit), Egypt (fresh oranges), Belize (orange juice), and Mexico (orange juice).

23.8 Marketing and promotion of citrus

Advertising and promotional activities have long served an important role in maintaining consumer demand for citrus products. Fresh and processed citrus products are considered cooperative goods as these commodities cannot be readily differentiated. Generic advertising and promotional activities have traditionally focused on the common attributes of citrus products that cannot be differentiated from one brand or another, such as nutritional content, origin, or safety (Forker and Ward, 1993; Ward, 2009). Generic promotional programs may increase total demand for a commodity and have been shown to provide positive net returns to industry stakeholders (Ward, 1997; Brown and Lee, 2002; Ward et al., 2015; Heng et al., 2018). Brand advertising, on the other hand, generally highlights many of the same generic attributes while seeking to differentiate the brand-specific products to compete within the category (Ward and Kilmer, 1989). Brand advertising may serve to increase demand, but also has an impact on capturing market share. Generic advertising can serve to collectively represent stakeholders of a homogenous commodity, such as producers of agricultural goods.

The structure and reach of advertising and promotional programs vary from market to market and vary across the fresh and processed sectors of the citrus industry. The introduction of new citrus varieties on the fresh market, for example, has reduced to a slight degree the homogeneity of the market for mandarin varieties, as indicative of the branding efforts associated with some new mandarin varieties, at least in the United States.

Advertising and promotion programs for 100% orange juice, while varying across regions, have served to enhance the category as a whole (Ward, 1997). In recent years, advertising and promotional activities for 100% orange juice and grapefruit juice have an increased focus on public relations and issues management activities to reflect the increase in consumer engagement across social media platforms. Historically, the Florida citrus industry's extensive investment in communicating the nutritional benefits and other attributes of citrus to consumers on both the domestic and international markets across several decades through television and print advertisements expanded the market for citrus products, especially in the United States. Educating and reinforcing consumers' knowledge of the health and wellness benefits of citrus products will remain an ongoing effort in an era where the information available is plentiful, but often, misleading, distorted, or altogether false. Two organizations heavily involved in generic promotional activities for orange juice are the FDOC and CitrusBR, based in São Paulo, Brazil. In recent years, both organizations have integrated proactive public relations and issues management programs into marketing activities.

The citrus industries in Florida and Brazil have undertaken proactive efforts to maintain the positive public opinion of citrus products in prominent markets around the world. Programs are designed to educate and reach key demographics and trusted influencers/nutrition experts who garner widespread reach alike. Programs are centered around reinforcing consumer awareness of the positive attributes of citrus, such as taste and nutritional benefits. Affinity for both raw and processed citrus product among younger consumers was identified as a critical issue to monitor and foster in the coming years. As with all industries, public relations activities have increased in relevance within the organizational structure over the last decade as consumers' access to media and information spans across more channels, including television, internet, and social networks. Although spending on social advertising has increased in recent years, investing in public relations activities has helped organizations with limited budgets reduce the need for paid media. Consumers are increasingly becoming exposed to new media across various platforms, such as television to digital media. In 2017, 81% of the US citizens had a social media profile, representing a 5% growth compared to the previous year (Kepios, 2017). The leading social network sites in 2017 were Facebook (2.06 billion active users), YouTube (1.5 billion active users), WhatsApp (1.3 billion active users), and Facebook Messenger (1.3 billion active users). Blogging services, such as Tumblr, also extend the reach of digital content (Kepios, 2017).

Consumers exposed to media across all platforms recall that the messages are positive, which can be attributed to the historical generic and brand advertising programs communicating the positive attributes of citrus, such as nutritional content or origin. Figure 23.6 illustrates the response of consumers' perception of the media they have seen, heard, or read about orange juice. The marginal decline in positive perception in November 2016 and March 2017 occurred following a 1-month lag when a spike in news media surrounds the release of the USDA citrus crop forecast.

Consumers who engage in social media were more likely to purchase 100% orange juice than those consumers who did not engage in social media, signifying a need to maintain a social media presence for both brand and generic programs (Figs. 23.6 and 23.7).

As the citrus industry grapples with the impact of HLB on the world supply of fresh and processed citrus products, maintaining consumers' affinity for citrus products has never been more crucial. The impact of a decline in supply in the United States includes a decline in distribution, higher prices at retail and food service, and a loss of shelf space at retail. Generic advertising and promotional programs have sought to provide perspective to consumers on the benefits of citrus and shore up demand for the long-term viability of the industry.

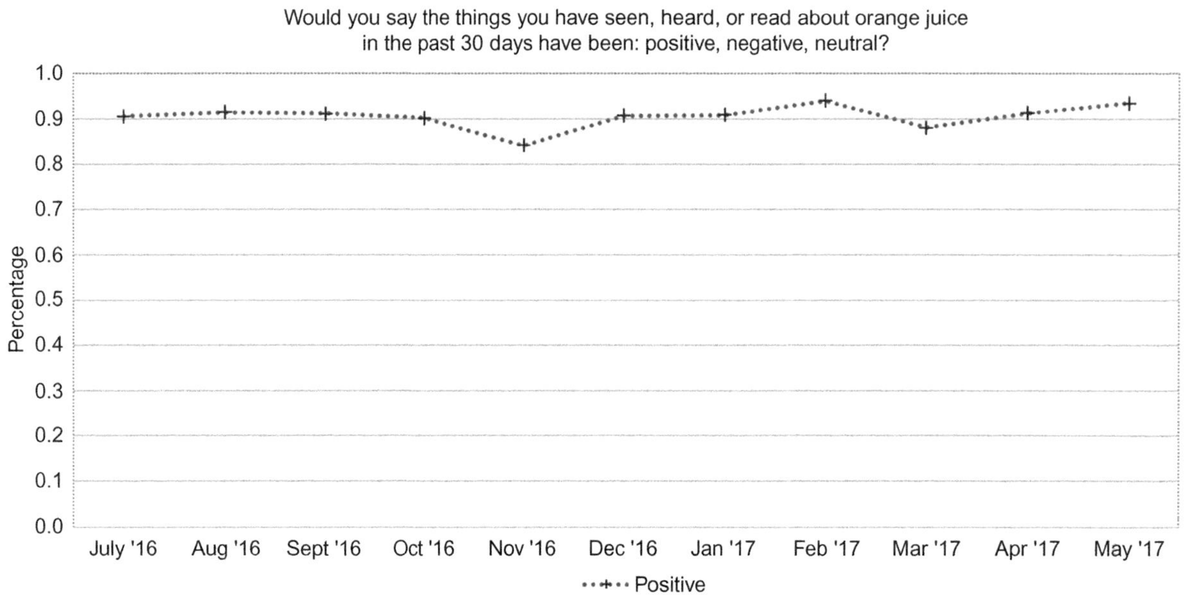

FIG. 23.6 Consumers' perception of media about orange juice is largely positive. *Florida Department of Citrus. n.d. OJ consumer tracking database. Unpublished.*

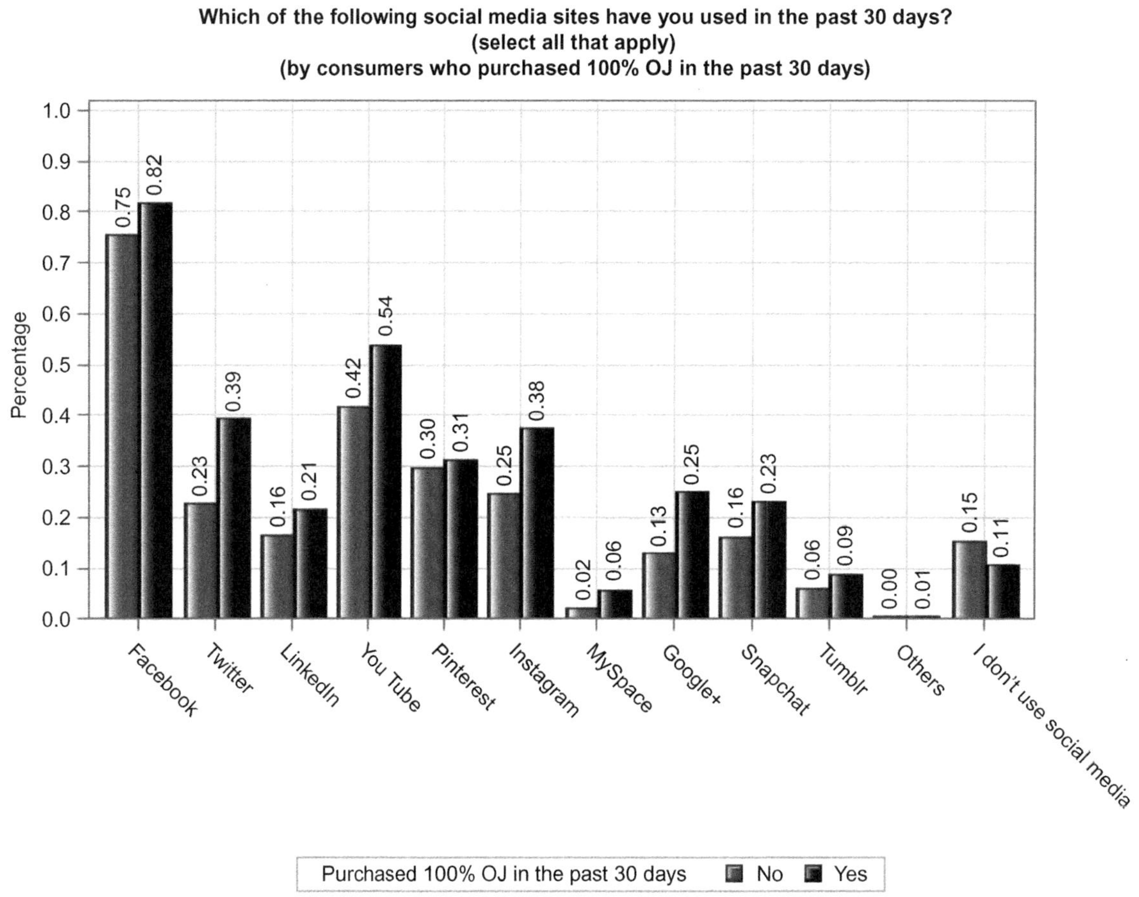

FIG. 23.7 Consumers engaged in social media were more likely to purchase 100% orange juice than consumers who did not engage in social media.

23.9 By-products from citrus processing

Several by-products are obtained through processing citrus fruit into juice. By weight, the most important by-product is pulp and peel. Depending on volume, the pulp and peel are further processed and transformed into a dehydrated product known as citrus pellets. Citrus pellets are used as animal feed. But surprisingly, the largest market for citrus pellets is Europe where they are used in rations for dairy cattle because of high fiber content. Citrus pellets enter the EU under a reduced tariff arrangement which makes them attractive relative to corn (either domestic or imported).

Analysis of citrus by-products is hampered by the lack of data. The FDOC collects data on the volume of citrus pellets and molasses produced in Florida, but does not collect data on peel oil or d-limonene, a high-valued solvent used in a variety of applications. Citrus essence production is also not reported.

Peel oil is collected from all varieties of citrus that are processed. The attributes of the peel oil are unique to the variety from which it was produced. Lemon oil is notable in that it is an important ingredient for soft drinks. In some countries, lemon oil is the primary product and lemon juice is the by-product. Citrus oils are used in a wide array of applications. For further discussion see Braddock (1999).

References

Braddock, R.J., 1999. Handbook of *Citrus* by-Products and Processing Technology. John Wiley & Sons, New York.

Brown, M.G., Lee, J.-Y., 2002. Brand and Generic Advertising Impacts in the Rotterdam Model. Working paper, Florida Department of Citrus, University of Florida.

Forker, O.D., Ward, R.W., 1993. Commodity Advertising: The Economics and Measurement of Generic Programs. Lexington Books. MacMillan, New York, NY.

GTA, 2017. Global Trade Atlas: International Import and Export Commodity Trade Data. Global Trade Information Services, Inc.

Heng, Y., Ward, R.W., House, L.A., Zansler, M.L., 2018. Assessing key factors influencing orange juice demand in the current market. Agribusiness, 1–15. https://doi.org/10.1002/agr.201596.

Kepios. 2017. Most famous social network sites worldwide as of September 2017, ranked by number of active users (in millions). In Statista—The Statistics Portal. Acessed November 9, 2017, Availabe from https://www.statista.com/statistics/272014/global-social-networks-ranked-by-number-of-users/.

USDA. 2017. Available from: https://www.fas.usda.gov/databases/production-supply-and-distribution-online-psd.

Ward, R.W., 1997. Advertising and promotions: Agro-food marketing. In: Padberg, D.I., Ritson, C., Albisu, L.M. (Eds.), CAB International in Association with CIHEAM. Zaragoza, Spain. CAB International, Wallingford, UK, ISBN: 0 85199 143 2.

Ward, R., 2009. Understanding the U.S. generic advertising system and its role in information management among commodities and food systems. Int. J. Food Syst. Dynam. 3, 204–211.

Ward, R.W., Kilmer, R.L., 1989. The Citrus Industry: A Domestic and International Economic Perspective. Iowa State Press, Ames, IA.

Ward, R.W., M. Peele, and L. Ortega. 2015. National Mango Board Consumer Research. Available from http://www.actahort.org/books/1075/1075_4.html.

Further reading

Florida Department of Citrus, 2018. OJ Consumer Tracking Database, unpublished.

Chapter 24

Citrus and health

Gang Ma[a], Lancui Zhang[a], Minoru Sugiura[b], Masaya Kato[a]
[a]*Department of Bioresource Sciences, Faculty of Agriculture, Shizuoka University, Shizuoka, Japan,* [b]*Department of Food Science and Nutrition, Doshisha Women's College of Liberal Arts, Kyoto, Japan*

24.1 Introduction

Citrus, which is widely cultivated in tropical and subtropical climates, is one of the most important crops in the world. Every year, approximately 130 million tons of citrus fruits are produced in >140 countries. Among them China, Brazil and the United States are the largest citrus producers, followed by Mexico, India and Spain. Citrus fruits have been consumed by humans since ancient times. As they are easily peelable and the flesh is soft and juicy with distinctive flavor, citrus fruits represent an important part of daily diet nowadays. Citrus cultivars such as oranges (*Citrus sinensis*), mandarins (*Citrus reticulata*, *Citrus clementina*, and *Citrus unshiu*), lemons (*Citrus limon*), and grapefruits (*Citrus* × *paradisi*) are popular all over the world, which are mainly consumed as fresh fruits or processed into juices and jams.

Citrus fruits not only taste good, but also are beneficial to human health (Gliszczyńska-Świgło et al., 2004; Lv et al., 2015; Economos and Clay, 1999) because of the wide range of compounds they contain. It is well recognized that citrus fruits, especially oranges, are an excellent source of ascorbic acid (vitamin C), which is a powerful antioxidant and good for human health. Citrus fruits are also rich in dietary fibers. Compared with other fruits and vegetables, the ratio of soluble to insoluble fiber is particularly high in citrus fruits. The soluble fiber in citrus fruits helps prevent diabetes and lower cholesterol levels. In addition to the macronutrients, citrus fruits are abundant in micronutrients, such as potassium, folate, vitamin B6, riboflavin, and calcium, which are essential for maintaining human health.

Citrus fruits have been used as herbal drugs in the traditional medicines for thousands of years. Citrus peel, known as *chen pi* or *ju pi*, is a nontoxic edible herbal medicine, and has been used to improve digestion, relieve intestinal gas and bloating, and resolve phlegm in traditional Chinese medicine. In the past decades, a growing number of epidemiologic and clinical studies have demonstrated that the consumption of citrus fruits is associated with the reduction of the risks of lifestyle-related diseases, such as cancers, cardiovascular diseases, osteoporosis, type-2 diabetes, and so on (Yuan et al., 2003; Sugiura et al., 2012, 2016a; Cirmi et al., 2016; Benavente-García and Castillo, 2008; Tripoli et al., 2007). In citrus fruits, phyotochemicals, such as flavonoids, carotenoids, and limonoids, are massively accumulated in the peel and pulp. The phyotochemicals of citrus fruits, especially flavonoids and carotenoids, have a broad range of pharmacological activities, and play important roles in the prevention and treatment of the lifestyle-related diseases.

In recent years, citrus fruits have attracted more and more attention because of the health-promoting benefits. To date, a number of bioactive compounds in citrus fruits have been isolated and identified, and their roles in maintaining human health have been extensively investigated. In this chapter, the roles of citrus fruits in human heath, the accumulation of major bioactive compounds, carotenoids, flavonoids, and ascorbic acid (AsA) in citrus fruits, and their health benefits are reviewed.

24.2 Carotenoid in citrus fruits

24.2.1 Carotenoid accumulation in citrus fruits

Carotenoids are a group of isoprenoid compounds widely distributed in nature. They are the main pigments in plants and contribute to the attractive yellow, orange, and red colors of flowers and fruits. Carotenoids not only fulfill multiple functions in regulating plant growth and development, but also have an immense value for human nutrition and health. To date, >800 carotenoids have been identified in nature of which lycopene, α-carotene, β-carotene, lutein, zeaxanthin, and β-cryptoxanthin are detected at a high level in human plasma. Among them, α-carotene, β-carotene, and β-cryptoxanthin, which can be converted to vitamin A in the body, are precursors of vitamin A. In addition to the pro-vitamin A activity, carotenoids also act as

The Genus Citrus. https://doi.org/10.1016/B978-0-12-812163-4.00024-3

antioxidants in the human body. In the recent years, more attention has been paid to the roles of carotenoids in the prevention of lifestyle-related diseases because of their high anti-oxidative activities. Many findings from the epidemiologic studies have demonstrated that dietary intakes of carotenoids reduces the risks of eye diseases, certain cancers, and inflammation (Sugiura et al., 2016a; Canfield et al., 2003; Männistö et al., 2004; Montonen et al., 2004; Cerhan et al., 2003).

Citrus fruits are a rich source of carotenoids. So far, approximately 115 carotenoids have been identified in citrus and among them β-cryptoxanthin and violaxanthin are commonly accumulated in most citrus fruits. The major carotenoids in citrus fruits are shown in Fig. 24.1. The basic structure of carotenoid is a C_{40} backbone skeleton containing a series

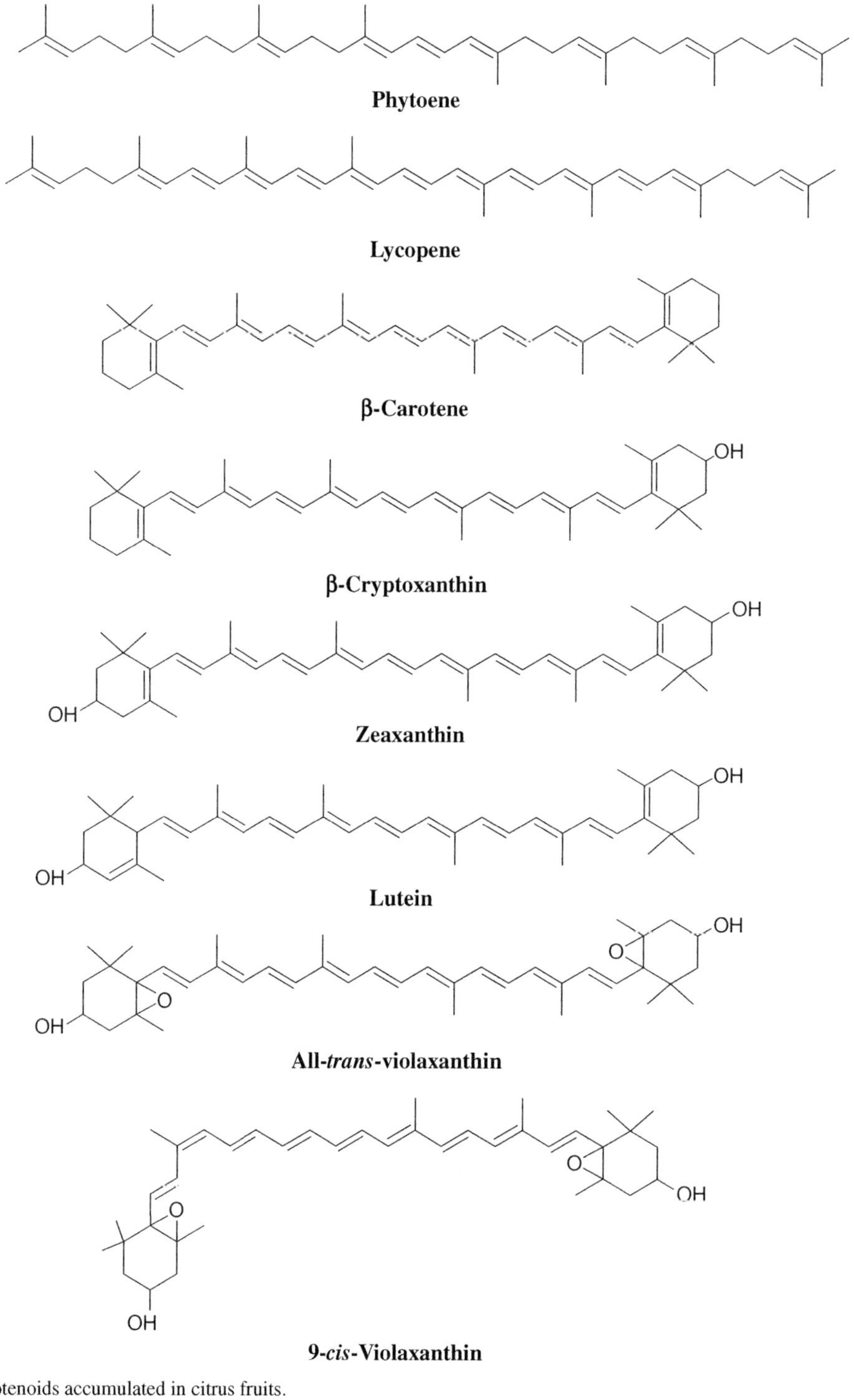

FIG. 24.1 Major carotenoids accumulated in citrus fruits.

of conjugated double bonds in the central chain. With the increases in number of conjugated double bonds in the polyene chain, carotenoids exhibit distinct colors, ranging from colorless (phytoene) to red (lycopene), orange, or yellow (β-carotene and lutein). In citrus fruits, carotenoids are the main pigments responsible for the appealing colors of peel and pulp (see Chapter 12), and their contents and compositions are key indexes of the commercial and nutrition values of citrus fruits.

As the accumulation of carotenoid in citrus fruits is influenced by growing conditions and geographical origins, the carotenoid profile varies among different citrus varieties. Ikoma et al. (2016) reported that according to the carotenoid composition in fruits, citrus cultivars could be roughly divided into three groups: β-cryptoxanthin abundant cultivars, violaxanthin abundant cultivars, and cultivars with low β-cryptoxanthin and violaxanthin contents. Mandarins, such as Satsuma and Ponkan, are β-cryptoxanthin-abundant cultivars, in which β-cryptoxanthin is massively accumulated as a major carotenoid in the flavedo and juice sacs. In contrast, most oranges are violaxanthin-abundant cultivars. In Valencia orange, violaxanthin is massively accumulated with 9-*cis*-violaxanthin as the principal carotenoid in the flavedo and juice sacs. Different from mandarins and most oranges, other species, such as lemon and grapefruit, accumulate low levels of β-cryptoxanthin and violaxanthin in the fruits. In lemon, a relatively low level of carotenoid accumulates in the mature fruits with the colorless carotenoids, phytoene and phytofluene, as the major carotenoids, while the contents of colored carotenoids, β-cryptoxanthin and violaxanthin, are very low. In red grapefruit, carotenes, such as phytoene, phytofluene, and lycopene, are predominantly accumulated, while the contents of xanthophylls are low in the mature fruits. In addition, different from most citrus cultivars which are in yellow or orange color, some grapefruits, oranges and pummelos exhibit red colors in the juice sacs because of the massive accumulation of lycopene.

24.2.2 The metabolism of β-cryptoxanthin in citrus fruits

β-Cryptoxanthin is one of the six main carotenoids found in human plasma, and beneficial to human health. In nature, the accumulation of β-cryptoxanthin is not a common event, and it is massively accumulated in only a few fruits, such as papaya, persimmon, citrus, and guava. Among them, citrus fruits are major sources of β-cryptoxanthin for human diet. In the past decade, the metabolism of β-cryptoxanthin in citrus fruits has been extensively elucidated (see Chapter 12); the key genes that regulate β-cryptoxanthin accumulation are identified. As shown in Fig. 24.2, β-cryptoxanthin is biosynthesized from geranylgeranyl pyrophosphate (GGPP) through a series of desaturations, cyclizations, and hydroxylation reactions. Phytoene synthase (PSY), phytoene desaturase (PDS), ξ-carotene desaturase (ZDS), lycopene β-cyclases (LCYb1 and LCYb2), and β-ring hydroxylase (HYb) are the key enzymes responsible for β-cryptoxanthin biosynthesis. In citrus fruits, β-cryptoxanthin is further converted to violaxanthin via zeaxanthin by β-ring hydroxylase (HYb) and zeaxanthin epoxidase (ZEP). Kato et al. (2004) reported that the higher expression levels of upstream synthesis genes (*CitPSY*, *CitPDS*, *CitZDS*, and *CitLCYb*) and low expression levels of downstream synthesis genes (*CitHYb* and *CitZEP*) led to the massive accumulation of β-cryptoxanthin in the juice sacs of mandarins.

β-Cryptoxanthin can be oxidatively cleaved in a site-specific manner by carotenoid cleavage dioxygenases (CCDs). In citrus, two CCDs, CitCCD1 and CitCCD4, have been isolated and the cleavage reactions of β-cryptoxanthin catalyzed by CitCCD1 and CitCCD4 are elucidated in citrus fruits. CitCCD1 is highly expressed in most citrus fruits, and it cleaves β-cryptoxanthin at the 9,10 and 9′,10′ positions, producing 3-hydroxy-β-ionone and β-ionone, which are aroma compounds, and C14 dialdehyde (Kato et al., 2006). In contrast to CitCCD1, CitCCD4 is specifically expressed in the flavedo of certain citrus varieties. In the case of CitCCD4, β-cryptoxanthin was cleaved at 7,8 or 7′,8′ position producing β-citraurin and *trans*-β-8′-carotenal (Ma et al., 2013). β-Citraurin is a color-imparting pigment responsible for the reddish color of citrus fruits. The citrus varieties accumulating β-citraurin are considered more attractive because of their red-orange color.

In citrus fruits, β-cryptoxanthin is present in free and esterified forms. In the esterified form, the hydroxyl groups of β-cryptoxanthin are acylated with different fatty acids. The esterification by fatty acids does not affect the antioxidant activity of β-cryptoxanthin, but instead improves its thermal stability (Fu et al., 2010). In the study by Ma et al. (2017), the results suggested that the esterification degree of β-cryptoxanthin increased gradually in the flavedos of two Satsuma mandarin cultivars (Miyagawa-wase and Yamashitabeni-wase) during the ripening process. In the mature fruits, >80% of β-cryptoxanthin was found to be present in esterified forms with β-cryptoxanthin laurate, myristate, and palmitate as the major β-cryptoxanthin esters. Since β-cryptoxanthin esters exhibited comparable bioavailability with the free β-cryptoxanthin, β-cryptoxanthin esters that accumulated in citrus fruits also serve as an important source of dietary xanthophylls for humans (Breithaupt et al., 2003).

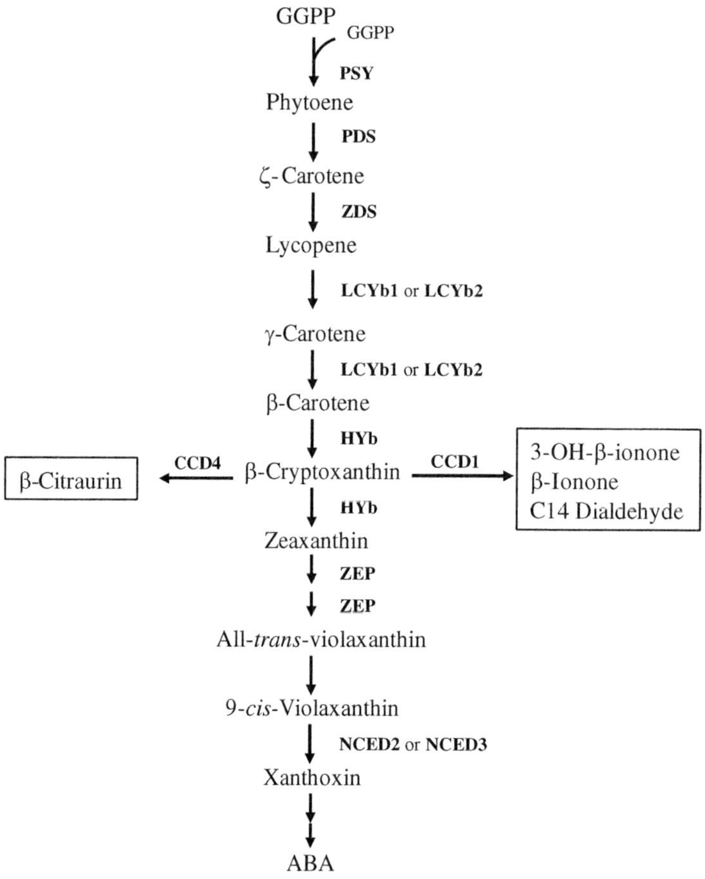

FIG. 24.2 Metabolic pathway of β-cryptoxanthin in citrus. GGPP, geranylgeranyl diphosphate. Enzymes are named according to the designation of their genes. PSY, phytoene synthase; PDS, phytoene desaturase; ZDS, ζ-carotene desaturase; LCYb, lycopene β-cyclase; HYb, β-ring hydroxylase; ZEP, zeaxanthin epoxidase; NCED, 9-*cis*-epoxycarotenoid dioxygenase; CCD, carotenoid cleavage dioxygenase.

24.2.3 The role of β-cryptoxanthin in human health

Of the six major carotenoids found in human serum, little is known about the physiological functions of β-cryptoxanthin, compared with those of β-carotene and lycopene. This may be due to the fact that the concentration of β-cryptoxanthin in human serum and breast milk is relatively low in many countries where people do not have a habit of eating citrus fruits containing high concentration of β-cryptoxanthin (Canfield et al., 2003; Michaud et al., 1998). It is therefore reasonable that β-cryptoxanthin has not received much attention compared to other types of carotenoids.

24.2.3.1 Serum β-cryptoxanthin levels

Findings from nutritional epidemiological studies published to date in Europe and the United States show some very interesting trends, namely serum β-cryptoxanthin is generally maintained at high levels despite the low β-cryptoxanthin intake (Albanes et al., 1997; Wahlqvist et al., 1994). For example, according to a report (Wahlqvist et al., 1994) published by a research group in Australia, although the daily β-cryptoxanthin intake in the population studied was 0.2 mg, which was one-tenth of the intakes of β-carotene, lycopene, or other carotenoids, the serum concentrations of these substances were found to be at the same levels. This finding led to the assumption that β-cryptoxanthin is easily absorbed and it can also survive relatively long in the body. Our survey, on the other hand, has revealed that serum β-cryptoxanthin levels rise dramatically in January, when mandarin fruits are in peak season, in a manner dependent on the frequency of mandarin fruit intake (Fig. 24.3), and that even during the preharvest month of September, serum β-cryptoxanthin levels are found at significantly higher levels in individuals with higher frequency of mandarin fruit intake during the winter season (Sugiura et al., 2002). This finding indicates that β-cryptoxanthin remained in the body over a relatively long period of time. Furthermore, a detailed follow-up of seasonal changes in serum β-cryptoxanthin levels over a year revealed Satsuma mandarin fruits to be the only food source affecting the serum concentrations of β-cryptoxanthin (Sugiura et al., 2004).

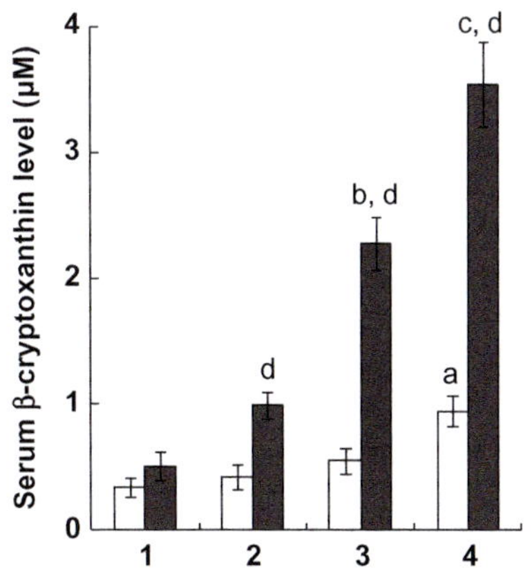

FIG. 24.3 Relationship of serum β-cryptoxanthin level with Satsuma mandarin intake. The levels of Satsuma mandarin consumption are as follows: Level 1: "I rarely eat Satsuma mandarin." Level 2: "Although I sometimes eat Satsuma mandarin, I eat fewer than three pieces of this fruit a week." Level 3: "I eat one to three Satsuma mandarins daily." Level 4: "I eat more than four Satsuma mandarins daily." Open columns show the results in September, closed columns show those in January. All data are represented as means ± SEM. Number of observations in group of levels 1, 2, 3, and 4 are 5, 11, 52, and 26, respectively. $^{a}P<0.05$ vs data of September in group of level 1, $^{b}P<0.05$, $^{c}P<0.01$ vs data of January in group of level 1 in Scheffe's test. $^{d}P<0.001$ vs data of September in the same group in paired *t*-test.

24.2.3.2 Characteristics of β-cryptoxanthin as identified in epidemiological studies

Epidemiological research on the association between fruit and vegetable intake and lifestyle-related diseases has witnessed remarkable progress in recent years, and carotenoids are regarded as one of the factors playing a key role in the prevention of lifestyle-related diseases such as cancer, type-2 diabetes, and so on. Many researchers have in fact focused their attention on carotenoids and have tried to identify and analyze the types of carotenoids that are most effective in preventing these types of diseases. Interestingly, according to the previous epidemiological studies on carotenoids, β-cryptoxanthin was found to be the only type of carotenoid that showed an association with lower rates of lung cancer, diabetes, and rheumatism (Männistö et al., 2004; Montonen et al., 2004; Cerhan et al., 2003). It is noteworthy that a statistically significant negative association between β-cryptoxanthin and lung cancer risks in smokers has been reported in several studies (Yuan et al., 2003). Furthermore, recently, we have found that serum β-cryptoxanthin is inversely associated with the risk for developing osteoporosis (Fig. 24.4), type-2 diabetes (Fig. 24.5), hyperlipidemia, liver dysfunction, and atherosclerosis, in a major investigation that made up the Mikkabi prospective cohort study (Sugiura et al., 2012, 2015a,b, 2016a,b; Nakamura et al., 2016). These findings suggest that β-cryptoxanthin may be physiologically superior to other carotenoids, but its underlying mechanism of action has not yet been fully elucidated.

β-Cryptoxanthin is a type of carotenoid that has a hydroxyl (–OH) group attached to one of the rings at the end of the carbon chain. Despite being a xanthophyll, it has pro-vitamin A activity as well as the ability to bind to a carotene binding protein and a retinoic acid receptor (Matsumoto et al., 2007; Rao et al., 1997). Moreover, due to the presence of the –OH

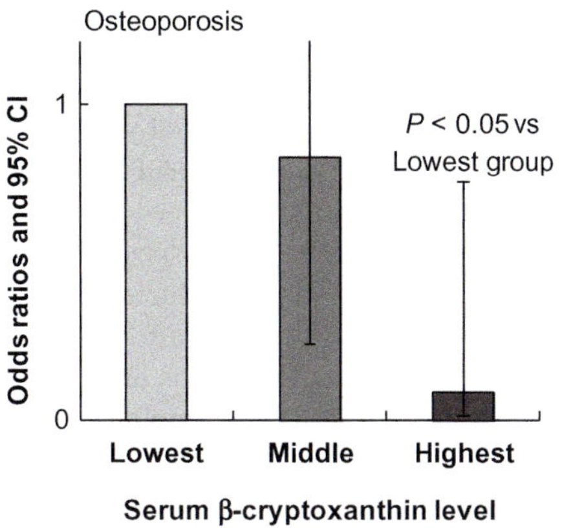

FIG. 24.4 Odds ratios and 95% confidence intervals of tertiles of baseline serum β-cryptoxanthin concentration on incidence of osteoporosis among postmenopausal Japanese female subjects.

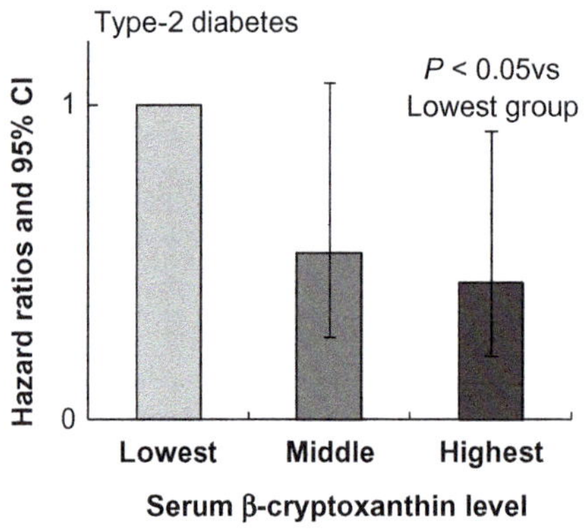

FIG. 24.5 Hazard ratios and 95% confidence intervals of tertiles of baseline serum β-cryptoxanthin concentration on incidence of type 2 diabetes among Japanese subjects.

group, β-cryptoxanthin has a higher polarity than β-carotene. However, its polarity is not quite as strong as lutein and zeaxanthin, which have –OH groups at both ends of their carbon chain. Due to these physicochemical properties specific to β-cryptoxanthin, the localization of β-cryptoxanthin in tissues or cells differs slightly from other types of carotenoids, which may be related to the diverse characteristics of β-cryptoxanthin. Further studies are necessary to provide more detailed information about the characteristics of β-cryptoxanthin.

24.3 Flavonoid in citrus fruits

24.3.1 Flavonoid composition in citrus fruits

Flavonoids are a group of polyphenolic secondary metabolites broadly expanded in nature, that fulfill a variety of important functions in plants including ultraviolet (UV) radiation protection, signal molecule, phytoalexins, and detoxifying agents, as well as defense against pathogen attacks (Falcone Ferreyra et al., 2012). In addition, flavonoids possess strong anti-oxidative and chelating activities, and have long been used as traditional medicines with a wide range of pharmacological benefits. Flavonoids, which are accumulated in almost all fruits and vegetables, are common substances in the daily diet. Citrus fruits are one of the richest sources of flavonoids, which are widely distributed in the seed, leaf, root, and fruits. To date, >100 flavonoids have been identified in citrus fruits, which are mainly present as sugar conjugates (Wang et al., 2017). According to the molecular structures, citrus flavonoids are primarily classified into four groups: flavanones, flavones, flavonols, and anthocyanins.

Flavanones are the major flavonoids massively accumulated in citrus fruits. Citrus flavanones are present in aglycone and glycoside forms. As shown in Fig. 24.6, eriodictyol, hesperetin, naringenin, and isosakuranetin are the main flavanone aglycones identified in citrus fruits. Compared with the flavanone aglycones, the flavonone glycosides are more abundant in most citrus fruits. In the glycoside forms, the flavanone aglycones are conjugated with rutinose and neohesperidose, respectively. The composition of flavanone glycoside varies among different citrus varieties. The two major rutinoside flavanones, hesperidin and narirutin, are accumulated in sweet orange, tangors, mandarins, and lemon. In contrast, naringin, neohesperidin, and neoeriocitrin, which are the major neohesperidoside flavanones, are predominantly accumulated in grapefruit, sour orange, and Natsudaidai. In addition, as the neohesperidoside flavanones have a bitter taste, the accumulation of naringin and neohesperidin gives a unique flavor to citrus fruits.

Flavones are a subgroup of flavonoids, which differ from flavanones by a C2–C3 double bond. Flavones are also present in aglycone and glycoside forms in citrus fruits. Flavone 7-*O*-rutinoside and flavone 7-*O*-neohesperidoside are the main flavone glycosides accumulated in citrus fruits. In addition, some of the flavones are highly methoxylated in citrus, and the polymethoxylated flavones (PMFs) (nobiletin, tangeretin, and sinensetin) are specifically accumulated in the flavedo of citrus fruits.

Flavonols are widely present in plants. In citrus, kaempferol, quercetin, and rutin are identified as the major citrus flavonols, which are preferentially accumulated in leaves of most citrus varieties. In citrus fruits, a very small amount of flavonols are specifically accumulated in a few species, such as lemon and lime.

Anthocyanins are water-soluble pigments synthesized via the phenylpropanoid pathway. In citrus, anthocyanins are specifically accumulated in the rind and juice vesicles of several cultivars of blood orange (see Chapter 12), such as Tarocco, Moro, and Sanguinello. The major anthocyanins of blood orange have been identified as cyanidin 3-glucoside and cyanidin 3-(6″-malonylglucoside) (Lo Piero, 2015). In citrus fruits, the accumulation of anthocyanin is highly regulated by the prevailing climatic conditions during fruit ripening.

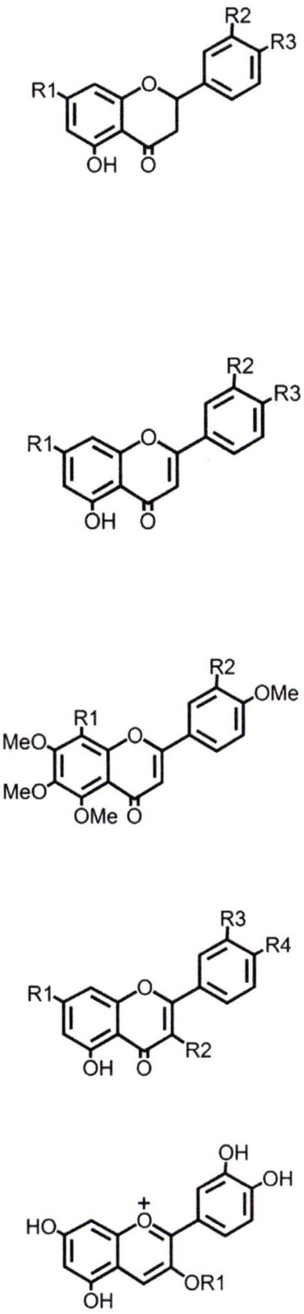

Flavanone

Eriodictyol	R1=OH, R2=OH, R3=OH
Eriocitrin	R1=O-Rutinose, R2=OH, R3=OH
Neoeriocitrin	R1=O-Neohesperidose, R2=OH, R3=OH
Hesperetin	R1=OH, R2=OH, R3=OMe
Hesperidin	R1=O-Rutinose, R2=OH, R3=OMe
Neohesperidin	R1=O-Neohesperidose, R2=OH, R3=OMe
Naringenin	R1=OH, R2=H, R3=OH
Narirutin	R1=O-Rutinose, R2=H, R3=OH
Naringin	R1=O-Neohesperidose, R2=H, R3=OH
Isosakuranetin	R1=OH, R2=H, R3=OMe
Neoponcirin	R1=O-Rutinose, R2=H, R3=OMe
Poncirin	R1=O-Neohesperidose, R2=OH, R3=OMe

Flavone

Apigenin	R1=OH, R2=H, R3=OH
Isorhoifolin	R1=O-Rutinose, R2=H, R3=OH
Rhoifolin	R1=O-Neohesperidose, R2=H, R3=OH
Diosmetin	R1=OH, R2=OH, R3=OMe
Diosmin	R1=O-Rutinose, R2=OH, R3=OMe
Neodiosmin	R1=O-Neohesperidose, R2=OH, R3=OMe
Chrysoeriol	R1=OH, R2=OMe, R3=OH
Chrysoeriol-7-O-Neohesperidose	R1=O-Neohesperidose, R2=OMe, R3=OH
Luteolin	R1=OH, R2=OH, R3=OH

Polymethoxylated flavone

Nobiletin	R1=OMe, R2=OMe
Sinensetin	R1=H, R2=OMe
Tangeretin	R1=OMe, R2=H

Flavonol

Quercetin	R1=OH, R2=OH, R3=OH, R4=OH
Rutin	R1=OH, R2=O-Rutinose, R3=OH, R4=OH
Kaempferol	R1=OH, R2=OH, R3=H, R4=OH

Anthocyanin

Cyanidin 3-glucoside	R1=glucosyl
Cyanidin 3-(6″-malonylglucoside)	R1=6″-malonylglucosyl

FIG. 24.6 Major flavonoids in citrus.

24.3.2 Heath benefits of citrus flavonoids

Flavonoids are one of the important compositions of herbal drugs, and have long been used in traditional medicine. Flavonoids have high antioxidant and radical scavenging activities, and dietary flavonoid protects the human body against free radicals and other pro-oxidative compounds. In addition, flavonoids are able to chelate with metals and bind with proteins. Recent research suggests the formation of complexes with metal ions and proteins may increase the potential of flavonoids to prevent some cardiovascular disorders (Selvaraj et al., 2014). Citrus fruits accumulate high amounts of flavonoids in the flavedo, albedo, seed, and juice sacs. In citrus, hesperidin, naringin, and PMFs are the characteristic flavonoids

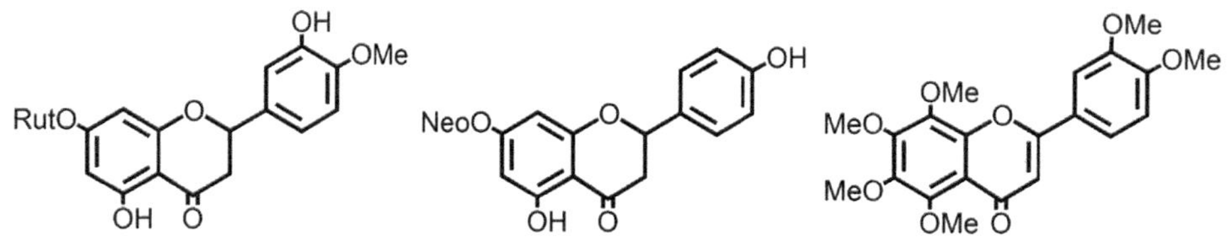

FIG. 24.7 Structures of hesperidin (left), naringin (middle), and nobiletin (right).

specifically accumulated in the fruits (Fig. 24.7). In the past decades, the roles of these citrus flavonoids in human health have been extensively investigated, and their pharmacological and therapeutic properties were well documented (Cirmi et al., 2016; Benavente-García and Castillo, 2008; Tripoli et al., 2007).

24.3.2.1 *Citrus flavanone hesperidin and its biological properties*

Hesperidin (7-*O*-rutinoside of hesperetin), a member of the flavanone group of flavonoids, can be isolated in large amounts from the fruit pulp and rind of some citrus species. Hesperidin is accumulated at a high level in various orange, mandarin, lemon, lime, and grapefruit varieties. Hesperidin was first isolated by Lebreton in 1927. In the early studies, hesperidin was reported to possess vitamin-like activity and to decrease capillary permeability, leakiness, and fragility (Scarborough and Bacharach, 1949). Hesperidin deficiency has been linked with abnormal capillary leakiness and pain in the hands, feet, and leg cramps at night. Over the past decades, extensive research was conducted on the bioactivity of hesperidin, and numerous biological properties, principally anticancer and anti-inflammatory activities have been identified (Garg et al., 2001; Roohbakhsh et al., 2015).

Hesperidin possesses significant anti-inflammatory activity. In the studies of Hirata et al. (2005) and Emim et al. (1994), it was reported that hesperidin diminished gene expression of cyclooxygenase-2 (*COX-2*) and provided anti-inflammatory effects in some experimental models (Hirata et al., 2005; Emim et al., 1994). Hesperidin is also a promising anticancer agent whose activity has been reported in multiple types of cancers, including gastric, colon, breast, lung, and liver cancers (Devi et al., 2015). Roohbakhsh found that the cancer chemopreventive properties of hesperidin were mainly attributed to its anti-oxidative and anti-inflammatory activities. Hesperidin inhibits tumor growth by increasing the anti-oxidant defense system, inducing apoptosis in cancerous cells, decreasing inflammatory cytokines and enzymes, and inhibiting angiogenesis and metastasis (Roohbakhsh et al., 2015). Moreover, hesperidin targets multiple cellular proteins at the same time in cancer cells. In addition to the anticancer and anti-inflammatory activities, there were also some reports on the antidiabetic effects of hesperidin (Akiyama et al., 2009; Jung et al., 2004). Feeding of 1% hesperidin prevented the elevation of blood glucose and serum insulin level in Goto-Kakizaki rats. The administration of hesperidin to this diabetic rat model reduced not only the levels of HbA1c (glycosylated hemoglobin) and glucose, but also total cholesterol and triglycerides (Kakadiya et al., 2010). In hesperidin-treated rats, LDL (low-density lipoprotein) level decreased, while HDL (high-density lipoprotein) level increased (Monforte et al., 1995). In the meanwhile, hesperidin activated peroxisome proliferator-activated receptor-γ in diabetic rat and caused a reduction in blood lipid peroxide levels (Agrawal et al., 2014).

To date, clinical studies using purified flavonoids are very limited. In hypertriglyceridemic subjects, the administration of glucosyl hesperidin (500 mg/day for 24 weeks) decreased plasma triglycerides and apoB levels (Miwa et al., 2005). In contrast, in moderately hypercholesterolemic men and women, supplementation with 800 mg/day of hesperidin for 4 weeks had no statistically different effects on the levels of total cholesterol or LDL cholesterol compared with control group (Demonty et al., 2010). From these results, it is conceivable that identification of the population groups that will benefit from treatment with purified citrus flavanones is not straightforward, and further studies are required. In addition, in overweight subjects, a polyphenol extract of citrus fruits (900 mg/day for 12 weeks) significantly decreased body weight and plasma glucose level (Dallas et al., 2014).

24.3.2.2 *Citrus flavanone naringin and its biological properties*

Naringin, 4′,5,7-trihydroxy flavanone 7-rhamnoglucoside, is a flavanone glycoside that is abundant in citrus fruits. Naringin was first discovered in the flowers of grapefruit by De Vry in 1857. Later researchers reported that naringin was accumulated at a high level in several citrus fruits, such as grapefruit, pummelo and sour orange. In grapefruit, the juice vesicles and seeds had the lowest concentration of naringin, while the segment membranes, albedo, and pith had the highest (Jourdan et al., 1985). Due to the bitterness of naringin, the accumulation of naringin imparts a distinct bitter taste to citrus juices. Naringin was one of the active compounds widely used in Chinese herbal medicines. In the recent years, the roles of naringin in human health have attracted interests from scientists, and significant progress has been made on the biological and pharmacological properties of naringin. In the study by Viswanatha et al. (2017), it was found that naringin was effective to ameliorate the neurobehavioral defects produced by diverse physical and chemical stimuli on oxidative stress in rodent brains. The meta-analysis that was performed to examine the effect of naringin on oxidative stress markers suggested that naringin restored the levels of all oxidative stress markers (oxidative, nitrosative, enzymes, and mitochondrial complexes) in different parts of the rodent brain. Alam et al. (2014) reported that naringin supplementation was useful for the treatment of obesity, diabetes, hypertension, and metabolic syndrome. The beneficial effects of naringin on metabolic syndrome were not only attributed to its high anti-oxidative and anti-inflammatory activities, but also closely related to the modification of molecular signal pathway by naringin. In metabolic syndrome, obesity, and related cardiovascular complications, naringin influences AMPK-, PPARα, and CPT-1-mediated fat utilization and preserves mitochondrial function. Recent fundamental studies have suggested that naringin has chemopreventive and anticancer activities in a various types of cancers, including oral, breast, colon, liver, lung, and ovarian cancer (Bharti et al., 2014). In addition, the supplementation of naringin was able to inhibit cell proliferation, induce apoptosis in tumor cell, and reduce the migration and invasion of some tumor cells (Chen et al., 2016).

24.3.2.3 Citrus polymethoxylated flavones and their biological properties

PMFs, including nobiletin, tangeretin, 3,3′,4′,5,6,7,8-heptamethoxyflavone, and sinensetin, are abundant in the peel tissues of certain types of citrus species. In citrus fruits, the PMFs are present as minor flavonoid components, and their contents are lower than hesperidin and naringin. However, as the polymethoxylated structure increases the metabolic stability and permeability of flavones, the PMFs are more easily absorbed and exhibited higher bioavailability than flavonoid glycosides.

Nobiletin (5,6,7,8,3′,4′-hexamethoxyflavone) is a major PMF that is specifically accumulated in citrus fruits (Green et al., 2007). Recently, extensive epidemiological studies have suggested that nobiletin was a promising candidate for development as a therapeutic agent with a broad range of pharmacological activities (Huang et al., 2016). In high-calorie diet-fed mice, nobiletin attenuated dyslipidemia through a reduction in VLDL (very low-density lipoprotein)-triglyceride secretion. Nobiletin did not activate any peroxisome proliferator-activated receptor (PPAR), indicating that the metabolic effects of nobiletin were PPAR independent (Mulvihill et al., 2011). Recently, anti-dementia activity of nobiletin has been reported. Nobiletin enhances protein kinase A (PKA)/extracellular signal-regulated kinase (ERK)/cAMP response element-binding protein (CREB) signaling in PC12D cells and cultured rat hippocampal neurons (Nagase et al., 2005a,b). In addition, nobiletin induces long-term potentiation, which is believed to be the cellular basis of learning and memory, via activating PKA-dependent phosphorylation of the alpha-amino-3-hydroxy-5-methyl-4-isoxazolepropionic acid (AMPA) receptor subunit GluR1 in hippocampus slices (Matsuzaki et al., 2008). These findings indicated that nobiletin exhibited memory-improving effects in various animal models and exerted a wide range of beneficial effects against pathological features of Alzheimer disease, oxidative stress, cholinergic neurodegeneration, and dysfunction of synaptic plasticity-related signaling.

24.3.3 The biosynthesis of flavonoids in citrus fruits

In citrus, some of the key enzymes were isolated and their roles in flavonoid accumulation were characterized (Lo Piero, 2015; Moriguchi et al., 2001; Khan and Dangles, 2014). As shown in Fig. 24.8, chalcone synthase (CHS), which is the first committed enzyme in the biosynthesis of flavonoids, condenses three malonyl-CoA molecules and one 4-coumaroyl-CoA to produce the chalcone scaffolds from which flavonoids derive. In the following step, the flavonoid chalcone is isomerized by chalcone isomerase (CHI) to form flavanone naringenin. The flavanone is further converted into flavone by flavone synthase (FNS), and into dihydroflavonol (dihydrokaempferol) by flavonone 3-hydroxylase (F3H). Dihydroflavonol, which is the substrate for the biosynthesis of flavonol and anthocyanin, lies at an important branch point in flavonoid biosynthesis. Flavonol synthase (FLS) catalyzes the oxidation of dihydroflavonol to produce flavonol. In the anthocyanin biosynthesis, dihydroflavonol reductase (DFR) competes for the substrate, dihydroflavonols, with FLS, and reduces the dihydroflavonol to its corresponding leucoanthocyanidin. Leucoanthocyanidin is further converted into anthocyanin catalyzed by leucoanthocyanidin dioxygenase (LDOX) and anthocyanidin reductase (ANR). In higher plants, the biosynthesis of flavonoids is not only controlled by the structural genes, but also regulated by MYB transcription factors. In citrus, a R2R3-MYB transcription factor, *Ruby*, was isolated by Butelli et al. (2012). It is responsible for anthocyanin biosynthesis, and the insertion of a Copia-like retrotransposon adjacent to Ruby activated the production of anthocyanin in blood orange. In citrus fruits, flavonoids are synthesized in the early ripening stage, and the massive accumulation of flavonoids along with the high gene expression of *CHS*, *CHI*, and *F3H* was observed in the young tissues.

Citrus flavonoids are a diverse group of compounds, which are derived from flavanone naringenin, and modified by hydroxylation, methylation, glucosylation, and rhamnosylation to form various derivatives. To date, although some genes encoding the enzymes that biosynthesize the core structure have been characterized in citrus fruits, the information on the

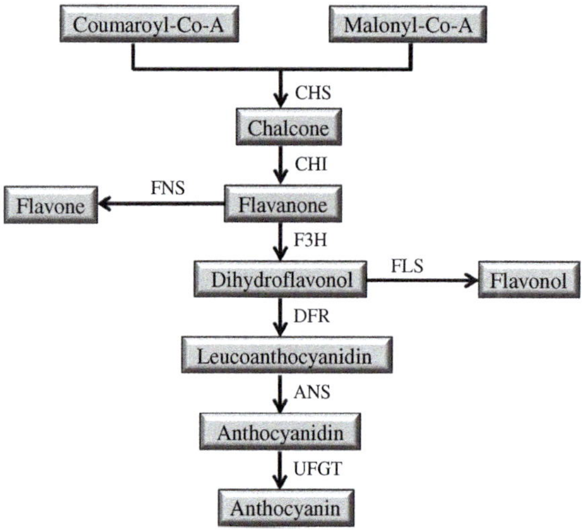

FIG. 24.8 Flavonoid biosynthetic pathway in citrus. CHS, chalcone synthase; CHI, chalcone isomerase; F3H, flavanone 3-hydroxylase; FNS, flavone synthase; FLS, flavonol synthase; DFR, dihydroflavonol 4-reductase; ANS, anthocyanidin synthase; UFGT, UDP-glucose flavonoid 3-O glucosyltransferase.

genes responsible for the process of hydroxylation, methylation, glucosylation, and rhamnosylation of flavonoids is still limited. In the future, further studies on citrus flavonoids biosynthesis will contribute to elucidating the mechanisms of flavonoid accumulation in citrus fruits, and facilitate the strategy to modify flavonoid content and composition in citrus. In addition, the improved flavonoid content and composition by manipulating the flavonoid biosynthetic pathway will be important for human health as well as for enhancing market value of citrus fruits.

24.4 Ascorbic acid in citrus fruits

24.4.1 The roles of ascorbic acid in human health

L-Ascorbic acid (AsA) also called vitamin C is one of the most important antioxidants accumulated in plants. AsA not only fulfills critical functions in regulating plant growth and development, but also is essential for maintaining human health (Moser and Chun, 2016; Al-Khudairy et al., 2017). The earliest recorded medical use of AsA was in the prevention of scurvy, which is a disease caused by the deficiency of vitamin C. During the 15th and 16th centuries, many sailors suffered from serious scurvy and died during the long sea voyages. In the mid-17th century, a British naval surgeon, Sir James Lind, found that the consumption of limes could control scurvy. However, the mechanisms responsible of the scurvy cure were completely unknown, until Dr. A. Szent-Gyorgyi isolated AsA from orange and cabbage juice, and proved that AsA was the active scurvy-preventing compound (Fig. 24.9). Dr. A. Szent-Gyorgyi was awarded the Nobel Prize in 1937 for the discovery of AsA (Hoffer, 1989).

Nowadays, although scurvy is rare in most countries, the roles of AsA in maintaining human health still attract interest from researchers. Extensive studies suggested that AsA was a powerful dietary anti-oxidant; daily intake of AsA could improve collagen formation, iron absorption, and immune function, and prevent the occurrences of cardiovascular diseases and age-related macular degeneration (Moser and Chun, 2016; Al-Khudairy et al., 2017; Hoffer, 1989; Pullar et al., 2017; Lane and Richardson, 2014; Wintergerst et al., 2006; Braakhuis et al., 2017). In the recent years, growing epidemiologic evidence showed that plasma concentration of AsA was inversely associated with the risks of certain cancers, such as lung, breast, stomach, and esophagus cancers (Mastrangelo et al., 2018). Generally, the plasma concentration of AsA in people

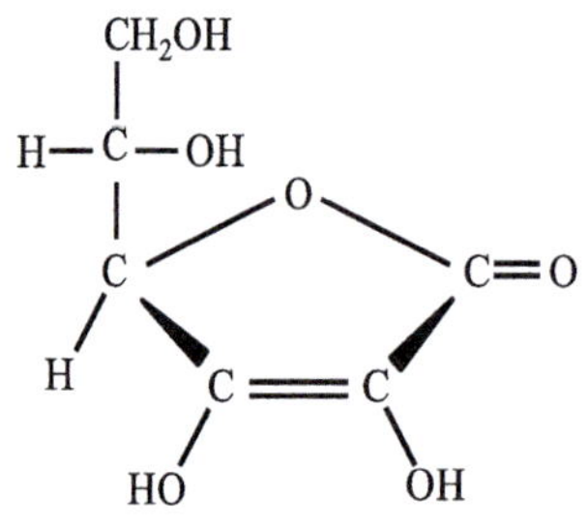

FIG. 24.9 Structure of L-Ascorbic acid (AsA).

with cancer was lower than that in healthy people. The application of high-dose AsA, which has been investigated in the cancer treatment since the 1970, was reported to be beneficial to cancer patients (Cameron and Campbell, 1974). Recently, new clinical trials suggested that the intake of high-dose AsA could improve quality of life and decrease cancer-related side effects in cancer patients. In addition to the cancer patients, vitamin C intake is also beneficial to smokers and nonsmokers who were exposed to secondhand smoke; supplementation with AsA reduced the oxidative stress caused by smoking and normalized DNA methylation (McEvoy et al., 2014; Shorey-Kendrick et al., 2017). To maintain health and reduce the risks of chronic diseases risks, the recommended dietary allowances of AsA are 75 mg/day (women) and 90 mg/day (men) for adults who do not smoke, while for smokers an additional 35 mg/day is recommended.

24.4.2 The accumulation of AsA in citrus fruits

AsA is biosynthesized and accumulated in plants and most animals. However, humans are unable to synthesize AsA due to the lack of L-gulonolactone oxidase, which is the enzyme in the final step of AsA biosynthesis. Thus, humans need to obtain AsA from the daily diet. Citrus fruits are important and rich sources of AsA. Gliszczynska-Swiglo reported that one glass (200 mL) of orange juice can deliver 30%–80% of recommended daily intake of AsA (Gliszczyńska-Świgło et al., 2004). In citrus fruits, the content of AsA varies greatly among different citrus varieties (Fig. 24.10). In general, orange

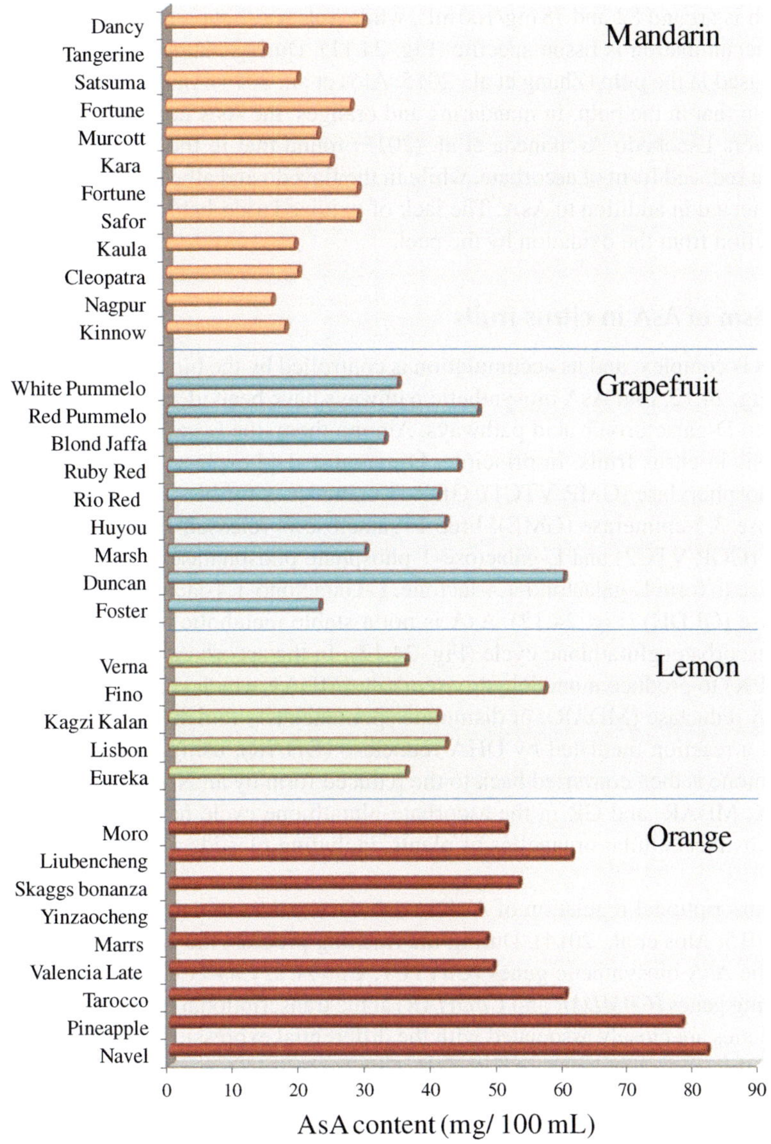

FIG. 24.10 Content of L-Ascorbic acid (AsA) among different citrus varieties.

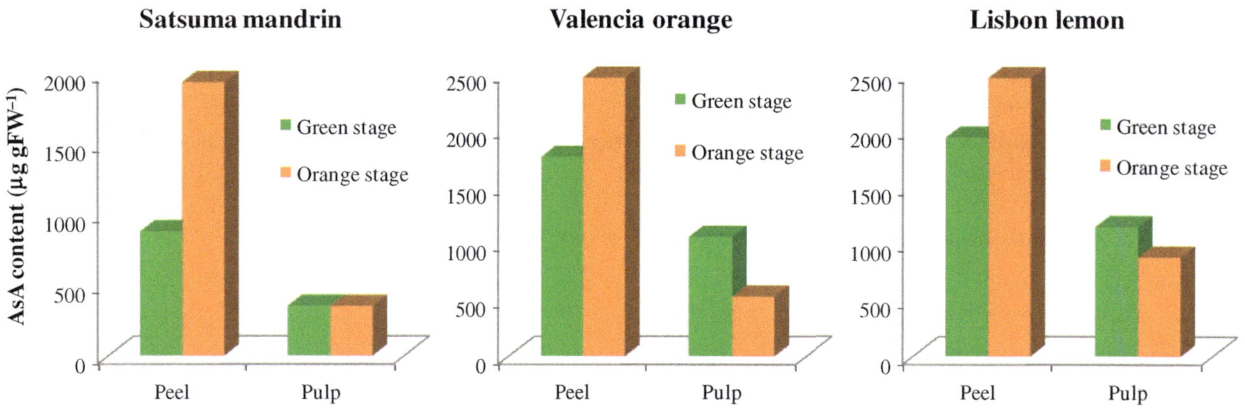

FIG. 24.11 Changes in L-Ascorbic acid (AsA) content in citrus fruits during maturation.

accumulates the highest AsA in the juice, followed by lemon, grapefruit, and mandarins. AsA content for instance, ranges from 29 to 82 mg/100 mL in the juice of orange. Among the orange varieties, Navel and Pineapple contain higher amounts of AsA in the juice, which is around 82 and 78 mg/100 mL, while Valencia Late contains lower amounts (29 mg/100 mL).

In citrus fruits, AsA accumulation is tissue-specific (Fig. 24.11). During maturation, AsA content increased gradually in the peel, while it decreased in the pulp (Zhang et al., 2015; Alós et al., 2014). In the mature fruits, the AsA content in the peel was much higher than that in the pulp. In mandarins and oranges, the AsA content in the peel was around four times that in the pulp. In addition, Escobedo-Avellaneda et al. (2013) found that in the pulp the ascorbate was predominantly present as AsA, which is a reduced form of ascorbate, while in the flavedo and albedo dehydroascorbic acid (oxidized form of ascorbate) was also detected in addition to AsA. The lack of oxidized ascorbate and low AsA content in the pulp might be attributed to the protection from the oxidation by the peel.

24.4.3 The metabolism of AsA in citrus fruits

AsA metabolism in citrus is complex, and its accumulation is controlled by the biosynthesis, breakdown, and regeneration processes. As shown in Fig. 24.12, four AsA biosynthetic pathways have been identified in citrus, including myo-inositol, L-gulose, L-galactose, and D-galacturonic acid pathways. Among them, the L-galactose pathway is proven to be a major route for AsA biosynthesis in citrus fruits. In principle, D-mannose-1-phosphate is converted into GDP-D-mannose by GDP-D-mannose pyrophosphorylase (GMP, VTC1). GDP-D-mannose is further converted into GDP-L-galactose via the action of GDP-D-mannose 3′5-epimerase (GME). Free L-galactose is released from GDP-L-galactose by the GDP-L-galactose phosphorylase (GGP, VTC2) and L-galactose-1-phosphate phosphatase (GPP, VTC4), and then is oxidized by L-galactose dehydrogenase to form L-galactono-1,4-lactone. L-Galactono-1,4-lactone is oxidized to AsA by L-galactono-1,4-lactone dehydrogenase (GLDH) (Fig. 24.12). AsA is not a stable metabolic product, and it can be further oxidized and regenerated via the ascorbate–glutathione cycle (Fig. 24.12). In the ascorbate–glutathione cycle, AsA is oxidized by ascorbate peroxidase (APX) to produce monodehydroascorbate (MDA), which can be reduced enzymatically to AsA by NADPH-dependent MDA reductase (MDAR), or dismutate spontaneously to dehydroascorbate (DHA). DHA is reduced to AsA enzymatically in a reaction mediated by DHA reductase (DHAR), using glutathione as an electron donor. The resulting oxidized glutathione is then converted back to the reduced form by an NADPH-dependent glutathione reductase (GR). The enzymes APX, MDAR, and GR in the ascorbate-glutathione cycle have been found to exist as isoenzymes, which are distributed in distinct cellular organelles of plants, including plastids, mitochondria, and peroxisomes (Zhang et al., 2015).

In citrus fruits, the transcriptional regulation of AsA metabolic genes is a major mechanism that regulates AsA accumulation (Zhang et al., 2015; Alós et al., 2014). During the ripening process, the accumulation of AsA in the flavedo and pulp is co-regulated by the AsA biosynthetic genes (*CitVTC1*, *CitVTC2*, *CitVTC4*, and *CitGLDH*), AsA oxidation genes *CitAPXs*, and AsA recycling genes (*CitMDAR* and *CitDHAR*) at the transcriptional level. In addition, the differences in AsA content among citrus varieties are closely associated with the differential expression levels of AsA metabolic genes. Zhang et al. (2015) reported that the expression levels of *CitVTC1*, *CitVTC2*, *CitVTC4*, and *CitGLDH* in the L-galactose pathway and *CitMDAR1* and *CitDHAR* in the ascorbate-glutathione cycle were much higher in orange than those in Satsuma mandarin. The higher expression levels of AsA biosynthetic and regeneration genes contributed to the massive accumulation of AsA in the juice sacs of orange fruits (Zhang et al., 2015).

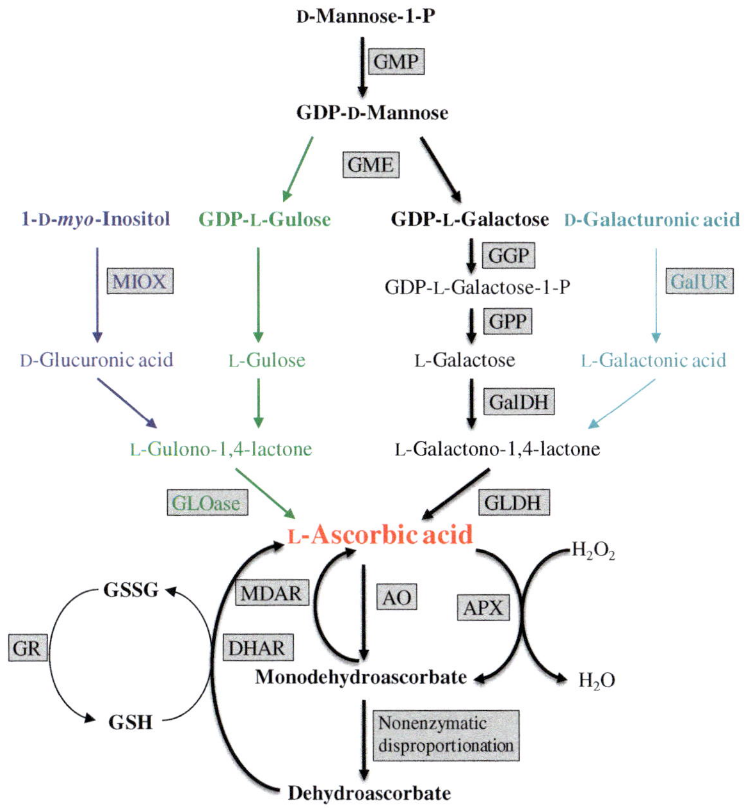

FIG. 24.12 Metabolic pathway of L-ascorbic acid (AsA) in citrus fruits. MIOX, myo-inositol oxygenase; GLOase, L-gulono-1,4-lactone oxidase; GMP, GDP-D-mannose pyrophosphorylase; GME, GDP-D-mannose 3′5-epimerase; GGP, GDP-L-galactose phosphorylase; GPP, L-galactose-1-phosphate phosphatase; GalDH, L-galactose dehydrogenase; GLDH, L-galactono-1,4-lactone dehydrogenase; APX, ascorbate peroxidase; AO, ascorbate oxidase; MDAR, monodehydroascorbate reductase; DHAR, dehydroascorbate reductase; GR, glutathione reductase; GSH, reduced glutathione; GSSG, oxidized glutathione; GalUR, D-galacturonic acid reductase.

24.4.4 The regulation of AsA in citrus fruits

In citrus fruits, the accumulation of AsA is affected by numerous factors, such as environmental conditions, harvesting procedures, and postharvest management (Mditshwa et al., 2017; Magwaza et al., 2017). Light, which is one of the most essential environmental factors in plant growth and development, has been proven to have a significant effect on AsA accumulation in citrus fruits. Lado et al. (2015) reported that shading fruits did not affect the accumulation of AsA in the pulp, but significantly reduced AsA level in the flavedo as compared with light-exposed fruits. Moreover, gene expression results suggested that the D-galacturonic pathway has a relevant contribution to AsA accumulation under light irradiation in the flavedo of citrus fruits (Lado et al., 2015). In addition to light quantity, AsA accumulation is also affected by light quality. In the study of the effects of red and blue LED (light-emitting diode) lights on AsA accumulation in citrus fruits, the results showed that red LED light did not affect the AsA metabolism in the juice sacs, while continuous irradiation with blue LED light at the intensity of $100\,\mu mol\,m^{-2}\,s^{-1}$ was an effective method to enhance the content of AsA in the juice sacs (Zhang et al., 2015).

AsA metabolism in citrus fruits is also influenced by temperature. Yungyuen et al. (2017) found that 10°C was effective to induce AsA accumulation in the juice sacs cultured in vitro. As compared with the control at 20°C, the AsA content was significantly enhanced at 10°C, while it was not affected at 30°C. The enhancement of AsA content at 10°C was regulated by the expression of genes in AsA biosynthetic, breakdown, and regeneration pathway. Likewise, 10°C has been proven to be an optimal temperature to maintain AsA level in citrus fruits after harvest. In sweet orange Blood Red, fruits that were stored at 10°C demonstrated better vitamin C retention than those stored at 5°C or 20°C.

Generally, AsA content decreased rapidly after harvest in citrus fruits. Studies indicated that the postharvest treatments could delay the reduction of AsA in the fruits after harvest (Mditshwa et al., 2017). Shin et al. (1998) reported that coating fruits with carnauba wax enriched with grapefruit seed extract retained higher AsA content in citrus. In the studies by Chien and Chou (2006) and Eskandari et al. (2014), the treatments of chitosan and Arabic gum were found to be effective in maintaining

As A content in citrus fruits after harvest. In addition, the delay of AsA loss was observed in several citrus varieties by using controlled atmosphere technology and modified atmosphere packaging after harvest. In Clemenules Clementine, fruits segments stored in CA (11% O_2; 10% CO_2) had higher AsA content than the fruits stored in normal air (Del-Valle et al., 2009).

24.5 Conclusion

High amounts of bioactive compounds are accumulated in citrus fruits, and have multiple functions in human nutrition and health. Recently, extensive studies have demonstrated that carotenoids, flavonoids, and AsA in citrus fruits possess high anti-oxidative and anti-inflammatory activities, and are promising agents in the prevention and treatment of chronic diseases. To date, some progress has been made in understanding carotenoid and flavonoid accumulation in citrus fruits. However, the information on the metabolism of β-cryptoxanthin, hesperidin, naringin, and PMFs in citrus is still limited. In future studies, further elucidation of the mechanisms of β-cryptoxanthin, hesperidin, naringin, and PMFs accumulation will be of great importance to enhance citrus fruit nutritional and commercial values. Additionally, although the beneficial effects of bioactive compounds in citrus fruits are mainly investigated in in vitro and in vivo studies, their therapeutic use remains limited because of the lack of adequate clinical evidence. More clinical trials on the intake, metabolism, and cytotoxicity of these bioactive compounds are still required to confirm their efficacy in the treatment of human diseases.

References

Agrawal, Y.O., Sharma, P.K., Shrivastava, B., Ojha, S., Upadhya, H.M., Arya, D.S., et al., 2014. Hesperidin produces cardioprotective activity via PPAR-γ pathway in ischemic heart disease model in diabetic rats. PLoS ONE 9 (11), e111212. https://doi.org/10.1371/journal.pone.0111212.

Akiyama, S., Katsumata, S., Suzuki, K., Nakaya, Y., Ishimi, Y., Uehara, M., 2009. Hypoglycemic and hypolipidemic effects of hesperidin and cyclodextrin-clathrated hesperetin in Goto-Kakizaki rats with type 2 diabetes. Biosci. Biotechnol. Biochem. 73 (12), 2779–2782.

Alam, M.A., Subhan, N., Rahman, M.M., Uddin, S.J., Reza, H.M., Sarker, S.D., 2014. Effect of citrus flavonoids, naringin and naringenin, on metabolic syndrome and their mechanisms of action. Adv. Nutr. 5 (4), 404–417. https://doi.org/10.3945/an.113.005603.

Albanes, D., Virtamo, J., Taylor, P.R., Rautalahti, M., Pietinen, P., Heinonen, O.P., 1997. Effects of supplemental beta-carotene, cigarette smoking, and alcohol consumption on serum carotenoids in the alpha-tocopherol, beta-carotene cancer prevention study. Am. J. Clin. Nutr. 66 (2), 366–372.

Al-Khudairy, L., Flowers, N., Wheelhouse, R., Ghannam, O., Hartley, L., Stranges, S., et al., 2017. Vitamin C supplementation for the primary prevention of cardiovascular disease. Cochrane Database Syst. Rev. 3, CD011114. https://doi.org/10.1002/14651858.CD011114.pub2.

Alós, E., Rodrigo, M.J., Zacarías, L., 2014. Differential transcriptional regulation of L-ascorbic acid content in peel and pulp of citrus fruits during development and maturation. Planta 239, 1113–1128.

Benavente-García, O., Castillo, J., 2008. Update on uses and properties of citrus flavonoids: new findings in anticancer, cardiovascular, and anti-inflammatory activity. Agric. Food Chem. 56 (15), 6185–6205. https://doi.org/10.1021/jf8006568.

Bharti, S., Rani, N., Krishnamurthy, B., Arya, D.S., 2014. Preclinical evidence for the pharmacological actions of naringin: a review. Planta Med. 80 (6), 437–451. https://doi.org/10.1055/s-0034-1368351.

Braakhuis, A., Raman, R., Vaghefi, E., 2017. The association between dietary intake of antioxidants and ocular disease. Diseases 5 (1), E3. https://doi.org/10.3390/diseases5010003.

Breithaupt, D.E., Weller, P., Wolters, M., Hahn, A., 2003. Plasma response to a single dose of dietary β-cryptoxanthin esters from papaya (*Carica papaya* L.) or non-esterified β-cryptoxanthin in adult human subjects: a comparative study. Br. J. Nutr. 90 (4), 795–801.

Butelli, E., Licciardello, C., Zhang, Y., Liu, J., Mackay, S., Bailey, P., et al., 2012. Retrotransposons control fruit-specific, cold-dependent accumulation of anthocyanins in blood oranges. Plant Cell 24 (3), 1242–1255. https://doi.org/10.1105/tpc.111.095232.

Cameron, E., Campbell, A., 1974. The orthomolecular treatment of cancer. II. Clinical trial of high-dose ascorbic acid supplements in advanced human cancer. Chem. Biol. Interact. 9 (4), 285–315.

Canfield, L.M., Clandinin, M.T., Davies, D.P., Fernandez, M.C., Jackson, J., Hawkes, J., et al., 2003. Multinational study of major breast milk carotenoids of healthy mothers. Eur. J. Nutr. 42, 133–141.

Cerhan, J.R., Saag, K.G., Merlino, L.A., Mikuls, T.R., Criswell, L.A., 2003. Antioxidant micronutrients and risk of rheumatoid arthritis in a cohort of older women. Am. J. Epidemiol. 157 (4), 345–354.

Chen, R., Qi, Q.L., Wang, M.T., Li, Q.Y., 2016. Therapeutic potential of naringin: an overview. Pharm. Biol. 54 (12), 3203–3210.

Chien, P., Chou, C., 2006. Antifungal activity of chitosan and its application to control post-harvest quality and fungal rotting of tankan citrus fruit (*Citrus tankan* Hayata). J. Sci. Food Agric. 86, 1964–1969.

Cirmi, S., Ferlazzo, N., Lombardo, G.E., Maugeri, A., Calapai, G., Gangemi, S., et al., 2016. Chemopreventive agents and inhibitors of cancer hallmarks: may citrus offer new perspectives? Nutrients 8 (11), E698.

Dallas, C., Gerbi, A., Elbez, Y., Caillard, P., Zamaria, N., Cloarec, M., 2014. Clinical study to assess the efficacy and safety of a citrus polyphenolic extract of red orange, grapefruit, and orange (Sinetrol-XPur) on weight management and metabolic parameters in healthy overweight individuals. Phytother. Res. 28 (2), 212–218. https://doi.org/10.1002/ptr.4981.

Del-Valle, V., Hernández-Muñoz, P., Catalá, R., Gavara, R., 2009. Optimization of an equilibrium modified atmosphere packaging (EMAP) for minimally processed mandarin segments. J. Food Eng. 91 (3), 474–481.

Demonty, I., Lin, Y., Zebregs, Y.E., Vermeer, M.A., van der Knaap, H.C., Jäkel, M., et al., 2010. The citrus flavonoids hesperidin and naringin do not affect serum cholesterol in moderately hypercholesterolemic men and women. J. Nutr. 140 (9), 1615–1620. https://doi.org/10.3945/jn.110.124735.

Devi, K.P., Rajavel, T., Nabavi, S.F., Setzer, W.N., Ahmadi, A., Mansouri, K., et al., 2015. Hesperidin: a promising anticancer agent from nature. Ind. Crop. Prod. 76, 582–589.

Economos, C., Clay, W.D., 1999. Nutritional and health benefits of citrus fruits. Food Nutr. Agric. 24, 11–18.

Emim, J.A., Oliveira, A.B., Lapa, A.J., 1994. Pharmacological evaluation of the anti-inflammatory activity of a citrus bioflavonoid, hesperidin, and the isoflavonoids, duartin and claussequinone, in rats and mice. J. Pharm. Pharmacol. 46 (2), 118–122.

Escobedo-Avellaneda, Z., Gutiérrez-Uribe, J., Valdez-Fragoso, A., Torres, J.A., Welti-Chanes, J., 2013. Phytochemicals and antioxidant activity of juice, flavedo, albedo and comminuted orange. J. Funct. Foods 6, 470–481.

Eskandari, A., Heidari, M., Daneshvar, M.H., Taheri, S., 2014. Studying effects of edible coatings of Arabic gum and olive oil on the storage life and maintain quality of postharvest sweet lemon (*Citrus lemontta*). Int. J. Agric. Crop. Sci. 7, 207–213.

Falcone Ferreyra, M.L., Rius, S.P., Casati, P., 2012. Flavonoids: biosynthesis, biological functions, and biotechnological applications. Front. Plant Sci. 28 (3), 222. https://doi.org/10.3389/fpls.2012.00222.

Fu, H.F., Xie, B., Gan, F., Ma, S., Zhu, X.R., Pan, S.Y., 2010. Effect of esterification with fatty acid of β-cryptoxanthin on its thermal stability and antioxidant activity by chemiluminescence method. Food Chem. 122 (3), 602–609. https://doi.org/10.1016/j.foodchem.2010.03.019.

Garg, A., Garg, S., Zaneveld, L.J., Singla, A.K., 2001. Chemistry and pharmacology of the citrus bioflavonoid hesperidin. Phytother. Res. 15 (8), 655–669.

Gliszczyńska-Świgło, A., Wróblewska, J., Lemańska, K., Klimczak, I., Tyrakowska, B., 2004. The contribution of polyphenols and vitamin C to the antioxidant activity of commercial orange juices and drinks. In: Proceedings of 14th IGWT Symposium Focusing New Century: Commodity-Trade-Environment, China, pp. 121–126.

Green, C.O., Wheatley, A.O., Osagie, A.U., St, A., Morrison, E.Y., Asemota, H.N., 2007. Determination of polymethoxylated flavones in peels of selected Jamaican and Mexican citrus (*Citrus* spp.) cultivars by high-performance liquid chromatography. Biomed. Chromatogr. 21 (1), 48–54.

Hirata, A., Murakami, Y., Shoji, M., Kadoma, Y., Fujisawa, S., 2005. Kinetics of radical-scavenging activity of hesperetin and hesperidin and their inhibitory activity on COX-2 expression. Anticancer Res. 25 (5), 3367–3374.

Hoffer, A., 1989. The discovery of vitamin C. Albert Szent-Gyorgyi, M.D., Ph.D. 1893–1986 "to see what everyone else has seen, but think what no one else has thought.". J. Orthomol. Med. 4, 24–26.

Huang, H., Li, L., Shi, W., Liu, H., Yang, J., Yuan, X., Wu, L., 2016. The multifunctional effects of nobiletin and its metabolites in vivo and in vitro. Evid. Based Complement. Alternat. Med. 2016, 2918796.

Ikoma, Y., Matsumoto, H., Kato, M., 2016. Diversity in the carotenoid profiles and the expression of genes related to carotenoid accumulation among citrus genotypes. Breed. Sci. 66 (1), 139–147. https://doi.org/10.1270/jsbbs.66.139.

Jourdan, P.S., McIntosh, C.A., Mansell, R.L., 1985. Naringin levels in *Citrus* tissues: II. Quantitative distribution of Naringin in *Citrus paradisi* Macfad. Plant Physiol. 77 (4), 903–908.

Jung, U.J., Lee, M.K., Jeong, K.S., Choi, M.S., 2004. The hypoglycemic effects of hesperidin and naringin are partly mediated by hepatic glucose-regulating enzymes in C57BL/KsJ-db/db mice. J. Nutr. 134 (10), 2499–2503.

Kakadiya, J., Mulani, H., Shah, N., 2010. Protective effect of hesperidin on cardiovascular complication in experimentally induced myocardial infarction in diabetes in rats. J. Basic Clin. Pharm. 1 (2), 85–91.

Kato, M., Ikoma, Y., Matsumoto, H., Sugiura, M., Hyodo, H., Yano, M., 2004. Accumulation of carotenoids and expression of carotenoid biosynthetic genes during maturation in citrus fruit. Plant Physiol. 134 (2), 824–837.

Kato, M., Matsumoto, H., Ikoma, Y., Okuda, H., Yano, M., 2006. The role of carotenoid cleavage dioxygenases in the regulation of carotenoid profiles during maturation in citrus fruit. J. Exp. Bot. 57, 2153–2164.

Khan, M.K., Dangles, O., 2014. A comprehensive review on flavanones, the major citrus polyphenols. J. Food Compos. Anal. 33 (1), 85–104.

Lado, J., Alós, E., Rodrigo, M.J., Zacarias, L., 2015. Light avoidance reduces ascorbic acid accumulation in the peel of citrus fruit. Plant Sci. 231, 138–147. https://doi.org/10.1016/j.plantsci.2014.12.002.

Lane, D.J., Richardson, D.R., 2014. The active role of vitamin C in mammalian iron metabolism: much more than just enhanced iron absorption! Free Radic. Biol. Med. 75, 69–83. https://doi.org/10.1016/j.freeradbiomed.2014.07.007.

Lo Piero, A.R., 2015. The state of the art in biosynthesis of anthocyanins and its regulation in pigmented sweet oranges [(*Citrus sinensis*) *L.* Osbeck]. J. Agric. Food Chem. 63 (16), 4031–4041. https://doi.org/10.1021/acs.jafc.5b01123.

Lv, X., Zhao, S., Ning, Z., Zeng, H., Shu, Y., Tao, O., et al., 2015. Citrus fruits as a treasure trove of active natural metabolites that potentially provide benefits for human health. Chem. Cent. J. 9, 68. https://doi.org/10.1186/s13065-015-0145-9.

Ma, G., Zhang, L.C., Matsuta, A., Matsutani, K., Yamawaki, K., Yahata, M., et al., 2013. Enzymatic formation of β-citraurin from β-cryptoxanthin and zeaxanthin by carotenoid cleavage dioxygenase4 in the flavedo of citrus fruit. Plant Physiol. 163 (2), 682–695. https://doi.org/10.1104/pp.113.223297.

Ma, G., Zhang, L., Iida, K., Madono, Y., Yungyuen, W., Yahata, M., et al., 2017. Identification and quantitative analysis of β-cryptoxanthin and β-citraurin esters in Satsuma mandarin fruit during the ripening process. Food Chem. 234, 356–364. https://doi.org/10.1016/j.foodchem.2017.05.015.

Magwaza, L.S., Mditshwa, A., Tesfay, S.Z., Opara, U.L., 2017. An overview of preharvest factors affecting vitamin C content of citrus fruit. Sci. Hortic. (Amsterdam) 216 (14), 12–21.

Männistö, S., Smith-Warner, S.A., Spiegelman, D., Albanes, D., Anderson, K., van den Brandt, P.A., et al., 2004. Dietary carotenoids and risk of lung cancer in a pooled analysis of seven cohort studies. Cancer Epidemiol. Biomark. Prev. 13 (1), 40–48.

Mastrangelo, D., Pelosi, E., Castelli, G., Lo-Coco, F., Testa, U., 2018. Mechanisms of anti-cancer effects of ascorbate: cytotoxic activity and epigenetic modulation. Blood Cells Mol. Dis. 69, 57–64. https://doi.org/10.1016/j.bcmd.2017.09.005.

Matsumoto, A., Mizukami, H., Mizuno, S., Umegaki, K., Nishikawa, J., Shudo, K., et al., 2007. beta-Cryptoxanthin, a novel natural RAR ligand, induces ATP-binding cassette transporters in macrophages. Biochem. Pharmacol. 74 (2), 256–264.

Matsuzaki, K., Miyazaki, K., Sakai, S., Yawo, H., Nakata, N., Moriguchi, S., et al., 2008. Nobiletin, a citrus flavonoid with neurotrophic action, augments protein kinase A-mediated phosphorylation of the AMPA receptor subunit, GluR1, and the postsynaptic receptor response to glutamate in murine hippocampus. Eur. J. Pharmacol. 578 (2–3), 194–200.

McEvoy, C.T., Schilling, D., Clay, N., Jackson, K., Go, M.D., Spitale, P., et al., 2014. Vitamin C supplementation for pregnant smoking women and pulmonary function in their newborn infants: a randomized clinical trial. JAMA 311 (20), 2074–2082. https://doi.org/10.1001/jama.2014.5217.

Mditshwa, A., Magwaza, L.S., Tesfay, S.Z., Opara, U.L., 2017. Postharvest factors affecting vitamin C content of citrus fruits: a review. Sci. Hortic. (Amsterdam) 218 (14), 95–104.

Michaud, D.S., Giovannucci, E.L., Ascherio, A., Rimm, E.B., Forman, M.R., Sampson, L., et al., 1998. Associations of plasma carotenoid concentrations and dietary intake of specific carotenoids in samples of two prospective cohort studies using a new carotenoid database. Cancer Epidemiol. Biomark. Prev. 7 (4), 283–290.

Miwa, Y., Mitsuzumi, H., Sunayama, T., Yamada, M., Okada, K., Kubota, M., et al., 2005. Glucosyl hesperidin lowers serum triglyceride level in hypertriglyceridemic subjects through the improvement of very low-density lipoprotein metabolic abnormality. J. Nutr. Sci. Vitaminol. (Tokyo) 51 (6), 460–470.

Monforte, M.T., Trovato, A., Kirjavainen, S., Forestieri, A.M., Galati, E.M., Lo Curto, R.B., 1995. Biological effects of hesperidin, a *Citrus* flavonoid. (note II): hypolipidemic activity on experimental hypercholesterolemia in rat. Farmaco 50 (9), 595–599.

Montonen, J., Knekt, P., Järvinen, R., Reunanen, A., 2004. Dietary antioxidant intake and risk of type 2 diabetes. Diabetes Care 27 (2), 362–366.

Moriguchi, T., Kita, M., Tomono, Y., Endo-Inagaki, T., Omura, M., 2001. Gene expression in flavonoid biosynthesis: Correlation with flavonoid accumulation in developing citrus fruit. Physiol. Plant. 111, 66–74.

Moser, M.A., Chun, O.K., 2016. Vitamin C and heart health: a review based on findings from epidemiologic studies. Int. J. Mol. Sci. 17 (8), E1328. https://doi.org/10.3390/ijms17081328.

Mulvihill, E.E., Assini, J.M., Lee, J.K., Allister, E.M., Sutherland, B.G., Koppes, J.B., 2011. Nobiletin attenuates VLDL overproduction, dyslipidemia, and atherosclerosis in mice with diet-induced insulin resistance. Diabetes 60 (5), 1446–1457. https://doi.org/10.2337/db10-0589.

Nagase, H., Omae, N., Omori, A., Nakagawasai, O., Tadano, T., Yokosuka, A., 2005a. Nobiletin and its related flavonoids with CRE-dependent transcription-stimulating and neuritegenic activities. Biochem. Biophys. Res. Commun. 337 (4), 1330–1336.

Nagase, H., Yamakuni, T., Matsuzaki, K., Maruyama, Y., Kasahara, J., Hinohara, Y., et al., 2005b. Mechanism of neurotrophic action of nobiletin in PC12D cells. Biochemistry 44 (42), 13683–13691.

Nakamura, M., Sugiura, M., Ogawa, K., Ikoma, Y., Yano, M., 2016. Serum β-cryptoxanthin and risk of high brachial-ankle pulse wave velocity: the Mikkabi cohort study. Nutr. Metab. Cardiovasc. Dis. 26 (9), 808–814. https://doi.org/10.1016/j.numecd.2016.04.001.

Pullar, J.M., Carr, A.C., Vissers, M.C.M., 2017. The roles of vitamin C in skin health. Nutrients 9 (8), E866. https://doi.org/10.3390/nu9080866.

Rao, M.N., Ghosh, P., Lakshman, M.R., 1997. Purification and partial characterization of a cellular carotenoid-binding protein from ferret liver. J. Biol. Chem. 272 (39), 24455–24460.

Roohbakhsh, A., Parhiz, H., Soltani, F., Rezaee, R., Iranshahi, M., 2015. Molecular mechanisms behind the biological effects of hesperidin and hesperetin for the prevention of cancer and cardiovascular diseases. Life Sci. 124, 64–74. https://doi.org/10.1016/j.lfs.2014.12.030.

Scarborough, H., Bacharach, L., 1949. Vitamins. In: Harris, R.S., Thimann, K.V. (Eds.), Vitamins and Hormones. Academic Press, New York, pp. 1–55.

Selvaraj, S., Krishnaswamy, S., Devashya, V., Sethuraman, S., Krishnan, U.M., 2014. Flavonoid-metal ion complexes: a novel class of therapeutic agents. Med. Res. Rev. 34 (4), 677–702. https://doi.org/10.1002/med.21301.

Shin, D.H., An, D.S., Koh, J.S., Lee, D.S., 1998. Wax-coating combined with grapefruit seed extract to reduce, physiological quality changes and decay of Satsuma Mandarin. Food Sci. Biotechnol. 7, 214–220.

Shorey-Kendrick, L.E., McEvoy, C.T., Ferguson, B., Burchard, J., Park, B.S., Gao, L., et al., 2017. Vitamin C prevents offspring DNA methylation changes associated with maternal smoking in pregnancy. Am. J. Respir. Crit. Care Med. 196 (6), 745–755. https://doi.org/10.1164/rccm.201610-2141OC.

Sugiura, M., Kato, M., Matsumoto, H., Nagao, A., Yano, M., 2002. Serum concentration of β-cryptoxanthin in Japan reflects the frequency of Satsuma mandarin (*Citrus unshiu* Marc.) consumption. J. Health Sci. 48, 350–353.

Sugiura, M., Matsumoto, H., Kato, M., Ikoma, Y., Yano, M., Nagao, A., 2004. Multiple linear regression analysis of the seasonal changes in the serum concentration of beta-cryptoxanthin. J. Nutr. Sci. Vitaminol. (Tokyo) 50 (3), 196–202.

Sugiura, M., Nakamura, M., Ogawa, K., Ikoma, Y., Yano, M., 2012. High serum carotenoids associated with lower risk for bone loss and osteoporosis in post-menopausal Japanese female subjects: prospective cohort study. PLoS ONE 7 (12), e52643. https://doi.org/10.1371/journal.pone.0052643.

Sugiura, M., Nakamura, M., Ogawa, K., Ikoma, Y., Yano, M., 2015a. High serum carotenoids associated with lower risk for type 2 diabetes among Japanese subjects: Mikkabi prospective cohort study. BMJ Open Diabetes Res. Care 3 (1), e000147. https://doi.org/10.1136/bmjdrc-2015-000147.

Sugiura, M., Nakamura, M., Ogawa, K., Ikoma, Y., Yano, M., 2015b. High serum carotenoids associated with lower risk for metabolic syndrome and its components among Japanese subjects: Mikkabi prospective cohort study. Br. J. Nutr. 114 (10), 1674–1682. https://doi.org/10.1017/S0007114515003268.

Sugiura, M., Nakamura, M., Ogawa, K., Ikoma, Y., Yano, M., 2016a. High vitamin C intake with high serum β-cryptoxanthin associated with lower risk for osteoporosis in post-menopausal Japanese female subjects: Mikkabi cohort study. J. Nutr. Sci. Vitaminol. (Tokyo) 62 (3), 185–191. https://doi.org/10.3177/jnsv.62.185.

Sugiura, M., Nakamura, M., Ogawa, K., Ikoma, Y., Yano, M., 2016b. High serum carotenoids associated with lower risk for elevated serum alanine aminotransferase among Japanese subjects: Mikkabi prospective cohort study. Br. J. Nutr. 115 (8), 1462–1469. https://doi.org/10.1017/S0007114516000374.

Tripoli, E., Guardia, M.L., Giammanco, S., Majo, D.D., Giammanco, M., 2007. *Citrus* flavonoids: molecular structure, biological activity and nutritional properties: a review. Food Chem. 104 (2), 466–479. https://doi.org/10.1016/j.foodchem.2006.11.054.

Viswanatha, G.L., Shylaja, H., Moolemath, Y., 2017. The beneficial role of Naringin-a citrus bioflavonoid, against oxidative stress-induced neurobehavioral disorders and cognitive dysfunction in rodents: a systematic review and meta-analysis. Biomed. Pharmacother. 94, 909–929.

Wahlqvist, M.L., Wattanapenpaiboon, N., Macrae, F.A., Lambert, J.R., MacLennan, R., Hsu-Hage, B.H., 1994. Changes in serum carotenoids in subjects with colorectal adenomas after 24 mo of beta-carotene supplementation. Australian polyp prevention project investigators. Am. J. Clin. Nutr. 60 (6), 936–943.

Wang, S., Yang, C., Tu, H., Zhou, J., Liu, X., Cheng, Y., et al., 2017. Characterization and metabolic diversity of flavonoids in citrus species. Sci. Rep. 7 (1), 10549. https://doi.org/10.1038/s41598-017-10970-2.

Wintergerst, E.S., Maggini, S., Hornig, D.H., 2006. Immune-enhancing role of vitamin C and zinc and effect on clinical conditions. Ann. Nutr. Metab. 50, 85–94.

Yuan, J.M., Stram, D.O., Arakawa, K., Lee, H.P., Yu, M.C., 2003. Dietary cryptoxanthin and reduced risk of lung cancer: the Singapore Chinese Health Study. Cancer Epidemiol. Biomark. Prev. 12 (9), 890–898.

Yungyuen, W., Ma, G., Zhang, L., Yamawaki, K., Yahata, M., Ohta, S., et al., 2017. Regulation of ascorbic acid metabolism in response to different temperatures in citrus juice sacs *in vitro*. Sci. Hortic. (Amsterdam) 217 (15), 1–7.

Zhang, L., Ma, G., Yamawaki, K., Ikoma, Y., Matsumoto, H., Yoshioka, T., et al., 2015. Regulation of ascorbic acid metabolism by blue LED light irradiation in citrus juice sacs. Plant Sci. 233, 134–142. https://doi.org/10.1016/j.plantsci.2015.01.010.

Index

Note: Page numbers followed by *f* indicate figures and *t* indicate tables.

D

Q

R

S

T

U

V

W

X

Y

Z

9780128121634